Power Flow Control Solutions for a Modern Grid using SMART Power Flow Controllers

IEEE Press
445 Hoes Lane
Piscataway, NJ 08854

Power Flow Control Solutions for a Modern Grid using SMART Power Flow Controllers

Kalyan K. Sen, PhD, PE, MBA, FIEEE
Mey Ling Sen, MEE, MIEEE
Sen Engineering Solutions, Inc.

IEEE PRESS SERIES ON POWER ENGINEERING

IEEE PRESS

WILEY

Published by John Wiley & Sons, Inc., Hoboken, New Jersey.
Published simultaneously in Canada.

Library of Congress Cataloging-in-Publication Data

Names: Sen, Kalyan K., author. | Sen, Mey Ling, author.
Title: Power flow control solutions for a modern grid using SMART power flow controllers / Kalyan K. Sen, PhD, PE, MBA, FIEEE, Mey Ling Sen, MEE, MIEEE, Sen Engineering Solutions, Inc.
Description: First edition. | Hoboken, New Jersey : John Wiley & Sons, Inc., [2021] | Series: IEEE press series on power engineering | Includes bibliographical references and index.
Identifiers: LCCN 2021031547 (print) | LCCN 2021031548 (ebook) | ISBN 9781119824350 (hardback) | ISBN 9781119824367 (adobe pdf) | ISBN 9781119824381 (epub)
Subjects: LCSH: Electric current regulators. | Electric power systems–Control. | Smart power grids. | Electric power Transmission.
Classification: LCC TK2851 .S48 2021 (print) | LCC TK2851 (ebook) | DDC 621.31–dc23
LC record available at https://lccn.loc.gov/2021031547
LC ebook record available at https://lccn.loc.gov/2021031548

Cover Design: Wiley
Cover Images: (Center) Courtesy of Kalyan Sen; (top) © Sam Robinson/Getty Images

Set in 9.5/12.5pt STIXTwoText by Straive, Pondicherry, India

10 9 8 7 6 5 4 3 2 1

MIX
Paper from responsible sources
FSC
www.fsc.org
FSC® C013604

To our family, friends,
and all our gurus who brought us to this point.

Contents

Authors' Biographies

Kalyan K. Sen was born in Bankura, West Bengal, India. He received BEE (first class honors, 1982), MSEE (1983), and PhD degrees (1987), all in Electrical Engineering, from *Jadavpur University* (India), *Tuskegee University* (USA), and *Worcester Polytechnic Institute* (USA), respectively. He also received an MBA (2012) from *Robert Morris University* (USA). He is the President and Chief Technology Officer of Sen Engineering Solutions, Inc. (www.sentransformer.com). From 1987 to 1990, he was an Assistant Professor at Prairie View A&M University. From 1990 to 2020, he worked mostly at Westinghouse and its successor companies in the United States, except during 1999–2001 when he worked at ABB in Sweden. He was a key member of the Flexible Alternating Current Transmission Systems (FACTS) development team at Westinghouse Science & Technology Center for which he became a Westinghouse Fellow Engineer. He contributed to the concept development, simulation, design, and commissioning of FACTS projects at Westinghouse since their inceptions in the 1990s. He conceived some of the basic concepts in power flow control technology for which he was elevated to the Institute of Electrical and Electronics Engineers (IEEE) Fellow grade with the citation: **for the development and application of power flow control technology**.

Kalyan has authored or coauthored more than 25 peer-reviewed publications, 8 issued patents, 2 books, and 3 book chapters in the areas of power flow control and power electronics. He is the coauthor of the book titled, *Introduction to FACTS Controllers: Theory, Modeling, and Applications*, IEEE Press and John Wiley & Sons, Inc. 2009, which is also published in Chinese and Indian (English) paperback editions. This book is used in universities and industries worldwide. His interests are in power converters, control systems, electrical machines, and power system simulations and studies. He is a licensed Professional Engineer in Pennsylvania and New York. He also served as a Fulbright Specialist (sponsored by the U.S. Government) and Global Initiative of Academic Networks (GIAN) Scholar (sponsored by the Government of India). He is an individual member of CIGRE.

Kalyan has served many organizations. He has been serving as an IEEE Power & Energy Society (PES) Distinguished Lecturer since 2002. In that capacity, he has given presentations on power flow control technology more than 150 times in 15 countries. He is an AdCom Member of the Power Electronics Society (PELS) and serves as the PELS Regions 1-6 Chair. He is the IEEE Division II Representative to the Board of Governors of Society on Social Implications of Technology (SSIT) and serves as the Chapters Committee Chair. He also serves as the Chair of IEEE Pittsburgh SSIT

Chapter. In 2003, he reestablished the Pittsburgh Chapters of the PES and the Industry Applications Society (IAS). Both Chapters received the "Outstanding Large Chapter" awards for their activities in 2004. He served as the Founding Chair of IEEE Pittsburgh PELS Chapter that received the Best Chapter Award in 2015. Under his Chairmanship, the IEEE Pittsburgh Section received the "Outstanding Large Section" award for its activities in 2005. He served as an Editor of the *IEEE Transactions on Power Delivery* from 2002 to 2007. He served as the Technical Program Chair of the 2008 PES General Meeting in Pittsburgh, and the Chapters and Sections Activities Track Chair at the 2008 IEEE Sections Congress in Quebec City, Canada. He has served as the Special Events Chair of the IEEE Pittsburgh Section for a decade. He received the IEEE Pittsburgh Section **Outstanding Volunteer Service** Award and Power & Energy Society **Outstanding Engineer** Award (2004). He is a Distinguished Toastmaster (DTM) who led District 13 of Toastmasters International (TI) as its Governor to be the 10th-ranking District in the world in 2007–2008. He has been serving as a Boy Scouts of America Leader for almost a decade.

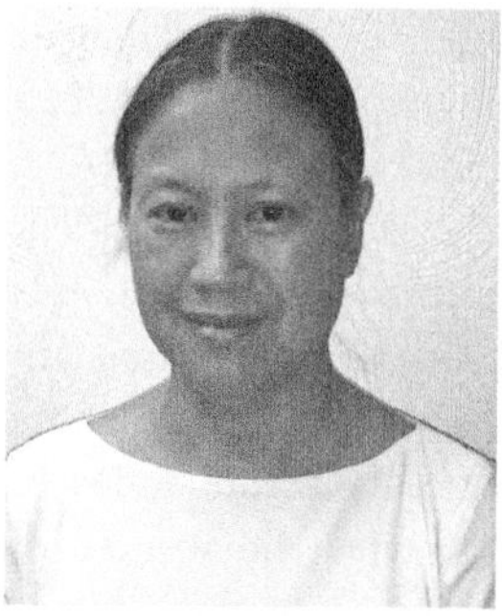

Mey Ling Sen was born in Aruba, Dutch Caribbean. She received BSEE (high distinction, 1988) and MEE (1990) degrees from *Worcester Polytechnic Institute* (USA) and *Rice University* (USA), respectively. As an Engineering Consultant, she worked at ABB and Westinghouse/CW. She is the Co-Founder and Chief Operating Officer of Sen Engineering Solutions, Inc. She is the co-inventor of the Sen Transformer, which is used as a Specific, Measurable, Attainable, Relevant, and Time-bound (SMART) Power Flow Controller that is based on functional requirements and a cost-effective solution. Her interests are in power electronics, electrical machines, and electric power engineering.

As a member of IEEE, she has served the Pittsburgh Chapters of PES and IAS in various positions, including Chapter Chair. Both Chapters received the "Outstanding Large Chapter" awards for their activities in 2008 and 2009, respectively. She also served IEEE Pittsburgh Section as the Treasurer in 2012 and Chair of Women in Engineering in 2016 and 2018–2019. She has been serving as the Special Events Chair of the IEEE Pittsburgh Section since 2020. She received IEEE Pittsburgh Section Power & Energy Society **Outstanding Engineer** Award (2018). She is a Distinguished Toastmaster (DTM). She served as the TI's President's Distinguished Area Governor in 2007–2008.

Technical Reviewers
J. M. DeSalvo
A. Parsotam
B. Shperling

Foreword

This book is a product of the authors' five decades of combined experiences in the research and development of power flow control technology. The traditional power grid as we know it is changing drastically. Mega-sized wind and solar projects are being integrated into the traditional majority carbon-based power grid in order to curb the production of greenhouse gases significantly.

Power systems of today were designed based on central generating stations and transmission and distribution lines to get the energy to the loads. However, with land-based and off-shore wind plants and distributed and utility-scale solar plants being connected to the grid, the old paradigm does not work since the geographic locations of the renewable resources do not in general coincide with the traditional generating plants. There is a need for the T&D systems to be revisited and modified/upgraded for the new power flow regimes. The line impedances that were tuned or optimized to serve certain flow patterns may now hinder delivery of the renewable energy to the desired destinations. The intermittent nature of the renewable energy sources brings additional challenges to system frequency and voltage control and to adopting the needed dynamic capability and the ability to control power flows bidirectionally at the right price. This can be mitigated with impedance regulation in strategically-selected transmission corridors. Furthermore, in many localities there are no new right-of-ways (ROWs), and rebuilding is limited to existing ones. Even though rebuild could be inevitable, flow control may help in some scenarios and may be much more economical.

The key to a clean energy transition depends on the electric grid's ability to generate and distribute renewable energy through the transmission and distribution system. The intermittency of supply and bidirectional flows, coupled with the remote locations of solar and wind projects, are challenging grid planners and operators. Even before we have reached large penetration of renewables, forecasters are factoring renewable curtailment as a major strategy to balance supply and demand, which adversely impacts the economics of the projects.

The concept of a SMART Power Flow Controller, developed in this book, is based on impedance management of the transmission line, which will be essential to (1) building the capacity to integrate and expand the use of clean distributed energy resources, (2) pursuing efficient asset utilization and reducing system losses, (3) facilitating greater transfer of clean energy from generation sites to load centers, and (4) improving grid reliability and resiliency. This technology can be customized, based on the required range and speed of operation, component non-obsolescence, ease of relocation, and interoperability.

This book starts with the derivation of the fundamentals of power flow in an AC transmission line and develops various solutions that can be used to enhance power flow while reducing the losses in AC transmission lines. The book builds on the evolution of power flow controllers in AC transmission systems covering the theory, modeling, and various applications. The subject is treated from the working engineers' point of view. After reading the appropriate parts of this book, students, teachers, and practicing engineers will be able to conduct studies of power system networks to mitigate their unique power flow problems.

The book's unique contribution is that it

- provides the basic theory and the step-by-step explanation of various power flow controllers;
- offers modeling techniques that are essential to electric utilities when conducting the needed studies and analysis;
- provides computer codes in the most widely-used Electro-Magnetic Transients Program (EMTP) formats;
- describes a new class of power flow controllers, based on the transformers/Load Tap Changers (LTCs) technology, developed by the authors and named the Sen Transformer (ST).

It is important to emphasize that the ST offers the equivalent control features of two devices – Phase Angle Regulator (PAR) and Voltage Regulator (VR) – for almost the price of one. If one purchases a PAR, which offers the phase angle or active power flow control only, the ST offers the added voltage or reactive power flow control capability with perhaps a small additional cost. The low-cost power flow control technology, such as ST, is of interest to utilities because of its simplicity, compared to power electronics inverter-based Unified Power Flow Controller.

I believe that the Sens' inventions are fundamental contributions toward the advancement of low-cost electric power flow control technology. A simulation model of the ST has already been developed in PSS/E, the most widely used load flow software, and the report is given in Appendix C. As an application example exercising the PSS/E model, it was verified that the ST performed as the most suitable candidate for power flow enhancement in a segment of the Chilean network. Also, a distribution-level Chinese demonstration of a 10-kV unit of ST confirmed the anticipated performance of the ST.

The topic of power flow control is of great interest to many power engineering professionals, utility engineers, large power equipment manufacturers, university professors, and students. The specialty of the book is that it develops the modern power flow control theories from the basics and supplements the theory with relevant computer models using the most widely used simulation software – EMTP and PSS/E. This book expands upon what the authors had presented in their last book, titled *Introduction to FACTS Controllers – Theory, Modeling, and Applications.*

In summary, the subject of power flow control cannot be overstressed; it is a very important topic to the electric power industry and electric utilities, particularly in today's environment. Due to the current need for integrating renewable energy sources into the grid reliably to reduce the carbon-based generation, electric utilities are seriously considering all available technical solutions. This is a timely book that gives the reader clear instructions on how to model, design, build, and evaluate power flow controllers. It supplements nicely the very few existing books. I realize that this book is practical, hands-on, and a true guide for the practicing engineers. The book gives significant amounts of detail in modeling and presentation that will be much appreciated by researchers/engineers in the field.

Since the 1990s, I have been interacting with Dr. Kalyan Sen on Flexible Alternating Current Transmission Systems (FACTS)-related projects. As the Lead Simulation Engineer at Westinghouse, Dr. Sen developed the FACTS models, which were essential for performing the feasibility study of the Convertible Static Compensator (CSC) FACTS project before its installation at the New York Power Authority (NYPA) Marcy 345 kV substation in central NY.

I have read this book with great interest. It is a work of love, written by two spouses who spent their entire careers in developing a much-needed power flow control technology that can help utilities worldwide to plan and operate their power grids. The specialty of the Sens' book is that it is coauthored by an engineer who actually designed and commissioned a number of power electronics-based FACTS controllers at Westinghouse since their inceptions in the 1990s. Therefore, the book includes a flavor of practical relevance. This book is going to aid the transformational change that is taking place in the electric utility industry worldwide.

White Plains, New York

Bruce B. Fardanesh Ph.D.,
IEEE Life Fellow
Vice President, System Planning & Analysis
New York Power Authority

Nomenclature

β	Relative phase angle of the series-compensating voltage with respect to the sending-end voltage
δ_s	Phase angle of the voltage at the sending end of a line
δ_r	Phase angle of the voltage at the receiving end of a line
δ	Power angle (difference of phase angles of the voltages at the two ends of a line)
δ'	Modified power angle (difference of phase angles of the voltages at the two ends of a line after compensation)
δ'_l	Lowest modified power angle
δ'_h	Highest modified power angle
ε	Least error of the calculated voltage and actually-tapped voltage in the Sen Transformer
θ	Phase-locked loop (PLL) angle
θ_I	Phase angle of the line current
θ_{VX}	Phase angle of the voltage across the line reactance
$^\circ$	Degree
φ	Power factor angle
ϕ	Phase angle between the voltage across the line reactance and the voltage difference between the sending and receiving ends
ω	Angular frequency
ψ	Phase-shift angle of the modified sending-end voltage with respect to the sending-end voltage
Ω	Unit of resistance, reactance, and impedance
a_1-b_1-c_1	Series-compensating windings in the *A* phase of ST
a_2-b_2-c_2	Series-compensating windings in the *B* phase of ST
a_3-b_3-c_3	Series-compensating windings in the *C* phase of ST
A-B-C	Exciter windings of ST
A_1-B_1-C_1	Shunt-compensating windings in the *A* phase of ST
A_2-B_2-C_2	Shunt-compensating windings in the *B* phase of ST
A_3-B_3-C_3	Shunt-compensating windings in the *C* phase of ST
A	Ampere (unit of current)
AC	Alternating Current
ANSI	American National Standards Institute
apr	Instantaneous apparent power rating
APR	Apparent Power Rating

ATC	Available Transfer Capability
b	Multiplier of V_{Xn} from 0 to 1
BPS	Bulk Power System
BTB-SSSC	Back-To-Back SSSC
BTB-STATCOM	Back-To-Back STATCOM
BYPBRK	Bypass breaker
C	Capacitance
$\cos(\varphi)$	Power factor
cp	Compensating points of the Sen Transformer
CSC	Convertible Static Compensator
c_r, $c_{s'}$, c_{se}	Intercept
d	Duty cycle
DC	Direct Current
DCLS	DC Link Switch
$\mathbf{E}$	Shunt- or series-connected voltage
EHV	Extra High Voltage
ES	Electronic Switch
F	Farad (unit of capacitance)
FACTS	Flexible Alternating Current Transmission Systems
FC	Fixed Capacitor
GaN	Gallium Nitride
GHG	Green-House Gas
GPFC	Generalized Power Flow Controller
GST	Generalized Sen Transformer
GTO	Gate-Turn Off
H	Henry (unit of inductance)
HV	High Voltage
Hz	Hertz (unit of frequency)
i	Instantaneous current, such as line current (i), exciting current (i_{ex}), sending-end current (i_s), source current (i_{src}), and so on
I	Line current magnitude
$\mathbf{I}$	Line current
IBR	Inverter-Based Resource
IEC	International Electrotechnical Commission
IEEE	Institute of Electrical and Electronics Engineers
$\mathbf{I_{ex}}$	Exciter current through the primary winding of the Sen Transformer
IPFC	Interline Power Flow Controller
I_n	Natural line current magnitude
IR	Impedance Regulator
$\mathbf{I_s}$	Current at the sending end of the line
$\mathbf{I_{src}}$	Source current
I_{TCR}	Current through Thyristor-Controlled Reactor
k	Number of TCSC sections
kHz	Kilo Hertz (unit of frequency)
k_R	Factor, representing the ratio of line resistance, R, when line current is I and the line resistance, R_n, when line current is I_n.

L	Inductance
LTC	Load Tap Changer
LV	Low Voltage
MC	Magnetic Circuit
m_r, $m_{s'}$, m_{se}	Slope
ms	Millisecond
MST	Multiline Sen Transformer
Mvar	Mega VAR (unit of reactive power)
n	(subscript) Natural
NC	Normally-Closed
NERC	North American Electric Reliability Corporation
NO	Normally-Open
p	Three-phase instantaneous active power
P	Active power
PAR	Phase Angle Regulator
P_{linen}	Power loss in the natural or uncompensated line
P_{link}	Active power on the common link
P_r	Active power at the receiving end of the line
P_{rh}	Highest active power at the receiving end of the line
P_{rl}	Lowest active power at the receiving end of the line
P_{rn}	Natural active power at the receiving end of the line
P_s	Active power at the sending end of the line
P_{se}	Exchanged active power by a Series Unit
P_{sh}	Exchanged active power by a Shunt Unit
P_{sn}	Natural active power at the sending end of the line
P_{src}	Active power at the source
$P_{s'}$	Active power at the modified sending end of the line
PFC	Power Flow Controller
POC	Point of Connection to the utility
PST	Phase-Shifting Transformer
pu	Per unit
q	Three-phase instantaneous quadrature power
Q	Quality factor
Q	Reactive power
QB	Quadrature Booster
Q_{linen}	Reactive power absorbed by the natural or uncompensated line
Q_{link}	Reactive power on the common link
Q_r	Reactive power at the receiving end of the line
Q_{rh}	Highest reactive power at the receiving end of the line
Q_{rl}	Lowest reactive power at the receiving end of the line
Q_{rn}	Natural reactive power at the receiving end of the line
Q_s	Reactive power at the sending end of the line
Q_{se}	Exchanged reactive power by a Series Unit
Q_{sh}	Exchanged reactive power by a Shunt Unit
Q_{sn}	Natural reactive power at the sending end of the line
Q_{src}	Reactive power at the source

$Q_{s'}$	Reactive power at the modified sending end of the line
r_k	Voltage error at a possible k^{th} operating point in the Sen Transformer
R	Line resistance
R'	Resistance of a section of a line
R_{eff}	Effective line resistance
ROM	Rough-Order Magnitude
RMS	Root Mean Square
RR	Reactance Regulator
R_{se}	Series-compensating resistance
R_{sh}	Shunt-compensating resistance
s	Second
S	Apparent power
SiC	Silicon Carbide
SMART	Specific, Measurable, Attainable, Relevant, and Time-bound
SPFC	SMART Power Flow Controller
S_r	Apparent power at the receiving end of the line
S_s	Apparent power at the sending end of the line
$_{se}$	(subscript) Series-exchanged, *i.e.* c_{se}, m_{se}, P_{se}, R_{se}, X_{se}, Q_{se}, Z_{se}
S_{se}	Exchanged apparent power by a Series Unit
$_{sh}$	(subscript) Shunt-exchanged, *i.e.* P_{sh}, R_{sh}, X_{sh}, Q_{sh}, Z_{sh}
S_{sh}	Exchanged apparent power by a Shunt Unit
SSSC	Static Synchronous Series Compensator
$S_{s'}$	Apparent power at the modified sending end of the line
ST	Sen Transformer
STATCOM	STATic synchronous COMpensator
SVC	Static Var Compensator
SynCon	Synchronous Condenser
t	Time
TCR	Thyristor-Controlled Reactor
THD	Total Harmonic Distortion
TNA	Transients Network Analyzer
TSC	Thyristor-Switched Capacitor
TSR	Thyristor-Switched Reactor
TTC	Transmission Transfer Capability
UHV	Ultra High Voltage
UPFC	Unified Power Flow Controller
UPS	Uninterruptible Power Supply
v	Instantaneous voltage
V	Phase voltage magnitude
$\mathbf{V}$	Phasor voltage
V	Volt (unit of voltage)
va	volt-ampere (unit of instantaneous apparent power)
VA	Volt-Ampere (unit of apparent power)
VAR	Volt-Ampere Reactive (unit of reactive power)
VR	Voltage Regulator
VRT	Voltage-Regulating Transformer

VSC	Voltage-Sourced Converter
$\mathbf{V_d}$	Voltage across the compensating resistance
$\mathbf{V_{dq}}$	Voltage across the compensating impedance
$\mathbf{V_q}$	Voltage across the compensating reactance
$\mathbf{V_r}$	Voltage at the receiving end of the line
$\mathbf{V_s}$	Voltage at the sending end of the line
$\mathbf{V_{s'}}$	Voltage at the modified sending end of the line
$\mathbf{V_{s'h}}$	Highest voltage at the modified sending end of the line
$\mathbf{V_{s'l}}$	Lowest voltage at the modified sending end of the line
$\mathbf{V_{s's}}$	Series-compensating voltage
$\mathbf{V_R}$	Voltage across the line resistance
$\mathbf{V_{Rn}}$	Natural voltage across the line resistance
$\mathbf{V_{R,X}}$	Voltage across the line impedance ($\mathbf{Z} = R + jX$)
$\mathbf{V_{Rn,Xn}}$	Natural voltage across the line impedance ($\mathbf{Z} = R + jX$)
$\mathbf{V_X}$	Voltage across the line reactance
$\mathbf{V_{Xn}}$	Natural voltage across the line reactance
W	Watt (unit of active power)
WECC	Western Electricity Coordinating Council
X	Line reactance (total)
X'_C	Capacitive reactance of a section of a line
X_{eff}	Effective line reactance
X'_L	Inductive reactance of a section of a line
X_{se}	Series-compensating reactance
X_{sh}	Shunt-compensating reactance
Z_{se}	Series-compensating impedance
Z_{sh}	Shunt-compensating impedance

Preface

Both authors have been involved in exploring various power flow controllers since the early 1990s. Kalyan Sen developed power electronics inverter-based Flexible Alternating Current Transmission Systems (FACTS) models while working at Westinghouse where pioneering development of FACTS products took place. Note that a forced-commutated inverter with a DC link capacitor is also referred to as a Voltage-Sourced Converter (VSC). Being an active contributor through patents, publications, design, and commissioning of much-advertised FACTS controllers since its inception in the 1990s, Kalyan has a first-hand knowledge of specific applications where the inverter-based controllers are the desirable solutions and where these solutions are not suitable at all. He has written an award-winning technical committee paper on the modeling of Unified Power Flow Controller (UPFC) in the *IEEE Transactions on Power Delivery*. Mey Ling Sen explored an alternate approach to the VSC-based FACTS Controllers that is cost effective while meeting functional requirements for most utility applications. This effort led to the concept of the Sen Transformer (ST). The ST is fundamentally different from the conventional transformer, in a sense that it uses three primary windings and nine secondary windings to create a compensating voltage that modifies the line voltage to be a specific magnitude and phase angle, whereas the conventional transformer only modifies the magnitude of the line voltage. As a result, by using an ST, the active and reactive power flows in the line can be regulated independently to maximize the revenue-generating active power flow and minimize the reactive power flow while maintaining the stability of the line voltage.

Since 2002, Kalyan has traveled around the world as an IEEE Distinguished Lecturer and lectured in more than 150 places in 15 countries. When he gives a presentation on power flow controllers, his approach is to start from the basics and lead up to the advanced concept of VSC-based FACTS Controllers and ST. His emphasis is based on real-world experience in modeling, simulation, design, and commissioning. He was requested in many places to compile his lecture material in the form of a book, which resulted in the publication of *Introduction to FACTS Controllers: Theory, Modeling, and Applications* in 2009. At the inception of the FACTS development in the 1990s, the main concerns were the high installation and operating costs of the FACTS Controllers. Over the decades, the list of drawbacks has expanded to include component obsolescence, costly maintenance, lack of trained-labor, impracticability of relocation and lack of interoperability. A desired feature of a Power Flow Controller (PFC) is that it is easily relocatable to wherever it is needed the most, since the need for power flow control may change with time due to new generation, load, and so on. Interoperability is desired so that components from various suppliers can be used, resulting in a global manufacturing standard, ease of maintenance, and ultimately lower cost to consumers.

The utilities are searching for a suitable power flow controller that offers its inherent features: simplicity, operational reliability, cost-effectiveness, component non-obsolescence, high efficiency, low maintenance, ease of relocation, and interoperability to meet their immediate needs to relieve grid congestion due to overload, peak load demand, and integration of renewable energy sources into the grid. The ST combines the best features of the FACTS controllers in terms of the ability to independently control active and reactive power flows while using time-tested and reliable transformer/Load Tap Changers (LTCs) technology that are familiar to the utilities worldwide for almost a century. More on LTCs can be found in the book, titled *On-Load Tap-Changers For Power Transformers: Operation, Principles, Applications and Selection*, by A. Krämer, Maschinenfabrik Reinhausen, 2000.

Power transformers are the workhorses that make transmission and distribution of AC electric power possible. Transformers step up the generator voltage (*e.g.* 25 kV) to the transmission level (*e.g.* 345 kV) and step down to distribution level (*e.g.* 13.8 kV) and, finally, to household utilization voltage (*e.g.* 120/240 V). With the addition of an LTC under load, transformers can easily regulate voltage. Specialty transformers, such as Phase Angle Regulators (PARs), can also regulate phase angle of the line voltage. The ST can regulate both the voltage magnitude and the phase angle simultaneously; as a result, the active and reactive power flows through the line can be controlled independently as desired.

The primary goal of this book is to present the fundamentals so that readers can retain the information clearly in their minds and provide a meaningful input in the selection process of adopting a particular solution. The book describes various concepts that are applicable to electric power industries. The concepts can be applied using traditional non-power electronics-based solutions and modern power electronics-based solutions or some hybrid of traditional-modern solutions. The reason for the primary goal is that a particular solution becomes obsolete as time progresses; however, the fundamental concepts remain the same.

Early power flow controllers employed basic technologies, such as transformers, capacitors, and reactors for the compensating voltage injection into the line. Later designs used power electronics to achieve much greater flexibility and optimization through an independent control of active and reactive power flows. When the first generation of power flow controllers, based on power electronics VSCs, were built in the 1990s, the Gate-Turn-Off thyristor (GTO) was the forced-commutated semiconductor switch of choice because of its availability in high power rating (4500 V, 4000 A) and its low forward voltage that resulted in low conduction loss. Early FACTS Controllers used VSCs with GTOs, switching once-per-cycle that resulted in the lowest switching loss and the lowest overall loss of about 1% of the rating of the VSC. These VSCs used special transformers to employ harmonic-neutralized techniques and produced high-quality AC waveforms without using filters. The inherent nature of a GTO is its relatively longer turn-on and turn-off times. More commonly used modern Pulse Width Modulation (PWM) techniques are based on instantaneous turn-on and turn-off of a switch. A voltage waveform that is created with a PWM technique consists of a fundamental component of interest and harmonic components, the dominant of which is related to the ratio of the switching and the fundamental frequencies. A higher switching frequency is desirable, because the higher dominant frequency requires a reduced-sized filter. To keep the sum of turn-on and turn-off times of a GTO to be less than 1% of the switching period, it would result in only several hundred Hz of switching frequency. This would require a fairly large-sized output filter to eliminate the unwanted low-order harmonic components, generated by a force-commutated inverter.

About a decade later, the VSC of choice started to use Insulated Gate Bipolar Transistor (IGBT)-based PWM techniques. An IGBT offers shorter turn-on and turn-off times, which is less than 1% of the switching period that results in a switching frequency of several kHz. A lower switching period means a higher switching frequency and higher order harmonic components that are not of significant interest, since they do not generate significant amount of harmonic currents for two reasons; first, higher order voltage harmonic components are lower in magnitudes and second, the higher order voltage harmonic components "see" higher reactances for a given inductance. However, some filtering may still be needed, since switching frequency could not be increased to the desired level in some cases due to generation of excessive losses (3–4% of the rating of the VSC) as a function of the increased switching frequency. Another decade later, the topology of choice has become multilevel VSCs that do not need any harmonic filtering. While the topologies of VSCs are changing, so are the semiconductor switching devices. The upcoming switches are based on silicon carbide (SiC) and gallium nitride (GaN) for desirable reasons, such as high-speed operation, which results in lower turn-on and turn-off times, thus lower switching loss, high-temperature operation, lower cooling requirement, and smaller circuits for the gate drive and the snubber. A higher switching frequency creates a higher Electro-Magnetic Interference (EMI), which requires the use of an additional EMI filter. The fact is that with various advances in the power electronics technology and semiconductor switches, the FACTS controllers become obsolete in a relatively few years and as a result, a one-to-one component replacement becomes impossible in 10–15 years. In the utility world where 45–50 years of equipment life is the norm, this means the entire power electronics inverter-based FACTS installation may need to be replaced several times in those 45- to 50-year period. In addition, simple maintenance requires highly skilled personnel that are not readily available. The global standard and interoperability do not exist due to a limited number of manufacturers. This is a highly expensive proposition perhaps two orders of magnitude more than a long-lived and easily maintained transformer/LTCs-based technology, such as ST.

Today's power grid has evolved into integration of inverter-based, renewable-sourced, electricity generation from solar and wind farms, which are intermittent in nature. Therefore, traditional steady-state power flow controllers, such as series-connected reactor/capacitor concepts, need to be updated with an improved dynamic response. Additionally, increasing installation of roof-top solar and its integration into a low-voltage distribution network has altered the traditional voltage profile in the distribution network and increased the need for a bidirectional power flow controller when the renewable generation is not available. Therefore, all available solutions need to be considered for future needs, which has led to the concept of SMART Controllers.

A considerable amount of effort has been put into modeling various controllers. Modeling is the only approach, before any hardware construction, for the verification of the performance of any concept. The book includes models of many controllers, developed using a freely available Electro-Magnetic Transients Program (EMTP) software package.

The book is divided into six chapters and three appendices. Chapter 1 presents the origin of power flow controllers and guides the reader to the selection process of a SMART Power Flow Controller (SPFC).

Chapter 2 is for anyone who would like to become familiar with the subject. It discusses various topics of the book in simple electrical engineering terms and corroborates the theory with relevant mathematics. The characteristic equations of various power flow controllers, including their equivalent compensating impedances, are developed. Using these equations, a set of example problems is given, which gives a quick back-of-the-envelope calculation results without much effort. A figure of merit, called *Sen Index*, is defined for all the Power Flow Controllers (PFCs).

Chapter 3 presents the fundamentals of modeling in EMTP and explains the basic differences of modeling various PFCs, such as the Voltage-Regulating Transformer (VRT), Phase Angle Regulator (PAR), Unified Power Flow Controller (UPFC), and Sen Transformer (ST). Following the Rough-Order Magnitude (ROM) calculations performed in Chapter 2, using simple equations to characterize a power flow solution, the ROM results may need to be refined by employing the modeling techniques developed in Chapter 3. An example simulation of a series-compensating voltage is shown to emulate a VRT, a PAR, and an Impedance Regulator (IR).

Chapter 4 presents the transformer-based PFCs and gives some baseline examples for comparison with other PFCs in the following chapters. It is shown how a VRT and a PAR may be modeled by using a series-compensating voltage.

Chapter 5 presents some early PFCs that use mechanical switches and set some baselines for comparison in the following chapters. It is shown how to model a virtual impedance that is equivalent to shunt-connected and/or series-connected inductive and/or capacitive compensators.

Chapter 6 presents the evolution of an ST and its wide variety of applications. The most up-to-date advancements in ST are described in this chapter. This includes various forms of two-core designs. Also included is a new factory-test method under full power.

Appendix A describes the operation of various items, such as (1) three-phase balanced and unbalanced voltage, current, and power; (2) symmetrical components; (3) d-q transformation; and (4) Fourier analysis. The reader will find it useful to see the industry techniques and the relevance of the theory and applications.

Appendix B presents the power flow control equations in a lossy line and compares the derived results from those in Chapter 2 for lossless lines. Simpler versions of these equations are derived in Chapter 2, considering the line resistance (R) is zero. These examples will be used as future references for those involved with PFCs. For the readers to recognize the importance of the equations and example solutions presented in Chapter 2, a list of all the "Examples" is placed at the end of Appendix B. Using the information received from Supervisory Control And Data Acquisition (SCADA) about the sending- and receiving-end voltages (V_s and V_r) and active and reactive power flows (P_r and Q_r), other power flow variables, such as the power angle (δ), can be calculated for a known line impedance ($\mathbf{Z} = R + jX$).

Appendix C presents a load flow study of the Chilean network, integrated with Sen Transformer, performed by Siemens PTI and sponsored by New York Power Authority.

Pittsburgh, Pennsylvania

Kalyan K. Sen
Mey Ling Sen

Acknowledgments

We both would like to thank our former colleagues at Westinghouse where the pioneering development of FACTS controllers started in 1990 under the technical leadership of Dr. L. Gyugyi. We are very grateful for all those who dedicated their time to review the manuscript thoroughly and provide valuable feedback. Special gratitude goes to Mr. R. Alexander, Dr. S. Behzadirafi, Mr. J. M. DeSalvo, Dr. M. Haj-Maharsi, Dr. A. J, F. Keri, Dr. T. Manna, Mr. A. Parsotam, Mr. G. Pedrick, Dr. B. Shperling, and Mr. R. Subramanian. We acknowledge the continuous effort to promote Sen Transformer – a SMART Power Flow Controller technology by Dr. S. Behzadirafi, Mr. A. Ettlinger, Dr. B. Fardanesh, and Mr. G. Pedrick of New York Power Authority, late Mr. M. Henderson of New England ISO, Mr. A. Parsotam of Newcastle, New South Wales, Australia, Prof. V. Dinavahi of University of Alberta, Prof. R. Gokaraju of University of Saskatchewan, Mr. R. Alexander, and Mr. J. M. DeSalvo. We thank American Public Power Association and Abby Anaday on Unsplash for using their respective photos of Hydropower and Windmill on the coverpage. After performing decades of research on various power flow control technologies, we feel that "the more we learn, the more we realize how little we know." We appreciate feedback from the readers. Our email addresses are senkk@ieee.org and senml@ieee.org, respectively.

K.K.S.
M.L.S.

I would like to acknowledge my three professors, in particular, late S. K. Dutta of Jadavpur University for teaching me the basics of power electronics – a new and upcoming subject during 1970–1980, Dr. P. K. Ray of Tuskegee University for giving me a thorough training on how to conduct a basic research, and Dr. A. E. Emanuel of Worcester Polytechnic Institute for perfecting the mold as my PhD thesis Advisor.

Very special thanks go to my two Westinghouse mentors – late T. Heinrich and late M. Brennen whose insights in power electronics were unparalleled. Proper thanks go to Dr. L. Gyugyi who convinced me to change my job from academia to industry, a change that I never regretted. He is the inspiration for my life-long passion to develop power flow control technologies. I feel fortunate to have worked in the development of FACTS products under his supervision.

Throughout my tenure at Westinghouse Science & Technology Center in Pittsburgh, I had a rare opportunity of working with extraordinary people from all over the world. Not only were they a fine class of engineers, they also had hobbies that could be considered as careers in their own rights. During the long hours of commissioning of the world's first UPFC in Kentucky and a STATCOM in Texas, and the TNA test in Montreal, the team of engineers kept everyone amused with their life stories. I thank the fine colleagues at the Tennessee Valley Authority, American Electric Power, New York Power Authority, Bonneville Power Administration, Western Area Power Administration, and the Electric Power Research Institute.

K.K.S.

About the Companion Website

This book is accompanied by a companion website:

www.wiley.com/go/sen/powerflowcontrol

All EMTP models that are discussed throughout the book are available in the companion website.

1

Smart Controllers

The future transmission and distribution system is on the way to be dramatically different from what it is today. Today's Bulk Power System (BPS), referred to as grid, that encompasses mainly electromechanical devices is continuously integrating Inverter-Based Resources (IBRs) to convert renewable energy sources into electricity. However, most of these sources are solar and wind, which are intermittent in nature. Commercial nuclear-powered generators, once turned on, deliver electric power continuously for the next 18 to 24 months before stopping for scheduled maintenance and refueling. This difference alone, in two different types of electricity generation, creates a need for a SMART controller that is capable of managing power flow dynamically. A SMART controller will be essential to increase the transmission capacity of the existing system with line impedance management and the needed dynamic performance. This technology will help operators optimize power flows across the grid to reduce voltage stress on the transmission network, reduce line loss, as well as reduce Green-House Gas (GHG) emissions from traditional carbon-based generation.

SMART is an acronym that stands for Specific, Measurable, Attainable, Relevant, and Time-bound. SMART solutions, which have been used for decades in many fields, are well suited for applications in electrical engineering. A controller, also referred to as a compensator, is a general term to describe a regulator that regulates voltage, current, phase angle of a line voltage, resistance, reactance, impedance, and so on, directly or indirectly in an electric circuit. A controller can be designed to perform a task (*e.g.* voltage regulation) and/or multiple tasks with varying cost and complexity. In the simplest term, a SMART Controller is what one procures is based on what one needs. A SMART Controller is an evolving technology that uses the best technical features from all previous concepts to meet the present need by blending functional requirements with the most cost-effective solution. A SMART Power Flow Controller (SPFC) enhances the power flow controllability in electric power transmission and distribution systems. It is recommended that utilities choose a solution that meets their **need** in terms of reliability, robustness, cost-effectiveness, component non-obsolescence, efficiency, ease of relocation, and interoperability.

1.1 Why is a Power Flow Controller Needed?

The electricity generation is typically located a distance away from the industrial and population, *i.e.* load centers. Therefore, electricity is transmitted from the points of generation to the points of utilization using transmission and distribution lines. One of the major achievements of the late 19th century was to implement long-distance alternating current (AC) transmission of electric power. For a given amount of transmitted power, a higher transmission voltage provides less footprint

Power Flow Control Solutions for a Modern Grid using SMART Power Flow Controllers, First Edition.
Kalyan K. Sen and Mey Ling Sen.

Companion website: www.wiley.com/go/sen/powerflowcontrol

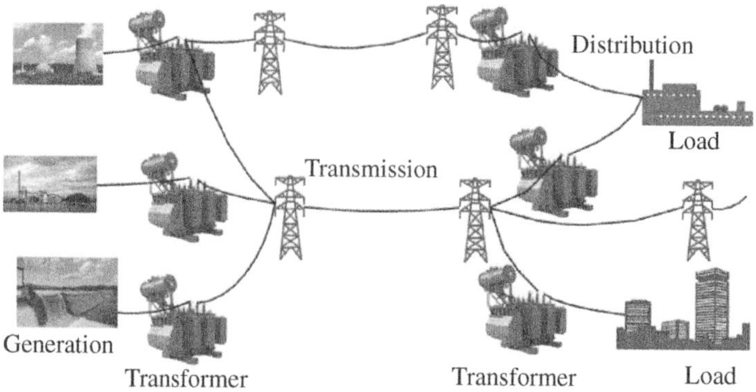

Figure 1-1 Part of a large interconnected transmission system supplying electric power from the generating point to the loads.

and less current in the line. Since loss in the line is proportional to the square of the current, it is most desirable to transmit electricity from generation to load centers through the transmission lines at higher voltage and lower current. Over the next century, various types of generation, such as coal, natural gas, nuclear, hydro, renewables, and so on, and loads were integrated through interconnected transmission and distribution lines to form the grid as shown in Figure 1-1.

Electricity flows freely from a higher potential to a lower potential. If two or more lines with different impedances are connected in parallel, the current in each line is, in some form, inversely proportional to the respective line impedance. This free flow of electricity might take longer paths to reach its destination and cause unwanted power losses in the lines. Additionally, this free flow may cause some transmission lines to be overloaded or underloaded. If the impedance of a line is larger compared to that of the lines connected in parallel, the current and the resulting power flow through the higher-impedance line is lower compared to that in the neighboring lines, and vice versa. Therefore, power flow in a line is, in some form, inversely proportional to the impedance of the line.

Transmission Reliability Margin (TRM) – the amount of Transmission Transfer Capability (TTC) needed to ensure that the interconnected transmission network is secure under a reasonable range of uncertainties in system conditions. These uncertainties may result from the following:

1) Simultaneous limitations with a parallel path.
2) Reservations for unscheduled flow, *i.e.* loop flow.
3) Reservations for unplanned transmission outages, *i.e.* for paths in which contingencies have not already been included in the calculation of TTC.

TRM does not include reservations for planned outages and other known transmission conditions, which have been included in the calculation of TTC (Minimum of {Thermal Limit, Voltage Limit, and Stability Limit}).

Capacity Benefit Margin (CBM) – the amount of TTC reserved by the load-serving entities to ensure access to generation from interconnected systems to meet generation reliability requirements. These reservations may include the following:

1) Transmission reserved by the Control Area Operator to accommodate operating reserves (spinning and supplemental). Such operating reserves may not exceed NERC and WECC applicable pool requirements or individual members' reliability requirements.

2) Transmission reserved for the import of ancillary services (such as spinning reserves) from another control area.
3) Transmission reserved for generation patterns and generation contingencies. These patterns and contingencies must be based upon reasonable, publicly available assumptions subject to evaluation in applicable dispute resolution proceedings.

According to "Available Transfer Capability Definitions and Determination," North American Electric Reliability Council, June 1996, TTC is defined as

$$\text{TTC} = \text{Minimum of } \{\text{Thermal Limit, Voltage Limit, and Stability Limit}\}. \tag{1-1}$$

Available Transfer Capability (ATC) is a measure of the transfer capability remaining in the physical transmission network for further commercial activity over and above already committed uses. Therefore,

$$\text{ATC} = \text{TTC} - \text{TRM} - \text{CBM} - \text{existing transmission commitments}. \tag{1-2}$$

It is often desirable to utilize the ATC to increase the power flow through a line as much as possible according to Equation (1-2), which results in a higher line utilization, meets greater customer needs, integrates new sources of energy, and avoids or defers building of new transmission lines, at least in the short term. Sometimes, it is the opposite when it may be desirable to lower the power flow of the line so that the power flow can be redirected to a desired transmission line that may include high-voltage and low-current and thus, low-loss lines. This is particularly important when a line becomes overloaded with a level of current that can trip the line or a fault current that must be limited. If an overloaded line trips, its current will be redirected in the available lines inversely proportional, depending on the lines' impedances. This may cause a previously underloaded line to become overloaded and tripped, which may create a possible cascaded failure of the grid, resulting in a blackout. This type of cascaded failure is described in the "Final Report on the August 14, 2003 Blackout in the United States and Canada: Causes and Recommendations" by U.S.-Canada Power System Outage Task Force, April 2004.

Inadequate management of reactive power and line voltage caused the blackout. Most of the recent power network blackouts have been related to line voltage collapses, which tend to occur from lack of reactive power supports in heavily stressed conditions, usually triggered by system faults. One of the ways to improve the reliability of the line when faced with similar occurrences would be to regulate the effective impedance of a line between its two ends with an Impedance Regulator (IR), which is also known as a Power Flow Controller (PFC).

It is also desirable to use a PFC to enhance the power flow in the line to maximize the revenue-generating active power flow and minimize the reactive power flow within a permissible limit, while maintaining the voltage stability. The flow of electricity in a particular line of a transmission system can be controlled by using a PFC as shown in Figure 1-2. Since the power flow in a line is inversely proportional to the impedance of the line and assuming the line to be an inductive impedance, the power flow can be decreased by inserting an additional inductive impedance in series with the line, thereby increasing the effective impedance of the line between its two ends. Also, the power flow can be increased by inserting an additional capacitive impedance in series with the line, thereby decreasing the effective impedance of the line between its two ends. Therefore, the regulation of the effective line impedance is the key to maintaining desired power flow in a line. Note that each of a Voltage Regulator (VR) and a Phase Angle Regulator (PAR) also regulates the effective line impedance indirectly. However, the Reactance Regulator (RR) regulates only the effective line reactance directly as discussed in Chapter 2, Section 2.2.2.6 (Representation of a Series-Compensating Voltage as a Series-Compensating Impedance).

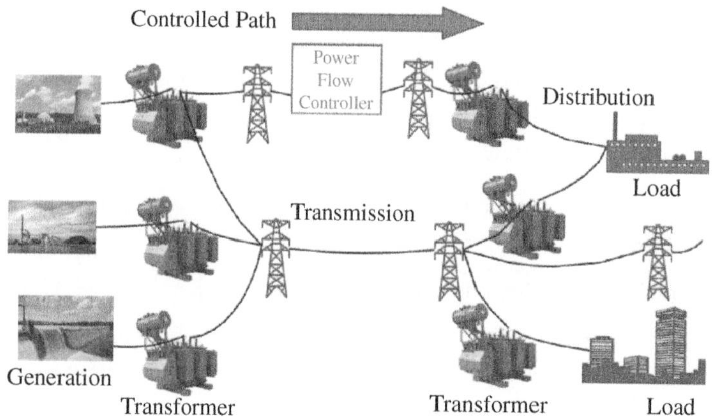

Figure 1-2 Power flow along a controlled path.

The demand for electrical energy around the world increases continuously; so does the use of various sources of energy from traditional synchronous generators, used at coal/natural gas/nuclear/hydro power stations to modern IBRs that convert renewable wind and solar energy into usable AC electricity. Often, the available sources of energy generation are far away from the load centers. The ever-growing need for transmitting more electricity can be met either by installing new transmission lines, characterized by a lengthy and costly process and/or by harnessing the dormant capacity of the underutilized transmission lines with a quicker and much less-costly option. The challenge is how to harness this dormant capacity in the most cost-effective way. Any investment alternative to harness the dormant capacity of the underutilized transmission lines should be supported by comparing the investment relative to other competing options with all cost/benefit considerations being evaluated for all tangible and non-tangible factors over the total life cycle.

The free flow of electricity from one particular point to another might not take the shortest path. Any unwanted path along the way causes extra power loss, loop flow of power, and reduced stability with increased voltage variation in the line. The power industry constantly searches for the most economical ways to transfer bulk power along a desired path. Before considering new transmission lines, it is desirable to explore all the options to increase the loadability of existing transmission lines. The free flow of power through unwanted longer paths, which causes extra losses in the lines can be mitigated with the use of a PFC. The optimum power flow through the lines will enhance the loadability of the lines in the most efficient way. This will reduce the carbon-based generation that is equal to the unwanted losses in the lines due to free flow of power, which will reduce GHG emissions and contribute to a reduction in global warming.

Traditional regulators, such as VR, PAR, and RR, regulate one of the three power flow control parameters (line voltage magnitude, its phase angle, and line reactance) and, in turn, control active and reactive power flows (P and Q) simultaneously, meaning both P and Q either increase or decrease. Since the effect of line reactance regulation is equivalent to essentially the combined effects of voltage regulation (using a VR) and phase angle regulation (using a PAR), the two main power flow control parameters are the line voltage magnitude and its phase angle. An IR is functionally equivalent to the combined effects of a VR and a PAR.

The optimization of P and Q flows in the transmission line requires an independent control of P and Q flows, which requires a simultaneous regulation of the line voltage magnitude and its phase angle. This is functionally equivalent to regulating the effective four-quadrant impedance (discussed in Section 1.2) of the line between its two ends. Early PFCs employed basic technologies, such as transformers, capacitors, and reactors for the compensating voltage injection in the line. Modern-day PFCs emulate a reactor or a capacitor by creating a compensating voltage whose phase angle is either leading or lagging the current that is passing through the compensating voltage. The advantage of using an emulated series capacitor as compared to an actual series-connected capacitor is the avoidance of creating any type of resonance with the inductance in the line and the synchronous machine in the form of sub-synchronous resonance. The voltage across the emulator is limited by the rating of the PFC whereas the voltage across a series-connected capacitor may be excessively high during resonance. In addition, just by changing the control algorithm, the same power hardware may be used to emulate a series-compensating capacitor as well as a reactor, instead of using a separate reactor.

An IR creates a series-connected virtual impedance that modifies the effective impedance of a line between its two ends. As a result, it is possible to increase power flow in an underloaded line, decrease power flow in an overloaded line, and control the flows of P and Q independently as desired. If deployed in critical locations, the IR will maximize the P flow that will generate the most revenue and minimize the Q flow, reducing unwanted power losses in transmission lines. This will increase the efficiency of the power grid, increase voltage stability margins, and may avoid a cascaded blackout as described in the "Final Report on the August 14, 2003 Blackout in the United States and Canada: Causes and Recommendations" by U.S.-Canada Power System Outage Task Force, April 2004.

The consequence of free flow of electricity following the Hanna-Juniper Line Loss on August 14 2003 contributed to a massive blackout of the North-Eastern USA and Canada. When a tree and the 345-kV line touched each other, the line tripped; as a result, the load got redistributed among the available transmission paths, which overloaded some lines that then tripped. These events normally would set off alarms in a local utility's control room and alert operators to activate controllers in neighboring regions to reroute power flows around the affected site. However, the alarm software failed, and thus the local operators were unaware of the problem. Transmission lines surrounding the failure spot were forced to shoulder more than their safe quota of electricity. Also, at this time, the reactive power supply was at a minimum and when plant operators tried to increase the reactive power flow, the generating plant shut down. This further destabilized the system's equilibrium, leading to additional lines and generators dropping out of the grid as the cascade continued. Within 8 minutes, 50 million people were experiencing a blackout.

The final report on the blackout stated that more high-voltage lines must be built and perhaps even more important, the power grid must be made SMARTER. A self-healing SMART grid is needed to be able to recognize the problem and then reconfigure the power grid. If IRs were strategically located, these overloaded lines would have their power flows controlled to be within their ATC limits and would neither trip the line nor contribute to the blackout. Hence, the IR improves grid reliability and resiliency.

1.2 Traditional Power Flow Control Concepts

The flow of AC power, irrespective of its source, has two components: active power and reactive power. A transmission line consists of electrical conductors that may be composed of many sections. Each section consists of a resistance (R'), inductive reactance (X'_L), and line-to-ground (shunt) capacitive reactance (X'_C) as shown in Figure 1-3. Since there is no significant storage of

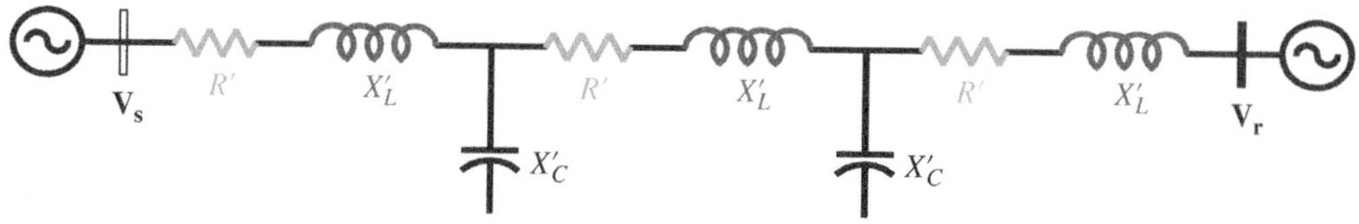

Figure 1-3 Representation of a transmission line between sending and receiving ends.

electric power at the utility scale, the active power, except for the loss in the resistance of the conductor, reaches from one end of the line to the other end. This active power can be used for lighting, heating, cooling, motion force in electric motors, and so on. The distributed series inductive and shunt capacitive reactances of the line absorb and generate reactive power, respectively. This reactive power flow causes an extra loss in the resistance of the line.

The symbols shown in the figure are

$\mathbf{V_s}$ is the sending-end voltage with a magnitude (V_s) and a phase angle (δ_s),
$\mathbf{V_r}$ is the receiving-end voltage with a magnitude (V_r) and a phase angle (δ_r),
R' is the line resistance in each section,
X'_L is the inductive reactance in each section of the line, and
X'_C is the line-to-ground (shunt) capacitive reactance in each section.

Transmission lines with lengths less than 50 miles (80.5 km) are classified as being short lines; lines of lengths between 50 and 150 miles (80.5 and 241.4 km) are classified as medium-length lines and lines above 150 miles (241.4 km) are considered long lines. Consider a line in the interconnected transmission system, connecting sources and loads as shown in Figure 1-1 as a relatively short line where the capacitive shunt reactance from the line to ground and among the lines can be ignored as shown in Figure 1-4. The resistances and inductive reactances from all the line sections are lumped together as shown in the figure. The natural power flow in an AC transmission line depends on (1) magnitudes of the sending and receiving-end voltages, (2) phase angle between these voltages, and (3) line impedance.

The additional symbols shown in the figure are

$\mathbf{V_{Xn}}$ is the natural voltage across the line reactance with a magnitude (V_{Xn}) and a phase angle (θ_{VXn}),
$\mathbf{V_{Rn}}$ is the natural voltage across the line resistance with a magnitude (V_{Rn}) and a phase angle (θ_{VRn}),
$\mathbf{I_n}$ is the natural line current with a magnitude (I_n) and a phase angle (θ_{In}),
P_{sn} is the natural active power flow at the sending end,
Q_{sn} is the natural reactive power flow at the sending end,

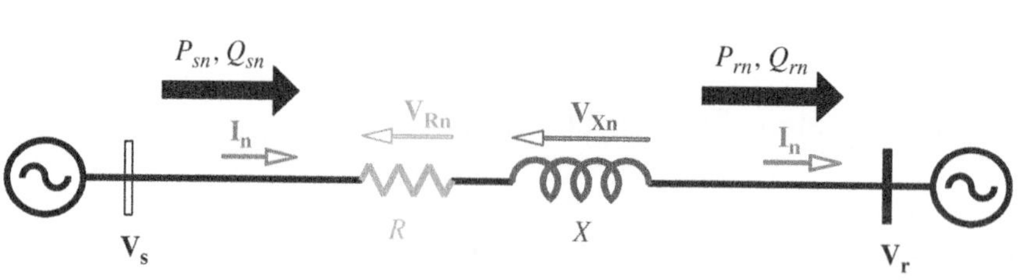

Figure 1-4 Power flow along a transmission line between sending and receiving ends.

P_{rn} is the natural active power flow at the receiving end,
Q_{rn} is the natural reactive power flow at the receiving end,
R is the line resistance ($R > 0$ and represents a positive resistance), and
X is the line reactance ($X > 0$ and represents an inductive reactance).

The natural active and reactive power flows (P_{sn} and Q_{sn}) at the sending end are derived in Appendix B as

$$P_{sn} = \frac{A}{\sqrt{R^2 + X^2}} \left\{ R\left(\frac{V_s}{V_r} - \cos\delta\right) + X \sin\delta \right\} \tag{B-12}$$

and

$$Q_{sn} = \frac{A}{\sqrt{R^2 + X^2}} \left\{ -R\sin\delta + X\left(\frac{V_s}{V_r} - \cos\delta\right) \right\} \tag{B-14}$$

where

$$A = \frac{V_s V_r}{\sqrt{R^2 + X^2}} \tag{B-13}$$

and the power angle is given in Chapter 2 as

$$\delta = \delta_s - \delta_r. \tag{2-27}$$

The natural active and reactive power flows (P_{rn} and Q_{rn}) at the receiving end are

$$P_{rn} = \frac{A}{\sqrt{R^2 + X^2}} \left\{ R\left(\cos\delta - \frac{V_r}{V_s}\right) + X \sin\delta \right\} \tag{B-21}$$

and

$$Q_{rn} = \frac{A}{\sqrt{R^2 + X^2}} \left\{ -R\sin\delta + X\left(\cos\delta - \frac{V_r}{V_s}\right) \right\}. \tag{B-22}$$

Ignoring the line resistance as shown in Figure 1-5a, the natural active and reactive power flows (P_{sn} and Q_{sn}) at the sending end and the natural active and reactive power flows (P_{rn} and Q_{rn}) at the receiving end for a relatively short lossless line are

$$P_{sn} = a \times \sin\delta, \tag{2-40}$$

$$Q_{sn} = a\left(\frac{V_s}{V_r} - \cos\delta\right), \tag{2-43}$$

$$P_{rn} = a \times \sin\delta \tag{2-46}$$

and

$$Q_{rn} = a\left(\cos\delta - \frac{V_r}{V_s}\right) \tag{2-48}$$

where

$$a = \frac{V_s V_r}{X}. \tag{2-41}$$

In addition to using these formulae to characterize a two-generator/single-line power system network, they may be used when designing an electrical generator where the $\mathbf{V_s}$ and $\mathbf{V_r}$ are the

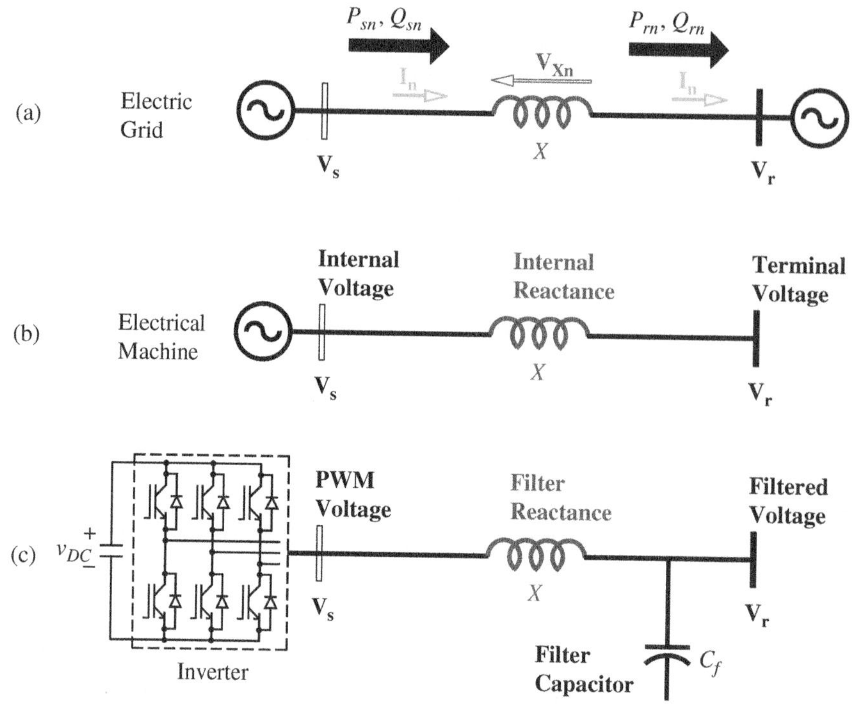

Figure 1-5 (a) Electric grid: power flow along a lossless transmission line between sending and receiving ends; (b) equivalent representation of an electrical machine; (c) equivalent representation of an inverter.

generator's internal voltage and terminal voltage, respectively, and X is the internal reactance of the generator as shown in Figure 1-5b. When designing an inverter, $\mathbf{V_s}$ represents the inverter's output voltage, which is typically created using a Pulse-Width Modulation (PWM) technique and passed through a filter that consists of an inductor with a reactance (X) and a capacitor (C_f) to create a filtered voltage, $\mathbf{V_r}$, as shown in Figure 1-5c.

The direct way to modify the effective line reactance (jX_{eff}) between its two ends is to connect a compensating reactance ($-jX_{se}$) in series with the line as shown in Figure 1-6. The active and reactive power flows (P_r and Q_r) at the receiving end of the line are given by the following equations:

$$P_r = a_{eff} \sin\delta \tag{2-207}$$

and

$$Q_r = a_{eff}\left(\cos\delta - \frac{V_r}{V_s}\right) \tag{2-208}$$

where

$$a_{eff} = \frac{V_s V_r}{X_{eff}}, \tag{2-209}$$

$$X_{eff} = X - X_{se}. \tag{2-210a}$$

Note that $X_{se} > 0$ represents a capacitive compensating reactance and $X_{se} < 0$ represents an inductive compensating reactance, respectively. However, $X_{eff} > 0$ represents an effective inductive reactance and $X_{eff} < 0$ represents an effective capacitive reactance, respectively.

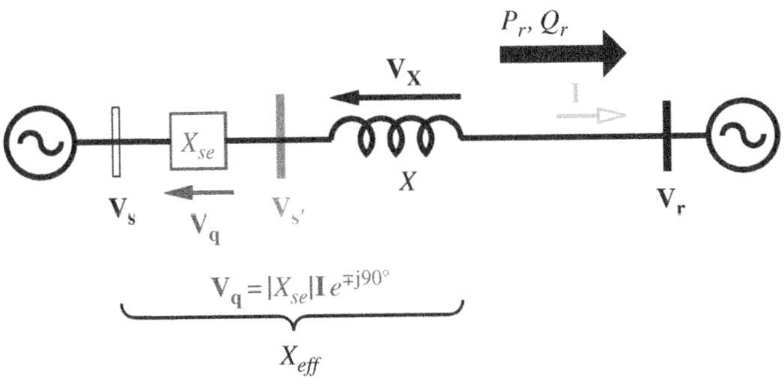

Figure 1-6 Power flow in a lossless line with a series-compensating reactance (X_{se}).

Depending on whether the compensating reactance ($-jX_{se}$) is capacitive or inductive, the voltage ($\mathbf{V_q} = jV_q$) across the compensating reactance lags or leads the prevailing line current (**I**) by 90°. This leads to the concept of an emulated reactance, which is defined as

$$X_{se} = -\frac{V_q}{I} \tag{1-3a}$$

or

$$X_{se} = \frac{V_{s's}}{I}, \tag{1-3b}$$

and is shown in Figure 1-7. In this concept, a compensating voltage $\left(\mathrm{V}_{s's} = -\mathbf{V_q}\right)$ is created, maintaining the quadrature relationship with the prevailing line current, and is connected in series with the line. Note that $(V_{s's} > 0)$ is defined to be a capacitive compensation and $V_{s's} < 0$ is defined to be an inductive compensation, respectively.

The concept of an emulated reactance can be further extended to represent an emulated impedance when the compensating voltage is not restricted to be in quadrature, but at any phase angle with respect to the prevailing line current (**I**). That means the compensating voltage $\left(\mathrm{V}_{s's} = -\mathbf{V_{dq}}\right)$ can be made to look like a virtual, four-quadrant, compensating impedance ($\mathbf{Z_{se}} = R_{se} - jX_{se}$) that consists of a resistance ($R_{se} = +R$ or $-R$) and a reactance ($X_{se} = X_C$ or $-X_L$) in series with the line without any discrete circuit component as shown in Figure 1-8b. This two-parameter, resistive (R_{se}) and reactive (X_{se}), control makes it possible to modify the magnitude and the phase angle of the line

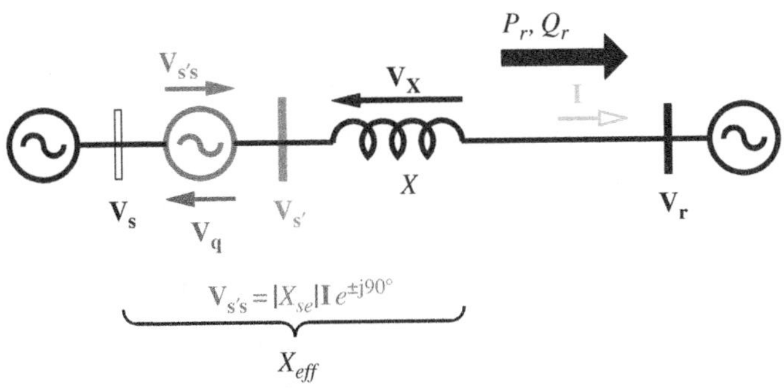

Figure 1-7 Power flow in a lossless line with a series-compensating voltage ($\mathbf{V_{s's}}$).

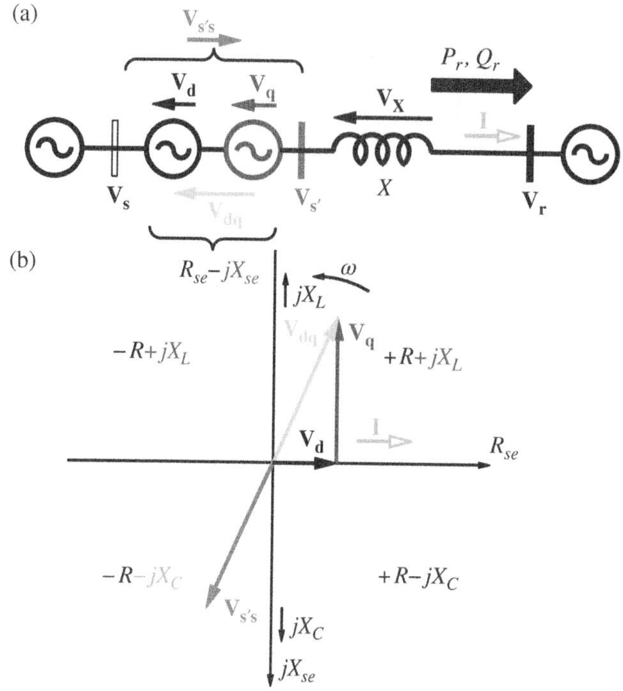

Figure 1-8 (a) Power transmission system with a series-compensating voltage ($\mathbf{V_{s's}}$); (b) four-quadrant emulated impedance.

voltage simultaneously, which results in an independent control of active and reactive power flows in the line.

The series-compensating voltage ($\mathbf{V_{s's}}$) is related to ($\mathbf{V_{dq}}$), such that

$$V_{s's} = -\mathbf{V_{dq}} \tag{1-4}$$

and

$$\mathbf{V_{dq}} = \mathbf{V_d} + \mathbf{V_q} = V_d + jV_q \tag{1-5}$$

where $\mathbf{V_d} = V_d$ and $\mathbf{V_q} = jV_q$ are the respective active or direct and reactive or quadrature components of the compensating voltage with load convention, meaning the line current (**I**) enters at the higher potential terminal of the voltages ($\mathbf{V_d}$ and $\mathbf{V_q}$) as shown in Figure 1-8a.

The natural or uncompensated power flow through a transmission line in a power system network is, in general, not optimal. Any of the power flow control parameters (line voltage magnitude, its phase angle, and line reactance) can be regulated with the use of the following equipment:

- Voltage-Regulating Transformer (VRT), shunt or parallel-connected switched reactor/capacitor, also known as Shunt Reactor (SR)/Shunt Capacitor (SC), Static Var Compensator (SVC), or STATic synchronous COMpensator (STATCOM) for voltage regulation as shown in Figure 1-9
- PAR or Phase-Shifting Transformer (PST) for phase angle regulation as shown in Figure 1-10
- Thyristor-Controlled Series Capacitor (TCSC) or Static Synchronous Series Compensator (SSSC) for series reactance regulation as shown in Figure 1-11.

The dynamic performance of a VRT is limited by the speed of operation of the mechanical Load Tap Changers (LTCs), which respond in seconds; this level of response time is acceptable in most

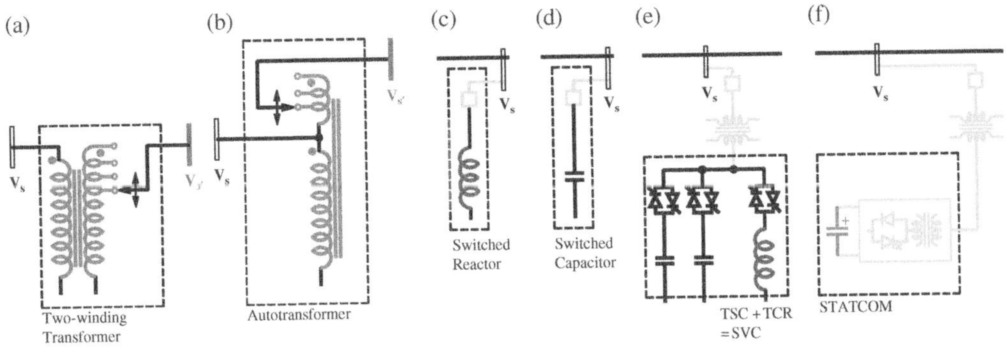

Figure 1-9 Transmission line Voltage Regulators: (a) Two-winding Transformer, (b) Autotransformer, (c) Switched Reactor, (d) Switched Capacitor, (e) TSC + TCR = SVC, and (f) STATCOM.

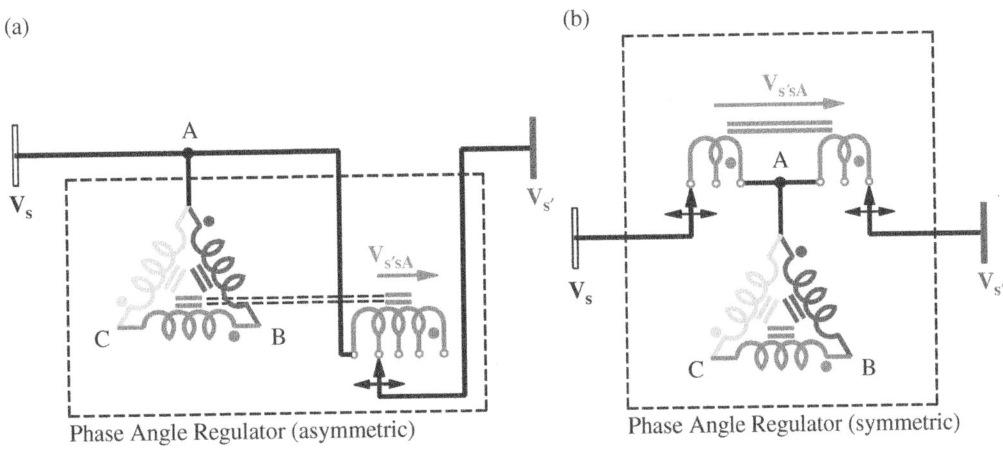

Figure 1-10 Transmission line voltage Phase Angle Regulators: (a) asymmetric and (b) symmetric.

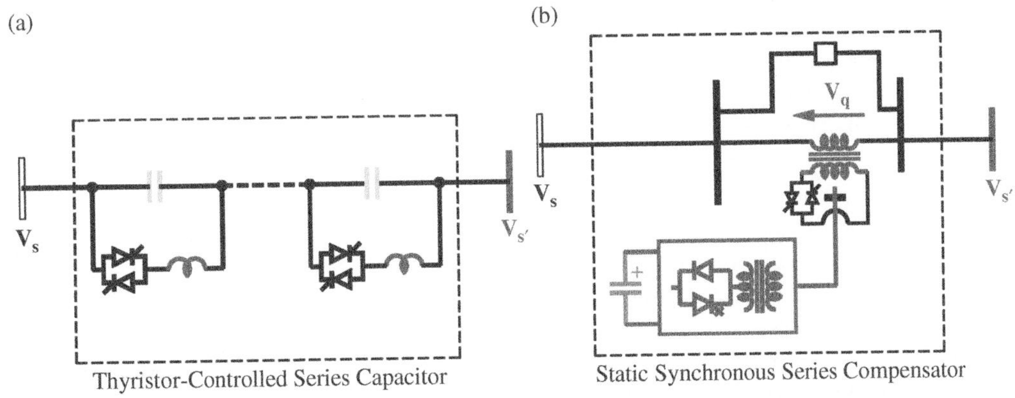

Figure 1-11 Transmission line Reactance Regulators: (a) Thyristor-Controlled Series Capacitor (TCSC) and (b) Static Synchronous Series Compensator (SSSC).

utility applications. However, if a faster response is desired, the mechanical LTCs can be upgraded with power electronics-based LTCs, called Thyristor-Controlled (TC) LTCs as discussed in Chapter 2. More on LTCs can be found in the book, titled "On-Load Tap-Changers For Power Transformers: Operation, Principles, Applications and Selection," by A. Krämer, Maschinenfabrik Reinhausen, 2000. The power electronics-based solutions can be divided into two categories, based on the type of semiconductor switches: naturally commutated switch, such as a thyristor and forced-commutated switch, such as Insulated Gate Bipolar Transistor (IGBT). Each of these solutions is based on engineering trade-offs.

When discussing dynamic and transient events, mechanical LTCs react in $\approx$ 3–5 s (3,000–5,000 ms); TC LTCs react in $<$ 1 s (1,000 ms) and IGBT-based converters react in $<$ 0.010 s (10 ms). These different technologies are referred to as slow, medium speed, and fast, respectively. Note that as the response time of a particular solution increases from slow using mechanical LTCs to medium speed using TC LTCs to fast using IGBTs, there is a corresponding increase in the solution's life-cycle costs (installation, operation, and maintenance), complexity, and impracticability of relocation. Other important features to consider are reliability, efficiency, component non-obsolescence, and interoperability.

For more than a century, the transmission line voltage magnitude has been regulated with transformers and tap changers. They are referred to, in this book, as the VRT in the form of a two-winding transformer with galvanically isolated primary and secondary windings, called a Shunt–Shunt configuration, and an autotransformer with an electrical connection between the primary and secondary windings, called a Shunt–Series configuration. In both types of transformers, the line voltage is applied to the primary windings. In the two-winding transformer, the full line voltage is induced in the secondary windings, whereas, in the autotransformer, only a fraction of the line voltage is induced in the secondary windings that are connected to the primary windings to produce the full line voltage. In both cases, the magnitude of the line voltage is regulated. The secondary voltage is varied with the use of LTCs. An LTC can step up/down the voltage without interruption of the load current. Both primary and secondary windings in the two-winding transformer carry the full transmitted power. Both primary and secondary windings in the autotransformer carry only a fraction of the full transmitted power. Therefore, if the galvanic isolation is not needed, the rating of the transformer can be significantly reduced with a Shunt–Series configuration as compared to a Shunt–Shunt configuration. Regardless of which configuration is used, the voltages at the input (primary) and output (secondary) terminals of both a two-winding transformer and an autotransformer are identical as discussed in Chapter 4, Section 4.1.

The primary reason for voltage regulation is due to the exchange of reactive power at the Point of Connection (POC) to the utility. However, a transformer neither generates nor absorbs reactive power. If a transformer delivers reactive power at one side (primary or secondary), it absorbs the same amount of reactive power on the other side (secondary or primary). Therefore, in the process of increasing voltage on the secondary side, it reduces voltage on the primary side. The opposite is true as well when, in the process of decreasing voltage on the secondary side, it increases voltage on the primary side. Figure 1-12 shows that a compensating voltage of $\pm$15% of the natural primary voltage (V_{sn}) of 0.988 pu results in a secondary voltage ($V_{s'}$) in the range of 0.872 to 1.095 pu. In the process, the primary voltage (V_s) varies in the range of 1.022 to 0.945 pu. Therefore, a desired 15% change in voltage at the secondary terminal may result in a net 10.7% increase due to the reduction of voltage at the primary terminal and a net 11.6% decrease due to the increase of voltage at the primary terminal as discussed in Chapter 4, Section 4.1.

The indirect way to regulate the magnitude of the line voltage is to connect a reactor or a capacitor in shunt with the line. A shunt-connected reactor absorbs reactive power from the line and lowers

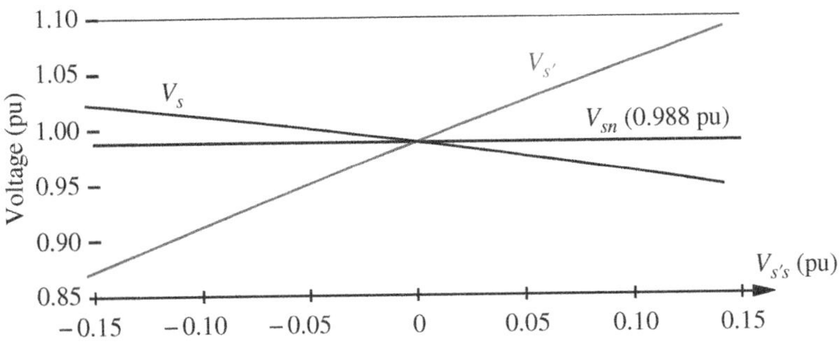

Figure 1-12 Ranges of voltages (V_S and $V_{S'}$) at the primary and secondary sides of a Voltage-Regulating Transformer.

the line voltage, whereas a shunt-connected capacitor raises the line voltage with its generated reactive power as discussed in Chapter 2. With a series-connected switch, such as back-to-back thyristors (triac), whose duty cycle can be varied, the shunt reactor can be made to operate as a variable reactor, which is called a Thyristor-Controlled Reactor (TCR). A Thyristor-Switched Capacitor (TSC) connects fixed capacitors in a step-like manner in shunt with the line through triacs. Therefore, a combination of the variable reactor and a parallel capacitor acts as a variable compensating reactor or capacitor, which is called SVC.

Voltage regulation can also be achieved by the field control of a synchronous motor (Synchronous Condenser or SynCon) that generates or absorbs var as in the cases of a shunt-connected capacitor or a shunt-connected reactor. Voltage regulation can also be achieved when the back emf of the SynCon is replaced with a power electronics-based Voltage-Sourced Converter (VSC), which is called STATCOM as discussed in Chapter 2, Section 2.3.1.2. More discussion on this topic is given in "Introduction to FACTS Controllers: Theory, Modeling, and Applications," by Sen and Sen, IEEE Press and John Wiley & Sons, 2009, Chapter 8, Section 8.1.

The power flow in a transmission line has traditionally been regulated with the use of a PAR. The line voltage is applied to the primary windings and the induced secondary voltage, called a compensating voltage that is varied with the use of LTCs is connected in series with the line. This compensating voltage is in quadrature with the phase-to-neutral voltage and as a result, the phase angle of the line voltage is regulated as discussed in Chapters 2 and 4. The PAR is configured in two forms – PAR asymmetric (asym) and symmetric (sym). In the process of varying the phase angle of the line voltage, a PAR (asym) also increases the magnitude of the line voltage. In a PAR (sym), while the phase angle is varied, the magnitude of the line voltage stays unchanged. When a high power flow enhancement is desired, the application of a PAR (sym) becomes limited, because of the need for a large amount of reactive power flow through the line. This large amount of reactive power flow creates significant additional losses, because of a large line current. Also, a larger-than-necessary rating of the PAR results when a large increase in active power flow is desired as discussed in Chapter 2, Section 2.5.2. Also as discussed in Chapter 2, Section 2.2.2.6, a PAR emulates an impedance in series with the line; however, this emulated impedance is not an independently controlled resistance and reactance; therefore, a PAR cannot control the active and reactive power flows in the line independently, whereas an IR offers an independent control of active and reactive power flows in the line as desired.

If the variable capacitor/reactor is connected in series with the line, the effective line reactance between the two ends of the line is regulated by the additional variable capacitor/reactor, which is called TCSC. The functionality of a TCSC can be realized with a series-compensating voltage as in the case of a SSSC. The SSSC maintains the compensating voltage almost in quadrature with the prevailing line current. A leading voltage emulates a reactor; a lagging voltage emulates a capacitor. A TCSC or SSSC modifies the magnitude and phase angle of the line voltage, which are the combined functionalities of a VR and a PAR as shown in Figure 2-28. Since, the series reactance compensation technique does not change the effective line resistance, it cannot control the active and reactive power flows in the line independently as shown in Figure 1-30.

In a lightly loaded line, the reactive power absorbed by the series reactance of the line may be much less in comparison to the reactive power generated by the line-to-ground, shunt capacitance of the line. The resulting voltage increase in the line may reach or exceed the allowable limits for other loads that are connected to the grid. In a heavily loaded transmission line, the reactive power needed by the series reactance of the line may be much more in comparison to the reactive power generated by the shunt capacitance of the line. The resulting voltage along the line may decrease to a point that is below an acceptable limit when the full performance of the load is not possible. If the voltage along the transmission line is increased to be regulated at its nominal value by using a VR, the active power flow increases over the natural flow as discussed in Chapter 2, Section 2.6.1 (Shunt-Compensating Reactance). If the phase angle between the voltages at the two ends of the transmission line is increased by using a PAR, the active power flow also increases. The unintended consequence of increasing active power flow by voltage regulation or phase angle regulation is that the reactive power flow in the line is also affected. When the line reactance is regulated, both the active and reactive power flows in the transmission line are varied simultaneously.

If the reactive power along the line is reduced, the freed-up capacity of the line can be used to increase the revenue-generating active power flow. As a result, the connected-generators will be required to supply less reactive power. Furthermore, the efficiencies of the generators and step-up transformers under a reduced reactive power condition will also increase. Since the loss in the line decreases due to less reactive power flow, the grid becomes more efficient. Therefore, it is desirable to compensate the lines to operate under independent, not simultaneous, control of the active and reactive power flows so that the line can facilitate the delivery of active power with the greatest value. The active and reactive power flows in a line can be regulated independently with a UPFC or an ST that controls the effective impedance of the line between its two ends, which is functionally equivalent to regulating both the magnitude and phase angle of the line voltage simultaneously.

1.3 Modern Power Flow Control Concepts

In the early 1990s, there was a renewed interest to experiment with novel power electronics VSCs-based PFCs due to the availability of high-power, forced-commutated, semiconductor switches, such as 4500 V, 4000 A-rated gate-turn-off (GTO) thyristors. A new definition, namely Flexible Alternating Current Transmission Systems (FACTS), was proposed as "alternating current transmission systems incorporating power electronic-based and other static controllers to enhance controllability and increased power transfer capability." The purpose of the FACTS solution was to completely overhaul and improve the power delivery techniques that were developed, since the introduction of free flow of electricity in the late 19th century. Under this program, Westinghouse installed a ±100-MVA-rated STATCON (STATic CONdenser as discussed in the literature) at the Tennessee Valley Authority's (TVA) Sullivan substation in Tennessee, USA. This new equipment was fundamentally different from the conventional thyristor-controlled SVC. This equipment was

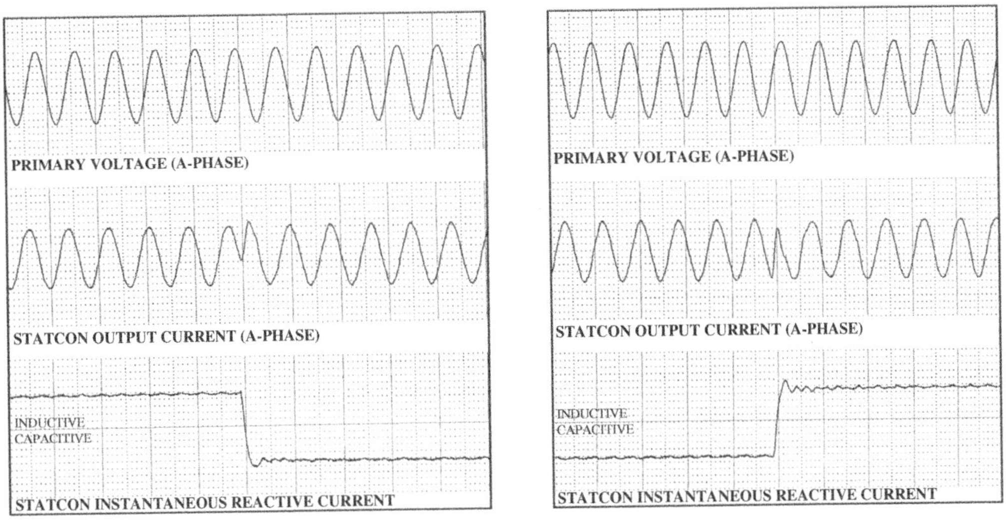

Figure 1-13 Response time of the first commercial STATCON for 100 Mvar capacitive step (left) and 100 Mvar inductive step (right) (field performance) (Westinghouse).

called STATCON because its steady-state output characteristics are similar to those of the rotating synchronous condenser. This was the world's first commercial installation of a STATCON, which was later renamed as STATCOM. This project was based on a previously demonstrated project that was documented in a report, titled *1 MVAR Advanced Static VAR Generator Development Program*, ESEERCO Report, EP 84-30, April 1987. The TVA-STATCOM demonstrated (i) realization of a ±100 MVA, 161 kV-rated VSC, based on a harmonic-neutralization technique that did not require the use of any filter at the output of the VSC, (ii) viability of GTO-based VSCs at high power for transmission applications and (iii) fast control response of a VSC-based compensator as shown in Figure 1-13. This STATCOM was built using MSDOS (Microsoft Disk Operating System), which became obsolete before its installation in 1995 due to the arrival of MS Windows operating system in 1993. The STATCOM was retired from service due to component obsolescence. This demonstrates the consequential risks associated with the new power electronics technologies in regard to the life expectancy under utility conventional standards. Rapid advances in semiconductor technology and computer operating systems do not align with the utility business model and quickly outpace the utility adoption standards. This is a prime example of technology obsolescence that should be considered when making a technical evaluation.

The capability of providing a 100-Mvar step change in 2 ms by a VSC-based STATCOM, using forced-commutated GTOs, was a major improvement in response time than what was obtained in a naturally commutated, thyristor-based SVC. The SVC uses thyristors, which have a natural transport delay of half of a power cycle, meaning when a thyristor turns on in a positive half cycle, it turns off naturally in the next negative half cycle following the zero-crossing of the voltage. Figure 1-14 shows a two-month voltage profile without an SVC (left) and with an SVC (right). This is a significant improvement over no compensation. The experience of the last five decades has shown that the response time of several cycles using an SVC is quite adequate in most utility applications. This is primarily why the higher cost of the VSC technology with component obsolescence issues have prevented its widespread use/adoption in utility applications.

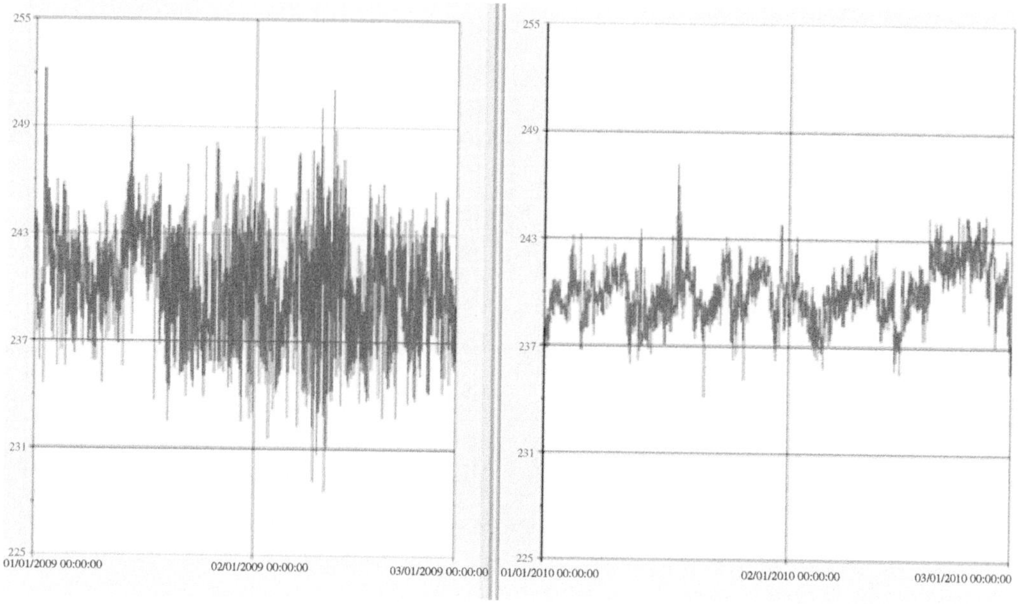

Figure 1-14 Voltage profile without an SVC (left) and with an SVC (right) (field performance) (Barot et al. 2014).

However, the fast response from a VSC may be just the right solution to address various issues in addition to var compensation, such as

- Unbalanced voltage
- Harmonic voltage and current
- Voltage spikes
- Voltage flicker.

One such application is shown in Figure 1-15 where a voltage source ($\mathbf{V_{source}}$) supplies power to an electric arc furnace load through a network whose impedance is represented as Thèvenin impedance ($\mathbf{Z_{Th}}$). The bus voltage ($\mathbf{V_{plant}}$) at the input of the plant is stepped down twice through, first, the Main Transformer and, then the Arc Furnace Transformer.

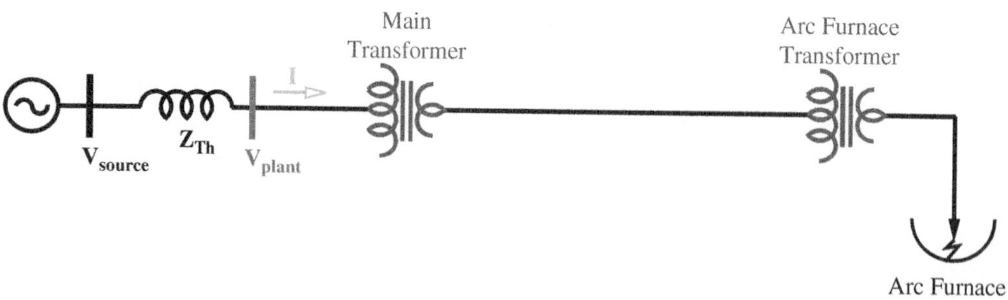

Figure 1-15 A single-line diagram of an electric arc furnace.

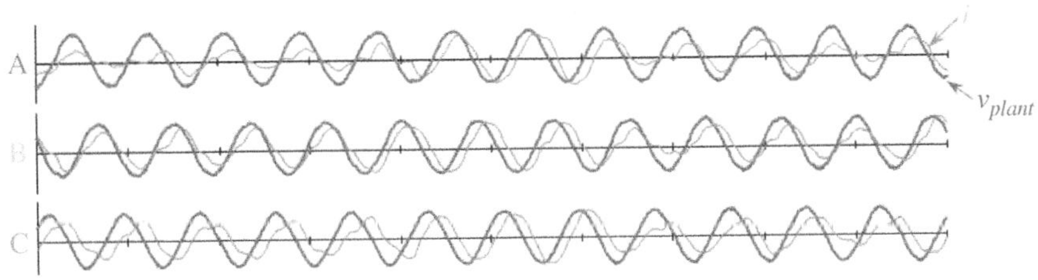

Figure 1-16 Instantaneous plant input bus voltage (v_{plant}) and plant current (i), drawn by a typical electric arc furnace (field performance) (Sen 2015).

A typical instantaneous plant input bus voltage (v_{plant}) and current (i), drawn by an electric arc furnace in three phases (A, B, and C) without any compensation, are shown in Figure 1-16. During the operation of the arc furnace, the random nature of the load current (*i.e.* electric arc) creates proportionally rapid and random voltage changes across $\mathbf{Z_{Th}}$. When this random voltage across $\mathbf{Z_{Th}}$ is subtracted from the regulated source voltage ($\mathbf{V_{source}}$), a varying voltage, known as voltage flicker, is created at the input voltage bus of the plant. However, the fast-acting STATCOM is capable of providing a unique solution to reduce this voltage flicker by supplying the fluctuating active power (Figure 1-19) and reactive power (Figure 1-18) needs of the rapidly-changing load of the electric arc furnace. Since a typical load is more inductive than capacitive in nature, the compensation scheme is designed to be more capacitive than inductive in nature. Since STATCOM operates both in capacitive and inductive modes in near-equal ranges, a STATCOM, combined with an optional Fixed Capacitor (FC) may be an appropriate solution in this application. In 1998, Westinghouse installed a +140/−20-MVA-rated power electronics VSC-based shunt compensator, consisting of a ±80-MVA-rated STATCOM and a +60-MVA-rated FC at a steel plant as shown in Figure 1-17.

The voltage ($\mathbf{V_{comp}}$) at the compensation bus is regulated to maintain the voltage flicker at an acceptable level. The control philosophy in this project is given by the following equations:

$$P_{util} = P_{furn} + P_{comp} \tag{1-6}$$

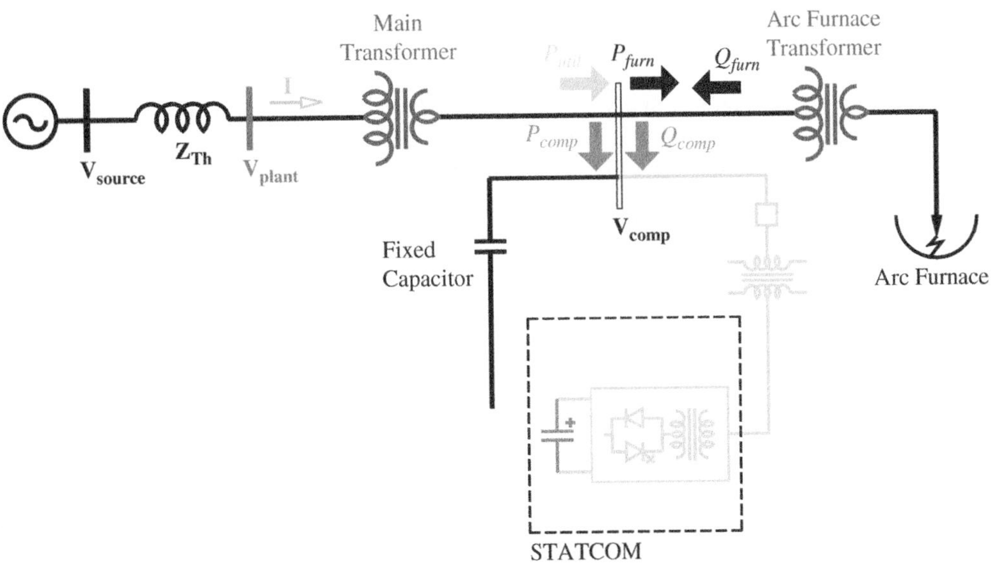

Figure 1-17 Use of a STATCOM, combined with a Fixed Capacitor (FC) as an arc furnace compensator.

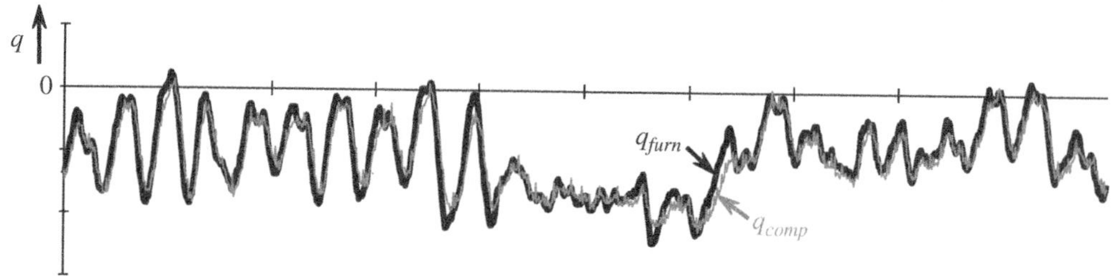

Figure 1-18 Instantaneous reactive power drawn by the furnace and supplied by the compensator (STATCOM and FC) (field performance) (Sen 2015).

where

P_{util} is the active power supplied by the utility
P_{furn} is the active power used by the arc furnace
P_{comp} is the active power exchanged by the compensator

and

$$Q_{furn} = Q_{comp} \tag{1-7}$$

where

Q_{furn} is the reactive power used by the arc furnace
Q_{comp} is the reactive power exchanged by the compensator.

As shown in Figure 1-18, the compensation scheme supplies nearly all the instantaneous reactive power (q) drawn by the furnace.

This preferred solution for a utility application is to use a larger-than-usual-sized DC capacitor, which provides a sink and a source of fluctuating active power for the load, thereby leveling off the active power drawn from the utility source. The instantaneous active power (p) drawn by the compensator, which is composed of a STATCOM and a FC, and the furnace and supplied by the utility are shown in Figure 1-19. The negative active power drawn by the compensator indicates that a part of the furnace power is supplied by the compensator. If the size of the DC capacitor was further increased to support a larger amount of fluctuating active power, the active power drawn from the utility could be made equivalent to a fixed resistive load.

A comparison of voltage flicker (termed as Level) without and with a compensation, provided by the STATCOM and the FC, is shown in Figure 1-20. The flicker measurements without and with a

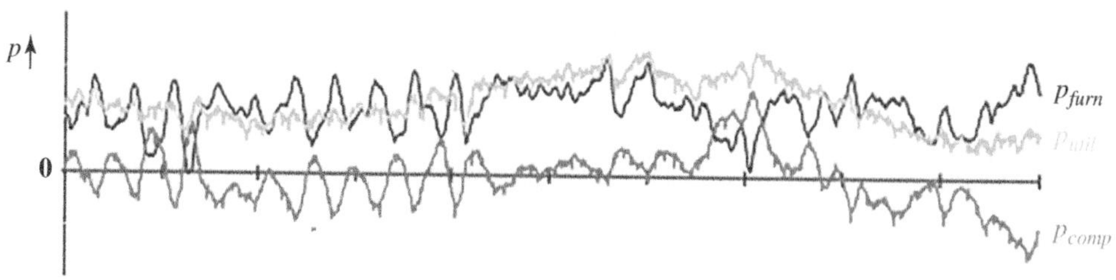

Figure 1-19 Instantaneous active power drawn by a compensator (STATCOM and FC) and a furnace and supplied by a utility (field performance) (Sen 2015).

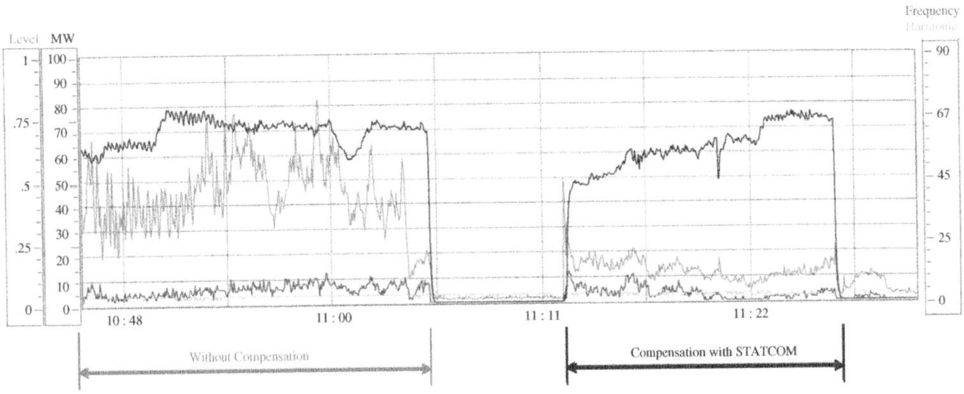

Figure 1-20 Flicker measurements without and with a compensation provided by a STATCOM (field performance) (Sen 2015).

compensation show that the STATCOM was designed correctly to reduce the flicker to a desired level of below 0.25% at 75 MW of load. Therefore, it was not commercially needed to increase the size of the DC capacitor any further. Even though this is the most useful application of a VSC-based STATCOM, it was also retired from service due to component obsolescence within a few years after commissioning and a less perfect, but adequate solution – naturally commutated, thyristor-based SVC was installed in its place.

The first generation of FACTS Controllers consisted of GTO-based VSCs. Since the commissioning of the world's first commercial STATCON at TVA in 1995, nine GTO-based VSCs were built. During the same time, several IGBT-based VSCs were also built. During the past two decades, the IGBT technology has advanced by several generations. So, what about the GTO technology? Unfortunately, it is no longer manufactured. While the IGBT is the industry workhorse at the present time, the future trend is to develop wide bandgap-type Silicon Carbide (SiC)- and Gallium Nitride (GaN)-based switches due to the several attractive attributes: higher-temperature operation, smaller turn-on and turn-off times that result in a lower switching loss, smaller gate-drive circuit, smaller snubber circuit, lower cooling requirement, and so on. So how can one source spare parts for the GTO-based FACTS Controllers? That is not an easy question to answer. Most of the installations with GTO-based VSCs were dismantled after a short life span, which was unforeseen at the inception of FACTS development. Based on the facts about FACTS Controllers of the last three decades, component obsolescence is the primary reason for a 10–15 years of life span of power electronics-based VSCs.

In this context, one might ask why not use the latest high-power electronics switches to replace the aging outdated switches? The answer is simple – it is impossible. Even though there is no available switch, which can be considered as a perfect switch, meaning zero forward voltage across the switch during conduction and zero transition time from on-to-off and vice versa, the snubber circuit, gate-drive circuit, and cooling requirement vary from a VSC that is made with one type of switch to another VSC that is made with a different type of switch. This fact alone forces the aging VSCs to retire when spare parts are unavailable.

Moreover, the control cabinet becomes completely outdated in a decade or so, requiring an upgrade. There is no initiative to keep any commercially used, legacy power electronics system alive after it passes its natural longevity. These facts need to be taken into account to calculate the true cost of power electronics VSC-based FACTS Controllers. In fact, the same is true for any power electronics-based system, including an IBR.

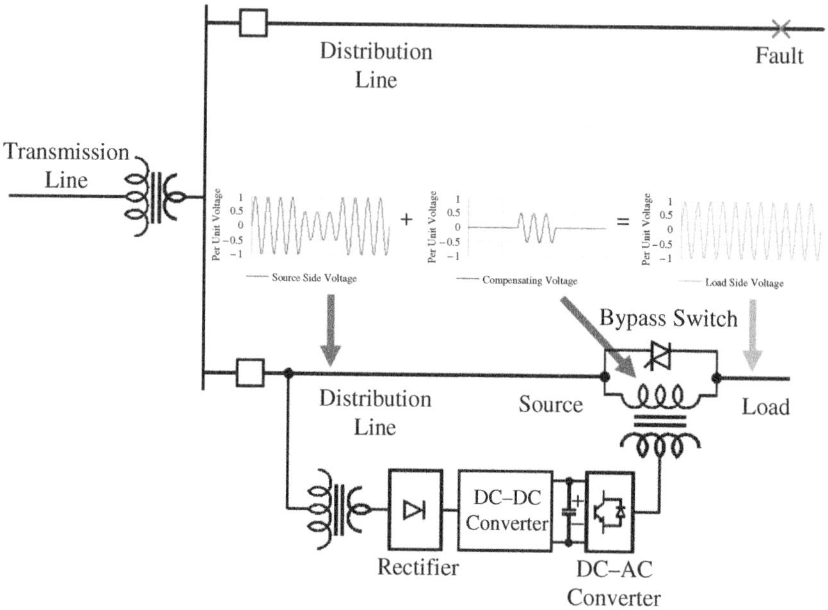

Figure 1-21 Single-line diagram of a Dynamic Voltage Restorer (DVR).

Even with the abovementioned challenges about the power electronics VSC-based solutions, there are applications where inverters are just the right solutions. One such application is the Dynamic Voltage Restorer (DVR). The purpose of a DVR, as its name suggests, is to restore the voltage of a critical load if a sag or swell occurs on any phase of the critical load, such as plants where good voltage regulation is essential to avoid a shutdown of production. DVR is a flexible voltage compensator that is needed for sensitive loads that requires a good voltage quality. IEEE 1159 defines a voltage sag as "the decrease in the rms voltage level to 10% – 90% (1–90% for EN 50160 standard) of nominal, at the power frequency for durations of ½ cycle to one (1) minute." Note that the equivalent terminology from the IEC is "dip."

Figure 1-21 shows a single-line diagram of a DVR that feeds a critical load, which is located on a power distribution line that is one of several lines supplied by a transmission line through a stepdown transformer. The heart of a DVR is a DC–AC converter (also called an inverter or VSC), which is normally bypassed with a thyristor-based bypass-switch. The DC capacitor is trickle-charged through a stepdown transformer, a rectifier, and a DC–DC converter.

During a voltage sag or swell due to a short-circuit fault, lightning or capacitor bank switching on an adjacent distribution line, the bypass-switch opens and a voltage that is generated by the VSC is placed in series with the load through a coupling transformer to restore the load's nominal voltage while the conventional system protection equipment clears the fault.

Since a DVR is installed upstream of a sensitive load, it senses the sagged or swelled voltage on its input terminals. The DVR control system is such that it continually compares the input voltage waveform with an internal reference voltage signal and determines the proper voltage that must be added or subtracted to restore the nominal voltage at the load terminals. Figure 1-22 shows a sag correction by a DVR during a field test. The sag occurred in one phase, which was corrected within a few ms.

A DVR is generally designed to store a fixed amount of energy that can be used for voltage compensation during a designated amount of time, for example, 30 fundamental frequency cycles

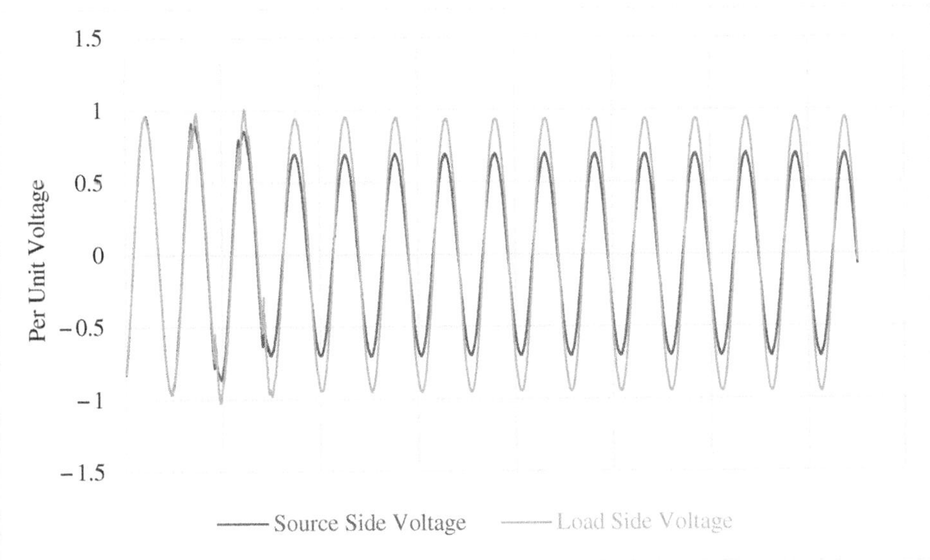

Figure 1-22 Sag correction by a DVR (field performance) (Sen 2015).

during a voltage sag or swell. This designated amount of time may be adequate to protect a sensitive load, such as a semiconductor manufacturing plant from a voltage sag or swell. However, in some applications where the input voltage to a critical load requires a continuous regulation within a specified range, the guaranteed voltage support for 30 fundamental cycles may be a limitation. This deficiency can be mitigated by a more general controller, which was designed in the 1980s and called an Active Power Line Conditioner (APLC). This electronic circuit topology was patented by Westinghouse in 1987 (U.S. patent number 4,651,265, titled "Active Power Conditioner System"). The APLC introduced the concept of a common DC link between two forced-commutated switch-based VSCs, which are connected back-to-back at their common DC link. One VSC is connected in shunt and another is connected in series with the line that supplies a load. These VSCs exchange active power (P_{link}) between the shunt- and series-connected VSCs for continuous regulation of line voltage in distribution-level applications as shown in Figure 1-23a.

The APLC extends the concept of an autotransformer, which is also a Shunt–Series configuration, meaning one unit is connected in shunt and the other unit is connected in series with the line. The APLC configuration, shown in Figure 1-23a, is identical to a stepdown autotransformer that supplies power to a load on the low-voltage side. The major difference between an APLC and an autotransformer is that the Shunt and the Series Units in an autotransformer exchange active power as well as reactive power. However, in an APLC, only active power is exchanged between the Shunt and the Series Units, since reactive power cannot flow through the common DC link capacitor. The same Shunt–Series VSCs concept was used later in the 1990s in the design of the Unified Power Flow Controller (UPFC) for regulation of line power in transmission-level applications as shown in Figure 1-23b.

The lessons learned from the installations of the first-generation FACTS controllers, such as ±160 MVA-rated UPFC at American Electric Power (AEP), ±100 MVA-rated CSC at New York Power Authority (NYPA), and ±80 MVA-rated UPFC at Korea Electric Power Corporation (KEPCO), are that FACTS controllers have limited applications due to their high life-time costs, which include installation, operation, and maintenance (specialized equipment and trained labor). The main

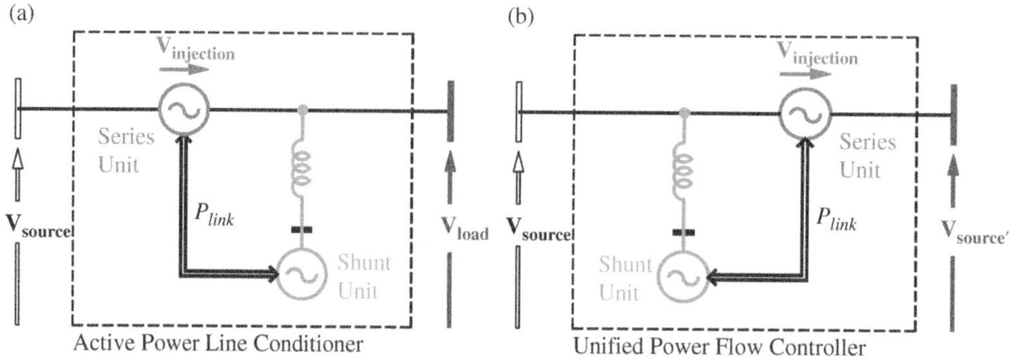

Figure 1-23 (a) Basic circuits for Active Power Line Conditioner and (b) Unified Power Flow Controller.

feature of very fast (millisecond-range) response time, offered by the power electronics inverter-based FACTS controllers is not needed in most utility applications. In search for the right PFC at an affordable price, the Shunt–Series configuration is used to create the Sen Transformer, which can provide a solution to meet the majority of power flow control needs for the utilities worldwide.

1.4 Cost of a Solution

The costs presented in this section are Rough Order Magnitude (ROM) that is based on authors' past three decades of experience on various technologies. This section provides an illustration/methodology for economic appraisal of different technologies that offer similar outcomes.

1.4.1 Defining a Cost-Effective Solution

The voltage/power flow compensation in the transmission/distribution network results in a higher asset utilization. The types of solutions may vary from using transformer and LTCs to power electronics inverters. Each of these solutions is based on engineering trade-offs. In particular, as the response times of various solutions increase from slow (3–5 s) to medium speed (<1 s) to fast (<0.010 s), there is a corresponding increase in the solution's life-cycle costs (installation, operation, and maintenance), complexity, and impracticability of relocation. Other important features to consider are reliability, robustness, efficiency, component non-obsolescence, and interoperability.

It is recognized that the superior response capability of a power electronics inverter-based solution may be beneficial in applications where voltage flicker is caused by an electric arc furnace load and dynamic voltage restoration is required for critical loads. The final selection of a solution, however, depends on knowing the functional requirements and analyzing the cost and benefit of each available solution that means the most features at the least total cost. In the case of a simple voltage regulation at a utility bus, an SC may be an adequate solution, whereas for an arc furnace load, the power electronics inverter may be the best solution.

Consider the three cases (Case 1: "do nothing;" Case 2: "do something;" and Case 3: "go above-and-beyond") of solutions for voltage regulation.

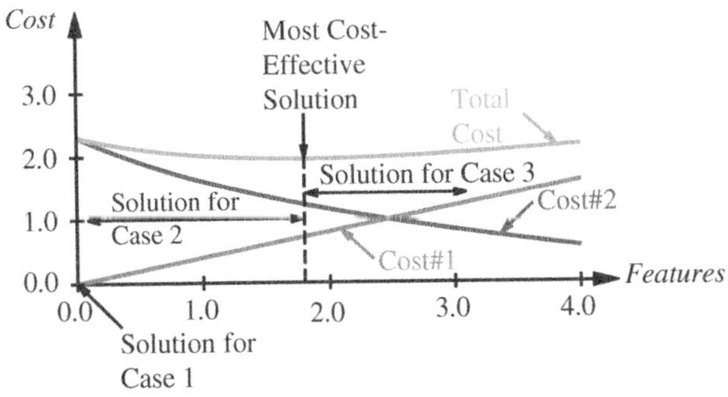

Figure 1-24 Cost versus features in various solutions (Case 1: "do nothing;" Case 2: "do something;" and Case 3: "go above and beyond.")

Case 1 represents a "do nothing" approach where the solution cost (*i.e.* cost #1) is zero; but the lost opportunity cost (*i.e.* cost #2), which is the cost of not providing a solution, such as penalty for causing a voltage flicker may be the highest as shown in Figure 1-24.

Case 2 represents a "do something" approach where the solution cost (cost #1) increases as the number of features in the solution increases. For example, a shunt-connected reactor or capacitor with a breaker may seem to be the simplest solution where the solution cost (cost #1) is greater than zero; however, the lost opportunity cost (cost #2) that accounts for the (1) penalty for not providing var support and (2) penalty for creating voltage flicker may be less than that in Case 1.

Case 3 represents a "go above-and-beyond" approach where the solution cost (cost #1) increases further as the number of features in the solution increases. When various other solutions options, such as SynCon, SVC, STATCOM, and so on with faster response times are considered, the solution cost (cost #1) goes up further; however, the rate of decrease in lost opportunity cost (cost #2) may be less than the rate of increase of the solution cost (cost #1). So, the optimum cost is where the total cost is the minimum. This is the intersection of the regions where Case 2 ends and Case 3 starts.

Better solutions provide primarily better regulation of voltage, thus create less flicker and less penalty. However, in the case of a STATCOM, the voltage is so well regulated that a secondary benefit emerges; the wear-and-tear on the electrodes in an electric arc furnace application becomes more uniform than in any other solution; this results in less frequent replacement of the electrodes, which reduces the overall operating cost of the plant. When all the costs, benefits, and penalties are taken into account, there may be a case where the cost of the added features of a particular solution outweighs the benefits. In between, there lies the cost-effective solution that provides the most features at the least total cost.

Consider the solution cost (C_1) is a combination of a fixed cost (FC_1) and a variable cost (m_1x), which is given by

$$C_1 = m_1x + FC_1 \tag{1-8}$$

where m_1 is the slope 1 and assumed to be 0.4, x is the number of features, and FC_1 is the fixed cost 1 and assumed to be 0.

Consider the opportunity cost (C_2) is a combination of a fixed cost (FC_2) and two variable costs (m_2x and 2^{-wx}), which is given by

$$C_2 = m_2x + FC_2 + 2^{-wx} \tag{1-9}$$

where m_2 is the slope 2 and assumed to be –0.2, x is the number of features, FC_2 is the fixed cost 2 and assumed to be 1.3, and w is the weighting factor and assumed to be 1.

Therefore, the total cost (C) is given by

$$C = C_1 + C_2 \tag{1-10}$$

where C_1 and C_2 are given in Equations (1-8) and (1-9), respectively. All the parameters may be assigned suitable values to represent cost versus features of a solution realistically.

1.4.2 Payback Time

Consider a transmission line with an ATC of 500 MW during the peak load of 200 hours per month. A \$10 M solution utilizes 50% of the ATC and its payback time is 10 months. What is the cost of power delivery in \$/hour/MW?

If an ST or a UPFC is used, what would be the payback time in each case? Assume the costs of a 100 MVA-rated ST and a 160 MVA-rated UPFC are \$10 M and \$80 M, respectively. It is calculated in Chapter 3 that a 30 MVA-rated ST or UPFC enhances the power flow in a particular line about 60 MW.

Solution Cost = \$10 M

Power flow enhancement = 0.5 × ATC = 0.5 × 500 = 250 MW

250 MW × 200 hours/month × Cost of power delivery × 10 months = \$10 M

Cost of power delivery = \$20/MWh

ST Data:

100 MVA rating Cost = \$10 M

Power flow enhancement using an ST of 30 MVA rating = 60 MW

60 MW × 200 hours/month × \$20/MWh × payback time = \$10 M × 30/100

Payback time for an ST = 12.5 months

UPFC Data:

160 MVA rating Cost = \$80 M

Power flow enhancement using a UPFC of 30 MVA rating = 60 MW

60 MW × 200 hours/month × \$20/MWh × payback time = \$80 M × 30/160

Payback time for a UPFC = 62.5 months

For a 60 MW of power flow enhancement, both the ST and the UPFC will generate a yearly revenue of 60 × 200 × 20 × 12 = \$2.88 M.

Note that the base case is related to a solution of voltage regulation where the power flow in the line is enhanced by 250 MW. Since the voltage regulation in the line cannot be increased beyond the statutory limitations, the 100% of the ATC cannot be realized using the voltage regulation solution. However, the full ATC can be achieved by using an impedance regulation method, offered by an ST or a UPFC. If the slow response time in 3–5 s is acceptable, an ST is a preferred solution over a UPFC, since the fast response time in <0.010 s from a power electronics inverter is not required in most utility applications and the payback time of the ST is 20% of the cost of a UPFC.

1.4.3 Economic Analysis

A comparison of the sizes and footprints of the world's first UPFC and an equivalent ST is shown in Figure 1-25. The UPFC applications estimate the project cost to be twice the equipment cost. Conservatively, the assumed project cost of the ST is also estimated to be twice the equipment cost, even

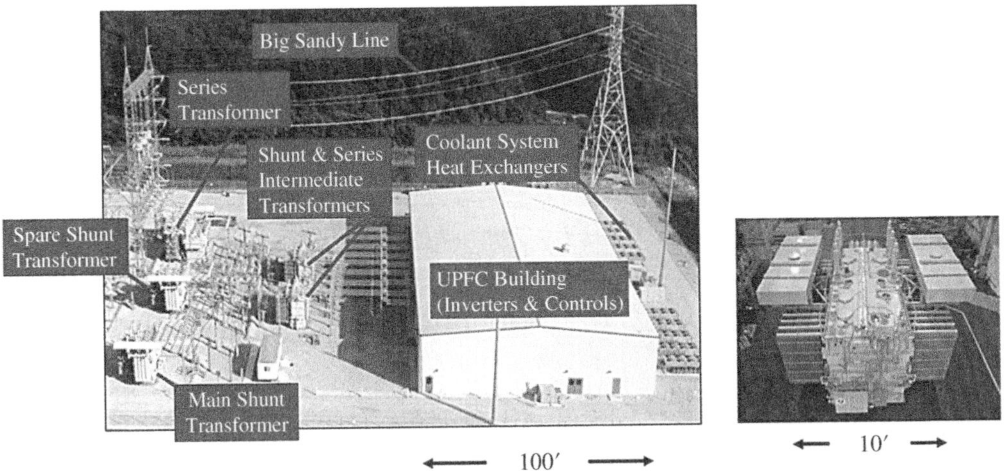

Figure 1-25 World's first UPFC at the AEP Inez substation (left) versus a comparably rated ST (right).

though the footprint of the ST is much smaller than that of the UPFC. The Operation & Maintenance (O&M) costs, $O\&M_{UPFC}$, for UPFC are easily an order of magnitude higher than the (O&M) costs, $O\&M_{ST}$, of the ST. Since there is very little published data on actual annual operating costs of GTO switches and required maintenance, the O&M costs are ignored for this analysis.

The life-cycle cost of a given type of solution (ST or UPFC) is the present value, determined from all cost categories for each design option over a time cycle that is equivalent for all alternatives. Appropriate discount rate for interest is applied. When comparing the ST and the UPFC, one should estimate the life of power electronics inverter-based UPFC as 15 years and that of the transformer/LTCs-based ST to be 45 years, although 50 years is a typical life span of power transformers. With alternatives that have unequal lives, it is customary to compare the alternatives over a period of time equal to the Least Common Multiple (LCM) of their lives using the best estimate for future renewal of shorter life alternative life cycle costs. In the case of the ST versus the UPFC, it is estimated that the UPFC costs for years 16–30 and years 31–45 to be same as the costs from years 1–15.

The equivalent future amount (F) at time, $t = n$, of any present amount (P) at time, $t = 0$, are related by

$$F = P(1 + i)^n = P(F/P, i\%, n) \tag{1-11}$$

where i is the discount rate or interest rate per period and n is the number of compounding periods. The notation, $(F/P, i\%, n)$, is interpreted as "Find F, given P, using an interest of $i\%$ over n years." The factor, $(1+i)^n$, referred to as single payment compound amount, converts P to F.

The uniform series compound amount converts an annual amount (A) to future amount (F) and is related by

$$F = A\frac{(1 + i)^n - 1}{i} = A(F/A, i\%, n). \tag{1-12}$$

Therefore, the uniform series present worth that converts an annual amount (A) to present amount (P) is related by

$$P = A\frac{(1 + i)^n - 1}{i(1 + i)^n} = A(P/A, i\%, n) = A(F/A, i\%, n)/(F/P, i\%, n). \tag{1-13}$$

Using Equation (1-13), (P/A, 10%, 45) = 9.8628 and (P/A, 10%, 15) = 7.6061. Therefore, (A/P, 10%, 45) = 1/9.8628 = 0.10139 and (A/P, 10%, 15) = 1/7.6061 = 0.13147.

Consider the baseline designs of an ST and a UPFC of 100 MVA rating, each of which costs \$10 M and \$50 M, respectively. Therefore, the first cost of the project for the ST is \$20 M and for the UPFC is 100 M.

The Annual Cost of UPFC for 15 years with a first-year installed cost of \$100 M

$$= \$100\text{ M} \times (A/P, 10\%, 15) + \$\text{X for Annual O\&M Costs of UPFC}$$

$$= \$100\text{ M} \times 0.13147 + \$\text{X for Annual O\&M Costs of UPFC}$$

$$= \$13.147\text{ M} + \$\text{X for Annual O\&M Costs of UPFC}$$

= the Annual Cost of UPFC for 45 years, since a second UPFC will be operational during 16–30 years and a third UPFC will be operational during 31–45 years.

The Annual Cost of ST for 45 years with a first-year installed cost of \$20 M

$$= \$20\text{ M} \times (A/P, 10\%, 45) + 0.1 \times \$\text{X for Annual O\&M Costs of ST}$$

$$= \$20\text{ M} \times 0.10139 + 0.1 \times \$\text{X for Annual O\&M Costs of ST}$$

$$= \$2.0278\text{ M} + 0.1 \times \$\text{X for Annual O\&M Costs of ST}$$

Annual Savings per year for using an ST, instead of a UPFC

$$= \$13.147\text{ M for Annual Cost of UPFC} - \$2.0278\text{ M for Annual Cost of ST}$$

$$= \$11.1192\text{ M}$$

Equivalent Present Value of UPFC for 45-year period (ignoring O&M costs)

$$= \$13.147\text{ M for Annual Cost of UPFC} \times (P/A, 10\%, 45)$$

$$= \$13.147\text{ M for Annual Cost of UPFC} \times 9.8628$$

$$= \$129.66\text{ M}$$

Present Value of Savings of ST over UPFC = \$129.66 M − \$20 M = \$109.66 M

The Economic Analysis of ST versus UPFC over a 45-year time period is shown in Table 1-1.

1.5 Independent Active and Reactive PFCs

The independent control of active and reactive power flows requires the simultaneous control of the magnitude and phase angle of the transmission line voltage, which can be achieved by a shunt-compensating voltage, using a Shunt–Shunt Compensator as shown in Figure 1-26. This concept dates back to the time when rectifiers and inverters were introduced to convert AC power from one voltage and frequency level to another with active power (P_{link}) transfer through a DC link. The most frequently-used topology is an AC–DC rectifier followed by a DC–AC inverter for variable speed motor drives and, if combined with a local energy storage, an Uninterruptible Power Supply (UPS). To improve the power quality at the rectifier's AC terminal and to accomplish a bidirectional power flow, two DC–AC inverters are connected back-to-back via their DC links as shown in Figure 1-26. This configuration in electric utility applications is known as Back-To-Back STATic synchronous COMpensator (BTB-STATCOM).

Table 1-1 Economic Analysis of ST versus UPFC over a 45-year time period.

Compensators	UPFC	ST
Base Equipment size	100 MVA	100 MVA
Life	15 years	45 years
Equipment First Cost	\$50 M	\$10 M
Project First Cost	\$100 M	\$20 M
Annual Operation & Maintenance (O&M) Costs	\$X	\$0.1 X
Discount Rate	10%	10%
Equivalent Present project Cost over 45-year life	\$129.66 M	\$20 M
Equivalent Present Project Cost over 45-year life ratio	6.483	1
Equivalent Annual Cost over 45-year life	\$13.147 M	\$2.0278 M
Equivalent Annual Cost over 45-year life ratio	6.483	1
Annual Savings of ST over UPFC		\$11.1192M
Present Value of Savings of ST over UPFC		\$109.66 M

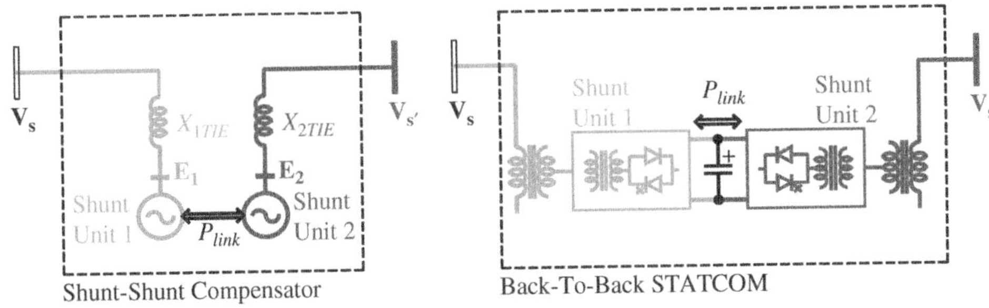

Figure 1-26 Point-to-point transfer of power with local reactive power compensation using a Shunt–Shunt Compensator-based BTB-STATCOM.

Assuming that the sending-end voltage ($\mathbf{V_s} = V_s \angle \delta_s$), receiving-end voltage ($\mathbf{V_r} = V_r \angle \delta_r$), and the line reactance (X) remain unchanged, the compensated active and reactive power flows (P_r and Q_r) at the receiving end are

$$P_r = a' \sin \delta' \tag{2-74}$$

and

$$Q_r = a' \left(\cos \delta' - \frac{V_r}{V_{s'}} \right) \tag{2-75}$$

where the magnitude of the voltage ($\mathbf{V_{s'}}$) at the modified sending end is $V_{s'}$ and the corresponding phase angle is $\delta_{s'}$,

$$a' = \frac{V_{s'} V_r}{X} \tag{2-61}$$

and the modified power angle is given by

$$\delta' = \delta_{s'} - \delta_r. \tag{2-62}$$

Combining Equations (2-74) and (2-75) and using Equation (2-61), it can be found that for given active and reactive power flows (P_r and Q_r) at the receiving end, a PFC with a Shunt–Shunt configuration applies a voltage ($\mathbf{V}_{s'} = V_{s'} \angle \delta'$) at the modified sending end, such that

$$V_{s'} = \frac{X}{V_r}\sqrt{P_r^2 + \left(Q_r + \frac{V_r^2}{X}\right)^2} \tag{2-81}$$

and

$$\delta' = \tan^{-1}\frac{P_r}{Q_r + \frac{V_r^2}{X}}. \tag{2-82a}$$

Therefore, a simultaneous regulation of the magnitude ($V_{s'}$) and the modified power angle (δ') of the line voltage is needed for an independent control of active and reactive power flows in the line.

The transfer of power flow from one line to another can be achieved with the use of a BTB-STATCOM that consists of at least two VSCs, each of which is connected in shunt (parallel) with the transmission line through a coupling transformer. All the VSCs are connected at their common DC link. The shunt-compensating voltage is of variable magnitude and phase angle, and it is also at any phase angle with the prevailing line current. Accordingly, it exchanges active and reactive powers with the line. The exchanged active power by one VSC flows bidirectionally through the common DC link to another VSC. Each shunt-connected VSC can also provide an independent shunt reactive power compensation at its AC terminal and, thereby, regulates the voltage at the POC. Each shunt-connected VSC is rated for the full line voltage and carries the full line current and, therefore, is rated for its full transmitted power. As an example, in order to increase the power flow in a line from its natural flow of 800 MVA to 1000 MVA, the rating of each of the two units of the Shunt–Shunt Compensator would be 1000 MVA. In some special cases for point-to-point transfer of power between two isolated networks with different voltages, phase angles, or frequencies, the use of the Shunt–Shunt Compensator may be the preferred solution.

The Shunt–Shunt Compensator may also be implemented, using a common electromechanical or magnetic link in which case, the compensating voltage is generated from either an electrical machine or a transformer with LTCs. In the case of a magnetic link, both active and reactive powers flow through the link. The point-to-point transfer of power from one line to another with different voltages, phase angles, or frequencies can be accomplished with the use of shunt–shunt-connected electrical machines. The Sen Transformer can also generate a shunt-compensating voltage for the interconnection of two nearby grids with different voltages and phase angles, but the frequencies of the two grids must be the same.

The Shunt–Shunt Compensator is capable of controlling the modified power angle (the phase angle between the modified sending-end voltage and the receiving-end voltage of the transmission line) in the range of $0° \le \delta' \le 360°$. The range of maximum transfer of power along a lossless transmission line (with a quality factor $Q = X/R = \infty$) between the modified sending and receiving ends takes place at $\delta' = 90°$ as shown in Figure 1-27. At the same time, the actual δ' is significantly lower and depends on the line length, system characteristics, and load flows. A transmission line with the natural (uncompensated) power angle (δ) in the range of 15 to 20° may have a possible range of compensation of an additional 5 to 10° as discussed in Chapter 2, Section 2.2.2. As an example, using a series-compensating voltage of 0.2 pu, a compensation range (ψ_{max}) of ±11.54° with an

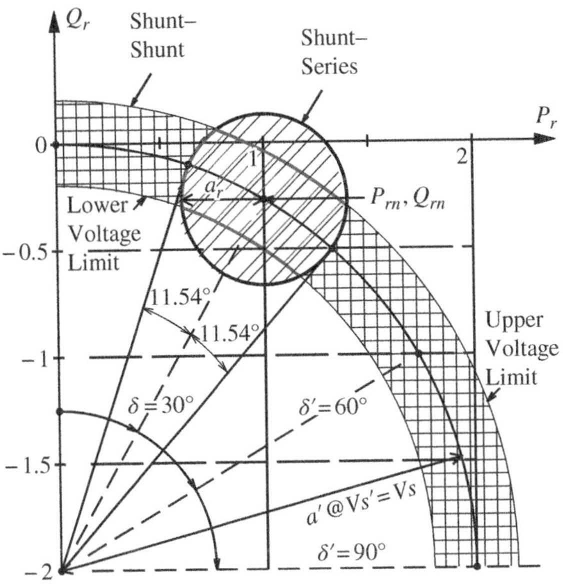

Figure 1-27 Ranges of Q_r versus P_r at the receiving end of the transmission line for the range of modified power angle ($\delta' = \delta_{s'} - \delta_r$) from 0° to 90° when $V_s = V_r = 1$ pu, $X = 0.5$ pu, and $R = 0$ ($X/R = \infty$) using Shunt–Shunt and Shunt–Series configurations, respectively.

uncompensated power angle (δ) of 30° is shown in the shaded area in Figure 1-27. Therefore, the Shunt–Shunt Compensator is severely restricted to operate within the first quadrant when used as a PFC. Then the question arises: "Is it possible to design a new PFC that can control the same amount of transmitted power that is controlled by the Shunt–Shunt Compensator and operate within the allowable range of power angles, while the power rating of this new PFC is a fraction of the conventional Shunt–Shunt Compensator?" For the same example of increasing the power flow in the line from its natural flow of 800 MVA to 1000 MVA, the rating of each of the two units of the Shunt–Series Compensator might be only 200 MVA, resulting in a saving of 80% of the power rating, compared to a Shunt–Shunt Compensator.

In an alternate approach to the Shunt–Shunt compensation, the same active and reactive power flows (P_r and Q_r) can be obtained using a PFC with a Shunt–Series configuration that applies a series-compensating voltage ($\mathbf{V_{s's}} = V_{s's} \angle (\delta_s + \beta)$), such that

$$\mathbf{V_{s's}} = \mathbf{V_{s'}} - \mathbf{V_s} \quad \text{or} \quad V_{s's} \angle (\delta_s + \beta) = V_{s'} \angle (\delta_s + \psi) - V_s \angle \delta_s \tag{2-94b}$$

or when $\delta_s = 0°$,

$$V_{s's} \angle \beta = V_{s'} \angle \psi - V_s \angle 0°$$

where the phase-shift angle,

$$\psi = \delta' - \delta \tag{2-51b}$$

or

$$\psi = (\delta_{s'} - \delta_r) - (\delta_s - \delta_r) = \delta_{s'} - \delta_s. \tag{2-51c}$$

The magnitude ($V_{s's}$) and the relative phase angle (β) of the series-compensating voltage ($\mathbf{V_{s's}}$) are

$$V_{s's} = \sqrt{V_{s'}^2 + V_s^2 - 2V_{s'}V_s \cos\psi} \tag{2-100}$$

and

$$\beta = \tan^{-1}\frac{V_{s'}\sin\psi}{V_{s'}\cos\psi - V_s}. \tag{2-101}$$

For a Shunt–Shunt Compensator, the relationship between the active and reactive power flows (P_r and Q_r) at the receiving end is shown by a quarter circle, the radius (a') of which is varied within the upper and lower limits by the modified sending-end voltage ($V_{s'}$) as shown in Figure 1-27. The relationship is given by the following equation:

$$P_r^2 + \left(Q_r + \frac{V_r^2}{X}\right)^2 = a'^2 \tag{2-79}$$

where

$$a' = \frac{V_{s'}V_r}{X}. \tag{2-61}$$

Equation (2-79) represents the locus of a circle, centered at $\left[0, -\frac{V_r^2}{X}\right]$ with a radius of a'.

For a Shunt–Series Compensator, the relationship between active and reactive power flows (P_r and Q_r) at the receiving end is shown by the small circle in Figure 1-27 and is given by the following equation:

$$(P_r - P_{rn})^2 + (Q_r - Q_{rn})^2 = a_r^2 \tag{2-136}$$

where P_{rn} and Q_{rn} are the natural active and reactive power flows at the receiving end of the line and

$$a_r = \frac{V_r V_{s's}}{X}. \tag{2-132}$$

Equation (2-136) represents the locus of a circle, centered at $[P_{rn}, Q_{rn}]$, with a radius of a_r that is given in Equation (2-132).

An ideal PFC controls the values of the power flow control parameters (line voltage magnitude, its phase angle, and line reactance) to regulate the magnitude and the phase angle of the line voltage simultaneously by adding a series-compensating voltage to the original voltage with the use of a Shunt–Series Compensator as shown in Figure 1-28. The compensating voltage is variable in magnitude and phase angle with respect to the line voltage. The series-connected VSC is rated for a fraction of the line voltage, but carries the full line current. The shunt-connected VSC is rated for the full line voltage, but carries only a fraction of the line current. Therefore, each VSC carries only a fraction of the full transmitted power. In an actual installation, both shunt- and series-connected VSCs are designed to be of the same voltage and current ratings, which reduces the inventory of spare parts. Therefore, their different interface voltages with the line are accomplished with selection of proper turns-ratios of the respective coupling transformers. For example, in the world' first UPFC at the AEP Inez substation, the two VSCs were designed identically to be rated at ±160 MVA and 37 kV phase-to-phase voltage. However, the shunt VSC was connected to a 138 kV-line through a coupling transformer and the series VSC was designed to inject 13.33 kV (~16% of the phase voltage) in series with the line through a coupling transformer.

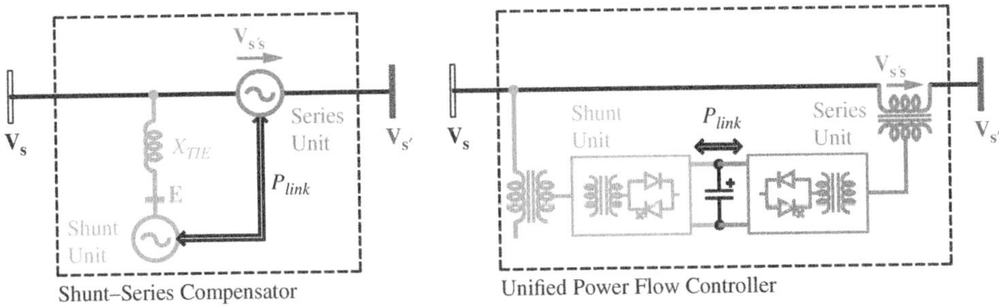

Figure 1-28 Independent active and reactive power flow controller with local reactive power compensation using a Shunt–Series Compensator-based UPFC.

The Shunt–Series Compensator connects a compensating voltage in series with the line at any relative phase angle in the range of $0° \leq \beta \leq 360°$ with respect to the line voltage at the POC. Figure 1-27 shows that a series-compensating voltage of 0.2 pu modifies the power angle by 11.54°, which may be near the allowable limit. The most important and unique feature of the Shunt–Series configuration is that for a given amount of transmission line power, the series-connected VSC has a large leverage between its own rating and the controlled transmission line power. The series-compensating voltage needs to be rated for only a fractional amount of transmitted power, whereas the shunt-connected VSC in the Shunt–Shunt configuration has no such leverage and it needs to be rated for the full amount of transmitted power. Because of this uniqueness, the Shunt–Series configuration is a preferred topology for a PFC. However, in some special cases for point-to-point transfer of power between two isolated networks with POC voltages (V_s and $V_{s'}$) as shown in Figure 1-26 or interconnection of two transmission lines with different voltages or phase angles (or frequencies), Shunt–Shunt configuration may be the preferred solution. One such system, called the North American Electric Reliability Corporation (NERC) Interconnections, consists of Eastern Interconnection, Western Interconnection, and Electric Reliability Council of Texas (ERCOT) Interconnection, which are three separate systems of 60 Hz frequency that are asynchronous to each other. Another such system exists in Japan, connecting a 50 Hz frequency system in the North and the East with a 60 Hz frequency system in the South and the West, that is asynchronous to each other.

The UPFC consists of two VSCs with a common DC link capacitor. The two VSCs are connected to the same transmission line through two coupling transformers: one is connected in shunt and the other is connected in series with the line. The series-compensating voltage is of variable magnitude and phase angle and it is also at any phase angle with the prevailing line current. Therefore, it exchanges active and reactive powers with the line. The exchanged active power (P_{link}) flows bidirectionally through the common DC link to and from the same line under compensation. Both shunt- and series-connected VSCs can also provide independent reactive power compensation at their respective AC terminals. The compensating voltage, being at any phase angle with the prevailing line current, emulates a four-quadrant, series-compensating impedance ($\mathbf{Z_{se}} = R_{se} - jX_{se}$) that consists of a resistance ($R_{se} = +R$ or $-R$) and a reactance ($X_{se} = X_C$ or $-X_L$) in series with the line. Therefore, the series-compensating voltage ($\mathbf{V_{s's}}$) acts as an IR. A positive resistance ($+R$) absorbs active power from the line; a negative resistance ($-R$) delivers active power to the line. The quadrature component of the compensating voltage with respect to the line current emulates a capacitive reactance if the compensating voltage lags the prevailing line current or an inductive reactance if the compensating voltage leads the prevailing line current. A capacitive reactance

(X_C) decreases the effective line reactance between its two ends and, in the process, increases the power flow in the line; an inductive reactance (X_L) increases the effective line reactance between its two ends and decreases the power flow in the line; since the power flow in the line is inversely proportional to the effective line impedance, which is assumed to be inductive.

As a special case, when the DC link capacitors of the two VSCs are not connected together, each of the shunt-connected VSC (STATCOM) and the series-connected VSC (SSSC) provides only a reactive power compensation that is independent of each other. Since there is no exchange of active power between the STATCOM and the SSSC, they act as RRs (X_{sh} or $X_{se} = X_C$ or $-X_L$).

In 1998, Westinghouse installed a ±160 MVA, 138 kV-rated FACTS Controller, namely UPFC, at the AEP's Inez substation in Kentucky, USA. This UPFC demonstrated for the first time that an impedance could be emulated by a series-compensating voltage whose phase is at any angle with respect to the prevailing line current. As a result, the magnitude and phase angle of the line voltage at the modified sending end could be regulated to the desired values by the UPFC, which is an IR; active and reactive power flows in a transmission line could be regulated independently while maintaining a fixed line voltage at the POC. While maintaining a unity power factor load, the active power flow in the line at the modified sending end is varied at different levels, such as 145 MW, 65 MW, 240 MW, and 145 MW, respectively, as shown in Figure 1-29. The active power flow in a line is reduced by making the effective line impedance higher than its natural value and increased by making the effective line impedance lower than its natural value while maintaining the reactive power flow to zero. Note that the utility application allowed the response time to be adjusted in seconds, even though the power electronics-based VSC is capable of providing faster responses in ms.

As a special case, the IR can be reconfigured to operate as a RR by connecting the SSSC only. The reactance emulation technique changes the active and reactive power flows simultaneously, meaning both powers either increase or decrease as shown in Figure 1-30; therefore, the line cannot be optimized for the highest amount of active power flow that generates the most revenue at the lowest amount of reactive power flow by using a RR alone.

It was demonstrated in the TVA-STATCON project (shown in Figure 1-13) that the line voltage can be regulated with a response time in ms; however, the fast response in ms cannot be utilized in power flow control in the AEP UPFC project in order to assure continued operation under various contingencies (*i.e.* all the possible variations in the number of lines connected together as a network at different times of a day, week, month or a year). Nevertheless, the cost of a FACTS controller is

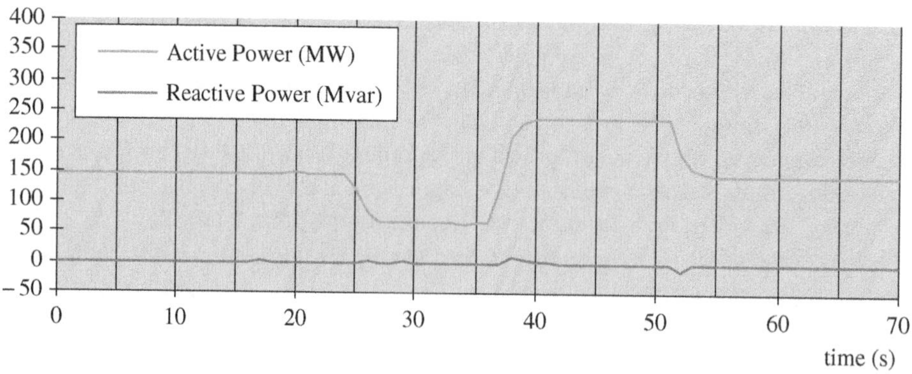

Figure 1-29 Independent power flow control by impedance regulation (field performance) (Sen and Keri 2003).

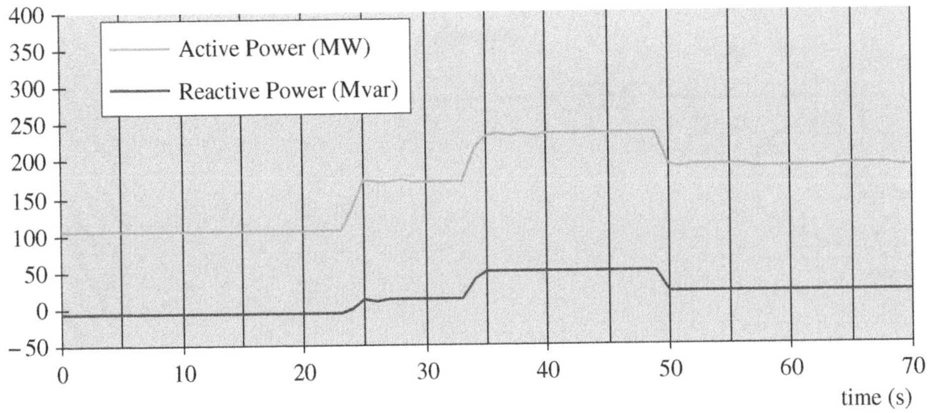

Figure 1-30 Simultaneous power flow control by reactance regulation (field performance) (Sen and Keri 2003).

about the same, whether it is used in slow-response or fast-response applications. This was the motivation to develop the Sen Transformer that meets the functional requirements to provide independent control of active and reactive power flows with responses in seconds and at a fractional amount of the cost of a VSC-based FACTS controller.

In 1998, a patent was granted to General Electric Company, which proposed to implement the independent control of active and reactive power flows such that the compensating voltage was generated using electrical machines (U.S. patent number 5,841,267, titled "Power Flow Control with Rotary Transformers").

The Sens (Kalyan and Mey Ling) proposed the idea of independent control of active and reactive power flows, using an IR, called the Sen Transformer, in a radically low-cost way by using redesigned transformer/LTC technology. The reason is that the transformer/LTC technology has been proven to be efficient, simple, and reliable in utility applications for decades. This implementation of an IR is completely different from the original Westinghouse and the GE concepts. The Sens were awarded five U.S. patents (four patents in 2002, all titled "Versatile Power Flow Transformers for Compensating Power Flow in a Transmission Line" and numbered 6,335,613, 6,384,581, 6,396,248, and 6,420,856, and one patent in 2005, titled "Multiline Power Flow Transformer for Compensating Power Flow Among Transmission Lines," and numbered 6,841,976). The Sen Transformer is fundamentally different from the conventional transformer, in a sense that it modifies both the magnitude and the phase angle of the line voltage while the conventional transformer only modifies the magnitude of the line voltage. Using a Sen Transformer, the active and reactive power flows in the line can be regulated independently as desired.

The Sen Transformer, shown in Figure 1-31, uses a Shunt Unit (Exciter Unit) and a Series Unit (Compensating-Voltage Unit) and creates a series-compensating voltage ($\mathbf{V}_{s's}$) that is variable in magnitude and phase angle to modify the sending-end voltage ($\mathbf{V}_s$) of the line to the modified sending-end voltage ($\mathbf{V}_{s'}$) and, in turn, controls both the magnitude and phase angle simultaneously; as a result, the active and reactive power flows in the line are controlled independently as desired. The compensating voltage of the Series Unit can be made to look like the effect of a positive resistance or a negative resistance and a capacitive reactance or an inductive reactance in each phase. The series-compensating voltage ($\mathbf{V}_{s's}$) is at any phase angle with the prevailing line current ($\mathbf{I}$) through the line reactance (X). Therefore, it exchanges active and reactive powers with the line, which is

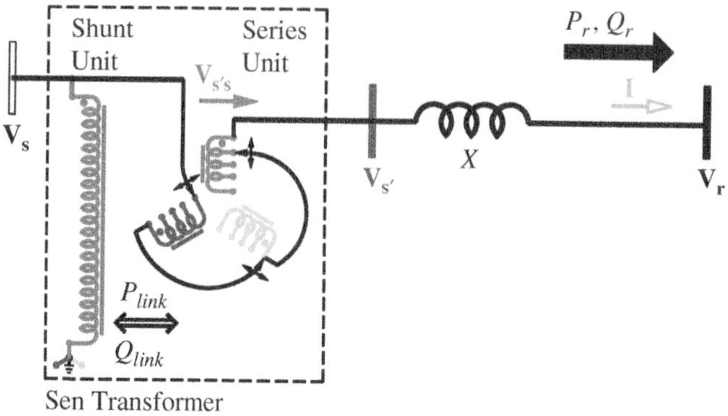

Figure 1-31 Sen Transformer (ST).

equivalent to emulating a four-quadrant, series-compensating impedance ($\mathbf{Z_{se}} = R_{se} - jX_{se}$) that consists of a resistance ($R_{se} = +R$ or $-R$) and a reactance ($X_{se} = X_C$ or $-X_L$) in series with the line. Therefore, the series-compensating voltage ($\mathbf{V_{s's}}$) acts as an IR. The ratio of the compensating voltage ($\mathbf{V_{s's}}$) and the prevailing line current (**I**) is a measure of a virtual four-quadrant emulated impedance. Note that these exchanged powers pass through the magnetic link as (P_{link} and Q_{link}).

The LTCs are preferably mechanical with vacuum or oil-immersed taps. These taps can respond in seconds, which is usually fast enough for utility power flow control needs. If a faster response is needed, the taps can be based on power electronics thyristors, which once turned on in a positive half-cycle of the voltage across it, commutate naturally in the negative half-cycle of the voltage. These taps can respond in a few power cycles, which is a 50-fold decrease in response time. Note that the thyristor technology also faces component obsolescence, albeit with a life cycle of 25–30 years, which is a decade or more longer than the life cycle of the VSC-based FACTS Controllers. The response time can be further reduced to < 0.010 s if a power electronics inverter-based FACTS controller is used. However, this type of fast response is almost never needed in utility applications. Besides, as the response speed of the solution increases from slow (3–5 s) to medium speed (< 1 s) to fast (< 0.010 s), there is a corresponding increase in the solution's life-cycle costs (installation, operation, and maintenance), complexity, and impracticability of relocation and decrease in the reliability significantly.

The VSC-based technology has the capability of providing fast (sub-cycle) dynamic response for a given transmission line impedance, although in a PFC the dynamic response of at least a few cycles of power supply frequency is necessary to operate safely under various contingencies. Most utility applications in the AC system allow regulation of the power flow in the line(s) in a "slow" manner as permitted by the speed of operation of the mechanical LTCs. If faster response is needed, the mechanical LTCs can be replaced with faster TC LTCs. The ST, shown in Figure 1-31, provides simultaneous voltage regulation at the POC and almost the same independent control of active and reactive power flows as the UPFC, albeit at a reduced dynamic rate, which is acceptable in most utility applications.

The STs with both types of LTCs (mechanical and TC) cover a wide range of requirements for power flow control in electric transmission lines. If the LTCs are too coarse for an IR, the number of taps on the winding may be increased. In comparison to a UPFC, which uses power electronics-based VSCs, the ST uses reliable and proven transformer and LTC-based technology that results in an order of magnitude less in operational/maintenance cost and equipment cost.

It is well established that the UPFC is the most versatile PFC that was ever developed. A detailed comparative analysis of the ST and UPFC is given in Chapter 6, Section 6.3 (Comparison Among the VRT, PAR, UPFC, and ST). The life-cycle costs (installation, operation, and maintenance) of the ST are less than the competing FACTS Controller, such as UPFC for the most utility applications due to the following reasons:

- For a one per unit (pu) power through the ST, the installed transformer rating may be as much as two pu, whereas the "all electronic" UPFC requires more than a four pu-rated transformer and more than eight pu of installed power electronics, which translates into a higher installation cost for the UPFC.
- The ST rides through the fault current, but the UPFC requires a protection scheme with an additional electronic bypass-switch, which translates into a higher installation cost for the UPFC.
- The power loss in the ST is less than 1% of its rated power whereas the power loss in the two coupling transformers and two intermediate transformers of a UPFC are 1–2% of the power flowing through the UPFC, which translates into a higher operating cost for the UPFC.
- There is no switching loss in the ST, whereas the switching and conduction losses in the two inverters of the UPFC can be 2–6% of the power flowing through the UPFC, depending on their configuration, which translates into a higher operating cost for the UPFC.
- The ST requires the use of LTCs whose contacts are immersed in the transformer oil. The maintenance expertise for ST is readily available in the industry. However, the power electronics inverter-based UPFC consists of semiconductor switches with appropriate snubber circuits that create power loss. To remove the heat generated from this loss, deionized water cooling and heat exchangers are needed. The failed switches need to be replaced, requiring specialized expertise. Therefore, the operational/maintenance cost of the UPFC is much higher than that of ST.
- The ST uses traditional but redesigned transformer and LTCs technology that has been proven to be efficient, simple, reliable, and robust in utility applications for decades. The UPFC uses thousands of electronic components that are constantly becoming obsolete. Therefore, the cost due to component obsolescence in a UPFC is far greater than that in an ST.
- The footprint of an ST is a fraction of that of a UPFC. Therefore, the ST is relocatable as the system needs change. The power electronics inverter-based UPFC is not practical to be relocated.
- The ST uses off-the-shelf transformer/LTC technology from any manufacturer and therefore, it is interoperable. The ST can be manufactured and serviced anywhere in the world. In contrast, there is no manufacturing standard established for the VSC-based FACTS Controllers. Since each manufacturer establishes its own unique design, the VSC-based FACTS Controllers are not interoperable. Its maintenance depends on the expertise of a specific manufacturer.

Impedance Regulators, such as the UPFC and ST, are capable of injecting a compensating voltage in series with the line in the entire range of 360°. However, in many instances, the capability of connecting a compensating voltage in series with a line within its entire range of 360° is not needed. The active power flow can be increased to the maximum possible level within the first 120° of the 360°-range of the relative phase angle. The active power flow can be decreased to the minimum possible level within the next 120° of 360°-range of the relative phase angle. The cost of the ST can be further reduced with its simpler design per the functional requirements to operate in a "limited angle" configuration, instead of the full 360°-range of operational configuration. There is no such cost advantage in the design of a UPFC. Hence, the ST is adequate and economically attractive to meet most of the utility's present need for independent control of the active and reactive power flows in the transmission lines.

In another unique operation, the ST with an autotransformer is the most cost-effective option that allows interfacing of two transmission systems with different voltage levels and implementing independent power flow control as shown in Figure 6-86.

The ST, in its basic design, uses three primary windings and nine secondary windings with either nine single-phase LTCs or three three-phase LTCs that are in direct contact with the transmission line. Therefore, the LTCs, in the basic design, are required to carry a high line current as well as even a higher fault current. The readily available LTCs may be challenging for use in Extra High Voltage (EHV) and Ultra High Voltage (UHV) applications. In these cases, the applications with greater than 230-kV voltage level require a two-core design where the taps are not exposed to high voltages as shown in Figure 6-72. A comparison of the sizes and footprints of the world's first UPFC and a comparably rated prototypic ST is shown in Figure 1-25.

The compensating voltage in an autotransformer is in-phase (0°) or out-of-phase (180°) with the line voltage and, therefore, regulates the magnitude of the transmission line voltage. The compensating voltage in the PAR is in quadrature (90° or −90°) with the line voltage and, therefore, regulates the phase angle of the transmission line voltage. The ST creates a series-compensating voltage that is variable in magnitude and phase angle and can control the transmission line voltage in both magnitude and phase angle simultaneously in order to achieve independent control of active and reactive power flows in the line. This compensating voltage may be thought of as two separate orthogonal compensating voltages of an autotransformer and a PAR (asym). Therefore, in the ST, the functions of the autotransformer and the PAR (asym) are combined in a single unit that results in a reduced amount of hardware from what is required separately in an autotransformer and in a PAR as shown in Figure 1-32.

Both the ST and UPFC are suitable for independent control of active and reactive power flows in a single transmission line in which they are installed. However, several transmission lines in close proximity may be connected to a common voltage bus. Therefore, any change in the power flow in one line will affect the power flows in the other lines as well. Thus, the excess power from one specific line cannot be transferred directly to another specific line. In a multiline transmission network, it would be advantageous to be able to transfer power from an overloaded to an underloaded line with minimum undesirable impact on the power flows in the other uncompensated lines.

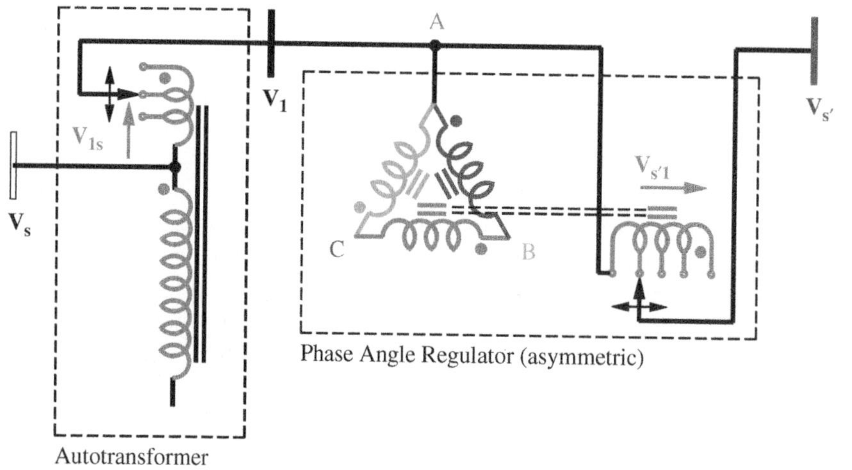

Figure 1-32 Autotransformer/PAR (asym).

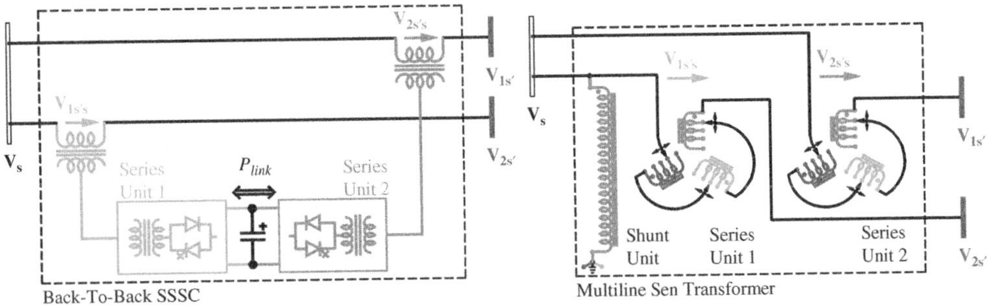

Figure 1-33 Multiline power flow concepts.

The common DC-link concept can be extended for power exchange between transmission lines with series–series-connected VSCs. The BTB-SSSC, also called interline power flow controller (IPFC), shown in Figure 1-33, consists of at least two VSCs; each VSC is connected in series with a transmission line. All the VSCs are connected at their common DC link. The BTB-SSSC transfers active power from one or more transmission lines, referred to as "leader" lines, to the others, referred to as "follower" lines, and provides an independent series reactive power compensation in each line. A BTB-SSSC selectively controls the active and reactive power flows in each line in a multiline transmission system and provides a power flow management for the transmission system by decreasing the power flow in an overloaded line and increasing the power flow in an underloaded line. The Multiline Sen Transformer, shown in Figure 1-33, provides the same functionality.

In summary, mechanically or electronically switched compensators are used as PFCs, but each of these compensators can control only one of the three power flow control parameters: line voltage magnitude, its phase angle, and line reactance. Although the active power flow in the line is regulated, the undesirable reactive power flow is also affected simultaneously, but the optimization of power flow that generates the most revenue can be achieved through independent control of active and reactive power flows in the transmission line. Therefore, the power industry's present need requires the use of PFCs that can independently control both active and reactive power flows in a transmission line, decrease the power flow in an overloaded line, and increase it in an underloaded line, while at the same time keeping the system voltage within the allowable upper and lower limits. The summary of choices for transmission line power flow control equipment is shown in Figure 1-34 in chronological order of their introduction in the industry.

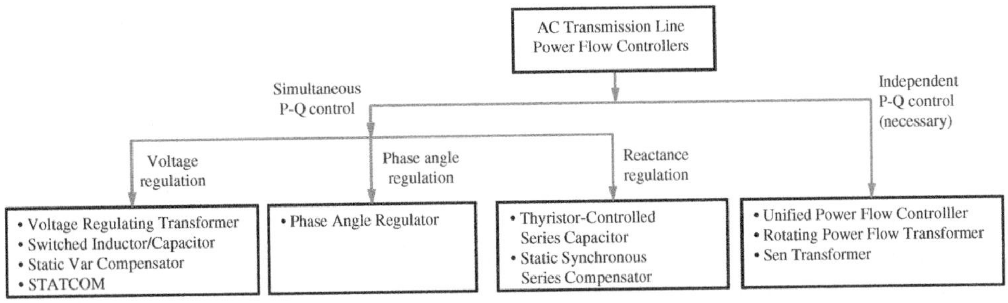

Figure 1-34 Choices for transmission line control equipment.

Table 1-2 Various features of all Shunt–Shunt and Shunt–Series configurations.

Features	VSC	EM	Transformer/LTCs
	Shunt–Shunt configuration		
Independent P-Q flow control	Yes	Yes	Yes
Different frequency system	Yes	Yes	No
Different phase angle system	Yes	Yes	Yes
Intermediate DC transmission	Yes	No	No
	Shunt–Series configuration		
Independent P-Q flow control	Yes	Yes	Yes
One frequency system	Yes	Yes	Yes
Cost	High	Medium	Low

The various features of all Shunt–Shunt and Shunt–Series configurations are summarized in Table 1-2. The VSC-based solutions provide more features than the electrical machine (EM)-based solutions, which provide more features than the transformer/LTC-based solutions. However, most of the power flow control needs are in synchronous networks and the use of the Shunt–Series configuration of the ST that uses transformer/LTC-based solutions is sufficient to meet the utilities' power flow control needs in the most cost-effective way.

It is shown in Chapter 3, Section 3.4 that the active and reactive power flow control area is virtually the same for the ST and UPFC. While the UPFC is a FACTS Controller that uses power electronics-based VSC, the ST, in its preferred form, uses only transformers and LTCs. The reference, titled "Comparison of the 'Sen' Transformer with the Unified Power Flow Controller," *IEEE Trans. on Power Delivery*, vol. 18, no. 4, pp. 1523–1533, Oct. 2003, states that "At the present time, two major drawbacks of all VSC-based FACTS Controllers are their high installation and operating costs." Over the decades, the list of drawbacks has expanded to include component obsolescence, impracticability of relocation, and lack of interoperability. Since the commissioning of the first commercial VSC-based FACTS Controller more than two decades ago, a great deal has been learned about the true needs of a utility for its everyday use and they are

- high reliability, requiring the lowest number of components
- low installation and operating costs
- component non-obsolescence
- interoperability (components usage from various suppliers), and
- easy relocation to adapt to changing power system's needs

while providing an independent control of active and reactive power flows. In response to these requirements, the novel impedance regulation method of a UPFC and the proven and reliable transformer/LTC technology that is used in a PAR for almost a century are combined to create the ST. This low-cost form of an IR can improve the fault level in otherwise weak networks, thereby making it possible to have a widely connected grid-scale renewable generation in weaker and isolated parts of the transmission networks as outlined in PES-TR-77, titled "Stability Definitions and Characterization of Dynamic Behavior in Systems with High Penetration of Power Electronic Interfaced Technologies," which is available at https://resourcecenter.ieee-pes.org/technical-publications/technical-reports/PES_TP_TR77_PSDP_stability_051320.html.

1.6 SMART Power Flow Controller (SPFC)

Today's power grid, shown in Figure 1-35, has integrated 100s of MW-range renewable generation. The primary driving factor for integrating more clean, renewable energy sources to the grid is the desire to reduce the use of carbon-based generation, which will lower GHG emissions. However, a typical characteristic of renewable energy, *i.e.* solar photovoltaic and offshore or land-based wind, is the intermittent nature and the resulting unsteady power flow to the load centers. To avoid resulting brown-outs and/or black-outs from this intermittency, the power flow in a transmission line needs to be controlled by using a SMART PFC (SPFC) that provides the needed dynamic response. So, traditional steady-state PFCs, such as series-connected reactor and capacitor need to be updated with an improved dynamic response. Additionally, there is an increased need for a bidirectional power flow when the renewable generation is unavailable.

Free flow of electricity may cause some transmission lines to be overloaded or underloaded. The congestions due to overloaded lines exist throughout the transmission network. The congestion points, referred to as bottlenecks, prevent desirable clean energy to reach desired load centers. When a line is congested, the present practice is to curtail renewable generation in front of the emerging bottleneck and turn on local, possibly, nongreen generation behind the bottleneck to meet the demand for electricity.

Successful transition to clean energy depends on the ability to generate and distribute renewable energy through the transmission and distribution networks. The intermittency of energy sources, need for bidirectional flows, remote locations of solar and wind projects, and so on, are challenging grid planners and operators to integrate more renewable energy into the grid. Even before the meaningful penetration of renewables is reached, forecasters are factoring renewable curtailment as a major strategy to balance supply and demand.

An SPFC, such as an IR with a proper dynamic response capability, can greatly aid in fulfilling these new requirements. The SPFC creates a variable virtual impedance that can be connected in series with the line, to maintain steady power flows to the load centers. The SPFC can limit the power flow in congested lines to be within their ATC so that the renewable generation does not need to be curtailed.

As discussed in great detail in Chapter 2, any PFC that connects a compensating voltage in series with the line is actually an IR, except an SSSC, which is a RR. The PAR acts as an IR in an indirect way with no capability to select the compensating resistance (R_{se}) and reactance

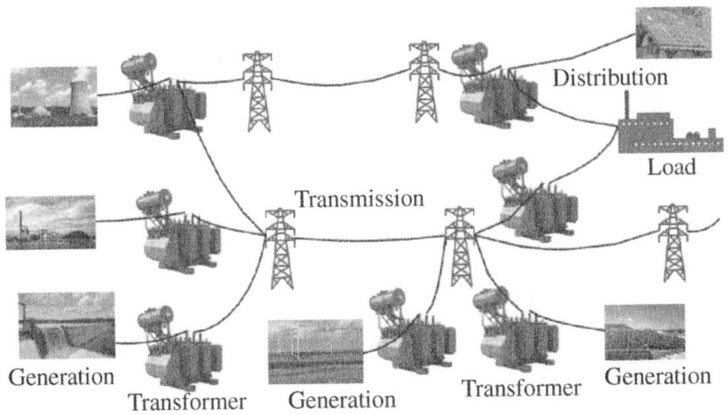

Figure 1-35 Today's grid with traditional generation and integrated renewable generation.

(X_{se}) independently as discussed in Chapter 2, Section 2.2.2.6. Therefore, a true IR that emulates a four-quadrant, series-compensating impedance ($\mathbf{Z_{se}} = R_{se} - jX_{se}$) that consists of a resistance ($R_{se} = +R$ or $-R$) and a reactance ($X_{se} = X_C$ or $-X_L$) in series with the line is really needed. Additionally, increasing the installation of roof-top solar as generation and its integration into a low voltage distribution network alters the traditional feeder voltage profile in the distribution networks, which can be mitigated, using an IR.

Advances in power electronics have made it possible to develop the UPFC, which is an IR. The VSC-based UPFC is capable of providing responses in the range of ms as shown in Figure 1-13 in the demonstration of the first commercial STATCOM at TVA Sullivan substation in 1995. However, the experiences from the last three decades show that the needed response time is in seconds in most utility applications as shown in Figures 1-29 and 1-30 in the demonstration of world's first UPFC at AEP Inez substation. Nevertheless, the cost of a UPFC is about the same, whether it is used in slow-response or fast-response applications. Therefore, it is desirable to explore the alternate designs of an IR that meet the functional requirements to provide independent control of active and reactive power flows with responses in seconds and at a fractional amount of the cost of VSC-based FACTS Controllers. This was the motivation to develop an SPFC whose objectives are as follows:

S – specific (design a power flow controller that meets utilities' needs)
M – measurable (high reliability, high efficiency, cost-effectiveness, component non-obsolescence, and ease of relocation)
A – attainable (realistic expectation about the outcome)
R – relevant (efficient power grid)
T – time-bound (delivery milestones).

The SPFC can control a bidirectional and independent active and reactive power flows dynamically as shown in Figure 1-36.

1.6.1 Example of an SPFC

In a particular application, the functional requirement of an SPFC may be written as follows:

The SPFC shall provide voltage regulation, phase angle regulation, positive and negative resistance regulation, capacitive and inductive reactance regulation, four-quadrant impedance regulation, and

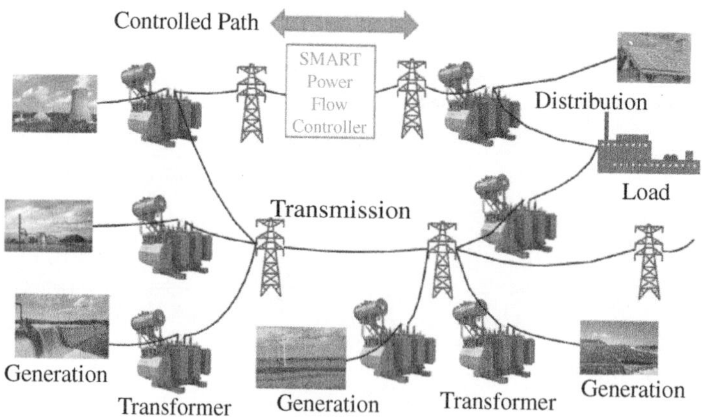

Figure 1-36 Interconnected transmission system, integrated with a SMART Power Flow Controller (SPFC).

independent regulation of active and reactive power flows in a line by connecting in series with the line a compensating voltage that has the following characteristics:

a) variable in magnitude within its design limit
b) variable in phase angle with respect to the line voltage or the prevailing line current
c) response time in the range of operation in less than 30 seconds
d) availability of 99.9999% of the time.

1.6.2 Justification

The natural power flow in an AC transmission line depends on (i) line voltage magnitude, (ii) its phase angle, and (iii) line impedance. The power flow in a line may be controlled by regulating any of these three parameters to optimize the voltage profile and the power flow in the line while maintaining the voltage stability and minimum power loss in the line.

1.6.3 Additional Information

The desired features of an SPFC are as follows:

- High reliability with the lowest number of components used
- Impedance control feature using a Shunt–Series configuration
- Lowest installation cost
- Lowest operating cost with minimum maintenance and losses
- Practically relocatable when the system needs change
- Free from component obsolescence for at least three decades, and
- Interoperability so that components from various suppliers can be used, resulting in a global manufacturing standard, ease of maintenance, and ultimately lower cost to consumers.

To meet the above-mentioned functional requirements, thyristor-based SVC and transformer/LTCs-based ST may be a suitable solution as shown in Figure 1-37. Since the proposed solution is based on functional requirements and the lowest cost, it is considered to be an SPFC.

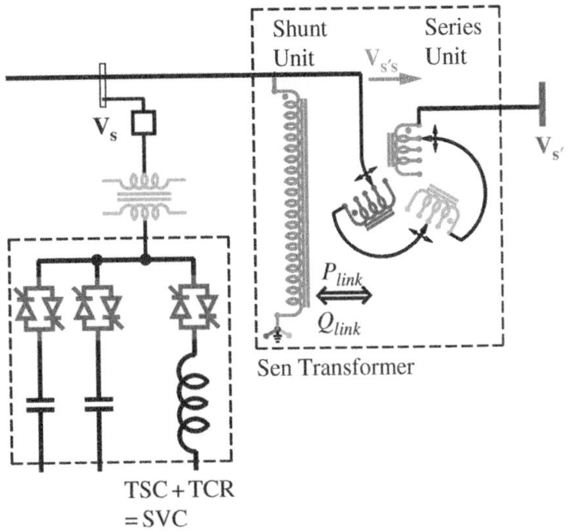

Figure 1-37 Voltage regulation with an SVC and independent power flow regulation with an ST.

1.7 Discussion

Various compensators for utility applications are summarized in Table 1-3. The features, advantages, and benefits of various solutions are listed in Table 1-4. The objective of using any of these solutions is to increase utility asset utilization.

It is recognized that the superior response capability of a power electronics inverter-based solution may be beneficial in applications where a voltage flicker, caused by an electric arc furnace load, needs to be reduced and dynamic voltage stability is required for critical loads. The final selection of a solution, however, depends on knowing the functional requirements and analyzing the cost and benefit of each available solution to determine the cost-effective solution that provides the most features at the least total cost. In the case of a simple voltage regulation at a utility bus, a SC may be an adequate solution, whereas for an arc furnace type of constantly variable load, the power electronics VSC-based STATCOM may be the best solution.

With the introduction of the first commercial STATCOM at TVA in 1995, it was anticipated that the power electronics VSC-based voltage regulation technique would be the new wave of the future. Instead, rotating machinery-based SynCons are being installed to provide some inertia to the power grid, since the grid is now becoming heavily loaded with IBRs that convert renewable wind and solar energy into usable AC electricity. Note that the IBRs do not have any mechanical inertia that is available in rotating machinery-based conventional power grid. In the meantime, many of the

Table 1-3 Various compensators for utility applications.

Compensators	Commercial names
Non-power electronics-based technology	Transformer and LTCs-based Voltage-Regulating Transformer (VRT) and Phase Angle Regulator (PAR), reactors/capacitors, Synchronous Condenser (SynCon), motor/generator, and Sen Transformer (ST)
Power electronics thyristor-based technology	Static Var Compensator (SVC) and Thyristor-Controlled Series Capacitor (TCSC)
Power electronics VSC-based technology	STATic synchronous COMpensator (STATCOM), Static Synchronous Series Compensator (SSSC), and Unified Power Flow Controller (UPFC)

Table 1-4 Features, advantages, and benefits of various solutions.

<table>
<tr><th>Solutions</th><th>Feature(s)</th><th>Advantage</th><th>Benefit</th></tr>
<tr><td>Shunt Reactor/Shunt Capacitor, VRT, SynCon, TCR, TSC, SVC, STATCOM</td><td>Voltage regulation</td><td>Meets operating voltage requirement of the load</td><td rowspan="4">Higher asset utilization</td></tr>
<tr><td>PAR</td><td>Phase angle regulation</td><td rowspan="3">Power flow magnitude and direction control</td></tr>
<tr><td>Series Reactor/Series Capacitor, TCSC, SSSC</td><td>Reactance regulation</td></tr>
<tr><td>ST, UPFC</td><td>Voltage regulation, phase angle regulation, and impedance regulation</td></tr>
</table>

nine inverters from the first-generation FACTS Controllers were decommissioned prematurely due to component obsolescence and without much payback.

The power industry's pressing need for the most economical ways to transfer bulk power along a desired path may relieve grid congestion in some markets during peak hours and integrate renewable energy from wind and solar sources. Apart from building new transmission lines that may take a decade or longer, there may be quicker and inexpensive options to utilize the existing transmission system infrastructure and harness the dormant capacity of the underutilized lines. This can only be possible through a dynamic line impedance management, which results in independent control of active and reactive power flows in the transmission lines. Independent control of active and reactive power flows leads to

- Reduction in reactive power flow, resulting in a reduction of losses in generators, transformers, and transmission lines, which increases the overall efficiency of the grid, thus lowering GHG emissions and reducing global warming
- Freeing up the generators, transformers, and transmission lines to carry more active power
- Power flow through the desired transmission paths that have high impedances, low power flow, and low line utilization
- Avoidance of grid congestion by redirecting excess power flow from an overloaded line to underloaded lines, instead of tripping the overloaded line and creating possible blackouts when power flow is needed the most, thus improving grid reliability and resiliency, and
- Delayed construction of new, expensive, high-voltage electric transmission lines.

The SPFC is proposed to enhance the controllability of the power flow in the power grid on the basis of functional requirements and cost-effective solutions. The SPFC is derived from utilizing the best features of all the technical concepts that are developed in the power flow control area until now. In the simplest term, a SMART Controller is what one procures is based on what one needs. Utilities are recommended to choose a solution that meets their **need** in terms of reliability, cost-effectiveness, component non-obsolescence, efficiency, ease of relocation, and interoperability.

The Sen Transformer technology meets the immediate need of the utility in terms of maximizing the revenue-generating active power flow while providing the highest efficiency. The ST uses transformers and mechanical LTCs and offers high reliability, high efficiency, low cost, component non-obsolescence, high power density, small footprint, and ease of relocation. The ST allows the utilities to avoid the initial purchase of an expensive PFC, along with the equally high cost of replacing a power electronics inverter-based system as it becomes obsolete. In addition to being a smaller, more dynamic, and customizable system that is less costly to maintain, the ST can be relocated to adapt to changing power system needs. Since ST uses existing transformer/LTCs-based technology, it offers a high level of interoperability, enabling components from various suppliers to be used for its manufacturing, operation, and maintenance. This also ensures a baseline global manufacturing standard, which increases the confidence of the consumer about the end product.

The power electronics inverter-based technology has the capability of providing fast (sub-cycle) dynamic response for a given transmission line impedance, although in a PFC, the dynamic response of at least a few line cycles is necessary to operate safely under various contingencies. Most utility applications allow regulation of the power flow in the line(s) in a "slow" manner as permitted by the speed of operation of the mechanical LTCs. Applications that require faster response times can make use of TC LTCs, instead of mechanical LTCs as shown in Figure 2-51. The STs with both types of LTCs (mechanical and TC) cover a wide range of requirements for power flow control in electric transmission lines. If the LTCs are too coarse for an IR, the number of taps on the winding may be increased.

The ST can be customized to operate in a limited-angle range that is suitable for a particular application. This reduces the number of secondary windings from nine to six and the number of three-phase LTCs from three to two. Consequently, the cost of an ST can be reduced due to its simpler design, whereas there is no such option for the reduction of cost in a power electronics inverter-based FACTS controller. The inverter is always designed to operate in the entire 360°-range of series-injection of a compensating voltage whether a particular application needs it or not.

Compared to the UPFC, the ST has inherent advantages: less costly, component non-obsolescence, portable, and reduced and easy maintenance along with many other features. The value proposition to the customer is that in comparison to a FACTS controller, a 5:1 reduction in equipment cost and a 10:1 reduction in operational/maintenance cost of an ST are expected. The ST provides 21st-century power flow control solutions with an impedance regulation using 20th-century power hardware, such as transformer/LTCs, which are proven to be reliable to the utilities worldwide and reduce the likelihood of obsolescence.

Modeling, also referred to as the simulation of an actual installation, is essential prior to the realization of a concept in the form of a full-scale implementation. Modeling is the key to understand the subject of an SPFC in the most cost-effective way. Various modeling techniques, which were successfully used during the development of the first generation of FACTS Controllers and numerous additional power system applications, are discussed throughout the book. It is important to note that a model is just an approximation of the actual equipment. A model, in its simplest form, can be of a first-order approximation to provide a Rough-Order Magnitude data of interest. A design-query can be answered by increasing the level of details in a model.

2

Power Flow Control Concepts

This chapter builds on the material that was covered in Chapter 1 where the need for Power Flow Controllers (PFCs) was reviewed. The PFCs included basic technologies, such as transformers, capacitors, and reactors, and modern-day Voltage-Sourced Converters (VSCs) that provide the regulations of the line voltage magnitude, its phase angle, and line reactance. Also covered were various transmission line regulators, including numerous combinations of VSCs in Shunt–Shunt and Shunt–Series configurations. The application of a compensating voltage that is variable in magnitude and phase angle, using an Impedance Regulator (IR) such as VSC-based Unified Power Flow Controller (UPFC), Electrical Machine (EM)-based Rotating Power Flow Transformer (RPFT), and transformer and Load Tap Changers (LTCs)-based Sen Transformer (ST), allows an independent control of active and reactive power flows in the line.

In those situations where several transmission lines are connected to a common bus and making a change in the power flow in one line could affect the power flows in other lines, the concept of a Series–Series configuration was discussed to transfer power from one specific line to another specific line without altering the power flows in the remaining lines. This concept may utilize series-connected VSCs, each of which is connected at their common DC link, as an Interline Power Flow Controller (IPFC). An alternate solution is a Multiline Sen Transformer (MST). The development of a SMART Power Flow Controller (SPFC) is also covered.

With the basic understanding of how PFCs were developed over the decades, Chapter 2 derives equations that will allow the reader to study uncompensated or compensated active and reactive power flows in a transmission line. The coverage of impedance emulation techniques gives the readers a clear understanding of various PFCs that create virtual reactors and capacitors, using VSCs, electrical machines, and transformer/LTCs-based technologies. The characteristic equations of various PFCs and numerous examples, showing how to use them to calculate a rough-order magnitude (ROM) power rating of a particular solution, enable the reader to understand the differences between various PFCs. Single-line applications include Voltage Regulator (VR) – STATic synchronous COMpensator (STATCOM), Static Var Compensator (SVC), Synchronous Condenser (SynCon), and shunt-connected capacitor/reactor; Phase Angle Regulator (PAR) – symmetric and asymmetric; Reactance Regulator (RR) – Static Synchronous Series Compensator (SSSC), Thyristor-Controlled Series Capacitor (TCSC) and series-connected capacitor/reactor; and IR – UPFC and ST. Parallel-line applications include Back-To-Back (BTB) operations of PFCs, such as BTB-STATCOM, BTB-SSSC, and MST.

Lastly, a comparison of installation and operational costs of various PFCs is presented in the form of Reactive Power Index (*RPI*), Loss Index (*LI*), and Apparent Power Rating (*APR*). The unique

Power Flow Control Solutions for a Modern Grid using SMART Power Flow Controllers, First Edition.
Kalyan K. Sen and Mey Ling Sen.

Companion website: www.wiley.com/go/sen/powerflowcontrol

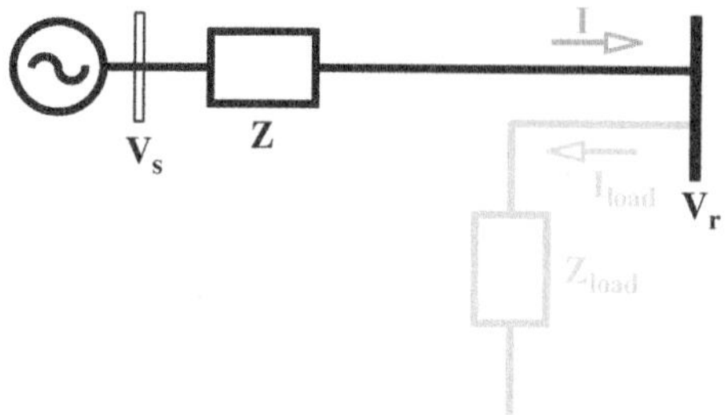

Figure 2-1 Representation of a power system network, consisting of a voltage source ($\mathbf{V_s}$), line impedance (**Z**), and load impedance ($\mathbf{Z_{load}}$).

concept of the *Sen Index* for comparing the costs of various PFCs is introduced to define the relationship between the increased cost of power delivery per unit incremental power delivered.

Consider the interconnected transmission system, connecting sources and loads as shown in Figure 1-1, in its simplest form as shown in Figure 2-1 where the sending-end voltage ($\mathbf{V_s}$) is supplying power to a load impedance ($\mathbf{Z_{load}}$) through a line with an impedance (**Z**), which is composed of a resistance (R) and an inductive reactance (X). Then, the line current (**I**) is given by

$$\mathrm{I} = \frac{\mathrm{V_S}}{\mathbf{Z} + \mathbf{Z}_{\mathrm{load}}} \tag{2-1}$$

The load voltage ($\mathbf{V_r}$) at the receiving end of the line is given by

$$\mathrm{V_r} = \mathbf{Z}_{\mathrm{load}}\mathrm{I}. \tag{2-2}$$

Substituting **I** from Equation (2-1) into Equation (2-2), the voltage divider equation can be written as

$$\mathrm{V_r} = \frac{\mathbf{Z}_{\mathrm{load}}}{\mathbf{Z} + \mathbf{Z}_{\mathrm{load}}}\mathrm{V_S}. \tag{2-3}$$

From Equation (2-3), it can be written that the magnitude of the load voltage is given by

$$V_r = \left|\frac{\mathbf{Z}_{\mathbf{load}}}{\mathbf{Z}_{\mathbf{load}} + \mathbf{Z}}\right| V_s \tag{2-4a}$$

or

$$V_r = \frac{1}{\left|1 + \frac{\mathbf{Z}}{\mathbf{Z}_{\mathbf{load}}}\right|} V_s. \tag{2-4b}$$

Considering a resistive load (R_{load}) and a highly inductive high-voltage transmission line with a reactance (X) as shown in Figure 2-2a, Equation (2-4a) becomes

$$V_r = \frac{R_{load}}{|R_{load} + jX|} V_s \quad \text{or} \quad V_r = \frac{R_{load}}{\sqrt{R_{load}^2 + X^2}} V_s. \tag{2-5}$$

From Ohm's law, the voltage across the line reactance (X) is given by

$$V_X = XI \tag{2-6}$$

and the voltage across the load resistance (R_{load}) is given by

$$V_r = R_{load} I_{load} \tag{2-7}$$

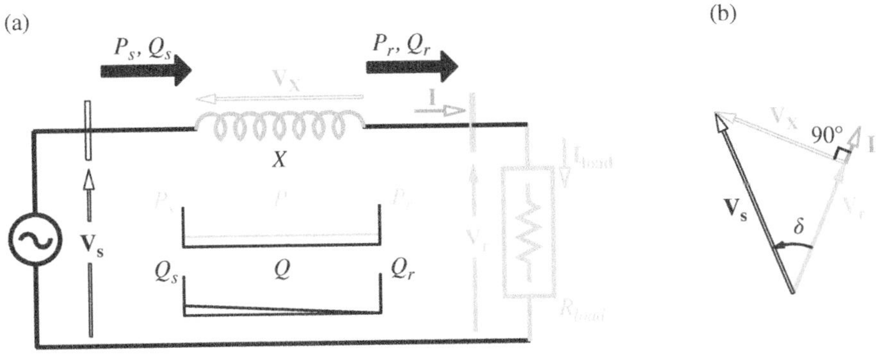

Figure 2-2 (a) One-generator/one-line power system network and (b) phasor diagram.

where

$$I = I_{load}. \tag{2-8}$$

The load power that is the power consumed at the receiving end of the line is given by

$$P_r = V_r I_{load}. \tag{2-9}$$

Equations (2-7) and (2-9) can be combined as

$$P_r = V_r I_{load} = V_r\left(\frac{V_r}{R_{load}}\right) = \frac{V_r^2}{R_{load}} \tag{2-10a}$$

or using Equation (2-5)

$$P_r = \frac{V_r^2}{R_{load}} = \left(\frac{V_r}{R_{load}}\right)^2 R_{load} = \frac{V_s^2}{R_{load}^2 + X^2} R_{load}. \tag{2-10b}$$

The line reactive power that is the reactive power absorbed by the line is given by

$$Q_{line} = V_X I \tag{2-11a}$$

or using Equation (2-6)

$$Q_{line} = I^2 X \tag{2-11b}$$

or using Equations (2-8), (2-7), and (2-5)

$$Q_{line} = \left(\frac{V_r}{R_{load}}\right)^2 X = \frac{V_s^2}{R_{load}^2 + X^2} X. \tag{2-11c}$$

In this case of a radial transmission system, the generator is not only supplying active power to the load at unity power factor, it is also delivering reactive power to the line. Figure 2-3 shows the variations of load voltage (V_r) and load power (P_r) as a function of load resistance (R_{load}).

At very low load resistance, the load voltage is low and so is the load power. At very high load resistance, the load current is low and so is the load power. In the middle range of load resistance, a desired amount of power can be delivered to the load; however, the load voltage is always lower than the nominal one per unit (pu) voltage. The condition for maximum power flow can be derived from Equation (2-10b) as

$$\frac{d}{dR_{load}} P_r = 0 \quad \text{or} \quad 0 = \frac{V_s^2}{R_{load}^2 + X^2} - V_s^2 R_{load} \frac{2R_{load}}{\left(R_{load}^2 + X^2\right)^2} \quad \text{or} \quad R_{load} = X. \tag{2-12}$$

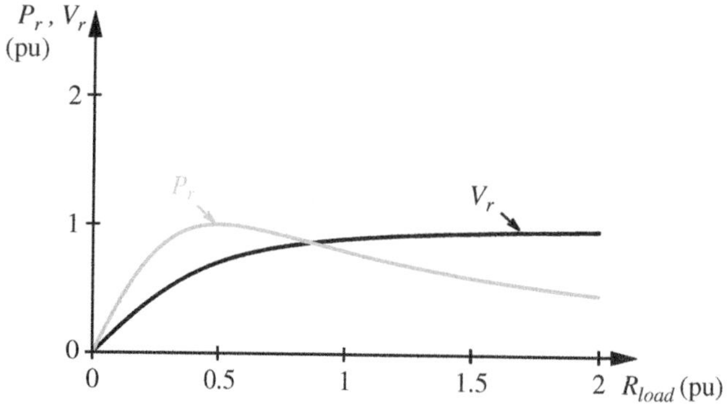

Figure 2-3 Variations of load voltage (V_r) and load power (P_r) as a function of load resistance (R_{load}) when V_s = 1 pu and X = 0.5 pu.

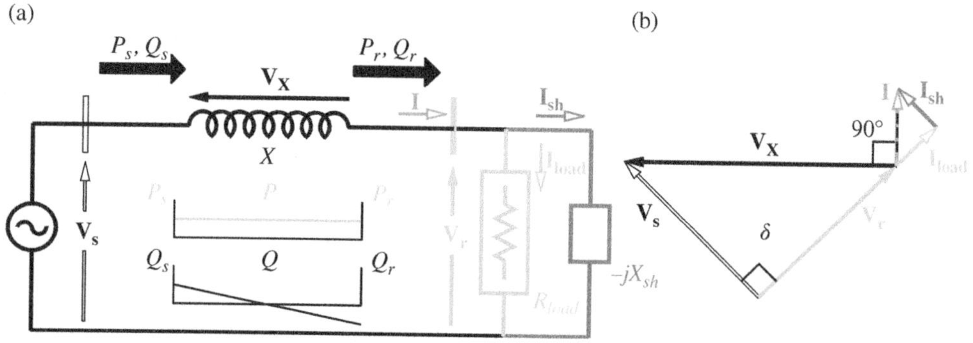

Figure 2-4 (a) One-generator/one-line power system network with a shunt voltage compensation and (b) phasor diagram.

The phasor diagram for a load resistance of R_{load} = 0.5 pu, shown in Figure 2-2b, indicates that the load voltage may be too low to be useful, even though this operating point delivers the maximum power to the load. Since most loads require their terminal voltages to be within some percentage of the nominal value per their design specifications, there needs to be a solution to restore the load voltage to its nominal value. The shunt-connected solution, shown in Figure 2-4a, restores the load voltage to its nominal value. Since this solution compensates the voltage deficiency, it is called a voltage compensator. The shunt-compensating reactance ($-jX_{sh}$) can be capacitive ($-jX_C = 1/j\omega C$) or inductive ($jX_L = j\omega L$); it is defined such that $X_{sh} > 0$ represents $-jX_{sh}$ to be capacitive and $X_{sh} < 0$ represents $-jX_{sh}$ to be inductive.

A parallel combination of load resistance (R_{load}) and shunt-compensating reactance ($-jX_{sh}$) is equivalent to

$$\mathbf{Z_{load}} = R_{load} \parallel (-jX_{sh}) = \frac{-jR_{load}X_{sh}}{R_{load} - jX_{sh}} = \frac{-jR_{load}X_{sh}}{R_{load} - jX_{sh}} \frac{R_{load} + jX_{sh}}{R_{load} + jX_{sh}} = \frac{R_{load}X_{sh}^2}{R_{load}^2 + X_{sh}^2} - j\frac{R_{load}^2 X_{sh}}{R_{load}^2 + X_{sh}^2}. \tag{2-13}$$

When combined with the line reactance, the total impedance in the circuit is given by

$$\mathbf{Z_{load}} + \mathbf{Z} = \frac{R_{load}X_{sh}^2}{R_{load}^2 + X_{sh}^2} - j\frac{R_{load}^2 X_{sh}}{R_{load}^2 + X_{sh}^2} + jX. \tag{2-14}$$

Similar to Equation (2-12), the condition for a maximum power transfer is

$$\frac{R_{load}X_{sh}^2}{R_{load}^2+X_{sh}^2} = -\frac{R_{load}^2X_{sh}}{R_{load}^2+X_{sh}^2}+X \quad \text{or} \quad R_{load}X_{sh}^2 = X\left(R_{load}^2+X_{sh}^2\right)-R_{load}^2X_{sh} \quad \text{or}$$
$$(R_{load}-X)X_{sh}^2+R_{load}^2X_{sh}-R_{load}^2X=0. \tag{2-15}$$

or

$$X_{sh} = \frac{-R_{load}^2 \pm \sqrt{\left(R_{load}^2\right)^2+4(R_{load}-X)R_{load}^2X}}{2(R_{load}-X)}. \tag{2-16}$$

For a given R_{load} and X, X_{sh} can be determined from Equation (2-16) for a maximum power transfer to the load.

For $R_{load}=X$, it is evident from Equation (2-15) that

$$X_{sh}=X. \tag{2-17}$$

From Equation (2-4a), the magnitude of the load voltage is given by

$$V_r = \frac{\left|\frac{R_{load}X_{sh}^2}{R_{load}^2+X_{sh}^2}-j\frac{R_{load}^2X_{sh}}{R_{load}^2+X_{sh}^2}\right|}{\left|\frac{R_{load}X_{sh}^2}{R_{load}^2+X_{sh}^2}-j\frac{R_{load}^2X_{sh}}{R_{load}^2+X_{sh}^2}+jX\right|}V_s \quad \text{or} \quad V_r = \frac{\left|R_{load}X_{sh}^2-jR_{load}^2X_{sh}\right|}{\left|R_{load}X_{sh}^2-jR_{load}^2X_{sh}+jX\left(R_{load}^2+X_{sh}^2\right)\right|}V_s$$

or

$$V_r = \frac{R_{load}|X_{sh}||X_{sh}-jR_{load}|}{\left|R_{load}X_{sh}^2+j\left\{X\left(R_{load}^2+X_{sh}^2\right)-R_{load}^2X_{sh}\right\}\right|}V_s \quad \text{or}$$
$$V_r = \frac{R_{load}|X_{sh}|\sqrt{R_{load}^2+X_{sh}^2}}{\sqrt{\left(R_{load}X_{sh}^2\right)^2+\left\{X\left(R_{load}^2+X_{sh}^2\right)-R_{load}^2X_{sh}\right\}^2}}V_s. \tag{2-18}$$

Since the load current (I_{load}) is resistive and the compensator current (I_{sh}) is reactive, the line current is defined by

$$I=\sqrt{I_{load}^2+I_{sh}^2}. \tag{2-19}$$

From Ohm's law, the load voltage is given by

$$V_r = X_{sh}I_{sh} \tag{2-20a}$$

or the compensator current is given by

$$I_{sh}=\frac{V_r}{X_{sh}}. \tag{2-20b}$$

Using Equations (2-20a) and (2-20b), the compensator reactive power that is the reactive power absorbed by the shunt compensator is given by

$$Q_{sh} = -V_rI_{sh} = -I_{sh}^2X_{sh} = -\frac{V_r^2}{X_{sh}}. \tag{2-21}$$

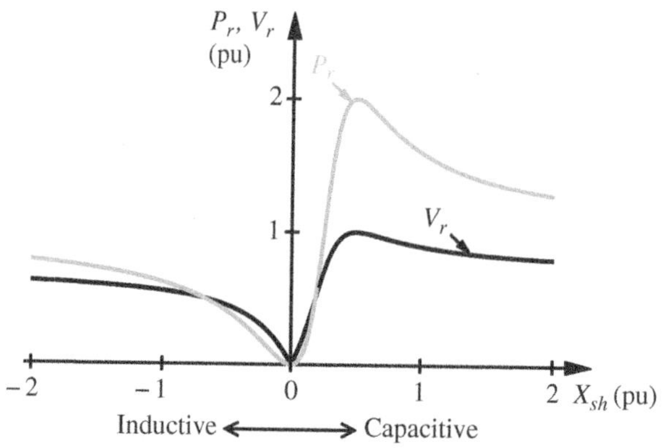

Figure 2-5 Variations of load voltage (V_r) and load power (P_r) as a function of compensating reactance (X_{sh}) when V_s = 1 pu, X = 0.5 pu, and R_{load} = 0.5 pu.

If $X_{sh} > 0$, Q_{sh} represents a generated reactive power and if $X_{sh} < 0$, Q_{sh} represents an absorbed reactive power.

Figure 2-5 shows the variations of load voltage (V_r) and load power (P_r) as a function of shunt-compensating reactance (X_{sh}). Note that $X_{sh} > 0$ means the shunt-compensating reactance is capacitive and $X_{sh} < 0$ means the shunt-compensating reactance is inductive. When inductively compensated, the load voltage is always less than its uncompensated value. However, for a particular capacitive compensation, the load voltage can be restored to its nominal value as shown in Figure 2-4b. The uncompensated line current, which is in phase with the resistive load voltage (shown in Figure 2-2b) now leads the load voltage because a portion of the line current that flows through the capacitive shunt compensator leads the load voltage by 90°.

The voltage regulation is essential to maintain the load voltage within its allowable limits at the load terminal. The increase of voltage at the load terminal also increases the load power, which is discussed in the next section. The load power (power transfer through the transmission line) is proportional to the sine of power angle (δ), which is the phase angle between the voltages at the sending and receiving ends of the line. As the power flow to the load increases so does the power angle (δ) as shown in Figure 2-4, compared to Figure 2-2.

Equation (2-4b) can be rewritten for an inductive-load condition as

$$V_r = \frac{1}{\left|1 + \frac{jX}{jX_L}\right|} V_s \quad \text{or} \quad V_r = \frac{1}{1 + \frac{X}{X_L}} V_s. \tag{2-22}$$

Considering the dominant parameter of a transmission line being an inductive reactance, adding a shunt reactor results in lowering the line voltage, since the denominator of the above equation is always greater than one.

A capacitive load results in increasing the line voltage, since the denominator of the following equation is always less than one.

$$V_r = \frac{1}{\left|1 + \frac{jX}{-jX_C}\right|} V_s \quad \text{or} \quad V_r = \frac{1}{1 - \frac{X}{X_C}} V_s. \tag{2-23}$$

Equation (2-5) can be rearranged as

$$V_r = \frac{1}{\sqrt{1 + \left(\frac{X}{R_{load}}\right)^2}} V_S. \tag{2-24}$$

The denominator of Equation (2-24) is always greater than one, hence $V_r < V_S$. Therefore, a resistive load results in lowering the line voltage. Adding a shunt-compensating resistance is not a preferred method for lowering the line voltage, since the shunt-compensating resistor absorbs active power, thus increases system losses.

The preferred method for lowering the load voltage is to use a shunt-compensating inductive reactance, which does not absorb any active power, but only reactive power. The preferred method for increasing the load voltage is to use a shunt-compensating capacitive reactance, which also does not absorb any active power, but generates reactive power.

During an early morning, the power consumption is low and as a result, the line carries a smaller amount of current. The var needed by the series-inductive reactance of the line may be smaller than the var generated by the line-to-ground, shunt-capacitive reactance. As a result, the line voltage may reach beyond its allowable upper limit. Therefore, a shunt-inductive compensation may be needed to absorb the extra var, generated by the line-to-ground, shunt-capacitive reactance and to decrease the line voltage to keep it within its allowable limits.

During the course of the day, as the power consumption increases, the line carries a larger amount of current. The var needed by the series-inductive reactance of the line may become larger than the var generated by the line-to-ground, shunt-capacitive reactance. The voltage across the line impedance increases, which lowers the voltage at any point along the line. The load voltage may dip below its allowable lower limit. Therefore, a shunt-capacitive compensation may be needed to supply extra var to the series-inductive reactance and to restore the line voltage within its allowable limits.

Example 2-1

In Figures 2-2a and 2-2b, the magnitude (V_s) of the sending-end voltage and the line reactance (X) are 1 pu and 0.5 pu, respectively. For maximum power transfer, (a) what is the load resistance (R_{load})? (b) what is the load voltage (V_r)? (c) what is the line current (I)? (d) what is the load power (P_r)? and (e) what is the line reactive power (Q_{line})?

Answer 2-1

a) From Equation (2-12), $R_{load} = X = 0.5$ pu.

b) From Equation (2-5), $V_r = \frac{R_{load}}{\sqrt{R_{load}^2 + X^2}} V_s = \frac{0.5 \times 1}{\sqrt{0.5^2 + 0.5^2}} = \frac{1}{\sqrt{2}}$ pu.

c) From Equations (2-8) and (2-7), $I = I_{load} = \frac{V_r}{R_{load}} = \frac{1/\sqrt{2}}{1/2} = \sqrt{2}$ pu.

d) From Equations (2-9) and (2-8), the load power, $P_r = V_r I_{load} = V_r I = \frac{1}{\sqrt{2}}\sqrt{2} = 1$ pu.

e) From Equation (2-11b), the line reactive power, $Q_{line} = I^2 X = 2 \times 0.5 = 1$ pu.

Example 2-2

In Figures 2-4a and 2-4b, (a) what type and how much of shunt compensation (X_{sh}) is needed for maximum amount of power transfer to the resistive load ($R_{load} = 0.5$ pu), assuming the line reactance, $X = 0.5$ pu? (b) what is the load voltage (V_r)? (c) what is the line current (I_{load})? (d) what is the load power (P_r)? (e) what is the line reactive power (Q_{line})? and (f) what is the rating of the shunt compensator (Q_{sh})?

Answer 2-2

a) From Equation (2-17), $X_{sh} = X = 0.5$ pu (capacitive).
b) From Equation (2-18),

$$V_r = \frac{R_{load}|X_{sh}|\sqrt{R_{load}^2 + X_{sh}^2}}{\sqrt{\left(R_{load}X_{sh}^2\right)^2 + \left\{X\left(R_{load}^2 + X_{sh}^2\right) - R_{load}^2 X_{sh}\right\}^2}}$$

$$V_s = \frac{0.5 \times 0.5\sqrt{0.5^2 + 0.5^2} \times 1}{\sqrt{(0.5 \times 0.5^2)^2 + \{0.5(0.5^2 + 0.5^2) - 0.5^2 \times 0.5\}^2}} = 1 \text{ pu.}$$

c) From Equation (2-7), $I_{load} = \dfrac{V_r}{R_{load}} = \dfrac{1}{1/2} = 2$ pu.

d) From Equation (2-10a), $P_r = \dfrac{V_r^2}{R_{load}} = \dfrac{1}{1/2} = 2$ pu.

e) From Equation (2-20b), $I_{sh} = \dfrac{V_r}{X_{sh}} = \dfrac{1}{1/2} = 2$ pu (capacitive).

From Equation (2-19), $I = \sqrt{I_{load}^2 + I_{sh}^2} = \sqrt{2^2 + 2^2} = 2\sqrt{2}$ pu.

From Equation (2-11b), the line reactive power, $Q_{line} = I^2 X = 8 \times 0.5 = 4$ pu.

f) From Equation (2-21), $Q_{sh} = -\dfrac{V_r^2}{X_{sh}} = -\dfrac{1}{1/2} = -2$ pu.

Additional Discussion

Since, the line is considered to be lossless, $P_s = P_r = 2$ pu. The apparent power (S_s) at the sending end, $V_s I = 1 \times 2\sqrt{2} = 2\sqrt{2}$ pu. Therefore, the reactive power at the sending end is $Q_s = \sqrt{S_s^2 - P_s^2} = \sqrt{8 - 4} = 2$ pu.

Note that the reactive power ($Q_{line} = 4$ pu) absorbed by the transmission line is supplied half and half by the sending-end generator and the receiving-end compensator.

From Equation (2-13), $\mathbf{Z_{load}} = R_{load} \parallel (-jX_{sh}) = \dfrac{-jR_{load}X_{sh}}{R_{load} - jX_{sh}} = \dfrac{R_{load}X_{sh}^2}{R_{load}^2 + X_{sh}^2} - j\dfrac{R_{load}^2 X_{sh}}{R_{load}^2 + X_{sh}^2} =$ $0.25 - j0.25$ pu.

From Equation (2-14), $\mathbf{Z_{load}} + \mathbf{Z} = \dfrac{R_{load}X_{sh}^2}{R_{load}^2 + X_{sh}^2} - j\dfrac{R_{load}^2 X_{sh}}{R_{load}^2 + X_{sh}^2} + jX = 0.25 + j0.25$ pu.

Figure 2-5 confirms that the chosen circuit parameters provide maximum power transfer from the source to the load.

Example 2-3

Consider a near no-load line, the magnitude (V_s) of the sending-end voltage ($\mathbf{V_s}$) in Figure 2-1 is 0.99 pu and $\mathbf{Z} = jX = j\,0.045$ pu; (a) what is the magnitude (V_r) of the receiving-end voltage ($\mathbf{V_r}$) when $\mathbf{Z_{load}} = jX_L$ where $X_L = 0.945$ pu? (b) what is the voltage magnitude (V_r) when $\mathbf{Z_{load}} = -jX_C$ where $X_C = 1.035$ pu? (c) what is the rating of the shunt compensator?

Answer 2-3

a) From Equation (2-22), $V_r = \dfrac{1}{1 + \dfrac{X}{X_L}} V_s = \dfrac{1}{1 + \dfrac{0.045}{0.945}} \times 0.99 = 0.945$ pu.

b) From Equation (2-23), $V_r = \dfrac{1}{1 - \dfrac{X}{X_C}} V_s = \dfrac{1}{1 - \dfrac{0.045}{1.035}} \times 0.99 = 1.035$ pu.

c) When the shunt compensator is inductive, its rating is $\dfrac{V_r^2}{X_L} = \dfrac{0.945^2}{0.945} = 0.945$ pu.

When the shunt compensator is capacitive, its rating is $\dfrac{V_r^2}{X_C} = \dfrac{1.035^2}{1.035} = 1.035$ pu.

Considering a resistive load (R_{load}) and a very low X/R ratio, *i.e.* a highly resistive low voltage distribution line with a resistance (R), Equation (2-4b) becomes

$$V_r = \frac{1}{\left|1 + \dfrac{R}{R_{load}}\right|} V_s \quad \text{or} \quad V_r = \frac{1}{1 + \dfrac{R}{R_{load}}} V_s. \tag{2-25}$$

For a positive load resistor, the voltage at the Point of Connection (POC) to the utility decreases, since the denominator of the Equation (2-25) is always more than one. In Equation (2-10a), $P_r > 0$ when the load resistor, $R_{load} > 0$.

Considering the modern-day photovoltaic solar installations in distribution system being a negative load, typically with unity power factor, *i.e.* a negative load resistor that feeds electrical energy to the grid. Recall in Equation (2-10a), $P_r < 0$ when the load resistor, $R_{load} < 0$. For a negative load resistor, the voltage at the POC increases, since the denominator of the Equation (2-25) is always less than one. This rise in voltage may become too large to cause voltage requirement violations to consumers who are close to the generation source on a distribution feeder.

The generation on the distribution circuit can also cause flicker. A flicker occurs when the voltage repeatedly changes by more than the allowable limit. This can happen when clouds pass over a solar farm or the wind changes its speed at the wind farm. The sudden change in voltage can cause lighting to change intensity and Uninterruptable Power Supply (UPS) systems to turn on and off.

According to American National Standards Institute (ANSI) standard C84.1, the voltage at the electric utility service, referred to as Service Voltage, which is usually the voltage at the meter socket or entrance switch is allowed to vary +5 to −5% for systems that are operating at 600 V and below. For systems that are operating above 600 V, this range is +5 to −2.5%.

The voltage at the user load, referred to as utilization voltage, which is generally considered to be the voltage at the terminals of the device or appliance or the voltage at the convenience outlet to

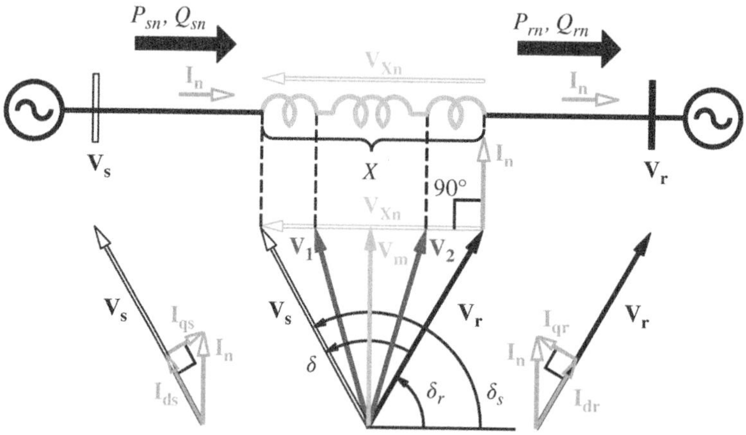

Figure 2-6 Two-generator/one-line power system network and the related phasor diagrams.

which these terminals are connected is allowed to vary +5 to −10% while providing full performance of the load.

Next, consider an interconnected transmission system with sources at both the sending and receiving ends of the line. A two-generator/one-line power system network with a sending-end voltage ($\mathbf{V_s} = V_s\angle\delta_s$) and a receiving-end voltage ($\mathbf{V_r} = V_r\angle\delta_r$), connected by a line reactance (X) and the related phasor diagrams are shown in Figure 2-6. The magnitudes of the voltages at the sending and receiving ends are V_s and V_r, respectively, and their corresponding phase angles are δ_s and δ_r, respectively.

Considering the natural voltage ($\mathbf{V_{Xn}}$) across the line reactance (X) to be the difference between the sending- and receiving-end voltages, it can be written from Figure 2-6 that

$$\mathbf{V_s} = \mathbf{V_r} + \mathbf{V_{Xn}} \quad \text{or} \quad V_s\angle\delta_s = V_r\angle\delta_r + V_{Xn}\angle\theta_{VXn} \tag{2-26a}$$

$$\mathbf{V_{Xn}} = \mathbf{V_s} - \mathbf{V_r} \quad \text{or} \quad V_{Xn}\angle\theta_{VXn} = V_s\angle\delta_s - V_r\angle\delta_r \tag{2-26b}$$

where $\mathbf{V_{Xn}}$, the difference between the sending-end voltage (V_s) and receiving-end voltage (V_r), is the natural voltage across the line reactance (X); θ_{VXn} is the phase angle of the voltage ($\mathbf{V_{Xn}}$) across the line reactance (X).

Equation (2-26a) can be rewritten as

$$V_s\angle\delta_s = V_r\angle(\delta_s - \delta) + V_{Xn}\angle(\delta_s + \theta_{VXns}) \quad \text{using Equation (2-27)} \tag{2-26c}$$

where θ_{VXns} is the phase angle of the voltage ($\mathbf{V_{Xn}}$) across the line reactance (X) with respect to the sending-end voltage, or

$$V_{Xn}\angle(\delta_s + \theta_{VXns}) = V_s\angle\delta_s - V_r\angle(\delta_s - \delta) \tag{2-26d}$$

and

$$V_s\angle(\delta_r + \delta) = V_r\angle\delta_r + V_{Xn}\angle(\delta_r + \theta_{VXnr}) \quad \text{using Equation (2-27)} \tag{2-26e}$$

where θ_{VXnr} is the phase angle of the voltage ($\mathbf{V_{Xn}}$) across the line reactance (X) with respect to the receiving-end voltage, or

$$V_{Xn}\angle(\delta_r + \theta_{VXnr}) = V_s\angle(\delta_r + \delta) - V_r\angle\delta_r \tag{2-26f}$$

where

$$\delta = \delta_s - \delta_r. \tag{2-27}$$

Since the line reactance (X) is inductive, the resulting natural line current ($\mathbf{I_n}$) lags the voltage ($\mathbf{V_{Xn}}$) by 90°. The voltage changes its magnitude and phase angle at any point along a line due to the voltage across the line reactance (X). This voltage across the line reactance (X) is caused by the flow of current in the line. However, the voltage at the POC with a load must be maintained within the regulatory limits. The intermediate line voltages (*i.e.* $\mathbf{V_1}$, $\mathbf{V_2}$, etc.) are smaller in magnitude than the sending- and receiving-end voltages ($\mathbf{V_s}$ and $\mathbf{V_r}$). The smallest voltage ($\mathbf{V_m}$) is at the midpoint of the transmission line in this illustration.

The direct or active and quadrature or reactive components of the line current ($\mathbf{I}$) at the sending end are $\mathbf{I_{ds}}$ and $\mathbf{I_{qs}}$ and the same at the receiving end are $\mathbf{I_{dr}}$ and $\mathbf{I_{qr}}$. Since the line current ($\mathbf{I}$) lags the sending-end voltage ($\mathbf{V_s}$) and leads the receiving-end voltage ($\mathbf{V_r}$), the line receives reactive power from the source at the sending end and also from the load at the receiving end. The power factor at the midpoint of the line is unity.

Considering the sending-end voltage phasor as reference (*i.e.* $\delta_s = 0°$), the natural voltage ($\mathbf{V_{Xn}}$) from Equation (2-26d) can be redefined as

$$V_{Xn}\angle\theta_{VXns} = V_s\angle 0° - V_r\angle(-\delta). \tag{2-28}$$

Equation (2-28) can be rewritten as

$$V_{Xn}\cos\theta_{VXns} + jV_{Xn}\sin\theta_{VXns} = V_s - V_r\cos\delta + jV_r\sin\delta.$$

Equating the real and imaginary parts of both sides of the above equation, it can be written that

$$V_{Xn}\cos\theta_{VXns} = V_s - V_r\cos\delta \tag{2-29a}$$

and

$$V_{Xn}\sin\theta_{VXns} = V_r\sin\delta. \tag{2-29b}$$

Therefore, the magnitude of the natural voltage ($\mathbf{V_{Xn}}$) across the line reactance is derived from Equations (2-29a) and (2-29b) using the identity ($\cos^2\theta_{VXns} + \sin^2\theta_{VXns} = 1$) as

$$V_{Xn} = \sqrt{V_{Xn}^2\left(\cos^2\theta_{VXns} + \sin^2\theta_{VXns}\right)} = \sqrt{\left(V_{Xn}\cos\theta_{VXns}\right)^2 + \left(V_{Xn}\sin\theta_{VXns}\right)^2}$$

or

$$V_{Xn} = \sqrt{(V_s - V_r\cos\delta)^2 + (V_r\sin\delta)^2}$$

or

$$V_{Xn} = \sqrt{V_s^2 - 2V_sV_r\cos\delta + V_r^2\left(\cos^2\delta + \sin^2\delta\right)}$$

or

$$V_{Xn} = \sqrt{V_s^2 - 2V_sV_r\cos\delta + V_r^2} \tag{2-30}$$

using the identity $\left(\cos^2\delta + \sin^2\delta\right) = 1$.

The phase angle of the natural voltage ($\mathbf{V_{Xn}}$) across the line reactance with respect to the sending-end voltage ($\mathbf{V_s}$) is derived from Equations (2-29a) and (2-29b) as

$$\theta_{VXns} = \tan^{-1}(\tan\theta_{VXns}) = \tan^{-1}\frac{\sin\theta_{VXns}}{\cos\theta_{VXns}} = \tan^{-1}\frac{V_{Xn}\sin\theta_{VXns}}{V_{Xn}\cos\theta_{VXns}}$$

or

$$\theta_{VXns} = \tan^{-1}\frac{V_r\sin\delta}{V_s - V_r\cos\delta}. \tag{2-31}$$

The natural line current ($\mathbf{I_n}$) is defined as

$$\mathbf{I_n} = \frac{\mathbf{V_{Xn}}}{jX} \quad \text{or} \quad I_n\angle\theta_{In} = \frac{V_{Xn}\angle\theta_{VXn}}{jX} \quad \text{or} \quad I_n\angle(\delta_s + \theta_{Ins}) = \frac{V_{Xn}\angle(\delta_s + \theta_{VXns})}{jX}$$

where θ_{Ins} is the phase angle of the natural line current with respect to the sending-end voltage.

Considering the sending-end voltage phasor as reference (*i.e.* $\delta_s = 0°$), it can be written from the above equation that

$$I_n\angle\theta_{Ins} = \frac{V_{Xn}\angle\theta_{VXns}}{jX} \tag{2-32a}$$

or

$$I_n\angle\theta_{Ins} = \frac{V_{Xn}}{X}\angle(\theta_{VXns} - 90°) \tag{2-32b}$$

where the magnitude of the natural line current ($\mathbf{I_n}$) is given by

$$I_n = \frac{V_{Xn}}{X} \tag{2-33}$$

and the phase angle of the natural line current ($\mathbf{I_n}$) with respect to the sending-end voltage ($\mathbf{V_s}$) is given by

$$\theta_{Ins} = \theta_{VXns} - 90° \tag{2-34}$$

where θ_{VXns} is given in Equation (2-31).

Alternately, from Equation (2-32a), it can be written that

$$I_n\cos\theta_{Ins} + jI_n\sin\theta_{Ins} = \frac{V_{Xn}\cos\theta_{VXns} + jV_{Xn}\sin\theta_{VXns}}{jX} \quad \text{or}$$

$$I_n\cos\theta_{Ins} + jI_n\sin\theta_{Ins} = \frac{V_{Xn}\sin\theta_{VXns}}{X} - j\frac{V_{Xn}\cos\theta_{VXns}}{X} \quad \text{or}$$

$$I_n\cos\theta_{Ins} + jI_n\sin\theta_{Ins} = \frac{V_r\sin\delta}{X} - j\frac{V_s - V_r\cos\delta}{X} \quad \text{using Equations (2-29a) and (2-29b).}$$

Equating the real and imaginary parts of both sides of the above equation, it can be written that

$$I_n\cos\theta_{Ins} = \frac{V_r\sin\delta}{X} \tag{2-35a}$$

and

$$I_n\sin\theta_{Ins} = -\frac{V_s - V_r\cos\delta}{X}. \tag{2-35b}$$

Therefore, the magnitude of the natural line current ($\mathbf{I_n}$) is derived from Equations (2-35a) and (2-35b) using the identity $\left(\cos^2\theta_{Ins} + \sin^2\theta_{Ins} = 1\right)$ as

$$I_n = \sqrt{I_n^2\left(\cos^2\theta_{Ins} + \sin^2\theta_{Ins}\right)} = \sqrt{(I_n\cos\theta_{Ins})^2 + (I_n\sin\theta_{Ins})^2} \qquad \text{or}$$

$$I_n = \frac{1}{X}\sqrt{(V_r\sin\delta)^2 + (V_s - V_r\cos\delta)^2} \qquad \text{or}$$

$$I_n = \frac{1}{X}\sqrt{V_s^2 - 2V_sV_r\cos\delta + V_r^2\left(\cos^2\delta + \sin^2\delta\right)} \qquad \text{or}$$

$$I_n = \frac{1}{X}\sqrt{V_s^2 - 2V_sV_r\cos\delta + V_r^2} \quad \text{using the identity } \left(\cos^2\delta + \sin^2\delta\right) = 1, \text{or}$$

$$I_n = \frac{V_{Xn}}{X} \tag{2-33}$$

using Equation (2-30).

The phase angle of the natural line current ($\mathbf{I_n}$) with respect to the sending-end voltage is derived from Equations (2-35a) and (2-35b) as

$$\theta_{Ins} = \tan^{-1}(\tan\theta_{Ins}) = \tan^{-1}\frac{\sin\theta_{Ins}}{\cos\theta_{Ins}} = \tan^{-1}\frac{I_n\sin\theta_{Ins}}{I_n\cos\theta_{Ins}}$$

or

$$\theta_{Ins} = \tan^{-1}\frac{V_r\cos\delta - V_s}{V_r\sin\delta}. \tag{2-36}$$

By inspection of Figure 2-7, it can be written that

$$\frac{\sin B}{V_1} = \frac{\sin C}{V_s} = \frac{\sin(\pi - B - C)}{bV_{Xn}} = \frac{\sin(B + C)}{bV_{Xn}}$$

where b = multiplier of V_{Xn} from 0 to 1, or

$$V_1 = \frac{\sin B}{\sin C}V_s = \frac{\sin B}{\sin(B + C)}bV_{Xn} \tag{2-37a}$$

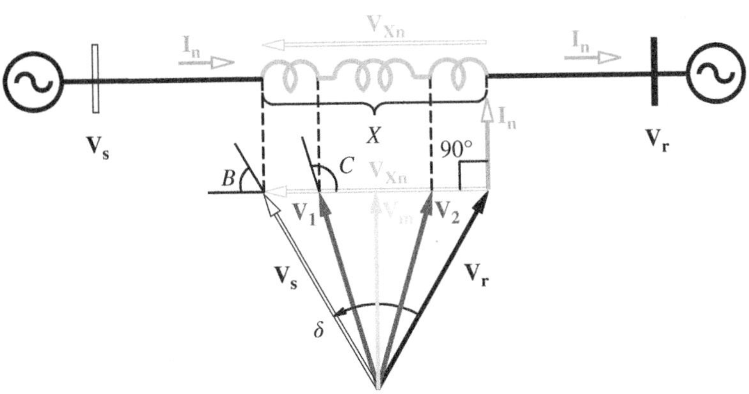

Figure 2-7 Two-generator/one-line power system network and the related phasor diagrams.

or

$$V_s \sin(B + C) = bV_{Xn} \sin C \qquad \text{or}$$

$$V_s \sin B \cos C = (bV_{Xn} - V_s \cos B) \sin C \qquad \text{or}$$

$$\tan C = \frac{V_s \sin B}{bV_{Xn} - V_s \cos B} = \frac{V_s \sin B}{bV_{Xn} - V_s\sqrt{1 - \sin^2 B}}. \tag{2-37b}$$

Also from Figure 2-7, it can be written that

$$\frac{\sin B}{V_r} = \frac{\sin \delta}{V_{Xn}} \qquad \text{or}$$

$$\sin B = \frac{V_r}{V_{Xn}} \sin \delta. \tag{2-37c}$$

Substituting Equation (2-37c) into Equation (2-37b), it can be written that

$$\tan C = \frac{V_s \sin B}{bV_{Xn} - V_s\sqrt{1 - \sin^2 B}} = \frac{V_s \frac{V_r}{V_{Xn}} \sin \delta}{bV_{Xn} - V_s\sqrt{1 - \frac{V_r^2}{V_{Xn}^2} \sin^2 \delta}} = \frac{V_s V_r \sin \delta}{bV_{Xn}^2 - V_s\sqrt{V_{Xn}^2 - V_r^2 \sin^2 \delta}}$$

or

$$\tan C = \frac{V_s V_r \sin \delta}{bV_{Xn}^2 - V_s\sqrt{V_s^2 - 2V_sV_r\cos\delta + V_r^2 - V_r^2 \sin^2 \delta}}$$

$$= \frac{V_s V_r \sin \delta}{bV_{Xn}^2 - V_s\sqrt{V_s^2 - 2V_sV_r\cos\delta + V_r^2 \cos^2 \delta}} = \frac{V_s V_r \sin \delta}{bV_{Xn}^2 - V_s\sqrt{(V_s - V_r\cos\delta)^2}}$$

using Equation (2-30), or

$$\tan C = \frac{V_s V_r \sin \delta}{bV_{Xn}^2 - V_s(V_s - V_r\cos\delta)}. \tag{2-37d}$$

Substituting Equations (2-37c) and (2-37d) into Equation (2-37a), it can be written that

$$V_1 = \frac{\sin B}{\sin C} V_s = \frac{\sin B}{\tan C}\sqrt{1 + \tan^2 C}\, V_s = \frac{V_s V_r \sin \delta}{V_{Xn}} \frac{\sqrt{1 + \tan^2 C}}{\tan C} = \frac{V_s V_r \sin \delta}{V_{Xn}} \sqrt{\left(\frac{1}{\tan C}\right)^2 + 1}$$

or

$$V_1 = \frac{V_s V_r \sin \delta}{V_{Xn}} \sqrt{1 + \left\{\frac{bV_{Xn}^2 - V_s(V_s - V_r\cos\delta)}{V_sV_r \sin\delta}\right\}^2}. \tag{2-37e}$$

In general, the voltage at any point along the line can be defined from Equation (2-37e) as

$$V_x = \frac{V_s V_r \sin \delta}{V_{Xn}} \sqrt{1 + \left\{\frac{bV_{Xn}^2 - V_s(V_s - V_r\cos\delta)}{V_sV_r \sin\delta}\right\}^2} \tag{2-38}$$

where $x = s, 1, m, 2, r,$ and so on as shown in Figure 2-7. Note that for $b = 0$ in Equation (2-38), $V_x = V_s$ and for $b = 1$, $V_x = V_r$.

Example 2-4

In Figure 2-6, the magnitudes (V_s and V_r) of the sending- and receiving-end voltages are one pu each, the power angle, $\delta = \delta_s - \delta_r = 30°$, and the line reactance, $X = 0.5$ pu; (a) what is the natural voltage ($\mathbf{V_{Xn}}$) across the line reactance (X)? (b) what is the resulting current ($\mathbf{I_n}$)? (c) plot the voltage magnitude along the line.

Answer 2-4

Electrical system data is given in Table 2-1.

a) From Equation (2-30), the magnitude of natural voltage is given by

$$V_{Xn} = \sqrt{V_s^2 - 2V_sV_r\cos\delta + V_r^2} = \sqrt{1^2 - 2\times 1\times 1\times\ \cos 30° + 1^2} = 0.517638 \text{ pu.}$$

b) From Equation (2-33), the natural line current, $I_n = \frac{V_{Xn}}{X} = \frac{0.517638}{0.5} = 1.035276$ pu. $\mathbf{I_n}$ lags $\mathbf{V_{Xn}}$ by 90° as shown in Figure 2-7.

c) Using Equation (2-38), the plot of the magnitude of voltage along the line is shown in Figure 2-8.

Note that these examples are worked out with 15 decimal digits for a greater accuracy, although the answers are shown with six decimal digits.

Table 2-1 Electrical system data.

Parameters	**Values**
Sending-end voltage (V_s) at reference phase angle	$1\angle 0°$ pu
Receiving-end voltage (V_r) and power angle ($-\delta$)	$1\angle -30°$ pu
Line reactance (X)	0.5 pu

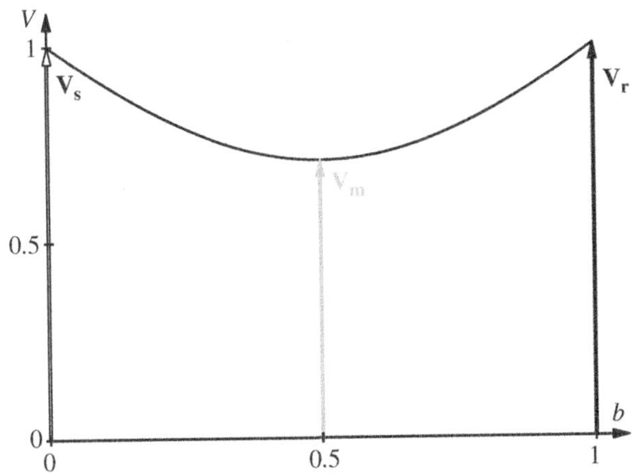

Figure 2-8 Voltage profile along a line when $V_s = V_r = 1$ pu and $\delta = 90°$.

2.1 Power Flow Equations for a Natural or Uncompensated Line

The natural or uncompensated active and reactive power flows (P_{sn} and Q_{sn}) at the sending end and (P_{rn} and Q_{rn}) at the receiving end of the line are derived as follows. The natural active power flow at the sending end is given by

$$P_{sn} = V_s I_n \cos(\delta_s - \theta_{In}) \qquad \text{or}$$

$P_{sn} = V_s I_n \cos(\delta_s - \{\delta_s + \theta_{Ins}\})$ where θ_{Ins} is the phase angle of the natural line current with respect to the sending-end voltage, or

$$P_{sn} = V_s I_n \cos(-\theta_{Ins}) \qquad \text{or}$$

$$P_{sn} = V_s I_n \cos\theta_{Ins}. \tag{2-39}$$

Equation (2-39) can be rewritten using Equation (2-35a) as

$$P_{sn} = V_s I_n \cos\theta_{Ins} = V_s \frac{V_r \sin\delta}{X} \qquad \text{or}$$

$$P_{sn} = a \times \sin\delta \tag{2-40}$$

where

$$a = \frac{V_s V_r}{X}. \tag{2-41}$$

The natural reactive power flow at the sending end is given by

$$Q_{sn} = V_s I_n \sin(\delta_s - \theta_{In}) \qquad \text{or}$$

$Q_{sn} = V_s I_n \sin(\delta_{se} - \{\delta_s + \theta_{Ins}\})$ where θ_{Ins} is the phase angle of the natural line current with respect to the sending-end voltage, or

$$Q_{sn} = V_s I_n \sin(-\theta_{Ins}) \qquad \text{or}$$

$$Q_{sn} = -V_s I_n \sin\theta_{Ins}. \tag{2-42}$$

Equation (2-42) can be rewritten using Equation (2-35b) as

$$Q_{sn} = -V_s I_n \sin\theta_{Ins} = -V_s \frac{-(V_s - V_r \cos\delta)}{X} \qquad \text{or}$$

$$Q_{sn} = a\left(\frac{V_s}{V_r} - \cos\delta\right) \tag{2-43}$$

where a is given in Equation (2-41).

Using Equation (2-41), Equations (2-40) and (2-43) can be combined as

$$\tan\delta = \frac{-P_{sn}}{Q_{sn} - \frac{V_s^2}{X}}. \tag{2-44}$$

The natural active power flow at the receiving end is given by

$$P_{rn} = V_r I_n \cos(\delta_r - \theta_{In}) \qquad \text{or}$$

$$P_{rn} = V_r I_n \cos(\{\delta_s - \delta\} - \{\delta_s + \theta_{Ins}\}) \qquad \text{using Equation (2-27)}$$

where θ_{Ins} is the phase angle of the natural line current with respect to the sending-end voltage, or

$$P_{rn} = V_r I_n \cos(-\delta - \theta_{Ins}) \quad \text{or}$$
$$P_{rn} = V_r I_n \cos(\delta + \theta_{Ins}) \quad \text{or}$$
$$P_{rn} = V_r \cos\delta I_n \cos\theta_{Ins} - V_r \sin\delta I_n \sin\theta_{Ins}. \tag{2-45}$$

Equation (2-45) can be rewritten using Equations (2-35a) and (2-35b) as

$$P_{rn} = V_r \cos\delta \frac{V_r \sin\delta}{X} - V_r \sin\delta \frac{-(V_s - V_r\cos\delta)}{X} \quad \text{or}$$
$$P_{rn} = \frac{V_r}{X}[\{V_r \sin\delta\}\cos\delta - \{-(V_s - V_r\cos\delta)\}\sin\delta] \quad \text{or}$$
$$P_{rn} = \frac{V_r}{X}[V_s \sin\delta] \quad \text{or}$$
$$P_{rn} = a \times \sin\delta \tag{2-46}$$

where a is given in Equation (2-41).

The natural reactive power flow at the receiving end is given by

$$Q_{rn} = V_r I_n \sin(\delta_r - \theta_{In}) \quad \text{or}$$
$$Q_{rn} = V_r I_n \sin(\{\delta_s - \delta\} - \{\delta_s + \theta_{Ins}\}) \quad \text{using Equation (2-27)}$$

where θ_{Ins} is the phase angle of the natural line current with respect to the sending-end voltage, or

$$Q_{rn} = V_r I_n \sin(-\delta - \theta_{Ins}) \quad \text{or}$$
$$Q_{rn} = -V_r I_n \sin(\delta + \theta_{Ins}) \quad \text{or}$$
$$Q_{rn} = -V_r \sin\delta I_n \cos\theta_{Ins} - V_r \cos\delta I_n \sin\theta_{Ins}. \tag{2-47}$$

Equation (2-47) can be rewritten using Equations (2-35a) and (2-35b) as

$$Q_{rn} = -V_r \sin\delta \frac{V_r \sin\delta}{X} - V_r \cos\delta \frac{-(V_s - V_r\cos\delta)}{X} \quad \text{or}$$
$$Q_{rn} = \frac{V_r}{X}[-V_r \sin\delta \sin\delta + (V_s - V_r\cos\delta)\cos\delta] \quad \text{or}$$
$$Q_{rn} = \frac{V_r}{X}[V_s\cos\delta - V_r] \quad \text{or}$$
$$Q_{rn} = a\left(\cos\delta - \frac{V_r}{V_s}\right) \tag{2-48}$$

where a is given in Equation (2-41).

Using Equation (2-41), Equations (2-46) and (2-48) can be combined as

$$\tan\delta = \frac{P_{rn}}{Q_{rn} + \frac{V_r^2}{X}} \tag{2-49}$$

where

$$\delta = \delta_s - \delta_r. \tag{2-27}$$

The equations for the natural active and reactive power flows (P_{sn} and Q_{sn}) at the sending end and (P_{rn} and Q_{rn}) at the receiving end are given in Table 2-2.

Table 2-2 Equations for natural active and reactive power flows (P_{sn} and Q_{sn}) at the sending end and (P_{rn} and Q_{rn}) at the receiving end.

$P_{sn} = a \times \sin\delta$ (2-40)	$Q_{sn} = a\left(\frac{V_s}{V_r} - \cos\delta\right)$ (2-43)	$a = \frac{V_s V_r}{X}$ (2-41)	$\delta = \delta_s - \delta_r$ (2-27)
$P_{rn} = a \times \sin\delta$ (2-46)	$Q_{rn} = a\left(\cos\delta - \frac{V_r}{V_s}\right)$ (2-48)	$a = \frac{V_s V_r}{X}$ (2-41)	$\delta = \delta_s - \delta_r$ (2-27)

Power flow is a function of three parameters: (1) magnitudes of line voltages at the sending and receiving ends, (2) phase angle between the line voltages at the sending and receiving ends, and (3) line reactance. Regulating one parameter only changes both active power and reactive power flows simultaneously; therefore, the line is not optimized for maximum active power flow with minimum reactive power flow. To achieve independent active and reactive power flows, two parameters need to be regulated simultaneously, which is discussed later in the chapter.

An additional reactor or capacitor, connected in series with the line, regulates the effective line reactance between its sending and receiving ends and the consequence is the simultaneous regulation of the active and reactive power flows in the line. The reactance regulation can be achieved with a combination of voltage regulation and phase angle regulation as explained in Section 2.2.2.5.2 (Reactance Regulator). Therefore, two primary parameters for regulating the active and reactive power flows in the line are (1) magnitude and (2) phase angle of the line voltage. When one or more power flow control parameters (voltage magnitude, its phase angle, and/or line reactance) are regulated, the line is considered to be "compensated."

Example 2-5

Consider the electrical system data as given in Table 2-1. What are the active and reactive power flows (P_{sn} and Q_{sn}) at the sending end, (P_{mn} and Q_{mn}) at the midpoint, and (P_{rn} and Q_{rn}) at the receiving end of the line?

Answer 2-5

From Equation (2-41): $a = \frac{V_s V_r}{X} = \frac{1 \times 1}{0.5} = 2.$

From Equation (2-40): $P_{sn} = a \times \sin\delta$ and (2-46): $P_{rn} = a \times \sin\delta = 2 \times \sin 30°$, the natural transmitted active power, $P_{sn} = P_{rn} = 1$ pu.

From Equation (2-43), the reactive power is $Q_{sn} = a\left(\frac{V_s}{V_r} - \cos\delta\right) = 2\left(\frac{1}{1} - \cos 30°\right) = 0.267649$ pu at the sending (source) end.

From Equation (2-48), the reactive power is $Q_{rn} = a\left(\cos\delta - \frac{V_r}{V_s}\right) = 1\left(\cos 30° - \frac{1}{1}\right) =$ -0.267949 pu at the receiving (load) end.

In the line shown in Figure 2-6, the natural active power flow is the same at the sending end ($P_{sn} = V_s I_{ds}$), at the midpoint ($P_{mn} = V_m I_n$), and at the receiving end ($P_{rn} = V_r I_{dr}$); it is one pu. However, the natural reactive power flow ($Q_{sn} = V_s I_{qs}$) at the sending end is 0.267949 pu (inductive); at the midpoint, it is $Q_{mn} = 0$; the natural reactive power flow ($Q_{rn} = V_r I_{qr}$) at the receiving end is 0.267949 pu (capacitive).

Additional Discussion

At the sending end, the direct component of current is $I_{ds} = P_{sn}/V_s = 1/1 = 1$ pu; the quadrature component of current is $I_{qs} = Q_{sn}/V_s = 0.267949/1 = 0.267949$ pu lagging; therefore, the natural line current is $I_n = \sqrt{I_{ds}^2 + I_{qs}^2} = \sqrt{1^2 + 0.267949^2} = 1.035276$ pu.

At the receiving end, the direct component of current is $I_{dr} = P_{rn}/V_r = 1/1 = 1$ pu; the quadrature component of current is $I_{qr} = Q_{rn}/V_r = 0.267949/1 = 0.267949$ pu leading; therefore, the natural line current is $I_n = \sqrt{I_{dr}^2 + I_{qr}^2} = \sqrt{1^2 + 0.267949^2} = 1.035276$ pu.

The voltage at the midpoint ($b = 0.5$) can be calculated by using Equation (2-38) as

$$V_m = \frac{V_s V_r \sin\delta}{V_{Xn}} \sqrt{1 + \left\{\frac{bV_{Xn}^2 - V_s(V_s - V_r \cos\delta)}{V_s V_r \sin\delta}\right\}^2}$$

$$= \frac{1 \times 1 \times 0.5}{0.517638} \sqrt{1 + \left\{\frac{0.5 \times 0.517638^2 - 1 \times (1 - 1 \times 0.866025)}{1 \times 1 \times 0.5}\right\}^2} = 0.965926 \text{ pu.}$$

where from Equation (2-30): $V_{Xn} = \sqrt{V_s^2 - 2V_s V_r \cos\delta + V_r^2}$, the magnitude of the natural voltage across the line reactance (X) is $V_{Xn} = 0.517638$ pu.

The direct component of current is $I_{md} = P_{mn}/V_m = 1/0.965926 = 1.035276$ pu $= I_n$; the quadrature component of current is $I_{mq} = \sqrt{I_n^2 - I_{md}^2} = 0$.

Therefore, there is only active, but no reactive power flow at the midpoint of the line. In this example of a two-bus network, the reactive power absorbed by the "left half" of the line is supplied by the sending-end voltage source (generator) and that absorbed by the "right half" of the line is supplied by the receiving-end voltage source (load) that can be thought of as a back emf of a motor.

2.2 Power Flow Equations for a Compensated Line

Consider the line is compensated at its sending end. Assuming that there are no changes in the line reactance (X) and the receiving-end voltage ($\mathbf{V_r} = V_r\angle\delta_r$), the power flow in the line can be regulated by controlling the magnitude and phase angle of the sending-end voltage so that it becomes the modified sending-end voltage ($\mathbf{V_{s'}} = V_{s'}\angle\delta_{s'}$). Controlling one parameter (voltage magnitude or its phase angle) using a PFC affects both the active and reactive power flows in the line simultaneously. The key to regulate the active and reactive power flows independently so that they become a particular pair of values is to control the sending-end voltage to a specific magnitude ($V_{s'}$) and phase angle ($\delta_{s'}$). There are two ways to accomplish this objective.

1) The modified sending-end voltage ($\mathbf{V_{s'}} = V_{s'}\angle\delta_{s'}$) can be obtained by using a shunt compensator, which is a voltage ($\mathbf{E} = E\angle(\delta_{s'} + \alpha)$) behind a tie-reactance (X_{TIE}) as shown in Figure 2-9a. Since the compensating voltage is connected in shunt with the line, it is controlled as a current source. When multiple shunt compensators are connected to a common POC, each one is controlled as a current source whose magnitude is the voltage across the tie reactor, which is the difference between $\mathbf{V_{s'}}$ and $\mathbf{E}$, divided by the tie-reactance (jX_{TIE}). Note that when a shunt compensator is used, the modified sending-end voltage ($\mathbf{V_{s'}}$) is the shunt-compensating voltage.
2) The modified sending-end voltage ($\mathbf{V_{s'}} = V_{s'}\angle\delta_{s'}$) can be obtained by using a series compensator, which adds a series-compensating voltage ($\mathbf{V_{s's}} = V_{s's}\angle(\delta_s + \beta)$) to the original sending-end

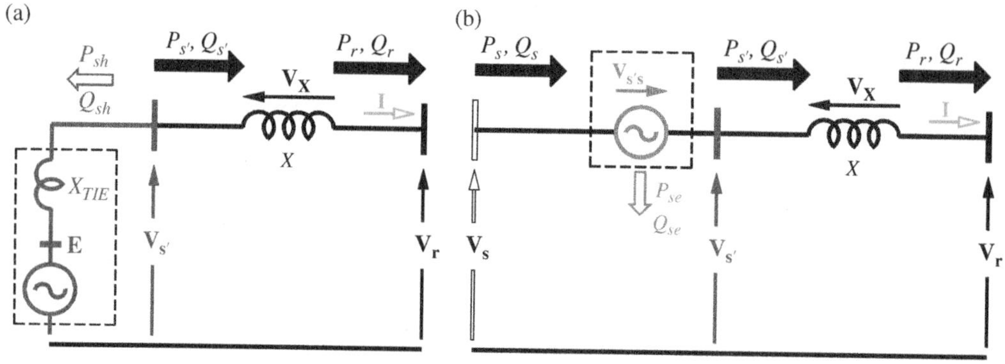

Figure 2-9 Two methods of controlling the line voltage: (a) with a shunt-compensating voltage and (b) with a series-compensating voltage.

voltage ($\mathbf{V_s} = V_s\angle\delta_s$) as shown in Figure 2-9b. Since the compensating voltage is connected in series, it is controlled as a voltage source.

Regardless of whether the compensating voltage is shunt-connected or series-connected, it is at any phase angle with respect to the current flowing through it and, therefore, it exchanges both active and reactive powers with the line at its terminal.

The additional symbols shown in the figure are

$\mathbf{V_{s'}}$ is the modified sending-end voltage with a magnitude ($V_{s'}$) and a phase angle ($\delta_{s'}$) or ($\delta_s + \psi$),
$\mathbf{V_{s's}}$ is the series-compensating voltage with a magnitude ($V_{s's}$) and a phase angle ($\delta_s + \beta$),
$\mathbf{E}$ is the shunt-connected voltage with a magnitude (E) and a phase angle ($\delta_{s'} + \alpha$),
X_{TIE} is the tie-reactance between the shunt-connected voltage ($\mathbf{E}$) and the modified sending-end voltage ($\mathbf{V_{s'}}$),
$\mathbf{I}$ is the line current with a magnitude (I) and a phase angle (θ_I),
P_s is the active power flow at the sending end,
Q_s is the reactive power flow at the sending end,
$P_{s'}$ is the active power flow at the modified sending end,
$Q_{s'}$ is the reactive power flow at the modified sending end,
P_r is the active power flow at the receiving end,
Q_r is the reactive power flow at the receiving end,
P_{sh} is the exchanged active power by the shunt-compensating unit,
Q_{sh} is the exchanged reactive power by the shunt-compensating unit,
P_{se} is the exchanged active power by the series-compensating unit, and
Q_{se} is the exchanged reactive power by the series-compensating unit.

In order to implement a shunt-compensating voltage with a specific magnitude and phase angle, it is necessary to provide a bidirectional path for the exchanged power of the shunt-compensating voltage with the line. To deliver or absorb the active power to and from the line, a source or a sink is required. To deliver or absorb the reactive power to and from the line, a capacitive or an inductive reactance is required. Figure 2-10 shows a point-to-point power transfer method with a Shunt–Shunt Compensator. The exchanged active powers (P_{1sh} and P_{2sh}) by the shunt-connected voltages ($\mathbf{E_1} = E_1\angle(\delta_s + \alpha_1)$ and $\mathbf{E_2} = E_2\angle(\delta_{s'} + \alpha_2)$) with their respective AC lines are maintained in balance. The losses in the compensators are usually supplied from the unit that delivers active power.

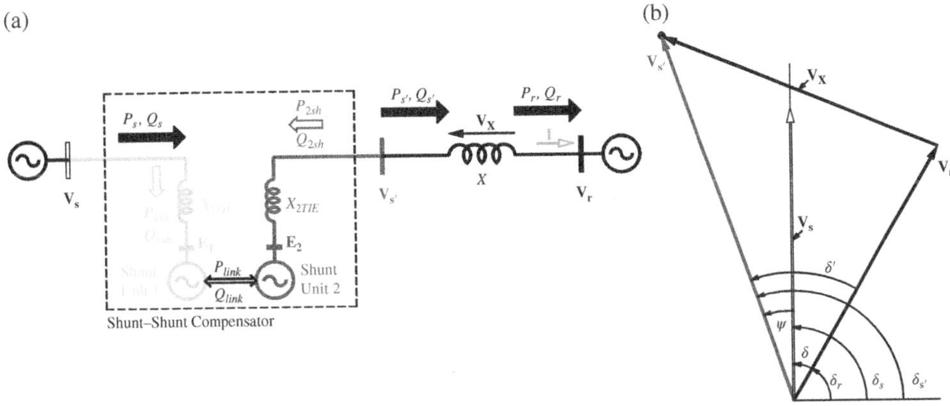

Figure 2-10 Scheme for implementing a shunt-compensating voltage.

The exchanged active power by one shunt-connected compensator with the line flows bidirectionally to the other shunt-connected compensator through the link, which can be electromechanical, DC or AC, or magnetic in nature. However, the exchanged reactive powers (Q_{1sh} and Q_{2sh}) by the shunt-connected voltages ($\mathbf{E_1} = E_1\angle(\delta_s + \alpha_1)$ and $\mathbf{E_2} = E_2\angle(\delta_{s'} + \alpha_2)$) with their respective AC lines are dependent on each other only in the case of a magnetic link as discussed next.

In order to implement a series-compensating voltage ($\mathbf{V_{s's}} = V_{s's}\angle(\delta_s + \beta)$) with a specific magnitude ($V_{s's}$) and operating in its entire controllable range of $0° \leq \beta \leq 360°$, it is also necessary to provide a bidirectional path for the exchanged power of the series-compensating voltage with the line. To deliver or absorb the active power to and from the line, a source or a sink is required. To deliver or absorb the reactive power to and from the line, a capacitive or an inductive reactance is required. Figure 2-11 shows a power flow control method with a Shunt–Series Compensator. The exchanged active power (P_{se}) by the series-compensating voltage ($\mathbf{V_{s's}}$) with the line flows bidirectionally to the shunt-connected voltage ($\mathbf{E} = E\angle(\delta_s + \alpha)$) through the link, which can be electromechanical, DC or AC, or magnetic in nature. The exchanged reactive power (Q_{se}) by the series-compensating voltage ($\mathbf{V_{s's}}$) with the line flows bidirectionally to the shunt-connected voltage

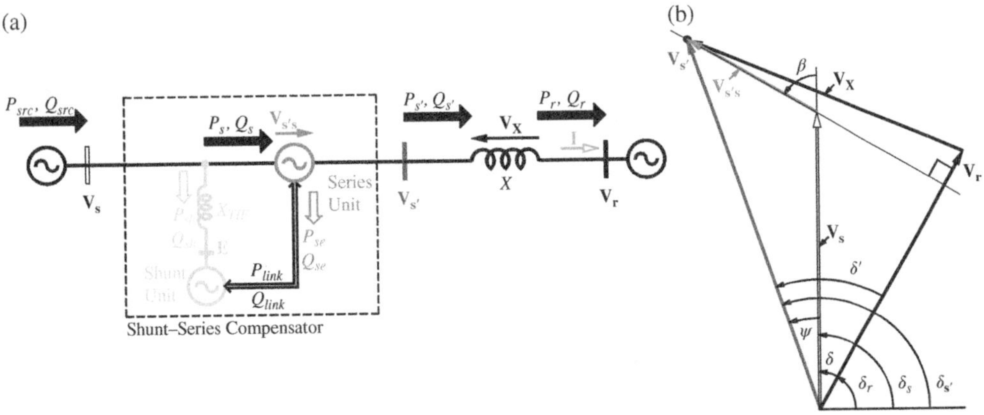

Figure 2-11 Scheme for implementing a series-compensating voltage.

($\mathbf{E} = E\angle(\delta_s + \alpha)$) only in the case of a magnetic link as discussed next. A series-compensating voltage ($\mathbf{V_{s's}}$) can modify the line voltage to a specific magnitude and phase angle, which results in an independent control of active and reactive power flows in the line. Note that the active and reactive power flows (P_{src} and Q_{src}) at the source are different than the active and reactive power flows (P_s and Q_s) at the sending end of the line. Their differences (P_{sh} and Q_{sh}) are the active and reactive power flows through the shunt-connected voltage (**E**).

This basic principle of a shunt- and series-compensating voltage can be implemented with the use of (1) electrical machines, (2) VSCs, and (3) transformer/LTCs-based technologies. The exchanged active power by the compensating Series Unit with the line passes through an electromechanical link, a DC or AC link, or a magnetic link, respectively. However, the exchanged reactive power by the respective Shunt and Series Units with the line are related only when they pass through the magnetic link as in the case of an ST. Moreover, when only the shunt- or series-reactance compensation is used, the exchanged active power with the line is limited to the losses in the compensating units.

1) Electric machine-based Shunt–Shunt and Shunt-Series Compensators are shown in Figure 2-12. The tie-reactance (X_{TIE}) may represent the internal reactance of the machine.

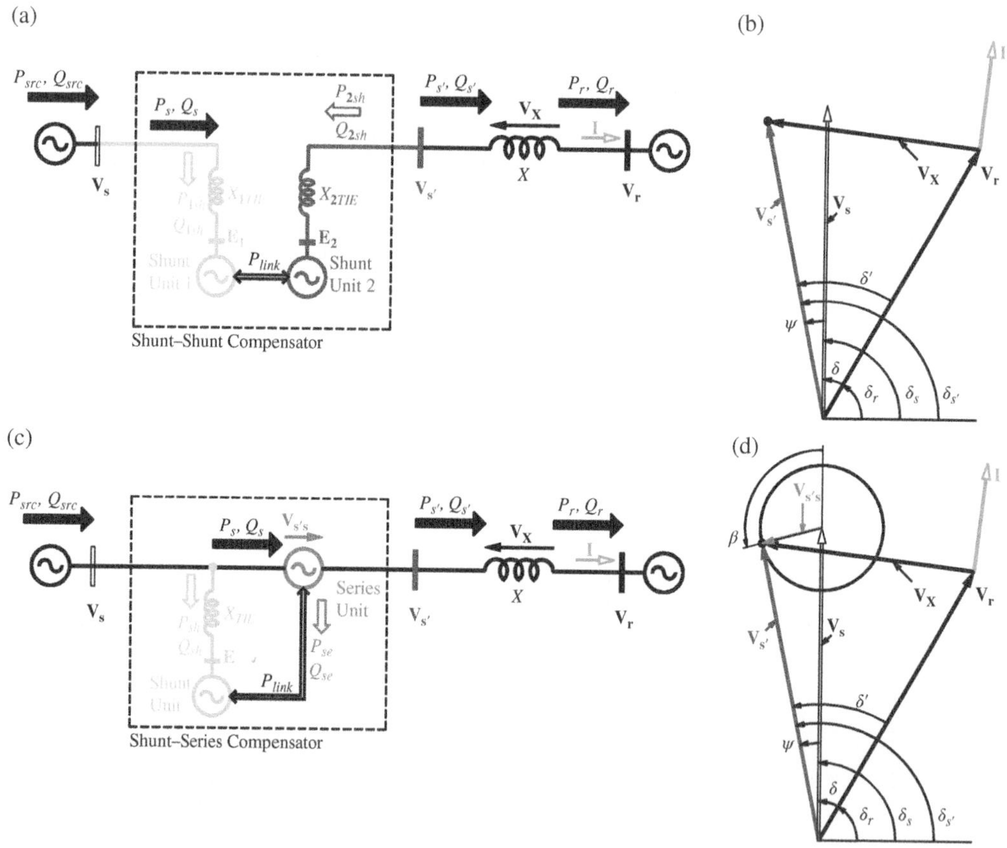

Figure 2-12 Electrical machine-based solutions for power flow controllers using (a) Shunt–Shunt Compensator and (b) its related phasor diagram; (c) Shunt–Series Compensator and (d) its related phasor diagram.

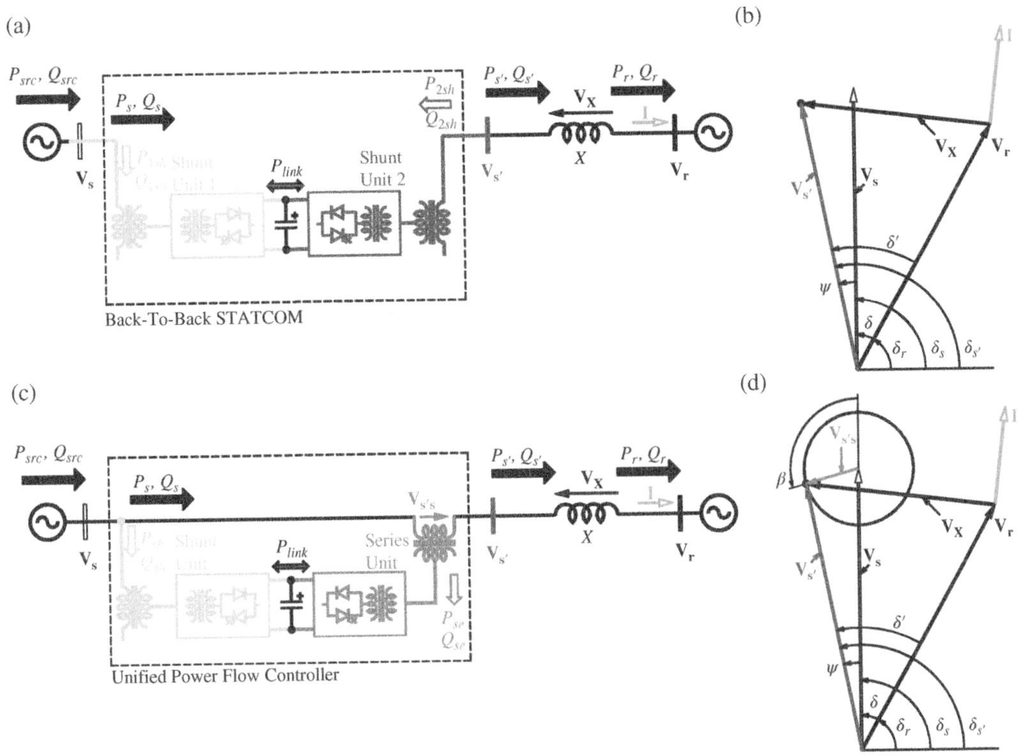

Figure 2-13 VSC-based solutions for power flow controllers using (a) Shunt–Shunt Compensator and (b) its related phasor diagram; (c) Shunt–Series Compensator and (d) its related phasor diagram.

2) VSC-based Shunt–Shunt and Shunt–Series Compensators are shown in Figure 2-13. The tie-reactance (X_{TIE}) may function as either the reactance of the magnetic circuit or the filter reactance at the output of the inverter. An energy storage may serve as a source or a sink as well.
3) Transformer/LTCs-based Shunt–Shunt and Shunt–Series Compensators are shown in Figure 2-14. The tie-reactance (X_{TIE}) may represent the transformer leakage reactance.

2.2.1 Shunt-Compensating Voltage

A shunt-compensating voltage can modify the sending-end voltage of the line. In some special cases for point-to-point transfer of power between two isolated networks or interconnection of two transmission lines with different voltages or phase angles (or frequencies), this scheme may be the preferred choice.

Referring to Figure 2-14b or Figure 2-14d, it can be written that

$$\mathbf{V_{s'}} = \mathbf{V_r} + \mathbf{V_X} \quad \text{or} \quad V_{s'}\angle\delta_{s'} = V_r\angle\delta_r + V_X\angle\theta_{VX} \tag{2-50a}$$

or

$$\mathbf{V_X} = \mathbf{V_{s'}} - \mathbf{V_r} \quad \text{or} \quad V_X\angle\theta_{VX} = V_{s'}\angle\delta_{s'} - V_r\angle\delta_r \tag{2-50b}$$

where $\mathbf{V_X}$, the difference between the modified sending-end voltage ($\mathbf{V_{s'}}$) and receiving-end voltage ($\mathbf{V_r}$), is the voltage ($\mathbf{V_X}$) across the line reactance (X) and θ_{VX} is the phase angle of the voltage ($\mathbf{V_X}$).

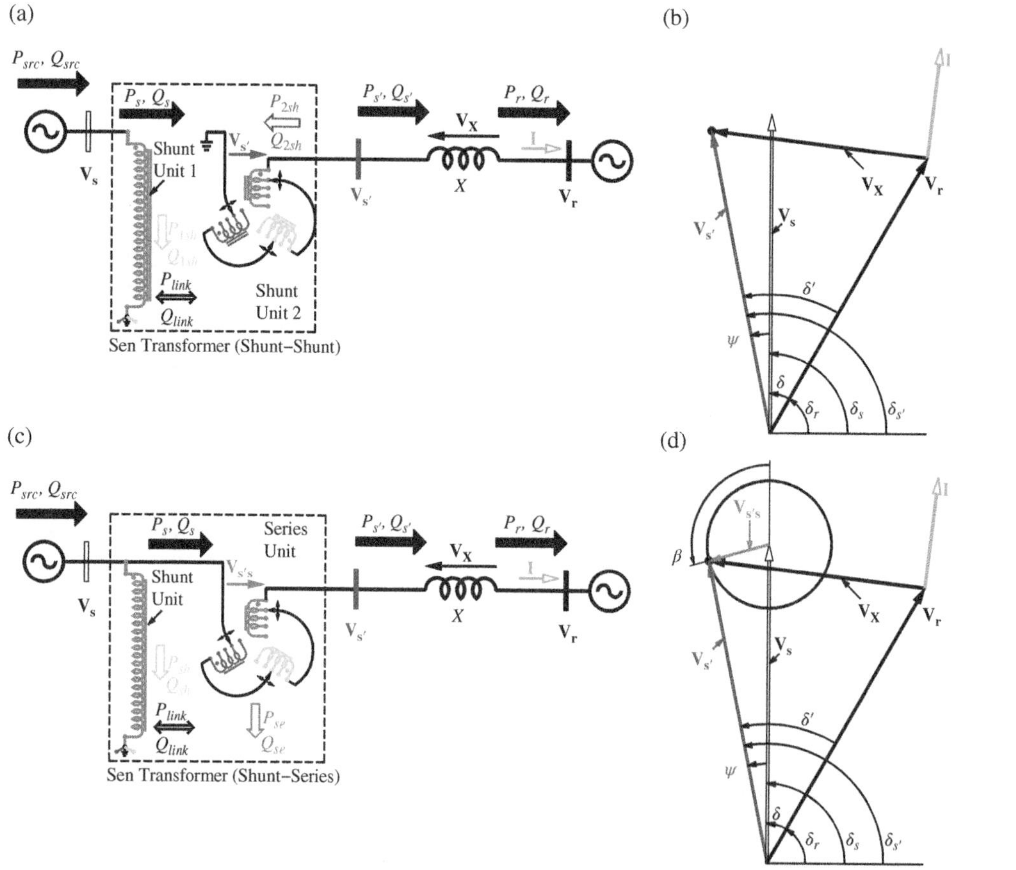

Figure 2-14 Transformer/LTCs-based solutions for power flow controllers using (a) Shunt–Shunt Compensator and (b) its related phasor diagram; (c) Shunt–Series Compensator and (d) its related phasor diagram.

It can also be written that

$$\delta' = \delta + \psi \tag{2-51a}$$

or

$$\psi = \delta' - \delta \tag{2-51b}$$

or

$$\psi = (\delta_{s'} - \delta_r) - (\delta_s - \delta_r) = \delta_{s'} - \delta_s \tag{2-51c}$$

or

$$\delta_{s'} = \delta_s + \psi. \tag{2-51d}$$

Using Equations (2-51d) and (2-27), the Equation (2-50a) can be rewritten as

$$V_{s'}\angle(\delta_s + \psi) = V_r\angle(\delta_s - \delta) + V_X\angle(\delta_s + \theta_{VXs}) \quad \text{or}$$

$$V_X\angle(\delta_s + \theta_{VXs}) = V_{s'}\angle(\delta_s + \psi) - V_r\angle(\delta_s - \delta) \tag{2-52a}$$

where θ_{VXs} is the phase angle of the voltage ($\mathbf{V_X}$) across the line reactance (X) with respect to the sending-end voltage and

$$V_{s'}\angle(\delta_r + \delta') = V_r\angle\delta_r + V_X\angle(\delta_r + \theta_{VXr}) \qquad \text{or}$$

$$V_X\angle(\delta_r + \theta_{VXr}) = V_{s'}\angle(\delta_r + \delta') - V_r\angle\delta_r \tag{2-52b}$$

where θ_{VXr} is the phase angle of the voltage ($\mathbf{V_X}$) across the line reactance (X) with respect to the receiving-end voltage.

Considering the sending-end voltage phasor as reference (*i.e.* $\delta_s = 0°$), it can be written from Equation (2-52a) that

$$V_X\angle\theta_{VXs} = V_{s'}\angle\psi - V_r\angle(-\delta)$$

or

$$V_X\cos\theta_{VXs} + jV_X\sin\theta_{VXs} = V_{s'}\cos\psi - V_r\cos\delta + jV_{s'}\sin\psi + jV_r\sin\delta.$$

Equating the real and imaginary parts of both sides of the above equation, it can be written that

$$V_X\cos\theta_{VXs} = V_{s'}\cos\psi - V_r\cos\delta \tag{2-53a}$$

and

$$V_X\sin\theta_{VXs} = V_{s'}\sin\psi + V_r\sin\delta. \tag{2-53b}$$

Therefore, the magnitude of the voltage ($\mathbf{V_X}$) across the line reactance is derived from Equations (2-53a) and (2-53b) using the identity $\left(\cos^2\theta_{VXs} + \sin^2\theta_{VXs} = 1\right)$ as

$$V_X = \sqrt{V_X^2\left(\cos^2\theta_{VXs} + \sin^2\theta_{VXs}\right)} = \sqrt{(V_X\cos\theta_{VXs})^2 + (V_X\sin\theta_{VXs})^2}$$

or

$$V_X = \sqrt{(V_{s'}\cos\psi - V_r\cos\delta)^2 + (V_{s'}\sin\psi + V_r\sin\delta)^2}$$

or

$$V_X = \sqrt{V_{s'}^2\left(\cos^2\psi + \sin^2\psi\right) - 2V_{s'}V_r\cos(\delta + \psi) + V_r^2\left(\cos^2\delta + \sin^2\delta\right)}$$

or

$$V_X = \sqrt{V_{s'}^2 - 2V_{s'}V_r\cos(\delta + \psi) + V_r^2} \tag{2-54a}$$

using the identities $\left(\cos^2\psi + \sin^2\psi = 1\right)$ and $\left(\cos^2\delta + \sin^2\delta = 1\right)$, or

$$V_X = \sqrt{V_{s'}^2 - 2V_{s'}V_r\cos\delta' + V_r^2} \tag{2-54b}$$

using Equation (2-51a).

The phase angle of the voltage ($\mathbf{V_X}$) across the line reactance with respect to the sending-end voltage is derived from Equations (2-53a) and (2-53b) as

$$\theta_{VXs} = \tan^{-1}\left(\tan\theta_{VXs}\right) = \tan^{-1}\frac{\sin\theta_{VXs}}{\cos\theta_{VXs}} = \tan^{-1}\frac{V_X\sin\theta_{VXs}}{V_X\cos\theta_{VXs}}$$

or

$$\theta_{VXs} = \tan^{-1}\frac{V_{s'}\sin\psi + V_r\sin\delta}{V_{s'}\cos\psi - V_r\cos\delta}. \tag{2-55}$$

The line current (**I**) is defined as

$$\mathbf{I} = \frac{\mathbf{V_X}}{jX} \quad \text{or} \quad I\angle\theta_I = \frac{V_X\angle\theta_{VX}}{jX} \quad \text{or} \quad I\angle(\delta_s + \theta_{Is}) = \frac{V_X\angle(\delta_s + \theta_{VXs})}{jX}$$

where θ_{Is} is the phase angle of the line current (**I**) with respect to the sending-end voltage ($\mathbf{V_s}$).

Considering the sending-end voltage phasor as reference (*i.e.* $\delta_s = 0°$), it can be written from the above equation that

$$I\angle\theta_{Is} = \frac{V_X\angle\theta_{VXs}}{jX} \tag{2-56a}$$

or

$$I\angle\theta_{Is} = \frac{V_X}{X}\angle(\theta_{VXs} - 90°) \tag{2-56b}$$

where the magnitude of the prevailing line current (**I**) is given by

$$I = \frac{V_X}{X} \tag{2-57}$$

and the phase angle of the line current (**I**) with respect to the sending-end voltage ($\mathbf{V_s}$) is given by

$$\theta_{Is} = \theta_{VXs} - 90° \tag{2-58}$$

where θ_{VXs} is given in Equation (2-55).

2.2.1.1 Power Flow at the Modified Sending End with a Shunt-Compensating Voltage

When the line is compensated, the active and reactive power flows ($P_{s'}$ and $Q_{s'}$) at the modified sending end are expressed by substituting V_s with $V_{s'}$, δ with δ', and a with a' in Equations (2-40), (2-43), and (2-41) as

$$P_{s'} = a'\sin\delta' \tag{2-59}$$

and

$$Q_{s'} = a'\left(\frac{V_{s'}}{V_r} - \cos\delta'\right) \tag{2-60}$$

where

$$a' = \frac{V_{s'}V_r}{X} \tag{2-61}$$

and the modified power angle is given by

$$\delta' = \delta_{s'} - \delta_r. \tag{2-62}$$

Using Equations (2-59) and (2-60), the apparent power ($S_{s'}$) at the modified sending end can be defined as

$$S_{s'} = \sqrt{P_{s'}^2 + Q_{s'}^2}. \tag{2-63}$$

Using Equation (2-61), Equations (2-59) and (2-60) can be combined as

$$P_{s'}^2 + \left(Q_{s'} - \frac{V_{s'}^2}{X}\right)^2 = a'^2 \tag{2-64}$$

and

$$\tan\delta' = \frac{P_{s'}}{\frac{V_{s'}^2}{X} - Q_{s'}}. \tag{2-65}$$

Equation (2-64) represents the locus of a circle, centered at $\left[0, \frac{V_{s'}^2}{X}\right]$ with a radius of a'.

The magnitude ($V_{s'}$) of the shunt-compensating voltage is expressed by combining Equations (2-61) and (2-64) as follows:

$$\left(\frac{V_{s'}^2}{X}\right)^2 - \left(\frac{V_r^2}{X^2} + \frac{2Q_{s'}}{X}\right)V_{s'}^2 + P_{s'}^2 + Q_{s'}^2 = 0$$

or

$$\left(V_{s'}^2\right)^2 - \left(V_r^2 + 2X\,Q_{s'}\right)V_{s'}^2 + X^2\left(P_{s'}^2 + Q_{s'}^2\right) = 0. \tag{2-66}$$

Solving the above quadratic equation results in

$$V_{s'}^2 = \frac{\left(V_r^2 + 2XQ_{s'}\right) \pm \sqrt{\left(V_r^2 + 2XQ_{s'}\right)^2 - 4X^2\left(P_{s'}^2 + Q_{s'}^2\right)}}{2}$$

or

$$V_{s'}^2 = \left(\frac{V_r^2}{2} + XQ_{s'}\right) \pm \sqrt{\left(\frac{V_r^2}{2} + XQ_{s'}\right)^2 - X^2\left(P_{s'}^2 + Q_{s'}^2\right)}$$

or

$$V_{s'} = +\sqrt{\left(\frac{V_r^2}{2} + XQ_{s'}\right) + \sqrt{\left(\frac{V_r^2}{2} + XQ_{s'}\right)^2 - X^2\left(P_{s'}^2 + Q_{s'}^2\right)}} \tag{2-67a}$$

or

$$V_{s'} = +\sqrt{\left(\frac{V_r^2}{2} + XQ_{s'}\right) - \sqrt{\left(\frac{V_r^2}{2} + XQ_{s'}\right)^2 - X^2\left(P_{s'}^2 + Q_{s'}^2\right)}} \tag{2-67b}$$

or

$$V_{s'} = -\sqrt{\left(\frac{V_r^2}{2} + XQ_{s'}\right) + \sqrt{\left(\frac{V_r^2}{2} + XQ_{s'}\right)^2 - X^2\left(P_{s'}^2 + Q_{s'}^2\right)}} \tag{2-67c}$$

or

$$V_{s'} = -\sqrt{\left(\frac{V_r^2}{2} + XQ_{s'}\right) - \sqrt{\left(\frac{V_r^2}{2} + XQ_{s'}\right)^2 - X^2\left(P_{s'}^2 + Q_{s'}^2\right)}}. \tag{2-67d}$$

The modified power angle (δ') of the shunt-compensating voltage ($\mathbf{V_{s'}}$) is derived from Equation (2-65) as

$$\delta' = \tan^{-1}\frac{P_{s'}}{\frac{V_{s'}^2}{X} - Q_{s'}}. \tag{2-68}$$

For independent control of active and reactive power flows ($P_{s'}$ and $Q_{s'}$), the magnitude ($V_{s'}$) of the modified sending-end voltage ($\mathbf{V_{s'}}$) and the modified power angle (δ') are determined from Equations (2-67a) and (2-68), respectively. For a desired amount of power flow, Equation (2-67a) results in a larger $V_{s'}$ and smaller δ' whereas Equation (2-67b) results in a smaller $V_{s'}$ and larger δ'. Equations (2-67c) and (2-67d) result in the same respective $V_{s'}$ as obtained in Equations (2-67a) and (2-67b); however, the corresponding δ' is 180° opposite. The final selection of $V_{s'}$ is based on the best design considerations.

Therefore, using Equations (2-63) and (2-67a), the line current (I) is given by

$$I = \frac{S_{s'}}{V_{s'}}. \tag{2-69}$$

Rearranging Equation (2-65), it can be written that

$$Q_{s'} = -\frac{1}{\tan\delta'}P_{s'} + \frac{V_{s'}^2}{X}; \tag{2-70a}$$

since for voltage regulation, only $V_{s'}$ changes and δ' remains the same as δ, Equation (2-70a) can be rewritten as

$$Q_{s'} = -\frac{1}{\tan\delta}P_{s'} + \frac{V_{s'}^2}{X} \tag{2-70b}$$

or substituting for tanδ from Equation (2-44), it can be written that

$$Q_{s'} = \frac{Q_{sn} - \frac{V_s^2}{X}}{P_{sn}}P_{s'} + \frac{V_{s'}^2}{X}. \tag{2-70c}$$

From Equations (2-70b) and (2-70c), the $P_{s'} - Q_{s'}$ relationship is defined with a slope ($m_{s'}$) and an intercept ($c_{s'}$) as

$$Q_{s'} = m_{s'}P_{s'} + c_{s'} \tag{2-71}$$

where

$$m_{s'} = -\frac{1}{\tan\delta} = \frac{Q_{sn} - \frac{V_s^2}{X}}{P_{sn}} \tag{2-72}$$

and

$$c_{s'} = \frac{V_{s'}^2}{X}. \tag{2-73}$$

In the case of a VR, the variations of reactive power ($Q_{s'}$) and the magnitude ($V_{s'}$) of the voltage at the modified sending end are shown in Figure 2-15 as functions of the active power ($P_{s'}$). When $P_{s'} = 0$ and $Q_{s'} = 0$, $V_{s'} = 0$; when $P_{s'} = P_{sn}$ and $Q_{s'} = Q_{sn}$, $V_{s'} = V_s$. The practical operating region of a VR is also shown. Note that the slope in the above equation may be constant for an electrical system, given in Table 2-1; however, its intercept keeps changing as the voltage ($V_{s'}$) at the modified sending end changes.

Figure 2-15 $Q_{s'}$ and $V_{s'}$ versus $P_{s'}$ at the modified sending end of the line for voltage regulation at power angle $\delta = 30°$ when $V_s = V_r = 1$ pu, $X = 0.5$ pu, and $R = 0$ ($X/R = \infty$), using a Shunt–Shunt configuration.

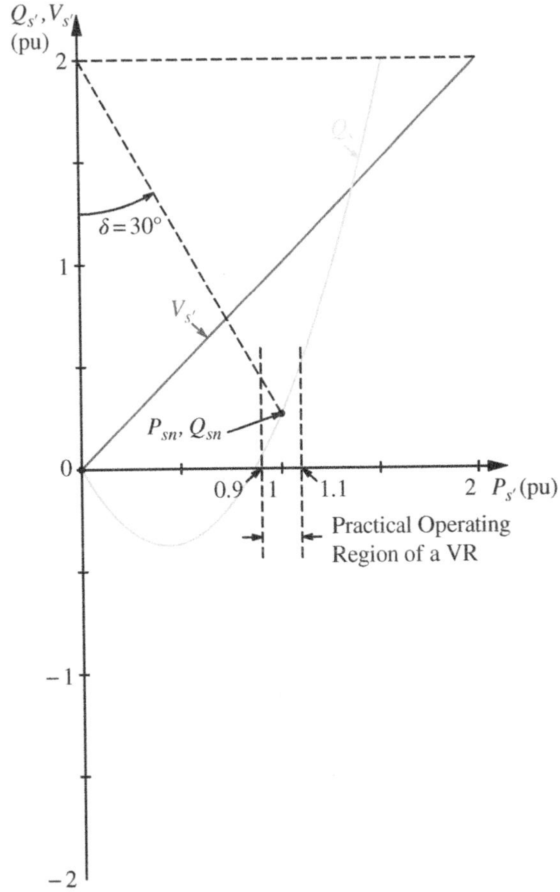

Example 2-6

Consider the electrical system data as given in Table 2-1. For $V_{s'} = 1.2$ pu and $V_{s'} = 0.8$ pu, calculate $m_{s'}$ and $c_{s'}$.

Answer 2-6

In Example 2-5, it was calculated that the natural transmitted active power, $P_{sn} = 1$ pu and the reactive power, $Q_{sn} = 0.267949$ pu at the source end. Therefore, from Equation (2-72),

$$m_{s'} = \frac{1}{P_{sn}}\left(Q_{sn} - \frac{V_S^2}{X}\right) = \frac{1}{1}\left(0.267949 - \frac{1^2}{0.5}\right) = -1.732051.$$

For $V_{s'} = 1.2$ pu, using Equation (2-73), $c_{s'} = \frac{V_{s'}^2}{X} = \frac{1.2^2}{0.5} = 2.88.$

For $V_{s'} = 0.8$ pu, using Equation (2-73), $c_{s'} = \frac{V_{s'}^2}{X} = \frac{0.8^2}{0.5} = 1.28.$

2.2.1.2 Power Flow at the Receiving End with a Shunt-Compensating Voltage

When the line is compensated, the active and reactive power flows (P_r and Q_r) at the receiving end are expressed by substituting V_s with $V_{s'}$, δ with δ', and a with a' in Equations (2-46) and (2-48) as

$$P_r = a' \sin\delta' \tag{2-74}$$

and

$$Q_r = a'\left(\cos\delta' - \frac{V_r}{V_{s'}}\right). \tag{2-75}$$

Using Equations (2-74) and (2-75), the apparent power (S_r) at the receiving end is given by

$$S_r = \sqrt{P_r^2 + Q_r^2}. \tag{2-76}$$

Therefore, using Equation (2-76), the line current (I) is given by

$$I = \frac{S_r}{V_r}. \tag{2-77}$$

The reactive power flow either at the modified sending end or at the receiving end can be determined by

$$Q_{s'} = Q_r + Q_{line} \tag{2-78a}$$

or

$$Q_r = Q_{s'} - Q_{line} \tag{2-78b}$$

where Q_{line} can be determined using Equation (2-11b).

Using Equation (2-61), Equations (2-74) and (2-75) can be combined as

$$P_r^2 + \left(Q_r + \frac{V_r^2}{X}\right)^2 = a'^2 \tag{2-79}$$

and

$$\tan\delta' = \frac{P_r}{Q_r + \frac{V_r^2}{X}}. \tag{2-80}$$

Equation (2-79) represents the locus of a circle, centered at $\left[0, -\frac{V_r^2}{X}\right]$ with a radius of a'.

Using Equation (2-61), Equation (2-79) can be rearranged as

$$V_{s'} = \frac{X}{V_r}\sqrt{P_r^2 + \left(Q_r + \frac{V_r^2}{X}\right)^2}. \tag{2-81}$$

From Equation (2-80), the modified power angle is given by

$$\delta' = \tan^{-1}\frac{P_r}{Q_r + \frac{V_r^2}{X}}, \tag{2-82a}$$

which can be rearranged to obtain the phase-shift angle as

$$\psi = \tan^{-1}\frac{P_r}{Q_r + \frac{V_r^2}{X}} - \delta \tag{2-82b}$$

using Equation (2-51b).

For independent control of active and reactive power flows (P_r and Q_r) at the receiving end, the required magnitude ($V_{s'}$) of the modified sending-end voltage ($\mathbf{V_{s'}}$) and the modified power angle (δ') are determined from Equations (2-81) and (2-82a) or (2-82b), respectively. Because of the prevailing line current being at any phase angle with the terminal voltage ($\mathbf{V_{s'}}$), the shunt-compensating voltage ($\mathbf{V_{s'}}$) exchanges bidirectional active and reactive powers with the line. This results in an emulation of a four-quadrant, shunt-compensating impedance ($\mathbf{Z_{sh}} = R_{sh} - jX_{sh}$) that consists of a resistance ($R_{sh} = +R$ or $-R$) and a reactance ($X_{sh} = X_C$ or $-X_L$) in shunt with the line. Therefore, the shunt-compensating voltage ($\mathbf{V_{s'}}$) acts as an IR. To deliver or absorb the active power to and from the line, the resistor acts as a source or a sink, depending on whether its sign is negative or positive. The capacitor and the reactor delivers and absorbs reactive power to and from the line, respectively. When only a shunt reactive compensation is used, the exchanged active power is just the losses in the shunt compensator, which is the functionality of either a SynCon or a STATCOM.

Rearranging Equation (2-80), it can be written that

$$Q_r = \frac{1}{\tan\delta'}P_r - \frac{V_r^2}{X}; \tag{2-83a}$$

since for voltage regulation, only $V_{s'}$ changes and δ' remains the same as δ, Equation (2-83a) can be rewritten as

$$Q_r = \frac{1}{\tan\delta}P_r - \frac{V_r^2}{X} \tag{2-83b}$$

or substituting for tanδ from Equation (2-49), it can be written that

$$Q_r = \frac{Q_{rn} + \frac{V_r^2}{X}}{P_{rn}}P_r - \frac{V_r^2}{X}. \tag{2-83c}$$

From Equations (2-83b) and (2-83c), the $P_r - Q_r$ relationship is defined with a slope (m_r) and an intercept (c_r) as

$$Q_r = m_r P_r + c_r \tag{2-84}$$

where

$$m_r = \frac{1}{\tan\delta} = \frac{Q_{rn} + \frac{V_r^2}{X}}{P_{rn}} \tag{2-85}$$

and

$$c_r = -\frac{V_r^2}{X}. \tag{2-86}$$

In the case of a VR, the variation of reactive power (Q_r) at the receiving end is shown in Figure 2-16 as a function of the active power (P_r). When $P_r = 0$, $Q_r = -V_r^2/X$; when $P_r = P_{rn}$, $Q_r = Q_{rn}$. The practical operating region of a VR is also shown.

The equations for active and reactive power flows ($P_{s'}$ and $Q_{s'}$) at the modified sending end and (P_r and Q_r) at the receiving end as a function of the magnitude ($V_{s'}$) and the modified power angle (δ') of the shunt-compensating voltage ($\mathbf{V_{s'}}$) are given in Table 2-3.

The equations relating the active and reactive power flows ($P_{s'}$ and $Q_{s'}$) at the modified sending end and (P_r and Q_r) at the receiving end are given in Table 2-4.

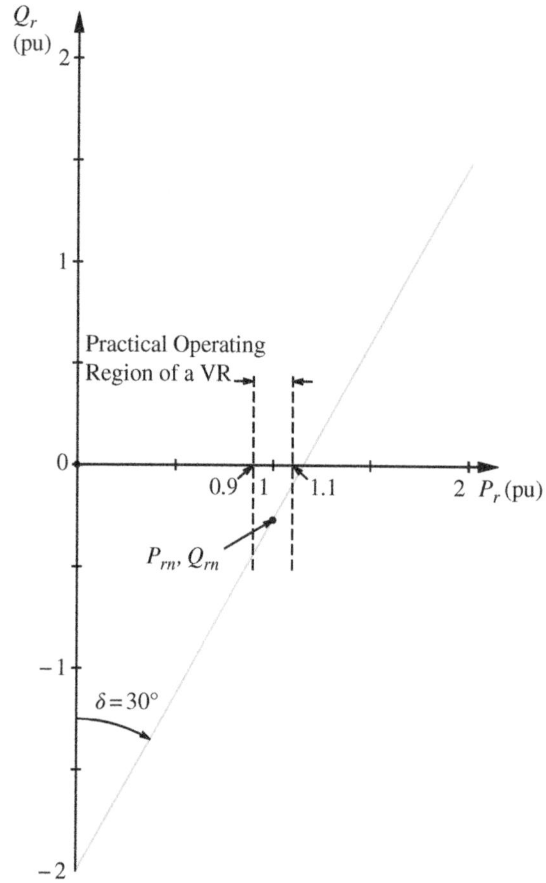

Figure 2-16 Q_r versus P_r at the receiving end of the line for voltage regulation at power angle $\delta = 30°$ when $V_s = V_r = 1$ pu, $X = 0.5$ pu, and $R = 0$ ($X/R = \infty$), using a Shunt–Shunt configuration.

Table 2-3 Equations for active and reactive power flows ($P_{s'}$ and $Q_{s'}$) at the modified sending end and (P_r and Q_r) at the receiving end as a function of the magnitude ($V_{s'}$) and the modified power angle (δ') of the shunt-compensating voltage ($\mathbf{V_{s'}}$).

$P_{s'} = a' \sin \delta'$ (2-59)	$Q_{s'} = a'\left(\frac{V_{s'}}{V_r} - \cos \delta'\right)$ (2-60)	$a' = \frac{V_{s'}V_r}{X}$ (2-61)	$\delta' = \delta_{s'} - \delta_r$ (2-62)
$P_r = a' \sin \delta'$ (2-74)	$Q_r = a'\left(\cos \delta' - \frac{V_r}{V_{s'}}\right)$ (2-75)	$a' = \frac{V_{s'}V_r}{X}$ (2-61)	$\delta' = \delta_{s'} - \delta_r$ (2-62)

Table 2-4 Equations relating the active and reactive power flows ($P_{s'}$ and $Q_{s'}$) at the modified sending end and (P_r and Q_r) at the receiving end.

$P_{s'}^2 + \left(Q_{s'} - \frac{V_{s'}^2}{X}\right)^2 = a'^2$ (2-64)	$a' = \frac{V_{s'}V_r}{X}$ (2-61)
$P_r^2 + \left(Q_r + \frac{V_r^2}{X}\right)^2 = a'^2$ (2-79)	$a' = \frac{V_{s'}V_r}{X}$ (2-61)

Example 2-7

Consider the electrical system data as given in Table 2-1. For $\beta = 0°$ or $\beta = 180°$, calculate m_r and c_r.

Answer 2-7

In Example 2-5, it was calculated that the natural transmitted active power, $P_{rn} = 1$ pu and the reactive power, $Q_{rn} = -0.267949$ pu at the load end. Therefore, from Equation (2-85),

$$m_r = \frac{1}{P_{rn}}\left(Q_{rn} + \frac{V_r^2}{X}\right) = \frac{1}{1}\left(-0.267949 + \frac{1^2}{0.5}\right) = 1.732051.$$

From Equation (2-86), $c_r = -\frac{V_r^2}{X} = -\frac{1^2}{0.5} = -2$ pu.

Example 2-8

Consider the electrical system data as given in Table 2-1. For a given $V_{s'} = 0.967715$ pu and $\delta' = 41.515336°$, calculate $P_{s'}$ and $Q_{s'}$.

Answer 2-8

From Equation (2-61), $a' = \frac{V_{s'}V_r}{X} = \frac{0.967715 \times 1}{0.5} = 1.935430.$

From Equation (2-59), $P_{s'} = a' \sin \delta' = 1.935430 \times \sin 41.515336° = 1.282843$ pu.

From Equation (2-60), $Q_{s'} = a'\left(\frac{V_{s'}}{V_r} - \cos \delta'\right) = 1.935430\left(\frac{0.967715}{1} - \cos 41.515336°\right) =$ 0.423737 pu.

Example 2-9

Consider the electrical system data as given in Table 2-1. For a given $V_{s'} = 0.967715$ pu and $\delta' = 41.515336°$, calculate P_r and Q_r.

Answer 2-9

From Equation (2-61), $a' = \frac{V_{s'}V_r}{X} = \frac{0.967715 \times 1}{0.5} = 1.935430.$

From Equation (2-74), $P_r = a' \times \sin \delta' = 1.935430 \times \sin 41.515336° = 1.282843$ pu.

From Equation (2-75), $Q_r = a'\left(\cos \delta' - \frac{V_r}{V_{s'}}\right) = 1.935430\left(\cos 41.515336° - \frac{1}{0.967715}\right) =$ -0.550792 pu.

For desired active and reactive power flows ($P_{s'}$ and $Q_{s'}$) at the modified sending end and (P_r and Q_r) at the receiving end, the equations for the magnitude ($V_{s'}$) and the modified power angle (δ') of the shunt-compensating voltage ($\mathbf{V}_{s'}$) are given in Table 2-5.

Table 2-5 Equations for magnitude ($V_{s'}$) and modified power angle (δ') of the shunt-compensating voltage ($\mathbf{V}_{s'}$) as a function of desired ($P_{s'}$ and $Q_{s'}$) at the modified sending end and (P_r and Q_r) the receiving end.

$V_{s'} = +\sqrt{\left(\frac{V_r^2}{2} + XQ_{s'}\right) + \sqrt{\left(\frac{V_r^2}{2} + XQ_{s'}\right)^2 - X^2(P_{s'}^2 + Q_{s'}^2)}}$ (2-67a)	$\delta' = \tan^{-1}\frac{P_{s'}}{\frac{V_{s'}^2}{X} - Q_{s'}}$ (2-68)
$V_{s'} = \frac{X}{V_r}\sqrt{P_r^2 + \left(Q_r + \frac{V_r^2}{X}\right)^2}$ (2-81)	$\delta' = \tan^{-1}\frac{P_r}{Q_r + \frac{V_r^2}{X}}$ (2-82a)

Example 2-10

Consider the electrical system data as given in Table 2-1. For a given $P_{s'} = 1.282843$ pu and $Q_{s'} = 0.423737$ pu, calculate $V_{s'}$ and δ'.

Answer 2-10

From Equation (2-67a), $V_{s'} = +\sqrt{\left(\frac{V_r^2}{2} + XQ_{s'}\right) + \sqrt{\left(\frac{V_r^2}{2} + XQ_{s'}\right)^2 - X^2(P_{s'}^2 + Q_{s'}^2)}} = \sqrt{\left(\frac{1^2}{2} + 0.5 \times 0.423737\right) + \sqrt{\left(\frac{1^2}{2} + 0.5 \times 0.423737\right)^2 - 0.5^2(1.282843^2 + 0.423737^2)}} = 0.967715$ pu.

From Equation (2-68), $\delta' = \tan^{-1}\frac{P_{s'}}{\frac{V_{s'}^2}{X} - Q_{s'}} = \tan^{-1}\frac{1.282843}{\frac{0.967715^2}{0.5} - 0.423737} = 41.515336°$.

These results confirm those obtained in Example 2-8.

Example 2-11

Consider the electrical system data as given in Table 2-1 For a given $P_r = 1.282843$ pu and $Q_r = -0.550792$ pu, calculate $V_{s'}$ and δ'.

Answer 2-11

From Equation (2-81), $V_{s'} = \frac{X}{V_r}\sqrt{P_r^2 + \left(Q_r + \frac{V_r^2}{X}\right)^2} = \frac{0.5}{1}\sqrt{1.282843^2 + \left(-0.550792 + \frac{1^2}{0.5}\right)^2} = 0.967715$ pu.

From Equation (2-82a), $\delta' = \tan^{-1}\frac{P_r}{Q_r + \frac{V_r^2}{X}} = \tan^{-1}\frac{1.282843}{-0.550792 + \frac{1^2}{0.5}} = 41.515336°$.

These results confirm those obtained in Example 2-9. Using the information about the receiving-end voltage (V_r) and active and reactive power flows (P_r and Q_r) from the Supervisory Control And Data Acquisition (SCADA) system, the line voltage ($V_{s'}$) and the power angle (δ') can be calculated for a known line reactance (X).

2.2.1.3 Exchanged Power by a Shunt-Compensating Voltage

Referring to Figure 2-9a, for any ($P_{s'}$ and $Q_{s'}$), the active power (P_{sh}) and reactive power (Q_{sh}) of the shunt compensator are defined as

$$P_{sh} = -P_{s'} \tag{2-87a}$$

and

$$Q_{sh} = -Q_{s'}. \tag{2-87b}$$

2.2.1.4 Representation of a Shunt-Compensating Voltage as a Shunt-Compensating Impedance

The active power (P_{sh}) and reactive power (Q_{sh}) of the shunt compensator are defined by

$$P_{sh} = I^2 R_{sh} \tag{2-88a}$$

and

$$Q_{sh} = -I^2 X_{sh} \tag{2-88b}$$

where the line current (I) is defined in Equation (2-57).

Using Equations (2-87a), (2-87b), (2-88a), and (2-88b), the shunt-compensating resistance (R_{sh}) and shunt-compensating reactance (X_{sh}) are given by

$$R_{sh} = \frac{P_{sh}}{I^2} \tag{2-89a}$$

or

$$R_{sh} = -\frac{P_{s'}}{I^2} \tag{2-89b}$$

and

$$X_{sh} = -\frac{Q_{sh}}{I^2} \tag{2-90a}$$

or

$$X_{sh} = \frac{Q_{s'}}{I^2}. \tag{2-90b}$$

The emulated shunt-compensating resistance (R_{sh}), reactance (X_{sh}), and impedance (Z_{sh}) are defined as

$$\mathbf{Z_{sh}} = Z_{sh}\angle\theta_{Zsh} = R_{sh} - jX_{sh} \tag{2-91a}$$

or

$$Z_{sh}\angle\theta_{Zsh} = \sqrt{R_{sh}^2 + X_{sh}^2}\angle\left(\tan^{-1}\frac{-X_{sh}}{R_{sh}}\right) \tag{2-91b}$$

where

$$Z_{sh}\angle\theta_{Zsh} = Z_{sh}\cos(\theta_{Zsh}) + jZ_{sh}\sin(\theta_{Zsh}). \tag{2-92}$$

Equating the real and imaginary parts of the Equations (2-91a) and (2-92), it can be written that

$$R_{sh} = Z_{sh}\cos(\theta_{Zsh}) \tag{2-93a}$$

and

$$X_{sh} = -Z_{sh}\sin(\theta_{Zsh}). \tag{2-93b}$$

Note that $X_{sh} > 0$ means the shunt-compensating reactance is capacitive and $X_{sh} < 0$ means the shunt-compensating reactance is inductive.

Example 2-12

Consider the electrical system data as given in Table 2-1. Using a Shunt–Shunt configuration, $V_{s'} = 0.967715$ pu, $\delta' = 41.515336°$, $P_{s'} = 1.282843$ pu, and $Q_{s'} = 0.423737$. Calculate V_X, I, R_{sh}, and X_{sh}.

Answer 2-12

From Equation (2-54a),

$$V_X = \sqrt{V_{s'}^2 - 2V_{s'}V_r \cos(\delta + \psi) + V_r^2} = \sqrt{0.967715^2 - 2 \times 0.967715 \times 1 \times \cos 41.515336° + 1^2}$$
$$= 0.698043 \text{ pu.}$$

From Equation (2-57), $I = \frac{V_X}{X} = \frac{0.698043}{0.5} = 1.396086$ pu.

From Equation (2-89b), $R_{sh} = -\frac{P_{s'}}{I^2} = -\frac{1.282843}{1.396086^2} = -0.658186$ pu. Note that the negative shunt-compensating resistance delivers active power.

From Equation (2-90b), $X_{sh} = \frac{Q_{s'}}{I^2} = \frac{0.423737}{1.396086^2} = 0.217406$ pu. Note that the positive shunt-compensating reactance delivers reactive power.

2.2.2 Series-Compensating Voltage as an Impedance Regulator, Voltage Regulator, and Phase Angle Regulator (Asymmetric)

Consider a two-generator/one-line power system network with a compensating voltage ($\mathbf{V_{s's}} = V_{s's}\angle(\delta_s + \beta)$) as shown in Figure 2-17a. When the compensating voltage ($\mathbf{V_{s's}}$) is added in series with the line at the sending-end voltage ($\mathbf{V_s}$), the modified sending-end voltage becomes ($\mathbf{V_{s'}} = \mathbf{V_s} + \mathbf{V_{s's}}$). Figure 2-17b shows the phasor diagram that is related to a series-compensating voltage with a fixed magnitude of 0.2 pu and its entire controllable range of $0° \leq \beta \leq 360°$. The difference voltage ($\mathbf{V_{s'}} - \mathbf{V_r}$) provides the voltage ($\mathbf{V_X}$) across the line reactance (X). As the relative phase angle (β) is varied over its entire controllable range of $0° \leq \beta \leq 360°$, the end of phasor ($\mathbf{V_{s's}}$) moves along a circle with its center located at the end of phasor ($\mathbf{V_s}$). The rotation of phasor ($\mathbf{V_{s's}}$) with an angle (β) modulates both the magnitude and phase angle of the phasor ($\mathbf{V_X}$). In a special operation, when $\beta = 0°$ or $180°$, the series-compensating voltage ($\mathbf{V_{s's}}$) controls the magnitude ($V_{s'}$) of the line voltage ($\mathbf{V_{s'}}$) as in the VR. In another special operation, when $\beta = \pm 90°$, the series-compensating voltage ($\mathbf{V_{s's}}$) controls mainly the phase angle (ψ) of the line voltage ($\mathbf{V_{s'}}$) as in the asymmetric PAR, called a PAR (asym).

In general, the compensating voltage ($\mathbf{V_{s's}}$) is also at any phase angle with the prevailing line current ($\mathbf{I}$). Therefore, it emulates in series with the line a positive resistance ($+R$) that absorbs active power from the line or a negative resistance ($-R$) that delivers active power to the line, and a capacitive reactance (X_C) that decreases the effective reactance between the two ends of the line, thus increases the power flow in the line or an inductive reactance (X_L) that increases the effective reactance between the two ends of the line, thus decreases the power flow in the line. Therefore, the desired series-compensating voltage is actually an impedance emulator.

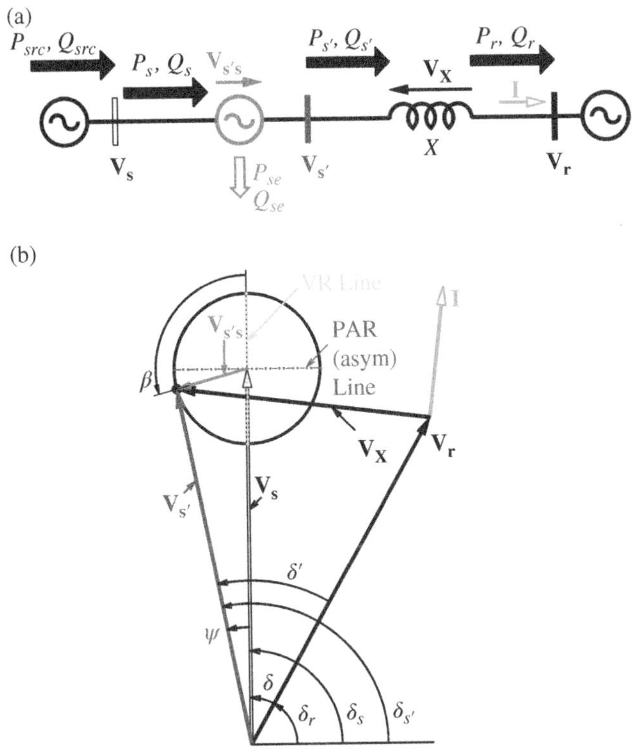

Figure 2-17 (a) Power transmission system with a series-compensating voltage ($\mathbf{V_{s's}}$); (b) phasor diagram.

From Figure 2-17b, it can be written that

$$\mathbf{V_{s'}} = \mathbf{V_s} + \mathbf{V_{s's}} \quad \text{or} \quad V_{s'}\angle\delta_{s'} = V_s\angle\delta_s + V_{s's}\angle(\delta_s + \beta)$$
$$\text{or} \quad V_{s'}\angle(\delta_s + \psi) = V_s\angle\delta_s + V_{s's}\angle(\delta_s + \beta) \tag{2-94a}$$

or

$$\mathbf{V_{s's}} = \mathbf{V_{s'}} - \mathbf{V_s} \quad \text{or} \quad V_{s's}\angle(\delta_s + \beta) = V_{s'}\angle\delta_{s'} - V_s\angle\delta_s$$
$$\text{or} \quad V_{s's}\angle(\delta_s + \beta) = V_{s'}\angle(\delta_s + \psi) - V_s\angle\delta_s \tag{2-94b}$$

where $\mathbf{V_{s's}}$, the difference between the modified sending-end voltage ($\mathbf{V_{s'}}$) and the sending-end voltage ($\mathbf{V_s}$), is the compensating voltage, β is the relative phase angle of the compensating voltage ($\mathbf{V_{s's}}$) with respect to the sending-end voltage ($\mathbf{V_s}$), and ψ is the phase-shift angle of the modified sending end voltage ($\mathbf{V_{s'}}$) with respect to the sending-end voltage ($\mathbf{V_s}$).

Alternatively, the above equations can be rewritten as

$$V_{s'}\angle(\delta_r + \delta') = V_s\angle(\delta_r + \delta) + V_{s's}\angle(\delta_r + \delta + \beta) \tag{2-95a}$$

or

$$V_{s's}\angle(\delta_r + \delta + \beta) = V_{s'}\angle(\delta_r + \delta') - V_s\angle(\delta_r + \delta). \tag{2-95b}$$

Considering the sending-end voltage phasor as reference (*i.e.* $\delta_s = 0°$), it can be written from Equation (2-94a) that

$$V_{s'}\angle\psi = V_s\angle 0° + V_{s's}\angle\beta$$

or

$$V_{s'}\cos\psi + jV_{s'}\sin\psi = V_s + V_{s's}\cos\beta + jV_{s's}\sin\beta.$$

Equating the real and imaginary parts of both sides of the above equation, it can be written that

$$V_{s'}\cos\psi = V_s + V_{s's}\cos\beta \tag{2-96a}$$

and

$$V_{s'}\sin\psi = V_{s's}\sin\beta. \tag{2-96b}$$

Therefore, the magnitude of the modified sending-end voltage ($\mathbf{V_{s'}}$) is derived from Equations (2-96a) and (2-96b) using the identity ($\cos^2\psi + \sin^2\psi = 1$) as

$$V_{s'} = \sqrt{V_{s'}^2\left(\cos^2\psi + \sin^2\psi\right)} = \sqrt{(V_{s'}\cos\psi)^2 + (V_{s'}\sin\psi)^2}$$

or

$$V_{s'} = \sqrt{(V_s + V_{s's}\cos\beta)^2 + (V_{s's}\sin\beta)^2}$$

or

$$V_{s'} = \sqrt{V_s^2 + V_{s's}^2\left(\cos^2\beta + \sin^2\beta\right) + 2V_sV_{s's}\cos\beta}$$

or

$$V_{s'} = \sqrt{V_s^2 + V_{s's}^2 + 2V_sV_{s's}\cos\beta} \tag{2-97}$$

using the identity $\left(\cos^2\beta + \sin^2\beta = 1\right)$.

The phase-shift angle of the compensating voltage ($\mathbf{V_{s'}}$) with respect to the sending-end voltage is derived from Equations (2-96a) and (2-96b) as

$$\psi = \tan^{-1}(\tan\psi) = \tan^{-1}\frac{\sin\psi}{\cos\psi} = \tan^{-1}\frac{V_{s'}\sin\psi}{V_{s'}\cos\psi}$$

or

$$\psi = \tan^{-1}\frac{V_{s's}\sin\beta}{V_s + V_{s's}\cos\beta}. \tag{2-98}$$

Considering the sending-end voltage phasor as reference (*i.e.* $\delta_s = 0°$), it can be written from Equation (2-94b) that

$$V_{s's}\angle\beta = V_{s'}\angle\psi - V_s\angle 0°$$

or

$$V_{s's}\cos\beta + jV_{s's}\sin\beta = V_{s'}\cos\psi - V_s + jV_{s'}\sin\psi.$$

Equating the real and imaginary parts of both sides of the above equation, it can be written that

$$V_{s's}\cos\beta = V_{s'}\cos\psi - V_s \tag{2-99a}$$

and

$$V_{s's}\sin\beta = V_{s'}\sin\psi. \tag{2-99b}$$

Therefore, the magnitude of the series-compensating voltage ($\mathbf{V}_{s's}$) is derived from Equations (2-99a) and (2-99b) using the identity $\left(\cos^2\beta + \sin^2\beta = 1\right)$ as

$$V_{s's} = \sqrt{V_{s's}^2\left(\cos^2\beta + \sin^2\beta\right)} = \sqrt{\left(V_{s's}\cos\beta\right)^2 + \left(V_{s's}\sin\beta\right)^2}$$

or

$$V_{s's} = \sqrt{\left(V_{s'}\cos\psi - V_s\right)^2 + \left(V_{s'}\sin\psi\right)^2}$$

or

$$V_{s's} = \sqrt{V_{s'}^2\left(\cos^2\psi + \sin^2\psi\right) + V_s^2 - 2V_{s'}V_s\cos\psi}$$

or

$$V_{s's} = \sqrt{V_{s'}^2 + V_s^2 - 2V_{s'}V_s\cos\psi} \tag{2-100}$$

using the identity $\left(\cos^2\psi + \sin^2\psi = 1\right)$.

The relative phase angle of the compensating voltage ($\mathbf{V}_{s's}$) with respect to the sending-end voltage is derived from Equations (2-99a) and (2-99b) as

$$\beta = \tan^{-1}\left(\tan\beta\right) = \tan^{-1}\frac{\sin\beta}{\cos\beta} = \tan^{-1}\frac{V_{s's}\sin\beta}{V_{s's}\cos\beta}$$

or

$$\beta = \tan^{-1}\frac{V_{s'}\sin\psi}{V_{s'}\cos\psi - V_s}. \tag{2-101}$$

Alternately, considering the receiving-end voltage phasor as reference (*i.e.* $\delta_r = 0°$), it can be written from Equation (2-95a) that

$$V_{s'}\angle\delta' = V_s\angle\delta + V_{s's}\angle(\delta + \beta)$$

or

$$V_{s'}\cos\delta' + jV_{s'}\sin\delta' = V_s\cos\delta + V_{s's}\cos\left(\delta + \beta\right) + jV_s\sin\delta + jV_{s's}\sin\left(\delta + \beta\right).$$

Equating the real and imaginary parts of both sides of the above equation, it can be written that

$$V_{s'}\cos\delta' = V_s\cos\delta + V_{s's}\cos\left(\delta + \beta\right) \tag{2-102a}$$

and

$$V_{s'}\sin\delta' = V_s\sin\delta + V_{s's}\sin\left(\delta + \beta\right). \tag{2-102b}$$

Therefore, the magnitude of the modified sending-end voltage ($\mathbf{V}_{s'}$) is derived from Equations (2-102a) and (2-102b) using the identity $(\cos^2\delta' + \sin^2\delta' = 1)$ as

$$V_{s'} = \sqrt{V_{s'}^2\left(\cos^2\delta' + \sin^2\delta'\right)} = \sqrt{\left(V_{s'}\cos\delta'\right)^2 + \left(V_{s'}\sin\delta'\right)^2}$$

or

$$V_{s'} = \sqrt{\left\{V_s\cos\delta + V_{s's}\cos\left(\delta + \beta\right)\right\}^2 + \left\{V_s\sin\delta + V_{s's}\sin\left(\delta + \beta\right)\right\}^2}$$

or

$$V_{s'} = \sqrt{V_s^2(\cos^2\delta + \sin^2\delta) + V_{s's}^2\{\cos^2(\delta + \beta) + \sin^2(\delta + \beta)\} + 2V_s V_{s's}\cos\beta}$$

or

$$V_{s'} = \sqrt{V_s^2 + V_{s's}^2 + 2V_s V_{s's}\cos\beta} \tag{2-103}$$

using the identities $(\cos^2\delta + \sin^2\delta = 1)$ and $\{\cos^2(\delta + \beta) + \sin^2(\delta + \beta) = 1\}$.

The phase angle of the voltage ($\mathbf{V_{s'}}$) at the modified sending end with respect to the receiving-end voltage is derived from Equations (2-102a) and (2-102b) as

$$\delta' = \tan^{-1}(\tan\delta') = \tan^{-1}\frac{\sin\delta'}{\cos\delta'} = \tan^{-1}\frac{V_{s'}\sin\delta'}{V_{s'}\cos\delta'}$$

or

$$\delta' = \tan^{-1}\frac{V_s\sin\delta + V_{s's}\sin(\delta + \beta)}{V_s\cos\delta + V_{s's}\cos(\delta + \beta)}. \tag{2-104}$$

Considering the receiving-end voltage phasor as reference (*i.e.* $\delta_r = 0°$), it can be written from Equation (2-95b) that

$$V_{s's}\angle(\delta + \beta) = V_{s'}\angle\delta' - V_s\angle\delta$$

or

$$V_{s's}\cos(\delta + \beta) + jV_{s's}\sin(\delta + \beta) = V_{s'}\cos\delta' - V_s\cos\delta + jV_{s'}\sin\delta' - jV_s\sin\delta.$$

Equating the real and imaginary parts of both sides of the above equation, it can be written that

$$V_{s's}\cos(\delta + \beta) = V_{s'}\cos\delta' - V_s\cos\delta \tag{2-105a}$$

and

$$V_{s's}\sin(\delta + \beta) = V_{s'}\sin\delta' - V_s\sin\delta. \tag{2-105b}$$

Therefore, the magnitude ($V_{s's}$) and the relative phase angle (β) of the series-compensating voltage ($\mathbf{V_{s's}}$) are derived from Equations (2-105a) and (2-105b) using the identity $\{\cos^2(\delta + \beta) + \sin^2(\delta + \beta) = 1\}$ as

$$V_{s's} = \sqrt{V_{s's}^2\{\cos^2(\delta + \beta) + \sin^2(\delta + \beta)\}} = \sqrt{\{V_{s's}\cos(\delta + \beta)\}^2 + \{V_{s's}\sin(\delta + \beta)\}^2}$$

or

$$V_{s's} = \sqrt{\{V_{s'}\cos\delta' - V_s\cos\delta\}^2 + \{V_{s'}\sin\delta' - V_s\sin\delta\}^2}$$

or

$$V_{s's} = \sqrt{V_{s'}^2(\cos^2\delta' + \sin^2\delta') + V_s^2(\cos^2\delta + \sin^2\delta) - 2V_{s'}V_s\cos(\delta' - \delta)}$$

or

$$V_{s's} = \sqrt{V_{s'}^2 + V_s^2 - 2V_{s'}V_s\cos(\delta' - \delta)} \tag{2-106}$$

using the identities $(\cos^2\delta' + \sin^2\delta' = 1)$ and $(\cos^2\delta + \sin^2\delta = 1)$.

The phase angle of the compensating voltage ($\mathbf{V_{s's}}$) with respect to the receiving-end voltage ($\mathbf{V_r}$) is derived from Equations (2-105a) and (2-105b) as

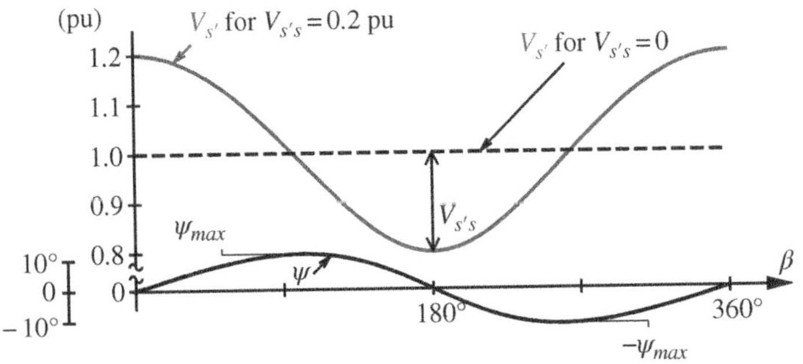

Figure 2-18 Variations of the magnitude ($V_{s'}$) and phase-shift angle (ψ) of the modified sending-end voltage as a function of the relative phase angle (β) of the series-compensating voltage ($\mathbf{V_{s's}}$) with a fixed magnitude of 0.2 pu.

$$\delta + \beta = \tan^{-1}\{\tan(\delta + \beta)\} = \tan^{-1}\frac{\sin(\delta + \beta)}{\cos(\delta + \beta)} = \tan^{-1}\frac{V_{s's}\sin(\delta + \beta)}{V_{s's}\cos(\delta + \beta)}$$

$$= \tan^{-1}\frac{V_{s'}\sin\delta' - V_s\sin\delta}{V_{s'}\cos\delta' - V_s\cos\delta}.$$

Therefore, the relative phase angle is given by

$$\beta = \tan^{-1}\frac{V_{s'}\sin\delta' - V_s\sin\delta}{V_{s'}\cos\delta' - V_s\cos\delta} - \delta. \quad \text{(2-107)}$$

In the transmission system considered above, the magnitude ($V_{s'}$) and phase-shift angle (ψ) of the modified sending-end voltage ($\mathbf{V_{s'}}$) are modulated between $\mathbf{V_s} + \mathbf{V_{s's}}$ at $\beta = 0°$ and $\mathbf{V_s} - \mathbf{V_{s's}}$ at $\beta = 180°$ and between $\delta_s + \psi_{max}$ and $\delta_s - \psi_{max}$, respectively, where $V_{s's} = 0.2$ pu and $\psi_{max} = 11.54°$ as shown in Figure 2-18.

For a desired series-compensating voltage ($\mathbf{V_{s's}} = V_{s's}\angle(\delta_s + \beta)$), the equations for magnitude ($V_{s'}$), modified power angle (δ'), and phase-shift angle (ψ) of the shunt-compensating voltage ($\mathbf{V_{s'}}$) are given in Table 2-6.

Example 2-13

Consider the electrical system data as given in Table 2-1. For a given $V_{s's} = 0.2$ pu and $\beta = 105°$, calculate $V_{s'}$, ψ, and δ'.

Answer 2-13

From Equation (2-97), $V_{s'} = \sqrt{V_s^2 + V_{s's}^2 + 2V_sV_{s's}\cos\beta} = \sqrt{1^2 + 0.2^2 + 2 \times 1 \times 0.2 \times \cos 105°} =$ 0.967715 pu.

Table 2-6 Equations for magnitude ($V_{s'}$), modified power angle (δ'), and phase-shift angle (ψ) of the modified sending-end voltage ($\mathbf{V_{s'}}$) as a function of desired series-compensating voltage ($\mathbf{V_{s's}}$) with magnitude ($V_{s's}$) and relative phase angle (β).

$V_{s'} = \sqrt{V_s^2 + V_{s's}^2 + 2V_sV_{s's}\cos\beta}$	(2-103)	$\delta' = \tan^{-1}\frac{V_s\sin\delta + V_{s's}\sin(\delta + \beta)}{V_s\cos\delta + V_{s's}\cos(\delta + \beta)}$	(2-104)
$V_{s'} = \sqrt{V_s^2 + V_{s's}^2 + 2V_sV_{s's}\cos\beta}$	(2-97)	$\psi = \tan^{-1}\frac{V_{s's}\sin\beta}{V_s + V_{s's}\cos\beta}$	(2-98)

Table 2-7 Equations for magnitude ($V_{s's}$) and relative phase angle (β) of the series-compensating voltage ($\mathbf{V_{s's}}$) as a function of desired modified sending-end voltage ($\mathbf{V_{s'}}$) with magnitude ($V_{s'}$) and modified power angle (δ') or phase-shift angle (ψ).

$V_{s's} = \sqrt{V_{s'}^2 + V_s^2 - 2V_{s'}V_s \cos(\delta' - \delta)}$ (2-106)	$\beta = \tan^{-1}\dfrac{V_{s'}\sin\delta' - V_s\sin\delta}{V_{s'}\cos\delta' - V_s\cos\delta} - \delta$ (2-107)
$V_{s's} = \sqrt{V_{s'}^2 + V_s^2 - 2V_{s'}V_s \cos\psi}$ (2-100)	$\beta = \tan^{-1}\dfrac{V_{s'}\sin\psi}{V_{s'}\cos\psi - V_s}$ (2-101)

From Equation (2-98), $\psi = \tan^{-1}\dfrac{V_{s's}\sin\beta}{V_s + V_{s's}\cos\beta} = \tan^{-1}\dfrac{0.2 \times \sin 105^\circ}{1 + 0.2 \times \cos 105^\circ} = 11.515336^\circ$.

From Equation (2-51a), $\delta' = \delta + \psi = 30^\circ + 11.515336^\circ = 41.515336^\circ$.

Alternately, from Equation (2-103), which is the same as Equation (2-97), $V_{s'} = 0.967715$ pu.

From Equation (2-104), $\delta' = \tan^{-1}\dfrac{V_s\sin\delta + V_{s's}\sin(\delta+\beta)}{V_s\cos\delta + V_{s's}\cos(\delta+\beta)} = \tan^{-1}\dfrac{1\times\sin 30^\circ + 0.2\times\sin(30^\circ+105^\circ)}{1\times\cos 30^\circ + 0.2\times\cos(30^\circ+105^\circ)}$
$= 41.515336^\circ$.

From Equation (2-51b), $\psi = \delta' - \delta = 41.515336^\circ - 30^\circ = 11.515336^\circ$.

For independent control of active and reactive power flows either ($P_{s'}$ and $Q_{s'}$) at the modified sending end or (P_r and Q_r) at the receiving end, the required magnitude ($V_{s'}$) and phase angle ($\delta_{s'}$) of the modified sending-end voltage ($\mathbf{V_{s'}}$) are determined either from Equations (2-67a) and (2-68) or from Equations (2-81) and (2-82a), respectively. Once the modified sending-end voltage ($\mathbf{V_{s'}}$) is determined, the magnitude ($V_{s's}$) and the relative phase angle (β) of the series-compensating voltage ($\mathbf{V_{s's}}$) can be determined following the equations given in Table 2-7.

Example 2-14

Consider the electrical system data as given in Table 2-1. For a given $V_{s'} = 0.967715$ pu, $\psi = 11.515336^\circ$, and $\delta' = 41.515336^\circ$, calculate $V_{s's}$ and β.

Answer 2-14

From Equation (2-100),

$$V_{s's} = \sqrt{V_{s'}^2 + V_s^2 - 2V_{s'}V_s\cos\psi} = \sqrt{0.967715^2 + 1^2 - 2\times 0.967715 \times 1 \times \cos 11.515336^\circ} = 0.2 \text{ pu}.$$

From Equation (2-101), $\beta = \tan^{-1}\dfrac{V_{s'}\sin\psi}{V_{s'}\cos\psi - V_s} = \tan^{-1}\dfrac{0.967715\times\sin 11.515336^\circ}{0.967715\times\cos 11.515336^\circ - 1} = 105^\circ$.

Alternately, from Equation (2-106),

$$V_{s's} = \sqrt{V_{s'}^2 + V_s^2 - 2V_{s'}V_s\cos(\delta' - \delta)}$$
$$= \sqrt{0.967715^2 + 1^2 - 2\times 0.967715\times 1\times\cos(41.515336^\circ - 30^\circ)}$$
$$= 0.2 \text{ pu}.$$

From Equation (2-107),

$$\beta = \tan^{-1}\frac{V_{s'}\sin\delta' - V_s\sin\delta}{V_{s'}\cos\delta' - V_s\cos\delta} - \delta = \tan^{-1}\frac{0.967715\times\sin 41.515336^\circ - 1\times\sin 30^\circ}{0.967715\times\cos 41.515336^\circ - 1\times\cos 30^\circ} - 30^\circ = 105^\circ.$$

These results confirm those obtained in Example 2-13.

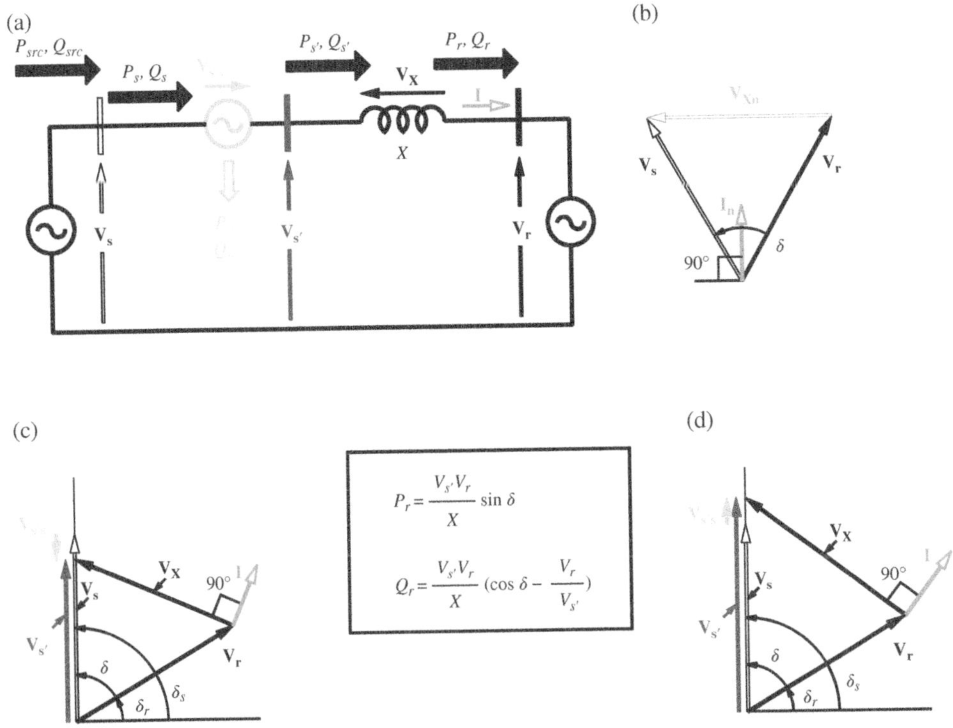

Figure 2-19 (a) Two-generator/one-line power system network and its series-compensating voltage to implement a VR and phasor diagrams for (b) an uncompensated line, (c) decreasing power flow in the line, and (d) increasing power flow in the line.

Figure 2-19a shows a two-generator/one-line power system network with a series-compensating voltage to implement a VR. Figure 2-19b shows the phasor diagram without any compensation. Figure 2-19c shows that the line current (**I**) decreases as the magnitude of the voltage between the modified sending end and receiving end across the line reactance (X) decreases from its uncompensated value. The line current (**I**) increases when the magnitude of the voltage between the modified sending end and receiving end across the line reactance (X) increases from its uncompensated value as shown in Figure 2-19d. The expressions for power flows at the receiving end of the line are shown in Figure 2-19, considering that the modified sending-end voltage is $V_{s'} \angle \delta_S$, *i.e.* the voltage magnitude at the modified sending end changes while the power angle remains unchanged.

Figure 2-20a shows a two-generator/one-line power system network with a series-compensating voltage to implement a PAR (asym). Figure 2-20b shows the phasor diagram without any compensation. Figure 2-20c shows that the line current (**I**) decreases as the phase angle of the voltage between the modified sending end and receiving end decreases from its uncompensated value. The line current (**I**) increases when the phase angle of the voltage between the modified sending end and receiving end increases from its uncompensated value as shown in Figure 2-20d. The expressions for power flows at the receiving end of the line are shown in Figure 2-20, considering that the modified sending-end voltage is $V_{s'} \angle \delta_{S'}$, *i.e.* both the voltage magnitude and the power angle at the modified sending end change.

Next, the active and reactive power flows (P_s and Q_s) at the sending end, (P_r and Q_r) at the receiving end, ($P_{s'}$ and $Q_{s'}$) at the modified sending end, and the exchanged active and reactive powers (P_{se} and Q_{se}) by the series-compensating voltage with the line are studied.

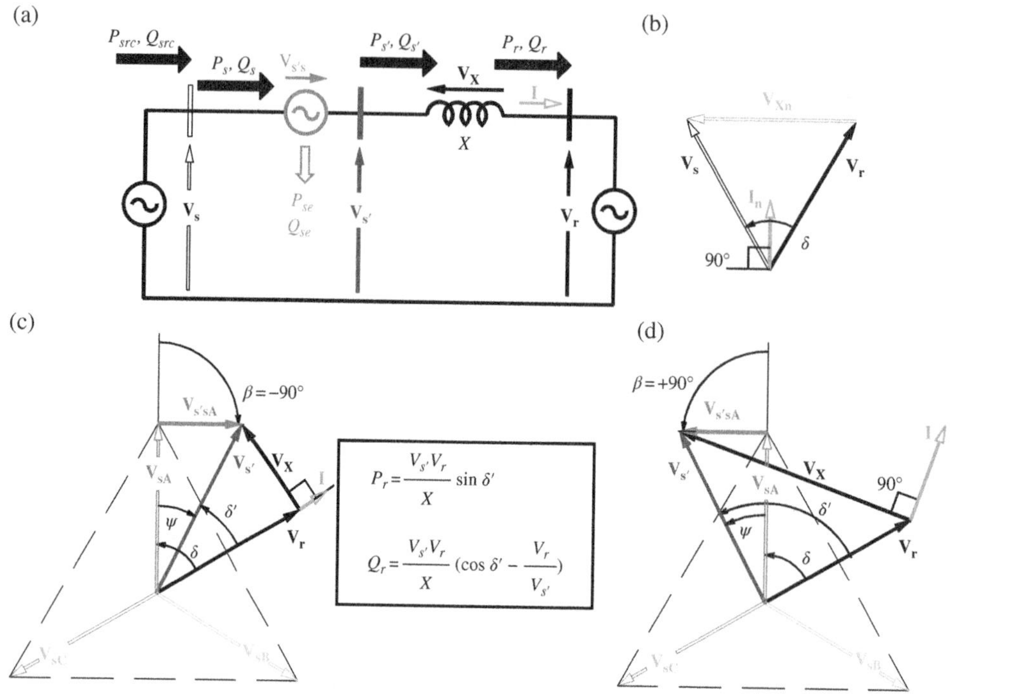

Figure 2-20 (a) Two-generator/one-line power system network and its series-compensating voltage to implement a PAR (asym) and phasor diagrams for (b) an uncompensated line, (c) decreasing power flow in the line, and (d) increasing power flow in the line.

From Equation (2-50b), it can be written that

$$\mathbf{V_X} = \mathbf{V_{s'}} - \mathbf{V_r} \qquad \text{or}$$

$$\mathbf{V_X} = \mathbf{V_s} + \mathbf{V_{s's}} - \mathbf{V_r} \quad \text{using Equation (2-94a)} \qquad \text{or}$$

$$V_X\angle\theta_{VX} = V_s\angle\delta_s + V_{s's}\angle(\delta_s + \beta) - V_r\angle\delta_r \qquad \text{or}$$

$$V_X\angle(\delta_s + \theta_{VXs}) = V_s\angle\delta_s + V_{s's}\angle(\delta_s + \beta) - V_r\angle(\delta_s - \delta) \quad \text{using Equation (2-27)}$$

where θ_{VXs} is the phase angle of the voltage across the line reactance with respect to the sending-end voltage.

Considering the sending-end voltage phasor as reference (*i.e.* $\delta_s = 0°$), it can be written from the above equation that

$$V_X\angle\theta_{VXs} = V_s\angle 0° + V_{s's}\angle\beta - V_r\angle(-\delta) \qquad \text{or}$$

$$V_X\cos\theta_{VXs} + jV_X\sin\theta_{VXs} = V_s + V_{s's}\cos\beta - V_r\cos\delta + jV_{s's}\sin\beta + jV_r\sin\delta.$$

Equating the real and imaginary parts of both sides of the above equation, it can be written that

$$V_X\cos\theta_{VXs} = V_s - V_r\cos\delta + V_{s's}\cos\beta \tag{2-108a}$$

and

$$V_X\sin\theta_{VXs} = V_r\sin\delta + V_{s's}\sin\beta. \tag{2-108b}$$

Therefore, the magnitude of the voltage ($\mathbf{V_X}$) across the transmission line is derived from Equations (2-108a) and (2-108b) using the identity $\left(\cos^2\theta_{VXs} + \sin^2\theta_{VXs} = 1\right)$ as

$$V_X = \sqrt{V_X^2\left(\cos^2\theta_{VXs} + \sin^2\theta_{VXs}\right)} = \sqrt{(V_X\cos\theta_{VXs})^2 + (V_X\sin\theta_{VXs})^2}$$

or

$$V_X = \sqrt{(V_s - V_r\cos\delta + V_{s's}\cos\beta)^2 + (V_r\sin\delta + V_{s's}\sin\beta)^2}$$

or

$$V_X = \sqrt{V_s^2 - 2V_sV_r\cos\delta + V_r^2\left(\cos^2\delta + \sin^2\delta\right) + V_{s's}^2\left(\cos^2\beta + \sin^2\beta\right) + 2V_sV_{s's}\cos\beta - 2V_{s's}V_r\cos(\delta+\beta)}$$

using the identity $[\cos\delta\cos\beta - \sin\delta\sin\beta = \cos(\delta+\beta)]$, or

$$V_X = \sqrt{V_s^2 - 2V_sV_r\cos\delta + V_r^2 + V_{s's}^2 + 2V_sV_{s's}\cos\beta - 2V_{s's}V_r\cos(\delta+\beta)} \quad (2\text{-}109)$$

using the identities $(\cos^2\delta + \sin^2\delta = 1)$ and $(\cos^2\beta + \sin^2\beta = 1)$.

Using Equation (2-30), Equation (2-109) can be rewritten as

$$V_X = \sqrt{V_{Xn}^2 + V_{s's}[V_{s's} + 2\{V_s\cos\beta - V_r\cos(\delta+\beta)\}]}. \quad (2\text{-}110)$$

The phase angle of the voltage ($\mathbf{V_X}$) across the line reactance with respect to the sending-end voltage is derived from Equations (2-108a) and (2-108b) as

$$\theta_{VXs} = \tan^{-1}(\tan\theta_{VXs}) = \tan^{-1}\frac{\sin\theta_{VXs}}{\cos\theta_{VXs}} = \tan^{-1}\frac{V_X\sin\theta_{VXs}}{V_X\cos\theta_{VXs}}$$

or

$$\theta_{VXs} = \tan^{-1}\frac{V_r\sin\delta + V_{s's}\sin\beta}{V_s - V_r\cos\delta + V_{s's}\cos\beta}. \quad (2\text{-}111)$$

The line current (**I**) is defined as

$$\mathbf{I} = \frac{\mathbf{V_X}}{jX} \quad \text{or} \quad I\angle\theta_I = \frac{V_X\angle\theta_{VX}}{jX} \quad \text{or} \quad I\angle(\delta_s + \theta_{Is}) = \frac{V_X\angle(\delta_s + \theta_{VXs})}{jX}.$$

Considering the sending-end voltage phasor as reference (*i.e.* $\delta_s = 0°$), it can be written from the above equation that

$$I\angle\theta_{Is} = \frac{V_X\angle\theta_{VXs}}{jX} \quad (2\text{-}56a)$$

or

$$I\angle\theta_{Is} = \frac{V_X}{X}\angle(\theta_{VXs} - 90°) \quad (2\text{-}56b)$$

where the magnitude of the prevailing line current (**I**) is given by

$$I = \frac{V_X}{X} \quad (2\text{-}57)$$

and the phase angle of the line current (**I**) with respect to the sending-end voltage ($\mathbf{V_s}$) is given by

$$\theta_{Is} = \theta_{VXs} - 90° \quad (2\text{-}58)$$

where θ_{VXs} is given in either Equation (2-55) or Equation (2-111).

Equations (2-57) and (2-110) can be combined as

$$I = \frac{\sqrt{V_{Xn}^2 + V_{s's}[V_{s's} + 2\{V_s\cos\beta - V_r\cos(\delta+\beta)\}]}}{X} \quad (2\text{-}112)$$

or

$$I = \sqrt{\frac{V_{Xn}^2}{X^2} + \frac{V_{s's}[V_{s's} + 2\{V_s \cos\beta - V_r \cos(\delta + \beta)\}]}{X^2}}$$

or

$$I = \sqrt{I_n^2 + \frac{V_{s's}[V_{s's} + 2\{V_s \cos\beta - V_r \cos(\delta + \beta)\}]}{X^2}} \tag{2-113}$$

using Equation (2-33).

Equation (2-56a) can be expanded as

$$I\angle\theta_{Is} = \frac{V_X\angle\theta_{VXs}}{jX} = \frac{V_X \cos\theta_{VXs} + jV_X \sin\theta_{VXs}}{jX} = \frac{V_X \sin\theta_{VXs}}{X} - j\frac{V_X \cos\theta_{VXs}}{X} \quad \text{or}$$

$$I\cos\theta_{Is} + jI\sin\theta_{Is} = \frac{V_r \sin\delta + V_{s's} \sin\beta}{X} - j\frac{V_s - V_r \cos\delta + V_{s's} \cos\beta}{X}$$

using Equations (2-108a) and (2-108b).

Equating the real and imaginary parts of both sides of the above equation, it can be written that

$$I\cos\theta_{Is} = \frac{V_r \sin\delta + V_{s's} \sin\beta}{X} \tag{2-114a}$$

and

$$I\sin\theta_{Is} = -\frac{V_s - V_r \cos\delta + V_{s's} \cos\beta}{X}. \tag{2-114b}$$

Therefore, the magnitude of the line current (**I**) is derived from Equations (2-114a) and (2-114b) using the identity $(\cos^2\theta_{Is} + \sin^2\theta_{Is} = 1)$ as

$$I = \sqrt{I^2(\cos^2\theta_{Is} + \sin^2\theta_{Is})} = \sqrt{(I\cos\theta_{Is})^2 + (I\sin\theta_{Is})^2} \quad \text{or}$$

$$I = \sqrt{\frac{\{V_r \sin\delta + V_{s's} \sin\beta\}^2 + \{V_s - V_r \cos\delta + V_{s's} \cos\beta\}^2}{X^2}}$$

$$I = \sqrt{\frac{V_s^2 - 2V_sV_r\cos\delta + V_r^2(\cos^2\delta + \sin^2\delta) + V_{s's}^2(\cos^2\beta + \sin^2\beta) + 2V_sV_{s's}\cos\beta - 2V_{s's}V_r\cos(\delta + \beta)}{X^2}}$$

using the identity $[\cos\delta\cos\beta - \sin\delta\sin\beta = \cos(\delta + \beta)]$, or

$$I = \frac{1}{X}\sqrt{V_s^2 - 2V_sV_r\cos\delta + V_r^2 + V_{s's}^2 + 2V_sV_{s's}\cos\beta - 2V_{s's}V_r\cos(\delta + \beta)}$$

using the identities $(\cos^2\delta + \sin^2\delta = 1)$ and $(\cos^2\beta + \sin^2\beta = 1)$, or

$$I = \frac{V_X}{X} \tag{2-57}$$

using Equation (2-109).

The phase angle of the line current (**I**) with respect to the sending-end voltage ($\mathbf{V_s}$) is given by

$$\theta_{Is} = \tan^{-1}\frac{V_r \sin\delta + V_{s's} \sin\beta}{V_s - V_r \cos\delta + V_{s's} \cos\beta} - 90° = \tan^{-1}\frac{V_r\cos\delta - V_s - V_{s's}\cos\beta}{V_r \sin\delta + V_{s's} \sin\beta} \tag{2-115}$$

Table 2-8 Line current for a VR and a PAR (asym).

	VR	PAR (asym)
I	$\sqrt{I_n^2 + \frac{V_{s's}[V_{s's} \pm 2\{V_s - V_r\cos\delta\}]}{X^2}}$	$\sqrt{I_n^2 + \frac{V_{s's}[V_{s's} \pm 2V_r \sin\delta]}{X^2}}$
θ_{Is}	$\tan^{-1}\frac{V_r\cos\delta - V_s \mp V_{s's}}{V_r \sin\delta}$	$\tan^{-1}\frac{V_r\cos\delta - V_s}{V_r\sin\delta \pm V_{s's}}$

using Equations (2-58) and (2-111), or

$$\theta_{Is} = \tan^{-1}(\tan\theta_{Is}) = \tan^{-1}\frac{\sin\theta_{Is}}{\cos\theta_{Is}} = \tan^{-1}\frac{I\sin\theta_{Is}}{I\cos\theta_{Is}}$$

or

$$\theta_{Is} = \tan^{-1}\frac{V_r\cos\delta - V_s - V_{s's}\cos\beta}{V_r\sin\delta + V_{s's}\sin\beta} \tag{2-115}$$

using Equations (2-114a) and (2-114b).

The magnitude and phase angle for the line current (**I**) are derived in Equations (2-113) and (2-115), respectively. They are shown in Table 2-8 for a VR and a PAR (asym), respectively.

Example 2-15

Consider the electrical system data as given in Table 2-1. For a given $V_{s's} = 0.2$ pu and $\beta = 105°$, calculate V_X and I.

Answer 2-15

In Example 2-3, it was calculated that the magnitude of the natural voltage, $V_{Xn} = 0.517638$ pu. Therefore, from Equation (2-110), $V_X = \sqrt{V_{Xn}^2 + V_{s's}[V_{s's} + 2\{V_s\cos\beta - V_r\cos(\delta+\beta)\}]}$

$$V_X = \sqrt{0.517638^2 + 0.2[0.2 + 2\{1\times\cos 105° - 1\times\cos(30° + 105°)\}]} = 0.698043 \text{ pu.}$$

From Equation (2-57), $I = \frac{V_X}{X} = \frac{0.698043}{0.5} = 1.396086$ pu.

Alternately, from Equation (2-112),

$$I = \frac{\sqrt{V_{Xn}^2 + V_{s's}[V_{s's} + 2\{V_s\cos\beta - V_r\cos(\delta+\beta)\}]}}{X}$$

$$= \sqrt{\frac{0.517638^2 + 0.2[0.2 + 2\{1\times\cos 105° - 1\times\cos(30° + 105°)\}]}{0.5^2}} = 1.396086 \text{ pu.}$$

In Example 2-4, it was calculated that the magnitude of the natural line current, $I_n = 1.035276$ pu. Therefore, from Equation (2-113),

$$I = \sqrt{I_n^2 + \frac{V_{s's}[V_{s's} + 2\{V_s\cos\beta - V_r\cos(\delta+\beta)\}]}{X^2}}$$

$$= \sqrt{1.035276^2 + \frac{0.2[0.2 + 2\{1\times\cos 105° - 1\times\cos(30° + 105°)\}]}{0.5^2}} = 1.396086 \text{ pu.}$$

2.2.2.1 Power Flow at the Sending End with a Series-Compensating Voltage

The active power flow at the sending end of the line when the line is series-compensated is given by

$$P_s = V_s I \cos(\delta_s - \theta_I) \qquad \text{or}$$
$$P_s = V_s I \cos(\delta_s - \{\delta_s + \theta_{Is}\})$$

where θ_{Is} is the phase angle of the line current (**I**) with respect to the sending-end voltage ($\mathbf{V_s}$), or

$$P_s = V_s I \cos(-\theta_{Is}) = V_s I \cos\theta_{Is} \qquad \text{or}$$
$$P_s = V_s I \cos\theta_{Is}. \tag{2-116}$$

Equation (2-116) can be rewritten using Equation (2-114a) as

$$P_s = \frac{V_s V_r}{X} \sin\delta + \frac{V_s V_{s's}}{X} \sin\beta \tag{2-117}$$

or

$$P_s = P_{sn} + a_s \sin\beta \tag{2-118}$$

where P_{sn} is given in Equation (2-40) and

$$a_s = \frac{V_s V_{s's}}{X}. \tag{2-119}$$

The reactive power flow at the sending end of the line when the line is series-compensated is given by

$$Q_s = V_s I \sin(\delta_s - \theta_I) \qquad \text{or}$$
$$Q_s = V_s I \sin(\delta_s - \{\delta_s + \theta_{Is}\})$$

where θ_{Is} is the phase angle of the line current (**I**) with respect to the sending-end voltage ($\mathbf{V_s}$), or

$$Q_s = V_s I \sin(-\theta_{Is}) = -V_s I \sin\theta_{Is} \qquad \text{or}$$
$$Q_s = -V_s I \sin\theta_{Is}. \tag{2-120}$$

Equation (2-120) can be rewritten using Equation (2-114b) as

$$Q_s = V_s \frac{V_s - V_r \cos\delta + V_{s's} \cos\beta}{X} \qquad \text{or} \qquad Q_s = \frac{V_s V_r}{X}\left(\frac{V_s}{V_r} - \cos\delta\right) + \frac{V_s V_{s's}}{X} \cos\beta \tag{2-121}$$

or

$$Q_s = Q_{sn} + a_s \cos\beta \tag{2-122}$$

where Q_{sn} and a_s are given in Equations (2-43) and (2-119), respectively.

Equations (2-118) and (2-122) can be combined as

$$(P_s - P_{sn})^2 + (Q_s - Q_{sn})^2 = a_s^2(\cos^2\beta + \sin^2\beta)$$

or

$$(P_s - P_{sn})^2 + (Q_s - Q_{sn})^2 = a_s^2 \tag{2-123}$$

using the identity $\left(\cos^2\beta + \sin^2\beta = 1\right)$ and

$$\tan\beta = \frac{P_s - P_{sn}}{Q_s - Q_{sn}}. \tag{2-124}$$

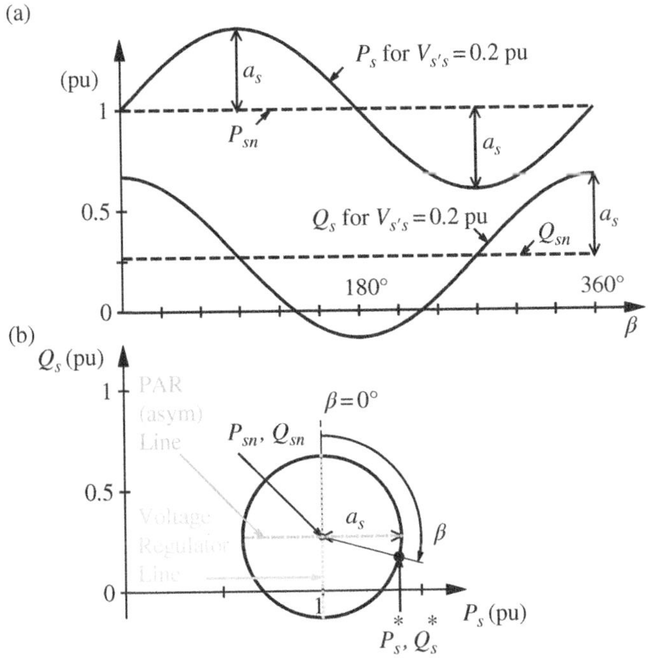

Figure 2-21 (a) Variation of the sending-end active and reactive power flows (P_s and Q_s) as a function of the relative phase angle (β) of the series-compensating voltage ($\mathbf{V_{s's}}$) with a fixed magnitude of 0.2 pu; (b) sending-end Q_s versus P_s.

Equation (2-123) defines the relationship between P_s and Q_s as a circle, centered at (P_{sn}, Q_{sn}) with a radius of a_s. The active and reactive power flows (P_s and Q_s) at the sending end are plotted in Figure 2-21a as functions of the relative phase angle (β). The active and reactive power flows at the sending end of the line can be regulated within the range defined by the *P-Q* plot in Figure 2-21b by controlling the magnitude ($V_{s's}$) of the series-compensating voltage ($\mathbf{V_{s's}}$) within its rated value and the relative phase angle of $0° \leq \beta \leq 360°$.

Equations (2-119) and (2-123) can be rearranged as

$$V_{s's} = \frac{X}{V_s}\sqrt{(P_s - P_{sn})^2 + (Q_s - Q_{sn})^2}. \tag{2-125}$$

Equation (2-124) can be rearranged as

$$\beta = \tan^{-1}\frac{P_s - P_{sn}}{Q_s - Q_{sn}}. \tag{2-126}$$

For the desired active and reactive power flows (P_s^* and Q_s^*) within the allowable range, the required magnitude ($V_{s's}$) and the relative phase angle (β) of the series-compensating voltage can be determined by substituting $P_s = P_s^*$ and $Q_s = Q_s^*$ in Equations (2-125) and (2-126). The relative phase angle (β) is the angle between the VR line (*i.e.* $\beta = 0°$ line) and the line that passes through (P_{sn}, Q_{sn}) and (P_s^*, Q_s^*).

For implementation of the VR, the relative phase angle is set to $\beta = 0°$ or $\beta = 180°$ in Equations (2-118) and (2-122). For $\beta = 0°$, the variations of P_s and Q_s are given by

$$P_s = P_{sn} + a_s \sin 0° \qquad \text{or} \qquad P_s = P_{sn}$$

and

$$Q_s = Q_{sn} + a_s \cos 0° \qquad \text{or} \qquad Q_s = Q_{sn} + a_s.$$

For $\beta = 180°$, the variations of P_s and Q_s are given by

$$P_s = P_{sn} + a_s \sin 180° \quad \text{or} \quad P_s = P_{sn}$$

and

$$Q_s = Q_{sn} + a_s \cos 180° \quad \text{or} \quad Q_s = Q_{sn} - a_s.$$

For $\beta = 0°$ or $\beta = 180°$, the variations of P_s and Q_s are then given by

$$P_s = P_{sn} \tag{2-127a}$$

and

$$Q_s = Q_{sn} \pm a_s \tag{2-127b}$$

or

$$V_{s's} = \pm X \frac{Q_s - Q_{sn}}{V_s} \quad \text{or} \quad V_{s's} = \frac{X}{V_s} |Q_s - Q_{sn}| \tag{2-127c}$$

using Equation (2-119). For a desired Q_s, the required $V_{s's}$ can be determined using Equation (2-127c).

For implementation of the PAR (asym), the relative phase angle is set to $\beta = 90°$ or $\beta = -90°$ in Equations (2-118) and (2-122). For $\beta = 90°$, the variations of P_s and Q_s are given by

$$P_s = P_{sn} + a_s \sin 90° \quad \text{or} \quad P_s = P_{sn} + a_s$$

and

$$Q_s = Q_{sn} + a_s \cos 90° \quad \text{or} \quad Q_s = Q_{sn}.$$

For $\beta = -90°$, the variations of P_s and Q_s are given by

$$P_s = P_{sn} + a_s \sin(-90°) \quad \text{or} \quad P_s = P_{sn} - a_s$$

and

$$Q_s = Q_{sn} + a_s \cos(-90°) \quad \text{or} \quad Q_s = Q_{sn}.$$

For $\beta = \pm 90°$, the variations of P_s and Q_s are then given by

$$P_s = P_{sn} \pm a_s \tag{2-128a}$$

or

$$V_{s's} = \pm X \frac{P_s - P_{sn}}{V_s} \quad \text{or} \quad V_{s's} = \frac{X}{V_s} |P_s - P_{sn}| \tag{2-128b}$$

using Equation (2-119) and

$$Q_s = Q_{sn}. \tag{2-128c}$$

For a desired P_s, the required $V_{s's}$ can be determined using Equation (2-128b).

In the case of $\beta = 0°$ or $\beta = 180°$, only the reactive power (Q_s), and, in the case of $\beta = \pm 90°$, only the active power (P_s) at the sending end vary with the magnitude ($V_{s's}$) of the series-compensating voltage ($\mathbf{V_{s's}}$).

Example 2-16

Consider the electrical system data as given in Table 2-1. Using a VR, (a) for $P_s = 1.4$ pu, what is $V_{s's}$? (b) for $P_s = 0.6$ pu, what is $V_{s's}$? Using a PAR (asym), (c) for $Q_s = 0.667949$ pu, what is $V_{s's}$? (d) for $Q_s = -0.132051$ pu, what is $V_{s's}$?

Answer 2-16

In Example 2-5, it was calculated that the natural transmitted active power, $P_{sn} = 1$ pu and the reactive power, $Q_{sn} = 0.267949$ pu at the source end. Therefore, from Equation (2-127c),

$$\text{a) } V_{s's} = \frac{X}{V_s}|Q_s - Q_{sn}| = \frac{0.5}{1}|0.667949 - 0.267949| = 0.2,$$

$$\text{b) } V_{s's} = \frac{X}{V_s}|Q_s - Q_{sn}| = \frac{0.5}{1}|-0.132051 - 0.267949| = 0.2;$$

from Equation (2-128b),

$$\text{c) } V_{s's} = \frac{X}{V_s}|P_s - P_{sn}| = \frac{0.5}{1}|1.4 - 1| = 0.2,$$

$$\text{d) } V_{s's} = \frac{X}{V_s}|P_s - P_{sn}| = \frac{0.5}{1}|0.6 - 1| = 0.2.$$

2.2.2.2 Power Flow at the Receiving End with a Series-Compensating Voltage

The active power flow at the receiving end of the line when the line is series-compensated is given by

$$P_r = V_r I \cos(\delta_r - \theta_I) \qquad \text{or}$$

$P_r = V_r I \cos(\{\delta_s - \delta\} - \{\delta_s + \theta_{Is}\})$ using Equation (2-27) where θ_{Is} is the phase angle of the line current ($\mathbf{I}$) with respect to the sending-end voltage ($\mathbf{V_s}$), or

$$P_r = V_r I \cos(-\delta - \theta_{Is}) = V_r I \cos(\delta + \theta_{Is}) \qquad \text{or}$$

$$P_r = V_r \cos\delta I \cos\theta_{Is} - V_r \sin\delta I \sin\theta_{Is}. \tag{2-129}$$

Equation (2-129) can be rewritten using Equations (2-114a) and (2-114b) as

$$P_r = V_r \cos\delta \frac{V_r \sin\delta + V_{s's}\sin\beta}{X} + V_r \sin\delta \frac{V_s - V_r\cos\delta + V_{s's}\cos\beta}{X}$$
$$= \frac{V_r V_s \sin\delta}{X} + \frac{V_r V_{s's}\cos\delta\sin\beta}{X} + \frac{V_r V_{s's}\sin\delta\cos\beta}{X}$$

or

$$P_r = \frac{V_s V_r}{X}\sin\delta + \frac{V_r V_{s's}}{X}\sin(\delta + \beta) \tag{2-130}$$

or

$$P_r = P_{rn} + a_r \sin(\delta + \beta) \tag{2-131}$$

where P_{rn} is given in Equation (2-46) and

$$a_r = \frac{V_r V_{s's}}{X}. \tag{2-132}$$

The reactive power flow at the receiving end of the line when the line is series-compensated is given by

$$Q_r = V_r I \sin(\delta_r - \theta_I) \qquad \text{or}$$

$Q_r = V_r I \sin(\{\delta_s - \delta\} - \{\delta_s + \theta_{Is}\})$ using Equation (2-27)

where θ_{Is} is the phase angle of the line current ($\mathbf{I}$) with respect to the sending-end voltage ($\mathbf{V_s}$), or

$$Q_r = V_r I \sin(-\delta - \theta_{Is}) = -V_r I \sin(\delta + \theta_{Is}) \qquad \text{or}$$
$$Q_r = -V_r \sin\delta\, I \cos\theta_{Is} - V_r \cos\delta\, I \sin\theta_{Is}. \tag{2-133}$$

Equation (2-133) can be rewritten using Equations (2-114a) and (2-114b) as

$$Q_r = -V_r \sin\delta \frac{V_r \sin\delta + V_{s's}\sin\beta}{X} + V_r \cos\delta \frac{V_s - V_r\cos\delta + V_{s's}\cos\beta}{X} \qquad \text{or}$$

$$Q_r = -\frac{V_r^2 \sin^2\delta}{X} - \frac{V_r V_{s's}\sin\delta\sin\beta}{X} + \frac{V_s V_r \cos\delta}{X} - \frac{V_r^2\cos^2\delta}{X} + \frac{V_r V_{s's}\cos\delta\cos\beta}{X} \qquad \text{or}$$

$$Q_r = \frac{V_s V_r}{X}\cos\delta - \frac{V_r^2}{X} + \frac{V_r V_{s's}}{X}\cos(\delta+\beta) \qquad \text{or}$$

$$Q_r = \frac{V_s V_r}{X}\left(\cos\delta - \frac{V_r}{V_s}\right) + \frac{V_r V_{s's}}{X}\cos(\delta+\beta) \tag{2-134}$$

or

$$Q_r = Q_{rn} + a_r\cos(\delta+\beta) \tag{2-135}$$

where Q_{rn} and a_r are given in Equations (2-48) and (2-132), respectively.

Equations (2-131) and (2-135) can be combined as

$$(P_r - P_{rn})^2 + (Q_r - Q_{rn})^2 = a_r^2\{\cos^2(\delta+\beta) + \sin^2(\delta+\beta)\}$$

or

$$(P_r - P_{rn})^2 + (Q_r - Q_{rn})^2 = a_r^2 \tag{2-136}$$

using the identity $\{\cos^2(\delta+\beta) + \sin^2(\delta+\beta) = 1\}$ and

$$\tan(\delta+\beta) = \frac{P_r - P_{rn}}{Q_r - Q_{rn}}. \tag{2-137}$$

Equation (2-136) defines the relationship between P_r and Q_r as a circle, centered at (P_{rn}, Q_{rn}) with a radius of a_r. The active and reactive power flows (P_r and Q_r) at the receiving end are plotted in Figure 2-22a as functions of the relative phase angle (β). The active and reactive power flows at the receiving end of the line can be regulated within the range defined by the *P-Q* plot in Figure 2-22b by controlling the magnitude ($V_{s's}$) of the series-compensating voltage ($\mathbf{V_{s's}}$) within its rated value and the relative phase angle of $0° \le \beta \le 360°$.

Equations (2-132) and (2-136) can be rearranged as

$$V_{s's} = \frac{X}{V_r}\sqrt{(P_r - P_{rn})^2 + (Q_r - Q_{rn})^2}. \tag{2-138}$$

Equation (2-137) can be rearranged as

$$\beta = \tan^{-1}\frac{P_r - P_{rn}}{Q_r - Q_{rn}} - \delta. \tag{2-139}$$

For the desired active and reactive power flows (P_r^* and Q_r^*) within the allowable range, the required magnitude ($V_{s's}$) and the relative phase angle (β) of the series-compensating voltage ($\mathbf{V_{s's}}$) can be determined by substituting $P_r = P_r^*$ and $Q_r = Q_r^*$ in Equations (2-138) and (2-139). The relative phase angle (β) is the angle between the VR line (*i.e.* $\beta = 0°$ line) and the line that passes through (P_{rn}, Q_{rn}) and (P_r^*, Q_r^*).

For implementation of the VR, the relative phase angle is set to $\beta = 0°$ or $\beta = 180°$ in Equations (2-131) and (2-135). For $\beta = 0°$, the variations of P_r and Q_r are given by

$$P_r = P_{rn} + a_r\sin(\delta + 0°) \qquad \text{or} \qquad P_r = P_{rn} + a_r\sin\delta$$

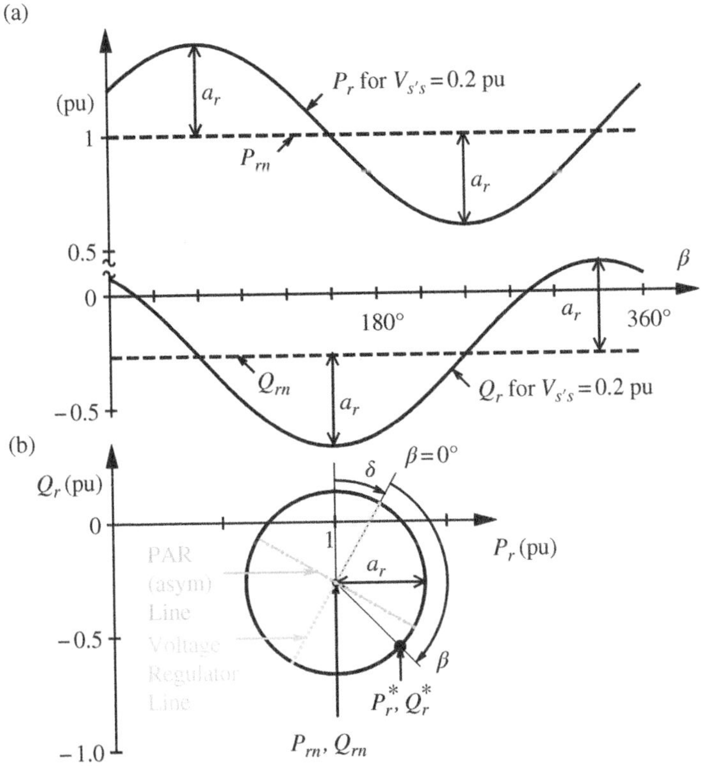

Figure 2-22 (a) Variation of the receiving-end active and reactive power flows (P_r and Q_r) as a function of the relative phase angle (β) of the series-compensating voltage ($\mathbf{V_{s's}}$) with a fixed magnitude of 0.2 pu; (b) receiving-end Q_r versus P_r.

and

$$Q_r = Q_{rn} + a_r \cos(\delta + 0°) \qquad \text{or} \qquad Q_r = Q_{rn} + a_r \cos\delta.$$

For $\beta = 180°$, the variations of P_r and Q_r are given by

$$P_r = P_{rn} + a_r \sin(\delta + 180°) \qquad \text{or} \qquad P_r = P_{rn} - a_r \sin\delta$$

and

$$Q_r = Q_{rn} + a_r \cos(\delta + 180°) \qquad \text{or} \qquad Q_r = Q_{rn} - a_r \cos\delta.$$

For $\beta = 0°$ or $\beta = 180°$, the variations of P_r and Q_r are then given by

$$P_r = P_{rn} \pm a_r \sin\delta \tag{2-140a}$$

or

$$V_{s's} = \pm X \frac{P_r - P_{rn}}{V_r \sin\delta} \qquad \text{or} \qquad V_{s's} = \frac{X}{V_r \sin\delta}|P_r - P_{rn}| \tag{2-140b}$$

using Equation (2-132) and

$$Q_r = Q_{rn} \pm a_r \cos\delta \tag{2-140c}$$

or

$$V_{s's} = \pm X \frac{Q_r - Q_{rn}}{V_r \cos\delta} \qquad \text{or} \qquad V_{s's} = \frac{X}{V_r \cos\delta}|Q_r - Q_{rn}| \tag{2-140d}$$

using Equation (2-132). For a desired P_r, the required $V_{s's}$ can be determined using Equation (2-140b); for a desired Q_r, the required $V_{s's}$ can be determined using Equation (2-140d).

Example 2-17

Consider the electrical system data as given in Table 2-1. Using a VR, (a) for $P_r = 1.2$ pu, what is $V_{s's}$? (b) for $P_r = 0.8$ pu, what is $V_{s's}$? (c) for $Q_r = 0.078461$ pu, what is $V_{s's}$? (d) for $Q_r = -0.614359$ pu, what is $V_{s's}$?

Answer 2-17

In Example 2-5, it was calculated that the natural transmitted active power, $P_{rn} = 1$ pu and the reactive power, $P_{rn} = -0.267949$ pu at the load end. Therefore, from Equation (2-140b),

$$\text{a) } V_{s's} = \frac{X}{V_r \sin\delta}|P_r - P_{rn}| = \frac{0.5}{1 \times \sin 30^\circ}|1.2 - 1| = 0.2,$$

$$\text{b) } V_{s's} = \frac{X}{V_r \sin\delta}|P_r - P_{rn}| = \frac{0.5}{1 \times \sin 30^\circ}|0.8 - 1| = 0.2;$$

from Equation (2-140d),

$$\text{c) } V_{s's} = \frac{X}{V_r \cos\delta}|Q_r - Q_{rn}| = \frac{0.5}{1 \times \cos 30^\circ}|0.078461 + 0.267949| = 0.2,$$

$$\text{d) } V_{s's} = \frac{X}{V_r \cos\delta}|Q_r - Q_{rn}| = \frac{0.5}{1 \times \cos 30^\circ}|-0.614359 + 0.267949| = 0.2.$$

Substituting $\beta = 0^\circ$ or $\beta = 180^\circ$ in Equation (2-137) or combining Equations (2-140a) and (2-140c) leads to

$$\tan\delta = \frac{P_r - P_{rn}}{Q_r - Q_{rn}} \quad \text{or} \quad Q_r = \frac{P_r - P_{rn}}{\tan\delta} + Q_{rn} \tag{2-141a}$$

or substituting for $\tan\delta$ from Equation (2-49)

$$Q_r = \frac{P_r - P_{rn}}{P_{rn}}\left(Q_{rn} + \frac{V_r^2}{X}\right) + Q_{rn} \quad \text{or} \quad Q_r = \left(\frac{P_r}{P_{rn}} - 1\right)\left(Q_{rn} + \frac{V_r^2}{X}\right) + Q_{rn}$$

or

$$Q_r = \frac{P_r}{P_{rn}}\left(Q_{rn} + \frac{V_r^2}{X}\right) - \frac{V_r^2}{X}, \tag{2-141b}$$

which is same as Equation (2-83c).

For implementation of the PAR (asym), the relative phase angle is set to $\beta = 90^\circ$ or $\beta = -90^\circ$ in Equations (2-131) and (2-135). For $\beta = 90^\circ$, the variations of P_r and Q_r are given by

$$P_r = P_{rn} + a_r \sin(\delta + 90^\circ) \quad \text{or} \quad P_r = P_{rn} + a_r \cos\delta$$

and

$$Q_r = Q_{rn} + a_r \cos(\delta + 90^\circ) \quad \text{or} \quad Q_r = Q_{rn} - a_r \sin\delta.$$

For $\beta = -90^\circ$, the variations of P_r and Q_r are given by

$$P_r = P_{rn} + a_r \sin(\delta - 90^\circ) \quad \text{or} \quad P_r = P_{rn} - a_r \cos\delta$$

and

$$Q_r = Q_{rn} + a_r \cos(\delta - 90^\circ) \quad \text{or} \quad Q_r = Q_{rn} + a_r \sin\delta.$$

For $\beta = \pm 90°$, the variations of P_r and Q_r are then given by

$$P_r = P_{rn} \pm a_r \cos\delta \tag{2-142a}$$

or

$$V_{s's} = +X\frac{P_r - P_{rn}}{V_r \cos\delta} \quad \text{or} \quad V_{s's} = \frac{X}{V_r \cos\delta}|P_r - P_{rn}| \tag{2-142b}$$

using Equation (2-132) and

$$Q_r = Q_{rn} \mp a_r \sin\delta \tag{2-142c}$$

or

$$V_{s's} = \mp X\frac{Q_r - Q_{rn}}{V_r \sin\delta} \quad \text{or} \quad V_{s's} = \frac{X}{V_r \sin\delta}|Q_r - Q_{rn}| \tag{2-142d}$$

using Equation (2-132). For a desired P_r, the required $V_{s's}$ can be determined using Equation (2-142b); for a desired Q_r, the required $V_{s's}$ can be determined using Equation (2-142d).

Example 2-18

Consider the electrical system data as given in Table 2-1. Using a PAR (asym), (a) for $P_r = 1.34641$ pu, what is $V_{s's}$? (b) for $P_r = 0.65359$ pu, what is $V_{s's}$? (c) for $Q_r = -0.467949$ pu, what is $V_{s's}$? (d) for $Q_r = -0.067949$ pu, what is $V_{s's}$?

Answer 2-18

In Example 2-5, it was calculated that the natural transmitted active power, $P_{rn} = 1$ pu and the reactive power, $Q_{rn} = -0.267949$ pu at the load end. Therefore, from Equation (2-142b),

a) $V_{s's} = \dfrac{X}{V_r \cos\delta}|P_r - P_{rn}| = \dfrac{0.5}{1 \times \cos 30°}|1.34641 - 1| = 0.2,$

b) $V_{s's} = \dfrac{X}{V_r \cos\delta}|P_r - P_{rn}| = \dfrac{0.5}{1 \times \cos 30°}|0.65359 - 1| = 0.2;$

from Equation (2-142d),

c) $V_{s's} = \dfrac{X}{V_r \sin\delta}|Q_r - Q_{rn}| = \dfrac{0.5}{1 \times \sin 30°}|-0.467949 + 0.267949| = 0.2,$

d) $V_{s's} = \dfrac{X}{V_r \sin\delta}|Q_r - Q_{rn}| = \dfrac{0.5}{1 \times \sin 30°}|-0.067949 + 0.267949| = 0.2.$

Substituting $\beta = 90°$ or $\beta = -90°$ in Equation (2-137) or combining Equations (2-142a) and (2-142c) leads to

$$\frac{P_r - P_{rn}}{Q_r - Q_{rn}} = -\cot\delta \quad \text{or} \quad Q_r = -\tan\delta(P_r - P_{rn}) + Q_{rn} \tag{2-143a}$$

or substituting for $\tan\delta$ from Equation (2-49)

$$Q_r = -\frac{P_{rn}}{Q_{rn} + \frac{V_r^2}{X}}(P_r - P_{rn}) + Q_{rn} \quad \text{or} \quad Q_r = \frac{-P_{rn}}{Q_{rn} + \frac{V_r^2}{X}}P_r + \frac{P_{rn}^2}{Q_{rn} + \frac{V_r^2}{X}} + Q_{rn}. \tag{2-143b}$$

The $P_r - Q_r$ relationship is defined with a slope (m_r) and an intercept (c_r) as

$$Q_r = m_r P_r + c_r \tag{2-144}$$

where

$$m_r = \frac{-P_{rn}}{Q_{rn} + \frac{V_r^2}{X}} \tag{2-145}$$

and

$$c_r = \frac{P_{rn}^2}{Q_{rn} + \frac{V_r^2}{X}} + Q_{rn}. \tag{2-146}$$

Therefore, in the case of $\beta = 0°$, $180°$or $\pm 90°$, both the active and reactive power flows (P_r and Q_r) at the receiving end vary with the magnitude ($V_{s's}$) of the series-compensating voltage ($\mathbf{V_{s's}}$). For a given $V_{s's}$, P_r and Q_r are determined for $\beta = 0°$ and $180°$, using Equations (2-140a) and (2-84); the same are determined for $\beta = \pm 90°$ using Equations (2-142a) and (2-144).

Example 2-19

Consider the electrical system data as given in Table 2-1. For $\beta = 90°$ or $\beta = -90°$, calculate m_r and c_r.

Answer 2-19

In Example 2-5, it was determined that the natural transmitted active power, $P_{rn} = 1$ pu and the reactive power, $Q_{rn} = -0.267949$ pu at the load end. Therefore, from Equation (2-145),

$$m_r = \frac{-P_{rn}}{Q_{rn} + \frac{V_r^2}{X}} = \frac{-1}{-0.267949 + \frac{1^2}{0.5}} = -0.577350.$$

From Equation (2-146), $c_r = \dfrac{P_{rn}^2}{Q_{rn} + \frac{V_r^2}{X}} + Q_{rn} = \dfrac{1^2}{-0.267949 + \frac{1^2}{0.5}} - 0.267949 = 0.309401$ pu.

Additional Discussion

Note that the slope (m_r) in Example 2-7 is 1.732051 and that in Example 2-19 is −0.577350; therefore, the product of these slopes is −1, meaning the $P_r - Q_r$ characteristic equations for the VR and the PAR (asym) are perpendicular, which is shown in Figure 2-22.

2.2.2.3 Power Flow at the Modified Sending End with a Series-Compensating Voltage

The active power flow at the modified sending end of the line when the line is series-compensated is given by

$$P_{s'} = V_{s'} I \cos(\delta_{s'} - \theta_I) \qquad \text{or}$$

$$P_{s'} = V_{s'} I \cos(\{\delta_s + \psi\} - \{\delta_s + \theta_{Is}\})$$

where θ_{Is} is the phase angle of the line current ($\mathbf{I}$) with respect to the sending-end voltage ($\mathbf{V_s}$), or

$$P_{s'} = V_{s'} I \cos(\psi - \theta_{Is}) \qquad \text{or}$$

$$P_{s'} = V_{s'} \cos\psi I \cos\theta_{Is} + V_{s'} \sin\psi I \sin\theta_{Is}. \tag{2-147}$$

Equation (2-147) can be rewritten using Equations (2-96a), (2-96b), (2-114a), and (2-114b) as

$$P_{s'} = (V_s + V_{s's}\cos\beta)\frac{V_r\sin\delta + V_{s's}\sin\beta}{X} + V_{s's}\sin\beta\frac{V_r\cos\delta - V_s - V_{s's}\cos\beta}{X} \quad \text{or}$$

$$P_{s'} = \frac{V_sV_r\sin\delta}{X} + \frac{V_{s's}V_r\cos\beta\sin\delta}{X} + \frac{V_{s's}V_r\sin\beta\cos\delta}{X} \quad \text{or}$$

$$P_{s'} = \frac{V_sV_r}{X}\sin\delta + \frac{V_rV_{s's}}{X}\sin(\delta+\beta) \tag{2-148}$$

or

$$P_{s'} = P_{sn} + a_r\sin(\delta+\beta) \tag{2-149}$$

where P_{sn} and a_r are given in Equations (2-40) and (2-132), respectively.

The reactive power flow at the modified sending end of the line when the line is series-compensated is given by

$$Q_{s'} = V_{s'}I\sin(\delta_{s'} - \theta_I) \quad \text{or}$$

$$Q_{s'} = V_{s'}I\sin(\{\delta_s + \psi\} - \{\delta_s + \theta_{Is}\})$$

where θ_{Is} is the phase angle of the line current (**I**) with respect to the sending-end voltage ($\mathbf{V_s}$), or

$$Q_{s'} = V_{s'}I\sin(\psi - \theta_{Is}) \quad \text{or}$$

$$Q_{s'} = V_{s'}\sin\psi I\cos\theta_{Is} - V_{s'}\cos\psi I\sin\theta_{Is}. \tag{2-150}$$

Equation (2-150) can be rewritten using Equations (2-96a), (2-96b), (2-114a), and (2-114b) as

$$Q_{s'} = V_{s's}\sin\beta\frac{V_r\sin\delta + V_{s's}\sin\beta}{X} + (V_s + V_{s's}\cos\beta)\frac{V_s - V_r\cos\delta + V_{s's}\cos\beta}{X} \quad \text{or}$$

$$Q_{s'} = \frac{V_rV_{s's}\sin\beta\sin\delta}{X} + \frac{V_{s's}^2\sin^2\beta}{X} + \frac{V_s^2}{X} + \frac{V_sV_{s's}\cos\beta}{X} - \frac{V_sV_r\cos\delta}{X} - \frac{V_rV_{s's}\cos\beta\cos\delta}{X} + \frac{V_sV_{s's}\cos\beta}{X} + \frac{V_{s's}^2\cos^2\beta}{X}$$

or

$$Q_{s'} = \frac{V_sV_r}{X}\left(\frac{V_s}{V_r} - \cos\delta\right) + \frac{V_{s's}^2}{X} + \frac{2V_sV_{s's}}{X}\cos\beta - \frac{V_rV_{s's}}{X}\cos(\delta+\beta) \tag{2-151}$$

or

$$Q_{s'} = Q_{sn} + \frac{V_{s's}^2}{X} + 2a_s\cos\beta - a_r\cos(\delta+\beta) \tag{2-152}$$

where Q_{sn}, a_s, and a_r are given in Equations (2-43), (2-119), and (2-132), respectively.

Equations (2-149) and (2-152) can be combined as

$$(P_{s'} - P_{sn})^2 + \left(Q_{s'} - Q_{sn} - \frac{V_{s's}^2}{X}\right)^2 = 4a_s^2\cos^2\beta - 4a_sa_r\cos\beta\cos(\delta+\beta) + a_r^2 \tag{2-153a}$$

or

$$(P_{s'} - P_{sn})^2 + (Q_{s'} - Q_{sn})^2 = a_r^2\sin^2(\delta+\beta) + \left[\frac{V_{s's}^2}{X} + 2a_s\cos\beta - a_r\cos(\delta+\beta)\right]^2 \tag{2-153b}$$

or substituting for a_s and a_r from Equations (2-119) and (2-132),

$$(P_{s'} - P_{sn})^2 + (Q_{s'} - Q_{sn})^2 = \left(\frac{V_{s's}V_r}{X}\right)^2 \sin^2(\delta + \beta) + \left[\frac{V_{s's}^2}{X} + 2\frac{V_{s's}V_s}{X}\cos\beta - \frac{V_{s's}V_r}{X}\cos(\delta + \beta)\right]^2$$

or

$$(P_{s'} - P_{sn})^2 + (Q_{s'} - Q_{sn})^2 = \frac{V_{s's}^2}{X^2}V_r^2\sin^2(\delta + \beta) + \frac{V_{s's}^2}{X^2}[V_{s's} + 2V_s\cos\beta - V_r\cos(\delta + \beta)]^2$$

or

$$(P_{s'} - P_{sn})^2 + (Q_{s'} - Q_{sn})^2 =$$
$$\frac{V_{s's}^2}{X^2}[V_{s's}^2 + 2V_{s's}\{2V_s\cos\beta - V_r\cos(\delta + \beta)\} + 4V_s^2\cos^2\beta - 4V_sV_r\cos\beta\cos(\delta + \beta) + V_r^2]$$

or

$$V_{s's}^4 + 2\{2V_s\cos\beta - V_r\cos(\delta + \beta)\}V_{s's}^3 +$$
$$\{4V_s^2\cos^2\beta - 4V_sV_r\cos\beta\cos(\delta + \beta) + V_r^2\}V_{s's}^2 - \{(P_{s'} - P_{sn})^2 + (Q_{s'} - Q_{sn})^2\}X^2 = 0 \tag{2-154}$$

and

$$\frac{P_{s'} - P_{sn}}{Q_{s'} - Q_{sn} - \dfrac{V_{s's}^2}{X}} = \frac{V_r\sin(\delta + \beta)}{2V_s\cos\beta - V_r\cos(\delta + \beta)}$$

or

$$(P_{s'} - P_{sn})\{2V_s\cos\beta - V_r\cos(\delta + \beta)\} = \left(Q_{s'} - Q_{sn} - \frac{V_{s's}^2}{X}\right)V_r\sin(\delta + \beta)$$

or

$$(P_{s'} - P_{sn})\{2V_s\cos\beta - V_r(\cos\delta\cos\beta - \sin\delta\sin\beta)\} = \left(Q_{s'} - Q_{sn} - \frac{V_{s's}^2}{X}\right)V_r(\sin\delta\cos\beta + \cos\delta\sin\beta)$$

or

$$(P_{s'} - P_{sn})V_r\sin\delta\sin\beta - \left(Q_{s'} - Q_{sn} - \frac{V_{s's}^2}{X}\right)V_r(\cos\delta\sin\beta)$$
$$= -(P_{s'} - P_{sn})\{2V_s\cos\beta - V_r(\cos\delta\cos\beta)\} + \left(Q_{s'} - Q_{sn} - \frac{V_{s's}^2}{X}\right)V_r(\sin\delta\cos\beta)$$

or

$$\left[(P_{s'} - P_{sn})V_r\sin\delta - \left(Q_{s'} - Q_{sn} - \frac{V_{s's}^2}{X}\right)V_r\cos\delta\right]\sin\beta =$$
$$\left[-(P_{s'} - P_{sn})(2V_s - V_r\cos\delta) + \left(Q_{s'} - Q_{sn} - \frac{V_{s's}^2}{X}\right)V_r\sin\delta\right]\cos\beta$$

or

$$\tan\beta = \frac{-(P_{s'} - P_{sn})\{2V_s - V_r\cos\delta\} + \left(Q_{s'} - Q_{sn} - \dfrac{V_{s's}^2}{X}\right)V_r\sin\delta}{(P_{s'} - P_{sn})V_r\sin\delta - \left(Q_{s'} - Q_{sn} - \dfrac{V_{s's}^2}{X}\right)V_r\cos\delta}; \tag{2-155a}$$

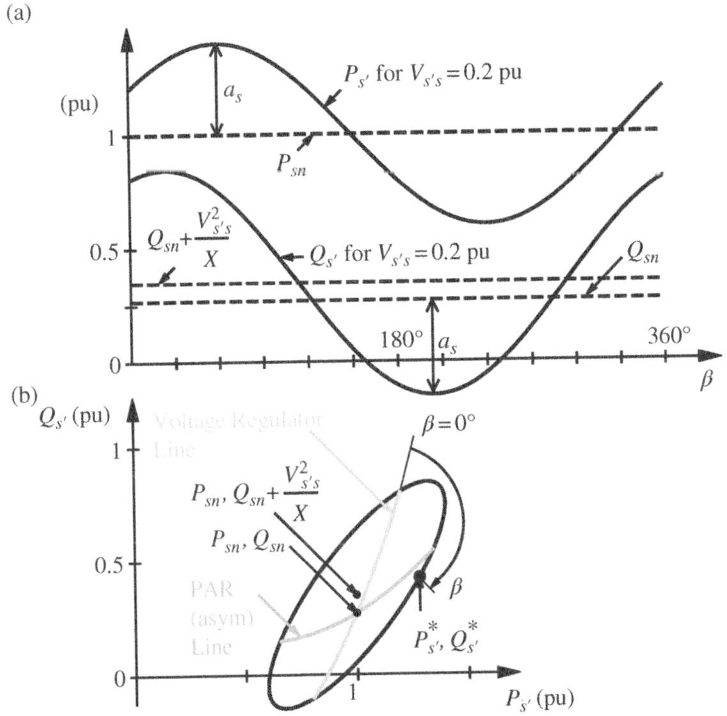

Figure 2-23 (a) Variation of the modified sending-end active and reactive power flows ($P_{s'}$ and $Q_{s'}$) as a function of the relative phase angle (β) of the series-compensating voltage ($\mathbf{V_{s's}}$) with a fixed magnitude of 0.2 pu; (b) modified sending-end $Q_{s'}$ versus $P_{s'}$.

it can also be written from Equations (2-149) and (2-152) that

$$\tan(\delta+\beta) = -\frac{P_{s'} - P_{sn}}{Q_{s'} - Q_{sn} - \frac{V^2_{s's}}{X} - 2a_s\cos\beta}. \tag{2-155b}$$

Equations (2-153a) and (2-153b) define the relationship between $P_{s'}$ and $Q_{s'}$ as an ellipse. The active and reactive power flows ($P_{s'}$ and $Q_{s'}$) at the modified sending end are plotted in Figure 2-23a as functions of the relative phase angle (β). The active and reactive power flows at the modified sending end of the line can be regulated within the range defined by the *P*-*Q* plot in Figure 2-23b by controlling the magnitude ($V_{s's}$) of the series-compensating voltage ($\mathbf{V_{s's}}$) within its rated value and the relative phase angle of $0° \leq \beta \leq 360°$.

For the desired active and reactive power flows ($P_{s'}{}^*$ and $Q_{s'}{}^*$) at the modified sending end, the required magnitude ($V_{s's}$) and the relative phase angle (β) of the series-compensating voltage ($\mathbf{V_{s's}}$) can be determined by substituting $P_{s'}$ with $P_{s'}{}^*$ and $Q_{s'}$ with $Q_{s'}{}^*$ in Equations (2-154) and (2-155a), respectively. Note that the Equation (2-154) is a fourth-order equation that may need to be solved by using a numerical method to determine the magnitude ($V_{s's}$) of the series-compensating voltage ($\mathbf{V_{s's}}$).

Alternately, a two-step method may be applied where the first step is to determine the magnitude ($V_{s'}$) and phase-shift angle (ψ) of the modified sending-end voltage ($\mathbf{V_{s'}}$) using Equations (2-67a), (2-68), and (2-51b) after substituting $P_{s'}$ with $P_{s'}{}^*$ and $Q_{s'}$ with $Q_{s'}{}^*$ and the second step is for a desired modified sending-end voltage ($\mathbf{V_{s'}}$), the magnitude ($V_{s's}$) and the relative phase angle (β) of the series-compensating voltage ($\mathbf{V_{s's}}$) are determined using Equations (2-100) and (2-101) as given in Table 2-11.

In a third method, which is a three-step method where the first step is to determine the active and reactive power flows (P_s and Q_s) at the sending end. If the desired active and reactive power flows at the modified sending end are $P_{s'}{}^*$ and $Q_{s'}{}^*$, the required exchanged active and reactive powers ($P_{se}{}^*$ and $Q_{se}{}^*$) are determined in the second step by using the following equations:

$$P_{se}^* = P_s - P_{s'}^* \tag{2-156a}$$

and

$$Q_{se}^* = Q_s - Q_{s'}^*. \tag{2-156b}$$

In the third step, the required magnitude ($V_{s's}$) and the relative phase angle (β) of the compensating voltage ($\mathbf{V_{s's}}$) are determined as shown in the next section.

The active and reactive power flows at any point in the line can also be controlled independently by using a closed-loop control scheme. The automatic power flow control mode of operation of the UPFC for achieving independent control of active and reactive power flows ($P_{s'}^*$ and $Q_{s'}^*$) at the modified sending end has been implemented in the world's first UPFC. The field results from this ±160 MVA-rated UPFC are shown in "Comparison of field results and digital simulation results of voltage-sourced converter-based FACTS controllers," by K. K. Sen and A. J. F. Keri, *IEEE Trans. on Power Delivery*, vol. 18, no. 1, pp. 300–306, Jan 2003. In 1998, this UPFC demonstrated for the first time that the active and reactive power flows in a line could be regulated independently while maintaining a fixed line voltage at the POC at Inez substation in Kentucky, USA.

The parameters a_s in Equation (2-119) and a_r in Equation (2-132) are the maximum power flow increases and are proportional to the magnitude ($V_{s's}$) of the series-compensating voltage ($\mathbf{V_{s's}}$). When the magnitude of the compensating voltage increases, so does the maximum power flow and vice versa. The same applies to the controllable range of the power flow in the *P-Q* plane. When the controllable range extends to all four quadrants, the bidirectional and independent control of active and reactive power flows is achieved.

For implementation of the VR, the relative phase angle is set to $\beta = 0°$ or $\beta = 180°$ in Equations (2-149) and (2-152). For $\beta = 0°$, the variations of $P_{s'}$ and $Q_{s'}$ are given by

$$P_{s'} = P_{sn} + a_r \sin(\delta + 0°) \qquad \text{or} \qquad P_{s'} = P_{sn} + a_r \sin\delta$$

and

$$Q_{s'} = Q_{sn} + \frac{V_{s's}^2}{X} + 2a_s \cos 0° - a_r \cos(\delta + 0°) \qquad \text{or} \qquad Q_{s'} = Q_{sn} + \frac{V_{s's}^2}{X} + 2a_s - a_r \cos\delta.$$

For $\beta = 180°$, the variations of $P_{s'}$ and $Q_{s'}$ are given by

$$P_{s'} = P_{sn} + a_r \sin(\delta + 180°) \qquad \text{or} \qquad P_{s'} = P_{sn} - a_r \sin\delta$$

and

$$Q_{s'} = Q_{sn} + \frac{V_{s's}^2}{X} + 2a_s \cos 180° - a_r \cos(\delta + 180°) \qquad \text{or} \qquad Q_{s'} = Q_{sn} + \frac{V_{s's}^2}{X} - 2a_s + a_r \cos\delta.$$

For $\beta = 0°$ or $\beta = 180°$, the variations of $P_{s'}$ and $Q_{s'}$ are then given by

$$P_{s'} = P_{sn} \pm a_r \sin\delta \tag{2-157a}$$

or

$$V_{s's} = \pm X\frac{P_{s'} - P_{sn}}{V_r \sin\delta} \quad \text{or} \quad V_{s's} = \frac{X}{V_r \sin\delta}|P_{s'} - P_{sn}| \tag{2-157b}$$

using Equation (2-132) and

$$Q_{s'} = Q_{sn} + \frac{V_{s's}^2}{X} \pm 2a_s \mp a_r \cos\delta \tag{2-157c}$$

or

$$V_{s's}^2 \pm (2V_s - V_r\cos\delta)V_{s's} - XQ_{s'} + XQ_{sn} = 0 \quad \text{using Equations (2-119) and (2-132)}$$

or

$$V_{s's} = \frac{\mp(2V_s - V_r\cos\delta) \pm \sqrt{(2V_s - V_r\cos\delta)^2 + 4X(Q_{s'} - Q_{sn})}}{2}$$

or

$$V_{s's} = \frac{1}{2}\left|(2V_s - V_r\cos\delta) - \sqrt{(2V_s - V_r\cos\delta)^2 + 4X(Q_{s'} - Q_{sn})}\right|. \tag{2-157d}$$

For a desired $P_{s'}$, the required $V_{s's}$ can be determined using Equation (2-157b); for a desired $Q_{s'}$, the required $V_{s's}$ can be determined using Equation (2-157d).

Example 2-20

Consider the electrical system data as given in Table 2-1. Using a VR, (a) for $P_{s'} = 1.2$ pu, what is $V_{s's}$? (b) for $P_{s'} = 0.8$ pu, what is $V_{s's}$? (c) for $Q_{s'} = 0.801539$ pu, what is $V_{s's}$? (d) for $Q_{s'} = -0.105641$ pu, what is $V_{s's}$?

Answer 2-20

In Example 2-5, it was calculated that the natural transmitted active power, $P_{sn} = 1$ pu and the reactive power, $Q_{sn} = 0.267949$ pu at the source end. Therefore, from Equation (2-157b),

$$\text{a)}\ V_{s's} = \frac{X}{V_r \sin\delta}|P_{s'} - P_{sn}| = \frac{0.5}{1 \times \sin 30^\circ}|1.2 - 1| = 0.2,$$

$$\text{b)}\ V_{s's} = \frac{X}{V_r \sin\delta}|P_{s'} - P_{sn}| = \frac{0.5}{1 \times \sin 30^\circ}|0.8 - 1| = 0.2;$$

from Equation (2-157d),

$$\text{c)}\ V_{s's} = \frac{1}{2}\left|(2V_s - V_r\cos\delta) - \sqrt{(2V_s - V_r\cos\delta)^2 + 4X(Q_{s'} - Q_{sn})}\right|$$

$$= \frac{1}{2}\left|(2\times1 - 1\times\cos 30^\circ) - \sqrt{(2\times1 - 1\times\cos 30^\circ)^2 + 4\times0.5(0.801539 - 0.267949)}\right| = 0.2,$$

d) from Equation (2-157d),

$$V_{s's} = \frac{1}{2}\left|(2V_s - V_r\cos\delta) - \sqrt{(2V_s - V_r\cos\delta)^2 + 4X(Q_{s'} - Q_{sn})}\right|$$

$$= \frac{1}{2}\left|(2\times1 - 1\times\cos 30^\circ) - \sqrt{(2\times1 - 1\times\cos 30^\circ)^2 + 4\times0.5(-0.105641 - 0.267949)}\right| = 0.2.$$

Substituting $\beta = 0°$ or $\beta = 180°$ in Equation (2-155b) or combining Equations (2-157a) and (2-157c) leads to

$$\tan\delta = -\frac{P_{s'} - P_{sn}}{Q_{s'} - Q_{sn} - \frac{V_{s's}^2}{X} \mp 2a_s} \quad \text{or} \quad Q_{s'} = -\frac{P_{s'} - P_{sn}}{\tan\delta} + Q_{sn} + \frac{V_{s's}^2}{X} \pm 2a_s; \tag{2-158a}$$

or substituting for tanδ from Equation (2-44),

$$Q_{s'} = -\frac{P_{s'}}{\tan\delta} + \frac{Q_{sn} - \frac{V_s^2}{X}}{-P_{sn}} P_{sn} + Q_{sn} + \frac{V_{s's}^2}{X} \pm 2a_s \quad \text{or} \quad Q_{s'} = -\frac{P_{s'}}{\tan\delta} + \frac{V_s^2}{X} + \frac{V_{s's}^2}{X} \pm 2a_s;$$

or substituting for a_s from Equation (2-119),

$$Q_{s'} = -\frac{P_{s'}}{\tan\delta} + \frac{(V_s \pm V_{s's})^2}{X}. \tag{2-158b}$$

Then,

$$Q_{s'} = -\frac{P_{s'}}{\tan\delta} + \frac{V_{s'}^2}{X} \tag{2-159}$$

since

$$V_{s'} = V_s \pm V_{s's}. \tag{2-160}$$

Note that Equation (2-159) is same as Equation (2-70b). Therefore, the $P_{s'}$ - $Q_{s'}$ relationship is defined with a slope ($m_{s'}$) and an intercept ($c_{s'}$) as

$$Q_{s'} = m_{s'} P_{s'} + c_{s'} \tag{2-71}$$

where

$$m_{s'} = -\frac{1}{\tan\delta} = \frac{1}{P_{sn}}\left(Q_{sn} - \frac{V_s^2}{X}\right) \tag{2-72}$$

and

$$c_{s'} = \frac{(V_s \pm V_{s's})^2}{X}. \tag{2-161}$$

Example 2-21

Consider the electrical system data as given in Table 2-1. For $\beta = 0°$ or $\beta = 180°$ and $V_{s's} = 0.2$ pu, calculate $m_{s'}$ and $c_{s'}$.

Answer 2-21

In Example 2-5, it was calculated that the natural transmitted active power, $P_{sn} = 1$ pu and the reactive power, $Q_{sn} = 0.267949$ pu at the source end. Therefore, from Equation (2-72),

$$m_{s'} = \frac{1}{P_{sn}}\left(Q_{sn} - \frac{V_s^2}{X}\right) = \frac{1}{1}\left(0.267949 - \frac{1^2}{0.5}\right) = -1.732051.$$

From Equation (2-161), $c_{s'} = \frac{(V_s \pm V_{s's})^2}{X} = \frac{(1 \pm 0.2)^2}{0.5} = 2.88, 1.28$. Note that for $\beta = 0°$, $c_{s'} = 2.88$ and for $\beta = 180°$, $c_{s'} = 1.28$.

For implementation of the PAR (asym), the relative phase angle is set to $\beta = 90°$ or $\beta = -90°$ in Equations (2-149) and (2-152). For $\beta = 90°$, the variations of $P_{s'}$ and $Q_{s'}$ are given by

$$P_{s'} = P_{sn} + a_r \sin(\delta + 90°) \qquad \text{or} \qquad P_{s'} = P_{sn} + a_r \cos\delta$$

and

$$Q_{s'} = Q_{sn} + \frac{V_{s's}^2}{X} + 2a_s \cos 90° - a_r \cos(\delta + 90°) \qquad \text{or} \qquad Q_{s'} = Q_{sn} + \frac{V_{s's}^2}{X} + a_r \sin\delta.$$

For $\beta = -90°$, the variations of $P_{s'}$ and $Q_{s'}$ are given by

$$P_{s'} = P_{sn} + a_r \sin(\delta - 90°) \qquad \text{or} \qquad P_{s'} = P_{sn} - a_r \cos\delta$$

and

$$Q_{s'} = Q_{sn} + \frac{V_{s's}^2}{X} + 2a_s \cos(-90°) - a_r \cos(\delta - 90°) \qquad \text{or} \qquad Q_{s'} = Q_{sn} + \frac{V_{s's}^2}{X} - a_r \sin\delta.$$

For $\beta = \pm 90°$, the variations of $P_{s'}$ and $Q_{s'}$ are then given by

$$P_{s'} = P_{sn} \pm a_r \cos\delta \tag{2-162a}$$

or

$$V_{s's} = \pm X \frac{P_{s'} - P_{sn}}{V_r \cos\delta} \qquad \text{or} \qquad V_{s's} = \frac{X}{V_r \cos\delta} |P_{s'} - P_{sn}| \tag{2-162b}$$

using Equation (2-132) and

$$Q_{s'} = Q_{sn} + \frac{V_{s's}^2}{X} \pm a_r \sin\delta \tag{2-162c}$$

or

$$V_{s's}^2 \pm V_r \sin\delta V_{s's} - XQ_{s'} + XQ_{sn} = 0 \quad \text{using Equation (2-132)}$$

or

$$V_{s's} = \frac{\mp V_r \sin\delta \pm \sqrt{(V_r \sin\delta)^2 + 4X(Q_{s'} - Q_{sn})}}{2} \qquad \text{or}$$

$$V_{s's} = \frac{1}{2}\left| V_r \sin\delta - \sqrt{(V_r \sin\delta)^2 + 4X(Q_{s'} - Q_{sn})} \right|. \tag{2-162d}$$

For a desired $P_{s'}$, the required $V_{s's}$ can be determined using Equation (2-162b); for a desired $Q_{s'}$, the required $V_{s's}$ can be determined using Equation (2-162d).

Example 2-22

Consider the electrical system data as given in Table 2-1. Using a PAR (asym), (a) for $P_{s'} = 1.34641$ pu, what is $V_{s's}$? (b) for $P_{s'} = 0.65359$ pu, what is $V_{s's}$? (c) for $Q_{s'} = 0.547949$ pu, what is $V_{s's}$? (d) for $Q_{s'} = 0.147949$ pu, what is $V_{s's}$?

Answer 2-22

In Example 2-5, it was calculated that the natural transmitted active power, $P_{sn} = 1$ pu and the reactive power $Q_{sn} = 0.267949$ pu at the source end. Therefore, from Equation (2-162b),

a) $V_{s's} = \dfrac{X}{V_r \cos\delta}|P_{s'} - P_{sn}| = \dfrac{0.5}{1 \times \cos 30°}|1.34641 - 1| = 0.2,$

b) $V_{s's} = \dfrac{X}{V_r \cos\delta}|P_{s'} - P_{sn}| = \dfrac{0.5}{1 \times \cos 30°}|0.65359 - 1| = 0.2;$

from Equation (2-162d),

$$\text{c)}\ V_{s's} = \frac{1}{2}\left|V_r \sin\delta - \sqrt{(V_r \sin\delta)^2 + 4X(Q_{s'} - Q_{sn})}\right|$$

$$= \frac{1}{2}\left|1 \times \sin 30^\circ - \sqrt{(1 \times \sin 30^\circ)^2 + 4 \times 0.5(0.547949 - 0.267949)}\right| = 0.2,$$

$$\text{d)}\ V_{s's} = \frac{1}{2}\left|V_r \sin\delta - \sqrt{(V_r \sin\delta)^2 + 4X(Q_{s'} - Q_{sn})}\right|$$

$$= \frac{1}{2}\left|1 \times \sin 30^\circ - \sqrt{(1 \times \sin 30^\circ)^2 + 4 \times 0.5(0.147949 - 0.267949)}\right| = 0.2.$$

Substituting $\beta = 90^\circ$ or $\beta = -90^\circ$ in Equation (2-155) or combining Equations (2-162a) and (2-162c) leads to

$$-\frac{P_{s'} - P_{sn}}{Q_{s'} - Q_{sn} - \dfrac{V_{s's}^2}{X}} = -\cot\delta \quad \text{or} \quad Q_{s'} = \tan\delta(P_{s'} - P_{sn}) + Q_{sn} + \frac{V_{s's}^2}{X} \tag{2-163a}$$

or substituting for $\tan\delta$ from Equation (2-44)

$$Q_{s'} = \frac{-P_{sn}}{Q_{sn} - \dfrac{V_S^2}{X}}(P_{s'} - P_{sn}) + Q_{sn} + \frac{V_{s's}^2}{X} \quad \text{or} \quad Q_{s'} = \frac{-P_{sn}}{Q_{sn} - \dfrac{V_S^2}{X}}P_{s'} + \frac{P_{sn}^2}{Q_{sn} - \dfrac{V_S^2}{X}} + Q_{sn} + \frac{V_{s's}^2}{X}. \tag{2-163b}$$

From Equation (2-163b), the $P_{s'}$ – $Q_{s'}$ relationship is defined with a slope ($m_{s'}$) and an intercept ($c_{s'}$) as

$$Q_{s'} = m_{s'}P_{s'} + c_{s'} \tag{2-164}$$

where

$$m_{s'} = \frac{-P_{sn}}{Q_{sn} - \dfrac{V_S^2}{X}} \tag{2-165}$$

and

$$c_{s'} = \frac{P_{sn}^2}{Q_s - \dfrac{V_S^2}{X}} + Q_{sn} + \frac{V_{s's}^2}{X}. \tag{2-166}$$

Therefore, in the case of $\beta = 0^\circ, 180^\circ$ or $\pm 90^\circ$, both the active and reactive power flows ($P_{s'}$ and $Q_{s'}$) at the modified sending end vary with the magnitude ($V_{s's}$) of the series-compensating voltage ($\mathbf{V}_{s's}$). For a given $V_{s's}$, ($P_{s'}$ and $Q_{s'}$) are determined for $\beta = 0^\circ, 180^\circ$ using Equations (2-157a) and (2-71) and the same are determined for $\beta = \pm 90^\circ$ using Equations (2-162a) and (2-164).

Example 2-23

Consider the electrical system data as given in Table 2-1. For $\beta = 90^\circ$ or $\beta = -90^\circ$ and $V_{s's} = 0.2$, calculate $m_{s'}$ and $c_{s'}$.

Answer 2-23

In Example 2-5, it was calculated that the natural transmitted active power, $P_{sn} = 1$ pu and the reactive power, $Q_{sn} = 0.267949$ pu at the source end.

From Equation (2-165), $m_{s'} = \dfrac{-P_{sn}}{Q_{sn} - \dfrac{V_S^2}{X}} = \dfrac{-1}{0.267949 - \dfrac{1^2}{0.5}} = 0.577350$ pu.

From Equation (2-166), $c_{s'} = \dfrac{P_{sn}^2}{Q_{sn} - \dfrac{V_S^2}{X}} + Q_{sn} + \dfrac{V_{s's}^2}{X} = \dfrac{1^2}{0.267949 - \dfrac{1^2}{0.5}} + 0.267949 + \dfrac{0.2^2}{0.5}$

$= -0.229401$ pu.

Additional Discussion

Note that the slope ($m_{s'}$) in Example 2-21 is −1.732051 and that in Example 2-23 is 0.577350; therefore, the product of these slopes is −1, meaning the $P_{s'} - Q_{s'}$ characteristic equations for the VR and the PAR (asym) are perpendicular, which is shown in Figure 2-23.

2.2.2.4 Exchanged Power by a Series-Compensating Voltage

The series-compensating voltage is an AC voltage source that carries a current, which is at any phase angle with its terminal voltage. Therefore, the compensating voltage, essentially, emulates a four-quadrant, series-compensating impedance at its terminals. This virtual impedance ($\mathbf{Z_{se}} = R_{se} - jX_{se}$) consists of a resistance ($R_{se} = +R$ or $-R$) and a reactance ($X_{se} = X_C$ or $-X_L$) in series with the line. A $X_{se} > 0$ means the compensating reactance is capacitive and $X_{se} < 0$ means the compensating reactance is inductive.

The series-compensating voltage ($\mathbf{V_{s's}}$) is related to ($\mathbf{V_{dq}}$), such that

$$\mathbf{V_{s's}} = -\mathbf{V_{dq}} \tag{1-4}$$

and

$$\mathbf{V_{dq}} = \mathbf{V_d} + \mathbf{V_q} = V_d + jV_q \tag{1-5}$$

where $\mathbf{V_d} = V_d$ and $\mathbf{V_q} = jV_q$ are the respective active or direct and reactive or quadrature components of the compensating voltage with load convention, meaning the line current (**I**) enters at the higher potential terminal of the voltages ($\mathbf{V_d}$ and $\mathbf{V_q}$) as shown in Figure 2-24a.

The series-compensating voltage ($\mathbf{V_{s's}}$) is at any phase angle with the prevailing line current (**I**) and, therefore, it exchanges with the line both active and reactive powers (P_{se} and Q_{se}), which are defined as

$$P_{se} = \mathbf{V_{dq}} \bullet \mathbf{I} = -\mathbf{V_{s's}} \bullet \mathbf{I} \qquad \text{or} \qquad P_{se} = V_d I \tag{2-167}$$

and

$$Q_{se} = |\mathbf{V_{dq}} \times \mathbf{I}| = |-\mathbf{V_{s's}} \times \mathbf{I}| \qquad \text{or} \qquad Q_{se} = V_q I \tag{2-168}$$

where V_d and V_q are the respective active or direct and reactive or quadrature components with respect to the prevailing line current (**I**).

Figure 2-24b shows the phasor diagram related to a series-compensating voltage with a fixed magnitude of 0.2 pu and its entire controllable range of $0° \leq \beta \leq 360°$. The compensating voltage ($\mathbf{V_{s's}}$) is added to the fixed sending-end voltage ($\mathbf{V_s}$) to produce the effective sending-end voltage ($\mathbf{V_{s'}} = \mathbf{V_s} + \mathbf{V_{s's}}$). Just for comparison, the characteristics of a VR that regulates the magnitude of the line voltage, a PAR (both types - symmetric and asymmetric) that regulates the phase angle of the line voltage and a RR that regulates the line reactance are also shown in the figure. In these special cases, the following occur.

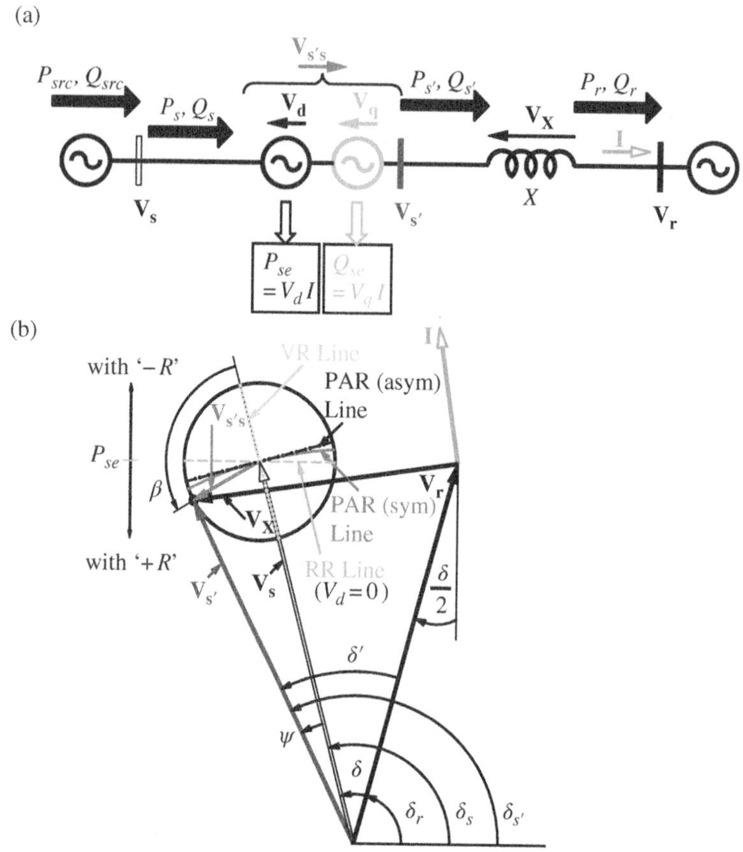

Figure 2-24 (a) Power transmission system with a series-compensating voltage ($\mathbf{V_{s's}}$); (b) phasor diagram.

- A series-compensating voltage can be in- or out-of-phase with the phase-to-neutral voltage of the line to implement a VR.
- A series-compensating voltage can be in quadrature with the phase-to-neutral voltage of the line to implement a PAR (both types - symmetric and asymmetric). Further details on a PAR (sym) and a PAR (asym) are given in Section 2.3.2 (Phase Angle Regulation).
- A series-compensating voltage can be such that it provides series reactance compensation because of being in quadrature with the prevailing line current. If the circular controllable area is equally divided by the series reactance compensator line ($V_d = 0$ or $P_{se} = 0$), the upper and lower halves represent P_{se} due to '$-R$' and '$+R$', respectively.

Any compensator, which provides compensation for one of the parameters (voltage, phase angle, or reactance) operates on a set of linear operating points. For example, Figure 2-22b shows that the VR and the PAR (asym) operating characteristic lines are neither horizontal nor vertical in the P-Q plane. Therefore, changing one parameter (voltage magnitude or phase angle) with the use of these regulators changes both the active and the reactive power flows in the line simultaneously.

The exchanged active power by the series-compensating voltage with the line can be defined as

$$P_{se} = P_s - P_{s'}, \tag{2-169a}$$

which can also be derived from Equations (2-118) and (2-149) as

$$P_{se} = a_s \sin\beta - a_r \sin(\delta + \beta) \tag{2-169b}$$

or

$$P_{se} = \frac{V_{s's}}{X}\{V_s \sin\beta - V_r(\sin\delta\cos\beta + \cos\delta\sin\beta)\} = \frac{V_{s's}}{X}\{(V_s - V_r\cos\delta)\sin\beta - V_r\sin\delta\cos\beta\}$$

with substitutions for a_s and a_r from Equations (2-119) and (2-132) or

$$P_{se} = \frac{V_{s's}V_r}{X}\left\{\left(\frac{V_s}{V_r} - \cos\delta\right)\sin\beta - \sin\delta\cos\beta\right\}. \tag{2-170}$$

Equations (2-40), (2-41), and (2-43) can be rearranged as

$$\frac{V_r\sin\delta}{X} = \frac{P_{sn}}{V_s} \tag{2-171}$$

and

$$\frac{V_r}{X}\left(\frac{V_s}{V_r} - \cos\delta\right) = \frac{Q_{sn}}{V_s}. \tag{2-172}$$

Substituting Equations (2-171) and (2-172) into Equation (2-170) leads to

$$P_{se} = \frac{V_{s's}}{V_s}(Q_{sn}\sin\beta - P_{sn}\cos\beta) \qquad \text{or} \qquad P_{se} = \frac{V_{s's}}{V_s}D\sin(\beta - E) \tag{2-173}$$

where

$$P_{sn} = D\sin(E) \tag{2-174a}$$

$$Q_{sn} = D\cos(E) \tag{2-174b}$$

$$D = \sqrt{P_{sn}^2 + Q_{sn}^2} \tag{2-174c}$$

$$E = \tan^{-1}\left(\frac{P_{sn}}{Q_{sn}}\right). \tag{2-174d}$$

Equations (2-171) and (2-172) can be combined as

$$P_{sn}^2 + Q_{sn}^2 = \left(\frac{V_sV_r}{X}\right)^2\left\{1 + \left(\frac{V_s}{V_r}\right)^2 - 2\frac{V_s}{V_r}\cos\delta\right\} = \left(\frac{V_sV_r}{X}\right)^2\frac{1}{V_r^2}(V_r^2 + V_s^2 - 2V_sV_r\cos\delta) \qquad \text{or}$$

$$P_{sn}^2 + Q_{sn}^2 = \frac{V_s^2}{X^2}V_{Xn}^2 \qquad \text{or} \qquad P_{sn}^2 + Q_{sn}^2 = \left(\frac{V_sV_{Xn}}{X}\right)^2 \tag{2-175}$$

where the natural voltage (V_{Xn}) across the line is given in Equation (2-30).

Equation (2-174c) is then

$$D = \sqrt{P_{sn}^2 + Q_{sn}^2} = \frac{V_sV_{Xn}}{X} \tag{2-176}$$

where the natural line current (I_n) is given in Equation (2-33).

By substituting D from Equation (2-176) and E from Equation (2-174d) into Equation (2-173), it can be written that

$$P_{se} = a_{se}\sin\left(\beta - \tan^{-1}\frac{P_{sn}}{Q_{sn}}\right) \tag{2-177}$$

where

$$a_{se} = \frac{V_{s's}V_{Xn}}{X}. \tag{2-178}$$

The exchanged reactive power by the series-compensating voltage with the line can be defined as

$$Q_{se} = Q_s - Q_{s'}, \tag{2-179a}$$

which can also be derived from Equations (2-122) and (2-152) as

$$Q_{se} = a_s \cos\beta - \frac{V_{s's}^2}{X} - 2a_s \cos\beta + a_r \cos(\delta + \beta)$$

$$Q_{se} = -\frac{V_{s's}^2}{X} - a_s \cos\beta + a_r \cos(\delta + \beta) \tag{2-179b}$$

or

$$Q_{se} = -\frac{V_{s's}^2}{X} + \frac{V_{s's}}{X}\{-V_s \cos\beta + V_r(\cos\delta\cos\beta - \sin\delta\sin\beta)\}$$

with substitutions for a_s and a_r from Equations (2-119) and (2-132) or

$$Q_{se} = -\frac{V_{s's}^2}{X} - \frac{V_{s's}V_r}{X}\left\{\left(\frac{V_s}{V_r} - \cos\delta\right)\cos\beta + \sin\delta\sin\beta\right\}$$

or using Equations (2-171) and (2-172), it can be written that

$$Q_{se} = -\frac{V_{s's}^2}{X} - \frac{V_{s's}}{V_s}(Q_{sn}\cos\beta + P_{sn}\sin\beta)$$

or

$$Q_{se} = -\frac{V_{s's}^2}{X} - \frac{V_{s's}}{V_s} D\cos(\beta - E) \tag{2-180}$$

with substitutions from Equations (2-174a) and (2-174b).

By substituting D from Equation (2-176) and E from Equation (2-174d) into Equation (2-180), it can be written that

$$Q_{se} = -\frac{V_{s's}^2}{X} - a_{se}\cos\left(\beta - \tan^{-1}\frac{P_{sn}}{Q_{sn}}\right) \tag{2-181}$$

where a_{se} is given in Equation (2-178).

The exchanged apparent power (S_{se}) is given by

$$S_{se} = \sqrt{P_{se}^2 + Q_{se}^2}. \tag{2-182}$$

Using Equation (2-178), Equations (2-177) and (2-181) can be combined as

$$P_{se}^2 + \left(Q_{se} + \frac{V_{s's}^2}{X}\right)^2 = a_{se}^2 \tag{2-183}$$

and

$$\tan\left(\beta - \tan^{-1}\frac{P_{sn}}{Q_{sn}}\right) = \frac{-P_{se}}{Q_{se} + \frac{V_{s's}^2}{X}}. \tag{2-184a}$$

Alternately, from Equations (2-169b) and (2-179b), it can be written that

$$\frac{P_{se}}{Q_{se}+\frac{V_{s's}^2}{X}} = \frac{a_s \sin\beta - a_r \sin(\delta+\beta)}{-a_s\cos\beta + a_r\cos(\delta+\beta)} = \frac{a_s\sin\beta - a_r\sin\delta\cos\beta - a_r\cos\delta\sin\beta}{-a_s\cos\beta + a_r\cos\delta\cos\beta - a_r\sin\delta\sin\beta}$$

$$= \frac{(a_s - a_r\cos\delta)\sin\beta - a_r\sin\delta\cos\beta}{-(a_s - a_r\cos\delta)\cos\beta - a_r\sin\delta\sin\beta} \quad \text{or}$$

$$P_{se}\{-(a_s - a_r\cos\delta)\cos\beta - a_r\sin\delta\sin\beta\} = \left(Q_{se}+\frac{V_{s's}^2}{X}\right)\{(a_s - a_r\cos\delta)\sin\beta - a_r\sin\delta\cos\beta\} \quad \text{or}$$

$$-P_{se}a_r\sin\delta\sin\beta - \left(Q_{se}+\frac{V_{s's}^2}{X}\right)(a_s - a_r\cos\delta)\sin\beta = P_{se}(a_s - a_r\cos\delta)\cos\beta$$

$$-\left(Q_{se}+\frac{V_{s's}^2}{X}\right)a_r\sin\delta\cos\beta$$

or

$$\tan\beta = \frac{P_{se}(a_s - a_r\cos\delta) - \left(Q_{se}+\frac{V_{s's}^2}{X}\right)a_r\sin\delta}{-P_{se}a_r\sin\delta - \left(Q_{se}+\frac{V_{s's}^2}{X}\right)(a_s - a_r\cos\delta)}. \tag{2-184b}$$

Equation (2-183) defines the relationship between P_{se} and Q_{se} as a circle centered at $(0, -V_{s's}^2/X)$ with a radius of a_{se}. The exchanged active and reactive powers (P_{se} and Q_{se}) are plotted in Figure 2-25a as functions of the relative phase angle (β). For a given magnitude of the series-compensating voltage, the exchanged capacitive power (Q_{se}) is larger than its inductive counterpart due to the fact that the capacitive compensation produces a larger line current. The series-compensating voltage, being at any phase angle with the prevailing line current, emulates a four-quadrant, series-compensating impedance ($\mathbf{Z_{se}} = R_{se} - jX_{se}$) that consists of a resistance ($R_{se} = +R$ or $-R$) and a reactance ($X_{se} = X_C$ or $-X_L$) in series with the line. Therefore, the

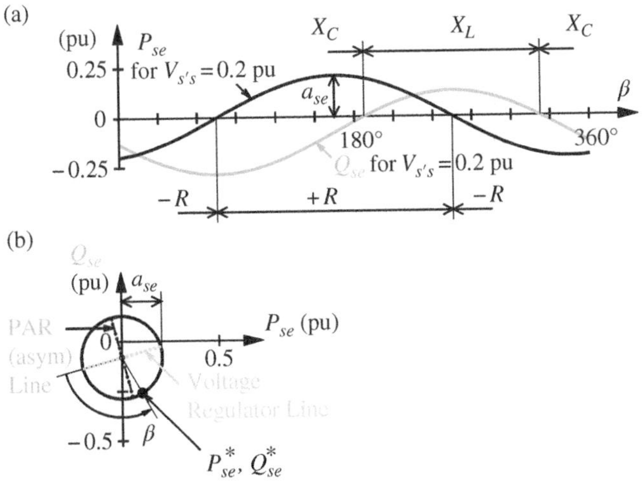

Figure 2-25 (a) Variation of the exchanged active and reactive powers (P_{se} and Q_{se}) as a function of the relative phase angle (β) of the series-compensating voltage ($\mathbf{V_{s's}}$) with a fixed magnitude of 0.2 pu; (b) exchanged reactive power (Q_{se}) versus active power (P_{se}).

series-compensating voltage ($\mathbf{V_{s's}}$) acts as an IR. Figure 2-25a shows that the series inductive-reactance compensation can be emulated with a series-compensating voltage at $\beta = 75°$ or $\beta = 255°$ in this particular example. The exchanged active and reactive powers can be regulated within the range defined by the *P-Q* plot in Figure 2-25b by controlling the magnitude ($V_{s's}$) of the series-compensating voltage ($\mathbf{V_{s's}}$) within its rated value and the relative phase angle of $0° \le \beta \le 360°$.

The magnitude ($V_{s's}$) of the series-compensating voltage is expressed by combining Equations (2-178) and (2-183) as follows:

$$P_{se}^2 + Q_{se}^2 + 2Q_{se}\frac{V_{s's}^2}{X} + \left(\frac{V_{s's}^2}{X}\right)^2 - \left(\frac{V_{s's}V_{Xn}}{X}\right)^2 = 0$$

or

$$\left(V_{s's}^2\right)^2 - \left(V_{Xn}^2 - 2XQ_{se}\right) V_{s's}^2 + X^2\left(P_{se}^2 + Q_{se}^2\right) = 0. \tag{2-185}$$

Solving the above quadratic equation results in

$$V_{s's}^2 = \frac{\left(V_{Xn}^2 - 2XQ_{se}\right) \pm \sqrt{\left(V_{Xn}^2 - 2XQ_{se}\right)^2 - 4X^2\left(P_{se}^2 + Q_{se}^2\right)}}{2}$$

or

$$V_{s's}^2 = \left(\frac{V_{Xn}^2}{2} - XQ_{se}\right) \pm \sqrt{\left(\frac{V_{Xn}^2}{2} - XQ_{se}\right)^2 - X^2\left(P_{se}^2 + Q_{se}^2\right)}$$

or

$$V_{s's} = +\sqrt{\left(\frac{V_{Xn}^2}{2} - XQ_{se}\right) + \sqrt{\left(\frac{V_{Xn}^2}{2} - XQ_{se}\right)^2 - X^2\left(P_{se}^2 + Q_{se}^2\right)}} \tag{2-186a}$$

or

$$V_{s's} = +\sqrt{\left(\frac{V_{Xn}^2}{2} - XQ_{se}\right) - \sqrt{\left(\frac{V_{Xn}^2}{2} - XQ_{se}\right)^2 - X^2\left(P_{se}^2 + Q_{se}^2\right)}} \tag{2-186b}$$

or

$$V_{s's} = -\sqrt{\left(\frac{V_{Xn}^2}{2} - XQ_{se}\right) + \sqrt{\left(\frac{V_{Xn}^2}{2} - XQ_{se}\right)^2 - X^2\left(P_{se}^2 + Q_{se}^2\right)}} \tag{2-186c}$$

or

$$V_{s's} = -\sqrt{\left(\frac{V_{Xn}^2}{2} - XQ_{se}\right) - \sqrt{\left(\frac{V_{Xn}^2}{2} - XQ_{se}\right)^2 - X^2\left(P_{se}^2 + Q_{se}^2\right)}}. \tag{2-186d}$$

The relative phase angle of the series-compensating voltage ($\mathbf{V_{s's}}$) is derived from Equation (2-184a) as

$$\beta = \tan^{-1}\frac{P_{sn}}{Q_{sn}} - \tan^{-1}\frac{P_{se}}{Q_{se} + \frac{V_{s's}^2}{X}}. \tag{2-187a}$$

Alternately, from Equation (2-184b), it can be written that

$$\beta = \tan^{-1}\frac{P_{se}(a_s - a_r\cos\delta) - \left(Q_{se} + \frac{V_{s's}^2}{X}\right)a_r\sin\delta}{-P_{se}a_r\sin\delta - \left(Q_{se} + \frac{V_{s's}^2}{X}\right)(a_s - a_r\cos\delta)}. \tag{2-187b}$$

For independent control of exchanged active and reactive powers (P_{se} and Q_{se}), the magnitude ($V_{s's}$) of the compensating voltage ($\mathbf{V_{s's}}$) and the relative phase angle (β) are determined from Equations (2-186b) and (2-187a) or (2-187b), respectively. For a given amount of exchanged power, it results in a larger $V_{s's}$ and smaller β per Equation (2-186a); it results in a smaller $V_{s's}$ and larger β per Equation (2-186b). Equations (2-186c) and (2-186d) result in the same respective $V_{s'}$ as obtained in Equations (2-186a) and (2-186b); however, the corresponding β is 180° opposite.

For the desired exchanged active and reactive powers (P_{se}^*, Q_{se}^*) within the allowable control range, the magnitude ($V_{s's}$) and the relative phase angle (β) of the series-compensating voltage ($\mathbf{V_{s's}}$) are determined by substituting P_{se} with P_{se}^* and Q_{se} with Q_{se}^* in Equations (2-186b) and (2-187a) or (2-187b), respectively.

For implementation of the VR, the relative phase angle is set to $\beta = 0°$ or $\beta = 180°$ in Equations (2-177) and (2-181). For $\beta = 0°$, the variations of P_{se} and Q_{se} are given by

$$P_{se} = a_{se}\sin\left(0° - \tan^{-1}\frac{P_{sn}}{Q_{sn}}\right) \quad \text{or} \quad P_{se} = -a_{se}\sin\left(\tan^{-1}\frac{P_{sn}}{Q_{sn}}\right)$$

and

$$Q_{se} = -\frac{V_{s's}^2}{X} - a_{se}\cos\left(0° - \tan^{-1}\frac{P_{sn}}{Q_{sn}}\right) \quad \text{or} \quad Q_{se} = -\frac{V_{s's}^2}{X} - a_{se}\cos\left(\tan^{-1}\frac{P_{sn}}{Q_{sn}}\right).$$

For $\beta = 180°$, the variations of P_{se} and Q_{se} are given by

$$P_{se} = a_{se}\sin\left(180° - \tan^{-1}\frac{P_{sn}}{Q_{sn}}\right) \quad \text{or} \quad P_{se} = a_{se}\sin\left(\tan^{-1}\frac{P_{sn}}{Q_{sn}}\right)$$

and

$$Q_{se} = -\frac{V_{s's}^2}{X} - a_{se}\cos\left(180° - \tan^{-1}\frac{P_{sn}}{Q_{sn}}\right) \quad \text{or} \quad Q_{se} = -\frac{V_{s's}^2}{X} + a_{se}\cos\left(\tan^{-1}\frac{P_{sn}}{Q_{sn}}\right).$$

For $\beta = 0°$ or $\beta = 180°$, the variations of P_{se} and Q_{se} are then given by

$$P_{se} = \mp a_{se}\sin\left(\tan^{-1}\frac{P_{sn}}{Q_{sn}}\right) \tag{2-188a}$$

or using Equation (2-178), it can be written that

$$V_{s's} = \mp X\frac{P_{se}}{V_{Xn}\mathrm{Sin}\left(\tan^{-1}\frac{P_{sn}}{Q_{sn}}\right)} \quad \text{or} \quad V_{s's} = \frac{X}{V_{Xn}\mathrm{Sin}\left(\tan^{-1}\frac{P_{sn}}{Q_{sn}}\right)}|P_{se}| \tag{2-188b}$$

and

$$Q_{se} = -\frac{V_{s's}^2}{X} \mp a_{se}\cos\left(\tan^{-1}\frac{P_{sn}}{Q_{sn}}\right) \tag{2-188c}$$

or

$$V_{s's}^2 \pm V_{Xn} \cos\left(\tan^{-1}\frac{P_{sn}}{Q_{sn}}\right) V_{s's} + XQ_{se} = 0 \quad \text{using Equation (2-178).}$$

Solving the above quadratic equation results in

$$V_{s's} = \frac{\mp V_{Xn} \cos\left(\tan^{-1}\frac{P_{sn}}{Q_{sn}}\right) \pm \sqrt{\left(V_{Xn} \cos\left(\tan^{-1}\frac{P_{sn}}{Q_{sn}}\right)\right)^2 - 4XQ_{se}}}{2} \quad \text{or}$$

$$V_{s's} = \frac{1}{2}\left| V_{Xn} \cos\left(\tan^{-1}\frac{P_{sn}}{Q_{sn}}\right) \mp \sqrt{\left(V_{Xn} \cos\left(\tan^{-1}\frac{P_{sn}}{Q_{sn}}\right)\right)^2 - 4XQ_{se}}\right|. \tag{2-188d}$$

For a desired P_{se}, the required $V_{s's}$ can be determined using Equation (2-188b) or

$$V_{s's} = \mp X\frac{P_{se}}{V_r \sin\delta} \quad \text{or} \quad V_{s's} = \frac{X}{V_r \sin\delta}|P_{se}| \tag{B-109e}$$

For a desired Q_{se}, the required $V_{s's}$ can be determined using Equation (2-188d) or

$$V_{s's} = \frac{\mp(V_s - V_r\cos\delta) \pm \sqrt{(V_s - V_r\cos\delta)^2 - 4XQ_{se}}}{2} \quad \text{or}$$

$$V_{s's} = \frac{1}{2}\left|V_s - V_r\cos\delta \mp \sqrt{(V_s - V_r\cos\delta)^2 - 4XQ_{se}}\right| \tag{B-109f}$$

Example 2-24

Consider the electrical system data as given in Table 2-1. Using a VR, (a) for $P_{se} = -0.2$ pu, what is $V_{s's}$? (b) for $P_{se} = 0.2$ pu, what is $V_{s's}$? (c) for $Q_{se} = -0.13359$ pu, what is $V_{s's}$? (d) for $Q_{se} = -0.02641$ pu, what is $V_{s's}$?

Answer 2-24

From Equation (B-109e),

a) $V_{s's} = \dfrac{X}{V_r \sin\delta}|P_{se}| = \dfrac{0.5}{1 \times \sin 30^\circ}|-0.2| = 0.2,$

b) $V_{s's} = \dfrac{X}{V_r \sin\delta}|P_{se}| = \dfrac{0.5}{1 \times \sin 30^\circ}|0.2| = 0.2;$

from Equation (B-109f),

c) $V_{s's} = \dfrac{1}{2}\left|V_s - V_r\cos\delta - \sqrt{(V_s - V_r\cos\delta)^2 - 4XQ_{se}}\right|$

$$= \frac{1}{2}\left|(1 - 1 \times \cos 30^\circ) - \sqrt{(1 - 1 \times \cos 30^\circ)^2 - 4 \times 0.5(-0.13359)}\right| = 0.2,$$

d) $V_{s's} = \dfrac{1}{2}\left|V_s - V_r\cos\delta + \sqrt{(V_s - V_r\cos\delta)^2 - 4XQ_{se}}\right|$

$$= \frac{1}{2}\left|(1 - 1 \times \cos 30^\circ) + \sqrt{(1 - 1 \times \cos 30^\circ)^2 - 4 \times 0.5(-0.02641)}\right| = 0.2.$$

Substituting $\beta = 0°$ or $\beta = 180°$ in Equation (2-187a) or combining Equations (2-188a) and (2-188c) leads to

$$\frac{P_{se}}{Q_{se} + \frac{V_{s's}^2}{X}} = \frac{P_{sn}}{Q_{sn}} \quad \text{or} \quad Q_{se} = \frac{Q_{sn}}{P_{sn}} P_{se} - \frac{V_{s's}^2}{X}. \tag{2-189}$$

From Equation (2-189), the $P_{se} - Q_{se}$ relationship is defined with a slope (m_{se}) and an intercept (c_{se}) as

$$Q_{se} = m_{se} P_{se} + c_{se} \tag{2-190}$$

where

$$m_{se} = \frac{Q_{sn}}{P_{sn}} \tag{2-191}$$

and

$$c_{se} = -\frac{V_{s's}^2}{X}. \tag{2-192}$$

Example 2-25

Consider the electrical system data as given in Table 2-1. For $\beta = 0°$ or $\beta = 180°$, calculate m_{se} and c_{se}.

Answer 2-25

In Example 2-5, it was calculated that the natural transmitted active power, $P_{sn} = 1$ pu and the reactive power, $Q_{sn} = 0.267949$ pu at the source end. Therefore, from Equation (2-191),

$$m_{se} = \frac{Q_{sn}}{P_{sn}} = \frac{0.267949}{1} = 0.267949 \text{ pu}.$$

From Equation (2-192), $c_{se} = -\frac{V_{s's}^2}{X} = -\frac{0.2^2}{0.5} = -0.08$ pu.

For implementation of the PAR (asym), the relative phase angle is set to $\beta = 90°$ or $\beta = -90°$ in Equations (2-177) and (2-181). For $\beta = 90°$, the variations of P_{se} and Q_{se} are given by

$$P_{se} = a_{se} \sin\left(90° - \tan^{-1}\frac{P_{sn}}{Q_{sn}}\right) \quad \text{or} \quad P_{se} = a_{se} \cos\left(\tan^{-1}\frac{P_{sn}}{Q_{sn}}\right)$$

and

$$Q_{se} = -\frac{V_{s's}^2}{X} - a_{se} \cos\left(90° - \tan^{-1}\frac{P_{sn}}{Q_{sn}}\right) \quad \text{or} \quad Q_{se} = -\frac{V_{s's}^2}{X} - a_{se} \sin\left(\tan^{-1}\frac{P_{sn}}{Q_{sn}}\right).$$

For $\beta = -90°$, the variations of P_{se} and Q_{se} are given by

$$P_{se} = a_{se} \sin\left(-90° - \tan^{-1}\frac{P_{sn}}{Q_{sn}}\right) \quad \text{or} \quad P_{se} = -a_{se} \cos\left(\tan^{-1}\frac{P_{sn}}{Q_{sn}}\right)$$

and

$$Q_{se} = -\frac{V_{s's}^2}{X} - a_{se}\cos\left(-90° - \tan^{-1}\frac{P_{sn}}{Q_{sn}}\right) \quad \text{or} \quad Q_{se} = -\frac{V_{s's}^2}{X} + a_{se}\sin\left(\tan^{-1}\frac{P_{sn}}{Q_{sn}}\right).$$

For $\beta = \pm 90°$, the variations of P_{se} and Q_{se} are then given by

$$P_{se} = \pm a_{se}\cos\left(\tan^{-1}\frac{P_{sn}}{Q_{sn}}\right) \tag{2-193a}$$

or using Equation (2-178)

$$V_{s's} = \pm X\frac{P_{se}}{V_{Xn}\cos\left(\tan^{-1}\frac{P_{sn}}{Q_{sn}}\right)} \quad \text{or} \quad V_{s's} = \frac{X}{V_{Xn}\cos\left(\tan^{-1}\frac{P_{sn}}{Q_{sn}}\right)}|P_{se}| \tag{2-193b}$$

and

$$Q_{se} = -\frac{V_{s's}^2}{X} \mp a_{se}\sin\left(\tan^{-1}\frac{P_{sn}}{Q_{sn}}\right) \tag{2-193c}$$

or

$$V_{s's}^2 \pm V_{Xn}\sin\left(\tan^{-1}\frac{P_{sn}}{Q_{sn}}\right)V_{s's} + XQ_{se} = 0 \quad \text{using Equation (2-178).}$$

Solving the above quadratic equation results in

$$V_{s's} = \frac{\mp V_{Xn}\sin\left(\tan^{-1}\frac{P_{sn}}{Q_{sn}}\right) \pm \sqrt{\left(V_{Xn}\sin\left(\tan^{-1}\frac{P_{sn}}{Q_{sn}}\right)\right)^2 - 4XQ_{se}}}{2}$$

or

$$V_{s's} = \frac{1}{2}\left|V_{Xn}\sin\left(\tan^{-1}\frac{P_{sn}}{Q_{sn}}\right) - \sqrt{\left(V_{Xn}\sin\left(\tan^{-1}\frac{P_{sn}}{Q_{sn}}\right)\right)^2 - 4XQ_{se}}\right|. \tag{2-193d}$$

For a desired P_{se}, the required $V_{s's}$ can be determined using Equation (2-193b) or

$$V_{s's} = \mp X\frac{P_{se}}{V_s - V_r\cos\delta} \quad \text{or} \quad V_{s's} = \frac{X}{V_s - V_r\cos\delta}|P_{se}|. \tag{B-110e}$$

For a desired Q_{se}, the required $V_{s's}$ can be determined using Equation (2-193d) or

$$V_{s's} = \frac{\mp V_r\sin\delta \pm \sqrt{(V_r\sin\delta)^2 - 4XQ_{se}}}{2} \quad \text{or} \quad V_{s's} = \frac{1}{2}\left|V_r\sin\delta - \sqrt{(V_r\sin\delta)^2 - 4XQ_{se}}\right|. \tag{B-110f}$$

Example 2-26

Consider the electrical system data as given in Table 2-1. Using a PAR (asym), (a) for $P_{se} = 0.05359$ pu, what is $V_{s's}$? (b) for $P_{se} = -0.05359$ pu, what is $V_{s's}$? (c) for $Q_{se} = -0.28$ pu, what is $V_{s's}$? (d) for $Q_{se} = 0.12$ pu, what is $V_{s's}$?

Answer 2-26

From Equation (B-110e),

$$\text{a) } V_{s's} = \frac{X}{V_s - V_r\cos\delta}|P_{se}| = \frac{0.5}{1 - 1\times\cos 30^\circ}|0.05359| = 0.2,$$

$$\text{b) } V_{s's} = \frac{X}{V_s - V_r\cos\delta}|P_{se}| = \frac{0.5}{1 - 1\times\cos 30^\circ}|-0.05359| = 0.2;$$

from Equation (B-110f),

$$\text{c) } V_{s's} = \frac{1}{2}\left|V_r\sin\delta - \sqrt{(V_r\sin\delta)^2 - 4XQ_{se}}\right|$$
$$= \frac{1}{2}\left|(1\times\sin 30^\circ) - \sqrt{(1\times\sin 30^\circ)^2 - 4\times 0.5(-0.28)}\right| = 0.2,$$

$$\text{d) } V_{s's} = \frac{1}{2}\left|V_r\sin\delta - \sqrt{(V_r\sin\delta)^2 - 4XQ_{se}}\right|$$
$$= \frac{1}{2}\left|(1\times\sin 30^\circ) - \sqrt{(1\times\sin 30^\circ)^2 - 4\times 0.5(0.12)}\right| = 0.2.$$

Substituting $\beta = 90^\circ$ or $\beta = -90^\circ$ in Equation (2-187a) or combining Equations (2-193a) and (2-193c) leads to

$$\frac{P_{se}}{Q_{se} + \frac{V_{s's}^2}{X}} = -\frac{Q_{sn}}{P_{sn}} \quad \text{or} \quad Q_{se} = -\frac{P_{sn}}{Q_{sn}}P_{se} - \frac{V_{s's}^2}{X}. \tag{2-194}$$

From Equation (2-194), the P_{se} - Q_{se} relationship is defined with a slope (m_{se}) and an intercept (c_{se}) as

$$Q_{se} = m_{se}P_{se} + c_{se} \tag{2-195}$$

where

$$m_{se} = -\frac{P_{sn}}{Q_{sn}} \tag{2-196}$$

and

$$c_{se} = -\frac{V_{s's}^2}{X}. \tag{2-197}$$

Example 2-27

Consider the electrical system data as given in Table 2-1. For $\beta = 90^\circ$ or $\beta = -90^\circ$, calculate m_{se} and c_{se}.

Answer 2-27

In Example 2-5, it was calculated that the natural transmitted active power, $P_{sn} = 1$ pu and the reactive power, $Q_{sn} = 0.267949$ pu at the source end. Therefore, from Equation (2-196),

$$m_{se} = -\frac{P_{sn}}{Q_{sn}} = -\frac{1}{0.267949} = -3.732051 \text{ pu}.$$

From Equation (2-197), $c_{se} = -\dfrac{V_{s's}^2}{X} = -\dfrac{0.2^2}{0.5} = -0.08$ pu.

Therefore, in the case of $\beta = 0°, 180°$, or $\pm 90°$, both the exchanged active and reactive powers (P_{se} and Q_{se}) vary with the magnitude ($V_{s's}$) of the series-compensating voltage ($\mathbf{V_{s's}}$). For a given $V_{s's}$, P_{se} and Q_{se} are determined for $\beta = 0°$and 180°, using Equations (2-188a) and (2-190) and the same are determined for $\beta = \pm 90°$ using Equations (2-193a) and (2-195).

Additional Discussion

Note that the slope (m_{se}) in Example 2-25 is 0.267949 and that in Example 2-27 is −1/0.267949; therefore, the product of these slopes is −1, meaning the $P_{se} - Q_{se}$ characteristics for the VR and the PAR (asym) are perpendicular, which is shown in Figure 2-25.

The relationships between active and reactive power flows (P_s and Q_s) at the sending end, (P_r and Q_r) at the receiving end, ($P_{s'}$ and $Q_{s'}$) at the modified sending end, and exchanged active and reactive powers (P_{se} and Q_{se}) by the series-compensating voltage ($\mathbf{V_{s's}}$) as a function of the modified power angle (δ') are given in Table 2-9.

The equations for active and reactive power flows (P_s and Q_s) at the sending end, (P_r and Q_r) at the receiving end, ($P_{s'}$ and $Q_{s'}$) at the modified sending end, and the exchanged active and reactive powers (P_{se} and Q_{se}) by the series-compensating voltage ($\mathbf{V_{s's}}$) as a function of the magnitude ($V_{s's}$) of the series-compensating voltage ($\mathbf{V_{s's}}$) are given in Table 2-10.

Example 2-28

Consider the electrical system data as given in Table 2-1. For a given $V_{s's} = 0.2$ pu and $\beta = 105°$, calculate P_s and Q_s.

Table 2-9 Relationships between active and reactive power flows (P_s and Q_s) at the sending end, (P_r and Q_r) at the receiving end, ($P_{s'}$ and $Q_{s'}$) at the modified sending end, and exchanged (P_{se} and Q_{se}) by the series-compensating voltage ($\mathbf{V_{s's}}$).

$(P_s - P_{sn})^2 + (Q_s - Q_{sn})^2 = a_s^2$	(2-123)	$a_s = \dfrac{V_s V_{s's}}{X}$ (2-119)
$(P_r - P_{rn})^2 + (Q_r - Q_{rn})^2 = a_r^2$	(2-136)	$a_r = \dfrac{V_r V_{s's}}{X}$ (2-132)
$(P_{s'} - P_{sn})^2 + \left(Q_{s'} - Q_{sn} - \dfrac{V_{s's}^2}{X}\right)^2 = 4a_s^2\cos^2\beta - 4a_s a_r \cos\beta\cos(\delta+\beta) + a_r^2$	(2-153a)	For a_s and a_r, see above.
$(P_{s'} - P_{sn})^2 + (Q_{s'} - Q_{sn})^2 = a_r^2\sin^2(\delta+\beta) + \left[\dfrac{V_{s's}^2}{X} + 2a_s\cos\beta - a_r\cos(\delta+\beta)\right]^2$	(2-153b)	
$P_{se}^2 + \left(Q_{se} + \dfrac{V_{s's}^2}{X}\right)^2 = a_{se}^2$	(2-183)	$a_{se} = \dfrac{V_{s's} V_{Xn}}{X}$ (2-178)

Table 2-10 Equations for active and reactive power flows (P_s and Q_s), (P_r and Q_r), ($P_{s'}$ and $Q_{s'}$), and (P_{se} and Q_{se}).

$P_s = P_{sn} + a_s \sin\beta$	(2-118)	$Q_s = Q_{sn} + a_s \cos\beta$	(2-122)	$a_s = \dfrac{V_s V_{s's}}{X}$	(2-119)
$P_r = P_{rn} + a_r \sin(\delta+\beta)$	(2-131)	$Q_r = Q_{rn} + a_r \cos(\delta+\beta)$	(2-135)	$a_r = \dfrac{V_r V_{s's}}{X}$	(2-132)
$P_{s'} = P_{sn} + a_r \sin(\delta+\beta)$	(2-149)	$Q_{s'} = Q_{sn} + \dfrac{V_{s's}^2}{X} + 2a_s \cos\beta - a_r \cos(\delta+\beta)$	(2-152)	For a_s and a_r, see above.	
$P_{se} = a_{se} \sin\left(\beta - \tan^{-1}\dfrac{P_{sn}}{Q_{sn}}\right)$	(2-177)	$Q_{se} = -\dfrac{V_{s's}^2}{X} - a_{se} \cos\left(\beta - \tan^{-1}\dfrac{P_{sn}}{Q_{sn}}\right)$	(2-181)	$a_{se} = \dfrac{V_{s's} V_{Xn}}{X}$	(2-178)

Answer 2-28

In Example 2-5, it was calculated that the natural transmitted active power, $P_{sn} = 1$ pu and the reactive power, $Q_{sn} = 0.267949$ pu at the source end.

From Equation (2-119), $a_s = \dfrac{V_s V_{s's}}{X} = \dfrac{1 \times 0.2}{0.5} = 0.4$ pu.

From Equation (2-118), $P_s = P_{sn} + a_s \sin\beta = 1 + 0.4 \times \sin 105° = 1.386370$ pu.

From Equation (2-122), $Q_s = Q_{sn} + a_s \cos\beta = 0.267949 + 0.4 \times \cos 105° = 0.164422$ pu.

Example 2-29

Consider the electrical system data as given in Table 2-1. For a given $V_{s's} = 0.2$ pu and $\beta = 105°$, calculate P_r and Q_r.

Answer 2-29

In Example 2-5, it was calculated that the natural transmitted active power, $P_{rn} = 1$ pu and the reactive power, $Q_{rn} = -0.267949$ pu at the load end.

From Equation (2-132), $a_r = \dfrac{V_r V_{s's}}{X} = \dfrac{1 \times 0.2}{0.5} = 0.4$ pu.

From Equation (2-131), $P_r = P_{rn} + a_r \sin(\delta+\beta) = 1 + 0.4 \times \sin(30° + 105°) = 1.282843$ pu.

From Equation (2-135), $Q_r = Q_{rn} + a_r \cos(\delta+\beta) = -0.267949 + 0.4 \times \cos(30° + 105°) = -0.550792$ pu.

Note that the values of P_r and Q_r, calculated in this example, are the same as those calculated in Example 2-9 with a shunt-compensating voltage.

Example 2-30

Consider the electrical system data as given in Table 2-1. For a given $V_{s's} = 0.2$ pu and $\beta = 105°$, calculate $P_{s'}$ and $Q_{s'}$.

Answer 2-30

In Example 2-5, it was calculated that the natural transmitted active power, $P_{sn} = 1$ pu and the reactive power, $Q_{sn} = 0.267949$ pu at the source end.

From Equation (2-119), $a_s = \dfrac{V_s V_{s's}}{X} = \dfrac{1 \times 0.2}{0.5} = 0.4$ pu.

From Equation (2-132), $a_r = \dfrac{V_r V_{s's}}{X} = \dfrac{1 \times 0.2}{0.5} = 0.4$ pu.

From Equation (2-149), $P_{s'} = P_{sn} + a_r \sin(\delta + \beta) = 1 + 0.4 \times \sin(30° + 105°) = 1.282843$ pu.

From Equation (2-152), $Q_{s'} = Q_{sn} + \dfrac{V_{s's}^2}{X} + 2a_s \cos\beta - a_r \cos(\delta + \beta) = 0.267949 + \dfrac{0.2^2}{0.5} + 2 \times 0.4 \times \cos 105° - 0.4 \times \cos(30° + 105°) = 0.423737$ pu.

Note that the values of $P_{s'}$ and $Q_{s'}$, calculated in this example, are the same as those calculated in Example 2-8 with a shunt-compensating voltage.

Additional Discussion

The total reactive power needed by the line, $Q_{total} = |Q_{s'}| + |Q_r| = 0.550792 + 0.423737 = 0.974529$ pu. Therefore, the line current, $I = \sqrt{Q_{total}/X} = \sqrt{0.974529/0.5} = 1.396086$ pu as calculated in Example 2-12.

Example 2-31

Consider the electrical system data as given in Table 2-1. For a given $V_{s's} = 0.2$ pu and $\beta = 105°$, calculate P_{se} and Q_{se}.

Answer 2-31

In Example 2-5, it was calculated that the magnitude of the natural voltage, $V_{Xn} = 0.517638$ pu, the natural transmitted active power, $P_{sn} = 1$ pu, and the reactive power, $Q_{sn} = 0.267949$ pu at the source end.

From Equation (2-119), $a_s = \dfrac{V_s V_{s's}}{X} = \dfrac{1 \times 0.2}{0.5} = 0.4$ pu.

From Equation (2-132), $a_r = \dfrac{V_r V_{s's}}{X} = \dfrac{1 \times 0.2}{0.5} = 0.4$ pu.

From Equation (2-178), $a_{se} = \dfrac{V_{s's} V_{Xn}}{X} = \dfrac{0.2 \times 0.517638}{0.5} = 0.207055$ pu.

From Equation (2-177), $P_{se} = a_{se} \sin\left(\beta - \tan^{-1}\dfrac{P_{sn}}{Q_{sn}}\right) = 0.207055 \times \sin\left(105° - \tan^{-1}\dfrac{1}{0.267949}\right) = 0.103528$ pu.

From Equation (2-181), $Q_{se} = -\dfrac{V_{s's}^2}{X} - a_{se} \cos\left(\beta - \tan^{-1}\dfrac{P_{sn}}{Q_{sn}}\right) = -\dfrac{0.2^2}{0.5} - 0.207055 \times \cos\left(105° - \tan^{-1}\dfrac{1}{0.267949}\right) = -0.259315$ pu.

For a desired active and reactive power flows (P_s and Q_s) at the sending end, (P_r and Q_r) at the receiving end, ($P_{s'}$ and $Q_{s'}$) at the modified sending end, and the exchanged active and reactive powers (P_{se} and Q_{se}) by the series-compensating voltage, the expressions for the magnitude ($V_{s's}$) and the relative phase angle (β) of the series-compensating voltage ($\mathbf{V_{s's}}$) are given in Table 2-11.

Table 2-11 Equations for magnitude ($V_{s's}$) and relative phase angle (β) of the series-compensating voltage ($\mathbf{V}_{s's}$) as a function of (P_s and Q_s) at the sending end, (P_r and Q_r) at the receiving end, and exchanged (P_{se} and Q_{se}).

$V_{s's} = \frac{X}{V_s}\sqrt{(P_s - P_{sn})^2 + (Q_s - Q_{sn})^2}$	(2-125)	$\beta = \tan^{-1}\frac{P_s - P_{sn}}{Q_s - Q_{sn}}$	(2-126)
$V_{s's} = \frac{X}{V_r}\sqrt{(P_r - P_{rn})^2 + (Q_r - Q_{rn})^2}$	(2-138)	$\beta = \tan^{-1}\frac{P_r - P_{rn}}{Q_r - Q_{rn}} - \delta$	(2-139)
$V_{s'} = +\sqrt{\left(\frac{V_r^2}{2} + XQ_{s'}\right) + \sqrt{\left(\frac{V_r^2}{2} + XQ_{s'}\right)^2 - X^2(P_{s'}^2 + Q_{s'}^2)}}$	(2-67a)	$\delta' = \tan^{-1}\frac{P_{s'}}{\frac{V_{s'}^2}{X} - Q_{s'}}$	(2-68)
$V_{s's} = \sqrt{V_{s'}^2 + V_s^2 - 2V_{s'}V_s\cos\psi}$	(2-100)	$\psi = \delta' - \delta$	(2-51b)
		$\beta = \tan^{-1}\frac{V_{s'}\sin\psi}{V_{s'}\cos\psi - V_s}$	(2-101)
$V_{s's} = +\sqrt{\left(\frac{V_{Xn}^2}{2} - XQ_{se}\right) - \sqrt{\left(\frac{V_{Xn}^2}{2} - XQ_{se}\right)^2 - X^2(P_{se}^2 + Q_{se}^2)}}$	(2-186b)	$\beta = \tan^{-1}\frac{P_{sn}}{Q_{sn}} - \tan^{-1}\frac{P_{se}}{Q_{se} + \frac{V_{s's}^2}{X}}$	(2-187a)
		$\beta = \tan^{-1}\frac{P_{se}(a_s - a_r\cos\delta) - \left(Q_{se} + \frac{V_{s's}^2}{X}\right)a_r\sin\delta}{-P_{se}a_r\sin\delta - \left(Q_{se} + \frac{V_{s's}^2}{X}\right)(a_s - a_r\cos\delta)}$	(2-187b)

Example 2-32

Consider the electrical system data as given in Table 2-1. For a given $P_s = 1.386370$ pu and $Q_s = 0.164422$ pu, calculate $V_{s's}$ and β.

Answer 2-32

In Example 2-5, it was calculated that the natural transmitted active power, $P_{sn} = 1$ pu and the reactive power, $Q_{sn} = 0.267949$ pu at the source end. Therefore, from Equation (2-125), the magnitude ($V_{s's}$) of the series-compensating voltage, $V_{s's} = \frac{X}{V_s}\sqrt{(P_s - P_{sn})^2 + (Q_s - Q_{sn})^2}$ $= \frac{0.5}{1}\sqrt{(1.386370 - 1)^2 + (0.164422 - 0.267944)^2} = 0.2$ pu.

From Equation (2-126), the relative phase angle (β) of the series-compensating voltage ($\mathbf{V}_{s's}$),

$$\beta = \tan^{-1}\frac{P_s - P_{sn}}{Q_s - Q_{sn}} = \tan^{-1}\frac{1.386370 - 1}{0.164422 - 0.267944} = 105°.$$

Example 2-33

Consider the electrical system data as given in Table 2-1. For a given $P_r = 1.282843$ pu and $Q_r = -0.550792$ pu, calculate $V_{s's}$ and β.

Answer 2-33

In Example 2-5, it was calculated that the natural transmitted active power, $P_{rn} = 1$ pu and the reactive power, $Q_{rn} = -0.267949$ pu at the load end. Therefore, from Equation (2-138), the magnitude ($V_{s's}$) of the series-compensating voltage, $V_{s's} = \frac{X}{V_r}\sqrt{(P_r - P_{rn})^2 + (Q_r - Q_{rn})^2} = \frac{0.5}{1}\sqrt{(1.282843 - 1)^2 + (-0.550792 + 0.267949)^2} = 0.2$ pu. From Equation (2-139), the relative phase angle (β) of the compensating voltage ($\mathbf{V_{s's}}$), $\beta = \tan^{-1}\frac{P_r - P_{rn}}{Q_r - Q_{rn}} - \delta = \tan^{-1}\frac{1.282843 - 1}{-0.550792 + 0.267943} - 30° = 105°$.

Example 2-34

Consider the electrical system data as given in Table 2-1. For a given $P_{s'} = 1.282843$ pu and $Q_{s'} = 0.423737$ pu, calculate $V_{s's}$ and β.

Answer 2-34

In Example 2-5, it was calculated that the natural transmitted active power, $P_{sn} = 1$ pu and the reactive power, $Q_{sn} = 0.267949$ pu at the source end. Therefore, from Equation (2-67a),

$$V_{s'} = +\sqrt{\left(\frac{V_r^2}{2} + XQ_{s'}\right) + \sqrt{\left(\frac{V_r^2}{2} + XQ_{s'}\right)^2 - X^2(P_{s'}^2 + Q_{s'}^2)}}$$

$$= \sqrt{\left(\frac{1^2}{2} + 0.5 \times 0.423737\right) + \sqrt{\left(\frac{1^2}{2} + 0.5 \times 0.423737\right)^2 - 0.5^2(1.282843^2 + 0.423737^2)}}$$

$$= 0.967715 \text{ pu.}$$

From Equation (2-68), $\delta' = \tan^{-1}\frac{P_{s'}}{\frac{V_{s'}^2}{X} - Q_{s'}} = \tan^{-1}\frac{1.282843}{\frac{0.967715^2}{0.5} - 0.423737} = 41.515336°$.

From Equation (2-51b), $\psi = \delta' - \delta = 41.515336° - 30° = 11.515336°$.

From Equation (2-100), $V_{s's} = \sqrt{V_{s'}^2 + V_s^2 - 2V_{s'}V_s\cos\psi} = \sqrt{0.967715^2 + 1^2 - 2 \times 0.967715 \times 1 \times \cos 11.515336°}$ $= 0.2$ pu.

From Equation (2-101), $\beta = \tan^{-1}\frac{V_{s'}\sin\psi}{V_{s'}\cos\psi - V_s} = \tan^{-1}\frac{0.967715 \times \sin 11.515336°}{0.967715 \times \cos 11.515336° - 1} = 105°$

Example 2-35

Consider the electrical system data as given in Table 2-1. For a given $P_{se} = 0.103528$ pu and $Q_{se} = -0.259315$ pu, calculate $V_{s's}$ and β.

Answer 2-35

In Example 2-5, it was calculated that the magnitude of the natural voltage, $V_{Xn} = 0.517638$ pu.

From Equation (2-186b), the magnitude ($V_{s's}$) of the series-compensating voltage ($\mathbf{V_{s's}}$),

$$V_{s's} = +\sqrt{\left(\frac{V_{Xn}^2}{2} - XQ_{se}\right) - \sqrt{\left(\frac{V_{Xn}^2}{2} - XQ_{se}\right)^2 - X^2\left(P_{se}^2 + Q_{se}^2\right)}}$$

$$= +\sqrt{\left(\frac{0.517638^2}{2} + 0.5 \times 0.259315\right) - \sqrt{\left(\frac{0.517638^2}{2} + 0.5 \times 0.259315\right)^2 - 0.5^2\left(0.103528^2 + 0.259315^2\right)}}$$

$$= 0.2 \text{ pu.}$$

From Equation (2-187a), the relative phase angle (β) of the series-compensating voltage ($\mathbf{V_{s's}}$),

$$\beta = \tan^{-1}\frac{P_{sn}}{Q_{sn}} - \tan^{-1}\frac{P_{se}}{Q_{se} + \frac{V_{s's}^2}{X}} = \tan^{-1}\frac{1}{0.267949} - \tan^{-1}\frac{0.103528}{-0.259315 + \frac{0.2^2}{0.5}} = 105°.$$

The equations for active and reactive power flows (P_s and Q_s) at the sending end, (P_r and Q_r) at the receiving end, ($P_{s'}$ and $Q_{s'}$) at the modified sending end, and the exchanged active and reactive powers (P_{se} and Q_{se}) by the series-compensating voltage as a function of the magnitude ($V_{s's}$) and the relative phase angle (β) of the series-compensating voltage ($\mathbf{V_{s's}}$) are given in Table 2-12 and Table 2-13 for a VR.

The equations relating active and reactive power flows (P_s and Q_s) at the sending end, (P_r and Q_r) at the receiving end, ($P_{s'}$ and $Q_{s'}$) at the modified sending end, and the exchanged active and reactive powers (P_{se} and Q_{se}) by the series-compensating voltage as a function of the magnitude ($V_{s's}$) and the relative phase angle (β) of the series-compensating voltage ($\mathbf{V_{s's}}$) are given in Table 2-14 and Table 2-15 for a PAR (asym).

Table 2-12 Equations for P_s, P_r, $P_{s'}$, and P_{se} when a VR is used.

$P_s = P_{sn}$	(2-127a)		
$P_r = P_{rn} \pm a_r \sin\delta$	(2-140a)	$a_r = \frac{V_r V_{s's}}{X}$	(2-132)
$P_{s'} = P_{sn} \pm a_r \sin\delta$	(2-157a)	$a_r = \frac{V_r V_{s's}}{X}$	(2-132)
$P_{se} = \mp a_{se} \sin\left(\tan^{-1}\frac{P_{sn}}{Q_{sn}}\right)$	(2-188a)	$a_{se} = \frac{V_{s's} V_{Xn}}{X}$ (2-178), $V_{Xn} = \sqrt{V_s^2 - 2V_s V_r \cos\delta + V_r^2}$	(2-30)

Table 2-13 Equations relating (P_s and Q_s), (P_r and Q_r), ($P_{s'}$ and $Q_{s'}$), and (P_{se} and Q_{se}) when a VR is used.

$Q_s = Q_{sn} \pm a_s$	(2-127b)	$a_s = \frac{V_s V_{s's}}{X}$	(2-119)		
$Q_r = m_r P_r + c_r$	(2-84)	$m_r = \frac{1}{P_{rn}}\left(Q_{rn} + \frac{V_r^2}{X}\right)$	(2-85)	$c_r = -\frac{V_r^2}{X}$	(2-86)
$Q_{s'} = m_{s'} P_{s'} + c_{s'}$	(2-71)	$m_{s'} = \frac{1}{P_{sn}}\left(Q_{sn} - \frac{V_s^2}{X}\right)$	(2-72)	$c_{s'} = \frac{(V_s \pm V_{s's})^2}{X}$	(2-161)
$Q_{se} = m_{se} P_{se} + c_{se}$	(2-190)	$m_{se} = \frac{Q_{sn}}{P_{sn}}$	(2-191)	$c_{se} = -\frac{V_{s's}^2}{X}$	(2-192)

Table 2-14 Equations for P_s, P_r, $P_{s'}$, and P_{se} when a PAR (asym) is used.

$P_s = P_{sn} \pm a_s$	(2-128a)	$a_s = \frac{V_s V_{s's}}{X}$	(2-119)		
$P_r = P_{rn} \pm a_r \cos\delta$	(2-142a)	$a_r = \frac{V_r V_{s's}}{X}$	(2-132)		
$P_{s'} = P_{sn} \pm a_r \cos\delta$	(2-162a)	$a_r = \frac{V_r V_{s's}}{X}$	(2-132)		
$P_{se} = \pm a_{se} \cos\left(\tan^{-1}\frac{P_{sn}}{Q_{sn}}\right)$	(2-193a)	$a_{se} = \frac{V_{s's} V_{Xn}}{X}$	(2-178)	$V_{Xn} = \sqrt{V_s^2 - 2V_s V_r \cos\delta + V_r^2}$	(2-30)

Table 2-15 Equations relating (P_s and Q_s), (P_r and Q_r), ($P_{s'}$ and $Q_{s'}$), and (P_{se} and Q_{se}) when a PAR (asym) is used.

$Q_s = Q_{sn}$	(2-128c)				
$Q_r = m_r P_r + c_r$	(2-144)	$m_r = \frac{-P_{rn}}{Q_{rn} + \frac{V_r^2}{X}}$	(2-145)	$c_r = \frac{P_{rn}^2}{Q_{rn} + \frac{V_r^2}{X}} + Q_{rn}$	(2-146)
$Q_{s'} = m_{s'} P_{s'} + c_{s'}$	(2-164)	$m_{s'} = \frac{-P_{sn}}{Q_{sn} - \frac{V_s^2}{X}}$	(2-165)	$c_{s'} = \frac{P_{sn}^2}{Q_{sn} - \frac{V_s^2}{X}} + Q_{sn} + \frac{V_{s's}^2}{X}$	(2-166)
$Q_{se} = m_{se} P_{se} + c_{se}$	(2-195)	$m_{se} = -\frac{P_{sn}}{Q_{sn}}$	(2-196)	$c_{se} = -\frac{V_{s's}^2}{X}$	(2-197)

2.2.2.5 Additional Series-Compensating Voltages

A series-compensating voltage can be used to implement (1) PAR (symmetric) and (2) RR.

The natural active power at the receiving end of the line is given by

$$P_{rn} = a \times \sin\delta \tag{2-46}$$

where

$$a = \frac{V_s V_r}{X}. \tag{2-41}$$

The power flow in a line can be reversed in one of the two ways, either by (1) modifying the power angle (δ') to be negative as shown in Figure 2-27 or (2) modifying the effective line reactance (X_{eff}) between its two ends to be capacitive as shown in Figures 2-35a and e and 2-38a and c, respectively.

The expressions for the magnitude and phase angle for the line current (**I**) are derived in Equations (2-113) and (2-115), respectively, where the relative phase angle (β) is as follows:

Equation (2-200): $\beta = \frac{\psi}{2} + 90°$ for $\psi > 0$ or $\beta = \frac{\psi}{2} + 270°$ for $\psi < 0$ for a PAR (sym)

Equation (2-227): $\beta = \tan^{-1}\frac{V_r \sin\delta}{V_s - V_r\cos\delta}$ for a RR.

2.2.2.5.1 Phase Angle Regulator (Symmetric)

A PAR (sym) connects a compensating voltage ($\mathbf{V_{s's}}$) in series with the line to regulate the phase-shift angle (ψ) of the line voltage ($\mathbf{V_s}$) at the sending end. The magnitude ($V_{s'}$) of the modified sending-end voltage ($\mathbf{V_{s'}}$) and the magnitude (V_s) of the sending-end voltage ($\mathbf{V_s}$) are the same. Therefore, substituting $V_{s'}$ with V_s in Equation (2-100), the expression for the magnitude ($V_{s's}$) of the series-compensating voltage ($\mathbf{V_{s's}}$) for a phase-shift angle (ψ) becomes

$$V_{s's} = \sqrt{V_s^2 + V_s^2 - 2V_sV_s\cos\psi} \quad \text{or} \quad V_{s's} = 2V_s\sin\left(\frac{\psi}{2}\right). \tag{2-198}$$

A $\psi > 0$ makes $V_{s's}$ positive and the power flow in the line increases; whereas, a $\psi < 0$ makes $V_{s's}$ negative and the power flow in the line decreases.

Substituting $V_{s'}$ with V_s, Equation (2-101) can be rewritten as

$$\beta = \tan^{-1}\frac{V_s\sin\psi}{V_s\cos\psi - V_s} \quad \text{or} \quad \beta = -\tan^{-1}\cot\frac{\psi}{2} \tag{2-199}$$

or

$$\beta = \frac{\psi}{2} + 90° \text{ for } \psi > 0 \quad \text{or} \quad \beta = \frac{\psi}{2} + 270° \text{ for } \psi < 0. \tag{2-200}$$

Figure 2-26a shows a two-generator/one-line power system network with a series-compensating voltage to implement a PAR (sym). Figure 2-26b shows the phasor diagram without any compensation. Figure 2-26c shows that the line current (**I**) decreases as the phase angle of the voltage between the modified sending end and receiving end decreases from its uncompensated value. The line current (**I**) increases when the phase angle of the voltage between the modified sending end and receiving end increases from its uncompensated value as shown in Figure 2-26d. Since for a phase-shift angle (ψ), the magnitude of the line voltage stays unchanged, it is referred to as PAR symmetric (sym). The expressions for power flows at the receiving end of the line are shown in Figure 2-26, considering that the modified sending-end voltage is $V_S \angle \delta_{S'}$, *i.e.* power angle changes while the voltage magnitude at the modified sending end remains unchanged.

The modified power angle (δ') and the phase-shift angle (ψ) are related by the following equation:

$$\delta' = \delta + \psi. \tag{2-51a}$$

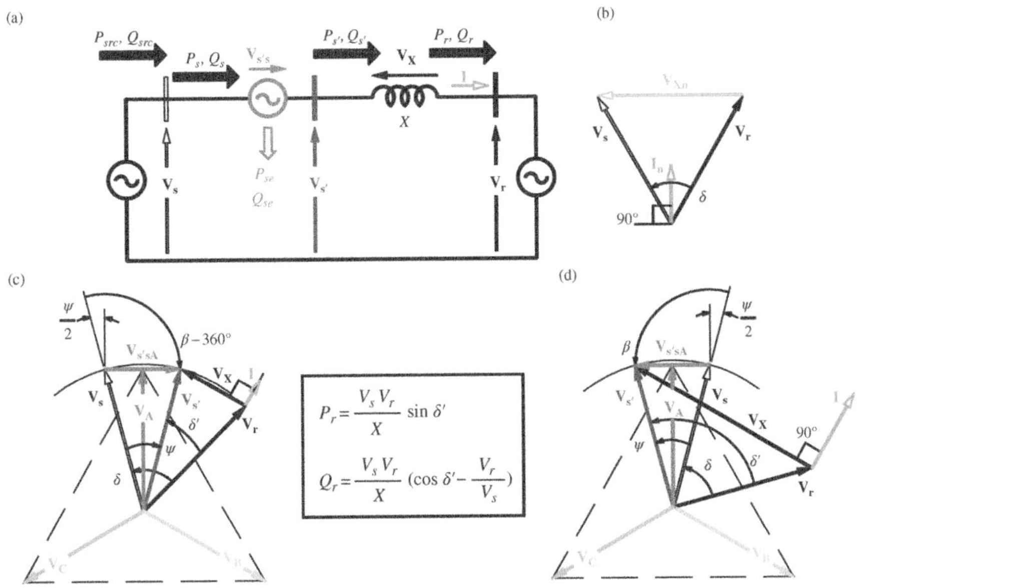

Figure 2-26 (a) Two-generator/one-line power system network and its series-compensating voltage to implement a PAR (sym) and phasor diagrams for (b) an uncompensated line, (c) decreasing power flow in the line, and (d) increasing power flow in the line.

The equations for active and reactive power flows ($P_{s'}$ and $Q_{s'}$) at the modified sending end of a line using a PAR (sym) are given below:

$$P_{s'} = a \times \sin\delta' \tag{2-201}$$

and

$$Q_{s'} = a\left(\frac{V_s}{V_r} - \cos\delta'\right) \tag{2-202}$$

where a is defined in Equation (2-41).

The equations for active and reactive power flows (P_r and Q_r) at the receiving end of a line using a PAR (sym) are given below:

$$P_r = a \times \sin\delta' \tag{2-203}$$

and

$$Q_r = a\left(\cos\delta' - \frac{V_r}{V_s}\right) \tag{2-204}$$

where a is defined in Equation (2-41).

Using Equation (2-41), Equations (2-201) and (2-202) can be combined as

$$P_{s'}^2 + \left(Q_{s'} - \frac{V_s^2}{X}\right)^2 = a^2\left\{(\sin\delta')^2 + (\cos\delta')^2\right\} = a^2 \quad \text{using the identity } \left(\cos^2\delta' + \sin^2\delta' = 1\right) \text{ or}$$

$$P_{s'}^2 + \left(Q_{s'} - \frac{V_s^2}{X}\right)^2 = a^2. \tag{2-205}$$

Using Equation (2-41), Equations (2-203) and (2-204) can be combined as

$$P_r^2 + \left(Q_r + \frac{V_r^2}{X}\right)^2 = a^2\left\{(\sin\delta')^2 + (\cos\delta')^2\right\} = a^2 \quad \text{using the identity } \left(\cos^2\delta' + \sin^2\delta' = 1\right) \text{ or}$$

$$P_r^2 + \left(Q_r + \frac{V_r^2}{X}\right)^2 = a^2. \tag{2-206}$$

The equations for active and reactive power flows ($P_{s'}$ and $Q_{s'}$) at the modified sending end and (P_r and Q_r) at the receiving end, using a PAR (sym), as a function of the modified power angle (δ') are given in Table 2-16.

The equations relating the active and reactive power flows ($P_{s'}$ and $Q_{s'}$) at the modified sending end and (P_r and Q_r) at the receiving end when a PAR (sym) is used are given in Table 2-17.

Table 2-16 Equations for active and reactive power flows ($P_{s'}$ and $Q_{s'}$) at the modified sending end and (P_r and Q_r) at the receiving end, using a PAR (sym), as a function of the modified power angle (δ').

$P_{s'} = a \times \sin\delta'$	(2-201)	$Q_{s'} = a\left(\frac{V_s}{V_r} - \cos\delta'\right)$	(2-202)	$a = \frac{V_sV_r}{X}$	(2-41)	$\delta' = \delta + \psi$	(2-51a)
$P_r = a \times \sin\delta'$	(2-203)	$Q_r = a\left(\cos\delta' - \frac{V_r}{V_s}\right)$	(2-204)	$a = \frac{V_sV_r}{X}$	(2-41)	$\delta' = \delta + \psi$	(2-51a)

Table 2-17 Equations relating the active and reactive power flows ($P_{s'}$ and $Q_{s'}$) at the modified sending end and (P_r and Q_r) at the receiving end when a PAR (sym) is used.

$P_{s'}^2 + \left(Q_{s'} - \frac{V_s^2}{X}\right)^2 = a^2$ (2-205)	$a = \frac{V_s V_r}{X}$ (2-41)
$P_r^2 + \left(Q_r + \frac{V_r^2}{X}\right)^2 = a^2$ (2-206)	$a = \frac{V_s V_r}{X}$ (2-41)

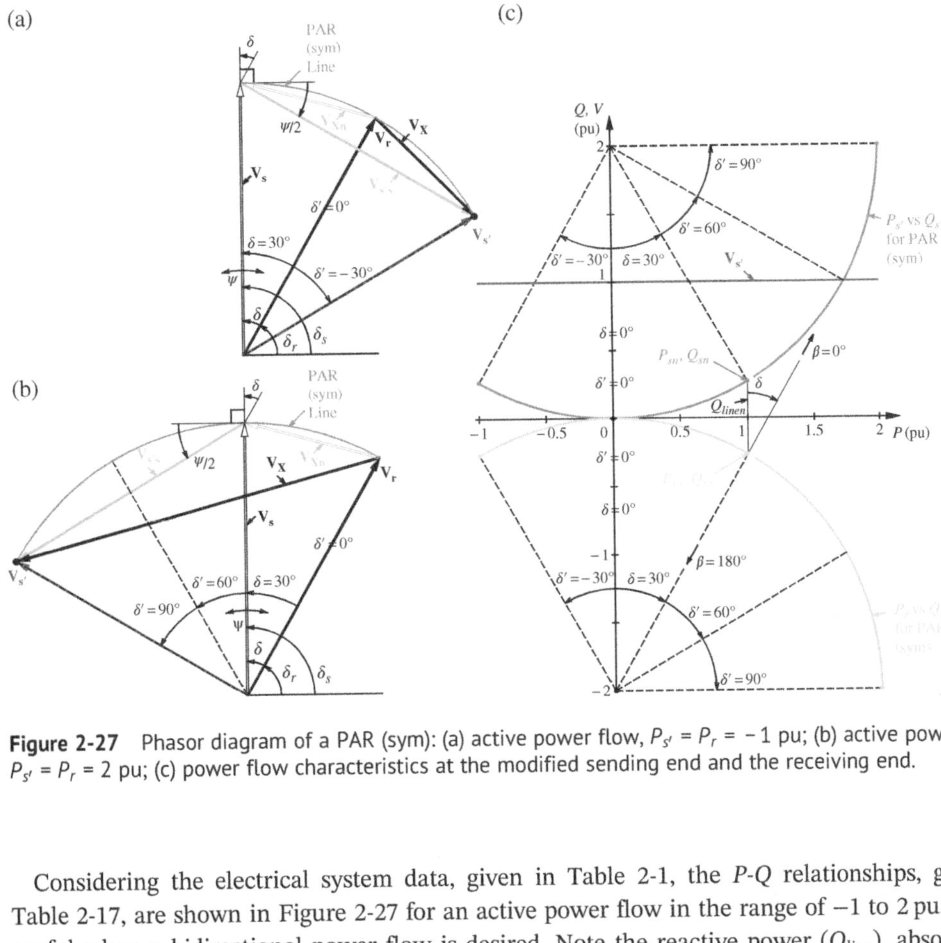

Figure 2-27 Phasor diagram of a PAR (sym): (a) active power flow, $P_{s'} = P_r = -1$ pu; (b) active power flow, $P_{s'} = P_r = 2$ pu; (c) power flow characteristics at the modified sending end and the receiving end.

Considering the electrical system data, given in Table 2-1, the *P-Q* relationships, given in Table 2-17, are shown in Figure 2-27 for an active power flow in the range of −1 to 2 pu. This is useful when a bidirectional power flow is desired. Note the reactive power (Q_{line}), absorbed by the line, is the sum of the absolute values of the reactive power at both ends of the line, since the power factors at both ends of the line are a mixture of leading and lagging power factors, otherwise it will be the difference of the absolute values of the reactive power at both ends of the line, according to Equations (2-320b) and (2-320a), discussed later in the chapter.

2.2.2.5.2 Reactance Regulator

A RR connects a variable capacitor/reactor with a compensating reactance (X_{se}) either directly or indirectly in series with the line to regulate the effective reactance of the line between its two ends. When a compensating reactance (X_{se}) is connected directly in series with a line, a voltage ($\mathbf{V_q}$) is

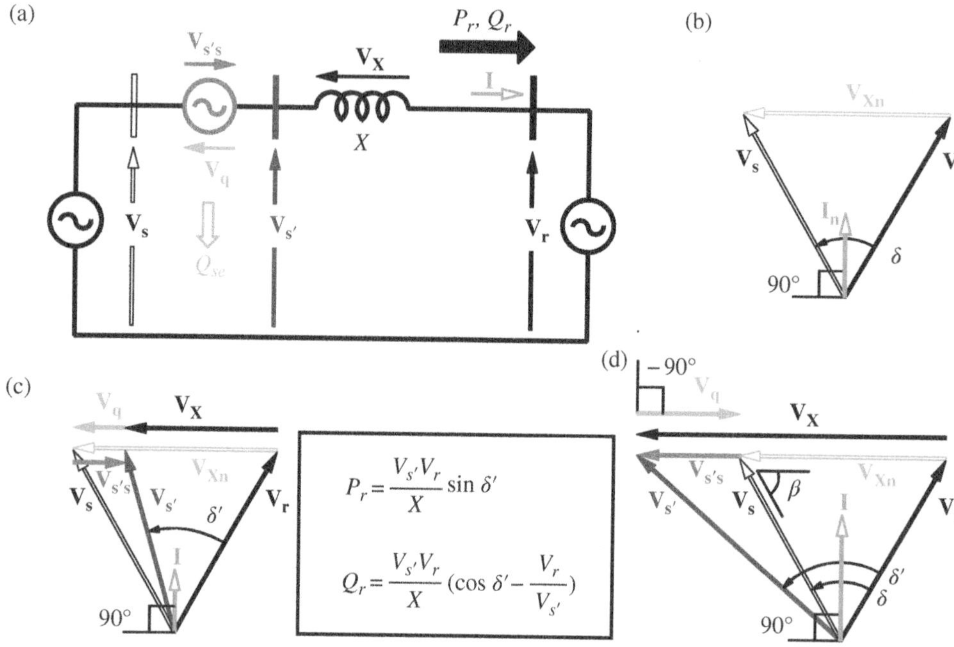

Figure 2-28 RR operating with a reactance control method: (a) two-generator/one-line power system network and its series-compensating voltage ($\mathbf{V_{s's}} = -\mathbf{V_q}$) to implement a RR and phasor diagrams for (b) an uncompensated line, (c) an inductively compensated line, and (d) a capacitively compensated line.

formed across the reactance as shown in Figure 2-28a. The voltage ($\mathbf{V_q} = jV_q$) is in quadrature with the prevailing line current (**I**). A compensating voltage ($\mathbf{V_{s's}} = -\mathbf{V_q}$) may also be used, maintaining the quadrature relationship with the prevailing line current (**I**). The compensating voltage modifies the magnitude ($V_{s'}$) and phase angle ($\delta_{s'}$) of the modified sending end voltage ($\mathbf{V_{s'}}$) simultaneously.

Substituting $V_{s'}$ with V_s, δ' with δ, and X with X_{eff}, the power flow Equations (2-74) and (2-75) can be rewritten using Equation (2-61) as

$$P_r = a_{eff} \sin \delta \tag{2-207}$$

and

$$Q_r = a_{eff}\left(\cos \delta - \frac{V_r}{V_s}\right) \tag{2-208}$$

where

$$a_{eff} = \frac{V_s V_r}{X_{eff}}, \tag{2-209}$$

$$X_{eff} = X - X_{se} \tag{2-210a}$$

or

$$X_{se} = X - X_{eff}, \tag{2-210b}$$

and

$$\delta = \delta_s - \delta_r. \tag{2-27}$$

X_{eff} is the effective reactance of the line between its two ends. The compensating reactance (X_{se}) is defined to be negative when the RR is operated in an inductive mode and positive when the RR is operated in a capacitive mode. The effective line reactance (X_{eff}) has a positive value when being inductive and a negative value when being capacitive.

Combining Equations (2-207) and (2-209), it can be written that

$$P_r = \frac{V_S V_r}{X_{eff}} \sin\delta$$

or

$$X_{eff} = \frac{V_S V_r}{P_r} \sin\delta. \tag{2-210c}$$

For a desired amount of active power flow (P_r), the required effective line reactance (X_{eff}) can be determined by using Equation (2-210c). Then, the series-compensating reactance (X_{se}) can be determined by using Equation (2-210b).

The line reactance (X) is inductive. The power flow can be decreased by inserting an additional inductive reactance in series with the line, thereby increasing the effective reactance of the line between its two ends. The power flow can also be increased by inserting an additional capacitive reactance in series with the line, thereby decreasing the effective reactance of the line between its two ends.

From Equation (2-210a), the normalized effective line reactance is given by

$$\frac{X_{eff}}{X} = \frac{X - X_{se}}{X}$$

or

$$\frac{X_{eff}}{X} = 1 - \frac{X_{se}}{X} \tag{2-211a}$$

or

$$\frac{X_{se}}{X} = 1 - \frac{X_{eff}}{X}. \tag{2-211b}$$

Using (2-46), (2-48), and (2-41), Equations (2-207), (2-208), and (2-209) can be rewritten as

$$P_r = \frac{X}{X_{eff}} \times \frac{V_S V_r}{X} \sin\delta \qquad \text{or} \qquad P_r = \frac{X}{X_{eff}} P_{rn} \tag{2-212}$$

and

$$Q_r = \frac{X}{X_{eff}} \times \frac{V_S V_r}{X}\left(\cos\delta - \frac{V_r}{V_S}\right) \qquad \text{or} \qquad Q_r = \frac{X}{X_{eff}} Q_{rn}. \tag{2-213}$$

From Equations (2-212) and (2-213), the expression for the normalized power flow in the line can be written as

$$\frac{P_r}{P_{rn}} = \frac{Q_r}{Q_{rn}} = \frac{X}{X_{eff}} = \frac{1}{1 - \dfrac{X_{se}}{X}} \tag{2-214}$$

using Equation (2-211a).

Figure 2-28a shows a two-generator/one-line power system network with a compensating voltage that emulates either a series-connected inductive reactance or a series-connected capacitive reactance. Figure 2-28b shows the phasor diagram without any compensation. Figure 2-28c shows that the line current (**I**) decreases from its natural or uncompensated value when the series-compensating voltage ($\mathbf{V_q}$), in load convention, leads the line current by 90° to provide an inductive reactance compensation. The line current (**I**) increases when the series-compensating voltage ($\mathbf{V_q}$) lags the prevailing line current (**I**) by 90° to provide a capacitive reactance compensation as shown in Figure 2-28d. The expressions for power flows at the receiving end of the line are shown in Figure 2-28, considering that the modified sending-end voltage is $V_{s'} \angle \delta_{s'}$, *i.e.* both the magnitude and the phase angle at the modified sending end change.

Figures 2-29a and 2-29b are identical to Figures 2-28a and 2-28b, respectively. Figure 2-29c shows that if the sending-end voltage ($\mathbf{V_s}$) is decreased with a VR and the phase angle is decreased with a PAR (sym), the equivalent modified sending-end voltage ($\mathbf{V_{s'}}$) can be achieved to mimic a series inductive reactance compensation. Figure 2-29d shows that if phase angle is increased with a PAR (sym) and the sending-end voltage ($\mathbf{V_s}$) is increased with a VR, the equivalent modified sending-end voltage ($\mathbf{V_{s'}}$) can be achieved to mimic a capacitive series reactance compensation. Therefore, a RR is indeed a combination of a VR and a PAR (sym). The same conclusion can be made if a combination of a VR and a PAR (asym) is used to mimic a RR.

Equations (2-212) and (2-213) can be combined as

$$Q_r = \frac{Q_{rn}}{P_{rn}} P_r, \tag{2-215}$$

which is a straight line with a fixed slope of Q_{rn}/P_{rn} as shown in Figure 2-38.

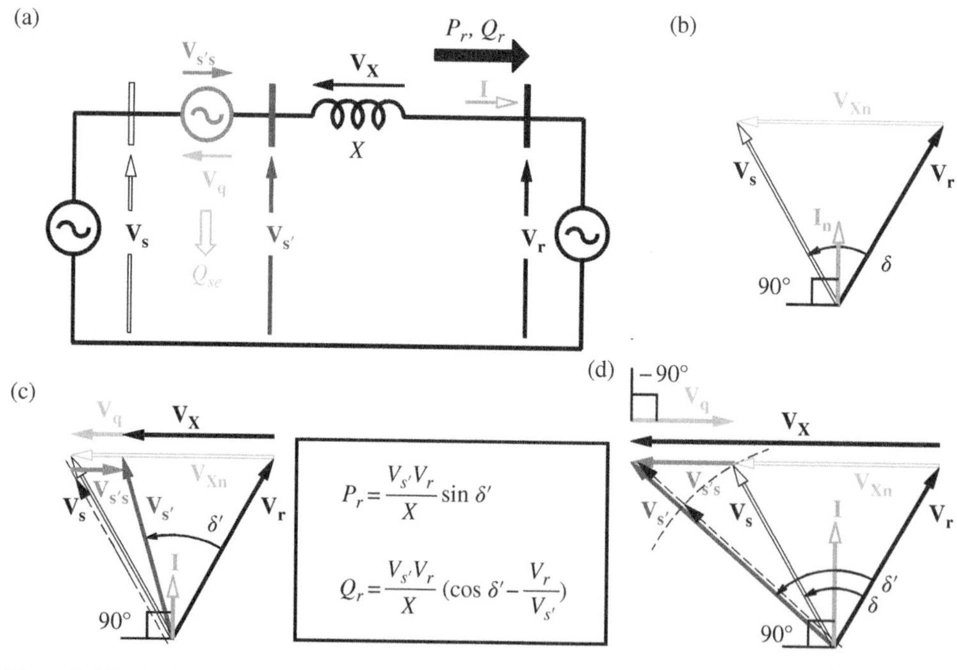

Figure 2-29 Series-reactance regulation as combined effects voltage regulation and phase angle regulation.

For a given amount of active power flow (P_r), the reactive power (Q_r) can be determined by using Equations (2-215). Then, the apparent power flow (S_r) and the prevailing line current (I) can be determined by using the following equations:

$$S_r = \sqrt{P_r^2 + Q_r^2} \tag{2-76}$$

$$I = \frac{S_r}{V_r}. \tag{2-77}$$

The active power flow ($P_{s'}$) at the modified sending end is given by

$$P_{s'} = P_r \tag{2-216a}$$

since the line is assumed to be lossless, or

$$P_{s'} = \frac{X}{X_{eff}} P_{rn} \tag{2-216b}$$

using Equation (2-212).

The reactive power flow ($Q_{s'}$) at the modified sending end is given by

$$Q_{s'} = Q_r + Q_{line}, \tag{2-78a}$$

which can be rewritten as

$$Q_{s'} = \frac{X}{X_{eff}} Q_{rn} + I^2 X \tag{2-217a}$$

using Equations (2-213) and (2-11b), or

$$Q_{s'} = \frac{P_{s'}}{P_{rn}} Q_{rn} + I^2 X \tag{2-217b}$$

with a substitution from Equation (2-216b).

The active power flow at the receiving end of the line when the line is series-compensated is given by

$$P_r = V_r I \cos(\delta_r - \theta_I) \qquad \text{or}$$
$$P_r = V_r I \cos(\delta_r - \{\delta_r + \theta_{Ir}\})$$

where θ_{Ir} is the phase angle of the line current with respect to the receiving-end voltage, or

$$P_r = V_r I \cos(-\theta_{Ir}) \qquad \text{or}$$
$$P_r = V_r I \cos\theta_{Ir} \qquad \text{or}$$
$$I \cos\theta_{Ir} = \frac{P_r}{V_r}. \tag{2-218}$$

The reactive power flow at the receiving end of the line when the line is series-compensated is given by

$$Q_r = V_r I \sin(\delta_r - \theta_I) \qquad \text{or}$$
$$Q_r = V_r I \sin(\delta_r - \{\delta_r + \theta_{Ir}\})$$

where θ_{Ir} is the phase angle of the line current with respect to the receiving-end voltage, or

$$Q_r = V_r I \sin(-\theta_{Ir}) \qquad \text{or}$$

$$Q_r = -V_r I \sin\theta_{Ir} \qquad \text{or}$$

$$I \sin\theta_{Ir} = -\frac{Q_r}{V_r}. \tag{2-219}$$

Using Equation (2-50a), the voltage at the modified sending end is given by

$$\mathbf{V_{s'}} = \mathbf{V_r} + \mathbf{V_X} \qquad \text{or}$$

$$V_{s'}\angle\delta_{s'} = V_r\angle\delta_r + V_X\angle\theta_{VX} \qquad \text{or}$$

$$V_{s'}\angle(\delta_r + \delta') = V_r\angle\delta_r + I\angle\theta_I(jX) \qquad \text{or}$$

$$V_{s'}\angle(\delta_r + \delta') = V_r\angle\delta_r + I\angle(\delta_r + \theta_{Ir})(jX).$$

Considering the receiving-end voltage phasor as reference (*i.e.* $\delta_r = 0°$), it can be written from the above equation that

$$V_{s'}\angle\delta' = \mathrm{V}_r\angle 0° + I\angle\theta_{Ir}(jX) \qquad \text{or}$$

$$V_{s'}\cos\delta' + jV_{s'}\sin\delta' = V_r + (I\cos\theta_{Ir} + jI\sin\theta_{Ir}) \times (jX) \qquad \text{or}$$

$$V_{s'}\cos\delta' + jV_{s'}\sin\delta' = V_r - XI\sin\theta_{Ir} + jXI\cos\theta_{Ir}.$$

Equating the real and imaginary parts of the both sides of the above equation, it can be written that

$$V_{s'}\cos\delta' = V_r - XI\sin\theta_{Ir} \tag{2-220a}$$

and

$$V_{s'}\sin\delta' = XI\cos\theta_{Ir}. \tag{2-220b}$$

Therefore, the magnitude of the modified sending-end voltage ($\mathbf{V_{s'}}$) is derived from Equations (2-220a) and (2-220b) using the identity $\left(\cos^2\delta' + \sin^2\delta' = 1\right)$ as

$$V_{s'} = \sqrt{V_{s'}^2\left(\cos^2\delta' + \sin^2\delta'\right)} = \sqrt{(V_{s'}\cos\delta')^2 + \left(V_{s'}\sin\delta'\right)^2} \qquad \text{or}$$

$$V_{s'} = \sqrt{(V_r - XI\sin\theta_{Ir})^2 + \left(XI\cos\theta_{Ir}\right)^2} \qquad \text{or}$$

$$V_{s'} = \sqrt{V_r^2 + (XI)^2\left(\cos^2\theta_{Ir} + \sin^2\theta_{Ir}\right) - 2V_r X(I\sin\theta_I)} \qquad \text{or}$$

$$V_{s'} = \sqrt{V_r^2 + (XI)^2 + 2XQ_r} \tag{2-221}$$

using the identity $\left(\cos^2\theta_{Ir} + \sin^2\theta_{Ir} = 1\right)$ and using Equation (2-219).

The phase angle of the modified sending-end voltage ($\mathbf{V_{s'}}$) is derived from Equations (2-220a) and (2-220b) as

$$\delta' = \tan^{-1}(\tan\delta') = \tan^{-1}\frac{\sin\delta'}{\cos\delta'} = \tan^{-1}\frac{V_{s'}\sin\delta'}{V_{s'}\cos\delta'} = \tan^{-1}\frac{XI\cos\theta_{Ir}}{V_r - XI\sin\theta_{Ir}}$$

$$= \tan^{-1}\frac{V_r XI\cos\theta_{Ir}}{V_r^2 - V_r XI\sin\theta_{Ir}}$$

or

$$\delta' = \tan^{-1}\frac{XP_r}{V_r^2 + XQ_r} \tag{2-222}$$

using Equations (2-218) and (2-219).

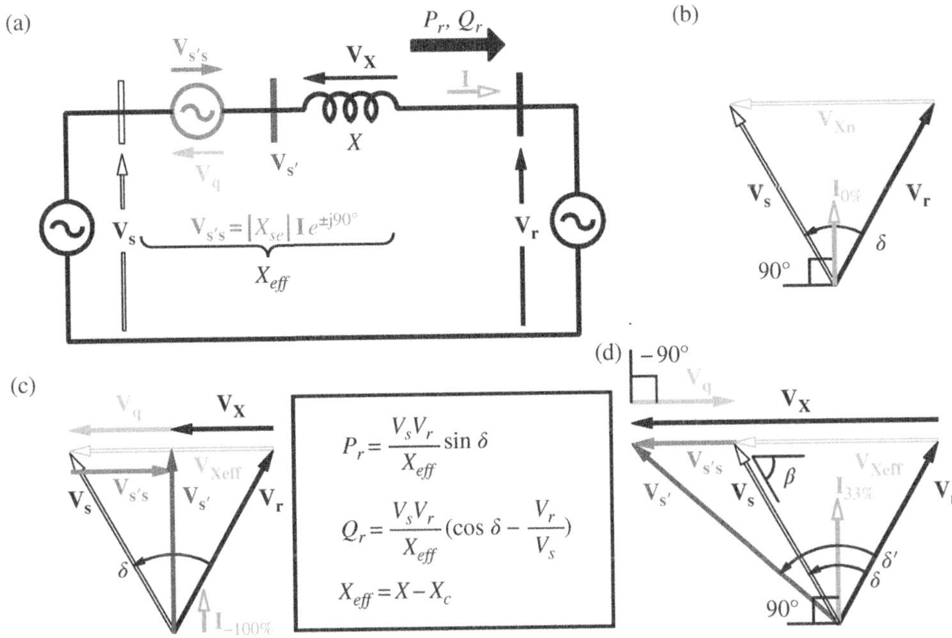

Figure 2-30 RR operating with a reactance control method: (a) two-generator/one-line power system network and its series-compensating voltage ($\mathbf{V_{s's}} = -\mathbf{V_q}$) to implement a RR and phasor diagrams for (b) an uncompensated line, (c) a 100% inductively compensated line, and (d) a 33% capacitively compensated line.

Figure 2-30 shows a two-generator/one-line power system network with the RR operated both in inductive and in capacitive modes and the related phasor diagrams. The line current decreases from $\mathbf{I_{0\%}}$, which is same as natural current ($\mathbf{I_n}$), to $\mathbf{I_{-100\%}}$, when the inductive reactance compensation ($-X_{se}/X$) increases from 0 to 100%. The line current increases from $\mathbf{I_{0\%}}$ to $\mathbf{I_{33\%}}$ when the capacitive reactance compensation (X_{se}/X) increases from 0 to 33%.

The effects of the compensating reactance (X_{se}) on the normalized effective line reactance and the normalized power flow in the line are shown in Figure 2-31. When the compensating reactance is inductive, the effective line reactance (X_{eff}) increases and the active and reactive power flows (P_r and Q_r) decrease as the reactance compensation ($-X_{se}/X$) increases. When the compensating reactance is capacitive, the effective line reactance (X_{eff}) decreases and the active and reactive power flows (P_r and Q_r) increase as the reactance compensation (X_{se}/X) increases.

The reactance control method, just implemented, is based on emulating a reactance in series with the line reactance. This control method is used to operate the RR, so that the series-compensating voltage ($V_{s's}$) is proportional to the prevailing line current (I) with the emulated reactance (X_{se}) being the constant of proportionality. For this control algorithm to work successfully, a line current must exist. Also, in the reactance control method, the polarity of the reactance is defined before the desired control action is to take place. If it is defined to be inductive, the most that can be achieved is the reduction of the line current to somewhat lower than the corresponding uncompensated value. The line current can never be brought to nearly zero by the controller since the successful operation of the control algorithm depends on the existence of the line current. If the emulated reactance is to be defined as capacitive, the line current, at first, increases. If the emulated capacitive reactance demand is higher than the line inductive reactance, the effective line reactance becomes capacitive and the power flow in the line reverses. But reversing the power flow while the line current is high, causes a number of problems. During the transition, when the power flow reverses, the line may

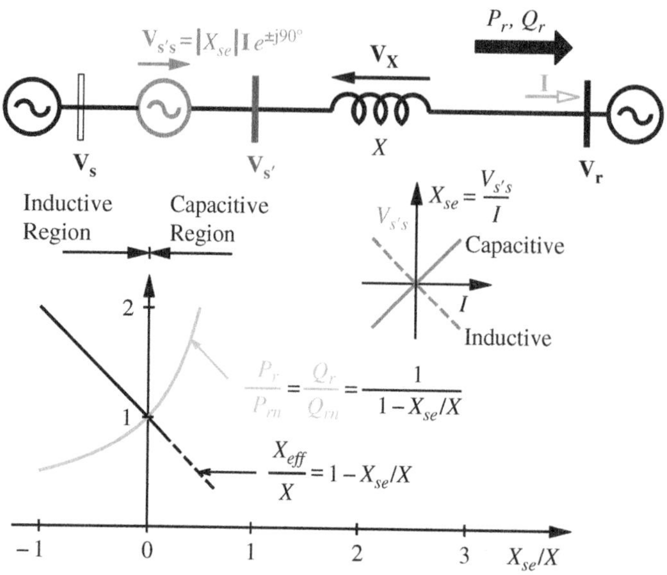

Figure 2-31 Effect of a compensating reactance on the power flow and the effective line reactance.

operate beyond its maximum thermal capability. Also, during the emulation of a higher capacitive reactance, there exists a point at which the inductive reactance of the line and the capacitive reactance emulated by the RR become equal, which can cause instability in the power flow of the line. In a practical implementation, if the current through the RR exceeds its rating, the VSC-based RR will be bypassed. The reactance control method provides the basic features of the RR in terms of controlling the power flow in the line. However, the undesirable feature of this control method appears during a reversal of power flow in the line, when the line current should be brought to nearly zero and be increased gradually in the opposite direction; instead, a high-current reverses, causing even higher transients.

The desirable way to reverse the power flow in a line is to implement a control method, which reduces the line current to nearly zero and reverses with absolute stability. That means, in the process of reversing the power flow in the line, this control method should first emulate an inductive reactance of higher and higher value, and then change to emulate a capacitive reactance of very high value, and gradually reduce the capacitive reactance value to increase the power flow in the reverse direction. The reactance control method does not allow a near-zero line current operating condition because this method relies primarily on the existence of the line current for its operation. The voltage control method is used to connect a compensating voltage in series with the line with any amount of bidirectional line current including, near zero. This compensating voltage is either in phase or in opposite phase with the voltage across the effective line reactance. This control method offers all the desirable features offered by the reactance control method in terms of controlling the power flow in the line. In addition, the voltage control method offers an absolute stability in the power flow, causing the power to go through near zero while changing its direction of flow.

Two control methods – reactance control and voltage control – are compared in the following section. The latter is shown to have a superior characteristic when a reversal of power flow in the line occurs.

2.2.2.5.2.1 Reactance Control Method Figure 2-32 shows a single-line diagram of a simple power transmission system with the RR operating with a capacitive reactance control method and the related phasor diagrams. The line current increases from an uncompensated value of $\mathbf{I}_{0\%}$ to a compensated value of $\mathbf{I}_{33\%}$ when the reactance compensation (X_{se}/X) increases from 0 to 33%. For a larger reactance compensation ($X_{se}/X = 300\%$), the line current ($\mathbf{I}_{300\%}$) flows in the reverse direction and decreases further as the compensation increases. Figure 2-33 shows the RR operating with an inductive reactance control method and the related phasor diagram. The line current decreases from $\mathbf{I}_{0\%}$ to $\mathbf{I}_{-100\%}$ when the reactance compensation ($-X_{se}/X$) increases from 0 to 100%.

The effects of the compensating reactance (X_{se}) on the normalized power flow in the line and the normalized effective line reactance are shown in Figure 2-34. In an operation of the RR with a capacitive reactance control method, the power flow (P_r) increases monotonically as the reactance compensation (X_{se}/X) increases during the period when $0 < X_{se}/X < 1$. When $X_{se}/X = 1$, the power flow (P_r) can possibly reach to an extremely high value limited by the line resistance only. Finally, the power flow (P_r) decreases monotonically in the reverse direction as the reactance compensation (X_{se}/X) increases during the period when $X_{se}/X > 1$. In an operation of the RR with an inductive reactance control method, the power flow (P_r) decreases monotonically as the reactance compensation ($-X_{se}/X$) increases when $X_{se}/X < 0$. The points to be made here are

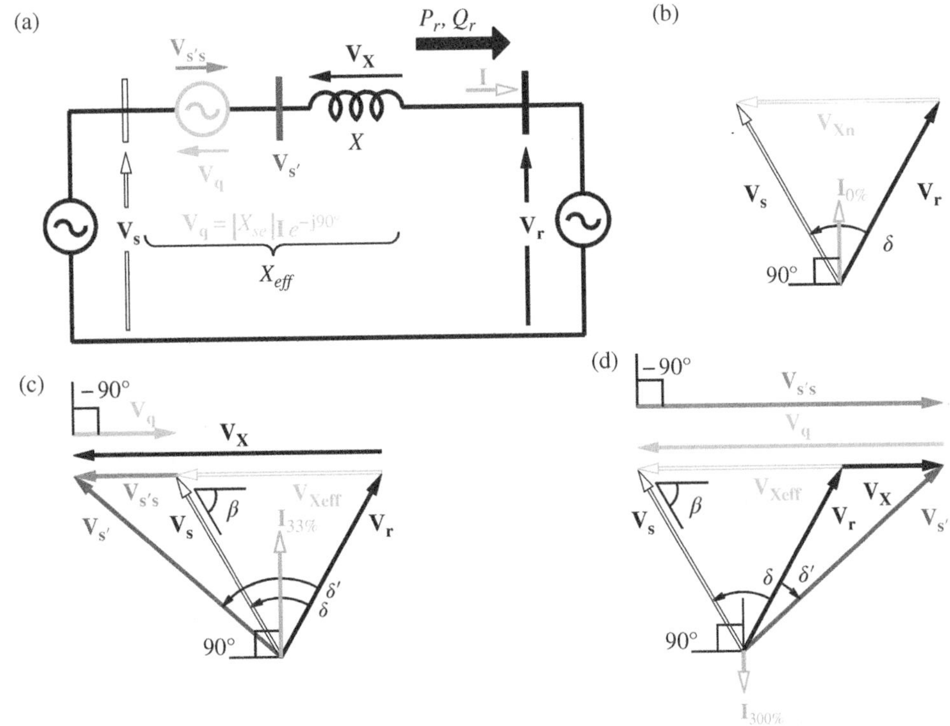

Figure 2-32 RR operating with a reactance control method: (a) two-generator/one-line power system network and its series-compensating voltage ($\mathbf{V}_{s's} = -\mathbf{V}_q$) to implement a RR and phasor diagrams for (b) an uncompensated line, (c) a 33% capacitively compensated line, and (d) a 300% capacitively compensated line.

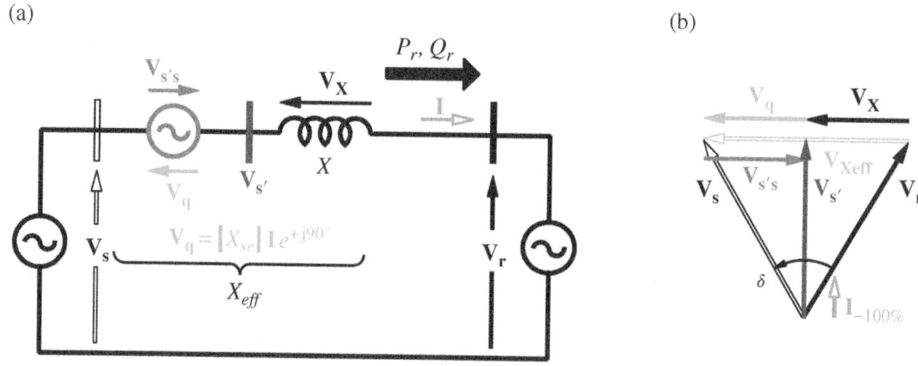

Figure 2-33 RR operating with a reactance control method: (a) two-generator/one-line power system network with a RR and (b) the phasor diagram for a 100% inductively compensated line.

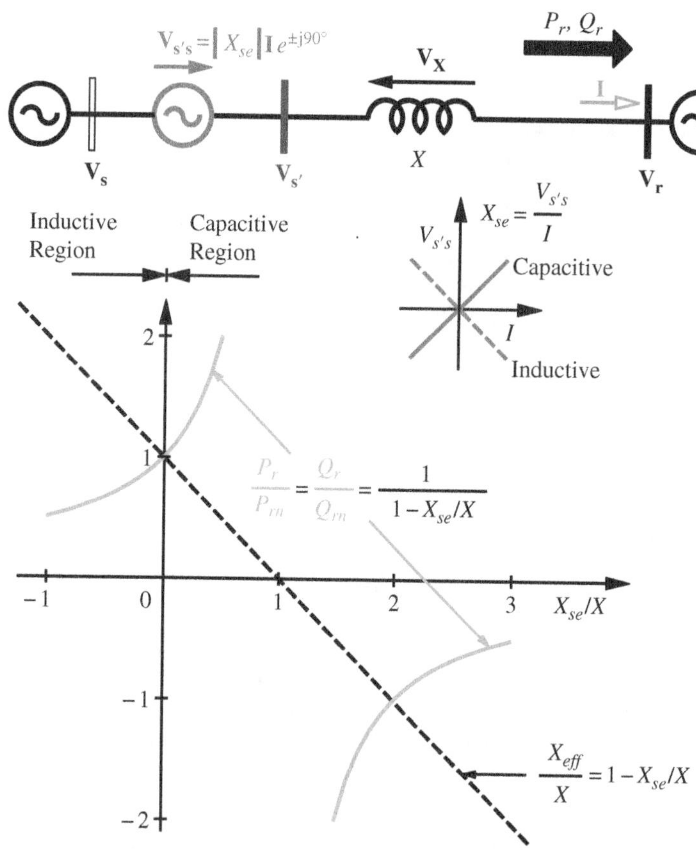

Figure 2-34 Effect of a compensating reactance on the power flow and the effective line reactance.

- The RR is capable of controlling the power flow in a line when operating with a reactance control method.
- In an operation with a capacitive reactance control method, this power flow can be reversed in theory, but not in practice.

However, while changing direction, the power flow (P_r) goes through two simultaneous extremes: one is a very high positive value when the effective line reactance (X_{eff}) is nearly zero inductive and the next one is a very high negative value when the effective line reactance (X_{eff}) is nearly zero capacitive. The fact that these two extremes are practically limited by the line resistance (R) only, which is an impractical operating point.

2.2.2.5.2.2 Voltage Control Method In an operation of the RR with a voltage control method, a compensating voltage ($\mathbf{V_{s's}}$) is connected in series with the line, either in phase or in opposite phase with the voltage ($\mathbf{V_{Xeff}} = \mathbf{V_{Xn}}$) across the line effective line reactance (X_{eff}). The compensating voltage ($\mathbf{V_{s's}}$) maintains its phase angle in quadrature with the prevailing line current ($\mathbf{I}$) and emulates a compensating reactance (X_{se}) in series with the line. In the process, the magnitude and phase angle of the line voltage at the modified sending end change simultaneously.

The series-compensating voltage ($\mathbf{V_{s's}}$) is related to ($\mathbf{V_q}$), such that

$$\mathbf{V_{s's}} = -\mathbf{V_q}. \tag{2-223}$$

The voltage, ($\mathbf{V_q}$), leading the line current, emulates an inductive reactance in series with the line, causing the power flow as well as the line current to decrease as the level of compensation increases, and the RR is considered to be operating in an inductive mode. The voltage, ($\mathbf{V_q}$), lagging the line current, emulates a capacitive reactance in series with the line, causing the power flow as well as the line current to increase as the level of compensation increases, and the RR is considered to be operating in a capacitive mode.

The compensating voltage ($\mathbf{V_{s's}}$) exchanges only reactive power (Q_{se}) with the line. This reactive power (Q_{se}) is defined as

$$Q_{se} = \left|\mathbf{V_q} \times \mathbf{I}\right| = \left|-\mathbf{V_{s's}} \times \mathbf{I}\right| \quad \text{or} \quad Q_{se} = V_q I. \tag{2-224}$$

By inspection of Figure 2-28d, it can be written that

$$\frac{V_r}{\sin\beta} = \frac{V_s}{\sin(\pi - \delta - \beta)}$$

or

$$V_s \sin\beta - V_r \sin(\delta + \beta) = 0. \tag{2-225}$$

Equation (2-225) can be rewritten as

$$V_s \sin\beta = V_r \sin(\delta + \beta) = V_r(\sin\delta\cos\beta + \cos\delta\sin\beta) \quad \text{or}$$

$$(V_s - V_r\cos\delta)\sin\beta = V_r \sin\delta\cos\beta \quad \text{or}$$

$$\tan\beta = \frac{V_r \sin\delta}{V_s - V_r\cos\delta}. \tag{2-226}$$

Therefore,

$$\beta = \tan^{-1}\frac{V_r \sin\delta}{V_s - V_r \cos\delta}. \tag{2-227}$$

From Equation (2-226), it can be written that,

$$\sin\beta = \frac{\tan\beta}{\sqrt{1+\tan^2\beta}} = \frac{V_r \sin\delta}{\sqrt{V_s^2 + V_r^2 - 2V_sV_r\cos\delta}} = \frac{V_r \sin\delta}{V_{Xn}} \tag{2-228a}$$

$$\cos\beta = \frac{1}{\sqrt{1+\tan^2\beta}} = \frac{V_s - V_r \cos\delta}{\sqrt{V_s^2 + V_r^2 - 2V_sV_r\cos\delta}} = \frac{V_s - V_r\cos\delta}{V_{Xn}} \tag{2-228b}$$

using Equation (2-30).

Substituting for cosβ from Equation (2-228b) in Equation (2-97), the magnitude of the modified sending-end voltage ($\mathbf{V}_{s'}$) is derived as

$$V_{s'} = \sqrt{V_s^2 + V_{s's}^2 + 2V_sV_{s's}\frac{V_s - V_r\cos\delta}{V_{Xn}}}. \tag{2-229}$$

Substituting for sinβ from Equation (2-228a) and cosβ from Equation (2-228b) in Equation (2-98), the phase-shift angle is derived as

$$\psi = \tan^{-1}\frac{V_{s's}\sin\beta}{V_s + V_{s's}\cos\beta} = \tan^{-1}\frac{V_{s's}\dfrac{V_r\sin\delta}{V_{Xn}}}{V_s + V_{s's}\dfrac{V_s - V_r\cos\delta}{V_{Xn}}} \quad \text{or}$$

$$\psi = \tan^{-1}\frac{V_{s's}V_r\sin\delta}{V_sV_{Xn} + V_{s's}(V_s - V_r\cos\delta)}. \tag{2-230}$$

As discussed in Section 2.2.2.6.4 (Equivalent Impedance of a Reactance Regulator), $V_{s's} > 0$ represents a series-capacitive compensating reactance and $V_{s's} < 0$ represents a series-inductive compensating reactance.

Alternately, for a given series-compensating reactance (X_{se}), the modified sending-end voltage ($\mathbf{V}_{s'}$) can be determined using the following equations:

$$X_{eff} = X - X_{se} \tag{2-210a}$$

$$\delta = \delta_s - \delta_r. \tag{2-27}$$

$$a_{eff} = \frac{V_sV_r}{X_{eff}} \tag{2-209}$$

$$P_r = a_{eff}\sin\delta \tag{2-207}$$

$$Q_r = a_{eff}\left(\cos\delta - \frac{V_r}{V_s}\right) \tag{2-208}$$

$$S_r = \sqrt{P_r^2 + Q_r^2} \tag{2-76}$$

$$I = \frac{S_r}{V_r} \tag{2-77}$$

$$V_{s'} = \sqrt{V_r^2 + (XI)^2 + 2XQ_r} \tag{2-221}$$

$$\delta' = \tan^{-1}\frac{XP_r}{V_r^2 + XQ_r} \tag{2-222}$$

$$\psi = \delta' - \delta. \tag{2-51b}$$

Figure 2-35 shows a single-line diagram of a simple power transmission system with the RR operating with a voltage control method when the compensating voltage is in phase with the voltage ($\mathbf{V_{Xeff}}$) across the effective line reactance (X_{eff}) and the related phasor diagrams. The line current decreases from $\mathbf{I_{0\%}}$ to $\mathbf{I_{-100\%}}$ when the voltage compensation ($-V_{s's}/V_{Xeff}$) increases from 0 to 50%, which is analogous to an increase in the inductive reactance compensation ($-X_{se}/X$) from 0 to 100%. The relationship between the voltage compensation ($V_{s's}/V_{Xeff}$) and the reactance compensation (X_{se}/X) is explained in the following section. When fully compensated (*i.e.* at $-V_{s's}/V_{Xeff} = 100\%$), the line current is zero. With further increase in the voltage compensation (*i.e.* at $-V_{s's}/V_{Xeff} = 150\%$), the line current ($\mathbf{I_{300\%}}$) flows in the reverse direction and any further increase in the voltage compensation ($-V_{s's}/V_{Xeff}$) only increases this reverse line current. Once again, a voltage compensation ($-V_{s's}/V_{Xeff}$) of 150% is analogous to a capacitive reactance compensation (X_{se}/X) of 300%, which is explained later in this section.

Figure 2-36 shows the RR operating with a voltage control method when the compensating voltage is in opposite phase with the voltage ($\mathbf{V_{Xeff}}$) across the effective line reactance (X_{eff}) and the

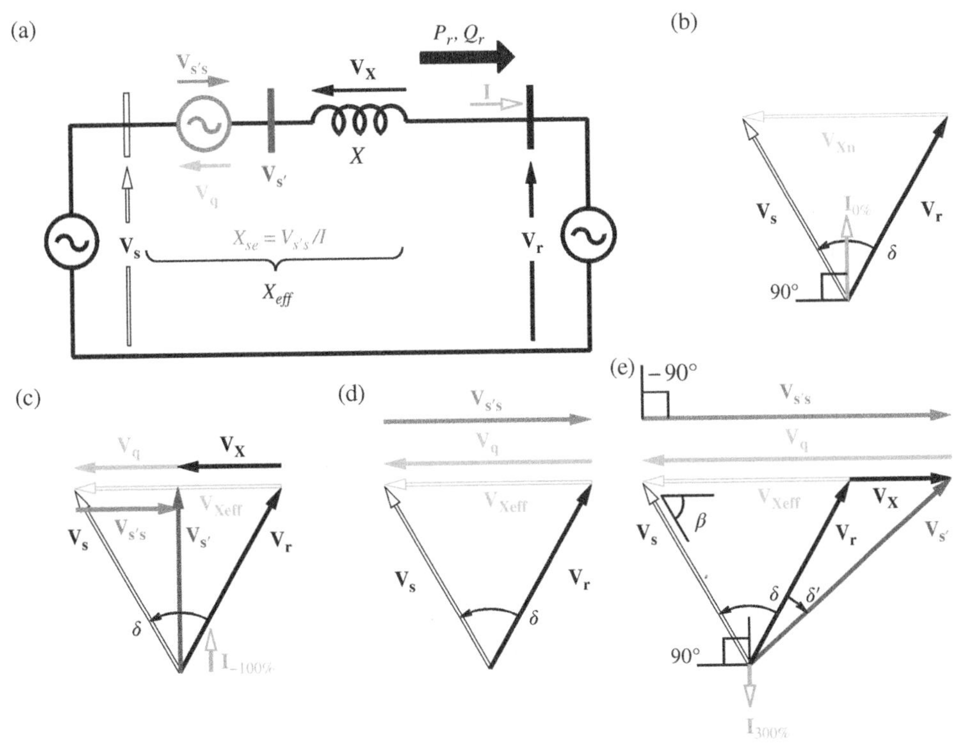

Figure 2-35 RR operating with a voltage control method with a compensating voltage in phase with the voltage across the effective line reactance: (a) two-generator/one-line power system network and its series-compensating voltage ($\mathbf{V_{s's}} = -\mathbf{V_q}$) to implement a RR and the phasor diagrams for (b) an uncompensated line, (c) a 100% inductively compensated line, (d) an infinite inductively compensated line, and (e) a 300% capacitively compensated line.

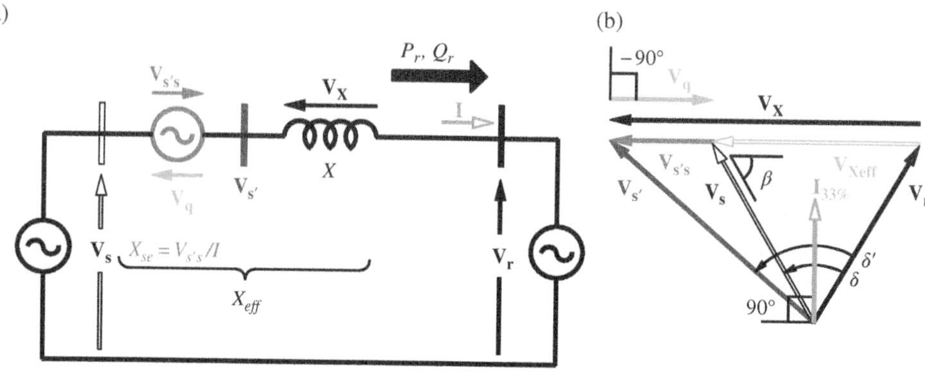

Figure 2-36 RR operating with a voltage control method when a compensating voltage is in opposite phase with the voltage across the effective line reactance: (a) two-generator/one-line power system network with a RR and (b) the phasor diagram for a 33% capacitively compensated line.

related phasor diagram. The line current increases from $\mathbf{I_{0\%}}$ to $\mathbf{I_{33\%}}$ when the voltage compensation ($V_{s's}/V_{Xeff}$) increases from 0 to 50%, which is analogous to an increase in the capacitive reactance compensation (X_{se}/X) from 0 to 33%. It can be noted from the phasor diagram that the voltage ($\mathbf{V_{Xeff}}$) across the effective line reactance (X_{eff}) is finite and is the difference between the voltage sources ($\mathbf{V_s}$ and $\mathbf{V_r}$). Also, this voltage phasor remains constant, whether or not there is any voltage compensation, in a system in which the sending- and receiving-end voltages are fixed. Moreover, this finite difference prevents the two opposing voltage phasors - one is the compensating voltage ($\mathbf{V_q}$) and the other one is the voltage ($\mathbf{V_X}$) across the inductive line reactance (X) - from being equal and making the line immune from resonance.

From the phasor diagram in Figure 2-36,

$$I_{33\%}X_{eff} = V_{Xeff} \tag{2-231}$$

$$I_{33\%}X = V_X = V_{Xeff} + V_{s's} = V_{Xeff}\left(1 + \frac{V_{s's}}{V_{Xeff}}\right) \tag{2-232}$$

Equations (2-231) and (2-232) can be combined as

$$\frac{X_{eff}}{X} = \frac{1}{1 + V_{s's}/V_{Xeff}} \tag{2-233a}$$

or

$$X_{eff} = \frac{1}{1 + V_{s's}/V_{Xeff}}X. \tag{2-233b}$$

The compensating voltage $V_{s's} > 0$ when it is in opposite phase with the voltage across the effective line reactance and $V_{s's} < 0$ when it is in phase with the voltage across the effective line reactance.

Combining Equations (2-211b) and (2-233a), the relationship between the compensating reactance and the compensating voltage can be found as

$$\frac{X_{se}}{X} = \frac{V_{s's}/V_{Xeff}}{1 + V_{s's}/V_{Xeff}} \tag{2-234a}$$

or

$$X_{se} = \frac{V_{s's}/V_{Xeff}}{1 + V_{s's}/V_{Xeff}} X \tag{2-234b}$$

or

$$X_{se} = \frac{V_{s's}}{V_{Xeff}} X_{eff} \tag{2-234c}$$

using Equation (2-233b), or

$$\frac{V_{Xeff}}{X_{eff}} = \frac{V_{s's}}{X_{se}} = I, \tag{2-234d}$$

which confirms Equation (1-3b).

Equations (2-212) and (2-213) can be rewritten with a substitution from Equation (2-233a) as

$$P_r = \left(1 + \frac{V_{s's}}{V_{Xeff}}\right) P_{rn} \tag{2-235}$$

and

$$Q_r = \left(1 + \frac{V_{s's}}{V_{Xeff}}\right) Q_{rn}. \tag{2-236}$$

From Equations (2-235) and (2-236), the expression for the normalized power flow in the line can be written as

$$\frac{P_r}{P_{rn}} = \frac{Q_r}{Q_{rn}} = 1 + \frac{V_{s's}}{V_{Xeff}}. \tag{2-237}$$

From Equations (2-214) and (2-237), it can be written that

$$\frac{X_{eff}}{X} = \frac{1}{1 + V_{s's}/V_{Xeff}}, \quad \text{which is Equation (2-233a).}$$

The expression for the normalized power flow in the line given in Equations (2-214) and (2-237) are basically the same. In an operation of the SSSC with a reactance control method, the control parameter is X_{se}/X, which is essentially $V_{s's}/V_X$. The power expression given by Equation (2-214) has a pole at $X_{se}/X = 1$, causing the power flow (P_r) to be unstable at resonance, whereas, in an operation of the RR with a voltage control method, the control parameter is $V_{s's}/V_{Xeff}$. The power expression given by Equation (2-237) has a zero at $V_{s's}/V_{Xeff} = -1$, causing an absolute stability at the instant when power flow (P_r) reverses.

The effects of the compensating voltage ($\mathbf{V_{s's}}$) on the normalized power flow in the line, the normalized effective line reactance, and the normalized compensating reactance of the line are shown in Figure 2-37. When the compensating voltage is in phase with the voltage across the effective line reactance, the effective line reactance increases monotonically as the voltage compensation ($-V_{s's}/V_{Xeff}$) increases during $0 > V_{s's}/V_{Xeff} > -1$. When fully compensated (*i.e.* at $V_{s's}/V_{Xeff} = -1$), the effective line reactance reaches infinity. Finally, the effective line reactance decreases monotonically in the reverse direction as the voltage compensation ($-V_{s's}/V_{Xeff}$) increases during $V_{s's}/V_{Xeff} < -1$. When the compensating voltage is in opposite phase with the voltage across the effective line

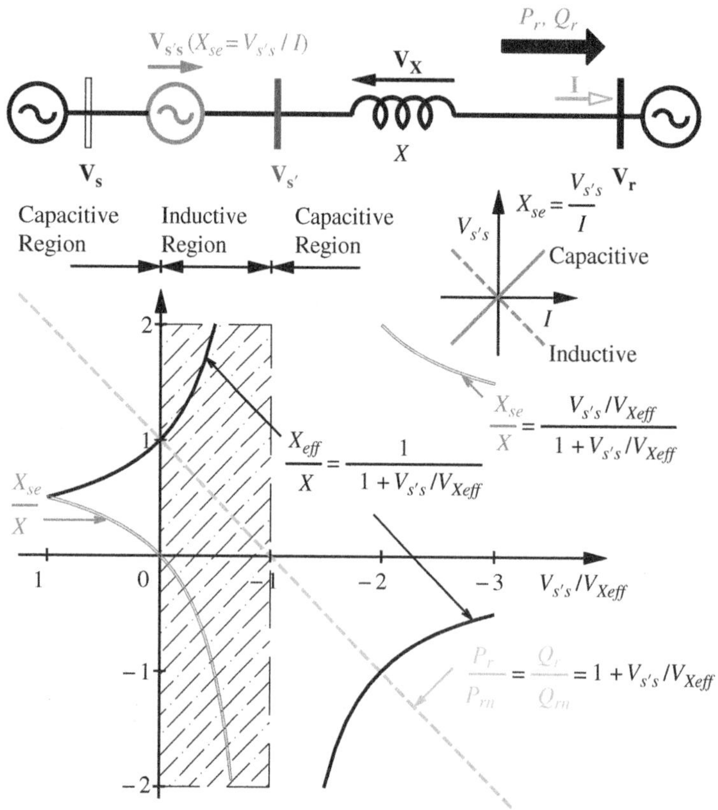

Figure 2-37 Effect of compensating voltage on power flow, effective line reactance, and compensating reactance.

reactance, the effective line reactance monotonically decreases as the voltage compensation ($V_{s's}/V_{Xeff}$) increases during $V_{s's}/V_{Xeff} > 0$. The points to be made here are

- The RR is capable of controlling the power flow in a line when operating with voltage control method.
- When the compensating voltage is in phase with the voltage across the effective line reactance, this power flow can be reversed.

However, while changing direction, the power flow (P_r) goes through zero, resulting in an absolute stability in the power flow (P_r) of the line. During the transition period when the power flow (P_r) reverses, the effective line reactance (X_{eff}) goes through two simultaneous extremes: one is a very high inductive value when the power flow (P_r) in the line is nearly zero positive and the next one is a very high capacitive value when the power flow (P_r) is nearly zero negative. The fact is that in the region where $0 > V_{s's}/V_{Xeff} > -1$, the RR acts like a reactance compensator operating with an inductive reactance control method. In the rest of the region, the RR acts like a reactance compensator operating with a capacitive reactance control method.

Equations (2-235) and (2-236) can be combined as

$$Q_r = \frac{Q_{rn}}{P_{rn}} P_r, \quad \text{which is Equation (2-215).}$$

Using Equations (2-235) and (2-236) for active and reactive power flows (P_r and Q_r), the apparent power flow (S_r) and the prevailing line current (I) are given by

$$S_r = \sqrt{P_r^2 + Q_r^2} \tag{2-76}$$

$$I = \frac{S_r}{V_r}. \tag{2-77}$$

The active power flow ($P_{s'}$) at the modified sending end is given by

$$P_{s'} = P_r \tag{2-216a}$$

since the line is assumed to be lossless, or

$$P_{s'} = \left(1 + \frac{V_{s's}}{V_{Xeff}}\right) P_{rn} \tag{2-238}$$

using Equation (2-235).

The reactive power flow ($Q_{s'}$) at the modified sending end is given by

$$Q_{s'} = Q_r + Q_{line}, \tag{2-78a}$$

which can be rewritten as

$$Q_{s'} = \left(1 + \frac{V_{s's}}{V_{Xeff}}\right) Q_{rn} + I^2 X \tag{2-239}$$

using Equations (2-236) and (2-11b), or

$$Q_{s'} = \frac{P_{s'}}{P_{rn}} Q_{rn} + I^2 X \tag{2-217b}$$

with a substitution from Equation (2-238).

From Equation (2-237), it can be determined that for $P_r/P_{rn} = -1$, $V_{s's}/V_{Xeff} = -2$ and for $P_r/P_{rn} = 2$, $V_{s's}/V_{Xeff} = 1$. Therefore, varying $V_{s's}/V_{Xeff}$ in the range of -2 to 1 and using Equations (2-235) and (2-215), the variation of active power (P_r) as a function of reactive power (Q_r) in the case of a RR is derived and shown in Figure 2-38. When $P_r = 0$, $Q_r = 0$; when $P_r = P_{rn}$, $Q_r = Q_{rn}$. The variation of active power ($P_{s'}$) as a function of reactive power ($Q_{s'}$) in the case of a RR is derived in Equation (2-217b) and shown in Figure 2-38. The figure also shows the variation of the magnitude of the modified sending-end voltage as given in Equation (2-229). When $P_r = 0$, $V_{s'} = V_r$; when $P_r = P_{rn}$, $V_{s'} = V_s$.

2.2.2.6 Representation of a Series-Compensating Voltage as a Series-Compensating Impedance

The series-compensating voltage ($\mathbf{V_{s's}}$) exchanges with the line both active and reactive powers (P_{se} and Q_{se}), which are defined as

$$P_{se} = \mathbf{V_{dq}} \bullet \mathbf{I} = -\mathbf{V_{s's}} \bullet \mathbf{I} \qquad \text{or} \qquad P_{se} = V_d I \tag{2-167}$$

and

$$Q_{se} = \left|\mathbf{V_{dq}} \times \mathbf{I}\right| = \left|-\mathbf{V_{s's}} \times \mathbf{I}\right| \qquad \text{or} \qquad Q_{se} = V_q I \tag{2-168}$$

where V_d and V_q are the respective active or direct and reactive or quadrature components with respect to the prevailing line current ($\mathbf{I}$).

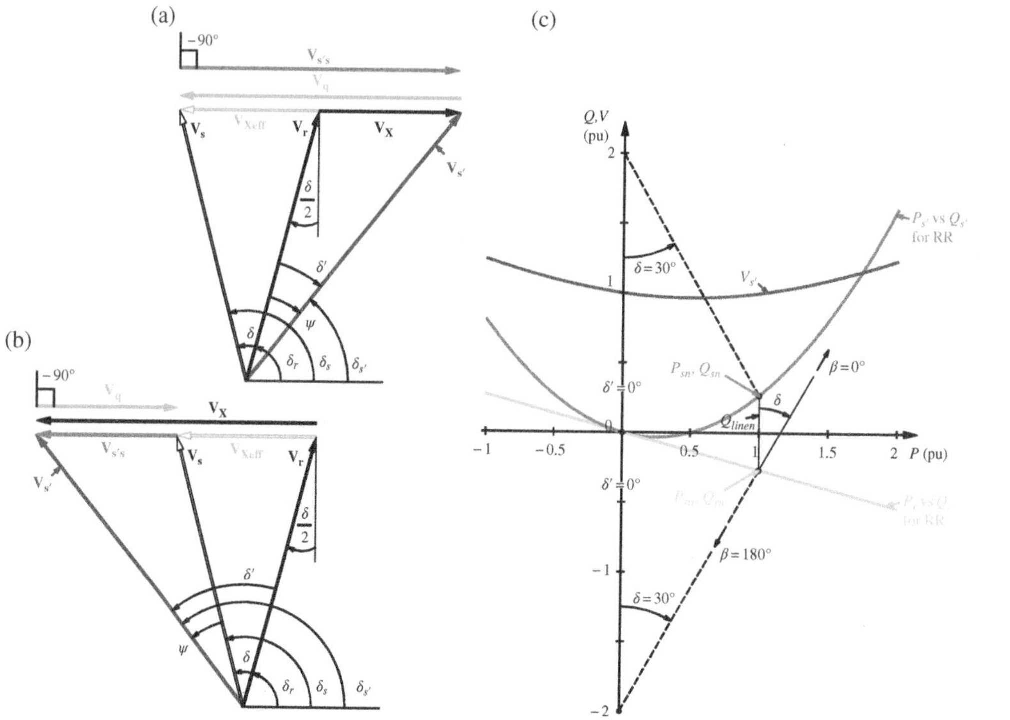

Figure 2-38 Power flow characteristics at the modified sending end and receiving end for reactance regulation and $V_{s'}$ at the modified sending end of the line at power angle $\delta = \delta_s - \delta_r = 30°$ when $V_s = V_r = 1$ pu, $X = 0.5$ pu, and $R = 0$ ($X/R = \infty$).

The exchanged active and reactive powers (P_{se} and Q_{se}) can also be defined as

$$P_{se} = P_s - P_{s'} \tag{2-169a}$$

$$Q_{se} = Q_s - Q_{s'} \tag{2-179a}$$

where the active and reactive power flows at the sending end are P_s and Q_s and the active and reactive power flows at the modified sending end are $P_{s'}$ and $Q_{s'}$, respectively.

For exchanged active and reactive powers (P_{se} and Q_{se}), the emulated series-compensating resistance (R_{se}) and reactance (X_{se}) can be determined using the current magnitude from Equation (2-57) and the following equations:

$$R_{se} = \frac{P_{se}}{I^2} \tag{2-240}$$

and

$$X_{se} = \frac{-Q_{se}}{I^2}. \tag{2-241}$$

The emulated series-compensating resistance (R_{se}), reactance (X_{se}), and impedance ($\mathbf{Z_{se}}$) are defined as

$$\mathbf{Z_{se}} = Z_{se}\angle\theta_{Zse} = R_{se} - jX_{se} \tag{2-242a}$$

or

$$Z_{se}\angle\theta_{Zse} = \sqrt{R_{se}^2 + X_{se}^2}\angle\left(\tan^{-1}\frac{-X_{se}}{R_{se}}\right) \tag{2-242b}$$

where

$$Z_{se}\angle\theta_{Zse} = Z_{se}\cos(\theta_{Zse}) + jZ_{se}\sin(\theta_{Zse}). \tag{2-243}$$

Equating the real and imaginary parts of the Equations (2-242a) and (2-243), it can be written that

$$R_{se} = Z_{se}\cos(\theta_{Zse}) \tag{2-244a}$$

and

$$X_{se} = -Z_{se}\sin(\theta_{Zse}). \tag{2-244b}$$

Note that $X_{se} > 0$ means the series-compensating reactance is capacitive and $X_{se} < 0$ means the series-compensating reactance is inductive.

Example 2-36

Consider the electrical system data as given in Table 2-1. For a given $V_{s's} = 0.2$ pu and $\beta = 105°$, calculate R_{se} and X_{se}.

Answer 2-36

In Example 2-12, it was calculated that the magnitude of line current, $I = 1.396086$ pu.
In Example 2-31, it was calculated that the exchanged active and reactive powers, $P_{se} = 0.103528$ pu and $Q_{se} = -0.259315$ pu.

From Equation (2-240), series-compensating resistance (R_{se}) can be calculated as

$$R_{se} = \frac{P_{se}}{I^2} = \frac{0.103528}{1.396086^2} = 0.053117 \text{ pu.}$$

From Equation (2-241), series-compensating reactance (X_{se}) can be calculated as

$$X_{se} = \frac{-Q_{se}}{I^2} = \frac{0.259315}{1.396086^2} = 0.133046 \text{ pu.}$$

Additional Discussion

In Example 2-28, P_s and Q_s are calculated as 1.386370 pu and 0.164422 pu, respectively.
In Example 2-30, $P_{s'}$ and $Q_{s'}$ are calculated as 1.282843 pu and 0.423737 pu, respectively.

Therefore, using Equations (2-169a) and (2-179a), the exchanged active and reactive powers (P_{se} and Q_{se}) can be calculated as $P_{se} = P_s - P_{s'} = 1.386370 - 1.282843 = 0.103528$ pu and $Q_{se} = Q_s - Q_{s'} = 0.164422 - 0.423737 = -0.259315$ pu, which were confirmed in Example 2-31.

For a series-compensating voltage, $\mathbf{V_{s's}} = V_{s's}\angle(\delta_s + \beta) = -\mathbf{V_{dq}}$, the equation defining the series-compensating impedance can be written as

$$Z_{se}\angle\theta_{Zse} = \frac{\mathbf{V_{dq}}}{I\angle(\theta_I)} \quad \text{or using Equation (1-4)} \quad Z_{se}\angle\theta_{Zse} = \frac{-\mathbf{V_{s's}}}{I\angle(\theta_I)} \quad \text{or}$$

$$Z_{se}\angle\theta_{Zse} = \frac{\mathbf{V_{s's}}\angle 180°}{I\angle(\theta_I)} \quad \text{or} \quad Z_{se}\angle\theta_{Zse} = \frac{V_{s's}\angle(\delta_s + \beta + 180°)}{I\angle(\delta_s + \theta_{Is})}$$

or

$$Z_{se}\angle\theta_{Zse} = \frac{V_{s's}}{I}\angle(\beta + 180° - \theta_{Is}), \tag{2-245a}$$

which can be further expanded using Equation (2-57) to obtain

$$Z_{se}\angle\theta_{Zse} = X\frac{V_{s's}}{V_X}\angle(\beta + 180° - \theta_{Is}) \tag{2-245b}$$

where θ_{Is} is the phase angle of the line current with respect to the sending-end voltage.

Equation (2-245b) can be expanded using Equation (2-115) as

$$Z_{se}\angle\theta_{Zse} = X\frac{V_{s's}}{V_X}\angle(\beta + 180° - \theta_{Is}) = X\frac{V_{s's}}{V_X}\angle\left(\beta + 180° - \tan^{-1}\frac{V_r\cos\delta - V_s - V_{s's}\cos\beta}{V_r\sin\delta + V_{s's}\sin\beta}\right)$$

or

$$Z_{se}\angle\theta_{Zse} = X\frac{V_{s's}}{V_X}\angle\left(\tan^{-1}\frac{-V_s\cos\beta + V_r\cos(\delta+\beta) - V_{s's}}{V_s\sin\beta - V_r\sin(\delta+\beta)}\right). \tag{2-246}$$

Digress:

$$F = \beta + 180° - \tan^{-1}\frac{V_r\cos\delta - V_s - V_{s's}\cos\beta}{V_r\sin\delta + V_{s's}\sin\beta} \quad \text{or}$$

$$\tan^{-1}\frac{V_r\cos\delta - V_s - V_{s's}\cos\beta}{V_r\sin\delta + V_{s's}\sin\beta} = \beta + 180° - F$$

or

$$\frac{V_r\cos\delta - V_s - V_{s's}\cos\beta}{V_r\sin\delta + V_{s's}\sin\beta} = \tan(\beta + 180° - F) \quad \text{or}$$

$$\frac{V_r\cos\delta - V_s - V_{s's}\cos\beta}{V_r\sin\delta + V_{s's}\sin\beta} = \tan(\beta - F)$$

or

$$\frac{V_r\cos\delta - V_s - V_{s's}\cos\beta}{V_r\sin\delta + V_{s's}\sin\beta} = \frac{\sin(\beta - F)}{\cos(\beta - F)} \quad \text{or}$$

$$\frac{V_r\cos\delta - V_s - V_{s's}\cos\beta}{V_r\sin\delta + V_{s's}\sin\beta} = \frac{\sin\beta\cos F - \cos\beta\sin F}{\cos\beta\cos F + \sin\beta\sin F} \quad \text{or}$$

$$(V_r\cos\delta - V_s - V_{s's}\cos\beta)(\cos\beta\cos F + \sin\beta\sin F) = (V_r\sin\delta + V_{s's}\sin\beta)(\sin\beta\cos F - \cos\beta\sin F)$$

or

$$(V_r\cos\delta - V_s - V_{s's}\cos\beta)\cos\beta\cos F + (V_r\cos\delta - V_s - V_{s's}\cos\beta)\sin\beta\sin F$$
$$= (V_r\sin\delta + V_{s's}\sin\beta)\sin\beta\cos F - (V_r\sin\delta + V_{s's}\sin\beta)\cos\beta\sin F$$

or

$$\{(V_r\cos\delta - V_s - V_{s's}\cos\beta)\sin\beta + (V_r\sin\delta + V_{s's}\sin\beta)\cos\beta\}\sin F$$
$$= \{(V_r\sin\delta + V_{s's}\sin\beta)\sin\beta - (V_r\cos\delta - V_s - V_{s's}\cos\beta)\cos\beta\}\cos F$$

or

$$\frac{\sin F}{\cos F} = \frac{(V_r \sin\delta + V_{s's} \sin\beta)\sin\beta - (V_r \cos\delta - V_s - V_{s's}\cos\beta)\cos\beta}{(V_r \cos\delta - V_s - V_{s's}\cos\beta)\sin\beta + (V_r \sin\delta + V_{s's}\sin\beta)\cos\beta}$$

or

$$\tan F = \frac{V_s \cos\beta - V_r \cos(\delta+\beta) + V_{s's}(\sin^2\beta + \cos^2\beta)}{-V_s \sin\beta + V_r \sin(\delta+\beta) + V_{s's}(\sin\beta\cos\beta - \sin\beta\cos\beta)} \quad \text{or}$$

$$\tan F = \frac{-V_s \cos\beta + V_r \cos(\delta+\beta) - V_{s's}}{V_s \sin\beta - V_r \sin(\delta+\beta)}$$

or

$$F = \tan^{-1}\frac{-V_s \cos\beta + V_r \cos(\delta+\beta) - V_{s's}}{V_s \sin\beta - V_r \sin(\delta+\beta)}.$$

From Equations (2-244a) and (2-244b), it can be written that

$$R_{se} = X\frac{V_{s's}}{V_X}\left\{\frac{V_s \sin\beta - V_r \sin(\delta+\beta)}{V_X}\right\} \tag{2-247a}$$

and

$$X_{se} = X\frac{V_{s's}}{V_X}\left\{\frac{V_s \cos\beta - V_r \cos(\delta+\beta) + V_{s's}}{V_X}\right\} \tag{2-247b}$$

where

$$V_X = \sqrt{\{V_s \sin\beta - V_r \sin(\delta+\beta)\}^2 + \{V_{s's} + V_s \cos\beta - V_r \cos(\delta+\beta)\}^2}$$

or

$$V_X = \sqrt{V_s^2(\cos^2\beta + \sin^2\beta) - 2V_sV_r\cos\delta + V_r^2\{\cos^2(\delta+\beta) + \sin^2(\delta+\beta)\} + V_{s's}^2 + 2V_{s's}V_s\cos\beta - 2V_{s's}V_r\cos(\delta+\beta)}$$

or

$$V_X = \sqrt{V_s^2 - 2V_sV_r\cos\delta + V_r^2 + V_{s's}^2 + 2V_{s's}V_s\cos\beta - 2V_{s's}V_r\cos(\delta+\beta)} \tag{2-109}$$

using the identities $(\cos^2\beta + \sin^2\beta = 1)$ and $\{\cos^2(\delta+\beta) + \sin^2(\delta+\beta) = 1\}$.

Using Equation (2-30), Equation (2-109) can be rewritten as

$$V_X = \sqrt{V_{Xn}^2 + V_{s's}[V_{s's} + 2\{V_s \cos\beta - V_r \cos(\delta+\beta)\}]}. \tag{2-110}$$

Note that $X_{se} > 0$ means the series-compensating reactance is capacitive and $X_{se} < 0$ means the series-compensating reactance is inductive.

Example 2-37

Consider the electrical system data as given in Table 2-1. For an IR operating at $V_{s's} = 0.2$ pu and $\beta = 105°$, calculate R_{se} and X_{se}.

Answer 2-37

In Example 2-12, it was calculated that the magnitude of the voltage (V_X) across the line $V_X = 0.698043$ pu.

Therefore, from Equation (2-247a), the series-compensating resistance (R_{se}) can be calculated as

$$R_{se} = X\frac{V_{s's}}{V_X}\left\{\frac{V_s \sin\beta - V_r \sin(\delta+\beta)}{V_X}\right\} = 0.5 \times \frac{0.2}{0.698043}\left\{\frac{1 \times \sin 105^\circ - 1 \times \sin(30^\circ + 105^\circ)}{0.698043}\right\}$$
$$= 0.053117 \text{ pu.}$$

From Equation (2-247b), the series-compensating reactance (X_{se}) can be calculated as

$$X_{se} = X\frac{V_{s's}}{V_X}\left\{\frac{V_s \cos\beta - V_r \cos(\delta+\beta) + V_{s's}}{V_X}\right\} = 0.5 \times \frac{0.2}{0.698043}$$
$$\left\{\frac{1 \times \cos 105^\circ - 1 \times \cos(30^\circ + 105^\circ) + 0.2}{0.698043}\right\} = 0.133046 \text{ pu.}$$

Additional Discussion

In Example 2-36, R_{se} and X_{se} were calculated as 0.053117 pu and 0.133046 pu, respectively.

For the electrical system data as given in Table 2-1 and an IR operating at $\beta = 60°$, the series-connected compensating voltage is varied from $V_{s's} = -1$ pu to 0.5 pu. Using Equations (2-247a) and (2-247b), variations of compensating resistance (R_{se}) and compensating reactance (X_{se}) are shown in Figure 2-39 as functions of the active power (P_r) at the receiving end of the line. Figure 2-40 shows the variations in compensating impedance magnitude (Z_{se}) and its phase angle (θ_{Zse}). Note that the active power flow increases from its nominal value of 1.0 pu when $X_{se} > 0$ and $V_{s's} > 0$; the active power flow decreases from its nominal value of 1.0 pu when $X_{se} < 0$ and $V_{s's} < 0$; the active power flow reverses when $X_{se} > 0$ and $V_{s's} < 0$. More on this point was presented

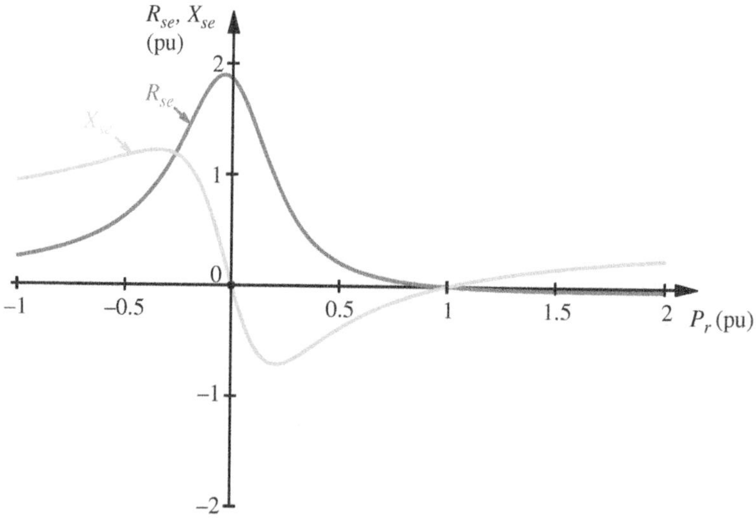

Figure 2-39 Variations of series-compensating resistance and reactance of an IR as functions of the active power at the receiving end of the line.

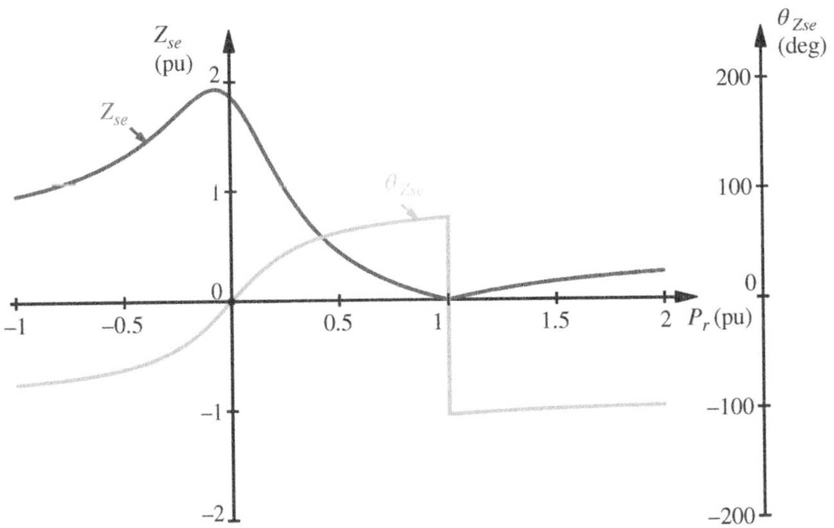

Figure 2-40 Variations of series-compensating impedance magnitude and its phase angle of an IR as functions of the active power at the receiving end of the line.

in "UPFC-unified power flow controller: theory, modeling, and applications," by K. K. Sen and E. J. Stacey at Power Engineering Society Winter Meeting, Tampa, USA, 1998 and published in *IEEE Trans. on Power Delivery*, vol. 13, no. 4, pp. 1453-1460, Oct. 1998.

Note that the phase angle of the compensating impedance changes from 75° to −105° (inductive to capacitive) instantly at the nominal power flow ($P_r = 1$ pu) at the receiving end, which reinforces the fact that an inductive compensation reduces power flow and a capacitive compensation increases power flow in an inductive line.

Selected operating points of the IR is given in Table 2-18. Note that $V_{s's} = -1$ pu and $\beta = 60°$ is actually an operating point of $V_{s's} = 1$ pu and $\beta = 240°$.

Table 2-18 Selected operating points of an IR and the corresponding series-compensating resistance, reactance, and impedance for the electrical system data as given in Table 2-1 and the IR operating at $\beta = 60°$.

$V_{s's}$ (pu)	$P_s = P_r$ (pu)	R_{se} (pu)	X_{se} (pu)	Z_{se} (pu)	θ_{Zse} (degrees)
−1.000000	−1.000000	0.250000	0.933013	0.965926	−75.000000
−0.750000	−0.500000	0.624499	1.165332	1.322119	−61.813215
−0.500000	0.000000	1.866025	0.000000	1.866025	0.000000
−0.250000	0.500000	0.208166	−0.388444	0.440706	61.813215
0.000000	1.000000	0.000000	0.000000	0.000000	0.000000
0.250000	1.500000	−0.028851	0.161513	0.164070	−100.128079
0.500000	2.000000	−0.032903	0.245592	0.247786	−97.630740

2.2.2.6.1 Equivalent Impedance of a Voltage Regulator (VR)

For a VR, the relative phase angle (β) is either 0° or 180°. Therefore, Equation (2-103) becomes

$$V_{s'} = V_s + V_{s's} \text{ for } \beta = 0°$$

or

$$V_{s'} = V_s - V_{s's} \text{ for } \beta = 180°.$$

In general,

$$V_{s'} = V_s + V_{s's} \tag{2-248}$$

where the regulation parameter is the series-compensating voltage ($\mathbf{V_{s's}}$), the magnitude of which $V_{s's} > 0$ represents the relative phase angle (β) equals 0° and $V_{s's} < 0$ represents $\beta = 180°$.

The phase-shift angle is given by Equation (2-98) as

$$\psi = 0° \quad \text{for} \quad \beta = 0° \text{ or } 180°.$$

For $\beta = 0°$, the series-compensating resistance (R_{se}) can be defined from Equation (2-247a) as

$$R_{se} = -X\frac{V_{s's}}{V_X}\frac{V_r \sin\delta}{V_X}$$

and the series-compensating reactance (X_{se}) can be defined from Equation (2-247b) as

$$X_{se} = X\frac{V_{s's}}{V_X}\frac{V_s - V_r\cos\delta + V_{s's}}{V_X}.$$

For $\beta = 180°$, the series-compensating resistance (R_{se}) can be defined from Equation (2-247a) as

$$R_{se} = X\frac{V_{s's}}{V_X}\frac{V_r \sin\delta}{V_X}.$$

and the series-compensating reactance (X_{se}) can be defined from Equation (2-247b) as

$$X_{se} = X\frac{V_{s's}}{V_X}\frac{-V_s + V_r\cos\delta + V_{s's}}{V_X}.$$

In general,

$$R_{se} = -X\frac{V_{s's}}{V_X}\frac{V_r \sin\delta}{V_X} \tag{2-249a}$$

and

$$X_{se} = X\frac{V_{s's}}{V_X}\frac{V_s - V_r\cos\delta + V_{s's}}{V_X} \tag{2-249b}$$

where $V_{s's} > 0$ represents $\beta = 0°$ and $V_{s's} < 0$ represents $\beta = 180°$.

Using Equation (2-242b), the series-compensating impedance ($Z_{se}\angle\theta_{Zse}$) can be written from Equations (2-249a) and (2-249b) as

$$Z_{se}\angle\theta_{Zse} = X\frac{V_{s's}}{V_X}\angle\left(\tan^{-1}\frac{V_s - V_r\cos\delta + V_{s's}}{V_r \sin\delta}\right) \tag{2-250}$$

where $V_{s's} > 0$ represents $\beta = 0°$ and $V_{s's} < 0$ represents $\beta = 180°$.

Example 2-38

Consider the electrical system data as given in Table 2-1. For a VR operating at $V_{s's} = 0.2$ pu and $\beta = 0°$, calculate $V_{s'}$, δ', R_{se}, and X_{se}.

Answer 2-38

From Equation (2-97), $V_{s'} = \sqrt{V_s^2 + V_{s's}^2 + 2V_sV_{s's}\cos\beta} = \sqrt{1^2 + 0.2^2 + 2\times 1\times 0.2\times \cos 0°} =$ 1.2 pu.

From Equation (2-98), $\psi = \tan^{-1}\dfrac{V_{s's}\sin\beta}{V_s + V_{s's}\cos\beta} = \tan^{-1}\dfrac{0.2\times \sin 0°}{1 + 0.2\times \cos 0°} = 0°$.

From Equation (2-51a), $\delta' = \delta + \psi = \delta + 0° = \delta = 30°$.

From Equation (2-54b), $V_X = \sqrt{V_{s'}^2 - 2V_{s'}V_r\cos\delta' + V_r^2} = \sqrt{1.2^2 - 2\times 1.2\times 1\times \cos 30° + 1^2}$ $= 0.601281$ pu.

From Equation (2-249a), the series-compensating resistance (R_{se}) can be calculated as

$$R_{se} = -X\frac{V_{s's}}{V_X}\frac{V_r\sin\delta}{V_X} = -0.5\times\frac{0.2}{0.601281}\frac{1\times \sin 30°}{0.601281} = -0.138298\text{ pu.}$$

From Equation (2-249b), the series-compensating reactance (X_{se}) can be calculated as

$$X_{se} = X\frac{V_{s's}}{V_X}\frac{V_s - V_r\cos\delta + V_{s's}}{V_X} = 0.5\times\frac{0.2}{0.601281}\frac{1 - 1\times \cos 30° + 0.2}{0.601281} = 0.092376\text{ pu.}$$

For the electrical system data as given in Table 2-1 and a VR operating at $\beta = 0°$, the series-compensating voltage is varied from $V_{s's} = -0.1$ pu to 0.1 pu. Using Equations (2-249a) and (2-249b), variations of compensating resistance (R_{se}) and compensating reactance (X_{se}) are shown in Figure 2-41 as functions of the active power (P_r) at the receiving end of the line. Figure 2-42 shows the variations in compensating impedance magnitude (Z_{se}) and its phase angle (θ_{Zse}).

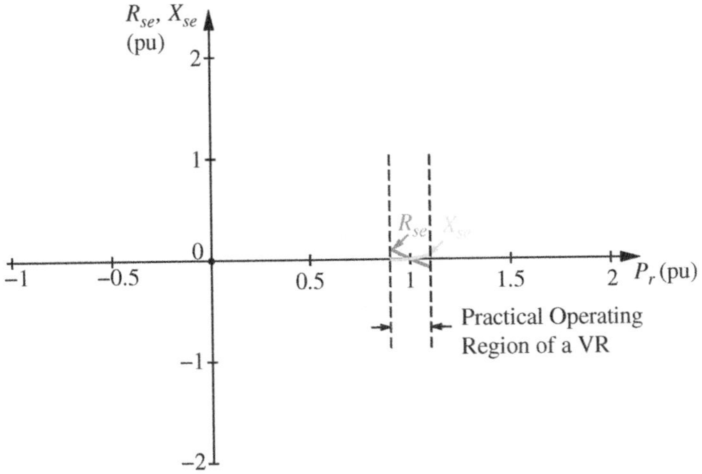

Figure 2-41 Variations of series-compensating resistance and reactance of a VR as functions of the active power at the receiving end of the line.

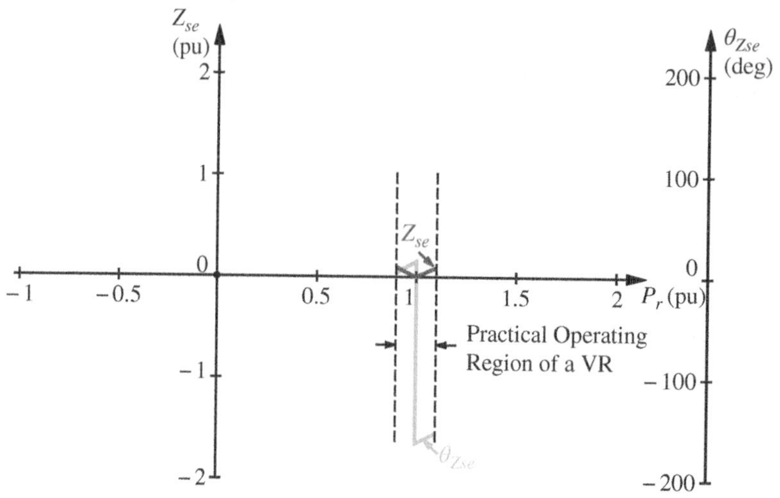

Figure 2-42 Variations of series-compensating impedance magnitude and its phase angle of a VR as functions of the active power at the receiving end of the line.

Table 2-19 Selected operating points of a VR and the corresponding series-compensating resistance, reactance, and impedance for the electrical system data as given in Table 2-1 and the VR operating at $\beta = 0°$.

$V_{s's}$ (pu)	$P_s = P_r$ (pu)	R_{se} (pu)	X_{se} (pu)	Z_{se} (pu)	θ_{Zse} (degrees)
−0.100000	0.900000	0.099540	−0.006764	0.099770	3.887227
−0.075000	0.925000	0.073971	−0.008725	0.074484	6.726911
−0.050000	0.950000	0.048628	−0.008167	0.049309	9.533805
−0.025000	0.975000	0.023866	−0.005202	0.024427	12.295292
0.000000	1.000000	0.000000	0.000000	0.000000	0.000000
0.025000	1.025000	−0.022705	0.007219	0.023825	−162.361980
0.050000	1.050000	−0.044038	0.016204	0.046924	−159.798949
0.075000	1.075000	−0.063847	0.026685	0.069199	−157.317551
0.100000	1.100000	−0.082036	0.038389	0.090574	−154.922794

Note that the phase angle of the compensating impedance changes from 15° to −165° (inductive to capacitive) instantly at the nominal power flow ($P_r = 1$ pu) at the receiving end, which reinforces the fact that an inductive compensation reduces power flow and a capacitive compensation increases power flow in an inductive line.

Selected operating points of the VR is given in Table 2-19. Note that $V_{s's} = -0.1$ pu and $\beta = 0°$ is actually an operating point of $V_{s's} = 0.1$ pu and $\beta = 180°$.

2.2.2.6.2 Equivalent Impedance of a Phase Angle Regulator (Asymmetric)

For a PAR (asym), the relative phase angle (β) is either −90° or 90°. The resulting effect is that the variable compensating voltage modifies mainly the phase angle of the line voltage in a PAR (asym) with some increase in the magnitude of the voltage as shown in the phasor diagrams Figures 2-20c and 2-20d, respectively.

For $\beta = 90°$, the series-compensating resistance (R_{se}) can be defined from Equation (2-247a) as

$$R_{se} = X\frac{V_{s's}}{V_X}\frac{V_s - V_r\cos\delta}{V_X}$$

and the series-compensating reactance (X_{se}) can be defined from Equation (2-247b) as

$$X_{se} = X\frac{V_{s's}}{V_X}\frac{V_r\sin\delta + V_{s's}}{V_X}.$$

For $\beta = -90°$, the series-compensating resistance (R_{se}) can be defined from Equation (2-247a) as

$$R_{se} = X\frac{V_{s's}}{V_X}\frac{-V_s + V_r\cos\delta}{V_X}$$

and the series-compensating reactance (X_{se}) can be defined from Equation (2-247b) as

$$X_{se} = X\frac{V_{s's}}{V_X}\frac{-V_r\sin\delta + V_{s's}}{V_X}.$$

In general,

$$R_{se} = X\frac{V_{s's}}{V_X}\frac{V_S - V_r\cos\delta}{V_X} \tag{2-251a}$$

and

$$X_{se} = X\frac{V_{s's}}{V_X}\frac{V_r\sin\delta + V_{s's}}{V_X} \tag{2-251b}$$

where $V_{s's} > 0$ represents $\beta = 90°$ and $V_{s's} < 0$ represents $\beta = -90°$.

Using Equation (2-242b), the series-compensating impedance ($Z_{se}\angle\theta_{Zse}$) can be written from Equations (2-251a) and (2-251b) as

$$Z_{se}\angle\theta_{Zse} = X\frac{V_{s's}}{V_X}\angle\left(\tan^{-1}\frac{-V_r\sin\delta - V_{s's}}{V_s - V_r\cos\delta}\right) \tag{2-252}$$

where $V_{s's} > 0$ represents $\beta = 90°$ and $V_{s's} < 0$ represents $\beta = -90°$.

Example 2-39

Consider the electrical system data as given in Table 2-1. For a PAR (asym) operating at $V_{s's} = 0.2$ pu and $\beta = 90°$, calculate $V_{s'}$, δ', R_{se}, and X_{se}.

Answer 2-39

From Equation (2-97), $V_{s'} = \sqrt{V_s^2 + V_{s's}^2 + 2V_sV_{s's}\cos\beta} = \sqrt{1^2 + 0.2^2 + 2\times 1\times 0.2\times\cos 90°} =$ 1.019804 pu.

From Equation (2-98), $\psi = \tan^{-1}\dfrac{V_{s's}\sin\beta}{V_s + V_{s's}\cos\beta} = \tan^{-1}\dfrac{0.2\times\sin 90°}{1 + 0.2\times\cos 90°} = 11.319933°$.

From Equation (2-51a), $\delta' = \delta + \psi = 30° + 11.319933° = 41.319933°$.

From Equation (2-54b),

$$V_X = \sqrt{V_{s'}^2 - 2V_{s'}V_r\cos\delta' + V_r^2} = \sqrt{1.019804^2 - 2\times 1.019804\times 1\times\cos 41.319933° + 1^2}$$
$$= 0.712706 \text{ pu.}$$

From Equation (2-251a), the series-compensating resistance (R_{se}) can be calculated as

$$R_{se} = X\frac{V_{s's}}{V_X}\frac{V_S - V_r\cos\delta}{V_X} = 0.5 \times \frac{0.2}{0.712706}\frac{1 - 1\times\cos 30^\circ}{0.712706} = 0.026376 \text{ pu}.$$

From Equation (2-251b), the series-compensating reactance (X_{se}) can be calculated as

$$X_{se} = X\frac{V_{s's}}{V_X}\frac{V_r\sin\delta + V_{s's}}{V_X} = 0.5 \times \frac{0.2}{0.712706}\frac{1\times\sin 30^\circ + 0.2}{0.712706} = 0.137809 \text{ pu}.$$

For the electrical system data as given in Table 2-1 and a PAR (asym) operating at $\beta = 90^\circ$, the series-compensating voltage is varied from $V_{s's} = -1.154701$ pu to 0.577350 pu. Using Equations (2-251a) and (2-251b), variations of compensating resistance (R_{se}) and compensating reactance (X_{se}) are shown in Figure 2-43 as functions of the active power (P_r) at the receiving end of the line. Figure 2-44 shows the variations in compensating impedance magnitude (Z_{se}) and its phase angle (θ_{Zse}).

Note that the phase angle of the compensating impedance changes from 105° to −75° (inductive to capacitive) instantly at the nominal power flow ($P_r = 1$ pu) at the receiving end, which reinforces the fact that an inductive compensation reduces power flow and a capacitive compensation increases power flow in an inductive line. Also, the phase angle of the compensating impedance changes from −180° to 180° (capacitive to inductive) instantly at $P_r = 0.133975$ pu at the receiving end.

Selected operating points of the PAR (asym) is given in Table 2-20. Note that $V_{s's} = -1.154701$ pu and $\beta = 90^\circ$ is actually an operating point of $V_{s's} = 1.154701$ pu and $\beta = 270^\circ$.

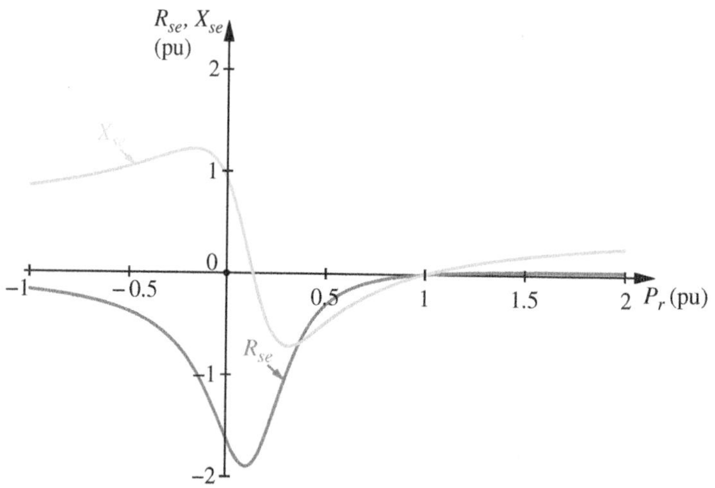

Figure 2-43 Variations of series-compensating resistance and reactance of a PAR (asym) as functions of the active power at the receiving end of the line.

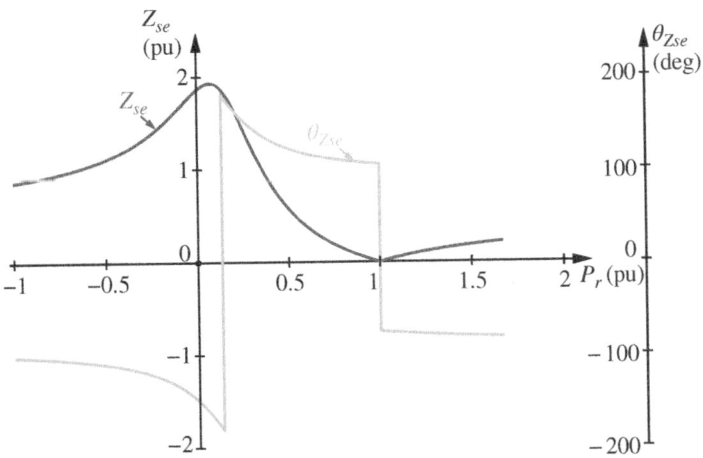

Figure 2-44 Variations of series-compensating impedance magnitude and its phase angle of a PAR (asym) as functions of the active power at the receiving end of the line.

Table 2-20 Selected operating points of a PAR (asym) and the corresponding series-compensating resistance, reactance, and impedance for the electrical system data as given in Table 2-1 and the PAR (asym) operating at $\beta = 90°$.

$V_{s's}$ (pu)	$P_s = P_r$ (pu)	R_{se} (pu)	X_{se} (pu)	Z_{se} (pu)	θ_{Zse} (degrees)
−1.154701	−1.000000	−0.173205	0.846410	0.863950	−101.565051
−0.866025	−0.500000	−0.381854	1.043244	1.110933	−110.103909
−0.577350	0.000000	−1.616025	0.933013	1.866025	−150.000000
−0.288675	0.500000	−0.308870	−0.487197	0.576855	122.373659
0.000000	1.000000	0.000000	0.000000	0.000000	0.000000
0.288675	1.500000	0.030217	0.177880	0.180428	−80.359028
0.577350	2.000000	0.032814	0.263869	0.265901	−82.911337

2.2.2.6.3 Equivalent Impedance of a Phase Angle Regulator (Symmetric)

For a PAR (sym), the regulation parameter is the phase-shift angle (ψ). For $\psi > 0, \beta = \frac{\psi}{2} + 90°$, the series-compensating resistance (R_{se}) can be defined from Equation (2-247a) as

$$R_{se} = X\frac{V_{s's}}{V_X}\left\{\frac{V_s\cos\left(\frac{\psi}{2}\right) - V_r\cos\left(\delta + \frac{\psi}{2}\right)}{V_X}\right\}$$

and the series-compensating reactance (X_{se}) can be defined from Equation (2-247b) as

$$X_{se} = X\frac{V_{s's}}{V_X}\left\{\frac{-V_s\sin\left(\frac{\psi}{2}\right) + V_r\sin\left(\delta + \frac{\psi}{2}\right) + V_{s's}}{V_X}\right\}.$$

For $\psi < 0, \beta = \frac{\psi}{2} + 270°$, the series-compensating resistance (R_{se}) can be defined from Equation (2-247a) as

$$R_{se} = X\frac{V_{s's}}{V_X}\left\{\frac{-V_s\cos\left(\frac{\psi}{2}\right) + V_r\cos\left(\delta + \frac{\psi}{2}\right)}{V_X}\right\}$$

and the series-compensating reactance (X_{se}) can be defined from Equation (2-247b) as

$$X_{se} = X\frac{V_{s's}}{V_X}\left\{\frac{V_s\sin\left(\frac{\psi}{2}\right) - V_r\sin\left(\delta + \frac{\psi}{2}\right) + V_{s's}}{V_X}\right\}.$$

In general,

$$R_{se} = X\frac{V_{s's}}{V_X}\left\{\frac{V_s\cos\left(\frac{\psi}{2}\right) - V_r\cos\left(\delta + \frac{\psi}{2}\right)}{V_X}\right\} \quad (2\text{-}253a)$$

and

$$X_{se} = X\frac{V_{s's}}{V_X}\left\{\frac{-V_s\sin\left(\frac{\psi}{2}\right) + V_r\sin\left(\delta + \frac{\psi}{2}\right) + V_{s's}}{V_X}\right\} \quad (2\text{-}253b)$$

where $V_{s's} > 0$ represents $\beta = \frac{\psi}{2} + 90°$ and $V_{s's} < 0$ represents $\beta = \frac{\psi}{2} + 270°$.

Therefore, the series-compensating impedance ($Z_{se}\angle\theta_{Zse}$) can be written from Equations (2-253a) and (2-253b) as

$$Z_{se}\angle\theta_{Zse} = X\frac{V_{s's}}{V_X}\angle\left(\tan^{-1}\frac{V_s\sin\left(\frac{\psi}{2}\right) - V_r\sin\left(\delta + \frac{\psi}{2}\right) - V_{s's}}{V_s\cos\left(\frac{\psi}{2}\right) - V_r\cos\left(\delta + \frac{\psi}{2}\right)}\right) \quad (2\text{-}254)$$

where $V_{s's} > 0$ represents $\beta = \frac{\psi}{2} + 90°$ and $V_{s's} < 0$ represents $\beta = \frac{\psi}{2} + 270°$.

Example 2-40

Consider the electrical system data as given in Table 2-1. For a PAR (sym) operating at $\psi = 11.478341°$, calculate $V_{s'}$, δ', R_{se}, and X_{se}.

Answer 2-40

For a PAR (sym), $V_{s'} = V_s$.

From Equation (2-198), $V_{s's} = 2V_s\sin\left(\frac{\psi}{2}\right) = 2\times 1\times\sin\left(\frac{11.478341°}{2}\right) = 0.2$ pu.

From Equation (2-200), $\beta = \frac{\psi}{2} + 90°$ for $\psi > 0$; $\beta = 95.739170°$.

From Equation (2-51a), $\delta' = \delta + \psi = 30° + 11.478341° = 41.478341°$.

From Equation (2-54b),

$$V_X = \sqrt{V_{s'}^2 - 2V_{s'}V_r\cos\delta' + V_r^2} = \sqrt{1^2 - 2\times 1\times 1\times\cos 41.478341° + 1^2} = 0.708229 \text{ pu.}$$

From Equation (2-253a), the series-compensating resistance (R_{se}) can be calculated as

$$R_{se} = X\frac{V_{s's}}{V_X}\left\{\frac{V_s\cos\left(\frac{\psi}{2}\right) - V_r\cos\left(\delta + \frac{\psi}{2}\right)}{V_X}\right\} = 0.5\times\frac{0.2}{0.708229}$$

$$\left\{\frac{\cos\left(\frac{11.478341°}{2}\right) - \cos\left(30° + \frac{11.478341°}{2}\right)}{0.708229}\right\} = 0.036545 \text{ pu.}$$

From Equation (2-253b), the series-compensating reactance (X_{se}) can be calculated as

$$X_{se} = X\frac{V_{s's}}{V_X}\left\{\frac{-V_s\sin\left(\frac{\psi}{2}\right) + V_r\sin\left(\delta + \frac{\psi}{2}\right) + V_{s's}}{V_X}\right\}$$

$$= 0.5 \times \frac{0.2}{V_X}\left\{\frac{-\sin\left(\frac{11.478341^\circ}{2}\right) + \sin\left(30^\circ + \frac{11.478341^\circ}{2}\right) + 0.2}{0.708229}\right\} = 0.136386 \text{ pu.}$$

For the electrical system data as given in Table 2-1, operating with a PAR (sym), the phase angle is varied from $\psi = -60^\circ$ to 60°. Using Equations (2-253a) and (2-253b), variations of compensating resistance (R_{se}) and compensating reactance (X_{se}) are shown in Figure 2-45 as functions of the active power (P_r) at the receiving end of the line. Figure 2-46 shows the variations in compensating impedance magnitude (Z_{se}) and its phase angle (θ_{Zse}).

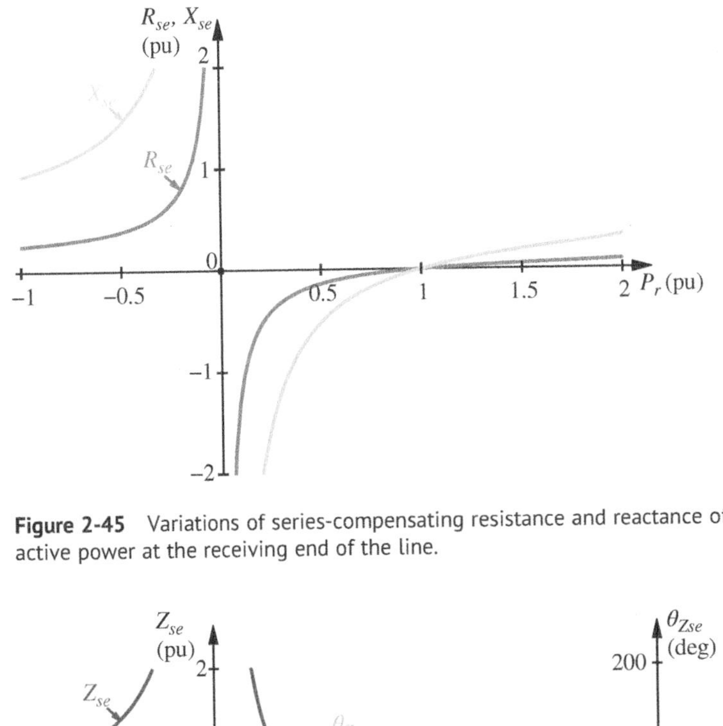

Figure 2-45 Variations of series-compensating resistance and reactance of a PAR (sym) as functions of the active power at the receiving end of the line.

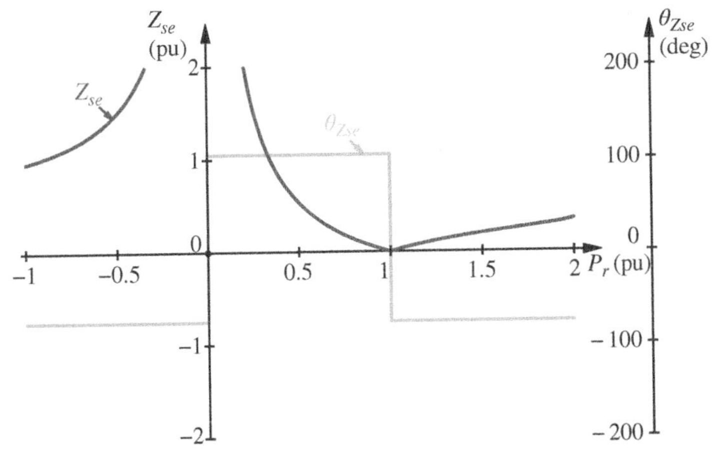

Figure 2-46 Variations of series-compensating impedance magnitude and its phase angle of a PAR (sym) as functions of the active power at the receiving end of the line.

Table 2-21 Selected operating points of a PAR (sym) and the corresponding series-compensating resistance, reactance, and impedance for the electrical system data as given in Table 2-1.

ψ	$P_s = P_r$ (pu)	R_{se} (pu)	X_{se} (pu)	Z_{se} (pu)	θ_{Zse} (degrees)
−60°	−1.000000	0.250000	0.933013	0.965926	−75°
−45°	−0.517638	0.379410	1.415976	1.465926	−75°
−30°	0.000000	∞	∞	∞	0°
−15°	0.517638	−0.129410	−0.482963	0.500000	105°
0°	1.000000	0.000000	0.000000	0.000000	0°
15°	1.414214	0.044139	0.164730	0.170541	−75°
30°	1.732051	0.066987	0.250000	0.258819	−75°
45°	1.931852	0.081350	0.303603	0.314313	−75°
60°	2.000000	0.091506	0.341506	0.353553	−75°

Note that the phase angle of the compensating impedance changes from 105° to −75° (inductive to capacitive) instantly at the nominal power flow ($P_r = 1$ pu) at the receiving end, which reinforces the fact that an inductive compensation reduces power flow and a capacitive compensation increases power flow in an inductive line. Also, the phase angle of the compensating impedance changes from −75° to 105° (capacitive to inductive) instantly at no power flow ($P_r = 0$) at the receiving end when both compensating resistance (R_{se}) and reactance (X_{se}) are ∞.

Selected operating points of the PAR (sym) is given in Table 2-21. Note that $V_{s's} = -1.154701$ pu and $\beta = 90°$ is actually an operating point of $V_{s's} = 1.154701$ pu and $\beta = 270°$.

2.2.2.6.4 Equivalent Impedance of a Reactance Regulator

The series-compensating voltage ($\mathbf{V_{s's}}$) is related to ($\mathbf{V_q}$), such that

$$\mathbf{V_{s's}} = -\mathbf{V_q} \tag{2-255}$$

where $\mathbf{V_q}$ is in quadrature with respect to the prevailing line current ($\mathbf{I}$) as shown in Figure 2-28.

The compensating voltage ($\mathbf{V_{s's}}$) exchanges with the line only the reactive power (Q_{se}), which is defined as

$$Q_{se} = |\mathbf{V_q} \times \mathbf{I}| = |-\mathbf{V_{s's}} \times \mathbf{I}| \qquad \text{or} \qquad Q_{se} = V_q I. \tag{2-256}$$

From Equations (2-247a) and (2-225), it can be written that

$$R_{se} = 0, \tag{2-257}$$

which confirms the fact that the exchanged active power (P_{se}) is zero.

From Equations (2-246) and (2-225), it can be written that

$$Z_{se}\angle\theta_{Zse} = X\frac{V_{s's}}{V_X}\angle(-90°). \tag{2-258}$$

Since the natural voltage ($\mathbf{V_{Xn}}$) across the line reactance (X) is given by $\mathbf{V_{Xn}} = \mathbf{V_s} - \mathbf{V_r}$, Equation (2-26b), it can be written by inspecting Figure 2-28d that the magnitude (V_{Xn}) of the natural voltage ($\mathbf{V_{Xn}}$) is

$$V_{Xn} = V_s \cos\beta + V_r \cos(\pi - \delta - \beta) = V_s \cos\beta - V_r \cos(\delta + \beta). \tag{2-259}$$

By inspection of Figure 2-28d, it can also be written that

$$V_X = V_{Xn} + V_{s's}. \tag{2-260}$$

From Equations (2-259) and (2-260), it can be written that

$$V_X = V_s \cos\beta - V_r \cos(\delta + \beta) + V_{s's}. \tag{2-261}$$

Therefore, from Equations (2-247b) and (2-261), it can be written as

$$X_{se} = X\frac{V_{s's}}{V_X} \tag{2-262}$$

where $V_{s's} > 0$ represents a series-capacitive compensating reactance and $V_{s's} < 0$ represents a series-inductive compensating reactance. Therefore, $X_{se} > 0$ means the series-compensating reactance is capacitive and $X_{se} < 0$ means the series-compensating reactance is inductive.

Equations (2-262) and (2-57) can be combined as

$$X_{se} = \frac{V_{s's}}{I} \tag{2-263a}$$

or

$$V_{s's} = IX_{se}. \tag{2-263b}$$

Note that only the absolute value is considered to determine the magnitude of the series-compensating voltage.

Example 2-41

Consider the magnitudes of the sending- and receiving-end voltages, $V_s = V_r = 1$ pu, the power angle, $\delta = \delta_s - \delta_r = 60°$, and the line reactance, $X = 0.5$. For a RR operating at $V_{s's} = 0.5$ pu (capacitive) and -0.5 pu (inductive), respectively, calculate $V_{s'}$, δ', and X_{se}.

Answer 2-41

From Equation (2-30), $V_{Xn} = \sqrt{V_s^2 - 2V_sV_r\cos\delta + V_r^2} = \sqrt{1^2 - 2\times 1\times 1\times \cos 60° + 1^2} = 1$ pu.

For a capacitive compensation of $V_{s's} = 0.5$ pu,

From Equation (2-229),

$$V_{s'} = \sqrt{V_s^2 + V_{s's}^2 + 2V_sV_{s's}\frac{V_s - V_r\cos\delta}{V_{Xn}}} = \sqrt{1^2 + 0.5^2 + 2\times 1\times 0.5\frac{1 - 1\times \cos 60°}{1}}$$

$$= 1.322876 \text{ pu.}$$

From Equation (2-230), $\psi = \tan^{-1}\dfrac{V_{s's}V_r\sin\delta}{V_sV_{Xn} + V_{s's}(V_s - V_r\cos\delta)} = \tan^{-1}\dfrac{0.5\times 1\times \sin 60°}{1\times 1 + 0.5(1 - 1\times \cos 60°)}$

$= 19.106605°$.

From Equation (2-51a), $\delta' = \delta + \psi = 60° + 19.106605° = 79.106605°$.

From Equation (2-260), $V_X = V_{Xn} + V_{s's} = 1 + 0.5 = 1.5$ pu.

From Equation (2-262), $X_{se} = X\dfrac{V_{s's}}{V_X} = 0.5 \times \dfrac{0.5}{1.5} = 0.166667$ pu.

For an inductive compensation of $V_{s's} = -0.5$ pu,

From Equation (2-229),

$$V_{s'} = \sqrt{V_s^2 + V_{s's}^2 + 2V_sV_{s's}\frac{V_s - V_r\cos\delta}{V_{Xn}}} = \sqrt{1^2 + 0.5^2 - 2\times 1\times 0.5\frac{1 - 1\times \cos 60°}{1}}$$

$$= \frac{\sqrt{3}}{2} = 0.866025 \text{ pu.}$$

From Equation (2-230),

$$\psi = \tan^{-1}\frac{V_{s's}V_r\sin\delta}{V_sV_{Xn}+V_{s's}(V_s-V_r\cos\delta)} = \tan^{-1}\frac{-0.5\times1\times\sin60^\circ}{1\times1-0.5(1-1\times\cos60^\circ)}$$
$$= \tan^{-1}\frac{-1}{\sqrt{3}} = -30^\circ.$$

From Equation (2-51a), $\delta' = \delta + \psi = 60^\circ - 30^\circ = 30^\circ$.
From Equation (2-260), $V_X = V_{Xn} + V_{s's} = 1 - 0.5 = 0.5$ pu.
From Equation (2-262), $X_{se} = X\frac{V_{s's}}{V_X} = 0.5 \times \frac{(-0.5)}{0.5} = -0.5$ pu.

Check:

From Equation (2-41) $a = \frac{V_sV_r}{X} = \frac{1\times1}{0.5} = 2.$
From Equation (2-27), $\delta = \delta_s - \delta_r = 60^\circ$.
From Equation (2-46), $P_{rn} = a\times\sin\delta = 2\times\sin60^\circ = \sqrt{3}$ pu.
From Equation (2-48), $Q_{rn} = a\left(\cos\delta - \frac{V_r}{V_s}\right) = 2\left(\cos60^\circ - \frac{1}{1}\right) = -1.$
From Equation (2-210a), $X_{eff} = X - X_{se} = 0.5 - (-0.5) = 1.$
From Equation (2-212), $P_r = \frac{X}{X_{eff}}P_{rn} = \frac{0.5}{1}\times\sqrt{3} = \frac{\sqrt{3}}{2}.$
From Equation (2-213), $Q_r = \frac{X}{X_{eff}}Q_{rn} = \frac{0.5}{1}\times(-1) = -\frac{1}{2}.$
From Equation (2-76), $S_r = \sqrt{P_r^2 + Q_r^2} = \sqrt{\frac{3}{4}+\frac{1}{4}} = 1.$
From Equation (2-77), $I = \frac{S_r}{V_r} = \frac{1}{1} = 1.$
From Equation (2-221), $V_{s'} = \sqrt{V_r^2 + (XI)^2 + 2XQ_r} = \sqrt{1^2 + (0.5\times1)^2 + 2\times0.5\times(-0.5)}$
$= \frac{\sqrt{3}}{2}.$

From Equation (2-222), $\delta' = \tan^{-1}\frac{XP_r}{V_r^2 + XQ_r} = \tan^{-1}\frac{0.5\times\frac{\sqrt{3}}{2}}{1+0.5\times(-0.5)} = \tan^{-1}\frac{1}{\sqrt{3}} = 30^\circ.$
From Equation (2-51b), $\psi = \delta' - \delta = 30^\circ - 60^\circ = -30^\circ$, which corroborates with the results above.

For the electrical system data as given in Table 2-1, operating with a RR, the series-compensating voltage is varied from $V_{s's} = -1$ pu to 1 pu. Using Equations (2-257) and (2-262), variations of compensating resistance (R_{se}) and compensating reactance (X_{se}) are shown in Figure 2-47 as functions of the active power (P_r) at the receiving end of the line. Figure 2-48 shows the variations in compensating impedance magnitude (Z_{se}) and its phase angle (θ_{Zse}).

Note that the phase angle of the compensating impedance changes from 90° to −90° (inductive to capacitive) instantly at the nominal power flow ($P_r = 1$ pu) at the receiving end, which reinforces the fact that an inductive compensation reduces power flow and a capacitive compensation increases power flow in an inductive line. Also, the phase angle of the compensating impedance magnitude changes from −90° to 90° (capacitive to inductive) instantly at no power flow ($P_r = 0$) at the receiving end when the compensating reactance (X_{se}) as well as the compensating impedance magnitude (Z_{se}) are ∞. An RR also reinforces the fact that the compensating resistance (R_{se}) is zero.

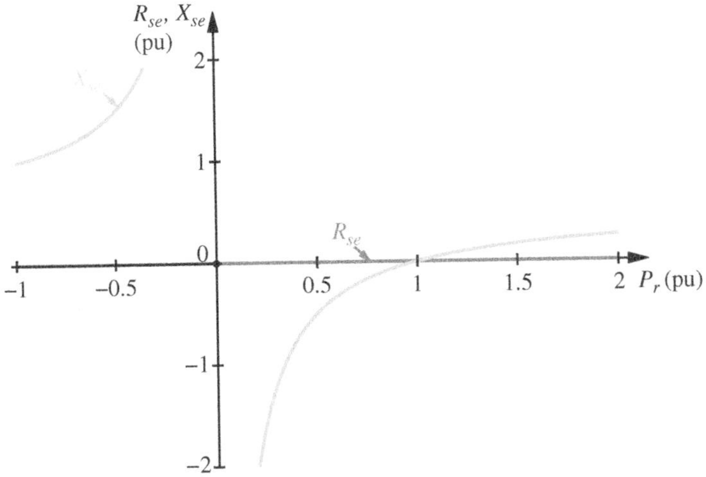

Figure 2-47 Variations of series-compensating resistance and reactance of a RR as functions of the active power at the receiving end of the line.

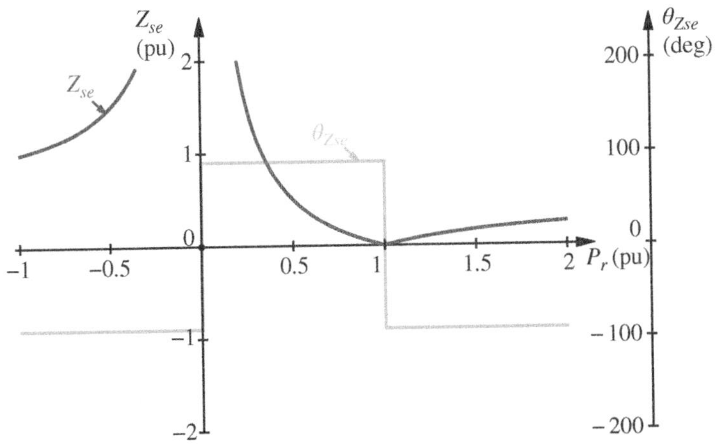

Figure 2-48 Variations of series-compensating impedance magnitude and its phase angle of a RR as functions of the active power at the receiving end of the line.

Selected operating points of the RR is given in Table 2-22.

The expressions for series-compensating resistance (R_{se}) and reactance (X_{se}) from each of an IR, a VR, a PAR (asym), a PAR (sym), and a RR are shown in Table 2-23.

The series-compensating resistance (R_{se}) and reactance (X_{se}) from each of an IR, a VR, a PAR (asym), a PAR (sym), and a RR are combined in the form of a series-compensating impedance ($Z_{se}\angle\theta_{Zse}$) and shown in Table 2-24.

Table 2-22 Selected operating points of a RR and the corresponding series-compensating resistance, reactance, and impedance for the electrical system data as given in Table 2-1.

$V_{s's}$ (pu)	$P_s = P_r$ (pu)	R_{se} (pu)	X_{se} (pu)	Z_{se} (pu)	θ_{Zse} (degrees)
−1.035276	−1.000000	0.000000	1.000000	1.000000	−90°
−0.776457	−0.500000	0.000000	1.500000	1.500000	−90°
−0.517638	0.000000	0.000000	∞	∞	±90°
−0.258819	0.500000	0.000000	−0.500000	0.500000	90°
0.000000	1.000000	0.000000	0.000000	0.000000	0°
0.258819	1.500000	0.000000	0.166667	0.166667	−90°
0.517638	2.000000	0.000000	0.250000	0.250000	−90°

Table 2-23 Series-compensating resistance (R_{se}) and reactance (X_{se}) for an IR, a VR, a PAR (asym), a PAR (sym), and a RR, respectively.

IR	$R_{se} = X\frac{V_{s's}}{V_X}\left\{\frac{V_s \sin\beta - V_r \sin(\delta+\beta)}{V_X}\right\}$	(2-247a)	$X_{se} = X\frac{V_{s's}}{V_X}\left\{\frac{V_s \cos\beta - V_r \cos(\delta+\beta) + V_{s's}}{V_X}\right\}$	(2-247b)
VR	$R_{se} = -X\frac{V_{s's}}{V_X}\frac{V_r \sin\delta}{V_X}$	(2-249a)	$X_{se} = X\frac{V_{s's}}{V_X}\frac{V_s - V_r \cos\delta + V_{s's}}{V_X}$	(2-249b)
PAR (asym)	$R_{se} = X\frac{V_{s's}}{V_X}\frac{V_s - V_r \cos\delta}{V_X}$	(2-251a)	$X_{se} = X\frac{V_{s's}}{V_X}\frac{V_r \sin\delta + V_{s's}}{V_X}$	(2-251b)
PAR (sym)	$R_{se} = X\frac{V_{s's}}{V_X}\left\{\frac{V_s \cos\left(\frac{\psi}{2}\right) - V_r \cos\left(\delta+\frac{\psi}{2}\right)}{V_X}\right\}$	(2-253a)	$X_{se} = X\frac{V_{s's}}{V_X}\left\{\frac{-V_s \sin\left(\frac{\psi}{2}\right) + V_r \sin\left(\delta+\frac{\psi}{2}\right) + V_{s's}}{V_X}\right\}$	(2-253b)
RR	$R_{se} = 0$	(2-257)	$X_{se} = X\frac{V_{s's}}{V_X}$	(2-262)

Table 2-24 Series-compensating impedance (Z_{se}) for an IR, a VR, a PAR (asym), a PAR (sym), and a RR, respectively.

IR	$Z_{se}\angle\theta_{Zse} = X\frac{V_{s's}}{V_X}\angle\left(\tan^{-1}\frac{-V_s \cos\beta + V_r \cos(\delta+\beta) - V_{s's}}{V_s \sin\beta - V_r \sin(\delta+\beta)}\right)$	(2-246)
VR	$Z_{se}\angle\theta_{Zse} = X\frac{V_{s's}}{V_X}\angle\left(\tan^{-1}\frac{V_s - V_r \cos\delta + V_{s's}}{V_r \sin\delta}\right)$	(2-250)
PAR (asym)	$Z_{se}\angle\theta_{Zse} = X\frac{V_{s's}}{V_X}\angle\left(\tan^{-1}\frac{-V_r \sin\delta - V_{s's}}{V_s - V_r \cos\delta}\right)$	(2-252)
PAR (sym)	$Z_{se}\angle\theta_{Zse} = X\frac{V_{s's}}{V_X}\angle\left(\tan^{-1}\frac{V_s \sin\left(\frac{\psi}{2}\right) - V_r \sin\left(\delta+\frac{\psi}{2}\right) - V_{s's}}{V_s \cos\left(\frac{\psi}{2}\right) - V_r \cos\left(\delta+\frac{\psi}{2}\right)}\right)$	(2-254)
RR	$Z_{se}\angle\theta_{Zse} = X\frac{V_{s's}}{V_X}\angle(-90°) = X_{se}\angle(-90°)$	(2-258, 2-262)

2.2.3 Comparison Between Series- and Shunt-Compensating Voltages

A compensating voltage for modifying the line voltage can be generated with one of three distinct methods by using (a) rotating machines, (b) power electronics converters, and (c) transformer/LTCs. A Shunt–Shunt configuration is used in some special cases for point-to-point transfer of power between two isolated networks with different voltages and/or phase angles. In an application in which power electronics-based VSCs or rotating machines are used, the frequencies of the two networks may be different and, therefore, it is referred to as an "asynchronous tie." A Shunt–Series configuration is used for power flow control. The final choice of the PFC topology is dependent on the system requirements and the life-cycle costs, involving installation, operation, and maintenance.

When a series-compensating voltage is used to modify the line voltage, the modified sending-end voltage is given by

$$\mathbf{V}_{s'} = \mathbf{V}_s + \mathbf{V}_{s's}. \tag{2-94a}$$

This equation may be rewritten in the case of an autotransformer, shown in Figure 2-49a, as

$$\mathbf{V}_{s'-se} = \mathbf{V}_s + \mathbf{V}_{comp} \tag{2-264}$$

where $\mathbf{V}_{s'\text{-}se}$, $\mathbf{V}_s$, and $\mathbf{V}_{comp}$ are modified sending-end, sending-end, and compensating voltages, respectively. When a shunt-compensating voltage is used, the modified sending-end voltage ($\mathbf{V}_{s'\text{-}sh}$) is the compensating voltage. The equation, in the case of a two-winding transformer, shown in Figure 2-49b, is given by

$$\mathbf{V}_{s'-sh} = \mathbf{V}_{comp}. \tag{2-265}$$

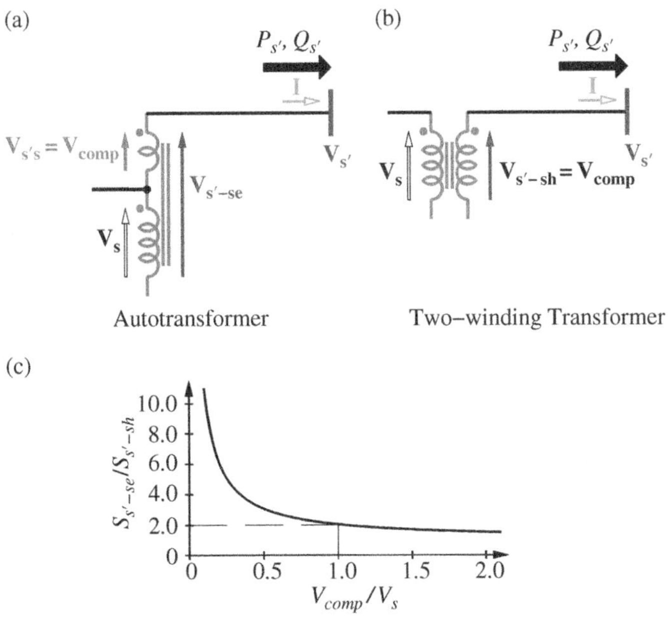

Figure 2-49 (a) Autotransformer. (b) Two-winding transformer. (c) Ratio of the transmitted apparent power with series-compensated and shunt-compensated voltages as a function of normalized compensating voltage.

Using Equations (2-264) and (2-265), the transmitted power at the modified sending end of a line is given by

$$S_{s'-se} = V_{s'-se} \times I \tag{2-266a}$$

when an autotransformer is used and

$$S_{s'-sh} = V_{s'-sh} \times I \tag{2-266b}$$

when a two-winding transformer is used where I is the line current.

Using Equations (2-266a) and (2-266b), the ratio of the transmitted apparent power ($S_{s'\text{-}se}$) by using a series-compensating voltage and ($S_{s'\text{-}sh}$) by using a shunt-compensating voltage is given by

$$\frac{S_{s'-se}}{S_{s'-sh}} = \frac{V_{s'-se} \times I}{V_{s'-sh} \times I} = \frac{V_{s'-se}}{V_{s'-sh}} = \frac{V_s + V_{comp}}{V_{comp}} = 1 + \frac{V_s}{V_{comp}} = 1 + \frac{1}{V_{comp}/V_s}. \tag{2-267}$$

The ratio ($S_{s'\text{-}se}/S_{s'\text{-}sh}$) is less than ∞ for V_{comp} greater than zero, but greater than one for V_{comp} less than ∞. Therefore, the PFC with the series-compensating voltage always offers a large leverage between its own rating and the transmitted apparent power in the line. The series-compensating voltage needs to be rated for only a fraction of the transmitted apparent power. The ratio of the transmitted apparent power with series- and shunt-compensating voltages as a function of the normalized compensating voltage is plotted in Figure 2-49c. When the compensating voltage is one pu, the series-compensating voltage transmits twice the apparent power that the shunt-compensating voltage transmits. When the series-compensating voltage is a fraction of the line voltage, the transmitted apparent power ratio becomes greater than two.

As an example, a 0.1-pu-rated series-compensating voltage in an autotransformer with one pu line current requires a Shunt–Series Compensator rating of 0.1 pu whereas, for the same outcome, a 1.1-pu-rated shunt-compensating voltage in a two-winding transformer with one pu line current requires a Shunt–Shunt Compensator rating of 1.1 pu. The rating and thus the cost advantage of the Shunt–Series Compensator makes it attractive for its use as a PFC. Note that the world's first UPFC used a Shunt–Series configuration of 160 MVA rating to regulate power flow in a 1000 MVA-rated transmission line. An alternate option of using a Shunt–Shunt configuration would require the use of two 1000 MVA-rated STATCOMs, connected back to back. A one-pu-power-rated Shunt–Series solution costs significantly less than a (1000/160) 6.25-pu-power-rated Shunt–Shunt solution. In conclusion, the series compensator in the Shunt–Series configuration is rated for the compensating (not transmitted) power in the AC line to which it is connected. However, the shunt compensator in the Shunt–Shunt configuration must be rated for the transmitted power, which is also the compensating power, in the AC line to which it is connected. Both the series and shunt compensators carry the same line current; however, the reduced voltage rating of the series compensator makes its power rating to be a fraction of the power rating of the shunt compensator.

The equations that describe the active and reactive power flows at the receiving end of a line with the use of a shunt-compensating voltage are given in Equations (2-74) and (2-75). The variation of active power as a function of reactive power is derived in Equation (2-79) and is shown in Figure 2-50. Modified power angle (δ') in the second and third quadrants ($+P_r/-Q_r$ and $-P_r/-Q_r$) are not used and, therefore, are not shown in Figure 2-50. If the fourth-quadrant ($-P_r/+Q_r$) operation of a Shunt–Shunt configuration is needed for reversal of active power flow, note that the Shunt–Series configuration with a series-compensating voltage can also be used for this purpose as shown by increasing the radius of the shaded circle in Figure 2-50. More discussion

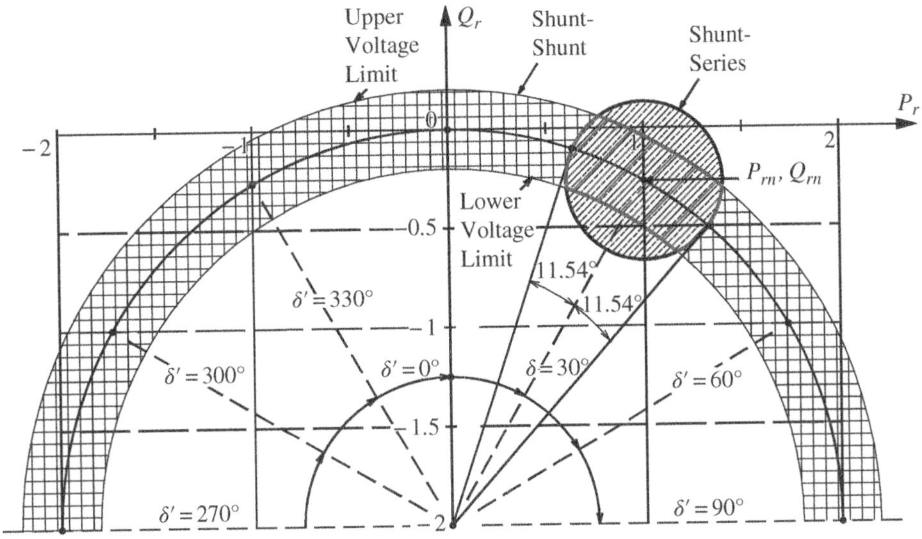

Figure 2-50 Q_r vs P_r at the receiving end of the line for the range of modified power angle ($\delta' = \delta_{s'} - \delta_r$) from 270° to 90° when $V_s = V_r = 1$ pu, $X = 0.5$ pu, and $R = 0$ ($X/R = \infty$).

on this topic is given in "Introduction to FACTS Controllers: Theory, Modeling, and Applications," by Sen and Sen, IEEE Press and John Wiley & Sons, 2009, Chapter 8, Section 8.3 (Shunt–Series Compensation Using a Unified Power Flow Controller (UPFC)).

For a Shunt–Shunt Compensator, the relationship between active and reactive power flows (P_r and Q_r) at the receiving end is shown by the half circle in Figure 2-50 as a function of the modified power angle (δ') and is given by the following equation:

$$P_r^2 + \left(Q_r + \frac{V_r^2}{X}\right)^2 = a'^2 \tag{2-79}$$

where

$$a' = \frac{V_{s'} V_r}{X}. \tag{2-61}$$

Equation (2-79) represents the locus of a circle, centered at $\left[0, -\frac{V_r^2}{X}\right]$ with a radius of a'.

For a Shunt–Series Compensator, the relationship between active and reactive power flows (P_r and Q_r) at the receiving end is shown by the small circle in Figure 2-50 and is given by the following equation:

$$(P_r - P_{rn})^2 + (Q_r - Q_{rn})^2 = a_r^2 \tag{2-136}$$

where

$$a_r = \frac{V_r V_{s's}}{X}. \tag{2-132}$$

The active and reactive power flows at the receiving end of the line can be regulated within the range defined by the *P-Q* plot in Figure 2-50 by controlling the magnitude ($V_{s's}$) of the series-compensating voltage ($\mathbf{V}_{s's}$) within its rated value and the relative phase angle of $0° \leq \beta \leq 360°$.

The Shunt–Shunt configuration is capable of controlling the power angle (the phase angle between the voltage, $V_{s'}$, at the POC, which is the modified sending end and the voltage, V_r, at the far end of the line, which is the receiving end) over its entire controllable range of $0° \le \delta' \le 360°$. The maximum transfer of active power along a lossless line (with quality factor $Q = X/R = \infty$) between the sending and receiving ends takes place at the 90° power angle. At the same time, the actual power angle is significantly lower and depends on the line length (*i.e.* line reactance, X), system characteristics, and load flows. As an example, a compensation range (ψ_{max}) of $\pm 11.54°$ with an uncompensated power angle of 30° is shown in the shaded area in Figure 2-50. In this application, the Shunt–Shunt configuration is severely restricted to operate within the first quadrant when used as a PFC. However, in some special cases for point-to-point transfer of power between two isolated networks or interconnection of two transmission lines with different voltages or phase angles (or frequencies), Shunt–Shunt compensation scheme may be the preferred option while selecting a SPFC.

2.3 Implementation of Power Flow Control Concepts

The various methods of power flow control that have been established so far are summarized in this section. In order to control the flow of power in the line, at least one parameter (voltage magnitude, its phase angle, or line reactance) needs to be regulated.

2.3.1 Voltage Regulation

The various methods of voltage regulation can be categorized as (a) direct method and (b) indirect method.

2.3.1.1 Direct Method

The direct method of voltage regulation of a line is to add a compensating voltage vectorially in or out of phase with the phase-to-neutral voltage of the line at the POC. The Voltage-Regulating Transformer (VRT), in its Shunt–Series configuration, generates a bipolar compensating voltage through transformer action and connects it in series with the line as shown in Figure 2-51a.

The VRT as shown in Figure 2-51a consists of an Exciter Unit and a Voltage-Regulating Unit. The Exciter Unit consists of a three-phase (A, B, and C), Y-connected primary windings and is connected in shunt with the line. The three-phase primary windings are excited from the three-phase line voltages ($\mathbf{V_{sA}}$, $\mathbf{V_{sB}}$, and $\mathbf{V_{sC}}$). A three-phase, bipolar compensating voltages ($\mathbf{V_{s'sA}}$, $\mathbf{V_{s'sB}}$, and $\mathbf{V_{s'sC}}$) that are either in phase or out of phase with the corresponding phase-to-neutral voltage are generated from the induced voltages in the secondary windings. The Voltage-Regulating Unit consists of a total of six secondary windings (two windings in each phase). The magnitude of the series-compensating voltage ($\mathbf{V_{s's}}$) is varied with the use of LTCs. The line is regulated at a voltage ($\mathbf{V_{s'}}$) from the uncompensated voltage ($\mathbf{V_s}$). The corresponding phasor diagram is shown in Figure 2-51b. This is an autotransformer-type VRT where the secondary winding is electrically connected with the primary winding. Note that Figure 2-51 is functionally the same as an autotransformer and is consistent with the modeling techniques used in Chapters 3 and 4. There are also other ways to implement a VRT, such as a two-winding transformer, which is a VRT in a Shunt–Shunt configuration.

Figure 2-52 shows the schematic diagram of a Thyristor-Controlled Load Tap Changer (LTC). A transformer winding is tapped at various places. Each of the tapped points is connected to

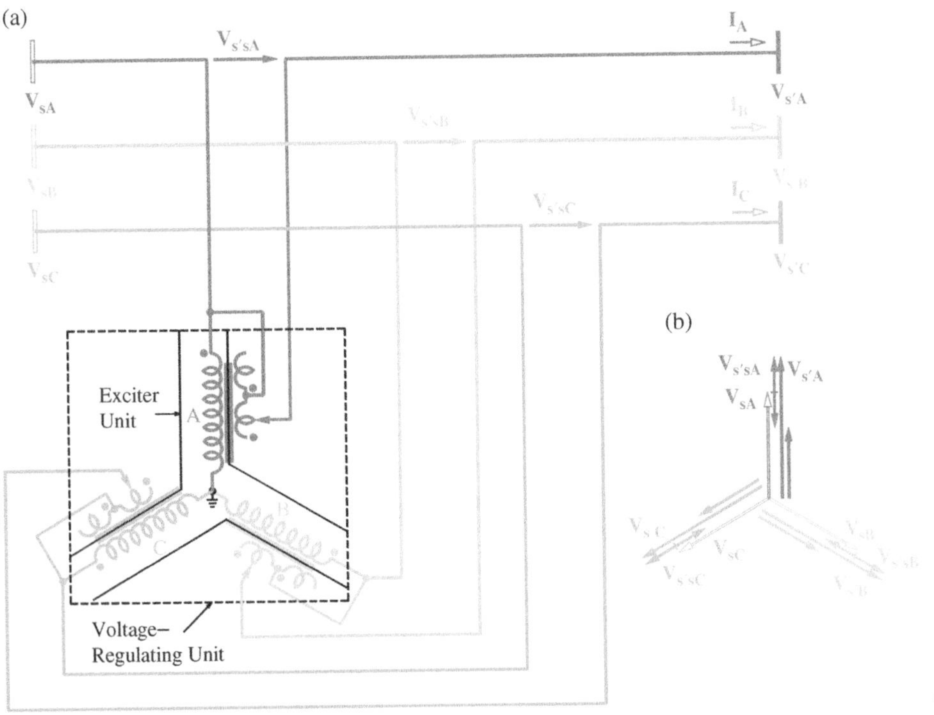

Figure 2-51 (a) Voltage-Regulating Transformer (Shunt–Series configuration); (b) phasor diagram.

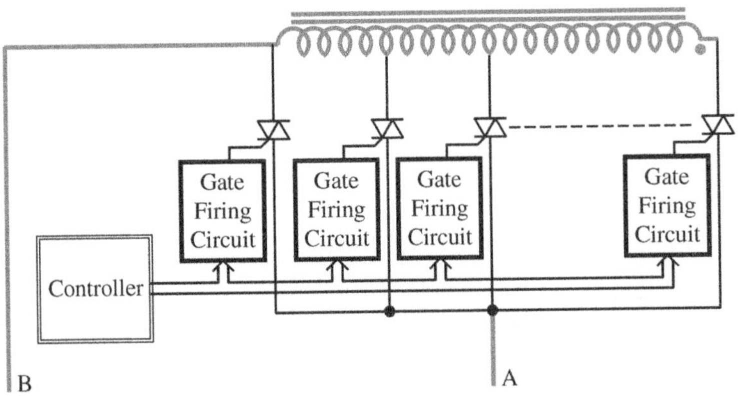

Figure 2-52 Thyristor-Controlled Load Tap Changer.

one side of a BTB thyristor (triac) switch. The other side of each triac switch is connected at point A. Depending on which thyristor is on, the voltage between points A and B can be varied between zero and the full-winding voltage with desired steps in between. In the mechanical contact version of this arrangement, the LTC connects with one of various taps to make a variable number of turns active between the connected tap and one end of the winding.

There are other ways to regulate the line voltage as described below.

2.3.1.2 Indirect Method

The indirect way to regulate the line voltage is to connect a variable capacitor or a variable reactor in shunt with the line. According to Equation (2-23), a shunt-connected capacitor raises the line voltage. According to Equation (2-22), a shunt-connected reactor lowers the line voltage. It is very common to connect a single or multiple banks of fixed shunt capacitors/reactors, using contactors (*i.e.* mechanical switches) to provide power factor correction/voltage regulation.

The shunt reactor or capacitor can also be connected through thyristor switches, which are power electronics-based naturally commutated switches that can be turned on during the positive half cycle of an AC voltage; however, these switches turn off naturally when the voltage reverses. During the evolution of this approach over the last five decades, the various electronic topologies and their control methods became known as Thyristor-Switched Capacitor (TSC), Thyristor-Switched Reactor (TSR), and Thyristor-Controlled Reactor (TCR). The TSC connects fixed capacitors in a step-like manner in shunt with the line. The TSR connects fixed reactors in a step-like manner in shunt with the line. However, it is rarely used. The TCR connects a reactor in shunt with the line through thyristor switches whose duty cycle can be varied, thereby controlling the flow of current through the reactor and making it appear like a variable reactor. A combination of TSC, TSR, and TCR is termed the SVC as shown in Figure 2-53.

A capacitive reactance ($-jX_C$) with a parallel inductive reactance ($+jX_L$) offers a shunt-compensating reactance of

$$-jX_{sh} = \frac{(+jX_L)(-jX_C)}{j(X_L - X_C)} = j\frac{X_L X_C}{(X_C - X_L)} \tag{2-268}$$

where $X_C > X_L$ makes X_{sh} inductive and $X_C < X_L$ makes X_{sh} capacitive.

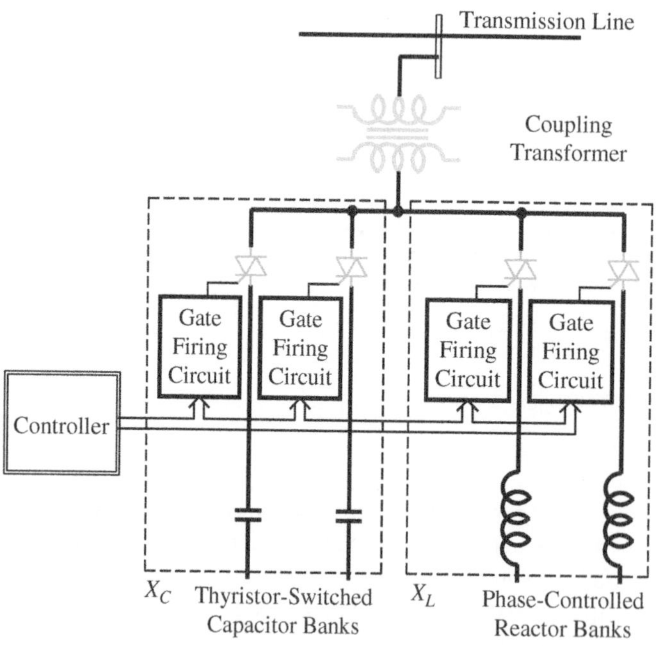

Figure 2-53 Static Var Compensator (SVC).

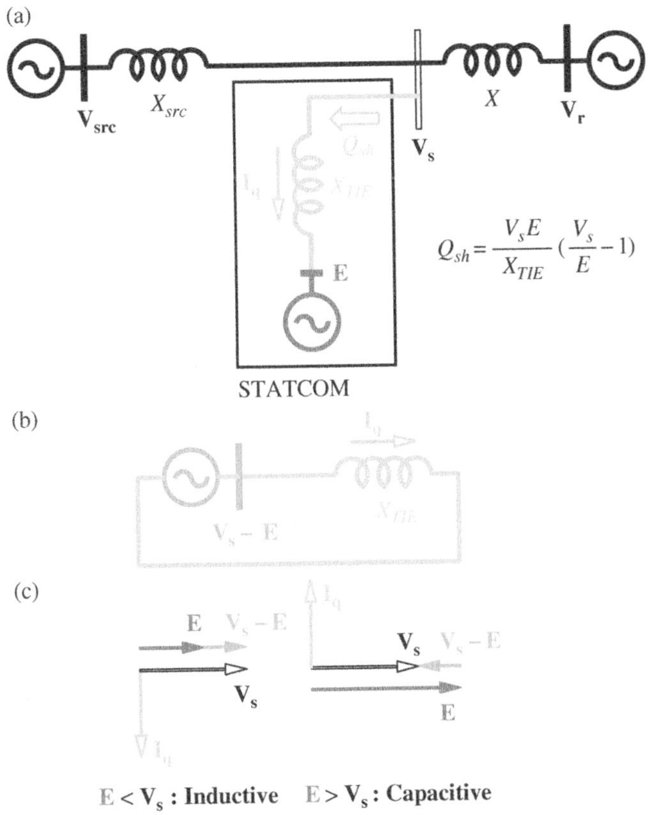

Figure 2-54 Concept of a STATCOM.

A two-generator/one-line power system network model is shown in the one-line diagram of Figure 2-54a. The network is driven by an ideal voltage source ($\mathbf{V_{src}}$) behind a Thèvenin source reactance (X_{src}). The line reactance is X and the sending- and receiving-end voltages are $\mathbf{V_s}$ and $\mathbf{V_r}$. The indirect way to implement a variable shunt capacitor or a variable shunt reactor is to use a shunt compensator, which generates a variable magnitude voltage (E) in phase with the line voltage (V_s) at the POC, and connects this voltage in shunt with the line through a tie-reactance (X_{TIE}). If the line voltage (V_s), compensator voltage (E), and tie-reactance (X_{TIE}) are thought of as the modified sending-end voltage ($V_{s'}$), receiving-end voltage (V_r), and line reactance (X), respectively, then the active power absorbed from the line at the POC is zero because the line voltage and the compensating voltage are in phase ($\delta' = 0°$). However, the compensating reactive power absorbed from the line can be expressed by rewriting Equations (2-60) and (2-61) as

$$Q_{sh} = \frac{V_s E}{X_{TIE}} \left(\frac{V_s}{E} - 1 \right). \tag{2-269}$$

The concept of connecting a variable amplitude synchronous AC voltage in shunt with the AC line through inductive impedance has been practiced since the introduction of the SynCon in which the internal voltage (E) of the synchronous machine is controlled by controlling the field current. The same concept has been implemented using power electronics VSC-based STATCOM. In contrast to SynCon, the power circuit of the VSC does not have any wear-and-tear because there is no moving parts in the main power circuit; also, the response time of a VSC is in ms compared to that in seconds in the case of a SynCon of a 100 Mvar rating. As shown in Chapter 1, the fast response of a STATCOM is beneficial in lowering voltage flicker caused by an electric arc furnace.

A STATCOM is implemented with the use of a variable magnitude, shunt compensator voltage as shown in Figure 2-54a. A net voltage ($\mathbf{V_s}$-$\mathbf{E}$) drives a current ($\mathbf{I_q}$) through the tie-reactance (X_{TIE}) as shown in Figure 2-54b. The current ($\mathbf{I_q}$) always lags the driving voltage ($\mathbf{V_s}$-$\mathbf{E}$) by 90°. Through control action, the magnitude of the shunt-compensator voltage (E) can be made higher or lower than the line voltage (V_S). When $E < V_S$ as shown in Figure 2-54c, the current ($\mathbf{I_q}$) through the STATCOM lags the terminal voltage (V_S) and the line "sees" an inductive reactance connected at its terminal. The STATCOM is considered to be operating in an inductive mode and reactive power is absorbed at the POC by the emulated reactor. Similarly, when $E > V_S$ as shown in Figure 2-54c, the current ($\mathbf{I_q}$) through the STATCOM leads the terminal voltage (V_S) and the line "sees" the STATCOM as a capacitive reactance. The STATCOM is considered to be operating in a capacitive mode and reactive power is delivered at the POC by the emulated capacitor.

Figure 2-55 shows a schematic diagram of the STATCOM. The STATCOM consists of a VSC (VSC1), a magnetic circuit (MC1), a shunt-coupling transformer (SHNTR), a shunt breaker (SHNBRK), current and voltage sensors, and a Control & Protection Unit. The operation of the STATCOM is such that the VSC produces a voltage that is almost in phase with the line voltage. Through control action, the VSC output voltage is maintained to be smaller or larger than the line voltage. Therefore, the STATCOM essentially injects an almost sinusoidal reactive current of variable magnitude at the POC. This reactive current, in turn, regulates the line voltage. Note that MC1 is typically a specialty transformer, such as an Interphase Transformer. However, the coupling transformer may

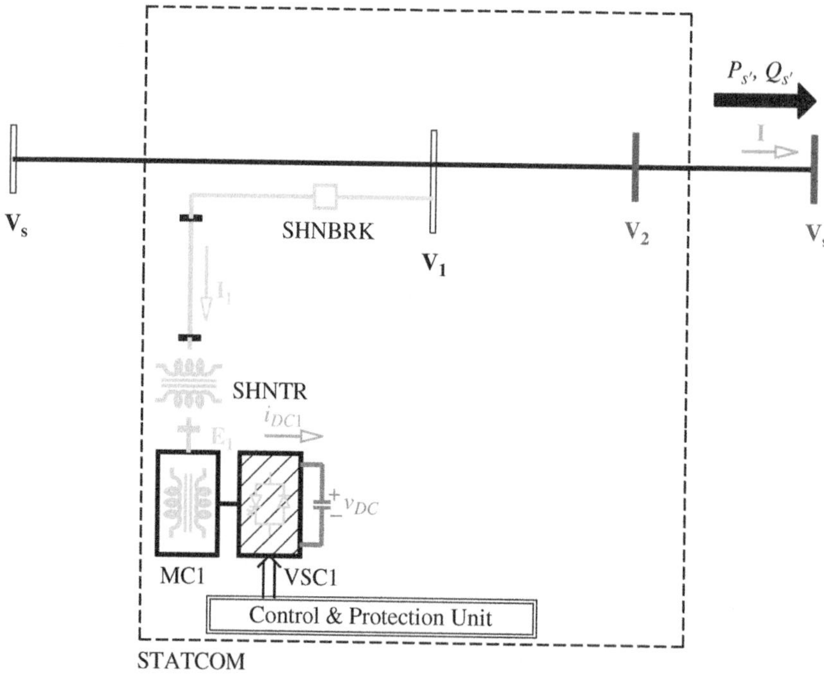

Figure 2-55 Static synchronous compensator (STATCOM).

be a conventional transformer. In the first generation of FACTS Controllers in 1990s and 2000s, in some cases, the specialty transformer and the coupling transformer were combined.

In one type of VSC, called Harmonic-Neutralized VSC (HN-VSC), a number of square wave voltages are generated at fundamental frequency by operating the controllable semiconductor switches once per cycle of the fundamental frequency. These square wave voltages are combined using a magnetic circuit in order to produce a high-quality sinusoidal voltage with acceptable harmonic content. Note that there are many circuit topologies, other than HN-VSC that can be used to create an electronic shunt compensator voltage, $\mathbf{E}_1$. It should be noted that VSC configurations and modulators are still evolving. Configurations, such as 2-level, 3-level, H-bridge, half bridge, and so on with modulators, such as HN, Pulse Width Modulation (PWM), and Space Vector (SV) have been used in high-power applications. It should also be noted that irrespective of configuration and modulation technique used in a VSC, all VSCs produce undesired harmonic voltage and current components that are injected into the power system network at the POC.

2.3.2 Phase Angle Regulation

The effective phase angle of the line voltage at the POC is varied by using a PAR, which is also known as a Phase-Shifting Transformer (PST). The PAR connects a voltage in series with the line and in quadrature with the phase-to-neutral voltage of the line. The PAR characteristics can be of asymmetric (asym) and symmetric (sym) types. Both PARs (asym and sym) may consist of a single-core or dual-core transformers. When a single-core transformer is used, the primary windings are always connected in Δ; however, when a dual-core transformer is used, the primary windings may be connected in Y or in Δ.

2.3.2.1 Single-core Phase Angle Regulator

Figures 2-56a, 2-56c, and 2-56e show one-line diagrams of a PAR (asym) for no operation, decreasing power flow, and increasing power flow in the line and the related phasor diagrams in Figures 2-56b, 2-56d, and 2-56f, respectively. In a PAR (asym), the regulation parameter is the magnitude ($V_{s's}$) of the series-compensating voltage ($\mathbf{V}_{s's}$), which is varied with LTCs. With proper polarities of the series-connected windings, the compensating voltage is either at −90° (Figure 2-56c) or 90° (Figure 2-56e) with respect to the primary phase-to-neutral voltage ($\mathbf{V}_s$).

Therefore, the magnitude and phase-shift angle of the modified sending-end voltage ($\mathbf{V}_{s'}$) are derived from Equations (2-97) and (2-98) as

$$V_{s'} = \sqrt{V_s^2 + V_{s's}^2} \tag{2-270}$$

and

$$\psi = \tan^{-1}\frac{V_{s's}}{V_s} \tag{2-271a}$$

where $V_{s's} > 0$ represents $\beta = 90°$ and $V_{s's} < 0$ represents $\beta = -90°$.

The series-compensating voltage introduces a phase-shift angle (ψ), shown in Figures 2-56d and 2-56f, whose magnitude (for a small change) in radians varies with the magnitude of the compensating voltage in pu where the phase-to-neutral voltage of the line is the base voltage. Therefore,

$$\psi \cong \tan\psi = \frac{V_{s's}}{V_s} \tag{2-271b}$$

using Equation (2-271a).

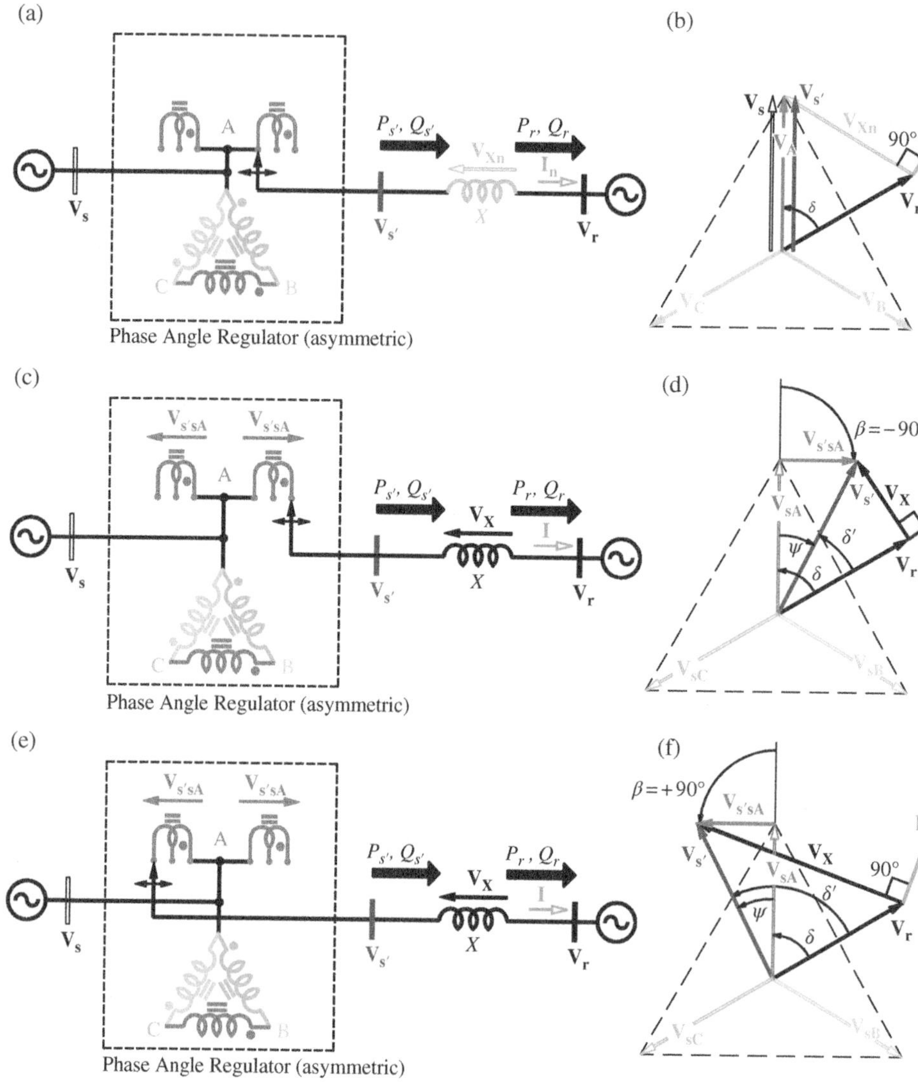

Figure 2-56 Two-generator/one-line power system network with (a) no compensation and (b) its related phasor diagram; (c) a series-compensating voltage, $\mathbf{V_{s's}}$ for decreasing power flow and (d) its related phasor diagram; (e) a series-compensating voltage, $\mathbf{V_{s's}}$ for increasing power flow and (f) its related phasor diagram using a PAR (asym).

The three-phase, series-compensating secondary windings are center-tapped at A, B, and C, respectively, when the compensating voltage is designed to be of equal magnitude for both $\beta = 90°$ and $\beta = -90°$. The three-phase line voltages ($\mathbf{V_{sA}}$, $\mathbf{V_{sB}}$, and $\mathbf{V_{sC}}$) are applied to Δ-connected primary windings. The series-compensating secondary voltage ($\mathbf{V_{s's}}$) that is in phase with the primary phase-to-phase voltage, but in quadrature with the primary phase-to-neutral voltage, is connected in series with the line. The magnitude ($V_{s's}$) of the series-compensating voltage ($\mathbf{V_{s's}}$) is varied with LTCs. The modified sending-end voltage ($\mathbf{V_{s'A}}$) is always larger than the sending-end voltage because of the addition of two orthogonal voltages: (1) sending-end voltage ($\mathbf{V_{sA}}$) and (2) series-compensating voltage ($\mathbf{V_{s'sA}}$).

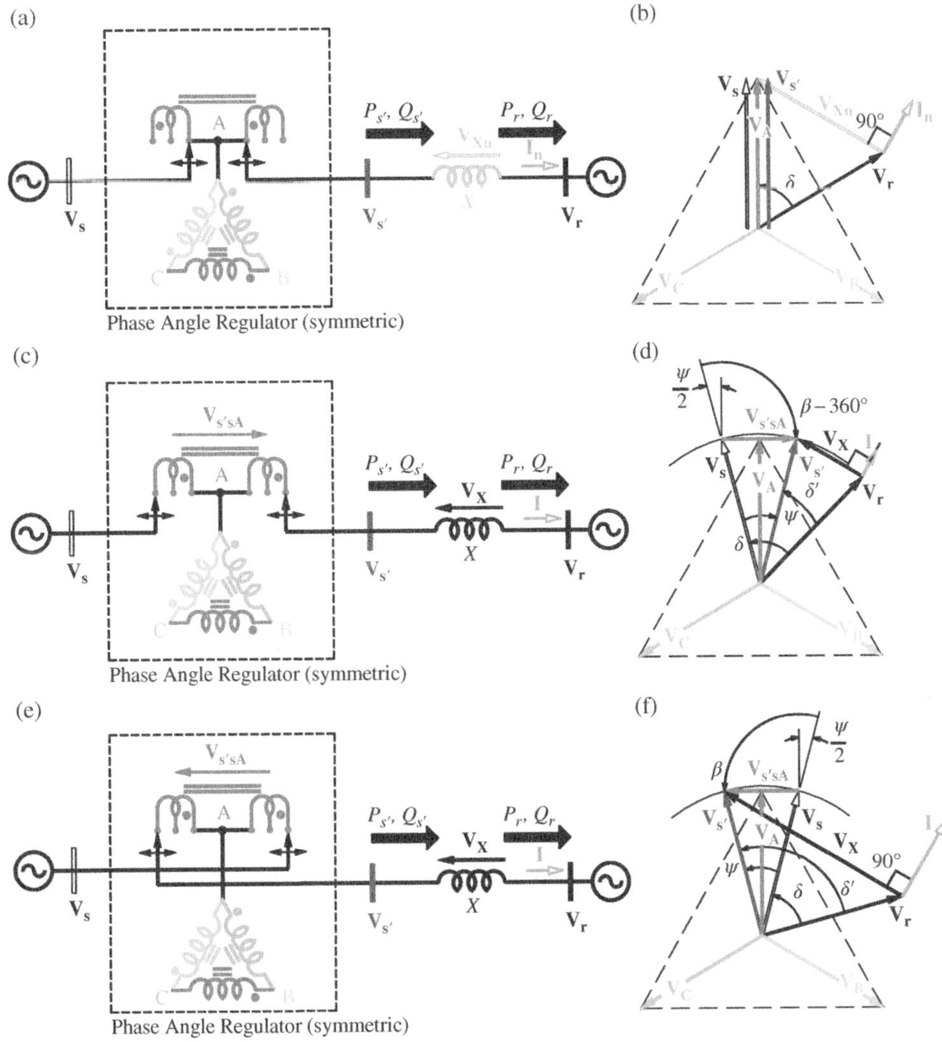

Figure 2-57 Two-generator/one-line power system network with (a) no compensation and (b) its related phasor diagram; (c) a series-compensating voltage, $\mathbf{V_{s's}}$, for decreasing power flow and (d) its related phasor diagram; (e) a series-compensating voltage, $\mathbf{V_{s's}}$, for increasing power flow and (f) its related phasor diagram using a PAR (sym).

Figures 2-57a, 2-57c, and 2-57e show one-line diagrams of a PAR (sym) for no operation, decreasing power flow, and increasing power flow in the line and the related phasor diagrams in Figures 2-57b, 2-57d, and 2-57f, respectively. In a PAR (sym), the regulation parameter is the phase-shift angle (ψ). With proper polarities of the series-connected windings, the compensating voltage is either at −90° (Figure 2-57c) or 90° (Figure 2-57e) with respect to the primary phase-to-neutral voltage ($\mathbf{V_s}$). The resulting effect is that only the phase angle of the line voltage between $\mathbf{V_s}$ and $\mathbf{V_{s'}}$ is changed, but the magnitude stays unchanged at V_s as shown in the phasor diagrams Figures 2-57d and 2-57f, respectively.

The three-phase series-compensating secondary windings are center-tapped at A, B, and C, respectively. The three-phase voltages ($\mathbf{V_A}$, $\mathbf{V_B}$, and $\mathbf{V_C}$) are applied to Δ-connected primary windings. The series-compensating secondary voltage ($\mathbf{V_{s's}}$) that is in phase with the primary phase-to-phase voltage, but in quadrature with the primary phase-to-neutral voltage, is connected in series with the line. The series-compensating voltage, $\mathbf{V_{s's}}$ (of magnitude, $V_{s's}$, and relative phase angle, β, with respect to $\mathbf{V_s}$), is varied with LTCs. As a result, the phase-shift angle (ψ) varies.

The difference between the PAR (asym) and PAR (sym) is that in the case of a PAR (asym), the line voltage is used to excite the primary windings, whereas in the case of a PAR (sym), the voltage at the midpoint of the series-compensating winding is used to excite the primary windings. In a PAR (asym), the variable compensating voltage modifies mainly the phase angle of the line voltage with some increase in the magnitude of the line voltage; in a PAR (sym), the variable compensating voltage modifies only the phase angle of the line voltage with no change in its magnitude.

2.3.2.2 Dual-core Phase Angle Regulator

In some configurations, the PAR consists of two transformers, referred to as dual-core PAR as shown in Figures 2-58 and 2-59. The first transformer (Exciter Unit) is called a regulating transformer. Its three-phase (A, B, and C), Y-connected primary windings are connected in shunt with the line. The three-phase primary windings are excited from the line voltages ($\mathbf{V_{sA}}$, $\mathbf{V_{sB}}$, and $\mathbf{V_{sC}}$) and three-phase voltages ($\mathbf{V_A}$, $\mathbf{V_B}$, and $\mathbf{V_C}$) are induced in the regulating transformer secondary windings. For series connection of these voltages, an electrical isolation is necessary. The second transformer (Series Unit) is called a series transformer, which is excited from the phase-to-phase voltages ($\mathbf{V_B} - \mathbf{V_C}$, $\mathbf{V_C} - \mathbf{V_A}$, and $\mathbf{V_A} - \mathbf{V_B}$) of the regulating transformer. Three-phase, bipolar compensating voltages ($\mathbf{V_{s'sA}}$, $\mathbf{V_{s'sB}}$, and $\mathbf{V_{s'sC}}$) that are in quadrature with the corresponding phase-to-neutral voltage are induced in the secondary windings of the series transformer. The magnitudes of the series-compensating voltages are varied using the LTCs in the secondary windings of the regulating transformer. The induced voltages of the series transformer are connected in series with the line. If the series transformer is a step-down transformer, the primary windings of the series

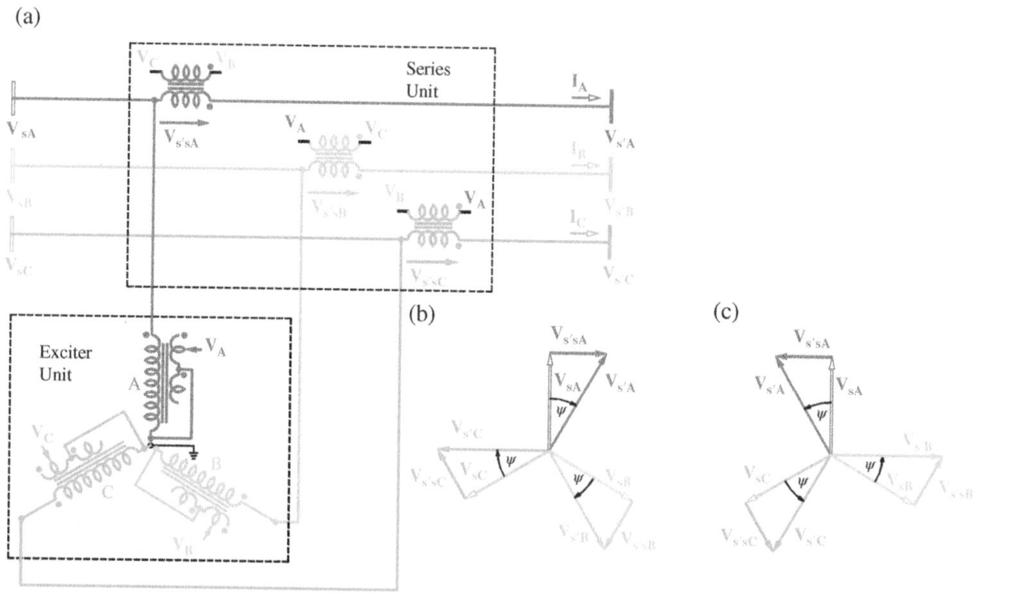

Figure 2-58 (a) Phase Angle Regulator (asym), (b) phasor diagram for decreasing power flow, and (c) phasor diagram for increasing power flow.

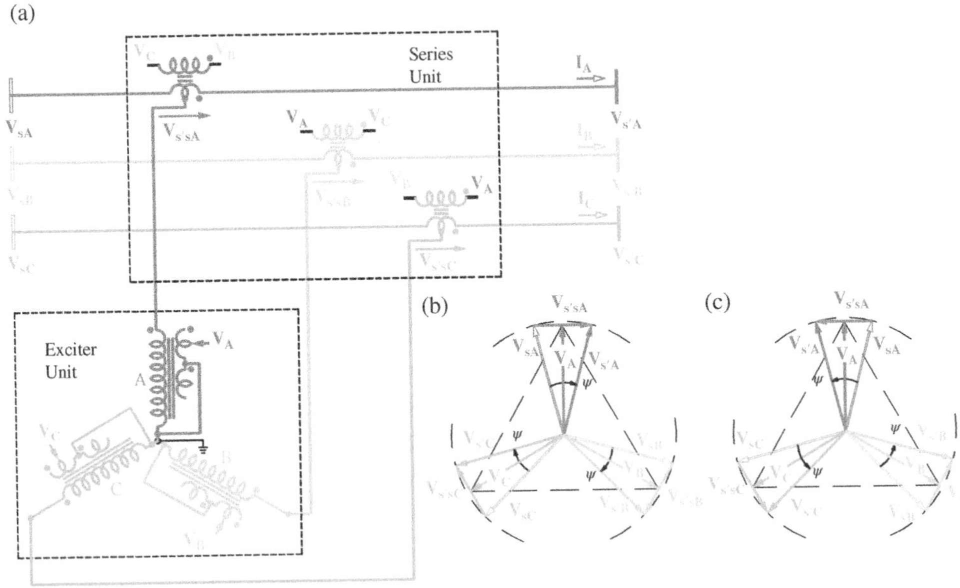

Figure 2-59 (a) Phase Angle Regulator (sym), (b) phasor diagram for decreasing power flow, and (c) phasor diagram for increasing power flow.

transformer as well as the secondary windings of the regulating transformer are high-voltage and low-current rated, so that the LTCs on the secondary side of the regulating transformer can operate at a low current and can ride through a high fault current. In contrast, when a PAR is realized with one transformer, the LTCs are always required to carry a high line current as well as even a higher fault current.

An asymmetrical, dual core design connects a voltage in series with the line and changes the phase angle of the line voltage with some increase of its magnitude. The primary windings of the regulating transformer are Y-connected as shown in Figure 2-58a. They are connected in shunt with the line and excited from the line voltages ($\mathbf{V_{sA}}$, $\mathbf{V_{sB}}$, and $\mathbf{V_{sC}}$). Three-phase voltages ($\mathbf{V_A}$, $\mathbf{V_B}$, and $\mathbf{V_C}$) are induced in the regulating transformer secondary windings. Three-phase, bipolar compensating voltages ($\mathbf{V_{s'sA}}$, $\mathbf{V_{s'sB}}$, and $\mathbf{V_{s'sC}}$) that are in quadrature with the corresponding phase-to-neutral voltage are induced in the secondary windings of the series transformer. The magnitudes of the series-compensating voltages are varied using the LTCs in the secondary windings of the regulating transformer. The induced voltages of the series transformer are connected in series with the line. The corresponding phasor diagrams for decreasing power flow and increasing power flow are shown in Figures 2-58b and 2-58c, respectively.

A symmetrical, dual core design connects a voltage in series with the line and changes the phase angle of the line voltage without changing its magnitude. The primary windings of the regulating transformer are Y-connected as shown in Figure 2-59a. They are connected in shunt with the line and excited from the voltage at the midpoints of the secondary windings of the series transformer. Three-phase voltages ($\mathbf{V_A}$, $\mathbf{V_B}$, and $\mathbf{V_C}$) are induced in the regulating transformer secondary windings. Three-phase, bipolar compensating voltages ($\mathbf{V_{s'sA}}$, $\mathbf{V_{s'sB}}$, and $\mathbf{V_{s'sC}}$) that are in quadrature with the corresponding phase-to-neutral voltage are induced in the secondary windings of the series transformer. The magnitudes of the series-compensating voltages are varied using the LTCs in the secondary windings of the regulating transformer. The induced voltages of the series transformer are connected in series with the line. The corresponding phasor diagrams for decreasing power flow and increasing power flow are shown in Figures 2-59b and 2-59c, respectively.

2.3.3 Series Reactance Regulation

In a special case, the magnitude of the line voltage and its phase angle can also be varied together in such a way that the effective line reactance (X_{eff}) between the sending and receiving ends is regulated.

2.3.3.1 Direct Method

The direct way to regulate the effective line reactance is to connect a capacitor or a reactor in series with the line. A series-connected capacitor increases the power flow in the line by decreasing the effective line reactance between its two ends. A series-connected reactor decreases the power flow in the line by increasing the effective line reactance between its two ends. Through the use of a TCSC, a series-connected variable capacitor or a variable reactor can be implemented as shown in Figure 2-60. Each section of a TCSC consists of a parallel combination of a fixed capacitor with a reactance (X_C) and a variable reactor with a reactance (X_L), emulated by a TCR. The equivalent series-compensating reactance (x_{se}) of each section of the TCSC can be determined during its inductive or capacitive operation. The total compensating reactance (X_{se}) across the TCSC is defined by

$$X_{se} = k \times x_{se} \tag{2-272a}$$

where k is the number of sections, connected in series; the series-compensating reactance (x_{se}) per section is determined by the parallel combination of a capacitive reactance ($-jX_C$) and an inductive reactance ($+jX_L$), which is given by

$$-jx_{se} = \frac{(+jX_L)(-jX_C)}{j(X_L - X_C)} = j\frac{X_L X_C}{(X_C - X_L)}. \tag{2-272b}$$

where $X_C > X_L$ makes X_{se} inductive and when $X_C < X_L$ makes X_{se} capacitive.

The total compensating voltage ($\mathbf{V_q}$) across the TCSC is the sum of voltages across k sections. The voltage across a section is the sum of two currents (the line current, $\mathbf{I}$, and the current, $\mathbf{I_{TCR}}$, through the TCR) multiplied by the series-connected capacitive reactance (X_C).

2.3.3.2 Indirect Method

The indirect way to emulate a series capacitor or a reactor is to connect a compensating voltage in series with the line and maintain its phase angle at 90° lagging or leading the prevailing line

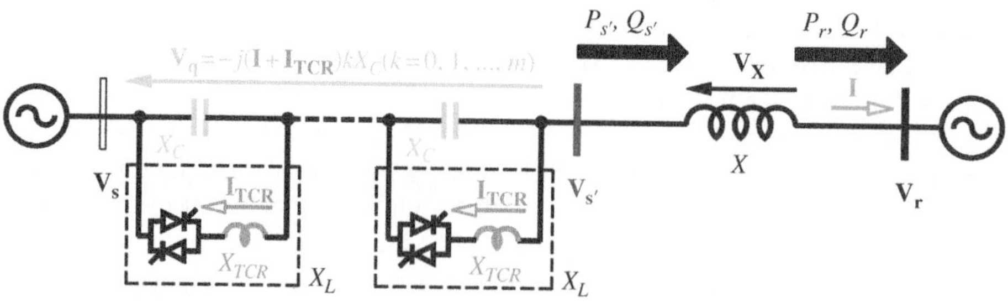

Figure 2-60 Thyristor-Controlled Series Capacitor (TCSC) (k is the number of sections, connected in series).

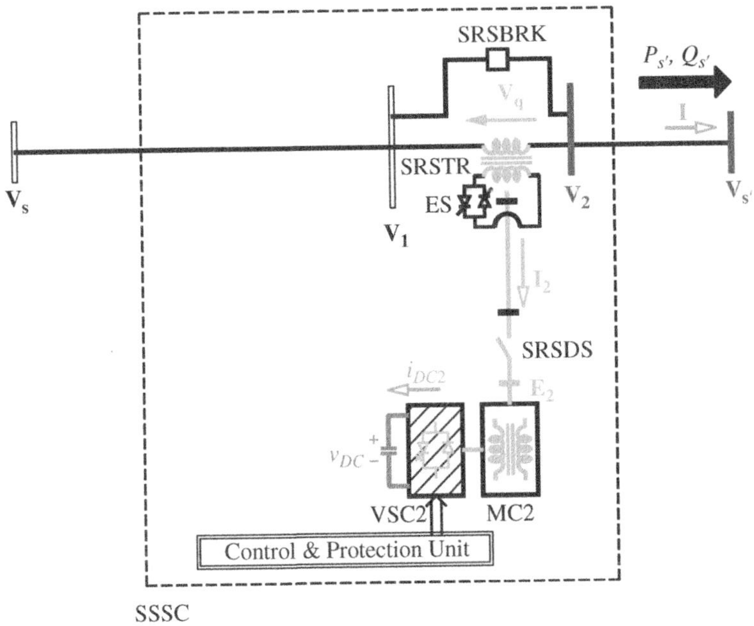

Figure 2-61 Static Synchronous Series Compensator (SSSC).

current. Through control action, the magnitude of the series-compensating voltage is varied in order to vary the emulated capacitor or reactor. The series-compensating voltage varies the magnitude and phase angle of the line voltage simultaneously. A variable magnitude, series-compensating voltage can also be implemented through the use of electrical machines, VSCs, or transformers/LTCs.

Figure 2-61 shows a schematic diagram of the SSSC. The SSSC consists of a HN-VSC (VSC2), a magnetic circuit (MC2), a series-coupling transformer (SRSTR), a series disconnect switch (SRSDS), an electronic bypass-switch (ES), a series bypass breaker (SRSBRK), current and voltage sensors, and a Control & Protection Unit. The operation of the SSSC is such that the VSC produces a voltage that is almost in quadrature with the prevailing line current. Through control action, the magnitude of the series-compensating voltage is varied to emulate a variable reactance in series with the line.

2.3.4 Impedance Regulation

In all the concepts presented thus far, only one parameter (voltage magnitude, its phase angle, or line reactance) is regulated one at a time and, therefore, both the active and reactive power flows are affected simultaneously, according to Equations (2-59), (2-60), and (2-61). In impedance regulation, an impedance that consists of two parameters – resistance and reactance are emulated by a series-compensating voltage. This compensating voltage is added to the original line voltage to modify both the magnitude and the phase angle of the line voltage. This modified line voltage, in turn, regulates the active and reactive power flows in the line independently as desired. There are several schemes to implement the concept of impedance regulation that results in an independent control of active and reactive power flows in a line.

Figure 2-62 shows the basic building blocks of VSC-based PFCs namely, STATCOM and SSSC and their combined topology called a UPFC. When the STATCOM and the SSSC operate as

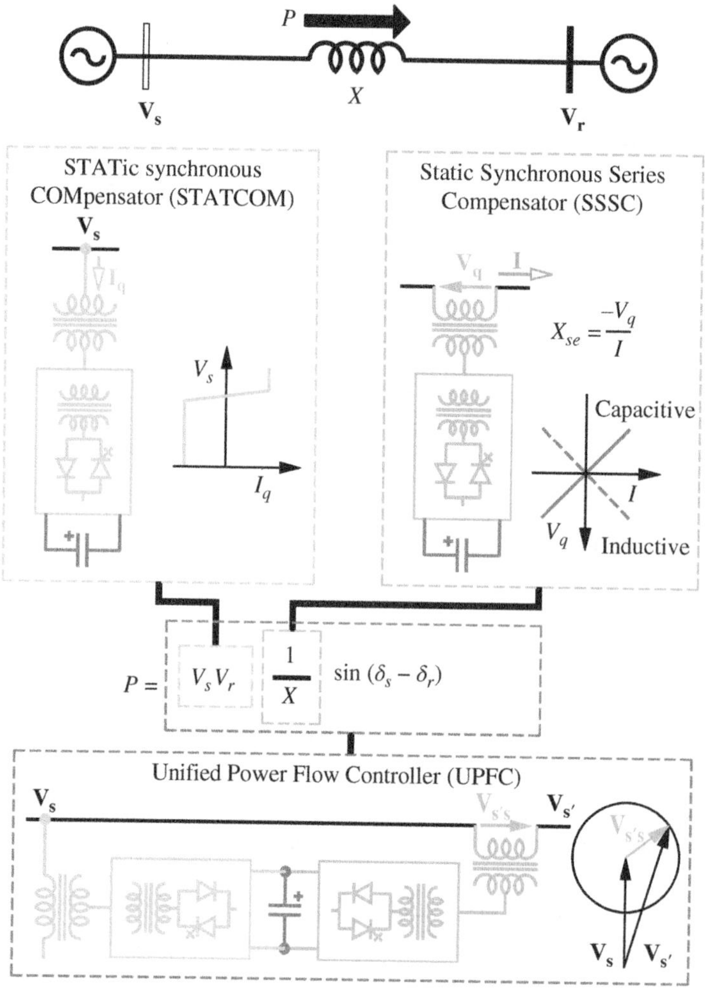

Figure 2-62 The UPFC (with a common DC link) and its building blocks: STATCOM and SSSC.

stand-alone compensators, they exchange mostly reactive power with the line and, in the process, control only one power flow control parameter – line voltage by the STATCOM and line reactance by the SSSC, respectively. In both cases, the compensating voltage and the current flowing through it are in quadrature. Of course, each of the STATCOM and SSSC also draws active power to provide its own losses that are a few percent of its rated power. When both the STATCOM and the SSSC are connected at their common DC link, the series-compensating voltage of the SSSC is no longer restricted to be in quadrature with the prevailing line current. This unique DC link connection adds the active power exchange capability to both the STATCOM and the SSSC. The series-compensating voltage can be at any phase angle with the prevailing line current, thereby exchanging active power with the line bidirectionally. This exchanged active power flows freely back and forth through their DC link. The result is an emulation of a four-quadrant, series-compensating impedance ($\mathbf{Z_{se}} = R_{se} - jX_{se}$) that consists of a resistance ($R_{se} = +R$ or $-R$) and a reactance ($X_{se} = X_C$ or $-X_L$) in series with the line as shown in Figure 1-8b. The modified line voltage

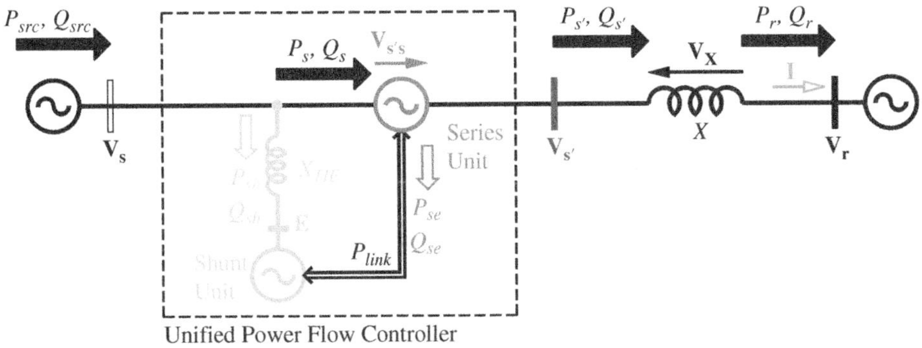

Figure 2-63 Two-generator/one-line power system network, integrated with a PFC – Unified Power Flow Controller (UPFC).

can be regulated to a desired magnitude and phase angle within the design limit to control the active and reactive power flows in the line independently in the P_r – Q_r circle as shown in Figure 2-22 for receiving-end power flows and in the $P_{s'}$ – $Q_{s'}$ ellipse as shown in Figure 2-23 for modified sending-end power flows. The hardware required for this three-decade old technology is rated for only a fraction of the transmitted power flow capability of the line. Note that the characteristics, shown in Figure 2-62, for various FACTS Controllers are using ideal specialty transformer and the coupling transformer without any leakage impedance and saturation. The actual characteristic differs, based on the transformer characteristics.

A two-generator/one-line power system network, integrated with a UPFC is shown in Figure 2-63. The VSC1 ($\mathbf{E} = E\angle(\delta_s + \alpha)$), shown as the Shunt Unit, is connected to the line voltage ($\mathbf{V_s} = V_s\angle\delta_s$) with a tie-reactance ($X_{TIE}$). The VSC2 ($\mathbf{V_{s's}} = V_{s's}\angle(\delta_s + \beta)$), shown as the Series Unit, is connected in series with the line. The active and reactive powers, exchanged by the shunt-connected VSC, are P_{sh} and Q_{sh}, respectively. The active and reactive powers, exchanged by the series-connected VSC, are P_{se} and Q_{se}, respectively. The active power (P_{link}) flows from one VSC to the other bidirectionally through their common DC link. However, the reactive powers (Q_{sh} and Q_{se}), exchanged by the two VSCs, are independent of each other.

2.3.4.1 Unified Power Flow Controller (UPFC)

Figure 2-64 shows the schematic diagram of the UPFC. The UPFC consists of two HN-VSCs (VSC1 and VSC2), two magnetic circuits (MC1 and MC2), a shunt-coupling transformer (SHNTR), a series-coupling transformer (SRSTR), a shunt breaker (SHNBRK), a series disconnect switch (SRSDS), an electronic bypass-switch (ES), a DC link switch (DCLS), a series bypass breaker (SRSBRK), current and voltage sensors, and a Control & Protection Unit. Each VSC is coupled with a transformer at its output. Both VSCs generate almost sinusoidal voltages.

When the DCLS is closed, the VSCs are connected at their common DC link. The series-connected, variable magnitude, compensating voltage that is at any phase angle with respect to the line current, exchanges an active power as well as a reactive power with the line. This exchanged active power at the AC terminal of the series-connected VSC flows to the AC terminal of the shunt-connected VSC through their common DC link. In addition, the shunt-connected VSC can deliver or absorb reactive power independently as desired. When the DCLS is open, the two VSCs can be operated as stand-alone, independent, reactive compensators, such as a shunt compensator

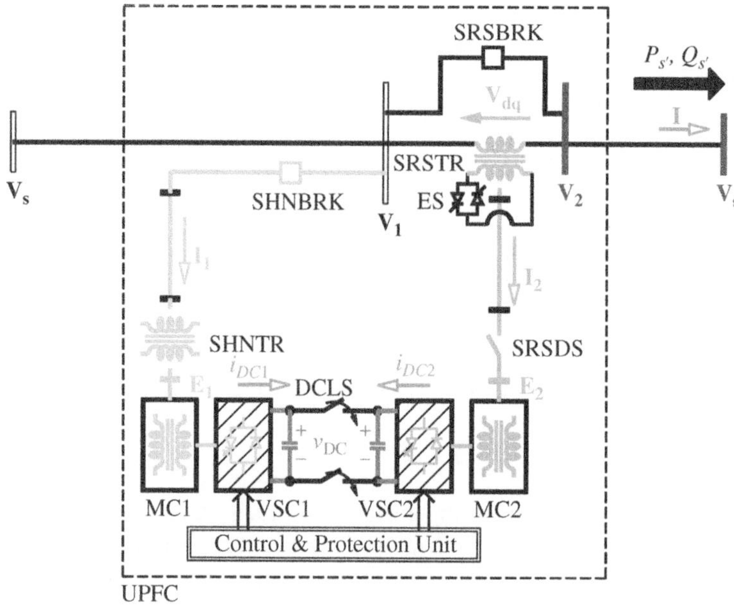

Figure 2-64 Unified Power Flow Controller (UPFC).

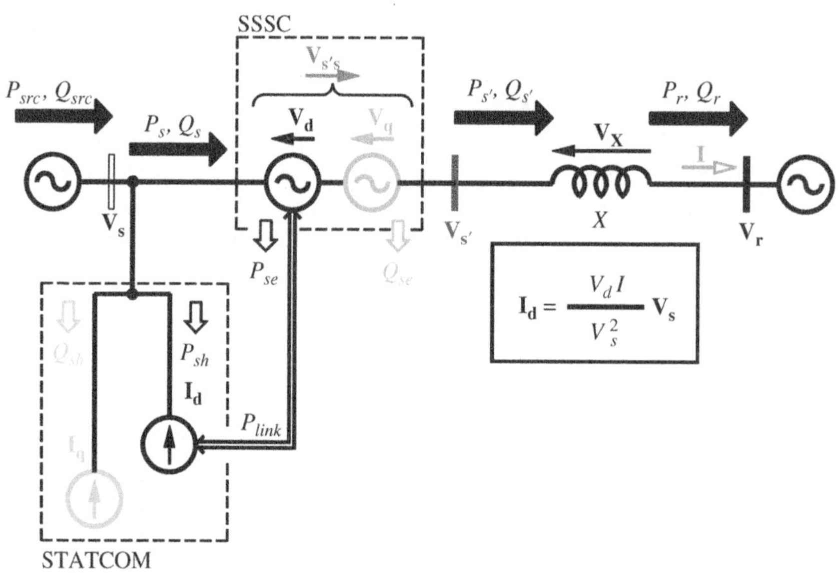

Figure 2-65 Basic UPFC model, integrated in a two-generator/one-line power system network.

(STATCOM) and a series compensator (SSSC) and, in turn, exchange almost exclusively reactive power with the line at their respective terminals.

A basic UPFC model, integrated in a two-generator/one-line power system network is shown in Figure 2-65. The figure shows the basic UPFC model, which consists of the STATCOM and the

SSSC that are connected at their common DC link. The SSSC connects a compensating voltage ($\mathbf{V_{s's}}$) in series with the line. The compensating voltage ($\mathbf{V_{s's}}$) consists of active and reactive components, which are $\mathbf{V_d}$ and $\mathbf{V_q}$ with load convention. The compensating voltage ($\mathbf{V_{s's}}$) is at any phase angle with the prevailing line current (**I**). The active component ($\mathbf{V_d} = V_d$) that is either in-phase or out-of-phase with the line current emulates a positive or a negative resistor in series with the line. The reactive component ($\mathbf{V_q} = jV_q$) that is in quadrature and either leading or lagging the line current emulates a reactor or a capacitor in series with the line.

The series-compensating voltage ($\mathbf{V_{s's}}$) exchanges active and reactive powers (P_{se} and Q_{se}) that are defined as

$$P_{se} = -\mathbf{V_{s's}} \bullet \mathbf{I} = \mathbf{V_{dq}} \bullet \mathbf{I} = V_d I = V_s I_d \tag{2-273}$$

and

$$Q_{se} = |-\mathbf{V_{s's}} \times \mathbf{I}| = |\mathbf{V_{dq}} \times \mathbf{I}| = V_q I. \tag{2-274}$$

The active power (P_{se}) that is exchanged with the line by the SSSC flows through the STATCOM and results in an active current flow ($\mathbf{I_d} = I_d$) through it. This current is either in phase or out of phase with the line voltage ($\mathbf{V_s}$) and emulates a positive or a negative resistor at the POC in shunt with the line. However, the reactive power (Q_{se}) that is exchanged with the line by the SSSC is independent of whether there is any reactive current ($\mathbf{I_q}$) flowing through the STATCOM. The loading effect (voltage change) of the power system network at the POC due to (1) active current flow through the STATCOM and (2) load change in the line may be compensated by the independent control of the reactive current flow through the STATCOM. This reactive component $(\mathbf{I_q} = jI_q)$ that is in quadrature with the line voltage emulates an inductive or a capacitive reactance at the POC in shunt with the line.

The fast-switching capability of the VSCs enables the UPFC to provide a fast dynamic response, if properly designed. Even though the Shunt–Series configuration is preferable for a PFC, instead of a Shunt–Shunt configuration in a synchronous system because of the advantages in lower rating, thus lower cost, the VSC-based UPFC started with two major drawbacks (high installation and operating costs) since its inception in the 1990s. Over the decades, the list of drawbacks has expanded to include component obsolescence, impracticability of relocation, and lack of interoperability. Moreover, the fast dynamic response in the range of ms is not required in most utility applications. Traditionally, a mechanical LTC is used since its speed of operation provides an adequate response time in most utility applications. If faster response is needed, the mechanical LTCs can be replaced with faster switches. This was the motivation to redesign the PFC as discussed next.

In the next section, a new SPFC is introduced to provide an independent control of active and reactive power flows in a line with the use of time-tested components, such as transformer/LTCs that are familiar to utilities worldwide for almost a century.

2.3.4.2 Sen Transformer (ST)

The schematic diagram of an ST and its operational phasor diagram are shown in Figure 2-66. The ST consists of an Exciter Unit and a Compensating-Voltage Unit that connects a compensating voltage in series with the line and, in turn, modifies the line voltage to a specific magnitude and phase angle within its design limit. While the series-compensating voltage ($\mathbf{V_{s's}}$) provides a compensation in its entire controllable range of $0° \leq \beta \leq 360°$, its exchanged active and reactive powers with the line flow bidirectionally through the magnetic link to the shunt-connected Exciter Unit as shown in Figure 2-11.

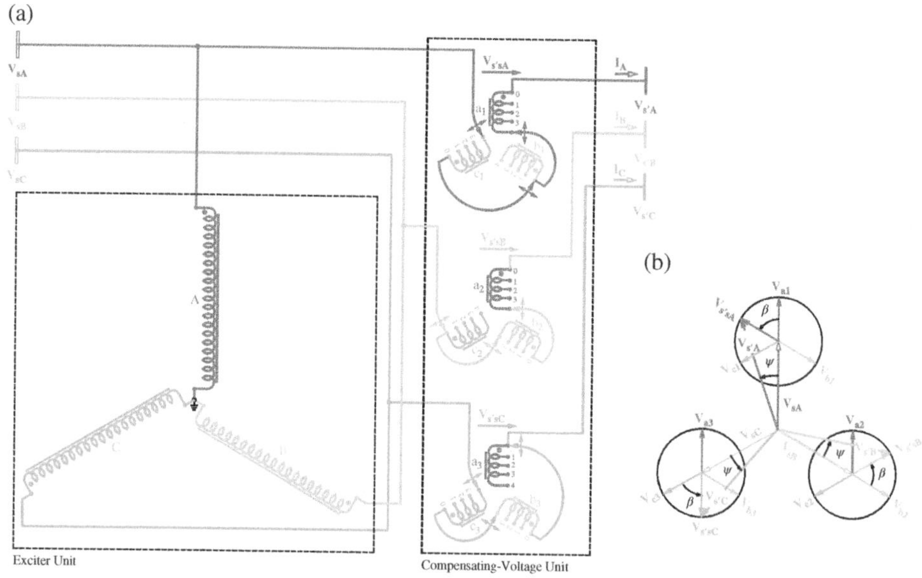

Figure 2-66 Sen Transformer (ST).

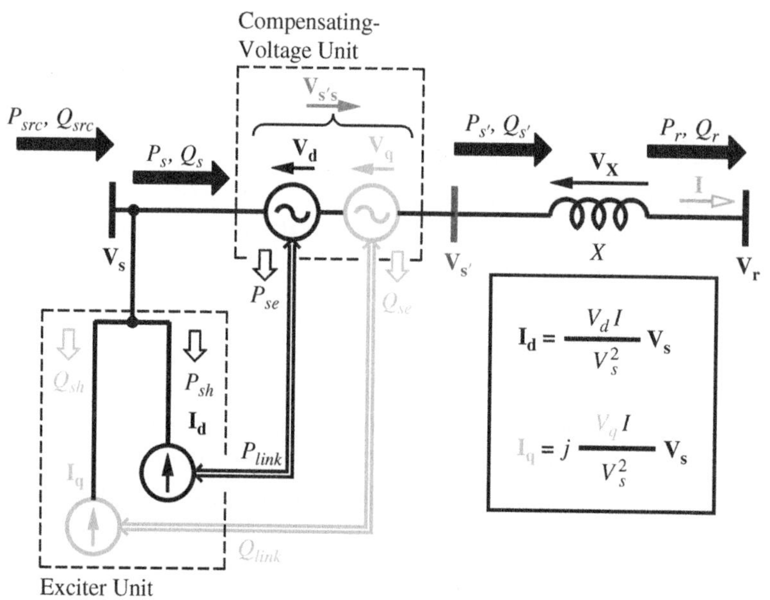

Figure 2-67 Basic ST model, integrated in a two-generator/one-line power system network.

A basic ST model, integrated in a two-generator/one-line power system network is shown in Figure 2-67. The figure shows the basic ST model, which consists of a shunt-connected Exciter Unit and a series-connected Compensating-Voltage Unit. The Compensating-Voltage Unit connects a compensating voltage ($\mathbf{V_{s's}}$) in series with the line. The compensating voltage ($\mathbf{V_{s's}}$) consists of active and reactive components, which are $\mathbf{V_d} = V_d$ and $\mathbf{V_q} = jV_q$ with load convention. The compensating voltage ($\mathbf{V_{s's}}$) is at any phase angle with the prevailing line current (**I**). The active component ($\mathbf{V_d}$) that is either in phase or out of phase with the line current, emulates a positive or a negative resistor in series with the line. The reactive component ($\mathbf{V_q}$) that is in quadrature either leading or lagging the line current, emulates a reactor or a capacitor in series with the line.

The series-compensating voltage ($\mathbf{V_{s's}}$) exchanges active and reactive powers (P_{se} and Q_{se}) that are defined as

$$P_{se} = -\mathbf{V_{s's}} \cdot \mathbf{I} = \mathbf{V_{dq}} \cdot \mathbf{I} = V_d I = V_s I_d \tag{2-275}$$

and

$$Q_{se} = |-\mathbf{V_{s's}} \times \mathbf{I}| = |\mathbf{V_{dq}} \times \mathbf{I}| = V_q I = V_s I_q. \tag{2-276}$$

The active and reactive powers (P_{se} and Q_{se}) that are exchanged with the line, flow through a magnetic link to the Exciter Unit and result in active and reactive currents ($\mathbf{I_d}$ or I_d and $\mathbf{I_q}$ or jI_q) flowing through it. The active current ($\mathbf{I_d}$) is either in phase or out of phase with the line voltage and emulates a positive or a negative resistor at the POC in shunt with the line. The reactive current ($\mathbf{I_q}$) is in quadrature with the line voltage and emulates an inductive or a capacitive reactance at the POC in shunt with the line.

The active and reactive powers exchanged by the UPFC and the ST with the line are the same. In both the UPFC and ST, the exchanged active power (P_{se}) contributes to the same amount of active current ($\mathbf{I_d}$) through the shunt connection. Therefore, when the series-connected Compensating-Voltage Unit emulates a $+R$, the shunt-connected Exciter Unit emulates a $-R$ and vice versa. In the case of the ST, the exchanged reactive power (Q_{se}) by the Compensating-Voltage Unit contributes to an amount of reactive current ($\mathbf{I_q}$) through the shunt connection. However, the reactive current ($\mathbf{I_q}$) through the STATCOM is independent of the reactive power (Q_{se}), exchanged by the SSSC with the line. Therefore, in the ST, when the series-connected Compensating-Voltage Unit emulates a capacitive reactance, the shunt-connected Exciter Unit emulates an inductive reactance and vice versa.

2.4 Interline Power Flow Concept

The ST or the UPFC is suitable for independent control of active and reactive power flows in one line to which it is connected. However, several lines are usually connected in close proximity to a common voltage bus. Any change in the power flow in one line will affect the power flows in the neighboring lines. Therefore, a compensation scheme is needed to transfer power directly from particular lines to other specific lines. In a multiline transmission network, it would be advantageous to be able to transfer power from an overloaded to an underloaded line with a minimum impact on the power flow in the other uncompensated lines.

A series-compensating voltage regulates the magnitude and phase angle of the line voltage independently and, therefore, controls the active and reactive power flows in a line independently as desired. In the process, it exchanges active and reactive powers at the POC with the line. The need for the exchanged active power demand can be met using a Series–Series configuration, which is composed of at least two DC-to-AC VSCs that are connected back to back at their common DC link. Hence, it is called a BTB-SSSC. Each SSSC is connected in series with an AC line. In a typical application, the BTB-SSSC can transfer active power from one or more lines, referred to as "leader" lines, to the others, referred to as "follower" lines. In addition, the BTB-SSSC can provide independent reactive power compensation in each line to which it is connected and thereby controls the effective reactance of each line between its two ends. In 2003, a VSC-based PFC (BTB-SSSC), also known as IPFC, was demonstrated to be able to provide this operation at Marcy substation in New York, USA. If the lines that are involved in the compensation scheme are not connected to a common voltage bus, they may be operated at different frequencies. This compensation scheme then becomes an "asynchronous tie."

Another PFC, based on transformer/LTCs, is the MST that provides independent control of active and reactive power flows in a line and also transfers active and reactive powers selectively from one line to another in a multiline transmission system without affecting the power flows in the uncompensated lines.

2.4.1 Back-to-Back SSSC

A basic BTB-SSSC that provides a series compensation in a three-generator/two-line power system network is shown in Figure 2-68. Both lines are connected together at the sending end with voltage ($\mathbf{V_s}$). The line 1 consists of a series-compensating voltage ($\mathbf{V_{1s's}}$), line reactance (X_1), and a receiving-end voltage ($\mathbf{V_{1r}}$). The line 2 consists of a series-compensating voltage ($\mathbf{V_{2s's}}$), line reactance (X_2), and a receiving-end voltage ($\mathbf{V_{2r}}$).

The modified sending-end voltages for lines 1 and 2 are

$$\mathbf{V_{1s'}} = \mathbf{V_s} + \mathbf{V_{1s's}} \tag{2-277}$$

$$\mathbf{V_{2s'}} = \mathbf{V_s} + \mathbf{V_{2s's}}. \tag{2-278}$$

Line 1 operates as the "leader" line where the series-compensating voltage ($\mathbf{V_{1s's}}$) is of any magnitude within its allowable limit and at any phase angle with the sending-end voltage as well as the prevailing line current ($\mathbf{I_1}$). Therefore, the series-compensating voltage ($\mathbf{V_{1s's}}$) exchanges active and reactive powers (P_{1se} and Q_{1se}) with the line and emulates a four-quadrant, series-compensating impedance ($\mathbf{Z_{1se}} = R_{1se} - jX_{1se}$) that consists of a compensating resistance ($R_{1se} = +R_1$ or $-R_1$) and a compensating reactance ($X_{1se} = X_{C1}$ or $-X_{L1}$) in series with line 1. A positive resistance ($+R_1$) absorbs active power from the line and a negative resistance ($-R_1$) delivers active power to the line; a capacitive reactance (X_{C1}) increases the power flow of the line and an inductive reactance (X_{L1}) decreases the power flow of the line.

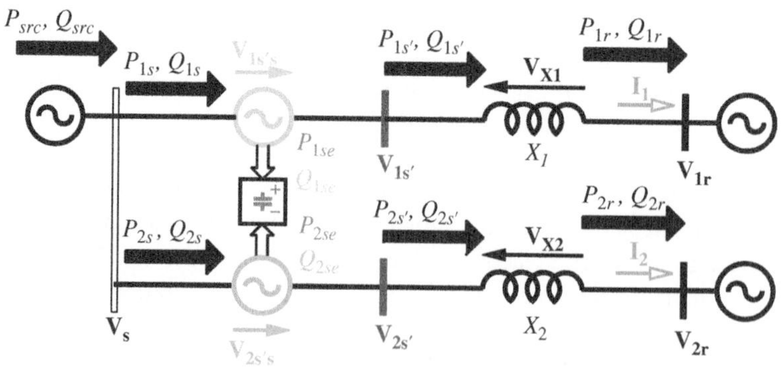

Figure 2-68 Basic BTB-SSSC, integrated in a three-generator/two-line power system network.

Line 2 operates as the "follower" line where the series-compensating voltage ($\mathbf{V}_{2s's}$) is of specific magnitude and phase angle with the prevailing line current ($\mathbf{I}_2$). The compensating voltage ($\mathbf{V}_{2s's}$), therefore, exchanges both active and reactive powers (P_{2se} and Q_{2se}) with the line and satisfies the active power and reactive power demands (both delivered and absorbed) by the series-compensating voltage ($\mathbf{V}_{1s's}$) in line 1. The compensating voltage ($\mathbf{V}_{2s's}$) emulates a series-compensating impedance ($\mathbf{Z}_{2se} = R_{2se} - jX_{2se}$) that consists of a resistance ($R_{2se} = +R_2$ or $-R_2$) and a reactance ($X_{2se} = X_{C2}$ or $-X_{L2}$) in series with line 2.

In an application, it may be desirable to provide series compensation to a number of lines in such a way that the active powers from specific "leader" lines flow through specific "follower" lines while the reactive compensation in each line is independently provided. Such a scheme can be implemented with several lines compensated by the SSSCs that are connected at their common DC link. Figure 2-69 shows a basic BTB-SSSC with two VSCs that are connected at their common DC link. The series-compensating voltage ($\mathbf{V}_{1s's}$) from VSC1 is connected in series with line 1. The series-compensating voltage ($\mathbf{V}_{2s's}$) from VSC2 is connected in series with line 2. The active power that flows through their common DC link is $P_{1se} = -P_{2se}$, assumimg the VSCs are lossless. In addition, the two VSCs exchange reactive power ($Q_{1se} \neq Q_{2se}$) with the respective line independently as desired. In an actual installation, if the "leader" line receives P_{1se}, the "follower" line supplies P_{2se}, which is the sum of P_{1se} and the losses in the VSCs.

The BTB-SSSC, shown in Figure 2-69, consists of two HN-VSCs (VSC1 and VSC2), two magnetic circuits (MC1 and MC2), two series-coupling transformers (SRSTR1 and SRSTR2), two series disconnect switches (SRSDS1 and SRSDS2), two electronic bypass-switches (ES1 and ES2), a DC link switch (DCLS), two series bypass breakers (SRSBK1 and SRSBK2), current and voltage sensors, and a Control & Protection Unit.

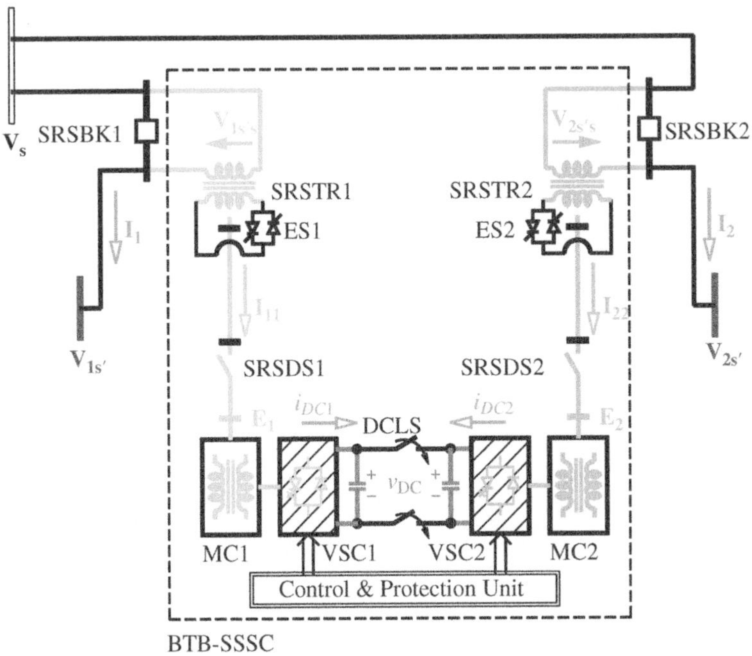

Figure 2-69 Implementation of a basic BTB-SSSC.

When the DCLS is open, the two VSCs can be used as independent reactive compensators, such as a series compensator, SSSC. Each compensator is coupled with a transformer at its output. The SSSC connects an almost sinusoidal voltage, of variable magnitude, in series with the line. When the SSSCs operate as stand-alone controllers, they exchange almost exclusively reactive power with the line at their respective AC terminals.

When the DCLS is closed, the VSCs are connected at their common DC link. While operating both the VSCs together as a BTB-SSSC, the exchanged power at the terminal of each VSC can be reactive as well as active. The series-compensating voltage can be at any phase angle with respect to the prevailing line current. The exchanged active power at the terminal of one VSC with the line flows to the terminal of the other VSC through the common DC link. Note that the BTB-SSSC power circuit must be rated for the compensating (not transmitted) power in the AC lines to which it is connected.

If the lines that are involved in the BTB-SSSC compensation scheme are not connected to a common voltage bus, they are not required to be operated at the same frequency. This compensation scheme, integrated in a four-generator/two-line power system network, is shown in Figure 2-70, then becomes an "asynchronous tie."

The modified sending-end voltages ($\mathbf{V}_{1s'}$ and $\mathbf{V}_{2s'}$) for lines 1 and 2 are

$$\mathbf{V}_{1s'} = \mathbf{V}_{1s} + \mathbf{V}_{1s'1s} \tag{2-279}$$

$$\mathbf{V}_{2s'} = \mathbf{V}_{2s} + \mathbf{V}_{2s'2s} \tag{2-280}$$

where $\mathbf{V}_{1s}$ and $\mathbf{V}_{2s}$ are the sending-end voltages for lines 1 and 2, respectively, and $\mathbf{V}_{1s'1s}$ and $\mathbf{V}_{2s'2s}$ are the series-compensating voltages for lines 1 and 2, respectively.

2.4.2 Multiline Sen Transformer (MST)

The alternative technique of power flow control in a multiline transmission system can be implemented using transformer/LTCs technology, called the MST as shown in Figure 2-71. The MST is based on the established technology of transformer/LTCs and is capable of providing an effective power flow management for multiple lines that are connected to a common voltage bus.

The general power flow equations in an uncompensated line are given by

$$P_{sn} = \frac{V_s V_r}{X} \sin(\delta_s - \delta_r) \tag{2-281}$$

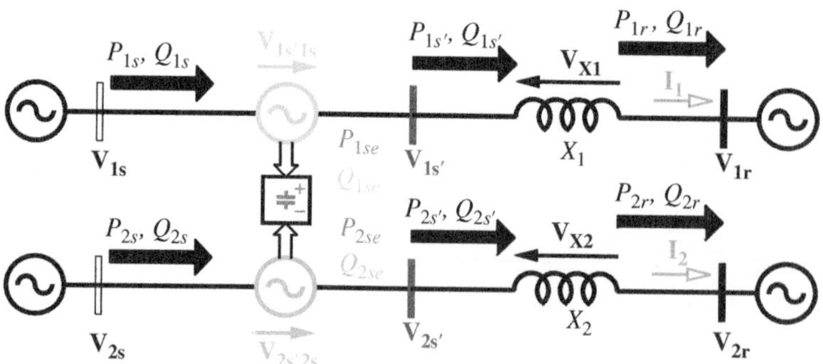

Figure 2-70 BTB-SSSC scheme as an asynchronous tie, integrated in a four-generator/two-line power system network.

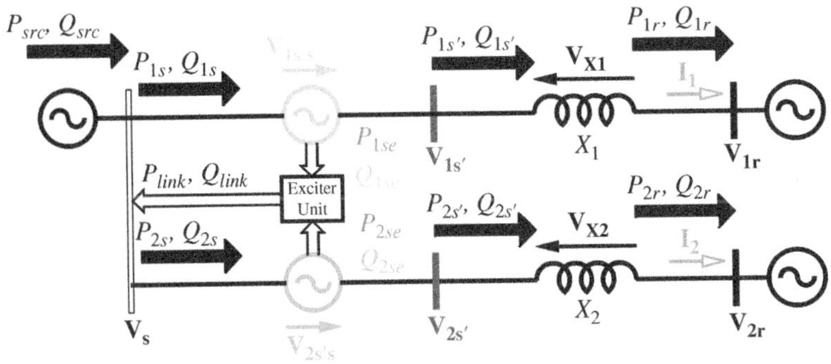

Figure 2-71 Basic MST, integrated in a three-generator/two-line power system network.

$$Q_{sn} = \frac{V_s V_r}{X}\left\{\frac{V_s}{V_r} - \cos(\delta_s - \delta_r)\right\} \tag{2-282}$$

$$P_{rn} = \frac{V_s V_r}{X}\sin(\delta_s - \delta_r) \tag{2-283}$$

and

$$Q_{rn} = \frac{V_s V_r}{X}\left\{\cos(\delta_s - \delta_r) - \frac{V_r}{V_s}\right\} \tag{2-284}$$

using Equations (2-40), (2-43), (2-46), (2-48), (2-41), and (2-27).

Therefore, the power flow equations for the receiving end in the compensated line 1 ("leader" line) can be rewritten by substituting P_{rn} with P_{1r}, Q_{rn} with Q_{1r}, V_s with $V_{1s'}$, V_r with V_{1r}, δ_s with $\delta_{1s'}$, δ_r with δ_{1r}, and X with X_1 in Equations (2-283) and (2-284) as

$$P_{1r} = \frac{V_{1s'} V_{1r}}{X_1}\sin(\delta_{1s'} - \delta_{1r}) \tag{2-285}$$

and

$$Q_{1r} = \frac{V_{1s'} V_{1r}}{X_1}\left\{\cos(\delta_{1s'} - \delta_{1r}) - \frac{V_{1r}}{V_{1s'}}\right\}. \tag{2-286}$$

Equations (2-285) and (2-286) can be combined as

$$P_{1r}^2 + \left(Q_{1r} + \frac{V_{1r}^2}{X_1}\right)^2 = \left(\frac{V_{1s'} V_{1r}}{X_1}\right)^2 \tag{2-287}$$

and

$$\tan(\delta_{1s'} - \delta_{1r}) = \frac{P_{1r}}{Q_{1r} + \frac{V_{1r}^2}{X_1}}. \tag{2-288}$$

The magnitude ($V_{1s'}$) of the modified sending-end voltage ($\mathbf{V_{1s'}}$) in line 1 is expressed by rearranging Equation (2-287) as

$$V_{1s'} = \frac{X_1}{V_{1r}}\sqrt{P_{1r}^2 + \left(Q_{1r} + \frac{V_{1r}^2}{X_1}\right)^2}. \tag{2-289}$$

The phase angle ($\delta_{1s'}$) of the modified sending-end voltage ($\mathbf{V_{1s'}}$) in line 1 is expressed by rearranging Equation (2-288) as

$$\delta_{1s'} = \tan^{-1}\frac{P_{1r}}{Q_{1r} + \dfrac{V_{1r}^2}{X_1}} + \delta_{1r}. \tag{2-290}$$

Considering the sending-end voltage phasor as reference ($\delta_s = 0°$), ($\psi_1 = \delta_{1s'} - \delta_s$) as the phase-shift angle of the modified sending-end voltage in line 1, and ($\delta_1 = \delta_s - \delta_{1r}$) as the natural power angle of line 1, Equation (2-290) can be rewritten as

$$\psi_1 = \tan^{-1}\frac{P_{1r}}{Q_{1r} + \dfrac{V_{1r}^2}{X_1}} - \delta_1. \tag{2-291}$$

Therefore, the magnitude ($V_{1s's}$) and the relative phase angle (β_1) of the series-compensating voltage ($\mathbf{V_{1s's}}$) in line 1 can be rewritten by substituting $V_{s'}$ with $V_{1s'}$ and ψ with ψ_1 in Equations (2-100) and (2-101) as

$$V_{1s's} = \sqrt{V_{1s'}^2 + V_s^2 - 2V_{1s'}V_s\cos\psi_1} \tag{2-292}$$

and

$$\beta_1 = \tan^{-1}\frac{V_{1s'}\sin\psi_1}{V_{1s'}\cos\psi_1 - V_s}. \tag{2-293}$$

The equations for the active and reactive powers (P_{1se} and Q_{1se}) exchanged in line 1 can be rewritten by substituting P_{sn} with P_{1sn}, Q_{sn} with Q_{1sn}, β with β_1, $V_{s's}$ with $V_{1s's}$, V_{Xn} with V_{1Xn}, X with X_1, and a_{se} with a_{1se} in Equations (2-177), (2-181), and (2-178) as

$$P_{1se} = a_{1se}\sin\left(\beta_1 - \tan^{-1}\frac{P_{1sn}}{Q_{1sn}}\right) \tag{2-294}$$

and

$$Q_{1se} = -\frac{V_{1s's}^2}{X_1} - a_{1se}\cos\left(\beta_1 - \tan^{-1}\frac{P_{1sn}}{Q_{1sn}}\right) \tag{2-295}$$

where

$$a_{1se} = \frac{V_{1s's}V_{1Xn}}{X_1}. \tag{2-296}$$

Note that the expression for the natural voltage (V_{1Xn}) across line 1 can be rewritten by substituting V_{Xn} with V_{1Xn}, V_r with V_{1r}, and δ with δ_1 in Equation (2-30) as

$$V_{1Xn} = \sqrt{V_s^2 - 2V_sV_{1r}\cos\delta_1 + V_{1r}^2}. \tag{2-297}$$

The natural power flow equations for the sending end in line 1 ("leader" line) can be rewritten by substituting P_{sn} with P_{1sn}, Q_{sn} with Q_{1sn}, V_r with V_{1r}, δ_r with δ_{1r}, and X with X_1 in Equations (2-281) and (2-282) as

$$P_{1sn} = \frac{V_sV_{1r}}{X_1}\sin(\delta_s - \delta_{1r}) \tag{2-298}$$

and

$$Q_{1sn} = \frac{V_s V_{1r}}{X_1}\left\{\frac{V_s}{V_{1r}} - \cos(\delta_s - \delta_{1r})\right\}. \tag{2-299}$$

If the multiline power flow concept is applied in two lines using series-compensating voltages generated by transformer/LTCs, it can be written from Figure 2-71 that

$$P_{1se} + P_{2se} = P_{link} \tag{2-300}$$

and

$$Q_{1se} + Q_{2se} = Q_{link}. \tag{2-301}$$

If the multiline power flow concept is applied in two lines using series-compensating voltages generated by VSCs or electrical machines, it can be written from Figure 2-68 that

$$P_{1se} + P_{2se} = 0 \tag{2-302}$$

while the reactive powers, exchanged with the leader and follower lines, are independent of each other. The active power exchanged with line 2 is the negative of the active power exchanged with line 1. In other words, if line 1 is delivering active power, line 2 is absorbing the same amount of active power that line 1 is delivering and vice versa, assuming the VSCs are lossless. Once the active and reactive powers (P_{2se} and Q_{2se}), which are exchanged with line 2 are known, the required series-compensating voltage ($\mathbf{V_{2s's}} = V_{2s's}\angle(\delta_s + \beta_2)$) can be determined as follows.

The expression for the magnitude ($V_{2s's}$) of the series-compensating voltage in line 2 can be rewritten by substituting $V_{s's}$ with $V_{2s's}$, P_{se} with P_{2se}, Q_{se} with Q_{2se}, V_{Xn} with V_{2Xn}, and X with X_2 in Equation (2-186b) as

$$V_{2s's} = +\sqrt{\left(\frac{V_{2Xn}^2}{2} - X_2 Q_{2se}\right) - \sqrt{\left(\frac{V_{2Xn}^2}{2} - X_2 Q_{2se}\right)^2 - X_2^2\left(P_{2se}^2 + Q_{2se}^2\right)}}. \tag{2-303}$$

The expression for the relative phase angle (β_2) of the series-compensating voltage ($\mathbf{V_{2s's}}$) can be rewritten by substituting $V_{s's}$ with $V_{2s's}$, P_{se} with P_{2se}, Q_{se} with Q_{2se}, P_{sn} with P_{2sn}, and Q_{sn} with Q_{2sn} in Equation (2-187a) as

$$\beta_2 = \tan^{-1}\frac{P_{2sn}}{Q_{2sn}} - \tan^{-1}\frac{P_{2se}}{Q_{2se} + \frac{V_{2s's}^2}{X_2}}. \tag{2-304}$$

Note that the expressions for the natural voltage (V_{2Xn}) across line 2 can be rewritten by substituting V_{Xn} with V_{2Xn}, V_r with V_{2r}, and δ with δ_2 in Equation (2-30) as

$$V_{2Xn} = \sqrt{V_s^2 - 2V_s V_{2r}\cos\delta_2 + V_{2r}^2}. \tag{2-305}$$

The natural power flow equations for the sending end in line 1 ("leader" line) can be rewritten by substituting P_{sn} with P_{2sn}, Q_{sn} with Q_{2sn}, V_r with V_{2r}, δ_r with δ_{2r}, and X with X_2 in Equations (2-281) and (2-282) as

$$P_{2sn} = \frac{V_s V_{2r}}{X_2}\sin(\delta_s - \delta_{2r}) \tag{2-306}$$

and

$$Q_{2sn} = \frac{V_s V_{2r}}{X_2}\left\{\frac{V_s}{V_{2r}} - \cos(\delta_s - \delta_{2r})\right\}. \tag{2-307}$$

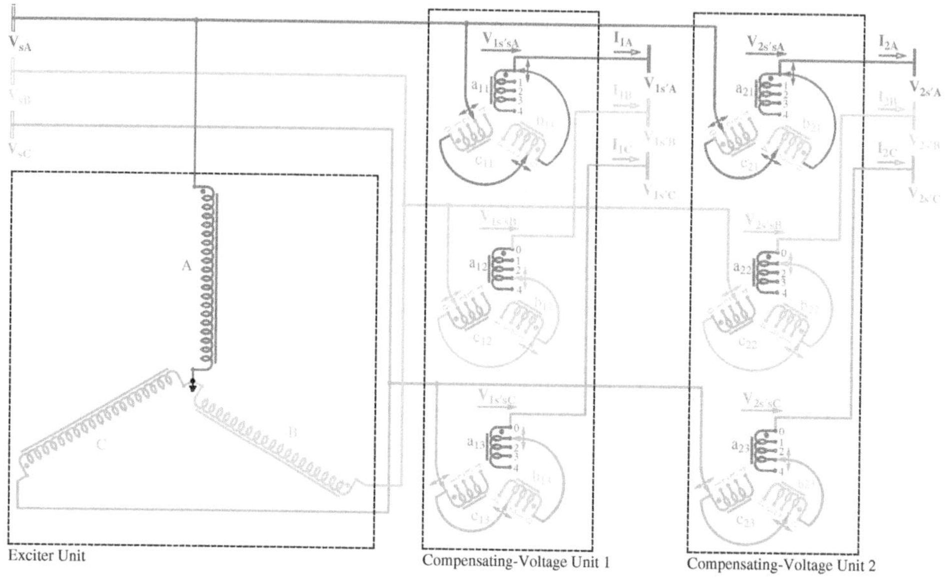

Figure 2-72 Multiline Sen transformer (MST).

Figure 2-72 shows a basic MST with two STs that are connected at their common magnetic link.

2.4.3 Back-to-Back STATCOM

For a point-to-point transfer of power between two isolated networks or interconnection of two lines with different voltages or phase angles (or frequencies), the magnitude and phase angle of the line voltage can be controlled to specific values by a shunt-compensating voltage. The compensating voltage exchanges active and reactive powers at the POC with the line. The need for the exchanged active power demand can be met using a Shunt–Shunt configuration, consisting of at least two DC-to-AC VSCs that are connected back to back at their common DC link. Hence, it is called a BTB STATCOM. Each STATCOM is connected in shunt to an AC line. In a typical application, the BTB-STATCOM transfers active power from one or more lines, referred to as "leader" lines, to the others, referred to as "follower" lines. In addition, the BTB-STATCOM provides independent reactive power compensation in each line to which it is connected and thereby regulates the voltage of the line at the POC. In 2000, the VSC-based BTB-STATCOM was demonstrated to be able to provide this operation as an "asynchronous tie" at Eagle Pass substation in Texas, USA, connecting American Electric Power (AEP) in the USA with Comisión Federal de Electricidad (CFE) in Mexico.

A two-generator/one-line power system network, integrated with a BTB STATCOM is shown in Figure 2-73. The two VSCs are connected at their common DC link. The VSC1 ($\mathbf{E_1} = E_1\angle(\delta_s + \alpha_1)$), shown as Shunt Unit 1, is connected to the line voltage ($\mathbf{V_s} = V_s\angle\delta_s$) with a tie-reactance ($X_{1TIE}$). The VSC2 ($\mathbf{E_2} = E_2\angle(\delta_{s'} + \alpha_2)$), shown as Shunt Unit 2, is connected to the line voltage ($\mathbf{V_{s'}} = V_{s'}\angle\delta_{s'}$) with a tie-reactance ($X_{2TIE}$). Shunt Unit 2 is operated as the "leader" unit in which the shunt-compensator voltage ($\mathbf{E_2}$) is of any magnitude within its allowable limit and at any phase angle with the modified sending-end voltage ($\mathbf{V_{s'}}$) to exchange active and reactive powers (P_{2sh} and Q_{2sh}) with the line. Shunt Unit 1 is operated as the "follower" unit in which the shunt-compensator

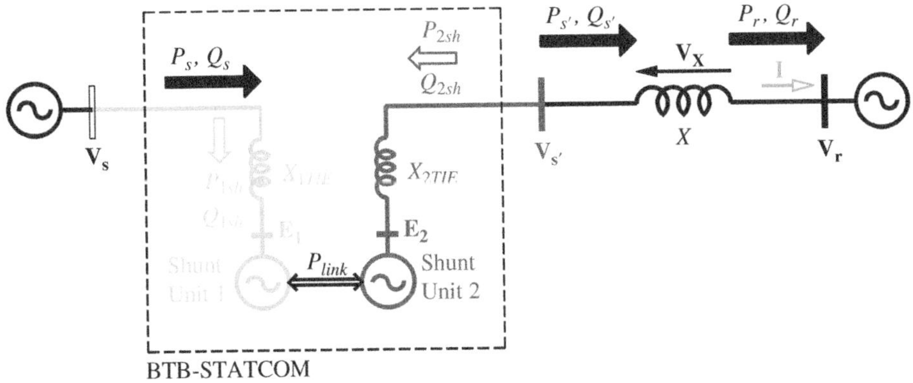

Figure 2-73 Back-To-Back STATCOM (BTB-STATCOM), integrated in a two-generator/one-line power system network.

voltage ($\mathbf{E}_1$) is of specific magnitude and phase angle with the sending-end voltage ($\mathbf{V}_s$) to exchange active and reactive powers (P_{1sh} and Q_{1sh}) at the POC. The net active power that flows through their common DC link is P_{link}. Each Shunt Unit exchanges reactive power ($Q_{1sh} \neq Q_{2sh}$) with its respective line independently as desired. In an actual installation, if the "leader" unit is designed to receive P_{2sh}, the "follower" unit is designed to supply P_{1sh}, which is the sum of P_{2sh} and the losses in both the units.

Figure 2-74 shows a basic BTB-STATCOM with two VSCs that are connected at their common DC link. The BTB-STATCOM consists of two HN-VSCs (VSC1 and VSC2), two magnetic circuits (MC1 and MC2), two shunt-coupling transformers (SHNTR1 and SHNTR2), two shunt breakers (SHNBK1 and SHNBK2), a DC link switch (DCLS), current and voltage sensors, and a Control & Protection Unit.

Figure 2-74 Back-to-Back STATCOM.

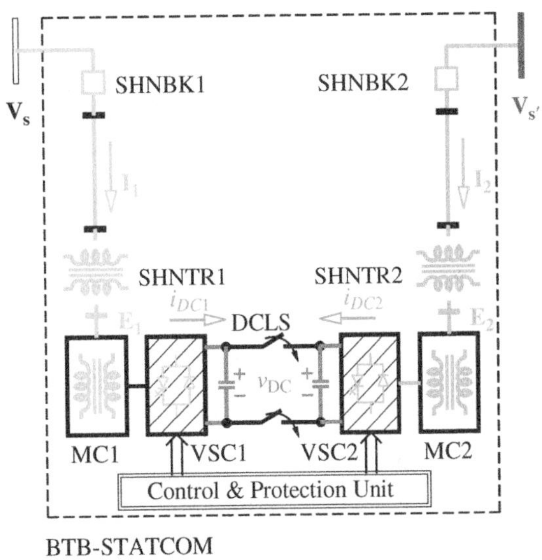

When the DCLS is open, the two VSCs can be used as independent reactive compensators (STATCOMs). Each VSC is coupled with a transformer at its output. The STATCOM injects an almost sinusoidal current, of variable magnitude, in shunt with the line. When the STATCOMs operate as stand-alone compensators, they exchange almost exclusively reactive power with the respective line.

When the DCLS is closed, the VSCs are connected at their common DC link. While operating both the VSCs together as a BTB-STATCOM, the exchanged power at the terminal of each VSC can be reactive as well as active. The shunt-compensating voltage can be at any phase angle with respect to the current flowing through it. The exchanged active power at the terminal of one VSC with the line flows to the terminal of the other VSC through their common DC link. However, the reactive power, exchanged with each line, is independent of each other. Note that the BTB-STATCOM's power circuit must be rated for the apparent power that is transmitted in the line.

2.4.4 Generalized Power Flow Controller

The VSC-based PFCs that have been mentioned thus far (UPFC, BTB-SSSC, and BTB-STATCOM) make use of at least two units and they are connected in Shunt–Series, Series–Series, and Shunt–Shunt configurations, respectively. In a multiline transmission network, it would be advantageous to extend the functions of these PFCs to compensate more than two lines in an area, so that the power flow management in all the compensated lines can be coordinated while interconnecting lines with different voltages, phase angles, or frequencies. The Generalized Power Flow Controller (GPFC), as shown in Figure 2-75, consists of a number of VSCs, each of which is connected in shunt

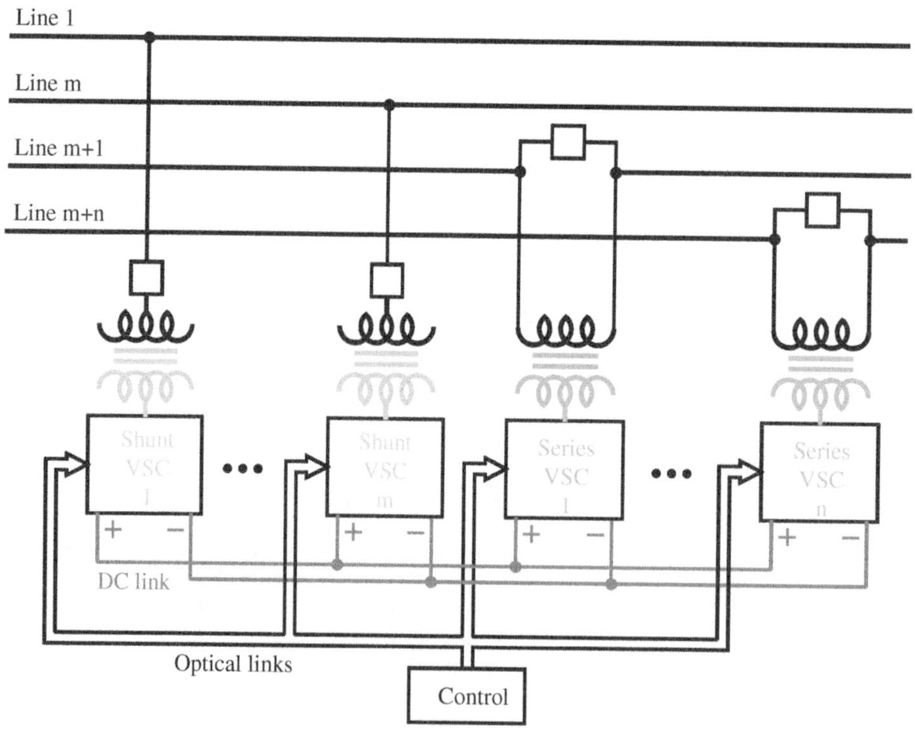

Figure 2-75 Generalized power flow controller with a DC link.

and/or series with a particular line through a coupling transformer. All the VSCs in the GPFC are connected at their common DC link. The compensating voltages are controlled in such a way that the active power can be transferred from one specific line to another specific line through their common DC link. In addition, each VSC provides an independent reactive compensation at its AC terminal.

The overall control scheme of a GPFC is such that if the active power exchanged by the mth shunt-connected VSC and nth series-connected VSC with their AC lines are P_{msh} and P_{nse}, respectively, the following active power balance is maintained, ignoring the losses in the VSCs:

$$P_{1sh} + P_{2sh} + \dots + P_{msh} + P_{1se} + P_{2se} + \dots + P_{nse} = 0. \tag{2-308}$$

Each shunt-connected VSC may be connected to a line through a tie reactor or coupling transformer. If a coupling transformer is used, its leakage reactance can replace the reactance of the tie reactor. In addition, a coupling transformer can keep the power electronics-based VSCs isolated from the high-voltage transmission line and also provide voltage matching between the line and the VSC, if needed. However, each series VSC in a DC link scheme must be connected through a coupling transformer. The main purpose of the series-coupling transformer is to convert a three-terminal (A-B-C) VSC into a six-terminal VSC with three isolated single-phase voltages that can be connected in series with a three-phase line.

In an alternate approach, the intermediate link between the Shunt and Series Units may be an AC link as shown in Figure 2-76.

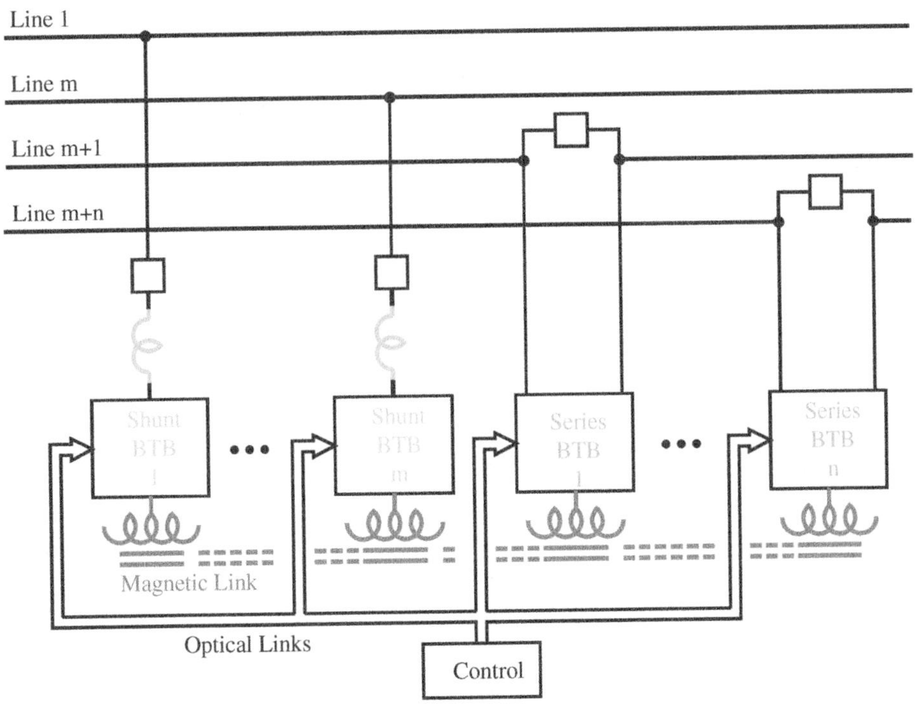

Figure 2-76 Generalized power flow controller with an AC link.

2.5 Figure of Merits Among Various PFCs

The two most commonly-used control techniques for a transmission line power flow are voltage regulation and phase angle regulation. The other control techniques, such as reactance regulation and impedance regulation, can be thought of as a combination of both voltage regulation and phase angle regulation. Using any regulation technique, except the impedance regulation, results in a control of active and reactive power flows simultaneously. However, the impedance regulation results in independent control of active and reactive power flows in the line. The merit of a PFC stems from the evaluations of *Reactive Power Index* (*RPI*), *Loss Index* (*LI*), *Apparent Power Rating* (*APR*), and so on, as described next. Each of these parameters is equivalent to a cost. The SMART PFC provides a solution that meets the functional requirements while keeping the overall cost at the lowest level.

A PFC, in its shunt–series form, modifies the sending-end voltage ($\mathbf{V_s} = V_s\angle\delta$) to a modified sending-end voltage ($\mathbf{V_{s'}} = V_{s'}\angle\delta'$) with a series-compensating voltage ($\mathbf{V_{s's}}$) whose magnitude is $V_{s's}$ and the relative phase angle with respect to the sending-end voltage is β. A PFC can be operated as a VR, a PAR (sym), a PAR (asym), a RR, and an IR. The electrical system data as given in Table 2-1 is used for the analysis in this section.

2.5.1 VR

A VR regulates the line voltage while keeping its phase angle unchanged. By varying the line voltage, the power flow characteristics ($P_{s'}$ versus $Q_{s'}$) and (P_r versus Q_r) are shown in Figure 2-77c. The limitation of the VR is that it is not designed to operate beyond the upper and lower voltage limits, for example ±10%. The resulting useful range of power flow due to the use of a VR is bound by the ($P_{s'}$ - $Q_{s'}$ and P_r - Q_r) characteristics in the practical operating region as shown in Figure 2-77c. Considering the base case active power flow of one pu for a system with parameters given in Table 2-1, the phasor diagrams for no power flow (Figure 2-77a), 100% power flow increase (Figure 2-77b), and the related *P*-*Q* diagrams (Figure 2-77c) are shown in the figure, using a VR.

2.5.2 PAR (sym)

A PAR (sym) changes the phase angle of the line voltage while keeping its magnitude unchanged, *i.e.* $V_{s'} = V_s$. Figures 2-78 through 2-80 show three cases (case 1: 40%, case 2: 70%, and case 3: 100%) of active power flow increase, which is defined as

$$\Delta P = P_r - P_{rn} = P_{s'} - P_{sn}. \tag{2-309}$$

The phasor diagram of a two-generator/one-line power system network, compensated with a PAR (sym), is shown in Figure 2-78a for Case 1 with an increased active power flow of 40%. Figure 2-78b shows the related *P*-*Q* characteristics. Note the reactive power (Q_{line1}) absorbed by the line for Case 1 is increased, compared to the reactive power (Q_{linen}) absorbed by an uncompensated line.

The phasor diagram of a two-generator/one-line power system network, compensated with a PAR (sym), is shown in Figure 2-79a for Case 2 with an increased active power flow of 70%. Figure 2-79b shows the related *P*-*Q* characteristics. Note the reactive power (Q_{line2}) absorbed by the line for Case 2 is significantly increased, compared to the reactive power (Q_{line1}) absorbed by the line for Case 1. This operating point may be chosen as the maximum allowable power flow

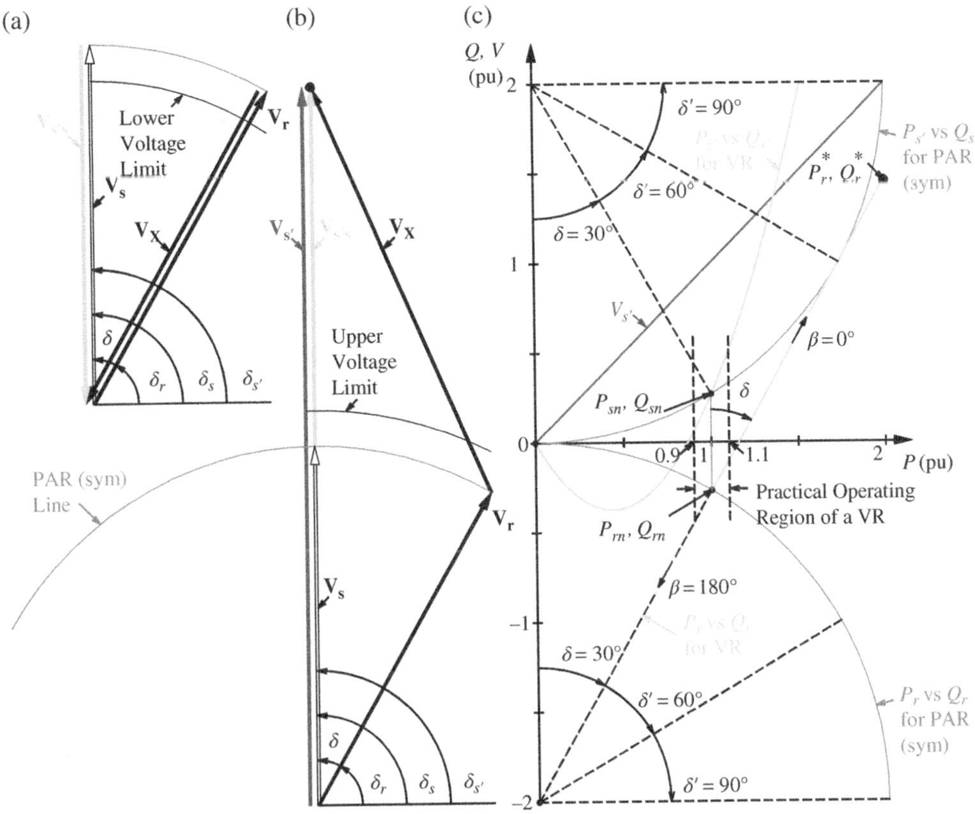

Figure 2-77 Phasor diagram of VR for (a) no power flow, (b) 100% power flow increase, and (c) related *P-Q* diagram, superimposed with the power flow characteristics of a PAR (sym).

limit due to the limitation of increased cost per unit of increased power delivery, called *Sen Index*, which is discussed later in the chapter.

Consider the electrical system data as given in Table 2-1. For a given $P_r = 1.7$ pu, calculate δ'_{max} and δ'_{min}.

From Equation (2-41), $a = \dfrac{V_s V_r}{X} = \dfrac{1 \times 1}{0.5} = 2$.

From Equation (2-46), $P_{rn} = a \times \sin\delta = 2 \times \sin 30° = 1$ pu.

From Equation (2-61), $a' = \dfrac{V_{s'} V_r}{X} = \dfrac{1 \times 1}{0.5} = 2$.

From Equation (2-74), $1.7 = P_r = a' \sin\delta' = 2 \times \sin\delta'$ pu.

Therefore, the maximum modified power angle, $\delta'_{\max} = \sin^{-1}\dfrac{1.7}{2} = 58.211669°$.

The phase-shift angle is given by

$$\psi = \delta'_{\max} - \delta, \tag{2-310}$$

which is $58.211669° - 30° = 28.211669°$.

In order to keep the symmetry in the design of the PAR (sym), the minimum modified power angle,

$$\delta'_{\min} = \delta - \psi, \tag{2-311}$$

which is $30° - 28.211669° = 1.788331°$.

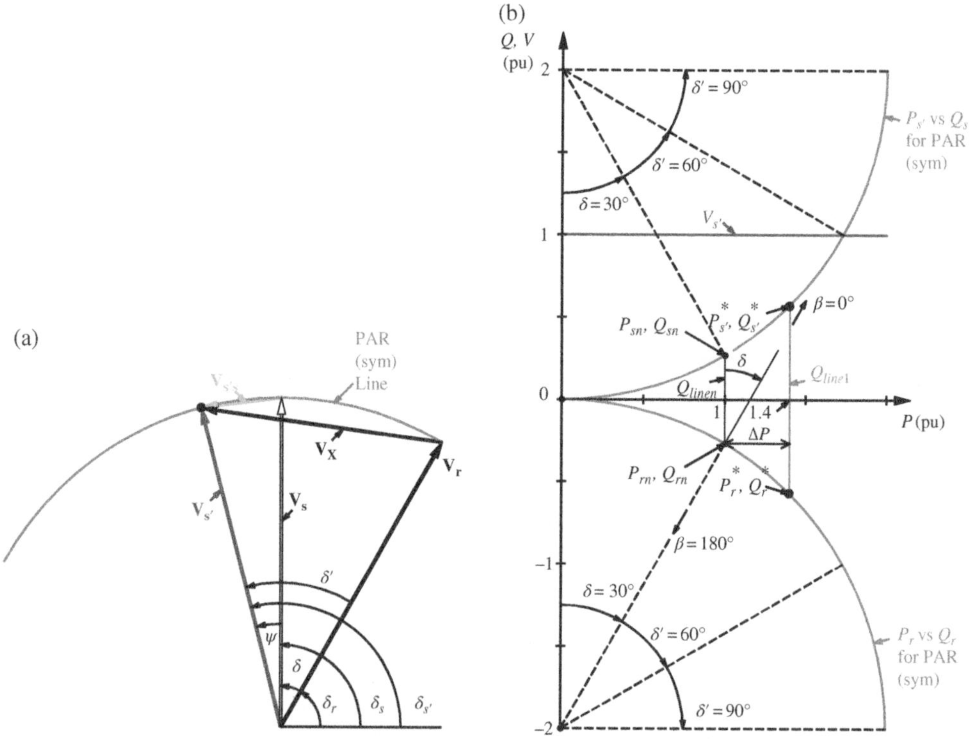

Figure 2-78 (a) Phasor diagram of a two-generator/one-line power system network with a PAR (sym) for Case 1 (40% enhancement of active power flow); (b) power flow characteristics at the modified sending end and the receiving end.

The phasor diagram of a two-generator/one-line power system network, compensated with a PAR (sym), is shown in Figure 2-80a for Case 3 with an increased active power flow of 100%. Figure 2-80b shows the related *P*-*Q* characteristics. Note the reactive power (Q_{line3}) absorbed by the line for Case 3 is significantly increased, compared to the reactive power (Q_{line2}) absorbed by the line for Case 2.

The limitation of the PAR (sym) is that it is not designed to operate beyond the maximum modified power angle (δ'_{max}), represented by the dashed line in the *P*-*Q* characteristic both at the sending and receiving ends. Therefore, if the thermal limit of the line is at 2 pu, as it is in this example, it is not possible to achieve the desired power flow using the PAR (sym). Note that the magnitude of the voltage ($V_{s'}$) at the modified sending end remains unchanged during the entire operating range of the PAR (sym).

2.5.3 PAR (asym)

A PAR (asym) changes primarily the phase angle of the line voltage with some increase in its magnitude. The phasor diagram of a two-generator/one-line power system network, compensated with a PAR (asym), is shown in Figure 2-81a for Case 1 with an increased active power flow of 40%.

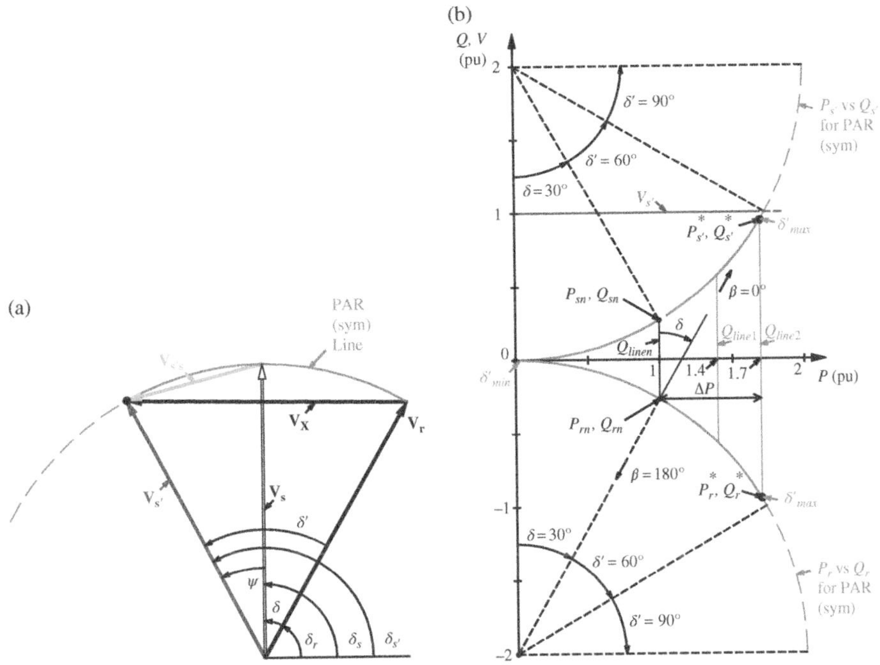

Figure 2-79 (a) Phasor diagram of a two-generator/one-line power system network with a PAR (sym) for Case 2 (70% enhancement of active power flow); (b) power flow characteristics at the modified sending end and the receiving end.

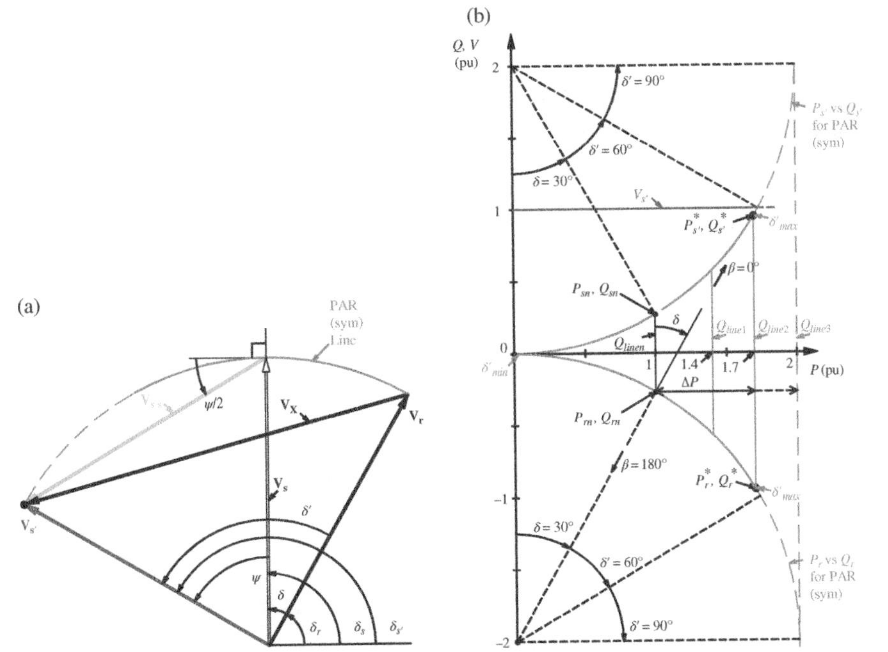

Figure 2-80 (a) Phasor diagram of a two-generator/one-line power system network with a PAR (sym) for Case 3 (100% enhancement of active power flow); (b) power flow characteristics at the modified sending end and the receiving end.

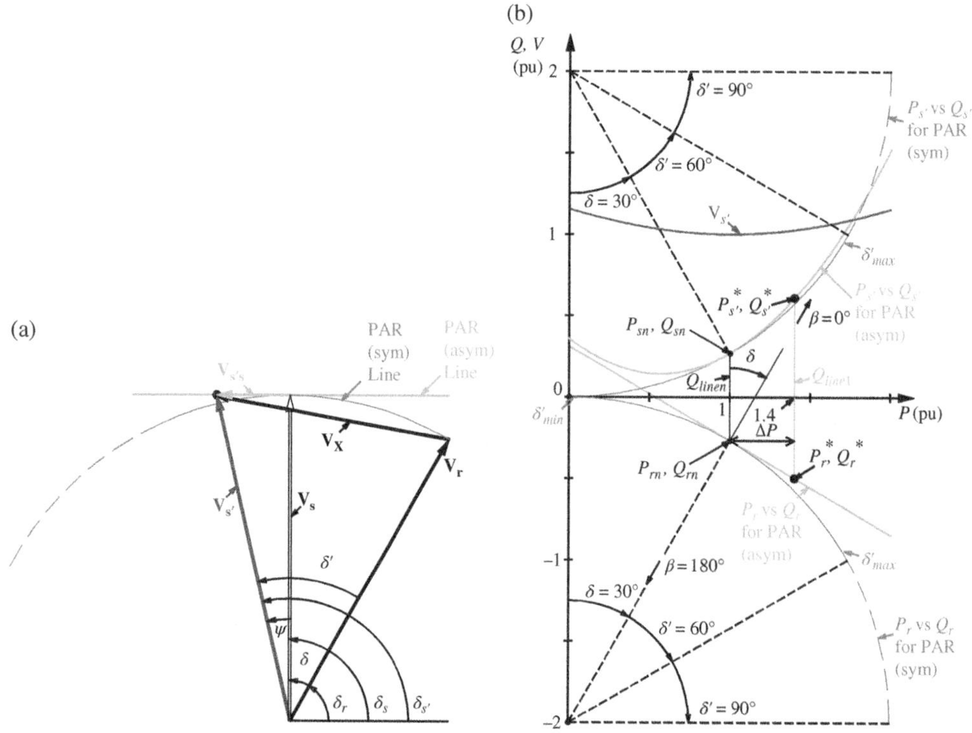

Figure 2-81 (a) Phasor diagram of a two-generator/one-line power system network with a PAR (asym) for Case 1 (40% enhancement of active power flow); (b) power flow characteristics at the modified sending end and the receiving end, superimposed with the power flow characteristics of a PAR (sym).

Figure 2-81b shows the related *P-Q* characteristics. Note the reactive power (Q_{line1}), absorbed by the line, for Case 1 is increased, compared to the reactive power (Q_{linen}) absorbed by an uncompensated line.

The phasor diagram of a two-generator/one-line power system network, compensated with a PAR (asym), is shown in Figure 2-82a for Case 2 with an increased active power flow of 70%. Figure 2-82b shows the related *P-Q* characteristics. Note the reactive power (Q_{line2}), absorbed by the line, for Case 2 is significantly increased, compared to the reactive power (Q_{line1}) absorbed by the line for Case 1.

The phasor diagram of a two-generator/one-line power system network, compensated with a PAR (asym), is shown in Figure 2-83a for Case 3 with an increased active power flow of 100%. Figure 2-83b shows the related *P-Q* characteristics. Note the reactive power (Q_{line3}), absorbed by the line, for Case 3 is significantly increased, compared to the reactive power (Q_{line2}) absorbed by the line for Case 2. However, the Q_{line3} in the case of a PAR (asym) is less than the Q_{line3} in the case of a PAR (sym). Therefore, if the active power flow needs to be increased by 100%, the use of a PAR (asym) is preferred over the use of a PAR (sym) in this example. Note that the magnitude of the voltage ($V_{s'}$) at the modified sending end always increases in the case of a PAR (asym) whether the power flow is increased or decreased.

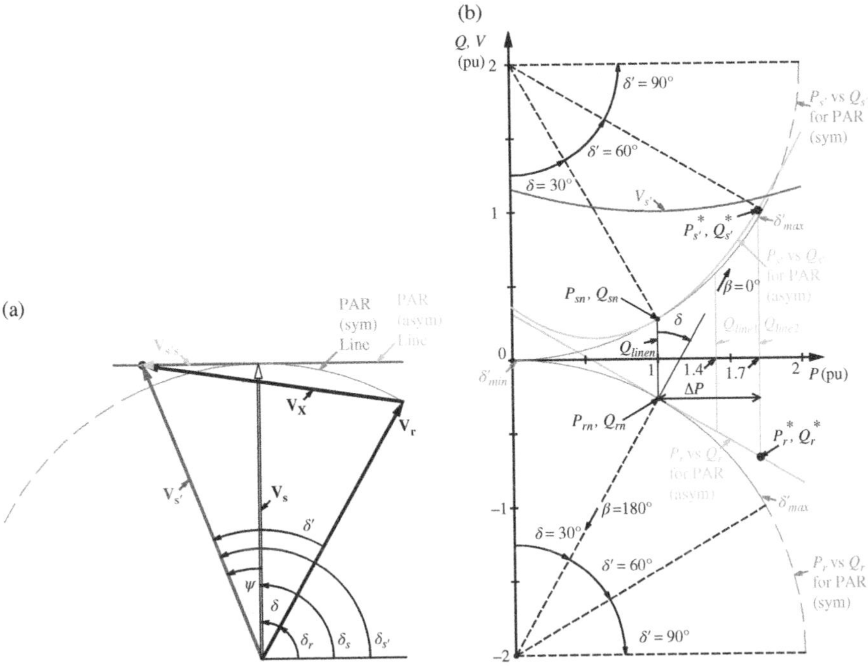

Figure 2-82 (a) Phasor diagram of a two-generator/one-line power system network with a PAR (asym) for Case 2 (70% enhancement of active power flow); (b) power flow characteristics at the modified sending end and the receiving end, superimposed with the power flow characteristics of a PAR (sym).

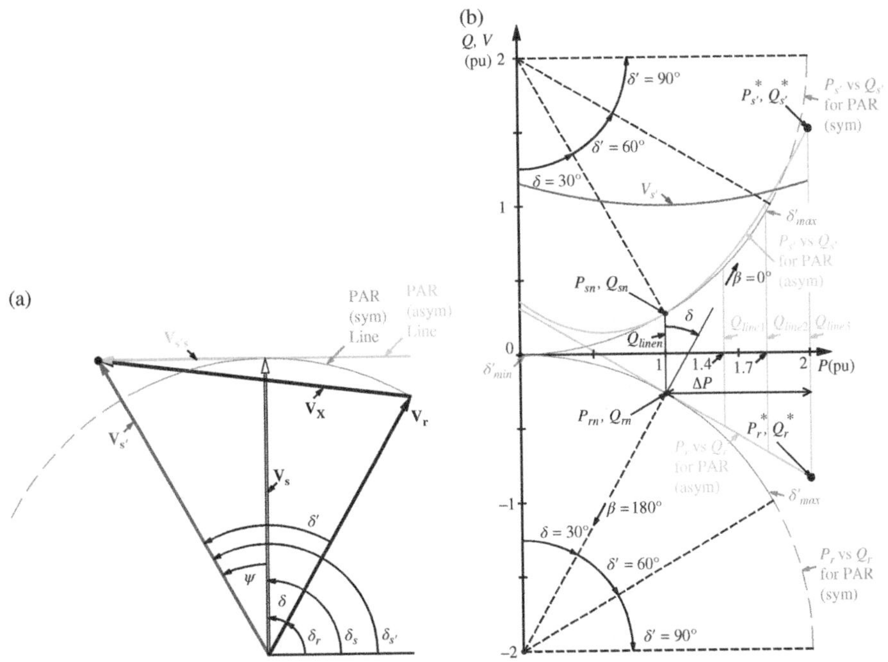

Figure 2-83 (a) Phasor diagram of a two-generator/one-line power system network with a PAR (asym) for Case 3 (100% enhancement of active power flow); (b) power flow characteristics at the modified sending end and the receiving end, superimposed with the power flow characteristics of a PAR (sym).

2.5.4 RR

A RR connects a series-compensating voltage ($V_{s's}$) that is in quadrature with the prevailing line current (I). The ratio of the compensating voltage and the prevailing line current is the compensating reactance (X_{se}). The phasor diagram of a two-generator/one-line power system network, compensated with a RR, is shown in Figure 2-84a for Case 1 with an increased active power flow of 40%. Figure 2-84b shows the related P-Q characteristics. Note the reactive power (Q_{line1}), absorbed by the line, for Case 1 is increased, compared to the reactive power (Q_{linen}) absorbed by an uncompensated line.

The phasor diagram of a two-generator/one-line power system network, compensated with a RR, is shown in Figure 2-85a for Case 2 with an increased active power flow of 70%. Figure 2-85b shows the related P-Q characteristics. Note the reactive power (Q_{line2}), absorbed by the line, for Case 2 is significantly increased, compared to the reactive power (Q_{line1}) absorbed by the line for Case 1.

The phasor diagram of a two-generator/one-line power system network, compensated with a RR, is shown in Figure 2-86a for Case 3 with an increased active power flow of 100%. Figure 2-86b shows the related P-Q characteristics. Note the reactive power (Q_{line3}), absorbed by the line, for Case 3 is significantly increased, compared to the reactive power (Q_{line2}) absorbed by the line for Case 2. However, the Q_{line3} in the case of a RR is less than the Q_{line3} in the case of a PAR (sym or asym). Therefore, if the active power flow needs to be increased by 100%, the use of a RR is preferred over the use of a PAR (sym or asym) in this example. Note that the magnitude of the voltage ($V_{s'}$) at the modified sending end increases in the case of a RR when power flow increases over the natural value and vice versa.

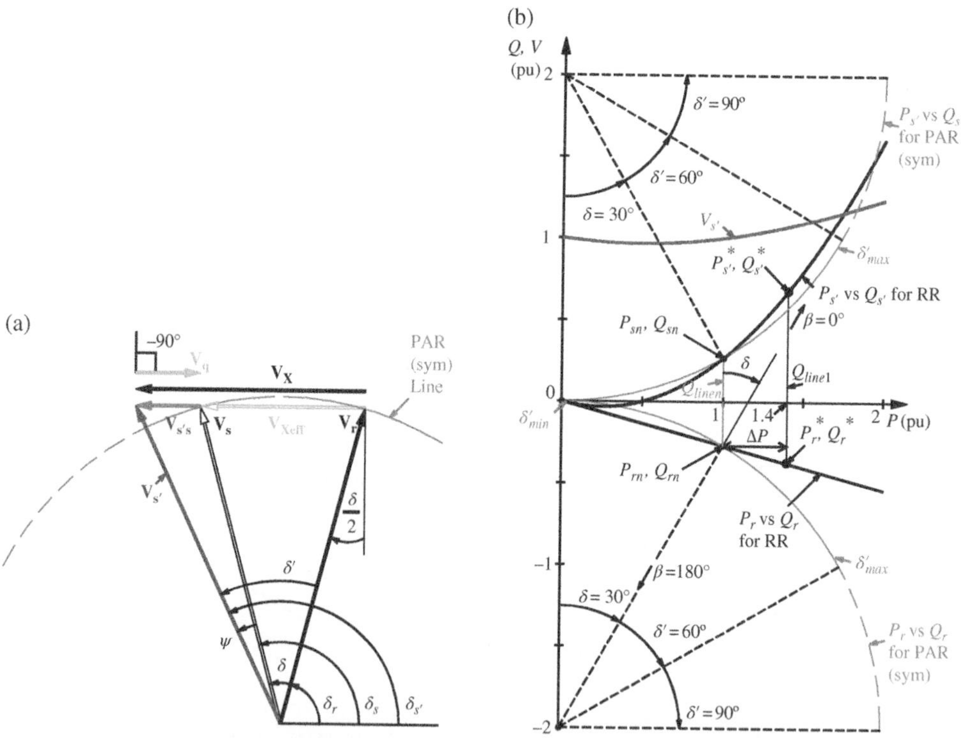

Figure 2-84 (a) Phasor diagram of a two-generator/one-line power system network with a RR for Case 1 (40% enhancement of active power flow); (b) power flow characteristics at the modified sending end and the receiving end, superimposed with the power flow characteristics of a PAR (sym).

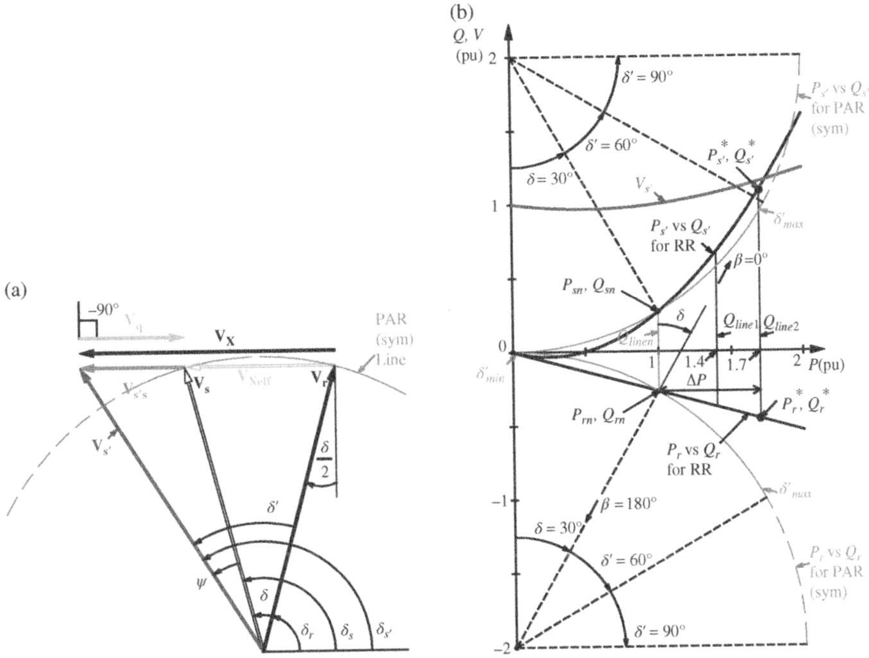

Figure 2-85 (a) Phasor diagram of a two-generator/one-line power system network with a RR for Case 2 (70% enhancement of active power flow); (b) power flow characteristics at the modified sending end and the receiving end, superimposed with the power flow characteristics of a PAR (sym).

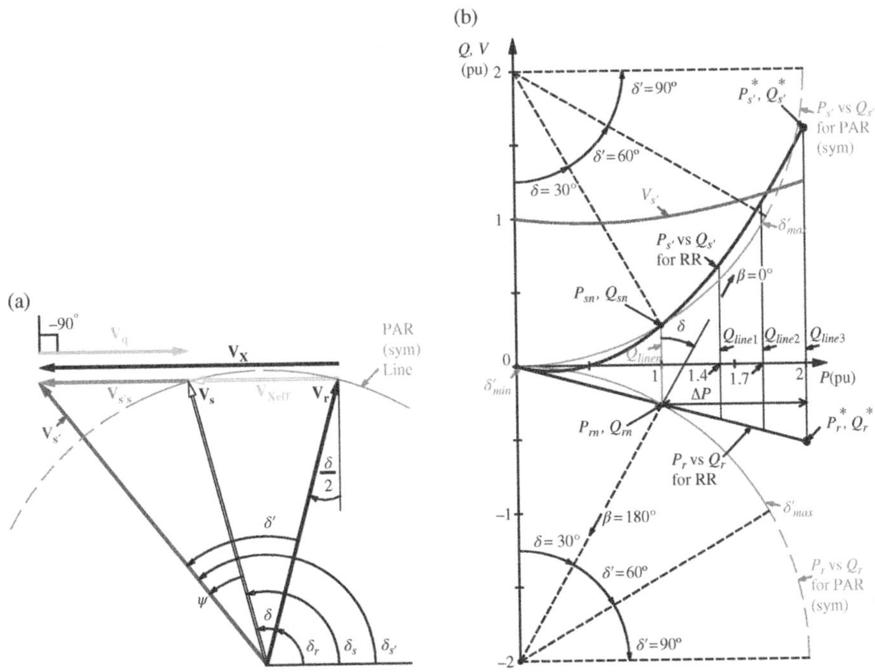

Figure 2-86 (a) Phasor diagram of a two-generator/one-line power system network with a RR for Case 3 (100% enhancement of active power flow); (b) power flow characteristics at the modified sending end and the receiving end, superimposed with the power flow characteristics of a PAR (sym).

2.5.5 IR

For the operation of an IR, a variable magnitude series-compensating voltage is applied during its entire controllable range of $0° \leq \beta \leq 360°$. The *P-Q* characteristic at the modified sending end as described in Section 2.2.2.3 (Power Flow at the Modified Sending End with a Series-Compensating Voltage) is elliptical; the *P-Q* characteristics at the receiving end as described in Section 2.2.2.2 (Power Flow at the Receiving End with a Series-Compensating Voltage) is circular. The corresponding phasor diagrams and the *P-Q* characteristics are as shown in Figures 2-87 through 2-89 for the three cases. The phasor diagram of a two-generator/one-line power system network, compensated with an IR, is shown in Figure 2-87a for Case 1 with an increased active power flow of 40%. Figure 2-87b shows the related *P-Q* characteristics. Note the reactive power (Q_{line1}), absorbed by the line, for Case 1 is increased, compared to the reactive power (Q_{linen}) absorbed by an uncompensated line; however, the reactive power flow at the receiving end remains unchanged as Q_{rn}.

The phasor diagram of a two-generator/one-line power system network, compensated with an IR, is shown in Figure 2-88a for Case 2 with an increased active power flow of 70%. Figure 2-88b shows the related *P-Q* characteristics. Note the reactive power (Q_{line2}), absorbed by the line, for Case 2 is increased, compared to the reactive power (Q_{line1}) absorbed by the line for Case 1; however, the reactive power flow at the receiving end remains unchanged as Q_{rn}.

The phasor diagram of a two-generator/one-line power system network, compensated with an IR, is shown in Figure 2-89a for Case 3 with an increased active power flow of 100%. The figure also shows the variation of voltage magnitude at the modified sending end. Figure 2-89b shows

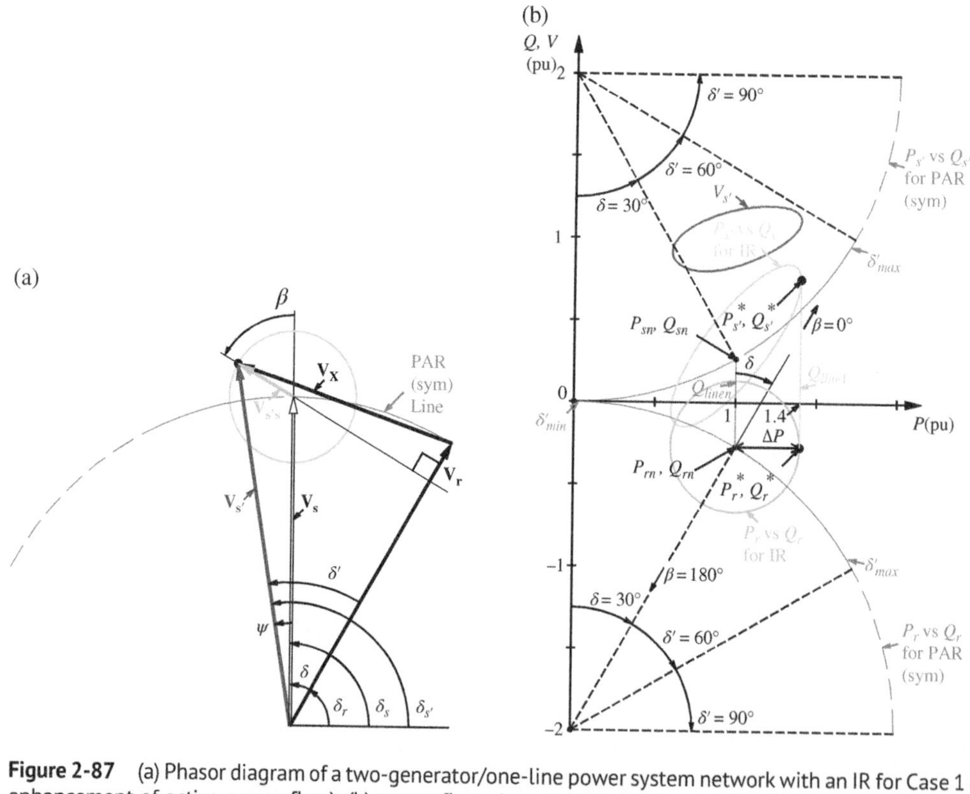

Figure 2-87 (a) Phasor diagram of a two-generator/one-line power system network with an IR for Case 1 (40% enhancement of active power flow); (b) power flow characteristics at the modified sending end and the receiving end, superimposed with the power flow characteristics of a PAR (sym).

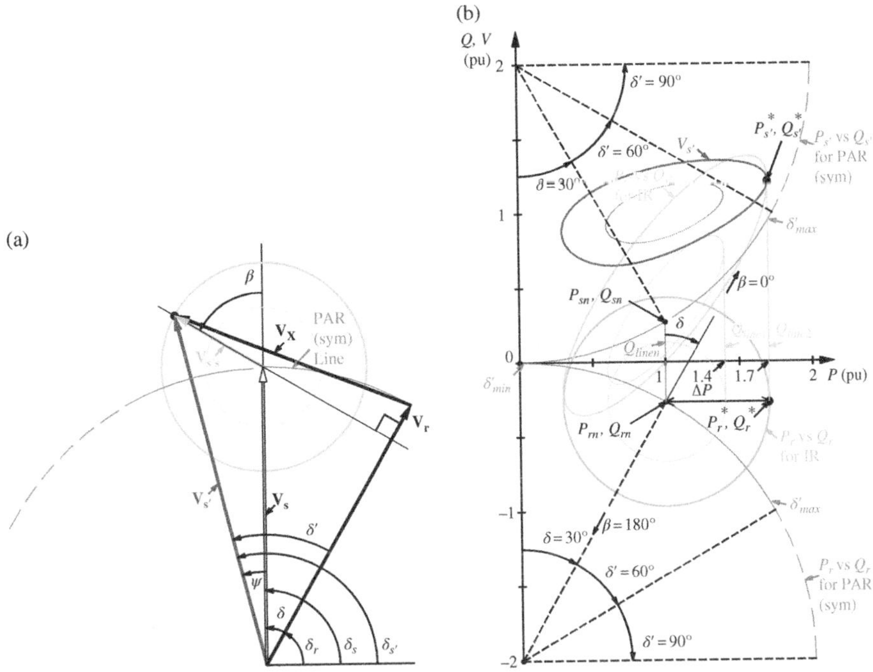

Figure 2-88 (a) Phasor diagram of a two-generator/one-line power system network with an IR for Case 2 (70% enhancement of active power flow); (b) power flow characteristics at the modified sending end and the receiving end, superimposed with the power flow characteristics of a PAR (sym).

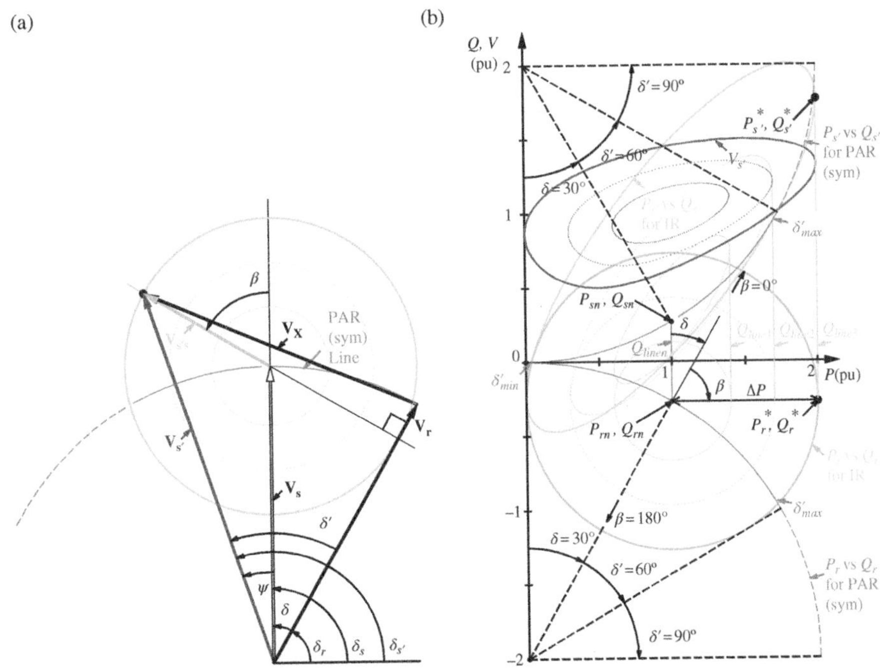

Figure 2-89 (a) Phasor diagram of a two-generator/one-line power system network with an IR for the Case 3 (100% enhancement of active power flow); (b) power flow characteristics at the modified sending end and the receiving end, superimposed with the power flow characteristics of a PAR (sym).

the related *P-Q* characteristics. Note the reactive power (Q_{line3}), absorbed by the line, for Case 3 is increased, compared to Q_{line2}; however, the reactive power flow at the receiving end remains unchanged as Q_{rn}. Figure 2-89b shows that the active power delivery at the load end can be nearly doubled from the natural power flow without any voltage compensation at the load end while maintaining a unity power factor. This is only possible to achieve by using an IR. Note that the magnitude of the voltage ($V_{s'}$) at the modified sending end may be increased or decreased during the entire operating range.

Note that the entire power flow control range, shown in Figures 2-87 through 2-89, of an IR may not be available for use due to system constraints of voltage limitation as shown in Figure 6.12. A VSC-based UPFC cannot be fully utilized due to these voltage constraints. However, an SPFC, such as an ST, can be designed for a limited voltage and limited angle (as shown in Figures 6-65–6-70) operations per functional requirements to maximize its use at the lowest *APR* that results in the lowest cost.

2.5.6 *RPI*, *LI*, and *APR* of a PFC

The cost of delivering active power through a line consists of (1) fixed costs, such as the cost of the line and maintenance cost and (2) load-dependent variable costs, such as need for reactive power in the line and power loss in the line. In order to increase power flow in the line, additional costs are incurred as follows:

a) Need for additional reactive power in the line.
b) Additional power loss in the line.
c) Cost of the PFC, which is directly proportional to the *APR* of the PFC.

For an IR, Equation (2-309) can be rewritten using Equations (2-131), (2-132), and (2-149) as

$$\Delta P = \frac{V_r V_{s's}}{X} \sin(\delta + \beta) \tag{2-312}$$

where ΔP is the change in power flow with respect to the natural power flow.

Therefore, the required rating of the series-compensating voltage for a desired power flow change (ΔP) can be written as

$$V_{s's} = \frac{X \times \Delta P}{V_r \sin(\delta + \beta)}. \tag{2-313}$$

The minimum voltage rating of the IR for a given power flow change occurs when $\delta + \beta = 90°$ or 270° and the resulting series-compensating voltage rating is

$$V_{s's} = \frac{X}{V_r} |\Delta P|. \tag{2-314}$$

For a given power angle (δ), the relative phase angle of the series-compensating voltage is given by

$$\beta = 90° - \delta \qquad \text{or} \qquad \beta = 270° - \delta. \tag{2-315}$$

From Equations (2-152), (2-135), and (2-315), the corresponding reactive power flow at the modified sending end and receiving end are given by

$$Q_{s'} = Q_{sn} + \frac{V_{s's}^2}{X} + 2a_s \sin\delta \tag{2-316}$$

and

$$Q_r = Q_{rn}. \tag{2-317}$$

Knowing the natural active power flow (P_{rn}) at the receiving end and the power flow change (ΔP), the active power flow (P_r) at the receiving end can be determined using Equation (2-309). Then, the apparent power (S_r) is given by

$$S_r = \sqrt{P_r^2 + Q_r^2}. \tag{2-76}$$

Therefore, the line current is given by

$$I = \frac{S_r}{V_r}. \tag{2-77}$$

When the current (I) flows through the line, the line experiences a power loss, which is given by

$$P_{line} = I^2 R \tag{2-318}$$

where R is the line resistance.

The reactive power, absorbed by the line, is given by

$$Q_{line} = I^2 X \tag{2-11b}$$

where X is the line reactance.

The reactive power absorbed by the line can also be given by

$$Q_{line} = Q_{s'} - Q_r. \tag{2-319}$$

Alternately,

$Q_{line} = |Q_{s'}| - |Q_r|$ if the power factors at both ends of the line are either lagging or leading (2-320a)

or

$Q_{line} = |Q_{s'}| + |Q_r|$ if the power factor at one end is lagging and that at the other end is leading. (2-320b)

The Reactive Power Index (RPI), which is a measure of total reactive power (Q_{line}), needed by the line for a unit of active power (P_r) transmitted, is defined as

$$RPI = \frac{Q_{line}}{P_r} = \frac{Q_{line}}{P_{rn} + \Delta P} \tag{2-321}$$

using Equation (2-309).

The power loss in the uncompensated line, carrying the natural current (I_n), is given as baseline loss by

$$P_{linen} = I_n^2 R_n \tag{2-322}$$

where R_n is the natural line resistance.

The reactive power absorbed by the line, carrying the natural current (I_n), is given as baseline reactive power by

$$Q_{linen} = I_n^2 X. \tag{2-323}$$

The Loss Index (LI), which is a measure of line loss as a multiple of baseline loss, is defined as

$$LI = k_R \frac{I^2}{I_n^2} \tag{2-324}$$

where

$$k_R = \frac{R}{R_n}, \tag{2-325}$$

the line resistance is R at line current (I) and the natural line resistance is R_n at natural line current (I_n). Note that as the line current (I) increases, the loss increases, which increases the temperature and, in turn, the resistance of the conductor.

The Apparent Power Rating (APR) of the PFC is defined as

$$APR = |V_{s's}|I \tag{2-326}$$

when the PFC is used to increase power flow with a prevailing line current (I);

$$APR = |V_{s's}|I_n \tag{2-327}$$

when the PFC is used to decrease power flow from a natural line current (I_n).

Therefore, Table 2-1 can be expanded to include the base case power flow data as shown in Table 2-25.

Considering the base case active power flow of one pu, the phasor diagrams for no power flow and 100% power flow increase, and the related P-Q diagrams are shown in Figures 2-90 through 2-93 using a PAR (sym), a PAR (asym), a RR, and an IR, respectively. In all cases, when $\psi = -\delta, \delta' = 0°$ and the active power flow stops. However, the reactive power only stops in the case of a PAR (sym) since $V_{s'}/V_r = V_r/V_{s'} = 1 = \cos\delta'$, but continues to flow from the modified sending end in the case of a PAR (asym) and from the receiving end in the case of an IR due to unequal voltages at the modified sending end and receiving end $(V_{s'}/V_r \neq 1, V_r/V_{s'} \neq 1)$, which are shown in

Table 2-25 Electrical system data and base power flow data.

Parameters	Values
Sending-end voltage (V_s) at reference phase angle	$1\angle 0°$ pu
Receiving-end voltage (V_r) and power angle ($-\delta$)	$1\angle -30°$ pu
Line reactance (X)	0.5 pu
Natural transmitted active power ($P_{sn} = P_{rn}$)	1 pu
Reactive power (Q_{sn}) at source end	0.267949 pu
Reactive power (Q_{rn}) at load end	−0.267949 pu
Natural line current (I_n)	1.035276 pu
Natural line current^2 (I_n^2)	1.071797 pu
Reactive power $(Q_{linen} = I_n^2 X)$, absorbed by an uncompensated line	0.535898 pu
Reactive Power Index, $RPI = Q_{linen}/(P_{rn} + \Delta P)$	0.535898
Loss Index, $LI = k_R \times I^2/I_n^2$ at natural line current (i.e. $I = I_n$); $k_R = 1$	1.0717097/1.0717097 = 1

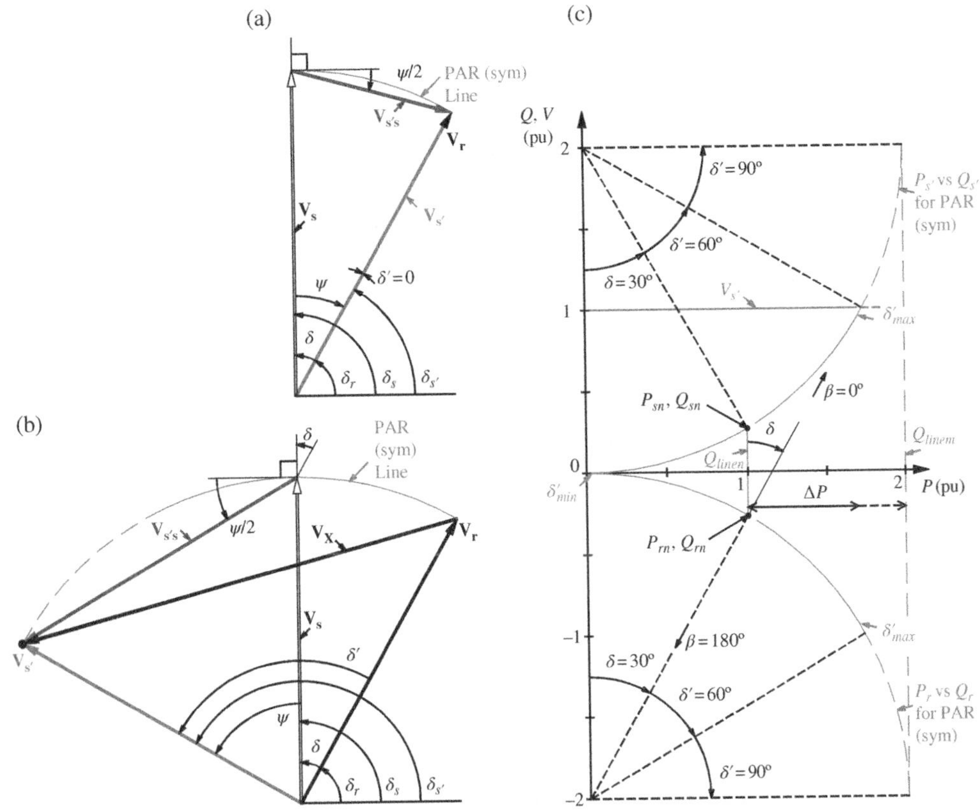

Figure 2-90 Phasor diagram of a PAR (sym) for (a) no power flow, (b) 100% power flow increase, and (c) related *P-Q* diagram.

Figures 2-91 and 2-93. The reactive power absorbed by the entire line is the difference between $Q_{s'}$ and Q_r. This is how a var compensator (SynCon, SVC, or STATCOM) operates.

Example 2-42

Consider the electrical system data as given in Table 2-1. For a PAR (sym) operating at $\psi = -30°$, calculate $P_{s'}$, $Q_{s'}$, P_r, Q_r, R_{se}, and X_{se}.

Answer 2-42

For a PAR (sym), $V_{s'} = V_s$.

From Equation (2-198), $V_{s's} = 2V_s \sin\left(\frac{\psi}{2}\right) = 2 \times 1 \times \sin\left(\frac{-30°}{2}\right) = -0.517638$ pu.

From Equation (2-200), $\beta = \frac{\psi}{2} + 270°$ for $\psi < 0$; $\beta = 255°$.

From Equation (2-51a), $\delta' = \delta + \psi = 30° - 30° = 0°$.

From Equation (2-41), $a = \frac{V_s V_r}{X} = \frac{1 \times 1}{0.5} = 2$.

From Equation (2-201), $P_{s'} = a \times \sin\delta' = 2 \times \sin 0° = 0$ pu.

From Equation (2-202), $Q_{s'} = a\left(\frac{V_s}{V_r} - \cos\delta'\right) = 2\left(\frac{1}{1} - \cos 0°\right) = 0$ pu.

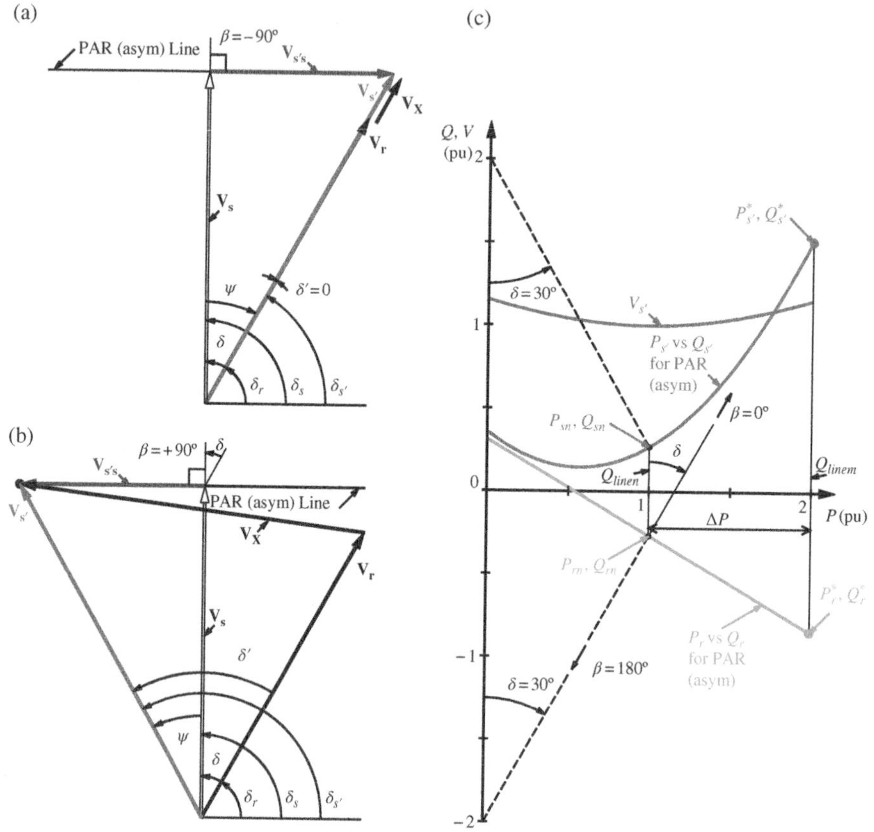

Figure 2-91 Phasor diagram of PAR (asym) for (a) no power flow, (b) 100% power flow increase, and (c) related *P-Q* diagram.

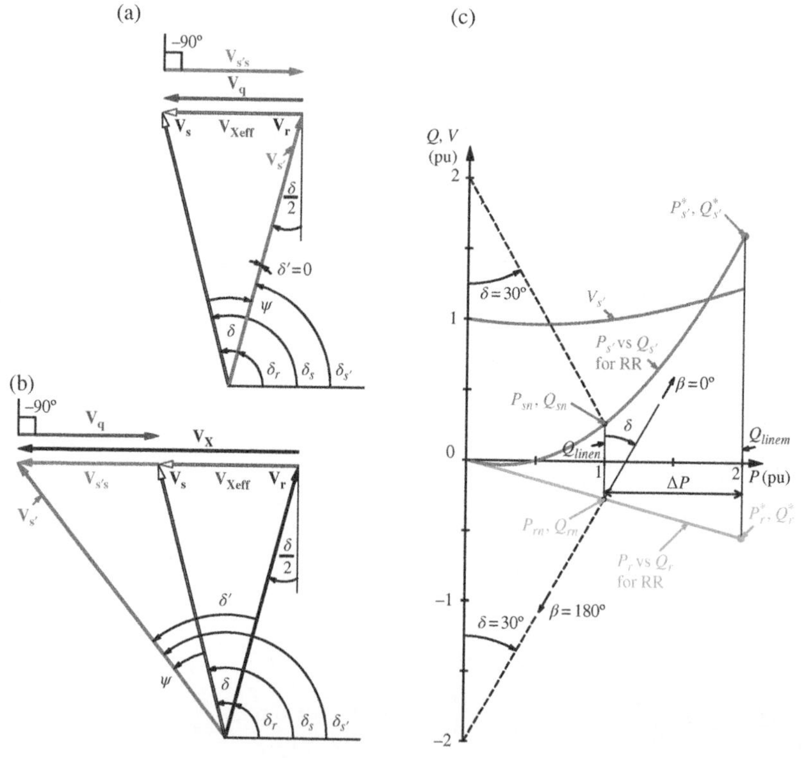

Figure 2-92 Phasor diagram of a RR for (a) no power flow, (b) 100% power flow increase, and (c) related *P-Q* diagram.

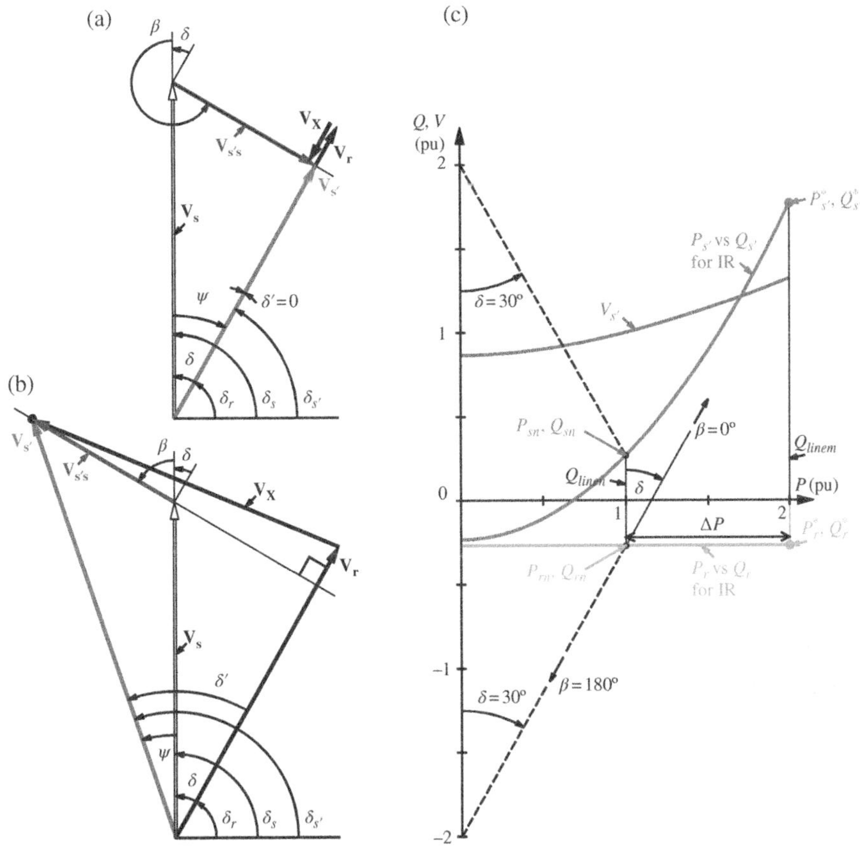

Figure 2-93 Phasor diagram of an IR for (a) no power flow, (b) 100% power flow increase, and (c) related *P-Q* diagram.

From Equation (2-203), $P_r = a \times \sin\delta' = 2 \times \sin 0° = 0$ pu.

From Equation (2-204), $Q_r = a\left(\cos\delta' - \frac{V_r}{V_s}\right) = 2\left(\cos 0° - \frac{1}{1}\right) = 0$ pu.

From Equation (2-54b), $V_X = \sqrt{V_{s'}^2 - 2V_{s'}V_r\cos\delta' + V_r^2} = \sqrt{1^2 - 2\times 1\times 1\times \cos 0° + 1^2} = 0$ pu.

From Equation (2-253a), the series-compensating resistance (R_{se}) can be calculated as

$$R_{se} = X\frac{V_{s's}}{V_X}\left\{\frac{V_s}{V_X}\cos\left(\frac{\psi}{2}\right) - \frac{V_r}{V_X}\cos\left(\delta + \frac{\psi}{2}\right)\right\} = 0.5 \times \frac{0.517638}{0}$$
$$\left\{\frac{1}{0}\cos\left(\frac{-30°}{2}\right) - \frac{1}{0}\cos\left(30° - \frac{30°}{2}\right)\right\} = \infty.$$

From Equation (2-253b), the series-compensating reactance (X_{se}) can be calculated as

$$X_{se} = X\frac{V_{s's}}{V_X}\left\{-\frac{V_s}{V_X}\sin\left(\frac{\psi}{2}\right) + \frac{V_r}{V_X}\sin\left(\delta + \frac{\psi}{2}\right) + \frac{V_{s's}}{V_X}\right\} = 0.5\frac{0.517638}{0}$$
$$\left\{-\frac{1}{0}\sin\frac{-30°}{2} + \frac{1}{0}\sin\left(30° - \frac{30°}{2}\right) + \frac{0.517638}{0}\right\} = \infty.$$

Example 2-43

Consider the electrical system data as given in Table 2-1. For a PAR (sym) operating at $\psi = 60°$, calculate $P_{s'}$, $Q_{s'}$, P_r, Q_r, R_{se}, X_{se}, *RPI*, *LI* for $k_R = 1$, and *APR*. Adjusting the control parameter (ψ), tabulate *RPI*, *LI*, and *APR* of a PAR (sym) for power flow increases in three cases: Case 1 (40%), Case 2 (70%), and Case 3 (100%).

Answer 2-43

For a PAR (sym), $V_{s'} = V_s$.

From Equation (2-198), $V_{s's} = 2V_s \sin\left(\frac{\psi}{2}\right) = 2 \times 1 \times \sin\left(\frac{60°}{2}\right) = 1$ pu.

From Equation (2-200), $\beta = \frac{\psi}{2} + 90°$ for $\psi > 0$; $\beta = 120°$.

From Equation (2-51a), $\delta' = \delta + \psi = 30° + 60° = 90°$.

From Equation (2-41), $a = \frac{V_s V_r}{X} = \frac{1 \times 1}{0.5} = 2$.

From Equation (2-201), $P_{s'} = a \times \sin\delta' = 2 \times \sin 90° = 2$ pu.

From Equation (2-202), $Q_{s'} = a\left(\frac{V_s}{V_r} - \cos\delta'\right) = 2\left(\frac{1}{1} - \cos 90°\right) = 2$ pu.

From Equation (2-203), $P_r = a \times \sin\delta' = 2 \times \sin 90° = 2$ pu.

From Equation (2-204), $Q_r = a\left(\cos\delta' - \frac{V_r}{V_s}\right) = 2\left(\cos 90° - \frac{1}{1}\right) = -2$ pu.

From Equation (2-76), $S_r = \sqrt{P_r^2 + Q_r^2} = \sqrt{2^2 + 2^2} = 2.828427$ pu.

From Equation (2-77), $I = \frac{S_r}{V_r} = \frac{2.828427}{1} = 2.828427$ pu.

From Equation (2-11b), $Q_{line} = I^2 X = 8 \times 0.5 = 4$ pu, which is the same as $Q_{s'} - Q_r = 2 - (-2) = 4$ pu, according to Equation (2-319).

From Equation (2-30), $V_{Xn} = \sqrt{V_s^2 - 2V_s V_r \cos\delta + V_r^2} = \sqrt{1^2 - 2 \times 1 \times 1 \times \cos 30° + 1^2}$ $= 0.517638$ pu.

From Equation (2-33), $I_n = \frac{V_{Xn}}{X} = \frac{0.517638}{0.5} = 1.035276$ pu.

From Equation (2-321), $RPI = \frac{Q_{line}}{P_r} = \frac{4}{2} = 2$ pu.

From Equation (2-324), $LI = k_R \frac{I^2}{I_n^2} = 1 \times \frac{2.828427^2}{1.035276^2} = 7.464102$.

From Equation (2-326), $APR = |V_{s's}| I = 1 \times 2.828427 = 2.828427$ pu.

From Equation (2-54b), $V_X = \sqrt{V_{s'}^2 - 2V_{s'} V_r \cos\delta' + V_r^2} = \sqrt{1^2 - 2 \times 1 \times 1 \times \cos 90° + 1^2}$ $= 1.414214$ pu.

From Equation (2-253a), the series-compensating resistance (R_{se}) can be calculated as

$$R_{se} = X \frac{V_{s's}}{V_X}\left\{\frac{V_s}{V_X}\cos\left(\frac{\psi}{2}\right) - \frac{V_r}{V_X}\cos\left(\delta + \frac{\psi}{2}\right)\right\} = 0.5 \times \frac{1}{1.414214}$$

$$\left\{\frac{1}{1.414214}\cos\left(\frac{60°}{2}\right) - \frac{1}{1.414214}\cos\left(30° + \frac{60°}{2}\right)\right\} = 0.091506 \text{ pu.}$$

Table 2-26 *RPI*, *LI*, and *APR* of a PAR (sym) for power flow increases in three cases: Case 1 (40%), Case 2 (70%), and Case 3 (100%).

Parameters	Base case	Case 1	Case 2	Case 3
Modified sending-end voltage ($V_{s'}$)	1 pu	1 pu	1 pu	1 pu
Phase-shift angle (ψ)	0°	14.427°	28.2117°	60°
Series-compensating voltage ($V_{s's}$)	0	0.2511 pu	0.4874 pu	1 pu
Relative phase angle (β)	0°	97.2135°	104.1058°	120°
Active power ($P_{s'} = P_r$)	1 pu	1.4 pu	1.7 pu	2 pu
Reactive power ($Q_{s'}$) at modified sending end	0.2679 pu	0.5717 pu	0.9464 pu	2 pu
Reactive power (Q_r) at receiving end	−0.2679 pu	−0.5717 pu	−0.9464 pu	−2 pu
Line current (I)	1.0353 pu	1.5122 pu	1.9457 pu	2.8284 pu
Line reactive power ($Q_{line} = I^2X$)	0.5359 pu	1.1434 pu	1.8929 pu	4 pu
Series-compensating Resistance (R_{se})	0	0.043 pu	0.0648 pu	0.0915 pu
Series-compensating Reactance (X_{se})	0	0.1604 pu	0.242 pu	0.3415 pu
Reactive Power Index, $RPI = Q_{line}/P_r$	0.5359	0.8167	1.1135	2
Loss Index, $LI = k_R \times I^2/I_n^2$ at natural line current (i.e. $I = I_n$); $k_R = 1$	1	2.1337	3.5321	7.4641
$APR = \|V_{s's}\|I$	0	0.3798 pu	0.9484 pu	2.8284 pu

From Equation (2-253b), the series-compensating reactance (X_{se}) can be calculated as

$$X_{se} = X\frac{V_{s's}}{V_X}\left\{-\frac{V_s}{V_X}\sin\left(\frac{\psi}{2}\right) + \frac{V_r}{V_X}\sin\left(\delta + \frac{\psi}{2}\right) + \frac{V_{s's}}{V_X}\right\}$$

$$= 0.5 \times \frac{1}{1.414214}\left\{-\frac{1}{1.414214}\sin\left(\frac{60^\circ}{2}\right) + \frac{1}{1.414214}\sin\left(30^\circ + \frac{60^\circ}{2}\right) + \frac{1}{1.414214}\right\}$$

$$= 0.341506 \text{ pu.}$$

Table 2-26 shows the *RPI*, *LI*, and *APR* of a PAR (sym) for power flow increases in three cases: Case 1 (40%), Case 2 (70%), and Case 3 (100%).

Example 2-44

Consider the electrical system data as given in Table 2-1. For a PAR (asym) operating at $V_{s's} = -0.577350$ pu, calculate $P_{s'}$, $Q_{s'}$, P_r, Q_r, R_{se}, and X_{se}.

Answer 2-44

From Equation (2-270), $V_{s'} = \sqrt{V_s^2 + V_{s's}^2} = \sqrt{1^2 + (-0.577350)^2} = 1.154701$ pu.

From Equation (2-271a), $\psi = \tan^{-1}\frac{V_{s's}}{V_s} = \tan^{-1}\frac{-0.577350}{1} = -30°$.

From Equation (2-51a), $\delta' = \delta + \psi = 30° - 30° = 0°$.

From Equation (2-61), $a' = \frac{V_{s'}V_r}{X} = \frac{1.154701 \times 1}{0.5} = 2.309402$.

From Equation (2-59), $P_{s'} = a' \sin\delta' = 2.309402 \times \sin 0° = 0$ pu.

From Equation (2-60), $Q_{s'} = a'\left(\frac{V_{s'}}{V_r} - \cos\delta'\right) = 2.309402\left(\frac{1.154701}{1} - \cos 0°\right) = 0.357266$ pu.

From Equation (2-74), $P_r = a' \sin\delta' = 2.309402 \times \sin 0° = 0$ pu.

From Equation (2-75), $Q_r = a'\left(\cos\delta' - \frac{V_r}{V_{s'}}\right) = 2.309402\left(\cos 0° - \frac{1}{1.154701}\right) = 0.309401$ pu.

From Equation (2-76), $S_r = \sqrt{P_r^2 + Q_r^2} = \sqrt{0^2 + 0.309401^2} = 0.309401$ pu.

From Equation (2-77), $I = \frac{S_r}{V_r} = \frac{0.309401}{1} = 0.309401$ pu.

From Equation (2-54b), $V_X = \sqrt{V_{s'}^2 - 2V_{s'}V_r\cos\delta' + V_r^2} = \sqrt{1.154701^2 - 2 \times 1.154701 \times 1 \times \cos 0° + 1^2}$ $= 0.154701$ pu.

From Equation (2-251a), the series-compensating resistance (R_{se}) can be calculated as

$$R_{se} = X\frac{V_{s's}}{V_X}\frac{V_s - V_r\cos\delta}{V_X} = 0.5 \times \frac{-0.577350}{0.154701}\frac{1 - 1 \times \cos 30°}{0.154701} = -1.616025 \text{ pu}.$$

From Equation (2-251b), the series-compensating reactance (X_{se}) can be calculated as

$$X_{se} = X\frac{V_{s's}}{V_X}\frac{V_r\sin\delta + V_{s's}}{V_X} = 0.5 \times \frac{-0.577350}{0.154701}\frac{1 \times \sin 30° - 0.577350}{0.154701} = 0.933013 \text{ pu}.$$

Additional Discussion

From Equation (2-11b), $Q_{line} = I^2X = 0.309401^2 \times 0.5 = 0.047865$ pu, which is the same as $Q_{line} = Q_{s'} - Q_r = 0.357266 - 0.309401 = 0.047865$ pu from Equation (2-319). Note that both $Q_{s'}$ and Q_r are positive numbers, meaning reactive power flows through the line from the higher voltage sending end to the lower voltage receiving end. This is how a STATCOM works, which is explained in Figure 2-54.

Example 2-45

Consider the electrical system data as given in Table 2-1. For a PAR (asym) operating at $V_{s's} = 0.577350$ pu, calculate $P_{s'}$, $Q_{s'}$, P_r, Q_r, R_{se}, X_{se}, *RPI*, *LI* for $k_R = 1$, and *APR*. Adjusting the control parameter, $V_{s's}$, tabulate *RPI*, *LI*, and *APR* of a PAR (asym) for power flow increases in three cases: Case 1 (40%), Case 2 (70%), and Case 3 (100%).

Answer 2-45

From Equation (2-270), $V_{s'} = \sqrt{V_s^2 + V_{s's}^2} = \sqrt{1^2 + 0.577350^2} = 1.154701$ pu.

From Equation (2-271a), $\psi = \tan^{-1}\frac{V_{s's}}{V_s} = \tan^{-1}\frac{0.577350}{1} = 30°$.

From Equation (2-51a), $\delta' = \delta + \psi = 30° + 30° = 60°$.

From Equation (2-61), $a' = \frac{V_{s'}V_r}{X} = \frac{1.154701 \times 1}{0.5} = 2.309402$.

From Equation (2-59), $P_{s'} = a' \sin\delta' = 2.309402 \times \sin 60° = 2$ pu.

From Equation (2-60), $Q_{s'} = a'\left(\frac{V_{s'}}{V_r} - \cos\delta'\right) = 2.309402\left(\frac{1.154701}{1} - \cos 60°\right) = 1.511967$ pu.

From Equation (2-74), $P_r = a' \sin \delta' = 2.309402 \times \sin 60° = 2$ pu.

From Equation (2-75), $Q_r = a'\left(\cos \delta' - \frac{V_r}{V_{s'}}\right) = 2.309402\left(\cos 60° - \frac{1}{1.154701}\right) = -0.845299$ pu.

From Equation (2-76), $S_r = \sqrt{P_r^2 + Q_r^2} = \sqrt{2^2 + 0.845299^2} = 2.171297$ pu.

From Equation (2-77), $I = \frac{S_r}{V_r} = \frac{2.171297}{1} = 2.171297$ pu.

From Equation (2-11b), $Q_{line} = I^2X = 2.171297^2 \times 0.5 = 2.357266$ pu, which is same as $Q_{s'} - Q_r = 1.511967 - (-0.845299) = 2.357266$ pu, according to Equation (3-19).

From Equation (2-30), $V_{Xn} = \sqrt{V_s^2 - 2V_sV_r \cos \delta + V_r^2} = \sqrt{1^2 - 2 \times 1 \times 1 \times \cos 30° + 1^2}$ $= 0.517638$ pu.

From Equation (2-33), $I_n = \frac{V_{Xn}}{X} = \frac{0.517638}{0.5} = 1.035276$ pu.

From Equation (2-321), $RPI = \frac{Q_{line}}{P_r} = \frac{2.357266}{2} = 1.178633$ pu.

From Equation (2-324), $LI = k_R \frac{I^2}{I_n^2} = 1 \times \frac{2.171297^2}{1.035276^2} = 4.398717$.

From Equation (2-326), $APR = |V_{s's}|I = 0.577350 \times 2.171297 = 1.253599$ pu.

From Equation (2-54b),

$$V_X = \sqrt{V_{s'}^2 - 2V_{s'}V_r \cos \delta' + V_r^2} = \sqrt{1.154701^2 - 2 \times 1.154701 \times 1 \times \cos 60° + 1^2} = 1.085649 \text{ pu.}$$

From Equation (2-251a), the series-compensating resistance (R_{se}) can be calculated as

$$R_{se} = X \frac{V_{s's}}{V_X} \frac{V_s - V_r \cos \delta}{V_X} = 0.5 \times \frac{0.577350}{1.085649} \frac{1 - 1 \times \cos 30°}{1.085649} = 0.032814 \text{ pu.}$$

From Equation (2-251b), the series-compensating reactance (X_{se}) can be calculated as

$$X_{se} = X \frac{V_{s's}}{V_X} \frac{V_r \sin \delta + V_{s's}}{V_X} = 0.5 \times \frac{0.577350}{1.085649} \frac{1 \times \sin 30° + 0.577350}{1.085649} = 0.263869 \text{ pu.}$$

Table 2-27 shows the *RPI*, *LI*, and *APR* of a PAR (asym) for power flow increases in three cases: Case 1 (40%), Case 2 (70%), and Case 3 (100%).

Example 2-46

Consider the electrical system data as given in Table 2-1. For a RR operating at $V_{s's} = -0.517638$ pu, calculate P_r, Q_r, $P_{s'}$, $Q_{s'}$, R_{se}, and X_{se}.

Answer 2-46

From Equation (2-229),

$$V_{s'} = \sqrt{V_s^2 + V_{s's}^2 + 2V_sV_{s's}\frac{V_s - V_r \cos \delta}{V_{Xn}}}$$
$$= \sqrt{1^2 + (-0.517638)^2 + 2 \times 1 \times (-0.517638)\frac{1 - 1 \times \cos 30°}{0.517638}} = 1.0 \text{ pu.}$$

Table 2-27 *RPI*, *LI*, and *APR* of a PAR (asym) for power flow increases in three cases: Case 1 (40%), Case 2 (70%), and Case 3 (100%).

Parameters	Base case	Case 1	Case 2	Case 3
Modified sending-end voltage ($V_{s'}$)	1 pu	1.0263 pu	1.0786 pu	1.1547 pu
Phase-shift angle (ψ)	0°	13°	22°	30°
Series-compensating voltage ($V_{s's}$)	0	0.2309 pu	0.4041 pu	0.5774 pu
Relative phase angle (β)	0°	90°	90°	90°
Active power ($P_{s'} = P_r$)	1 pu	1.4 pu	1.7 pu	2 pu
Reactive power ($Q_{s'}$) at modified sending end	0.2679 pu	0.6056 pu	0.9988 pu	1.512 pu
Reactive power (Q_r) at receiving end	−0.2679 pu	−0.4989 pu	−0.6721 pu	−0.8453 pu
Line current (I)	1.0353 pu	1.4862 pu	1.8280 pu	2.1713 pu
Line reactive power ($Q_{line} = I^2X$)	0.5359 pu	1.1044 pu	1.6709 pu	2.3573 pu
Series-compensating Resistance (R_{se})	0	0.0280 pu	0.0324 pu	0.0328 pu
Series-compensating Reactance (X_{se})	0	0.1528 pu	0.2187 pu	0.2639 pu
Reactive Power Index, $RPI = Q_{line}/P_r$	0.5359	0.7889	0.9829	1.1786
Loss Index, $LI = k_R \times I^2/I_n^2$ at natural line current (*i.e.* $I = I_n$); $k_R = 1$	1	2.0609	3.1179	4.3987
$APR = \|V_{s's}\|I$	0	0.3432 pu	0.7388 pu	1.2536 pu

From Equation (2-230),

$$\psi = \tan^{-1}\frac{V_{s's}V_r \sin\delta}{V_sV_{Xn} + V_{s's}(V_s - V_r\cos\delta)}$$
$$= \tan^{-1}\frac{(-0.517638)\times 1\times \sin 30^\circ}{1\times 0.517638 + (-0.517638)\times(1-1\times\cos 30^\circ)} = -30^\circ$$

From Equation (2-51a), $\delta' = \delta + \psi = 30^\circ - 30^\circ = 0^\circ$.

From Equation (2-235), $P_r = \left(1+\frac{V_{s's}}{V_{Xeff}}\right)P_{rn} = \left(1+\frac{-0.517638}{0.517638}\right)\times 1 = 0.$

From Equation (2-236), $Q_r = \left(1+\frac{V_{s's}}{V_{Xeff}}\right)Q_{rn} = \left(1+\frac{-0.517638}{0.517638}\right)\times(-0.267949) = 0.$

From Equation (2-76), $S_r = \sqrt{P_r^2 + Q_r^2} = \sqrt{0^2+0^2} = 0.$

From Equation (2-77), $I = \frac{S_r}{V_r} = \frac{0}{1} = 0.$

From Equation (2-11b), $Q_{line} = I^2X = 0^2 \times 0.5 = 0.$

From Equation (2-238), $P_{s'} = \left(1+\frac{V_{s's}}{V_{Xeff}}\right)P_{rn} = \left(1+\frac{-0.517638}{0.517638}\right)\times 1 = 0.$

From Equation (2-78a), $Q_{s'} = Q_r + Q_{line} = 0 + 0 = 0.$

From Equation (2-260), $V_X = V_{Xn} + V_{s's} = 0.517638 + (-0.517638) = 0.$

From Equation (2-257), $R_{se} = 0.$

From Equation (2-262), $X_{se} = X\frac{V_{s's}}{V_X} = 0.5\times\frac{-0.517638}{0} = \infty.$

Example 2-47

Consider the electrical system data as given in Table 2-1. For a RR operating at $V_{s's} = 0.5$ pu, calculate P_r, Q_r, $P_{s'}$, $Q_{s'}$, R_{se}, X_{se}, *RPI*, *LI* for $k_R = 1$, and *APR*. Adjusting the control parameter, $V_{s's}$, tabulate *RPI*, *LI*, and *APR* of a RR for power flow increases in three cases: Case 1 (40%), Case 2 (70%), and Case 3 (100%).

Answer 2-47

From Equation (2-229),

$$V_{s'} = \sqrt{V_s^2 + V_{s's}^2 + 2V_sV_{s's}\frac{V_s - V_r\cos\delta}{V_{Xn}}}$$

$$= \sqrt{1^2 + 0.517638^2 + 2 \times 1 \times 0.517638\frac{1 - 1 \times \cos 30°}{0.517638}} = 1.239314 \text{ pu.}$$

From Equation (2-230),

$$\psi = \tan^{-1}\frac{V_{s's}V_r\sin\delta}{V_sV_{Xn} + V_{s's}(V_s - V_r\cos\delta)}$$

$$= \tan^{-1}\frac{0.517638 \times 1 \times \sin 30°}{1 \times 0.517638 + 0.517638 \times (1 - 1 \times \cos 30°)} = 23.793977°.$$

From Equation (2-51a), $\delta' = \delta + \psi = 30° + 23.793977° = 53.793977°$.

From Equation (2-235), $P_r = \left(1 + \frac{V_{s's}}{V_{Xeff}}\right)P_{rn} = \left(1 + \frac{0.517638}{0.517638}\right) \times 1 = 2$ pu.

From Equation (2-236), $Q_r = \left(1 + \frac{V_{s's}}{V_{Xeff}}\right)Q_{rn} = \left(1 + \frac{0.517638}{0.517638}\right) \times (-0.267949) = -0.535898$ pu.

From Equation (2-76), $S_r = \sqrt{P_r^2 + Q_r^2} = \sqrt{2^2 + (-0.535898)^2} = 2.070552$ pu.

From Equation (2-77), $I = \frac{S_r}{V_r} = \frac{2.070552}{1} = 2.070552$ pu.

From Equation (2-11b), $Q_{line} = I^2X = 2.070552^2 \times 0.5 = 2.143594$ pu.

From Equation (2-238), $P_{s'} = \left(1 + \frac{V_{s's}}{V_{Xeff}}\right)P_{rn} = \left(1 + \frac{0.517638}{0.517638}\right) \times 1 = 2$ pu.

From Equation (2-78a), $Q_{s'} = Q_r + Q_{line} = -0.535898 + 2.143594 = 1.607695$ pu.

From Equation (2-260), $V_X = V_{Xn} + V_{s's} = 0.517638 + 0.517638 = 1.035276$, pu.

From Equation (2-257), $R_{se} = 0$.

From Equation (2-262), $X_{se} = X\frac{V_{s's}}{V_X} = 0.5 \times \frac{0.517638}{1.035276} = 0.25$ pu.

Table 2-28 shows the *RPI*, *LI*, and *APR* of a RR for power flow increases in three cases: Case 1 (40%), Case 2 (70%), and Case 3 (100%).

Table 2-28 *RPI*, *LI*, and *APR* of a RR for power flow increases in three cases: Case 1 (40%), Case 2 (70%), and Case 3 (100%).

Parameters	Base case	Case 1	Case 2	Case 3
Modified sending-end voltage ($V_{s'}$)	1 pu	1.0724 pu	1.1484 pu	1.2393 pu
Phase-shift angle (ψ)	0°	10.7484°	17.7442°	23.794°
Series-compensating voltage ($V_{s's}$)	0	0.207055 pu	0.362347 pu	0.517638 pu
Relative phase angle (β)	0°	75°	75°	75°
Active power ($P_{s'} = P_r$)	1 pu	1.4 pu	1.7 pu	2 pu
Reactive power ($Q_{s'}$) at modified sending end	0.2679 pu	0.6752 pu	1.0932 pu	1.6077 pu
Reactive power (Q_r) at receiving end	−0.2679 pu	−0.3751 pu	−0.4556 pu	−0.5359 pu
Line current (I)	1.0353 pu	1.4494 pu	1.76 pu	2.0706 pu
Line reactive power ($Q_{line} = I^2X$)	0.5359 pu	1.0504 pu	1.5487 pu	2.1436 pu
Series-compensating Resistance (R_{se})	0	0.0 pu	0.0 pu	0.0 pu
Series-compensating Reactance (X_{se})	0	0.1429 pu	0.2059 pu	0.25 pu
Reactive Power Index, $RPI = Q_{line}/P_r$	0.5359	0.7502	0.911	1.0718
Loss Index, $LI = k_R \times I^2/I_n^2$ at natural line current (*i.e.* $I = I_n$); $k_R = 1$	1	1.96	2.89	4
$APR = \lvert V_{s's}\rvert I$	0	0.300 pu	0.6377 pu	1.0718 pu

Example 2-48

Consider the electrical system data as given in Table 2-1. For an IR operating at $V_{s's} = 0.5$ pu and $\beta = 240°$, calculate $P_{s'}$, $Q_{s'}$, P_r, Q_r, R_{se}, and X_{se}.

Answer 2-48

From Equation (2-97),

$$V_{s'} = \sqrt{V_s^2 + V_{s's}^2 + 2V_sV_{s's}\cos\beta} = \sqrt{1^2 + 0.5^2 + 2 \times 1 \times 0.5 \times \cos 240°} = 0.866025 \text{ pu.}$$

From Equation (2-98), $\psi = \tan^{-1}\dfrac{V_{s's}\sin\beta}{V_s + V_{s's}\cos\beta} = \tan^{-1}\dfrac{0.5 \times \sin 240°}{1 + 0.5 \times \cos 240°} = -30°.$

From Equation (2-51a), $\delta' = \delta + \psi = 30° - 30° = 0°$.

From Equation (2-61), $a' = \dfrac{V_{s'}V_r}{X} = \dfrac{0.866025 \times 1}{0.5} = 1.732051.$

From Equation (2-59), $P_{s'} = a'\sin\delta' = 1.732051 \times \sin 0° = 0$ pu.

From Equation (2-60), $Q_{s'} = a'\left(\dfrac{V_{s'}}{V_r} - \cos\delta'\right) = 1.732051\left(\dfrac{0.866025}{1} - \cos 0°\right) = -0.232051$ pu.

From Equation (2-74), $P_r = a'\sin\delta' = 1.732051 \times \sin 0° = 0$ pu.

From Equation (2-75), $Q_r = a'\left(\cos\delta' - \dfrac{V_r}{V_{s'}}\right) = 1.732051\left(\cos 0° - \dfrac{1}{0.866025}\right) = -0.267949$ pu.

From Equation (2-76), $S_r = \sqrt{P_r^2 + Q_r^2} = \sqrt{0^2 + 0.267949^2} = 0.267949$ pu.

From Equation (2-77), $I = \frac{S_r}{V_r} = \frac{0.267949}{1} = 0.267949$ pu.

From Equation (2-54b),

$$V_X = \sqrt{V_{s'}^2 - 2V_{s'}V_r\cos\delta' + V_r^2} = \sqrt{0.866025^2 - 2 \times 0.866025 \times 1 \times \cos 0^\circ + 1^2} = 0.133975 \text{ pu.}$$

From Equation (2-247a), the series-compensating resistance (R_{se}) can be calculated as

$$R_{se} = X\frac{V_{s's}}{V_X}\left\{\frac{V_s\sin\beta - V_r\sin(\delta+\beta)}{V_X}\right\} = 0.5 \times \frac{0.5}{0.133975}\left\{\frac{1 \times \sin 240^\circ - 1 \times \sin(30^\circ + 240^\circ)}{0.133975}\right\}$$
$$= 1.866025 \text{ pu.}$$

From Equation (2-247b), the series-compensating reactance (X_{se}) can be calculated as

$$X_{se} = X\frac{V_{s's}}{V_X}\left\{\frac{V_s\cos\beta - V_r\cos(\delta+\beta) + V_{s's}}{V_X}\right\} = 0.5$$
$$\times \frac{0.5}{0.133975}\left\{\frac{1 \times \cos 240^\circ - 1 \times \cos(30^\circ + 240^\circ) + 0.5}{0.133975}\right\} = 0.$$

Additional Discussion

From Equation (2-11b), $Q_{line} = I^2X = 0.267949^2 \times 0.5 = 0.035898$ pu, which is same as $Q_{line} = Q_{s'} - Q_r = (-0.232051) - (-0.267949) = 0.035898$ pu from Equation (2-319). Note that both $Q_{s'}$ and Q_r are negative numbers, meaning reactive power flows through the line from the higher voltage receiving end to the lower voltage sending end. This is how a STATCOM works, which is explained in Figure 2-54.

Example 2-49

Consider the electrical system data as given in Table 2-1. For an IR operating at $V_{s's} = 0.5$ pu and $\beta = 60^\circ$, calculate $P_{s'}$, $Q_{s'}$, P_r, Q_r, R_{se}, X_{se}, *RPI*, *LI* for $k_R = 1$, and *APR*. Adjusting the control parameter ($V_{s's}$), while $\beta = 60^\circ$, tabulate *RPI*, *LI*, and *APR* of an IR for power flow increases in three cases: Case 1 (40%), Case 2 (70%), and Case 3 (100%).

Answer 2-49

From Equation (2-97), $V_{s'} = \sqrt{V_s^2 + V_{s's}^2 + 2V_sV_{s's}\cos\beta} = \sqrt{1^2 + 0.5^2 + 2 \times 1 \times 0.5 \times \cos 60^\circ}$ $= 1.322876$ pu.

From Equation (2-98), $\psi = \tan^{-1}\frac{V_{s's}\sin\beta}{V_s + V_{s's}\cos\beta} = \tan^{-1}\frac{0.5 \times \sin 60^\circ}{1 + 0.5 \times \cos 60^\circ} = 19.106605^\circ$.

From Equation (2-51a), $\delta' = \delta + \psi = 30^\circ + 19.106605^\circ = 49.106605^\circ$.

From Equation (2-61), $a' = \frac{V_{s'}V_r}{X} = \frac{1.322876 \times 1}{0.5} = 2.645751$ pu.

From Equation (2-59), $P_{s'} = a'\sin\delta' = 2.645751 \times \sin 49.106605^\circ = 2$ pu.

From Equation (2-60), $Q_{s'} = a'\left(\frac{V_{s'}}{V_r} - \cos\delta'\right) = 2.645751\left(\frac{1.322876}{1} - \cos 49.106605°\right)$ $= 1.767949$ pu.

From Equation (2-74), $P_r = a' \sin\delta' = 2.645751 \times \sin 49.106605° = 2$ pu.

From Equation (2-75), $Q_r = a'\left(\cos\delta' - \frac{V_r}{V_{s'}}\right) = 2.645751\left(\cos 49.106605° - \frac{1}{1.322876}\right)$ $= -0.267949$ pu.

From Equation (2-76), $S_r = \sqrt{P_r^2 + Q_r^2} = \sqrt{2^2 + 0.267949^2} = 2.017869$ pu.

From Equation (2-77), $I = \frac{S_r}{V_r} = \frac{2.017869}{1} = 2.017869$ pu.

From Equation (2-11b), $Q_{line} = I^2X = 2.017869^2 \times 0.5 = 2.035898$ pu, which is same as $Q_{line} = Q_{s'} - Q_r = 1.767949 - (-0.267949) = 2.035898$ pu from Equation (2-319). Note that $Q_{s'}$ is a positive number and Q_r is a negative number, meaning reactive power flows into the line from its both ends.

From Equation (2-30), $V_{Xn} = \sqrt{V_s^2 - 2V_sV_r\cos\delta + V_r^2} = \sqrt{1^2 - 2 \times 1 \times 1 \times \cos 30° + 1^2}$ $= 0.517638$ pu.

From Equation (2-33), $I_n = \frac{V_{Xn}}{X} = \frac{0.517638}{0.5} = 1.035276$ pu.

From Equation (2-321), $RPI = \frac{Q_{line}}{P_r} = \frac{2.035898}{2} = 1.017949$.

From Equation (2-324), $LI = k_R\frac{I^2}{I_n^2} = 1 \times \frac{2.017869^2}{1.035276^2} = 3.799038$.

From Equation (2-326), $APR = |V_{s's}|I = 0.5 \times 2.017869 = 1.008935$ pu.

From Equation (2-54b),

$$V_X = \sqrt{V_{s'}^2 - 2V_{s'}V_r\cos\delta' + V_r^2} = \sqrt{1.322875^2 - 2 \times 1.322875 \times 1 \times \cos 49.106605° + 1^2}$$
$$= 1.008935 \text{ pu.}$$

From Equation (2-247a), the series-compensating resistance (R_{se}) can be calculated as

$$R_{se} = X\frac{V_{s's}}{V_X}\left\{\frac{V_s\sin\beta - V_r\sin(\delta+\beta)}{V_X}\right\} = 0.5 \times \frac{0.5}{1.008935}\left\{\frac{1 \times \sin 60° - 1 \times \sin(30° + 60°)}{1.008935}\right\}$$
$$= -0.032903 \text{ pu.}$$

From Equation (2-247b), the series-compensating reactance (X_{se}) can be calculated as

$$X_{se} = X\frac{V_{s's}}{V_X}\left\{\frac{V_s\cos\beta - V_r\cos(\delta+\beta) + V_{s's}}{V_X}\right\} = 0.5 \times \frac{0.5}{1.008935}$$
$$\left\{\frac{1 \times \cos 60° - 1 \times \cos(30° + 60°) + 0.5}{1.008935}\right\} = 0.245592 \text{ pu.}$$

Table 2-29 shows the *RPI*, *LI*, and *APR* of an IR for power flow increases in three cases: Case 1 (40%), Case 2 (70%), and Case 3 (100%).

Figure 2-94 shows the voltage magnitude at the modified sending end for the range of active power flow from 0 to 2 pu in the cases of a PAR (sym), a PAR (asym), a RR, and an IR. While the voltage magnitude stays constant for a PAR (sym), it always increases from the uncompensated

Table 2-29 *RPI*, *LI*, and *APR* of an IR for power flow increases in three cases: Case 1 (40%), Case 2 (70%), and Case 3 (100%).

Parameters	Base case	Case 1	Case 2	Case 3
Modified sending-end voltage ($V_{s'}$)	1 pu	1.1136 pu	1.2135 pu	1.3229 pu
Phase-shift angle (ψ)	0°	8.9483°	14.465°	19.1066°
Series-compensating voltage ($V_{s's}$)	0	0.2 pu	0.35 pu	0.5 pu
Relative phase angle (β)	0°	60°	60°	60°
Active power ($P_{s'} = P_r$)	1 pu	1.4 pu	1.7 pu	2 pu
Reactive power ($Q_{s'}$) at modified sending end	0.2679 pu	0.7479 pu	1.213 pu	1.7679 pu
Reactive power (Q_r) at receiving end	−0.2679 pu	−0.2679 pu	−0.2679 pu	−0.2679 pu
Line current (I)	1.0353 pu	1.4254 pu	1.721 pu	2.0179 pu
Line reactive power ($Q_{line} = I^2X$)	0.5359 pu	1.0159 pu	1.4809 pu	2.0359 pu
Series-compensating Resistance (R_{se})	0	−0.0264 pu	−0.0317 pu	−0.0329 pu
Series-compensating Reactance (X_{se})	0	0.1378 pu	0.2009 pu	0.2456 pu
Reactive Power Index, $RPI = Q_{line}/P_r$	0.5359	0.7256	0.8711	1.0179
Loss Index, $LI = k_R \times I^2/I_n^2$ at natural line current (i.e. $I = I_n$); $k_R = 1$	1	1.8957	2.7634	3.799
$APR = \|V_{s's}\|I$	0	0.2851 pu	0.6023 pu	1.0089 pu

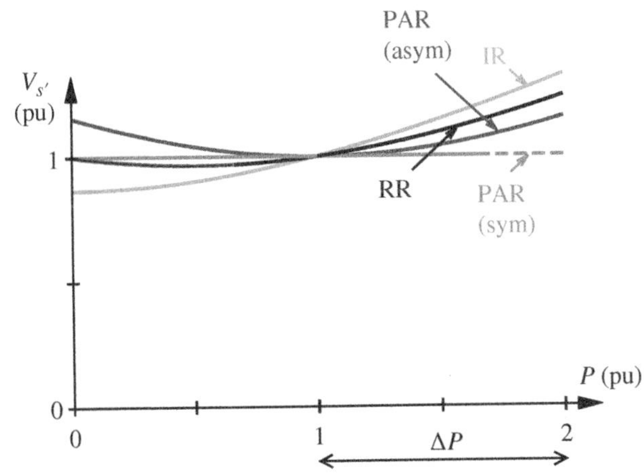

Figure 2-94 Voltage magnitude at the modified sending end for the range of active power flow from 0 to 2 pu in the cases of a PAR (sym), a PAR (asym), a RR, and an IR.

value for a PAR (asym); for a RR and an IR, the voltage magnitude is less than the uncompensated value when the power flow is less than the uncompensated value and vice versa.

The power flow characteristics at the modified sending end and receiving end for the range of active power flow from 0 to 2 pu in the cases of a PAR (sym), a PAR (asym), a RR, and an IR are shown in Figures 2-95 and 2-96, respectively. For a power flow from 1 to 2 pu, the characteristics

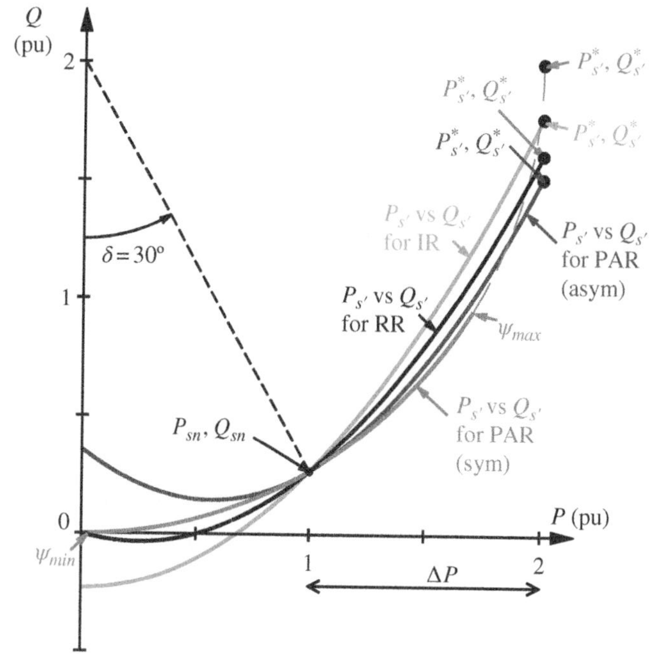

Figure 2-95 Power flow characteristics at the modified sending end for the range of active power flow from 0 to 2 pu in the cases of a PAR (sym), a PAR (asym), a RR, and an IR.

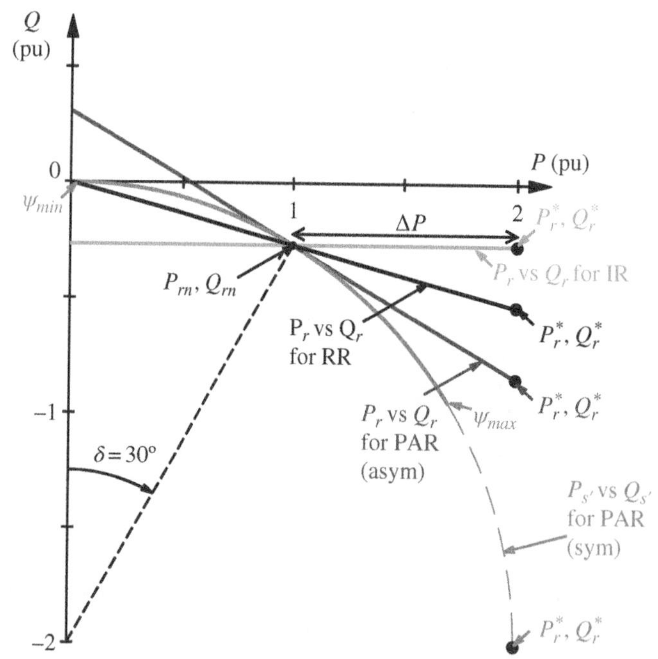

Figure 2-96 Power flow characteristics at the receiving end for the range of active power flow from 0 to 2 pu in the cases of a PAR (sym), a PAR (asym), a RR, and an IR.

are similar at the modified sending end as shown in Figure 2-95; however, they are very different at the receiving end as shown in Figure 2-96.

While the active power flow at the receiving end varies, the reactive power flow at the receiving end must vary when either a PAR (sym or asym) or a RR is used; however, the reactive power flow at the receiving end may be chosen to remain the same as the natural reactive power flow in the case of an IR. In fact, the reactive power flow at the receiving end may be set to zero, if so desired. The superior characteristics of the IR are quite evident when *RPI*, *LI*, and *APR* are examined in Figures 2-97 through 2-99.

Table 2-25 can be expanded to include the comparison of *APR* of a PAR (sym), a PAR (asym), a RR, and an IR for 100% power flow decrease as shown in Table 2-30.

Table 2-25 can be expanded to include the comparison of *RPI*, *LI*, and *APR* of a PAR (sym), a PAR (asym), a RR, and an IR for 100% power flow increase as shown in Table 2-31.

The phase angle regulation method, in symmetric form, regulates the phase angle of the line voltage without any change in its magnitude and, in asymmetric form, regulates mainly the phase angle of the line voltage with some increase in its magnitude. In both forms, a compensating voltage with variable magnitude and predetermined phase angle (90° or −90°) is connected in series with the line.

In contrast, in the impedance regulation method, a compensating voltage with variable magnitude and variable phase angle, in the entire controllable range of $0° \leq \beta \leq 360°$, is connected in series with the line. The prevailing line current, being at any phase angle with respect to the compensating voltage emulates a four-quadrant, compensating impedance ($\mathbf{Z_{se}} = R_{se} - jX_{se}$) that consists of a resistance ($R_{se} = +R$ or $-R$) and a reactance ($X_{se} = X_C$ or $-X_L$) in series with the line. By changing the emulated impedance within its design limit, it can be made to be resistive, reactive, or a combination of the two.

The emulated, four-quadrant impedance modifies the effective impedance (both resistance and reactance) of the line between its two ends, which, in turn, modifies the sending-end voltage to be of

Table 2-30 Comparison of *APR* of a PAR (sym), a PAR (asym), a RR, and an IR for 100% power flow decrease.

Parameters	Base case	PAR (sym)	PAR (asym)	RR	IR
Modified sending-end voltage ($V_{s'}$)	1 pu	1 pu	1.1547 pu	1 pu	0.8660 pu
Phase-shift angle (ψ)	0°	−30°	−30°	−30°	−30°
Series-compensating voltage ($V_{s's}$)	0	0.5176 pu	0.5774 pu	0.5176 pu	0.5 pu
Relative phase angle (β)	0°	255°	−90°	255°	240°
Active power ($P_{s'} = P_r$)	1 pu	0	0	0	0
Reactive power ($Q_{s'}$) at modified sending end	0.2679 pu	0	0.3573 pu	0	−0.2321 pu
Reactive power (Q_r) at receiving end	−0.2679 pu	0	0.3094 pu	0	−0.2679 pu
Line current (I)	1.0353 pu	0	0.3094 pu	0	0.2679 pu
Line reactive power ($Q_{line} = I^2X$)	0.5359 pu	0	0.0479 pu	0	0.0359 pu
Series-compensating Resistance (R_{se})	0	∞	−1.6160 pu	0	1.8660 pu
Series-compensating Reactance (X_{se})	0	∞	0.9330 pu	∞	0 pu
$APR = \|V_{s's}\|I_n$	0	0.5359 pu	0.5977 pu	0.5359 pu	0.5176 pu

Table 2-31 Comparison of *RPI*, *LI*, and *APR* of a PAR (sym), a PAR (asym), a RR, and an IR for 100% power flow increase.

Parameters	Base case	PAR (sym)	PAR (asym)	RR	IR
Modified sending-end voltage ($V_{s'}$)	1 pu	1 pu	1.1547 pu	1.2393 pu	1.3229 pu
Phase-shift angle (ψ)	0°	60°	30°	23.794°	19.1066°
Series-compensating voltage ($V_{s's}$)	0	1 pu	0.5774 pu	0.5176 pu	0.5 pu
Relative phase angle (β)	0°	120°	90°	75°	60°
Active power ($P_{s'} = P_r$)	1 pu	2 pu	2 pu	2 pu	2 pu
Reactive power ($Q_{s'}$) at modified sending end	0.2679 pu	2 pu	1.512 pu	1.6077 pu	1.7679 pu
Reactive power (Q_r) at receiving end	−0.2679 pu	−2 pu	−0.8453 pu	−0.5359 pu	−0.2679 pu
Line current (I)	1.0353 pu	2.8284 pu	2.1713 pu	2.0706 pu	2.0179 pu
Line reactive power ($Q_{line} = I^2X$)	0.5359 pu	4 pu	2.3573 pu	2.1436 pu	2.0359 pu
Series-compensating Resistance (R_{se})	0	0.0915 pu	0.0328 pu	0	−0.0329 pu
Series-compensating Reactance (X_{se})	0	0.3415 pu	0.2639 pu	0.25 pu	0.2456 pu
Reactive Power Index, $RPI = Q_{line}/P_r$	0.5359	2	1.1786	1.0718	1.0179
Loss Index, $LI = k_R \times I^2/I_n^2$ at natural line current (*i.e.* $I = I_n$); $k_R = 1$	1	7.4641	4.3987	4	3.799
$APR = \|V_{s's}\|I$	0	2.8284 pu	1.2536 pu	1.0718 pu	1.0089 pu

a specific magnitude and a phase angle that results in an independent control of active and reactive power flows in the line.

The direct benefit of independent control is to optimize both the active and reactive power flows so that the revenue-generating active power flow can be maximized. While maintaining the voltage stability, if the reactive power flow is minimized, it results in lower losses and higher efficiency in the electric transmission system, and lower wholesale electric market costs to loads.

Among all the PFCs, the IR requires lower overall reactive power, causes lower losses in the line, and requires lower *APR* than the PAR and the RR as shown in Figures 2-97 through 2-99.

Considering the line resistance (R) to be zero, the *RPI* per unit of active power (P_r) transmitted is given by

$$\frac{RPI}{P_r} = \frac{Q_{line}}{P_r^2} = \frac{I^2X}{P_r^2} = \frac{V_{Xn}^2X}{V_s^2V_r^2}\frac{R_{se}^2 + X_{eff}^2}{\left[R_{se}\left\{\cos\delta - \frac{V_r}{V_s}\right\} + X_{eff}\sin\delta\right]^2}. \tag{B-151b}$$

The *LI* per unit of active power (P_r) transmitted is given by

$$\frac{LI}{P_r} = k_R\frac{I^2}{I_n^2P_r} = k_R\frac{V_{Xn}^2}{I_n^2V_sV_r}\frac{1}{R_{se}\left\{\cos\delta - \frac{V_r}{V_s}\right\} + X_{eff}\sin\delta} = k_R\frac{1}{V_sV_r}\frac{X^2}{R_{se}\left\{\cos\delta - \frac{V_r}{V_s}\right\} + X_{eff}\sin\delta}. \tag{B-152b}$$

Figure 2-97 Reactive Power Index (*RPI*) versus active power (P_r) flow at the receiving end of the line.

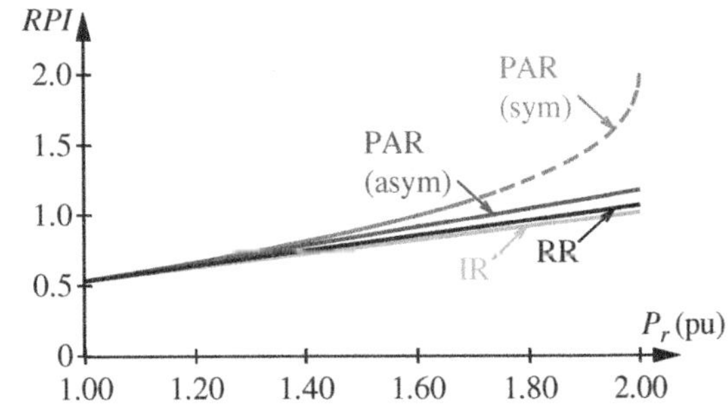

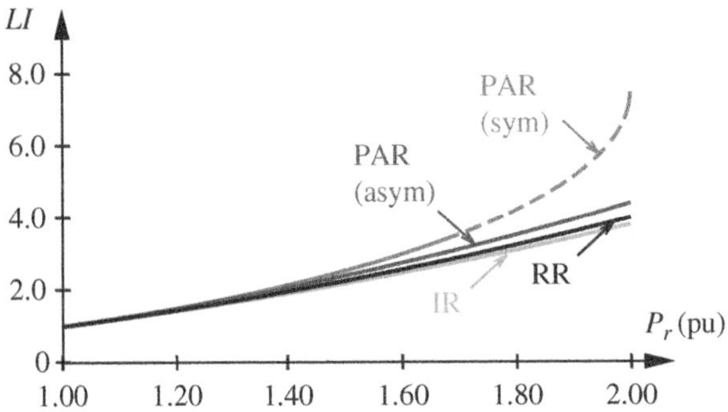

Figure 2-98 Loss Index (*LI*) versus active power (P_r) flow at the receiving end of the line.

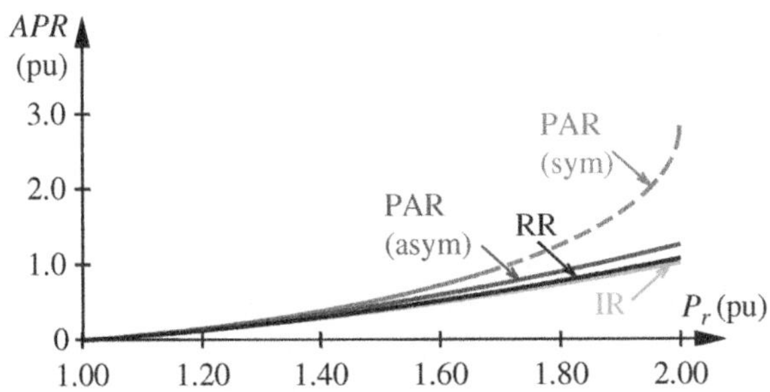

Figure 2-99 Apparent Power Rating (*APR*) versus active power (P_r) flow at the receiving end of the line.

The *APR* per unit of active power (P_r) transmitted is given by

$$\frac{APR}{P_r} = \frac{|V_{s's}|I}{P_r} = \frac{\sqrt{R_{se}^2 + X_{se}^2 I^2}}{P_r} = \frac{V_{Xn}^2}{V_s V_r} \frac{\sqrt{R_{se}^2 + X_{se}^2}}{R_{se}\left\{\cos\delta - \frac{V_r}{V_s}\right\} + X_{eff}\sin\delta}. \tag{B-153b}$$

Equations (B-151b), (B-152b), and (B-153b) apply to a PAR (sym), a PAR (asym), and an IR, since all three PFCs emulate a four-quadrant, series-compensating impedance ($\mathbf{Z_{se}} = R_{se} - jX_{se}$) that consists of a resistance ($R_{se} = +R$ or $-R$) and a reactance ($X_{se} = X_C$ or $-X_L$) in series with the line. However, for a RR, $R_{se} = 0$. Therefore, the Equations (B-151b), (B-152b), and (B-153b) can be rewritten as follows.

The *RPI* per unit of active power (P_r) transmitted is given by

$$\frac{RPI}{P_r} = \frac{Q_{line}}{P_r^2} = \frac{I^2X}{P_r^2} = \frac{V_{Xn}^2 X}{V_s^2 V_r^2}\frac{1}{[\sin\delta]^2} = \frac{V_{Xn}^2}{P_{rn}^2}\frac{1}{X} \tag{B-151c}$$

using Equations (2-41) and (2-46).

The *LI* per unit of active power (P_r) transmitted is given by

$$\frac{LI}{P_r} = k_R\frac{I^2}{I_n^2 P_r} = k_R\frac{V_{Xn}^2}{I_n^2 V_s V_r}\frac{1}{X_{eff}\sin\delta} = k_R\frac{1}{V_s V_r}\frac{X^2}{X_{eff}\sin\delta} = k_R\frac{1}{P_{rn}}\frac{X}{X_{eff}} \tag{B-152c}$$

using Equations (2-33), (2-41), and (2-46).

The *APR* per unit of active power (P_r) transmitted is given by

$$\frac{APR}{P_r} = \frac{|V_{s's}|I}{P_r} = \frac{X_{se}I^2}{P_r} = \frac{X_{se}V_{Xn}^2}{X_{eff}^2}\frac{X_{eff}}{V_s V_r\sin\delta} = \frac{V_{Xn}^2}{V_s V_r}\frac{X_{se}}{X_{eff}\sin\delta} \tag{B-153c}$$

using (2-234d), (2-207), and (2-209). Note that $V_{Xn} = V_{Xeff}$.

Table 2-25 can be expanded to include the comparison of *RPI*, *LI*, and *APR* per unit of active power (P_r) transmitted for a PAR (sym), a PAR (asym), a RR, and an IR in the case of a 100% power flow increase at the receiving end of the line is shown in Table 2-32. Among all the PFCs, the IR requires the lowest *RPI*, *LI*, and *APR* per unit of active power (P_r) transmitted.

The limitation of the PAR (sym) is that it is not designed to operate beyond the maximum modified power angle (δ'_{max}), represented by the dashed line. Therefore, if the thermal limit of the line is at 2 pu, as it is in this example, it is not possible to achieve the desired power flow using the PAR (sym). Although the IR being the best-performing PFC with the lowest increase of the *RPI*, *LI*, and *APR* as the active power flow in the line increases, its performance may be restricted up to 1.7 pu active power flow in the line, so the performance of the PAR (sym) may be compared up to its maximum modified power angle (δ'_{max}). For a 70% increase in active power flow (0.7 pu of active power flow enhancement over the natural active power flow of 1 pu) in the line, the *APR* of the IR is 0.6023 pu as shown in Table 2-29, whereas that of the PAR (sym) is 0.9484 pu as shown in

Table 2-32 Comparison of *RPI*, *LI*, and *APR* per unit of active power (P_r) transmitted for a PAR (sym), a PAR (asym), a RR, and an IR in the case of a 100% power flow increase at the receiving end of the line.

Parameters	Base case	PAR (sym)	PAR (asym)	RR	IR
Active power ($P_{s'} = P_r$)	1 pu	2 pu	2 pu	2 pu	2 pu
Series-compensating Resistance (R_{se})	0	0.0915 pu	0.0328 pu	0	−0.0329 pu
Series-compensating Reactance (X_{se})	0	0.3415 pu	0.2639 pu	0.25 pu	0.2456 pu
Effective Line Reactance, $X_{eff} = X - X_{se}$	0.5 pu	0.1585 pu	0.2361 pu	0.25 pu	0.2544 pu
RPI/P_r	0.5359	1	0.5893	0.5359	0.509
LI/P_r	1	3.7321	2.1994	2	1.9
APR/P_r	0	1.4142	0.6268	0.5359	0.5045

Table 2-26. For a 200 MW active power flow enhancement, the rating of the IR is $200/0.7 \times 0.6023 = 172$ MVA, whereas the rating of the PAR (sym) is $200/0.7 \times 0.9484 = 271$ MVA. Besides, the lowest-rated initial equipment cost, the IR also provides the lowest expenses for the reactive power support and the line loss during its decades of operation.

Example 2-50

Consider two identical lines that are connected in parallel as shown in Figure 2-68. Using the electrical system data for each line as given in Table 2-1, use a BTB-SSSC with a series-compensating voltage ($\mathbf{V_{1s's}}$) to make $P_{1r} = 2 \times P_{1rn}$ and $Q_{1r} = Q_{1rn}$. Calculate $V_{1s's}$, β_1, P_{1se}, and Q_{1se}. For the second line with a given series-compensating voltage ($\mathbf{V_{2s's}}$) when $P_{link} = 0$ and $Q_{2se} = 0$, calculate $V_{2s's}$ and β_2.

Answer 2-50

From Table 2-1 data, $V_s = 1$ pu, $V_{1r} = V_{2r} = 1$ pu, $X_1 = X_2 = 0.5$ pu, and $\delta_1 = \delta_2 = 30°$.

Applying Equation (2-46): $P_{rn} = a \times \sin\delta$ and Equation (2-48): $Q_{rn} = a\left(\cos\delta - \frac{V_r}{V_s}\right)$, the natural transmitted active power, $P_{1sn} = P_{1rn} = 1\text{ pu} = P_{2sn} = P_{2rn}$ and natural reactive powers at the sending end, $Q_{1sn} = 0.267949\text{ pu} = Q_{2sn}$ and that the receiving end $Q_{1rn} = -0.267949\text{ pu} = Q_{2rn}$ in the uncompensated lines 1 ("leader" line) and 2 ("follower" line).

With a given series-compensating voltage ($\mathbf{V_{1s's}}$), $P_{1r} = 2$ pu, $Q_{1r} = -0.267949$ pu.

From Equation (2-289), the magnitude ($V_{1s'}$) of the modified sending-end voltage ($\mathbf{V_{1s'}}$) in line 1,

$$V_{1s'} = \frac{X_1}{V_{1r}}\sqrt{P_{1r}^2 + \left(Q_{1r} + \frac{V_{1r}^2}{X_1}\right)^2} = \frac{0.5}{1}\sqrt{2^2 + \left(-0.267949 + \frac{1^2}{0.5}\right)^2} = 1.322876\text{ pu.}$$

From Equation (2-291), the phase-shift angle of the modified sending-end voltage in line 1,

$$\psi_1 = \tan^{-1}\frac{P_{1r}}{Q_{1r} + \frac{V_{1r}^2}{X_1}} - \delta_1 = \tan^{-1}\frac{2}{-0.267949 + \frac{1^2}{0.5}} - 30° = 19.106605°.$$

From Equation (2-292), the magnitude ($V_{1s's}$) and from Equation (2-293), the relative phase angle (β_1) of the series-compensating voltage in line 1,

$$V_{1s's} = \sqrt{V_{1s'}^2 + V_s^2 - 2V_{1s'}V_s\cos\psi_1} = \sqrt{1.322876^2 + 1^2 - 2 \times 1.322876 \times 1 \times \cos 19.106605°}$$
$$= 0.5\text{ pu and}$$

$$\beta_1 = \tan^{-1}\frac{V_{1s'}\sin\psi_1}{V_{1s'}\cos\psi_1 - V_s} = \tan^{-1}\frac{1.322876 \times \sin 19.106605°}{1.322876 \times \cos 19.106605° - 1} = 60°.$$

From Equation (2-297), the natural voltage (V_{1Xn}) across the line 1,

$$V_{1Xn} = \sqrt{V_s^2 - 2V_sV_{1r}\cos\delta_1 + V_{1r}^2} = \sqrt{1^2 - 2 \times 1 \times 1 \times \cos 30° + 1^2} = 0.517638\text{ pu.}$$

From Equation (2-296), $a_{1se} = \frac{V_{1s's}V_{1Xn}}{X_1} = \frac{0.5 \times 0.517638}{0.5} = 0.517638$ pu.

From Equation (2-294), the exchanged active power (P_{1se}) and from Equation (2-295), the exchanged reactive power (Q_{1se}) between line 1 to line 2,

$$P_{1se} = a_{1se}\sin\left(\beta_1 - \tan^{-1}\frac{P_{1sn}}{Q_{1sn}}\right) = 0.517638 \times \sin\left(60° - \tan^{-1}\frac{1}{0.267649}\right) = -0.133974\text{ pu.}$$

$$Q_{1se} = -\frac{V_{1s's}^2}{X_1} - a_{1se}\cos\left(\beta_1 - \tan^{-1}\frac{P_{1sn}}{Q_{1sn}}\right) = -\frac{0.5^2}{0.5} - 0.517638$$
$$\times \cos\left(60° - \tan^{-1}\frac{1}{0.267649}\right) = -1 \text{ pu}.$$

For an ideal BTB-SSSC (neglecting its internal power losses), $P_{link} = 0$. Also, Q_{1se} and Q_{2se} are independent of each other. It is also given that $Q_{2se} = 0$ pu.

From Equation (2-300), $P_{1se} + P_{2se} = P_{link} = 0$; therefore, $P_{2se} = -P_{1se} = +0.133974$ pu.

From Equation (2-305), the natural voltage (V_{2Xn}) across line 2, $V_{2Xn} = \sqrt{V_s^2 - 2V_sV_{2r}\cos\delta_2 + V_{2r}^2} = \sqrt{1^2 - 2\times1\times1\times\cos30° + 1^2} = 0.517638$ pu.

From Equation (2-303), the magnitude ($V_{2s's}$) and from Equation (2-304), the relative phase angle (β_2) of the series-compensating voltage in line 2,

$$V_{2s's} = +\sqrt{\left(\frac{V_{2Xn}^2}{2} - X_2Q_{2se}\right) - \sqrt{\left(\frac{V_{2Xn}^2}{2} - X_2Q_{2se}\right)^2 - X_2^2(P_{2se}^2 + Q_{2se}^2)}}$$
$$= +\sqrt{\left(\frac{0.517638^2}{2} - 0.5\times0\right) - \sqrt{\left(\frac{0.517638^2}{2} - 0.5\times0\right)^2 - 0.5^2(0.133974^2 + 0^2)}}$$
$$= 0.133974 \text{ pu}.$$

$$\beta_2 = \tan^{-1}\frac{P_{2sn}}{Q_{2sn}} - \tan^{-1}\frac{P_{2se}}{Q_{2se} + \frac{V_{2s's}^2}{X_2}} = \tan^{-1}\frac{1}{0.267649} - \tan^{-1}\frac{0.133974}{\frac{0.133974^2}{0.5}} = 0.$$

2.6 Comparison Between Shunt-Compensating Reactance and Series-Compensating Reactance

Consider a two-generator/one-line power system network, shown in Figure 2-100, with the magnitudes (V_s and V_r) of the sending- and receiving-end voltages to be equal to V (one pu), the power angle, $\delta = \delta_s - \delta_r = 90°$, and the line reactance, $X = 1$ pu.

Rewriting Equations (2-40), (2-46), (2-43), and (2-48) using Equation (2-41), the active and reactive power flows at the sending and receiving ends of the line are given by

$$P_{sn} = P_{rn} = \frac{V_sV_r}{X}\sin(\delta) = \frac{V^2}{X}\sin(\delta) \tag{2-328}$$

$$Q_{sn} = \frac{V_sV_r}{X}\{1 - \cos(\delta)\} = \frac{V^2}{X}\{1 - \cos(\delta)\} \tag{2-329}$$

$$Q_{rn} = \frac{V_sV_r}{X}\{\cos(\delta) - 1\} = \frac{V^2}{X}\{\cos(\delta) - 1\}. \tag{2-330}$$

Equation (2-319) can be rewritten for an uncompensated line as

$$Q_{linen} = Q_{sn} - Q_{rn} = 2\frac{V^2}{X}(1 - \cos\delta). \tag{2-331}$$

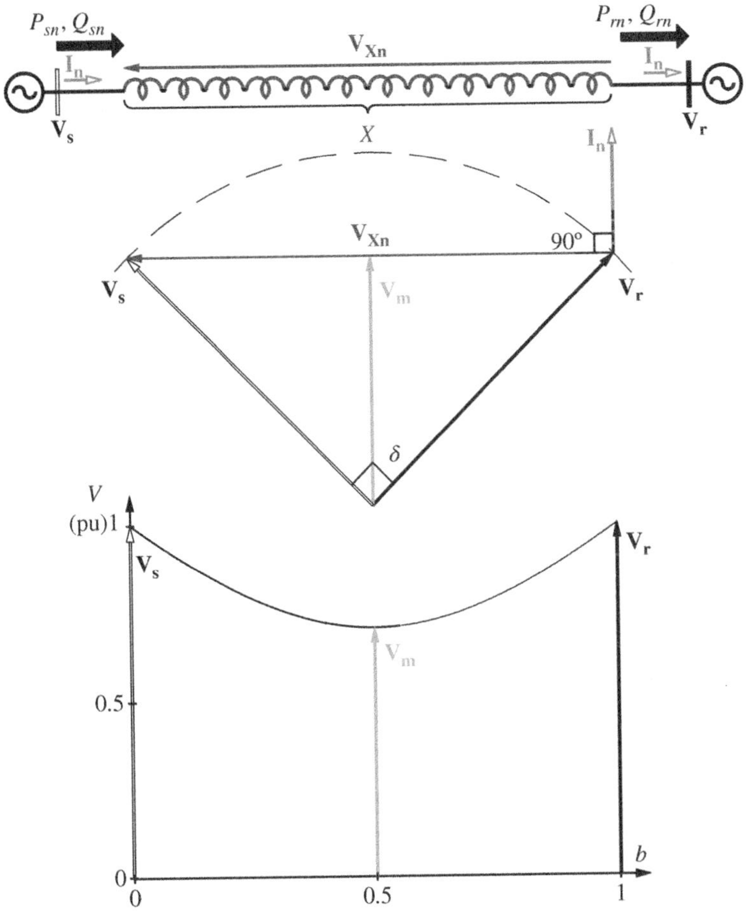

Figure 2-100 Two-generator/one-line power system network without any compensation and its related phasor diagram.

Equation (2-30) can be rewritten as

$$V_{Xn} = \sqrt{V_s^2 - 2V_sV_r\cos\delta + V_r^2} = \sqrt{V^2 - 2V^2\cos\delta + V^2}$$

$$V_{Xn} = 2V\sin\left(\frac{\delta}{2}\right). \tag{2-332}$$

From Equations (2-38) and (2-332), the voltage at any point in the line is given by

$$V_x = V\cos\left(\frac{\delta}{2}\right)\sqrt{1 + \left\{(2b-1)\tan\left(\frac{\delta}{2}\right)\right\}^2} \tag{2-333}$$

where b = multiplier of V_{Xn} from 0 to 1 and $x = s, 1, m, 2, r,$ and so on as shown in Figure 2-8.

Digress:

$$V_x = \frac{V_s V_r \sin\delta}{V_{Xn}} \sqrt{1 + \left\{\frac{bV_{Xn}^2 - V_s(V_s - V_r\cos\delta)}{V_s V_r \sin\delta}\right\}^2}$$

$$= \frac{V^2 2\sin\left(\frac{\delta}{2}\right)\cos\left(\frac{\delta}{2}\right)}{2V\sin\left(\frac{\delta}{2}\right)} \sqrt{1 + \left\{\frac{b \times 4V^2\sin^2\left(\frac{\delta}{2}\right) - V^2(1 - \cos\delta)}{V^2\sin\delta}\right\}^2}$$

$$= V\cos\left(\frac{\delta}{2}\right)\sqrt{1 + \left\{\frac{b \times 4\sin^2\left(\frac{\delta}{2}\right) - 2\sin^2\left(\frac{\delta}{2}\right)}{2\sin\left(\frac{\delta}{2}\right)\cos\left(\frac{\delta}{2}\right)}\right\}^2}$$

$$= V\cos\left(\frac{\delta}{2}\right)\sqrt{1 + \left\{\frac{2b \times \sin\left(\frac{\delta}{2}\right) - \sin\left(\frac{\delta}{2}\right)}{\cos\left(\frac{\delta}{2}\right)}\right\}^2} = V\cos\left(\frac{\delta}{2}\right)\sqrt{1 + \left\{(2b-1)\tan\left(\frac{\delta}{2}\right)\right\}^2}.$$

Using Equation (2-333), the plot of the magnitude of voltage (voltage profile) along the line is shown in Figure 2-100. If a load is to be connected along the midpoint in the line, the line voltage of 0.707 pu is not suitable for the load, since it is below the allowable lower limit.

2.6.1 Shunt-Compensating Reactance

The voltage at any point along the line can be restored to its desired nominal value of one pu by using a shunt-compensating reactance.

2.6.1.1 Restoration of Voltage at the Midpoint of the Line

Figure 2-101 shows a scheme that can be used to restore the line voltage at one pu. Following Figure 2-100, the left half of the line with reactance of $X/2$ may be treated as a line with its sending- and receiving-end voltages, $V_s = 1$ pu and $V_r = 1$ pu with a power angle of $\delta/2$. The line current ($\mathbf{I_s}$) lags the sending-end voltage ($\mathbf{V_s}$) by $\delta/4$ and leads the midpoint voltage ($\mathbf{V_m}$) by $\delta/4$. Similarly, the right half of the line with reactance of $X/2$ may be treated as a line with its sending- and receiving-end voltages to be one pu with a power angle of $\delta/2$. The line current ($\mathbf{I_r}$) lags the midpoint voltage ($\mathbf{V_m}$) by $\delta/4$ and leads the receiving-end voltage ($\mathbf{V_r}$) by $\delta/4$. The compensator current ($\mathbf{I_q}$) is the difference between the currents in the left half of the line and the right half of the line and is given by

$$\mathbf{I_q} = \mathbf{I_s} - \mathbf{I_r}. \tag{2-334}$$

Since the compensator current ($\mathbf{I_q}$) leads the midpoint voltage by 90°, the compensator must be capacitive, which was demonstrated at the beginning of this chapter. Using Equation (2-333), the plot of the magnitude of the voltage along the line is shown in Figure 2-101.

Rewriting Equations (2-328), (2-329), and (2-330), the active and reactive power flows at the sending and receiving ends of the line are given by

$$P_s = P_r = \frac{V^2}{X/2}\sin\left(\frac{\delta}{2}\right) = \frac{2V^2}{X}\sin\left(\frac{\delta}{2}\right) \tag{2-335}$$

$$Q_s = \frac{V^2}{X/2}\left\{1 - \cos\left(\frac{\delta}{2}\right)\right\} = \frac{2V^2}{X}\left\{1 - \cos\left(\frac{\delta}{2}\right)\right\} \tag{2-336}$$

$$Q_r = \frac{V^2}{X/2}\left\{\cos\left(\frac{\delta}{2}\right) - 1\right\} = \frac{2V^2}{X}\left\{\cos\left(\frac{\delta}{2}\right) - 1\right\}. \tag{2-337}$$

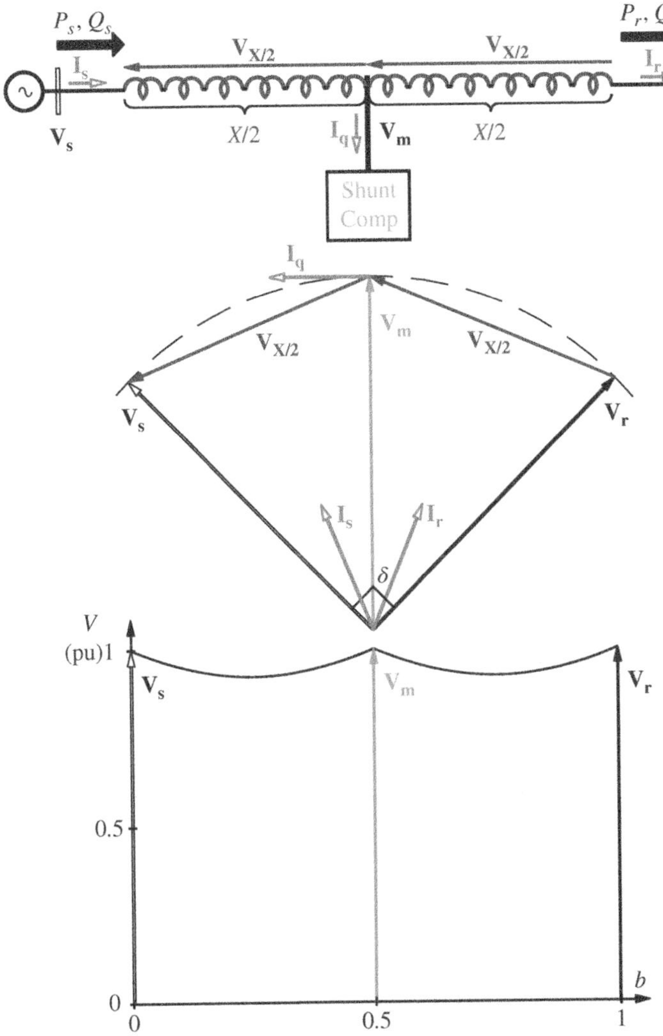

Figure 2-101 Two-generator/one-line power system network with shunt compensation and its related phasor diagram.

From Equation (2-331), the reactive power needed by each of the two segments of the line is given by

$$Q_{line} = Q_s - Q_r = 2 \times \frac{2V^2}{X}\left\{1 - \cos\left(\frac{\delta}{2}\right)\right\} = 4\frac{V^2}{X}\left\{1 - \cos\left(\frac{\delta}{2}\right)\right\}. \tag{2-338}$$

The shunt compensator provides reactive power to the right half of the left half of the line and to the left of the right half of the line. Therefore, the rating of the shunt compensator is given by

$$Q_{sh} = Q_{line} = 4\frac{V^2}{X}\left\{1 - \cos\left(\frac{\delta}{2}\right)\right\}. \tag{2-339}$$

Increasing the voltage along the line at various points increases the Available Transfer Capability (ATC) of the line. This is significant for lines that operate with a large power angle (δ). Considering

$V_s = V_r = 1$ pu and $X = 1$ pu, if V_m is raised to be one pu, the ATC changes from sin δ to $2 \times \sin(\delta/2)$. For a $\delta = 30°$, it is a 3.53% increase of active power flow. However, in an extreme case, $\delta = 90°$ would result in an increase of 41.4% in active power flow.

2.6.1.2 Restoration of Voltage at the One-Third and Two-Third Points of the Line

The voltage at one-third and two-third points of the line can be restored to its desired one pu value by using two shunt-connected reactance compensators as shown in Figure 2-102.

The sum of the compensator currents, $\mathbf{I_{q1}} + \mathbf{I_{q2}}$, is the difference between the currents in the left one-third of the line and the right one-third of the line and is given by

$$\mathbf{I_{q1}} + \mathbf{I_{q2}} = \mathbf{I_s} - \mathbf{I_r}. \tag{2-340}$$

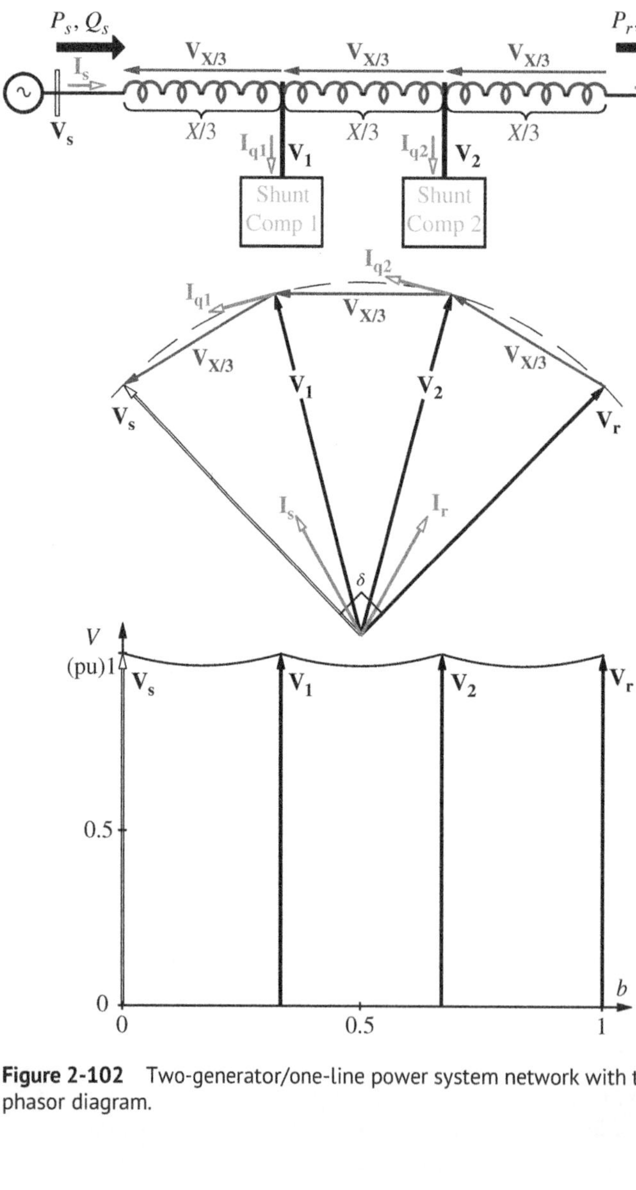

Figure 2-102 Two-generator/one-line power system network with two shunt compensations and its related phasor diagram.

Using Equation (2-333), the plot of the magnitude of voltage along the line is shown in Figure 2-102. Rewriting Equations (2-328), (2-329), and (2-330), the active and reactive power flows at the sending and receiving ends of the line are given by

$$P_s = P_r = \frac{V^2}{X/3}\sin\left(\frac{\delta}{3}\right) = \frac{3V^2}{X}\sin\left(\frac{\delta}{3}\right) \tag{2-341}$$

$$Q_s = \frac{V^2}{X/3}\left\{1 - \cos\left(\frac{\delta}{3}\right)\right\} = \frac{3V^2}{X}\left\{1 - \cos\left(\frac{\delta}{3}\right)\right\} \tag{2-342}$$

$$Q_r = \frac{V^2}{X/3}\left\{\cos\left(\frac{\delta}{3}\right) - 1\right\} = \frac{3V^2}{X}\left\{\cos\left(\frac{\delta}{3}\right) - 1\right\}. \tag{2-343}$$

From Equation (2-331), the reactive power needed by each of the three segments of the line is given by

$$Q_{line} = Q_s - Q_r = 2 \times \frac{3V^2}{X}\left\{1 - \cos\left(\frac{\delta}{3}\right)\right\} = 6\frac{V^2}{X}\left\{1 - \cos\left(\frac{\delta}{3}\right)\right\}. \tag{2-344}$$

Each shunt compensator provides reactive power to the right half of the left third of the line and to the left half of the right third of the line. Therefore, the rating of each shunt compensator is given by

$$Q_{sh} = Q_{line} = 6\frac{V^2}{X}\left\{1 - \cos\left(\frac{\delta}{3}\right)\right\}. \tag{2-345}$$

2.6.1.3 Restoration of Voltage at the One-Fourth, Half, and Three-Fourth Points of the Line

The voltage at one-fourth, half, and three-fourth points of the line can be restored to its desired one pu value by using three shunt-connected reactance compensators as shown in Figure 2-103.

The sum of the compensator currents, $\mathbf{I_{q1}} + \mathbf{I_{q2}} + \mathbf{I_{q3}}$, is the difference between the currents in the left one-fourth of the line and the right one-fourth of the line and is given by

$$\mathbf{I_{q1}} + \mathbf{I_{q2}} + \mathbf{I_{q3}} = \mathbf{I_s} - \mathbf{I_r}. \tag{2-346}$$

Using Equation (2-333), the plot of the magnitude of voltage along the line is shown in Figure 2-103. Rewriting Equations (2-328), (2-329), and (2-330), the active and reactive power flows at the sending and receiving ends of the line are given by

$$P_s = P_r = \frac{V^2}{X/4}\sin\left(\frac{\delta}{4}\right) = \frac{4V^2}{X}\sin\left(\frac{\delta}{4}\right) \tag{2-347}$$

$$Q_s = \frac{V^2}{X/4}\left\{1 - \cos\left(\frac{\delta}{4}\right)\right\} = \frac{4V^2}{X}\left\{1 - \cos\left(\frac{\delta}{4}\right)\right\} \tag{2-348}$$

$$Q_r = \frac{V^2}{X/4}\left\{\cos\left(\frac{\delta}{4}\right) - 1\right\} = \frac{4V^2}{X}\left\{\cos\left(\frac{\delta}{4}\right) - 1\right\}. \tag{2-349}$$

From Equation (2-331), the reactive power needed by each of the four segments of the line is given by

$$Q_{line} = Q_s - Q_r = 2 \times \frac{4V^2}{X}\left\{1 - \cos\left(\frac{\delta}{4}\right)\right\} = 8\frac{V^2}{X}\left\{1 - \cos\left(\frac{\delta}{4}\right)\right\}. \tag{2-350}$$

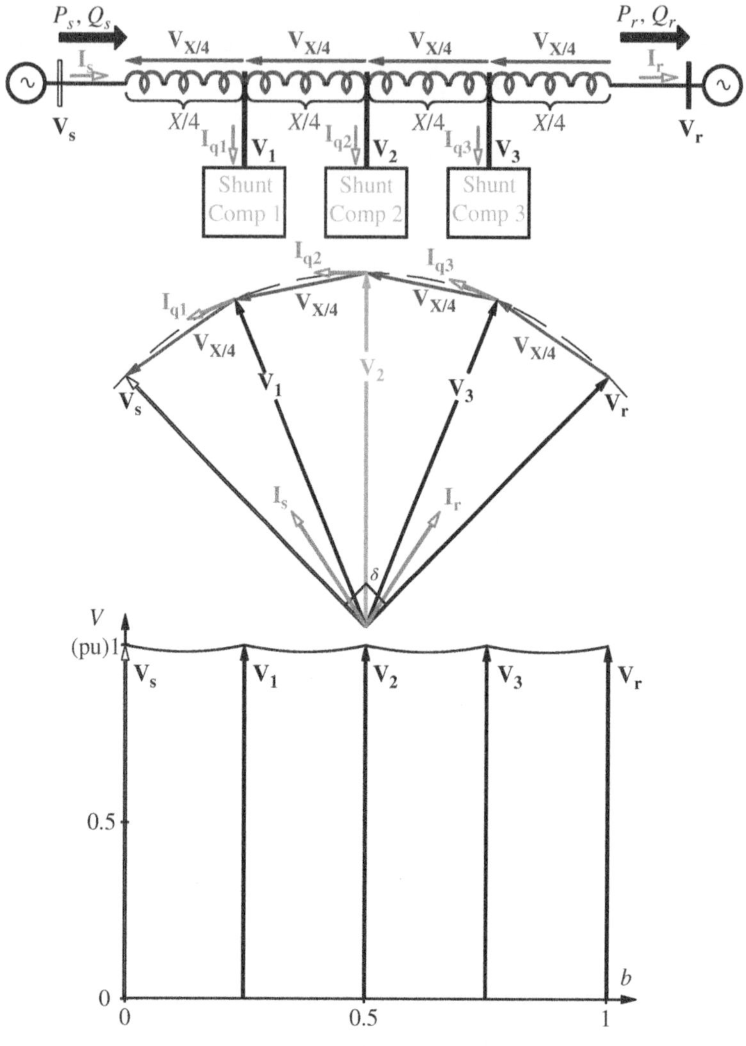

Figure 2-103 Two-generator/one-line power system network with three shunt compensations and its related phasor diagram.

Each shunt compensator provides reactive power to the right half of the left fourth of the line and to the left half of the right fourth of the line. Therefore, the rating of each shunt compensator is given by

$$Q_{sh} = Q_{line} = 8\frac{V^2}{X}\left\{1 - \cos\left(\frac{\delta}{4}\right)\right\}. \tag{2-351}$$

Each capacitive compensation unit at the POC with the line supplies reactive power to the right half of the left quarter of the line and to the left half of the right quarter of the line. In this case, the sending- and the receiving-end voltages supply reactive power to the left half of the first quarter and the right half of the fourth quarter of the line, respectively. Note that if a line is compensated at ¼, ½, and ¾ points to maintain the nominal voltage at each compensating point, the ATC changes from sinδ to 4×sin(δ/4). For a $\delta = 90°$, that is a 53% increase in active power flow.

2.6.1.4 Restoration of Voltage at *n* Points of the Line

In general, the active and reactive power flows at the sending and receiving ends of the line are given by

$$P_s = P_r = \frac{(n+1)V^2}{X}\sin\left(\frac{\delta}{n+1}\right) \quad (2\text{-}352)$$

$$Q_s = \frac{(n+1)V^2}{X}\left\{1-\cos\left(\frac{\delta}{n+1}\right)\right\} \quad (2\text{-}353)$$

$$Q_r = \frac{(n+1)V^2}{X}\left\{\cos\left(\frac{\delta}{n+1}\right)-1\right\} \quad (2\text{-}354)$$

$$Q_{sh} = Q_{line} = 2(n+1)\frac{V^2}{X}\left\{1-\cos\left(\frac{\delta}{n+1}\right)\right\} \quad (2\text{-}355)$$

where n is the number of shunt compensators, each placed at equal distance. Note that Q_{line} in Equation (2-355) refers to the reactive power in each segment of the line, thus the rating (Q_{sh}) of each shunt compensator. The combined *APR* (S_{sh}) of all the shunt compensators is given by

$$APR = S_{sh} = n \times Q_{sh} = 2n(n+1)\frac{V^2}{X}\left\{1-\cos\left(\frac{\delta}{n+1}\right)\right\}. \quad (2\text{-}356)$$

Using Equation (2-355), the *RPI*, which is a measure of total reactive power needed by the line for a unit of active power transmitted, is given by

$$RPI = (n+1)\frac{Q_{line}}{P_r} = \frac{2}{P_r}(n+1)^2\frac{V^2}{X}\left\{1-\cos\left(\frac{\delta}{n+1}\right)\right\}. \quad (2\text{-}357)$$

For a large value of n,

$$\sin\left(\frac{\delta}{n+1}\right) \cong \frac{\delta}{n+1} \quad (2\text{-}358a)$$

and

$$\cos\left(\frac{\delta}{n+1}\right) \cong \cos(0) = 1. \quad (2\text{-}358b)$$

Therefore, using Equations (2-358a) and (2-358b), (2-352), (2-353), and (2-354) can be approximated as

$$P_s = P_r = \frac{(n+1)V^2}{X}\frac{\delta}{n+1} \quad \text{or} \quad P_s = P_r = \frac{V^2}{X}\delta \quad (2\text{-}359)$$

and

$$Q_s = 0 = Q_r. \quad (2\text{-}360)$$

Using Equation (2-352), Table 2-33 shows the active power (P_r) at the sending and receiving ends of the line as a function of the number (n) of shunt compensators used for a power angle of $\delta = 30°$. Also included in Table 2-33 are the reactive power (Q_r) at the receiving end from Equation (2-354), line current (I) from Equations (2-76) and (2-77), reactive power (Q_{line}) needed by each segment of the line from Equation (2-355), *RPI* from Equation (2-357), *LI* from Equation (2-324), and *APR* (S_{sh}) from Equation (2-356).

Table 2-33 Figure of merits of shunt compensation with a power angle of δ = 30°.

n	$P_s = P_r$ (pu)	Q_r (pu)	I (pu)	Q_{line} (pu)	RPI	LI	APR (S_{sh})
0	0.500000	−0.133975	0.517638	0.267949	0.535898	1.000000	0.000000
1	0.517638	−0.068148	0.522105	0.136297	0.526610	1.017332	0.136297
2	0.520945	−0.045577	0.522934	0.091153	0.524932	1.020568	0.182307
3	0.522105	−0.034221	0.523225	0.068441	0.524348	1.021703	0.205323
4	0.522642	−0.027391	0.523360	0.054781	0.524078	1.022228	0.219124
5	0.522934	−0.022832	0.523433	0.045664	0.523931	1.022514	0.228318
∞	$\pi/6$	0	$\pi/6$	0			

Table 2-34 Figure of merits of shunt compensation with a power angle of δ = 60°.

n	$P_s = P_r$ (pu)	Q_r (pu)	I (pu)	Q_{line} (pu)	RPI	LI	APR (S_{sh})
0	0.866025	−0.500000	1.000000	1.000000	1.154701	1.000000	0.000000
1	1.000000	−0.267949	1.035276	0.535898	1.071797	1.071797	0.535898
2	1.026060	−0.180922	1.041889	0.361844	1.057962	1.085533	0.723689
3	1.035276	−0.136297	1.044210	0.272593	1.053220	1.090374	0.817780
4	1.039558	−0.109262	1.045285	0.218524	1.051042	1.092620	0.874096
5	1.041889	−0.091153	1.045869	0.182307	1.049864	1.093842	0.911535
∞	$\pi/3$	0	$\pi/3$	0			

Table 2-35 Figure of merits of shunt compensation with a power angle of δ = 90°.

n	$P_s = P_r$ (pu)	Q_r (pu)	I (pu)	Q_{line} (pu)	RPI	LI	APR (S_{sh})
0	1.000000	−1.000000	1.414214	2.000000	2.000000	1.000000	0.000000
1	1.414214	−0.585786	1.530734	1.171573	1.656854	1.171573	1.171573
2	1.500000	−0.401924	1.552914	0.803848	1.607695	1.205771	1.607695
3	1.530734	−0.304482	1.560723	0.608964	1.591229	1.217927	1.826891
4	1.545085	−0.244717	1.564345	0.489435	1.583844	1.223587	1.957739
5	1.552914	−0.204445	1.566314	0.408890	1.579830	1.226670	2.044450
∞	$\pi/2$	0	$\pi/2$	0			

Table 2-34 is a repetition of Table 2-33 with a power angle of $\delta = 60°$.

Table 2-35 is a repetition of Table 2-33 with a power angle of $\delta = 90°$.

The active power flow as a function of number of shunt compensators is shown in Figure 2-104. As the number of compensators increases, the power flow reaches to the theoretical limit of $\pi/6 = 0.5229$ when $\delta = 30°$, $\pi/3 = 1.0419$ when $\delta = 60°$, and $\pi/2 = 1.57079$ when $\delta = 90°$. The

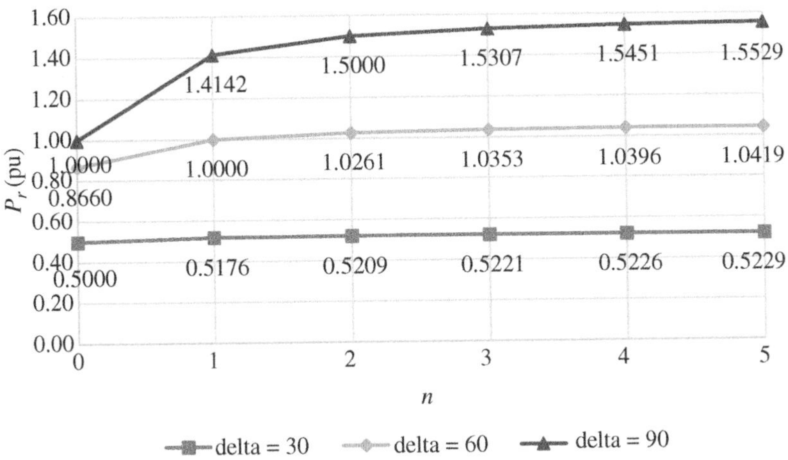

Figure 2-104 Active power at the sending and receiving ends versus number of shunt compensators.

justification of multiple shunt compensators can only be made if the power angle is large enough. Even with a large power angle, such as when $\delta = 90°$, Table 2-35 shows that the active power flow increases by 41.42% with just the use of first shunt compensator. With the use of the second shunt compensator, the active power flow increases only by another 8.58%. Therefore, care must be taken in selecting the number of shunt compensators for use.

In a shunt-compensated line, the *RPI* as a function of number of shunt compensators is shown in Figure 2-105.

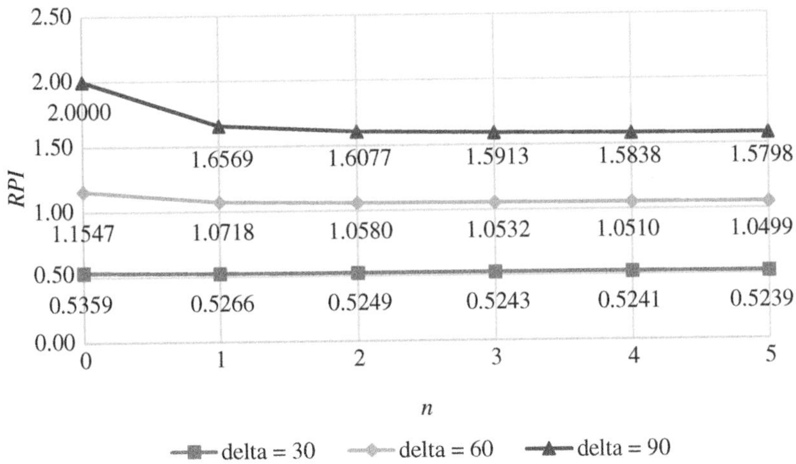

Figure 2-105 *RPI* versus number of shunt compensators.

In a shunt-compensated line, the *LI* as a function of number of shunt compensators is shown in Figure 2-106.

In a shunt-compensated line, the total *APR* of the shunt compensators as a function of number of shunt compensators is shown in Figure 2-107.

Even though the power flow in a line increases as the number of shunt compensators increases, there is a theoretical limit of power flow increase, which is related to the power angle of the line. However, the power flow of the line can be greatly increased with a series compensation up to the thermal limit of the line.

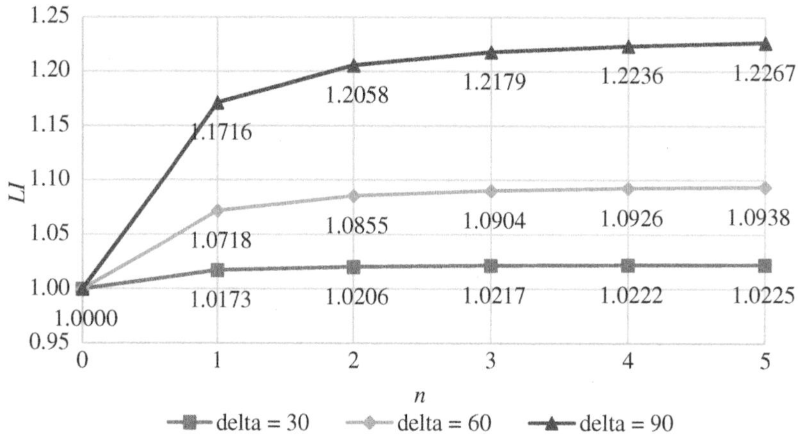

Figure 2-106 *LI* versus number of compensators.

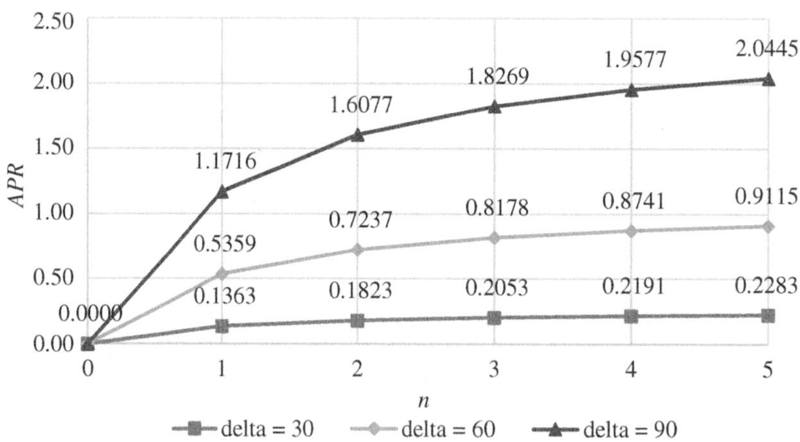

Figure 2-107 Total *APR* of shunt compensation versus number of shunt compensators.

2.6.2 Series-Compensating Reactance

The series-connected reactance compensator delivers or absorbs reactive power to or from the line. In the process, the active and reactive power flows in the line change. The active and reactive power flows at the sending and receiving ends of the series-compensated line are given by

$$P_s = P_r = \frac{V_s V_r}{X_{eff}} \sin\delta = \frac{V^2}{X_{eff}} \sin\delta \tag{2-361}$$

$$Q_s = \frac{V_s V_r}{X_{eff}} \left(\frac{V_s}{V_r} - \cos\delta \right) = \frac{V^2}{X_{eff}} (1 - \cos\delta) \tag{2-362}$$

$$Q_r = \frac{V_s V_r}{X_{eff}} \left(\cos\delta - \frac{V_r}{V_s} \right) = \frac{V^2}{X_{eff}} (\cos\delta - 1) \tag{2-363}$$

where

$$V_s = V_r = V$$

$$X_{eff} = X - X_{se} \tag{2-210a}$$

$$\delta = \delta_s - \delta_r. \tag{2-27}$$

Note that $X_{se} > 0$ means the series-compensating reactance is capacitive and $X_{se} < 0$ means the series-compensating reactance is inductive. Also, $X_{eff} > 0$ represents an effective inductive reactance and $X_{eff} < 0$ represents an effective capacitive reactance, respectively.

From Equations (2-352), (2-361), and (2-210a), it can be written that

$$P_s = P_r = \frac{(n+1)V^2}{X} \sin\left(\frac{\delta}{n+1} \right) = \frac{V^2}{X - X_{se}} \sin(\delta). \tag{2-364}$$

Rearranging Equation (2-364), the series-compensating reactance for the same active power flow (P_r) enhancement as in the case of a shunt-compensated line is given by

$$X_{se} = X \left\{ 1 - \frac{\sin(\delta)}{(n+1)\sin\left(\frac{\delta}{n+1} \right)} \right\}. \tag{2-365}$$

Using Equations (2-263a) and (2-326), the equivalent *APR* (S_{se}) of the series compensators for the same active power flow enhancement, shown in Tables 2-32, 2-33, and 2-34, are given by

$$APR = S_{se} = |V_{s's}| I = I|X_{se}| \times I$$

or

$$APR = S_{se} = I^2 |X_{se}|. \tag{2-366}$$

Table 2-36 shows the series-compensating reactances (X_{se}) for the same active power at the sending and receiving ends as just discussed with various number (n) of shunt compensators for a power angle of $\delta = 30°$. The active power (P_r) is determined using Equation (2-361) for verification. Also included in Table 2-36 are the reactive power (Q_r) at the receiving end from Equation (2-363), line current (I) from Equations (2-76) and (2-77), Q_{line} from Equation (2-11b), *RPI* from Equation (2-321), *LI* from Equation (2-324), and *APR* (S_{se}) from Equation (2-366).

Table 2-37 is a repetition of Table 2-36 with a power angle of $\delta = 60°$.

Table 2-38 is a repetition of Table 2-36 with a power angle of $\delta = 90°$.

Table 2-36 Figure of merits of series compensation with a power angle of $\delta = 30°$.

n	X_{se} (pu)	$P_s = P_r$ (pu)	Q_r (pu)	I (pu)	Q_{line} (pu)	*RPI*	*LI*	*APR* (S_{se})
0	0.000000	0.500000	−0.133975	0.517638	0.267949	0.535859	1.000000	0.000000
1	0.034074	0.517638	−0.138701	0.535898	0.287187	0.554803	1.071797	0.009786
2	0.040205	0.520945	−0.139587	0.539321	0.290868	0.558347	1.085533	0.011694
3	0.042338	0.522105	−0.139898	0.540523	0.292165	0.559590	1.090374	0.012370
4	0.043323	0.522642	−0.140042	0.541079	0.292767	0.560166	1.092620	0.012683
5	0.043857	0.522934	−0.140120	0.541382	0.293094	0.560479	1.093842	0.012854

Table 2-37 Figure of merits of series compensation with a power angle of $\delta = 60°$.

n	X_{se} (pu)	$P_s = P_r$ (pu)	Q_r (pu)	I (pu)	Q_{line} (pu)	*RPI*	*LI*	*APR* (S_{se})
0	0.000000	0.866025	−0.500000	1.000000	1.000000	1.154701	1.000000	0.000000
1	0.133975	1.000000	−0.577350	1.154701	1.333333	1.333333	1.333333	0.178633
2	0.155970	1.026060	−0.592396	1.184793	1.403733	1.368081	1.403733	0.218941
3	0.163484	1.035276	−0.597717	1.195434	1.429062	1.380368	1.429062	0.233628
4	0.166930	1.039558	−0.600189	1.200379	1.440909	1.386078	1.440909	0.240530
5	0.168793	1.041889	−0.601535	1.203070	1.447377	1.389185	1.447377	0.244307

Table 2-38 Figure of merits of series compensation with a power angle of $\delta = 90°$.

n	X_{se} (pu)	$P_s = P_r$ (pu)	Q_r (pu)	I (pu)	Q_{line} (pu)	*RPI*	*LI*	*APR* (S_{se})
0	0.000000	1.000000	−1.000000	1.414214	2.000000	2.000000	1.000000	0.000000
1	0.292893	1.414214	−1.414214	2.000000	4.000000	2.828428	2.000000	1.171573
2	0.333333	1.500000	−1.500000	2.121320	4.500000	3.000000	2.250000	1.500000
3	0.346719	1.530734	−1.530734	2.164784	4.686292	3.061467	2.343146	1.624824
4	0.352786	1.545085	−1.545085	2.185080	4.774575	3.090170	2.387288	1.684405
5	0.356049	1.552914	−1.552914	2.196252	4.823085	3.105829	2.411543	1.717257

The series-compensating reactance as a function of equivalent number of shunt compensators for the same active power flow (P_r) enhancement is shown in Figure 2-108.

In a series-compensated line, the *RPI* as a function of equivalent number of shunt compensators for the same active power flow (P_r) enhancement is shown in Figure 2-109.

In a series-compensated line, the *LI* as a function of equivalent number of shunt compensators for the same active power flow (P_r) enhancement is shown in Figure 2-110.

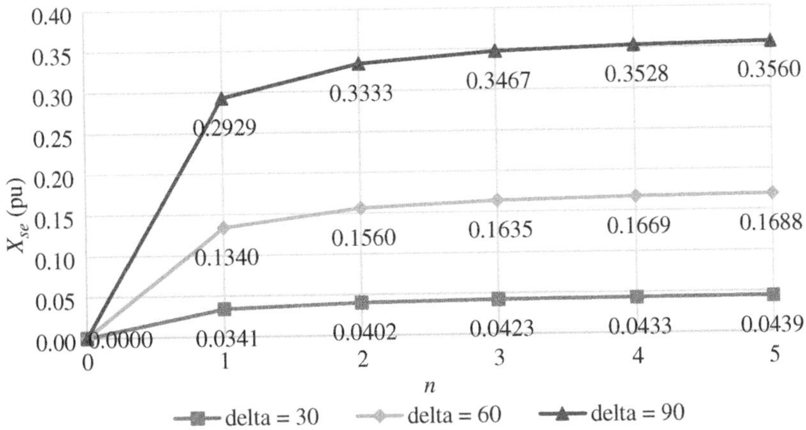

Figure 2-108 Series-compensating reactance versus equivalent shunt compensation for the same active power flow (P_r) enhancement as a function of a number of shunt compensators.

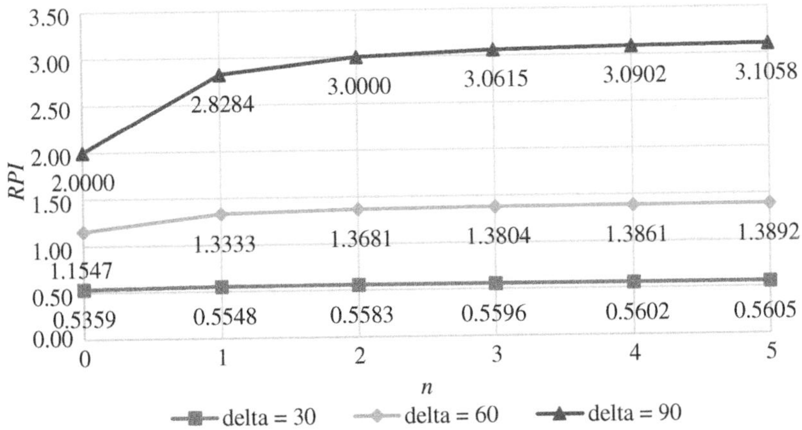

Figure 2-109 *RPI* of series compensation versus equivalent shunt compensation for the same active power flow (P_r) enhancement as a function of a number of shunt compensators.

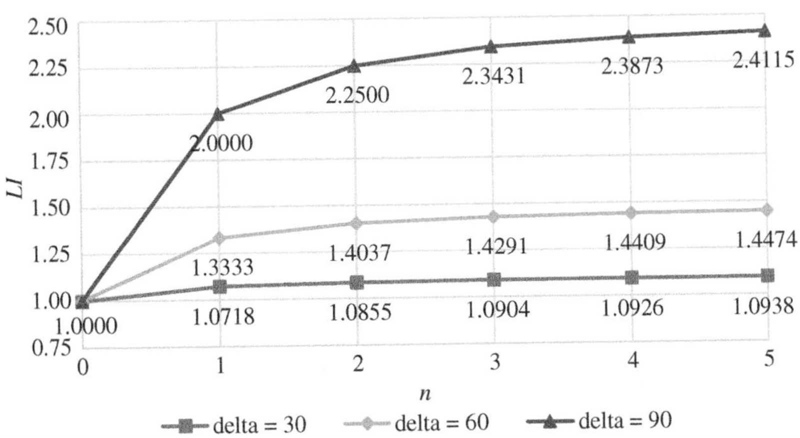

Figure 2-110 *LI* of series compensation versus equivalent shunt compensation for the same active power flow (P_r) enhancement as a function of a number of shunt compensators.

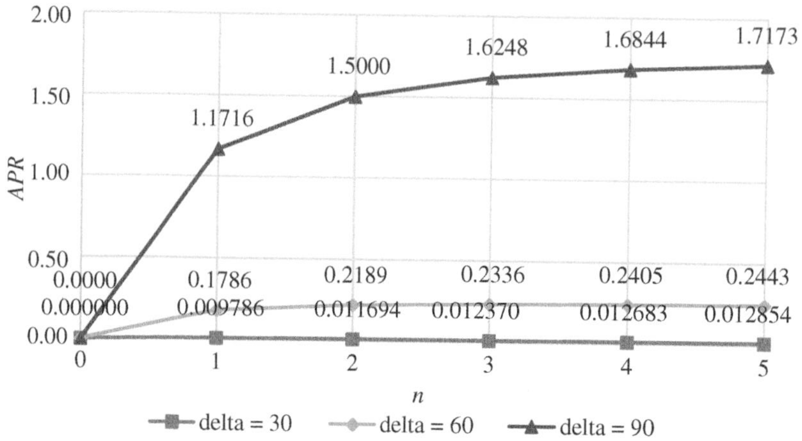

Figure 2-111 *APR* of series compensation versus equivalent shunt compensation for the same active power flow (P_r) enhancement as a function of a number of shunt compensators.

In a series-compensated line, the *APR* as a function of equivalent number of shunt compensators for the same active power flow (P_r) enhancement is shown in Figure 2-111.

It can be noted that for a given amount of active power flow enhancement in a line, the *RPI* and *LI* may be lower in the case of a shunt compensation than in a series compensation as shown in Figures 2-105 and 2-109 and Figures 2-106 and 2-110; however, the *APR* is higher in the case of a shunt compensation than in a series compensation as shown in Figures 2-107 and 2-111.

2.7 Calculation of *RPI*, *LI*, and *APR* for a PAR (sym), a PAR (asym), a RR, and an IR in a Lossy Line

Consider an AC line with a resistance (R), an inductive reactance (X), carrying power from the sending end with a voltage ($\mathbf{V_s} = V_s\angle\delta_s$) to the receiving end with a voltage ($\mathbf{V_r} = V_r\angle\delta_r$) as shown in Figure 2-112. The phasor diagram is drawn for a $X/R = \tan\varphi = \cot\phi = 3.7$. Note that φ is the power factor angle of the line impedance; however, ϕ is the phase angle between the reactive voltage ($\mathbf{V_{Xn}}$) across the line reactance and the voltage ($\mathbf{V_{Rn,Xn}}$) across the line impedance, which is the difference between the sending- and receiving-end voltages. This voltage causes a natural current ($\mathbf{I_n} = I_n\angle\theta_{In}$) to flow through the line impedance ($\mathbf{Z} = R + jX$) as shown in Figure 2-112.

When a RR is connected in series with the line, the effective line reactance (X_{eff}), given by Equation (2-210a), is shown in Figure 2-113.

Consider the electrical system data as given in Table 2-1 with an added line resistance (R) so that the $X/R = 10$. The natural active power flow at the receiving end of the line can be determined by using the following equation:

$$P_{rn} = \frac{A}{\sqrt{R^2 + X^2}}\left\{R\left(\cos\delta - \frac{V_r}{V_s}\right) + X\sin\delta\right\} \tag{B-21}$$

and it is 0.963 pu. It can be verified using the following equation:

$$P_r = \frac{A}{\sqrt{R^2 + X^2}}\left\{R\left(\cos\delta' - \frac{V_r}{V_s}\right) + X\sin\delta'\right\} \tag{B-117}$$

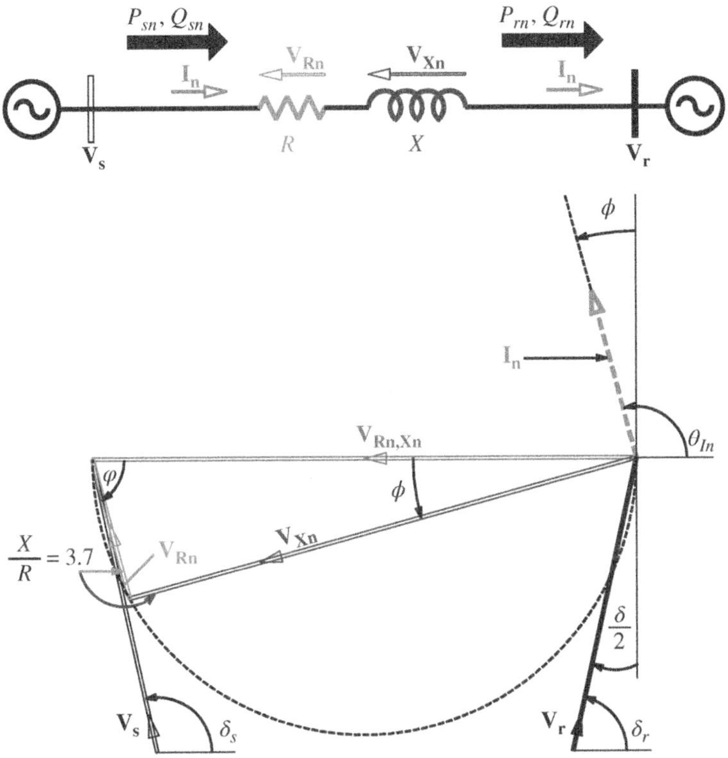

Figure 2-112 Two-generator/one-line power system network with a lossy line and its phasor diagram.

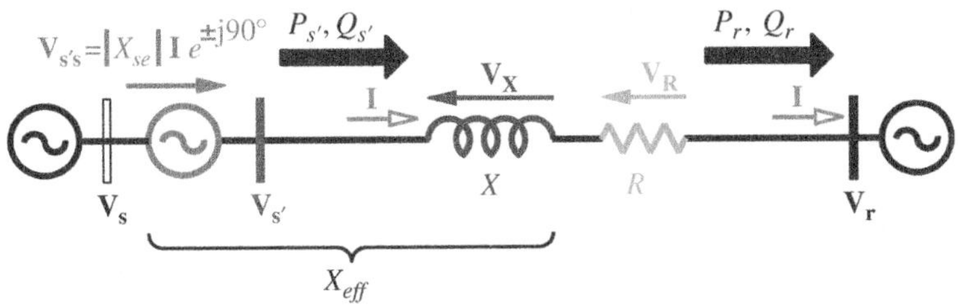

Figure 2-113 Two-generator/one-line power system network with a lossy line and a RR.

that the PAR (sym) is able to increase active power flow from its natural active power flow of 0.963 pu to about 1.786 pu at the receiving end of the line. Therefore, the same active power flow enhancement ($\Delta P = 1.786$ pu $- 0.963$ pu $= 0.823$ pu) is considered while studying other PFCs, such as a PAR (asym), a RR, and an IR. The *RPI*, *LI*, and *APR* are determined in each case. The control parameters and their effects on $V_{s's}$, β, $V_{s'}$, and ψ in a PAR (sym), a PAR (asym), a RR, and an IR are shown in Table 2-39. For a RR, the effects on β, $V_{s'}$, and ψ are dependent on the X/R ratio of the line. For $R = 0$, the expressions for β, $V_{s'}$, and ψ are given in Equations (2-227), (2-221) or (2-229), and (2-230), respectively. The range of a RR is $-X_L$ to $+X_C$.

Table 2-39 Control parameters and their effects on $V_{s's}$, β, $V_{s'}$, and ψ in a PAR (sym), a PAR (asym), a RR, and an IR.

	Control parameter(s)	$V_{s's}$	β	$V_{s'}$	Ψ
PAR (sym)	ψ	$V_{s's} = 2V_s \sin\left(\frac{\psi}{2}\right)$ (2-198)	$\beta = \frac{\psi}{2} + 90°$ for $\psi > 0$ or $\beta = \frac{\psi}{2} + 270°$ for $\psi < 0$ (2-200)	V_s	Range: $-\psi_{max}$ to $+\psi_{max}$
PAR (asym)	$V_{s's}$	Range: 0–$V_{s's}$	−90° or 90°	$V_{s'} = \sqrt{V_s^2 + V_{s's}^2}$ (2-270)	$\psi = \tan^{-1}\frac{V_{s's}}{V_s}$ (2-271a)
RR	X_{se} Range: $-X_L$ to $+X_C$	$V_{s's} = IX_{se}$ (2-263b)	$\beta = \tan^{-1}[\{-R(Q_r - Q_{rn}) + X(P_r - P_{rn})\}/\{R(P_r - P_{rn}) + X(Q_r - Q_{rn})\}] - \delta$ (B-82)	$V_{s'} = \sqrt{V_r^2 + (R^2 + X^2)I^2 + 2(RP_r + XQ_r)}$ (B-132)	$\psi = \tan^{-1}\frac{XP_r - RQ_r}{V_r^2 + RP_r + XQ_r} - \delta$ (B-134)
IR	$\mathbf{V_{s's}} = V_{s's}\angle(\delta_s + \beta)$	Range: 0–$V_{s's}$	Range: 0°–360°	$V_{s'} = \sqrt{V_s^2 + V_{s's}^2 + 2V_s V_{s's} \cos\beta}$ (2-97)	$\psi = \tan^{-1}\frac{V_{s's} \sin\beta}{V_s + V_{s's} \cos\beta}$ (2-98)

2.7.1 PAR (sym)

In the case of a PAR (sym), the following steps are needed.

Step 0: Determine the range of the control parameter.

It can be found from Equations (2-51a), (B-13), and (B-117) that for an active power flow ($P_r = 1.786$ pu), the control parameter of a PAR (sym) is $\psi = 50°$. Therefore, the phase-shift angle (ψ) is varied in the range of 0 to 50° and the results are shown in Figures 2-114 through 2-116.

Step 1: For a desired phase-shift angle (ψ), calculate the modified power angle (δ') using the following equations:

$$\delta = \delta_s - \delta_r \tag{2-27}$$

$$\delta' = \delta + \psi. \tag{2-51a}$$

Step 2: Calculate the active power (P_r) and reactive power (Q_r) at the receiving end using the following equations:

$$P_r = \frac{A}{\sqrt{R^2 + X^2}} \left\{ R\left(\cos\delta' - \frac{V_r}{V_s}\right) + X\sin\delta' \right\} \tag{B-117}$$

$$Q_r = \frac{A}{\sqrt{R^2 + X^2}} \left\{ -R\sin\delta' + X\left(\cos\delta' - \frac{V_r}{V_s}\right) \right\} \tag{B-118}$$

where

$$A = \frac{V_s V_r}{\sqrt{R^2 + X^2}}. \tag{B-13}$$

Step 3: Calculate the magnitude of the natural line current (I_n) using the following equations:

$$V_{Rn,Xn} = \sqrt{V_s^2 - 2V_sV_r\cos\delta + V_r^2} \tag{B-4}$$

$$I_n = \frac{V_{Rn,Xn}}{\sqrt{R^2 + X^2}}. \tag{B-8}$$

Step 4: Using P_r and Q_r in Step 2 and $V_{s'}$s from Equation (2-198), calculate line current (I), reactive power (Q_{line}) absorbed by the line, *RPI*, *LI* for $k_R = 1$, and *APR* using the following equations:

$$S_r = \sqrt{P_r^2 + Q_r^2} \tag{2-76}$$

$$I = \frac{S_r}{V_r} \tag{2-77}$$

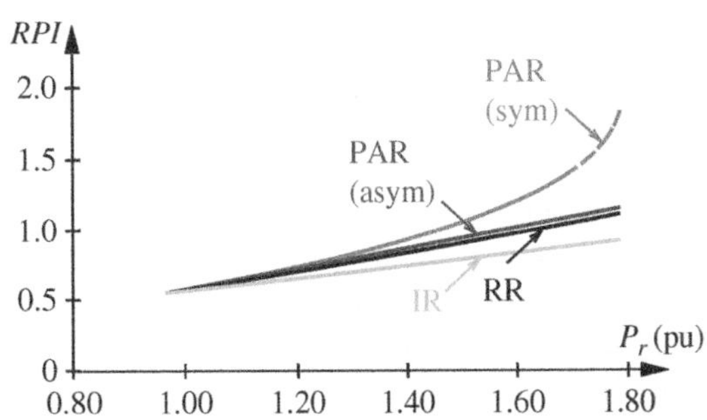

Figure 2-114 Reactive Power Index (*RPI*) versus active power flow (P_r) at the receiving end of the line.

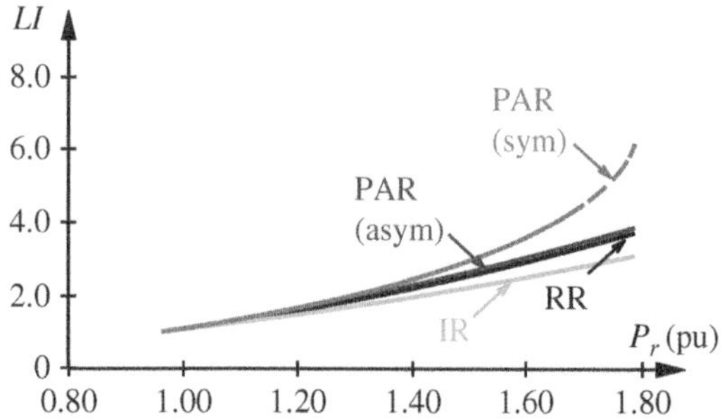

Figure 2-115 Loss Index (LI) versus active power flow (P_r) at the receiving end of the line.

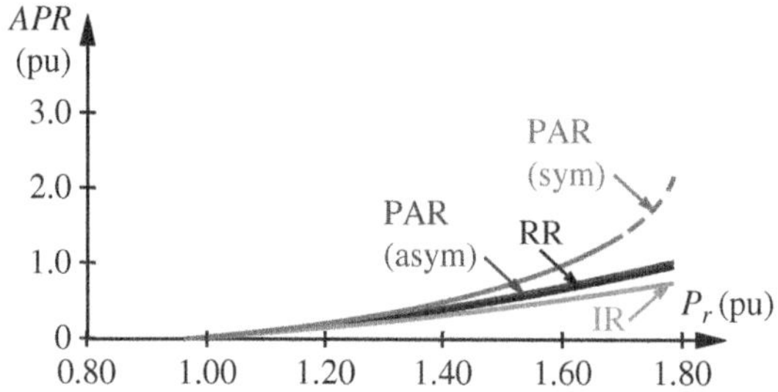

Figure 2-116 Apparent Power Rating (APR) versus active power flow (P_r) at the receiving end of the line.

$$Q_{line} = I^2X \tag{2-11b}$$

$$RPI = \frac{Q_{line}}{P_r} \tag{2-321}$$

$$LI = k_R\frac{I^2}{I_n^2} \tag{2-324}$$

$$V_{s's} = 2V_s \sin\left(\frac{\psi}{2}\right) \tag{2-198}$$

$$APR = |V_{s's}|I. \tag{2-326}$$

2.7.2 PAR (asym)

In the case of a PAR (asym), the following steps are needed.

Step 0: Determine the range of the control parameter.

It can be found from Equations (2-270), (2-271a), (2-51a), (B-34), and (B-43) that for an active power flow ($P_r = 1.786$ pu), the control parameter of a PAR (asym) is $V_{s's} = 0.509$ pu. Therefore, the magnitude of the compensating voltage ($V_{s's}$) is varied in the range of 0 to 0.509 pu and the results are shown in Figures 2-114 through 2-116.

Step 1: For a desired series-compensating voltage ($\mathbf{V_{s's}}$) with magnitude ($V_{s's}$), calculate the modified sending-end voltage ($V_{s'}$), phase-shift angle (ψ), and the modified power angle (δ') using the following equations:

$$V_{s'} = \sqrt{V_s^2 + V_{s's}^2} \quad (2\text{-}270)$$

$$\psi = \tan^{-1}\frac{V_{s's}}{V_s} \quad (2\text{-}271a)$$

$$\delta = \delta_s - \delta_r \quad (2\text{-}27)$$

$$\delta' = \delta + \psi. \quad (2\text{-}51a)$$

Step 2: Calculate the active power (P_r) and reactive power (Q_r) at the receiving end using the following equations:

$$P_r = \frac{A'}{\sqrt{R^2 + X^2}}\left\{R\left(\cos\delta' - \frac{V_r}{V_{s'}}\right) + X\sin\delta'\right\} \quad (B\text{-}43)$$

$$Q_r = \frac{A'}{\sqrt{R^2 + X^2}}\left\{-R\sin\delta' + X\left(\cos\delta' - \frac{V_r}{V_{s'}}\right)\right\} \quad (B\text{-}44)$$

where

$$A' = \frac{V_{s'}V_r}{\sqrt{R^2 + X^2}}. \quad (B\text{-}34)$$

Step 3: Calculate the magnitude of the natural line current (I_n) using the following equations:

$$V_{Rn,Xn} = \sqrt{V_s^2 - 2V_sV_r\cos\delta + V_r^2} \quad (B\text{-}4)$$

$$I_n = \frac{V_{Rn,Xn}}{\sqrt{R^2 + X^2}}. \quad (B\text{-}8)$$

Step 4: Using P_r and Q_r in Step 2, calculate the line current (I), reactive power absorbed by the line (Q_{line}), reactive power index (RPI), loss index (LI) for $k_R = 1$, and APR using the following equations:

$$S_r = \sqrt{P_r^2 + Q_r^2}. \quad (2\text{-}76)$$

$$I = \frac{S_r}{V_r} \quad (2\text{-}77)$$

$$Q_{line} = I^2X \quad (2\text{-}11b)$$

$$RPI = \frac{Q_{line}}{P_r} \quad (2\text{-}321)$$

$$LI = k_R\frac{I^2}{I_n^2} \quad (2\text{-}324)$$

$$APR = |V_{s's}|I. \quad (2\text{-}326)$$

2.7.3 RR

In the case of a RR, the following steps are needed.

Step 0: Determine the range of the control parameter.

It can be found from Equations (2-210a), (B-125), and (B-123) that for an active power flow ($P_r = 1.786$ pu), the control parameter of a RR is $X_{se} = 0.2445$ pu. Therefore, the compensating-reactance (X_{se}) is varied in the range of 0 to 0.2445 pu and the results are shown in Figures 2-114 through 2-116.

Step 1: For a desired series-compensating reactance (X_{se}), calculate the effective line reactance (X_{eff}) using the following equation:

$$X_{eff} = X - X_{se}. \tag{2-210a}$$

Step 2: Calculate the active power (P_r) and reactive power (Q_r) at the receiving end using the following equations:

$$P_r = \frac{A_{eff}}{\sqrt{R^2 + X_{eff}^2}} \left\{ R\left(\cos\delta - \frac{V_r}{V_s}\right) + X_{eff}\sin\delta \right\} \tag{B-123}$$

$$Q_r = \frac{A_{eff}}{\sqrt{R^2 + X_{eff}^2}} \left\{ -R\sin\delta + X_{eff}\left(\cos\delta - \frac{V_r}{V_s}\right) \right\} \tag{B-124}$$

where

$$A_{eff} = \frac{V_s V_r}{\sqrt{R^2 + X_{eff}^2}}. \tag{B-125}$$

Step 3: Calculate the magnitude of the natural line current (I_n) using the following equations:

$$V_{Rn,Xn} = \sqrt{V_s^2 - 2V_s V_r \cos\delta + V_r^2} \tag{B-4}$$

$$I_n = \frac{V_{Rn,Xn}}{\sqrt{R^2 + X^2}}. \tag{B-8}$$

Step 4: Using P_r and Q_r in Step 2, calculate the line current (I), reactive power absorbed by the line (Q_{line}), Reactive Power Index (RPI), Loss Index (LI) for $k_R = 1$, and Apparent Power Rating (APR) using the following equations:

$$S_r = \sqrt{P_r^2 + Q_r^2} \tag{2-76}$$

$$I = \frac{S_r}{V_r} \tag{2-77}$$

$$Q_{line} = I^2 X \tag{2-11b}$$

$$RPI = \frac{Q_{line}}{P_r} \tag{2-321}$$

$$LI = k_R \frac{I^2}{I_n^2} \tag{2-324}$$

$$V_{s's} = I X_{se} \tag{2-263b}$$

$$APR = |V_{s's}| I. \tag{2-326}$$

2.7.4 IR

In the case of an IR, the following steps are needed.

Step 0: Determine the range of the control parameter.

It can be found from Equations (2-97), (2-98), (2-51a), (B-34), and (B-43) that for an active power flow ($P_r = 1.786$ pu), the control parameters of an IR are $V_{s's} = 0.415$ pu and $\beta = 49°$. Keeping the relative phase angle (β) fixed at 49°, the magnitude ($V_{s's}$) is varied in the range of 0 to 0.415 pu and the results are shown in Figures 2-114 through 2-116.

Step 1: For a desired series-compensating voltage ($\mathbf{V_{s's}}$) with magnitude ($V_{s's}$) and relative phase angle (β), calculate the magnitude ($V_{s'}$) of the modified sending-end voltage ($\mathbf{V_{s'}}$), the phase-shift angle (ψ), and the modified power angle (δ') using the following equations:

$$V_{s'} = \sqrt{V_s^2 + V_{s's}^2 + 2V_sV_{s's}\cos\beta} \tag{2-97}$$

$$\psi = \tan^{-1}\frac{V_{s's}\sin\beta}{V_s + V_{s's}\cos\beta}. \tag{2-98}$$

$$\delta = \delta_s - \delta_r \tag{2-27}$$

$$\delta' = \delta + \psi. \tag{2-51a}$$

Step 2: Calculate the active power (P_r) and reactive power (Q_r) at the receiving end using the following equations:

$$P_r = \frac{A'}{\sqrt{R^2 + X^2}}\left\{R\left(\cos\delta' - \frac{V_r}{V_{s'}}\right) + X\sin\delta'\right\} \tag{B-43}$$

$$Q_r = \frac{A'}{\sqrt{R^2 + X^2}}\left\{-R\sin\delta' + X\left(\cos\delta' - \frac{V_r}{V_{s'}}\right)\right\} \tag{B-44}$$

where

$$A' = \frac{V_{s'}V_r}{\sqrt{R^2 + X^2}}. \tag{B-34}$$

Step 3: Calculate the magnitude of the natural line current (I_n) using the following equations:

$$V_{Rn,Xn} = \sqrt{V_s^2 - 2V_sV_r\cos\delta + V_r^2} \tag{B-4}$$

$$I_n = \frac{V_{Rn,Xn}}{\sqrt{R^2 + X^2}}. \tag{B-8}$$

Step 4: Using P_r and Q_r in Step 2, calculate the line current (I), reactive power absorbed by the line (Q_{line}), RPI, LI for $k_R = 1$, and APR using the following equations:

$$S_r = \sqrt{P_r^2 + Q_r^2} \tag{2-76}$$

$$I = \frac{S_r}{V_r} \tag{2-77}$$

$$Q_{line} = I^2X \tag{2-11b}$$

$$RPI = \frac{Q_{line}}{P_r} \tag{2-321}$$

$$LI = k_R\frac{I^2}{I_n^2} \tag{2-324}$$

$$APR = |V_{s's}|I. \tag{2-326}$$

In each case, the IR requires lower overall reactive power, causes lower losses in the line, and requires lower *APR* than the PAR (sym and asym) and the RR as shown in Figures 2-114 through 2-116.

Note that the PAR (sym) is not capable of reaching the thermal limit of the line at 2 pu, in this example, when the ideal line is replaced with a more realistic line with a $X/R = 10$.

The *RPI* per unit of active power (P_r) transmitted is given by

$$\frac{RPI}{P_r} = \frac{V_{Rn,Xn}^2 X}{V_s^2 V_r^2} \frac{R_{eff}^2 + X_{eff}^2}{\left[R_{eff}\left\{\cos\delta - \frac{V_r}{V_s}\right\} + X_{eff}\sin\delta\right]^2}. \tag{B-151}$$

The *LI* per unit of active power (P_r) is given by

$$\frac{LI}{P_r} = k_R \frac{1}{V_s V_r} \frac{R^2 + X^2}{R_{eff}\left\{\cos\delta - \frac{V_r}{V_s}\right\} + X_{eff}\sin\delta}. \tag{B-152}$$

The *APR* per unit of active power (P_r) is given by

$$\frac{APR}{P_r} = \frac{V_{Rn,Xn}^2}{V_s V_r} \frac{\sqrt{R_{se}^2 + X_{se}^2}}{R_{eff}\left\{\cos\delta - \frac{V_r}{V_s}\right\} + X_{eff}\sin\delta}. \tag{B-153}$$

Equations (B-151), (B-152), and (B-153) apply to a PAR (sym), a PAR (asym), and an IR, since all three PFCs emulate a four-quadrant, series-compensating impedance ($\mathbf{Z_{se}} = R_{se} - jX_{se}$) that consists of a resistance ($R_{se} = +R$ or $-R$) and a reactance ($X_{se} = X_C$ or $-X_L$) in series with the line. However, for a RR, $R_{se} = 0$. Therefore, Equations (B-151), (B-152), and (B-153) can be rewritten as follows.

The *RPI* per unit of active power (P_r) transmitted is given by

$$\frac{RPI}{P_r} = \frac{V_{Rn,Xn}^2 X}{V_s^2 V_r^2} \frac{R^2 + X_{eff}^2}{\left[R\left\{\cos\delta - \frac{V_r}{V_s}\right\} + X_{eff}\sin\delta\right]^2}. \tag{B-151a}$$

The *LI* per unit of active power (P_r) transmitted is given by

$$\frac{LI}{P_r} = k_R \frac{1}{V_s V_r} \frac{R^2 + X^2}{R\left\{\cos\delta - \frac{V_r}{V_s}\right\} + X_{eff}\sin\delta}. \tag{B-152a}$$

The *APR* per unit of active power (P_r) transmitted is given by

$$\frac{APR}{P_r} = \frac{V_{Rn,Xn}^2}{V_s V_r} \frac{X_{se}}{R\left\{\cos\delta - \frac{V_r}{V_s}\right\} + X_{eff}\sin\delta}. \tag{B-153a}$$

Table 2-27 can be expanded to include the quality factor of the line to be Q = 10, which modifies the natural active power flow, $P_{rn} = 0.963$ pu. The comparison of *RPI*, *LI*, and *APR* per unit of active power (P_r) transmitted for a PAR (sym), a PAR (asym), a RR, and an IR for an active power flow enhancement ($\Delta P = 1.786$ pu $- 0.963$ pu $= 0.823$ pu) at the receiving end of the line is shown in Table 2-40.

Table 2-40 Comparison of *RPI*, *LI*, and *APR* per unit of active power (P_r) transmitted for a PAR (sym), a PAR (asym), a RR, and an IR for an active power flow enhancement (ΔP = 1.786 pu − 0.963 pu = 0.823 pu) at the receiving end of the line.

Parameters	Base case	PAR (sym)	PAR (asym)	RR	IR
Active power (P_r) at receiving end	0.963 pu	1.786 pu	1.786 pu	1.786 pu	1.786 pu
Series-compensating Resistance (R_{se})	0	0.0533 pu	0.0081 pu	0	−0.791 pu
Series-compensating Reactance (X_{se})	0	0.326 pu	0.2511 pu	0.2445 pu	0.2153 pu
Effective Line Resistance ($R_{eff} = R + R_{se}$)	0	0.1033 pu	0.0581 pu	0.05 pu	−0.741 pu
Effective Line Reactance ($X_{eff} = X + X_{se}$)	0	0.174 pu	0.2489 pu	0.2555 pu	0.2847 pu
RPI/P_r	0.55	1.025	0.643	0.62	0.513
LI/P_r	1	3.453	2.165	2.086	1.727
APR/P_r	0	1.21	0.577	0.541	0.42

Note that the IR offers the lowest *RPI*, *LI*, and *APR* per unit of active power (P_r) transmitted, which result in the lowest need for reactive power, lowest loss in the line, and the lowest rating of the controller. The lowest *RPI*, *LI*, and *APR* contribute to the lowest overall cost of a solution.

The equivalent compensating resistance (R_{se}) and reactance (X_{se}) are determined in each case as follows.

PAR (sym):

Using Equation (2-200), the relative phase angle, $\beta = \frac{\psi}{2} + 90°$ for $\psi > 0$ or $\beta = \frac{50°}{2} + 90° = 115°$.

Using Equation (B-4), the magnitude ($V_{Rn,Xn}$) of the voltage ($\mathbf{V_{Rn,Xn}}$) across the transmission line is determined as

$$V_{Rn,Xn} = \sqrt{V_s^2 - 2V_sV_r\cos\delta + V_r^2} = \sqrt{1^2 - 2\times 1\times 1\times \cos 30°} = 0.517638 \text{ pu}.$$

Using Equation (B-53), the magnitude ($V_{R,X}$) of the voltage ($\mathbf{V_{R,X}}$) across the transmission line is determined as

$$V_{R,X} = \sqrt{V_{Rn,Xn}^2 + V_{s's}[V_{s's} + 2\{V_s\cos\beta - V_r\cos(\delta+\beta)\}]}$$
$$= \sqrt{0.517638^2 + 0.845\times[0.845 + 2\{1\times\cos 115° - 1\times\cos(30° + 115°)\}]} = 1.2856 \text{ pu}.$$

Using Equation (B-143a),

$$R_{se} = \sqrt{R^2+X^2}\frac{V_{s's}}{V_{R,X}}\left\{\frac{V_s\sin\left(\frac{\psi}{2} + \tan^{-1}\frac{X}{R}\right) - V_r\sin\left(\delta + \frac{\psi}{2} + \tan^{-1}\frac{X}{R}\right) - V_{s's}\cos\left(\tan^{-1}\frac{X}{R}\right)}{V_{R,X}}\right\}$$

$$= \sqrt{0.05^2+0.5^2}\frac{0.845}{1.2856}\left\{\frac{1\times\sin\left(\frac{50°}{2} + \tan^{-1}10\right) - 1\times\sin\left(30° + \frac{50°}{2} + \tan^{-1}10\right) - 0.845\times\cos(\tan^{-1}10)}{1.2856}\right\}$$

$$= 0.0533 \text{ pu}.$$

Using Equation (B-143b),

$$X_{se} = \sqrt{R^2+X^2}\frac{V_{s's}}{V_{R,X}}\left\{\frac{V_s\cos\left(\frac{\psi}{2}+\tan^{-1}\frac{X}{R}\right)-V_r\cos\left(\delta+\frac{\psi}{2}+\tan^{-1}\frac{X}{R}\right)+V_{s's}\sin\left(\tan^{-1}\frac{X}{R}\right)}{V_{R,X}}\right\}$$

$$= \sqrt{0.05^2+0.5^2}\frac{0.845}{1.2856}$$

$$\left\{\frac{1\times\cos\left(\frac{50^\circ}{2}+\tan^{-1}10\right)-1\times\cos\left(30^\circ+\frac{50^\circ}{2}+\tan^{-1}10\right)+0.845\times\sin(\tan^{-1}10)}{1.2856}\right\}$$

$$= 0.326\ \text{pu.}$$

PAR (asym):

Using Equation (B-53), the magnitude ($V_{R,X}$) of the voltage ($\mathbf{V_{R,X}}$) across the transmission line is calculated as

$$V_{R,X} = \sqrt{V_{Rn,Xn}^2+V_{s's}[V_{s's}+2\{V_s\cos\beta-V_r\cos(\delta+\beta)\}]}$$

$$= \sqrt{0.517638^2+0.509\times[0.509+2\{1\times\cos 90^\circ-1\times\cos(30^\circ+90^\circ)\}]} = 1.0179\ \text{pu.}$$

Using Equation (B-141a),

$$R_{se} = \sqrt{R^2+X^2}\frac{V_{s's}}{V_{R,X}}\left\{\frac{V_s\sin\left(\tan^{-1}\frac{X}{R}\right)-V_r\sin\left(\delta+\tan^{-1}\frac{X}{R}\right)-V_{s's}\cos\left(\tan^{-1}\frac{X}{R}\right)}{V_{R,X}}\right\}$$

$$= \sqrt{0.05^2+0.5^2}\frac{0.509}{1.0179}\left\{\frac{1\times\sin(\tan^{-1}10)-1\times\sin(30^\circ+\tan^{-1}10)-0.509\times\cos(\tan^{-1}10)}{1.0179}\right\}$$

$$= 0.0081\ \text{pu.}$$

Using Equation (B-141b),

$$X_{se} = \sqrt{R^2+X^2}\frac{V_{s's}}{V_{R,X}}\left\{\frac{V_s\cos\left(\tan^{-1}\frac{X}{R}\right)-V_r\cos\left(\delta+\tan^{-1}\frac{X}{R}\right)+V_{s's}\sin\left(\tan^{-1}\frac{X}{R}\right)}{V_{R,X}}\right\}$$

$$= \sqrt{0.05^2+0.5^2}\frac{0.509}{1.0179}\left\{\frac{1\times\cos(\tan^{-1}10)-1\times\cos(30^\circ+\tan^{-1}10)+0.509\times\sin(\tan^{-1}10)}{1.0179}\right\}$$

$$= 0.2511\ \text{pu.}$$

IR:

Using Equation (B-53), the magnitude ($V_{R,X}$) of the voltage ($\mathbf{V_{R,X}}$) across the transmission line is calculated as

$$V_{R,X} = \sqrt{V_{Rn,Xn}^2+V_{s's}[V_{s's}+2\{V_s\cos\beta-V_r\cos(\delta+\beta)\}]}$$

$$= \sqrt{0.517638^2+0.415\times[0.415+2\times\{1\times\cos 49^\circ-1\times\cos(30^\circ+49^\circ)\}]} = 0.909\ \text{pu.}$$

Using Equation (B-138a),

$$R_{se} = \sqrt{R^2+X^2}\frac{V_{s's}}{V_{R,X}}\left\{\frac{-V_s\cos\left(\beta+\tan^{-1}\frac{X}{R}\right)+V_r\cos\left(\delta+\beta+\tan^{-1}\frac{X}{R}\right)-V_{s's}\cos\left(\tan^{-1}\frac{X}{R}\right)}{V_{R,X}}\right\}$$

$$= \sqrt{0.05^2+0.5^2}\frac{0.415}{0.909}\left\{\frac{-1\times\cos(49^\circ+\tan^{-1}10)+1\times\cos(30^\circ+49^\circ+\tan^{-1}10)-0.415\times\cos(\tan^{-1}10)}{0.909}\right\}$$

$$= -0.791 \text{ pu.}$$

Using Equation (B-138b),

$$X_{se} = \sqrt{R^2+X^2}\frac{V_{s's}}{V_{R,X}}\left\{\frac{V_s\sin\left(\beta+\tan^{-1}\frac{X}{R}\right)-V_r\sin\left(\delta+\beta+\tan^{-1}\frac{X}{R}\right)+V_{s's}\sin\left(\tan^{-1}\frac{X}{R}\right)}{V_{R,X}}\right\}$$

$$= \sqrt{0.05^2+0.5^2}\frac{0.415}{0.909}\left\{\frac{1\times\sin(49^\circ+\tan^{-1}10)-1\times\sin(30^\circ+49^\circ+\tan^{-1}10)+0.415\times\sin(\tan^{-1}10)}{0.909}\right\}$$

$$= 0.2153 \text{ pu.}$$

2.8 *Sen Index* of a PFC

Consider the power flow in the line is increased by increasing the line current (I) from its natural (uncompensated) value (I_n). Therefore, using Equations (2-318): $P_{line} = I^2R$ and (2-322): $P_{linen} = I_n^2R_n$, the additional power loss in the line due to active power flow increase (ΔP) at the receiving end is given by

$$P_{additional} = P_{line} - P_{linen} \quad \text{or} \quad P_{additional} = I^2R - I_n^2R_n. \tag{2-367}$$

Using Equations (2-11b): $Q_{line} = I^2X$ and (2-323): $Q_{linen} = I_n^2X$, an additional reactive power in the line due to active power flow increase (ΔP) is given by

$$Q_{additional} = Q_{line} - Q_{linen} \quad \text{or} \quad Q_{additional} = (I^2 - I_n^2)X. \tag{2-368}$$

The installation cost of a PFC is directly related to the *APR* of the PFC. In addition, there are recurring costs in terms of power loss in the PFC, operational costs, maintenance costs, and so on.

Sen Index (*SI*) is defined as the increased cost of power delivery per unit of incremental power delivered, which is given by

$$SI = \frac{P_{additional} + Q_{additional} + APR}{\Delta P} \tag{2-369}$$

neglecting the recurring costs, or

$$SI = \frac{(I^2 - I_n^2)(R + X) + APR}{\Delta P} \tag{2-370}$$

where

$$\Delta P = P_r - P_{rn} \tag{2-309}$$

$$k_R = \frac{R}{R_n} = 1. \tag{2-325}$$

SI is unitless, since the increased cost of power delivery is related to the sum of the additionally -required active and reactive powers and the *APR* of the PFC for a unit of incremental power

Table 2-41 Comparison of *SI* of a PAR (sym), a PAR (asym), a RR, and an IR for an active power flow enhancement (ΔP = 1.786 pu − 0.963 pu = 0.823 pu) enhancement at the receiving end of the line.

Parameters	Base case	PAR (sym)	PAR (asym)	RR	IR
Modified sending-end voltage ($V_{s'}$)	1 pu	1 pu	1.122 pu	1.141	1.31 pu
Phase-shift angle (ψ)	0°	50°	26.976°	25.144°	13.83°
Series-compensating voltage ($V_{s's}$)	0	0.845 pu	0.509 pu	0.486 pu	0.415 pu
Relative phase angle (β)	0°	115°	90°	86.07°	49°
Active power ($P_{s'}$) at modified sending end	1.016 pu	2.114 pu	1.991 pu	1.984	1.95 pu
Active power (P_r) at receiving end	0.963 pu	1.786 pu	1.786 pu	1.786 pu	1.786 pu
Reactive power ($Q_{s'}$) at modified sending end	0.166 pu	1.441 pu	1.096 pu	1.103	1.348 pu
Reactive power (Q_r) at receiving end	−0.364 pu	−1.831 pu	−0.956 pu	−0.874 pu	−0.288 pu
Line current (I)	1.030 pu	2.558 pu	2.026 pu	1.988 pu	1.809 pu
Line reactive power ($Q_{line} = I^2X$)	0.530 pu	3.273 pu	2.052 pu	1.977 pu	1.636 pu
Reactive Power Index, $RPI = Q_{line}/P_r$	0.55	1.832	1.149	1.107	0.916
Loss Index, $LI = k_R \times I^2/I_n^2$ at natural line current (*i.e.* $I = I_n$); $k_R = 1$	1	6.168	3.867	3.725	3.084
$APR = \lvert V_{s's} \rvert I$	0	2.162 pu	1.031 pu	0.967 pu	0.751 pu
$Sen\ Index = \{(I^2 - I_n^2)(R + X) + APR\}/\Delta P$	0	6.293	3.288	3.11	2.392

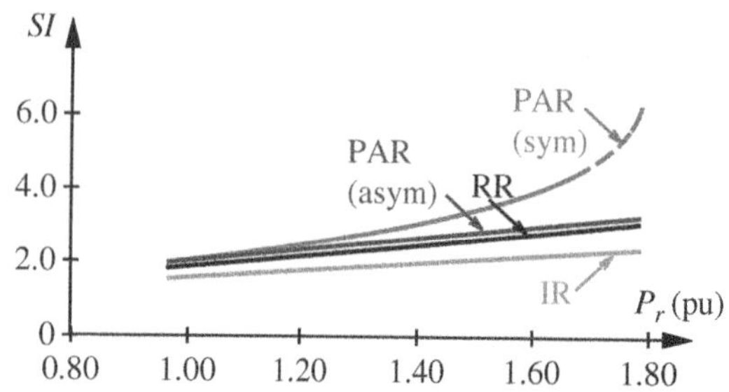

Figure 2-117 *Sen Index* (SI) versus active power flow (P_r) at the receiving end of the line using a PAR (sym), a PAR (asym), a RR, and an IR.

delivery. If desired, the unit of *SI* may be expressed in \$/MW delivered. In that case, the delivery cost may be determined, considering the actual cost of various types of solutions, such as VSC-based, transformer/LTC-based, and so on. Note that a transformer/LTCs-based solution does not generate or absorb reactive power; therefore, the additional reactive power needed by the line during a power flow enhancement must be supplied by another source. In contrast, a shunt or series capacitor, including a VSC-based or an electrical machine-based solution, generates reactive power that should be taken into account for determining the absolute reactive power requirement from other sources during a power flow enhancement. As a first order of approximation of *SI*, it is evident that the higher the SI, the poorer the investment is and the lower the SI, the better the investment is.

Table 2-31 can be expanded to include the comparison of *SI* of a PAR (sym), a PAR (asym), a RR, and an IR for an active power flow enhancement ($\Delta P = 1.786\ \text{pu} - 0.963\ \text{pu} = 0.823\ \text{pu}$) at the receiving end of the line as shown in Table 2-41.

The figure of merit, called *Sen Index*, for each of the most commonly used PFCs, such as a PAR (sym), a PAR (asym), a RR, and an IR is shown in Figure 2-117.

3

Modeling Principles

The best way to understand the subject matter of this book is to practice some mathematical modeling. There are many simulation software programs available for this purpose. However, the most widely used mathematical modeling software program in power system transient studies is based on Electromagnetic Transients Program (EMTP). All EMTP models that are discussed throughout the book are available at the following ftp address: www.wiley.com/go/sen/powerflowcontrol. Various modeling techniques that are presented throughout this book can be easily translated into other simulation software. Various sample models are discussed in this book to lay the foundation for a complete modeling experience, involving power system networks and Power Flow Controllers (PFCs) with the use of EMTP software. Models are created and developed incrementally. They show application of equations, developed in Chapter 2 and Appendices A and B.

While there are several different user interfaces available for an EMTP software, the datafiles of models in this book are created and edited, using a simple text editor, such as NOTEPAD. Each datafile name consists of a maximum of eight characters, followed by the .DAT extension, such as **301NTWK1.DAT**. According to the convention, followed in this book, the first character in the datafile name is reserved for Chapter number (3 in this case). The second and third characters are reserved for file number (01 in this case). The four characters from fourth through seventh are reserved for a description of the datafile network (NTWK) in this case, and the last character is optional; in this case, it represents network #1. In this chapter, the structure of the EMTP datafile and modeling techniques are presented in detail.

3.1 The Modeling in EMTP

Code 3-1 shows the EMTP template datafile with .DAT extension. This datafile is structured according to the flow diagram as shown in Figure 3-1. When the datafile is run in EMTP, a set of codes is executed. Each line of code, called a datacard, refers to the 1970s style of punchcard that contained one line of computer code on one datacard. The datafile may be sectionalized for modularity as

Power Flow Control Solutions for a Modern Grid using SMART Power Flow Controllers, First Edition.
Kalyan K. Sen and Mey Ling Sen.

Companion website: www.wiley.com/go/sen/powerflowcontrol

discussed in the next subsection. The not-yet-discussed entries are indicated in bold characters. The significance of the comment datacards that start with a "C" in column 1 is discussed in the next subsection.

Code 3-1 EMTP template datafile (TEMPLATE.DAT).

```
BEGIN NEW DATA CASE
C File Name: TEMPLATE.DAT
C 00000001111111111222222222233333333334444444444555555555566666666667777777777778
C 34567890123456789012345678901234567890123456789012345678901234567890123456789012345678901234567890
C ----dt<---tmax<---Xopt<---Copt<--Toler<        ><        >
C 00000001111111111222222222233333333334444444444555555555566666666667777777777778
C 34567890123456789012345678901234567890123456789012345678901234567890123456789012345678901234567890
C -Iprnt<--Iplot<-Idoubl<-KssOut<-MaxOut<        ><        ><---Icat<-Nenerg
TACS HYBRID
C 00000001111111111222222222233333333334444444444555555555566666666667777777777778
C 34567890123456789012345678901234567890123456789012345678901234567890123456789012345678901234567890
C TACS OUTPUT
BLANK RECORD ENDING TACS
C 00000001111111111222222222233333333334444444444555555555566666666667777777777778
C 34567890123456789012345678901234567890123456789012345678901234567890123456789012345678901234567890
C <---Nodes--><---Nodes--><----R<----L<----C                                  V
C <-Bus1<-Bus2<-Bus3<-Bus4<-Ohms<---mH<---uF                                  V
BLANK RECORD ENDING BRANCHES
C 00000001111111111222222222233333333334444444444555555555566666666667777777777778
C 34567890123456789012345678901234567890123456789012345678901234567890123456789012345678901234567890
C <-Bus1<-Bus2<---Tclose<----Topen<-------Ie<----Flash<--Request<-----Target<--O
BLANK RECORD ENDING SWITCHES
C 00000001111111111222222222233333333334444444444555555555566666666667777777777778
C 34567890123456789012345678901234567890123456789012345678901234567890123456789012345678901234567890
C <--Bus<I<-----Ampl<-----Freq<----Phase<-------A1<------T1><---Tstart<----Tstop
BLANK RECORD ENDING SOURCES
BLANK RECORD ENDING NODE VOLTAGE OUTPUT
BLANK RECORD ENDING PLOT
```

In the datafile, first, some general constants are defined. Next, the control or the Transient Analysis of Control Systems (TACS) section receives its input signals from the sensors or measuring switches in the electrical network model, described by the BRANCH and the SWITCH cards. The control operations and calculations are performed in this section. The SOURCE section contains TACS-controlled and independent voltage sources, which establish the power flow in a transmission line. The compensating voltages that are created in the TACS

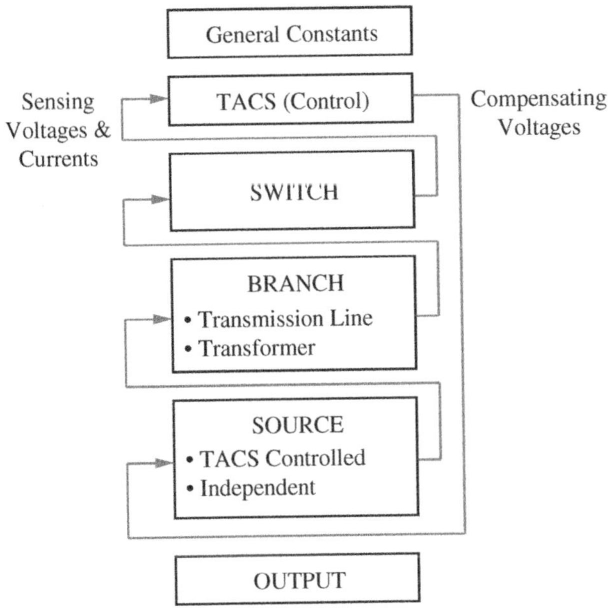

Figure 3-1 EMTP modeling structure.

section are fed to a magnetic circuit, which is located in the BRANCH section. The BRANCH section contains the transmission line, transformer, and so on. The effects of a nonideal magnetic circuit, which includes leakage reactance, magnetic saturation, and so on, can be modeled in this section. The measuring switches and the breakers are located in the SWITCH section. Finally, the OUTPUT section is defined.

The operation of various PFCs, described in the book, are verified in a power system network, considering a three-phase base power of 160 MVA and a phase-to-phase (*P–P*) base rms voltage, $V_{rms(P\text{-}P)}$, of 138 kV_{rms}. Given a system with a phase-to-phase rms voltage, $V_{rms(P\text{-}P)}$, of 138 kV_{rms}, the base voltage (V_{base}) is chosen to be the phase-to-neutral peak voltage (V_p) and calculated as

$$V_{base} = V_p = V_{rms(P-N)} \times \sqrt{2} = \frac{V_{rms(P-P)}}{\sqrt{3}} \times \sqrt{2} = \frac{138 \times 10^3}{\sqrt{3}}\sqrt{2} = 112{,}676.528\ \text{V}. \quad (3\text{-}1)$$

The base power (P_{base}) is defined as

$$P_{base} = \sqrt{3}V_{rms(P-P)}I_{rms} = 3V_{rms(P-N)}I_{rms} = 3 \times \frac{V_p}{\sqrt{2}} \times \frac{I_p}{\sqrt{2}} = 1.5V_pI_p = 1.5V_{base}I_{base} \quad (3\text{-}2)$$

where $V_{rms(P\text{-}N)}$ is the phase-to-neutral rms voltage and I_{rms} is the phase rms current.

Therefore, the base current (I_{base}), which is chosen to be the peak current (I_p), is calculated as

$$I_{base} = I_p = \sqrt{2} \times I_{rms} = \frac{P_{base}}{1.5 \times V_{base}} = \frac{160 \times 10^6}{1.5 \times 112{,}676.528} = 946.66\ \text{A}. \tag{3-3}$$

The base impedance (Z_{base}) is calculated as

$$Z_{base} = \frac{V_{base}}{I_{base}} = \frac{112{,}676.528}{946.66}\Omega = 119.025\ \Omega. \tag{3-4}$$

The voltage and current peak values (V_p and I_p) in their corresponding engineering units (V or A) can be converted into their corresponding per unit (pu) values as

$$V_{pu} = \frac{V_p}{V_{base}} \tag{3-5a}$$

and

$$I_{pu} = \frac{I_p}{I_{base}}. \tag{3-5b}$$

Therefore, the voltage and current pu values (V_{pu} and I_{pu}) can be converted into their corresponding engineering units (V or A) as

$$V_p = V_{pu} \times V_{base} \tag{3-6a}$$

and

$$I_p = I_{pu} \times I_{base}. \tag{3-6b}$$

The same rules apply for the instantaneous values. However, the voltage and current pu values are converted into their corresponding engineering rms values using the following equations:

$$V_{rms(P-P)} = V_{pu} \times V_{base} \times \sqrt{\frac{3}{2}} \tag{3-7a}$$

and

$$I_{rms} = I_{pu} \times I_{base} \times \frac{1}{\sqrt{2}}. \tag{3-7b}$$

Therefore, using the Equations (3-7a), (3-7b), and (3-2), the expression for Apparent Power Rating (APR) in an engineering unit (VA) is

$$\begin{aligned} APR &= \sqrt{3}V_{rms(P-P)}I_{rms} = \sqrt{3} \times V_{pu} \times V_{base} \times \sqrt{\frac{3}{2}} \times I_{pu} \times I_{base} \times \frac{1}{\sqrt{2}} \\ &= \frac{3}{2} \times V_{base} \times I_{base} \times V_{pu} \times I_{pu} \end{aligned}$$

or

$$APR = P_{base} \times V_{pu} \times I_{pu}. \tag{3-8}$$

Therefore, using Equation (3-8), the expression for APR in pu can be written as

$$APR_{pu} = \frac{APR}{P_{base}} = V_{pu} \times I_{pu}. \tag{3-9}$$

For instantaneous voltage and current (v_{pu} and i_{pu}), the instantaneous apparent power rating (apr) in pu can be defined as

$$apr_{pu} = v_{pu} \times i_{pu}. \tag{3-10}$$

The instantaneous apr in an engineering unit (va) can be defined as

$$apr = P_{base} \times v_{pu} \times i_{pu}. \tag{3-11}$$

3.1.1 A Single-Generator/Single-Line Model

Consider a single-line diagram, shown in Figure 3-2, that consists of a three-phase source voltage ($\mathbf{V_{src}}$) of 138 kV_{rms} (P–P) rms or $\sqrt{3}$ pu as defined in Equation (3-1) and a source impedance with a resistance (R_{src}) and reactance (X_{src}). The resistance (R_{src}) is connected between $\mathbf{V_{src}}$ (VSRC) and $\mathbf{V_{src'}}$ (VSRCP) nodes. The reactance (X_{src}) is connected between $\mathbf{V_{src'}}$ (VSRCP) and $\mathbf{V_s}$ (VS) nodes. The $\mathbf{V_s}$ (VS) node is connected to $\mathbf{V_1}$ (BUS01) node with a short, using a 1 mΩ resistance, which is not shown in the single-line diagram. The electrical system data is given in Table 3-1. The not-yet-discussed entries are indicated in bold characters.

The source impedance at the sending end (from the infinity node to the Point of Connection, POC, to the utility) is represented by a resistance (R_{src}) and an inductive reactance (X_{src}). The reactance (X_{src}) is 6.25% of the base impedance ($Z_{base} = 119.025\ \Omega$), which is $X_{src} = 7.43875\ \Omega$. This translates to a reactor, $L_{src} = X_{src}/\omega = 0.0625 \times Z_{base}/\omega = 19.73$ mH where the angular frequency $\omega = 2 \times \pi \times f$ (system frequency, $f = 60$ Hz) rad/s. A quality factor (Q) of 7.4 translates to a resistance, $R_{src} = X_{src}/Q = 7.43875/7.4 = 1.0053\ \Omega$.

An EMTP test datafile (with .DAT extension) is shown in Code 3-2 to measure an open-circuit voltage (v_{1Apu}) at BUS01A. The not-yet-discussed entries are indicated in bold characters. The A-phase voltage is shown in Figure 3-3. The lower case variables are defined as instantaneous values.

Figure 3-2 Single-line diagram of a three-phase voltage source with a source impedance.

Table 3-1 Electrical system data.

Parameters	Values
Source voltage ($\mathbf{V_{src}}$) phase-to-neutral	**112,676.528∠0° $\mathbf{V_{peak}}$**
Source resistance (R_{src})	**1.0053 Ω**
Source reactance (X_{src})	**7.43875 Ω (19.73 mH)**

Code 3-2 EMTP test datafile (TEST.DAT).

```
BEGIN NEW DATA CASE
C File Name: TEST.DAT
C A SINGLE-GENERATOR/SINGLE-LINE POWER SYSTEM NETWORK
C 00000001111111111222222222233333333334444444444555555555566666666667777777777778
C 34567890123456789012345678901234567890123456789012345678901234567890123456789 0
C ----dt<---tmax<---Xopt<---Copt<--Toler<        ><        >
16.666-6150.00-3
C 00000001111111111222222222233333333334444444444555555555566666666667777777777778
C 345678901234567890123456789012345678901234567890123456789012345678901234567890
C -Iprnt<--Iplot<-Idoubl<-KssOut<-MaxOut<        ><        ><---Icat<-Nenerg
      20      10                                                   1
TACS HYBRID
C BASE VOLTAGE
99Vbase    = 112676.528
90BUS01A
90BUS01B
90BUS01C
C 00000001111111111222222222233333333334444444444555555555566666666667777777777778
C 345678901234567890123456789012345678901234567890123456789012345678901234567890
C BUS01 PER-UNIT VOLTAGE COMPUTATION
99v1Apu    = BUS01A / Vbase
C TACS OUTPUT
33v1Apu
BLANK RECORD ENDING TACS
C 00000001111111111222222222233333333334444444444555555555566666666667777777777778
C 345678901234567890123456789012345678901234567890123456789012345678901234567890
C <---Nodes--><---Nodes--><----R<----L<----C                                    V
C <-Bus1<-Bus2<-Bus3<-Bus4<-Ohms<---mH<---uF                                    V
  VSRCA VSRCPA             1.0053
  VSRCB VSRCPBVSRCA VSRCPA
  VSRCC VSRCPCVSRCA VSRCPA
  VSRCPAVSA                      19.73
  VSRCPBVSB   VSRCPAVSA
  VSRCPCVSC   VSRCPAVSA
  VSA   BUS01A             0.001
  VSB   BUS01BVSA   BUS01A
  VSC   BUS01CVSA   BUS01A
BLANK RECORD ENDING BRANCHES
BLANK RECORD ENDING SWITCHES
```

(Continued)

Code 3-2 (Continued)

```
C 00000001111111111222222222233333333334444444444555555555566666666667777777778
C 34567890123456789012345678901234567890123456789012345678901234567890123456789 0
C <--Bus<I<-----Ampl<-----Freq<----Phase<-------A1<------T1><---Tstart<----Tstop
14VSRCA    112676.528      60.00       0.00                         -1.0
14VSRCB    112676.528      60.00    -120.00                         -1.0
14VSRCC    112676.528      60.00     120.00                         -1.0
BLANK RECORD ENDING SOURCES
BLANK RECORD ENDING NODE VOLTAGE OUTPUT
BLANK RECORD ENDING PLOT
```

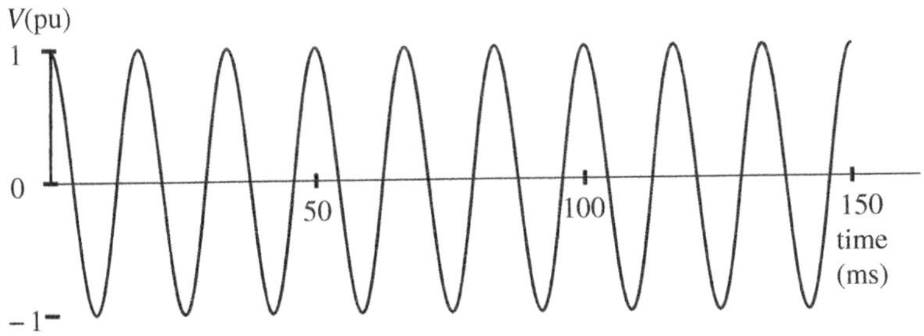

Figure 3-3 The *A*-phase voltage (v_{1Apu}).

The EMTP datafile is "column-specific." There are 80 columns in each line. To indicate a comment datacard, a "C" in column 1 and a blank column 2 must appear. The column numbers are kept in track by using two-line comment datacards and reading the first digit in the top line and the second digit in the bottom line. A reading of the EMTP program user's manual or ATP rule book (*ATP Rule Book* 1990) is recommended for further details.

The following datacard indicates the start of the datafile:

```
BEGIN NEW DATA CASE
```

In the next datacard, the time step ($\Delta t = 16.666$ μs) is entered in columns 1–8 and the simulation time, $T_{max} = 150.00$ ms, is entered in columns 9–16:

```
C File Name: TEST.DAT
C A SINGLE-GENERATOR/SINGLE-LINE POWER SYSTEM NETWORK
C 00000001111111111222222222233333333334444444444555555555566666666667777777778
C 34567890123456789012345678901234567890123456789012345678901234567890123456789 0
C ----dt<---tmax<---Xopt<---Copt<--Toler<        ><        >
16.666-6150.00-3
```

The next datacard indicates a miscellaneous datacard:

```
C 00000001111111111222222222233333333334444444444555555555566666666667777777777778
C 345678901234567890123456789012345678901234567890123456789012345678901234567890
C -Iprnt<--Iplot<-Idoubl<-KssOut<-MaxOut<       ><       ><---Icat<-Nenerg
      20      10                                               1
```

Iprnt 20 in columns 7–8 denotes “printout every 20th point.” Iplot 10 in columns 15–16 denotes “plot every 10th point.” Icat 1 in column 64 denotes a plot file to be generated and saved.

Next, in the TACS section, the datacards are located for implementing a control algorithm. An Itype of 99 is used in columns 1–2 preceding a variable name or a constant of maximum 6 characters in length in columns 3–8:

```
TACS HYBRID
C BASE VOLTAGE
99Vbase    = 112676.528
```

The next three datacards read the three-phase voltages at the BUS01 node, defined in the BRANCH section of the datacards:

```
90BUS01A
90BUS01B
90BUS01C
```

An Itype of 90 is used in columns 1–2, preceding a node name in columns 3–8 for a network node voltage measurement. Note that an Itype of 91 is used in columns 1–2, preceding a node name in columns 3–8 for a branch current measurement.

```
C 00000001111111111222222222233333333334444444444555555555566666666667777777777778
C 345678901234567890123456789012345678901234567890123456789012345678901234567890
C BUS01 NODE PER-UNIT VOLTAGE COMPUTATION
99v1Apu    = BUS01A / Vbase
C TACS OUTPUT
33v1Apu
```

An Itype of 33 is used in columns 1–2 that is followed by a variable name in columns 3–8 for the values to be sent to the output. Other variable names, each of maximum 6 characters long, can be listed in the designated columns, such as 9 through 14, 15 through 20, 21 through 26, and so on.

The next datacard indicates the end of the TACS section:

```
BLANK RECORD ENDING TACS
```

The branch datacards are located in the BRANCH section where columns 1–2 are left blank for the Itype. Columns 3–8 are designated for the first node name. Columns 9–14 are designated for the second node name. Designated parameters are placed within designated columns. The resistance values in Ohms (Ω) are entered in columns 27–32. Special attention is needed for right or left justification within the specified columns. Throughout the book, the constants and variable names are left-justified and the numbers are right-justified. The inductance values in milliHenries (mH) are entered in columns 33–38. The capacitance values in microFarads (μF) are entered in columns 39–44.

```
C 0000000111111111122222222223333333333444444444455555555556666666666777777777778
C 3456789012345678901234567890123456789012345678901234567890123456789012345678890
C <---Nodes--><---Nodes--><----R<----L<----C                                    V
C <-Bus1<-Bus2<-Bus3<-Bus4<-Ohms<---mH<---uF                                    V
  VSRCA VSRCPA                  1.0053
  VSRCB VSRCPBVSRCA VSRCPA
  VSRCC VSRCPCVSRCA VSRCPA
  VSRCPAVSA                           19.73
  VSRCPBVSB    VSRCPAVSA
  VSRCPCVSC    VSRCPAVSA
  VSA    BUS01A                  0.001
  VSB    BUS01BVSA    BUS01A
  VSC    BUS01CVSA    BUS01A
```

The source impedance for the *A* phase is entered in one line. The source impedances for the *b* and *c* phases are the same as that in the *A* phase and are copied by using *A* phase node names in Bus3 and Bus4 columns of the *b* and *c* phases. The impedances with proper node names and resistance and inductance values are entered into the appropriate columns. The next datacard indicates the end of the BRANCH section:

```
BLANK RECORD ENDING BRANCHES
```

There are no switches in this datafile. Therefore, the next datacard indicates the end of the SWITCH section:

```
BLANK RECORD ENDING SWITCHES
```

The SOURCE section contains the source datacards.

```
C 0000000111111111122222222223333333333444444444455555555556666666666777777777778
C 3456789012345678901234567890123456789012345678901234567890123456789012345678890
C <--Bus<I<-----Ampl<-----Freq<----Phase<-------A1<------T1><---Tstart<----Tstop
14VSRCA    112676.528      60.00       0.00                          -1.0
14VSRCB    112676.528      60.00    -120.00                          -1.0
14VSRCC    112676.528      60.00     120.00                          -1.0
```

An Itype of 14 in columns 1–2 is used for a sinusoidal voltage source. A node name is entered in columns 3–8. The peak value of the voltage source (112,676.528 V in the present case) is entered in columns 11–20. The value of the frequency (60 Hz in the present case) is entered in columns 21–30, and the phase angles of 0°, −120°, and 120° for the three phases of the sending-end voltage sources are entered in columns 31–40. A −1.0 in columns 67–70 is needed if an initial phasor solution is desired at the start of the simulation.

The next datacard indicates the end of the SOURCE section:

```
BLANK RECORD ENDING SOURCES
```

The following two datacards are required for a successful run of the simulation:

```
BLANK RECORD ENDING NODE VOLTAGE OUTPUT
BLANK RECORD ENDING PLOT
```

3.1.2 A Two-Generator/Single-Line Model

The network model (shown in Figure 3-2) is expanded to form a two-generator/single-line power system network, including a PFC as shown in the single-line diagram of Figure 3-4. The PFC modifies the sending-end voltage ($\mathbf{V_s}$) to become the modified sending-end voltage ($\mathbf{V_{s'}}$). The receiving-end voltage ($\mathbf{V_r}$) is set at 138 kV_{rms}, which is 112,676.528 V_{peak}, lagging the source voltage by 20°. The transmission line impedance in each phase is represented by a resistance (R) and a reactance (X). The resistance (R) is connected between the $\mathbf{V_{s'}}$ (VSP) and the $\mathbf{V_{r'}}$ (VRP) nodes. The reactance (X) is connected between the $\mathbf{V_{r'}}$ (VRP) and the $\mathbf{V_r}$ (VR) nodes. The inductive reactance (X) of 18.75% of the base impedance ($Z_{base} = 119.025\ \Omega$), which is $X = 0.1875 \times 119.025 = 22.31625\ \Omega$. This translates to a reactor, $L = X/\omega = 0.1875 \times Z_{base}/\omega = 22.31625/(2 \times \pi \times 60) = 59.19\ \text{mH}$. A quality factor (Q) of 7.4 translates to a resistance, $R = X/Q = 22.31625/7.4 = 3.0159\ \Omega$. The active and reactive power flows at the sending end are P_s and Q_s, respectively. The active and reactive power flows at the modified sending end are $P_{s'}$ and $Q_{s'}$, respectively. The active and reactive power flows at the receiving end are P_r and Q_r, respectively. The line current (I) is measured at the $\mathbf{V_{s'}}$ (VSP) node. The electrical system data is given in Table 3-2. The not-yet-discussed entries are indicated in bold characters.

The power system network model, shown in Figure 3-5, is used throughout this book unless otherwise specified. The $\mathbf{V_1}$ (BUS01) and $\mathbf{V_2}$ (BUS02) nodes are connected with a Normally-Closed, series breaker, SRSBRK, which is represented with a 1 mΩ resistance in the computer code.

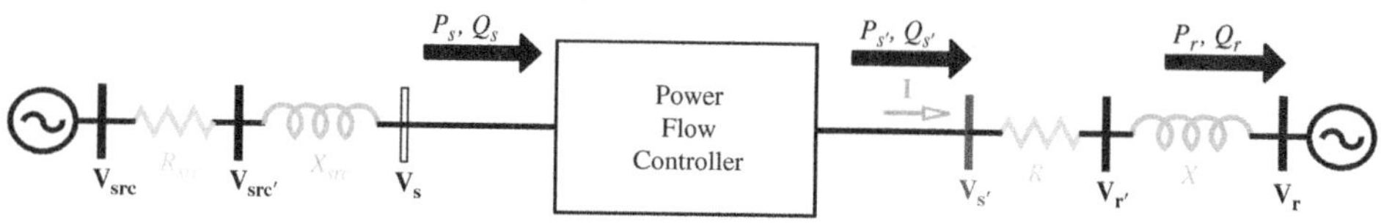

Figure 3-4 Single-line diagram of a two-generator/single-line power system network.

Table 3-2 Electrical system data.

Parameters	Values
Source voltage ($\mathbf{V_{src}}$) phase-to-neutral	$112{,}676.528\angle 0°\ V_{peak}$
Source resistance (R_{src})	1.0053 Ω
Source reactance (X_{src})	7.43875 Ω (19.73 mH)
Receiving-end voltage ($\mathbf{V_r}$) phase-to-neutral	**$112{,}676.528\angle -20°\ V_{peak}$**
Line resistance (R)	**3.0159 Ω**
Line reactance (X)	**22.31625 Ω (59.19 mH)**

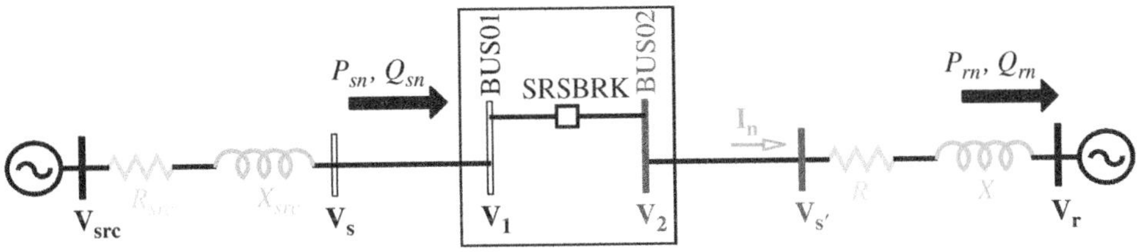

Figure 3-5 Single-line diagram of a two-generator/single-line power system with BUS01 and BUS02 nodes, connected with a Normally-Closed series breaker (SRSBRK).

The modified symbols shown in the figure are

- $\mathbf{I_n}$ is the natural current through the line,
- P_{sn} is the natural active power flow at the sending end,
- Q_{sn} is the natural reactive power flow at the sending end,
- P_{rn} is the natural active power flow at the receiving end, and
- Q_{rn} is the natural reactive power flow at the receiving end.

Note that $\mathbf{V_s} = \mathbf{V_{s'}} = \mathbf{V_{sn}}$, the natural voltage at the sending end when the series breaker (SRSBRK) is closed.

In the natural or uncompensated network, no PFC is connected. The natural active and reactive power flows (P_{rn} and Q_{rn}) at the receiving end of the line are 210 MW (1.31 pu) and −67 Mvar (−0.42 pu), respectively. The network is studied with opening the series breaker (SRSBRK) and placing a PFC between the BUS01 and BUS02 nodes. The simplified two-generator/single-line power system network reveals most of the PFC's performance under both dynamic and steady-state conditions. The network file **301NTWK1.DAT** is shown in Code 3-3.

Code 3-3 EMTP datafile for a two-generator/single-line power system network (301NTWK1.DAT).

```
BEGIN NEW DATA CASE
C File Name: 301NTWK1.DAT
C A TWO-GENERATOR/SINGLE-LINE POWER SYSTEM NETWORK
C 0000000111111111122222222223333333333444444444455555555556666666666777777777778
C 3456789012345678901234567890123456789012345678901234567890123456789012345678901234567890
C ----dt<---tmax<---Xopt<---Copt<--Toler<        ><        >
16.666-6150.00-3
C 0000000111111111122222222223333333333444444444455555555556666666666777777777778
C 3456789012345678901234567890123456789012345678901234567890123456789012345678901234567890
C -Iprnt<--Iplot<-Idoubl<-KssOut<-MaxOut<        ><        ><---Icat<-Nenerg
      20      10                                                          1
```

(Continued)

Code 3-3 (Continued)

```
TACS HYBRID
99TWOPI    = 2.0 * PI
C BASE VOLTAGE, CURRENT, AND POWER
99Vbase    = 112676.528
99Ibase    = 946.662704
99Zbase    = Vbase / Ibase
99Pbase    = 1.5 * Vbase * Ibase
C INPUTS FROM NODE VOLTAGES AND LINE CURRENTS
$INCLUDE 302MEAS1.SWT
C NORMALIZATION OF MEASURED VARIABLES AND COMPUTATION OF PEAK VOLTAGES/CURRENTS
$INCLUDE 303PTCT.SCL
C IDEAL PHASE-LOCKED LOOP
$INCLUDE 304IPLL.PLL
C COMPUTATION OF LINE RESISTANCE
$INCLUDE 305LINER.CMP
C TACS OUTPUT
33theta v1pu  v1Apu
BLANK RECORD ENDING TACS
C 00000001111111111222222222233333333334444444444555555555566666666667777777777 8
C 34567890123456789012345678901234567890123456789012345678901234567890123456789 0
C <---Nodes--><---Nodes--><----R<----L<----C                                     V
C <-Bus1<-Bus2<-Bus3<-Bus4<-Ohms<---mH<---uF                                     V
C REPRESENTATION OF A SHORT BETWEEN NODES VS AND BUS01
  VSA   BUS01A            0.001
  VSB   BUS01B            0.001
  VSC   BUS01C            0.001
C REPRESENTATION OF A SHORT BETWEEN NODES BUS01 AND BUS02
  BUS01ABUS02A            0.001
  BUS01BBUS02B            0.001
  BUS01CBUS02C            0.001
C CONNECTION OF NETWORK 1 BRANCHES (FIGURE 3-5)
$INCLUDE 306NTWK1.BRN
BLANK RECORD ENDING BRANCHES
C 00000001111111111222222222233333333334444444444555555555566666666667777777777 8
C 34567890123456789012345678901234567890123456789012345678901234567890123456789 0
C <-Bus1<-Bus2<---Tclose<----Topen<-------Ie<----Flash<--Request<-----Target<--O
C MEASUREMENT OF LINE CURRENTS
$INCLUDE 307MEAS2.SWT
BLANK RECORD ENDING SWITCHES
C 00000001111111111222222222233333333334444444444555555555566666666667777777777 8
```

(*Continued*)

Code 3-3 (Continued)

```
C 345678901234567890123456789012345678901234567890123456789012345678901234567890
C CONNECTION OF NETWORK 1 SOURCES (FIGURE 3-5)
$INCLUDE 308NTWK1.SRC
BLANK RECORD ENDING SOURCES
BLANK RECORD ENDING NODE VOLTAGE OUTPUT
BLANK RECORD ENDING PLOT
```

The datafile of the two-generator/single-line power system network with filename **301NTWK1.DAT** is described below.

The datafile may include various $INCLUDE files with different extensions, other than .DAT. The datacard "$INCLUDE 302MEAS1.SWT" attaches the $INCLUDE file 302MEAS1.SWT, which measures the voltages and currents as shown in Code 3-4. The instantaneous voltage measuring points are at the sending end ($\mathbf{V_s}$), the modified sending end ($\mathbf{V_{s'}}$), and the receiving end ($\mathbf{V_r}$). The instantaneous current measuring points are at the modified sending end ($\mathbf{V_{s'}}$).

Code 3-4 EMTP $INCLUDE file for inputs from node voltages and line currents (302MEAS1.SWT).

```
C File Name: 302MEAS1.SWT
C INPUTS FROM NODE VOLTAGES AND LINE CURRENTS
C 000000011111111112222222222333333333344444444445555555555666666666677777777778
C 345678901234567890123456789012345678901234567890123456789012345678901234567890
C INPUTS FROM NODE VOLTAGES
90BUS01A
90BUS01B
90BUS01C
90BUS02A
90BUS02B
90BUS02C
90VRA
90VRB
90VRC
C INPUTS FROM LINE CURRENTS
91VSPA
91VSPB
91VSPC
```

Code 3-5 EMTP $INCLUDE file for normalizing the measured voltages and currents (303PTCT.SCL).

```
C File Name: 303PTCT.SCL
C NORMALIZATION OF MEASURED VARIABLES AND COMPUTATION OF PEAK VOLTAGES/CURRENTS
C 00000001111111111222222222233333333334444444444555555555566666666667777777778
C 345678901234567890123456789012345678901234567890123456789012345678901234567890
C BUS01 NODE PER-UNIT VOLTAGE COMPUTATION
99v1Apu   = BUS01A / Vbase
99v1Bpu   = BUS01B / Vbase
99v1Cpu   = BUS01C / Vbase
C CALCULATION OF THE PEAK OF THE BUS01 NODE VOLTAGE PER EQUATION (3-15)
99v1pu    = SQRT((v1Apu * v1Apu + v1Bpu * v1Bpu + v1Cpu * v1Cpu) * 2 / 3)
C BUS02 NODE PER-UNIT VOLTAGE COMPUTATION
99v2Apu   = BUS02A / Vbase
99v2Bpu   = BUS02B / Vbase
99v2Cpu   = BUS02C / Vbase
C CALCULATION OF THE PEAK OF THE BUS02 NODE VOLTAGE PER EQUATION (3-15)
99v2pu    = SQRT((v2Apu * v2Apu + v2Bpu * v2Bpu + v2Cpu * v2Cpu) * 2 / 3)
C RECEIVING-END PER-UNIT VOLTAGE COMPUTATION
99vrApu   = VRA / Vbase
99vrBpu   = VRB / Vbase
99vrCpu   = VRC / Vbase
C CALCULATION OF THE PEAK OF THE VR NODE VOLTAGE PER EQUATION (3-15)
99vrpu    = SQRT((vrApu * vrApu + vrBpu * vrBpu + vrCpu * vrCpu) * 2 / 3)
C LINE PER-UNIT CURRENT COMPUTATION AT Vs' NODE
99iApu    = VSPA / Ibase
99iBpu    = VSPB / Ibase
99iCpu    = VSPC / Ibase
C CALCULATION OF THE PEAK OF THE LINE CURRENT PER EQUATION (3-15)
99ipu     = SQRT((iApu * iApu + iBpu * iBpu + iCpu * iCpu) * 2 / 3)
C SERIES-CONNECTED PER-UNIT VOLTAGE COMPUTATION
99v12Apu  = v1Apu - v2Apu
99v12Bpu  = v1Bpu - v2Bpu
99v12Cpu  = v1Cpu - v2Cpu
C CALCULATION OF THE PEAK OF THE V12 VOLTAGE PER EQUATION (3-15)
99v12pu   = SQRT((v12Apu * v12Apu + v12Bpu * v12Bpu + v12Cpu * v12Cpu) * 2 / 3)
```

The datacard "$INCLUDE 303PTCT.SCL" attaches the $INCLUDE file 303PTCT.SCL that scales or normalizes the measured voltages and currents as shown in Code 3-5.

The line voltage (V_1) at BUS01 node is expressed in terms of the three-phase instantaneous node voltages as follows:

$$v_A = V_1 \cos\theta \tag{3-12a}$$

$$v_B = V_1 \cos(\theta - 2\pi/3) \tag{3-12b}$$

$$v_C = V_1 \cos(\theta + 2\pi/3). \tag{3-12c}$$

Therefore,

$$\begin{aligned} & v_A^2 + v_B^2 + v_C^2 \\ &= V_1^2\left[\cos^2\theta + \cos^2(\theta - 2\pi/3) + \cos^2(\theta + 2\pi/3)\right] \\ &= \frac{V_1^2}{2}\left[2\cos^2\theta + 2\cos^2(\theta - 2\pi/3) + 2\cos^2(\theta + 2\pi/3)\right] \\ &= \frac{V_1^2}{2}[1 + \cos(2\theta) + 1 + \cos 2(\theta - 2\pi/3) + 1 + \cos 2(\theta + 2\pi/3)] \\ &= \frac{V_1^2}{2}[3 + \{\cos(2\theta) + \cos(2\theta + 2\pi/3) + \cos(2\theta - 2\pi/3)\}]. \end{aligned} \tag{3-13}$$

The three cosine waves that are of equal magnitude and phase-shifted from each other by $2\pi/3$ radians add up to be zero. Therefore,

$$v_A^2 + v_B^2 + v_C^2 = \frac{3}{2}V_1^2. \tag{3-14}$$

The expression for the phase-to-neutral peak voltage (V_1) at BUS01 node is

$$V_1 = \sqrt{\frac{2}{3}\left(v_A^2 + v_B^2 + v_C^2\right)}. \tag{3-15}$$

Therefore, the phase-to-phase rms voltage (V_{1rms}) at BUS01 node can be expressed in terms of the instantaneous *a*-*b*-*c* phase voltages as

$$V_{1rms} = \sqrt{\frac{3}{2}}V_1 \quad \text{or}$$

$$V_{1rms} = \sqrt{v_A^2 + v_B^2 + v_C^2} \tag{3-16}$$

using Equation (3-15).

The datacard "$INCLUDE 304IPLL.PLL" attaches the $INCLUDE file 304IPLL.PLL that calculates the phase angle, referred to as the Phase-Locked Loop (PLL) angle, of the *A*-phase voltage (v_{1A}) at the BUS01 node as shown in Code 3-6. The PLL angle keeps track of the voltage at the POC to the utility. This voltage keeps changing from its original (uncompensated) value because of the variation of line current and the resulting voltage across the line impedance between the POC and the generator node, also known as infinite node. The relative phase angle (β) of the compensating voltage ($\mathbf{V}_{s's}$) is always with respect to the PLL angle as described in the theory, developed in Chapter 2. This $INCLUDE file also calculates the modulus of the three-phase voltage (v_{1dqpu}), which is filtered with a first-order lag filter. A first-order lag filter with 4 ms time-constant may be implemented by setting all the coefficients in the numerator and denominator to be zero, except $b_1 = 0.004$ s in the following equation:

$$v_{1dqf} = \frac{1 + a_1 s + a_2 s^2 + a_3 s^3 + \ldots}{1 + b_1 s + b_2 s^2 + b_3 s^3 + \ldots} v_{1dqpu}. \tag{3-17}$$

For a balanced three-phase voltage, the modulus is the same as the peak of the sinusoidal voltage. Since the control algorithm is based on regulating the positive-sequence fundamental signals, all unwanted harmonic components that are present in the measured voltage and current signals must be filtered out before they are fed back to the control algorithm. This avoids unwanted signals from passing through the control algorithm and creating undesired output signals. Any negative sequence fundamental component that is present in the measured signal must also be filtered out.

The PLL angle is referenced to a cosine wave, meaning that the peak of the waveform is located at $\theta = 0$ radian; the phase angle (θ) linearly increases from 0 to 2π radians during a complete cycle of a cosine wave as shown in Figure 3-6. Transforming a three-phase (a-b-c) reference voltage into an equivalent two-phase d-q stationary reference frame (see Appendix A.5), the stationary-frame voltages (v_{ds} and v_{qs}) can be obtained, using Equation (A-75a) in Appendix A, as

$$v_{ds} = v_A = V_1 \cos\theta \tag{3-18}$$

$$\begin{aligned} v_{qs} &= -\frac{1}{\sqrt{3}}(v_A + 2v_C) = -\frac{V_1}{\sqrt{3}}(\cos\theta + 2\cos(\theta + 120^\circ)) \\ &= -\frac{V_1}{\sqrt{3}}\left(\cos\theta - \cos\theta - \sqrt{3}\sin\theta\right) = V_1 \sin\theta. \end{aligned} \tag{3-19}$$

Therefore, using Equations (3-18) and (3-19), the cosine-referenced phase angle of the A-phase voltage (v_{1A}) at the BUS01 node is given by

$$\theta = \tan^{-1}\left(\frac{\sin\theta}{\cos\theta}\right) = \tan^{-1}\left(\frac{V_1\sin\theta}{V_1\cos\theta}\right) = \tan^{-1}\left(\frac{v_{qs}}{v_{ds}}\right). \tag{3-20}$$

A small value of 10^{-8} is added to the denominator in the calculation of (v1qspu/v1dspu) to avoid a division by zero, if $v_{ds} = 0$. If v_{ds} is negative, the phase angle is corrected by adding π radians to the calculated angle. If the calculated angle is negative, 2π radians is added to make the phase angle positive; if the calculated angle is more than 2π radians, 2π radians is subtracted to keep the phase angle between 0 and 2π radians. If a sine-referenced phase angle is needed, $\pi/2$ radians is added to the above-mentioned cosine-referenced phase angle.

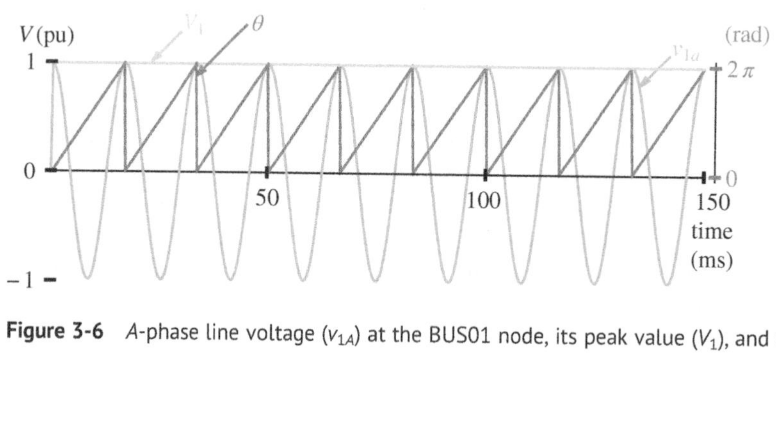

Figure 3-6 *A*-phase line voltage (v_{1A}) at the BUS01 node, its peak value (V_1), and the PLL angle (θ).

Code 3-6 EMTP $INCLUDE file for implementing an ideal PLL (304IPLL.PLL).

```
C File Name: 304IPLL.PLL
C IDEAL PHASE-LOCKED LOOP
C 00000001111111111222222222233333333334444444444555555555566666666667777777778
C 34567890123456789012345678901234567890123456789012345678901234567890123456789
C CONVERT FROM 3-PHASE TO 2-PHASE IN STATIONARY REF FRAME PER EQUATION (A-75a)
99v1dspu   = v1Apu
99v1qspu   = -(v1Apu + 2.0 * v1Cpu) / SQRT(3)
C PHASE-LOCKED LOOP (LOCKED TO A-PHASE VOLTAGE OF BUS01 NODE) PER EQUATION 3-20
99th       = ATAN(v1qspu / (v1dspu+1.0E-8))
99the      = th + PI * (v1dspu .LT. 0)
99theta    = the + TWOPI * (the .LT. 0) - TWOPI * (the .GE. TWOPI)
99COSPLL   = COS(theta)
99SINPLL   = SIN(theta)
C BUS01 NODE VOLTAGE MAGNITUDE COMPUTATION FROM STATIONARY TO
C ROTATING REFERENCE FRAME PER EQUATIONS A55-b
99v1dpu    =  v1dspu * COSPLL + v1qspu * SINPLL
99v1qpu    = -v1dspu * SINPLL + v1qspu * COSPLL
C CALCULATION OF THE PEAK OF THE VOLTAGE AT BUS01 NODE IN D-Q
C ROTATING REFERENCE PER EQUATION (3-23)
99v1dqpu   = SQRT(v1dpu * v1dpu + v1qpu * v1qpu)
C LOW-PASS FILTER INTRODUCES A RAMP-RATE AND REDUCES SYSTEM TRANSIENTS
 1v1dqf    +v1dqpu
        1.0
        1.0    0.0040
```

From Equations (3-18) and (3-19), it can be written that

$$v_{ds}^2 + v_{qs}^2 = V_1^2\left(\cos^2\theta + \sin^2\theta\right) = V_1^2 \tag{3-21}$$

using the identity ($\cos^2\theta + \sin^2\theta = 1$).

Therefore, the expression for the phase-to-neutral peak voltage (V_1) at BUS01 node is given by

$$V_1 = \sqrt{v_{ds}^2 + v_{qs}^2}. \tag{3-22}$$

The Equation (3-22) can be expanded further, using Equation (A-55a), as

$$V_1 = \sqrt{v_{ds}^2 + v_{qs}^2} = \sqrt{\left(v_d \cos\theta - v_q \sin\theta\right)^2 + \left(v_d \sin\theta + v_q \cos\theta\right)^2}$$

or

$$V_1 = \sqrt{v_d^2\left(\cos^2\theta + \sin^2\theta\right) + v_q^2\left(\cos^2\theta + \sin^2\theta\right)} = \sqrt{v_d^2 + v_q^2} \tag{3-23}$$

using the identity ($\cos^2\theta + \sin^2\theta = 1$).

The Equation (3-20) can be expanded further, using Equation (A-55a), as

$$\theta = \tan^{-1}\left(\frac{v_{qs}}{v_{ds}}\right) = \tan^{-1}\left(\frac{V_1 \sin\theta}{V_1 \cos\theta}\right) = \tan^{-1}\left(\frac{v_d \sin\theta + v_q \cos\theta}{v_d \cos\theta - v_q \sin\theta}\right) = \tan^{-1}\left(\frac{v_q}{v_d}\right) \tag{3-24}$$

with a substitution of $\theta = 0°$ to represent a stationary frame.

The datacard "$INCLUDE 305LINER.CMP" attaches the $INCLUDE file 305LINER.CMP that computes the modified sending-end power, receiving-end power, and line resistance as shown in Code 3-7. First, the magnitude (I_{dq}) and the relative phase angle (θ_{ir}) of the line current (i) are calculated with respect to the PLL angle (θ). Then the line resistance (R) is calculated as explained in Section 3.3.

Code 3-7 EMTP $INCLUDE file for computing the line resistance (305LINER.CMP).

```
C File Name: 305LINER.CMP
C LINE RESISTANCE COMPUTATION
C 00000001111111111222222222233333333334444444444555555555566666666667777777778
C 34567890123456789012345678901234567890123456789012345678901234567890123456789O
C LINE CURRENT MAGNITUDE COMPUTATION PER EQUATIONS (A-75a), (A-55b) and (3-23).
99idspu    =  iApu
99iqspu    = -(iApu + 2.0 * iCpu) / SQRT(3)
99idpu     =  idspu * COSPLL + iqspu * SINPLL
99iqpu     = -idspu * SINPLL + iqspu * COSPLL
99idqpu    = SQRT(idpu * idpu + iqpu * iqpu)
C RELATIVE PHASE ANGLE COMPUTATION PER EQUATION 3-24
99thetaa   = ATAN( iqpu / (idpu + 1.0E-8) )
99thetab   = thetaa + PI * (idpu .LT. 0.0)
99thetir   = thetab + TWOPI * (thetab .LT. 0.0) - TWOPI * (thetab .GE. TWOPI)
C MODIFIED SENDING-END POWER COMPUTATION
99Psppu    = (v2Apu*iApu + v2Bpu*iBpu + v2Cpu*iCpu) * Vbase * Ibase / Pbase
99Qsppu    = (v2Apu*iCpu - v2Cpu*iApu) * SQRT(3) * Vbase * Ibase / Pbase
C MODIFIED SENDING-END POWER COMPUTATION IN MW AND MVAR
99PspMW    = Psppu * Pbase / 1e6
99QspMV    = Qsppu * Pbase / 1e6
C RECEIVING-END POWER COMPUTATION
```

(Continued)

Code 3-7 (Continued)

```
99Prpu     = (vrApu*iApu + vrBpu*iBpu + vrCpu*iCpu) * Vbase * Ibase / Pbase
99Qrpu     = (vrApu*iCpu - vrCpu*iApu) * SQRT(3) * Vbase * Ibase / Pbase
C RECEIVING-END POWER COMPUTATION IN MW AND MVAR
99PrMW     = Prpu * Pbase / 1e6
99QrMVAR   = Qrpu * Pbase / 1e6
C RECEIVING-END MVA COMPUTATION PER EQUATION (2-76)
99SrMVA    = SQRT(PrMW * PrMW + QrMVAR * QrMVAR)
C LINE RESISTANCE COMPUTATION
99Rline    = (Psppu - Prpu) / ipu / ipu * Zbase
C LOW-PASS FILTER INTRODUCES A RAMP-RATE AND REDUCES SYSTEM TRANSIENTS
 1Rlinef   +Rline
        1.0
        1.0     0.0083
```

VS and BUS01 nodes are connected with a short that is represented by a resistance, $R = 0.001\ \Omega$, as shown in Code 3-8. This short may be replaced with a current-measuring sensor, if needed. BUS01 and BUS02 nodes are connected initially with a short that is represented by a resistance, $R = 0.001\ \Omega$. These two nodes are reserved for connecting the series-compensating voltage of the PFC.

Code 3-8 Representation of shorts between VS and BUS01 nodes and between BUS01 and BUS02 nodes.

```
C 00000001111111111222222222233333333334444444444555555555566666666667777777778
C 34567890123456789012345678901234567890123456789012345678901234567890123456789 0
C <---Nodes--><---Nodes--><----R<----L<----C                                  V
C <-Bus1<-Bus2<-Bus3<-Bus4<-Ohms<---mH<---uF                                   V
C REPRESENTATION OF A SHORT BETWEEN NODES VS AND BUS01
  VSA   BUS01A            0.001
  VSB   BUS01B            0.001
  VSC   BUS01C            0.001
C REPRESENTATION OF A SHORT BETWEEN NODES BUS01 AND BUS02
  BUS01ABUS02A            0.001
  BUS01BBUS02B            0.001
  BUS01CBUS02C            0.001
```

The datacard "$INCLUDE 306NTWK1.BRN" attaches the $INCLUDE file 306NTWK1.BRN that connects a single transmission line and a source impedance as shown in Code 3-9. Note that the resistance and reactance parameters are entered in the same line in their designated columns.

Code 3-9 EMTP $INCLUDE file for implementing the source impedance and the line impedance (306NTWK1.BRN).

```
C File Name: 306NTWK1.BRN
C SOURCE IMPEDANCE & LINE IMPEDANCE
C 00000001111111111222222222233333333334444444444555555555566666666667777777778
C 34567890123456789012345678901234567890123456789012345678901234567890123456789 0
C <---Nodes--><---Nodes--><----R<----L<----C                                  V
C <-Bus1<-Bus2<-Bus3<-Bus4<-Ohms<---mH<---uF                                   V
  VSRCA VSA                 1.0053 19.73
  VSRCB VSB   VSRCA VSA
  VSRCC VSC   VSRCA VSA
  VSPA  VRA                 3.0159 59.19
  VSPB  VRB   VSPA  VRA
  VSPC  VRC   VSPA  VRA
```

The SWITCH section contains the measuring switch datacards. There are different types of switches. The datacard "$INCLUDE 307MEAS2.SWT" attaches the $INCLUDE file 307MEAS2. SWT that measures branch currents as shown in Code 3-10. For a measuring switch, columns 1–2 are left blank. Node "1" name is typed in columns 3–8 and node "2" name in columns 9–14. The word MEASURING is typed in columns 55–63.

Code 3-10 EMTP $INCLUDE file for measuring the branch currents (307MEAS2.SWT).

```
C File Name: 307MEAS2.SWT
C MEASUREMENT OF LINE CURRENTS
C 00000001111111111222222222233333333334444444444555555555566666666667777777778
C 345678901234567890123456789012345678901234567890123456789012345678901234567890
C <-Bus1<-Bus2<---Tclose<----Topen<-------Ie<----Flash<--Request<-----Target<--O
  BUS02AVSPA                                            MEASURING
  BUS02BVSPB                                            MEASURING
  BUS02CVSPC                                            MEASURING
```

The SOURCE section contains the source datacards. The datacard "$INCLUDE 308NTWK1. SRC" attaches the $INCLUDE file 308NTWK1.SRC that connects source voltages as shown in Code 3-11. The value of the phase angles of −20°, −140°, and 100° for the three phases of the receiving-end voltage sources are entered in columns 31–40.

Code 3-11 EMTP $INCLUDE file for the source voltages and the receiving-end voltages (308NTWK1.SRC).

```
C File Name: 308NTWK1.SRC
C SOURCE VOLTAGES AND RECEIVING-END VOLTAGES
C 00000001111111111222222222233333333334444444444555555555566666666667777777778
C 3456789012345678901234567890123456789012345678901234567890123456789012345678901234567890
C <--Bus<I<-----Ampl<-----Freq<----Phase<-------A1<------T1><---Tstart<----Tstop
14VSRCA    112676.528     60.00      0.00                           -1.0
14VSRCB    112676.528     60.00   -120.00                           -1.0
14VSRCC    112676.528     60.00    120.00                           -1.0
14VRA      112676.528     60.00    -20.00                           -1.0
14VRB      112676.528     60.00   -140.00                           -1.0
14VRC      112676.528     60.00    100.00                           -1.0
```

The simulation results from the file **301NTWK1.DAT** are shown in Figure 3-6. The PLL signal (reference phase angle, θ) is locked to the A-phase voltage at the BUS01 node.

To study faults in power systems, an A-phase-to-ground fault is applied to the power system network. The measured current is sent to the plot file by setting Iout in column 80 with 1. The corresponding datafile **309NTWK2.DAT** is shown in Code 3-12.

Code 3-12 EMTP datafile for a two-generator/single-line faulted power system network (309NTWK2.DAT).

```
BEGIN NEW DATA CASE
C File Name: 309NTWK2.DAT
C A TWO-GENERATOR/SINGLE-LINE POWER SYSTEM NETWORK WITH LINE-TO-NEUTRAL FAULT
C 00000001111111111222222222233333333334444444444555555555566666666667777777778
C 3456789012345678901234567890123456789012345678901234567890123456789012345678901234567890
C ----dt<---tmax<---Xopt<---Copt<--Toler<        ><        >
16.666-6150.00-3
C 00000001111111111222222222233333333334444444444555555555566666666667777777778
C 3456789012345678901234567890123456789012345678901234567890123456789012345678901234567890
C -Iprnt<--Iplot<-Idoubl<-KssOut<-MaxOut<        ><        ><---Icat<-Nenerg
      20       10                                                  1
TACS HYBRID
99TWOPI    = 2.0 * PI
C BASE VOLTAGE, CURRENT, AND POWER
99Vbase    = 112676.528
99Ibase    = 946.662704
99Zbase    = Vbase / Ibase
99Pbase    = 1.5 * Vbase * Ibase
C INPUTS FROM NODE VOLTAGES AND LINE CURRENTS
```

(*Continued*)

Code 3-12 (Continued)

```
$INCLUDE 302MEAS1.SWT
C NORMALIZATION OF MEASURED VARIABLES AND COMPUTATION OF PEAK VOLTAGES/CURRENTS
$INCLUDE 303PTCT.SCL
C VECTOR PHASE-LOCKED LOOP
$INCLUDE 310VPLL.PLL
C COMPUTATION OF LINE RESISTANCE
$INCLUDE 305LINER.CMP
C TACS OUTPUT
33theta v1pu  v1Apu
BLANK RECORD ENDING TACS
C 000000011111111112222222222333333333344444444445555555555666666666677777777778
C 345678901234567890123456789012345678901234567890123456789012345678901234567890
C <---Nodes--><---Nodes--><----R<----L<----C                                   V
C <-Bus1<-Bus2<-Bus3<-Bus4<-Ohms<---mH<---uF                                   V
C REPRESENTATION OF A SHORT BETWEEN NODES VS AND BUS01
  VSA   BUS01A            0.001
  VSB   BUS01B            0.001
  VSC   BUS01C            0.001
C REPRESENTATION OF A SHORT BETWEEN NODES BUS01 AND BUS02
  BUS01ABUS02A            0.001
  BUS01BBUS02B            0.001
  BUS01CBUS02C            0.001
C CONNECTION OF NETWORK 2 BRANCHES WITH DAMPING RESISTANCES (FIGURE 3-5)
$INCLUDE 311NTWK2.BRN
BLANK RECORD ENDING BRANCHES
C 000000011111111112222222222333333333344444444445555555555666666666677777777778
C 345678901234567890123456789012345678901234567890123456789012345678901234567890
C <-Bus1<-Bus2<---Tclose<----Topen<-------Ie<----Flash<--Request<-----Target<--O
C MEASUREMENT OF LINE CURRENTS
$INCLUDE 307MEAS2.SWT
C FAULT STARTS
  BUS01A            .050      0.100       5.0                                   1
C FAULT ENDS
BLANK RECORD ENDING SWITCHES
C 000000011111111112222222222333333333344444444445555555555666666666677777777778
C 345678901234567890123456789012345678901234567890123456789012345678901234567890
C NETWORK 1 SOURCES (FIGURE 3-5)
$INCLUDE 308NTWK1.SRC
BLANK RECORD ENDING SOURCES
BLANK RECORD ENDING NODE VOLTAGE OUTPUT
BLANK RECORD ENDING PLOT
```

The additional datacard in the SWITCH section starts an *A*-phase-to-ground fault at BUS01A node at 50 ms by closing the switch and opening it 50 ms later to clear the fault when the current goes below 5 A. The Ideal PLL (IPLL), used here, cannot preserve the PLL angle in the event when there is a distorted line voltage at the point of measurement due to the line faults as shown in Figure 3-8. This issue can be overcome with the use of a Vector PLL (VPLL), which is discussed next.

3.2 Vector Phase-Locked Loop (VPLL)

Figure 3-7 shows the block diagram of the Vector PLL (VPLL). The following steps are needed to implement a successful VPLL.

Step 1: Using Equations (A-75a), the *d*-*q* stationary frame components (v_{1ds} and v_{1qs}) of the line voltage (v_1) at the BUS01 node are calculated.

Step 2: Using Equation (A-55b), the quadrature-axis rotating frame component (v_{1q}) is calculated.

Step 3: The quadrature-axis rotating frame component (v_{1q}), also called PLL error (PLLerr), is regulated to be zero with the use of a PI Regulator, also called an Error Regulator, with a proportional constant (K_{PLLp}) and an integral constant (K_{PLLi}). These constants are chosen as $K_{PLLp} = 100$ and the ratio (K_{PLLi}/K_{PLLp}) to be 2.5 as explained in "Operation of a Phase Locked Loop System Under Distorted Utility Conditions," by V. Kaura and V. Blasko in *IEEE Trans. on Industry Applications*, vol. 33, no. 1, pp. 58–63, Jan/Feb 1997.

Step 4: The output of the PI Regulator, which is the synchronously rotating angular speed (ω), is integrated to produce the PLL angle (θ).

If the quadrature-axis rotating frame component (v_{1q}) is regulated to be zero, the voltage vector, which is composed of all of the three-phase voltages, aligns itself with the direct-axis component (v_{1d}) that becomes the peak (V_1) of the phase-to-neutral voltage (v_1) at BUS01 node under a steady-state condition. In the EMTP datafile **301NTWK1.DAT**, the datacard "$INCLUDE 310VPLL.PLL" attaches the $INCLUDE file 310VPLL.PLL that calculates the phase angle of the voltage at BUS01 node, which is referred to the *A*-phase voltage (v_{1A}) as shown in Code 3-13 and presented in "SSSC-static synchronous series compensator: theory, modeling, and applications," by K. K. Sen at Power

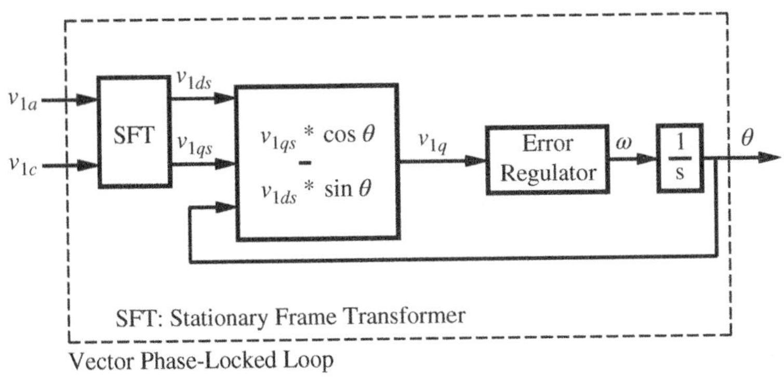

Figure 3-7 Block diagram of a Vector PLL.

Code 3-13 EMTP $INCLUDE file for a Vector Phase-Locked Loop (310VPLL.PLL).

```
C File Name: 310VPLL.PLL
C VECTOR PHASE-LOCKED LOOP
C 0000000111111111122222222223333333333444444444455555555556666666666777777777778
C 3456789012345678901234567890123456789012345678901234567890123456789012345678901234567890
C CONVERT FROM 3-PHASE TO 2-PHASE PER EQUATION (A-75a) (STEP 1)
99v1dspu  = v1Apu
99v1qspu  = -(v1Apu + 2.0 * v1Cpu) / SQRT(3)
C PHASE-LOCKED LOOP (LOCKED TO A-PHASE VOLTAGE OF BUS01 NODE)
99KPLLp   = 100.0
99KPLLi   = 250.0
C CALCULATION OF v1q (PLLerr) PER EQUATION (A-55b) (STEP 2)
99PLLerr  = v1qspu * COS(thetai) - v1dspu * SIN(thetai)
C INTEGRATION OF PLLerr PER EQUATION (3-26)
99PLLint  = PLLini + PLLerr * DELTAT
PLLini  +PLLint
C CALCULATION OF omega (rad/s) AT THE OUTPUT OF A PI REGULATOR (STEP 3)
99omega   = KPLLp * PLLerr + KPLLi * PLLint
C INTEGRATION OF omega TO CALCULATE theta PER EQUATION (3-26) (STEP 4)
99thetaf  = thetai + omega * DELTAT
  thetai  +thetaf
C theta IS WRAPPED TO BE BOUND BETWEEN 0 AND 2π RADIANS
99theta   = thetaf - TWOPI * TRUNC(thetaf/TWOPI)
99COSPLL  = COS(theta)
99SINPLL  = SIN(theta)
C ------TACS INITIAL CONDITIONS------
C -name>xx<--------|
77PLLint  1.5080
C BUS01 NODE VOLTAGE MAGNITUDE COMPUTATION PER EQUATIONS A48-b and 3-23
99v1dpu   =  v1dspu * COSPLL + v1qspu * SINPLL
99v1qpu   = -v1dspu * SINPLL + v1qspu * COSPLL
99v1dqpu  = SQRT(v1dpu * v1dpu + v1qpu * v1qpu)
C LOW-PASS FILTER INTRODUCES A RAMP-RATE AND REDUCES SYSTEM TRANSIENTS
 1v1dqf   +v1dqpu
       1.0
       1.0     0.0040
```

Engineering Society Summer Meeting, Berlin, Germany, 1997 and published in *IEEE Trans. on Power Delivery*, vol. 13, no. 1, pp. 241–246, Jan. 1998. Note that the numerical integration is implemented using first-order Euler method as

$$\omega = \dot{\theta} = \frac{\theta_f - \theta_i}{\Delta t} \tag{3-25}$$

where ω is the angular frequency, θ_i is the initial value, and θ_f is the final value of the PLL angle during the integration time step, Δt. Rearranging Equation (3-25), it can be written that

$$\theta_f = \theta_i + \omega \times \Delta t. \tag{3-26}$$

At the end of each time step (Δt), the final value of PLL angle (θ_f) is stored for use as the initial value of PLL angle (θ_i) at the next time step of integration. The same process of initialization is used for integration of any other variable.

Also note that if all the integrators in the VPLL is started from zero initial condition, the PLL angle (θ) will take some time to reach to its steady-state value. To reduce this settling time, the initial condition (*IC*) of the integrator in the PI Regulator is set at

$$IC = \frac{\omega}{K_{PLLi}} = \frac{2\pi f}{K_{PLLi}} = \frac{2 \times \pi \times 60}{250} = 1.5080. \tag{3-27}$$

Figure 3-8 shows that the VPLL is able to track the near original phase in the case of an *A*-phase-to-ground fault. Note that during a fault, the current through the line inductance changes rapidly, causing numerical instability in the simulation. To minimize this problem, a numerical snubber in the form of a resistance across each line inductance is used. The value of the snubber resistance should be high enough so that it does not cause spurious results in the simulation but facilitates the decaying of an oscillating current during a disturbance. The datacard "$INCLUDE 311NTWK2.BRN," shown in Code 3-14, attaches the $INCLUDE file 311NTWK2.BRN that connects a single transmission line and a source impedance with snubber resistances across all inductances.

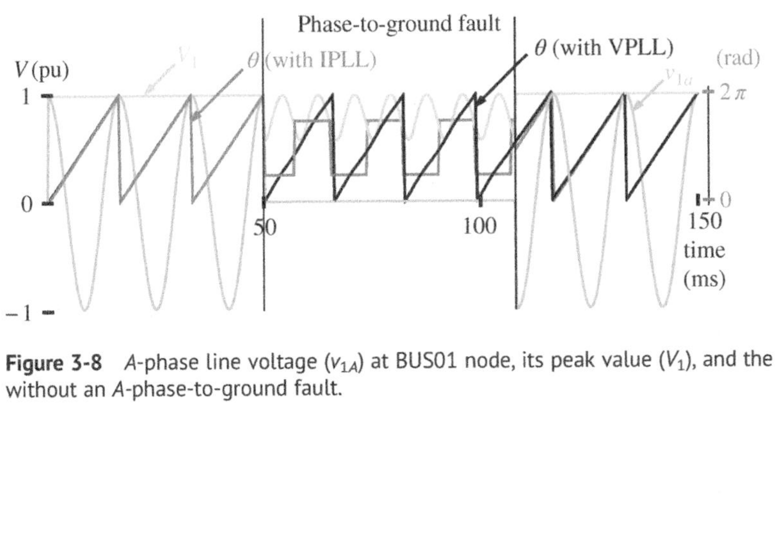

Figure 3-8 *A*-phase line voltage (v_{1A}) at BUS01 node, its peak value (V_1), and the PLL angle (θ) with and without an *A*-phase-to-ground fault.

Code 3-14 EMTP $INCLUDE file for implementing the source impedance and the line impedance (311NTWK2.BRN).

```
C File Name: 311NTWK2.BRN
C SOURCE/LINE IMPEDANCES WITH SNUBBER RESISTANCES ACROSS ALL INDUCTANCES
C 00000001111111111222222222233333333334444444444555555555566666666667777777778
C 3456789012345678901234567890123456789012345678901234567890123456789012345678901234567890
C <---Nodes--><---Nodes--><----R<----L<----C                                        V
C <-Bus1<-Bus2<-Bus3<-Bus4<-Ohms<---mH<---uF                                        V
  VSRCA VSRCPA            1.0053
  VSRCPAVSA                     19.73
  VSRCB VSRCPB            1.0053
  VSRCPBVSB                     19.73
  VSRCC VSRCPC            1.0053
  VSRCPCVSC                     19.73
  VSPA  VRPA              3.0159
  VRPA  VRA                     59.19
  VSPB  VRPB              3.0159
  VRPB  VRB                     59.19
  VSPC  VRPC              3.0159
  VRPC  VRC                     59.19
C HIGH SNUBBER RESISTANCES ACROSS ALL INDUCTANCES
  VSRCPAVSA               5.00+2
  VSRCPBVSB               5.00+2
  VSRCPCVSC               5.00+2
  VRPA  VRA               4.00+3
  VRPB  VRB               4.00+3
  VRPC  VRC               4.00+3
```

3.3 Transmission Line Steady-State Resistance Calculator

In some applications, it is necessary to know the value of the resistance (R) of the transmission line between two points. Figure 3-9 shows the block diagram of a scheme for the computation of transmission line resistance. The active power flow at the two points are measured, the difference of which can be assumed to be the loss across the transmission line resistance (R), since other active power losses are insignificant. The value of the resistance can be computed by dividing the power loss across the resistance by the square of the line current. This approach of calculating the line resistance is accurate only for a steady-state condition. Therefore, a low pass filter with a narrow bandwidth is used to eliminate any inaccuracy in the calculated resistance values. The low-pass filter reduces the changes in resistance values, calculated during short electrical transients.

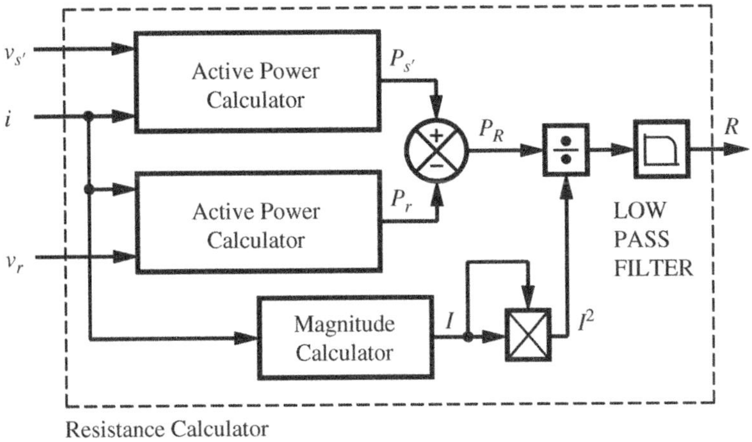

Figure 3-9 Block diagram of a steady-state resistance calculator.

3.4 Simulation of an Independent PFC, Integrated in a Two-Generator/Single-Line Power System Network

An independent PFC regulates the active and reactive power flows in the transmission line independently by connecting a compensating voltage in series with the line as shown in Figure 2-11. The series-compensating voltage exchanges both active and reactive powers with the transmission line. In the case of the Sen Transformer (ST), both the active and reactive powers, exchanged by the Series Unit, appear at the transmission line through the magnetic link to the Shunt Unit. However, in the case of the Unified Power Flow Controller (UPFC), only the exchanged active power by the Series Unit passes through the DC link to the Shunt Unit. Both the Shunt and Series Units exchange reactive power at their AC terminals independently as desired. The mathematical models of the ST and the UPFC are shown in Figure 3-10. The corresponding datafile **312PQPFC.DAT** is shown in Code 3-15.

Note that the high resistances are needed to avoid floating nodes in the circuit. Special attention is needed to select the high resistances across the current sources. These resistances are set at 100 MΩ, which may need to be adjusted for a stable simulation of a particular operating point. Refer to the note below Table 4-10.

In this section, the effects of the ST and the UPFC on power flow control in a transmission line are presented. The network model of Figure 3-5 is used with the short between BUS01 and BUS02 nodes, replaced, first, with the mathematical model of the ST and then with that of the UPFC. In each ST and UPFC mathematical model, the compensating voltage is connected in series with the transmission line through a coupling transformer (SRSTR). Table 3-2 can be expanded to include the series-coupling transformer as shown in bold in Table 3-3.

The leakage reactance of the series-coupling transformer of the PFC is represented by an inductive impedance of 15% of the base impedance ($Z_{base} = 119.025\ \Omega$) with a quality factor (Q) of 10 that translates into an inductance ($L = 47.358\ \text{mH}$) where $L = 0.15 \times Z_{base}/\omega = 0.15 \times 119.025/2/\pi/60$, and a resistance ($R = 1.7854\ \Omega$) where $R = 0.15 \times Z_{base}/Q = 0.15 \times 119.025/10$. The turns-ratio of the series-coupling transformer is considered to be 1:1, and the leakage reactance is distributed equally in both primary and secondary sides. The datacard "$INCLUDE 315TRAN2.TRN"

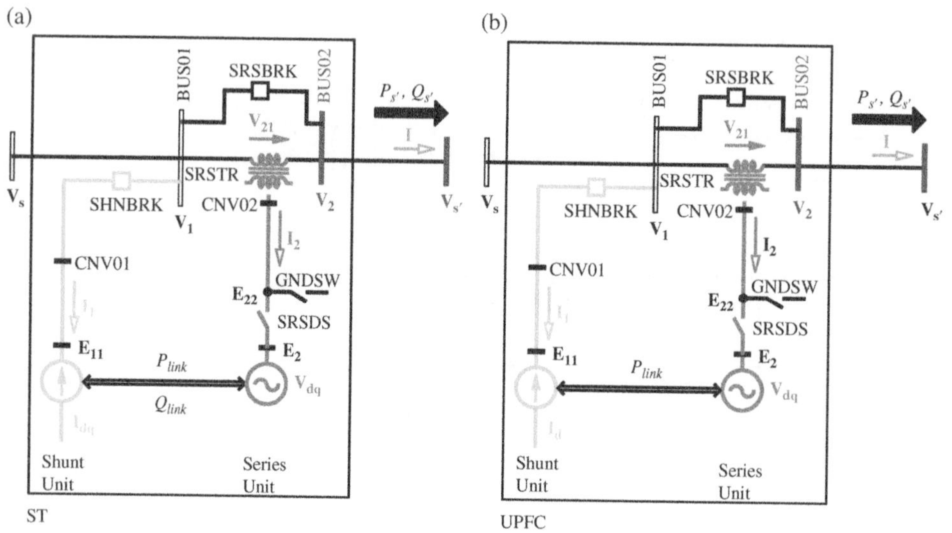

Figure 3-10 Mathematical models of the (a) ST and (b) UPFC. Note the difference in reactive power in the link between the Shunt Unit and the Series Unit.

Code 3-15 EMTP datafile for an independent PFC (Shunt-Series), integrated in a two-generator/single-line power system network (312PQPFC.DAT).

```
BEGIN NEW DATA CASE
C File Name: 312PQPFC.DAT
C PQ POWER FLOW CONTROLLER (PFC) WITH A SHUNT-CONNECTED CURRENT SOURCE (SHUNT
C UNIT) AND A SERIES-CONNECTED VOLTAGE SOURCE (SERIES UNIT)
C 00000001111111111222222222233333333334444444444555555555566666666667777777778
C 34567890123456789012345678901234567890123456789012345678901234567890123456789 0
C ----dt<---tmax<---Xopt<---Copt<--Toler<        ><        >
8.3333-6150.00-3
C 00000001111111111222222222233333333334444444444555555555566666666667777777778
C 34567890123456789012345678901234567890123456789012345678901234567890123456789 0
C -Iprnt<--Iplot<-Idoubl<-KssOut<-MaxOut<        ><        ><---Icat<-Nenerg
      30      10                                                        1
TACS HYBRID
99CRD      = 180 / PI
99CDR      = 1 / CRD
99TWOPI    = 2.0 * PI
C BASE VOLTAGE, CURRENT, AND POWER
99Vbase    = 112676.528
99Ibase    = 946.662704
99Zbase    = Vbase / Ibase
99Pbase    = 1.5 * Vbase * Ibase
C INPUTS FROM NODE VOLTAGES AND LINE CURRENTS
```

(Continued)

Code 3-15 (Continued)

```
$INCLUDE 302MEAS1.SWT
C MEASUREMENT OF CURRENTS FLOWING THROUGH THE SHUNT UNIT
91CNV01A
91CNV01B
91CNV01C
C MEASUREMENT OF CURRENTS FLOWING THROUGH THE SERIES UNIT
91CNV02A
91CNV02B
91CNV02C
C NORMALIZATION OF MEASURED VARIABLES AND COMPUTATION OF PEAK VOLTAGES/CURRENTS
$INCLUDE 303PTCT.SCL
C VECTOR PHASE-LOCKED LOOP
$INCLUDE 310VPLL.PLL
C USER'S INPUT FILE
$INCLUDE 313PQPFC.USR
C MATHEMATICAL MODEL OF A PFC WITH SHUNT-SERIES CONFIGURATION
$INCLUDE 314PQPFC.CON
C COMPUTATION OF LINE RESISTANCE
$INCLUDE 305LINER.CMP
C TACS OUTPUT
33VskV  PrMW  QrMVARP2seMWQ2seMVS2seMV
BLANK RECORD ENDING TACS
C 00000001111111111222222222233333333334444444444555555555566666666667777777778
C 34567890123456789012345678901234567890123456789012345678901234567890123456789 0
C <---Nodes--><---Nodes--><----R<----L<----C                                   V
C <-Bus1<-Bus2<-Bus3<-Bus4<-Ohms<---mH<---uF                                   V
C REPRESENTATION OF A SHORT BETWEEN NODES VS AND BUS01
  VSA   BUS01A              0.001
  VSB   BUS01B              0.001
  VSC   BUS01C              0.001
C CONNECTION OF NETWORK 2 BRANCHES WITH DAMPING RESISTANCES (FIGURE 3-5)
$INCLUDE 311NTWK2.BRN
C HIGH RESISTANCES TO AVOID FLOATING NODES IN THE CIRCUIT
  BUS01ACNV01A            1.00+8
  BUS01BCNV01B            1.00+8
  BUS01CCNV01C            1.00+8
  e2A   e22A              1.00+8
  e2B   e22B              1.00+8
  e2C   e22C              1.00+8
  e22A                    1.00+8
  e22B                    1.00+8
  e22C                    1.00+8
  i1A                     1.00+8
  i1B                     1.00+8
  i1C                     1.00+8
```

(Continued)

Code 3-15 (Continued)

```
C CONNECTION OF A 3-PHASE SERIES-COUPLING TRANSFORMER WITH LEAKAGE IMPEDANCE
$INCLUDE 315TRAN2.TRN
BLANK RECORD ENDING BRANCHES
C 00000001111111111222222222233333333334444444444555555555566666666667777777778
C 34567890123456789012345678901234567890123456789012345678901234567890123456789O
C <-Bus1<-Bus2<---Tclose<----Topen<-------Ie<----Flash<--Request<-----Target<--O
C SHUNT BREAKER TO CONNECT/DISCONNECT THE SHUNT UNIT TO THE CIRCUIT
13BUS01ACNV01A                                                        SHNBRK
13BUS01BCNV01B                                                        SHNBRK
13BUS01CCNV01C                                                        SHNBRK
C SERIES DISCONNECT SWITCH TO CONNECT/DISCONNECT THE SERIES UNIT TO THE CIRCUIT
13e2A   e22A                                                          SRSDS
13e2B   e22B                                                          SRSDS
13e2C   e22C                                                          SRSDS
C TECHNIQUE TO SHORT THE SERIES-COMPENSATING VOLTAGE TO CALCULATE THE
C INITIAL CONDITIONS CORRECTLY
13e22A                                                CLOSED          GNDSW
13e22B                                                CLOSED          GNDSW
13e22C                                                CLOSED          GNDSW
C MEASUREMENT OF LINE CURRENTS
$INCLUDE 307MEAS2.SWT
C MEASUREMENT OF CURRENTS FLOWING THROUGH THE SHUNT UNIT
  CNV01Ai1A                                           MEASURING
  CNV01Bi1B                                           MEASURING
  CNV01Ci1C                                           MEASURING
C MEASUREMENT OF CURRENTS FLOWING THROUGH THE SERIES UNIT
  CNV02Ae22A                                          MEASURING
  CNV02Be22B                                          MEASURING
  CNV02Ce22C                                          MEASURING
BLANK RECORD ENDING SWITCHES
C 00000001111111111222222222233333333334444444444555555555566666666667777777778
C 34567890123456789012345678901234567890123456789012345678901234567890123456789O
C CONNECTION OF SHUNT UNIT CURRENT SOURCES TO THE NETWORK
60i1A   -1
60i1B   -1
60i1C   -1
C CONNECTION OF SERIES UNIT VOLTAGE SOURCES TO THE NETWORK
60e2A
60e2B
60e2C
C CONNECTION OF NETWORK 1 SOURCES (FIGURE 3-5)
$INCLUDE 308NTWK1.SRC
```

(Continued)

Code 3-15 (Continued)

```
BLANK RECORD ENDING SOURCES
BLANK RECORD ENDING NODE VOLTAGE OUTPUT
BLANK RECORD ENDING PLOT
```

Table 3-3 Electrical system data.

Parameters	Values
Source voltage ($\mathbf{V_{src}}$) phase-to-neutral	112,676.528∠0° V_{peak}
Source resistance (R_{src})	1.0053 Ω
Source reactance (X_{src})	7.43875 Ω (19.73 mH)
Receiving-end voltage ($\mathbf{V_r}$) phase-to-neutral	112,676.528∠−20° V_{peak}
Line resistance (R)	3.0159 Ω
Line reactance (X)	22.31625 Ω (59.19 mH)
Series-coupling transformer (SRSTR) turns-ratio	**1:1**
Primary winding resistance	**0.8927 Ω**
Primary winding reactance	**23.679 mH**
Secondary winding resistance	**0.8927 Ω**
Secondary winding reactance	**23.679 mH**

attaches the $INCLUDE file 315TRAN2.TRN that connects a coupling transformer in series with the transmission line as shown in Code 3-16.

The term TRANSFORMER is a reserved word in EMTP for a physical model of a transformer. SRSTRA in columns 39–44 is the name for the internal node at the top of the magnetizing branch. This name uniquely identifies the transformer. The numbers 9999 in columns 13–16 terminate transformer magnetizing branch datacards, which do not exist in this model. The winding number is specified in columns 1–2. A "1" indicates the primary winding and a "2" indicates the secondary winding. BUS01A node in columns 3–8 indicates node "1" name. BUS02A node in columns 9–14 indicates node "2" name. A blank node name indicates a connection to the ground. The winding resistance in Ohms is specified in columns 27–32. The winding inductance in mH is specified in columns 33–38. Rated winding voltage in pu or number of turns is specified in columns 39–44.

The natural active and reactive power flows (P_{rn} and Q_{rn}) at the receiving end of the line are 132 MW (0.82 pu) and −41 Mvar (−0.25 pu), respectively. Note the reduction of the natural power flow at the receiving end in comparison to that calculated in Section 3.1.1 is due to the series-coupling transformer, which results in an additional inductive impedance in series with the line.

Code 3-16 EMTP $INCLUDE file for a three-phase series-coupling transformer (315TRAN2.TRN) with a leakage impedance.

```
C File Name: 315TRAN2.TRN
C A 3-PHASE SERIES-COUPLING TRANSFORMER WITH LEAKAGE IMPEDANCE
C 00000001111111111222222222233333333334444444444555555555566666666667777777777778
C 34567890123456789012345678901234567890123456789012345678901234567890123456789O
C ------SERIES-COUPLING TRANSFORMER FOR PHASE A------
C TRANSFORMER <--Refxxxxxx<--Iss<--Phi<-Name<-Rmag<---------------------------O
  TRANSFORMER                          SRSTRA
C NEXT LINE (9999 IN COLUMNS 13-16) INDICATES A LINEAR TRANSFORMER
           9999
C <---Nodes--><----------><---Rk<---Lk<Volt|---------------------------------->O
C <-Bus1<-Bus2<----------><-Ohms<---mH<---kV---------------------------------->O
 1CNV02A                   0.892723.679   1.0
 2BUS01ABUS02A             0.892723.679   1.0
C ------SERIES-COUPLING TRANSFORMER FOR PHASE B------
  TRANSFORMER                          SRSTRB
           9999
 1CNV02B                   0.892723.679   1.0
 2BUS01BBUS02B             0.892723.679   1.0
C ------SERIES-COUPLING TRANSFORMER FOR PHASE C------
  TRANSFORMER                          SRSTRC
           9999
 1CNV02C                   0.892723.679   1.0
 2BUS01CBUS02C             0.892723.679   1.0
```

The datacard "$INCLUDE 313PQPFC.USR" attaches the $INCLUDE file 313PQPFC.USR that defines the user's input as shown in Code 3-17. This file is used to define a compensating voltage that is connected in series with the line. The file shows that the Normally-Open (NO) shunt breaker is closed at 10 ms. The ST is represented by a series-compensating voltage of 12 kV_{rms} (0.15 pu) and a relative phase angle of 0°. The ST is activated at 20 ms. Keeping the magnitude fixed at 0.15 pu, the relative phase angle is varied within its entire controllable range of $0° \leq \beta \leq 360°$ through the user's input file.

The datacard "$INCLUDE 314PQPFC.CON" attaches the $INCLUDE file 314PQPFC.CON that defines the mathematical model of the shunt-connected sinusoidal current source (Shunt Unit) and the series-connected sinusoidal voltage source (Series Unit) as shown in Figure 3-10. The instantaneous current $\mathbf{i_{dq}} = i_{1d} + ji_{1q}$ through the Shunt Unit is calculated in d-q rotating reference frame as follows.

Step 1: Computation of Exchanged Power at the Series Converter

Substituting v_A, v_B, and v_C and i_A, i_B, and i_C in the following equations

$$p = v_A i_A + v_B i_B + v_C i_C \tag{A-16}$$

Code 3-17 EMTP $INCLUDE file for user's input (313PQPFC.USR).

```
C File Name: 313PQPFC.USR
C USER'S INPUT FILE FOR PQPFC
C 0000000111111111122222222223333333333444444444455555555556666666666777777777778
C 345678901234567890123456789012345678901234567890123456789012345678901234567890
C A SHUNT BREAKER IS CLOSED.
99SHNBRK   = TIMEX .GE. 0.01
C SET ST = 1.0 FOR ST OR ST = 0.0 FOR UPFC
99ST       = 1.0
C SETTING THE COMPENSATING VOLTAGE
99e2cml    = 0.15 * (TIMEX .GE. 0.020)
99e2dqcm   = e2cml
C LOW-PASS FILTER INTRODUCES A RAMP-RATE AND REDUCES SYSTEM TRANSIENTS
 1e2dqrf   +e2dqcm
         1.0
         1.0      0.0160
99betcml   = 0 * (TIMEX .GE. 0.020)
99betcmd   = betcml * CDR
C LOW-PASS FILTER INTRODUCES A RAMP-RATE AND REDUCES SYSTEM TRANSIENTS
 1beta     +betcmd
         1.0
         1.0      0.0020
```

$$q = \sqrt{3}(v_A i_C - v_C i_A) \tag{A-99}$$

with e_{2Apu}, e_{2Bpu}, and e_{2Cpu} and i_{2Apu}, i_{2Bpu}, and i_{2Cpu}, it can be rewritten that

$$p_{2sepu} = e_{2Apu} i_{2Apu} + e_{2Bpu} i_{2Bpu} + e_{2Cpu} i_{2Cpu} \tag{A-16a}$$

$$q_{2sepu} = \sqrt{3}\left(e_{2Apu} i_{2Cpu} - e_{2Cpu} i_{2Apu}\right) \tag{A-99a}$$

with proper scaling by using V_{base}, I_{base}, and P_{base}. The corresponding EMTP code is shown below.

```
C EXCHANGED POWER COMPUTATION AT THE SERIES CONVERTER
99p2sepu  = (e2Apu * i2Apu + e2Bpu * i2Bpu + e2Cpu * i2Cpu) *Vbase*Ibase / Pbase
99q2sepu  = (e2Apu * i2Cpu - e2Cpu * i2Apu) * SQRT(3) * Vbase * Ibase / Pbase
```

The instantaneous exchanged active and reactive powers (p_{2sepu} and q_{2sepu}) at the converter are calculated. It is important to note that the exchanged power with the line differs from the power at the converter due to the active power loss and reactive power stored in the circuit reactance.

Step 2: Shunt Unit Current Computation in d-q rotating frame

The instantaneous exchanged active and reactive powers (p_{2sepu} and q_{2sepu}) by the Series Unit at the converter are considered to be the same as those by the Shunt Unit at the converter when the power loss in the PFC is neglected as a first order of approximation. Substituting p_{pu}, q_{pu}, v_{dpu}, v_{qpu}, i_{dpu}, and i_{qpu} in the following equations

$$p_{pu} = v_{dpu} i_{dpu} + v_{qpu} i_{qpu} \tag{A-91}$$

$$q_{pu} = v_{qpu} i_{dpu} - v_{dpu} i_{qpu} \tag{A-92}$$

$$i_{dpu} = \frac{p_{pu} v_{dpu} + q_{pu} v_{qpu}}{v_{dpu}^2 + v_{qpu}^2} \tag{A-93}$$

$$i_{qpu} = \frac{p_{pu} v_{qpu} - q_{pu} v_{dpu}}{v_{dpu}^2 + v_{qpu}^2} \tag{A-94}$$

with p_{2sepu}, q_{2sepu}, v_{1dpu}, v_{1qpu}, i_{1dpu}, and i_{1qpu}, it can be rewritten that

$$p_{2sepu} = v_{1dpu} i_{1dpu} + v_{1qpu} i_{1qpu} \tag{A-91a}$$

$$q_{2sepu} = v_{1qpu} i_{1dpu} - v_{1dpu} i_{1qpu} \tag{A-92a}$$

$$i_{1dpu} = \frac{p_{2sepu} v_{1dpu} + q_{2sepu} v_{1qpu}}{v_{1dpu}^2 + v_{1qpu}^2} \tag{A-93a}$$

$$i_{1qpu} = \frac{p_{2sepu} v_{1qpu} - q_{2sepu} v_{1dpu}}{v_{1dpu}^2 + v_{1qpu}^2}. \tag{A-94a}$$

Note that v_{1dpu}, v_{1qpu}, i_{1dpu}, and i_{1qpu} are in per units. The corresponding EMTP code is shown below.

```
C SHUNT UNIT CURRENT COMPUTATION PER EQNS (A-93a) AND (A-94a) in D-Q ROTATING FRAME
99Denom    = v1dpu * v1dpu + v1qpu * v1qpu
99i1dpu    = (p2sepu * v1dpu + q2sepu * ST * v1qpu) / Denom
99i1qpu    = (p2sepu * v1qpu - q2sepu * ST * v1dpu) / Denom
```

The switch "ST" is set to be 1 for Sen Transformer (ST) and 0 for Unified Power Flow Controller (UPFC) in the $INCLUDE file 313PQPFC.USR.

Step 3: Shunt Unit Current Computation in A-B-C time domain

The d-q rotating reference frame components (i_{1dpu} and i_{1qpu}) are converted into the corresponding stationary reference frame using the phase-locked loop angle according to Equation (A-55a) as shown below.

```
C SHUNT UNIT CURRENT COMPUTATION PER EQUATIONS (A-55a) in D-Q STATIONARY FRAME
99i1dspu   = i1dpu * COSPLL - i1qpu * SINPLL
99i1qspu   = i1dpu * SINPLL + i1qpu * COSPLL
```

The stationary reference frame components (i_{1dspu} and i_{1qspu}) are converted into the corresponding A-B-C time domain with proper scaling by using I_{base} according to Equation (A-75b) as shown below.

```
C SHUNT UNIT CURRENT COMPUTATION PER EQUATIONS (A-75b) in a-b-c DOMAIN
99i1Apu   = i1dspu
99i1Cpu   = -0.5 * (i1dspu + i1qspu * SQRT(3))
99i1Bpu   = -(i1Apu + i1Cpu)

99i1A     = i1Apu * Ibase
99i1B     = i1Bpu * Ibase
99i1C     = i1Cpu * Ibase
```

The corresponding EMTP $INCLUDE file (314PQPFC.CON) is shown in Code 3-18.

Code 3-18 EMTP $INCLUDE file for *P-Q* power flow control, using the mathematical model of a PFC in a Shunt-Series configuration (314PQPFC.CON).

```
C File Name: 314PQPFC.CON
C PQPFC - MATHEMATICAL MODEL OF A PFC WITH A SHUNT-SERIES CONFIGURATION WITH
C OPEN-LOOP CONTROL OF THE SERIES UNIT
C 00000001111111111222222222233333333334444444444555555555566666666667777777777 8
C 34567890123456789012345678901234567890123456789012345678901234567890123456789 0
99tha2    = theta + PI / 2 + PI + beta
C This addition of PI/2 is because the compensating voltage is sine reference
C and the PLL is cosine reference.
C PI radian is added to reverse the polarity of the series-coupling transformer.
99theta2  = tha2 + TWOPI * (tha2 .LT. 0.0) - TWOPI * (tha2 .GE. TWOPI)
C SERIES UNIT VOLTAGE
99vsrspk  = e2dqrf * Vbase
99e2a     = vsrspk * SIN(theta2)
99e2b     = vsrspk * SIN(theta2 - 120 * CDR)
99e2c     = vsrspk * SIN(theta2 + 120 * CDR)
99e2Apu   = e2A / Vbase
99e2Bpu   = e2B / Vbase
99e2Cpu   = e2C / Vbase
99e2pu    = SQRT((e2Apu * e2Apu + e2Bpu * e2Bpu + e2Cpu * e2Cpu) * 2 /3)
C SERIES UNIT VOLTAGE MAGNITUDE COMPUTATION PER EQS. (A-75a), (A-55b) and (3-23)
99e2dspu  =  e2Apu
99e2qspu  = -(e2Apu + 2.0 * e2Cpu) / SQRT(3)
99e2dpu   =  e2dspu * COSPLL + e2qspu * SINPLL
99e2qpu   = -e2dspu * SINPLL + e2qspu * COSPLL
99e2dqpu  = SQRT(e2dpu * e2dpu + e2qpu * e2qpu)
C SERIES UNIT CURRENT COMPUTATION
99i2Apu   = CNV02A / Ibase
99i2Bpu   = CNV02B / Ibase
99i2Cpu   = CNV02C / Ibase
99i2pu    = SQRT((i2Apu * i2Apu + i2Bpu * i2Bpu + i2Cpu * i2Cpu) * 2 / 3)
C SERIES UNIT CURRENT MAGNITUDE COMPUTATION PER EQS. (A-75a), (A-55b) and (3-23)
99i2dspu  = i2Apu
```

(Continued)

Code 3-18 (Continued)

```
99i2qspu   = -(i2Apu + 2.0 * i2Cpu) / SQRT(3)
99i2dpu    =  i2dspu * COSPLL + i2qspu * SINPLL
99i2qpu    = -i2dspu * SINPLL + i2qspu * COSPLL
99i2dqpu   = SQRT(i2dpu * i2dpu + i2qpu * i2qpu)
C EXCHANGED POWER IN PU BY THE SERIES CONVERTER
99p2sepu   = (e2Apu * i2Apu + e2Bpu * i2Bpu + e2Cpu * i2Cpu) *Vbase*Ibase / Pbase
99q2sepu   = (e2Apu * i2Cpu - e2Cpu * i2Apu) * SQRT(3) * Vbase * Ibase / Pbase
C EXCHANGED MW AND MVAR BY THE SERIES CONVERTER
99p2seMW   = p2sepu * Pbase / 1e6
99q2seMV   = q2sepu * Pbase / 1e6
C EXCHANGED MVA BY THE SERIES CONVERTER PER EQUATION (2-182)
99s2seMV   = SQRT(p2seMW * p2seMW + q2seMV * q2seMV)
C SHUNT UNIT CURRENT COMPUTATION PER EQNS (A-93a) AND (A-94a) in D-Q ROTATING FRAME
99Denom    = v1dpu * v1dpu + v1qpu * v1qpu
99i1dpu    = (p2sepu * v1dpu + q2sepu * ST * v1qpu) / Denom
99i1qpu    = (p2sepu * v1qpu - q2sepu * ST * v1dpu) / Denom
99i1dqpu   = SQRT(i1dpu * i1dpu + i1qpu * i1qpu)
C SHUNT UNIT CURRENT COMPUTATION PER EQUATIONS (A-55a) in D-Q STATIONARY FRAME
99i1dspu   = i1dpu * COSPLL - i1qpu * SINPLL
99i1qspu   = i1dpu * SINPLL + i1qpu * COSPLL
C SHUNT UNIT CURRENT COMPUTATION PER EQUATIONS (A-75b) in a-b-c DOMAIN
99i1Apu    = i1dspu
99i1Cpu    = -0.5 * (i1dspu + i1qspu * SQRT(3))
99i1Bpu    = -(i1Apu + i1Cpu)
99i1pu     = SQRT((i1Apu * i1Apu + i1Bpu * i1Bpu + i1Cpu * i1Cpu) * 2 / 3)
99i1A      = i1Apu * Ibase
99i1B      = i1Bpu * Ibase
99i1C      = i1Cpu * Ibase
99i1AApu   = CNV01A / Ibase
99i1BBpu   = CNV01B / Ibase
99i1CCpu   = CNV01C / Ibase
C EXCHANGED POWER BY THE SHUNT UNIT AT THE POC
99Pshpu    = (v1Apu * i1AApu + v1Bpu * i1BBpu + v1Cpu * i1CCpu)*Vbase*Ibase/Pbase
99Qshpu    = (v1Apu * i1CCpu - v1Cpu * i1AApu) * SQRT(3) * Vbase * Ibase / Pbase
99PshMW    = Pshpu * Pbase / 1e6
99QshMV    = Qshpu * Pbase / 1e6
C EXCHANGED POWER BY THE SERIES UNIT AT THE POC
99Psepu    = (v12Apu * iApu + v12Bpu * iBpu + v12Cpu * iCpu) *Vbase*Ibase / Pbase
99Qsepu    = (v12Apu * iCpu - v12Cpu * iApu) * SQRT(3) * Vbase * Ibase / Pbase
99PseMW    = Psepu * Pbase / 1e6
99QseMV    = Qsepu * Pbase / 1e6
```

(Continued)

Code 3-18 (Continued)

```
C BUS01 NODE LINE-TO-LINE VOLTAGE COMPUTATION PER EQUATION (3-7A)
99Vs        = v1pu * Vbase * SQRT(3 / 2)
C Vs in kV
99VskV      = Vs / 1e3
C BUS02 NODE LINE-TO-LINE VOLTAGE COMPUTATION PER EQUATION (3-7A)
99Vsp       = v2pu * Vbase * SQRT(3 / 2)
C Vsp in kV
99VspkV     = Vsp / 1e3
C Setting up the disconnect switches
99GNDSW     = 0
99SRSDS     = .NOT. GNDSW
```

In order to run the mathematical model of either ST or UPFC, the selection parameter (ST) is set to either 1 or 0, respectively. In the $INCLUDE file for user's input (Code 3-17), a desired series-compensating voltage of e2cm1 = 0.15 and a relative phase angle of betcm1 = 0 are defined. Just letting the simulation run for a period of time, such as 150 ms, the last solution point is recorded, provided the solution settles down to a steady-state condition. Keeping the magnitude of the series-compensating voltage fixed at 0.15 pu, the relative phase angle is varied in one-degree interval; the mathematical model of the ST and the UPFC are rerun to obtain the steady-state results in the range of the relative phase angle $0° \leq \beta \leq 360°$.

Table 3-4 shows sample data points of variations of magnitude (V_s) of the sending-end voltage ($\mathbf{V_s}$), active power (P_r) and reactive power (Q_r) flows at the receiving end, exchanged active power (P_{se}), reactive power (Q_{se}), and apparent power (S_{se}) with the line by the Series Unit during the entire range of the relative phase angle of $0° \leq \beta \leq 360°$.

Table 3-5 shows the effects of the variation of the series-compensating voltage on the active power (P_r) and reactive power (Q_r) flows at the receiving end in the cases of a VR and a PAR (asym).

Figure 3-11 shows the active, reactive, and apparent powers (P_{se}, Q_{se}, and S_{se}) exchanged by the series-compensating voltages of the ST and the UPFC with the line. The maximum exchanged active, reactive, and apparent powers by the Series Unit are 21 MW, 29 Mvar, and 30 MVA, respectively. In the case of the ST, both the active and reactive powers flow bidirectionally through the Shunt Unit (Exciter Unit). In the case of the UPFC, only the active power flows bidirectionally through the Shunt Unit (STATCOM).

Consistent with Figure 2-25a, it is shown in Figure 3-11 that the reactive power exchanged by the series-compensating voltage (Series Unit) is capacitive over the majority of the control range of $0° \leq \beta \leq 360°$, except during $205° \leq \beta \leq 335°$ when the exchanged reactive power is inductive. The Shunt Unit absorbs inductive power when the Series Unit delivers capacitive power and vice versa. Also, it should be noted that the Series Unit absorbs from the line a maximum of 12 Mvar of inductive power and delivers a maximum of 29 Mvar of capacitive power. Therefore, the Shunt Unit absorbs from the line a maximum of 29 Mvar of inductive power and delivers a maximum of 12 Mvar of capacitive power. The Shunt Unit is inductive over the majority of the control range of $0° \leq \beta \leq 360°$, except during $205° \leq \beta \leq 335°$ when it is capacitive.

While the relative phase angle (β) of the series-compensating voltage ($\mathbf{V_{s's}}$) varies, the line current ($\mathbf{I}$) and the active and reactive power flows (P_r and Q_r) in the transmission line vary and so does the

Table 3-4 Sending-end voltage, power flows at the receiving end, and the exchanged powers with the line for the electrical system data, shown in Table 3-3, as the relative phase angle varies in the range of 0° $\leq \beta \leq$ 360°.

Case for Sen Transformer (set ST=1 in 313PQPFC.USR)						
β (deg)	V_s (kV)	P_r (MW)	Q_r (Mvar)	P_{se} (MW)	Q_{se} (Mvar)	S_{se} (MVA)
0	132.93	151.90	13.24	−21.25	−8.44	22.86
30	132.87	175.51	−7.19	−19.04	−18.20	26.34
60	133.70	187.74	−36.90	−12.32	−25.91	28.69
90	135.23	185.02	−68.34	−2.36	−29.48	29.58
120	137.06	167.56	−92.99	8.42	−27.47	28.74
150	138.70	139.83	−103.70	16.83	−19.96	26.11
180	139.70	109.58	−97.19	20.04	−8.99	21.96
210	139.75	85.43	−75.35	16.97	1.97	17.08
240	138.87	74.04	−44.55	8.78	9.53	12.96
270	137.29	78.15	−13.42	−1.78	11.76	11.89
300	135.47	96.15	9.80	−11.70	8.56	14.49
330	133.88	123.05	19.41	−18.64	1.19	18.68
360	132.93	151.91	13.22	−21.25	−8.43	22.87
Case for Unified Power Flow Controller (set ST=0 in 313PQPFC.USR)						
β (deg)	V_s (kV)	P_r (MW)	Q_r (Mvar)	P_{se} (MW)	Q_{se} (Mvar)	S_{se} (MVA)
0	133.33	152.24	14.57	−21.25	−8.63	22.93
30	133.73	176.27	−4.29	−18.83	−18.57	26.44
60	134.92	188.85	−32.79	−11.80	−26.21	28.74
90	136.60	186.31	−63.73	−1.68	−29.48	29.53
120	138.32	168.77	−88.76	8.96	−27.16	28.60
150	139.61	140.71	−100.69	17.04	−19.56	25.95
180	140.10	109.97	−95.86	20.03	−8.79	21.88
210	139.67	85.35	−75.64	16.99	1.93	17.10
240	138.44	73.65	−45.94	8.97	9.43	13.02
270	136.75	77.67	−15.17	−1.52	11.77	11.87
300	135.07	95.81	8.50	−11.53	8.66	14.42
330	133.82	123.00	19.23	−18.63	1.21	18.67
360	133.33	152.26	14.55	−21.25	−8.63	22.94

voltage across the source impedance. Therefore, the voltage phasor ($\mathbf{V_s}$) at the POC (*i.e.* at the node, $\mathbf{V_s}$) varies its magnitude and phase angle from the natural line voltage ($\mathbf{V_{sn}}$). This variation from the natural voltage is highly dependent on the short-circuit impedance of the transmission line at the POC and the load flow in the line.

Figure 3-12 shows the variations of the phase-to-phase voltage at the node ($\mathbf{V_s}$) from its natural value of 136.7 kV_{rms} (*P–P*) or ($\sqrt{3} \times 0.99$ pu) with the operation of the ST and the UPFC. If the POC

Table 3-5 Effects of the variation of series-compensating voltage on the power flows at the receiving end in the cases of a VR and a PAR (asym).

	VR		PAR (asym)	
$V_{s's}$ (pu)	P_r (MW)	Q_r (Mvar)	P_r (MW)	Q_r (Mvar)
−0.15	109.58	−97.19	78.11	−13.44
−0.10	117.13	−78.15	95.96	−22.56
−0.05	124.49	−59.37	113.81	−31.69
0.00	131.64	−40.84	131.64	−40.84
0.05	138.59	−22.55	149.46	−49.99
0.10	145.35	−4.53	167.25	−59.16
0.15	151.90	13.24	185.02	−68.34

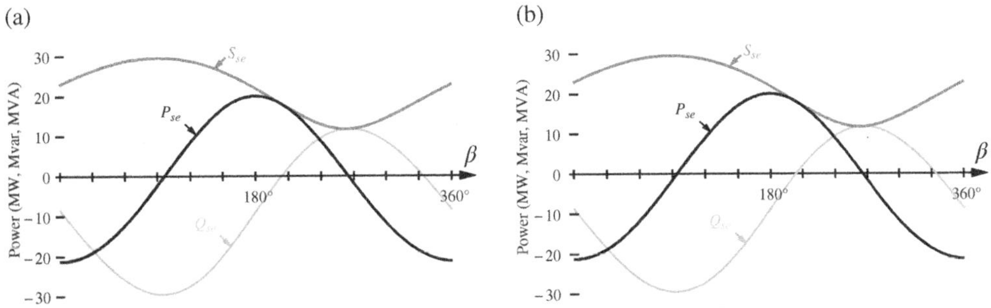

Figure 3-11 Variations of exchanged powers between the transmission line and the Series Unit of the (a) ST and (b) UPFC.

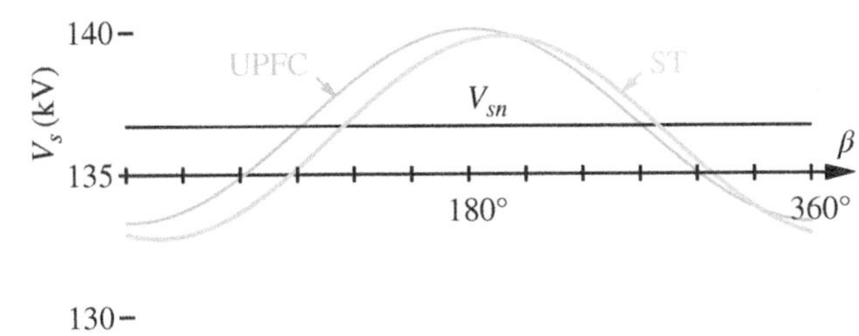

Figure 3-12 Variations of node voltage with the operation of the ST and UPFC.

is near a "stiff" node, the voltage variation at the POC (*i.e.* at the node, $\mathbf{V_s}$) is smaller and vice versa. In the case of the UPFC, the bidirectional flow of active power through the Shunt Unit emulates a $+R$ or a $-R$ and, in turn, causes a voltage sag or swell at the POC. In the case of the ST, the additional reactive power flow through the Shunt Unit emulates an inductance or a capacitance, thus causing

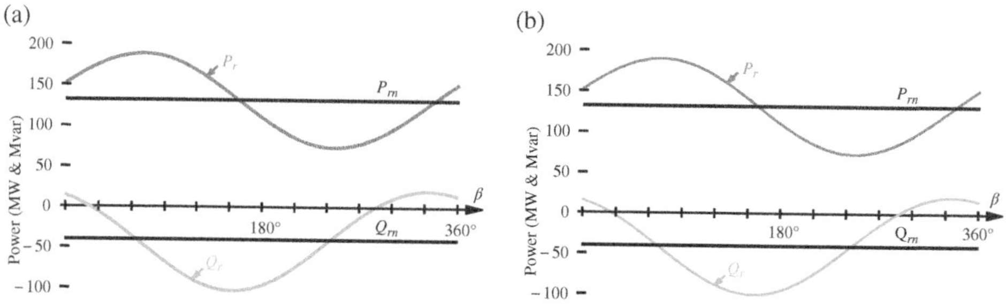

Figure 3-13 Variations of the receiving-end power flows with the operations of the (a) ST and (b) UPFC.

an additional voltage sag or swell at the POC. Note that in the range of the relative phase angle $205° \leq \beta \leq 335°$, the voltage ($\mathbf{V_s}$) is higher in the case of the ST than in the case of the UPFC. In the rest of the region, the voltage ($\mathbf{V_s}$) is higher in the case of the UPFC than in the case of the ST.

Figure 3-13a shows that with the use of the ST, both the active and reactive power flows at the receiving end of the line under consideration vary by 57 and −58 MW and by 60 and −63 Mvar, respectively, from the natural power flows (P_{rn} = 132 MW and Q_{rn} = − 41 Mvar). Figure 3-13b shows that with the use of the UPFC, both the active and reactive power flows at the receiving end of the same line vary by ±58 MW and by ±60 Mvar, respectively, from the natural power flows.

The controllable range of active and reactive power flows with the operation of the ST and UPFC are presented in Figure 3-14. The ST and UPFC offer circular controllable areas that are almost identical. The radius of the controllable area increases with larger series-compensating voltages. With a proper series-compensating voltage, the circular controllable area covers all four quadrants, resulting in a reversal of power flow from its natural direction. This phenomenon is presented in "Introduction to FACTS Controllers: Theory, Modeling, and Applications," by Sen and Sen, IEEE Press and John Wiley & Sons, 2009, Chapter 8, Section 8.3.2. Special operations of the ST with $\beta = 0°$ and $-0.15 \leq E_{2dq} \leq 0.15$ and $\beta = 90°$ and $-0.15 \leq E_{2dq} \leq 0.15$ show the functionalities of the VR and the PAR (asym), respectively. Whereas the ST offers a circular controllable area, the PAR (asym) and the VR offer linear control range.

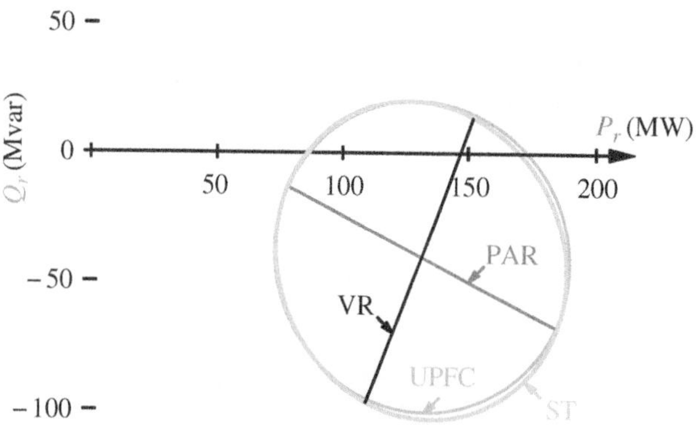

Figure 3-14 Comparison of the active and reactive power flows at the receiving end of the line with the operations of the VR, PAR (asym), UPFC, and ST.

The summary of the study results is given below.

- The variation of phase-to-phase voltage at the POC is between 132.8 kV and 139.8 kV in the case of the ST and between 133.3 kV and 140.1 kV in the case of the UPFC, whereas the natural voltage is 136.7 kV.
- Line active power flow changes from the natural flow of 132 MW by 57 and −58 MW in the case of the ST and by ±58 MW in the case of the UPFC.
- Line reactive power flow changes from the natural flow of −41 Mvar by 60 and −63 Mvar in the case of the ST and by ±60 Mvar in the case of the UPFC.
- Rating of the ST: 30 MVA.
- Rating of the UPFC: a Series Unit of 30 MVA and a Shunt Unit of 21 MVA.

Since the Shunt and the Series Units of the UPFC are designed to be of the same rating, a 30 MVA-rated Shunt Unit is able to provide a maximum of 21 Mvar of reactive power while delivering the required 21 MW of active power. If this capability of supplying shunt reactive power is really needed, particularly when the short-circuit impedance at the POC is high, a separate capacitor bank as a shunt reactive compensator may be used along with the ST as discussed in Section 1.6.

4

Transformer-Based Power Flow Controllers

The transformer-based Power Flow Controllers (PFCs) are composed of two fundamental building blocks: Voltage-Regulating Transformer (VRT) and Phase Angle Regulator (PAR). The VRT primarily regulates the magnitude of the line voltage (*i.e.* increase or decrease of the secondary voltage) without significantly changing its original phase angle. The PAR (sym) primarily regulates the phase angle of the line voltage with little or no change in its original voltage magnitude. The PAR (asym) also primarily regulates the phase angle of the line voltage with some increase in the voltage magnitude. The objective in this chapter is to study these traditional transformer-based PFCs. The operations of various PFCs are verified in a single-generator/single-line power system network, considering a three-phase base power of 160 MVA and a phase-to-phase (*P–P*) base rms voltage, $V_{rms(P\text{-}P)}$, of 138 kV_{rms} as defined in Chapter 3, Section 3.1.

4.1 Voltage-Regulating Transformer (VRT)

The VRT can be of two kinds: (a) autotransformer and (b) two-winding transformer as shown in Figure 4-1. In the VRT, the primary winding with n_1 turns is excited from an input voltage source and a voltage is induced in the secondary winding with n_2 turns. In an autotransformer, the secondary voltage is usually a fraction of the input (*i.e.* primary) voltage and the output voltage is the phasor sum of the primary and secondary voltages. The compensating secondary voltage is connected in series with the transmission line. There is no electrical isolation between the input and output voltages. In a two-winding transformer, the input and output voltages are electrically isolated. The compensating voltage, which is the induced voltage in the secondary winding is the output voltage and is connected in shunt with the line. In both the autotransformer and the two-winding transformer, the active number of turns in the secondary windings are varied with the use of load tap changers (LTCs) to vary the compensating voltage.

The output voltage is

$$V_2 = V_1 \times \left(1 + \frac{n_2}{n_1}\right) \text{ in the case of an autotransformer} \tag{4-1a}$$

$$V_2 = V_1 \times \frac{n_2}{n_1} \text{ in the case of a two-winding transformer.} \tag{4-1b}$$

Power Flow Control Solutions for a Modern Grid using SMART Power Flow Controllers, First Edition.
Kalyan K. Sen and Mey Ling Sen.

Companion website: www.wiley.com/go/sen/powerflowcontrol

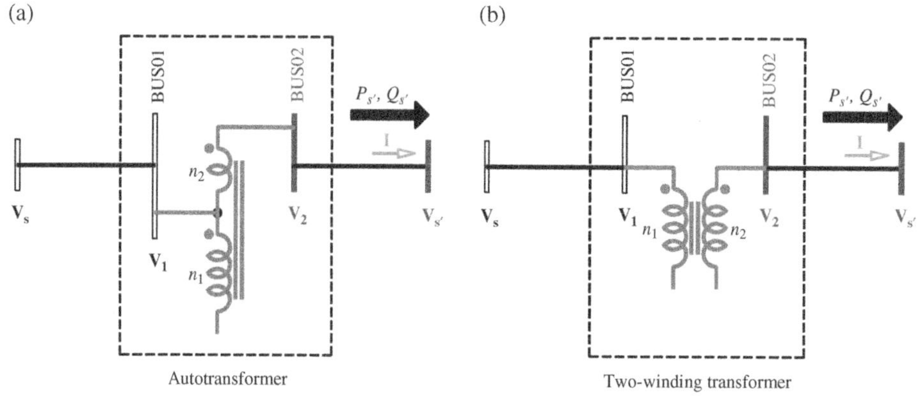

Figure 4-1 (a) Autotransformer. (b) two-winding transformer.

There are other variations of these circuit topologies. In sub-transmission (34.5–69 kV) and distribution (2.3–39 kV) category transformers, which are usually two-winding transformers, the taps are on the high-voltage (HV) primary windings. When the taps change their positions, the secondary low-voltage (LV) windings function as a VRT. While the sub-transmission transformers consist of on-load LTCs and distribution transformers consist of off-load LTCs.

In this chapter, two types of VRTs are studied, namely (1) an autotransformer-type VRT where the secondary winding is electrically connected with the primary winding and (2) two-winding VRT. The autotransformer-type VRT is a Shunt-Series configuration that generates a bipolar compensating voltage through transformer action and connects it in series with the line. The two-winding VRT is a Shunt-Shunt configuration that generates a compensating voltage through transformer action and connects it in shunt with the line.

4.1.1 Voltage Regulating Transformer (Shunt-Series Configuration)

The Voltage Regulating Transformer (VRT) in its Shunt-Series configuration can be studied with the use of the mathematical model of the ST as shown in Figure 3-10a. The model includes all the modes of operation, such as a VRT, a PAR (asym), and an IR as discussed in Chapter 3, Section 3.4. The network model of Figure 3-5 is used with the short between BUS01 and BUS02 nodes, which is replaced with the mathematical model of the ST. The corresponding Electromagnetic Transients Program (EMTP) datafile **401PQPFC.DAT**, which is the same as **312PQPFC.DAT**, except with a different series-coupling transformer without any leakage impedance, is shown in Code 4-1.

Code 4-1 EMTP datafile for mathematical model of ST, operating as a VRT in a Shunt-Series configuration, integrated in a two-generator/single-line power system network (401PQPFC.DAT).

```
BEGIN NEW DATA CASE
C File Name: 401PQPFC.DAT
C STUDY THE EFFECT OF AN AUTOTRANSFORMER, USING THE MATHEMATICAL MODEL OF ST
```

(Continued)

Code 4-1 (Continued)

```
C WITH A SHUNT-CONNECTED CURRENT SOURCE (SHUNT UNIT) AND A
C SERIES-CONNECTED VOLTAGE SOURCE (SERIES UNIT), USING AN
C IDEAL SERIES-COUPLING TRANSFORMER WITHOUT ANY LEAKAGE IMPEDANCE
C 00000001111111111222222222233333333334444444444555555555566666666667777777778
C 34567890123456789012345678901234567890123456789012345678901234567890123456789 0
C ----dt<---tmax<---Xopt<---Copt<--Toler<        ><        >
8.3333-6150.00-3
C 00000001111111111222222222233333333334444444444555555555566666666667777777778
C 34567890123456789012345678901234567890123456789012345678901234567890123456789 0
C -Iprnt<--Iplot<-Idoubl<-KssOut<-MaxOut<        ><        ><---Icat<-Nenerg
      30      10                                                1
TACS HYBRID
99CRD      = 180 / PI
99CDR      = 1 / CRD
99TWOPI    = 2.0 * PI
C BASE VOLTAGE, CURRENT, AND POWER
99Vbase    = 112676.528
99Ibase    = 946.662704
99Zbase    = Vbase / Ibase
99Pbase    = 1.5 * Vbase * Ibase
C INPUTS FROM NODE VOLTAGES AND LINE CURRENTS
$INCLUDE 302MEAS1.SWT
C MEASUREMENT OF CURRENTS FLOWING THROUGH THE SHUNT UNIT
91CNV01A
91CNV01B
91CNV01C
C MEASUREMENT OF CURRENTS FLOWING THROUGH THE SERIES UNIT
91CNV02A
91CNV02B
91CNV02C
C NORMALIZATION OF MEASURED VARIABLES AND COMPUTATION OF PEAK VOLTAGES/CURRENTS
$INCLUDE 303PTCT.SCL
C VECTOR PHASE-LOCKED LOOP
$INCLUDE 310VPLL.PLL
C USER'S INPUT FILE
$INCLUDE 313PQPFC.USR
C MATHEMATICAL MODEL OF ST
$INCLUDE 314PQPFC.CON
C COMPUTATION OF LINE RESISTANCE
$INCLUDE 305LINER.CMP
C TACS OUTPUT
33v12pu v1pu  v2pu  Prpu  Qrpu
```

(Continued)

Code 4-1 (Continued)

```
BLANK RECORD ENDING TACS
C 00000001111111111222222222233333333334444444444555555555566666666667777777778
C 345678901234567890123456789012345678901234567890123456789012345678901234567890
C <---Nodes--><---Nodes--><----R<----L<----C                                    V
C <-Bus1<-Bus2<-Bus3<-Bus4<-Ohms<---mH<---uF                                     V
C REPRESENTATION OF A SHORT BETWEEN NODES VS AND BUS01
  VSA   BUS01A            0.001
  VSB   BUS01B            0.001
  VSC   BUS01C            0.001
C CONNECTION OF NETWORK 2 BRANCHES WITH DAMPING RESISTORS (FIGURE 3-5)
$INCLUDE 311NTWK2.BRN
C HIGH RESISTANCES TO AVOID FLOATING NODES IN THE CIRCUIT
  BUS01ACNV01A            1.00+8
  BUS01BCNV01B            1.00+8
  BUS01CCNV01C            1.00+8
  e2A   e22A              1.00+8
  e2B   e22B              1.00+8
  e2C   e22C              1.00+8
  e22A                    1.00+8
  e22B                    1.00+8
  e22C                    1.00+8
  i1A                     1.00+8
  i1B                     1.00+8
  i1C                     1.00+8
C CONNECTION OF A 3-PHASE SERIES-COUPLING TRANSFORMER WITH NO LEAKAGE IMPEDANCE
$INCLUDE 402IDLT2.TRN
BLANK RECORD ENDING BRANCHES
C 00000001111111111222222222233333333334444444444555555555566666666667777777778
C 345678901234567890123456789012345678901234567890123456789012345678901234567890
C <-Bus1<-Bus2<---Tclose<----Topen<-------Ie<----Flash<--Request<-----Target<--O
C SHUNT BREAKER TO CONNECT/DISCONNECT THE SHUNT UNIT TO THE CIRCUIT
13BUS01ACNV01A                                                   SHNBRK
13BUS01BCNV01B                                                   SHNBRK
13BUS01CCNV01C                                                   SHNBRK
C SERIES DISCONNECT SWITCH TO CONNECT/DISCONNECT THE SERIES UNIT TO THE CIRCUIT
13e2A   e22A                                                     SRSDS
13e2B   e22B                                                     SRSDS
13e2C   e22C                                                     SRSDS
C TECHNIQUE TO SHORT THE SERIES-COMPENSATING VOLTAGE TO CALCULATE THE
C INITIAL CONDITIONS CORRECTLY
13e22A                                              CLOSED       GNDSW
13e22B                                              CLOSED       GNDSW
13e22C                                              CLOSED       GNDSW
```

(Continued)

Code 4-1 (Continued)

```
C MEASUREMENT OF LINE CURRENTS
$INCLUDE 307MEAS2.SWT
C MEASUREMENT OF CURRENTS FLOWING THROUGH THE SHUNT UNIT
  CNV01Ai1A                                               MEASURING
  CNV01Bi1B                                               MEASURING
  CNV01Ci1C                                               MEASURING
C MEASUREMENT OF CURRENTS FLOWING THROUGH THE SERIES UNIT
  CNV02Ae22A                                              MEASURING
  CNV02Be22B                                              MEASURING
  CNV02Ce22C                                              MEASURING
BLANK RECORD ENDING SWITCHES
C 00000001111111111222222222233333333334444444444555555555566666666667777777777778
C 34567890123456789012345678901234567890123456789012345678901234567890123456789 0
C CONNECTION OF SHUNT UNIT CURRENT SOURCES TO THE NETWORK
60i1A   -1
60i1B   -1
60i1C   -1
C CONNECTION OF SERIES UNIT VOLTAGE SOURCES TO THE NETWORK
60e2A
60e2B
60e2C
C CONNECTION OF NETWORK 1 SOURCES IN FIGURE 3-5
$INCLUDE 308NTWK1.SRC
BLANK RECORD ENDING SOURCES
BLANK RECORD ENDING NODE VOLTAGE OUTPUT
BLANK RECORD ENDING PLOT
```

The $INCLUDE file 402IDLT2.TRN attaches a series-coupling transformer without any leakage impedance in series with the transmission line as shown in Code 4-2. The leakage impedance of the series-coupling transformer is deliberately set to nearly zero (0.001 Ω) to preserve its ideal operation.

Code 4-2 EMTP $INCLUDE file for a three-phase, series-coupling transformer (402TRAN2.TRN) without any leakage impedance.

```
C File Name: 402IDLT2.TRN
C IDEAL 3-PHASE SERIES-COUPLING TRANSFORMER WITHOUT ANY LEAKAGE IMPEDANCE
C 00000001111111111222222222233333333334444444444555555555566666666667777777777778
```

(*Continued*)

Code 4-2 (Continued)

```
C 345678901234567890123456789012345678901234567890123456789012345678901234567890
C ------SERIES-COUPLING TRANSFORMER FOR PHASE A------
C TRANSFORMER <--Refxxxxxx<--Iss<--Phi<-Name<-Rmag<---------------------------O
  TRANSFORMER                                   SRSTRA
C NEXT LINE (9999 IN COLUMNS 13-16) INDICATES A LINEAR TRANSFORMER
            9999
C <---Nodes--><----------><---Rk<---Lk<Volt|---------------------------------->O
C <-Bus1<-Bus2<----------><-Ohms<---mH<---kV---------------------------------->O
 1CNV02A                          0.001 0.001   1.0
 2BUS01ABUS02A                    0.001 0.001   1.0
C ------SERIES-COUPLING TRANSFORMER FOR PHASE B------
  TRANSFORMER                                   SRSTRB
            9999
 1CNV02B                          0.001 0.001   1.0
 2BUS01BBUS02B                    0.001 0.001   1.0
C ------SERIES-COUPLING TRANSFORMER FOR PHASE C------
  TRANSFORMER                                   SRSTRC
            9999
 1CNV02C                          0.001 0.001   1.0
 2BUS01CBUS02C                    0.001 0.001   1.0
```

Table 4-1 Electrical system data.

Parameters	Values
Source voltage ($\mathbf{V}_{\mathbf{src}}$) phase-to-neutral	$112{,}676.528\angle 0°\ V_{peak}$
Source resistance (R_{src})	1.0053 Ω
Source reactance (X_{src})	7.43875 Ω (19.73 mH)
Receiving-end voltage ($\mathbf{V}_{\mathbf{r}}$) phase-to-neutral	$112{,}676.528\angle{-20°}\ V_{peak}$
Line resistance (R)	3.0159 Ω
Line reactance (X)	22.31625 Ω (59.19 mH)
Series-coupling transformer (SRSTR) turns-ratio	1:1
Primary winding resistance	0.001 Ω
Primary winding leakage inductance	0.001 mH
Secondary winding resistance	0.001 Ω
Secondary winding leakage inductance	0.001 mH

Table 4-1 shows the electrical system data for a two-generator/single-line power system network, compensated by the mathematical model of the ST, operating as an ideal VRT in a Shunt-Series configuration.

Table 4-2 Effects of the variation of series-compensating voltage ($V_{s's}$) on measured voltage (V_{21} or $V_{s's}$), the sending-end and the modified sending-end voltages (V_1 or V_s and V_2 or $V_{s'}$), and the power flows (P_r and Q_r) at the receiving end.

$V_{s's}$ (pu)	V_{21} or $V_{s's}$ (pu)	V_1 or V_s (pu)	V_2 or $V_{s'}$ (pu)	P_r (pu)	Q_r (pu)
−0.150	−0.150	1.022	0.872	1.11	−1.01
−0.100	−0.100	1.012	0.912	1.18	−0.81
−0.050	−0.050	1.000	0.950	1.24	−0.62
0.000	0.000	0.988	0.988	1.31	−0.43
0.050	0.050	0.975	1.025	1.37	−0.24
0.100	0.100	0.960	1.060	1.42	−0.06
0.150	0.150	0.945	1.095	1.48	0.12

The EMTP datafile **401PQPFC.DAT** is rerun with the relative phase angle of the series-compensating voltage ($\mathbf{V_{s's}}$) set to $\beta = 0°$ while its magnitude ($V_{s's}$) is varied from −0.15 pu to 0.15 pu in the $INCLUDE file 313PQPFC.USR that defines the user's input as shown in Code 3-17. Table 4-2 shows the variation of the series-compensating voltage magnitude ($V_{s's}$) and its effects on measured voltage (V_{21} or $V_{s's}$), the sending-end voltage magnitude (V_1 or V_s), the modified sending-end voltage magnitude (V_2 or $V_{s'}$), and the resulting active and reactive power flows (P_r and Q_r) at the receiving end.

The series-compensating voltage exchanges active and reactive powers with the transmission line. Both of these active and reactive powers appear at the transmission line through the magnetic link and the Shunt Unit as shown in Figure 3-10a. The simulation result is shown in Figure 4-2. Since the modified sending-end voltage, $\mathbf{V_{s'}}$ (same as $\mathbf{V_2}$ at the BUS02 node), and the sending-end voltage, $\mathbf{V_s}$ (same as $\mathbf{V_1}$ at the BUS01 node), are in phase in the case of a VRT, their scalar difference, $\mathbf{V_{21}}$, is the series-compensating voltage ($\mathbf{V_{s's}}$). A variable-magnitude, bipolar, series-compensating voltage ($\mathbf{V_{s's}}$) increases or decreases $\mathbf{V_{s'}}$ from its uncompensated value, $\mathbf{V_s}$. It is discussed in Chapter 2 that an increase in the line voltage is caused by delivering capacitive power to the line and a decrease in the line voltage is caused by absorbing inductive power from the line at the Point of Connection (POC) to the utility. Since a transformer neither generates nor absorbs reactive power, the primary winding must absorb inductive power from the line to deliver capacitive power to the line through the secondary winding and vice versa; however, absorbing var from the line at the BUS01 node reduces the line voltage and vice versa. The figure shows that the voltage ($\mathbf{V_{s'}}$) at the modified sending end is 0.15 pu more than the voltage ($\mathbf{V_s}$) at the sending end; however, the net increase is only 0.107 pu since the voltage ($\mathbf{V_s}$) decreases by 0.043 pu from its natural value of 0.988 pu to a new value of 0.945 pu. The figure also shows that with a compensating voltage ($\mathbf{V_{s's}}$) of 0.15 pu and a negative polarity, the voltage ($\mathbf{V_{s'}}$) at the modified sending end is 0.15 pu less than the voltage ($\mathbf{V_s}$) at the sending end; however, the net decrease is only 0.116 pu since the voltage ($\mathbf{V_s}$) increases by 0.034 pu from its natural value of 0.988 pu to a new value of 1.022 pu. Therefore, the range of $V_{s'}$ is $0.872 = 1.022 - 0.15$ to $1.095 = 0.945 + 0.15$ when the compensating voltage magnitude ($V_{s's}$) varies from −0.15 to 0.15 pu.

Changing the voltage at the modified sending end affects the active and reactive power flows (P_r and Q_r) at the receiving end. The reactive power flow (Q_r) at the receiving end varies from −1.01 to 0.12 pu during the range of compensating voltage magnitude ($V_{s's}$) from −0.15 to 0.15 pu, while the uncompensated reactive power flow (Q_{rn}) at the receiving end is −0.43 pu. However, the active

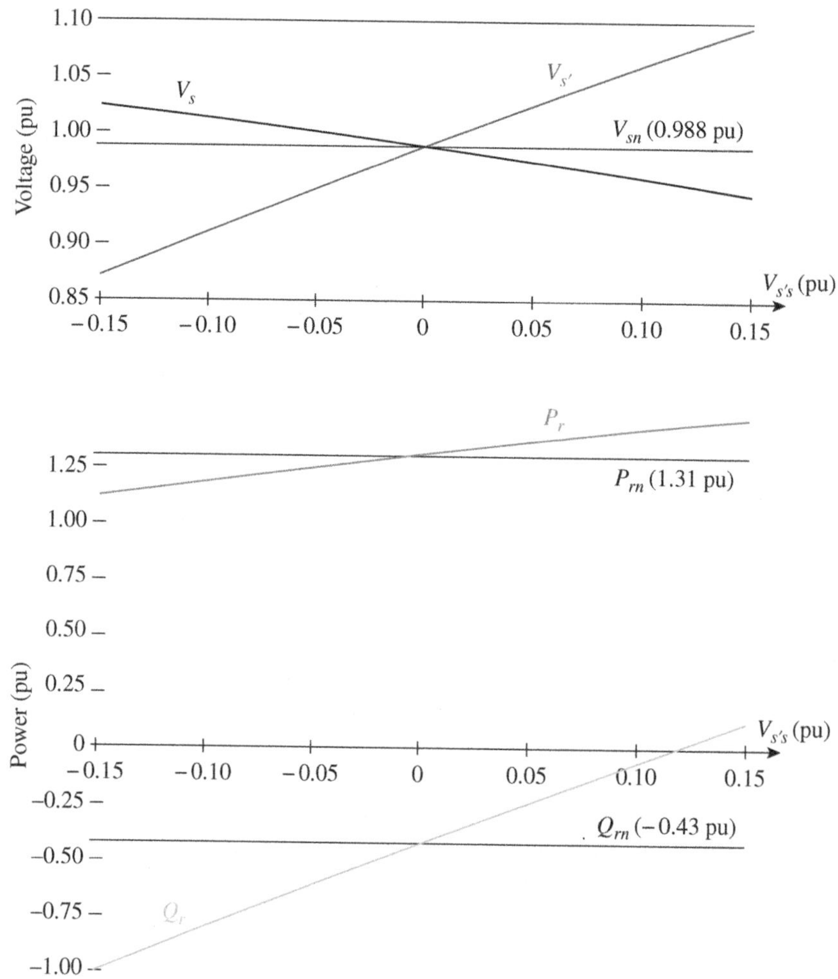

Figure 4-2 Ranges of voltage magnitudes (V_s and $V_{s'}$) at the sending and the modified sending ends and the active and reactive power flows (P_r and Q_r) at the receiving end. The VRT in a Shunt-Series configuration is implemented using the mathematical model of ST through a coupling transformer without any leakage impedance.

power flow (P_r) at the receiving end varies from 1.11 to 1.48 pu for the same range of variation in the compensating voltage magnitude ($V_{s's}$), while the uncompensated active power flow (P_{rn}) at the receiving end is 1.31 pu. Therefore, a compensating voltage magnitude ($V_{s's}$) in the range of −0.15 to 0.15 pu varies the reactive power flow (Q_r) by 1.13 pu and active power flow (P_r) by 0.37 pu.

For a VRT in a Shunt-Series configuration with a fixed turns-ratio (n_2/n_1), the series-compensating voltage magnitude ($V_{s's}$) is a multiplier (n_2/n_1) of the sending-end voltage at BUS01 node, according to Equation (4-1a), which is simulated below. The corresponding EMTP datafile **403PQPFC.DAT**, which is the same as **401PQPFC.DAT**, except with a different user's input file 404PQPFC.USR is shown in Code 4-3. In this user's input file, the series-compensating voltage magnitude (**99e2dqcm**) is defined as a multiplier (**e2cm1**) of the voltage (**v1pu**) at the Point of Connection (POC) to the utility.

Code 4-3 EMTP datafile for the mathematical model of the ST, operating as a VRT in a Shunt-Series configuration, integrated in a two-generator/single-line power system network (403PQPFC.DAT). The series-compensating voltage magnitude ($V_{s's}$) is a multiplier of the sending-end voltage magnitude (V_s).

```
BEGIN NEW DATA CASE
C File Name: 403PQPFC.DAT
C STUDY THE EFFECT OF A BETTER REPRESENTATIVE OF AN AUTOTRANSFORMER, USING THE
C MATHEMATICAL MODEL OF ST WITH A SHUNT-CONNECTED CURRENT SOURCE (SHUNT UNIT)
C AND A SERIES-CONNECTED VOLTAGE SOURCE (SERIES UNIT), USING AN
C IDEAL SERIES-COUPLING TRANSFORMER WITHOUT ANY LEAKAGE IMPEDANCE
C 0000000111111111122222222223333333333444444444455555555556666666666777777777778
C 3456789012345678901234567890123456789012345678901234567890123456789012345678901234567890
C ----dt<---tmax<---Xopt<---Copt<--Toler<        ><        >
8.3333-6150.00-3
C 0000000111111111122222222223333333333444444444455555555556666666666777777777778
C 3456789012345678901234567890123456789012345678901234567890123456789012345678901234567890
C -Iprnt<--Iplot<-Idoubl<-KssOut<-MaxOut<        ><        ><---Icat<-Nenerg
      30      10                                                        1
TACS HYBRID
99CRD     = 180 / PI
99CDR     = 1 / CRD
99TWOPI   = 2.0 * PI
C BASE VOLTAGE, CURRENT, AND POWER
99Vbase   = 112676.528
99Ibase   = 946.662704
99Zbase   = Vbase / Ibase
99Pbase   = 1.5 * Vbase * Ibase
C INPUTS FROM NODE VOLTAGES AND LINE CURRENTS
$INCLUDE 302MEAS1.SWT
C MEASUREMENT OF CURRENTS FLOWING THROUGH THE SHUNT UNIT
91CNV01A
91CNV01B
91CNV01C
C MEASUREMENT OF CURRENTS FLOWING THROUGH THE SERIES UNIT
91CNV02A
91CNV02B
91CNV02C
C NORMALIZATION OF MEASURED VARIABLES AND COMPUTATION OF PEAK VOLTAGES/CURRENTS
$INCLUDE 303PTCT.SCL
C VECTOR PHASE-LOCKED LOOP
$INCLUDE 310VPLL.PLL
C USER'S INPUT FILE
$INCLUDE 404PQPFC.USR
```

(*Continued*)

Code 4-3 (Continued)

```
C MATHEMATICAL MODEL OF ST
$INCLUDE 314PQPFC.CON
C COMPUTATION OF LINE RESISTANCE
$INCLUDE 305LINER.CMP
C TACS OUTPUT
33v12pu v1pu  v2pu  Prpu  Qrpu
BLANK RECORD ENDING TACS
C 0000000111111111122222222223333333333444444444455555555556666666666777777777778
C 3456789012345678901234567890123456789012345678901234567890123456789012345678901234567890
C <---Nodes--><---Nodes--><----R<----L<----C                                   V
C <-Bus1<-Bus2<-Bus3<-Bus4<-Ohms<---mH<---uF                                  V
C REPRESENTATION OF A SHORT BETWEEN NODES VS AND BUS01
  VSA   BUS01A            0.001
  VSB   BUS01B            0.001
  VSC   BUS01C            0.001
C CONNECTION OF NETWORK 2 BRANCHES WITH DAMPING RESISTORS (FIGURE 3-5)
$INCLUDE 311NTWK2.BRN
C HIGH RESISTANCES TO AVOID FLOATING NODES IN THE CIRCUIT
  BUS01ACNV01A           1.00+8
  BUS01BCNV01B           1.00+8
  BUS01CCNV01C           1.00+8
  e2A   e22A             1.00+8
  e2B   e22B             1.00+8
  e2C   e22C             1.00+8
  e22A                   1.00+8
  e22B                   1.00+8
  e22C                   1.00+8
  i1A                    1.00+8
  i1B                    1.00+8
  i1C                    1.00+8
C CONNECTION OF A 3-PHASE SERIES-COUPLING TRANSFORMER WITH NO LEAKAGE IMPEDANCE
$INCLUDE 402IDLT2.TRN
BLANK RECORD ENDING BRANCHES
C 0000000111111111122222222223333333333444444444455555555556666666666777777777778
C 3456789012345678901234567890123456789012345678901234567890123456789012345678901234567890
C <-Bus1<-Bus2<---Tclose<----Topen<-------Ie<----Flash<--Request<-----Target<--O
C SHUNT BREAKER TO CONNECT/DISCONNECT THE SHUNT UNIT TO THE CIRCUIT
13BUS01ACNV01A                                                          SHNBRK
13BUS01BCNV01B                                                          SHNBRK
13BUS01CCNV01C                                                          SHNBRK
C SERIES DISCONNECT SWITCH TO CONNECT/DISCONNECT THE SERIES UNIT TO THE CIRCUIT
13e2A   e22A                                                            SRSDS
13e2B   e22B                                                            SRSDS
```

(*Continued*)

Code 4-3 (Continued)

```
13e2C    e22C                                                              SRSDS
C TECHNIQUE TO SHORT THE SERIES-COMPENSATING VOLTAGE TO CALCULATE THE
C INITIAL CONDITIONS CORRECTLY
13e22A                                                    CLOSED           GNDSW
13e22B                                                    CLOSED           GNDSW
13e22C                                                    CLOSED           GNDSW
C MEASUREMENT OF LINE CURRENTS
$INCLUDE 307MEAS2.SWT
C MEASUREMENT OF CURRENTS FLOWING THROUGH THE SHUNT UNIT
  CNV01Ai1A                                               MEASURING
  CNV01Bi1B                                               MEASURING
  CNV01Ci1C                                               MEASURING
C MEASUREMENT OF CURRENTS FLOWING THROUGH THE SERIES UNIT
  CNV02Ae22A                                              MEASURING
  CNV02Be22B                                              MEASURING
  CNV02Ce22C                                              MEASURING
BLANK RECORD ENDING SWITCHES
C 00000001111111111222222222233333333334444444444555555555566666666667777777778
C 34567890123456789012345678901234567890123456789012345678901234567890123456789 0
C CONNECTION OF SHUNT UNIT CURRENT SOURCES TO THE NETWORK
60i1A    -1
60i1B    -1
60i1C    -1
C CONNECTION OF SERIES UNIT VOLTAGE SOURCES TO THE NETWORK
60e2A
60e2B
60e2C
C CONNECTION OF NETWORK 1 OR 2 SOURCES (FIGURE 3-5)
$INCLUDE 308NTWK1.SRC
BLANK RECORD ENDING SOURCES
BLANK RECORD ENDING NODE VOLTAGE OUTPUT
BLANK RECORD ENDING PLOT
```

Note that the sending-end voltage at BUS01 node also varies with the varying line current. The EMTP datafile **403PQPFC.DAT** is rerun with the relative phase angle of the series-compensating voltage set to $\beta = 0°$ while its magnitude is varied from −15 to 15% of the sending-end voltage magnitude (V_s) in the $INCLUDE file 404PQPFC.USR that defines the user's input as shown in Code 4-4. It is shown that the series-compensating voltage magnitude ($V_{s's}$) is set at 15% of the sending-end voltage magnitude (V_s).

Code 4-4 EMTP $INCLUDE file for user's input (404PQPFC.USR).

```
C File Name: 404PQPFC.USR
C USER'S INPUT FILE FOR PQPFC
C 00000001111111111222222222233333333334444444444555555555566666666667777777778
C 3456789012345678901234567890123456789012345678901234567890123456789012345678901234567890
C A SHUNT BREAKER IS CLOSED.
99SHNBRK   = TIMEX .GE. 0.01
C SET ST = 1.0 FOR ST OR ST = 0.0 FOR UPFC
99ST         = 1.0
C SETTING THE COMPENSATING VOLTAGE
99e2cm1    = 0.15 * (TIMEX .GE. 0.020)
99e2dqcm   = e2cm1 * v1pu
C LOW-PASS FILTER INTRODUCES A RAMP-RATE AND REDUCES SYSTEM TRANSIENTS
 1e2dqrf   +e2dqcm
       1.0
       1.0     0.0160
99betcm1   = 0 * (TIMEX .GE. 0.020)
99betcmd   = betcm1 * CDR
C LOW-PASS FILTER INTRODUCES A RAMP-RATE AND REDUCES SYSTEM TRANSIENTS
 1beta      +betcmd
       1.0
       1.0     0.0020
```

Table 4-3 Effects of the variation of series-compensating voltage ($V_{s's}$) on simulated voltage (V_{21} or $V_{s's}$), the sending-end and the modified sending-end voltages (V_1 or V_s and V_2 or $V_{s'}$), and the power flows (P_r and Q_r) at the receiving end.

$V_{s's}/V_s$	V_{21} or $V_{s's}$ (pu)	V_1 or V_s (pu)	V_2 or $V_{s'}$ (pu)	P_r (pu)	Q_r (pu)
−0.150	−0.153	1.023	0.870	1.10	−1.02
−0.100	−0.101	1.012	0.911	1.18	−0.82
−0.050	−0.050	1.000	0.950	1.24	−0.62
0.000	0.000	0.988	0.988	1.31	−0.43
0.050	0.049	0.975	1.024	1.37	−0.25
0.100	0.096	0.962	1.058	1.42	−0.08
0.150	0.142	0.948	1.090	1.47	0.09

Table 4-3 shows the variation of the normalized series-compensating voltage magnitude ($V_{s's}/V_s$) and its effects on simulated voltage (V_{21} or $V_{s's}$), the sending-end voltage magnitude (V_1 or V_s), the modified sending-end voltage magnitude (V_2 or $V_{s'}$), and the resulting active and reactive power flows (P_r and Q_r) at the receiving end. The magnitude of the varying compensating voltage ($V_{s's}$)

at various operating points can either be simulated or calculated from the turns-ratio between the effective compensating winding and the exciting winding, using the sending-end voltage (V_s), which varies with load. For example, $0.15 \times 0.948 = 0.142$, where 0.15 is the turns-ratio and 0.142 pu is the actual compensating voltage magnitude ($V_{s's}$), since the sending-end voltage magnitude (V_s) decreases from its natural value (V_{sn}) of 0.988 pu to 0.948 pu as shown in the last row of Tablc 4 3.

Figure 4-3 shows that the variation of the series-compensating voltage magnitude ($V_{s's}$) in the range of −15 to 15% of the sending-end voltage magnitude (V_s) results in a variation of the

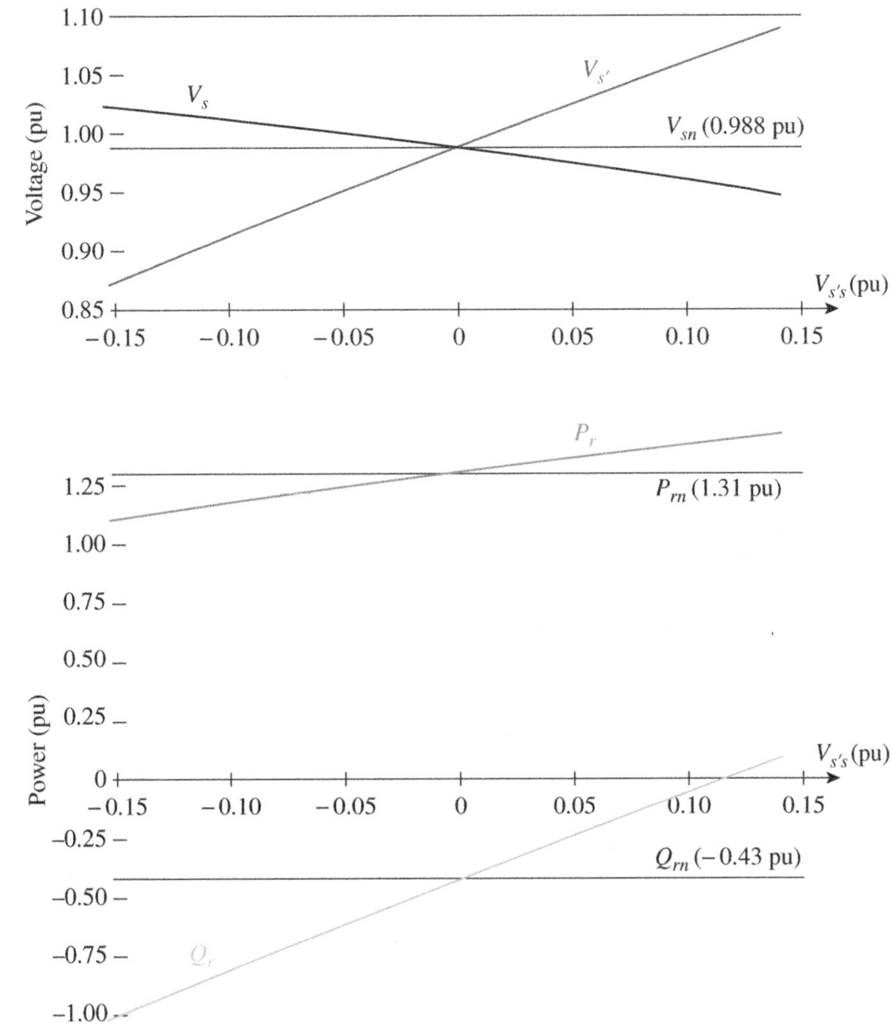

Figure 4-3 Ranges of voltage magnitudes (V_s and $V_{s'}$) at the sending and the modified sending ends and the active and reactive power flows (P_r and Q_r) at the receiving end. The VRT in a Shunt-Series configuration is implemented using the mathematical model of ST through a coupling transformer without any leakage impedance. The series-compensating voltage magnitude ($V_{s's}$) is a multiplier of the sending-end voltage magnitude (V_s).

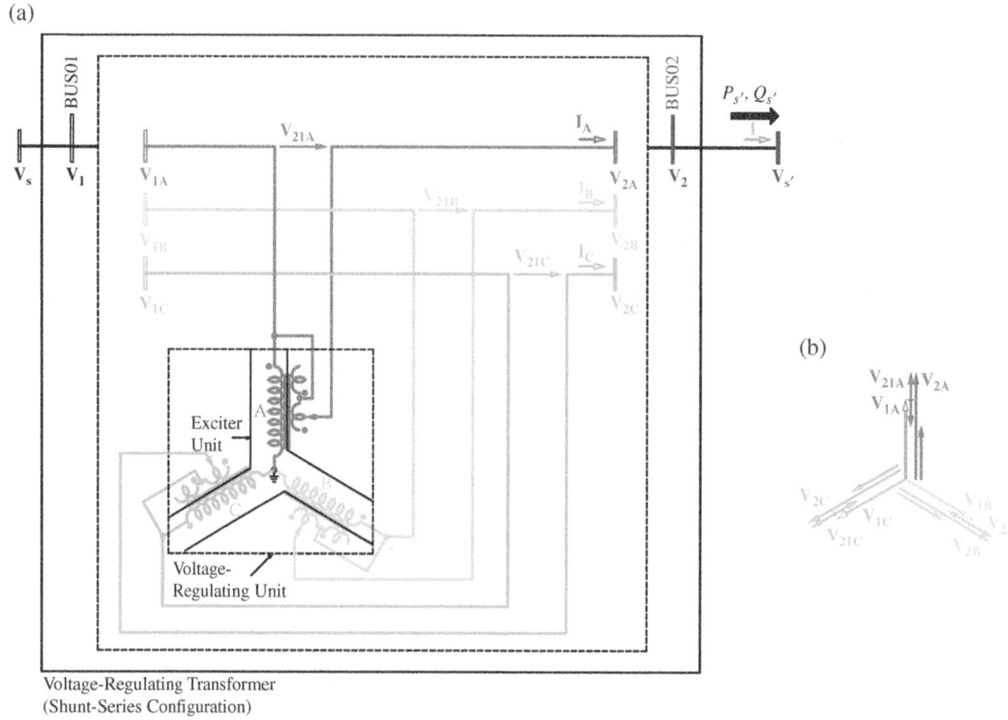

Figure 4-4 (a) Voltage-regulating transformer (shunt-series configuration); (b) phasor diagram.

modified sending-end voltage magnitude ($V_{s'}$) in the range of 0.870 to 1.090 pu. In the process, the sending-end voltage magnitude (V_s) varies in the range of 1.023 to 0.948 pu. The reactive power flow (Q_r) at the receiving end varies from −1.02 to 0.09 pu during the range of compensating voltage magnitude ($V_{s's}$) from −15 to 15% of the sending-end voltage magnitude (V_s). However, the active power flow (P_r) at the receiving end varies in the range of 1.10 to 1.47 pu for the same range of variation in the compensating voltage magnitude ($V_{s's}$). Therefore, a range of −15 to 15% in compensating voltage magnitude ($V_{s's}$) varies the reactive power flow (Q_r) by 1.11 pu and active power flow (P_r) by 0.37 pu.

Next, an ideal VRT model in a Shunt-Series configuration is studied, using a TRANSFORMER model in EMTP. The network model of Figure 3-5 is used with the short between BUS01 and BUS02 nodes, which is replaced with the VRT as shown in Figure 4-4. The corresponding EMTP datafile **405IDLAT.DAT** is shown in Code 4-5. Note that the implementation of a VRT in a Shunt-Series configuration in Figure 4-4 is different from the autotransformer, which is shown in Figure 4-1a; however, it is functionally the same and representative of the model in EMTP.

The VRT as shown in Figure 4-4a consists of an Exciter Unit and a Voltage-Regulating Unit. The Exciter Unit consists of a three-phase (A, B, and C), Y-connected primary windings and is connected in shunt with the line. The three-phase primary windings are excited from the three-phase line voltages ($\mathbf{V_{sA}}$, $\mathbf{V_{sB}}$, and $\mathbf{V_{sC}}$). A three-phase, bipolar compensating voltages ($\mathbf{V_{s'sA}}$, $\mathbf{V_{s'sB}}$, and $\mathbf{V_{s'sC}}$) that are either in phase or out of phase with the corresponding phase-to-neutral voltage are generated from the induced voltages in the secondary windings. The Voltage-Regulating Unit consists of a total of six secondary windings (two windings in each phase). The magnitude of the series-compensating voltage ($\mathbf{V}_{s's}$) is varied with the use of LTCs. The line is regulated at a voltage ($\mathbf{V}_{s'}$) from the uncompensated voltage ($\mathbf{V}_s$). The corresponding phasor diagram is shown in Figure 4-4b.

Code 4-5 EMTP datafile for an ideal VRT (Shunt-Series configuration), integrated in a two-generator/single-line power system network (405IDLAT.DAT).

```
BEGIN NEW DATA CASE
C File Name: 405IDLAT.DAT
C IDEAL 3-PHASE VRT (SHUNT-SERIES CONFIGURATION) WITHOUT ANY LEAKAGE IMPEDANCE
C 0000000111111111122222222223333333333444444444455555555556666666666777777777778
C 3456789012345678901234567890123456789012345678901234567890123456789012345678901234567890
C ----dt<---tmax<---Xopt<---Copt<--Toler<        ><        >
8.3333-6150.00-3
C 0000000111111111122222222223333333333444444444455555555556666666666777777777778
C 3456789012345678901234567890123456789012345678901234567890123456789012345678901234567890
C -Iprnt<--Iplot<-Idoubl<-KssOut<-MaxOut<        ><        ><---Icat<-Nenerg
      30      10                                                   1
TACS HYBRID
99CRD       = 180 / PI
99CDR       = 1 / CRD
99TWOPI     = 2.0 * PI
C BASE VOLTAGE, CURRENT, AND POWER
99Vbase     = 112676.528
99Ibase     = 946.662704
99Zbase     = Vbase / Ibase
99Pbase     = 1.5 * Vbase * Ibase
C INPUTS FROM NODE VOLTAGES AND LINE CURRENTS
$INCLUDE 302MEAS1.SWT
C NORMALIZATION OF MEASURED VARIABLES AND COMPUTATION OF PEAK VOLTAGES/CURRENTS
$INCLUDE 303PTCT.SCL
C IDEAL PHASE-LOCKED LOOP
$INCLUDE 304IPLL.PLL
C COMPUTATION OF LINE RESISTANCE
$INCLUDE 305LINER.CMP
C TACS OUTPUT
33v12pu v1pu  v2pu  Prpu  Qrpu
BLANK RECORD ENDING TACS
C 0000000111111111122222222223333333333444444444455555555556666666666777777777778
C 3456789012345678901234567890123456789012345678901234567890123456789012345678901234567890
C <---Nodes--><---Nodes--><----R<----L<----C                                      V
```

(Continued)

Code 4-5 (Continued)

```
C <-Bus1<-Bus2<-Bus3<-Bus4<-Ohms<---mH<---uF                                          V
C REPRESENTATION OF A SHORT BETWEEN NODES VS AND BUS01
  VSA   BUS01A                0.001
  VSB   BUS01B                0.001
  VSC   BUS01C                0.001
C CONNECTION OF NETWORK 2 BRANCHES WITH DAMPING RESISTORS (FIGURE 3-5)
$INCLUDE 311NTWK2.BRN
C CONNECTION OF AN IDEAL (NO LEAKAGE IMPEDANCE) 3-PHASE VRT IN A SHUNT-SERIES
C CONFIGURATION
$INCLUDE 406IDLAT.TRN
BLANK RECORD ENDING BRANCHES
C 00000001111111111222222222233333333334444444444555555555566666666667777777777778
C 34567890123456789012345678901234567890123456789012345678901234567890123456789890
C <-Bus1<-Bus2<---Tclose<----Topen<-------Ie<----Flash<--Request<-----Target<--O
C MEASUREMENT OF LINE CURRENTS
$INCLUDE 307MEAS2.SWT
BLANK RECORD ENDING SWITCHES
C 00000001111111111222222222233333333334444444444555555555566666666667777777777778
C 34567890123456789012345678901234567890123456789012345678901234567890123456789890
C CONNECTION OF NETWORK 1 OR 2 SOURCES (FIGURE 3-5)
$INCLUDE 308NTWK1.SRC
BLANK RECORD ENDING SOURCES
BLANK RECORD ENDING NODE VOLTAGE OUTPUT
BLANK RECORD ENDING PLOT
```

The datafile **405IDLAT.DAT** includes a datacard, which attaches the $INCLUDE file 406IDLAT.TRN that connects an ideal VRT (autotransformer without any leakage impedance) to the transmission line for increasing the modified sending-end voltage magnitude ($V_{S'}$) by 15% as shown in Code 4-6.

Code 4-6 EMTP $INCLUDE file for an ideal three-phase VRT in a Shunt-Series configuration for increasing 15% line voltage (406IDLAT.TRN).

```
C File Name: 406IDLAT.TRN
C IDEAL 3-PHASE VRT (SHUNT-SERIES CONFIGURATION) WITHOUT ANY LEAKAGE IMPEDANCE
C FOR INCREASING 15% LINE VOLTAGE
C 00000001111111111222222222233333333334444444444555555555566666666667777777777778
C 34567890123456789012345678901234567890123456789012345678901234567890123456789890
```

(*Continued*)

Code 4-6 (Continued)

```
C ------ SHUNT-SERIES CONFIGURATION FOR PHASE A------
C TRANSFORMER <--Refxxxxxx<--Iss<--Phi<-Name<-Rmag<---------------------------O
  TRANSFORMER                                    AUTOTA
C NEXT LINE (9999 IN COLUMNS 13-16) INDICATES A LINEAR TRANSFORMER
            9999
C <---Nodes--><---------><---Rk<---Lk<Volt|--------------------------------->O
C <-Bus1<-Bus2<----------><-Ohms<---mH<---kV---------------------------------->O
 1BUS01A                          0.001 0.001 1.000
 2BUS02ABUS01A                    0.001 0.001 0.150
C ------ SHUNT-SERIES CONFIGURATION FOR PHASE B------
  TRANSFORMER                                    AUTOTB
            9999
 1BUS01B                          0.001 0.001 1.000
 2BUS02BBUS01B                    0.001 0.001 0.150
C ------ SHUNT-SERIES CONFIGURATION FOR PHASE C------
  TRANSFORMER                                    AUTOTC
            9999
 1BUS01C                          0.001 0.001 1.000
 2BUS02CBUS01C                    0.001 0.001 0.150
```

Table 4-4 shows the electrical system data for a two-generator/single-line power system network, compensated by an ideal VRT in a Shunt-Series configuration for increasing line voltage.

Table 4-4 Electrical system data.

Parameters	Values
Source voltage ($\mathbf{V_{src}}$) phase-to-neutral	112,676.528∠0° V_{peak}
Source resistance (R_{src})	1.0053 Ω
Source reactance (X_{src})	7.43875 Ω (19.73 mH)
Receiving-end voltage ($\mathbf{V_r}$) phase-to-neutral	112,676.528∠−20° V_{peak}
Line resistance (R)	3.0159 Ω
Line reactance (X)	22.31625 Ω (59.19 mH)
Shunt-Series configuration for increasing line voltage	**1:(0–0.15)**
Primary winding resistance	0.001 Ω
Primary winding leakage inductance	0.001 mH
Secondary winding resistance	0.001 Ω
Secondary winding leakage inductance	0.001 mH

The $INCLUDE file 406IDLAT.TRN is modified for −15% of voltage compensation (for decreasing line voltage) as shown in Code 4-7.

Code 4-7 EMTP $INCLUDE file for an ideal three-phase VRT in a Shunt-Series configuration for decreasing 15% line voltage (406IDLAT.TRN).

```
C File Name: 406IDLAT.TRN
C IDEAL 3-PHASE VRT (SHUNT-SERIES CONFIGURATION) WITHOUT ANY LEAKAGE IMPEDANCE
C FOR DECREASING 15% LINE VOLTAGE
C 00000001111111111222222222233333333334444444444555555555566666666667777777778
C 34567890123456789012345678901234567890123456789012345678901234567890123456789 0
C ------ SHUNT-SERIES CONFIGURATION FOR PHASE A------
C TRANSFORMER <--Refxxxxxx<--Iss<--Phi<-Name<-Rmag<---------------------------O
  TRANSFORMER                                 AUTOTA
C NEXT LINE (9999 IN COLUMNS 13-16) INDICATES A LINEAR TRANSFORMER
            9999
C <---Nodes--><----------><---Rk<---Lk<Volt|--------------------------------->O
C <-Bus1<-Bus2<----------><-Ohms<---mH<---kV--------------------------------->O
 1BUS01A                        0.001 0.001 1.000
 2BUS01ABUS02A                  0.001 0.001 0.150
C ------ SHUNT-SERIES CONFIGURATION FOR PHASE B------
  TRANSFORMER                                 AUTOTB
            9999
 1BUS01B                        0.001 0.001 1.000
 2BUS01BBUS02B                  0.001 0.001 0.150
C ------ SHUNT-SERIES CONFIGURATION FOR PHASE C------
  TRANSFORMER                                 AUTOTC
            9999
 1BUS01C                        0.001 0.001 1.000
 2BUS01CBUS02C                  0.001 0.001 0.150
```

Table 4-5 shows the electrical system data for a two-generator/single-line power system network, compensated by an ideal VRT in a Shunt-Series configuration for decreasing line voltage.

Table 4-6 shows the variation of the normalized series-compensating voltage magnitude ($V_{s's}/V_s$) and its effects on simulated voltage (V_{21} or $V_{s's}$), the sending-end voltage magnitude (V_1 or V_s), the modified sending-end voltage magnitude (V_2 or $V_{s'}$), and the resulting active and reactive power flows (P_r and Q_r) at the receiving end.

Figure 4-5 shows the variations of the sending-end voltage magnitude (V_s) and the modified sending-end voltage magnitude ($V_{s'}$) as a function of the simulated voltage (V_{21} or $V_{s's}$) when the series-compensating voltage magnitude ($V_{s's}$) varies in the range of −15 to 15% of the sending-end voltage magnitude (V_s) using a TRANSFORMER model in EMTP to represent a VRT in

Table 4-5 Electrical system data.

Parameters	Values
Source voltage ($\mathbf{V_{src}}$) phase-to-neutral	112,676.528∠0° V_{peak}
Source resistance (R_{src})	1.0053 Ω
Source reactance (X_{src})	7.43875 Ω (19.73 mH)
Receiving-end voltage ($\mathbf{V_r}$) phase-to-neutral	112,676.528∠−20° V_{peak}
Line resistance (R)	3.0159 Ω
Line reactance (X)	22.31625 Ω (59.19 mH)
Shunt-Series configuration for **decreasing** line voltage	1:(0–0.15)
Primary winding resistance	0.001 Ω
Primary winding leakage inductance	0.001 mH
Secondary winding resistance	0.001 Ω
Secondary winding leakage inductance	0.001 mH

Table 4-6 Effects of the variation of series-compensating voltage ($V_{s's}$) on simulated voltage (V_{21} or $V_{s's}$), the sending-end and the modified sending-end voltages (V_1 or V_s and V_2 or $V_{s'}$), and the power flows (P_r and Q_r) at the receiving end.

$V_{s's}/V_s$	V_{21} or $V_{s's}$ (pu)	V_1 or V_s (pu)	V_2 or $V_{s'}$ (pu)	P_r (pu)	Q_r (pu)
−0.150	−0.153	1.023	0.869	1.10	−1.02
−0.100	−0.101	1.012	0.911	1.18	−0.82
−0.050	−0.050	1.000	0.950	1.24	−0.62
0.000	0.000	0.988	0.988	1.31	−0.43
0.050	0.049	0.975	1.024	1.37	−0.25
0.100	0.096	0.962	1.058	1.42	−0.08
0.150	0.142	0.948	1.090	1.47	0.09

a Shunt-Series configuration. The figure also shows the variation of active and reactive power flows (P_r and Q_r) at the receiving end. The results using a TRANSFORMER model in EMTP to represent a VRT in a Shunt-Series configuration are identical to what were shown in Figure 4-3, using the mathematical model of the ST, operated as a VRT in a Shunt-Series configuration (see Figure 3-10a), through a series-coupling transformer without any leakage impedance. The series-compensating voltage magnitude ($V_{s's}$) is a multiplier of the sending-end voltage magnitude (V_s).

4.1.2 Two-Winding Transformer

An ideal VRT model in EMTP in a Shunt-Shunt configuration is studied. The network model of Figure 3-5 is used with the short between BUS01 and BUS02 nodes, which is replaced with a

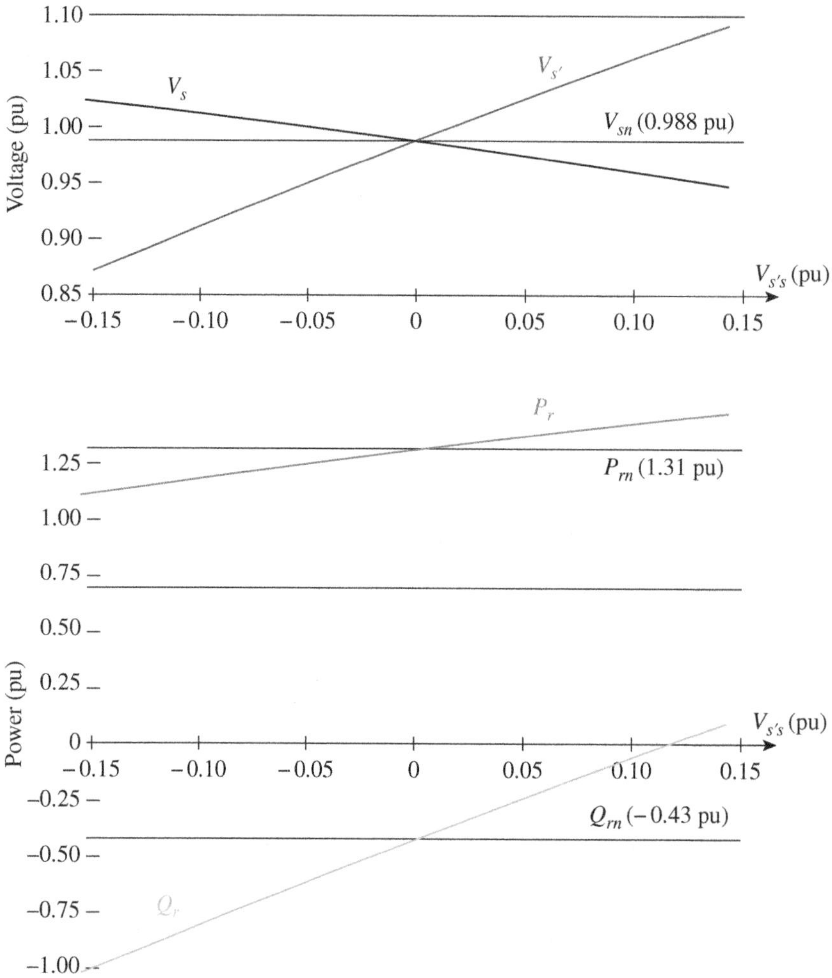

Figure 4-5 Ranges of voltage magnitudes (V_s and $V_{s'}$) at the sending and the modified sending ends and the active and reactive power flows (P_r and Q_r) at the receiving end with a TRANSFORMER model in EMTP to represent a VRT in a Shunt-Series configuration without any leakage impedance.

two-winding transformer as shown in Figure 4-6. The corresponding EMTP datafile **407IDLTT.DAT** is shown in Code 4-8. The datafile **407IDLTT.DAT** is the same as **405IDLAT.DAT**, except the VRT in a Shunt-Series configuration is replaced with a VRT in a Shunt-Shunt configuration.

The VRT consists of two parts: Exciter Unit and Voltage-Regulating Unit as shown in Figure 4-6a. The Exciter Unit consists of a three-phase (A, B, and C), Y-connected primary windings that are connected in shunt with the line. The three-phase primary windings are excited from the three-phase line voltages ($\mathbf{V_{sA}}$, $\mathbf{V_{sB}}$, and $\mathbf{V_{sC}}$). Three-phase compensating voltages ($\mathbf{V_{s'A}}$, $\mathbf{V_{s'B}}$, and $\mathbf{V_{s'C}}$) that are in phase with the corresponding phase-to-neutral primary voltages are generated from the induced voltages in the secondary windings. The Voltage-Regulating Unit consists of a total of three secondary windings (one winding in each phase). With the use of LTCs, the magnitude of the compensating voltage ($\mathbf{V_{s'}}$) is varied. The compensating voltage is connected in shunt with the line to increase or decrease the line voltage from an uncompensated voltage ($\mathbf{V_s}$) to a compensated voltage ($\mathbf{V_{s'}}$). Figure 4-6b shows the related phasor diagram.

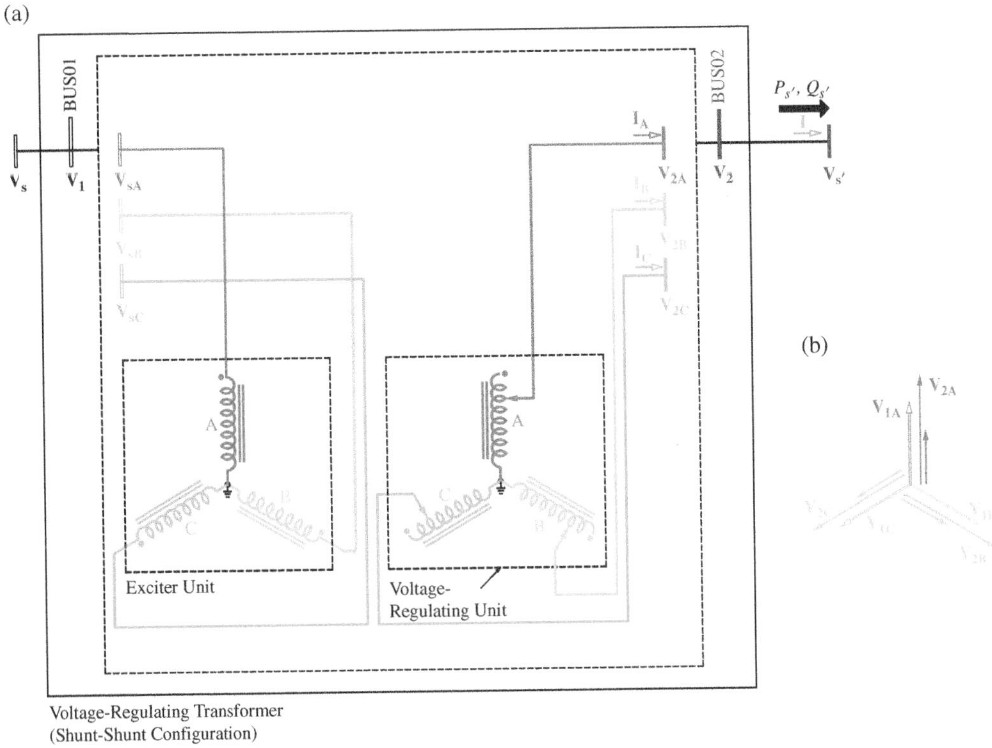

Figure 4-6 (a) Voltage-regulating transformer (two-winding transformer); (b) phasor diagram.

Code 4-8 EMTP datafile for an ideal VRT (Shunt-Shunt configuration), integrated in a two-generator/single-line power system network (407IDLTT.DAT).

```
BEGIN NEW DATA CASE
C File Name: 407IDLTT.DAT
C IDEAL 3-PHASE VRT (SHUNT-SHUNT CONFIGURATION) WITHOUT ANY LEAKAGE IMPEDANCE
C 0000000111111111122222222223333333333444444444455555555556666666666777777777778
C 3456789012345678901234567890123456789012345678901234567890123456789012345678901234567890
C ----dt<---tmax<---Xopt<---Copt<--Toler<        ><        >
8.3333-6150.00-3
C 0000000111111111122222222223333333333444444444455555555556666666666777777777778
C 3456789012345678901234567890123456789012345678901234567890123456789012345678901234567890
C -Iprnt<--Iplot<-Idoubl<-KssOut<-MaxOut<        ><        ><---Icat<-Nenerg
      30      10                                                       1
TACS HYBRID
99CRD      = 180 / PI
99CDR      = 1 / CRD
99TWOPI    = 2.0 * PI
```

(Continued)

Code 4-8 (Continued)

```
C BASE VOLTAGE, CURRENT, AND POWER
99Vbase    = 112676.528
99Ibase    = 946.662704
99Zbase    = Vbase / Ibase
99Pbase    = 1.5 * Vbase * Ibase
C INPUTS FROM NODE VOLTAGES AND LINE CURRENTS
$INCLUDE 302MEAS1.SWT
C NORMALIZATION OF MEASURED VARIABLES AND COMPUTATION OF PEAK VOLTAGES/CURRENTS
$INCLUDE 303PTCT.SCL
C IDEAL PHASE-LOCKED LOOP
$INCLUDE 304IPLL.PLL
C COMPUTATION OF LINE RESISTANCE
$INCLUDE 305LINER.CMP
C TACS OUTPUT
33v12pu v1pu  v2pu  Prpu  Qrpu
BLANK RECORD ENDING TACS
C 00000001111111111222222222233333333334444444444555555555566666666667777777777778
C 34567890123456789012345678901234567890123456789012345678901234567890123456789O
C <---Nodes--><---Nodes--><----R<----L<----C                                    V
C <-Bus1<-Bus2<-Bus3<-Bus4<-Ohms<---mH<---uF                                     V
C REPRESENTATION OF A SHORT BETWEEN NODES VS AND BUS01 NODES
  VSA   BUS01A             0.001
  VSB   BUS01B             0.001
  VSC   BUS01C             0.001
C CONNECTION OF NETWORK 2 BRANCHES WITH DAMPING RESISTORS (FIGURE 3-5)
$INCLUDE 311NTWK2.BRN
C CONNECTION OF IDEAL 3-PHASE 2-WINDING TRANSFORMER (WITH NO LEAKAGE IMPEDANCE)
$INCLUDE 408IDLTT.TRN
BLANK RECORD ENDING BRANCHES
C 00000001111111111222222222233333333334444444444555555555566666666667777777777778
C 34567890123456789012345678901234567890123456789012345678901234567890123456789O
C <-Bus1<-Bus2<---Tclose<----Topen<-------Ie<----Flash<--Request<-----Target<--O
C MEASUREMENT OF LINE CURRENTS
$INCLUDE 307MEAS2.SWT
BLANK RECORD ENDING SWITCHES
C 00000001111111111222222222233333333334444444444555555555566666666667777777777778
C 34567890123456789012345678901234567890123456789012345678901234567890123456789O
C CONNECTION OF NETWORK 1 OR 2 SOURCES (FIGURE 3-5)
$INCLUDE 308NTWK1.SRC
BLANK RECORD ENDING SOURCES
BLANK RECORD ENDING NODE VOLTAGE OUTPUT
BLANK RECORD ENDING PLOT
```

The datafile **407IDLTT.DAT** includes a datacard, which attaches the $INCLUDE file 408IDLTT.TRN that connects an ideal VRT (two-winding transformer without any leakage impedance) to the transmission line as shown in Code 4-9. For a 15% of voltage compensation, the turns-ratio of the secondary windings is set at 1.150. For a −15% of voltage compensation, the turns-ratio of the secondary windings is set at 0.850 as shown in Code 4-10.

Code 4-9 EMTP $INCLUDE file for an ideal three-phase VRT in a Shunt-Shunt configuration for increasing 15% line voltage (408IDLTT.TRN).

```
C File Name: 408IDLTT.TRN
C IDEAL 3-PHASE VRT IN A SHUNT-SHUNT CONFIGURATION WITHOUT ANY LEAKAGE IMPEDANCE
C FOR INCREASING 15% LINE VOLTAGE
C 000000011111111112222222222333333333344444444445555555555666666666677777777778
C 345678901234567890123456789012345678901234567890123456789012345678901234567890
C ------TWO-WINDING TRANSFORMER FOR PHASE A------
C TRANSFORMER <--Refxxxxxx<--Iss<--Phi<-Name<-Rmag<---------------------------O
  TRANSFORMER                             TWDGTA
C NEXT LINE (9999 IN COLUMNS 13-16) INDICATES A LINEAR TRANSFORMER
            9999
C <---Nodes--><----------><---Rk<---Lk<Volt|---------------------------------->O
C <-Bus1<-Bus2<----------><-Ohms<---mH<---kV---------------------------------->O
 1BUS01A                    0.001 0.001 1.000
 2BUS02A                    0.001 0.001 1.150
C ------TWO-WINDING TRANSFORMER FOR PHASE B------
  TRANSFORMER                             TWDGTB
            9999
 1BUS01B                    0.001 0.001 1.000
 2BUS02B                    0.001 0.001 1.150
C ------TWO-WINDING TRANSFORMER FOR PHASE C------
  TRANSFORMER                             TWDGTC
            9999
 1BUS01C                    0.001 0.001 1.000
 2BUS02C                    0.001 0.001 1.150
```

Code 4-10 EMTP $INCLUDE file for an ideal three-phase VRT in a Shunt-Shunt configuration for decreasing 15% line voltage (408IDLTT.TRN).

```
C File Name: 408IDLTT.TRN
C IDEAL 3-PHASE VRT IN A SHUNT-SHUNT CONFIGURATION WITHOUT ANY LEAKAGE IMPEDANCE
C FOR DECREASING 15% LINE VOLTAGE
C 000000011111111112222222222333333333344444444445555555555666666666677777777778
C 345678901234567890123456789012345678901234567890123456789012345678901234567890
```

(Continued)

Code 4-10 (Continued)

```
C ------TWO-WINDING TRANSFORMER FOR PHASE A------
C TRANSFORMER <--Refxxxxxx<--Iss<--Phi<-Name<-Rmag<----------------------------O
  TRANSFORMER                                       TWDGTA
C NEXT LINE (9999 IN COLUMNS 13-16) INDICATES A LINEAR TRANSFORMER
            9999
C <---Nodes--><----------><---Rk<---Lk<Volt|---------------------------------->O
C <-Bus1<-Bus2<----------><-Ohms<---mH<---kV---------------------------------->O
 1BUS01A                          0.001 0.001 1.000
 2BUS02A                          0.001 0.001 0.850
C ------TWO-WINDING TRANSFORMER FOR PHASE B------
  TRANSFORMER                                       TWDGTB
            9999
 1BUS01B                          0.001 0.001 1.000
 2BUS02B                          0.001 0.001 0.850
C ------TWO-WINDING TRANSFORMER FOR PHASE C------
  TRANSFORMER                                       TWDGTC
            9999
 1BUS01C                          0.001 0.001 1.000
 2BUS02C                          0.001 0.001 0.850
```

Table 4-7 Electrical system data.

Parameters	Values
Source voltage ($\mathbf{V_{src}}$) phase-to-neutral	$112{,}676.528\angle 0°\ V_{peak}$
Source resistance (R_{src})	1.0053 Ω
Source reactance (X_{src})	7.43875 Ω (19.73 mH)
Receiving-end voltage ($\mathbf{V_r}$) phase-to-neutral	$112{,}676.528\angle -20°\ V_{peak}$
Line resistance (R)	3.0159 Ω
Line reactance (X)	22.31625 Ω (59.19 mH)
Two-winding transformer (TWDGT) configuration	**1:(0.85–1.15)**
Primary winding resistance	0.001 Ω
Primary winding leakage inductance	0.001 mH
Secondary winding resistance	0.001 Ω
Secondary winding leakage inductance	0.001 mH

Table 4-7 shows the electrical system data for a two-generator/single-line power system network, compensated by an ideal VRT in a Shunt-Shunt configuration.

Table 4-8 shows the variation of the shunt-compensating voltage magnitude ($V_{s'}$) and its effects on the sending-end voltage magnitude (V_s) and the resulting active and reactive power flows (P_r and Q_r) at the receiving end.

Table 4-8 Effects of the variation of shunt-compensating voltage ($V_{s'}$) on simulated voltage (V_{21} or $V_{s's}$), the sending-end and the modified sending-end voltages (V_1 or V_s and V_2 or $V_{s'}$), and the power flows (P_r and Q_r) at the receiving end.

$V_{s'}/V_s$	V_{21} or $V_{s's}$ (pu)	V_1 or V_s (pu)	V_2 or $V_{s'}$ (pu)	P_r (pu)	Q_r (pu)
0.850	−0.153	1.023	0.869	1.10	−1.02
0.900	−0.101	1.012	0.911	1.18	−0.82
0.950	−0.050	1.000	0.950	1.24	−0.62
1.000	0.000	0.988	0.988	1.31	−0.43
1.050	0.049	0.975	1.024	1.37	−0.25
1.100	0.096	0.962	1.058	1.42	−0.08
1.150	0.142	0.948	1.090	1.47	0.09

Figure 4-7 shows the variations of the sending-end voltage magnitude (V_s) and the modified sending-end voltage magnitude ($V_{s'}$) as a function of the simulated voltage (V_{21} or $V_{s's}$) when the shunt-compensating voltage magnitude ($V_{s'}$) varies in the range of 85 to 115% of the

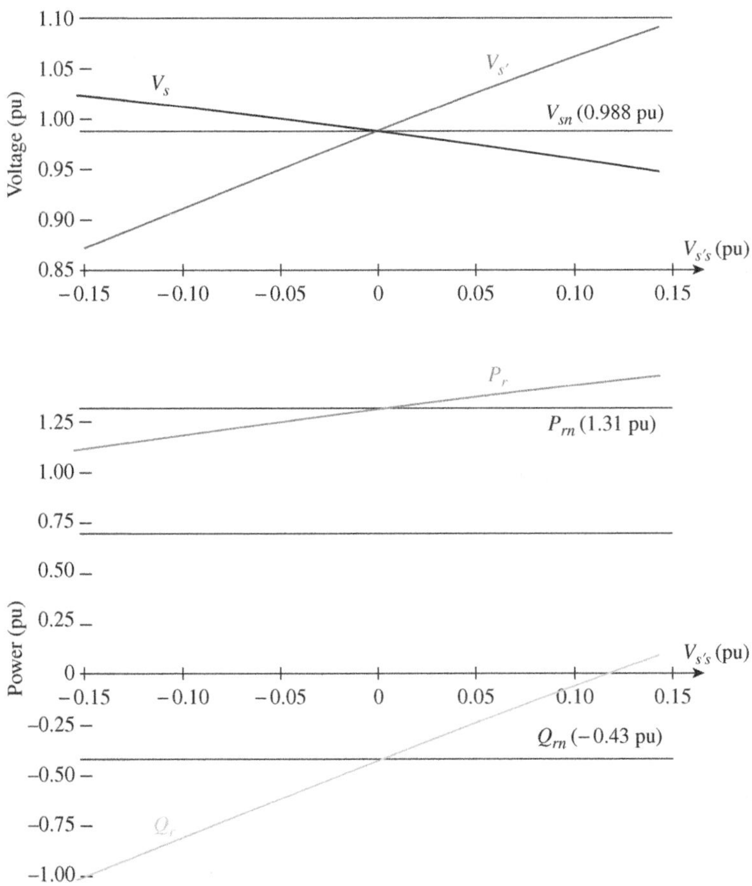

Figure 4-7 Ranges of voltage magnitudes (V_s and $V_{s'}$) at the sending and the modified sending ends and the active and reactive power flows (P_r and Q_r) at the receiving end with a TRANSFORMER model in EMTP to represent a VRT in a Shunt-Shunt configuration without any leakage impedance.

sending-end voltage magnitude (V_s) using a TRANSFORMER model in EMTP to represent a VRT in a Shunt-Shunt configuration. This is implemented by varying the turns-ratio (n_2/n_1) from 0.85 to 1.15 while keeping n_1 constant at 1.00. The figure also shows the variation of active and reactive power flows (P_r and Q_r) at the receiving end. The results are identical to what were shown using a VRT in Shunt-Series configurations [both series-compensating voltage, generated by the mathematical model of the ST (Figure 4-3) and TRANSFORMER (Figure 4-5) model in EMTP]. The results show that the VRT changes the reactive power flow at the receiving end of the line much more than it changes the active power flow, which is consistent with the Equations (2-140a) and (2-140c), developed in Chapter 2.

4.2 Phase Angle Regulator (PAR)

A PAR connects a compensating voltage in series with the transmission line. The series-compensating voltage is in quadrature with the phase-to-neutral voltage of the line. There are two types of PARs: (1) asymmetric and (2) symmetric. Both types are in a Shunt-Series configuration. While the shunt-exciting windings are connected to the line voltage in an asymmetric type as shown in Figure 2-56, the shunt-exciting windings are connected at the midpoints of the series-compensating windings in a symmetric type as shown in Figure 2-57.

4.2.1 PAR (Asymmetric)

The PAR (asym) can be studied with the use of the mathematical model of the ST as shown in Figure 3-10a. The network model of Figure 3-5 is used with the short between BUS01 and BUS02 nodes, which is replaced with the mathematical model of the ST. For a PAR (asym) with a given turns-ratio, the series-compensating voltage is a multiplier of the sending-end voltage at BUS01 node, because the sending-end voltage at BUS01 node varies with the varying line current. The corresponding EMTP datafile **403PQPFC.DAT**, shown in Code 4-3, is rerun with the relative phase angle of the series-compensating voltage set to $\beta = 90°$ while its magnitude ($V_{s's}$) is varied from −15 to 15% of the sending-end voltage magnitude (V_s). The datafile **403PQPFC.DAT** includes a datacard, which attaches the $INCLUDE file 404PQPFC.USR that defines the user's input as shown in Code 4-11. It is shown that the series-compensating voltage magnitude ($V_{s's}$) is set at 15% of the sending-end voltage magnitude (V_s).

Code 4-11 EMTP $INCLUDE file for user's input (404PQPFC.USR).

```
C File Name: 404PQPFC.USR
C USER'S INPUT FILE FOR PQPFC
C 00000001111111111222222222233333333334444444444555555555566666666667777777778
C 34567890123456789012345678901234567890123456789012345678901234567890123456789
C A SHUNT BREAKER IS CLOSED.
99SHNBRK   = TIMEX .GE. 0.01
C SET ST = 1.0 FOR ST OR ST = 0.0 FOR UPFC
```

(Continued)

Code 4-11 (Continued)

```
99ST        = 1.0
C SETTING THE COMPENSATING VOLTAGE
99e2cm1     = 0.15 * (TIMEX .GE. 0.020)
99e2dqcm    - e2cm1 * v1pu
C LOW-PASS FILTER INTRODUCES A RAMP-RATE AND REDUCES SYSTEM TRANSIENTS
 1e2dqrf   +e2dqcm
        1.0
        1.0     0.0160
99betcm1    = 90 * (TIMEX .GE. 0.020)
99betcmd    = betcm1 * CDR
C LOW-PASS FILTER INTRODUCES A RAMP-RATE AND REDUCES SYSTEM TRANSIENTS
 1beta      +betcmd
        1.0
        1.0     0.0020
```

Table 4-9 Electrical system data.

Parameters	Values
Source voltage ($\mathbf{V_{src}}$) phase-to-neutral	112,676.528∠0° V_{peak}
Source resistance (R_{src})	1.0053 Ω
Source reactance (X_{src})	7.43875 Ω (19.73 mH)
Receiving-end voltage ($\mathbf{V_r}$) phase-to-neutral	112,676.528∠−20° V_{peak}
Line resistance (R)	3.0159 Ω
Line reactance (X)	22.31625 Ω (59.19 mH)
Series-coupling transformer (SRSTR) turns-ratio	1:1
Primary winding resistance	0.001 Ω
Primary winding leakage inductance	0.001 mH
Secondary winding resistance	0.001 Ω
Secondary winding leakage inductance	0.001 mH

Table 4-9 shows the electrical system data for a two-generator/single-line power system network, compensated by the mathematical model of the ST, operating as an ideal PAR (asym).

Table 4-10 shows the variation of the normalized series-compensating voltage magnitude ($V_{s's}/V_s$) and its effects on simulated voltage (V_{21} or $V_{s's}$), the sending-end voltage magnitude (V_1 or V_s), the modified sending-end voltage magnitude (V_2 or $V_{s'}$), and the resulting active and reactive power flows (P_r and Q_r) at the receiving end.

Note the high resistances across the current sources need to be set at some lower value (for example, 3.4 kΩ) to avoid numerical oscillation while simulating the operating point of series-compensating voltage of 15% of the sending-end voltage magnitude (V_s).

Table 4-10 Effects of the variation of series-compensating voltage ($V_{s's}$) on simulated voltage (V_{21} or $V_{s's}$), the sending-end and the modified sending-end voltages (V_1 or V_s and V_2 or $V_{s'}$), and the power flows (P_r and Q_r) at the receiving end.

$V_{s's}/V_s$	V_{21} or $V_{s's}$ (pu)	V_1 or V_s (pu)	V_2 or $V_{s'}$ (pu)	P_r (pu)	Q_r (pu)
−0.150	−0.149	0.993	1.004	0.78	−0.15
−0.100	−0.099	0.993	0.997	0.96	−0.24
−0.050	−0.050	0.991	0.992	1.13	−0.33
0.000	0.000	0.988	0.988	1.31	−0.43
0.050	0.049	0.984	0.985	1.48	−0.52
0.100	0.098	0.979	0.984	1.65	−0.61
0.150	0.146	0.972	0.983	1.81	−0.70

Figure 4-8 shows that the variation of the series-compensating voltage magnitude ($V_{s's}$) in the range of −15 to 15% of the sending-end voltage magnitude (V_s) results in a variation of the modified sending-end voltage magnitude ($V_{s'}$) in the range of 1.004 to 0.983 pu. In the process, the sending-end voltage magnitude (V_s) varies in the range of 0.993 to 0.972 pu. The reactive power flow (Q_r) at

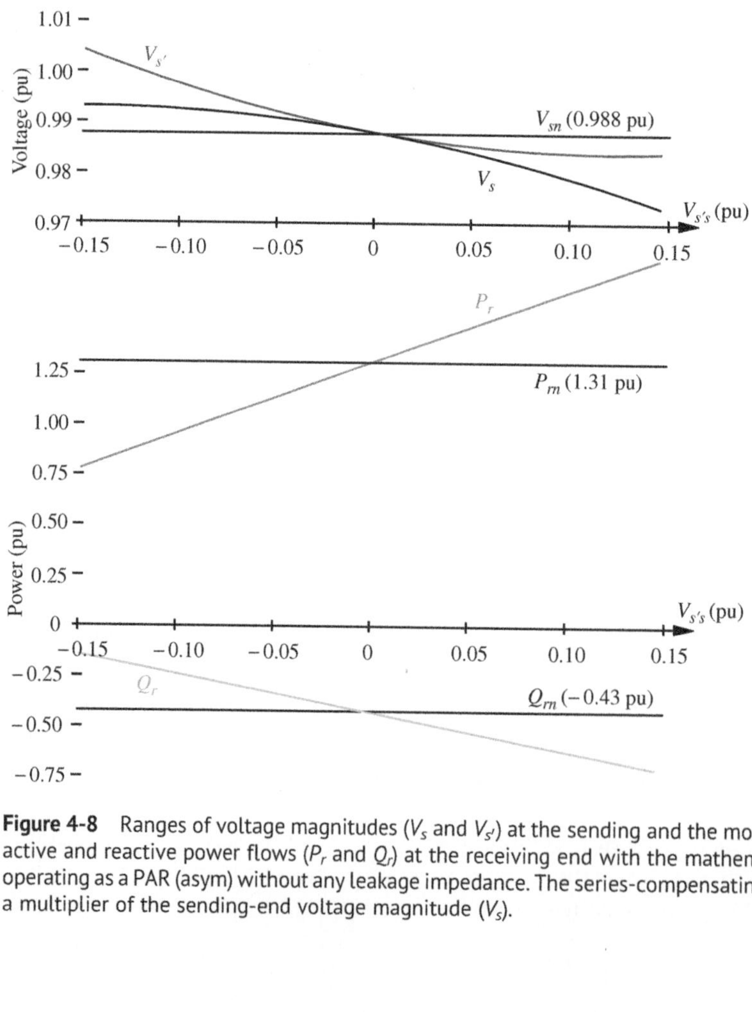

Figure 4-8 Ranges of voltage magnitudes (V_s and $V_{s'}$) at the sending and the modified sending ends and the active and reactive power flows (P_r and Q_r) at the receiving end with the mathematical model of the ST, operating as a PAR (asym) without any leakage impedance. The series-compensating voltage magnitude ($V_{s's}$) is a multiplier of the sending-end voltage magnitude (V_s).

the receiving end varies from −0.15 to −0.7 pu during the range of compensating voltage magnitude ($V_{s's}$) from −15 to 15% of the sending-end voltage magnitude (V_s). However, the active power flow (P_r) at the receiving end varies in the range of 0.78 to 1.81 pu for the same range of variation in the compensating voltage magnitude ($V_{s's}$). Therefore, a range of −15 to 15% in compensating voltage magnitude ($V_{s's}$), in quadrature with phase-to-neutral voltage, varies the reactive power flow (Q_r) by −0.55 pu and the active power flow (P_r) by 1.03 pu.

Figure 4-9 shows the PAR (asym) configuration that is bypassed and connected to the network model of Figure 3-5 with the short between BUS01 and BUS02 nodes, which is replaced with a TRANSFORMER model in EMTP to represent a PAR (asym). The three-phase voltages ($\mathbf{V_{1A}}$, $\mathbf{V_{1B}}$, and $\mathbf{V_{1C}}$) at BUS01 node are applied to Δ-connected primary windings with double-pole, double-through breakers that can be used to reverse the applied polarity. As shown in Figure 4-9a, the LTCs are connected to the left end of the secondary windings; therefore, at this neutral tap position, no compensating voltage is applied. The sending-end voltage ($\mathbf{V_s}$ or $\mathbf{V_1}$) is the same as the modified sending-end voltage ($\mathbf{V_{s'}}$ or $\mathbf{V_2}$) as shown in the phasor diagram (Figure 4-9b).

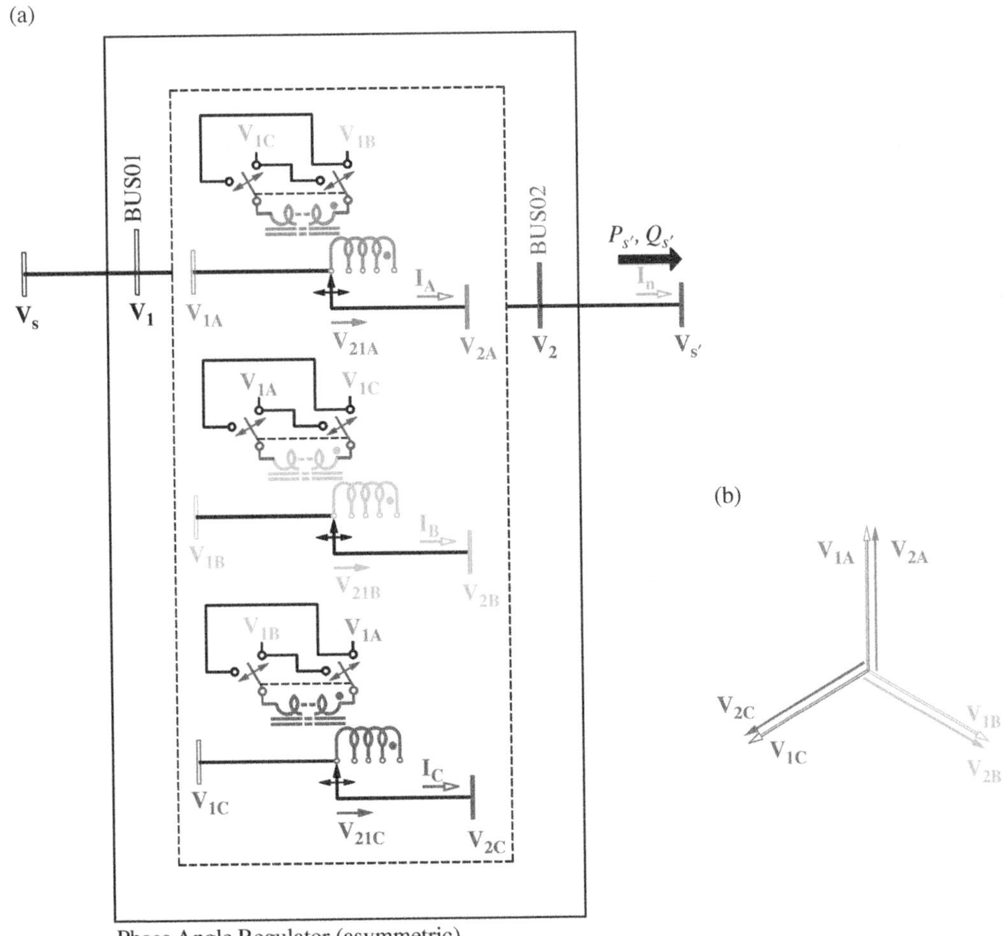

Figure 4-9 (a) PAR (asym) configuration in a bypass-mode of operation; (b) phasor diagram.

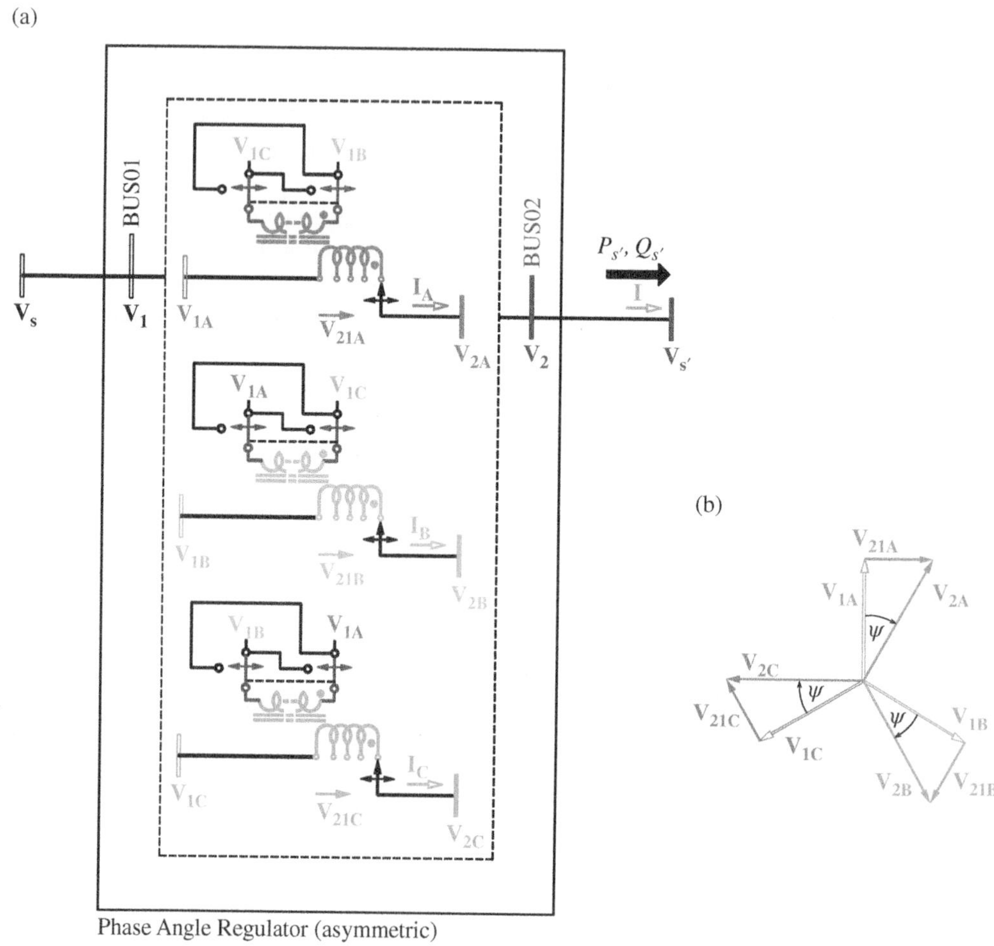

Figure 4-10 (a) PAR (asym) configuration for decreasing power flow; (b) phasor diagram.

Figure 4-10a shows the PAR (asym) configuration for decreasing power flow. The secondary voltage (**V**$_{21}$) that is in phase with the primary phase-to-phase voltage, but in quadrature with the primary phase-to-neutral voltage, is connected in series with the transmission line. The polarity of the series-connected windings is set such that the compensating secondary voltage is connected in series with the transmission line at −90° to the primary phase-to-neutral voltage. The resulting effect is that the phase angle of the line voltage between BUS01 and BUS02 nodes is decreased as shown in the phasor diagram Figure 4-10b. A decrease in phase angle of the line voltage results in a decrease in power flow in the line.

An ideal PAR (asym) is implemented in the datafile **409PAR1.DAT**, using a TRANSFORMER model in EMTP as shown in Code 4-12, which is the same as **405IDLAT.DAT**, except the VRT in a Shunt-Series configuration model is replaced with a PAR (asym) model. The datafile **409PAR1.DAT** includes a datacard, which attaches the $INCLUDE file 410PAR1D.TRN that connects an ideal PAR (asym) without any leakage impedance to the transmission line as shown in Code 4-13. Note that in order to keep the same volts per turn in both Y and Δ windings, the Δ-connected primary windings have $\sqrt{3}$ times the turns of the Y-connected primary windings.

Code 4-12 EMTP datafile for a PAR (asym), integrated in a two-generator/single-line power system network (409PAR1.DAT).

```
BEGIN NEW DATA CASE
C File Name: 409PAR1.DAT
C IDEAL PHASE ANGLE REGULATOR (ASYM) WITHOUT ANY LEAKAGE IMPEDANCE
C 0000000111111111122222222223333333333444444444455555555556666666666777777777778
C 3456789012345678901234567890123456789012345678901234567890123456789012345678901234567890
C ----dt<---tmax<---Xopt<---Copt<--Toler<        ><        >
8.3333-6150.00-3
C 0000000111111111122222222223333333333444444444455555555556666666666777777777778
C 3456789012345678901234567890123456789012345678901234567890123456789012345678901234567890
C -Iprnt<--Iplot<-Idoubl<-KssOut<-MaxOut<        ><        ><---Icat<-Nenerg
      30      10                                                  1
TACS HYBRID
99CRD      = 180 / PI
99CDR      = 1 / CRD
99TWOPI    = 2.0 * PI
C BASE VOLTAGE, CURRENT, AND POWER
99Vbase    = 112676.528
99Ibase    = 946.662704
99Zbase    = Vbase / Ibase
99Pbase    = 1.5 * Vbase * Ibase
C INPUTS FROM NODE VOLTAGES AND LINE CURRENTS
$INCLUDE 302MEAS1.SWT
C NORMALIZATION OF MEASURED VARIABLES AND COMPUTATION OF PEAK VOLTAGES/CURRENTS
$INCLUDE 303PTCT.SCL
C IDEAL PHASE-LOCKED LOOP
$INCLUDE 304IPLL.PLL
C TACS OUTPUT
33v12pu v1pu  v2pu  Prpu  Qrpu
C COMPUTATION OF LINE RESISTANCE
$INCLUDE 305LINER.CMP
BLANK RECORD ENDING TACS
C 0000000111111111122222222223333333333444444444455555555556666666666777777777778
C 3456789012345678901234567890123456789012345678901234567890123456789012345678901234567890
C <---Nodes--><---Nodes--><----R<----L<----C                                   V
C <-Bus1<-Bus2<-Bus3<-Bus4<-Ohms<---mH<---uF                                   V
C REPRESENTATION OF A SHORT BETWEEN NODES VS AND BUS01
  VSA   BUS01A              0.001
  VSB   BUS01B              0.001
  VSC   BUS01C              0.001
C CONNECTION OF NETWORK 2 BRANCHES WITH DAMPING RESISTORS (FIGURE 3-5)
$INCLUDE 311NTWK2.BRN
```

(*Continued*)

Code 4-12 (Continued)

```
C CONNECTION OF AN IDEAL (NO LEAKAGE IMPEDANCE) SINGLE-CORE PAR (ASYM)
$INCLUDE 410PAR1D.TRN
BLANK RECORD ENDING BRANCHES
C 00000001111111111222222222233333333334444444444555555555566666666667777777778
C 34567890123456789012345678901234567890123456789012345678901234567890123456789
C <-Bus1<-Bus2<---Tclose<----Topen<-------Ie<----Flash<--Request<-----Target<--O
C MEASUREMENT OF LINE CURRENTS
$INCLUDE 307MEAS2.SWT
BLANK RECORD ENDING SWITCHES
C 00000001111111111222222222233333333334444444444555555555566666666667777777778
C 34567890123456789012345678901234567890123456789012345678901234567890123456789
C CONNECTION OF NETWORK 1 OR 2 SOURCES (FIGURE 3-5)
$INCLUDE 308NTWK1.SRC
BLANK RECORD ENDING SOURCES
BLANK RECORD ENDING NODE VOLTAGE OUTPUT
BLANK RECORD ENDING PLOT
```

Code 4-13 EMTP $INCLUDE file for an ideal single-core PAR (asym) for decreasing power flow (410PAR1D.TRN).

```
C File Name: 410PAR1D.TRN
C IDEAL SINGLE-CORE PHASE ANGLE REGULATOR (ASYM) WITHOUT ANY LEAKAGE IMPEDANCE
C FOR DECREASING PHASE-SHIFT ANGLE
C 00000001111111111222222222233333333334444444444555555555566666666667777777778
C 34567890123456789012345678901234567890123456789012345678901234567890123456789
C ------PHASE ANGLE REGULATOR (ASYM) FOR PHASE A------
C TRANSFORMER <--Refxxxxxx<--Iss<--Phi<-Name<-Rmag<--------------------------O
  TRANSFORMER                       PARA
C NEXT LINE (9999 IN COLUMNS 13-16) INDICATES A LINEAR TRANSFORMER
            9999
C <---Nodes--><----------><---Rk<---Lk<Volt|--------------------------------->O
C <-Bus1<-Bus2<----------><-Ohms<---mH<---kV--------------------------------->O
 1BUS01BBUS01C            0.001 0.001 1.732
 2BUS02ABUS01A            0.001 0.001 0.150
C ------PHASE ANGLE REGULATOR (ASYM) FOR PHASE B------
  TRANSFORMER                       PARB
            9999
 1BUS01CBUS01A            0.001 0.001 1.732
 2BUS02BBUS01B            0.001 0.001 0.150
```

(Continued)

Code 4-13 (Continued)

```
C ------PHASE ANGLE REGULATOR (ASYM) FOR PHASE C------
 TRANSFORMER                                    PARC
           9999
1BUS01ABUS01B                 0.001 0.001 1.732
2BUS02CBUS01C                 0.001 0.001 0.150
```

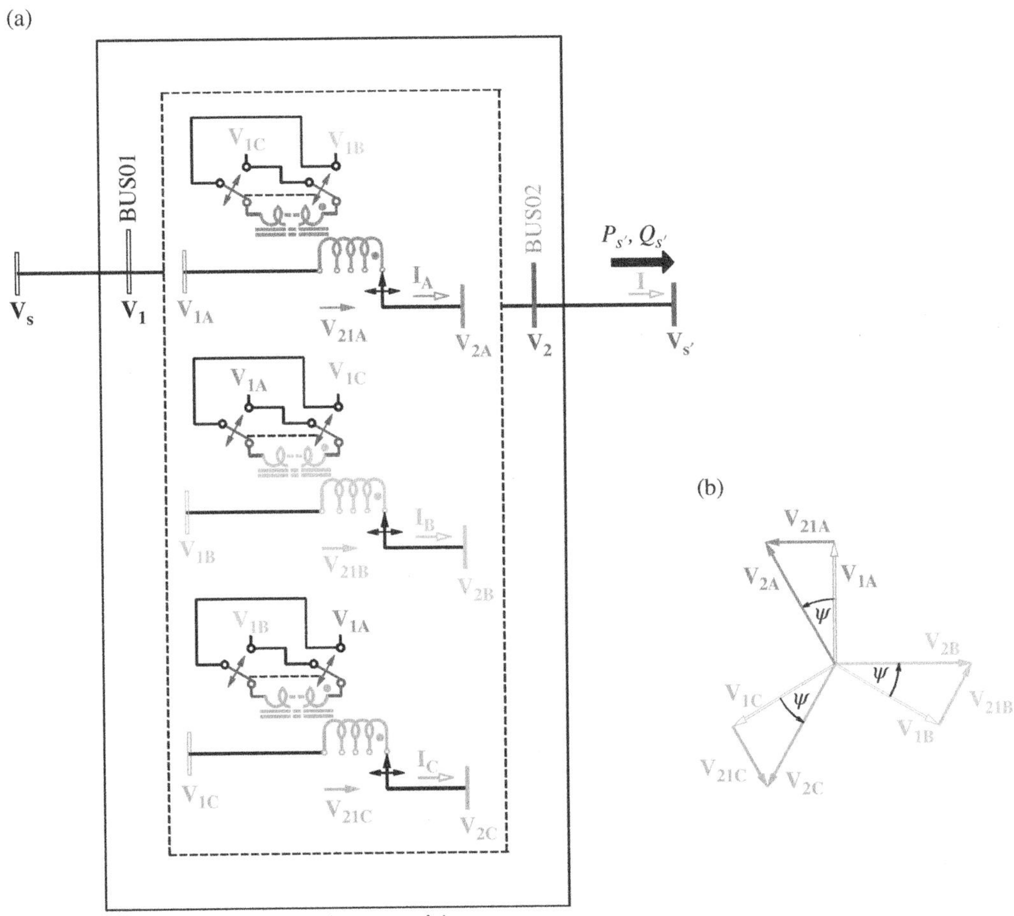

Figure 4-11 (a) PAR (asym) configuration for increasing power flow; (b) phasor diagram.

Figure 4-11a shows the PAR (asym) configuration for increasing power flow. The corresponding $INCLUDE file 410PAR1I.TRN is shown in Code 4-14. The polarity of the series-connected windings is set such that the compensating secondary voltage is connected in series with the transmission line at 90° to the primary phase-to-neutral voltage. The resulting effect is that only the phase angle of the line voltage between BUS01 and BUS02 nodes is increased as shown in the phasor diagram Figure 4-11b. An increase in phase angle of the line voltage results in an increase in power flow in the line. Note that the modified sending-end voltage (**V**$_2$) is always greater than the sending-end voltage (**V**$_1$), regardless of whether the phase-shift angle (ψ) is positive or negative, due to the quadrature addition of the compensating voltage (**V**$_{21}$) and the sending-end voltage (**V**$_1$).

Code 4-14 EMTP $INCLUDE file for an ideal single-core PAR (asym) for increasing power flow (410PAR1I.TRN).

```
C File Name: 410PAR1I.TRN
C IDEAL SINGLE-CORE PHASE ANGLE REGULATOR (ASYM) WITHOUT ANY LEAKAGE IMPEDANCE
C FOR INCREASING PHASE-SHIFT ANGLE
C 00000001111111111222222222233333333334444444444555555555566666666667777777778
C 34567890123456789012345678901234567890123456789012345678901234567890123456789
C ------PHASE ANGLE REGULATOR (ASYM) FOR PHASE A------
C TRANSFORMER <--Refxxxxxx<--Iss<--Phi<-Name<-Rmag<----------------------------O
  TRANSFORMER                                   PARA
C NEXT LINE (9999 IN COLUMNS 13-16) INDICATES A LINEAR TRANSFORMER
            9999
C <---Nodes--><---------><---Rk<---Lk<Volt|---------------------------------->O
C <-Bus1<-Bus2<---------><-Ohms<---mH<---kV---------------------------------->O
 1BUS01CBUS01B                0.001 0.001 1.732
 2BUS02ABUS01A                0.001 0.001 0.150
C ------PHASE ANGLE REGULATOR (ASYM) FOR PHASE B------
  TRANSFORMER                                   PARB
            9999
 1BUS01ABUS01C                0.001 0.001 1.732
 2BUS02BBUS01B                0.001 0.001 0.150
C ------PHASE ANGLE REGULATOR (ASYM) FOR PHASE C------
  TRANSFORMER                                   PARC
            9999
 1BUS01BBUS01A                0.001 0.001 1.732
 2BUS02CBUS01C                0.001 0.001 0.150
```

Table 4-11 shows the electrical system data for a two-generator/single-line power system network, compensated by an ideal PAR (asym).

Table 4-12 shows the variation of the normalized series-compensating voltage magnitude ($V_{s's}/V_s$) and its effects on simulated voltage (V_{21} or $V_{s's}$), the sending-end voltage magnitude (V_1 or V_s), the modified sending-end voltage magnitude (V_2 or $V_{s'}$), and the resulting active and reactive power flows (P_r and Q_r) at the receiving end.

The results using a TRANSFORMER model in EMTP to represent a PAR (asym) are shown in Figure 4-12. The results are identical to what were shown in Figure 4-8 where the mathematical model of the ST (see Figure 3-10a) generated a series-compensating voltage through a

Table 4-11 Electrical system data.

Parameters	Values
Source voltage ($\mathbf{V}_{\mathbf{src}}$) phase-to-neutral	112,676.528∠0° V_{peak}
Source resistance (R_{src})	1.0053 Ω
Source reactance (X_{src})	7.43875 Ω (19.73 mH)
Receiving-end voltage ($\mathbf{V}_{\mathbf{r}}$) phase-to-neutral	112,676.528∠−20° V_{peak}
Line resistance (R)	3.0159 Ω
Line reactance (X)	22.31625 Ω (59.19 mH)
Phase Angle Regulator, PAR (asym) turns-ratio	**1.732:(−0.15 to 0.15)**
Primary winding resistance	0.001 Ω
Primary winding leakage inductance	0.001 mH
Secondary winding resistance	0.001 Ω
Secondary winding leakage inductance	0.001 mH

Table 4-12 Effects of the variation of series-compensating voltage ($V_{s's}$) on simulated voltage (V_{21} or $V_{s's}$), the sending-end and the modified sending-end voltages (V_1 or V_s and V_2 or $V_{s'}$), and the power flows (P_r and Q_r) at the receiving end.

$V_{s's}/V_s$	V_{21} or $V_{s's}$ (pu)	V_1 or V_s (pu)	V_2 or $V_{s'}$ (pu)	P_r (pu)	Q_r (pu)
−0.150	−0.149	0.993	1.004	0.78	−0.15
−0.100	−0.099	0.992	0.997	0.96	−0.24
−0.050	−0.050	0.991	0.992	1.13	−0.33
0.000	0.000	0.988	0.988	1.31	−0.43
0.050	0.049	0.984	0.985	1.48	−0.52
0.100	0.098	0.979	0.984	1.65	−0.61
0.150	0.146	0.973	0.984	1.82	−0.70

series-coupling transformer without any leakage impedance to simulate a PAR (asym). The results show that the PAR (asym) changes the active power flow at the receiving end of the line much more than it changes the reactive power flow, which is consistent with the Equations (2-142a) and (2-142c), developed in Chapter 2.

Note that in addition to changing the phase angle of the line voltage, the PAR (asym) also increases the magnitude of the line voltage. There are two contributing factors to this effect. First, the operation of the PAR (sym) or PAR (asym) changes the power flow of the line. This loading effect coupled with the Thèvenin impedance of the network changes the voltage at the POC. Second, when the compensating voltage is added or subtracted in quadrature with the phase-to-neutral voltage, the resultant voltage is always larger than the original voltage. In order to obtain the same voltage magnitude at the sending and the modified sending ends, the PAR (sym) may be implemented with the following configuration.

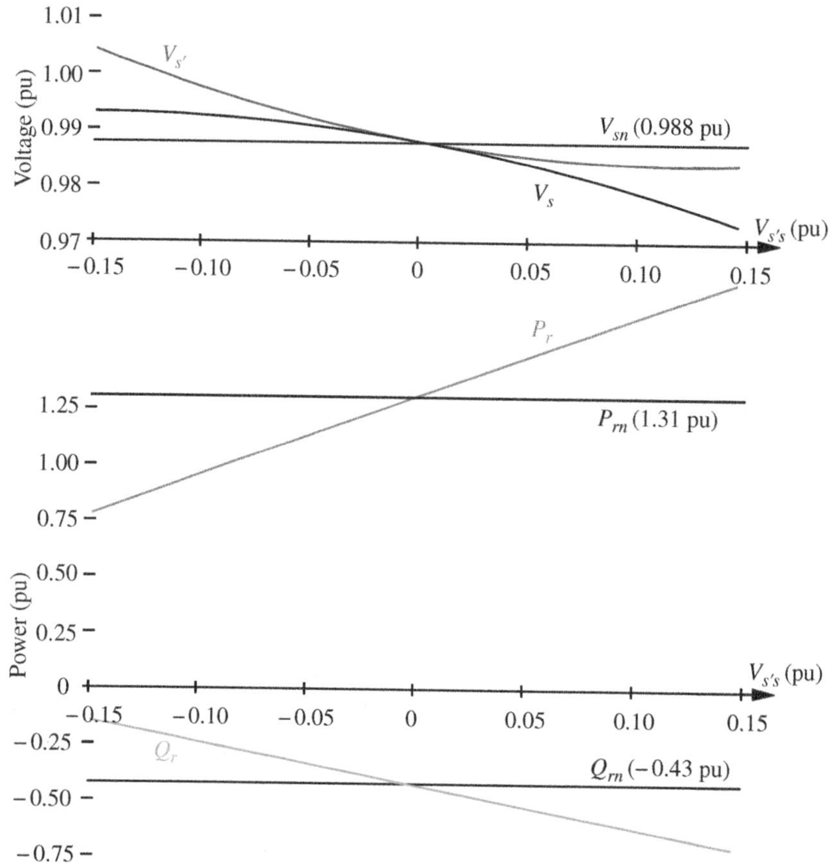

Figure 4-12 Ranges of voltage magnitudes (V_s and $V_{s'}$) at the sending and the modified sending ends and the active and reactive power flows (P_r and Q_r) at the receiving end with a TRANSFORMER model in EMTP to represent a PAR (asym) without any leakage impedance.

4.2.2 PAR (Symmetric)

Figure 4-13 shows the PAR (sym) configuration that is bypassed and connected to the network model of Figure 3-5 with a short between BUS01 and BUS02 nodes, which is replaced with a TRANSFORMER model in EMTP to represent a PAR (sym). The three-phase compensating secondary windings are center-tapped at A, B, and C, respectively. The three-phase voltages ($\mathbf{V_A}$, $\mathbf{V_B}$, and $\mathbf{V_C}$) are applied to Δ-connected primary windings with double-pole, double-throw breakers that can be used to reverse the applied polarity. The secondary voltage ($\mathbf{V_{21}}$) that is in phase with the primary phase-to-phase voltage, but in quadrature with the primary phase-to-neutral voltage, is connected in series with the transmission line. With proper polarities of the series-connected windings, the compensating secondary voltage is connected in series with the transmission line at 90° or −90° to the primary phase-to-neutral voltage at the midpoints of the secondary windings. As shown in Figure 4-13a, the LTCs are connected to the center taps; therefore, no compensating voltage is applied. The sending-end voltage ($\mathbf{V_s}$ or $\mathbf{V_1}$) is the same as the modified sending-end voltage ($\mathbf{V_{s'}}$ or $\mathbf{V_2}$) as shown in the phasor diagram Figure 4-13b.

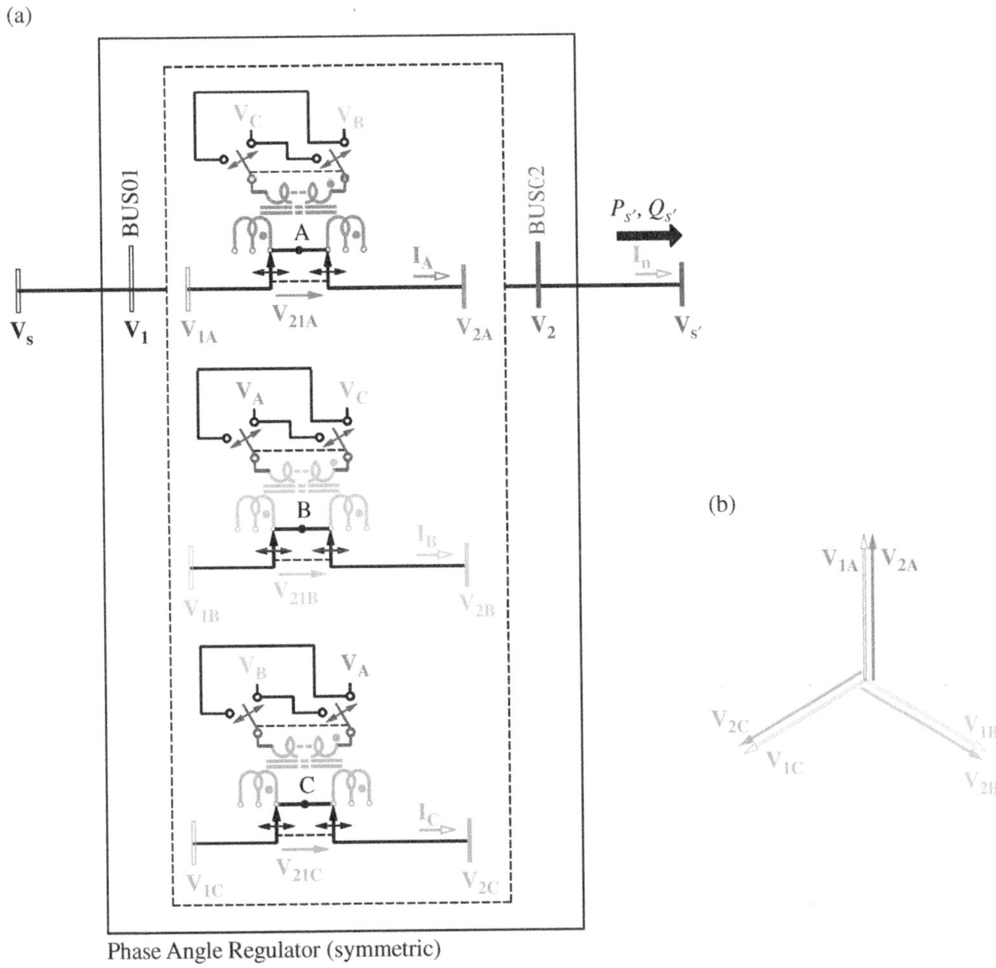

Figure 4-13 (a) PAR (sym) configuration in a bypass-mode of operation; (b) phasor diagram.

Figure 4-14a shows the PAR (sym) configuration for decreasing power flow. The secondary voltage ($\mathbf{V}_{21}$) that is in phase with the primary phase-to-phase voltage, but in quadrature with the primary phase-to-neutral voltage at the midpoints of the winding, is connected in series with the transmission line. The polarity of the series-connected windings is set such that the compensating secondary voltage is connected in series with the transmission line at −90° to the primary phase-to-neutral voltage at the midpoints of the secondary windings. The resulting effect is that only the phase angle of the line voltage between BUS01 and BUS02 nodes is decreased as shown in the phasor diagram Figure 4-14b. A decrease in phase angle of the line voltage results in a decrease in power flow in the line. Note that the magnitude of the voltage stays unchanged as shown in the phasor diagram Figure 4-14b.

An ideal PAR (sym) is implemented in the datafile **411PAR2.DAT**, using a TRANSFORMER model in EMTP as shown in Code 4-15, which is the same as **404IDLAT.DAT**, except the VRT in a Shunt-Series configuration is replaced with a PAR (sym) model. The datafile **411PAR2.DAT** includes a datacard, which attaches the $INCLUDE file 412PAR2D.TRN that connects an ideal PAR (sym) without any leakage impedance to the transmission line as shown in Code 4-16.

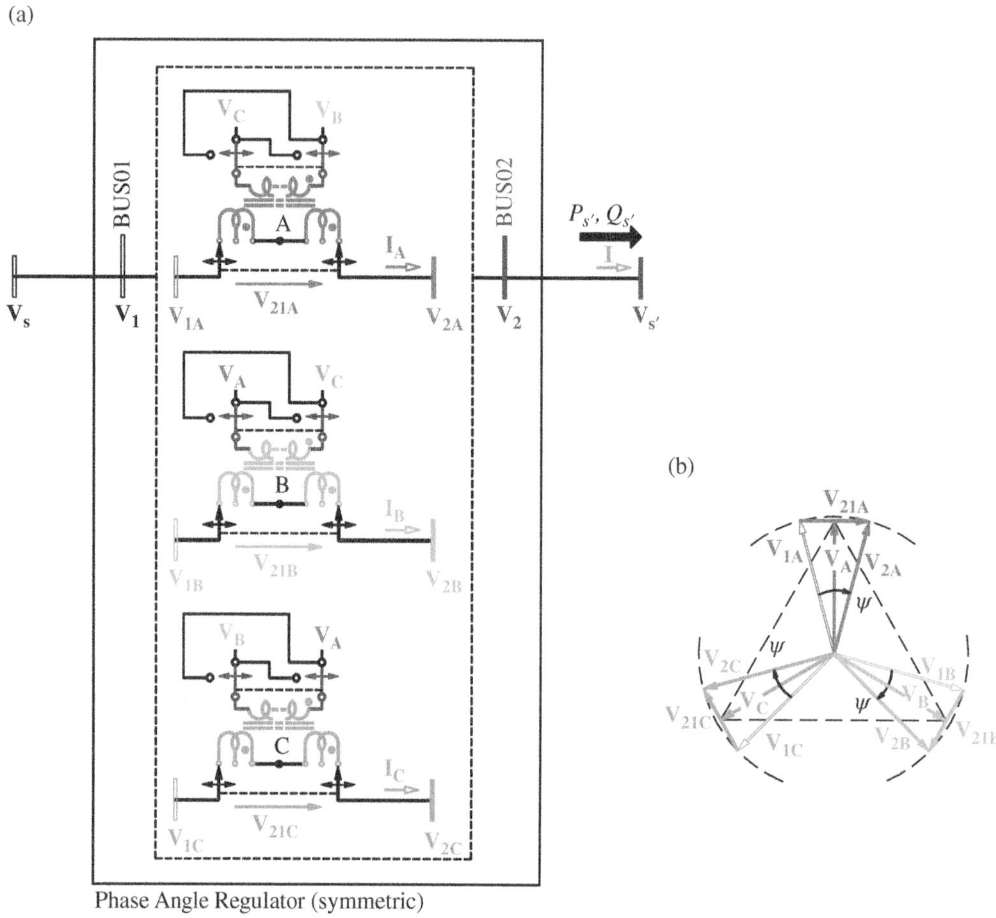

Figure 4-14 (a) PAR (sym) configuration for decreasing power flow; (b) phasor diagram.

Code 4-15 EMTP datafile for a PAR (sym), integrated in a two-generator/single-line power system network (411PAR2.DAT).

```
BEGIN NEW DATA CASE
C File Name: 411PAR2.DAT
C IDEAL PHASE ANGLE REGULATOR (SYM) WITHOUT ANY LEAKAGE IMPEDANCE
C 00000001111111111222222222233333333334444444444555555555566666666667777777778
C 34567890123456789012345678901234567890123456789012345678901234567890123456789 0
C ----dt<---tmax<---Xopt<---Copt<--Toler<        ><        >
8.3333-6150.00-3
C 00000001111111111222222222233333333334444444444555555555566666666667777777778
C 34567890123456789012345678901234567890123456789012345678901234567890123456789 0
```

(*Continued*)

Code 4-15 (Continued)

```
C -Iprnt<--Iplot<-Idoubl<-KssOut<-MaxOut<       ><       ><---Icat<-Nenerg
      30      10                                                 1
TACS HYBRID
99CRD      = 180 / PI
99CDR      = 1 / CRD
99TWOPI    = 2.0 * PI
C BASE VOLTAGE, CURRENT, AND POWER
99Vbase    = 112676.528
99Ibase    = 946.662704
99Zbase    = Vbase / Ibase
99Pbase    = 1.5 * Vbase * Ibase
C INPUTS FROM NODE VOLTAGES AND LINE CURRENTS
$INCLUDE 302MEAS1.SWT
C NORMALIZATION OF MEASURED VARIABLES AND COMPUTATION OF PEAK VOLTAGES/CURRENTS
$INCLUDE 303PTCT.SCL
C IDEAL PHASE-LOCKED LOOP
$INCLUDE 304IPLL.PLL
C COMPUTATION OF LINE RESISTANCE
$INCLUDE 305LINER.CMP
C TACS OUTPUT
33v12pu v1pu  v2pu  Prpu  Qrpu
BLANK RECORD ENDING TACS
C 0000000111111111122222222223333333333444444444455555555556666666666777777777778
C 3456789012345678901234567890123456789012345678901234567890123456789012345678901234567890
C <---Nodes--><---Nodes--><----R<----L<----C                                    V
C <-Bus1<-Bus2<-Bus3<-Bus4<-Ohms<---mH<---uF                                    V
C REPRESENTATION OF A SHORT BETWEEN NODES VS AND BUS01
  VSA   BUS01A            0.001
  VSB   BUS01B            0.001
  VSC   BUS01C            0.001
C CONNECTION OF NETWORK 2 BRANCHES WITH DAMPING RESISTORS (FIGURE 3-5)
$INCLUDE 311NTWK2.BRN
C CONNECTION OF AN IDEAL (NO LEAKAGE IMPEDANCE) SINGLE-CORE PAR (SYM)
$INCLUDE 412PAR2D.TRN
BLANK RECORD ENDING BRANCHES
C 0000000111111111122222222223333333333444444444455555555556666666666777777777778
C 3456789012345678901234567890123456789012345678901234567890123456789012345678901234567890
C <-Bus1<-Bus2<---Tclose<----Topen<-------Ie<----Flash<--Request<-----Target<--O
C MEASUREMENT OF LINE CURRENTS
$INCLUDE 307MEAS2.SWT
BLANK RECORD ENDING SWITCHES
C 0000000111111111122222222223333333333444444444455555555556666666666777777777778
C 3456789012345678901234567890123456789012345678901234567890123456789012345678901234567890
```

(Continued)

Code 4-15 (Continued)

```
C CONNECTION OF NETWORK 1 OR 2 SOURCES (FIGURE 3-5)
$INCLUDE 308NTWK1.SRC
BLANK RECORD ENDING SOURCES
BLANK RECORD ENDING NODE VOLTAGE OUTPUT
BLANK RECORD ENDING PLOT
```

Code 4-16 EMTP $INCLUDE file for an ideal single-core PAR (sym) for decreasing power flow (412PAR2D.TRN).

```
C File Name: 412PAR2D.TRN
C IDEAL SINGLE-CORE PHASE ANGLE REGULATOR (SYM) WITHOUT ANY LEAKAGE IMPEDANCE
C FOR DECREASING PHASE-SHIFT ANGLE
C 00000001111111111222222222233333333334444444444555555555566666666667777777778
C 34567890123456789012345678901234567890123456789012345678901234567890123456789 0
C ------PHASE ANGLE REGULATOR (SYM) FOR PHASE A------
C TRANSFORMER <--Refxxxxxx<--Iss<--Phi<-Name<-Rmag<----------------------------O
  TRANSFORMER                          PARA
C NEXT LINE (9999 IN COLUMNS 13-16) INDICATES A LINEAR TRANSFORMER
            9999
C <---Nodes--><----------><---Rk<---Lk<Volt|---------------------------------->O
C <-Bus1<-Bus2<----------><-Ohms<---mH<---kV---------------------------------->O
 1BUSB  BUSC                0.001 0.001 1.732
 2BUS02ABUSA                0.001 0.001 0.075
 3BUSA  BUS01A              0.001 0.001 0.075
C ------PHASE ANGLE REGULATOR (SYM) FOR PHASE B------
  TRANSFORMER                          PARB
            9999
 1BUSC  BUSA                0.001 0.001 1.732
 2BUS02BBUSB                0.001 0.001 0.075
 3BUSB  BUS01B              0.001 0.001 0.075
C ------PHASE ANGLE REGULATOR (SYM) FOR PHASE C------
  TRANSFORMER                          PARC
            9999
 1BUSA  BUSB                0.001 0.001 1.732
 2BUS02CBUSC                0.001 0.001 0.075
 3BUSC  BUS01C              0.001 0.001 0.075
```

Figure 4-15a shows the PAR (sym) configuration for increasing power flow. The corresponding $INCLUDE file 412PAR2I.TRN is shown in Code 4-17. The polarity of the series-connected windings is set such that the compensating secondary voltage is connected in series with the transmission line at 90° to the primary phase-to-neutral voltage at the midpoints of the secondary

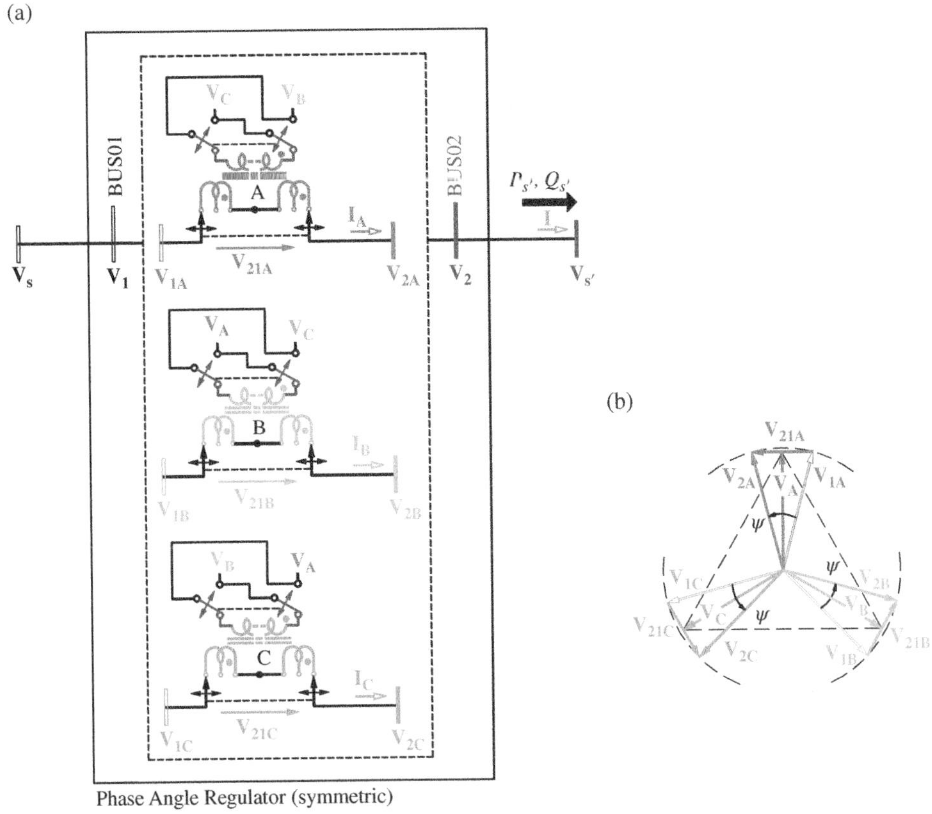

Figure 4-15 (a) PAR (sym) configuration for increasing power flow; (b) phasor diagram.

windings. The resulting effect is that only the phase angle of the line voltage between BUS01 and BUS02 nodes is increased as shown in the phasor diagram (Figure 4-15b). An increase in phase angle of the line voltage results in an increase in power flow in the line.

Code 4-17 EMTP $INCLUDE file for an ideal single-core PAR (sym) for increasing power flow (412PAR2I.TRN).

```
C File Name: 412PAR2I.TRN
C IDEAL SINGLE-CORE PHASE ANGLE REGULATOR (SYM) WITHOUT ANY LEAKAGE IMPEDANCE
C FOR INCREASING PHASE-SHIFT ANGLE
C 0000000111111111122222222223333333333444444444455555555556666666666777777777778
C 345678901234567890123456789012345678901234567890123456789012345678901234567890
C ------PHASE ANGLE REGULATOR (SYM) FOR PHASE A------
C TRANSFORMER <--Refxxxxxx<--Iss<--Phi<-Name<-Rmag<--------------------------O
  TRANSFORMER                            PARA
C NEXT LINE (9999 IN COLUMNS 13-16) INDICATES A LINEAR TRANSFORMER
            9999
```

(Continued)

Code 4-17 (Continued)

```
C <---Nodes--><----------><---Rk<---Lk<Volt|--------------------------------->0
C <-Bus1<-Bus2<----------><-Ohms<---mH<---kV--------------------------------->0
 1BUSC  BUSB                 0.001 0.001 1.732
 2BUS02ABUSA                 0.001 0.001 0.075
 3BUSA  BUS01A               0.001 0.001 0.075
C ------PHASE ANGLE REGULATOR (SYM) FOR PHASE B------
  TRANSFORMER                           PARB
          9999
 1BUSA  BUSC                 0.001 0.001 1.732
 2BUS02BBUSB                 0.001 0.001 0.075
 3BUSB  BUS01B               0.001 0.001 0.075
C ------PHASE ANGLE REGULATOR (SYM) FOR PHASE C------
  TRANSFORMER                           PARC
          9999
 1BUSB  BUSA                 0.001 0.001 1.732
 2BUS02CBUSC                 0.001 0.001 0.075
 3BUSC  BUS01C               0.001 0.001 0.075
```

Table 4-13 Electrical system data.

Parameters	Values
Source voltage ($\mathbf{V_{src}}$) phase-to-neutral	$112{,}676.528\angle 0°$ V_{peak}
Source resistance (R_{src})	1.0053 Ω
Source reactance (X_{src})	7.43875 Ω (19.73 mH)
Receiving-end voltage ($\mathbf{V_r}$) phase-to-neutral	$112{,}676.528\angle -20°$ V_{peak}
Line resistance (R)	3.0159 Ω
Line reactance (X)	22.31625 Ω (59.19 mH)
PAR (sym) turns-ratio	1.732:(−0.15 to 0.15)
	Secondary (center-tapped)
Primary winding resistance	0.001 Ω
Primary winding leakage inductance	0.001 mH
Secondary winding resistance	0.001 Ω
Secondary winding leakage inductance	0.001 mH

Table 4-13 shows the electrical system data for a two-generator/single-line power system network, compensated by an ideal PAR (sym).

Table 4-14 shows the variation of the normalized series-compensating voltage magnitude ($V_{s's}/V_s$) and its effects on simulated voltage (V_{21} or $V_{s's}$), the sending-end voltage magnitude (V_1 or V_s), the modified sending-end voltage magnitude (V_2 or $V_{s'}$), and the resulting active and reactive power flows (P_r and Q_r) at the receiving end. Note that $0.15 \times 0.976/\sqrt{\{1^2 + 0.075^2\}} = 0.146$ where 0.15 is the turns-ratio and 0.146 pu is the actual compensating voltage magnitude ($V_{s's}$), since the

Table 4-14 Effects of the variation of series-compensating voltage ($V_{s's}$) on simulated voltage (V_{21} or $V_{s's}$), the sending-end and the modified sending-end voltages (V_1 or V_s and V_2 or $V_{s'}$), and the power flows (P_r and Q_r) at the receiving end.

$V_{s's}/V_s$	V_{21} or $V_{s's}$ (pu)	V_1 or V_s (pu)	V_2 or $V_{s'}$ (pu)	P_r (pu)	Q_r (pu)
−0.150	−0.149	0.996	0.996	0.77	−0.19
−0.100	−0.099	0.994	0.994	0.95	−0.26
−0.050	−0.050	0.991	0.991	1.13	−0.34
0.000	0.000	0.988	0.988	1.31	−0.43
0.050	0.049	0.984	0.984	1.48	−0.53
0.100	0.098	0.980	0.980	1.65	−0.63
0.150	0.146	0.976	0.976	1.81	−0.75

sending-end voltage magnitude (V_s) decreases from its natural value (V_{sn}) of 0.988 to 0.976 pu as shown in the last row of Table 4-14.

Figure 4-16 shows that the variation of the series-compensating voltage magnitude ($V_{s's}$) in the range of −15 to 15% of the voltage ($\mathbf{V_{A\text{-}B\text{-}C}}$) at the midpoints of the compensating voltage

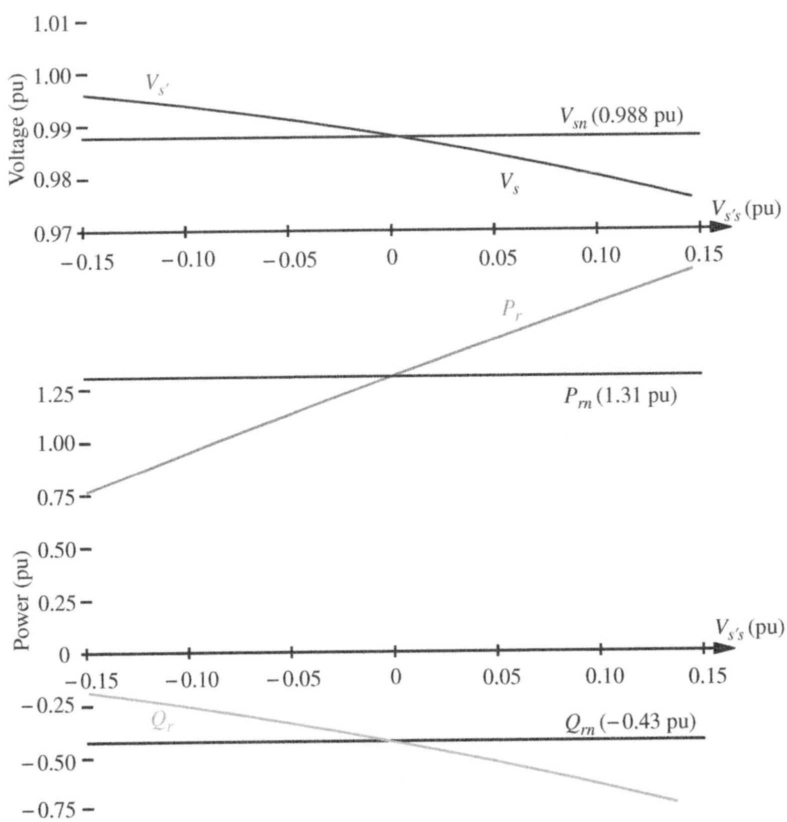

Figure 4-16 Ranges of voltage magnitudes (V_s and $V_{s'}$) at the sending and the modified sending ends and the active and reactive power flows (P_r and Q_r) at the receiving end with a TRANSFORMER model in EMTP to represent a PAR (sym) without any leakage impedance. Note that the Δ-connected primary windings are excited from the center-tapped series compensating windings.

(secondary) windings results in a variation of the modified sending-end voltage magnitude ($V_{s'}$) in the range of 0.996 to 0.976 pu and an identical voltage (V_s) at the sending end. The change in voltage at the POC is solely due to the change in the loading effect of the line. The reactive power flow (Q_r) at the receiving end varies from −0.19 to −0.75 pu during the range of compensating voltage magnitude ($V_{s's}$) from −15 to 15% of the voltage (**V_{A-B-C}**) at the midpoints of the compensating voltage (secondary) windings. However, the active power flow (P_r) at the receiving end varies in the range of 0.77 to 1.81 pu for the same range of variation in the compensating voltage magnitude ($V_{s's}$). Therefore, a range of −15 to 15% in compensating voltage magnitude ($V_{s's}$), in quadrature with phase-to-neutral voltage varies the reactive power flow (Q_r) by −0.56 pu and the active power flow (P_r) by 1.04 pu.

5

Mechanically-Switched Voltage Regulators and Power Flow Controllers

This chapter builds on the material that was presented in Chapter 2 where it was shown that a shunt-connected capacitor increases the line voltage, whereas a shunt-connected reactor lowers the line voltage at the Point of Connection (POC) to the utility. It was also shown that a series-connected reactor lowers the active and reactive power flows in the line simultaneously, whereas a series-connected capacitor increases the active and reactive power flows in the line simultaneously. The objective in this chapter is to show that the mechanically-switched shunt- and series-connected capacitors and reactors can be used as Voltage Regulators (VRs) and Power Flow Controllers (PFCs), respectively. The operation of various VRs and PFCs are verified in a single-generator/single-line power system network, considering a three-phase base power of 160 MVA and a phase-to-phase (*P–P*) base rms voltage, $V_{rms(P\text{-}P)}$, of 138 kV_{rms} as defined in Chapter 3, Section 3.1. It is also shown how to model a virtual impedance using a series-connected compensator.

5.1 Shunt Compensation

Both mechanically-switched capacitors and reactors are studied in this section.

5.1.1 Mechanically-Switched Capacitor (MSC)

It is shown in Chapters 2 and 4 that the effect of voltage regulation results in a change of active and reactive power flows in the line. For an increase in the line voltage, the increase in the reactive power flow is significantly more than the increase in the active power flow as described in Figures 4-2, 4-3, 4-5, and 4-7. As an example, the value of the shunt-connected capacitor is adjusted until the reactive current (I_{1q}) drawn from the line at the POC is 1 pu. The resulting line voltage (V_1) is 1.035 pu. Therefore, the shunt-compensating reactance is $X_{sh} = V_1/I_{1q} = 1.035/1 = 1.035\,\text{pu}$, which is $1.035 \times Z_{base} = 1.035 \times 119.025 = 123.19\,\Omega$. Refer to Equation (3-4) in Chapter 3, Section 3.1 for the definition of base impedance, Z_{base}. Therefore, the shunt-connected capacitor, $C_{sh} = 1/(\omega X_{sh}) = 1/(2 \times \pi \times 60 \times 123.19) = 21.532\,\mu\text{F}$.

A shunt-compensating, mechanically-switched capacitor (C_{sh}) is implemented in the Electromagnetic Transients Program (EMTP) datafile **501SHREA.DAT** as shown in Code 5-1. The datacard "$INCLUDE 502SHREA.USR" attaches the $INCLUDE file 502SHREA.USR that defines the user's input as shown in Code 5-2. The datacard "$INCLUDE 503SHREA.CON" attaches the $INCLUDE file 503SHREA.CON that performs mathematical calculations as shown in Code 5-3. The corresponding shunt-compensating capacitor branch file is 504SHRE1.BRN as shown in Code 5-4.

Power Flow Control Solutions for a Modern Grid using SMART Power Flow Controllers, First Edition.
Kalyan K. Sen and Mey Ling Sen.

Companion website: www.wiley.com/go/sen/powerflowcontrol

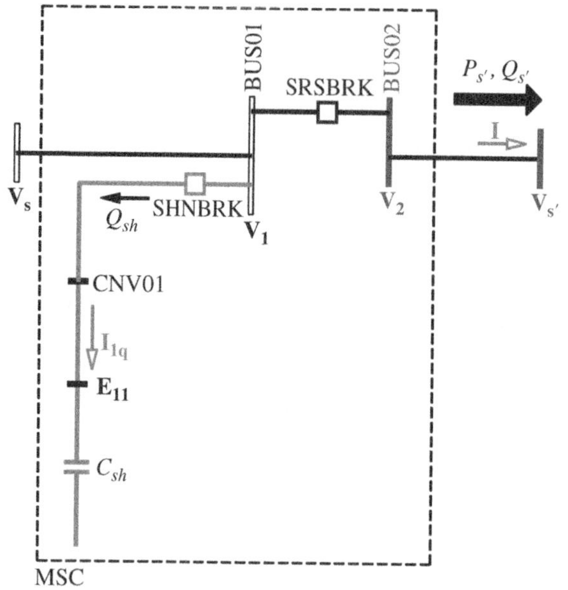

Figure 5-1 Shunt-compensating, mechanically-switched capacitor, connected to a line.

Code 5-1 EMTP datafile for a shunt-compensating, mechanically-switched capacitor, integrated in a two-generator/single-line power system network (501SHREA.DAT).

```
BEGIN NEW DATA CASE
C File Name: 501SHREA.DAT
C OPERATION OF A SHUNT-COMPENSATING, MECHANICALLY-SWITCHED CAPACITOR
C 00000001111111111222222222233333333334444444444555555555566666666667777777778
C 34567890123456789012345678901234567890123456789012345678901234567890123456789 0
C ----dt<---tmax<---Xopt<---Copt<--Toler<        ><        >
16.666-6300.00-3
C 00000001111111111222222222233333333334444444444555555555566666666667777777778
C 34567890123456789012345678901234567890123456789012345678901234567890123456789 0
C -Iprnt<--Iplot<-Idoubl<-KssOut<-MaxOut<        ><        ><---Icat<-Nenerg
       5      10                                                  1
TACS HYBRID
99TWOPI    = 2.0 * PI
C BASE VOLTAGE, CURRENT, IMPEDANCE, AND POWER
99Vbase    = 112676.528
99Ibase    = 946.662704
```

(Continued)

Code 5-1 (Continued)

```
99Zbase    = Vbase / Ibase
99Pbase    = 1.5 * Vbase * Ibase
C INPUTS FROM NODE VOLTAGES AND LINE CURRENTS
$INCLUDE 302MEAS1.CWT
C MEASUREMENT OF CURRENTS FLOWING THROUGH THE SHUNT-CONNECTED CAPACITOR
91CNV01A
91CNV01B
91CNV01C
C NORMALIZATION OF MEASURED VARIABLES AND COMPUTATION OF PEAK VOLTAGES/CURRENTS
$INCLUDE 303PTCT.SCL
C VECTOR PHASE-LOCKED LOOP
$INCLUDE 310VPLL.PLL
C USER'S INPUT FILE
$INCLUDE 502SHREA.USR
$INCLUDE 503SHREA.CON
C COMPUTATION OF LINE RESISTANCE
$INCLUDE 305LINER.CMP
C TACS OUTPUT
33v1pu  v1Apu i1Apu i1q    Qshpu
BLANK RECORD ENDING TACS
C 0000000111111111122222222223333333333444444444455555555556666666666777777777778
C 3456789012345678901234567890123456789012345678901234567890123456789012345678901234567890
C <---Nodes--><---Nodes--><----R<----L<----C                                        V
C <-Bus1<-Bus2<-Bus3<-Bus4<-Ohms<---mH<---uF                                          V
C REPRESENTATION OF SHORTED VS AND BUS01 NODES
  VSA   BUS01A           0.001
  VSB   BUS01B           0.001
  VSC   BUS01C           0.001
C CONNECTION OF NETWORK 2 BRANCHES WITH DAMPING RESISTORS (FIGURE 3-5)
$INCLUDE 311NTWK2.BRN
C REPRESENTATION OF SHORTED BUS01 AND BUS02 NODES
  BUS01ABUS02A           0.001
  BUS01BBUS02B           0.001
  BUS01CBUS02C           0.001
C HIGH RESISTANCES TO AVOID FLOATING NODES IN THE CIRCUIT
  BUS01ACNV01A          1.00+8
  BUS01BCNV01B          1.00+8
  BUS01CCNV01C          1.00+8
```

(Continued)

Code 5-1 (Continued)

```
C CONNECTION OF SHUNT-CONNECTED CAPACITOR BRANCHES (FIGURE 5-1)
$INCLUDE 504SHRE1.BRN
BLANK RECORD ENDING BRANCHES
C <-Bus1<-Bus2<---Tclose<----Topen<-------Ie<----Flash<--Request<-----Target<--O
C SHUNT BREAKER TO CONNECT/DISCONNECT THE SHUNT CAPACITOR/REACTOR TO THE CIRCUIT
13BUS01ACNV01A                                                         SHNBRK
13BUS01BCNV01B                                                         SHNBRK
13BUS01CCNV01C                                                         SHNBRK
C MEASUREMENT OF LINE CURRENTS
$INCLUDE 307MEAS2.SWT
C MEASUREMENT OF CURRENTS FLOWING THROUGH THE SHUNT CAPACITOR/REACTOR
  CNV01Ae11A                                            MEASURING
  CNV01Be11B                                            MEASURING
  CNV01Ce11C                                            MEASURING
BLANK RECORD ENDING SWITCHES
C 00000001111111111222222222233333333334444444444555555555566666666667777777778
C 34567890123456789012345678901234567890123456789012345678901234567890123456789 0
C CONNECTION OF NETWORK 1 SOURCES (FIGURE 3-5)
$INCLUDE 308NTWK1.SRC
BLANK RECORD ENDING SOURCES
BLANK RECORD ENDING NODE VOLTAGE OUTPUT
BLANK RECORD ENDING PLOT
```

Code 5-2 EMTP $INCLUDE file for user's input (502SHREA.USR).

```
C File Name: 502SHREA.USR
C USER'S INPUT FILE FOR A SHUNT-COMPENSATING, MECHANICALLY-SWITCHED CAPACITOR
C 00000001111111111222222222233333333334444444444555555555566666666667777777778
C 345678901234567890123456789012345678901234567890123456789012345678901234567890
C A SHUNT BREAKER IS CLOSED.
99SHNBRK  = (TIMEX .GE. 0.05) .AND. (TIMEX .LE. 0.25)
```

Table 5-1 shows the electrical system data for a two-generator/single-line power system network, compensated by a shunt-connected capacitor.

Figure 5-2 shows the EMTP simulation results from connecting a three-phase, shunt-compensating capacitor (C_{sh}) to the transmission line. Between 0 and 50 ms, the shunt breaker (SHNBRK) stays open and the natural voltage at the POC is 0.99 pu. At 50 ms, the SHNBRK closes, and the shunt-compensating capacitor is connected to the line. The inrush A-phase current (i_{LA}) exceeds 4 pu. The instantaneous A-phase voltage exceeds 0.75 pu over the nominal value. Under

Code 5-3 EMTP $INCLUDE file for calculation of exchanged power by a shunt compensator (503SHREA.CON).

```
C File Name: 503SHREA.CON
C CALCULATION OF EXCHANGED POWER BY A SHUNT COMPENSATOR
C 00000001111111111222222222233333333334444444444555555555566666666667777777778
C 34567890123456789012345678901234567890123456789012345678901234567890123456789 0
C SHUNT CURRENT COMPUTATION
99i1Apu    = CNV01A / Ibase
99i1Bpu    = CNV01B / Ibase
99i1Cpu    = CNV01C / Ibase
C LINE CURRENT MAGNITUDE COMPUTATION PER EQUATION (3-15)
99i1pu     = SQRT((i1Apu * i1Apu + i1Bpu * i1Bpu + i1Cpu * i1Cpu) * 2 / 3)
C LINE CURRENT MAGNITUDE COMPUTATION PER EQUATIONS (A-75a), (A-55b) and (3-23)
99i1ds     = i1Apu
99i1qs     = -(i1Apu + 2.0 * i1Cpu) / SQRT(3)
99i1d      =  i1ds * COSPLL + i1qs * SINPLL
99i1q      = -i1ds * SINPLL + i1qs * COSPLL
99i1dq     = SQRT(i1d * i1d + i1q * i1q)
C EXCHANGED POWER BY THE SHUNT UNIT AT THE POC
99Pshpu    = (v1Apu * i1Apu + v1Bpu * i1Bpu + v1Cpu * i1Cpu) *Vbase*Ibase / Pbase
99Qshpu    = (v1Apu * i1Cpu - v1Cpu * i1Apu) * SQRT(3) * Vbase * Ibase / Pbase
```

Code 5-4 EMTP $INCLUDE file for implementing a three-phase, shunt-compensating capacitor (504SHRE1.BRN).

```
C File Name: 504SHRE1.BRN
C THREE-PHASE, SHUNT-COMPENSATING CAPACITOR TO DRAW 1 PU CURRENT
C 00000001111111111222222222233333333334444444444555555555566666666667777777778
C 345678901234567890123456789012345678901234567890123456789012345678901234567890
C <---Nodes--><---Nodes--><----R<----L<----C                                   V
C <-Bus1<-Bus2<-Bus3<-Bus4<-Ohms<---mH<---uF                                   V
  e11A                                 21.532
  e11B                                 21.532
  e11C                                 21.532
```

a steady-state condition, the A-phase current (i_{1A}) through the shunt-compensating capacitor leads the A-phase line voltage (v_{1A}) of by 90° with a magnitude of 1 pu. The line voltage becomes 1.03 pu at the POC. Therefore, the compensating reactive power becomes −1.03 pu (capacitive). At 250 ms, the SHNBRK opens to disconnect the shunt-compensating capacitor from the line, the current through the shunt-compensating capacitor reduces to zero, and the line voltage at the POC becomes

Table 5-1 Electrical system data.

Parameters	Values
Source voltage ($\mathbf{V}_{\mathbf{src}}$) phase-to-neutral	112,676.528∠0° V_{peak}
Source resistance (R_{src})	1.0053 Ω
Source reactance (X_{src})	7.43875 Ω (19.73 mH)
Receiving-end voltage ($\mathbf{V}_{\mathbf{r}}$) phase-to-neutral	112,676.528∠−20° V_{peak}
Line resistance (R)	3.0159 Ω
Line reactance (X)	22.31625 Ω (59.19 mH)
Shunt-compensating capacitor (C_{sh})	**21.532 μF**

the original 0.99 pu. When the shunt-compensating capacitor disconnects, a spike in voltage occurs at the POC due to the sudden change in flow of current through the line reactor.

Just prior to closing the switch, the voltage on one side of the contact of the shunt breaker (SHNBRK) is the shunt-compensating capacitor voltage, which is zero. On the other side of the contact is the line voltage and it is different instantaneously on three different phases. The worst inrush current will occur in any phase if the contact closes on a peak voltage. The most desirable scenario is to close the contact on zero voltage.

The inrush current through the shunt-compensating capacitor (C_{sh}) can be limited by connecting the shunt-compensating capacitor through a series bypass reactor (L_{byp}) as shown in Figure 5-3. After some time, the series bypass reactor is shorted out by closing a bypass breaker (BYPBRK) and leaving only the shunt-compensating capacitor in the circuit. The series bypass reactor with an inductive reactance (X_{byp}) is chosen to be 20% of the shunt-compensating reactance (X_{sh}) with a quality factor, Q = 10. An inductive reactance of 20% of X_{sh} translates to a reactor, $L_{byp} = X_{byp}/\omega = 0.2 \times X_{sh}/\omega = 0.2 \times 123.19/(2 \times \pi \times 60) = 65.355$ mH and a resistance, $R_{byp} = X_{byp}/Q = 0.2 \times 123.19/10 = 2.4638\ \Omega$.

A shunt-compensating, mechanically-switched capacitor (C_{sh}) with a bypass series-connected resistor and reactor (R_{byp} and L_{byp}) is implemented in the datafile **505SHREA.DAT**, which is the same as **501SHREA.DAT**, except with a change in the shunt-compensating reactance file as shown in Code 5-5. The datacard "$INCLUDE 506SHREA.USR" attaches the $INCLUDE file 506SHREA.USR that defines the user's input as shown in Code 5-6. The corresponding shunt-compensating capacitor branch file is 507SHRE2.BRN as shown in Code 5-7.

Table 5-2 shows the electrical system data for a two-generator/single-line power system network, compensated by a shunt-connected capacitor with a series reactor.

Figure 5-4 shows the EMTP simulation results when a three-phase shunt-compensating capacitor is connected to the transmission line through a three-phase series reactor that is bypassed after a designated time. Between 0 and 50 ms, the shunt breaker (SHNBRK) stays open and the natural voltage at the POC is 0.99 pu. At 50 ms, the SHNBRK closes and the shunt capacitor with the series reactor is connected to the line. The inrush A-phase current (i_{1A}) barely exceeds 3.2 pu. The instantaneous A-phase voltage (v_{1A}) reaches 0.18 pu over the nominal value. At 150 ms, the BYPBRK closes to short out the series reactor, causing some transients to occur in the currents and the

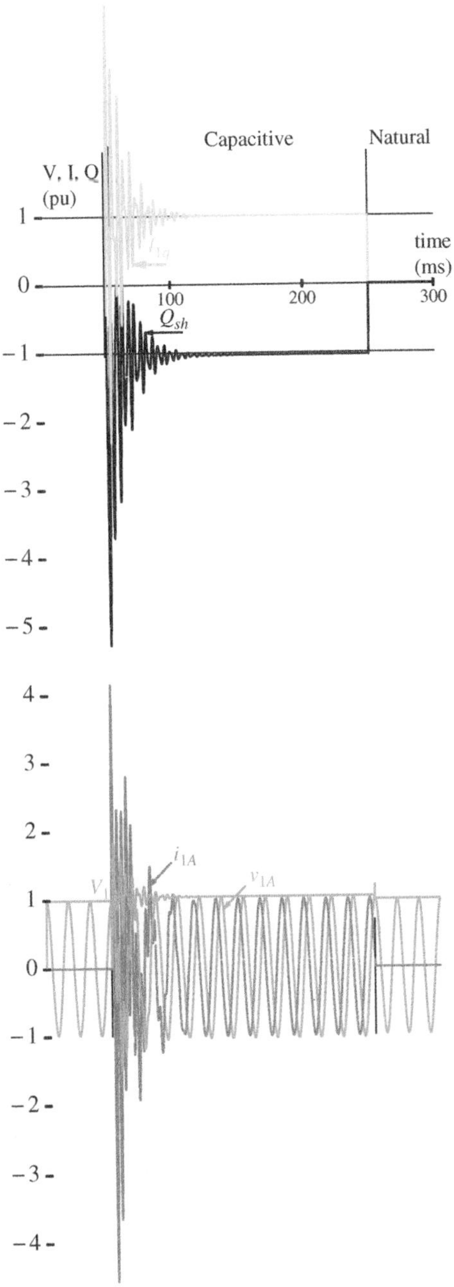

Figure 5-2 Exchanged reactive power (Q_{sh}), the resulting voltages (v_{1A} and V_1), and currents (i_{1A} and I_{1q}) due to a shunt-compensating, mechanically-switched capacitor, connected to a line.

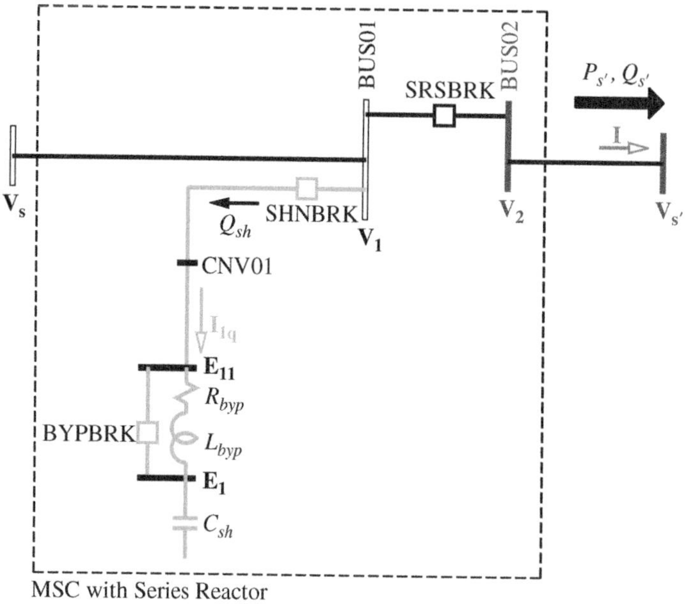

Figure 5-3 Shunt-compensating, mechanically-switched capacitor with a series bypass reactor, connected to a line.

Code 5-5 EMTP datafile for a shunt-compensating, mechanically-switched capacitor with a series bypass reactor, integrated in a two-generator/single-line power system network (505SHREA.DAT).

```
BEGIN NEW DATA CASE
C File Name: 505SHREA.DAT
C OPERATION OF A SHUNT-COMPENSATING, MECHANICALLY-SWITCHED CAPACITOR WITH A
C SERIES BYPASS REACTOR
C 0000000111111111122222222223333333333444444444455555555556666666666777777777778
C 3456789012345678901234567890123456789012345678901234567890123456789012345678901234567890
C ----dt<---tmax<---Xopt<---Copt<--Toler<        ><        >
16.666-6300.00-3
C 0000000111111111122222222223333333333444444444455555555556666666666777777777778
C 3456789012345678901234567890123456789012345678901234567890123456789012345678901234567890
C -Iprnt<--Iplot<-Idoubl<-KssOut<-MaxOut<        ><        ><---Icat<-Nenerg
       5      10                                                   1
TACS HYBRID
99TWOPI    = 2.0 * PI
C BASE VOLTAGE, CURRENT, IMPEDANCE, AND POWER
99Vbase    = 112676.528
99Ibase    = 946.662704
```

(Continued)

Code 5-5 (Continued)

```
99Zbase    = Vbase / Ibase
99Pbase    = 1.5 * Vbase * Ibase
C INPUTS FROM NODE VOLTAGES AND LINE CURRENTS
$INCLUDE 302MEAS1.SWT
C MEASUREMENT OF CURRENTS FLOWING THRU SHUNT-CONNECTED CAPACITOR WITH REACTOR
91CNV01A
91CNV01B
91CNV01C
C NORMALIZATION OF MEASURED VARIABLES AND COMPUTATION OF PEAK VOLTAGES/CURRENTS
$INCLUDE 303PTCT.SCL
C VECTOR PHASE-LOCKED LOOP
$INCLUDE 310VPLL.PLL
C USER'S INPUT FILE
$INCLUDE 506SHREA.USR
$INCLUDE 503SHREA.CON
C COMPUTATION OF LINE RESISTANCE
$INCLUDE 305LINER.CMP
C TACS OUTPUT
33v1pu  v1Apu i1Apu i1q   Qshpu
BLANK RECORD ENDING TACS
C 00000001111111111222222222233333333334444444444555555555566666666667777777777 8
C 34567890123456789012345678901234567890123456789012345678901234567890123456789 0
C <---Nodes--><---Nodes--><----R<----L<----C                                   V
C <-Bus1<-Bus2<-Bus3<-Bus4<-Ohms<---mH<---uF                                   V
C REPRESENTATION OF SHORTED VS AND BUS01 NODES
  VSA   BUS01A            0.001
  VSB   BUS01B            0.001
  VSC   BUS01C            0.001
C CONNECTION OF NETWORK 2 BRANCHES WITH DAMPING RESISTORS (FIGURE 3-5)
$INCLUDE 311NTWK2.BRN
C REPRESENTATION OF SHORTED BUS01 AND BUS02 NODES
  BUS01ABUS02A            0.001
  BUS01BBUS02B            0.001
  BUS01CBUS02C            0.001
C HIGH RESISTANCES TO AVOID FLOATING NODES IN THE CIRCUIT
  BUS01ACNV01A           1.00+8
  BUS01BCNV01B           1.00+8
  BUS01CCNV01C           1.00+8
  e11A  e1A              1.00+8
  e11B  e1B              1.00+8
  e11C  e1C              1.00+8
```

(*Continued*)

Code 5-5 (Continued)

```
C SHUNT-CONNECTED CAPACITOR WITH SERIES REACTOR BRANCHES (FIGURE 5-3)
$INCLUDE 507SHRE2.BRN
BLANK RECORD ENDING BRANCHES
C <-Bus1<-Bus2<---Tclose<----Topen<-------Ie<----Flash<--Request<-----Target<--O
C SHUNT BREAKER TO CONNECT/DISCONNECT THE SHUNT CAPACITOR/REACTOR TO THE CIRCUIT
13BUS01ACNV01A                                                            SHNBRK
13BUS01BCNV01B                                                            SHNBRK
13BUS01CCNV01C                                                            SHNBRK
C MEASUREMENT OF LINE CURRENTS
$INCLUDE 307MEAS2.SWT
C MEASUREMENT OF CURRENTS FLOWING THROUGH SHUNT CAPACITOR WITH SERIES REACTOR
  CNV01Ae11A                                            MEASURING
  CNV01Be11B                                            MEASURING
  CNV01Ce11C                                            MEASURING
13e11A  e1A                                                               BYPBRK
13e11B  e1B                                                               BYPBRK
13e11C  e1C                                                               BYPBRK
BLANK RECORD ENDING SWITCHES
C 00000001111111111222222222233333333334444444444555555555566666666667777777778
C 34567890123456789012345678901234567890123456789012345678901234567890123456789 0
C CONNECTION OF NETWORK 1 SOURCES (FIGURE 3-5)
$INCLUDE 308NTWK1.SRC
BLANK RECORD ENDING SOURCES
BLANK RECORD ENDING NODE VOLTAGE OUTPUT
BLANK RECORD ENDING PLOT
```

Code 5-6 EMTP $INCLUDE file for user's input (506SHREA.USR).

```
C File Name: 506SHREA.USR
C USER'S INPUT FILE FOR A SHUNT-COMPENSATING, MECHANICALLY-SWITCHED CAPACITOR
C WITH A SERIES BYPASS REACTOR
C 00000001111111111222222222233333333334444444444555555555566666666667777777778
C 345678901234567890123456789012345678901234567890123456789012345678901234567890
C A SHUNT BREAKER IS CLOSED.
99SHNBRK  = (TIMEX .GE. 0.05) .AND. (TIMEX .LE. 0.25)
99BYPBRK  = (TIMEX .GE. 0.15)
```

Code 5-7 EMTP $INCLUDE file for implementing a three-phase, shunt-compensating capacitor with a series bypass reactor (507SHRE2.BRN).

```
C File Name: 507SHRE2.BRN
C SHUNT-COMPENSATING CAPACITOR (WITH SERIES BYPASS REACTOR) TO DRAW 1 PU CURRENT
C 00000001111111111222222222233333333334444444444555555555566666666667777777778
C 34567890123456789012345678901234567890123456789012345678901234567890123456789
C <---Nodes--><---Nodes--><----R<----L<----C                                   V
C <-Bus1<-Bus2<-Bus3<-Bus4<-Ohms<---mH<---uF                                   V
  e11A  e1A               2.463865.355
  e11B  e1B               2.463865.355
  e11C  e1C               2.463865.355
  e1A                                 21.532
  e1B                                 21.532
  e1C                                 21.532
```

Table 5-2 Electrical system data.

Parameters	Values
Source voltage ($\mathbf{V_{src}}$) phase-to-neutral	112,676.528∠0° V_{peak}
Source resistance (R_{src})	1.0053 Ω
Source reactance (X_{src})	7.43875 Ω (19.73 mH)
Receiving-end voltage ($\mathbf{V_r}$) phase-to-neutral	112,676.528∠−20° V_{peak}
Line resistance (R)	3.0159 Ω
Line reactance (X)	22.31625 Ω (59.19 mH)
Shunt-compensating capacitor, C_{sh} **(with a series bypass resistor and reactor: R_{byp} and L_{byp})**	21.532 μF **(2.4638 Ω, 65.355 mH)**

voltages. After the transients die out, the A-phase current (i_{1A}) through the shunt capacitor leads the A-phase line voltage (v_{1A}) by 90° with a magnitude of 1 pu. From this point onward, the circuit operation is the same as what is shown in Figure 5-2.

The chosen shunt-compensating capacitor of 21.532 μF at $V_{rms(P\text{-}P)}$ of 138 kV_{rms} results in a ~155 Mvar rating. Switching the capacitor in one step would likely cause too much change in voltage, which is >2.5%. Often the planners like to keep the change in voltage <1.5%. The solution is to switch the capacitor in two or three steps, which lessens the impact of the switching transients on the network.

There are two commercially-available controlled-switching techniques:

1) A two-step switching – a closing impedance (R_{byp}, L_{byp}), termed as a "lossy reactor," is first inserted in series with the shunt-compensating capacitor (C_{sh}) as shown in Figure 5-3 and then shorted out some time (8–20 ms) later.
2) Capacitor-switching at voltage zero-crossing – the capacitor is energized on each phase when the voltage across the capacitor closely matches the instantaneous system voltage, meaning the voltage across the closing pole being close to zero.

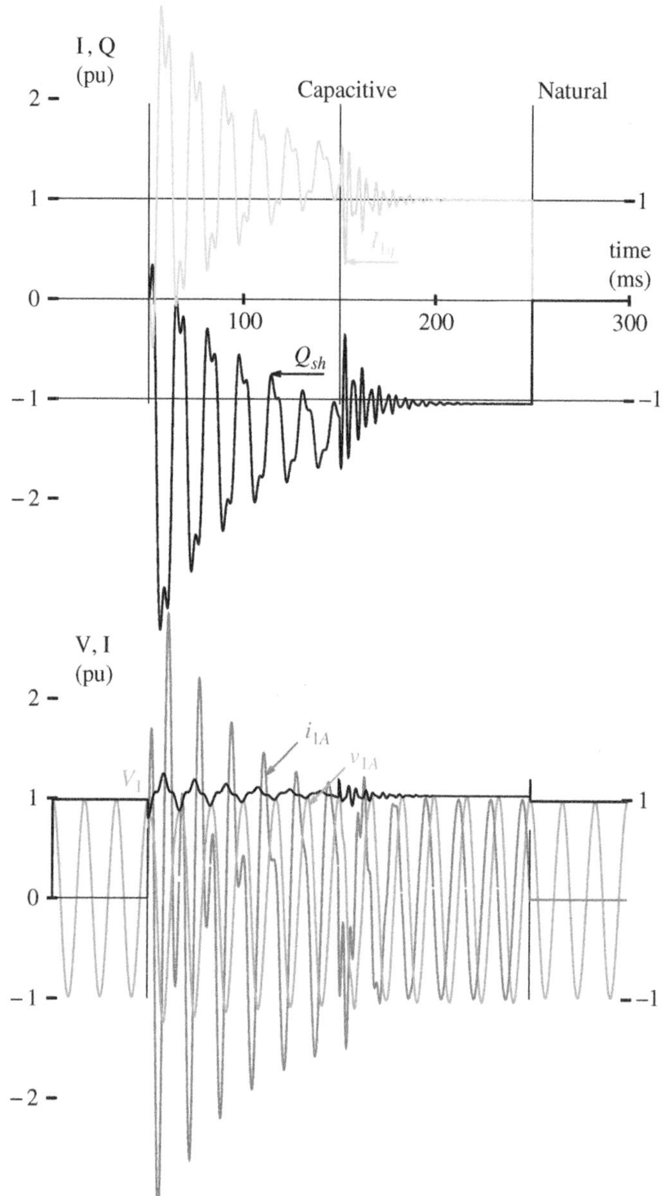

Figure 5-4 Exchanged reactive power (Q_{sh}), the resulting voltages (v_{1A} and V_1), and currents (i_{1A} and I_{1q}) due to a shunt-compensating, mechanically-switched capacitor (with a series bypass reactor), connected to a line.

Both of these controlled-switching techniques have similar performance in terms of reducing inrush current and overvoltage. However, it is more challenging to determine when the voltage across the capacitor is zero in the second technique. In the first technique, the closing resistors may be an option as well. All of these commercially available controlled switching devices reduce the inrush current by at least a factor of 3. Note that the chosen base power of 160 MVA is arbitrary; so is the value of 1 pu capacitive current in this example. In reality, the compensating current and, therefore, the generated var are based on the system's need.

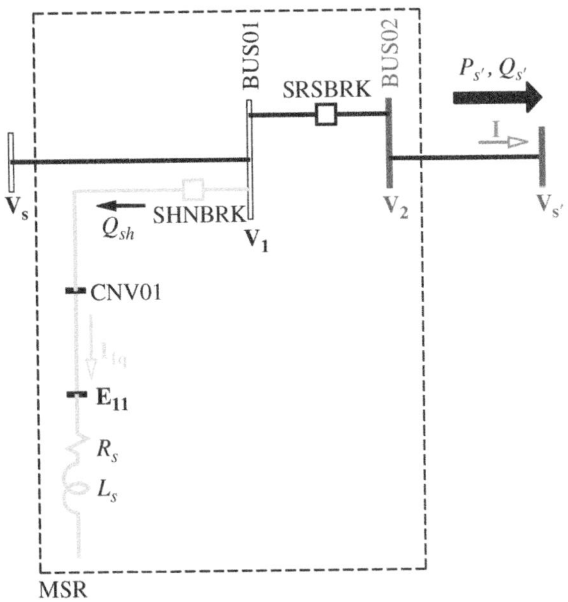

Figure 5-5 Shunt-compensating, mechanically-switched reactor, connected to a line.

5.1.2 Mechanically-Switched Reactor (MSR)

It is discussed in Chapter 2 that a shunt-compensating reactor, shown in Figure 5-5, decreases the line voltage at the POC and also decreases the delivery of reactive power to the receiving end of the transmission line significantly more than the delivery of active power.

The shunt-connected reactor is adjusted until the reactive current (I_1) drawn from the line at the POC is 1 pu. The resulting line voltage (V_1) is 0.94 pu. Therefore, the shunt-compensating reactance is $X_{sh} = V_1/I_{1q} = 0.94/1 = 0.94$ pu, which is $0.94 \times Z_{base} = 0.94 \times 119.025 = 111.8835\,\Omega$. Therefore, the shunt-connected reactor $L_p = X_{sh}/\omega = 111.8835/(2 \times \pi \times 60) = 296.78$ mH.

A resistor (R_p), connected in parallel to the reactor, is $R_p = \mathrm{Q} \times X_{sh} = 10 \times 111.8835 = 1118.835\,\Omega$, where the quality factor $\mathrm{Q} = R_p/X_{sh} = 10$. The equivalent series combination of a resistor and a reactor is

$$R_s = R_p\frac{1}{1+\mathrm{Q}^2} = \frac{1118.835}{101} = 11.077\ \Omega \quad \text{and} \quad L_s = L_p\frac{\mathrm{Q}^2}{1+\mathrm{Q}^2} = 296.78\frac{100}{101} = 293.84\ \text{mH}$$

as shown in Code 5-8.

A shunt-compensating, mechanically-switched reactor is implemented in the datafile **501SHREA.DAT**. The corresponding shunt inductive reactance branch file is 508SHRE3.BRN as shown in Code 5-8.

Table 5-3 shows the electrical system data for a two-generator/single-line power system network, compensated by a shunt reactor.

Figure 5-6 shows the EMTP simulation results when a three-phase shunt-compensating inductive reactance (X_{sh}) is connected to the transmission line. Between 0 and 50 ms, the shunt breaker (SHNBRK) stays open and the natural voltage at the POC is 0.99 pu. At 50 ms, the SHNBRK closes and the shunt reactor is connected to the line. After the closing of the breaker, the current through

Code 5-8 EMTP $INCLUDE file for implementing a three-phase, shunt-compensating reactor (508SHRE3.BRN).

```
C File Name: 508SHRE3.BRN
C THREE-PHASE, SHUNT-COMPENSATING REACTOR TO DRAW 1 PU CURRENT
C 00000001111111111222222222233333333334444444444555555555566666666667777777778
C 34567890123456789012345678901234567890123456789012345678901234567890123456789 0
C <---Nodes--><---Nodes--><----R<----L<----C                                  V
C <-Bus1<-Bus2<-Bus3<-Bus4<-Ohms<---mH<---uF                                  V
  e11A                    11.077293.84
  e11B                    11.077293.84
  e11C                    11.077293.84
```

Table 5-3 Electrical system data.

Parameters	Values
Source voltage ($\mathbf{V}_{\mathbf{src}}$) phase-to-neutral	112,676.528∠0° V_{peak}
Source resistance (R_{src})	1.0053 Ω
Source reactance (X_{src})	7.43875 Ω (19.73 mH)
Receiving-end voltage ($\mathbf{V}_{\mathbf{r}}$) phase-to-neutral	112,676.528∠−20° V_{peak}
Line resistance (R)	3.0159 Ω
Line reactance (X)	22.31625 Ω (59.19 mH)
Shunt-compensating resistor and reactor (R_s, L_s)	**11.077 Ω, 293.84 mH**

the reactor builds up from its zero initial condition. Under steady-state conditions, the A-phase current (i_{1A}) through the shunt reactor lags the A-phase line voltage (v_{1A}) by 90° with a magnitude of 1 pu. The line voltage becomes 0.94 pu at the POC. Therefore, the compensating reactive power becomes 0.94 pu. At 250 ms, the SHNBRK opens to disconnect the shunt reactor from the line. The current through the shunt reactor becomes zero, causing a spike in voltage across the reactor. There is also a spike in voltage at the POC due to the sudden change in flow of current through the line inductance. The line voltage at the POC becomes the original 0.99 pu.

5.2 Series Compensation

Both mechanically-switched reactors and capacitors are studied in this section.

5.2.1 Mechanically-Switched Reactor (MSR)

It is discussed in Chapter 2 that a series-compensating reactor, shown in Figure 5-7, modifies the effective reactance of the line between its two ends and regulates the active and reactive power flows in the line simultaneously.

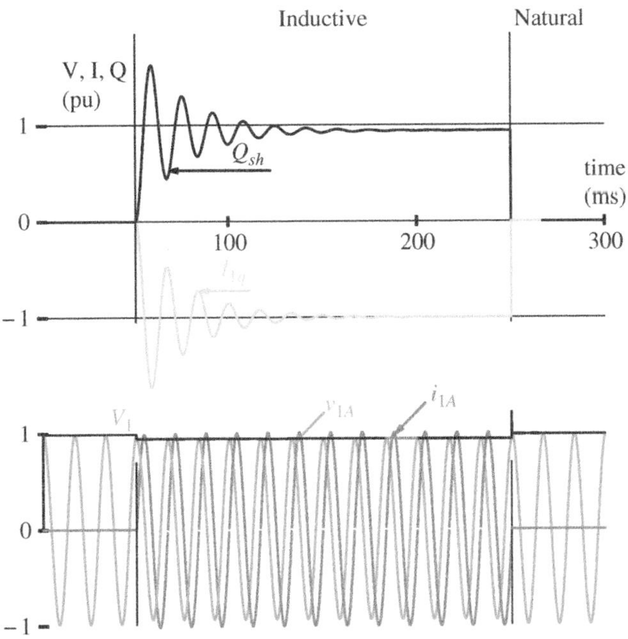

Figure 5-6 Exchanged reactive power (Q_{sh}), the resulting voltages (v_{1A} and V_1), and currents (i_{1A} and I_{1q}) due to a shunt-compensating, mechanically-switched reactor, connected to a line.

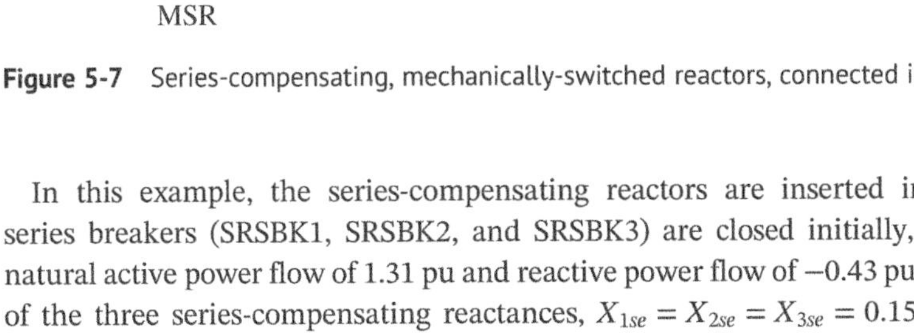

Figure 5-7 Series-compensating, mechanically-switched reactors, connected in series with the line.

In this example, the series-compensating reactors are inserted in three stages. All three series breakers (SRSBK1, SRSBK2, and SRSBK3) are closed initially, leaving the line with its natural active power flow of 1.31 pu and reactive power flow of −0.43 pu at the receiving end. Each of the three series-compensating reactances, $X_{1se} = X_{2se} = X_{3se} = 0.15 \times Z_{base} = 0.15 \times 119.02 = 17.854\,\Omega$. Therefore, each compensating reactor, $L_{1se} = L_{2se} = L_{3se} = X_{1se}/\omega = 17.854/(2 \times \pi \times 60) = 47.358$ mH. The first compensating reactance (X_{1se}) has a quality factor (Q) of 10, whereas the other two reactors (X_{2se} and X_{3se}) have a quality factor of ∞. Therefore, the series resistance, $R_{1se} = X_{1se}/10 = 1.7854\,\Omega$ and that of X_{2se} and X_{3se} are $R_{2se} = X_{2se}/\infty = 0$ and are $R_{3se} = X_{3se}/\infty = 0$. If the same operating points are studied using a VSC-based series compensator, it may be considered that the compensating reactance (X_{1se}) includes the leakage reactance of the series-coupling transformer that is an integral part of the SSSC.

A series-compensating reactor is implemented in the datafile **509SEREA.DAT** as shown in Code 5-9. The datacard "$INCLUDE 510SEREA.USR" attaches the $INCLUDE file 510SEREA.USR that defines the user's input as shown in Code 5-10. The datacard "$INCLUDE 511SEREA.CON" attaches the $INCLUDE file 511SEREA.CON that performs mathematical calculations as shown in Code 5-11. The corresponding series reactor branch file is 512SERE1.BRN as shown in Code 5-12.

Code 5-9 EMTP datafile for series-compensating, mechanically-switched reactors, integrated in a two-generator/single-line power system network (509SEREA.DAT).

```
BEGIN NEW DATA CASE
C File Name: 509SEREA.DAT
C OPERATION OF SERIES-COMPENSATING, MECHANICALLY-SWITCHED REACTORS
C 000000011111111112222222222333333333344444444445555555555666666666677777777778
C 345678901234567890123456789012345678901234567890123456789012345678901234567890
C ----dt<---tmax<---Xopt<---Copt<--Toler<      ><      >
16.666-6600.00-3
C 000000011111111112222222222333333333344444444445555555555666666666677777777778
C 345678901234567890123456789012345678901234567890123456789012345678901234567890
C -Iprnt<--Iplot<-Idoubl<-KssOut<-MaxOut<      ><      ><---Icat<-Nenerg
      40      40                                                1
TACS HYBRID
99CRD     = 180 / PI
99CDR     = 1 / CRD
99TWOPI   = 2.0 * PI
C BASE VOLTAGE, CURRENT, AND POWER
99Vbase   = 112676.528
99Ibase   = 946.662704
99Zbase   = Vbase / Ibase
99Pbase   = 1.5 * Vbase * Ibase
C INPUTS FROM NODE VOLTAGES AND LINE CURRENTS
$INCLUDE 302MEAS1.SWT
C NORMALIZATION OF MEASURED VARIABLES AND COMPUTATION OF PEAK VOLTAGES/CURRENTS
$INCLUDE 303PTCT.SCL
C VECTOR PHASE-LOCKED LOOP
$INCLUDE 310VPLL.PLL
C USER'S INPUT FILE
$INCLUDE 510SEREA.USR
$INCLUDE 511SEREA.CON
C COMPUTATION OF LINE RESISTANCE
$INCLUDE 305LINER.CMP
C COMPENSATING VOLTAGE MAGNITUDE AND PHASE ANGLE COMPUTATION
99v12dsp  = v12Apu
99v12qsp  = -(v12Apu + 2.0 * v12Cpu) / SQRT(3)
```

(Continued)

Code 5-9 (Continued)

```
99v12dpu   =  v12dsp * COSPLL + v12qsp * SINPLL
99v12qpu   = -v12dsp * SINPLL + v12qsp * COSPLL
99v12dqp   = SQRT(v12dpu * v12dpu + v12qpu * v12qpu)
C LOW-PASS FILTER INTRODUCES A RAMP-RATE AND REDUCES SYSTEM TRANSIENTS
 1v12dqf   +v12dqp
        1.0
        1.0     0.0040
99zse      = v12dqf / idqpu
99thv12a   = ATAN( v12qpu / (v12dpu + 1.0E-8) )
99thv12b   = thv12a + PI * (v12dpu .LT. 0.0)
99thv12    = thv12b + TWOPI * (thv12b .LT. 0.0) - TWOPI * (thv12b .GE. TWOPI)
C LOW-PASS FILTER INTRODUCES A RAMP-RATE AND REDUCES SYSTEM TRANSIENTS
 1thv12f   +thv12
        1.0
        1.0     0.0040
99tv12dq   = thv12f * CRD
99theti    = thetir + TWOPI * (thetir .LT. -PI) - TWOPI * (thetir .GE. PI)
99thidq    = theti * CRD
99zsetha   = thv12f - thetir
99zsethb   = zsetha + TWOPI * (zsetha .LT. -PI) - TWOPI * (zsetha .GE. PI)
99thzse    = zsethb * CRD
99beta     = (thv12f + PI) * CRD
C The PI is added, since the polarity of V21 is opposite of the polarity of V12.
99tha2     = theta + PI / 2 + PI + beta * CDR
C This addition of PI/2 is because the compensating voltage is sine reference
C and the PLL is cosine reference.
C PI radian is added to reverse the polarity of the series-coupling transformer.
99theta2   = tha2 + TWOPI * (tha2 .LT. 0.0) - TWOPI * (tha2 .GE. TWOPI)
99vsrspk   = v12dqf * Vbase
99e2A      = vsrspk * SIN(theta2)
99e2B      = vsrspk * SIN(theta2 - 120 * CDR)
99e2C      = vsrspk * SIN(theta2 + 120 * CDR)
99e2Apu    = e2A / Vbase
33v12ApuiApu  Qsepu Prpu  Qrpu  v12dqftv12dqidqpu thidq zse thzse beta  e2Apu
BLANK RECORD ENDING TACS
C 00000001111111111222222222233333333334444444444555555555566666666667777777778
C 34567890123456789012345678901234567890123456789012345678901234567890123456789 0
C <---Nodes--><---Nodes--><----R<----L<----C                                   V
C <-Bus1<-Bus2<-Bus3<-Bus4<-Ohms<---mH<---uF                                   V
C REPRESENTATION OF SHORTED VS AND BUS01 NODES
  VSA   BUS01A             0.001
```

(*Continued*)

Code 5-9 (Continued)

```
  VSB   BUS01B             0.001
  VSC   BUS01C             0.001
C CONNECTION OF NETWORK 2 BRANCHES WITH DAMPING RESISTORS (FIGURE 3-5)
$INCLUDE 311NTWK2.BRN
C HIGH RESISTANCES TO AVOID FLOATING NODES IN THE CIRCUIT
  BUS01ABUS02A            1.00+8
  BUS01BBUS02B            1.00+8
  BUS01CBUS02C            1.00+8
C SERIES-CONNECTED REACTOR (FIGURE 5-7)
$INCLUDE 512SERE1.BRN
BLANK RECORD ENDING BRANCHES
C <-Bus1<-Bus2<--Vig---|<-Ihold--|<-Tdeion-|xxxxxxxxxxCLOSEDSAMEGRID->SIGNAL  oo
C CLOSED SERIES BREAKERS AT START OF THE SIMULATION
13BUS01ABUS02A                                          CLOSED          SRSBK1
13BUS01BBUS02B                                          CLOSED          SRSBK1
13BUS01CBUS02C                                          CLOSED          SRSBK1
13BUS11ABUS12A                                          CLOSED          SRSBK2
13BUS11BBUS12B                                          CLOSED          SRSBK2
13BUS11CBUS12C                                          CLOSED          SRSBK2
13BUS12ABUS02A                                          CLOSED          SRSBK3
13BUS12BBUS02B                                          CLOSED          SRSBK3
13BUS12CBUS02C                                          CLOSED          SRSBK3
C MEASUREMENT OF LINE CURRENTS
$INCLUDE 307MEAS2.SWT
BLANK RECORD ENDING SWITCHES
C 0000000111111111122222222223333333333444444444455555555556666666666777777777778
C 3456789012345678901234567890123456789012345678901234567890123456789012345678901234567890
C CONNECTION OF NETWORK 1 SOURCES (FIGURE 3-5)
$INCLUDE 308NTWK1.SRC
BLANK RECORD ENDING SOURCES
BLANK RECORD ENDING NODE VOLTAGE OUTPUT
BLANK RECORD ENDING PLOT
```

Code 5-10 EMTP $INCLUDE file for user's input (510SEREA.USR).

```
C File Name: 510SEREA.USR
C USER'S INPUT FILE FOR SERIES-COMPENSATING, MECHANICALLY-SWITCHED REACTORS
C 0000000111111111122222222223333333333444444444455555555556666666666777777777778
C 3456789012345678901234567890123456789012345678901234567890123456789012345678901234567890
C NORMALLY-CLOSED SERIES BREAKERS ARE OPENED AND RECLOSED.
99SRSBK1  = 1 - ((TIMEX .GE. 0.025) .AND. (TIMEX .LE. 0.30))
99SRSBK2  = 1 - ((TIMEX .GE. 0.050) .AND. (TIMEX .LE. 0.30))
99SRSBK3  = 1 - ((TIMEX .GE. 0.175) .AND. (TIMEX .LE. 0.30))
```

Code 5-11 EMTP $INCLUDE file for calculation of exchanged-power by a series-compensator (511SEREA.CON).

```
C File Name: 511SEREA.CON
C CALCULATION OF EXCHANGED POWER BY A SERIES COMPENSATOR
C 00000001111111111222222222233333333334444444444555555555566666666667777777778
C 34567890123456789012345678901234567890123456789012345678901234567890123456789 0
C EXCHANGED POWER BY THE SERIES UNIT WITH THE LINE
99Psepu    = (v12Apu * iApu + v12Bpu * iBpu + v12Cpu * iCpu) *Vbase*Ibase / Pbase
99Qsepu    = (v12Apu * iCpu - v12Cpu * iApu) * SQRT(3) * Vbase * Ibase / Pbase
```

Code 5-12 EMTP $INCLUDE file for implementing three-phase, series-compensating reactors (512SERE1.BRN).

```
C File Name: 512SERE1.BRN
C THREE-PHASE, SERIES-COMPENSATING REACTORS
C 00000001111111111222222222233333333334444444444555555555566666666667777777778
C 34567890123456789012345678901234567890123456789012345678901234567890123456789 0
C <---Nodes--><---Nodes--><----R<----L<----C                                   V
C <-Bus1<-Bus2<-Bus3<-Bus4<-Ohms<---mH<---uF                                   V
  BUS01ABUS11A            1.785447.358
  BUS01BBUS11B            1.785447.358
  BUS01CBUS11C            1.785447.358
  BUS11ABUS12A                47.358
  BUS11BBUS12B                47.358
  BUS11CBUS12C                47.358
  BUS12ABUS02A                47.358
  BUS12BBUS02B                47.358
  BUS12CBUS02C                47.358
```

Table 5-4 shows the electrical system data for a two-generator/single-line power system network, compensated by series-connected reactors.

Figure 5-8 shows the EMTP simulation results when a three-phase series reactor bank is connected to the line. The normally-closed series breaker 1 (SRSBK1) opens at 25 ms and stays open for the next 275 ms to connect the reactance (R_{1se}, L_{1se}) in series with the transmission line between the BUS01 and BUS02 nodes. A voltage (v_{12}) develops between the BUS01 and BUS02 nodes. The A-phase voltage (v_{12A}) leads the line current (i_A) in the corresponding phase by almost 90° because the effective impedance of the line is inductive. Both the active and reactive power flows (P_r and Q_r) in the line decrease. Further increases in the inductive compensation by opening the series breakers SRSBK2 (*i.e.* inserting L_{2se}) at 50 ms and opening SRSBK3 (*i.e.* inserting L_{3se}) at 175 ms decrease the active and reactive power flows in the line simultaneously. These operating points are verified in series compensation with VSCs Sen and Sen (2009). The compensating reactive power (Q_{se})

Table 5-4 Electrical system data.

Parameters	Values
Source voltage ($\mathbf{V_{src}}$) phase-to-neutral	112,676.528∠0° V_{peak}
Source resistance (R_{src})	1.0053 Ω
Source reactance (X_{src})	7.43875 Ω (19.73 mH)
Receiving-end voltage ($\mathbf{V_r}$) phase-to-neutral	112,676.528∠−20° V_{peak}
Line resistance (R)	3.0159 Ω
Line reactance (X)	22.31625 Ω (59.19 mH)
Series-compensating resistor and reactor 1 (R_{1se}, L_{1se})	**1.7854 Ω, 47.358 mH**
Series-compensating reactor 2 (L_{2se})	**47.358 mH**
Series-compensating reactor 3 (L_{3se})	**47.358 mH**

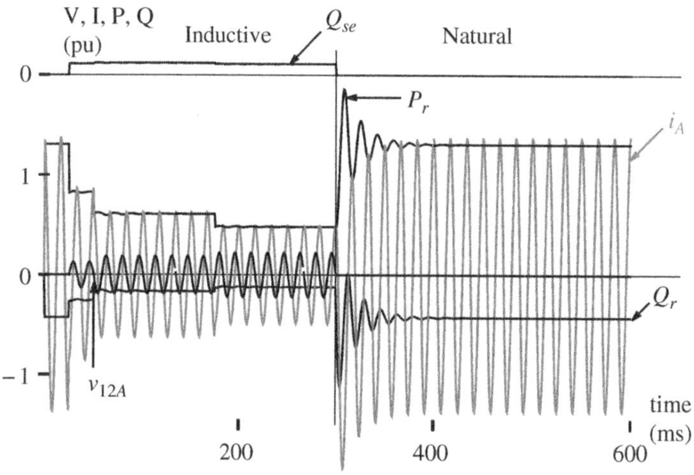

Figure 5-8 Exchanged reactive power (Q_{se}), the resulting voltage (v_{12A}), current (i_A), and the active and reactive power flows (P_r and Q_r) at the receiving end of a line due to series-compensating, mechanically-switched reactors, connected to the line.

exchanged with the line is shown. At 300 ms, all the series breakers are closed, bypassing the series inductive compensation. The line current and power flow return to the uncompensated levels.

Next, the impedance between BUS01 and BUS02 nodes is calculated from measuring the voltages between the BUS01 and BUS02 nodes and the line current, and taking their ratio. This compensating voltage is connected in series with the line, replacing the series-compensating reactor to demonstrate that the effect of a virtual reactor is the same as that of an actual reactor, described in Section 5.2.3. Even though only the effect of a compensating reactor is studied in this section, the steps for calculating the virtual impedance is the same as that in the case of a virtual reactance. The instantaneous filtered magnitude (v_{1dqf}) and the phase angle (θ_{v12dq}) of the voltage across the series-compensated reactors with respect to the PLL are shown in Figures 5-9 and 5-10, respectively. The corresponding datacards are shown in bold in Code 5-9.

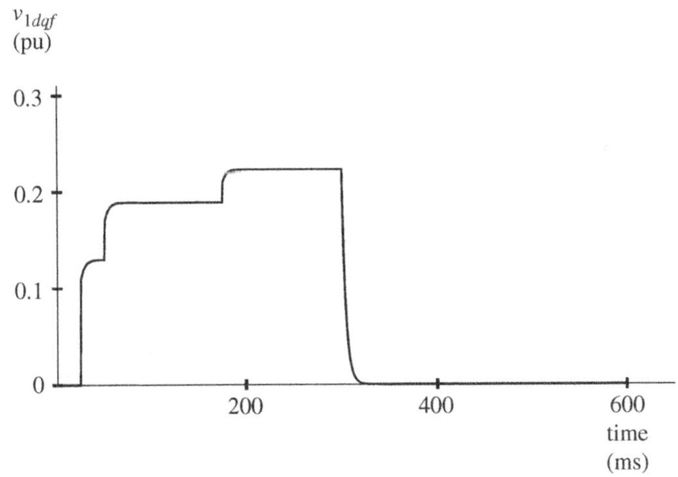

Figure 5-9 Instantaneous filtered magnitude of the voltage across the series-compensating reactors.

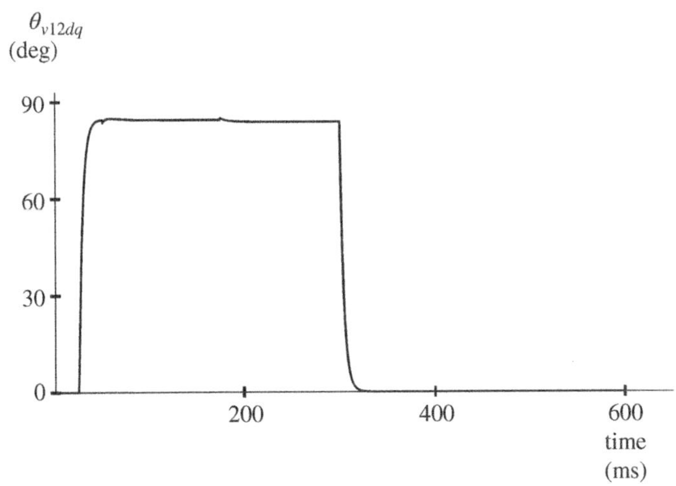

Figure 5-10 Instantaneous filtered phase angle of the voltage across the series-compensating reactors with respect to the PLL angle.

The instantaneous magnitude (i_{dq}) in pu of the line current, which is the current through the series-connected reactor, and its phase angle (θ_{idq}) in degrees are shown in Figures 5-11 and 5-12, respectively. The corresponding datacards are shown in bold in Code 5-9.

The instantaneous magnitude (z_{se}) and phase angle (θ_{zse}) of the impedance of the series-compensating reactors are shown in Figures 5-13 and 5-14, respectively. The corresponding datacards are shown in bold in Code 5-9. Note that the calculated magnitudes are the same (0.15, 0.30, and 0.45 pu) as those that were used in this study. Also note that when various reactors are connected in series with the line during the period from 25 to 300 ms, the voltage leads the line current by nearly 90°. The exact phase angle depends on the quality factor (Q) of the total amount of series-compensating reactor.

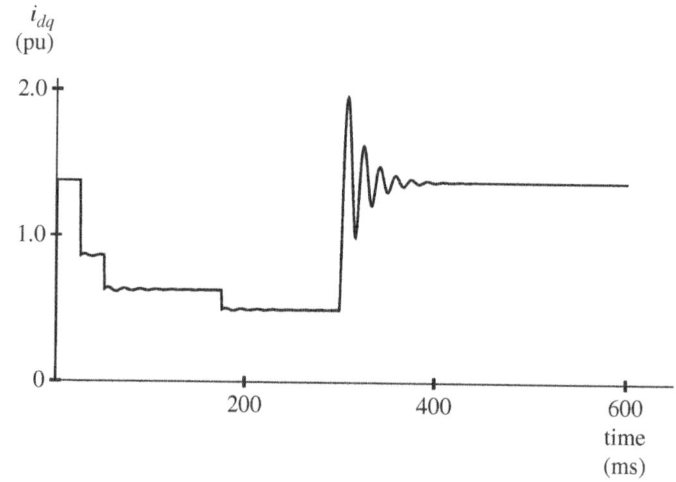

Figure 5-11 Instantaneous magnitude (i_{dq}) of the current through the series-compensating reactors.

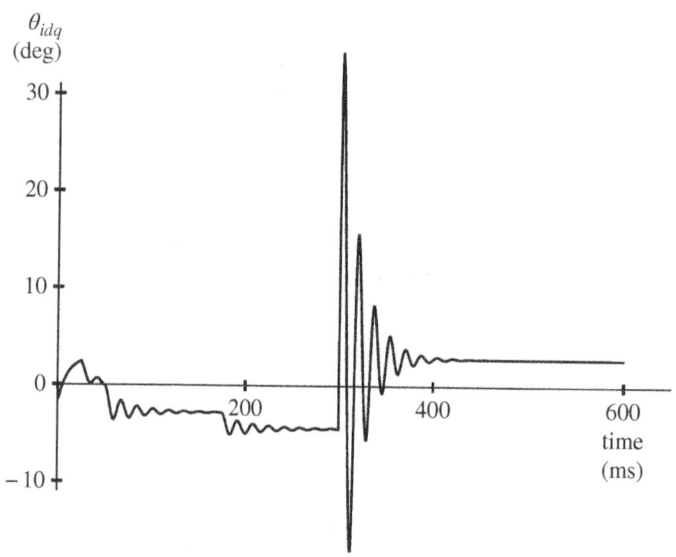

Figure 5-12 Instantaneous phase angle (θ_{idq}) of the current through the series-compensating reactors with respect to the PLL angle.

Since $\mathbf{V}_{21} = -\mathbf{V}_{12}$, an angle of π radians needs to be added to the phase angle, shown in Figure 5-10, to calculate the instantaneous relative phase angle (β) of the compensating voltage ($\mathbf{V}_{s's}$) as shown in Figure 5-15. This addition of π radians or 180° is also present in the calculation of the relative phase angle (β) during the time when the line is uncompensated, which is of no concern, since during no compensation, the magnitude of the series-compensating voltage is zero.

From the magnitude (v_{1dqf}) of the compensating voltage, shown in Figure 5-9, and the relative phase angle (β), shown in Figure 5-15, the three-phase compensating voltage in time domain is calculated as shown in bold in Code 5-9. The *A* phase of the calculated compensating voltage

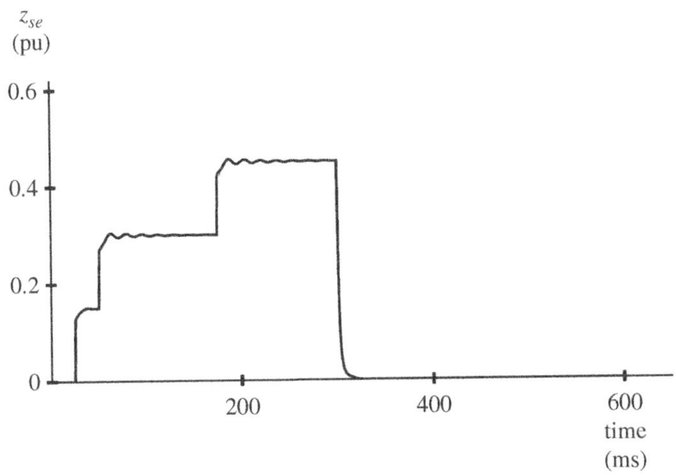

Figure 5-13 Instantaneous magnitude (z_{se}) of the impedance of the series-compensating reactors.

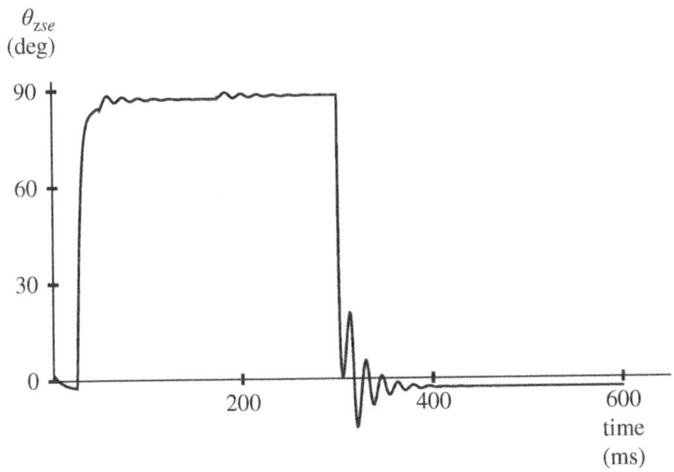

Figure 5-14 Instantaneous phase angle (θ_{zse}) of the voltage across the series-compensating reactors with respect to the prevailing line current.

(v_{12Apu}) and the A phase of the measured voltage (e_{2Apu}) between BUS01 and BUS02 nodes are superimposed in Figure 5-16 and found to be identical. Therefore, the process of generating the compensating voltage is validated.

5.2.2 Mechanically-Switched Capacitor (MSC) with a Reactor

In this example, the series reactances from a three-phase reactor and a three-phase capacitor, shown in Figure 5-17, are inserted in two stages. The two series breakers (SRSBK1 and SRSBK2)

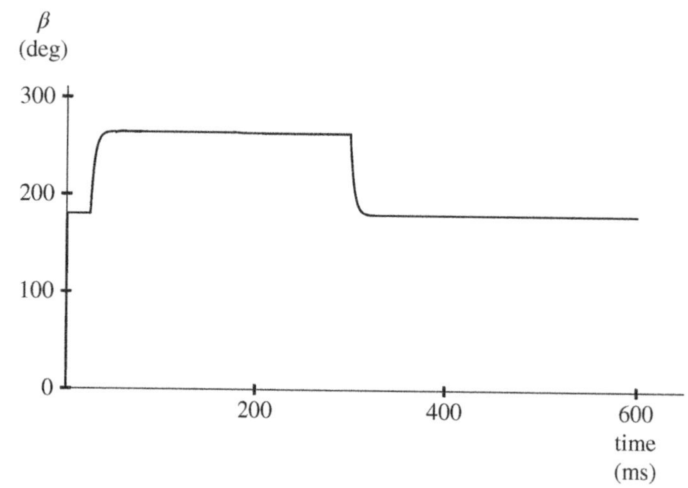

Figure 5-15 Instantaneous relative phase angle (β) of the series-compensating voltage with respect to the PLL angle.

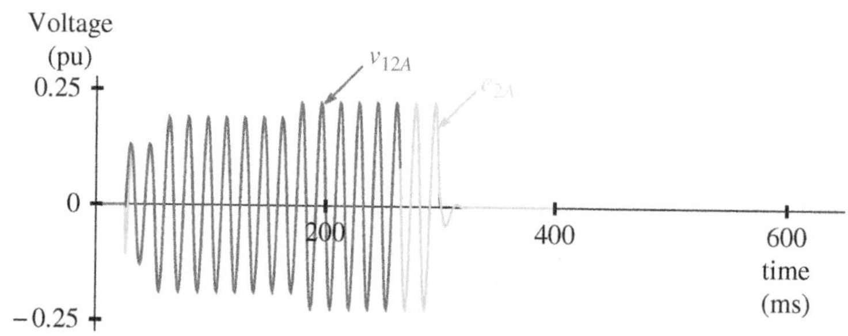

Figure 5-16 Two superimposed voltages: (1) *A* phase of the calculated compensating voltage (v_{12Apu}) and (2) *A* phase of the measured voltage (e_{2Apu}) between BUS01 and BUS02.

are closed initially, leaving the line with its natural active power flow of 1.31 pu and reactive power flow of −0.43 pu at the receiving end. The first series-compensating reactance, $X_{1se} = 0.15 \times Z_{base} = 0.15 \times 119.025 = 17.854\,\Omega$ and corresponding reactor, $L_{1se} = X_{1se}/\omega = 17.854/(2 \times \pi \times 60) = 47.358$ mH. Therefore, the series resistor, $R_{1se} = X_{1se}/10 = 1.7854\,\Omega$, assuming the quality factor (Q) is 10. The second series-compensating reactance, $X_{2se} = 0.15 \times Z_{base} = 0.15 \times 119.025 = 17.854\,\Omega$ and the corresponding capacitor, $C_{2se} = 1/(\omega X_{2se}) = 1/(2 \times \pi \times 60 \times 17.854) = 148.57\mu$F.

A series-compensating reactor/capacitor is implemented in the datafile **513SEREA.DAT**, which is the same as **509SEREA.DAT**, except with a change in the series reactance file as shown in Code 5-13. The datacard "$INCLUDE 514SERE2.BRN" attaches the $INCLUDE file 514SERE2.BRN that connects a branch file as shown in Code 5-14.

Table 5-5 shows the electrical system data for a two-generator/single-line power system network, compensated by a series-connected reactor/capacitor.

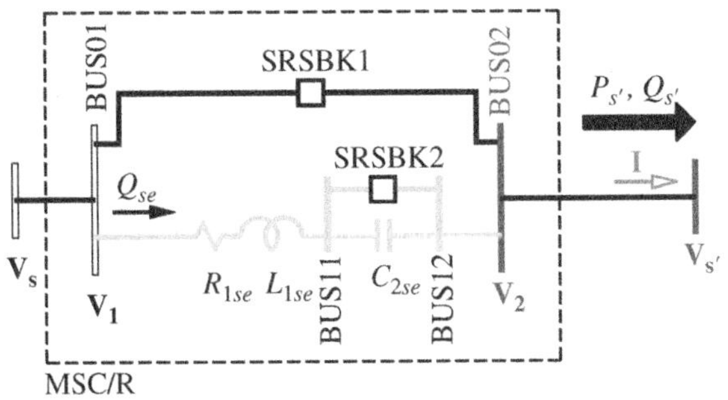

Figure 5-17 Series-compensating, mechanically-switched capacitor with a reactor, connected to the line.

Figure 5-18 shows the EMTP simulation results when three-phase series reactor/capacitor banks are connected to the transmission line. The normally-closed series breaker 1 (SRSBK1) opens at 25 ms and stays open for the next 275 ms to connect the inductive reactance (X_L) in series with the transmission line between the BUS01 and BUS02 nodes. A voltage (v_{12}) develops between BUS01 and BUS02 nodes. The A-phase voltage (v_{12A}) leads or lags the line current (i_A) when the effective impedance of the line is inductive or capacitive, respectively. Both active and reactive power flows (P_r and Q_r) in the line decrease. At 50 ms, series breaker 2 (SRSBK2) is opened, adding

Code 5-13 EMTP datafile for a series-compensating, mechanically-switched capacitor with a reactor, integrated in a two-generator/single-line power system network (513SEREA.DAT).

```
BEGIN NEW DATA CASE
C File Name: 513SEREA.DAT
C OPERATION OF A SERIES-COMPENSATING, MECHANICALLY-SWITCHED REACTOR/CAPACITOR
C 000000011111111112222222222333333333344444444445555555555666666666677777777778
C 345678901234567890123456789012345678901234567890123456789012345678901234567890
C ----dt<---tmax<---Xopt<---Copt<--Toler<        ><        >
16.666-6600.00-3
C 000000011111111112222222222333333333344444444445555555555666666666677777777778
C 345678901234567890123456789012345678901234567890123456789012345678901234567890
C -Iprnt<--Iplot<-Idoubl<-KssOut<-MaxOut<        ><        ><---Icat<-Nenerg
      40      10                                                        1
TACS HYBRID
99CRD       = 180 / PI
99CDR       = 1 / CRD
99TWOPI     = 2.0 * PI
C BASE VOLTAGE, CURRENT, AND POWER
99Vbase     = 112676.528
99Ibase     = 946.662704
99Zbase     = Vbase / Ibase
99Pbase     = 1.5 * Vbase * Ibase
```

(Continued)

Code 5-13 (Continued)

```
C INPUTS FROM NODE VOLTAGES AND LINE CURRENTS
$INCLUDE 302MEAS1.SWT
C NORMALIZATION OF MEASURED VARIABLES AND COMPUTATION OF PEAK VOLTAGES/CURRENTS
$INCLUDE 303PTCT.SCL
C VECTOR PHASE-LOCKED LOOP
$INCLUDE 310VPLL.PLL
C USER'S INPUT FILE
$INCLUDE 510SEREA.USR
$INCLUDE 511SEREA.CON
C COMPUTATION OF LINE RESISTANCE
$INCLUDE 305LINER.CMP
C COMPENSATING VOLTAGE MAGNITUDE AND PHASE ANGLE COMPUTATION
99v12dsp  = v12Apu
99v12qsp  = -(v12Apu + 2.0 * v12Cpu) / SQRT(3)
99v12dpu  =  v12dsp * COSPLL + v12qsp * SINPLL
99v12qpu  = -v12dsp * SINPLL + v12qsp * COSPLL
99v12dqp  = SQRT(v12dpu * v12dpu + v12qpu * v12qpu)
C LOW-PASS FILTER INTRODUCES A RAMP-RATE AND REDUCES SYSTEM TRANSIENTS
 1v12dqf  +v12dqp
      1.0
      1.0    0.0040
99zse     = v12dqf / idqpu
99thv12a  = ATAN( v12qpu / (v12dpu + 1.0E-8) )
99thv12b  = thv12a + PI * (v12dpu .LT. 0.0)
99thv12   = thv12b + TWOPI * (thv12b .LT. 0.0) - TWOPI * (thv12b .GE. TWOPI)
C LOW-PASS FILTER INTRODUCES A RAMP-RATE AND REDUCES SYSTEM TRANSIENTS
 1thv12f  +thv12
      1.0
      1.0    0.0040
99tv12dq  = thv12f * CRD
99theti   = thetir + TWOPI * (thetir .LT. -PI) - TWOPI * (thetir .GE. PI)
99thidq   = theti * CRD
99zsetha  = thv12f - thetir
99zsethb  = zsetha + TWOPI * (zsetha .LT. -PI) - TWOPI * (zsetha .GE. PI)
99thzse   = zsethb * CRD
99beta    = (thv12f + PI) * CRD
C The PI is added, since the polarity of V21 is opposite of the polarity of V12.
99tha2    = theta + PI / 2 + PI + beta * CDR
C This addition of PI/2 is because the compensating voltage is sine reference
C and the PLL is cosine reference.
C PI radian is added to reverse the polarity of the series-coupling transformer.
99theta2  = tha2 + TWOPI * (tha2 .LT. 0.0) - TWOPI * (tha2 .GE. TWOPI)
99vsrspk  = v12dqf * Vbase
99e2a     = vsrspk * SIN(theta2)
```

(*Continued*)

Code 5-13 (Continued)

```
99e2b       = vsrspk * SIN(theta2 - 120 * CDR)
99e2c       = vsrspk * SIN(theta2 + 120 * CDR)
99e2Apu     = e2A / Vbase
33v12ApuiApu  Qsepu Prpu  Qrpu
BLANK RECORD ENDING TACS
C 0000000111111111122222222223333333333444444444455555555556666666666777777777778
C 3456789012345678901234567890123456789012345678901234567890123456789012345678901234567890
C <---Nodes--><---Nodes--><----R<----L<----C                                          V
C <-Bus1<-Bus2<-Bus3<-Bus4<-Ohms<---mH<---uF                                          V
C REPRESENTATION OF SHORTED VS AND BUS01 NODES
  VSA   BUS01A             0.001
  VSB   BUS01B             0.001
  VSC   BUS01C             0.001
C CONNECTION OF NETWORK 2 BRANCHES WITH DAMPING RESISTORS (FIGURE 3-5)
$INCLUDE 311NTWK2.BRN
C HIGH RESISTANCES TO AVOID FLOATING NODES IN THE CIRCUIT
  BUS01ABUS02A            1.00+8
  BUS01BBUS02B            1.00+8
  BUS01CBUS02C            1.00+8
C SERIES-CONNECTED REACTOR (FIGURE 5-7)
$INCLUDE 514SERE2.BRN
BLANK RECORD ENDING BRANCHES
C <-Bus1<-Bus2<--Vig---|<-Ihold--|<-Tdeion-|xxxxxxxxxxCLOSEDSAMEGRID->SIGNAL  oo
C CLOSED SERIES BREAKERS AT START OF THE SIMULATION
13BUS01ABUS02A                                                 CLOSED          SRSBK1
13BUS01BBUS02B                                                 CLOSED          SRSBK1
13BUS01CBUS02C                                                 CLOSED          SRSBK1
13BUS11ABUS12A                                                 CLOSED          SRSBK2
13BUS11BBUS12B                                                 CLOSED          SRSBK2
13BUS11CBUS12C                                                 CLOSED          SRSBK2
13BUS12ABUS02A                                                 CLOSED          SRSBK3
13BUS12BBUS02B                                                 CLOSED          SRSBK3
13BUS12CBUS02C                                                 CLOSED          SRSBK3
C MEASUREMENT OF LINE CURRENTS
$INCLUDE 307MEAS2.SWT
BLANK RECORD ENDING SWITCHES
C 0000000111111111122222222223333333333444444444455555555556666666666777777777778
C 3456789012345678901234567890123456789012345678901234567890123456789012345678901234567890
C CONNECTION OF NETWORK 1 SOURCES (FIGURE 3-5)
$INCLUDE 308NTWK1.SRC
BLANK RECORD ENDING SOURCES
BLANK RECORD ENDING NODE VOLTAGE OUTPUT
BLANK RECORD ENDING PLOT
```

Code 5-14 EMTP $INCLUDE file for implementing a three-phase series-connected capacitor with a reactor (514SERE2.BRN).

```
C File Name: 514SERE2.BRN
C THREE-PHASE, SERIES-COMPENSATING CAPACITOR WITH A REACTOR
C 00000001111111111222222222233333333334444444444555555555566666666667777777778
C 34567890123456789012345678901234567890123456789012345678901234567890123456789 0
C <---Nodes--><---Nodes--><----R<----L<----C                                  V
C <-Bus1<-Bus2<-Bus3<-Bus4<-Ohms<---mH<---uF                                  V
  BUS01ABUS11A            1.785447.358
  BUS01BBUS11B            1.785447.358
  BUS01CBUS11C            1.785447.358
  BUS11ABUS12A                        148.57
  BUS11BBUS12B                        148.57
  BUS11CBUS12C                        148.57
  BUS12ABUS02A             0.001
  BUS12BBUS02B             0.001
  BUS12CBUS02C             0.001
```

Table 5-5 Electrical system data.

Parameters	Values
Source voltage ($\mathbf{V_{src}}$) phase-to-neutral	112,676.528∠0° V_{peak}
Source resistance (R_{src})	1.0053 Ω
Source reactance (X_{src})	7.43875 Ω (19.73 mH)
Receiving-end voltage ($\mathbf{V_r}$) phase-to-neutral	112,676.528∠−20° V_{peak}
Line resistance (R)	3.0159 Ω
Line reactance (X)	22.31625 Ω (59.19 mH)
Series-connected resistor/reactor (R_{1se}, L_{1se})	1.7854 Ω, 47.358 mH
Series-compensating capacitor (X_{2se})	**148.57 μF**

15% capacitive compensation that nullifies the inductive compensation that is already in place. This is exactly what happens when a VSC-based SSSC is used. A part of the VSC's capacitive rating is used to nullify the effect of the added series inductive reactance, contributed by the series-coupling transformer leakage reactance, which is typically 15% of the full power rating of the SSSC. These operating points are verified in series compensation with VSCs in Sen and Sen (2009). The compensating reactive power (Q_{se}) exchanged with the line is shown. At 300 ms, all the series breakers are closed, bypassing the series inductive and capacitive compensations. The line current and power flow return to the uncompensated or natural levels.

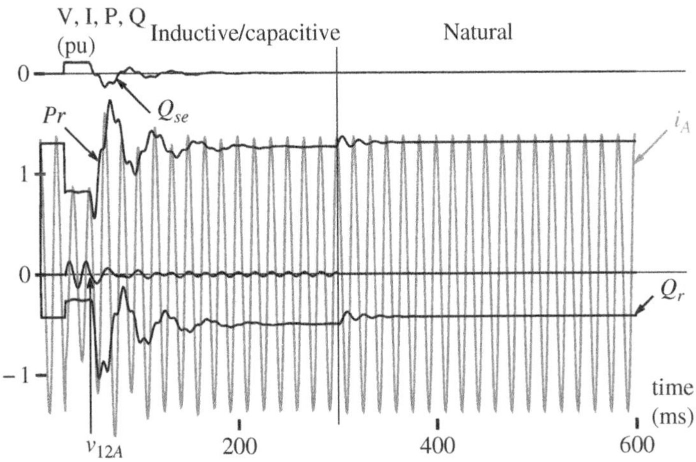

Figure 5-18 Exchanged reactive power (Q_{se}), the resulting voltage (v_{12A}), current (i_A), and the active and reactive power flows (P_r and Q_r) at the receiving end of a line due to a series-compensating, mechanically-switched reactor/capacitor, connected to the line.

5.2.3 Series Reactance Emulator

The compensating voltage ($\mathbf{V}_{s's}$), calculated in Section 5.2.1, is used to replace the series-connected reactor between the BUS01 and BUS02 nodes as shown in Figure 5-19. The corresponding EMTP datafile **515PQPFC.DAT** and the user's input file are shown in Code 5-15 and Code 5-16, respectively.

Table 5-6 shows the electrical system data for a two-generator/single-line power system network, compensated by an ideal series-connected PFC, operating as a virtual reactor.

Figure 5-20 shows the EMTP simulation results when the compensating voltage is integrated in a two-generator/single-line power system network during the time from 75 to 300 ms. The normally-

Figure 5-19 Series-compensating voltage to emulate a reactor in series with the line.

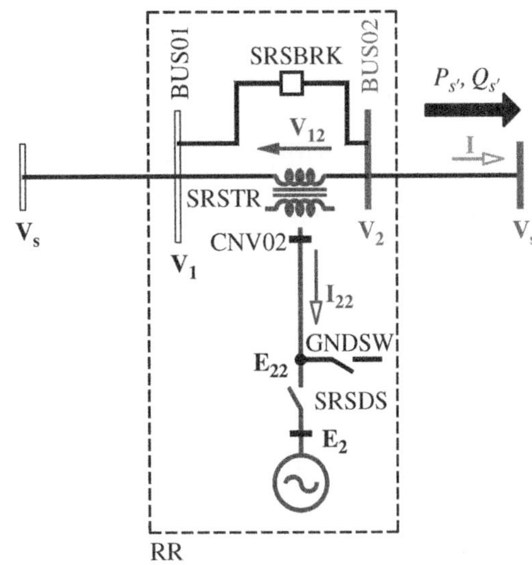

Code 5-15 EMTP datafile for a series-compensating reactor, integrated in a two-generator/single-line power system network (515PQPFC.DAT).

```
BEGIN NEW DATA CASE
C File Name: 515PQPFC.DAT
C STUDY OF THE EMULATION OF A SERIES-COMPENSATING REACTOR, USING THE MATHEMATICAL
C MODEL OF ST WITH A SHUNT-CONNECTED CURRENT SOURCE (SHUNT UNIT) AND
C A SERIES-CONNECTED VOLTAGE SOURCE (SERIES UNIT),USING AN IDEAL
C SERIES-COUPLING TRANSFORMER WITHOUT ANY LEAKAGE IMPEDANCE
C 0000000111111111122222222223333333333444444444455555555556666666666777777777778
C 345678901234567890123456789012345678901234567890123456789012345678901234567890
C ----dt<---tmax<---Xopt<---Copt<--Toler<       ><       >
16.666-6600.00-3
C 0000000111111111122222222223333333333444444444455555555556666666666777777777778
C 345678901234567890123456789012345678901234567890123456789012345678901234567890
C -Iprnt<--Iplot<-Idoubl<-KssOut<-MaxOut<       ><       ><---Icat<-Nenerg
      40      10                                                 1
TACS HYBRID
99CRD      = 180 / PI
99CDR      = 1 / CRD
99TWOPI    = 2.0 * PI
C BASE VOLTAGE, CURRENT, AND POWER
99Vbase    = 112676.528
99Ibase    = 946.662704
99Zbase    = Vbase / Ibase
99Pbase    = 1.5 * Vbase * Ibase
C INPUTS FROM NODE VOLTAGES AND LINE CURRENTS
$INCLUDE 302MEAS1.SWT
C MEASUREMENT OF CURRENTS FLOWING THROUGH THE SERIES UNIT
91CNV02A
91CNV02B
91CNV02C
C NORMALIZATION OF MEASURED VARIABLES AND COMPUTATION OF PEAK VOLTAGES/CURRENTS
$INCLUDE 303PTCT.SCL
C VECTOR PHASE-LOCKED LOOP
$INCLUDE 310VPLL.PLL
C USER'S INPUT FILE
$INCLUDE 516PQPFC.USR
C COMPUTATION OF LINE RESISTANCE
$INCLUDE 305LINER.CMP
C SERIES-COMPENSATING VOLTAGE
99tha2     = theta + PI / 2 + PI + beta
C This addition of PI/2 is because the compensating voltage is sine reference
C and the PLL is cosine reference.
C PI radian is added to reverse the polarity of the series-coupling transformer.
```

(Continued)

Code 5-15 (Continued)

```
99theta2   = tha2 + TWOPI * (tha2 .LT. 0.0) - TWOPI * (tha2 .GE. TWOPI)
99vsrspk   = e2dqrf * Vbase
99e2a      = vsrspk * SIN(theta2)
99e2b      = vsrspk * SIN(theta2 - 120 * CDR)
99e2c      = vsrspk * SIN(theta2 + 120 * CDR)
C Setting up the disconnect switches
99GNDSW    = 0
99SRSDS    = .NOT. GNDSW
33Prpu  Qrpu
BLANK RECORD ENDING TACS
C 00000001111111111222222222233333333334444444444555555555566666666667777777778
C 34567890123456789012345678901234567890123456789012345678901234567890123456789O
C <---Nodes--><---Nodes--><----R<----L<----C                                  V
C <-Bus1<-Bus2<-Bus3<-Bus4<-Ohms<---mH<---uF                                  V
C REPRESENTATION OF SHORTED VS AND BUS01 NODES
  VSA   BUS01A             0.001
  VSB   BUS01B             0.001
  VSC   BUS01C             0.001
C CONNECTION OF NETWORK2 BRANCHES WITH DAMPING RESISTORS (FIGURE 3-5)
$INCLUDE 311NTWK2.BRN
C HIGH RESISTANCES TO AVOID FLOATING NODES IN THE CIRCUIT
  e2a   e22a              1.00+8
  e2b   e22b              1.00+8
  e2c   e22c              1.00+8
  e22a                    1.00+8
  e22b                    1.00+8
  e22c                    1.00+8
C CONNECTION OF A 3-PHASE SERIES-COUPLING TRANSFORMER WITH NO LEAKAGE IMPEDANCE
$INCLUDE 402IDLT2.TRN
BLANK RECORD ENDING BRANCHES
C 00000001111111111222222222233333333334444444444555555555566666666667777777778
C 34567890123456789012345678901234567890123456789012345678901234567890123456789O
C <-Bus1<-Bus2<---Tclose<----Topen<-------Ie<----Flash<--Request<-----Target<--O
C SERIES DISCONNECT SWITCH TO CONNECT/DISCONNECT THE SERIES UNIT TO THE CIRCUIT
13e2a   e22a                                                        SRSDS
13e2b   e22b                                                        SRSDS
13e2c   e22c                                                        SRSDS
C TECHNIQUE TO SHORT THE SERIES-COMPENSATING VOLTAGE TO CALCULATE THE
C INITIAL CONDITIONS CORRECTLY
13e22a                                          CLOSED              GNDSW
13e22b                                          CLOSED              GNDSW
13e22c                                          CLOSED              GNDSW
```

(Continued)

Code 5-15 (Continued)

```
C MEASUREMENT OF LINE CURRENTS
$INCLUDE 307MEAS2.SWT
C MEASUREMENT OF CURRENTS FLOWING THROUGH THE SERIES UNIT
  CNV02Ae22a                                               MEASURING
  CNV02Be22b                                               MEASURING
  CNV02Ce22c                                               MEASURING
BLANK RECORD ENDING SWITCHES
C 00000001111111111222222222233333333334444444444555555555566666666667777777778
C 34567890123456789012345678901234567890123456789012345678901234567890123456789 0
C CONNECTION OF SERIES UNIT VOLTAGE SOURCES TO THE NETWORK
60e2a
60e2b
60e2c
C CONNECTION OF NETWORK1 SOURCES (FIGURE 3-5)
$INCLUDE 308NTWK1.SRC
BLANK RECORD ENDING SOURCES
BLANK RECORD ENDING NODE VOLTAGE OUTPUT
BLANK RECORD ENDING PLOT
```

Code 5-16 EMTP $INCLUDE file for user's input (516PQPFC.USR).

```
C File Name: 516PQPFC.USR
C USER'S INPUT FILE FOR PQPFC
C 00000001111111111222222222233333333334444444444555555555566666666667777777778
C 34567890123456789012345678901234567890123456789012345678901234567890123456789 0
C SETTING THE COMPENSATING VOLTAGE
99e2cm1    = 0.1889 * (TIMEX .GE. 0.050)
99e2cm2    = 0.0339 * (TIMEX .GE. 0.175)
99e2cm3    = 0.2228 * (TIMEX .GE. 0.300)
99e2dqcm   = e2cm1 + e2cm2 - e2cm3
C LOW-PASS FILTER INTRODUCES A RAMP-RATE AND REDUCES SYSTEM TRANSIENTS
 1e2dqrf  +e2dqcm
       1.0
       1.0    0.0160
99betcm1   = 264.31 * (TIMEX .GE. 0.050)
99betcm2   = 0.68 * (TIMEX .GE. 0.175)
99betcmd   = (betcm1 - betcm2) * CDR
C LOW-PASS FILTER INTRODUCES A RAMP-RATE AND REDUCES SYSTEM TRANSIENTS
 1beta     +betcmd
       1.0
       1.0    0.0020
```

closed series breaker (SRSBRK) opens at 75 ms and stays open for the next 225 ms to connect the compensating voltage ($\mathbf{V_{s's}}$) whose magnitude and relative phase angle are shown in Figures 5-9 and 5-15, respectively. Both the active and reactive power flows (P_r and Q_r) at the receiving end of the line decrease. Further increase in the inductive compensation at 175 ms decreases the active and reactive power flows at the receiving end of the line simultaneously. At 300 ms, the series-compensating voltage is set to 0, and the line current and power flow return to the uncompensated levels. The active and reactive power flows (P_r and Q_r) at the receiving end of the line, shown in Figure 5-8, due to a series inductive compensation are superimposed in Figure 5-20 and found to be identical under steady-state operating conditions. Therefore, the series-compensating voltage demonstrates that the effect of a virtual reactor is the same as that of an actual reactor as described in Section 5.2.1.

Table 5-6 Electrical system data.

Parameters	Values
Source voltage ($\mathbf{V_{src}}$) phase-to-neutral	112,676.528∠0° V_{peak}
Source resistance (R_{src})	1.0053 Ω
Source reactance (X_{src})	7.43875 Ω (19.73 mH)
Receiving-end voltage ($\mathbf{V_r}$) phase-to-neutral	112,676.528∠−20° V_{peak}
Line resistance (R)	3.0159 Ω
Line reactance (X)	22.31625 Ω (59.19 mH)
Series-coupling transformer (SRSTR) turns-ratio	1:1
Primary winding resistance	0.001 Ω
Primary winding leakage inductance	0.001 mH
Secondary winding resistance	0.001 Ω
Secondary winding leakage inductance	0.001 mH

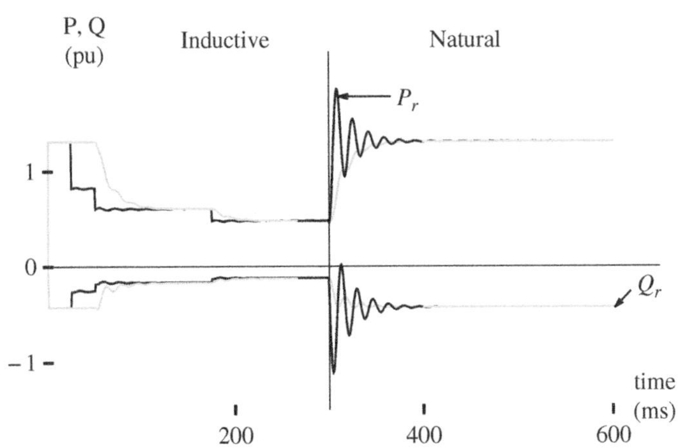

Figure 5-20 Validation of a series-compensating voltage as an emulated reactor.

This technique of creating a series-compensating voltage may be extended further to make the compensating voltage of any magnitude within the design limit and at any phase angle with respect to the prevailing line current, which, in turn, emulates a four-quadrant, series-compensating impedance ($\mathbf{Z_{se}} = R_{se} - jX_{se}$) that consists of a resistance ($R_{se} = +R$ or $-R$) and a reactance ($X_{se} = X_C$ or $-X_L$) in series with the line.

6

Sen Transformer

Sen Transformer (ST), a power-regulating transformer, was presented in *Introducing the Family of Sen Transformers: A Set of Power Flow Controlling Transformers*, by K. K. Sen and M. L. Sen at Power Engineering Society Summer Meeting, Chicago, USA, 2002, and published in *IEEE Trans. on Power Delivery*, vol. 18, no. 1, pp. 149–157, January 2003. The ST is proposed as part of a broader SMART Power Flow Controller (SPFC) concept that enhances the controllability in an electric power transmission system by using functional requirements and cost-effective solutions. An SPFC is an evolving technology that uses the best technical features from all previous concepts that have been developed in the power flow control area until now. It uses any solution that is power electronics-based, non-power electronics-based or a combination of both, irrespective of being stationary or rotating, to meet the needs of utilities worldwide. The features of a Unified Power Flow Controller (UPFC) – voltage regulation, resistance regulation, reactance regulation, impedance regulation, phase angle regulation, and independent control of active and reactive power flows – were demonstrated at American Electric Power's Inez substation in 1998. The ST, a member of an SPFC family, adopts the best features of a UPFC and the proven low-cost hardware, used in a Phase Angle Regulator (PAR), to create a viable Power Flow Controller (PFC) that can be affordable to utilities worldwide.

In this chapter, the discussion topics are as follows:

- Power flows in uncompensated and compensated lines
- Shunt-Shunt and Shunt-Series configurations and the associated Transformer/Load Tap Changers (LTCs)-based solutions
- ST model using the Electro-Magnetic Transients Program (EMTP) software for the
  - regulation of compensating voltage magnitude and phase angle
  - selective programming of the Tap Control Unit to emulate an impedance in series with the line
  - factory tests and power flow tests required for the verification of ST performance
- Multiline ST (MST) with either shunt-connected or series-connected Compensating-Voltage Units
- Generalized ST (GST) with both shunt-connected and series-connected Compensating-Voltage Units
- Summary of the comparison of the VRT, PAR, UPFC, and ST characteristics.

Power Flow Control Solutions for a Modern Grid using SMART Power Flow Controllers, First Edition.
Kalyan K. Sen and Mey Ling Sen.

Companion website: www.wiley.com/go/sen/powerflowcontrol

Today's power grid has evolved into a massive integration of renewable generation, which is intermittent in nature. The intermittency stems from clouds passing over a solar farm or the wind changing its speed at the wind farm. Therefore, traditional steady-state power flow controllers, such as series-connected reactor/capacitor concepts need to be updated with an improved dynamic response. Additionally, increasing installation of roof-top solar and its integration into a low-voltage distribution network has altered the traditional voltage profile in the distribution network and increased the need for a bidirectional power flow controller when the renewable generation is not available as outlined in PES-TR-77, titled "Stability Definitions and Characterization of Dynamic Behavior in Systems with High Penetration of Power Electronic Interfaced Technologies," which is available at https://resourcecenter.ieee-pes.org/technical-publications/technical-reports/PES_TP_TR77_PSDP_stability_051320.html. The ST is capable of regulating a bidirectional power flow and maintaining the desired voltage profile.

In Chapter 2, it was derived that any regulation technique whether it is voltage regulation, power flow regulation, resistance regulation, reactance regulation, impedance regulation or phase angle regulation is actually a direct or indirect implementation of an impedance regulation technique. The ST can regulate the line impedance directly, which is beneficial in stabilizing Inverter-Based Resources (IBRs), such as solar farms where numerous inverters are optimized to operate with their control gains, which are tuned to match the parameters of the plant that the inverters are connected to. In an event when a line trips, the impedance of the plant changes and the control parameters become out-of-tune, causing the IBRs' operation to become unstable and the IBRs to trip as outlined in PES-TR-77. Therefore, utilizing the impedance regulation feature of ST, the IBRs can be operated with stability during various network contingencies.

It is widely accepted that the active power flow in a line is primarily influenced by the phase angle difference of the voltages at both ends of the line and reactive power flow in a line primarily influenced by the magnitude difference of the voltages at both ends of the line. This is true when the line reactance is more dominant than the line resistance. However, in low-voltage distribution lines, it is opposite where line resistance is more dominant than the line reactance. Therefore, both the line resistance and reactance must be considered in determining the accurate power flow in a line. If the line impedance needs to be altered, an Impedance Regulator (IR), such as ST, is needed.

Modeling is essential for a successful implementation of a concept. Before the realization of a utility project, a load flow and voltage stability study are usually performed. In these studies, the power system network may consist of tens of thousands of voltage busses. However, a transient study typically involves the simulation of a reduced power system network that consists of only a few to tens of voltage busses around the installation site. The transients study can be performed using a mathematical model of a particular controller, such as an IR. This model is also suitable for a transient study during a fault and to formulate a protection scheme.

In this chapter, an ST model is developed using EMTP. The emphasis of this model is to preserve the details of an ST configuration that cannot be obtained from a load flow model, since a load flow model generates steady-state results, using a positive-sequence single-phase equivalent circuit. The operation of the EMTP model is verified with the model connected to several different power system networks that can easily be upgraded to a particular utility's more representative power system network. The model is run for a sufficiently long time until steady-state conditions are reached for each operating point under consideration. These results give the transmission network planners important information before proceeding further in the realization of an actual installation.

6.1 Existing Solutions

As described in detail in Chapter 2, the electric power flow through an AC transmission line is a function of the line impedance ($\mathbf{Z} = R + jX$), the magnitudes of the sending-end voltage ($\mathbf{V_s} = V_s \angle \delta_s$) and the receiving-end voltage ($\mathbf{V_r} = V_r \angle \delta_r$), and the phase angle ($\delta$) between these two voltages. Figure 6-1 shows a two-generator/one-line power system network with only a line reactance (X), connecting the sending-end and receiving-end voltage sources. The difference between the sending-end voltage ($\mathbf{V_s}$) and the receiving-end voltage ($\mathbf{V_r}$) determines the voltage ($\mathbf{V_{Xn}}$) across the line reactance (X) when the line resistance (R) is neglected. The voltage ($\mathbf{V_{Xn}}$) determines the line current ($\mathbf{I_n}$) and, in turn, the active and reactive power flows (P_{sn} and Q_{sn}) at the sending end and (P_{rn} and Q_{rn}) at the receiving end of the line. The suffix "n" indicates the natural power flow condition when the line is uncompensated.

The natural active and reactive power flows (P_{sn} and Q_{sn}) at the sending end and the natural active and reactive power flows (P_{rn} and Q_{rn}) at the receiving end are defined in Chapter 2 as

$$P_{sn} = a \times \sin\delta \tag{2-40}$$

where

$$a = \frac{V_s V_r}{X} \tag{2-41}$$

and

$$\delta = \delta_s - \delta_r, \tag{2-27}$$

$$Q_{sn} = a\left(\frac{V_s}{V_r} - \cos\delta\right), \tag{2-43}$$

$$P_{rn} = a \times \sin\delta, \tag{2-46}$$

and

$$Q_{rn} = a\left(\cos\delta - \frac{V_r}{V_s}\right). \tag{2-48}$$

The natural or uncompensated power flow through a transmission line in a power network is, in general, not optimal. It can be improved with a PFC by regulating the power flow control

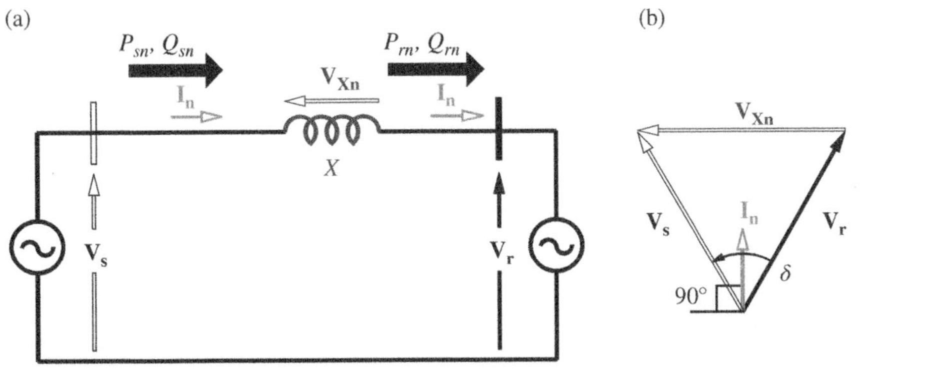

Figure 6-1 (a) Two-generator/one-line uncompensated power system network; (b) phasor diagram.

parameters, such as line voltage magnitude, its phase angle, and line reactance. Any of the power flow control parameters can be regulated with the use of the following equipment:

- Voltage-Regulating Transformer (VRT), shunt or parallel-connected switched reactor/capacitor, also known as Shunt Reactor (SR)/Shunt Capacitor (SC), Static Var Compensator (SVC), or STATic synchronous COMpensator (STATCOM) for voltage regulation
- Phase Angle Regulator (PAR) or Phase-Shifting Transformer (PST) for phase angle regulation
- Thyristor-Controlled Series Capacitor (TCSC) or Static Synchronous Series Compensator (SSSC) for series reactance regulation.

By regulating any one of the power flow control parameters using one of the above equipment, both the active and reactive power flows (P and Q) in the line are affected simultaneously.

Consider that the Point of Connection (POC) to the utility is at its sending end. Assuming that there are no changes in the line impedance and the receiving-end voltage, a PFC can control the active and reactive power flows to be a particular pair of values by modifying the sending-end voltage to a specific magnitude and phase angle ($\mathbf{V_{s'}} = V_{s'}\angle\delta_{s'}$).

This can be accomplished by connecting a compensating voltage in shunt with the line, using a Shunt-Shunt configuration. In 2000, a Back-To-Back STATic synchronous COMpensator (BTB-STATCOM) in the form of a Shunt-Shunt configuration was commissioned at the Eagle Pass substation in Texas, USA, to connect two asynchronous networks – Comisión Federal de Electricidad (CFE) in Mexico and American Electric Power (AEP) in USA. A ±36 MVA-rated BTB-STATCOM asynchronous tie demonstrated for the first time that the active power could be transferred from one power grid to another through the BTB-STATCOM while at the same time, independent shunt reactive compensations could be provided at the points of connection with both power grids – CFE and AEP. In some special cases for point-to-point transfer of power between two isolated networks or interconnection of two power grids with different voltages or phase angles (or frequencies), this scheme may be the preferred choice.

In a Shunt-Series configuration, referred to as an IR, a series-compensating voltage ($\mathbf{V_{s's}} = V_{s's}\angle(\delta_s + \beta)$) is connected in series with the line as shown in Figure 6-2a. When the

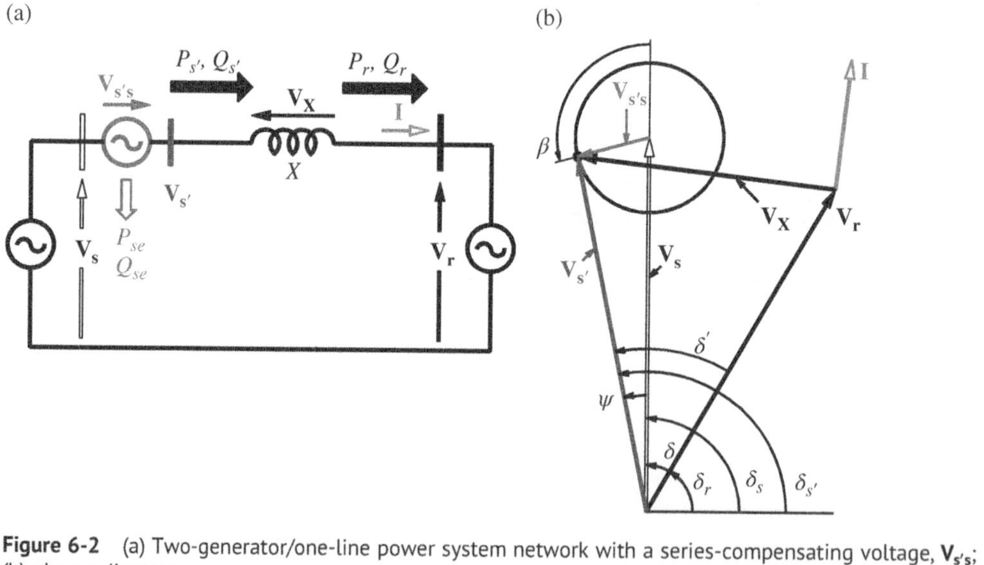

Figure 6-2 (a) Two-generator/one-line power system network with a series-compensating voltage, $\mathbf{V_{s's}}$; (b) phasor diagram.

series-compensating voltage is added to the sending-end voltage ($\mathbf{V_s}$), the modified sending-end voltage becomes $\mathbf{V_{s'}}$. The difference between the modified sending-end voltage ($\mathbf{V_{s'}}$) and the receiving-end voltage ($\mathbf{V_r}$) determines the voltage ($\mathbf{V_X}$) across the line reactance (X). The voltage ($\mathbf{V_X}$) determines the line current (**I**) and, in turn, the active and reactive power flows (P and Q) in the line. The related phasor diagram is shown in Figure 6-2b. The circle, formed around the tip of phasor ($\mathbf{V_s}$) with radius ($\mathbf{V_{s's}}$), is the locus of $\mathbf{V_{s'}}$. The circle defines the controllable area of an IR. Several PFCs with a Shunt-Series configuration have been installed in USA, South Korea, and China for use as UPFCs. In 1998, a ±160 MVA-rated UPFC, installed at AEP's Inez substation in Kentucky, USA, demonstrated for the first time that the active and reactive power flows in a line could be regulated independently while maintaining a fixed line voltage at the POC.

Regardless of whether the compensating voltage is series- or shunt-connected, it is at any phase angle with the prevailing line current; therefore, it exchanges active and reactive powers with the line at the POC and emulates a four-quadrant, compensating impedance that consists of a positive resistance ($+R$) or a negative resistance ($-R$) and a capacitive reactance (X_C) or an inductive reactance (X_L) in series with the line. When a VSC-based Shunt-Series configuration, such as a UPFC, is used, the exchanged active power by the Series Unit flows bidirectionally through the DC link to the Shunt Unit. When a magnetic link-based Shunt-Series configuration, such as an ST, is used, the exchanged active and reactive powers by the Series Unit flow bidirectionally through the link to the Shunt Unit. Alternately, the exchanged active and reactive powers by the Series Unit flow bidirectionally through the link to another Series Unit in a Series-Series configuration. In a VSC-based Series-Series configuration, such as an Interline Power Flow Controller (IPFC), only the exchanged active power flows through the DC link. However, in a magnetic link-based Series-Series configuration, such as a Multi-line Sen Transformer (MST), the exchanged active and reactive powers flow bidirectionally through the magnetic link.

In 2003, a ±100 MVA-rated IPFC, installed at New York Power Authority's Marcy substation in New York, USA, demonstrated for the first time that the active and reactive (P and Q) powers in a transmission line could be transferred from one specific line to another specific line without altering the power flows in the remaining lines where several transmission lines are connected to a common bus. Note that in a similar situation, regulating power flow in one line using a Shunt-Series configuration can affect the power flows in other lines.

When a VSC-based Shunt-Shunt configuration, such as a BTB-STATCOM, is used, the exchanged active power by one Shunt Unit flows bidirectionally through the DC link to another Shunt Unit. When a magnetic link-based Shunt-Shunt configuration, such as an ST, is used, the exchanged active and reactive powers by one Shunt Unit flow bidirectionally through the link to the other Shunt Unit as discussed in Section 6.8.

The active and reactive power flows ($P_{s'}$ and $Q_{s'}$) at the modified sending end and active and reactive power flows (P_r and Q_r) at the receiving end for a modified power angle ($\delta' = \delta + \psi$) are defined as

$$P_{s'} = a' \sin \delta' \tag{2-59}$$

where

$$a' = \frac{V_{s'} V_r}{X} \tag{2-61}$$

and

$$\delta' = \delta_{s'} - \delta_r \tag{2-62}$$

or

$$\delta' = \delta + \psi, \tag{2-51a}$$

$$Q_{s'} = a'\left(\frac{V_{s'}}{V_r} - \cos\delta'\right), \tag{2-60}$$

$$P_r = a' \sin\delta' \tag{2-74}$$

and

$$Q_r = a'\left(\cos\delta' - \frac{V_r}{V_{s'}}\right). \tag{2-75}$$

From Equations (2-74), (2-75), and (2-61), it can be found that for given active and reactive power flows (P_r and Q_r) at the receiving end, a PFC with a Shunt-Shunt configuration applies a voltage ($\mathbf{V_{s'}}$), such that

$$V_{s'} = \frac{X}{V_r}\sqrt{P_r^2 + \left(Q_r + \frac{V_r^2}{X}\right)^2} \tag{2-81}$$

and

$$\delta' = \tan^{-1}\frac{P_r}{Q_r + \dfrac{V_r^2}{X}}, \tag{2-82a}$$

which can be rearranged to obtain the phase-shift angle as

$$\psi = \tan^{-1}\frac{P_r}{Q_r + \dfrac{V_r^2}{X}} - \delta \tag{2-82b}$$

using Equation (2-51b).

In an alternate approach, the same active and reactive power flows (P_r and Q_r) can be obtained using a PFC with a Shunt-Series configuration that applies a series-compensating voltage ($\mathbf{V_{s's}}$) such that

$$V_{s's}\angle\beta = V_{s'}\angle\psi - V_s\angle 0° \tag{2-94b}$$

with $\delta_s = 0°$ where

$$\psi = \delta' - \delta. \tag{2-51b}$$

The magnitude ($V_{s's}$) and relative phase angle (β) of the series-compensating voltage ($\mathbf{V_{s's}}$) are

$$V_{s's} = \sqrt{V_{s'}^2 + V_s^2 - 2V_{s'}V_s\cos\psi} \tag{2-100}$$

and

$$\beta = \tan^{-1}\frac{V_{s'}\sin\psi}{V_{s'}\cos\psi - V_s}. \tag{2-101}$$

Transformer/Load Tap Changers (LTCs)-based solutions for PFCs, using Shunt-Shunt and Shunt-Series configurations, integrated with a two-generator/one-line power system network, and their phasor diagrams are shown in Figure 6-3. In Figure 6-3a, the primary three-phase windings (A, B, and C) are excited from the line voltage in Shunt Unit 1; the modified sending-end

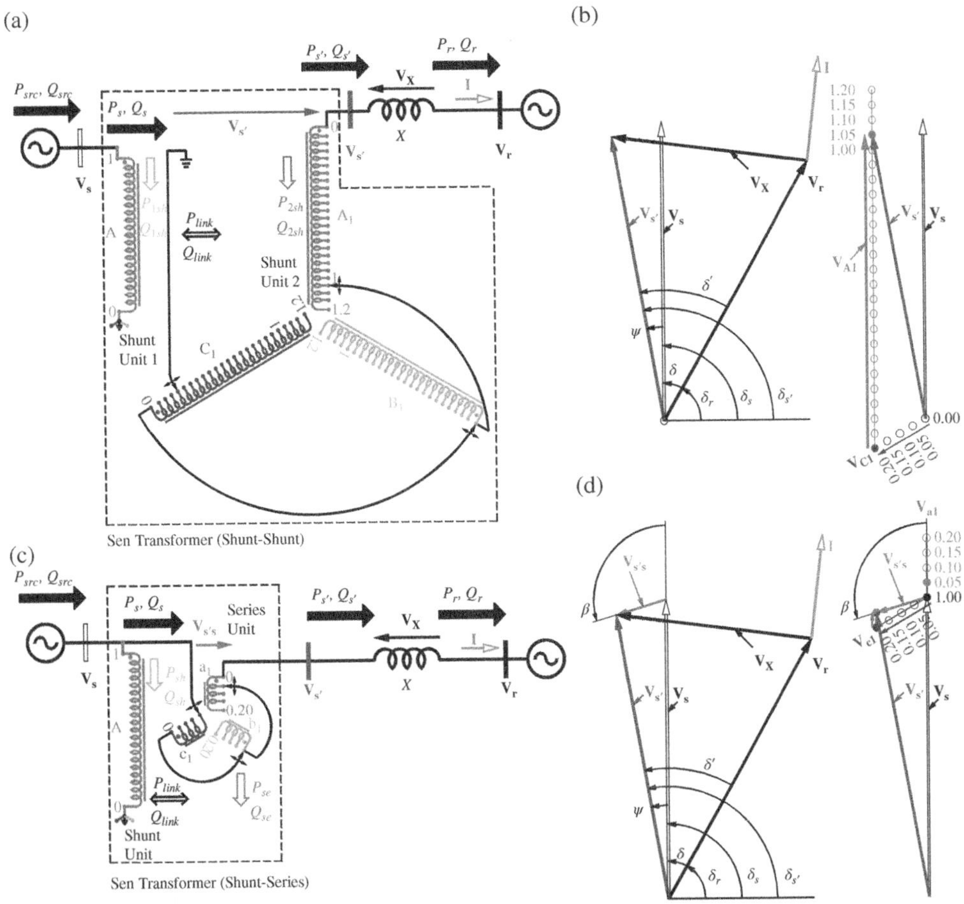

Figure 6-3 Transformer/LTCs-based solutions for power flow controllers using (a) Shunt-Shunt configuration and (b) its related phasor diagram; (c) Shunt-Series configuration and (d) its related phasor diagram.

voltage ($\mathbf{V}_{s'}$), which is the shunt-compensating voltage with magnitude ($V_{s'}$) and phase-shift angle (ψ), is generated for the A phase in Shunt Unit 2 by implementing the equation:

$$\mathbf{V}_{s'A} = \mathbf{V}_{A1} + \mathbf{V}_{B1} + \mathbf{V}_{C1}$$

or

$$V_{s'A}\angle\psi_A = K_{A1}V_{sA} + K_{B1}V_{sB} + K_{C1}V_{sC} = K_{A1}V_{sA} + K_{B1}e^{-j2\pi/3}V_{sA} + K_{C1}e^{j2\pi/3}V_{sA} \quad (6\text{-}1)$$

where

$\mathbf{V}_{A1}$, $\mathbf{V}_{B1}$, and $\mathbf{V}_{C1}$ are active voltages in the shunt-compensating windings (A_1, B_1, and C_1) of the A phase,

$K_{A1} = 1.05$, $K_{B1} = 0$, and $K_{C1} = 0.20$ are the active turns-ratios of the secondary, shunt-compensating windings in the A phase and the corresponding primary windings,

$V_{s'A}$ is the magnitude of the A-phase shunt-compensating voltage, which is also the voltage at the modified sending end of the line,

ψ_A is the phase-shift angle for the A-phase voltage at the modified sending end of the line, and

V_{sA} is the magnitude of the A-phase voltage at the sending end of the line.

Note that the A-phase shunt-compensating winding is 105% active; the B-phase shunt-compensating winding is inactive, since it is bypassed; the C-phase shunt-compensating winding is 20% active at the particular operating point, shown in the Figure 6-3a. Figure 6-3b shows the corresponding phasor diagram. Note that the filled dots in the phasor diagram (Figures 6-3b and 6-3d) represent the end of the A and C compensating phasors, respectively.

The same modified sending-end voltage ($\mathbf{V_{s'}}$) can be generated by connecting the compensating voltage ($\mathbf{V_{s's}}$) with magnitude ($V_{s's}$) and the relative phase angle (β) in series with the sending-end voltage ($\mathbf{V_s}$). Figure 6-3c shows the implementation of the Equation (2-94a) with $\delta_s = 0°$ for the A phase:

$$V_{s'A}\angle\psi_A = V_{sA}\angle 0° + V_{s'sA}\angle\beta_A \tag{6-2}$$

where

$$V_{s'sA}\angle\beta_A = \mathbf{V_{s'sA}} = \mathbf{V_{a1}} + \mathbf{V_{b1}} + \mathbf{V_{c1}}$$

or

$$V_{s'sA}\angle\beta_A = k_{a1}V_{sA} + k_{b1}V_{sB} + k_{c1}V_{sC} = k_{a1}V_{sA} + k_{b1}e^{-j2\pi/3}V_{sA} + k_{c1}e^{j2\pi/3}V_{sA} \tag{6-3}$$

where

$\mathbf{V_{a1}}$, $\mathbf{V_{b1}}$, and $\mathbf{V_{c1}}$ are active voltages in the series-compensating windings (a_1, b_1, and c_1) of the A phase,
$k_{a1} = 0.05$, $k_{b1} = 0$, and $k_{c1} = 0.20$ are the active turns-ratios of the secondary, series-compensating windings in the A phase and the corresponding primary windings,
$V_{s'sA}$ is the magnitude of the series-compensating voltage in the A phase, and
β_A is the relative phase angle of the series-compensating voltage in the A phase.

Note that the A-phase series-compensating winding is 5% active; the B-phase, series-compensating winding is inactive, since it is bypassed; the C-phase series-compensating winding is 20% active at the particular operating point, shown in Figure 6-3c. Figure 6-3d shows the corresponding phasor diagram.

It is important to note that the compensating voltages in both the series- and shunt-connected Compensating-Voltage Units are composed of three series-connected voltages from the three phases: A, B, and C; however, the series-compensating voltage is connected in series with the line, while the shunt-compensating voltage is connected in shunt with the line.

It is also important to note that in the Shunt-Series configuration, the series-compensating voltage carries the full line current and its terminal voltage rating is a fractional amount of the line voltage. Therefore, the rating of the Shunt-Series configuration is only a fraction of the transmitted power of the line. In the Shunt-Shunt configuration, the shunt-compensating voltage also carries the full line current, but its terminal voltage rating is the same as the full line voltage. Therefore, the rating of the Shunt-Shunt configuration is the same as the transmitted power of the line. For the same amount of controllable transmitted power, the Shunt-Shunt configuration requires higher-power-rated hardware than what is required in a Shunt-Series configuration. For the same amount of transmitted power, the Shunt-Shunt configuration costs more than the Shunt-Series configuration. Therefore, the preferred topology of a PFC uses a Shunt-Series configuration. A Shunt-Shunt configuration may, however, be the preferred choice for the interconnection of two nearby grids with different voltages or phase angles.

The Voltage-Sourced Converter (VSC)-based technology has the capability of providing a fast dynamic response, but this capability is not required in most utility applications, where the need is to regulate the line voltage and the power flow in the line(s) in a relatively slow manner.

Although the UPFC offers a more versatile power flow control capability than any previously developed PFC, it has two major drawbacks: its high installation and operating costs. Over the decades, the list of drawbacks has expanded to include component obsolescence, impracticability of relocation and, lack of interoperability. In order to address these issues, the ST is proposed as part of a broader SPFC concept. The ST is based on the proven transformers/LTCs technology that is known to the power industry worldwide for almost a century as reliable and cost-effective.

It is well established that for an optimum operation of a line, independent control of active and reactive power flows in the line is necessary. These require the line voltage to be modified to a specific magnitude and phase angle. For many decades, the regulations of voltage magnitude and its phase angle have been accomplished using the VRT and PAR, respectively. The ST performs both voltage and phase angle regulations in one unit that is the most cost-effective solution.

The ST is a new family of power flow controlling transformers that provides the same independent control of active and reactive power flows as the UPFC, albeit at a reduced dynamic rate, but with an estimated 5:1 reduction in equipment cost and a 10:1 reduction in operational cost. The objective in this chapter is to describe the evolution of the ST and introduce it as the most cost-effective and independent active and reactive PFC in a utility application.

6.1.1 Voltage Regulation

The VRT, in a Shunt-Series configuration, consists of two parts: Exciter Unit and Voltage-Regulating unit as shown in Figure 6-4a. The Exciter Unit consists of three-phase (A, B, and C),

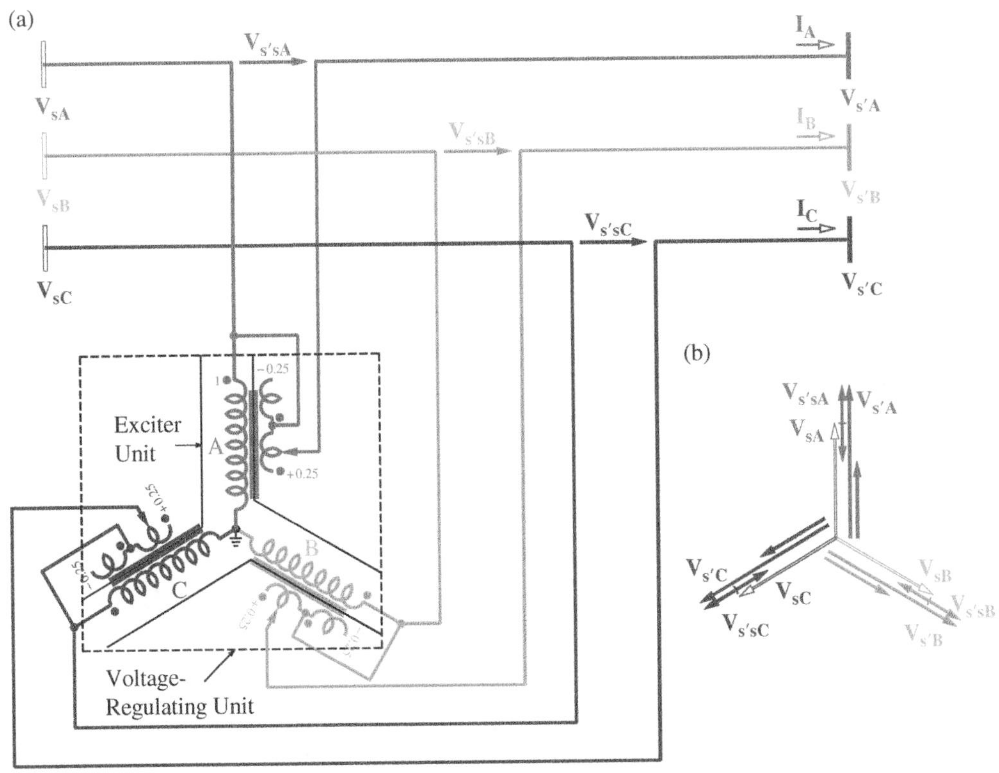

Figure 6-4 (a) Voltage-Regulating Transformer (Shunt-Series configuration); (b) phasor diagram.

Y-connected primary windings that are connected in shunt with the line. The three-phase primary windings are excited from the three-phase line voltages ($\mathbf{V_{sA}}$, $\mathbf{V_{sB}}$, and $\mathbf{V_{sC}}$). A three-phase set of bipolar compensating voltages ($\mathbf{V_{s'sA}}$, $\mathbf{V_{s'sB}}$, and $\mathbf{V_{s'sC}}$) in the range of 0.25 pu that are either in phase or out of phase with the corresponding phase-to-neutral voltage are generated from the induced secondary voltages. The Voltage-Regulating Unit consists of a total of six secondary windings (two windings in each phase). With the use of LTCs, the magnitude of the compensating voltage ($\mathbf{V_{s's}}$) is varied. The compensating voltage is connected in series with the line to increase or decrease the line voltage from an uncompensated voltage ($\mathbf{V_s}$) to a compensated voltage ($\mathbf{V_{s'}}$). Figure 6-4b shows the related phasor diagram. Figure 6-4 shows that the line voltage can be regulated between 0.75 and 1.25 pu. Note that the LTC taps are set to regulate the line voltage at 1.125 pu.

The VRT, in a Shunt-Shunt configuration, consists of two parts: Exciter Unit and Voltage-Regulating Unit as shown in Figure 6-5a. The Exciter Unit consists of a three-phase (A, B, and C), Y-connected primary windings that are connected in shunt with the line. The three-phase primary windings are excited from the three-phase line voltages ($\mathbf{V_{sA}}$, $\mathbf{V_{sB}}$, and $\mathbf{V_{sC}}$). Three-phase compensating voltages ($\mathbf{V_{s'A}}$, $\mathbf{V_{s'B}}$, and $\mathbf{V_{s'C}}$) in the range of 1.25 pu that are in phase with the corresponding phase-to-neutral primary voltages are generated from the induced voltages in the secondary windings. The Voltage-Regulating Unit consists of a total of three secondary windings (one winding in each phase). With the use of LTCs, the magnitude of the compensating voltage ($\mathbf{V_{s'}}$) is varied. The compensating voltage is connected in shunt with the line to increase or decrease the line voltage from an uncompensated voltage ($\mathbf{V_s}$) to a compensated voltage ($\mathbf{V_{s'}}$). Figure 6-5b shows the related phasor diagram. Figure 6-5 shows that the line voltage can be regulated up to 1.25 pu. Note that the LTC taps are set to regulate the line voltage at 1.125 pu.

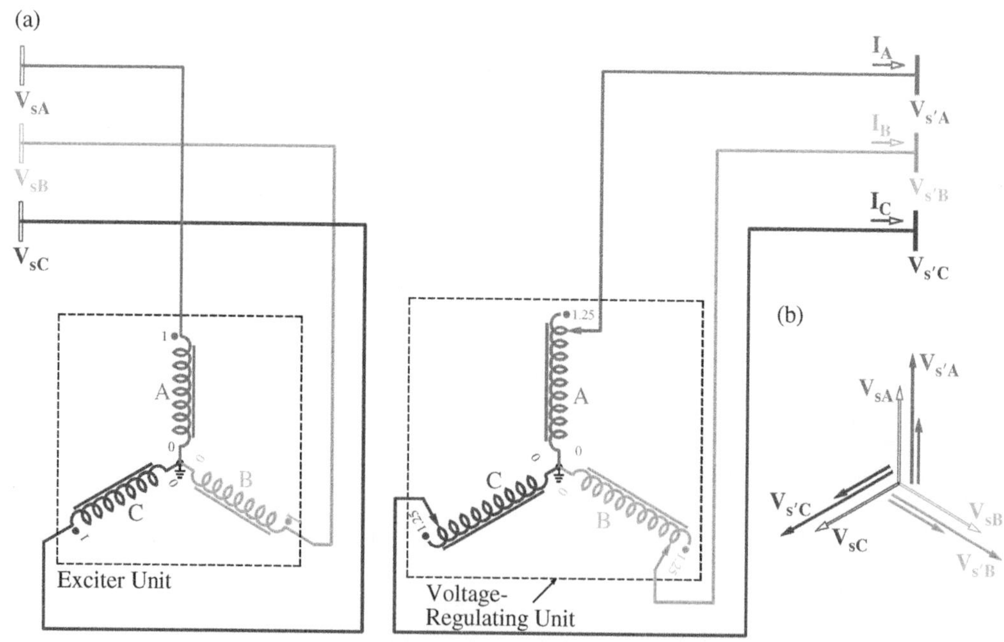

Figure 6-5 (a) Voltage-Regulating Transformer (Shunt-Shunt configuration); (b) phasor diagram.

6.1.2 Phase Angle Regulation

The effective phase angle of the line voltage is varied by using the PAR (asym). The PAR (asym) – a Shunt-Series configuration gets excited from the shunt-connected line voltage and connects a compensating voltage in series with the line; the compensating voltage is in quadrature with the phase-to-neutral line voltage as shown in Figure 6-6a. In a typical configuration, the PAR (asym) consists of two transformers. The first transformer (Exciter Unit) is called a regulating transformer. Its three-phase (A, B, and C), Y-connected primary windings are connected in shunt with the line. The three-phase primary windings are excited from the line voltages ($\mathbf{V_{sA}}$, $\mathbf{V_{sB}}$, and $\mathbf{V_{sC}}$) and three-phase voltages ($\mathbf{V_A}$, $\mathbf{V_B}$, and $\mathbf{V_C}$) in the range of 0 to 0.25 pu are induced in the regulating transformer secondary windings. For a series connection of this voltage to the line, an electrical isolation is necessary. The second transformer (Series Unit) is called a series transformer, which is connected in series with the line. The secondary voltages of the regulating transformer are the primary voltages of the series transformer. These voltages are phase-to-phase voltages ($\mathbf{V_B} - \mathbf{V_C}$, $\mathbf{V_C} - \mathbf{V_A}$, and $\mathbf{V_A} - \mathbf{V_B}$) in the range of 0 to $0.25 \times \sqrt{3}$ pu and in quadrature with the phase-to-neutral voltages of the line. Three-phase, bipolar compensating voltages ($\mathbf{V_{s'sA}}$, $\mathbf{V_{s'sB}}$, and $\mathbf{V_{s'sC}}$) that are in quadrature with the phase-to-neutral voltages of the line are induced in the secondary windings of the series transformer. The magnitude of the series-compensating voltage is varied using the LTCs in the secondary windings of the regulating transformer. The induced voltage of the series transformer is connected in series with the line. Figure 6-6b shows the related

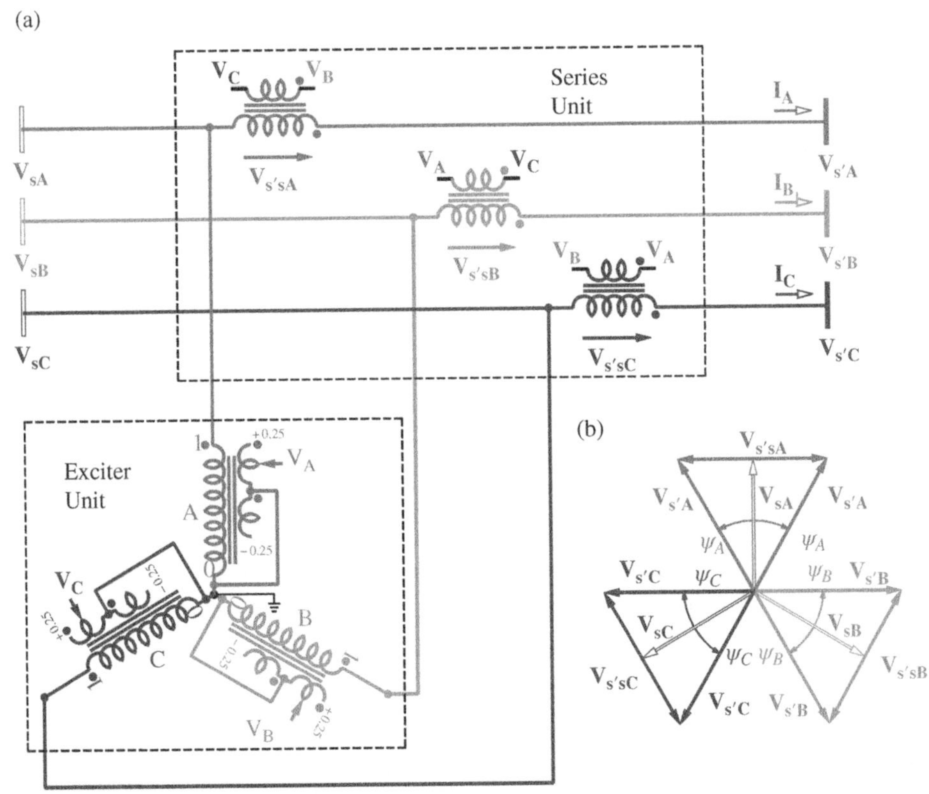

Figure 6-6 (a) Phase Angle Regulator (asymmetric); (b) phasor diagram.

phasor diagram. A positive compensating voltage creates a positive phase-shift angle (ψ) and vice versa. Figure 6-6 shows that the compensating voltage can be regulated in the range of $\pm 0.25 \times \sqrt{3}$ pu, followed by a gain of 5:3 from the turns-ratio of the series transformer. Note that the LTC taps are set to regulate the compensating voltage at $-0.625/\sqrt{3}$ pu, which is $-0.125 \times \sqrt{3} \times 5/3$ pu.

The VRT, in either a Shunt-Series configuration or a Shunt-Shunt configuration, modifies the line voltage by increasing or decreasing the magnitude, but not its phase angle. The phase angle remains almost the same as that of the original voltage before compensation. The PAR (asym) modifies, primarily, the phase angle of the line voltage and, secondarily, increases the magnitude of the voltage as discussed in Chapters 1 through 4. But, for a regulation of the active and reactive power flows in the line independently, the PFC needs to modify the line voltage ($\mathbf{V_s} = V_s \angle \delta_s$) to a specific magnitude ($V_{s'}$) and phase angle ($\delta_{s'}$) as desired.

6.2 Desired Solution

A series-compensating voltage ($\mathbf{V_{s's}}$), shown in Figure 6-7a, modifies the magnitude and phase angle of the line voltage independently as desired. For the desired amount of active and reactive power flows in the line, the compensating voltage needs to be a specific magnitude ($V_{s's}$) and at a relative phase angle (β) with respect to the line voltage.

The desired series-compensating voltage can be composed of two orthogonal voltages: one regulates the magnitude of the line voltage and the other regulates the phase angle of the line voltage as can be implemented with the use of the VRT and PAR, respectively. The ST combines the functions of the VRT and the PAR (asym) in a single unit that results in a reduced amount of hardware from what is required separately for a VRT and a PAR.

Consider a two-generator/one-line power system network with a sending-end voltage ($\mathbf{V_s} = V_s \angle \delta_s$), a receiving-end voltage ($\mathbf{V_r} = V_r \angle \delta_r$), the voltage ($\mathbf{V_X} = V_X \angle \theta_{VX}$) across the line reactance (X), and a compensating voltage ($\mathbf{V_{s's}}$) as shown in Figure 6-7a. When the compensating voltage ($\mathbf{V_{s's}} = V_{s's} \angle (\delta_s + \beta)$) is added in series with the line, the modified sending-end voltage becomes $\mathbf{V_{s'}} = \mathbf{V_s} + \mathbf{V_{s's}}$. Figure 6-7b shows the phasor diagram related to the series-compensating voltage with a fixed magnitude of 0.2 pu and its entire controllable range of the relative phase angle of $0° \le \beta \le 360°$. The difference voltage ($\mathbf{V_X} = \mathbf{V_{s'}} - \mathbf{V_r}$) provides the voltage ($\mathbf{V_X}$) across the line reactance (X). As the relative phase angle (β) is varied over its full 360° range, the end of phasor ($\mathbf{V_{s's}}$) moves along a circle with its center located at the end of phasor ($\mathbf{V_s}$). The rotation of phasor ($\mathbf{V_{s's}}$) modulates the magnitude and phase angle of phasor ($\mathbf{V_X}$). The figure also shows the characteristics of a Voltage Regulator (VR), a PAR (symmetric and asymmetric), and a Reactance Regulator (RR).

In the transmission system considered above, the magnitude ($V_{s'}$) and phase angle ($\delta_{s'}$) of the modified sending-end voltage ($\mathbf{V_{s'}}$) are modulated between $\mathbf{V_{s'}} = \mathbf{V_s} + \mathbf{V_{s's}}$ and $\mathbf{V_{s'}} = \mathbf{V_s} - \mathbf{V_{s's}}$ and between $\delta_{s'} = \delta_s + \psi_{max}$ and $\delta_{s'} = \delta_s - \psi_{max}$, respectively, where $V_{s's} = 0.2$ pu and ψ_{max} = 11.54° during the relative phase angle ranges β_{12} and β_{34} as shown in Figure 6-7c. Note that the line voltage is allowed to be controlled within a narrow band, for example 10% of the nominal value.

The compensating voltage ($\mathbf{V_{s's}}$) is at any phase angle with the prevailing line current ($\mathbf{I}$) and, therefore, exchanges with the line both active power ($P_{se} = V_d I$) and reactive power ($Q_{se} = V_q I$), where $\mathbf{V_d} = V_d$ and $\mathbf{V_q} = jV_q$ are the respective active or direct and reactive or quadrature components of the compensating voltage with load convention, meaning the line current ($\mathbf{I}$) enters at the higher potential terminal of the compensating voltage as shown in Figure 6-7a. These exchanged active and reactive powers (P_{se} and Q_{se}) are also sinusoidal functions of the relative phase angle (β)

as shown in Figure 6-7d. For a given magnitude of a compensating voltage, the exchanged capacitive power (Q_{se}) by the compensating voltage ($\mathbf{V_{s's}}$) is larger than its inductive counterpart due to the fact that the capacitive compensation produces a larger line current. The compensating voltage, being at any phase angle with the prevailing line current, emulates a four-quadrant, series-compensating impedance ($\mathbf{Z} = R_{se} - jX_{se}$) that consists of a resistance ($R_{se} = +R$ or $-R$) and a reactance ($X_{se} - X_C$ or $-X_L$) in series with the line. Therefore, the series-compensating voltage ($\mathbf{V_{s's}}$) acts as an IR.

The active and reactive power flows (P_r and Q_r) at the receiving end vary with the relative phase angle (β) in a sinusoidal manner as shown in Figure 6-7e. The relationship between Q_r and P_r is a circle centered at (P_{rn}, Q_{rn}) with a radius of a_r, where P_{rn} and Q_{rn} are the natural active and reactive power flows at the receiving end of the line without any compensation, and a_r is the maximum power flow enhancement. The active and reactive power flows (P_r and Q_r) at the receiving end

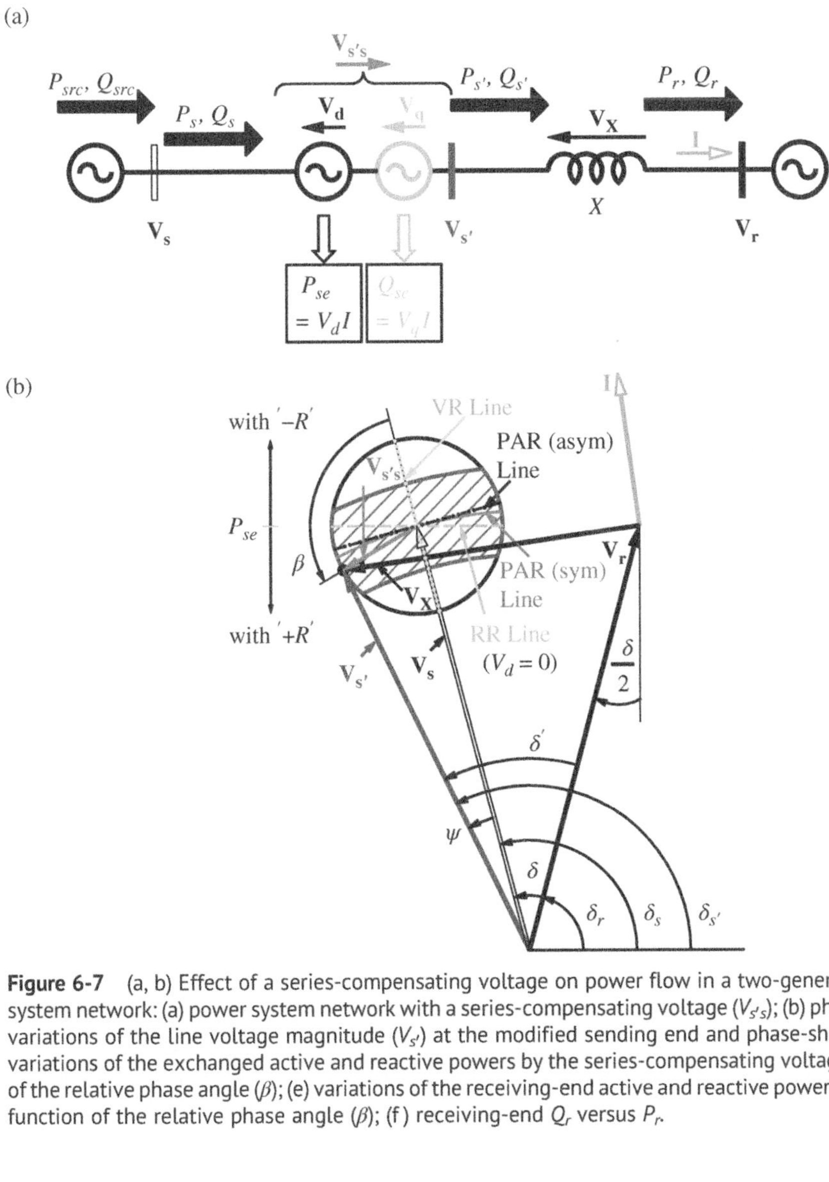

Figure 6-7 (a, b) Effect of a series-compensating voltage on power flow in a two-generator/one-line power system network: (a) power system network with a series-compensating voltage ($V_{s's}$); (b) phasor diagram; (c) the variations of the line voltage magnitude ($V_{s'}$) at the modified sending end and phase-shift angle (ψ); (d) the variations of the exchanged active and reactive powers by the series-compensating voltage ($V_{s's}$) as a function of the relative phase angle (β); (e) variations of the receiving-end active and reactive power flows (P_r and Q_r) as a function of the relative phase angle (β); (f) receiving-end Q_r versus P_r.

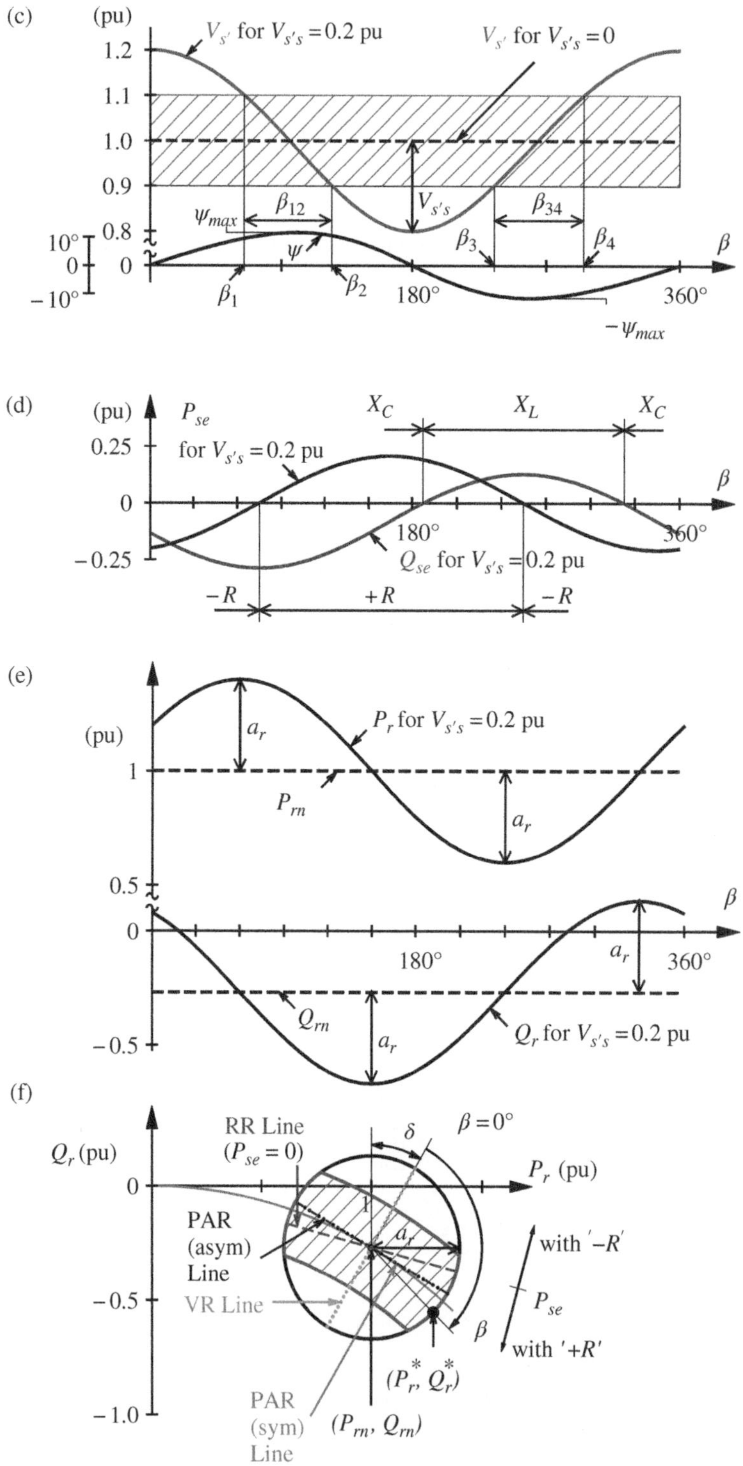

Figure 6-7 (Continued)

of the line can be controlled within the range defined by the *P-Q* plot in Figure 6-7f by choosing the magnitude ($V_{s's}$) of the compensating voltage ($\mathbf{V_{s's}}$) within its rated value and the relative phase angle of $0° \leq \beta \leq 360°$. If the magnitude ($V_{s's}$) of the compensating voltage is increased, the controllable range of the power flows in the *P-Q* plane is also increased. When the controllable range extends to all four quadrants, bidirectional and independent control of active and reactive power flows are achieved. For the desired active and reactive power flows (P_r^* and Q_r^*), within the allowable control range, the required magnitude ($V_{s's}$) and the phase angle (β) between the voltage regulator line (*i.e.* $\beta = 0°$ line) and the line that passes through (P_{rn}, Q_{rn}) and (P_r^*, Q_r^*) can be determined from Equations (2-138) and (2-139) as

$$V_{s's} = \frac{X}{V_r}\sqrt{\left(P_r^* - P_{rn}\right)^2 + \left(Q_r^* - Q_{rn}\right)^2}$$

and

$$\beta = \tan^{-1}\frac{P_r^* - P_{rn}}{Q_r^* - Q_{rn}} - \delta.$$

When a series-compensating voltage is used to modify the line voltage, the following occur.

- The compensating voltage can be either in phase or out of phase with the phase-to-neutral voltage of the line to implement the function of the VRT.
- The compensating voltage can be in quadrature with the phase-to-neutral voltage of the line to implement the function of the PAR (asym).
- The compensating voltage can be such that it provides a series reactance compensation because it is in quadrature with the prevailing line current. If the circular controllable area is equally divided by the reactance compensator line ($V_d = 0$ or $P_{se} = 0$), the upper and lower halves represent P_{se} due to $-R$ and $+R$, respectively.

Any compensator that provides compensation for one power flow control parameter (voltage magnitude, its phase angle, or line reactance) operates on a set of linear operating points inside the *P-Q* circle as defined above. Since these operating characteristic lines are neither horizontal nor vertical in the *P-Q* plane at the receiving end, controlling one parameter with the use of these compensators affects both the active and reactive power flows in the line simultaneously as shown in Figure 6-7f. The same conclusion can be drawn in regard to the active and reactive power flows at the modified sending end.

6.2.1 ST as a New Voltage Regulator

Figure 6-8a shows a new voltage regulator with a Shunt-Series configuration, which is a three-phase transformer with Y-connected primary windings (A, B, and C), called the Exciter Unit, and three secondary windings (one winding in each phase: a_1 on the core of the *A* phase, b_2 on the core of the *B* phase, and c_3 on the core of the *C* phase), called the Compensating-Voltage Unit. The three-phase sending-end voltages ($\mathbf{V_{sA}}$, $\mathbf{V_{sB}}$, and $\mathbf{V_{sC}}$) are applied to the Exciter Unit. The in-phase component of the compensating voltage for any phase is induced in a secondary winding that is placed on the corresponding phase of the transformer core. By choosing the number of turns of the secondary windings with the use of LTCs, the magnitude of the compensating voltage ($\mathbf{V_{s's}}$) is varied in the range of 0 to 0.2 pu. Note that the three compensating windings (a_1, b_2, and c_3) have the same number of turns. The compensating voltage is connected in series with the line to increase the line voltage from an uncompensated voltage ($\mathbf{V_s}$) to a compensated voltage ($\mathbf{V_{s'}}$). Figure 6-8b shows the related phasor diagram. Figure 6-8 shows that the line voltage can be regulated between 1 and 1.2 pu. Note that the LTC taps are set to regulate the line voltage at 1.1 pu.

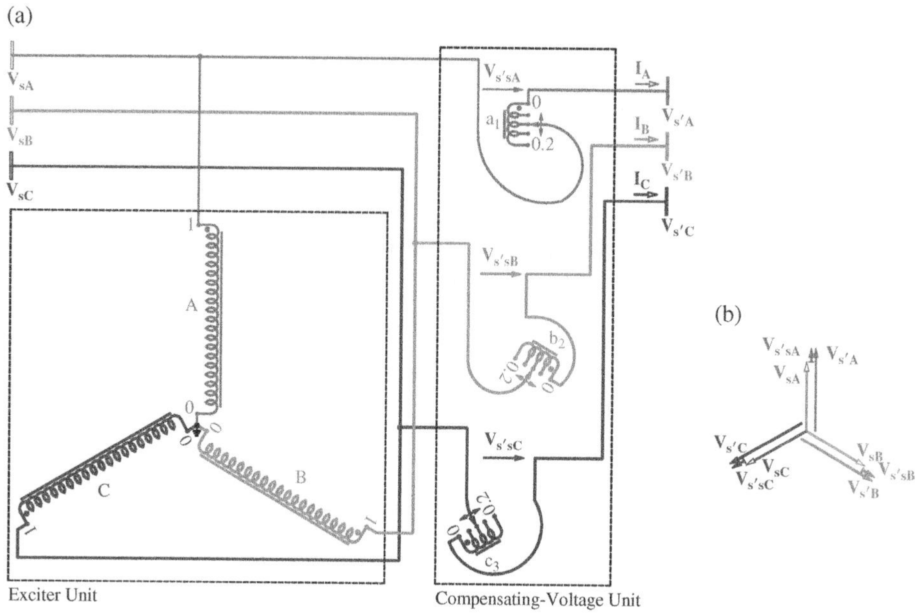

Figure 6-8 (a) Voltage-Regulating Transformer for increasing line voltage; (b) phasor diagram.

Figure 6-9a shows a new voltage regulator with a Shunt-Series configuration, which is a three-phase transformer with Y-connected primary windings (A, B, and C), called the Exciter Unit, and six secondary windings (two windings in each phase: a_2 and a_3 on the core of the *A* phase; b_1 and b_3 on the core of the *B* phase; and c_1 and c_2 on the core of the *C* phase), called the Compensating-Voltage Unit. The three-phase sending-end voltages ($\mathbf{V_{sA}}$, $\mathbf{V_{sB}}$, and $\mathbf{V_{sC}}$) are applied to the Exciter Unit. The out-of-phase component of the compensating voltage for any phase is derived from the phasor sum of the voltages induced in two equal-turn windings that are placed on the remaining two phases of the transformer core. By choosing the number of turns of the secondary windings with the use of LTCs, the magnitude of the compensating voltage ($\mathbf{V_{s's}}$) is varied in the range of 0 to 0.2 pu. Note that the six compensating windings (a_2, a_3, b_1, b_3, c_1, and c_2) have the same number of turns. The compensating voltage is connected in series with the line to decrease the line voltage from an uncompensated voltage ($\mathbf{V_s}$) to a compensated voltage ($\mathbf{V_{s'}}$). Figure 6-9b shows the related phasor diagram. Figure 6-9 shows that the line voltage can be regulated between 0.8 and 1 pu. Note that the LTC taps are set to regulate the line voltage at 0.9 pu.

Figure 6-10a shows the ST, which is a three-phase transformer with Y-connected primary windings (A, B, and C), called the Exciter Unit, and nine secondary windings (three windings in each phase: a_1, a_2, and a_3 on the core of the *A* phase; b_1, b_2, and b_3 on the core of the *B* phase; and c_1, c_2, and c_3 on the core of the *C* phase), called the Compensating-Voltage Unit. The three-phase sending-end voltages ($\mathbf{V_{sA}}$, $\mathbf{V_{sB}}$, and $\mathbf{V_{sC}}$) are applied to the Exciter Unit. The compensating voltage is connected in series with the line to increase or decrease the line voltage from an uncompensated voltage ($\mathbf{V_s}$) to a compensated voltage ($\mathbf{V_{s'}}$). The in-phase component of the compensating voltage for any phase is induced in a secondary winding that is placed on the corresponding phase of the transformer core. The out-of-phase component of the compensating voltage for any phase is derived from the phasor sum of the voltages induced in two equal-turn windings that are placed on the remaining two phases of the transformer core. By choosing the number of turns of the secondary

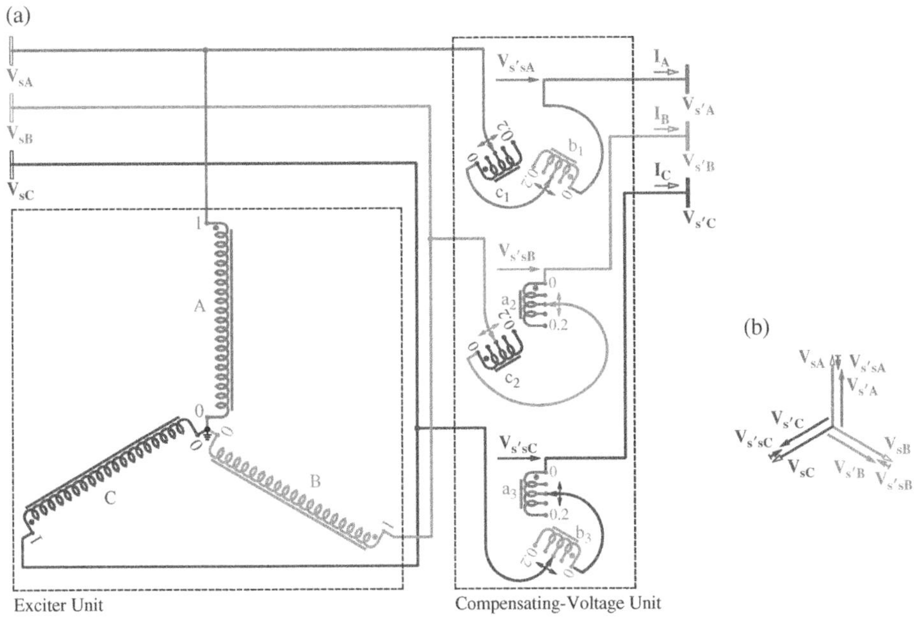

Figure 6-9 (a) Voltage-Regulating Transformer for decreasing line voltage; (b) phasor diagram.

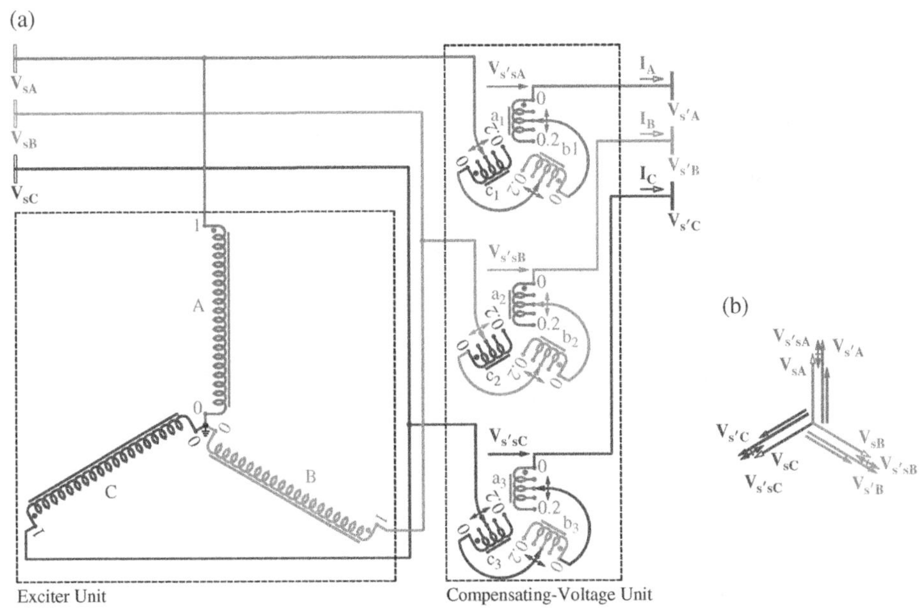

Figure 6-10 (a) ST for voltage regulation; (b) phasor diagram.

windings with the use of LTCs, the magnitude of the compensating voltage ($\mathbf{V}_{\mathbf{s's}}$) is varied in the range of 0 to 0.2 pu. Figure 6-10 shows that the line voltage can be regulated between 0.8 and 1.2 pu. Note that the LTC taps are set to regulate the line voltage at 1 pu, since the sum of the three equal magnitude and 120° phase-shifted compensating voltages is zero.

At this point, one might ask why should one use the ST as a voltage regulator when the VRT, as shown in Figure 6-4a, offers the same functionality with a simpler configuration? This is described in the next section.

6.2.2 ST as an Independent PFC

The circuit configuration, shown in Figure 6-10a, can also be used to form an IR. The effect of impedance emulation is such that the series-compensating voltage is at any phase angle with respect to the line voltage as shown in Figure 6-11a; therefore, the voltage ($\mathbf{V}_{s'}$) at the modified sending end can be regulated to a desired magnitude ($V_{s'}$) and phase-shift angle (ψ) and, in turn, the active and reactive power flows in a line can be controlled independently as desired. The power circuit is identical for both the VR and the IR. Figure 6-11b shows the related phasor diagram.

By choosing the number of turns from each of the three windings through three LTCs (LTC_1, LTC_2, and LTC_3) and, therefore, the magnitudes of the components of the three 120° phase-shifted induced voltages, the compensating voltage ($\mathbf{V}_{s's}$) in any phase is derived from the phasor sum of the voltages induced in a three-phase secondary winding set (a_1, b_1, and c_1 for compensation in the A phase; a_2, b_2, and c_2 for compensation in the B phase; and a_3, b_3, and c_3 for compensation in the C phase).

Since the LTCs that are used in the ST change their positions in steps, the compensating points in the ST are discrete in the allowable control range of the relative phase angle of $0° \leq \beta \leq 360°$. The magnitude of the induced voltage (V), the voltage (x) across each tap of the LTC, and the number (N) of taps in the LTC are related by the following equation:

$$N = \frac{V}{x}. \tag{6-4}$$

Assuming the sending-end voltage, $V_s = 1$ pu, the allowable magnitude (V) of the induced voltage ($\mathbf{V}$) to be 0.2 pu, and the voltage (x) across each tap of an LTC is 0.05 pu, there are $N = 4$ taps

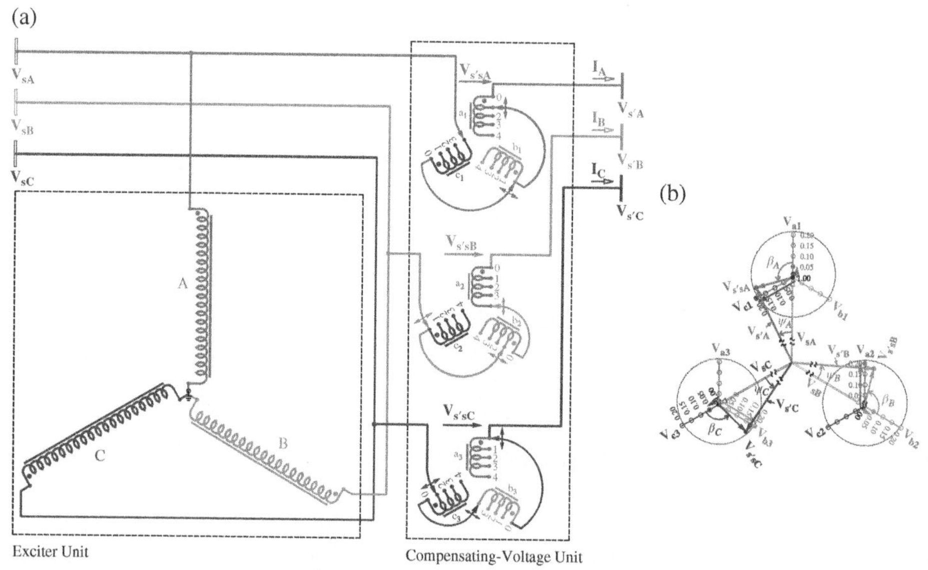

Figure 6-11 (a) ST for voltage compensation with $V_{s's} = 0.18$ pu and $\beta = 106.1°$; (b) phasor diagram; note that the circle represents the entire range of the relative phase angle of $0° \leq \beta \leq 360°$.

associated with each secondary winding. Therefore, each of the taps in LTC_1, LTC_2, and LTC_3 are set at 0, 1, 2, 3, or 4; the corresponding secondary winding is active by 0, 5, 10, 15, and 20%, respectively, and contributes 0.0, 0.05, 0.10, 0.15, and 0.20 pu toward forming the compensating voltage. The total number of possible compensating points (cp) in this case is defined by

$$cp = 3N(N + 1) \tag{6-5}$$

and it is 60. The compensating secondary windings are labeled in Figure 6-11a in terms of taps at a 5% interval.

Note that for any relative phase angle (β), except $\beta = 0°$, 120°, and 240°, a compensating voltage is composed of two neighboring available induced voltages. For example, for a compensating voltage ($\mathbf{V_{s'sA}}$) with $V_{s'sA} = 0.18$ pu and $\beta_A = 106.1°$ (see later in Table 6-1), the A-phase series-compensating winding is 5% active; the B-phase series-compensating winding is inactive, since it is bypassed; the C-phase series-compensating winding is 20% active as shown in Figure 6-11. It should also be noted that each of the windings (a_1, b_2, and c_3) is tapped at the same number of turns through LTC_1; each of the windings (b_1, c_2, and a_3) is tapped at the same number of turns through LTC_2; and each of the windings (c_1, a_2, and b_3) is tapped at the same number of turns through LTC_3. However, the number of active turns in the a_1-b_2-c_3 set, b_1-c_2-a_3 set, and c_1-a_2-b_3 set can be different from each other with one exception – when the ST is used as a voltage regulator to decrease the modified sending-end voltage. Therefore, for the relative phase angle, $\beta = 180°$, there are the same number of turns in the two compensating windings in each phase as shown in Figure 6-9. Note that under a balanced condition, the relative phase angle $\beta = \beta_A = \beta_B = \beta_C$ and the phase-shift angle $\psi = \psi_A = \psi_B = \psi_C$.

The compensating voltage ($\mathbf{V_{s's}} = -\mathbf{V_{dq}}$) is also at any phase angle with respect to the prevailing line current ($\mathbf{I}$). The active or direct component ($\mathbf{V_d} = V_d$) of the compensating voltage provides the series resistance emulation ($R_{se} = +R$ or $-R$). The positive resistance ($+R$) absorbs active power from the line and the negative resistance ($-R$) delivers active power to the line. The reactive or quadrature component ($\mathbf{V_q} = jV_q$) of the compensating voltage provides the series reactance emulation ($X_{se} = X_C$ or $-X_L$). The capacitive reactance (X_C) reduces the effective reactance of the line between its two ends and, in the process, increases the power flow of the line; the inductive reactance (X_L) increases the effective reactance of the line between its two ends, and in the process, decreases the power flow of the line. Therefore, the desired compensating voltage is actually an impedance emulator.

A comprehensive study of the power system network, integrated with the PFC can only show how many taps is the right number. Figure 6-12 shows that the theoretical circular control area is actually a hexagon. A larger number of taps in the LTCs make the compensating points closer to each other and vice versa.

Figure 6-12a shows the theoretically possible compensating points of the modified sending-end voltage with the use of the ST. Figure 6-12b shows the corresponding active and reactive power flows at the receiving end within the entire range of the relative phase angle of $0° \leq \beta \leq 360°$. Figures 6-12c and 6-12d show the practical control range of the ST. The filled dots, in Figure 6-12c, show a total of 36 compensating points of the ST within the allowable voltage range, which is 10% of the nominal voltage in the relative phase angle range of $0° \leq \beta \leq 360°$. Figure 6-12d shows that the maximum active power flow enhancement at the receiving end, within the allowable voltage range, is actually 0.35 pu, instead of the theoretically possible 0.40 pu as shown in Figure 6-12b.

In Chapter 2, Section 2.2.2.6.1, it was shown that the compensating voltage needed for a voltage regulation is equivalent to a four-quadrant, series-compensating impedance ($\mathbf{Z} = R_{se} - jX_{se}$). Therefore, both functions of voltage regulation and independent control of active and reactive power

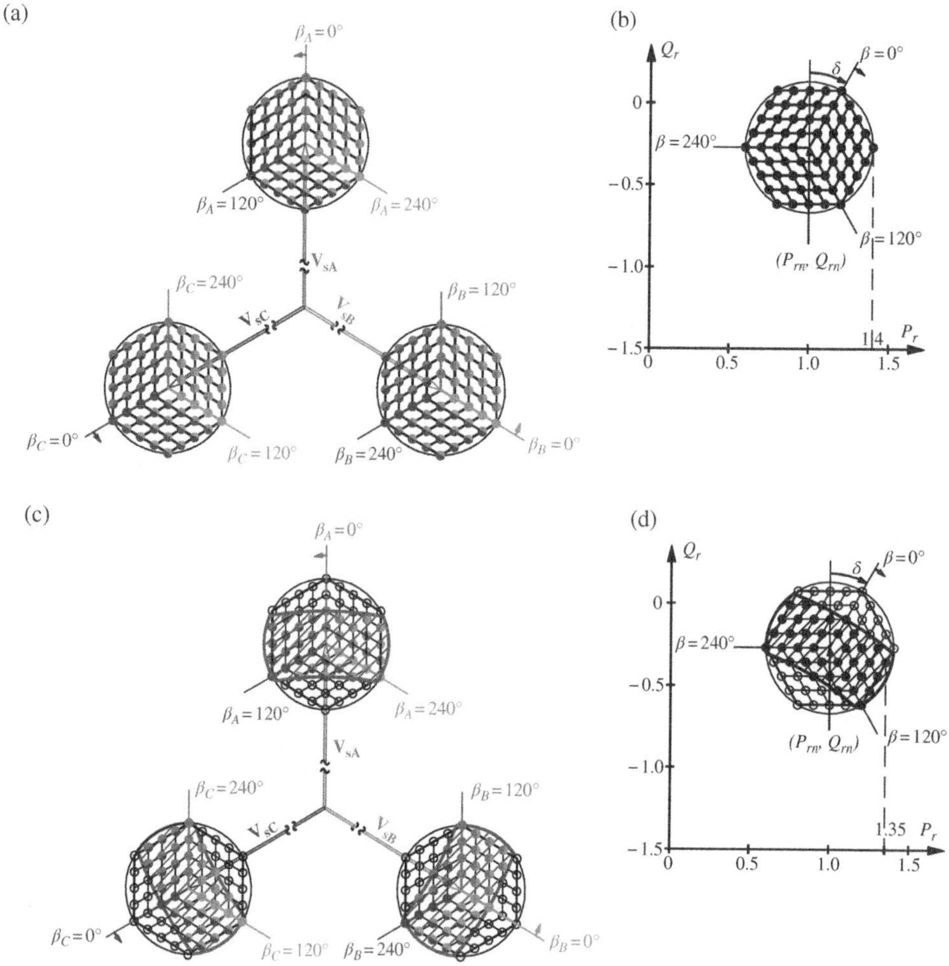

Figure 6-12 Compensating points with the use of the ST within the entire range of the relative phase angle of $0° \leq \beta \leq 360°$. Theoretically possible (a) modified sending-end voltage and (b) active and reactive power flows at the receiving end; practically possible (c) modified sending-end voltage and (d) active and reactive power flows at the receiving end.

flows can be implemented in just one unit by proper programming of the Tap Control Unit for a desired emulation of impedance $\left(\mathbf{Z}^*_{\mathbf{se}} = R^*_{se} - jX^*_{se}\right)$ as shown in Figure 6-13.

6.2.3 Control of ST

The series-compensating voltage ($\mathbf{V}_{\mathbf{s's}}$) emulates a resistance ($R_{se} = +R$ or $-R$) and a reactance ($X_{se} = X_C$ or $-X_L$) in series with the line without using any physical circuit component, except with a transformer and LTCs. The control of the ST can be in several categories as described below. In each category, first the compensating voltage magnitude ($V_{s's}$) and the relative phase angle (β) with respect to the exciter voltage ($\mathbf{V}_\mathbf{s}$) are calculated. Depending on the operating range of the relative phase angle of β ($0° \leq \beta \leq 120°$, $120° \leq \beta \leq 240°$, or $240° \leq \beta \leq 360°$), the two neighboring active compensating windings in each phase are selected. At $\beta = 0°$, 120°, and 240°, only one of the three secondary windings in each phase is active. Then, from the measured magnitude of the exciter voltage (V_s), the Tap Control Unit determines the number of turns necessary on each winding of the Compensating-Voltage Unit. Based on this calculation, the appropriate taps are switched on to connect the required number of turns in series with the line. Since every turn of the compensating winding is not likely to be connected to a tap, there will be a discrepancy between the required number of active turns and the actual number of active turns through the connected tap. A special control algorithm with a hysteresis must be implemented to avoid any possible jittering effect between the neighboring taps.

6.2.3.1 Impedance Emulation

Figure 6-13 shows the general control block diagram of the series impedance emulation using the ST. An instantaneous three-phase set of line voltages (v_s) is used to calculate the Phase-Locked Loop (PLL) angle (θ), which is phase locked to the A-phase line voltage (v_{sA}). From an instantaneous three-phase set of measured line currents (i), the magnitude (I) and its relative phase angle (θ_{ir}) with respect to the PLL angle (θ) are calculated. From the compensating resistance demand ($R_{se}*$) and the compensating reactance demand ($X_{se}*$), the magnitude ($Z_{se}*$) and phase angle ($\theta_{Zse}*$) of the compensating impedance demand are calculated. The magnitude (I) of the line current multiplied by the magnitude of the compensating impedance ($Z_{se}*$) demand is the magnitude ($V_{dq}*$) of the combined V_d* and V_q*, which is same as the magnitude of the compensating voltage ($V_{s's}*$) demand. The relative phase angle (β) of the compensating voltage demand is $\theta_{ir}+\theta_{Zse}*+\pi$. The addition of π is due to the fact that $\mathbf{V_{s's}} = -\mathbf{V_{dq}}$. Once the desired magnitude of the compensating voltage ($V_{s's}*$) demand and the relative phase angle (β) are defined, the Tap Control Unit figures out the

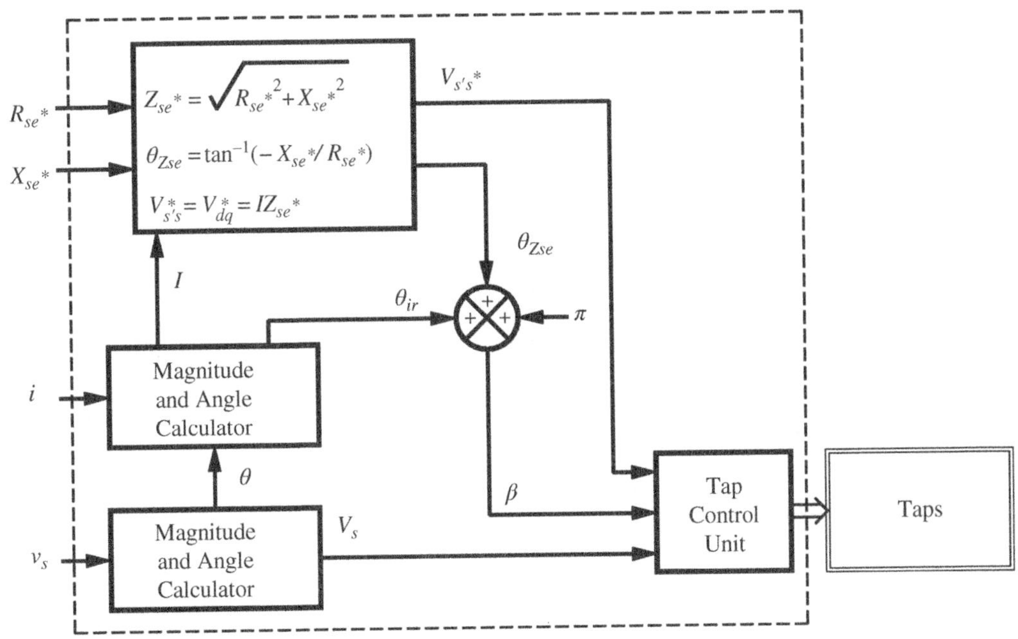

Figure 6-13 Series impedance emulation control block diagram of the ST.

contribution from each winding of a three-phase set (a_1, b_1, and c_1 for compensation in the *A* phase; a_2, b_2, and c_2 for compensation in the *B* phase; and a_3, b_3, and c_3 for compensation in the *C* phase) to constitute ($V_{s's}*$).

6.2.3.2 Resistance Emulation

As a special case, the control block diagram for the series resistance emulation (positive or negative) is shown in Figure 6-14. The calculated magnitude (I) of the line current multiplied by the compensating resistance ($R_{se}*$) demand is the compensating voltage magnitude (V_d*) demand. The compensating resistance ($R_{se}*$) is defined to be positive when it is a positive resistance ($+R$) and negative when it is a negative resistance ($-R$). The relative phase angle (β) of this compensating voltage demand is either $\theta_{ir}+\pi$ if the demanding compensating resistance is positive or $\theta_{ir}+2\pi$ if the demanding compensating resistance is negative. Once the desired compensating voltage $\left(V^*_{s's} = \mid V^*_d \mid\right)$ demand and the relative phase angle (β) are defined, the Tap Control Unit figures out the contribution from each winding of a three-phase set (a_1, b_1, and c_1 for compensation in the *A* phase; a_2, b_2, and c_2 for compensation in the *B* phase; and a_3, b_3, and c_3 for compensation in the *C* phase) to constitute ($V_{s's}*$).

6.2.3.3 Reactance Emulation

A simple modification to the control block diagram, shown in Figure 6-14, allows the emulation of a series reactance. The desired block diagram is shown in Figure 6-15. The calculated magnitude (I) of the line current multiplied by the compensating reactance demand ($X_{se}*$) is the compensating voltage magnitude (V_q*) demand. The compensating reactance ($X_{se}*$) is defined to be positive when it is a capacitive reactance and negative when it is an inductive reactance. The relative phase angle (β) of this compensating voltage demand is either $\theta_{ir} - \pi/2 + \pi$ if the demanding compensating reactance is capacitive or $\theta_{ir}+\pi/2+\pi$ if the demanding compensating reactance is inductive. Once the desired compensating voltage $\left(V^*_{s's} = \mid V^*_q \mid\right)$ demand and the relative phase angle (β) are defined, the Tap Control Unit figures out the contribution from each winding of a three-phase set (a_1, b_1, and c_1 for

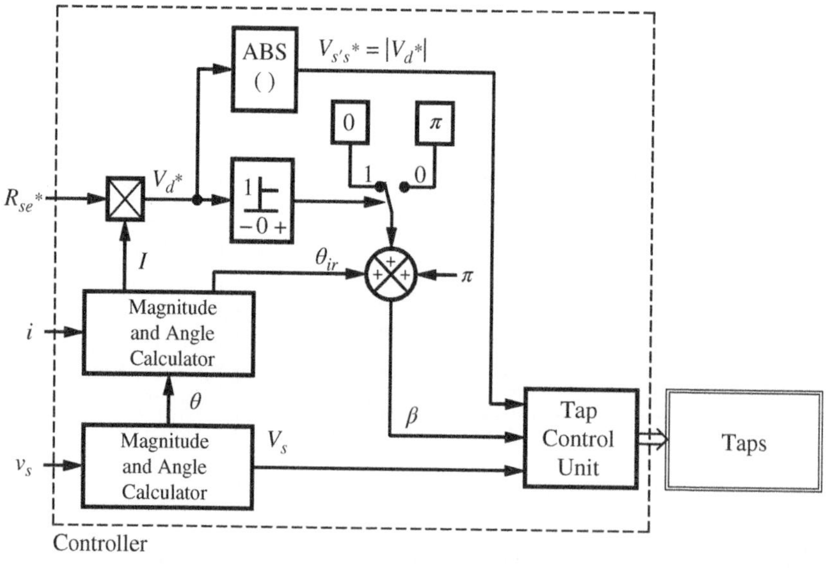

Figure 6-14 Series resistance emulation control block diagram of the ST.

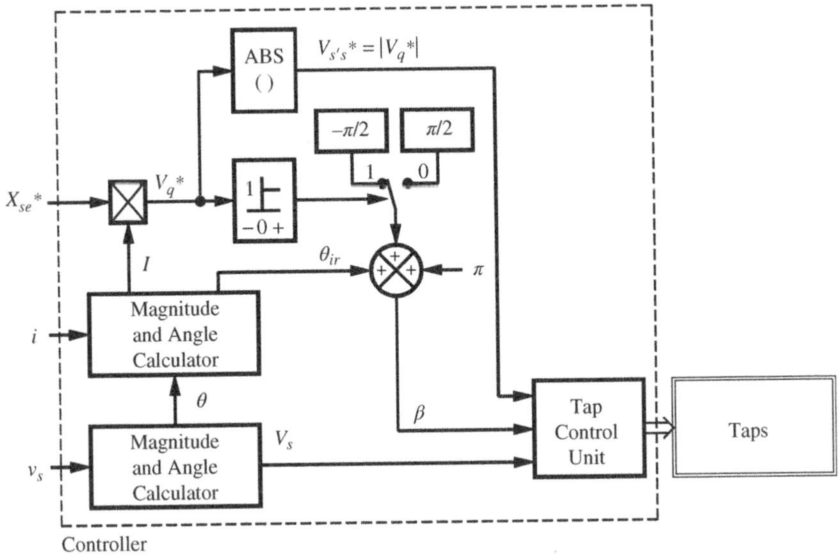

Figure 6-15 Series reactance emulation control block diagram of the ST.

compensation in the A phase; a_2, b_2, and c_2 for compensation in the B phase; and a_3, b_3, and c_3 for compensation in the C phase) to constitute ($V_{s's}*$).

6.2.3.4 Closed-Loop Power Flow Control

The closed-loop automatic power flow control block diagram is shown in Figure 6-16. The automatic power flow control algorithm works in such a way that the desired active and reactive power flows ($P_{s'}*$ and $Q_{s'}*$) are compared with the measured active and reactive power flows ($P_{s'}$ and $Q_{s'}$)

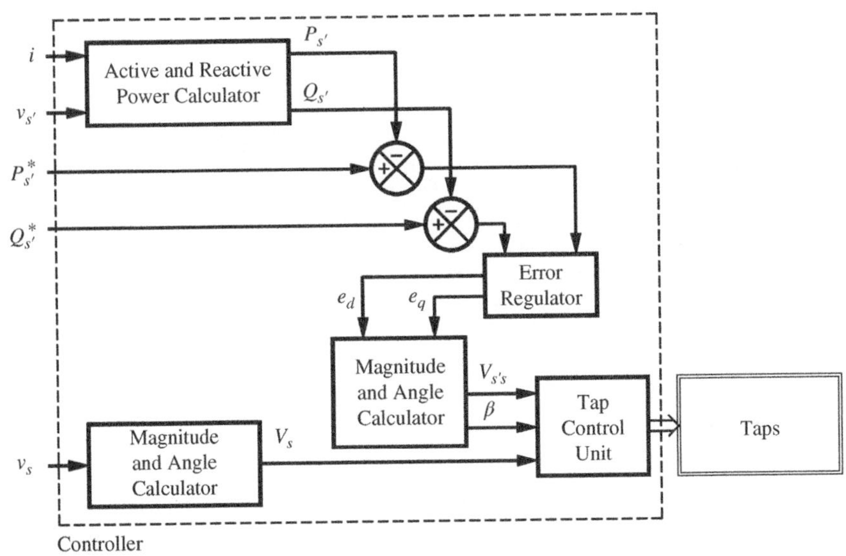

Figure 6-16 Closed-loop automatic power flow control block diagram of the ST.

and the errors are passed through an error regulator, which produces, with respect to the exciter voltage ($\mathbf{V_s}$), the required direct and quadrature components (e_d and e_q) of the series-compensating voltage. Next, the magnitude ($V_{s's}$) of the compensating voltage ($v_{s's} = -e_{dq}$) and the relative phase angle (β) with respect to the phase angle of the exciter voltage are calculated. Note that for desired active and reactive power flows ($P_{s'}^*$ and $Q_{s'}^*$), if the magnitude ($V_{s's}$) of the compensating voltage exceeds the rated voltage of the Compensating-Voltage Unit, then $V_{s's}$ is limited to its design rating and the active and reactive power flows are regulated to an intermediate pair of values.

6.2.3.5 Open-Loop Power Flow Control

The controllable range of the active and reactive power flows can easily be determined with an open-loop control by specifying to the Tap Control Unit the rated compensating voltage ($V_{s's}^*$) and varying the relative phase angle (β) within its entire 360° range as shown in Figure 6-17.

Once the compensating voltage magnitude ($V_{s's}$) and the relative phase angle (β) are determined, the tip of phasor ($\mathbf{V_{s's}}$) falls within four possible operating points as shown in Figure 6-18.

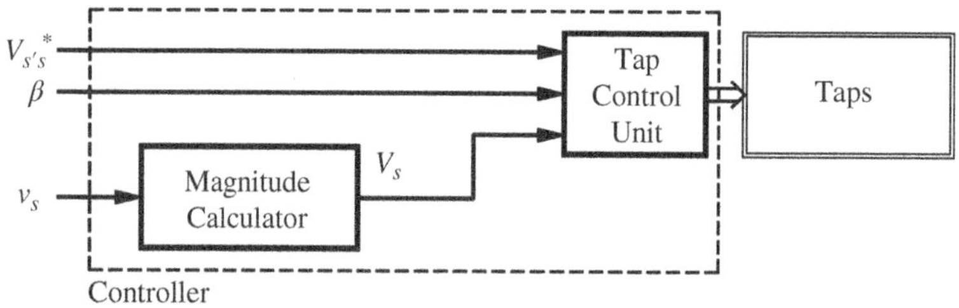

Figure 6-17 Open-loop Compensating-Voltage Unit control block diagram of the ST.

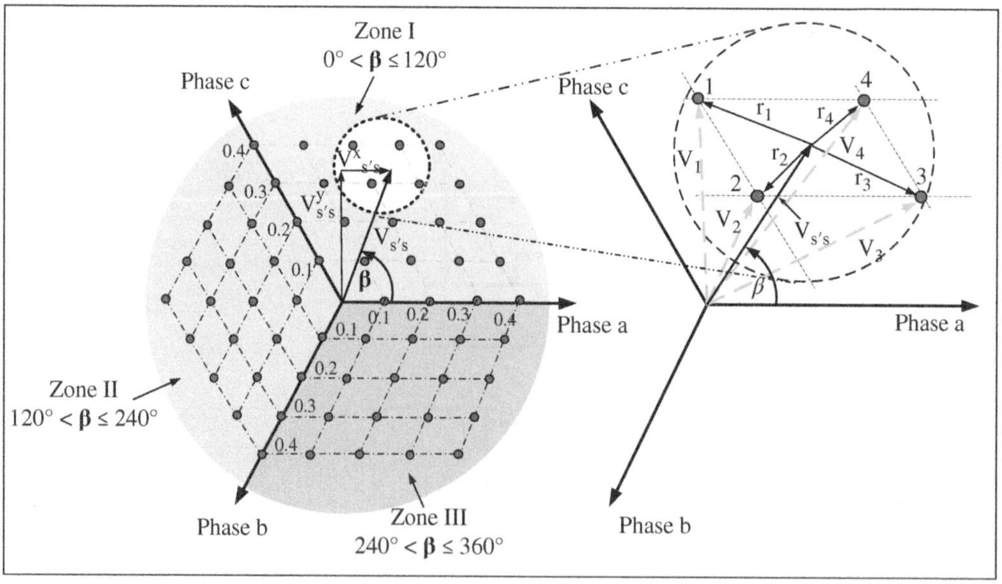

Figure 6-18 Selection of tap positions (Faruque and Dinavahi-2007).

The selected operating point is based on producing the least error given by the following equation:

$$\varepsilon = \|r_k\|_2 = \sqrt{\left(V_k^x - V_{s's}^x\right)^2 + \left(V_k^y - V_{s's}^y\right)^2} \tag{6-6}$$

where ε is the least error and r_k is the error at a possible operating point for $k = 1, 2, 3$, and 4. V_k^x and $V_{s's}^x$ are the projections of voltages of possible operating points and the compensating voltage on Phase a (*i.e.* x-axis). V_k^y and $V_{s's}^y$ are the projections of voltages of possible operating points and the compensating voltage on the perpendicular to Phase a (*i.e.* y-axis). A hysteresis loop must be implemented to eliminate jittering effect among the four neighboring taps that surround the tip of phasor ($\mathbf{V_{s's}}$). More discussion on this topic is given in "A tap-changing algorithm for the implementation of Sen transformer," by Faruque and Dinavahi (2007).

The magnitude ($V_{s's}$) and the relative phase angle (β) of the compensating voltage ($\mathbf{V_{s's}}$) can be calculated in terms of taps associated with a particular phase as shown in Figure 6-12. Note that the windings (a_1, b_1, and c_1) are associated with the compensating voltage for the A phase. Similarly, the windings (a_2, b_2, and c_2) are associated with the compensating voltage for the B phase; the windings (a_3, b_3, and c_3) are associated with the compensating voltage for the C phase. It should also be noted that each of the windings (a_1, b_2, and c_3) is tapped at the same number of turns through LTC_1; each of the windings (b_1, c_2, and a_3) is tapped at the same number of turns through LTC_2; and each of the windings (c_1, a_2, and b_3) is tapped at the same number of turns through LTC_3. However, the number of turns in the a_1-b_2-c_3 set, b_1-c_2-a_3 set, and c_1-a_2-b_3 set can be different from each other. The compensating voltage for the A phase and the relative phase angle in the range of $0° \le \beta_A \le 360°$ is derived as

$$V_{s'sA}\angle\beta_A = k_{a1}V_{sA} + k_{b1}e^{-j2\pi/3}V_{sA} + k_{c1}e^{j2\pi/3}V_{sA} \tag{6-3}$$

where
k_{a1}, k_{b1}, and k_{c1} are the active turns-ratios of the secondary, series-compensating windings in the A phase and the corresponding primary windings.

Therefore,

$$V_{s'sA}\cos\beta_A + jV_{s'sA}\sin\beta_A = k_{a1}V_{sA} + k_{b1}V_{sA}\left(-1/2 - j\sqrt{3}/2\right) + k_{c1}V_{sA}\left(-1/2 + j\sqrt{3}/2\right)$$

or

$$V_{s'sA}\cos\beta_A + jV_{s'sA}\sin\beta_A = (k_{a1} - k_{b1}/2 - k_{c1}/2)V_{sA} + j(\sqrt{3}/2)(k_{c1} - k_{b1})V_{sA}.$$

Equating the real and imaginary parts of both sides of the above equation, it can be written that

$$V_{s'sA}\cos\beta_A = (k_{a1} - k_{b1}/2 - k_{c1}/2)V_{sA} \tag{6-7a}$$

and

$$V_{s'sA}\sin\beta_A = \left(\sqrt{3}/2\right)(k_{c1} - k_{b1})V_{sA}. \tag{6-7b}$$

Equations (6-7a) and (6-7b) can be converted into a polar form as

$$V_{s'sA}\angle\beta_A = \sqrt{V_{s'sA}^2\left(\cos^2\beta_A + \sin^2\beta_A\right)}\angle\tan^{-1}\frac{\sin\beta_A}{\cos\beta_A}$$ using the identity ($\cos^2\beta_A + \sin^2\beta_A = 1$) or

$$V_{s'sA}\angle\beta_A = \sqrt{(V_{s'sA}\cos\beta_A)^2 + (V_{s'sA}\sin\beta_A)^2}\angle\tan^{-1}\frac{V_{s'sA}\sin\beta_A}{V_{s'sA}\cos\beta_A}$$ or

$$V_{s'sA}\angle\beta_A = V_{sA}\sqrt{k_{a1}^2 + k_{b1}^2 + k_{c1}^2 - k_{a1}k_{b1} - k_{b1}k_{c1} - k_{c1}k_{a1}}\angle\tan^{-1}\frac{\sqrt{3}(k_{c1} - k_{b1})}{2k_{a1} - k_{b1} - k_{c1}}. \tag{6-8}$$

For $\beta_A = 0°$, the winding (a_1) is active while the windings (b_1 and c_1) are inactive; for $\beta_A = 120°$, the winding (c_1) is active while the windings (a_1 and b_1) are inactive; for $\beta_A = 240°$, the winding (b_1) is active while the windings (c_1 and a_1) are inactive.

For $0° < \beta_A < 120°$, the windings (a_1 and c_1) are active while the winding (b_1) is inactive. Therefore, setting $k_{b1} = 0$, Equations (6-7a) and (6-7b) can be written as

$$V_{s'sA} \cos\beta_A = (k_{a1} - k_{c1}/2)V_{sA} \tag{6-9a}$$

and

$$V_{s'sA} \sin\beta_A = \left(\sqrt{3}/2\right)(k_{c1})V_{sA}. \tag{6-9b}$$

Solving Equations (6-9a) and (6-9b), it can be written that

$$k_{a1} = \frac{V_{s'sA}}{V_{sA}} \cos\beta_A + \frac{V_{s'sA}}{\sqrt{3}V_{sA}} \sin\beta_A \tag{6-10a}$$

and

$$k_{c1} = \frac{2V_{s'sA}}{\sqrt{3}V_{sA}} \sin\beta_A. \tag{6-10b}$$

As a special case of $\beta_A = 0°$ when the line voltage increases, it can be seen from Equation (6-9b) that $k_{c1} = 0$. Therefore, from Equation (6-9a), it can be written that

$$k_{a1} = \frac{V_{s'sA}}{V_{sA}}. \tag{6-10c}$$

For $120° < \beta_A < 240°$, the windings (b_1 and c_1) are active while the winding (a_1) is inactive. Therefore, setting $k_{a1} = 0$, Equations (6-7a) and (6-7b) can be written as

$$V_{s'sA} \cos\beta_A = (-k_{b1}/2 - k_{c1}/2)V_{sA} \tag{6-11a}$$

and

$$V_{s'sA} \sin\beta_A = \left(\sqrt{3}/2\right)(k_{c1} - k_{b1})V_{sA}. \tag{6-11b}$$

Solving Equations (6-11a) and (6-11b), it can be written that

$$k_{b1} = -\frac{V_{s'sA}}{\sqrt{3}V_{sA}} \sin\beta_A - \frac{V_{s'sA}}{V_{sA}} \cos\beta_A \tag{6-12a}$$

and

$$k_{c1} = \frac{V_{s'sA}}{\sqrt{3}V_{sA}} \sin\beta_A - \frac{V_{s'sA}}{V_{sA}} \cos\beta_A. \tag{6-12b}$$

As a special case of $\beta_A = 180°$ when the line voltage decreases, it can be seen from Equation (6-11b) that $k_{b1} = k_{c1}$. Therefore, from Equation (6-11a), it can be written that

$$k_{b1} = k_{c1} = \frac{V_{s'sA}}{V_{sA}}. \tag{6-10c}$$

The same conclusion can be derived from the Equations (6-12a) and (6-12b).

For $240° < \beta_A < 360°$, the windings (a_1 and b_1) are active while the winding (c_1) is inactive. Therefore, setting $k_{c1} = 0$, Equations (6-7a) and (6-7b) can be written as

$$V_{s'sA}\cos\beta_A = (k_{a1} - k_{b1}/2)V_{sA} \tag{6-13a}$$

and

$$V_{s'sA}\sin\beta_A = \left(\sqrt{3}/2\right)(-k_{b1})V_{sA}. \tag{6-13b}$$

Solving Equations (6-13a) and (6-13b), it can be written that

$$k_{a1} = \frac{V_{s'sA}}{V_{sA}}\cos\beta_A - \frac{V_{s'sA}}{\sqrt{3}V_{sA}}\sin\beta_A \tag{6-14a}$$

and

$$k_{b1} = -\frac{2V_{s'sA}}{\sqrt{3}V_{sA}}\sin\beta_A. \tag{6-14b}$$

Equation (6-2) can be further derived as

$$V_{s'A}\angle\psi_A = V_{sA} + V_{s'sA}\angle\beta_A \tag{6-2}$$

or

$$V_{s'A}\cos\psi_A + jV_{s'A}\sin\psi_A = V_{sA} + V_{s'sA}\cos\beta_A + jV_{s'sA}\sin\beta_A.$$

Equating the real and imaginary parts of both sides of the above equation, it can be written that

$$V_{s'A}\cos\psi_A = V_{sA} + V_{s'sA}\cos\beta_A \tag{6-15a}$$

and

$$V_{s'A}\sin\psi_A = V_{s'sA}\sin\beta_A. \tag{6-15b}$$

Using Equations (6-7a) and (6-7b), Equations (6-15a) and (6-15b) can be combined as

$$V_{s'A}\cos\psi_A = V_{sA} + (k_{a1} - k_{b1}/2 - k_{c1}/2)V_{sA} \tag{6-16a}$$

and

$$V_{s'A}\sin\psi_A = \left(\sqrt{3}/2\right)(k_{c1} - k_{b1})V_{sA}. \tag{6-16b}$$

Equations (6-16a) and (6-16b) can be converted into a polar form as

$$V_{s'A}\angle\psi_A = \sqrt{V_{s'A}^2\left(\cos^2\psi_A + \sin^2\psi_A\right)}\angle\tan^{-1}\frac{\sin\psi_A}{\cos\psi_A} \text{ using the identity } \left(\cos^2\psi_A + \sin^2\psi_A = 1\right)$$

or

$$V_{s'A}\angle\psi_A = \sqrt{(V_{s'A}\cos\psi_A)^2 + (V_{s'A}\sin\psi_A)^2}\angle\tan^{-1}\frac{V_{s'A}\sin\psi_A}{V_{s'A}\cos\psi_A} \quad \text{or}$$

$$V_{s'A}\angle\psi_A = V_{sA}\sqrt{1 + k_{a1}^2 + k_{b1}^2 + k_{c1}^2 - k_{a1}k_{b1} - k_{b1}k_{c1} - k_{c1}k_{a1} + 2k_{a1} - k_{b1} - k_{c1}}\ \angle\tan^{-1}\frac{\sqrt{3}(k_{c1} - k_{b1})}{2 + 2k_{a1} - k_{b1} - k_{c1}} \tag{6-17a}$$

or

$$V_{s'A}\angle\psi_A = V_{sA}\sqrt{(1 + k_{a1})^2 + k_{b1}^2 + k_{c1}^2 - (1 + k_{a1})k_{b1} - k_{b1}k_{c1} - k_{c1}(1 + k_{a1})}\ \angle\tan^{-1}\frac{\sqrt{3}(k_{c1} - k_{b1})}{2(1 + k_{a1}) - k_{b1} - k_{c1}}. \tag{6-17b}$$

For an equivalent shunt compensation, the active turns-ratios (K_{A1}, K_{B1}, and K_{C1}) of the shunt-compensating windings and the active turns-ratios (k_{a1}, k_{b1}, and k_{c1}) of the series-compensating windings, shown in Figure 6-3, may be related as

$$K_{A1} = 1 + k_{a1} \quad (6\text{-}18a)$$

$$K_{B1} = k_{b1} \quad (6\text{-}18b)$$

$$K_{C1} = k_{c1}. \quad (6\text{-}18c)$$

Therefore, Equations (6-16a) and (6-16b) can be rewritten as

$$V_{s'A} \cos \psi_A = (K_{A1} - K_{B1}/2 - K_{C1}/2) V_{sA} \quad (6\text{-}19a)$$

and

$$V_{s'A} \sin \psi_A = \left(\sqrt{3}/2\right)(K_{C1} - K_{B1}) V_{sA}. \quad (6\text{-}19b)$$

Either combining Equations (6-19a) and (6-19b) as shown later in the chapter or substituting Equations (6-18a), (6-18b), and (6-18c) into Equation (6-17b), it can be written that

$$V_{s'A} \angle \psi_A = V_{sA} \sqrt{K_{A1}^2 + K_{B1}^2 + K_{C1}^2 - K_{A1}K_{B1} - K_{B1}K_{C1} - K_{C1}K_{A1}} \angle \tan^{-1} \frac{\sqrt{3}(K_{C1} - K_{B1})}{2K_{A1} - K_{B1} - K_{C1}}. \quad (6\text{-}20)$$

For $\psi_A = 0°$, the winding (A_1) is active while the windings (B_1 and C_1) are inactive; for $\psi_A = 120°$, the winding (C_1) is active while the windings (A_1 and B_1) are inactive; for $\psi_A = 240°$, the winding (B_1) is active while the windings (C_1 and A_1) are inactive. However, only the cases $-90° < \psi_A < 90°$ are considered in this book.

For $0° < \psi_A < 120°$, the windings (A_1 and C_1) are active while the winding (B_1) is inactive. Therefore, setting $K_{B1} = 0$, Equations (6-19a) and (6-19b) can be written as

$$V_{s'A} \cos \psi_A = (K_{A1} - K_{C1}/2) V_{sA} \quad (6\text{-}21a)$$

and

$$V_{s'A} \sin \psi_A = \left(\sqrt{3}/2\right)(K_{C1}) V_{sA}. \quad (6\text{-}21b)$$

Solving Equations (6-21a) and (6-21b), it can be written that

$$K_{A1} = \frac{V_{s'A}}{V_{sA}} \cos \psi_A + \frac{V_{s'A}}{\sqrt{3} V_{sA}} \sin \psi_A \quad (6\text{-}22a)$$

and

$$K_{C1} = \frac{2V_{s'A}}{\sqrt{3} V_{sA}} \sin \psi_A. \quad (6\text{-}22b)$$

As a special case of $\psi_A = 0°$, it can be seen from Equation (6-21b) that $k_{C1} = 0$. Therefore, from Equation (6-21a), it can be written that

$$k_{A1} = \frac{V_{s'A}}{V_{sA}}. \quad (6\text{-}22c)$$

For $240° < \psi_A < 360°$, the windings (A_1 and B_1) are active while the winding (C_1) is inactive. Therefore, setting $K_{C1} = 0$, Equations (6-19a) and (6-19b) can be written as

$$V_{s'A}\cos\psi_A = (K_{A1} - K_{B1}/2)V_{sA} \tag{6-23a}$$

and

$$V_{s'A}\sin\psi_A = \left(\sqrt{3}/2\right)(-K_{B1})V_{sA}. \tag{6-23b}$$

Solving Equations (6-23a) and (6-23b), it can be written that

$$K_{A1} = \frac{V_{s'A}}{V_{sA}}\cos\psi_A - \frac{V_{s'A}}{\sqrt{3}V_{sA}}\sin\psi_A \tag{6-24a}$$

and

$$K_{B1} = -\frac{2V_{s'A}}{\sqrt{3}V_{sA}}\sin\psi_A. \tag{6-24b}$$

Once the magnitude ($V_{s'}$) and the phase-shift angle (ψ) of the modified sending-end voltage ($\mathbf{V_{s'}}$) are known, the compensated active and reactive power flows (P_r and Q_r) at the receiving end of the line can be determined using the following equations:

$$P_r = \frac{A'}{\sqrt{R^2+X^2}}\left\{\left(R\cos\delta' - \frac{V_r}{V_{s'}}\right) + \sin\delta'\right\} \tag{B-43}$$

and

$$Q_r = \frac{A'}{\sqrt{R^2+X^2}}\left\{-R\sin\delta' + X\left(\cos\delta' - \frac{V_r}{V_{s'}}\right)\right\} \tag{B-44}$$

where the magnitude of the voltage at the modified sending end is $V_{s'}$ and the modified power angle is given by

$$\delta' = \delta + \psi \tag{2-51a}$$

and

$$A' = \frac{V_{s'}V_r}{\sqrt{R^2+X^2}}. \tag{B-34}$$

Note that $V_{s'}$ and ψ in Equations (B-43), (B-44), and (2-51a) can be substituted using $V_{s'A}$ and ψ_A from Equation (6-20) when assuming a balanced three-phase system.

Using Equations (B-43) and (B-44), the apparent power (S_r) at the receiving end and the line current (I) can be determined using the following equations:

$$S_r = \sqrt{P_r^2 + Q_r^2}. \tag{2-76}$$

$$I = \frac{S_r}{V_r}. \tag{2-77}$$

For a series-compensating voltage ($\mathbf{V_{s's}} = V_{s's}\angle\beta$), the exchanged active and reactive powers (P_{se} and Q_{se}) by the Compensating-Voltage Unit (Series Unit) can be determined by using the following equations:

$$P_{se} = -\frac{RV_{s's}^2}{R^2+X^2} + \frac{1}{\sqrt{R^2+X^2}}[-R\{A_s\cos\beta - A_r\cos(\delta+\beta)\} + X\{A_s\sin\beta - A_r\sin(\delta+\beta)\}] \tag{B-101}$$

and

$$Q_{se} = -\frac{XV_{s's}^2}{R^2 + X^2} + \frac{1}{\sqrt{R^2 + X^2}}[-R\{A_s \sin\beta - A_r \sin(\delta + \beta)\} - X\{A_s \cos\beta - A_r \cos(\delta + \beta)\}] \tag{B-102}$$

where

$$A_s = \frac{V_s V_{s's}}{\sqrt{R^2 + X^2}} \tag{B-62}$$

$$A_r = \frac{V_r V_{s's}}{\sqrt{R^2 + X^2}}. \tag{B-76}$$

Using Equations (B-101) and (B-102), the exchanged apparent power (S_{se}) by the Compensating-Voltage Unit (Series Unit) can be determined by using the following equation:

$$S_{se} = \sqrt{P_{se}^2 + Q_{se}^2}. \tag{2-182}$$

The effect of implementing an open-loop control in voltage-injection mode, traveling from one operating point to the next in one particular hexagon in each phase, is studied, using an electrical system data as shown in Table 2-1, which is reproduced below.

Table 6-1 shows the active turns-ratios of the three secondary, series-compensating windings in the A phase for the entire range of the relative phase angle of $0° \leq \beta_A \leq 360°$. The table shows the variations of the relative phase angle (β_A) and the phase-shift angle (ψ_A) due to the change in turns-ratios. The theoretical values of the compensating voltage ($V_{s'sA}$) and the modified sending-end voltage ($V_{s'A}$) are calculated using Equations (6-8) and (6-17a) or (6-17b), respectively, considering the sending-end voltage (V_{sA}) to be 1 pu during the entire range of the relative phase angle of $0° \leq \beta_A \leq 360°$ and shown in Table 6-1. The table is expanded for each hexagon, related to 5, 10, 15, and 20% compensating voltages.

To generate the compensating voltage ($V_{s'sB}$) and the modified sending-end voltage ($V_{s'B}$) for the B phase, Equations (6-2) and (6-3) may be rewritten as

$$V_{s'B}\angle\psi_B = V_{sB}\angle 0° + V_{s'sB}\angle\beta_B \tag{6-25}$$

where

$$V_{s'sB}\angle\beta_B = \mathbf{V_{s'sB}} = \mathbf{V_{b2}} + \mathbf{V_{c2}} + \mathbf{V_{a2}} \quad \text{or}$$

$$V_{s'sB}\angle\beta_B = k_{b2}V_{sB} + k_{c2}V_{sC} + k_{a2}V_{sA} = k_{b2}e^{-j2\pi/3}V_{sA} + k_{c2}e^{j2\pi/3}V_{sA} + k_{a2}V_{sA} \tag{6-26}$$

Table 2-1 Electrical system data.

Parameters	Values
Sending-end voltage (V_s) at reference phase angle	$1\angle 0°$ pu
Receiving-end voltage (V_r) and power angle ($-\delta$)	$1\angle -30°$ pu
Line resistance (R)	0 pu
Line reactance (X)	0.5 pu

Table 6-1 Active turns-ratios of the secondary, series-compensating windings in the *a* phase, corresponding series-compensating voltage, and modified sending-end voltage.

Case for natural power flow (center point of the hexagon characteristic)						
Turns-ratios			**$V_{s'sA}$ (pu)**	**β_A (deg)**	**$V_{s'A}$ (pu)**	**ψ_A (deg)**
k_{a1}	k_{b1}	k_{c1}				
0.00	0.00	0.00	0.00	0.00	1.00	0.00
0.00	0.00	0.00	0.00	360.00	1.00	0.00
Case for 0.05 pu compensating voltage (smallest hexagon)						
Turns-ratios			**$V_{s'sA}$ (pu)**	**β_A (deg)**	**$V_{s'A}$ (pu)**	**ψ_A (deg)**
k_{a1}	k_{b1}	k_{c1}				
0.05	0.00	0.00	0.05	0.00	1.05	0.00
0.05	0.00	0.05	0.05	60.00	1.03	2.42
0.00	0.00	0.05	0.05	120.00	0.98	2.54
0.00	0.05	0.05	0.05	180.00	0.95	0.00
0.00	0.05	0.00	0.05	240.00	0.98	−2.54
0.05	0.05	0.00	0.05	300.00	1.03	−2.42
0.05	0.00	0.00	0.05	360.00	1.05	0.00
Case for 0.10 pu compensating voltage (second smallest hexagon)						
Turns-ratios			**$V_{s'sA}$ (pu)**	**β_A (deg)**	**$V_{s'A}$ (pu)**	**ψ_A (deg)**
k_{a1}	k_{b1}	k_{c1}				
0.10	0.00	0.00	0.10	0.00	1.10	0.00
0.10	0.00	0.05	0.09	30.00	1.08	2.31
0.10	0.00	0.10	0.10	60.00	1.05	4.72
0.05	0.00	0.10	0.09	90.00	1.00	4.95
0.00	0.00	0.10	0.10	120.00	0.95	5.21
0.00	0.05	0.10	0.09	150.00	0.93	2.68
0.00	0.10	0.10	0.10	180.00	0.90	0.00
0.00	0.10	0.05	0.09	210.00	0.93	−2.68
0.00	0.10	0.00	0.10	240.00	0.95	−5.21

(Continued)

0.05	0.10	0.00	0.09	270.00	1.00	−4.95
0.10	0.10	0.00	0.10	300.00	1.05	−4.72
0.10	0.05	0.00	0.09	330.00	1.08	−2.31
0.10	0.00	0.00	0.10	360.00	1.10	0.00
Case for 0.15 pu compensating voltage (second largest hexagon)						
Turns-ratios			**$V_{s'sA}$ (pu)**	**β_A (deg)**	**$V_{s'A}$ (pu)**	**ψ_A (deg)**
k_{a1}	k_{b1}	k_{c1}				
0.15	0.00	0.00	0.15	0.00	1.15	0.00
0.15	0.00	0.05	0.13	19.11	1.13	2.20
0.15	0.00	0.10	0.13	40.89	1.10	4.50
0.15	0.00	0.15	0.15	60.00	1.08	6.89
0.10	0.00	0.15	0.13	79.11	1.03	7.22
0.05	0.00	0.15	0.13	100.89	0.98	7.59
0.00	0.00	0.15	0.15	120.00	0.93	7.99
0.00	0.05	0.15	0.13	139.11	0.90	5.50
0.00	0.10	0.15	0.13	160.89	0.88	2.83
0.00	0.15	0.15	0.15	180.00	0.85	0.00
0.00	0.15	0.10	0.13	199.11	0.88	−2.83
0.00	0.15	0.05	0.13	220.89	0.90	−5.50
0.00	0.15	0.00	0.15	240.00	0.93	−7.99
0.05	0.15	0.00	0.13	259.11	0.98	−7.59
0.10	0.15	0.00	0.13	280.89	1.03	−7.22
0.15	0.15	0.00	0.15	300.00	1.08	−6.89
0.15	0.10	0.00	0.13	319.11	1.10	−4.50
0.15	0.05	0.00	0.13	340.89	1.13	−2.20
0.15	0.00	0.00	0.15	360.00	1.15	0.00

Table 6-1 (Continued)

Case for 0.20 pu compensating voltage (largest hexagon)						
Turns-ratios			$V_{s'sA}$ (pu)	β_A (deg)	$V_{s'A}$ (pu)	ψ_A (deg)
k_{a1}	k_{b1}	k_{c1}				
0.20	0.00	0.00	0.20	0.00	1.20	0.00
0.20	0.00	0.05	0.18	13.90	1.18	2.11
0.20	0.00	0.10	0.17	30.00	1.15	4.31
0.20	0.00	0.15	0.18	46.10	1.13	6.59
0.20	0.00	0.20	0.20	60.00	1.11	8.95
0.15	0.00	0.20	0.18	73.90	1.06	9.37
0.10	0.00	0.20	0.17	90.00	1.01	9.83
0.05	0.00	0.20	0.18	106.10	0.97	10.33
0.00	0.00	0.20	0.20	120.00	0.92	10.89
0.00	0.05	0.20	0.18	133.90	0.88	8.44
0.00	0.10	0.20	0.17	150.00	0.85	5.82
0.00	0.15	0.20	0.18	166.10	0.83	3.00
0.00	0.20	0.20	0.20	180.00	0.80	0.00
0.00	0.20	0.15	0.18	193.90	0.83	−3.00
0.00	0.20	0.10	0.17	210.00	0.85	−5.82
0.00	0.20	0.05	0.18	226.10	0.88	−8.44
0.00	0.20	0.00	0.20	240.00	0.92	−10.89
0.05	0.20	0.00	0.18	253.90	0.97	−10.33
0.10	0.20	0.00	0.17	270.00	1.01	−9.83
0.15	0.20	0.00	0.18	286.10	1.06	−9.37
0.20	0.20	0.00	0.20	300.00	1.11	−8.95
0.20	0.15	0.00	0.18	313.90	1.13	−6.59
0.20	0.10	0.00	0.17	330.00	1.15	−4.31
0.20	0.05	0.00	0.18	346.10	1.18	−2.11
0.20	0.00	0.00	0.20	360.00	1.20	0.00

where

$\mathbf{V_{b2}}$, $\mathbf{V_{c2}}$, and $\mathbf{V_{a2}}$ are active voltages in the series-compensating windings (b_2, c_2, and a_2) of the *B* phase,
$k_{b2} = 0.05$, $k_{c2} = 0$, and $k_{a2} = 0.20$ are the active turns-ratios of the secondary, series-compensating windings in the *B* phase and the corresponding primary windings,
$V_{s'sB}$ is the magnitude of the series-compensating voltage in the *B* phase, and
β_B is the relative phase angle of the series-compensating voltage in the *B* phase.

Using Equations (6-8) and (6-17a) or (6-17b) and replacing k_{a1}, k_{b1}, and k_{c1} with k_{b2}, k_{c2}, and k_{a2}, respectively, it can be written that

$$V_{s'sB}\angle\beta_B = V_{sB}\sqrt{k_{b2}^2 + k_{c2}^2 + k_{a2}^2 - k_{b2}k_{c2} - k_{c2}k_{a2} - k_{a2}k_{b2}}\angle\tan^{-1}\frac{\sqrt{3}(k_{a2} - k_{c2})}{2k_{b2} - k_{c2} - k_{a2}} \tag{6-27}$$

and

$$V_{s'B}\angle\psi_B = V_{sB}\sqrt{1 + k_{b2}^2 + k_{c2}^2 + k_{a2}^2 - k_{b2}k_{c2} - k_{c2}k_{a2} - k_{a2}k_{b2} + 2k_{b2} - k_{c2} - k_{a2}}\angle\tan^{-1}\frac{\sqrt{3}(k_{a2} - k_{c2})}{2 + 2k_{b2} - k_{c2} - k_{a2}} \tag{6-28a}$$

or

$$V_{s'B}\angle\psi_B = V_{sB}\sqrt{(1 + k_{b2})^2 + k_{c2}^2 + k_{a2}^2 - (1 + k_{b2})k_{c2} - k_{c2}k_{a2} - k_{a2}(1 + k_{b2})}\angle\tan^{-1}\frac{\sqrt{3}(k_{a2} - k_{c2})}{2(1 + k_{b2}) - k_{c2} - k_{a2}}. \tag{6-28b}$$

Table 6-2 shows the active turns-ratios of the three secondary, series-compensating windings in the *B* phase for the entire range of the relative phase angle of $0° \leq \beta_B \leq 360°$. The table shows the variations of the relative phase angle (β_B) and the phase-shift angle (ψ_B) due to the change in turns-ratios. The theoretical values of the compensating voltage ($V_{s'sB}$) and the modified sending-end voltage ($V_{s'B}$) are calculated using Equations (6-27) and (6-28a) or (6-28b), respectively, considering the sending-end voltage (V_{sB}) to be 1 pu during the entire range of the relative phase angle of $0° \leq \beta_B \leq 360°$ and shown in Table 6-2. The table is expanded for each hexagon, related to 5, 10, 15, and 20% compensating voltages.

To generate the compensating voltage ($\mathbf{V_{s'sC}}$) and the modified sending-end voltage ($\mathbf{V_{s'C}}$) for the C phase, Equations (6-2) and (6-3) may be rewritten as

$$V_{s'C}\angle\psi_C = V_{sC}\angle 0° + V_{s'sC}\angle\beta_C \tag{6-29}$$

where

$$V_{s'sC}\angle\beta_C = \mathbf{V_{s'sC}} = \mathbf{V_{c3}} + \mathbf{V_{a3}} + \mathbf{V_{b3}}$$

or

$$V_{s'sC}\angle\beta_C = k_{c3}V_{sC} + k_{a3}V_{sA} + k_{b3}V_{sB} = k_{c3}e^{j2\pi/3}V_{sA} + k_{a3}V_{sA} + k_{b3}e^{-j2\pi/3}V_{sA} \tag{6-30}$$

where

$\mathbf{V_{c3}}$, $\mathbf{V_{a3}}$, and $\mathbf{V_{b3}}$ are active voltages in the series-compensating windings (c_3, a_3, and b_3) of the *C* phase,
$k_{c3} = 0.05$, $k_{a3} = 0$, and $k_{b3} = 0.20$ are the active turns-ratios of the secondary, series-compensating windings in the *C* phase and the corresponding primary windings,

Table 6-2 Active turns-ratios of the secondary, series-compensating windings in the *B* phase, corresponding series-compensating voltage, and modified sending-end voltage.

Case for natural power flow (center point of the hexagon characteristic)						
Turns-ratios			$V_{s'sB}$ (pu)	β_B (deg)	$V_{s'B}$ (pu)	ψ_B (deg)
k_{b2}	k_{c2}	k_{a2}				
0.00	0.00	0.00	0.00	0.00	1.00	0.00
0.00	0.00	0.00	0.00	360.00	1.00	0.00

Case for 0.05 pu compensating voltage (smallest hexagon)						
Turns-ratios			$V_{s'sB}$ (pu)	β_B (deg)	$V_{s'B}$ (pu)	ψ_B (deg)
k_{b2}	k_{c2}	k_{a2}				
0.05	0.00	0.00	0.05	0.00	1.05	0.00
0.05	0.00	0.05	0.05	60.00	1.03	2.42
0.00	0.00	0.05	0.05	120.00	0.98	2.54
0.00	0.05	0.05	0.05	180.00	0.95	0.00
0.00	0.05	0.00	0.05	240.00	0.98	−2.54
0.05	0.05	0.00	0.05	300.00	1.03	−2.42
0.05	0.00	0.00	0.05	360.00	1.05	0.00

Case for 0.10 pu compensating voltage (second smallest hexagon)						
Turns-ratios			$V_{s'sB}$ (pu)	β_B (deg)	$V_{s'B}$ (pu)	ψ_B (deg)
k_{b2}	k_{c2}	k_{a2}				
0.10	0.00	0.00	0.10	0.00	1.10	0.00
0.10	0.00	0.05	0.09	30.00	1.08	2.31
0.10	0.00	0.10	0.10	60.00	1.05	4.72
0.05	0.00	0.10	0.09	90.00	1.00	4.95
0.00	0.00	0.10	0.10	120.00	0.95	5.21
0.00	0.05	0.10	0.09	150.00	0.93	2.68
0.00	0.10	0.10	0.10	180.00	0.90	0.00
0.00	0.10	0.05	0.09	210.00	0.93	−2.68

(*Continued*)

Table 6-2 (Continued)

Case for 0.10 pu compensating voltage (second smallest hexagon)						
Turns-ratios			$V_{s'sB}$ (pu)	β_B (deg)	$V_{s'B}$ (pu)	ψ_B (deg)
k_{b2}	k_{c2}	k_{a2}				
0.00	0.10	0.00	0.10	240.00	0.95	−5.21
0.05	0.10	0.00	0.09	270.00	1.00	−4.95
0.10	0.10	0.00	0.10	300.00	1.05	−4.72
0.10	0.05	0.00	0.09	330.00	1.08	−2.31
0.10	0.00	0.00	0.10	360.00	1.10	0.00

Case for 0.15 pu compensating voltage (second largest hexagon)						
Turns-ratios			$V_{s'sB}$ (pu)	β_B (deg)	$V_{s'B}$ (pu)	ψ_B (deg)
k_{b2}	k_{c2}	k_{a2}				
0.15	0.00	0.00	0.15	0.00	1.15	0.00
0.15	0.00	0.05	0.13	19.11	1.13	2.20
0.15	0.00	0.10	0.13	40.89	1.10	4.50
0.15	0.00	0.15	0.15	60.00	1.08	6.89
0.10	0.00	0.15	0.13	79.11	1.03	7.22
0.05	0.00	0.15	0.13	100.89	0.98	7.59
0.00	0.00	0.15	0.15	120.00	0.93	7.99
0.00	0.05	0.15	0.13	139.11	0.90	5.50
0.00	0.10	0.15	0.13	160.89	0.88	2.83
0.00	0.15	0.15	0.15	180.00	0.85	0.00
0.00	0.15	0.10	0.13	199.11	0.88	−2.83
0.00	0.15	0.05	0.13	220.89	0.90	−5.50
0.00	0.15	0.00	0.15	240.00	0.93	−7.99
0.05	0.15	0.00	0.13	259.11	0.98	−7.59
0.10	0.15	0.00	0.13	280.89	1.03	−7.22
0.15	0.15	0.00	0.15	300.00	1.08	−6.89
0.15	0.10	0.00	0.13	319.11	1.10	−4.50

Table 6-2 (Continued)

Case for 0.15 pu compensating voltage (second largest hexagon)						
Turns-ratios			$V_{s'sB}$ (pu)	β_B (deg)	$V_{s'B}$ (pu)	ψ_B (deg)
k_{b2}	k_{c2}	k_{a2}				
0.15	0.05	0.00	0.13	340.89	1.13	−2.20
0.15	0.00	0.00	0.15	360.00	1.15	0.00
Case for 0.20 pu compensating voltage (largest hexagon)						
Turns-ratios			$V_{s'sB}$ (pu)	β_B (deg)	$V_{s'B}$ (pu)	ψ_B (deg)
k_{b2}	k_{c2}	k_{a2}				
0.20	0.00	0.00	0.20	0.00	1.20	0.00
0.20	0.00	0.05	0.18	13.90	1.18	2.11
0.20	0.00	0.10	0.17	30.00	1.15	4.31
0.20	0.00	0.15	0.18	46.10	1.13	6.59
0.20	0.00	0.20	0.20	60.00	1.11	8.95
0.15	0.00	0.20	0.18	73.90	1.06	9.37
0.10	0.00	0.20	0.17	90.00	1.01	9.83
0.05	0.00	0.20	0.18	106.10	0.97	10.33
0.00	0.00	0.20	0.20	120.00	0.92	10.89
0.00	0.05	0.20	0.18	133.90	0.88	8.44
0.00	0.10	0.20	0.17	150.00	0.85	5.82
0.00	0.15	0.20	0.18	166.10	0.83	3.00
0.00	0.20	0.20	0.20	180.00	0.80	0.00
0.00	0.20	0.15	0.18	193.90	0.83	−3.00
0.00	0.20	0.10	0.17	210.00	0.85	−5.82
0.00	0.20	0.05	0.18	226.10	0.88	−8.44
0.00	0.20	0.00	0.20	240.00	0.92	−10.89
0.05	0.20	0.00	0.18	253.90	0.97	−10.33
0.10	0.20	0.00	0.17	270.00	1.01	−9.83
0.15	0.20	0.00	0.18	286.10	1.06	−9.37

(Continued)

Table 6-2 (Continued)

Case for 0.15 pu compensating voltage (second largest hexagon)						
Turns-ratios			$V_{s'sB}$ (pu)	β_B (deg)	$V_{s'B}$ (pu)	ψ_B (deg)
k_{b2}	k_{c2}	k_{a2}				
0.20	0.20	0.00	0.20	300.00	1.11	−8.95
0.20	0.15	0.00	0.18	313.90	1.13	−6.59
0.20	0.10	0.00	0.17	330.00	1.15	−4.31
0.20	0.05	0.00	0.18	346.10	1.18	−2.11
0.20	0.00	0.00	0.20	360.00	1.20	0.00

$V_{s'sC}$ is the magnitude of the series-compensating voltage in the C phase, and
β_C is the relative phase angle of the series-compensating voltage in the C phase.

Using Equations (6-8) and (6-17a) or (6-17b) and replacing k_{a1}, k_{b1}, and k_{c1} with k_{c3}, k_{a3}, and k_{b3}, respectively, it can be written that

$$V_{s'sC}\angle\beta_C = V_{sC}\sqrt{k_{c3}^2 + k_{a3}^2 + k_{b3}^2 - k_{c3}k_{a3} - k_{a3}k_{b3} - k_{b3}k_{c3}}\angle\tan^{-1}\frac{\sqrt{3}(k_{b3} - k_{a3})}{2k_{c3} - k_{a3} - k_{b3}} \tag{6-31}$$

and

$$V_{s'C}\angle\psi_C = V_{sC}\sqrt{1 + k_{c3}^2 + k_{a3}^2 + k_{b3}^2 - k_{c3}k_{a3} - k_{a3}k_{b3} - k_{b3}k_{c3} + 2k_{c3} - k_{a3} - k_{b3}}\ \angle\tan^{-1}\frac{\sqrt{3}(k_{b3} - k_{a3})}{2 + 2k_{c3} - k_{a3} - k_{b3}} \tag{6-32a}$$

or

$$V_{s'C}\angle\psi_C = V_{sC}\sqrt{(1 + k_{c3})^2 + k_{a3}^2 + k_{b3}^2 - (1 + k_{c3})k_{a3} - k_{a3}k_{b3} - k_{b3}(1 + k_{c3})}\ \angle\tan^{-1}\frac{\sqrt{3}(k_{b3} - k_{a3})}{2(1 + k_{c3}) - k_{a3} - k_{b3}}. \tag{6-32b}$$

Table 6-3 shows the active turns-ratios of the three secondary, series-compensating windings in the C phase for the entire range of the relative phase angle of $0° \leq \beta_C \leq 360°$. The table shows the variations of the relative phase angle (β_C) and the phase-shift angle (ψ_C) due to the change in turns-ratios. The theoretical values of the compensating voltage ($V_{s'sC}$) and the modified sending-end voltage ($V_{s'C}$) are calculated using Equations (6-31) and (6-32a) or (6-32b), respectively, considering the sending-end voltage (V_{sC}) to be 1 pu during the entire range of the relative phase angle of $0° \leq \beta_C \leq 360°$ and shown in Table 6-3. The table is expanded for each hexagon, related to 5, 10, 15, and 20% compensating voltages.

It should be noted that each of the windings (a_1, b_2, and c_3) is tapped at the same number of turns through LTC_1; each of the windings (b_1, c_2, and a_3) is tapped at the same number of turns through LTC_2; and each of the windings (c_1, a_2, and b_3) is tapped at the same number of turns through LTC_3. However, the number of turns in the a_1-b_2-c_3 set, b_1-c_2-a_3 set, and c_1-a_2-b_3 set can be different from each other.

Table 6-3 Active turns-ratios of the secondary, series-compensating windings in the C phase, corresponding series-compensating voltage, and modified sending-end voltage.

Case for natural power flow (center point of the hexagon characteristic)						
Turns-ratios			$V_{s'sC}$ (pu)	β_C (deg)	$V_{s'C}$ (pu)	ψ_C (deg)
k_{c3}	k_{a3}	k_{b3}				
0.00	0.00	0.00	0.00	0.00	1.00	0.00
0.00	0.00	0.00	0.00	360.00	1.00	0.00
Case for 0.05 pu compensating voltage (smallest hexagon)						
Turns-ratios			$V_{s'sC}$ (pu)	β_C (deg)	$V_{s'C}$ (pu)	ψ_C (deg)
k_{c3}	k_{a3}	k_{b3}				
0.05	0.00	0.00	0.05	0.00	1.05	0.00
0.05	0.00	0.05	0.05	60.00	1.03	2.42
0.00	0.00	0.05	0.05	120.00	0.98	2.54
0.00	0.05	0.05	0.05	180.00	0.95	0.00
0.00	0.05	0.00	0.05	240.00	0.98	−2.54
0.05	0.05	0.00	0.05	300.00	1.03	−2.42
0.05	0.00	0.00	0.05	360.00	1.05	0.00
Case for 0.10 pu compensating voltage (second smallest hexagon)						
Turns-ratios			$V_{s'sC}$ (pu)	β_C (deg)	$V_{s'C}$ (pu)	ψ_C (deg)
k_{c3}	k_{a3}	k_{b3}				
0.10	0.00	0.00	0.10	0.00	1.10	0.00
0.10	0.00	0.05	0.09	30.00	1.08	2.31
0.10	0.00	0.10	0.10	60.00	1.05	4.72
0.05	0.00	0.10	0.09	90.00	1.00	4.95
0.00	0.00	0.10	0.10	120.00	0.95	5.21
0.00	0.05	0.10	0.09	150.00	0.93	2.68
0.00	0.10	0.10	0.10	180.00	0.90	0.00
0.00	0.10	0.05	0.09	210.00	0.93	−2.68

(Continued)

Table 6-3 (Continued)

Case for 0.10 pu compensating voltage (second smallest hexagon)						
Turns-ratios			$V_{s'sC}$ (pu)	β_C (deg)	$V_{s'C}$ (pu)	ψ_C (deg)
k_{c3}	k_{a3}	k_{b3}				
0.00	0.10	0.00	0.10	240.00	0.95	−5.21
0.05	0.10	0.00	0.09	270.00	1.00	−4.95
0.10	0.10	0.00	0.10	300.00	1.05	−4.72
0.10	0.05	0.00	0.09	330.00	1.08	−2.31
0.10	0.00	0.00	0.10	360.00	1.10	0.00
Case for 0.15 pu compensating voltage (second largest hexagon)						
Turns-ratios			$V_{s'sC}$ (pu)	β_C (deg)	$V_{s'C}$ (pu)	ψ_C (deg)
k_{c3}	k_{a3}	k_{b3}				
0.15	0.00	0.00	0.15	0.00	1.15	0.00
0.15	0.00	0.05	0.13	19.11	1.13	2.20
0.15	0.00	0.10	0.13	40.89	1.10	4.50
0.15	0.00	0.15	0.15	60.00	1.08	6.89
0.10	0.00	0.15	0.13	79.11	1.03	7.22
0.05	0.00	0.15	0.13	100.89	0.98	7.59
0.00	0.00	0.15	0.15	120.00	0.93	7.99
0.00	0.05	0.15	0.13	139.11	0.90	5.50
0.00	0.10	0.15	0.13	160.89	0.88	2.83
0.00	0.15	0.15	0.15	180.00	0.85	0.00
0.00	0.15	0.10	0.13	199.11	0.88	−2.83
0.00	0.15	0.05	0.13	220.89	0.90	−5.50
0.00	0.15	0.00	0.15	240.00	0.93	−7.99
0.05	0.15	0.00	0.13	259.11	0.98	−7.59
0.10	0.15	0.00	0.13	280.89	1.03	−7.22
0.15	0.15	0.00	0.15	300.00	1.08	−6.89

Table 6-3 (Continued)

Case for 0.15 pu compensating voltage (second largest hexagon)						
Turns-ratios			$V_{s'sC}$ (pu)	β_C (deg)	$V_{s'C}$ (pu)	ψ_C (deg)
k_{c3}	k_{a3}	k_{b3}				
0.15	0.10	0.00	0.13	319.11	1.10	−4.50
0.15	0.05	0.00	0.13	340.89	1.13	−2.20
0.15	0.00	0.00	0.15	360.00	1.15	0.00
Case for 0.20 pu compensating voltage (largest hexagon)						
Turns-ratios			$V_{s'sC}$ (pu)	β_C (deg)	$V_{s'C}$ (pu)	ψ_C (deg)
k_{c3}	k_{a3}	k_{b3}				
0.20	0.00	0.00	0.20	0.00	1.20	0.00
0.20	0.00	0.05	0.18	13.90	1.18	2.11
0.20	0.00	0.10	0.17	30.00	1.15	4.31
0.20	0.00	0.15	0.18	46.10	1.13	6.59
0.20	0.00	0.20	0.20	60.00	1.11	8.95
0.15	0.00	0.20	0.18	73.90	1.06	9.37
0.10	0.00	0.20	0.17	90.00	1.01	9.83
0.05	0.00	0.20	0.18	106.10	0.97	10.33
0.00	0.00	0.20	0.20	120.00	0.92	10.89
0.00	0.05	0.20	0.18	133.90	0.88	8.44
0.00	0.10	0.20	0.17	150.00	0.85	5.82
0.00	0.15	0.20	0.18	166.10	0.83	3.00
0.00	0.20	0.20	0.20	180.00	0.80	0.00
0.00	0.20	0.15	0.18	193.90	0.83	−3.00
0.00	0.20	0.10	0.17	210.00	0.85	−5.82
0.00	0.20	0.05	0.18	226.10	0.88	−8.44
0.00	0.20	0.00	0.20	240.00	0.92	−10.89
0.05	0.20	0.00	0.18	253.90	0.97	−10.33

(Continued)

Table 6-3 (Continued)

Case for 0.20 pu compensating voltage (largest hexagon)						
Turns-ratios			$V_{s'sC}$ (pu)	β_C (deg)	$V_{s'C}$ (pu)	ψ_C (deg)
k_{c3}	k_{a3}	k_{b3}				
0.10	0.20	0.00	0.17	270.00	1.01	−9.83
0.15	0.20	0.00	0.18	286.10	1.06	−9.37
0.20	0.20	0.00	0.20	300.00	1.11	−8.95
0.20	0.15	0.00	0.18	313.90	1.13	−6.59
0.20	0.10	0.00	0.17	330.00	1.15	−4.31
0.20	0.05	0.00	0.18	346.10	1.18	−2.11
0.20	0.00	0.00	0.20	360.00	1.20	0.00

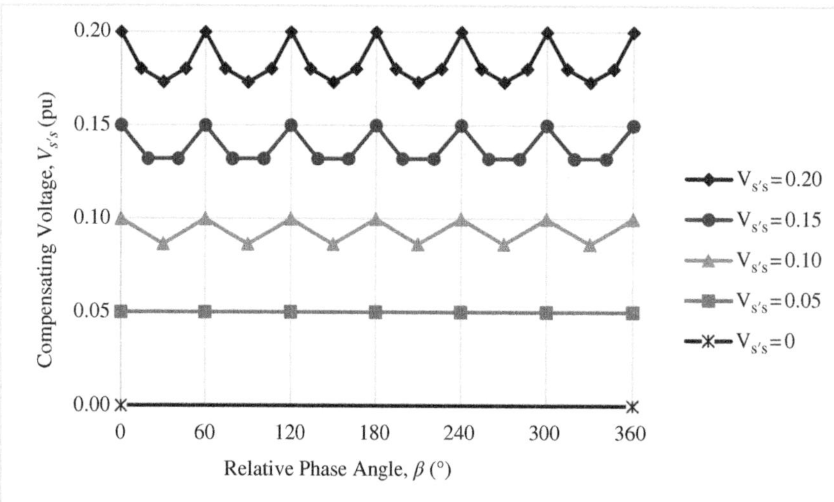

Figure 6-19 Compensating voltage ($V_{s's}$) in pu during the entire range of the relative phase angle of 0° $\leq \beta \leq$ 360°.

The compensating voltage ($V_{s's}$), given in Table 6-1 through Table 6-3, during the entire range of the relative phase angle of 0° $\leq \beta \leq$ 360° is shown in Figure 6-19.

The modified sending-end voltage ($V_{s'}$), given in Table 6-1 through Table 6-3, during the entire range of the relative phase angle of 0° $\leq \beta \leq$ 360° is shown in Figure 6-20.

The phase-shift angle (ψ), given in Table 6-1 through Table 6-3, during the entire range of the relative phase angle 0° $\leq \beta \leq$ 360° is shown in Figure 6-21.

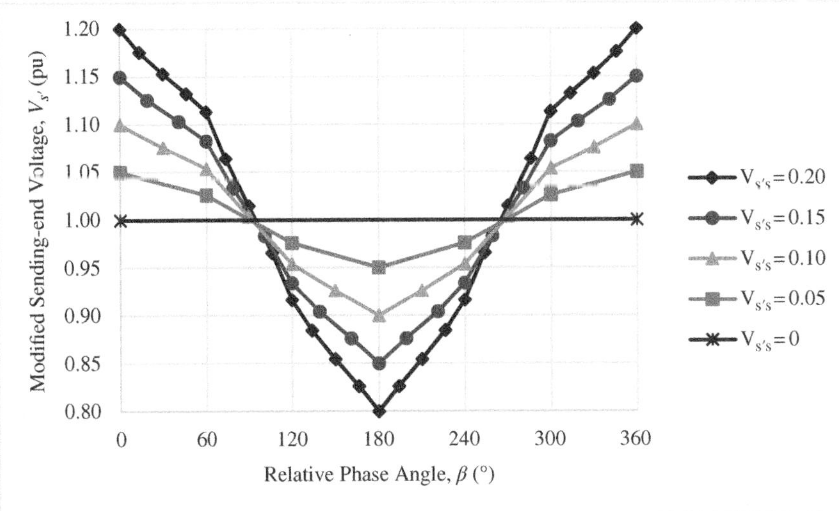

Figure 6-20 Modified sending-end voltage ($\mathbf{V_{s'}}$) in pu during the entire range of the relative phase angle of 0° $\leq \beta \leq$ 360°.

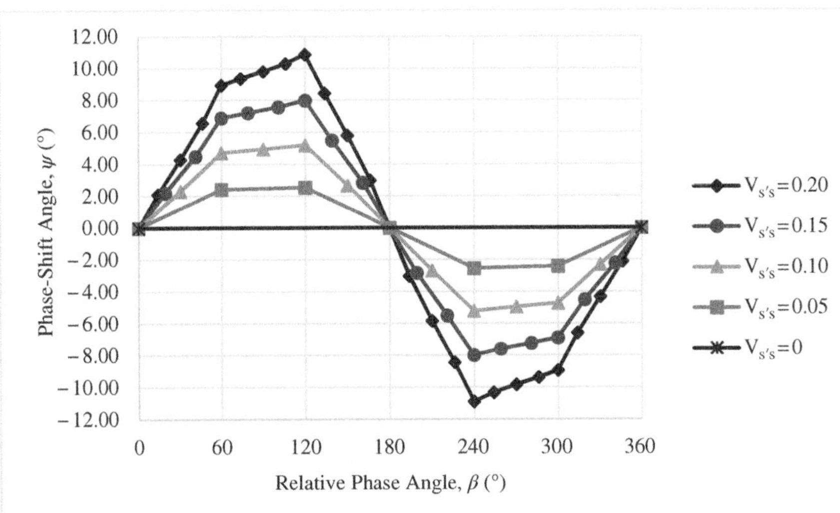

Figure 6-21 Phase-shift angle (ψ) in degrees during the entire range of the relative phase angle of 0° $\leq \beta \leq$ 360°.

Table 6-4 shows the variations of the theoretical magnitude ($V_{s's}$) of the compensating voltage ($\mathbf{V_{s's}}$) and the theoretical magnitude ($V_{s'}$) of the modified sending-end voltage ($\mathbf{V_{s'}}$) as a function of the relative phase angle (β), considering the sending-end voltage (V_s) to be 1 pu. The resulting active power (P_r), reactive power (Q_r), and apparent power (S_r) flows at the receiving end, exchanged active power (P_{se}), reactive power (Q_{se}), and apparent power (S_{se}) by the Compensating-Voltage Unit (Series Unit), and the prevailing line current (I) are also shown during the entire range of the relative phase angle of 0° $\leq \beta \leq$ 360° in Table 6-4 for the k_{a1}-k_{b1}-k_{c1} active turns-ratios from Table 6-1, k_{b2}-k_{c2}-k_{a2} active turns-ratios from Table 6-2, and k_{c3}-k_{a3}-k_{b3} active turns-ratios from Table 6-3. The table is expanded for each hexagon, related to 5, 10, 15, and 20% compensating voltages.

Table 6-4 Compensating voltage, modified sending-end voltage, power flows at the receiving end, exchanged powers by the Series Unit and the line current for electrical system data, shown in Table 2-1, and the active turns-ratios of the secondary, series-compensating windings in the *A*, *B*, and *C* phases per respective Tables 6-1, 6-2, and 6-3.

$V_{s's}$ (pu)	β (deg)	$V_{s'}$ (pu)	P_r (pu)	Q_r (pu)	S_r (pu)	P_{se} (pu)	Q_{se} (pu)	S_{se} (pu)	I (pu)
Case for natural power flows (center point of the hexagon characteristic)									
0.00	0.00	1.00	1.0000	−0.2679	1.0353	0.0000	0.0000	0.0000	1.0353
0.00	360.00	1.00	1.0000	−0.2679	1.0353	0.0000	0.0000	0.0000	1.0353
Case for 0.05 pu compensating voltage (smallest hexagon)									
$V_{s's}$ (pu)	β (deg)	$V_{s'}$ (pu)	P_r (pu)	Q_r (pu)	S_r (pu)	P_{se} (pu)	Q_{se} (pu)	S_{se} (pu)	I (pu)
0.05	0.00	1.05	1.0500	−0.1813	1.0655	−0.0500	−0.0184	0.0533	1.0655
0.05	60.00	1.03	1.1000	−0.2679	1.1322	−0.0134	−0.0550	0.0566	1.1322
0.05	120.00	0.98	1.0500	−0.3546	1.1082	0.0366	−0.0416	0.0554	1.1082
0.05	180.00	0.95	0.9500	−0.3546	1.0140	0.0500	0.0084	0.0507	1.0140
0.05	240.00	0.98	0.9000	−0.2679	0.9390	0.0134	0.0450	0.0470	0.9390
0.05	300.00	1.03	0.9500	−0.1813	0.9672	−0.0366	0.0316	0.0484	0.9672
0.05	360.00	1.05	1.0500	−0.1813	1.0655	−0.0500	−0.0184	0.0533	1.0655
Case for 0.10 pu compensating voltage (second smallest hexagon)									
$V_{s's}$ (pu)	β (deg)	$V_{s'}$ (pu)	P_r (pu)	Q_r (pu)	S_r (pu)	P_{se} (pu)	Q_{se} (pu)	S_{se} (pu)	I (pu)
0.10	0.00	1.10	1.1000	−0.0947	1.1041	−0.1000	−0.0468	0.1104	1.1041
0.09	30.00	1.08	1.1500	−0.1813	1.1642	−0.0634	−0.0784	0.1008	1.1642
0.10	60.00	1.05	1.2000	−0.2679	1.2296	−0.0268	−0.1200	0.1230	1.2296
0.09	90.00	1.00	1.1500	−0.3546	1.2034	0.0232	−0.1016	0.1042	1.2034
0.10	120.00	0.95	1.1000	−0.4412	1.1852	0.0732	−0.0932	0.1185	1.1852
0.09	150.00	0.93	1.0000	−0.4412	1.0930	0.0866	−0.0382	0.0947	1.0930
0.10	180.00	0.90	0.9000	−0.4412	1.0023	0.1000	0.0068	0.1002	1.0023
0.09	210.00	0.93	0.8500	−0.3546	0.9210	0.0634	0.0484	0.0798	0.9210
0.10	240.00	0.95	0.8000	−0.2679	0.8437	0.0268	0.0800	0.0844	0.8437
0.09	270.00	1.00	0.8500	−0.1813	0.8691	−0.0232	0.0716	0.0753	0.8691

Table 6-4 (Continued)

Case for 0.10 pu compensating voltage (second smallest hexagon)									
$V_{s's}$ (pu)	β (deg)	$V_{s'}$ (pu)	P_r (pu)	Q_r (pu)	S_r (pu)	P_{se} (pu)	Q_{se} (pu)	S_{se} (pu)	I (pu)
0.10	300.00	1.05	0.9000	−0.0947	0.9050	−0.0732	0.0532	0.0905	0.9050
0.09	330.00	1.08	1.0000	−0.0947	1.0045	−0.0866	0.0082	0.0870	1.0045
0.10	360.00	1.10	1.1000	−0.0947	1.1041	−0.1000	−0.0468	0.1104	1.1041
Case for 0.15 pu compensating voltage (second smallest hexagon)									
$V_{s's}$ (pu)	β (deg)	$V_{s'}$ (pu)	P_r (pu)	Q_r (pu)	S_r (pu)	P_{se} (pu)	Q_{se} (pu)	S_{se} (pu)	I (pu)
0.15	0.00	1.15	1.1500	−0.0081	1.1500	−0.1500	−0.0852	0.1725	1.1500
0.13	19.11	1.13	1.2000	−0.0947	1.2037	−0.1134	−0.1118	0.1592	1.2037
0.13	40.89	1.10	1.2500	−0.1813	1.2631	−0.0768	−0.1484	0.1671	1.2631
0.15	60.00	1.08	1.3000	−0.2679	1.3273	−0.0402	−0.1950	0.1991	1.3273
0.13	79.11	1.03	1.2500	−0.3546	1.2993	0.0098	−0.1716	0.1719	1.2993
0.13	100.89	0.98	1.2000	−0.4412	1.2785	0.0598	−0.1582	0.1691	1.2785
0.15	120.00	0.93	1.1500	−0.5278	1.2653	0.1098	−0.1548	0.1898	1.2653
0.13	139.11	0.90	1.0500	−0.5278	1.1752	0.1232	−0.0948	0.1555	1.1752
0.13	160.89	0.88	0.9500	−0.5278	1.0868	0.1366	−0.0448	0.1438	1.0868
0.15	180.00	0.85	0.8500	−0.5278	1.0005	0.1500	−0.0048	0.1501	1.0005
0.13	199.11	0.88	0.8000	−0.4412	0.9136	0.1134	0.0418	0.1209	0.9136
0.13	220.89	0.90	0.7500	−0.3546	0.8296	0.0768	0.0784	0.1097	0.8296
0.15	240.00	0.93	0.7000	−0.2679	0.7495	0.0402	0.1050	0.1124	0.7495
0.13	259.11	0.98	0.7500	−0.1813	0.7716	−0.0098	0.1016	0.1021	0.7716
0.13	280.89	1.03	0.8000	−0.0947	0.8056	−0.0598	0.0882	0.1066	0.8056
0.15	300.00	1.08	0.8500	−0.0081	0.8500	−0.1098	0.0648	0.1275	0.8500
0.13	319.11	1.10	0.9500	−0.0081	0.9500	−0.1232	0.0248	0.1257	0.9500
0.13	340.89	1.13	1.0500	−0.0081	1.0500	−0.1366	−0.0252	0.1389	1.0500
0.15	360.00	1.15	1.1500	−0.0081	1.1500	−0.1500	−0.0852	0.1725	1.1500

(Continued)

Table 6-4 (Continued)

Case for 0.20 pu compensating voltage (second largest hexagon)									
$V_{s's}$ (pu)	β (deg)	$V_{s'}$ (pu)	P_r (pu)	Q_r (pu)	S_r (pu)	P_{se} (pu)	Q_{se} (pu)	S_{se} (pu)	I (pu)
0.20	0.00	1.20	1.2000	0.0785	1.2026	−0.2000	−0.1336	0.2405	1.2026
0.18	13.90	1.18	1.2500	−0.0081	1.2500	−0.1634	−0.1552	0.2254	1.2500
0.17	30.00	1.15	1.3000	−0.0947	1.3034	−0.1268	−0.1868	0.2258	1.3034
0.18	46.10	1.13	1.3500	−0.1813	1.3621	−0.0902	−0.2284	0.2456	1.3621
0.20	60.00	1.11	1.4000	−0.2679	1.4254	−0.0536	−0.2800	0.2851	1.4254
0.18	73.90	1.06	1.3500	−0.3546	1.3958	−0.0036	−0.2516	0.2516	1.3958
0.17	90.00	1.01	1.3000	−0.4412	1.3728	0.0464	−0.2332	0.2378	1.3728
0.18	106.10	0.97	1.2500	−0.5278	1.3568	0.0964	−0.2248	0.2446	1.3568
0.20	120.00	0.92	1.2000	−0.6144	1.3481	0.1464	−0.2264	0.2696	1.3481
0.18	133.90	0.88	1.1000	−0.6144	1.2599	0.1598	−0.1614	0.2271	1.2599
0.17	150.00	0.85	1.0000	−0.6144	1.1736	0.1732	−0.1064	0.2033	1.1736
0.18	166.10	0.83	0.9000	−0.6144	1.0897	0.1866	−0.0614	0.1964	1.0897
0.20	180.00	0.80	0.8000	−0.6144	1.0087	0.2000	−0.0264	0.2017	1.0087
0.18	193.90	0.83	0.7500	−0.5278	0.9171	0.1634	0.0252	0.1653	0.9171
0.17	210.00	0.85	0.7000	−0.4412	0.8274	0.1268	0.0668	0.1433	0.8274
0.18	226.10	0.88	0.6500	−0.3546	0.7404	0.0902	0.0984	0.1335	0.7404
0.20	240.00	0.92	0.6000	−0.2679	0.6571	0.0536	0.1200	0.1314	0.6571
0.18	253.90	0.97	0.6500	−0.1813	0.6748	0.0036	0.1216	0.1217	0.6748
0.17	270.00	1.01	0.7000	−0.0947	0.7064	−0.0464	0.1132	0.1223	0.7064
0.18	286.10	1.06	0.7500	−0.0081	0.7500	−0.0964	0.0948	0.1352	0.7500
0.20	300.00	1.11	0.8000	0.0785	0.8038	−0.1464	0.0664	0.1608	0.8038
0.18	313.90	1.13	0.9000	0.0785	0.9034	−0.1598	0.0314	0.1629	0.9034
0.17	330.00	1.15	1.0000	0.0785	1.0031	−0.1732	−0.0136	0.1737	1.0031
0.18	346.10	1.18	1.1000	0.0785	1.1028	−0.1866	−0.0686	0.1988	1.1028
0.20	360.00	1.20	1.2000	0.0785	1.2026	−0.2000	−0.1336	0.2405	1.2026

Note that as the compensating voltages increases from inner hexagons to outer ones, the ratio (S_r/S_{se}) decreases, which corroborates Figure 2-49.

Figure 6-22 shows the active power (P_r), reactive power (Q_r), and apparent power (S_r) flows at the receiving end, given in Table 6-4, during the entire range of the relative phase angle of $0° \leq \beta \leq 360°$.

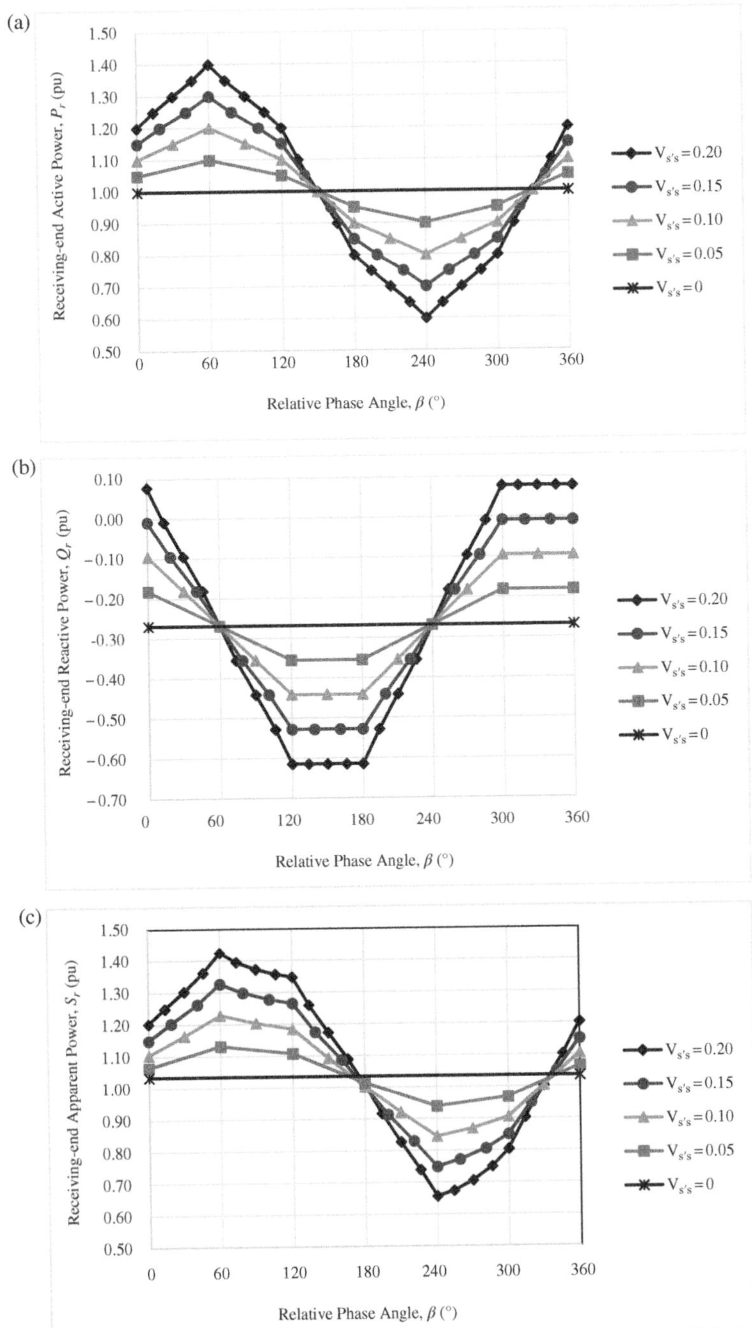

Figure 6-22 (a) Active power (P_r), (b) reactive power (Q_r), and (c) apparent power (S_r) flows in pu at the receiving end during the entire range of the relative phase angle of $0° \leq \beta \leq 360°$.

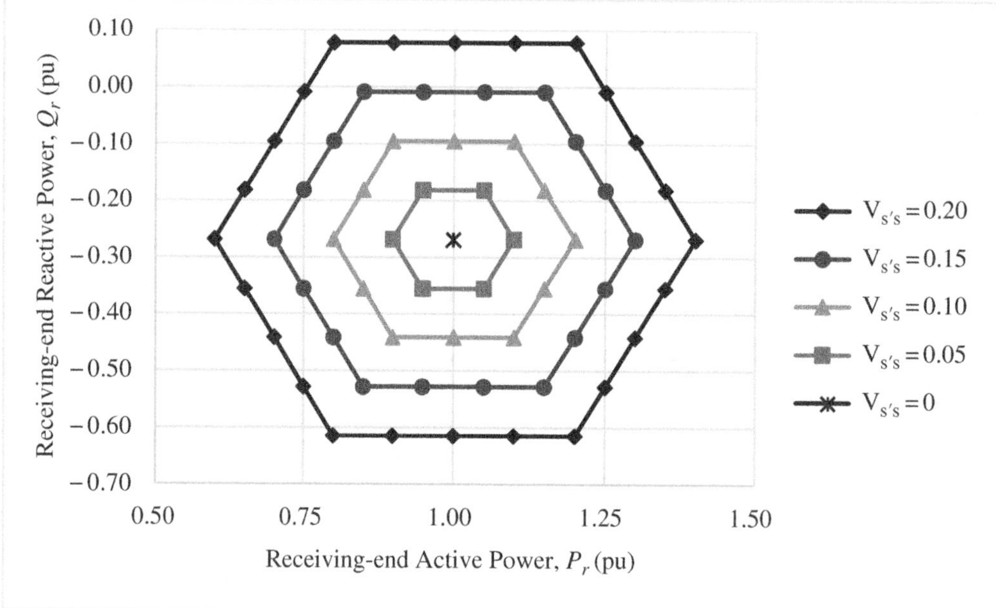

Figure 6-23 Active power (P_r) versus reactive power (Q_r) in pu at the receiving end during the entire range of the relative phase angle of $0° \leq \beta \leq 360°$.

Figure 6-23 shows the active power (P_r) versus reactive power (Q_r) at the receiving end, given in Table 6-4, during the entire range of the relative phase angle of $0° \leq \beta \leq 360°$.

Figure 6-24 shows the exchanged active power (P_{se}), reactive power (Q_{se}), and apparent power (S_{se}) by the Compensating-Voltage Unit (Series Unit), given in Table 6-4, during the entire range of the relative phase angle of $0° \leq \beta \leq 360°$.

Figure 6-25 shows the exchanged active power (P_{se}) versus reactive power (Q_{se}) by the Compensating-Voltage Unit (Series Unit), given in Table 6-4, during the entire range of the relative phase angle of $0° \leq \beta \leq 360°$.

Figure 6-26 shows the prevailing line current (I), given in Table 6-4, during the entire range of the relative phase angle of $0° \leq \beta \leq 360°$.

The summary of the study results for a 20% compensating voltage is given below.

- The line voltage at the POC is fixed at one pu.
- Active power flow at the receiving end changes from the natural flow of one pu by 0.4 and −0.4 pu.
- Reactive power flow at the receiving end changes from the natural flow of −0.2679 pu by 0.3464 and −0.3465 pu.
- Rating of the ST: 0.2851 pu.

Note that a 0.2851 pu-rated ST regulates 1.4254 pu apparent power in the line as shown in Figures 6-24c and 6-22c, respectively.

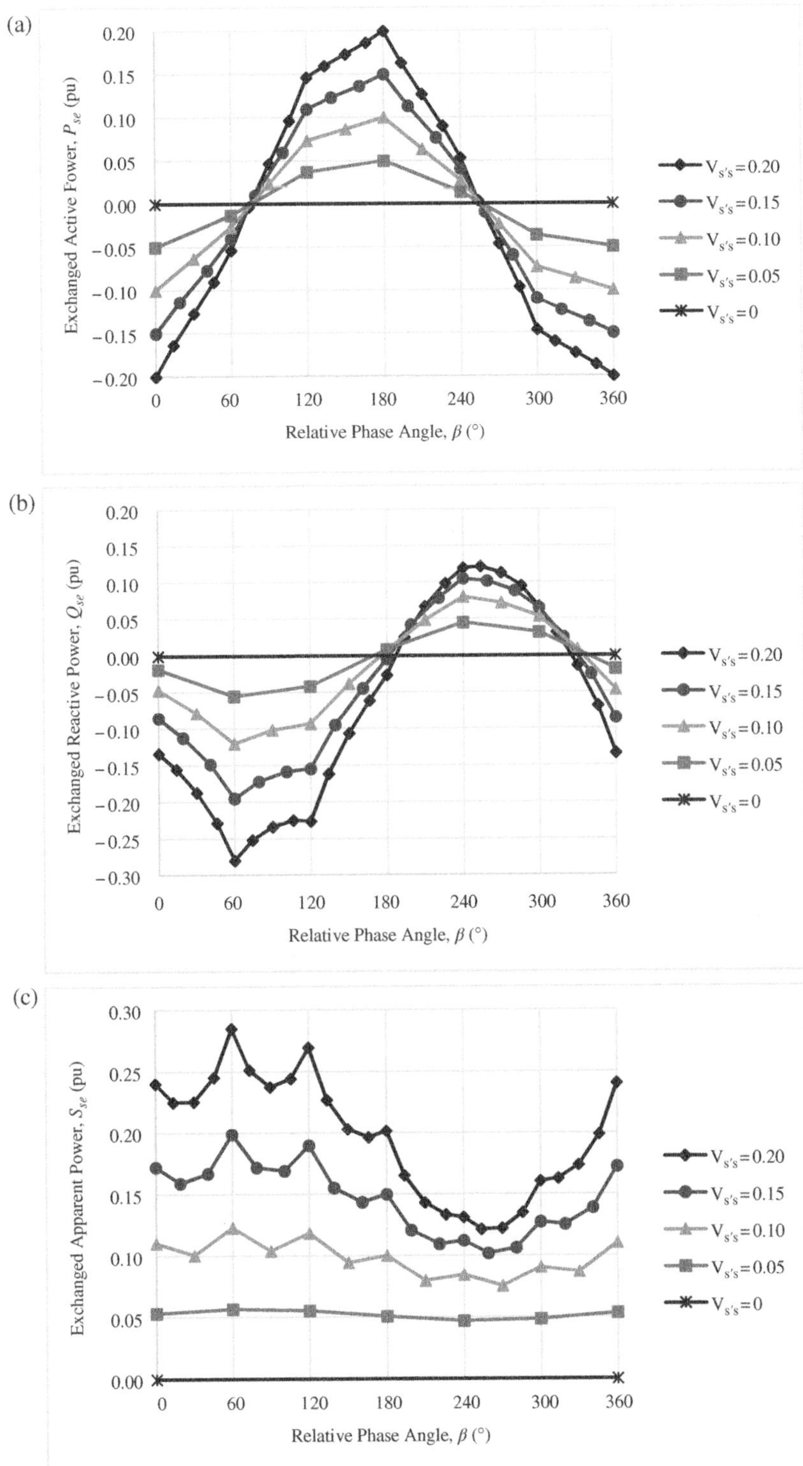

Figure 6-24 (a) Exchanged active power (P_{se}), (b) reactive power (Q_{se}), and (c) apparent power (S_{se}) by the Series Unit in pu during the entire range of the relative phase angle of $0° \leq \beta \leq 360°$.

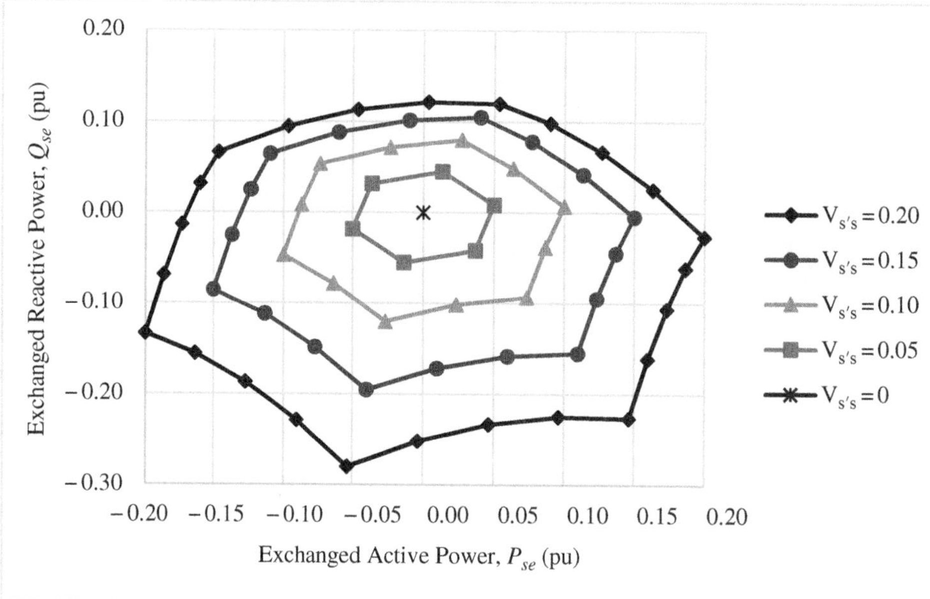

Figure 6-25 Exchanged active power (P_{se}) versus reactive power (Q_{se}) in pu by the Series Unit during the entire range of the relative phase angle of $0° \leq \beta \leq 360°$.

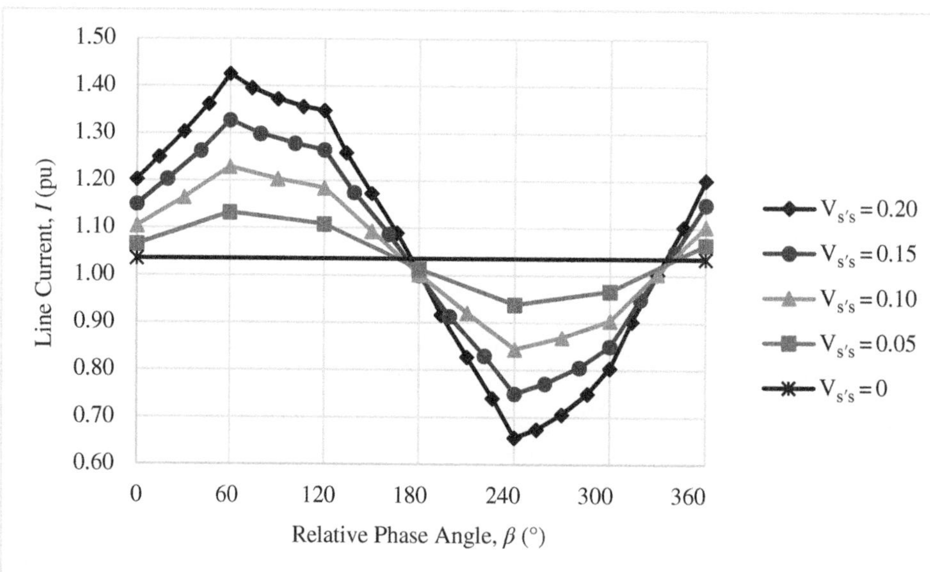

Figure 6-26 Line current (I) in pu during the entire range of the relative phase angle of $0° \leq \beta \leq 360°$.

6.2.4 Simulation of ST Integrated in a Two-Generator/One-Line Power System Network

A basic simulation of an ideal ST, without any leakage reactance, was carried out and the results are shown in this section. The operation of the ST is verified in a two-generator/one-line power system network, considering a three-phase base power of 160 MVA and a phase-to-phase (*P–P*) base rms voltage, $V_{rms(P\text{-}P)}$, of 138 kV_{rms} as defined in Chapter 3, Section 3.1. The network model of Figure 3-5 is used with the short between BUS01 and BUS02, which is replaced with the ST as shown in Figure 6-27. The corresponding EMTP datafile **601IDLST.DAT** is shown in Code 6-1. The EMTP code in bold is the additional code that is not discussed yet.

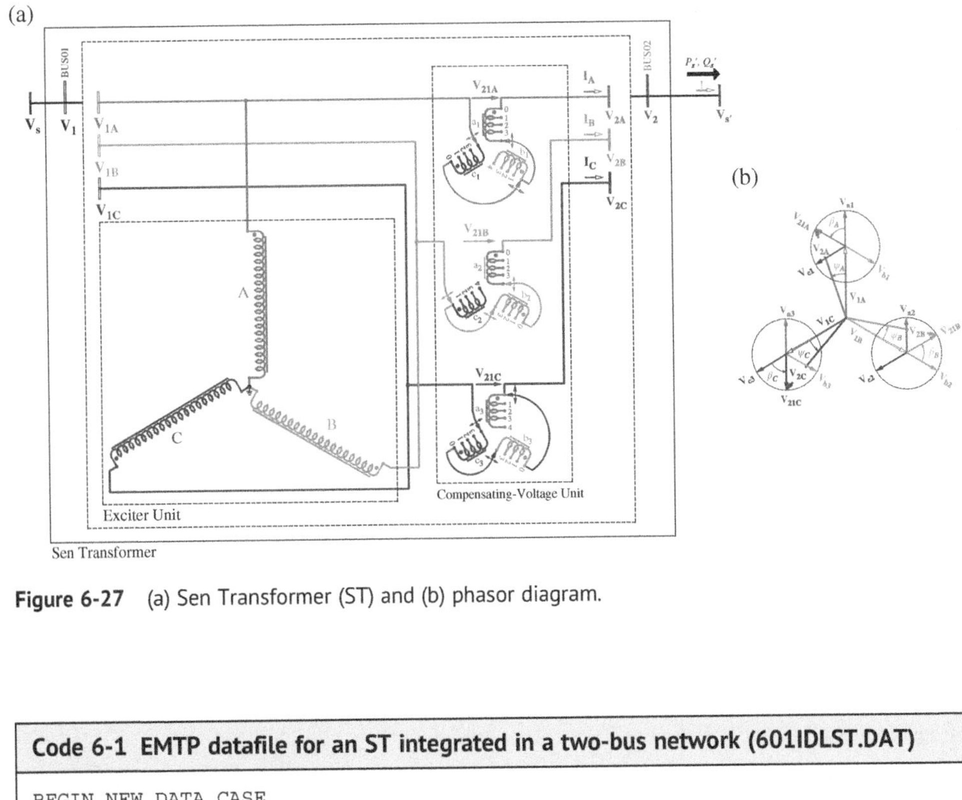

Figure 6-27 (a) Sen Transformer (ST) and (b) phasor diagram.

Code 6-1 EMTP datafile for an ST integrated in a two-bus network (601IDLST.DAT)

```
BEGIN NEW DATA CASE
C File Name: 601IDLST.DAT
C IDEAL SEN TRANSFORMER WITHOUT ANY LEAKAGE IMPEDANCE
C 0000000111111111122222222223333333333444444444455555555556666666666777777777778
C 3456789012345678901234567890123456789012345678901234567890123456789012345678901234567890
C ----dt<---tmax<---Xopt<---Copt<--Toler<        ><        >
16.666-6150.00-3
C 0000000111111111122222222223333333333444444444455555555556666666666777777777778
C 3456789012345678901234567890123456789012345678901234567890123456789012345678901234567890
C -Iprnt<--Iplot<-Idoubl<-KssOut<-MaxOut<        ><        ><---Icat<-Nenerg
      20      10                                                  1
```

(Continued)

Code 6-1 (Continued)

```
TACS HYBRID
99TWOPI    = 2.0 * PI
C BASE VOLTAGE, CURRENT, AND POWER
99Vbase    = 112676.528
99Ibase    = 946.662704
99Zbase    = Vbase / Ibase
99Pbase    = 1.5 * Vbase * Ibase
$INCLUDE 302MEAS1.SWT
$INCLUDE 303PTCT.SCL
$INCLUDE 304IPLL.PLL
$INCLUDE 305LINER.CMP
C EXCHANGED POWER IN PU BY THE SERIES UNIT WITH THE LINE
99P2sepu   = (v12Apu * iApu + v12Bpu * iBpu + v12Cpu * iCpu) *Vbase*Ibase / Pbase
99Q2sepu   = (v12Apu * iCpu - v12Cpu * iApu) * SQRT(3) * Vbase * Ibase / Pbase
C EXCHANGED MW AND MVAR BY THE SERIES UNIT WITH THE LINE
99P2seMW   = P2sepu * Pbase / 1e6
99Q2seMV   = Q2sepu * Pbase / 1e6
C EXCHANGED MVA WITH THE LINE PER EQUATION (2-182)
99S2seMV   = SQRT(P2seMW * P2seMW + Q2seMV * Q2seMV)
C RMS LINE CURRENT
99Irms     = ipu * Ibase / SQRT(2)
C TACS OUTPUT
33v1pu  v12pu v2pu  PrMW  QrMVARSrMVA P2seMWQ2seMVS2seMVIrms
BLANK RECORD ENDING TACS
C 0000000111111111122222222223333333333444444444455555555556666666666777777777778
C 3456789012345678901234567890123456789012345678901234567890123456789012345678901234567890
C <-Bus1<-Bus2<-Bus3<-Bus4<----R<----L<----C                                          V
  VSA   BUS01A            0.0001
  VSB   BUS01BVSA   BUS01A
  VSC   BUS01CVSA   BUS01A
$INCLUDE 306NTWK1.BRN
$INCLUDE 602IDLST.TRN
BLANK RECORD ENDING BRANCHES
C 0000000111111111122222222223333333333444444444455555555556666666666777777777778
C 3456789012345678901234567890123456789012345678901234567890123456789012345678901234567890
C <-Bus1<-Bus2<---Tclose<----Topen<-------Ie<----Flash<--Request<-----Target<--O
$INCLUDE 307MEAS2.SWT
BLANK RECORD ENDING SWITCHES
C 0000000111111111122222222223333333333444444444455555555556666666666777777777778
C 3456789012345678901234567890123456789012345678901234567890123456789012345678901234567890
$INCLUDE 308NTWK1.SRC
BLANK RECORD ENDING SOURCES
BLANK RECORD ENDING NODE VOLTAGE OUTPUT
BLANK RECORD ENDING PLOT
```

The datafile **601IDLST.DAT** includes a card, which attaches the $INCLUDE file 602IDLST.TRN that connects an ST, without any leakage impedance, to the transmission line as shown in Code 6-2. The EMTP code is set up corresponding to the operating point when no compensating voltage ($V_{s's}$ = 0.0 pu and β = 0°) is used. Therefore, all nine secondary windings of the ST were set to a turns-ratio of a near-zero value by using a small number, *i.e.* 1.0×10^{-6}.

Code 6-2 EMTP $INCLUDE file for an ST with an operating point, $V_{s's}$ = 0.0 pu and β = 0° (602IDLST.TRN).

```
C File Name: 602IDLST.TRN
C 3-PHASE IDEAL SEN TRANSFORMER WITHOUT ANY LEAKAGE IMPEDANCE
C 00000001111111111222222222233333333334444444444555555555566666666667777777777778
C 34567890123456789012345678901234567890123456789012345678901234567890123456789O
C ------SERIES COMPENSATING TRANSFORMER FOR THE A PHASE------
C TRANSFORMER <--Refxxxxxx<--Iss<--Phi<-Name<-Rmag<--------------------------O
  TRANSFORMER                       SRSTRA
C NEXT LINE (9999 IN COLUMNS 13-16) INDICATES A LINEAR TRANSFORMER
            9999
C <---Nodes--><----------><-Ohms<---mH<---kV
C <-Bus1<-Bus2<----------><---Rk<---Lk<Volt|---------------------------------->O
 1BUS01A                    1.0E-61.0E-6    1.0
 2A1HEADA1TAIL              1.0E-61.0E-61.0E-6
 3A2HEADA2TAIL              1.0E-61.0E-61.0E-6
 4A3HEADA3TAIL              1.0E-61.0E-61.0E-6
C 00000001111111111222222222233333333334444444444555555555566666666667777777777778
C 34567890123456789012345678901234567890123456789012345678901234567890123456789O
C <-Bus1<-Bus2<-Bus3<-Bus4<----R<----L<----C                                    V
  A1HEADBUS02A              1.0E-6
  A2HEADBUS02B              1.0E-6
  A3HEADBUS02C              1.0E-6
C ------SERIES COMPENSATING TRANSFORMER FOR THE B PHASE------
  TRANSFORMER                       SRSTRB
            9999
 1BUS01B                    1.0E-61.0E-6    1.0
 2B1HEADB1TAIL              1.0E-61.0E-61.0E-6
 3B2HEADB2TAIL              1.0E-61.0E-61.0E-6
 4B3HEADB3TAIL              1.0E-61.0E-61.0E-6
C 00000001111111111222222222233333333334444444444555555555566666666667777777777778
C 34567890123456789012345678901234567890123456789012345678901234567890123456789O
C <-Bus1<-Bus2<-Bus3<-Bus4<----R<----L<----C                                    V
  B1HEADA1TAIL              1.0E-6
  B2HEADA2TAIL              1.0E-6
  B3HEADA3TAIL              1.0E-6
```

(Continued)

Code 6-2 (Continued)

```
C ------SERIES COMPENSATING TRANSFORMER FOR THE C PHASE------
  TRANSFORMER                                    SRSTRC
            9999
 1BUS01C                    1.0E-61.0E-6    1.0
 2C1HEADC1TAIL              1.0E-61.0E-61.0E-6
 3C2HEADC2TAIL              1.0E-61.0E-61.0E-6
 4C3HEADC3TAIL              1.0E-61.0E-61.0E-6
C 00000001111111111222222222233333333334444444444555555555566666666667777777778
C 34567890123456789012345678901234567890123456789012345678901234567890123456789
C <-Bus1<-Bus2<-Bus3<-Bus4<----R<----L<----C                                    V
  C1HEADB1TAIL              1.0E-6
  C2HEADB2TAIL              1.0E-6
  C3HEADB3TAIL              1.0E-6
C 00000001111111111222222222233333333334444444444555555555566666666667777777778
C 34567890123456789012345678901234567890123456789012345678901234567890123456789
C <-Bus1<-Bus2<-Bus3<-Bus4<----R<----L<----C                                    V
  C1TAILBUS01A              1.0E-6
  C2TAILBUS01B              1.0E-6
  C3TAILBUS01C              1.0E-6
```

The ST consists of three primary windings and nine secondary windings (three for series connection in each of the three phases). The primary and secondary windings are considered to have one pu and 0.2 pu voltage ratings, respectively. Considering an LTC contact located at every 5% tap setting, each secondary winding consists of four LTC taps. The leakage resistance and reactance of the ST were set to 1.0×10^{-6} Ω and 1.0×10^{-6} mH, respectively, since the secondary effects are not of interest for a preliminary study. Using the above datacase with zero turns-ratio, the uncompensated or natural active and reactive power flows (P_{rn} and Q_{rn}) at the receiving end of the line are simulated to be 209.87 MW (1.3117 pu) and −66.97 Mvar (−0.4185 pu), respectively. To study the effects of other operating points of an ST, the active turns-ratios are set according to Table 6-1 through Table 6-3, which show the active turns-ratios of the three secondary, compensating-windings in the A, B, and C phases for the entire range of the relative phase angle of $0° \leq \beta \leq 360°$. The effect of implementing an open-loop control in voltage-injection mode, traveling from one operating point to the next in four hexagons, related to 5, 10, 15, and 20% compensating voltages, is studied, using an electrical system data as shown in Table 3-2.

Table 3-2 Electrical system data.

Parameters	Values
Source voltage ($\mathbf{V}_{src}$) phase-to-neutral	1 pu (112,676.528 V_{peak})∠0°
Source resistance (R_{src})	0.0084 pu (1.0053 Ω)
Source reactance (X_{src})	0.0625 pu (19.73 mH)
Receiving-end voltage ($\mathbf{V}_r$) phase-to-neutral	1 pu (112,676.528 V_{peak})∠−20°
Line resistance (R)	0.0253 pu (3.0159 Ω)
Line reactance (X)	0.1875 pu (59.19 mH)

It is recommended to create a $INCLUDE file, such as, ST4060P0.TRN, for each of the operating points of ST, following Code 6-2. Code 6-3 shows the setup of LTCs for a series-compensating voltage ($\mathbf{V_{s's}}$) with magnitude, $V_{s's}$ = 0.20 pu, and the relative phase angle, β = 60°. These $INCLUDE files can be created and edited, using a text editor, such as NOTEPAD. Each $INCLUDE file name consists of eight characters, followed by the .TRN extension, such as ST4060P0.TRN. According to the convention, followed in this book, the first two characters (ST) in the $INCLUDE file name are reserved for Sen Transformer. The third character is reserved for the highest tap setting, such as 4, which represents $V_{s's}$ = 0.20 pu, assuming 0.05 pu across each LTC tap. The five characters from fourth through eighth are reserved for the relative phase angle (β), such as 060P0 to represent β = 60.0°. Note that the turns-ratios for the appropriate secondary windings in this $INCLUDE file have changed from 1.0E-6, shown in Code 6-2, to 0.20 pu.

Code 6-3 EMTP $INCLUDE file for an ST with an operating point, $V_{s's}$ = 0.20 pu and β = 60.0° (ST4060P0.TRN).

```
C File Name: ST4060P0.TRN
C 3-PHASE IDEAL SEN TRANSFORMER WITHOUT ANY LEAKAGE IMPEDANCE
C OPERATING POINT: Vs's = 0.20 pu and beta = 60.0 degrees.
C 0000000111111111122222222223333333333444444444455555555556666666666777777777778
C 3456789012345678901234567890123456789012345678901234567890123456789012345678901234567890
C ------SERIES COMPENSATING TRANSFORMER FOR THE A PHASE------
C TRANSFORMER <--Refxxxxxx<--Iss<--Phi<-Name<-Rmag<---------------------------O
  TRANSFORMER                         SRSTRA
C NEXT LINE (9999 IN COLUMNS 13-16) INDICATES A LINEAR TRANSFORMER
            9999
C <---Nodes--><----------><-Ohms<---mH<---kV
C <-Bus1<-Bus2<----------><---Rk<---Lk<Volt|--------------------------------->O
 1BUS01A                  1.0E-61.0E-6   1.0
 2A1HEADA1TAIL            1.0E-61.0E-6  0.20
 3A2HEADA2TAIL            1.0E-61.0E-6  0.20
 4A3HEADA3TAIL            1.0E-61.0E-61.0E-6
C 0000000111111111122222222223333333333444444444455555555556666666666777777777778
C 3456789012345678901234567890123456789012345678901234567890123456789012345678901234567890
C <-Bus1<-Bus2<-Bus3<-Bus4<----R<----L<----C                                    V
  A1HEADBUS02A            1.0E-6
  A2HEADBUS02B            1.0E-6
  A3HEADBUS02C            1.0E-6
C ------SERIES COMPENSATING TRANSFORMER FOR THE B PHASE------
  TRANSFORMER                         SRSTRB
            9999
 1BUS01B                  1.0E-61.0E-6   1.0
 2B1HEADB1TAIL            1.0E-61.0E-61.0E-6
 3B2HEADB2TAIL            1.0E-61.0E-6  0.20
 4B3HEADB3TAIL            1.0E-61.0E-6  0.20
```

(Continued)

Code 6-3 (Continued)

```
C 0000000111111111122222222223333333333444444444455555555556666666666777777777778
C 345678901234567890123456789012345678901234567890123456789012345678901234567890
C <-Bus1<-Bus2<-Bus3<-Bus4<----R<----L<----C                                    V
  B1HEADA1TAIL            1.0E-6
  B2HEADA2TAIL            1.0E-6
  B3HEADA3TAIL            1.0E-6
C ------SERIES COMPENSATING TRANSFORMER FOR THE C PHASE------
  TRANSFORMER                         SRSTRC
            9999
 1BUS01C                  1.0E-61.0E-6   1.0
 2C1HEADC1TAIL            1.0E-61.0E-6  0.20
 3C2HEADC2TAIL            1.0E-61.0E-61.0E-6
 4C3HEADC3TAIL            1.0E-61.0E-6  0.20
C 0000000111111111122222222223333333333444444444455555555556666666666777777777778
C 345678901234567890123456789012345678901234567890123456789012345678901234567890
C <-Bus1<-Bus2<-Bus3<-Bus4<----R<----L<----C                                    V
  C1HEADB1TAIL            1.0E-6
  C2HEADB2TAIL            1.0E-6
  C3HEADB3TAIL            1.0E-6
C 0000000111111111122222222223333333333444444444455555555556666666666777777777778
C 345678901234567890123456789012345678901234567890123456789012345678901234567890
C <-Bus1<-Bus2<-Bus3<-Bus4<----R<----L<----C                                    V
  C1TAILBUS01A            1.0E-6
  C2TAILBUS01B            1.0E-6
  C3TAILBUS01C            1.0E-6
```

Table 6-5 shows the variations of theoretical magnitude ($V_{s's}$) of the compensating voltage ($\mathbf{V_{s's}}$) and the theoretical magnitude ($V_{s'}$) of the modified sending-end voltage ($\mathbf{V_{s'}}$) as a function of the relative phase angle (β), considering the sending-end voltage (V_s) is 1 pu. The table also shows the corresponding values of simulated magnitude ($V_{s's}$) of the compensating voltage ($\mathbf{V_{s's}}$), the simulated magnitude (V_s) of sending-end voltage ($\mathbf{V_s}$), and the simulated magnitude ($V_{s'}$) of modified sending-end voltage ($\mathbf{V_{s'}}$). The resulting active power (P_r), reactive power (Q_r), and apparent power (S_r) flows at the receiving end, exchanged active power (P_{se}), reactive power (Q_{se}), and apparent power (S_{se}) by the Compensating-Voltage Unit (Series Unit), and the prevailing line current (I) are also shown during the entire range of the relative phase angle of $0° \leq \beta \leq 360°$ in Table 6-5 for the k_{a1}-k_{b1}-k_{c1} active turns-ratios from Table 6-1, k_{b2}-k_{c2}-k_{a2} active turns-ratios from Table 6-2, and k_{c3}-k_{a3}-k_{b3} active turns-ratios from Table 6-3. The table is expanded for each hexagon, related to 5, 10, 15, and 20% compensating voltages.

Figure 6-28 shows the magnitude of the sending-end voltage (V_s), given in Table 6-5, during the entire range of the relative phase angle of $0° \leq \beta \leq 360°$. The voltage ($\mathbf{V_s}$) keeps changing from its original (uncompensated) value because of the variation of line current and the resulting voltage across the line impedance between the POC and the infinite bus. Due to this variation in the

Table 6-5 Compensating voltage, modified sending-end voltage, power flows at the receiving end, exchanged powers by the Series Unit and the line current for electrical system data, shown in Table 3-2, and the active turns-ratios of the secondary, series-compensating windings in the *A*, *B*, and *C* phases per respective Tables 6-1, 6-2, and 6-3.

Case for natural power flows (center point of the hexagon characteristic)												
Theoretical $V_{s's}$ (pu)	β (deg)	Theoretical $V_{s'}$ (pu)	Simulated V_s (pu)	Simulated $V_{s's}$ (pu)	Simulated $V_{s'}$ (pu)	Simulated P_r (MW)	Simulated Q_r (Mvar)	Simulated S_r (MVA)	Simulated P_{se} (MW)	Simulated Q_{se} (Mvar)	Simulated S_{se} (MVA)	Simulated I (A)
0.00	0.00	1.00	0.9886	0.0000	0.9886	209.8713	−66.9674	220.2966	0.0000	0.0000	0.0000	921.6545
0.00	360.00	1.00	0.9886	0.0000	0.9886	209.8713	−66.9674	220.2966	0.0000	0.0000	0.0000	921.6545
Case for 0.05 pu compensating voltage (smallest hexagon)												
Theoretical $V_{s's}$ (pu)	β (deg)	Theoretical $V_{s'}$ (pu)	Simulated V_s (pu)	Simulated $V_{s's}$ (pu)	Simulated $V_{s'}$ (pu)	Simulated P_r (MW)	Simulated Q_r (Mvar)	Simulated S_r (MVA)	Simulated P_{se} (MW)	Simulated Q_{se} (Mvar)	Simulated S_{se} (MVA)	Simulated I (A)
0.05	0.00	1.05	0.9757	0.0488	1.0245	219.0008	−37.9146	222.2586	−10.8009	−0.9508	10.8427	929.8627
0.05	60.00	1.03	0.9791	0.0490	1.0045	238.4317	−64.8420	247.0914	−6.2068	−10.3819	12.0958	1033.756
0.05	120.00	0.98	0.9916	0.0496	0.9678	229.5456	−94.9013	248.3896	4.6252	−11.4133	12.3149	1039.187
0.05	180.00	0.95	1.0010	0.0500	0.9509	200.0051	−97.4233	222.4710	10.9389	−2.0753	11.1340	930.7513
0.05	240.00	0.98	0.9973	0.0499	0.9733	180.3229	−69.0618	193.0955	5.5525	7.8665	9.6287	807.8534
0.05	300.00	1.03	0.9846	0.0492	1.0101	190.4236	−39.6402	194.5058	−5.3248	7.9581	9.5752	813.7535
0.05	360.00	1.05	0.9757	0.0488	1.0245	219.0008	−37.9146	222.2586	−10.8009	−0.9508	10.8427	929.8627

(Continued)

Table 6-5 (Continued)

Case for 0.10 pu compensating voltage (second smallest hexagon)												
Theoretical $V_{s's}$ (pu)	B (deg)	Theoretical $V_{s'}$ (pu)	Simulated V_s (pu)	Simulated $V_{s's}$ (pu)	Simulated $V_{s'}$ (pu)	Simulated P_r (MW)	Simulated Q_r (Mvar)	Simulated S_r (MVA)	Simulated P_{se} (MW)	Simulated Q_{se} (Mvar)	Simulated S_{se} (MVA)	Simulated I (A)
0.10	0.00	1.10	0.9623	0.0962	1.0585	227.4150	−10.2654	227.6466	−21.4199	−4.5869	21.9056	952.4044
0.09	30.00	1.08	0.9660	0.0837	1.0393	246.5931	−36.1864	249.2340	−16.9077	−12.2011	20.8503	1042.72
0.10	60.00	1.05	0.9687	0.0969	1.0206	265.9381	−62.6947	273.2283	−13.0807	−23.0101	26.4683	1143.105
0.09	90.00	1.00	0.9814	0.0850	0.9851	258.0261	−92.3275	274.0471	−2.3813	−23.1693	23.2913	1146.53
0.10	120.00	0.95	0.9935	0.0993	0.9477	249.3795	−123.3573	278.2215	8.3926	−26.3347	27.6397	1163.994
0.09	150.00	0.93	1.0035	0.0869	0.9292	219.9132	−126.3575	253.6297	15.5381	−15.6326	22.0411	1061.11
0.10	180.00	0.90	1.0126	0.1013	0.9113	189.3842	−129.2707	229.2974	21.9676	−7.5183	23.2185	959.3109
0.09	210.00	0.93	1.0094	0.0874	0.9347	169.4773	−99.8812	196.7201	16.5770	4.5746	17.1966	823.0175
0.10	240.00	0.95	1.0051	0.1005	0.9588	149.8632	−71.1161	165.8809	10.4785	12.9679	16.6723	693.9957
0.09	270.00	1.00	0.9926	0.0860	0.9963	160.9309	−41.3573	166.1601	−0.4562	14.2760	14.2833	695.1635
0.10	300.00	1.05	0.9795	0.0980	1.0320	171.2644	−12.9956	171.7568	−11.1929	12.5605	16.8240	718.5783
0.09	330.00	1.08	0.9713	0.0841	1.0450	199.7975	−11.6211	200.1352	−16.0133	5.1947	16.8348	837.305
0.10	360.00	1.10	0.9623	0.0962	1.0585	227.4150	−10.2654	227.6466	−21.4199	−4.5869	21.9056	952.4044

Table 6-5 (Continued)

Case for 0.15 pu compensating voltage (second largest hexagon)												
Theoretical $V_{s's}$ (pu)	β (deg)	Theoretical $V_{s'}$ (pu)	Simulated V_s (pu)	Simulated $V_{s's}$ (pu)	Simulated $V_{s'}$ (pu)	Simulated P_r (MW)	Simulated Q_r (Mvar)	Simulated S_r (MVA)	Simulated P_{se} (MW)	Simulated Q_{se} (Mvar)	Simulated S_{se} (MVA)	Simulated I (A)
0.15	0.00	1.15	0.9484	0.1423	1.0907	235.1379	15.9884	235.6808	−31.8173	−10.5745	33.5285	986.0172
0.13	19.11	1.13	0.9525	0.1260	1.0723	254.0543	−8.9321	254.2113	−27.4311	−16.5373	32.0304	1063.543
0.13	40.89	1.10	0.9556	0.1264	1.0544	273.1410	−34.4630	275.3065	−23.6532	−25.5273	34.8011	1151.799
0.15	60.00	1.08	0.9577	0.1436	1.0370	292.3349	−60.5339	298.5365	−20.6196	−37.6019	42.8844	1248.986
0.13	79.11	1.03	0.9704	0.1284	1.0026	285.3844	−89.7123	299.1530	−10.0826	−37.0558	38.4030	1251.566
0.13	100.89	0.98	0.9826	0.1300	0.9665	277.7151	−120.2854	302.6454	0.5251	−39.3359	39.3394	1266.177
0.15	120.00	0.93	0.9941	0.1491	0.9286	269.3012	−152.2419	309.3554	11.1448	−44.7649	46.1314	1294.249
0.13	139.11	0.90	1.0048	0.1329	0.9085	239.9764	−155.7705	286.0999	19.1842	−32.8363	38.0296	1196.955
0.13	160.89	0.88	1.0146	0.1342	0.8889	209.5153	−159.1908	263.1318	26.4771	−23.3729	35.3176	1100.863
0.15	180.00	0.85	1.0235	0.1535	0.8699	177.9953	−162.4845	241.0053	33.0339	−16.6634	36.9988	1008.293
0.13	199.11	0.88	1.0208	0.1350	0.8943	157.8746	−132.0820	205.8398	27.7033	−2.2607	27.7954	861.171
0.13	220.89	0.90	1.0169	0.1345	0.9194	138.0478	−102.2630	171.7990	21.5727	8.2893	23.1105	718.7553
0.15	240.00	0.93	1.0119	0.1518	0.9452	118.5807	−73.1211	139.3129	14.8211	15.0811	21.1448	582.843
0.13	259.11	0.98	0.9997	0.1322	0.9833	130.6034	−43.0590	137.5185	3.8420	17.7758	18.1862	575.3357
0.13	280.89	1.03	0.9869	0.1306	1.0196	141.8894	−14.3846	142.6167	−6.9268	17.2822	18.6187	596.6652
0.15	300.00	1.08	0.9735	0.1460	1.0542	152.4510	12.8982	152.9956	−17.4530	13.9477	22.3416	640.0873

(Continued)

Table 6-5 (Continued)

Case for 0.15 pu compensating voltage (second largest hexagon)												
Theoretical $V_{s's}$ (pu)	**β (deg)**	**Theoretical $V_{s'}$ (pu)**	**Simulated V_s (pu)**	**Simulated $V_{s's}$ (pu)**	**Simulated $V_{s'}$ (pu)**	**Simulated P_r (MW)**	**Simulated Q_r (Mvar)**	**Simulated S_r (MVA)**	**Simulated P_{se} (MW)**	**Simulated Q_{se} (Mvar)**	**Simulated S_{se} (MVA)**	**Simulated I (A)**
0.13	319.11	1.10	0.9660	0.1278	1.0659	180.8804	13.9692	181.4190	−21.6963	8.1669	23.1825	759.0017
0.13	340.89	1.13	0.9576	0.1267	1.0781	208.4650	15.0000	209.0039	−26.4755	−0.0828	26.4756	874.4089
0.15	360.00	1.15	0.9484	0.1423	1.0907	235.1379	15.9884	235.6808	−31.8173	−10.5745	33.5285	986.0172
Case for 0.20 pu compensating voltage (largest hexagon)												
Theoretical $V_{s's}$ (pu)	**β (deg)**	**Theoretical $V_{s'}$ (pu)**	**Simulated V_s (pu)**	**Simulated $V_{s's}$ (pu)**	**Simulated $V_{s'}$ (pu)**	**Simulated P_r (MW)**	**Simulated Q_r (Mvar)**	**Simulated S_r (MVA)**	**Simulated P_{se} (MW)**	**Simulated Q_{se} (Mvar)**	**Simulated S_{se} (MVA)**	**Simulated I (A)**
0.20	0.00	1.20	0.9342	0.1868	1.1211	242.1959	40.8642	245.6191	−41.9581	−18.5920	45.8927	1027.596
0.18	13.90	1.18	0.9386	0.1692	1.1035	260.8427	16.9331	261.3917	−37.7361	−23.0660	44.2273	1093.584
0.17	30.00	1.15	0.9420	0.1632	1.0863	279.6624	−7.6260	279.7664	−34.0514	−30.3967	45.6449	1170.458
0.18	46.10	1.13	0.9444	0.1703	1.0696	298.5951	−32.7497	300.3857	−31.0325	−40.6536	51.1443	1256.723
0.20	60.00	1.11	0.9459	0.1892	1.0534	317.5775	−58.3680	322.8967	−28.8081	−53.8688	61.0881	1350.902
0.18	73.90	1.06	0.9587	0.1728	1.0203	311.5698	−87.0678	323.5067	−18.4657	−52.7767	55.9138	1353.454
0.17	90.00	1.01	0.9710	0.1682	0.9855	304.8625	−117.1550	326.5983	−8.0549	−54.3341	54.9279	1366.389
0.18	106.10	0.97	0.9827	0.1772	0.9489	297.4265	−148.6235	332.4928	2.3651	−58.8547	58.9022	1391.05
0.20	120.00	0.92	0.9936	0.1987	0.9107	289.2355	−181.4549	341.4426	12.7300	−66.6496	67.8544	1428.492
0.18	133.90	0.88	1.0050	0.1812	0.8890	260.1203	−185.5608	319.5237	21.7143	−53.6609	57.8879	1336.791

Table 6-5 (Continued)

Case for 0.20 pu compensating voltage (largest hexagon)												
Theoretical $V_{s's}$ (pu)	β (deg)	Theoretical $V_{s'}$ (pu)	Simulated V_s (pu)	Simulated $V_{s's}$ (pu)	Simulated $V_{s'}$ (pu)	Simulated P_r (MW)	Simulated Q_r (Mvar)	Simulated S_r (MVA)	Simulated P_{se} (MW)	Simulated Q_{se} (Mvar)	Simulated S_{se} (MVA)	Simulated I (A)
0.17	150.00	0.85	1.0154	0.1759	0.8676	229.7973	−189.5392	297.8790	29.9330	−42.9949	52.3884	1246.236
0.18	166.10	0.83	1.0249	0.1848	0.8467	198.3388	−193.3682	277.0010	37.3840	−34.9565	51.1812	1158.888
0.20	180.00	0.80	1.0334	0.2067	0.8268	165.8300	−197.0263	257.5247	44.0829	−29.8296	53.2270	1077.406
0.18	193.90	0.83	1.0313	0.1859	0.8520	145.5061	−165.6337	220.4689	38.8826	−12.9714	40.9892	922.3753
0.17	210.00	0.85	1.0279	0.1780	0.8782	125.4775	−134.7765	184.1448	32.7845	−0.1007	32.7846	770.4061
0.18	226.10	0.88	1.0233	0.1845	0.9052	105.8126	−104.5569	148.7563	25.9755	8.8532	27.4428	622.3515
0.20	240.00	0.92	1.0176	0.2035	0.9327	86.5736	−75.0674	114.5866	18.6386	14.0168	23.3210	479.3958
0.18	253.90	0.97	1.0058	0.1813	0.9713	99.5315	−44.7385	109.1240	7.6215	18.2597	19.7865	456.5419
0.17	270.00	1.01	0.9933	0.1720	1.0081	111.7553	−15.7833	112.8644	−3.1695	19.1569	19.4173	472.1904
0.18	286.10	1.06	0.9802	0.1767	1.0431	123.2523	11.7896	123.8149	−13.7057	17.0541	21.8790	518.0041
0.20	300.00	1.11	0.9666	0.1933	1.0764	134.0345	37.9830	139.3125	−23.9618	12.2955	26.9323	582.8413
0.18	313.90	1.13	0.9597	0.1730	1.0869	162.3042	38.7963	166.8766	−27.7082	8.1147	28.8721	698.1614
0.17	330.00	1.15	0.9520	0.1649	1.0979	189.7989	39.5479	193.8754	−31.9354	1.4333	31.9676	811.1159
0.18	346.10	1.18	0.9435	0.1701	1.1093	216.4496	40.2373	220.1579	−36.6780	−7.5435	37.4457	921.0739
0.20	360.00	1.20	0.9342	0.1868	1.1211	242.1959	40.8642	245.6191	−41.9581	−18.5920	45.8927	1027.596

magnitude of the sending-end voltage (V_s), the magnitudes of the actual compensating voltage ($V_{s's}$) and the modified sending-end voltage ($V_{s'}$) will vary from their theoretical values for a fixed sending-end voltage of one pu as listed in Tables 6-1, 6-2, and 6-3, and shown in Figures 6-19 and 6-20, respectively. While the magnitude of the sending-end voltage is dependent on line current and line impedance, the relative phase angle (β) and the phase-shift angle (ψ) are not affected, since these angles are only dependent on turns-ratios, according to Equations (6-8) and (6-17a) or (6-17b) for the *A* phase, (6-27) and (6-28a) or (6-28b) for the *B* phase, and (6-31) and (6-32a) or (6-32b) for the *C* phase, respectively.

The magnitude of the varying compensating voltage ($V_{s's}$) at various operating points can either be simulated or calculated from Equation (6-8, 6-27, or 6-31), taking into account that the sending-end voltage (V_s) varies as shown in Figure 6-28. The simulated magnitude ($V_{s's}$) of the compensating voltage in pu, given in Table 6-5, during the entire range of the relative phase angle of $0° \leq \beta \leq 360°$ is shown in Figure 6-29.

Similarly, the varying modified sending-end voltage ($V_{s'}$) at various operating points can either be simulated or calculated from Equation (6-17a or 6-17b, 6-28a or 6-28b, or 6-32a or 6-32b), taking into account that the sending-end voltage (V_s) varies as shown in Figure 6-28. The simulated magnitude ($V_{s'}$) of the modified sending-end voltage in pu, given in Table 6-5, during the entire range of the relative phase angle of $0° \leq \beta \leq 360°$ is shown in Figure 6-30.

Figure 6-31 shows the active power (P_r), reactive power (Q_r), and apparent power (S_r) flows at the receiving end, given in Table 6-5, during the entire range of the relative phase angle of $0° \leq \beta \leq 360°$. Figure 6-32 shows the active power (P_r) versus reactive power (Q_r) flow at the receiving end, given in Table 6-5, during the entire range of the relative phase angle of $0° \leq \beta \leq 360°$. The *P-Q* plot is somewhat different from an ideal hexagon (shown in Figure 6-23) due to the fact that the primary voltage ($\mathbf{V_s}$) keeps changing from its original (uncompensated) value because of the change of line current and the resulting voltage across the line impedance between the POC

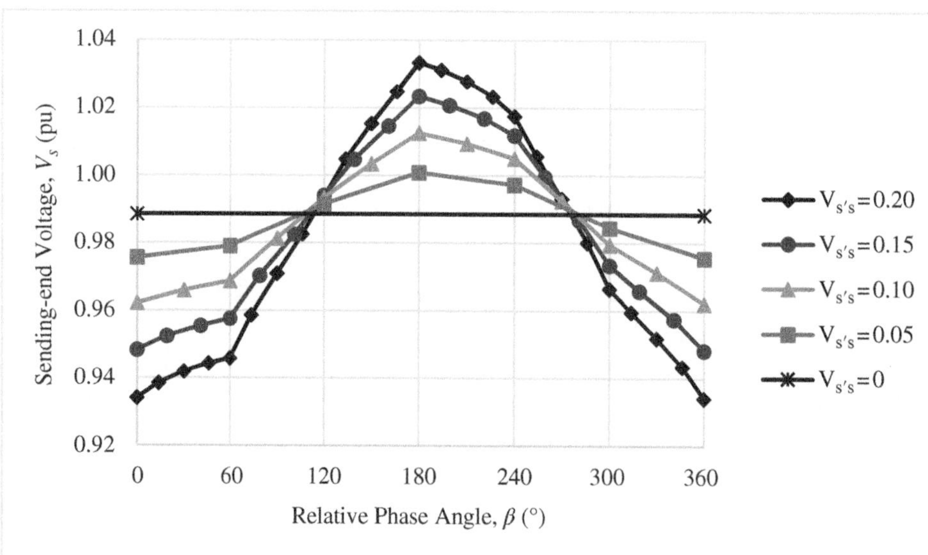

Figure 6-28 Magnitude of the simulated sending-end voltage (V_s) in pu during the entire range of the relative phase angle of $0° \leq \beta \leq 360°$.

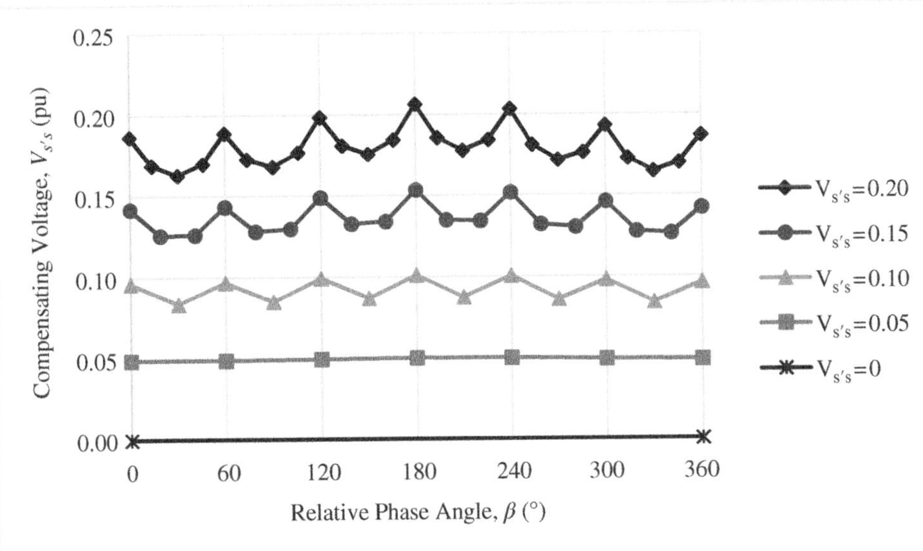

Figure 6-29 Magnitude of the simulated compensating voltage ($V_{s's}$) in pu during the entire range of the relative phase angle of $0° \leq \beta \leq 360°$.

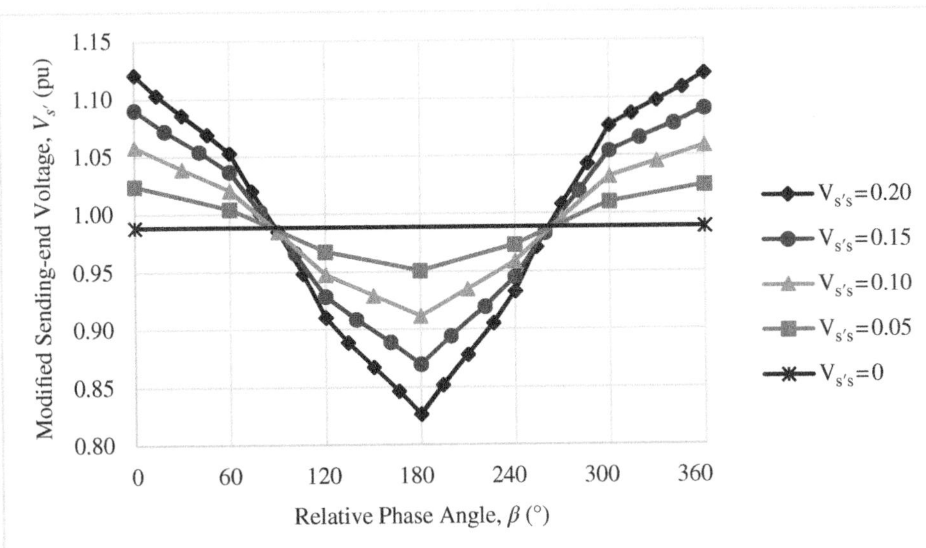

Figure 6-30 Magnitude of the simulated modified sending-end voltage ($V_{s'}$) in pu during the entire range of the relative phase angle of $0° \leq \beta \leq 360°$.

and the infinite bus on the source side. A more ideal hexagon-like characteristic results if the POC is closer to an infinite bus.

Figure 6-33 shows the exchanged active power (P_{se}), reactive power (Q_{se}), and apparent power (S_{se}) by the Compensating-Voltage Unit (Series Unit), given in Table 6-5, during the entire range of the relative phase angle of $0° \leq \beta \leq 360°$.

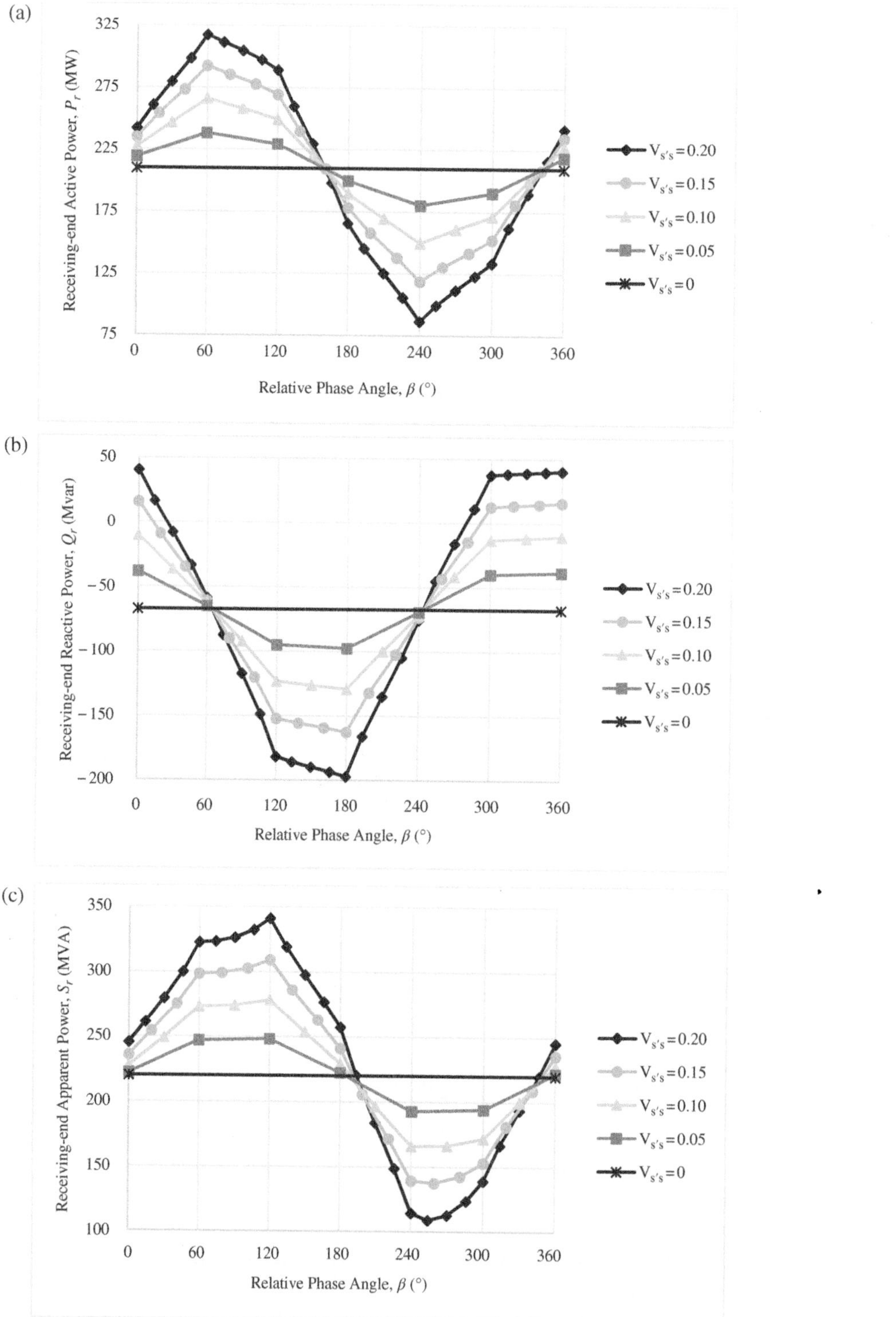

Figure 6-31 (a) Active power (P_r) in MW, (b) reactive power (Q_r) in Mvar, and (c) apparent Power (S_r) in MVA at the receiving end during the entire range of the relative phase angle of $0° \leq \beta \leq 360°$.

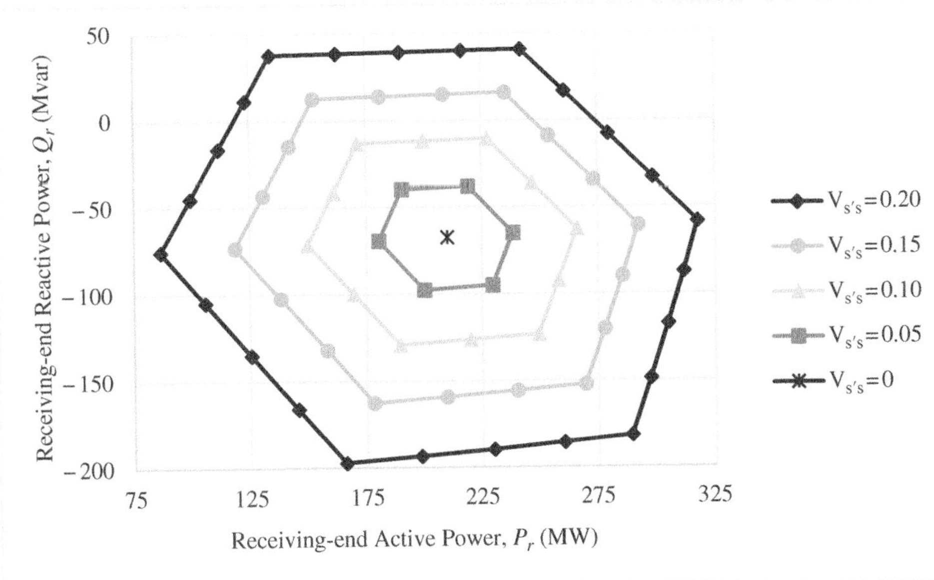

Figure 6-32 Active power (P_r) in MW versus reactive power (Q_r) in Mvar at the receiving end during the entire range of the relative phase angle of $0° \leq \beta \leq 360°$.

Figure 6-34 shows the exchanged active power (P_{se}) versus reactive power (Q_{se}) by the Compensating-Voltage Unit (Series Unit), given in Table 6-5, during the entire range of the relative phase angle of $0° \leq \beta \leq 360°$.

Figure 6-35 shows the prevailing line current (I), given in Table 6-5, during the entire range of the relative phase angle of $0° \leq \beta \leq 360°$.

The summary of the study results for a 20% compensating voltage is given below.

- The variation of line voltage at the POC is between 0.9342 and 1.0334 pu, whereas the natural voltage is 0.9866 pu.
- Active power flow at the receiving end changes from the natural flow of 210 MW (1.31 pu) by 108 and −123 MW.
- Reactive power flow at the receiving end changes from the natural flow of −67 Mvar (−0.42 pu) by 59 and −130 Mvar.
- Rating of the ST: 68 MVA.

Note that a 68 MVA-rated ST regulates 342 MVA in the line as shown in Figures 6-33c and 6-31c, respectively.

6.2.5 Simulation of ST Integrated in a Three-Generator/Four-Line Power System Network

Since the power system network is more complex than what is considered so far in this book, the simulation of ST was repeated with a more realistic network as shown in Figure 6-36, using an electrical system data as shown in Table 6-6. The corresponding EMTP datafile **603IDLST.DAT** is shown in Code 6-4. The difference between this file and **601IDLST.DAT** is that they have different branch and source files. The updated branches are defined in the $INCLUDE file 604NTWK4.BRN as shown in Code 6-5. The updated sources are defined in the $INCLUDE file 605NTWK4.SRC as

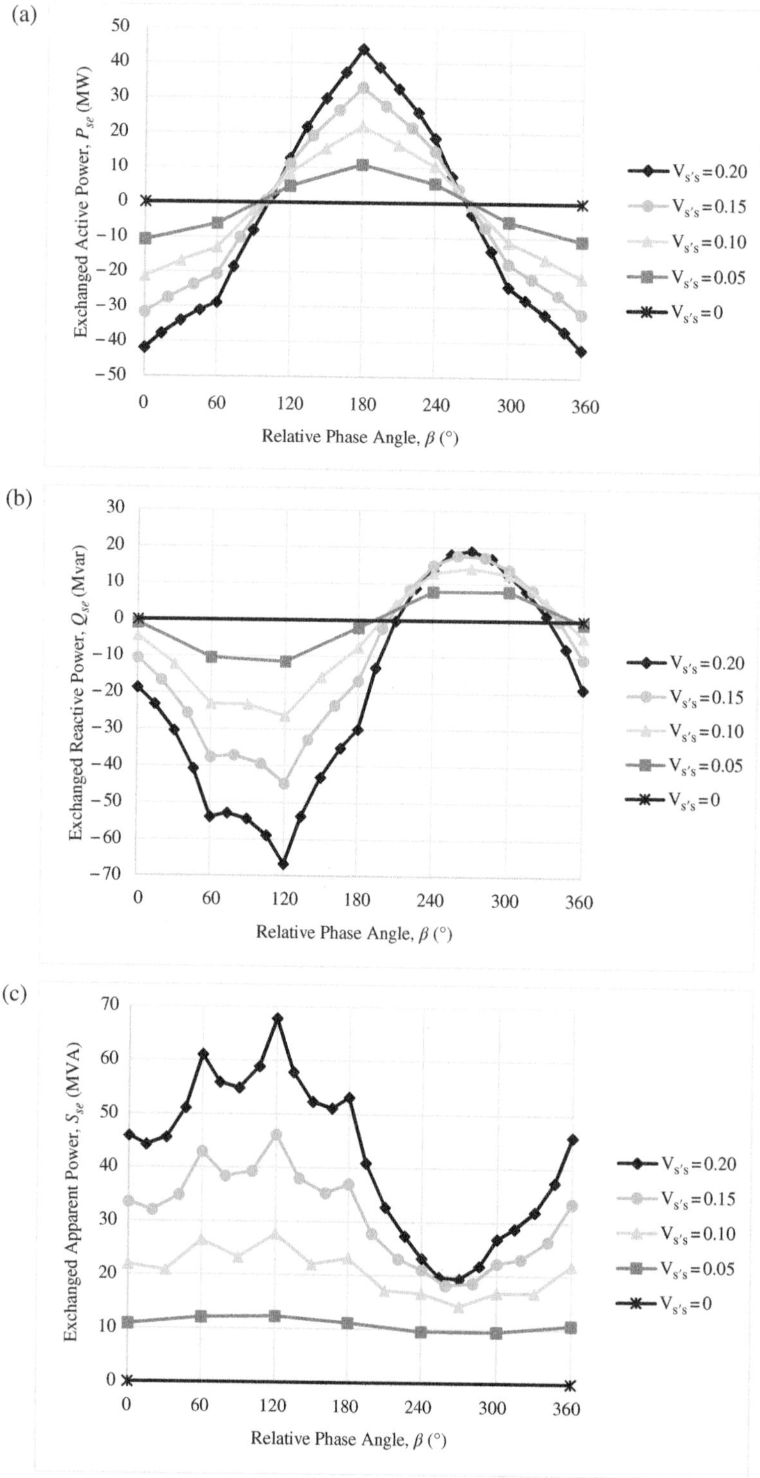

Figure 6-33 (a) Exchanged active power (P_{se}) in MW, (b) reactive power (Q_{se}) in Mvar, and (c) apparent power (S_{se}) in MVA by the Series Unit during the entire range of the relative phase angle of $0° \leq \beta \leq 360°$.

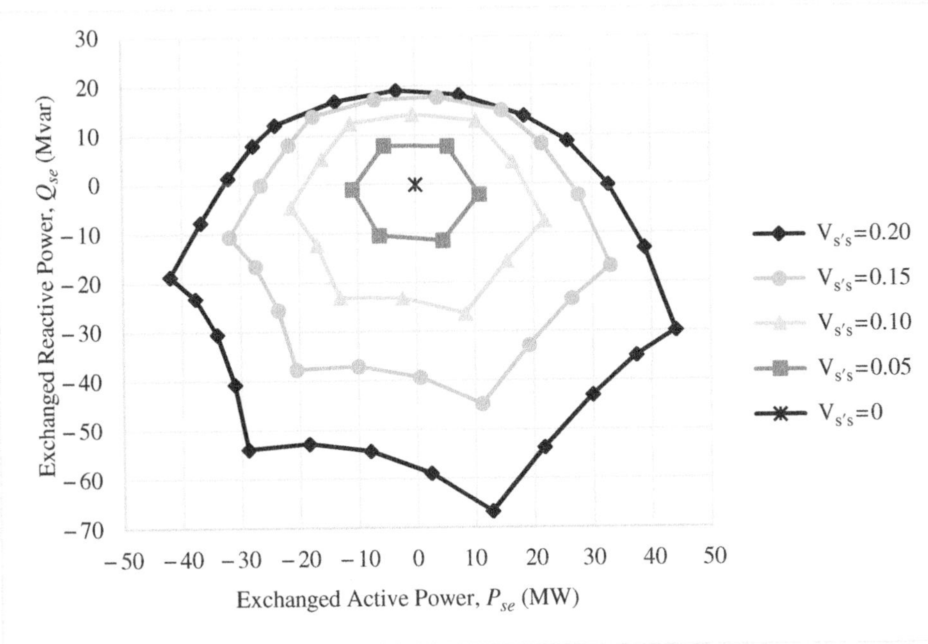

Figure 6-34 Exchanged active power (P_{se}) in MW versus reactive power (Q_{se}) in Mvar by the Series Unit during the entire range of the relative phase angle of $0° \leq \beta \leq 360°$.

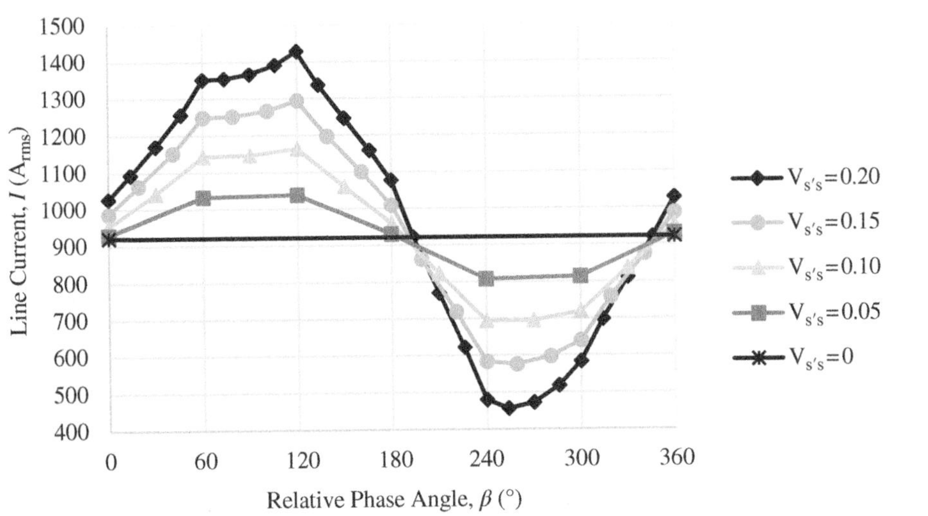

Figure 6-35 Line current (I) in A_{rms} during the entire range of the relative phase angle of $0° \leq \beta \leq 360°$.

shown in Code 6-6. The short between BUS01 and BUS02 was replaced with the ST. The uncompensated or natural active and reactive power flows (P_{rn} and Q_{rn}) at the receiving end of the line are simulated to be 167.89 MW (1.0494 pu) and −53.57 Mvar (−0.3348 pu), respectively. To study the effects of other operating points of an ST, the active turns-ratios are set according to Table 6-1 through Table 6-3, which show the active turns-ratios of the three secondary, series-compensating windings in the A, B, and C phases for the entire range of the relative phase angle of $0° \leq \beta \leq 360°$.

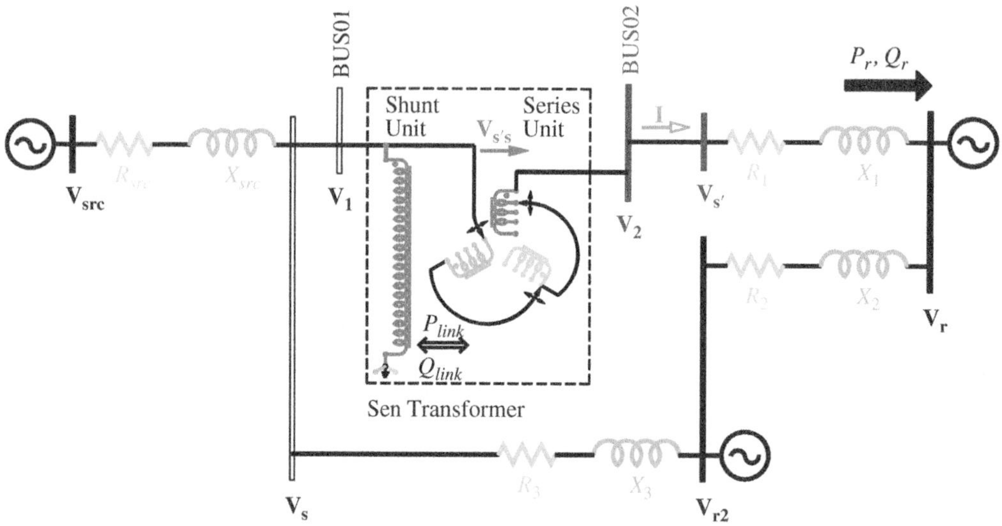

Figure 6-36 Three-generator/four-line power system network, integrated with a Sen Transformer.

Table 6-6 Electrical system data.

Parameters	**Values**
Source voltage ($\mathbf{V_{src}}$) phase-to-neutral	112,676.528∠0° V_{peak}
Source resistance (R_{src})	1.0053 Ω
Source reactance (X_{src})	19.73 mH
Receiving-end voltage ($\mathbf{V_r}$) phase-to-neutral	112,676.528∠−20° V_{peak}
Line 1 resistance (R_1)	3.0159 Ω
Line 1 reactance (X_1)	59.19 mH
Line 2 resistance (R_2)	**3.0159 Ω**
Line 2 reactance (X_2)	**59.19 mH**
Line 3 resistance (R_3)	**3.0159 Ω**
Line 3 reactance (X_3)	**59.19 mH**
Receiving-end 2 voltage ($\mathbf{V_{r2}}$) phase-to-neutral	**112,676.528∠−20° V_{peak}**

Table 6-7 shows the variations of theoretical magnitude ($V_{s's}$) of the compensating voltage ($\mathbf{V_{s's}}$) and the theoretical magnitude ($V_{s'}$) of the modified sending-end voltage ($\mathbf{V_{s'}}$) as a function of the relative phase angle (β), considering the sending-end voltage (V_s) is 1 pu. The table also shows corresponding values of simulated magnitude ($V_{s's}$) of compensating voltage ($\mathbf{V_{s's}}$), simulated magnitude (V_s) of sending-end voltage ($\mathbf{V_s}$), and simulated magnitude ($V_{s'}$) of modified sending-end voltage ($\mathbf{V_{s'}}$). The resulting active power (P_r), reactive power (Q_r), and apparent power (S_r) flows

Code 6-4 EMTP datafile for an ST integrated in a three-generator/four-line power system network (603IDLST.DAT).

```
BEGIN NEW DATA CASE
C File Name: 603IDLST.DAT
C IDEAL SEN TRANSFORMER WITHOUT ANY LEAKAGE IMPEDANCE
C 00000001111111111222222222233333333334444444444555555555566666666667777777778
C 34567890123456789012345678901234567890123456789012345678901234567890123456789 0
C ----dt<---tmax<---Xopt<---Copt<--Toler<        ><        >
16.666-6150.00-3
C 00000001111111111222222222233333333334444444444555555555566666666667777777778
C 34567890123456789012345678901234567890123456789012345678901234567890123456789 0
C -Iprnt<--Iplot<-Idoubl<-KssOut<-MaxOut<        ><        ><---Icat<-Nenerg
      20        10                                                        1
TACS HYBRID
99TWOPI    = 2.0 * PI
C BASE VOLTAGE, CURRENT, AND POWER
99Vbase    = 112676.528
99Ibase    = 946.662704
99Zbase    = Vbase / Ibase
99Pbase    = 1.5 * Vbase * Ibase
$INCLUDE 302MEAS1.SWT
$INCLUDE 303PTCT.SCL
$INCLUDE 304IPLL.PLL
$INCLUDE 305LINER.CMP
C EXCHANGED POWER IN PU BY THE SERIES UNIT WITH THE LINE
99P2sepu   = (v12Apu * iApu + v12Bpu * iBpu + v12Cpu * iCpu) *Vbase*Ibase / Pbase
99Q2sepu   = (v12Apu * iCpu - v12Cpu * iApu) * SQRT(3) * Vbase * Ibase / Pbase
C EXCHANGED MW AND MVAR BY THE SERIES UNIT WITH THE LINE
99P2seMW   = P2sepu * Pbase / 1e6
99Q2seMV   = Q2sepu * Pbase / 1e6
C EXCHANGED MVA WITH THE LINE PER EQUATION (2-182)
99S2seMV   = SQRT(P2seMW * P2seMW + Q2seMV * Q2seMV)
C RMS LINE CURRENT
99Irms     = ipu * Ibase / SQRT(2)
C TACS OUTPUT
33v1pu  v12pu v2pu  PrMW  QrMVARSrMVA P2seMWQ2seMVS2seMVIrms
BLANK RECORD ENDING TACS
C 00000001111111111222222222233333333334444444444555555555566666666667777777778
C 34567890123456789012345678901234567890123456789012345678901234567890123456789 0
C <-Bus1<-Bus2<-Bus3<-Bus4<----R<----L<----C                                 V
  VSA   BUS01A            0.0001
  VSB   BUS01BVSA   BUS01A
  VSC   BUS01CVSA   BUS01A
$INCLUDE 604NTWK4.BRN
$INCLUDE 602IDLST.TRN
BLANK RECORD ENDING BRANCHES
C 00000001111111111222222222233333333334444444444555555555566666666667777777778
```

(Continued)

Code 6-4 (Continued)

```
C 34567890123456789012345678901234567890123456789012345678901234567890123456789
C <-Bus1<-Bus2<---Tclose<----Topen<-------Ie<----Flash<--Request<-----Target<--O
$INCLUDE 307MEAS2.SWT
BLANK RECORD ENDING SWITCHES
C 00000001111111111222222222233333333334444444444555555555566666666667777777778
C 34567890123456789012345678901234567890123456789012345678901234567890123456789
$INCLUDE 605NTWK4.SRC
BLANK RECORD ENDING SOURCES
BLANK RECORD ENDING NODE VOLTAGE OUTPUT
BLANK RECORD ENDING PLOT
```

Code 6-5 EMTP $INCLUDE file for implementing four-line branches (604NTWK4.BRN).

```
C File Name: 604NTWK4.BRN
C SOURCE IMPEDANCES & LINE IMPEDANCES
C 00000001111111111222222222233333333334444444444555555555566666666667777777778
C 34567890123456789012345678901234567890123456789012345678901234567890123456789
C <-Bus1<-Bus2<-Bus3<-Bus4<----R<----L<----C                                   V
  VSRCA VSA                1.0053 19.73
  VSRCB VSB   VSRCA VSA
  VSRCC VSC   VSRCA VSA
  VSPA  VRA                3.0159 59.19
  VSPB  VRB   VSPA  VRA
  VSPC  VRC   VSPA  VRA
  VSA   VR2A               3.0159 59.19
  VSB   VR2B  VSA   VR2A
  VSC   VR2C  VSA   VR2A
  VR2A  VRA                3.0159 59.19
  VR2B  VRB   VR2A  VRA
  VR2C  VRC   VR2A  VRA
```

Code 6-6 EMTP $INCLUDE file for implementing three-generator sources (605NTWK4.SRC).

```
C File Name: 605NTWK4.SRC
C SOURCE VOLTAGES AND RECEIVING-END VOLTAGES
C 00000001111111111222222222233333333334444444444555555555566666666667777777778
C 34567890123456789012345678901234567890123456789012345678901234567890123456789
C <--Bus<I<-----Ampl<-----Freq<----Phase<-------A1<------T1><---Tstart<----Tstop
14VSRCA    112676.528     60.00      0.00                               -1.0
14VSRCB    112676.528     60.00   -120.00                               -1.0
14VSRCC    112676.528     60.00    120.00                               -1.0
14VRA      112676.528     60.00    -20.00                               -1.0
14VRB      112676.528     60.00   -140.00                               -1.0
14VRC      112676.528     60.00    100.00                               -1.0
14VR2A     112676.528     60.00    -20.00                               -1.0
14VR2B     112676.528     60.00   -140.00                               -1.0
14VR2C     112676.528     60.00    100.00                               -1.0
```

Table 6-7 Compensating voltage, modified sending-end voltage, power flows at the receiving end, exchanged powers by the Series Unit and the line current for electrical system data, shown in Table 6-6, and the active turns-ratios of the secondary, series-compensating windings in the *A*, *B*, and *C* phases per respective Tables 6-1, 6-2, and 6-3.

Case for natural power flows (left point of the hexagon characteristic)												
Theoretical $V_{s's}$ (pu)	β (deg)	Theoretical $V_{s'}$ (pu)	Simulated V_s (pu)	Simulated $V_{s's}$ (pu)	Simulated $V_{s'}$ (pu)	Simulated P_r (MW)	Simulated Q_r (Mvar)	Simulated S_r (MVA)	Simulated P_{se} (MW)	Simulated Q_{se} (Mvar)	Simulated S_{se} (MVA)	Simulated I (A)
0.00	0.00	1.00	0.9854	0.0000	0.9854	167.8971	−53.5739	176.2373	0.0000	0.0000	0.0000	737.3236
0.00	360.00	1.00	0.9854	0.0000	0.9854	167.8971	−53.5739	176.2373	0.0000	0.0000	0.0000	737.3236
Case for 0.05 pu compensating voltage (smallest hexagon)												
Theoretical $V_{s's}$ (pu)	β (deg)	Theoretical $V_{s'}$ (pu)	Simulated V_s (pu)	Simulated $V_{s's}$ (pu)	Simulated $V_{s'}$ (pu)	Simulated P_r (MW)	Simulated Q_r (Mvar)	Simulated S_r (MVA)	Simulated P_{se} (MW)	Simulated Q_{se} (Mvar)	Simulated S_{se} (MVA)	Simulated I (A)
0.05	0.00	1.05	0.9752	0.0488	1.0240	177.1905	−22.2712	178.5847	−8.6780	−0.7189	8.7077	747.1443
0.05	60.00	1.03	0.9783	0.0489	1.0036	198.9843	−48.8321	204.8885	−5.3581	−8.4691	10.0217	857.1917
0.05	120.00	0.98	0.9881	0.0494	0.9644	189.9523	−80.9602	206.4859	3.2371	−9.6744	10.2017	863.8745
0.05	180.00	0.95	0.9950	0.0498	0.9453	158.0533	−86.0990	179.9830	8.5884	−2.5338	8.9544	752.9944
0.05	240.00	0.98	0.9918	0.0496	0.9680	136.0127	−58.3573	148.0035	4.7809	5.5686	7.3394	619.2018
0.05	300.00	1.03	0.9818	0.0491	1.0073	146.1122	−26.6767	148.5275	−3.8748	6.1765	7.2913	621.3942
0.05	360.00	1.05	0.9752	0.0488	1.0240	177.1905	−22.2712	178.5847	−8.6780	−0.7189	8.7077	747.1443

(*Continued*)

Table 6-7 (Continued)

Case for 0.10 pu compensating voltage (second smallest hexagon)												
Theoretical $V_{s's}$ (pu)	β (deg)	Theoretical $V_{s'}$ (pu)	Simulated V_s (pu)	Simulated $V_{s's}$ (pu)	Simulated $V_{s'}$ (pu)	Simulated P_r (MW)	Simulated Q_r (Mvar)	Simulated S_r (MVA)	Simulated P_{se} (MW)	Simulated Q_{se} (Mvar)	Simulated S_{se} (MVA)	Simulated I (A)
0.10	0.00	1.10	0.9645	0.0964	1.0609	185.9395	7.7919	186.1027	−17.4020	−4.3975	17.9491	778.5975
0.09	30.00	1.08	0.9679	0.0838	1.0413	207.4645	−17.9332	208.2381	−14.0464	−10.3610	17.4543	871.2054
0.10	60.00	1.05	0.9704	0.0970	1.0224	229.2082	−44.1446	233.4205	−11.3192	−19.6206	22.6515	976.5608
0.09	90.00	1.00	0.9805	0.0849	0.9842	220.9932	−75.8488	233.6473	−2.7308	−19.6506	19.8394	977.5095
0.10	120.00	0.95	0.9900	0.0990	0.9444	212.2188	−108.7666	238.4680	5.7143	−22.9052	23.6072	997.6779
0.09	150.00	0.93	0.9974	0.0864	0.9236	180.3627	−114.2952	213.5277	11.6920	−14.2640	18.4435	893.3353
0.10	180.00	0.90	1.0040	0.1004	0.9036	147.6561	−119.8201	190.1557	17.0423	−8.6056	19.0918	795.5538
0.09	210.00	0.93	1.0012	0.0867	0.9271	125.3670	−91.2498	155.0594	13.3534	1.5577	13.4439	648.7218
0.10	240.00	0.95	0.9974	0.0997	0.9514	103.4047	−63.1691	121.1729	9.0218	8.0415	12.0855	506.9507
0.09	270.00	1.00	0.9877	0.0855	0.9914	114.2978	−31.1392	118.4636	0.3959	10.1251	10.1329	495.616
0.10	300.00	1.05	0.9774	0.0977	1.0298	124.6529	−0.3318	124.6534	−8.2655	8.9512	12.1837	521.512
0.09	330.00	1.08	0.9713	0.0841	1.0450	155.6689	3.7713	155.7145	−12.5727	3.6733	13.0983	651.4626
0.10	360.00	1.10	0.9645	0.0964	1.0609	185.9395	7.7919	186.1027	−17.4020	−4.3975	17.9491	778.5975

Case for 0.15 pu compensating voltage (second largest hexagon)												
Theoretical $V_{s's}$ (pu)	β (deg)	Theoretical $V_{s'}$ (pu)	Simulated V_s (pu)	Simulated $V_{s's}$ (pu)	Simulated $V_{s'}$ (pu)	Simulated P_r (MW)	Simulated Q_r (Mvar)	Simulated S_r (MVA)	Simulated P_{se} (MW)	Simulated Q_{se} (Mvar)	Simulated S_{se} (MVA)	Simulated I (A)
0.15	0.00	1.15	0.9533	0.1430	1.0963	194.1522	36.6049	197.5728	−26.1303	−10.7403	28.2515	826.5848
0.13	19.11	1.13	0.9570	0.1266	1.0774	215.4017	11.7217	215.7204	−22.7820	−15.0583	27.3088	902.5089
0.13	40.89	1.10	0.9598	0.1270	1.0591	236.8728	−13.6741	237.2672	−20.0013	−22.5294	30.1268	992.6542
0.15	60.00	1.08	0.9619	0.1443	1.0416	258.5094	−39.5231	261.5132	−17.8977	−33.2174	37.7323	1094.092
0.13	79.11	1.03	0.9721	0.1286	1.0044	251.1119	−70.7780	260.8959	−9.3340	−32.2267	33.5512	1091.51
0.13	100.89	0.98	0.9818	0.1299	0.9657	243.1563	−103.2514	264.1701	−0.9135	−34.2982	34.3104	1105.208
0.15	120.00	0.93	0.9909	0.1486	0.9255	234.6331	−136.9178	271.6601	7.3108	−39.7090	40.3764	1136.544
0.13	139.11	0.90	0.9988	0.1321	0.9031	202.8785	−142.8702	248.1362	13.9654	−29.6624	32.7855	1038.127
0.13	160.89	0.88	1.0059	0.1331	0.8813	170.2111	−148.8040	226.0850	19.9579	−22.5131	30.0859	945.8714
0.15	180.00	0.85	1.0123	0.1518	0.8604	136.7063	−154.6995	206.4475	25.3156	−18.4869	31.3472	863.7139
0.13	199.11	0.88	1.0098	0.1336	0.8847	114.1773	−125.3236	169.5360	21.8002	−6.1377	22.6477	709.2873
0.13	220.89	0.90	1.0064	0.1331	0.9100	91.9740	−96.3968	133.2350	17.5738	2.4122	17.7386	557.4149
0.15	240.00	0.93	1.0021	0.1503	0.9361	70.1552	−67.9951	97.6990	12.7707	7.2518	14.6860	408.7429
0.13	259.11	0.98	0.9927	0.1313	0.9765	81.8239	−35.6467	89.2516	4.1771	10.9516	11.7212	373.4016
0.13	280.89	1.03	0.9828	0.1300	1.0154	92.9631	−4.5073	93.0723	−4.4423	11.2549	12.0999	389.3865
0.15	300.00	1.08	0.9722	0.1458	1.0527	103.5710	25.4023	106.6406	−13.0528	8.4539	15.5514	446.1521

(Continued)

Table 6-7 (Continued)

Case for 0.15 pu compensating voltage (second largest hexagon)												
Theoretical $V_{s's}$ (pu)	β (deg)	Theoretical $V_{s'}$ (pu)	Simulated V_s (pu)	Simulated $V_{s's}$ (pu)	Simulated $V_{s'}$ (pu)	Simulated P_r (MW)	Simulated Q_r (Mvar)	Simulated S_r (MVA)	Simulated P_{se} (MW)	Simulated Q_{se} (Mvar)	Simulated S_{se} (MVA)	Simulated I (A)
0.13	319.11	1.10	0.9666	0.1279	1.0666	134.4718	29.2360	137.6133	−16.9250	4.8154	17.5967	575.7323
0.13	340.89	1.13	0.9603	0.1270	1.0811	164.6877	32.9728	167.9561	−21.2729	−1.6443	21.3364	702.6775
0.15	360.00	1.15	0.9533	0.1430	1.0963	194.1522	36.6049	197.5728	−26.1303	−10.7403	28.2515	826.5848
Case for 0.20 pu compensating voltage (largest hexagon)												
Theoretical $V_{s's}$ (pu)	β (deg)	Theoretical $V_{s'}$ (pu)	Simulated V_s (pu)	Simulated $V_{s's}$ (pu)	Simulated $V_{s'}$ (pu)	Simulated P_r (MW)	Simulated Q_r (Mvar)	Simulated S_r (MVA)	Simulated P_{se} (MW)	Simulated Q_{se} (Mvar)	Simulated S_{se} (MVA)	Simulated I (A)
0.20	0.00	1.20	0.9417	0.1883	1.1300	201.8392	64.1662	211.7932	−34.8236	−19.4541	39.8892	886.0788
0.18	13.90	1.18	0.9456	0.1705	1.1119	222.8070	40.1260	226.3914	−31.5216	−22.2686	38.5941	947.1531
0.17	30.00	1.15	0.9488	0.1643	1.0942	243.9989	15.5515	244.4940	−28.7301	−28.0882	40.1792	1022.889
0.18	46.10	1.13	0.9512	0.1715	1.0772	265.3611	−9.5029	265.5312	−26.5539	−36.9876	45.5323	1110.902
0.20	60.00	1.11	0.9528	0.1906	1.0609	286.8369	−34.9785	288.9618	−25.0974	−49.0096	55.0619	1208.929
0.18	73.90	1.06	0.9631	0.1736	1.0249	280.2524	−65.7616	287.8645	−16.5793	−47.1514	49.9812	1204.338
0.17	90.00	1.01	0.9730	0.1685	0.9874	273.1143	−97.7646	290.0850	−8.2023	−48.1920	48.8851	1213.628
0.18	106.10	0.97	0.9822	0.1771	0.9485	265.4100	−130.9660	295.9638	−0.0206	−52.4061	52.4061	1238.223
0.20	120.00	0.92	0.9908	0.1982	0.9081	257.1292	−165.3348	305.6976	7.9102	−60.0587	60.5774	1278.946
0.18	133.90	0.88	0.9992	0.1801	0.8839	225.5353	−171.7440	283.4822	15.2854	−48.7250	51.0663	1186.004

Table 6-7 (Continued)

Case for 0.20 pu compensating voltage (largest hexagon)												
Theoretical $V_{s's}$ (pu)	β (deg)	Theoretical $V_{s'}$ (pu)	Simulated V_s (pu)	Simulated $V_{s's}$ (pu)	Simulated $V_{s'}$ (pu)	Simulated P_r (MW)	Simulated Q_r (Mvar)	Simulated S_r (MVA)	Simulated P_{se} (MW)	Simulated Q_{se} (Mvar)	Simulated S_{se} (MVA)	Simulated I (A)
0.17	150.00	0.85	1.0069	0.1744	0.8603	192.9676	−178.1208	262.6091	21.9709	−40.1849	45.7990	1098.677
0.18	166.10	0.83	1.0138	0.1828	0.8375	159.4984	−184.4436	243.8425	27.9858	−34.6804	44.5639	1020.163
0.20	180.00	0.80	1.0198	0.2040	0.8158	125.2081	−190.6902	228.1224	33.3621	−32.4283	46.5256	954.3951
0.18	193.90	0.83	1.0177	0.1835	0.8408	102.4474	−160.5369	190.4405	30.0783	−17.7817	34.9413	796.7454
0.17	210.00	0.85	1.0148	0.1758	0.8670	80.0115	−130.7877	153.3208	26.0126	−7.0378	26.9478	641.4478
0.18	226.10	0.88	1.0108	0.1822	0.8942	57.9611	−101.5238	116.9041	21.3031	−0.1267	21.3034	489.0913
0.20	240.00	0.92	1.0060	0.2012	0.9220	36.3524	−72.8206	81.3900	16.0863	3.0650	16.3757	340.5112
0.18	253.90	0.97	0.9970	0.1797	0.9627	48.7734	−40.1867	63.1966	7.5224	8.5104	11.3584	264.3954
0.17	270.00	1.01	0.9873	0.1710	1.0020	60.6769	−8.7449	61.3038	−1.0545	10.4299	10.4831	256.4767
0.18	286.10	1.06	0.9770	0.1761	1.0397	72.0575	21.4803	75.1910	−9.6127	9.1100	13.2438	314.5763
0.20	300.00	1.11	0.9663	0.1933	1.0760	82.9132	50.4736	97.0680	−18.1219	4.8448	18.7583	406.103
0.18	313.90	1.13	0.9612	0.1733	1.0885	113.6491	54.0705	125.8560	−21.6215	2.8460	21.8080	526.5433
0.17	330.00	1.15	0.9554	0.1655	1.1018	143.7593	57.5559	154.8529	−25.5472	−1.9841	25.6241	647.8576
0.18	346.10	1.18	0.9489	0.1711	1.1157	173.1765	60.9231	183.5803	−29.9380	−9.4797	31.4031	768.0446
0.20	360.00	1.20	0.9417	0.1883	1.1300	201.8392	64.1662	211.7932	−34.8236	−19.4541	39.8892	886.0788

at the receiving end, exchanged active power (P_{se}), reactive power (Q_{se}), and apparent power (S_{se}) by the Compensating-Voltage Unit (Series Unit), and the prevailing line current (I) are also shown during the entire range of the relative phase angle of $0° \leq \beta \leq 360°$ in Table 6-7 for the k_{a1}-k_{b1}-k_{c1} active turns-ratios from Table 6-1, k_{b2}-k_{c2}-k_{a2} active turns-ratios from Table 6-2, and k_{c3}-k_{a3}-k_{b3} active turns-ratios from Table 6-3. The table is expanded for each hexagon, related to 5, 10, 15, and 20% compensating voltages.

Figure 6-37 shows the magnitude of the sending-end voltage (V_s), given in Table 6-7, during the entire range of the relative phase angle of $0° \leq \beta \leq 360°$.

The simulated magnitude ($V_{s's}$) of the compensating voltage, given in Table 6-7, during the entire range of the relative phase angle of $0° \leq \beta \leq 360°$ is shown in Figure 6-38.

The simulated magnitude ($V_{s'}$) of the modified sending-end voltage, given in Table 6-7, during the entire range of the relative phase angle of $0° \leq \beta \leq 360°$ is shown in Figure 6-39.

Figure 6-40 shows the active power (P_r), reactive power (Q_r), and apparent power (S_r) flows at the receiving end, given in Table 6-7, during the entire range of the relative phase angle of $0° \leq \beta \leq 360°$. Figure 6-41 shows the active power (P_r) versus reactive power (Q_r) flow at the receiving end, given in Table 6-7, during the entire range of the relative phase angle of $0° \leq \beta \leq 360°$.

Figure 6-42 shows the exchanged active power (P_{se}), reactive power (Q_{se}), and apparent power (S_{se}) by the Compensating-Voltage Unit (Series Unit), given in Table 6-7, during the entire range of the relative phase angle of $0° \leq \beta \leq 360°$.

Figure 6-43 shows the exchanged active power (P_{se}) versus reactive power (Q_{se}) by the Compensating-Voltage Unit (Series Unit), given in Table 6-7, during the entire range of the relative phase angle of $0° \leq \beta \leq 360°$.

Figure 6-44 shows the prevailing line current (I), given in Table 6-7, during the entire range of the relative phase angle of $0° \leq \beta \leq 360°$.

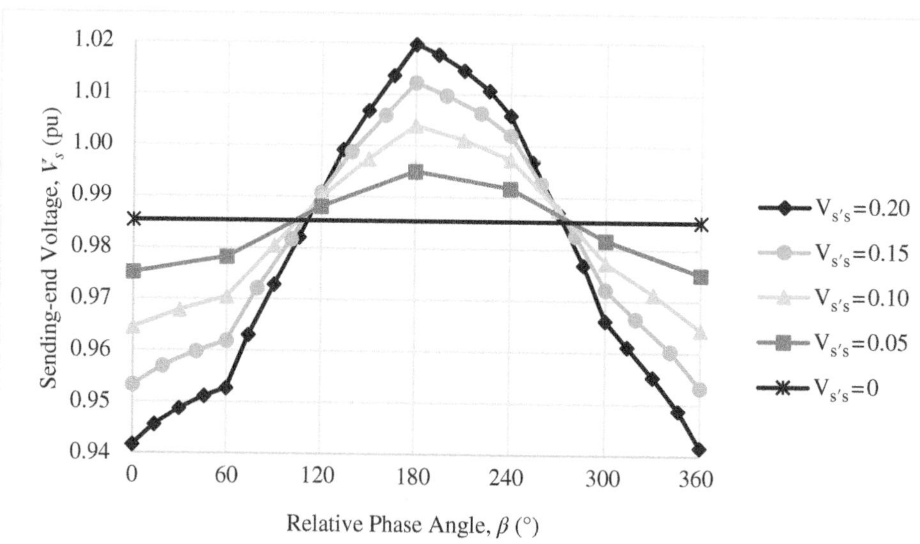

Figure 6-37 Magnitude of the simulated sending-end voltage (V_s) in pu during the entire range of the relative phase angle of $0° \leq \beta \leq 360°$.

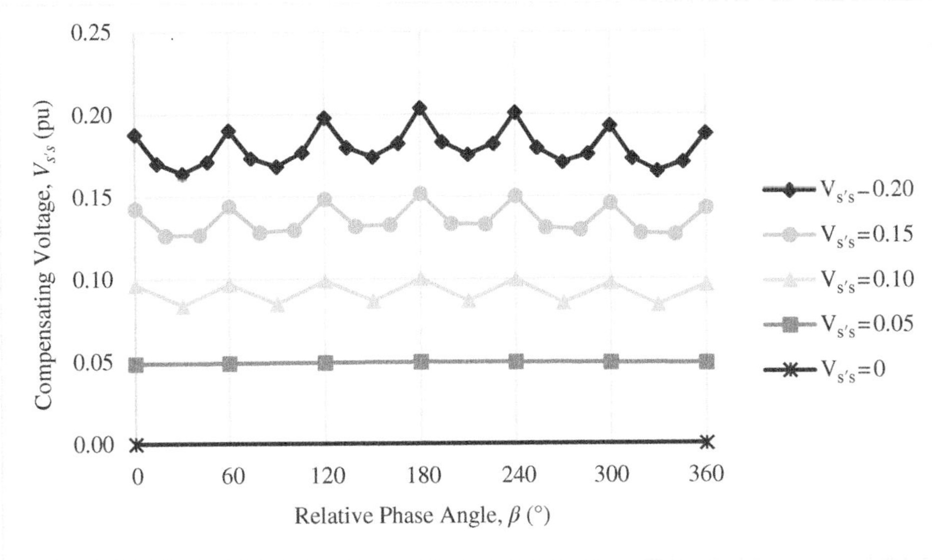

Figure 6-38 Magnitude of the simulated compensating voltage ($\mathbf{V_{s's}}$) in pu during the entire range of the relative phase angle of $0° \leq \beta \leq 360°$.

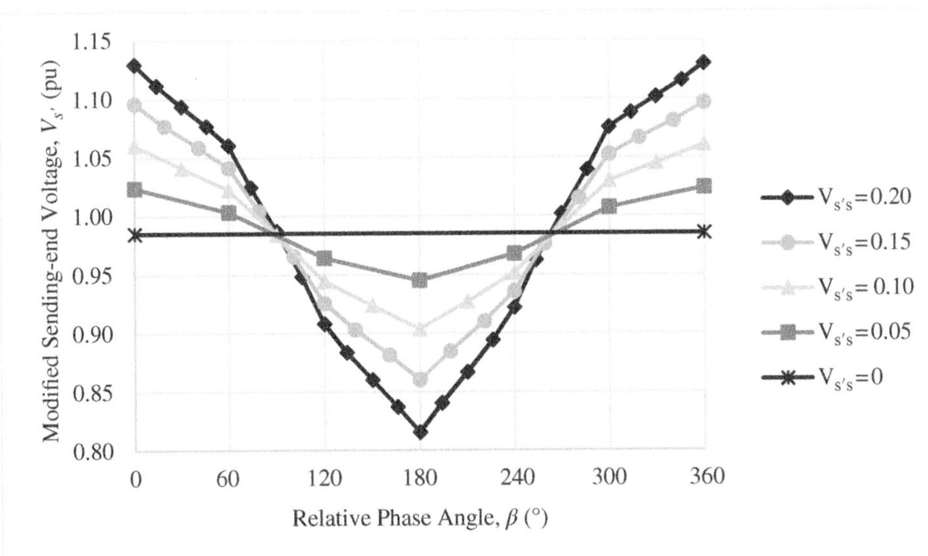

Figure 6-39 Magnitude of the simulated modified sending-end voltage ($\mathbf{V_{s'}}$) in pu during the entire range of the relative phase angle of $0° \leq \beta \leq 360°$.

The summary of the study results for a 20% compensating voltage is given below.

- The variation of line voltage at the POC is between 0.9417 and 1.0198 pu, whereas the natural voltage is 0.9854 pu.
- Active power flow at the receiving end changes from the natural flow of 168 MW (1.0494 pu) by 119 and −132 MW.

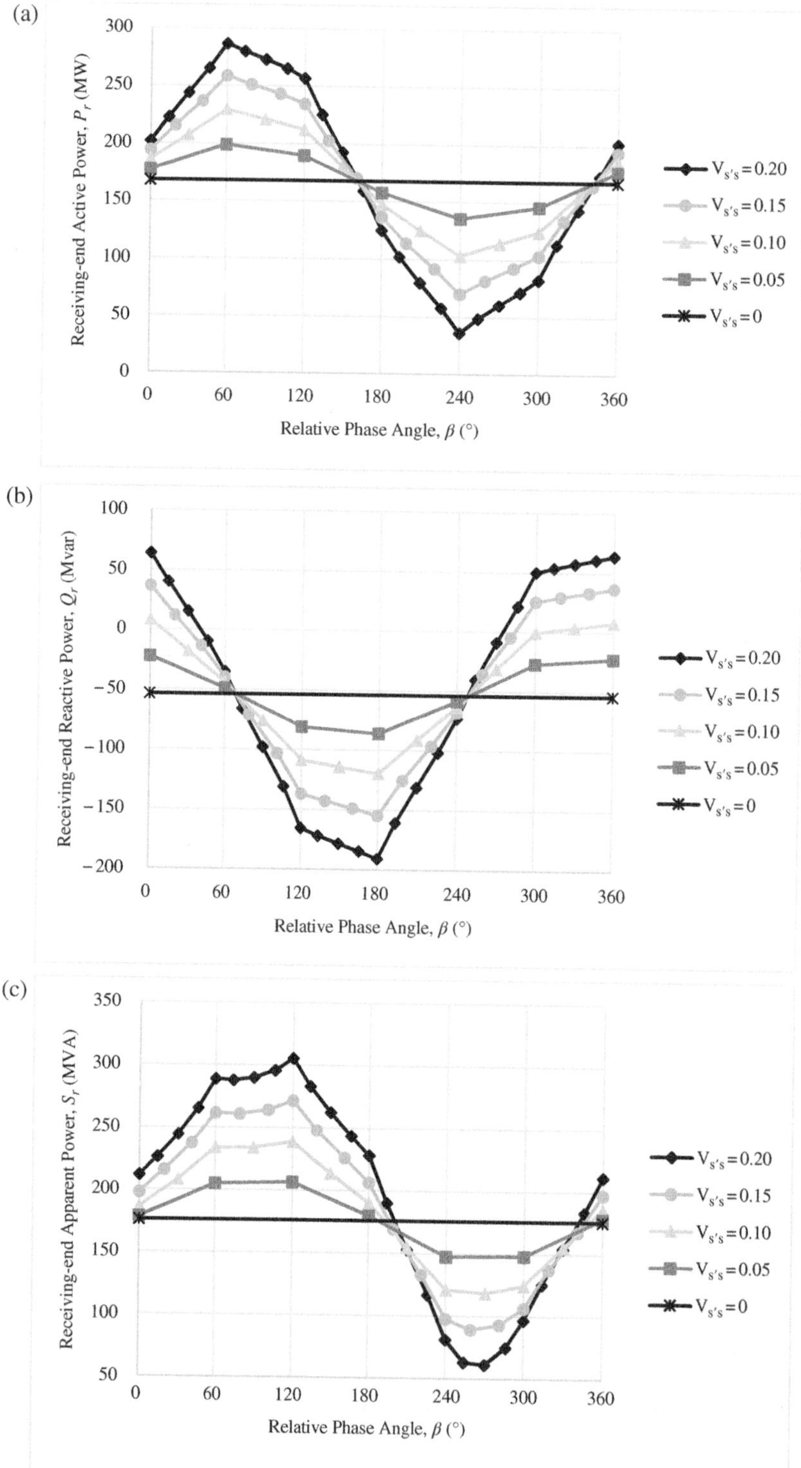

Figure 6-40 (a) Active power (P_r) in MW, (b) reactive power (Q_r) in Mvar, and (c) apparent power (S_r) in MVA at the receiving end during the entire range of the relative phase angle of $0° \leq \beta \leq 360°$.

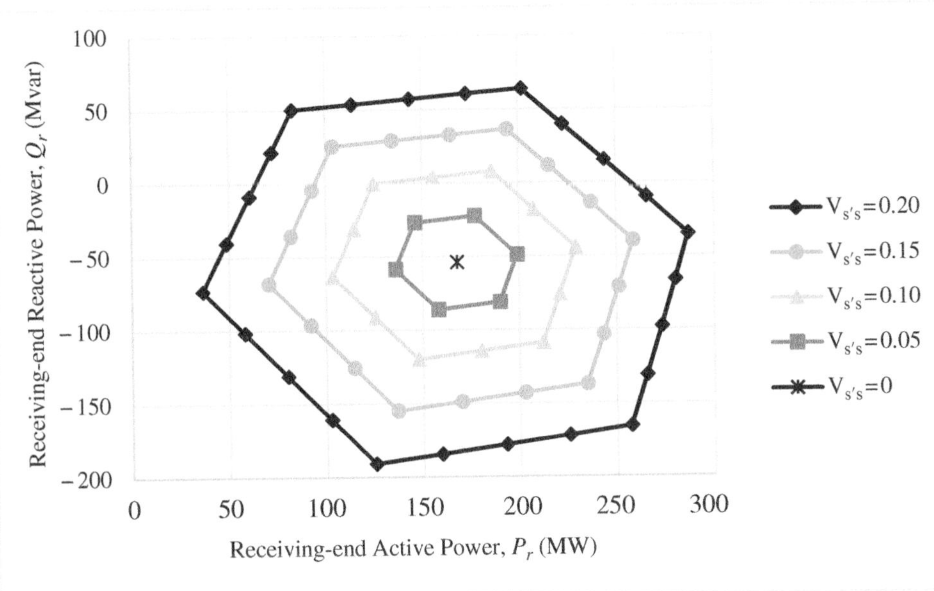

Figure 6-41 Active power (P_r) in MW versus reactive power (Q_r) in Mvar at the receiving end during the entire range of the relative phase angle of 0° $\leq \beta \leq$ 360°.

- Reactive power flow at the receiving end changes from the natural flow of −54 Mvar (−0.3348 pu) by 44 and −137 Mvar.
- Rating of the ST: 61 MVA.

Note that a 61 MVA-rated ST regulates 306 MVA in the line as shown in Figures 6-42c and 6-40c, respectively.

A comparison of the simulation results between Figures 6-28 through 6-35 and Figures 6-37 through 6-44 reveals that the actual numeric values for any particular parameter, for example, line current (I) may be different in the two networks; however, its characteristic due to the operation of the ST is similar. The same is true for all other comparative parameters, such as $P_r - Q_r$ characteristic, $P_{se} - Q_{se}$ characteristic, and so on.

6.2.6 Testing of ST

There are two types of tests that need to be performed to certify an ST for its integration in a power system network. These tests are (1) factory tests and (2) power flow tests. The factory tests include standard transformer tests, such as resistance, no-load (core loss), short-circuit (copper loss), impedance, dielectric, surge, heat run, and so on. For the power flow test, it is desirable to test the ST while it is integrated in an actual power system network. If an energized network is unavailable, the ST may be tested while integrated in a custom-made network as shown in Figure 6-45, which ensures that the ST carries its rated power according to its design specification as confirmed by the simulation results.

The additional symbols shown in the figure are

$\mathbf{I_s}$ is the current at the sending end of the line,
$\mathbf{I_{ex}}$ is the exciter current through the primary winding of the Sen Transformer, and
$\mathbf{I_{src}}$ is the source current.

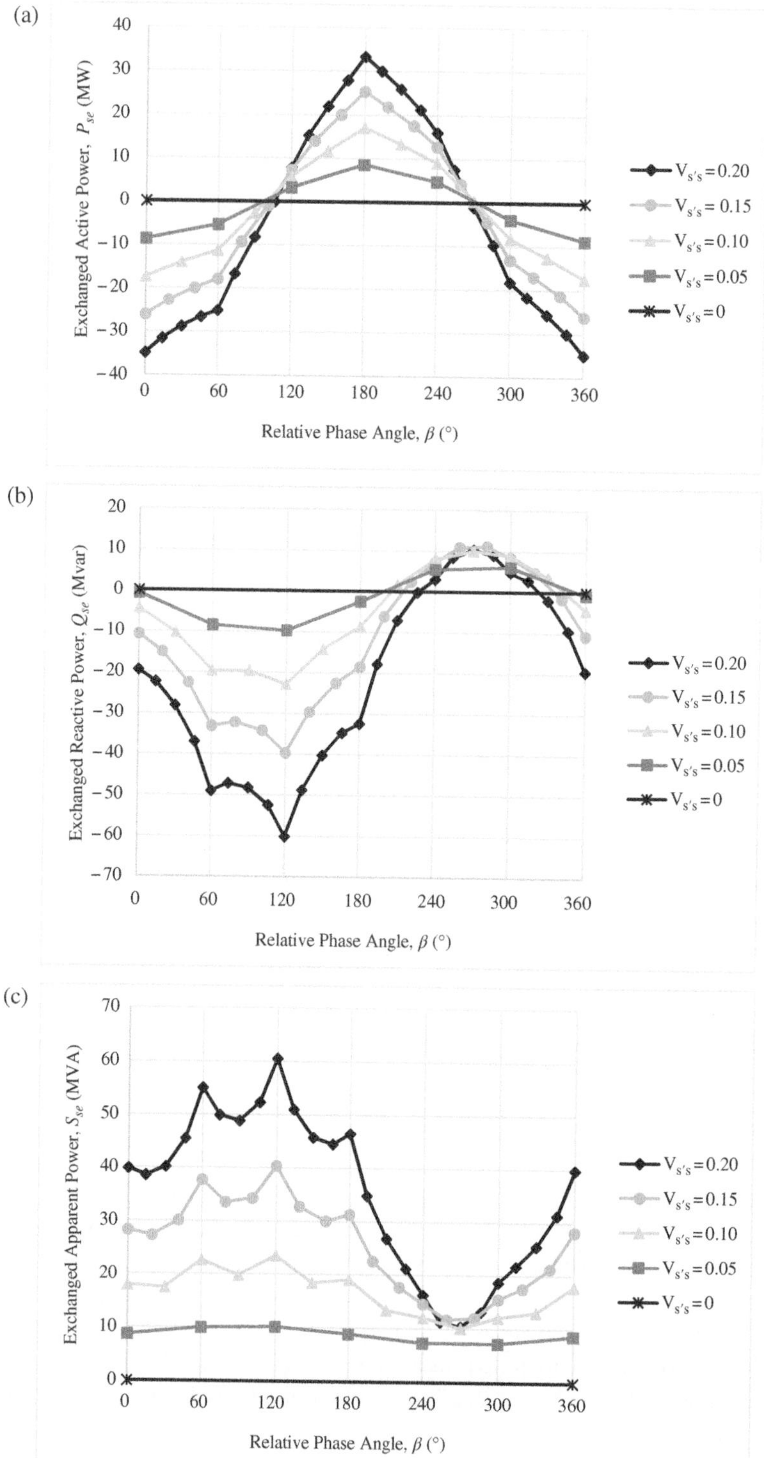

Figure 6-42 (a) Exchanged active power (P_{se}) in MW, (b) reactive power (Q_{se}) in Mvar, and (c) apparent power (S_{se}) in MVA by the Series Unit during the entire range of the relative phase angle of $0° \leq \beta \leq 360°$.

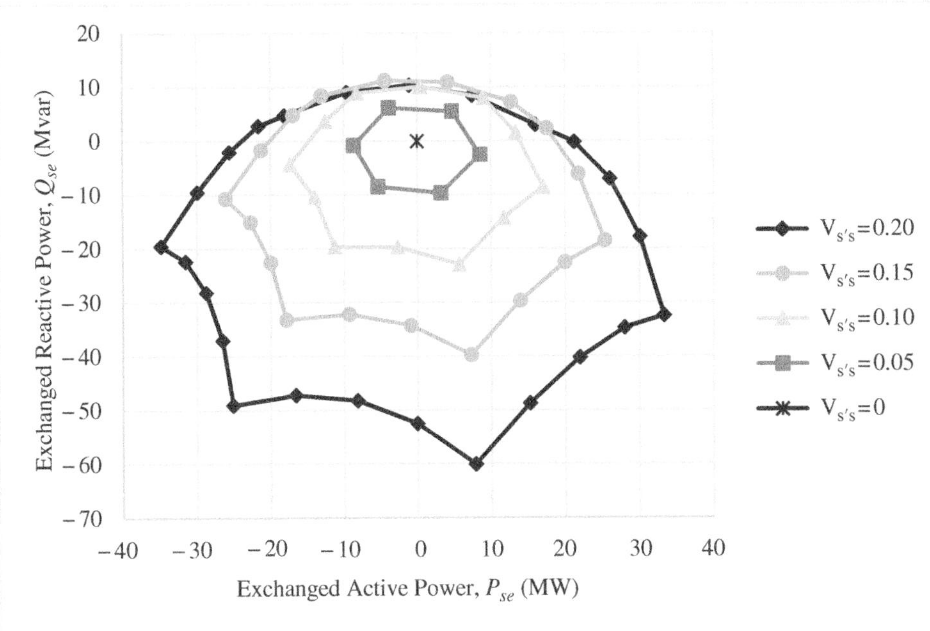

Figure 6-43 Exchanged active power (P_{se}) in MW versus reactive power (Q_{se}) in Mvar by the Series Unit during the entire range of the relative phase angle of $0° \leq \beta \leq 360°$.

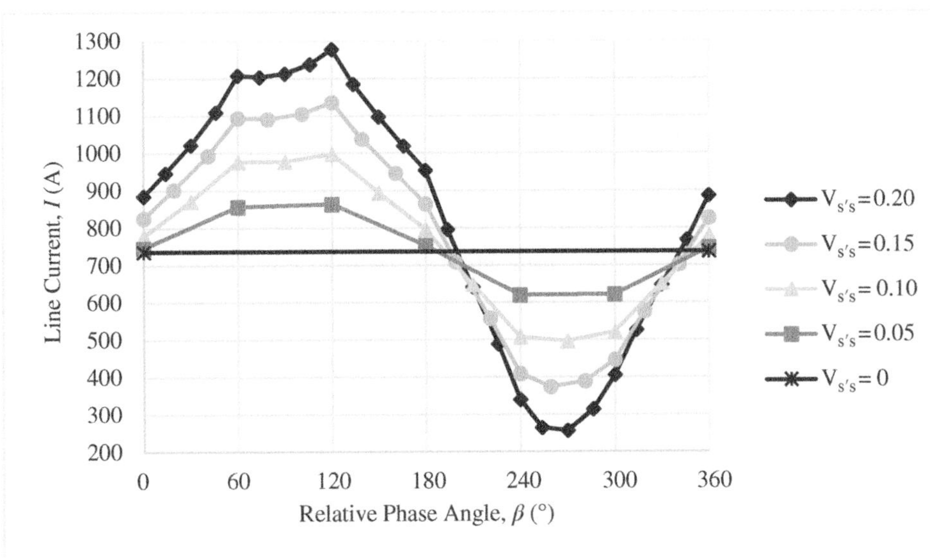

Figure 6-44 Line current (I) in A_{rms} during the entire range of the relative phase angle of $0° \leq \beta \leq 360°$.

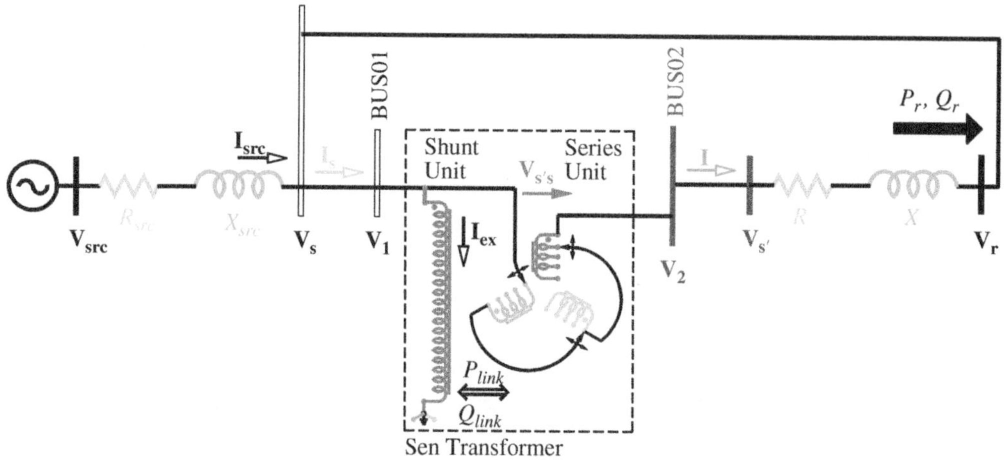

Figure 6-45 Factory test circuit of an ST.

The relationships between the source current ($\mathbf{I_{src}}$), sending-end current ($\mathbf{I_s}$), exciter current ($\mathbf{I_{ex}}$), and line current ($\mathbf{I}$) are as follows:

$$\mathbf{I_s} = \mathbf{I} + \mathbf{I_{ex}} \tag{6-33a}$$

$$\mathbf{I_s} = \mathbf{I} + \mathbf{I_{src}}. \tag{6-33b}$$

Therefore,

$$\mathbf{I_{src}} = \mathbf{I_{ex}}. \tag{6-34}$$

Note that the source is only required to supply the exciter current. The electrical system data is given in Table 6-8. The corresponding EMTP datafile **606IDLST.DAT** is shown in Code 6-7. The difference between this file and **601IDLST.DAT** is that they have different branch and source files. The short between BUS01 and BUS02 is replaced with the ST. The uncompensated or natural active and reactive power flows (P_{rn} and Q_{rn}) at the receiving end ($\mathbf{V_r}$ bus) of the line are simulated to be 0 MW (0 pu) and 0 Mvar (0 pu), respectively. To study the effects of other operating points of an ST, the active turns-ratios are set according to Table 6-1 through Table 6-3, which show the active turns-ratios of the three secondary, series-compensating windings in the A, B, and C phases for the entire range of the relative phase angle of $0° \leq \beta \leq 360°$.

Table 6-8 Electrical system data.

Parameters	Values
Source voltage ($\mathbf{V_{src}}$) phase-to-neutral	1 pu (112,676.528 V_{peak})$\angle 0°$
Source resistance (R_{src})	0.0084 pu (1.0053 Ω)
Source reactance (X_{src})	0.0625 pu (19.73 mH)
Line resistance (R)	0.0253 pu (3.0159 Ω)
Line reactance (X)	0.1875 pu (59.19 mH)

Code 6-7 EMTP datafile for an ST integrated in a one-generator/one-line power system network (606IDLST.DAT).

```
BEGIN NEW DATA CASE
C File Name: 606IDLST.DAT
C IDEAL SEN TRANSFORMER WITHOUT ANY LEAKAGE IMPEDANCE
C 00000001111111111222222222233333333334444444444555555555566666666667777777778
C 34567890123456789012345678901234567890123456789012345678901234567890123456789
C ----dt<---tmax<---Xopt<---Copt<--Toler<        ><        >
16.666-6150.00-3
C 00000001111111111222222222233333333334444444444555555555566666666667777777778
C 34567890123456789012345678901234567890123456789012345678901234567890123456789
C -Iprnt<--Iplot<-Idoubl<-KssOut<-MaxOut<        ><        ><---Icat<-Nenerg
      20      10                                                   1
TACS HYBRID
99TWOPI    = 2.0 * PI
C BASE VOLTAGE, CURRENT, AND POWER
99Vbase    = 112676.528
99Ibase    = 946.662704
99Zbase    = Vbase / Ibase
99Pbase    = 1.5 * Vbase * Ibase
$INCLUDE 302MEAS1.SWT
91VSA
99isA      = VSA
99iA       = iApu * Ibase
99iexA     = isA - iA
$INCLUDE 303PTCT.SCL
$INCLUDE 304IPLL.PLL
$INCLUDE 305LINER.CMP
C EXCHANGED POWER IN PU BY THE SERIES UNIT WITH THE LINE
99P2sepu   = (v12Apu * iApu + v12Bpu * iBpu + v12Cpu * iCpu) *Vbase*Ibase / Pbase
99Q2sepu   = (v12Apu * iCpu - v12Cpu * iApu) * SQRT(3) * Vbase * Ibase / Pbase
C EXCHANGED MW AND MVAR BY THE SERIES UNIT WITH THE LINE
99P2seMW   = P2sepu * Pbase / 1e6
99Q2seMV   = Q2sepu * Pbase / 1e6
C EXCHANGED MVA WITH THE LINE PER EQUATION (2-182)
99S2seMV   = SQRT(P2seMW * P2seMW + Q2seMV * Q2seMV)
C RMS LINE CURRENT
99Irms     = ipu * Ibase / SQRT(2)
C TACS OUTPUT
33v1pu  v12pu v2pu  PrMW  QrMVARSrMVA P2seMWQ2seMVS2seMVIrms  isA   iA    iexA
BLANK RECORD ENDING TACS
C 00000001111111111222222222233333333334444444444555555555566666666667777777778
C 34567890123456789012345678901234567890123456789012345678901234567890123456789
C <-Bus1<-Bus2<-Bus3<-Bus4<----R<----L<----C                                  V
  VSRCA VSA                 1.0053 19.73
  VSRCB VSB   VSRCA VSA
  VSRCC VSC   VSRCA VSA
```

(Continued)

Code 6-7 (Continued)

```
  VSPA   VRA                     3.0159 59.19
  VSPB   VRB    VSPA   VRA
  VSPC   VRC    VSPA   VRA
  VRA    VSA                     0.0001
  VRB    VSB                     0.0001
  VRC    VSC                     0.0001
$INCLUDE ST4000P0.TRN
BLANK RECORD ENDING BRANCHES
C 00000001111111111222222222233333333334444444444555555555566666666667777777778
C 34567890123456789012345678901234567890123456789012345678901234567890123456789 0
C <-Bus1<-Bus2<---Tclose<----Topen<-------Ie<----Flash<--Request<-----Target<--O
$INCLUDE 307MEAS2.SWT
  VSA    BUS01A                                                   MEASURING
  VSB    BUS01B                                                   MEASURING
  VSC    BUS01C                                                   MEASURING
BLANK RECORD ENDING SWITCHES
C 00000001111111111222222222233333333334444444444555555555566666666667777777778
C 34567890123456789012345678901234567890123456789012345678901234567890123456789 0
C <--Bus<I<-----Ampl<-----Freq<----Phase<-------A1<------T1><---Tstart<----Tstop
14VSRCA    112676.528     60.00       0.00                              -1.0
14VSRCB    112676.528     60.00    -120.00                              -1.0
14VSRCC    112676.528     60.00     120.00                              -1.0
BLANK RECORD ENDING SOURCES
BLANK RECORD ENDING NODE VOLTAGE OUTPUT
BLANK RECORD ENDING PLOT
```

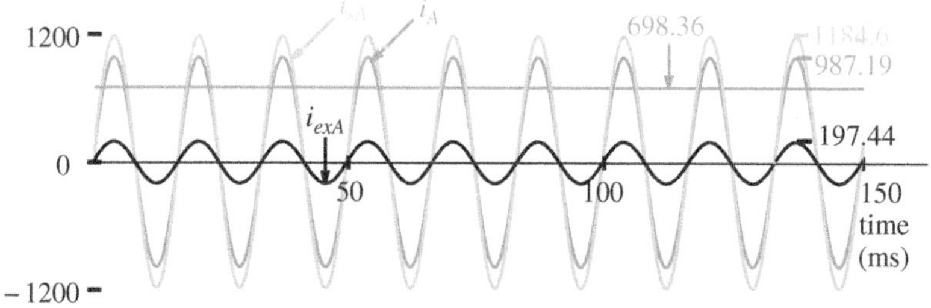

Figure 6-46 The instantaneous *A*-phase, sending-end current (i_{sA}), line current (i_A), and exciting current (i_{exA}).

Figure 6-46 shows the results from running Code 6-7 for a series-compensating voltage ($\mathbf{V}_{s's}$) of 0.2 pu at the relative phase angle, $\beta = 0°$. The rms line current (I_{rms}) is simulated to be 698.36 A_{rms}, shown in Table 6-9, which is equivalent to 987 A_{peak}. Therefore, the primary exciting current ($I_{ex\text{-}rms}$) is $0.2 \times I_{rms} = 0.2 \times 698.36 = 139.67$ A_{rms}, which is equivalent to 197 A_{peak} and is the same as the source current (I_{src}) per Equation (6-34). Since the line current, which is also the secondary compensating winding current, is in phase with the primary exciting current, the sending-end current is the sum of the line current and the exciting current, *i.e.*, $I_{s-rms} = I_{rms} + I_{ex-rms} = 698.36 + 139.67 = 838$ A_{rms}, which is equivalent to 1185 A_{peak}.

Table 6-9 Compensating voltage, modified sending-end voltage, power flows at the receiving end, exchanged powers by the Series Unit and the line current for electrical system data, shown in Table 6-8, and the active turns-ratios of the secondary, compensating windings in the *A*, *B*, and *C* phases per respective Tables 6-1, 6-2, and 6-3.

Case for natural power flows (center point of the hexagon characteristic)												
Theoretical $V_{s's}$ (pu)	β (deg)	Theoretical $V_{s'}$ (pu)	Simulated V_s (pu)	Simulated $V_{s's}$ (pu)	Simulated $V_{s'}$ (pu)	Simulated P_r (MW)	Simulated Q_r (Mvar)	Simulated S_r (MVA)	Simulated P_{se} (MW)	Simulated Q_{se} (Mvar)	Simulated S_{se} (MVA)	Simulated I (A)
0.00	0.00	1.00	1.0000	0.0000	1.0000	0.0000	0.0000	0.0000	0.0000	0.0000	0.0000	0.0000
0.00	360.00	1.00	1.0000	0.0000	1.0000	0.0000	0.0000	0.0000	0.0000	0.0000	0.0000	0.0000
Case for 0.05 pu compensating voltage (smallest hexagon)												
Theoretical $V_{s's}$ (pu)	β (deg)	Theoretical $V_{s'}$ (pu)	Simulated V_s (pu)	Simulated $V_{s's}$ (pu)	Simulated $V_{s'}$ (pu)	Simulated P_r (MW)	Simulated Q_r (Mvar)	Simulated S_r (MVA)	Simulated P_{se} (MW)	Simulated Q_{se} (Mvar)	Simulated S_{se} (MVA)	Simulated I (A)
0.05	0.00	1.05	0.9992	0.0500	1.0491	5.6545	41.8362	42.2166	−0.2827	−2.0918	2.1108	176.7689
0.05	60.00	1.03	0.9992	0.0500	1.0251	39.0585	16.0212	42.2167	−0.2827	−2.0918	2.1108	176.7689
0.05	120.00	0.98	0.9992	0.0500	0.9751	33.4041	−25.8151	42.2167	−0.2827	−2.0918	2.1108	176.7689
0.05	180.00	0.95	0.9992	0.0500	0.9492	−5.6545	−41.8363	42.2167	−0.2827	−2.0918	2.1108	176.7689
0.05	240.00	0.98	0.9992	0.0500	0.9751	−39.0585	−16.0212	42.2167	−0.2827	−2.0918	2.1108	176.7689
0.05	300.00	1.03	0.9992	0.0500	1.0251	−33.4040	25.8150	42.2166	−0.2827	−2.0918	2.1108	176.7689
0.05	360.00	1.05	0.9992	0.0500	1.0491	5.6545	41.8362	42.2166	−0.2827	−2.0918	2.1108	176.7689

(*Continued*)

Table 6-9 (Continued)

Case for 0.10 pu compensating voltage (second smallest hexagon)												
Theoretical $V_{s's}$ (pu)	β (deg)	Theoretical $V_{s'}$ (pu)	Simulated V_s (pu)	Simulated $V_{s's}$ (pu)	Simulated $V_{s'}$ (pu)	Simulated P_r (MW)	Simulated Q_r (Mvar)	Simulated S_r (MVA)	Simulated P_{se} (MW)	Simulated Q_{se} (Mvar)	Simulated S_{se} (MVA)	Simulated I (A)
0.10	0.00	1.10	0.9967	0.0997	1.0963	11.2528	83.2568	84.0139	−1.1253	−8.3256	8.4013	352.6605
0.09	30.00	1.08	0.9975	0.0864	1.0732	44.5645	57.6661	72.8792	−0.8454	−6.2546	6.3115	305.6668
0.10	60.00	1.05	0.9967	0.0997	1.0501	77.7289	31.8833	84.0139	−1.1253	−8.3256	8.4013	352.6605
0.09	90.00	1.00	0.9975	0.0864	1.0012	72.2224	−9.7618	72.8792	−0.8454	−6.2546	6.3115	305.6668
0.10	120.00	0.95	0.9967	0.0997	0.9508	66.4762	−51.3737	84.0140	−1.1253	−8.3256	8.4013	352.6605
0.09	150.00	0.93	0.9975	0.0864	0.9237	27.6580	−67.4270	72.8792	−0.8454	−6.2546	6.3115	305.6668
0.10	180.00	0.90	0.9967	0.0997	0.8970	−11.2528	−83.2569	84.0139	−1.1253	−8.3256	8.4013	352.6605
0.09	210.00	0.93	0.9975	0.0864	0.9237	−44.5652	−57.6656	72.8792	−0.8454	−6.2546	6.3115	305.6668
0.10	240.00	0.95	0.9967	0.0997	0.9508	−77.7290	−31.8833	84.0139	−1.1253	−8.3256	8.4013	352.6605
0.09	270.00	1.00	0.9975	0.0864	1.0012	−72.2225	9.7610	72.8791	−0.8454	−6.2546	6.3115	305.6668
0.10	300.00	1.05	0.9967	0.0997	1.0501	−66.4762	51.3736	84.0138	−1.1253	−8.3256	8.4013	352.6605
0.09	330.00	1.08	0.9975	0.0864	1.0732	−27.6572	67.4273	72.8791	−0.8454	−6.2546	6.3115	305.6668
0.10	360.00	1.10	0.9967	0.0997	1.0963	11.2528	83.2568	84.0139	−1.1253	−8.3256	8.4013	352.6605

Case for 0.15 pu compensating voltage (second largest hexagon)												
Theoretical $V_{s's}$ (pu)	β (deg)	Theoretical $V_{s'}$ (pu)	Simulated V_s (pu)	Simulated $V_{s's}$ (pu)	Simulated $V_{s'}$ (pu)	Simulated P_r (MW)	Simulated Q_r (Mvar)	Simulated S_r (MVA)	Simulated P_{se} (MW)	Simulated Q_{se} (Mvar)	Simulated S_{se} (MVA)	Simulated I (A)
0.15	0.00	1.15	0.9926	0.1489	1.1414	16.7399	123.8548	124.9810	−2.5110	−18.5781	18.7470	526.8048
0.13	19.11	1.13	0.9942	0.1315	1.1193	49.8681	98.7067	110.5886	−1.9595	−14.4976	14.6294	465.3685
0.13	40.89	1.10	0.9942	0.1315	1.0970	82.9415	73.1472	110.5884	−1.9595	−14.4975	14.6294	465.3678
0.15	60.00	1.08	0.9926	0.1489	1.0748	115.6314	47.4304	124.9811	−2.5110	−18.5781	18.7470	526.8048
0.13	79.11	1.03	0.9942	0.1315	1.0272	110.4164	6.1655	110.5885	−1.9595	−14.4976	14.6294	465.3678
0.13	100.89	0.98	0.9942	0.1315	0.9779	104.8179	−35.2567	110.5887	−1.9595	−14.4976	14.6294	465.3685
0.15	120.00	0.93	0.9926	0.1489	0.9271	98.8917	−76.4247	124.9812	−2.5110	−18.5781	18.7470	526.8048
0.13	139.11	0.90	0.9942	0.1315	0.8989	60.5484	−92.5404	110.5887	−1.9595	−14.4976	14.6294	465.3685
0.13	160.89	0.88	0.9942	0.1315	0.8710	21.8765	−108.4031	110.5885	−1.9595	−14.4975	14.6294	465.3678
0.15	180.00	0.85	0.9926	0.1489	0.8437	−16.7398	−123.8549	124.9811	−2.5110	−18.5781	18.7470	526.8048
0.13	199.11	0.88	0.9942	0.1315	0.8710	−49.8688	−98.7062	110.5884	−1.9595	−14.4975	14.6294	465.3678
0.13	220.89	0.90	0.9942	0.1315	0.8989	−82.9422	−73.1467	110.5886	−1.9595	−14.4976	14.6294	465.3685
0.15	240.00	0.93	0.9926	0.1489	0.9271	−115.6315	−47.4303	124.9810	−2.5110	−18.5781	18.7470	526.8048
0.13	259.11	0.98	0.9942	0.1315	0.9779	−110.4165	−6.1663	110.5885	−1.9595	−14.4976	14.6294	465.3685
0.13	280.89	1.03	0.9942	0.1315	1.0272	−104.8180	35.2558	110.5883	−1.9595	−14.4975	14.6294	465.3678
0.15	300.00	1.08	0.9926	0.1489	1.0748	−98.8915	76.4245	124.9809	−2.5110	−18.5781	18.7470	526.8048

(Continued)

Table 6-9 (Continued)

Case for 0.15 pu compensating voltage (second largest hexagon)												
Theoretical $V_{s's}$ (pu)	β (deg)	Theoretical $V_{s'}$ (pu)	Simulated V_s (pu)	Simulated $V_{s's}$ (pu)	Simulated $V_{s'}$ (pu)	Simulated P_r (MW)	Simulated Q_r (Mvar)	Simulated S_r (MVA)	Simulated P_{se} (MW)	Simulated Q_{se} (Mvar)	Simulated S_{se} (MVA)	Simulated I (A)
0.13	319.11	1.10	0.9942	0.1315	1.0970	−60.5476	92.5406	110.5884	−1.9595	−14.4975	14.6294	465.3678
0.13	340.89	1.13	0.9942	0.1315	1.1193	−21.8757	108.4033	110.5886	−1.9595	−14.4976	14.6294	465.3685
0.15	360.00	1.15	0.9926	0.1489	1.1414	16.7399	123.8548	124.9810	−2.5110	−18.5781	18.7470	526.8048
Case for 0.20 pu compensating voltage (largest hexagon)												
Theoretical $V_{s's}$ (pu)	β (deg)	Theoretical $V_{s'}$ (pu)	Simulated V_s (pu)	Simulated $V_{s's}$ (pu)	Simulated $V_{s'}$ (pu)	Simulated P_r (MW)	Simulated Q_r (Mvar)	Simulated S_r (MVA)	Simulated P_{se} (MW)	Simulated Q_{se} (Mvar)	Simulated S_{se} (MVA)	Simulated I (A)
0.20	0.00	1.20	0.9868	0.1974	1.1842	22.0636	163.2443	164.7286	−4.4128	−32.6487	32.9456	698.3642
0.18	13.90	1.18	0.9893	0.1783	1.1632	54.9193	138.7460	149.2200	−3.6032	−26.6585	26.9009	631.0545
0.17	30.00	1.15	0.9901	0.1715	1.1418	87.8110	113.6260	143.6024	−3.3315	−24.6484	24.8725	606.7971
0.18	46.10	1.13	0.9893	0.1783	1.1203	120.4133	88.1315	149.2198	−3.6032	−26.6584	26.9008	631.0536
0.20	60.00	1.11	0.9868	0.1974	1.0989	152.4056	62.5146	164.7288	−4.4128	−32.6487	32.9456	698.3642
0.18	73.90	1.06	0.9893	0.1783	1.0528	147.6171	21.8108	149.2198	−3.6032	−26.6584	26.9008	631.0536
0.17	90.00	1.01	0.9901	0.1715	1.0048	142.3084	−19.2343	143.6025	−3.3315	−24.6484	24.8726	606.7971
0.18	106.10	0.97	0.9893	0.1783	0.9553	136.5306	−60.2160	149.2201	−3.6032	−26.6585	26.9009	631.0545
0.20	120.00	0.92	0.9868	0.1974	0.9045	130.3421	−100.7300	164.7289	−4.4128	−32.6487	32.9456	698.3642
0.18	133.90	0.88	0.9893	0.1783	0.8751	92.6980	−116.9345	149.2200	−3.6032	−26.6585	26.9009	631.0545

Table 6-9 (Continued)

Case for 0.20 pu compensating voltage (largest hexagon)												
Theoretical $V_{s's}$ (pu)	β (deg)	Theoretical $V_{s'}$ (pu)	Simulated V_s (pu)	Simulated $V_{s's}$ (pu)	Simulated $V_{s'}$ (pu)	Simulated P_r (MW)	Simulated Q_r (Mvar)	Simulated S_r (MVA)	Simulated P_{se} (MW)	Simulated Q_{se} (Mvar)	Simulated S_{se} (MVA)	Simulated I (A)
0.17	150.00	0.85	0.9901	0.1715	0.8459	54.4975	−132.8596	143.6024	−3.3315	−24.6484	24.8725	606.7971
0.18	166.10	0.83	0.9893	0.1783	0.8173	16.1174	−148.3468	149.2198	−3.6032	−26.6584	26.9008	631.0536
0.20	180.00	0.80	0.9868	0.1974	0.7895	−22.0636	−148.3468	164.7286	−4.4128	−32.6487	32.9456	698.3642
0.18	193.90	0.83	0.9893	0.1783	0.8173	−54.9199	−138.7455	149.2197	−3.6032	−26.6584	26.9008	631.0536
0.17	210.00	0.85	0.9901	0.1715	0.8459	−87.8116	−113.6255	143.6023	−3.3315	−24.6484	24.8725	606.7971
0.18	226.10	0.88	0.9893	0.1783	0.8751	−120.4140	−88.1310	149.2198	−3.6032	−26.6585	26.9009	631.0545
0.20	240.00	0.92	0.9868	0.1974	0.9045	−152.4057	−62.5146	164.7286	−4.4128	−32.6487	32.9456	698.3642
0.18	253.90	0.97	0.9893	0.1783	0.9553	−147.6172	−21.8115	149.2198	−3.6032	−26.6585	26.9009	631.0545
0.17	270.00	1.01	0.9901	0.1715	1.0048	−142.3085	19.2335	143.6022	−3.3315	−24.6484	24.8725	606.7971
0.18	286.10	1.06	0.9893	0.1783	1.0528	−136.5307	60.2151	149.2196	−3.6032	−26.6584	26.9008	631.0536
0.20	300.00	1.11	0.9868	0.1974	1.0989	−130.3419	100.7298	164.7284	−4.4128	−32.6487	32.9455	698.3642
0.18	313.90	1.13	0.9893	0.1783	1.1203	−92.6971	116.9347	149.2195	−3.6032	−26.6584	26.9008	631.0536
0.17	330.00	1.15	0.9901	0.1715	1.1418	−54.4967	132.8599	143.6023	−3.3315	−24.6484	24.8725	606.7971
0.18	346.10	1.18	0.9893	0.1783	1.1632	−16.1166	148.3470	149.2199	−3.6032	−26.6585	26.9009	631.0545
0.20	360.00	1.20	0.9868	0.1974	1.1842	22.0636	163.2443	164.7286	−4.4128	−32.6487	32.9456	698.3642

Table 6-9 shows the variations of the theoretical magnitude ($V_{s's}$) of the compensating voltage ($\mathbf{V_{s's}}$) and the theoretical magnitude ($V_{s'}$) of the modified sending-end voltage ($\mathbf{V_{s'}}$) as a function of the relative phase angle (β), considering the sending-end voltage (V_s) is 1 pu. The table also shows corresponding values of the simulated magnitude ($V_{s's}$) of the compensating voltage ($\mathbf{V_{s's}}$), the simulated magnitude (V_s) of the sending-end voltage ($\mathbf{V_s}$), and the simulated magnitude ($V_{s'}$) of the modified sending-end voltage ($\mathbf{V_{s'}}$). The resulting active power (P_r), reactive power (Q_r), and apparent power (S_r) flows at the receiving end, exchanged active power (P_{se}), reactive power (Q_{se}), and apparent power (S_{se}) by the Compensating-Voltage Unit (Series Unit), and the prevailing line current (I) are also shown during the entire range of the relative phase angle of $0° \leq \beta \leq 360°$ in Table 6-9 for the k_{a1}-k_{b1}-k_{c1} active turns-ratios from Table 6-1, k_{b2}-k_{c2}-k_{a2} active turns-ratios from Table 6-2, and k_{c3}-k_{a3}-k_{b3} active turns-ratios from Table 6-3. The table is expanded for each hexagon, related to 5, 10, 15, and 20% compensating voltages.

Figure 6-47 shows the magnitude of the sending-end voltage (V_s), given in Table 6-9, during the entire range of the relative phase angle of $0° \leq \beta \leq 360°$.

The simulated magnitude ($V_{s's}$) of the compensating voltage, given in Table 6-9, during the entire range of the relative phase angle of $0° \leq \beta \leq 360°$, is shown in Figure 6-48.

The simulated magnitude ($V_{s'}$) of the modified sending-end voltage, given in Table 6-9, during the entire range of the relative phase angle of $0° \leq \beta \leq 360°$, is shown in Figure 6-49.

Figure 6-50 shows the active power (P_r), reactive power (Q_r), and apparent power (S_r) flows at the receiving end, given in Table 6-9, during the entire range of the relative phase angle of $0° \leq \beta \leq 360°$. Figure 6-51 shows the active power (P_r) versus reactive power (Q_r) flow at the receiving end, given in Table 6-9, during the entire range of the relative phase angle of $0° \leq \beta \leq 360°$. Note that the center of the hexagon, located at (0, 0), confirms that the initial power flow in the test network is zero due

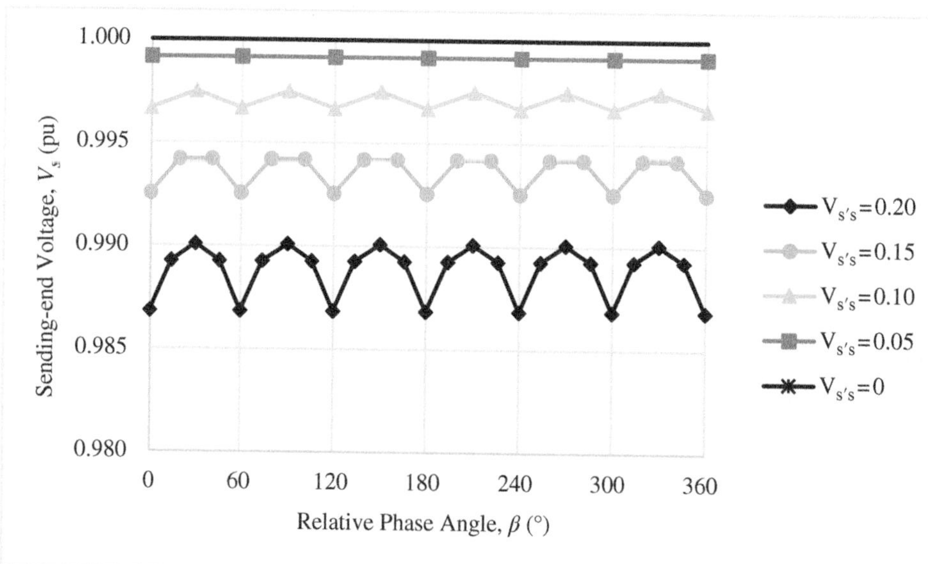

Figure 6-47 Magnitude of the simulated sending-end voltage ($\mathbf{V_s}$) in pu during the entire range of the relative phase angle of $0° \leq \beta \leq 360°$.

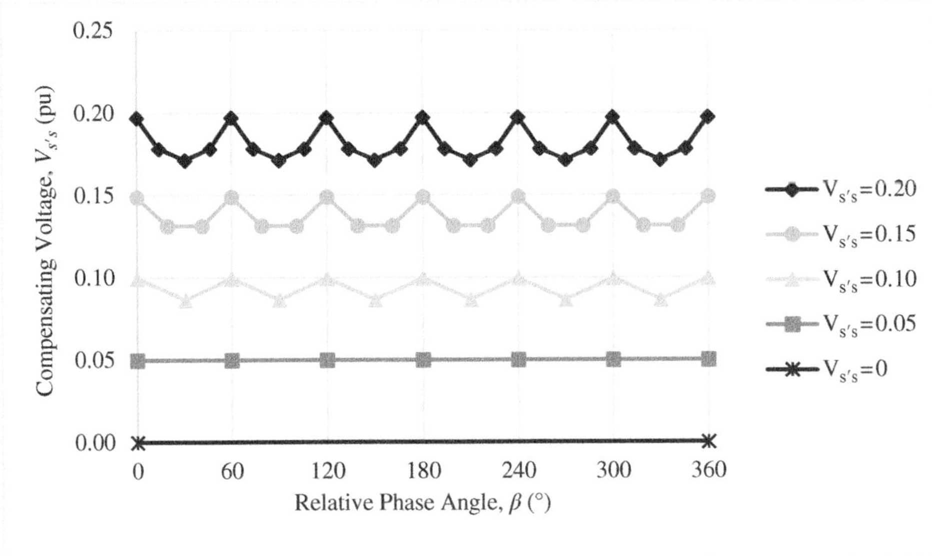

Figure 6-48 Magnitude of the simulated compensating voltage ($\mathbf{V_{s's}}$) in pu during the entire range of the relative phase angle of 0° $\leq \beta \leq$ 360°.

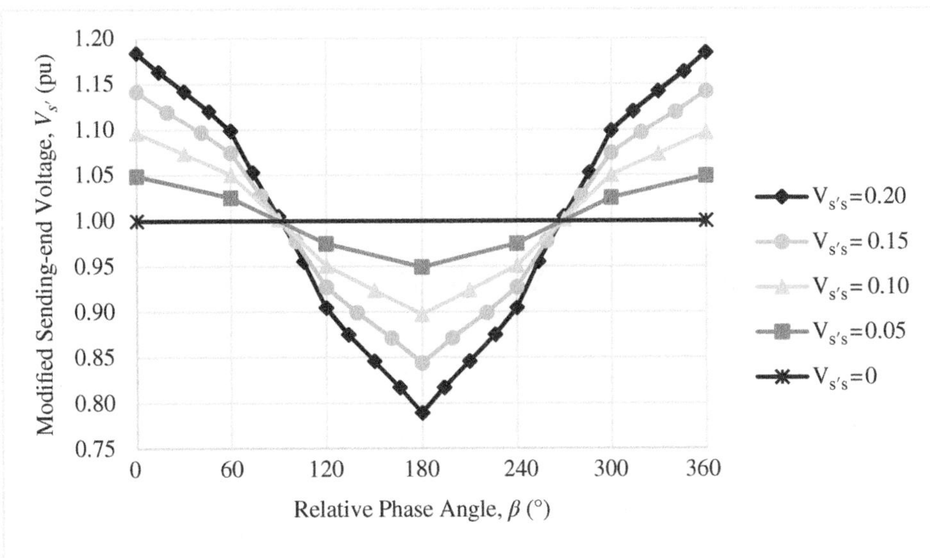

Figure 6-49 Magnitude of the simulated modified sending-end voltage ($\mathbf{V_{s'}}$) in pu during the entire range of the relative phase angle of 0° $\leq \beta \leq$ 360°.

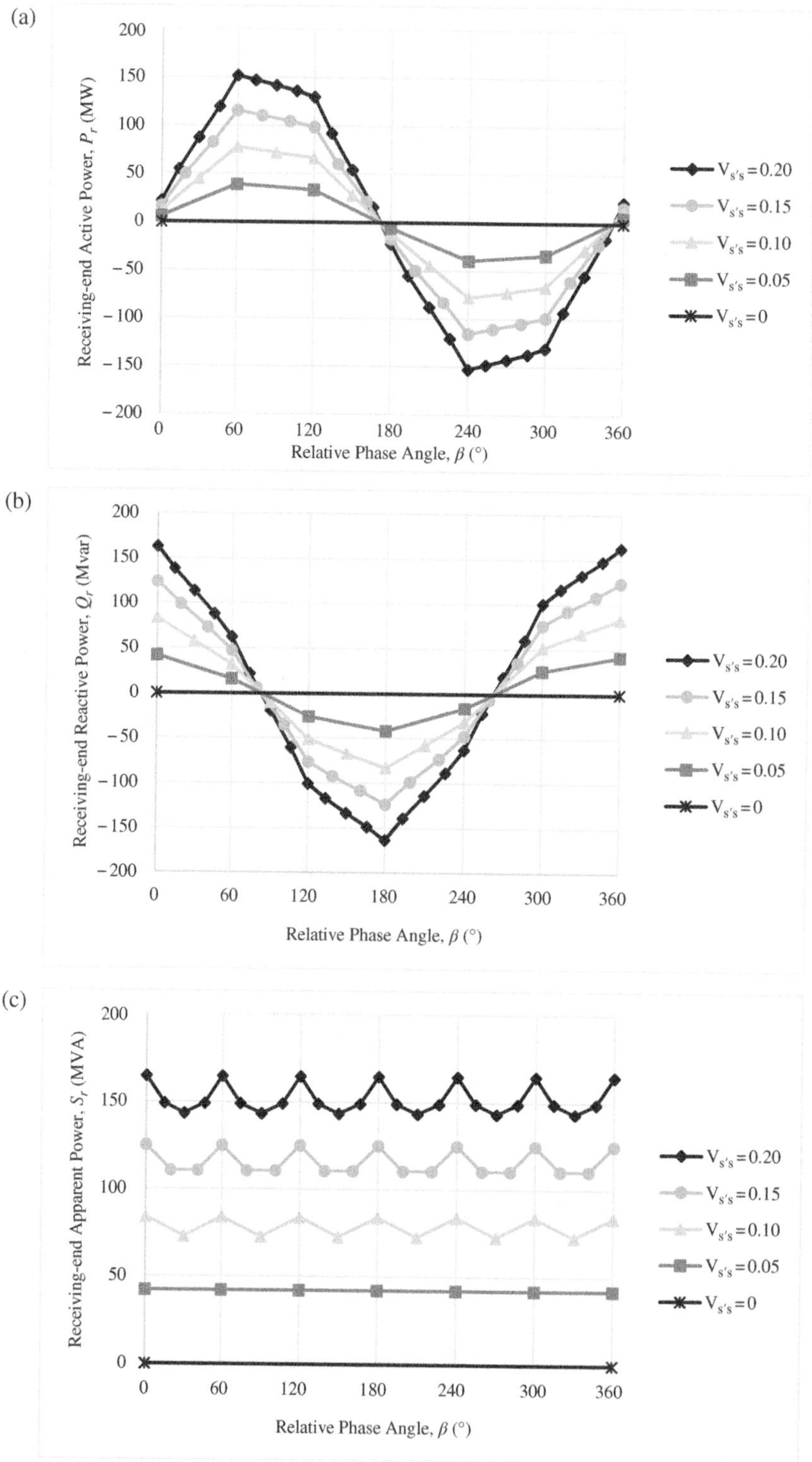

Figure 6-50 (a) Active power (P_r) in MW, (b) reactive power (Q_r) in Mvar, and (c) apparent power (S_r) in MVA flows at the receiving end during the entire range of the relative phase angle of $0° \leq \beta \leq 360°$.

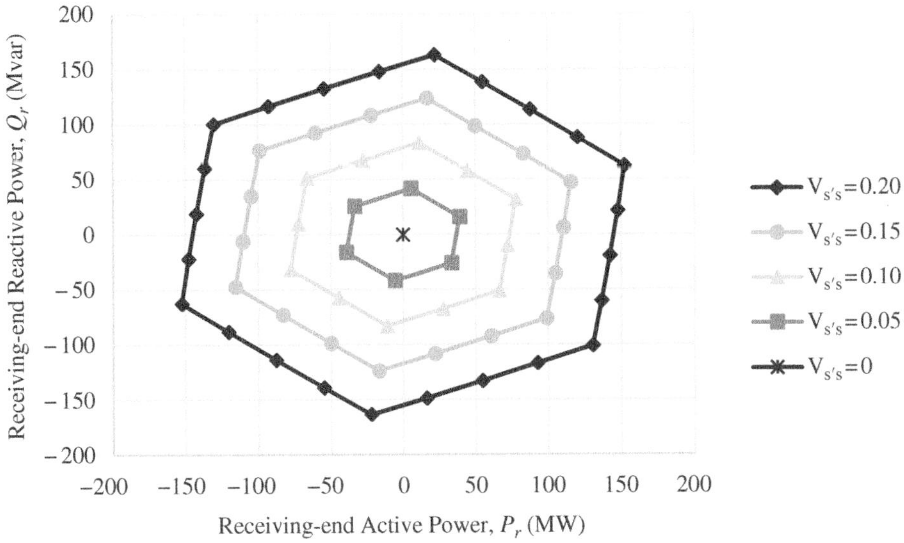

Figure 6-51 Active power (P_r) in MW versus reactive power (Q_r) in Mvar flows at the receiving end during the entire range of the relative phase angle of $0° \leq \beta \leq 360°$.

to having the same voltages at both ends of the test network's line reactor. The power flow in the test network occurs, since the ST modifies the voltage at the modified sending-end of the test network's line reactor while the receiving end is shorted with the sending end of the test network as shown in Figure 6-45.

Figure 6-52 shows the exchanged active power (P_{se}), reactive power (Q_{se}), and apparent power (S_{se}) by the Compensating-Voltage Unit (Series Unit), given in Table 6-9, during the entire range of the relative phase angle of $0° \leq \beta \leq 360°$.

Figure 6-53 shows the exchanged active power (P_{se}) versus reactive power (Q_{se}) by the Compensating-Voltage Unit (Series Unit), given in Table 6-9, during the entire range of the relative phase angle of $0° \leq \beta \leq 360°$.

Figure 6-54 shows the prevailing line current (I), given in Table 6-9, during the entire range of the relative phase angle of $0° \leq \beta \leq 360°$.

The summary of the study results for a 20% compensating voltage is given below.

- The line voltage at the POC is fixed at one pu.
- Active power flow at the receiving end changes from the natural flow of 0 MW by 152 and −152 MW.
- Reactive power flow at the receiving end changes from the natural flow of 0 Mvar by 163 and −148 Mvar.
- Rating of the ST: 33 MVA.

Note that a 33 MVA-rated ST regulates 165 MVA in the line as shown in Figures 6-52c and 6-50c, respectively.

Many a times, the source impedance (R_{src} and X_{src}) between the infinite bus on the source side and the POC where the ST is tested is not known. A realistic source impedance of 6.25%, shown in

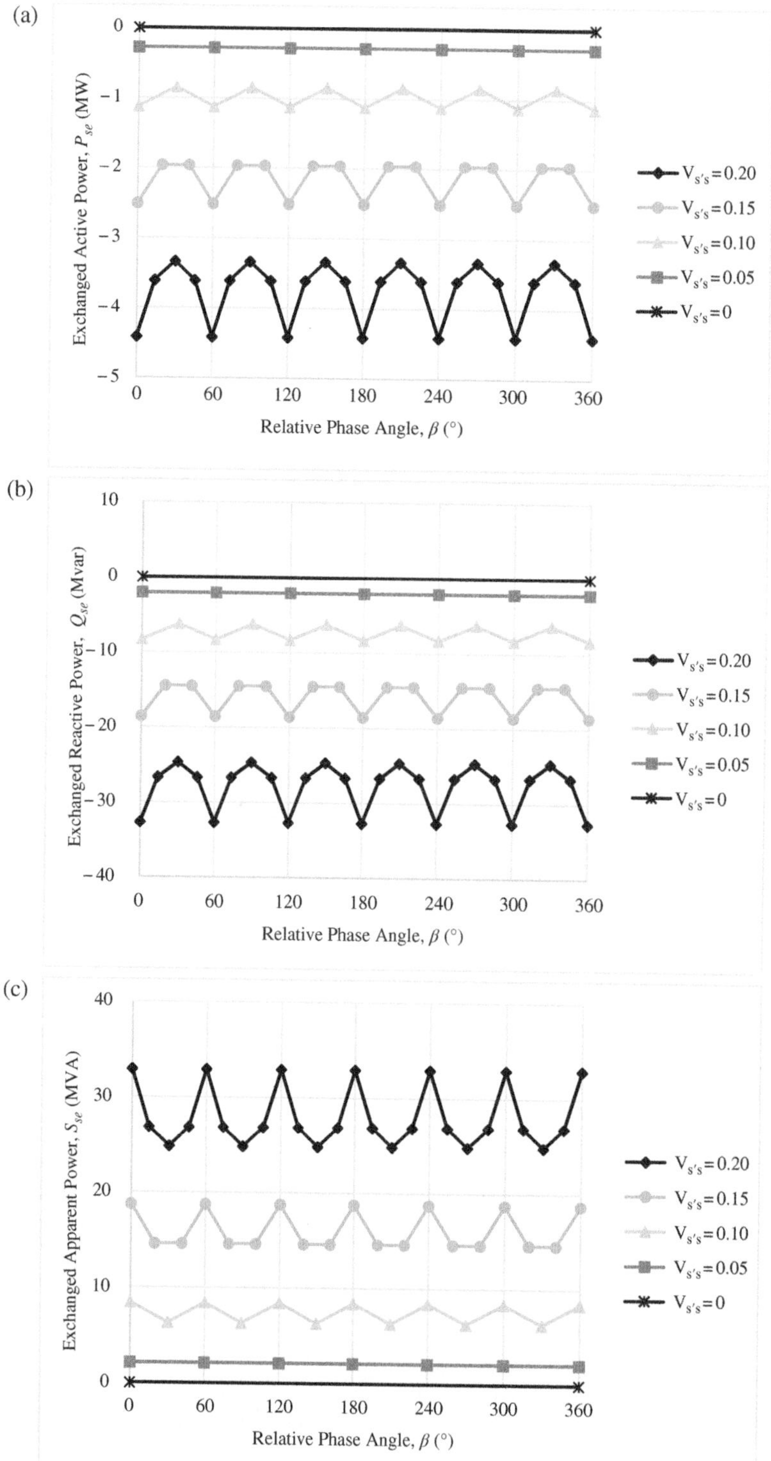

Figure 6-52 (a) Exchanged active power (P_{se}) in MW, (b) reactive power (Q_{se}) in Mvar, and (c) apparent power (S_{se}) in MVA by the Series Unit during the entire range of the relative phase angle of $0° \leq \beta \leq 360°$.

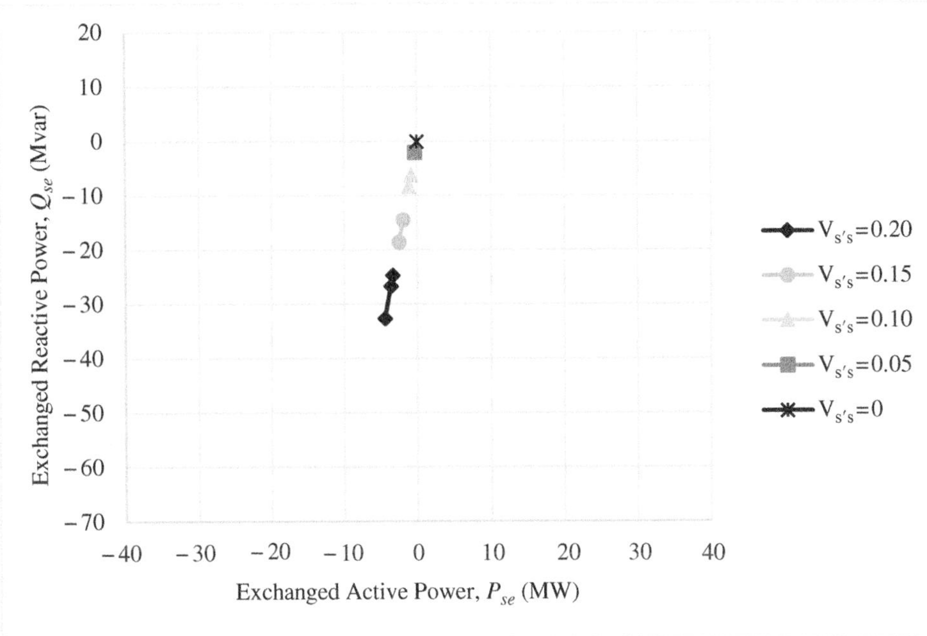

Figure 6-53 Exchanged active power (P_{se}) in MW versus reactive power (Q_{se}) in Mvar by the Series Unit during the entire range of the relative phase angle of $0° \leq \beta \leq 360°$.

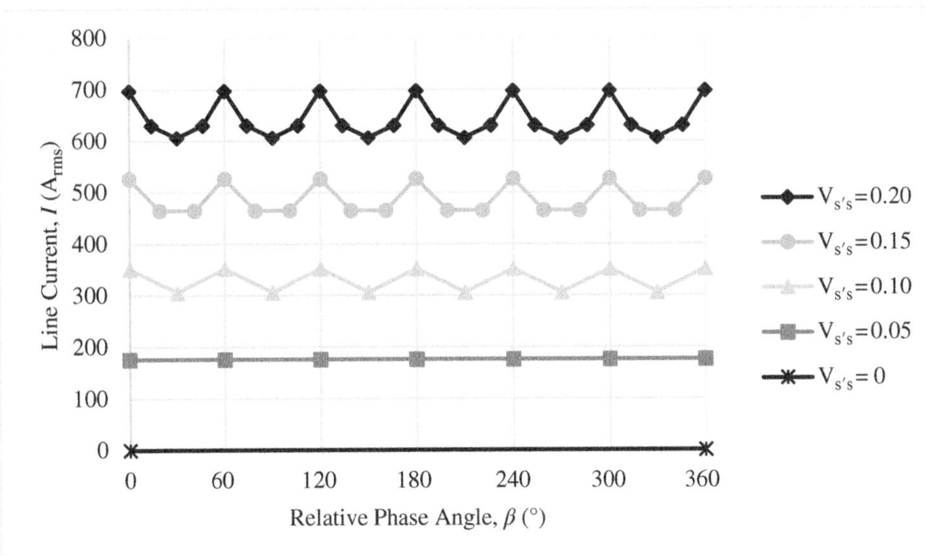

Figure 6-54 Line current (I) in A_{rms} during the entire range of the relative phase angle of $0° \leq \beta \leq 360°$.

Figure 6-45, considering the base impedance (Z_{base}) is 119.025 Ω, shows in Figure 6-47 that the variation of voltage at the POC is ~1%. Neglecting the source impedance and, therefore, assuming the voltage at the POC to be one pu, the power flow test of the ST is rerun. The electrical system data is given in Table 6-10. Figure 6-45 can be modified to model the test circuit as shown in Figure 6-55.

Note that the source is only required to supply the exciter current as defined by

$$\mathbf{I}_{\mathbf{src}} = \mathbf{I}_{\mathbf{ex}}. \tag{6-34}$$

The corresponding EMTP datafile **607IDLST.DAT** is shown in Code 6-8. The difference between this file and **606IDLST.DAT** is that they have a different branch file. The short between BUS01 and BUS02 is replaced with the ST. The uncompensated or natural active and reactive power flows (P_{rn} and Q_{rn}) at the receiving end of the line are simulated to be 0 MW (0 pu) and 0 Mvar (0 pu), respectively.

Note that the voltage at the right terminal of the line impedance is fixed at the line voltage ($\mathbf{V_r}$), which is same as the line voltage ($\mathbf{V_s}$), *i.e.* 138 kV in this example. At minimum turns-ratio (*i.e.* 0%), the voltages at both sides of the line impedance is the same and, therefore, the line current and power is zero. However, at maximum turns-ratio (*i.e.* 20%), the voltage at the left terminal of the line impedance can be as much as 20% of 138 kV/ $\sqrt{3}$ = 15.934 kV more than the voltage

Table 6-10 Electrical system data.

Parameters	Values
Source voltage ($\mathbf{V_{src}}$) phase-to-neutral	1 pu (112,676.528 V_{peak})∠0°
Line resistance (*R*)	0.0253 pu (3.0159 Ω)
Line reactance (*X*)	0.1875 pu (59.19 mH)

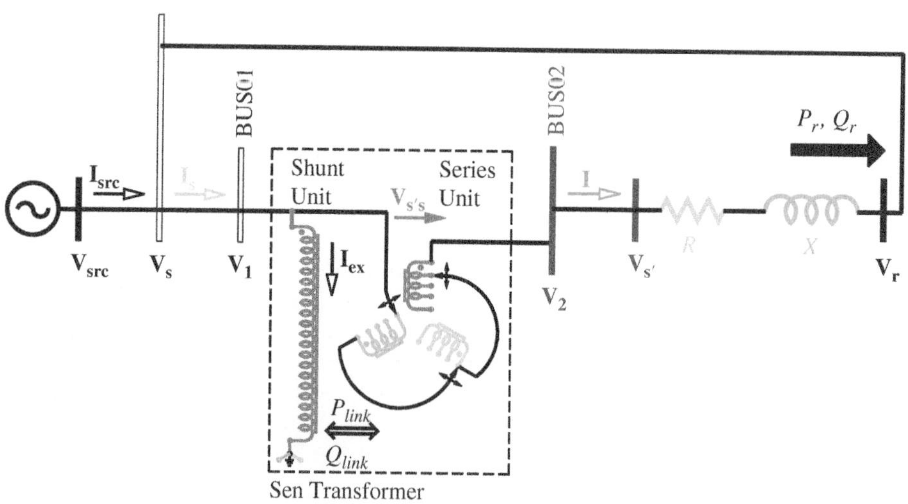

Figure 6-55 Modified factory test circuit of an ST.

at the right terminal, because of the series-compensating voltage of the ST that would create a current flow of 707.68 A through an impedance of 3.0159 Ω and 59.19 mH.

To study the effects of other operating points of an ST, the active turns-ratios are set according to Table 6-1 through Table 6-3, which show the active turns-ratios of the three secondary, series-compensating windings in the A, B, and C phases for the entire range of the relative phase angle of $0° \leq \beta \leq 360°$.

Figure 6-56 shows the results from running Code 6-8 for a series-compensating voltage ($\mathbf{V}_{s's}$) of 0.2 pu at the relative phase angle, $\beta = 180°$. The rms line current (I_{rms}) is simulated to be

Code 6-8 EMTP datafile for an ST integrated in a one-generator/one-line power system network (607IDLST.DAT).

```
BEGIN NEW DATA CASE
C File Name: 607IDLST.DAT
C IDEAL SEN TRANSFORMER WITHOUT ANY LEAKAGE IMPEDANCE
C 00000001111111111222222222233333333334444444444555555555566666666667777777778
C 34567890123456789012345678901234567890123456789012345678901234567890
C ----dt<---tmax<---Xopt<---Copt<--Toler<        ><        >
16.666-6150.00-3
C 00000001111111111222222222233333333334444444444555555555566666666667777777778
C 34567890123456789012345678901234567890123456789012345678901234567890
C -Iprnt<--Iplot<-Idoubl<-KssOut<-MaxOut<        ><        ><---Icat<-Nenerg
      20        10                                                1
TACS HYBRID
99TWOPI    = 2.0 * PI
C BASE VOLTAGE, CURRENT, AND POWER
99Vbase    = 112676.528
99Ibase    = 946.662704
99Zbase    = Vbase / Ibase
99Pbase    = 1.5 * Vbase * Ibase
$INCLUDE 302MEAS1.SWT
91VSA
99isA      = VSA
99iA       = iApu * Ibase
99IexA     = isA - iA
$INCLUDE 303PTCT.SCL
$INCLUDE 304IPLL.PLL
$INCLUDE 305LINER.CMP
C EXCHANGED POWER IN PU BY THE SERIES UNIT WITH THE LINE
99P2sepu   = (v12Apu * iApu + v12Bpu * iBpu + v12Cpu * iCpu) *Vbase*Ibase / Pbase
99Q2sepu   = (v12Apu * iCpu - v12Cpu * iApu) * SQRT(3) * Vbase * Ibase / Pbase
C EXCHANGED MW AND MVAR BY THE SERIES UNIT WITH THE LINE
99P2seMW   = P2sepu * Pbase / 1e6
```

(Continued)

Code 6-8 (Continued)

```
99Q2seMV   = Q2sepu * Pbase / 1e6
C EXCHANGED MVA WITH THE LINE PER EQUATION (2-182)
99S2seMV   = SQRT(P2seMW * P2seMW + Q2seMV * Q2seMV)
C RMS LINE CURRENT
99Irms     = ipu * Ibase / SQRT(2)
C TACS OUTPUT
33v1pu  v12pu v2pu  PrMW  QrMVARSrMVA P2seMWQ2seMVS2seMVIrms  isA   iA    iexA
BLANK RECORD ENDING TACS
C 0000000111111111122222222223333333333444444444455555555556666666666777777777778
C 3456789012345678901234567890123456789012345678901234567890123456789012345678890
C <-Bus1<-Bus2<-Bus3<-Bus4<----R<----L<----C                                    V
  VSRCA VSA               0.0001
  VSRCB VSB   VSRCA VSA
  VSRCC VSC   VSRCA VSA
  VSPA  VRA               3.0159 59.19
  VSPB  VRB   VSPA  VRA
  VSPC  VRC   VSPA  VRA
  VRA   VSA               0.0001
  VRB   VSB               0.0001
  VRC   VSC               0.0001
$INCLUDE ST4180P0.TRN
BLANK RECORD ENDING BRANCHES
C 0000000111111111122222222223333333333444444444455555555556666666666777777777778
C 3456789012345678901234567890123456789012345678901234567890123456789012345678890
C <-Bus1<-Bus2<---Tclose<----Topen<-------Ie<----Flash<--Request<-----Target<--O
$INCLUDE 307MEAS2.SWT
  VSA   BUS01A                                           MEASURING
  VSB   BUS01B                                           MEASURING
  VSC   BUS01C                                           MEASURING
BLANK RECORD ENDING SWITCHES
C 0000000111111111122222222223333333333444444444455555555556666666666777777777778
C 3456789012345678901234567890123456789012345678901234567890123456789012345678890
C <--Bus<I<-----Ampl<-----Freq<----Phase<-------A1<------T1><---Tstart<----Tstop
14VSRCA   112676.528     60.00      0.00                              -1.0
14VSRCB   112676.528     60.00   -120.00                              -1.0
14VSRCC   112676.528     60.00    120.00                              -1.0
BLANK RECORD ENDING SOURCES
BLANK RECORD ENDING NODE VOLTAGE OUTPUT
BLANK RECORD ENDING PLOT
```

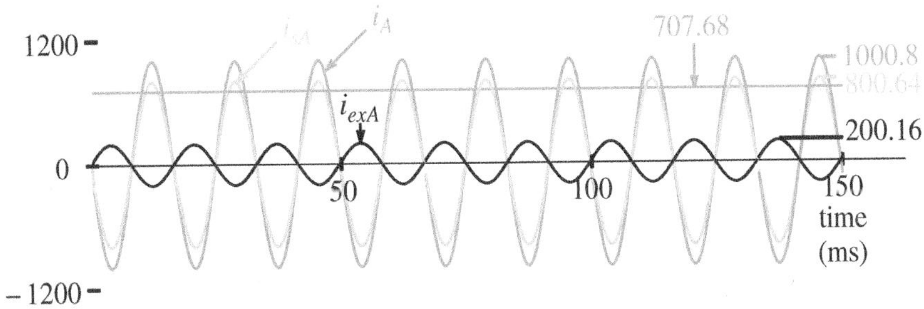

Figure 6-56 The instantaneous *A*-phase, sending-end current (i_{sA}), line current (i_A), and exciting current (i_{exA}).

707.68 A_{rms}, shown in Table 6-11, which is equivalent to 1000 A_{peak}. Therefore, the primary exciting current ($I_{ex\text{-}rms}$) is $0.2 \times I_{rms} = 0.2 \times 707.68 = 141.54$ A_{rms}, which is equivalent to 200 A_{peak} and is the same as the source current (I_{src}) per Equation (6-34). Since the line current, which is also the secondary compensating winding current, is in opposite phase with the primary exciting current, the sending-end current is the difference of the line current and the exciting current, *i.e.* $I_{S-rms} = I_{rms} - I_{ex-rms} = 707.68 - 141.54 = 566$ A_{rms}, which is equivalent to 800 A_{peak}. If the relative phase angle (β) is changed from 180° to 0°, the line current, which is also the secondary compensating winding current, will be in phase with the primary exciting current; the sending-end current will be the sum of the line current and the exciting current, which is 1200 A_{peak}. Therefore, the range of the sending-end current varies between 800 A_{peak} and 1200 A_{peak}, in the example, and is based on the phasor sum of the two currents (line current, i and exciting current, i_{ex}), which stems from the relative phase angle (β).

Table 6-11 shows the variations of the theoretical magnitude ($V_{s's}$) of the compensating voltage ($\mathbf{V_{s's}}$) and the theoretical magnitude ($V_{s'}$) of the modified sending-end voltage ($\mathbf{V_{s'}}$) as a function of the relative phase angle (β), considering the sending-end voltage, $V_s = 1$ pu. The table also shows the corresponding values of the simulated magnitude ($V_{s's}$) of the compensating voltage ($\mathbf{V_{s's}}$), the simulated magnitude (V_s) of the sending-end voltage ($\mathbf{V_s}$), and the simulated magnitude ($V_{s'}$) of the modified sending-end voltage ($\mathbf{V_{s'}}$). The resulting active power (P_r), reactive power (Q_r), and apparent power (S_r) flows at the receiving end, exchanged active power (P_{se}), reactive power (Q_{se}), and apparent power (S_{se}) by the Compensating-Voltage Unit (Series Unit), and the prevailing line current (I) are also shown during the entire range of the relative phase angle of $0° \leq \beta \leq 360°$ in Table 6-11 for the k_{a1}-k_{b1}-k_{c1} active turns-ratios from Table 6-1, k_{b2}-k_{c2}-k_{a2} active turns-ratios from Table 6-2, and k_{c3}-k_{a3}-k_{b3} active turns-ratios from Table 6-3. The table is expanded for each hexagon, related to 5, 10, 15, and 20% compensating voltages.

A comparison of the results in Tables 6-9 and 6-11 reveals that their difference is less than 1%, which may be acceptable. This difference is due to the fact that the simulation results in Table 6-9 included a source impedance (R_{src} and X_{src}) between the infinite bus on the source side and the POC where the ST is tested. However, the calculated results in Table 6-11 ignore the source impedance and the resulting ~1% voltage variation, shown in Figure 6-47, and assume that the voltage at the POC is to be 1 pu.

Figure 6-57 shows the magnitude of the sending-end voltage (V_s), given in Table 6-11, during the entire range of the relative phase angle of $0° \leq \beta \leq 360°$.

The simulated magnitude ($V_{s's}$) of the compensating voltage, given in Table 6-11, during the entire range of the relative phase angle of $0° \leq \beta \leq 360°$, is shown in Figure 6-58.

Table 6-11 Compensating voltage, modified sending-end voltage, power flows at the receiving end, exchanged powers by the Series Unit and the line current for electrical system data, shown in Table 6-10, and the active turns-ratios of the secondary, series compensating windings in the *A*, *B*, and *C* phases per respective Tables 6-1, 6-2, and 6-3.

Case for natural power flows (center point of the hexagon characteristic)												
Theoretical $V_{s's}$ (pu)	**β (deg)**	**Theoretical $V_{s'}$ (pu)**	**Simulated V_s (pu)**	**Simulated $V_{s's}$ (pu)**	**Simulated $V_{s'}$ (pu)**	**Simulated P_r (MW)**	**Simulated Q_r (Mvar)**	**Simulated S_r (MVA)**	**Simulated P_{se} (MW)**	**Simulated Q_{se} (Mvar)**	**Simulated S_{se} (MVA)**	**Simulated I (A)**
0.00	0.00	1.00	1.0000	0.0000	1.0000	0.0000	0.0000	0.0000	0.0000	0.0000	0.0000	0.0000
0.00	360.00	1.00	1.0000	0.0000	1.0000	0.0000	0.0000	0.0000	0.0000	0.0000	0.0000	0.0000
The table is expanded for each hexagon, related to 5, 10, 15, and 20% compensating voltages. Case for 0.05 pu compensating voltage (smallest hexagon)												
Theoretical $V_{s's}$ (pu)	**β (deg)**	**Theoretical $V_{s'}$ (pu)**	**Simulated V_s (pu)**	**Simulated $V_{s's}$ (pu)**	**Simulated $V_{s'}$ (pu)**	**Simulated P_r (MW)**	**Simulated Q_r (Mvar)**	**Simulated S_r (MVA)**	**Simulated P_{se} (MW)**	**Simulated Q_{se} (Mvar)**	**Simulated S_{se} (MVA)**	**Simulated I (A)**
0.05	0.00	1.05	1.0000	0.0500	1.0500	5.6643	41.9060	42.2870	−0.2832	−2.0953	2.1143	176.9161
0.05	60.00	1.03	1.0000	0.0500	1.0259	39.1238	16.0476	42.2870	−0.2832	−2.0953	2.1143	176.9161
0.05	120.00	0.98	1.0000	0.0500	0.9760	33.4595	−25.8583	42.2870	−0.2832	−2.0953	2.1143	176.9161
0.05	180.00	0.95	1.0000	0.0500	0.9500	−5.6642	−41.9060	42.2870	−0.2832	−2.0953	2.1143	176.9161
0.05	240.00	0.98	1.0000	0.0500	0.9760	−39.1237	−16.0476	42.2870	−0.2832	−2.0953	2.1143	176.9161
0.05	300.00	1.03	1.0000	0.0500	1.0259	−33.4595	25.8584	42.2870	−0.2832	−2.0953	2.1143	176.9161
0.05	360.00	1.05	1.0000	0.0500	1.0500	5.6643	41.9060	42.2870	−0.2832	−2.0953	2.1143	176.9161

Case for 0.10 pu compensating voltage (second smallest hexagon)												
Theoretical $V_{s's}$ (pu)	β (deg)	Theoretical $V_{s'}$ (pu)	Simulated V_s (pu)	Simulated $V_{s's}$ (pu)	Simulated $V_{s'}$ (pu)	Simulated P_r (MW)	Simulated Q_r (Mvar)	Simulated S_r (MVA)	Simulated P_{se} (MW)	Simulated Q_{se} (Mvar)	Simulated S_{se} (MVA)	Simulated I (A)
0.10	0.00	1.10	1.0000	0.1000	1.1000	11.3287	83.8127	84.5749	−1.1328	−8.3812	8.4574	353.8358
0.09	30.00	1.08	1.0000	0.0866	1.0759	44.7882	57.9544	73.2440	−0.8496	−6.2859	6.3431	306.4308
0.10	60.00	1.05	1.0000	0.1000	1.0536	78.2483	32.0955	84.5749	−1.1328	−8.3812	8.4574	353.8358
0.09	90.00	1.00	1.0000	0.0866	1.0037	72.5839	−9.8113	73.2440	−0.8496	−6.2859	6.3431	306.4308
0.10	120.00	0.95	1.0000	0.1000	0.9539	66.9197	−51.7172	84.5749	−1.1328	−8.3812	8.4574	353.8358
0.09	150.00	0.93	1.0000	0.0866	0.9260	27.7960	−67.7648	73.2440	−0.8496	−6.2859	6.3431	306.4308
0.10	180.00	0.90	1.0000	0.1000	0.9000	−11.3285	−83.8128	84.5749	−1.1328	−8.3812	8.4574	353.8359
0.09	210.00	0.93	1.0000	0.0866	0.9260	−44.7887	−57.9539	73.2440	−0.8496	−6.2859	6.3431	306.4308
0.10	240.00	0.95	1.0000	0.1000	0.9539	−78.2482	−32.0956	84.5749	−1.1328	−8.3812	8.4574	353.8358
0.09	270.00	1.00	1.0000	0.0866	1.0037	−72.5840	9.8104	73.2440	−0.8496	−6.2859	6.3431	306.4308
0.10	300.00	1.05	1.0000	0.1000	1.0536	−66.9196	51.7173	84.5749	−1.1328	−8.3812	8.4574	353.8358
0.09	330.00	1.08	1.0000	0.0866	1.0759	−27.7951	67.7652	73.2440	−0.8496	−6.2859	6.3431	306.4308
0.10	360.00	1.10	1.0000	0.1000	1.1000	11.3287	83.8127	84.5749	−1.1328	−8.3812	8.4574	353.8358

(Continued)

Table 6-11 (Continued)

Case for 0.15 pu compensating voltage (second largest hexagon)												
Theoretical $V_{s's}$ (pu)	β (deg)	Theoretical $V_{s'}$ (pu)	Simulated V_s (pu)	Simulated $V_{s's}$ (pu)	Simulated $V_{s'}$ (pu)	Simulated P_r (MW)	Simulated Q_r (Mvar)	Simulated S_r (MVA)	Simulated P_{se} (MW)	Simulated Q_{se} (Mvar)	Simulated S_{se} (MVA)	Simulated I (A)
0.15	0.00	1.15	1.0000	0.1500	1.1500	16.9932	125.7195	126.8628	−2.5489	−18.8578	19.0293	530.7554
0.13	19.11	1.13	1.0000	0.1323	1.1258	50.4526	99.8611	111.8826	−1.9825	−14.6672	14.8006	468.0827
0.13	40.89	1.10	1.0000	0.1323	1.1034	83.9127	74.0022	111.8824	−1.9825	−14.6672	14.8005	468.0821
0.15	60.00	1.08	1.0000	0.1500	1.0828	117.3729	48.1434	126.8628	−2.5489	−18.8578	19.0293	530.7554
0.13	79.11	1.03	1.0000	0.1323	1.0332	111.7085	6.2366	111.8825	−1.9825	−14.6672	14.8005	468.0821
0.13	100.89	0.98	1.0000	0.1323	0.9836	106.0442	−35.6701	111.8826	−1.9825	−14.6672	14.8006	468.0828
0.15	120.00	0.93	1.0000	0.1500	0.9341	100.3800	−77.5760	126.8629	−2.5489	−18.8578	19.0293	530.7555
0.13	139.11	0.90	1.0000	0.1323	0.9042	61.2563	−93.6237	111.8826	−1.9825	−14.6672	14.8006	468.0828
0.13	160.89	0.88	1.0000	0.1323	0.8761	22.1318	−109.6716	111.8824	−1.9825	−14.6672	14.8005	468.0822
0.15	180.00	0.85	1.0000	0.1500	0.8500	−16.9927	−125.7196	126.8628	−2.5489	−18.8578	19.0293	530.7556
0.13	199.11	0.88	1.0000	0.1323	0.8761	−50.4530	−99.8607	111.8824	−1.9825	−14.6672	14.8005	468.0822
0.13	220.89	0.90	1.0000	0.1323	0.9042	−83.9132	−74.0019	111.8825	−1.9825	−14.6672	14.8006	468.0828
0.15	240.00	0.93	1.0000	0.1500	0.9341	−117.3727	−48.1435	126.8627	−2.5489	−18.8578	19.0293	530.7555
0.13	259.11	0.98	1.0000	0.1323	0.9836	−111.7085	−6.2375	111.8825	−1.9825	−14.6672	14.8006	468.0828
0.13	280.89	1.03	1.0000	0.1323	1.0332	−106.0441	35.6693	111.8823	−1.9825	−14.6672	14.8006	468.0821
0.15	300.00	1.08	1.0000	0.1500	1.0828	−100.3797	77.5762	126.8627	−2.5489	−18.8578	19.0293	530.7554

Table 6-11 (Continued)

0.13	319.11	1.10	1.0000	0.1323	1.1034	−61.2552	93.6241	111.8823	−1.9825	−14.6672	14.8005	468.0821
0.13	340.89	1.13	1.0000	0.1323	1.1258	−22.1306	109.6719	111.8825	−1.9825	−14.6672	14.8006	468.0827
0.15	360.00	1.15	1.0000	0.1500	1.1500	16.9932	125.7195	126.8628	−2.5489	−18.8578	19.0293	530.7554
Case for 0.20 pu compensating voltage (largest hexagon)												
Theoretical $V_{s's}$ (pu)	**β (deg)**	**Theoretical $V_{s'}$ (pu)**	**Simulated V_s (pu)**	**Simulated $V_{s's}$ (pu)**	**Simulated $V_{s'}$ (pu)**	**Simulated P_r (MW)**	**Simulated Q_r (Mvar)**	**Simulated S_r (MVA)**	**Simulated P_{se} (MW)**	**Simulated Q_{se} (Mvar)**	**Simulated S_{se} (MVA)**	**Simulated I (A)**
0.20	0.00	1.20	1.0000	0.2000	1.2000	22.6578	167.6262	169.1506	−4.5315	−33.5251	33.8299	707.675
0.18	13.90	1.18	1.0000	0.1803	1.1758	56.1171	141.7679	152.4705	−3.6818	−27.2392	27.4869	637.8904
0.17	30.00	1.15	1.0000	0.1732	1.1533	89.5772	115.9090	146.4888	−3.3986	−25.1438	25.3725	612.8646
0.18	46.10	1.13	1.0000	0.1803	1.1325	123.0374	90.0501	152.4703	−3.6818	−27.2391	27.4868	637.8894
0.20	60.00	1.11	1.0000	0.2000	1.1136	156.4975	64.1912	169.1508	−4.5314	−33.5251	33.8299	707.6751
0.18	73.90	1.06	1.0000	0.1803	1.0642	150.8331	22.2845	152.4704	−3.6818	−27.2391	27.4868	637.8895
0.17	90.00	1.01	1.0000	0.1732	1.0149	145.1687	−19.6222	146.4889	−3.3986	−25.1438	25.3724	612.8647
0.18	106.10	0.97	1.0000	0.1803	0.9657	139.5044	−61.5290	152.4706	−3.6818	−27.2392	27.4869	637.8905
0.20	120.00	0.92	1.0000	0.2000	0.9165	133.8403	−103.4349	169.1508	−4.5314	−33.5251	33.8299	707.6752
0.18	133.90	0.88	1.0000	0.1803	0.8846	94.7165	−119.4825	152.4706	−3.6818	−27.2392	27.4869	637.8906
0.17	150.00	0.85	1.0000	0.1732	0.8544	55.5920	−135.5304	146.4888	−3.3986	−25.1438	25.3725	612.8648
0.18	166.10	0.83	1.0000	0.1803	0.8261	16.4676	−151.5784	152.4703	−3.6818	−27.2391	27.4868	637.8896

(Continued)

Table 6-11 (Continued)

Case for 0.20 pu compensating voltage (largest hexagon)												
Theoretical $V_{s's}$ (pu)	β (deg)	Theoretical $V_{s'}$ (pu)	Simulated V_s (pu)	Simulated $V_{s's}$ (pu)	Simulated $V_{s'}$ (pu)	Simulated P_r (MW)	Simulated Q_r (Mvar)	Simulated S_r (MVA)	Simulated P_{se} (MW)	Simulated Q_{se} (Mvar)	Simulated S_{se} (MVA)	Simulated I (A)
0.20	180.00	0.80	1.0000	0.2000	0.8000	−22.6569	−167.6264	169.1507	−4.5314	−33.5251	33.8300	707.6753
0.18	193.90	0.83	1.0000	0.1803	0.8261	−56.1172	−141.7676	152.4703	−3.6818	−27.2391	27.4868	637.8896
0.17	210.00	0.85	1.0000	0.1732	0.8544	−89.5774	−115.9087	146.4887	−3.3986	−25.1438	25.3725	612.8648
0.18	226.10	0.88	1.0000	0.1803	0.8846	−123.0376	−90.0499	152.4704	−3.6818	−27.2392	27.4869	637.8906
0.20	240.00	0.92	1.0000	0.2000	0.9165	−156.4971	−64.1915	169.1505	−4.5314	−33.5251	33.8300	707.6752
0.18	253.90	0.97	1.0000	0.1803	0.9657	−150.8330	−22.2855	152.4704	−3.6818	−27.2392	27.4869	637.8905
0.17	270.00	1.01	1.0000	0.1732	1.0149	−145.1686	19.6214	146.4886	−3.3986	−25.1438	25.3725	612.8647
0.18	286.10	1.06	1.0000	0.1803	1.0642	−139.5042	61.5282	152.4701	−3.6818	−27.2391	27.4868	637.8895
0.20	300.00	1.11	1.0000	0.2000	1.1136	−133.8397	103.4351	169.1505	−4.5314	−33.5251	33.8300	707.6751
0.18	313.90	1.13	1.0000	0.1803	1.1325	−94.7152	119.4830	152.4702	-3.6818	−27.2391	27.4868	637.8894
0.17	330.00	1.15	1.0000	0.1732	1.1533	−55.5906	135.5309	146.4887	−3.3315	−24.6484	24.8725	606.7971
0.18	346.10	1.18	1.0000	0.1803	1.1758	-16.4661	151.5787	152.4705	−3.6818	−27.2392	27.4869	637.8904
0.20	360.00	1.20	1.0000	0.2000	1.2000	22.6578	167.6262	169.1506	−4.5315	−33.5251	33.8299	707.675

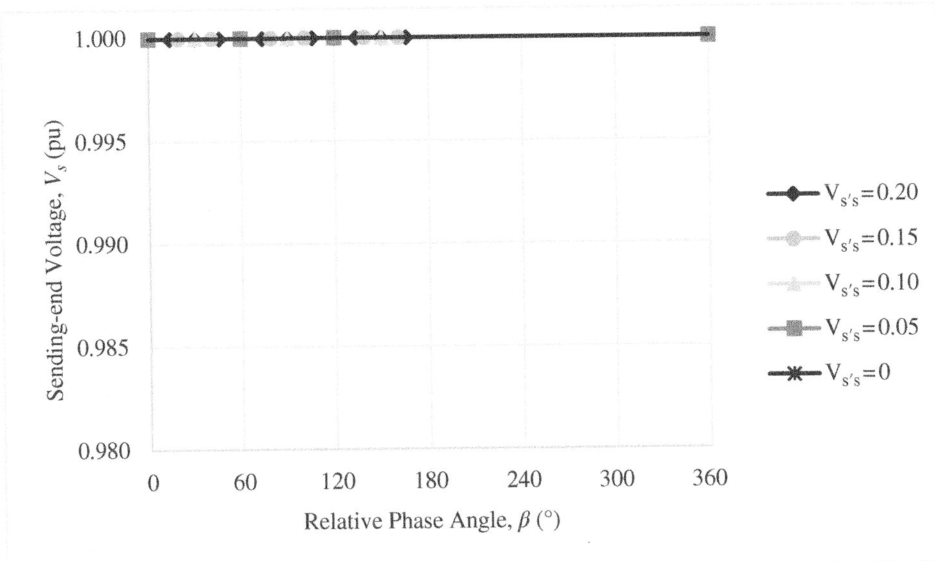

Figure 6-57 Magnitude of the simulated sending-end voltage ($\mathbf{V_s}$) during the entire range of the relative phase angle of $0° \leq \beta \leq 360°$.

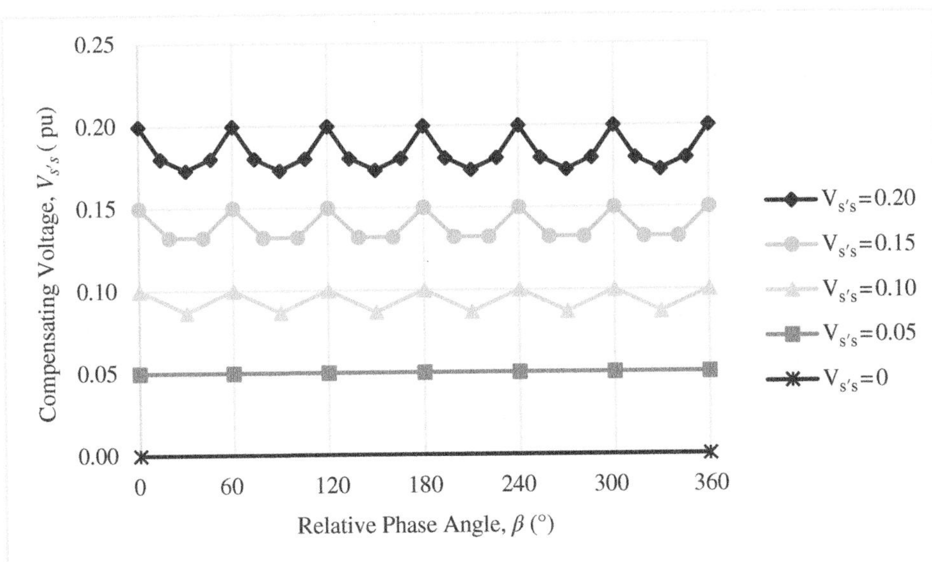

Figure 6-58 Magnitude of the simulated compensating voltage ($\mathbf{V_{s's}}$) during the entire range of the relative phase angle of $0° \leq \beta \leq 360°$.

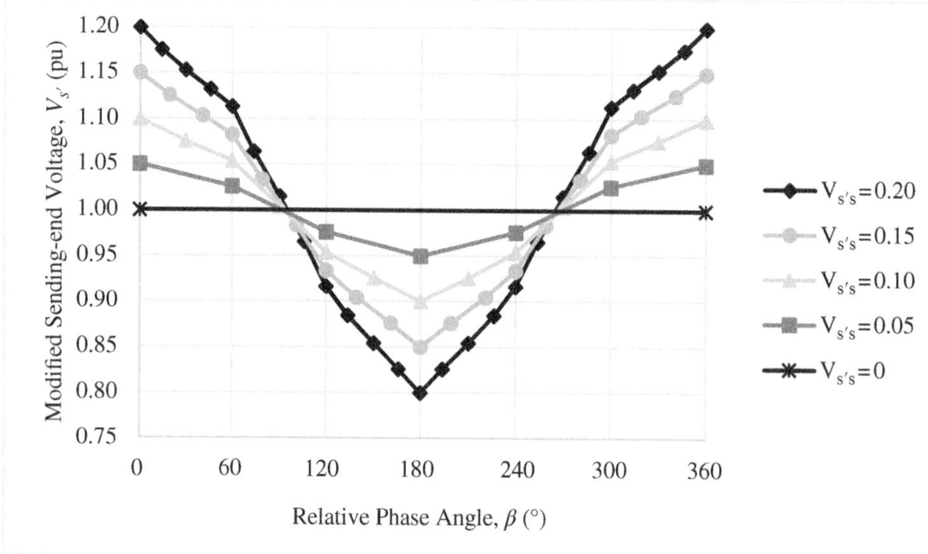

Figure 6-59 Magnitude of the simulated modified sending-end voltage ($\mathbf{V}_{s'}$) during the entire range of the relative phase angle of 0° $\leq \beta \leq$ 360°.

The simulated magnitude ($V_{s'}$) of the modified sending-end voltage, given in Table 6-11, during the entire range of the relative phase angle of 0° $\leq \beta \leq$ 360°, is shown in Figure 6-59.

Figure 6-60 shows the active power (P_r), reactive power (Q_r), and apparent power (S_r) flows at the receiving end, given in Table 6-11, during the entire range of the relative phase angle of 0° $\leq \beta \leq$ 360°. Figure 6-61 shows the active power (P_r) versus the reactive power (Q_r) flow at the receiving end, given in Table 6-11, during the entire range of the relative phase angle of 0° $\leq \beta \leq$ 360°. Note that the center of the hexagon, located at (0, 0), confirms that the initial power flow in the test network is zero due to having the same voltages at both ends of the test network's line reactor. The power flow in the test network occurs, since the ST modifies the voltage at the modified sending-end of the test network's line reactor when the receiving end is shorted with the sending end of the test network as shown in Figure 6-55.

Figure 6-62 shows the exchanged active power (P_{se}), reactive power (Q_{se}), and apparent power (S_{se}) by the Compensating-Voltage Unit (Series Unit), given in Table 6-11, during the entire range of the relative phase angle of 0° $\leq \beta \leq$ 360°.

Figure 6-63 shows the exchanged active power (P_{se}) versus reactive power (Q_{se}) by the Compensating-Voltage Unit (Series Unit), given in Table 6-11, during the entire range of the relative phase angle of 0° $\leq \beta \leq$ 360°.

Figure 6-64 shows the prevailing line current (I), given in Table 6-11, during the entire range of the relative phase angle of 0° $\leq \beta \leq$ 360°.

The summary of the study results for a 20% compensating voltage is given below.

- The line voltage at the POC is fixed at one pu.
- Active power flow at the receiving end changes from the natural flow of 0 MW by 157 and −157 MW.

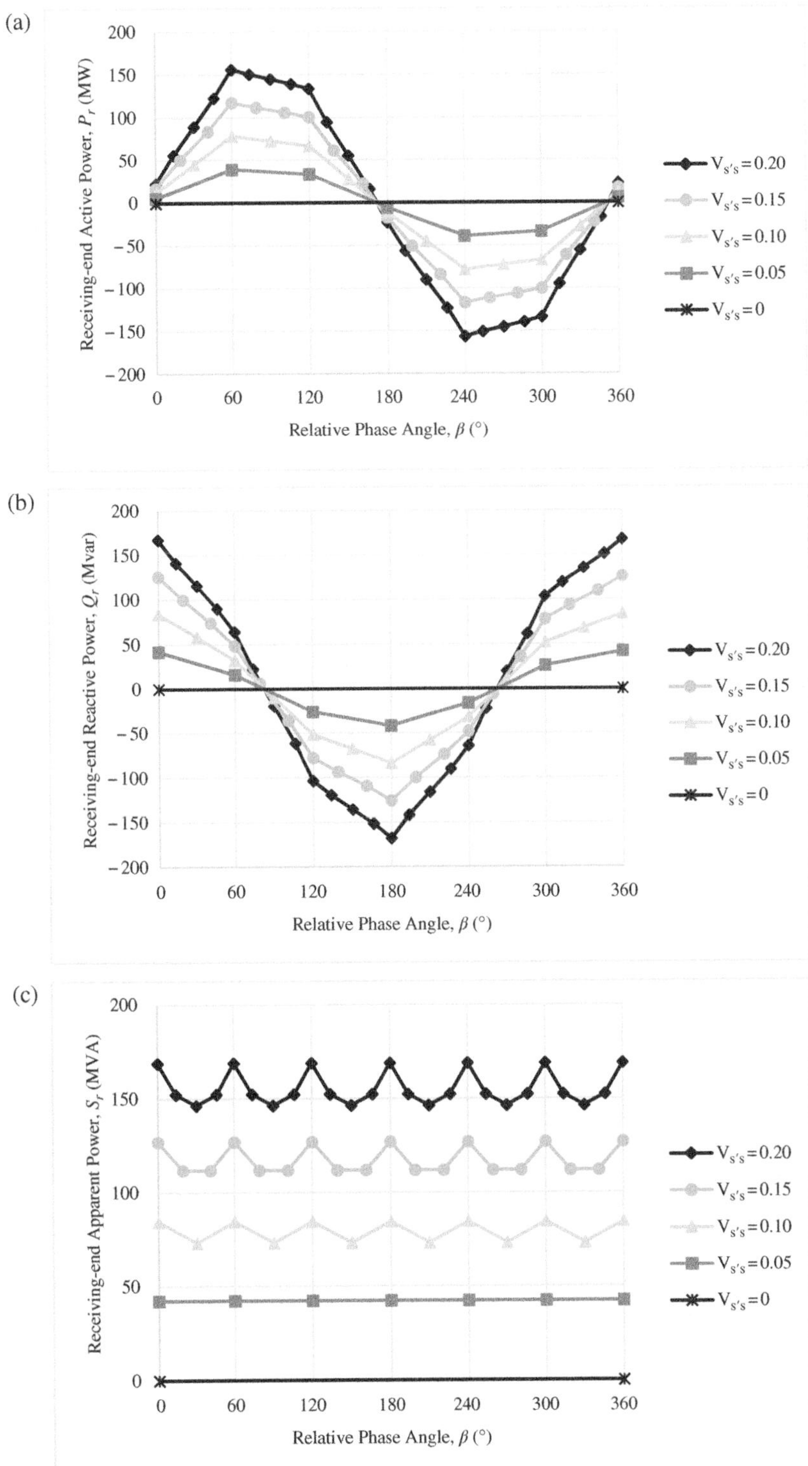

Figure 6-60 (a) Active power (P_r), (b) reactive power (Q_r), and (c) apparent power (S_r) flows at the receiving end during the entire range of the relative phase angle of $0° \leq \beta \leq 360°$.

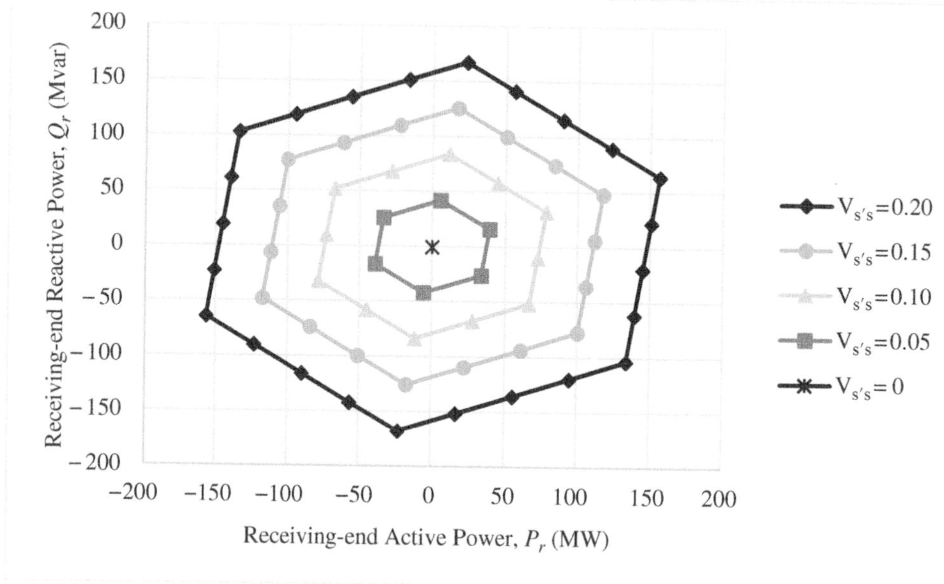

Figure 6-61 Active power (P_r) versus reactive power (Q_r) flows at the receiving end during the entire range of the relative phase angle of $0° \leq \beta \leq 360°$.

- Reactive power flow at the receiving end changes from the natural flow of 0 Mvar by 168 and −168 Mvar.
- Rating of the ST: 34 MVA.

Note that a 34 MVA-rated ST regulates 169 MVA in the line as shown in Figures 6-62c and 6-60c, respectively.

Since the source impedance and the leakage impedance are neglected in the previous example, an alternate approach is explored for an expedited determination of the power flow characteristics with the ST, integrated in a power system network. For the electrical system data, given in Table 6-10 and considering the sending-end voltage (V_s) is 1 pu, perform the following calculations as a function of the relative phase angle (β) during the entire range of the relative phase angle of $0° \leq \beta \leq 360°$ by using

Equation (6-17a) for the theoretical magnitude ($V_{s'}$) and the phase-shift angle (ψ) of the modified sending-end voltage ($\mathbf{V_{s'}}$),
Equation (2-51a) for the modified power angle (δ'),
Equation (B-43) for the active power (P_r),
Equation (B-44) for the reactive power (Q_r),
Equation (2-76) for the apparent power (S_r),
Equation (2-77) for the line current (I),
Equation (6-8) for the theoretical magnitude ($V_{s's}$) and the relative phase angle (β) of the compensating voltage ($\mathbf{V_{s's}}$),
Equation (B-101) for the exchanged active power (P_{se}) by the Compensating-Voltage Unit (Series Unit),
Equation (B-102) for the exchanged reactive power (Q_{se}) by the Compensating-Voltage Unit (Series Unit), and

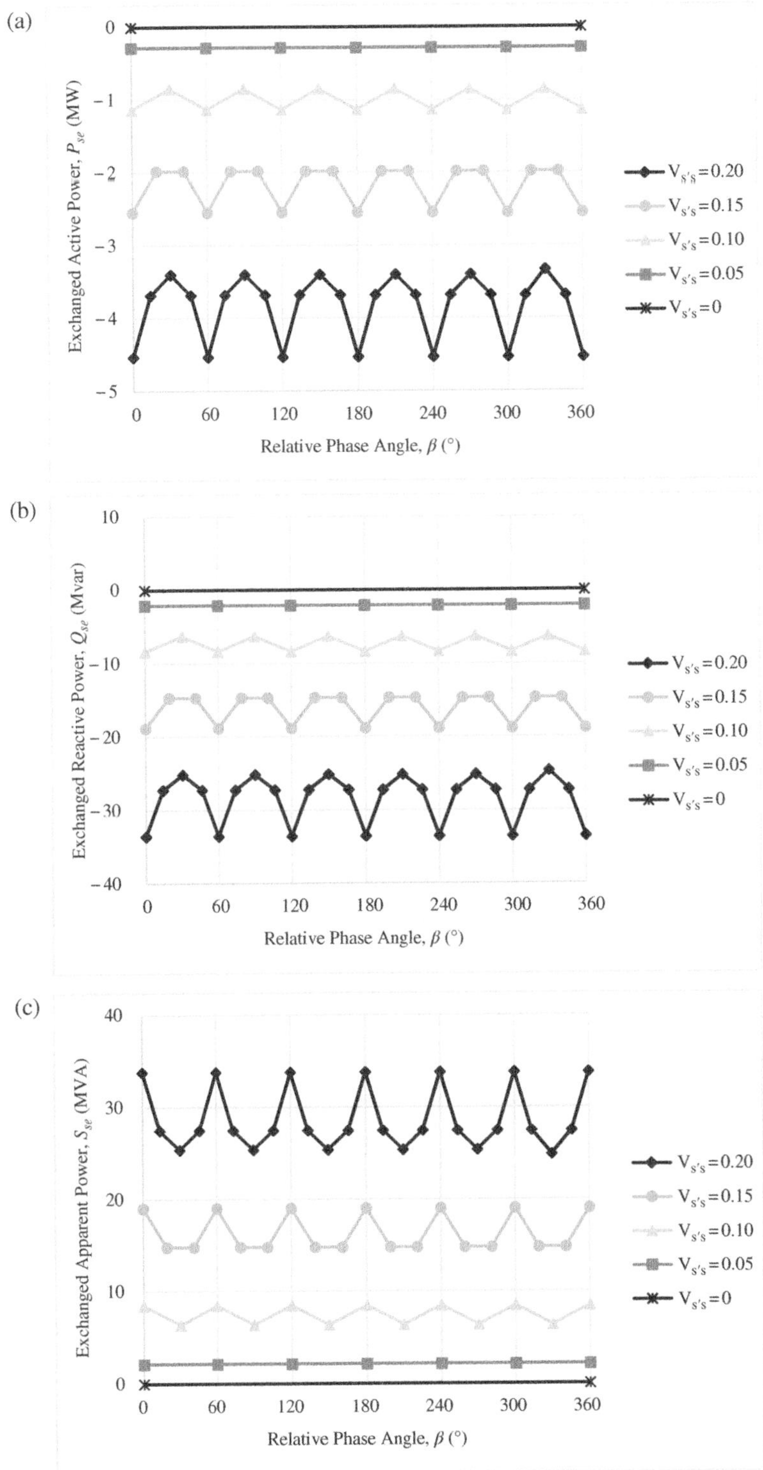

Figure 6-62 (a) Exchanged active power (P_{se}), (b) reactive power (Q_{se}), and (c) apparent power (S_{se}) by the Series Unit during the entire range of the relative phase angle of $0° \leq \beta \leq 360°$.

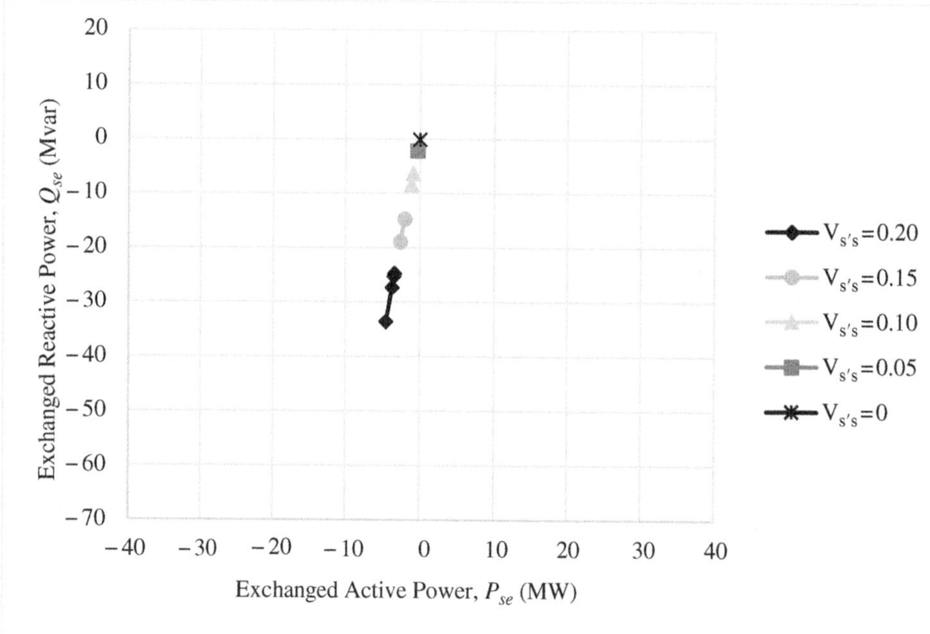

Figure 6-63 Exchanged active power (P_{se}) versus reactive power (Q_{se}) by the Series Unit during the entire range of the relative phase angle of 0° $\leq \beta \leq$ 360°.

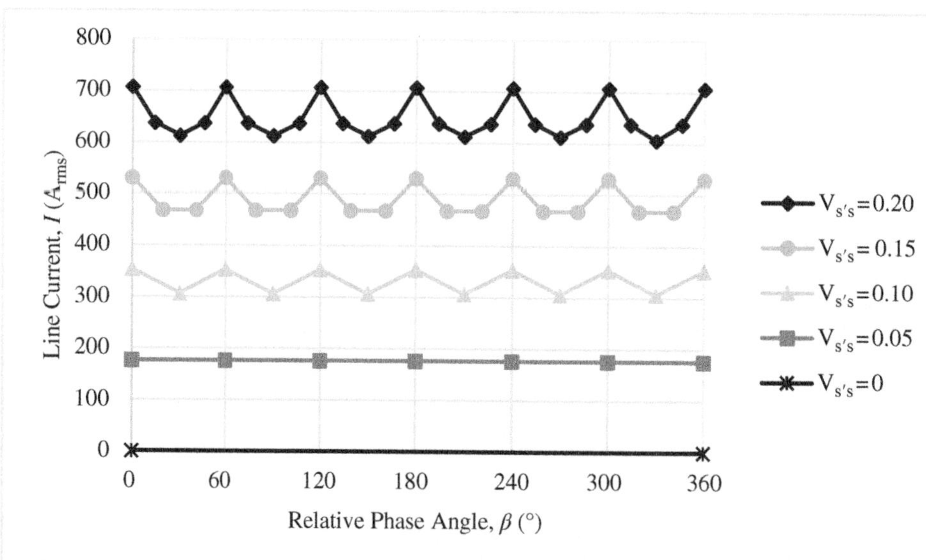

Figure 6-64 Line current (I) in A_{rms} during the entire range of the relative phase angle of 0° $\leq \beta \leq$ 360°.

Equation (2-182) for the exchanged apparent power (S_{se}) by the Compensating-Voltage Unit (Series Unit), respectively. The results are shown in Table 6-12.

A comparison of the results in Tables 6-11 and 6-12 reveals that their difference is virtually none. A slight discrepancy in the results is due to accounting, in the EMTP model, the fractional milliohm impedance in ST as its leakage impedance and in the test network. The table is expanded for each hexagon, related to 5, 10, 15, and 20% compensating voltages.

6.2.7 Limited-Angle Operation of ST

The ST requires the use of a three-phase transformer with three primary windings and nine secondary windings (or three single-phase transformers, each with one primary winding and three secondary windings). The compensating voltage ($\mathbf{V}_{s's}$) is of variable magnitude and at a variable phase angle. However, in many instances, the capability of connecting a compensating voltage in series with a line within its entire range of $0° \leq \beta \leq 360°$ is not needed. The active power flow can be increased to the maximum possible level within the first 120° of the possible range of the relative phase angle of $0° \leq \beta \leq 360°$. The active power flow can be decreased to the minimum possible level within the next 120° of the possible $0° \leq \beta \leq 360°$. In this case, the circuit configuration for the full $0° \leq \beta \leq 360°$ range of operation can be simplified to the "limited angle" configuration.

At $\beta = 0°$, 120°, and 240°, only one of the three secondary windings in each phase is active. This active winding in each phase carries the full line current at the rated winding voltage that is a fraction of the line voltage. Therefore, during the $0° \leq \beta \leq 360°$ range of operation, each of the nine secondary windings takes its turn to provide solely the full compensating voltage and carry the line current. Therefore, each of the nine secondary windings must be designed to carry the rated line current at rated winding voltage. This requires the magnetic rating in Volt-second of the 12 windings (three primary and nine secondary) of the ST to be $12/6 = 2$ pu, considering a transformer with six windings (three primary and three secondary) to be of one pu rating.

The ST with $0° \leq \beta \leq 360°$ range of operation capability uses only three primary windings and at the most six secondary windings at any given operating point. The remaining three secondary windings stay inactive. Therefore, it is possible to achieve a $0° \leq \beta \leq 360°$ operating range just by using six secondary windings instead of nine secondary windings with hardware reconfiguration in every 120° range of operation. In this way, the number of windings required in the ST is reduced at the expense of additional disconnect switches. Considering the three primary windings, the total number of windings required is $3 + 6 = 9$, instead of $3 + 9 = 12$. Therefore, the practical magnetic rating in Volt-second of the ST is only $9/6 = 1.5$ pu for the 120° range of operation, instead of two pu for $0° \leq \beta \leq 360°$ range of operation. Note that by using six secondary windings, one of the six operating regions (0° to 120°, 120° to 240°, 240° to 360°, 300° to 60°, 60° to 180°, and 180° to 300°) can be selected.

Figures 6-65 through 6-70 show, through transformer action, the generation of a compensating voltage ($\mathbf{V}_{s's}$), which is of variable magnitude and operated within a limited angle. Figure 6-65a shows that for the limited angle operation between 0° and 120°, the line voltage is applied to a shunt-connected, three-phase transformer's primary windings in the Exciter Unit. A total of six secondary windings (a_1 and a_2 on the core of the *A* phase; b_2 and b_3 on the core of the *B* phase; and c_3 and c_1 on the core of the *C* phase) constitute the limited angle Compensating-Voltage Unit. By choosing the number of turns from each of the two windings, the compensating voltage in any phase is derived from the phasor sum of the voltages induced in a two-phase winding set (a_1 and c_1 for connection in the *A* phase; a_2 and b_2 for connection in the *B* phase; and b_3 and c_3 for connection in the *C* phase). The relative phase angle (β) of the series-compensating voltage ($\mathbf{V}_{s's}$)

Table 6-12 Compensating voltage, modified sending-end voltage, power flows at the receiving end, exchanged powers by the Series Unit and the line current for electrical system data, shown in Table 6-10, and the active turns-ratios of the secondary, series-compensating windings in the *A*, *B*, and *C* phases per respective Tables 6-1, 6-2, and 6-3 using mathematical equations.

Case for natural power flows (center point of the hexagon characteristic)									
Theoretical $V_{s's}$ (pu)	β (deg)	Theoretical $V_{s'}$ (pu)	Calculated P_r (MW)	Calculated Q_r (Mvar)	Calculated S_r (MVA)	Simulated P_{se} (MW)	Simulated Q_{se} (Mvar)	Simulated S_{se} (MVA)	Calculated I (A)
0.00	0.00	1.00	0.0000	0.0000	0.0000	0.0000	0.0000	0.0000	0.0000
0.00	360.00	1.00	0.0000	0.0000	0.0000	0.0000	0.0000	0.0000	0.0000

Case for 0.05 pu compensating voltage (smallest hexagon)									
Theoretical $V_{s's}$ (pu)	β (deg)	Theoretical $V_{s'}$ (pu)	Calculated P_r (MW)	Calculated Q_r (Mvar)	Calculated S_r (MVA)	Simulated P_{se} (MW)	Simulated Q_{se} (Mvar)	Simulated S_{se} (MVA)	Calculated I (A)
0.05	0.00	1.05	5.6624	41.9015	42.2823	−0.2831	−2.0951	2.1141	176.8960
0.05	60.00	1.03	39.1189	16.0470	42.2823	−0.2831	−2.0951	2.1141	176.8960
0.05	120.00	0.98	33.4566	−25.8545	42.2823	−0.2831	−2.0951	2.1141	176.8960
0.05	180.00	0.95	−5.6624	−41.9015	42.2823	−0.2831	−2.0951	2.1141	176.8960
0.05	240.00	0.98	−39.1189	−16.0470	42.2823	−0.2831	−2.0951	2.1141	176.8960
0.05	300.00	1.03	−33.4566	25.8545	42.2823	−0.2831	−2.0951	2.1141	176.8960
0.05	360.00	1.05	5.6524	41.9015	42.2823	−0.2831	−2.0951	2.1141	176.8960

Case for 0.10 pu compensating voltage (second smallest hexagon)									
Theoretical $V_{s's}$ (pu)	β (deg)	Theoretical $V_{s'}$ (pu)	Calculated P_r (MW)	Calculated Q_r (Mvar)	Calculated S_r (MVA)	Simulated P_{se} (MW)	Simulated Q_{se} (Mvar)	Simulated S_{se} (MVA)	Calculated I (A)
0.10	0.00	1.10	11.3247	83.8030	84.5647	−1.1325	−8.3803	8.4565	353.7921
0.09	30.00	1.08	44.7813	57.9485	73.2352	−0.8494	−6.2852	6.3424	306.3929
0.10	60.00	1.05	78.2379	32.0940	84.5647	−1.1325	−8.3803	8.4565	353.7921
0.09	90.00	1.00	72.5755	−9.8075	73.2352	−0.8494	−6.2852	6.3424	306.3929
0.10	120.00	0.95	66.9131	−51.7090	84.5647	−1.1325	−8.3803	8.4565	353.7921
0.09	150.00	0.93	27.7942	−67.7560	73.2352	−0.8494	−6.2852	6.3424	306.3929
0.10	180.00	0.90	−11.3247	−83.8030	84.5647	−1.1325	−8.3803	8.4565	353.7921
0.09	210.00	0.93	−44.7813	−57.9485	73.2352	−0.8494	−6.2852	6.3424	306.3929
0.10	240.00	0.95	−78.2379	−32.0940	84.5647	−1.1325	−8.3803	8.4565	353.7921
0.09	270.00	1.00	−72.5755	9.8075	73.2352	−0.8494	−6.2852	6.3424	306.3929
0.10	300.00	1.05	−66.9131	51.7090	84.5647	−1.1325	−8.3803	8.4565	353.7921
0.09	330.00	1.08	−27.7942	67.7560	73.2352	−0.8494	−6.2852	6.3424	306.3929
0.10	360.00	1.10	11.3247	83.8030	84.5647	−1.1325	−8.3803	8.4565	353.7921

(Continued)

Table 6-12 (Continued)

Case for 0.15 pu compensating voltage (second largest hexagon)									
Theoretical $V_{s's}$ (pu)	β (deg)	Theoretical $V_{s'}$ (pu)	Calculated P_r (MW)	Calculated Q_r (Mvar)	Calculated S_r (MVA)	Simulated P_{se} (MW)	Simulated Q_{se} (Mvar)	Simulated S_{se} (MVA)	Calculated I (A)
0.15	0.00	1.15	16.9871	125.7044	126.8470	−2.5481	−18.8557	19.0271	530.6881
0.13	19.11	1.13	50.4437	99.8500	111.8686	−1.9818	−14.6655	14.7988	468.0229
0.13	40.89	1.10	83.9002	73.9955	111.8686	−1.9818	−14.6655	14.7988	468.0229
0.15	60.00	1.08	117.3568	48.1410	126.8470	−2.5481	−18.8557	19.0271	530.6881
0.13	79.11	1.03	111.6944	6.2395	111.8686	−1.9818	−14.6655	14.7988	468.0229
0.13	100.89	0.98	106.0321	−35.6620	111.8686	−1.9818	−14.6655	14.7988	468.0229
0.15	120.00	0.93	100.3697	−77.5635	126.8470	−2.5481	−18.8557	19.0271	530.6881
0.13	139.11	0.90	61.2508	−93.6105	111.8686	−1.9818	−14.6655	14.7988	468.0229
0.13	160.89	0.88	22.1318	−109.6575	111.8686	−1.9818	−14.6655	14.7988	468.0229
0.15	180.00	0.85	−16.9871	−125.7044	126.8470	−2.5481	−18.8557	19.0271	530.6881
0.13	199.11	0.88	−50.4437	−99.8500	111.8686	−1.9818	−14.6655	14.7988	468.0229
0.13	220.89	0.90	−83.9002	−73.9955	111.8686	−1.9818	−14.6655	14.7988	468.0229
0.15	240.00	0.93	−117.3568	−48.1410	126.8470	−2.5481	−18.8557	19.0271	530.6881
0.13	259.11	0.98	−111.6944	−6.2395	111.8686	−1.9818	−14.6655	14.7988	468.0229
0.13	280.89	1.03	−106.0321	35.6620	111.8686	−1.9818	−14.6655	14.7988	468.0229
0.15	300.00	1.08	−100.3697	77.5635	126.8470	−2.5481	−18.8557	19.0271	530.6881

Table 6-12 (Continued)

Case for 0.15 pu compensating voltage (second largest hexagon)									
Theoretical $V_{s's}$ (pu)	β (deg)	Theoretical $V_{s'}$ (pu)	Calculated P_r (MW)	Calculated Q_r (Mvar)	Calculated S_r (MVA)	Simulated P_{se} (MW)	Simulated Q_{se} (Mvar)	Simulated S_{se} (MVA)	Calculated I (A)
0.13	319.11	1.10	−61.2508	93.6105	111.8686	−1.9818	−14.6655	14.7988	468.0229
0.13	340.89	1.13	−22.1318	109.6575	111.8686	−1.9818	−14.6655	14.7988	468.0229
0.15	360.00	1.15	16.9871	125.7044	126.8470	−2.5481	−18.8557	19.0271	530.6881

Case for 0.20 pu compensating voltage (largest hexagon)									
Theoretical $V_{s's}$ (pu)	β (deg)	Theoretical $V_{s'}$ (pu)	Calculated P_r (MW)	Calculated Q_r (Mvar)	Calculated S_r (MVA)	Simulated P_{se} (MW)	Simulated Q_{se} (Mvar)	Simulated S_{se} (MVA)	Calculated I (A)
0.20	0.00	1.20	22.6495	167.6059	169.1294	−4.5299	−33.5212	33.8259	707.5841
0.18	13.90	1.18	56.1060	141.7514	152.4512	−3.6805	−27.2360	27.4835	637.8077
0.17	30.00	1.15	89.5626	115.8969	146.4703	−3.3974	−25.1409	25.3694	612.7858
0.18	46.10	1.13	123.0192	90.0425	152.4512	−3.6805	−27.2360	27.4835	637.8077
0.20	60.00	1.11	156.4757	64.1880	169.1294	−4.5299	−33.5212	33.8259	707.5841
0.18	73.90	1.06	150.8134	22.2865	152.4512	−3.6805	−27.2360	27.4835	637.8077
0.17	90.00	1.01	145.1510	−19.6150	146.4703	−3.3974	−25.1409	25.3694	612.7858
0.18	106.10	0.97	139.4886	−61.5165	152.4512	−3.6805	−27.2360	27.4835	637.8077
0.20	120.00	0.92	133.8263	−103.4180	169.1294	−4.5299	−33.5212	33.8259	707.5841
0.18	133.90	0.88	94.7073	−119.4650	152.4512	−3.6805	−27.2360	27.4835	637.8077

(Continued)

Table 6-12 (Continued)

Case for 0.20 pu compensating voltage (largest hexagon)									
Theoretical $V_{s's}$ (pu)	β (deg)	Theoretical $V_{s'}$ (pu)	Calculated P_r (MW)	Calculated Q_r (Mvar)	Calculated S_r (MVA)	Simulated P_{se} (MW)	Simulated Q_{se} (Mvar)	Simulated S_{se} (MVA)	Calculated I (A)
0.17	150.00	0.85	55.5884	−135.5119	146.4703	−3.3974	−25.1409	25.3694	612.7858
0.18	166.10	0.83	16.4695	−151.5589	152.4512	−3.6805	−27.2360	27.4835	637.8077
0.20	180.00	0.80	−22.6495	−167.6059	169.1294	−4.5299	−33.5212	33.8259	707.5841
0.18	193.90	0.83	−56.1060	−141.7514	152.4512	−3.6805	−27.2360	27.4835	637.8077
0.17	210.00	0.85	−89.5626	−115.8969	146.4703	−3.3974	−25.1409	25.3694	612.7858
0.18	226.10	0.88	−123.0192	−90.0425	152.4512	−3.6805	−27.2360	27.4835	637.8077
0.20	240.00	0.92	−156.4757	−64.1880	169.1294	−4.5299	−33.5212	33.8259	707.5841
0.18	253.90	0.97	−150.8134	−22.2865	152.4512	−3.6805	−27.2360	27.4835	637.8077
0.17	270.00	1.01	−145.1510	19.6150	146.4703	−3.3974	−25.1409	25.3694	612.7858
0.18	286.10	1.06	−139.4886	61.5165	152.4512	−3.6805	−27.2360	27.4835	637.8077
0.20	300.00	1.11	−133.8263	103.4180	169.1294	−4.5299	−33.5212	33.8259	707.5841
0.18	313.90	1.13	−94.7073	119.4650	152.4512	−3.6805	−27.2360	27.4835	637.8077
0.17	330.00	1.15	−55.5884	135.5119	146.4703	−3.3974	−25.1409	25.3694	612.7858
0.18	346.10	1.18	−16.4695	151.5589	152.4512	−3.6805	−27.2360	27.4835	637.8077
0.20	360.00	1.20	22.6495	167.6059	169.1294	−4.5299	−33.5212	33.8259	707.5841

with respect to the line voltage ($\mathbf{V}_s$) can vary between 0° and 120°. Figure 6-65b shows the related phasor diagram.

The series-compensating voltages in the A, B, and C phases are defined from Equations (6-8), (6-27), and (6-31) as

$$V_{s'sA}\angle\beta_A = V_{sA}\sqrt{k_{a1}^2 + k_{c1}^2 - k_{c1}k_{a1}}\angle\tan^{-1}\frac{\sqrt{3}(k_{c1})}{2k_{a1} - k_{c1}} \tag{6-8a}$$

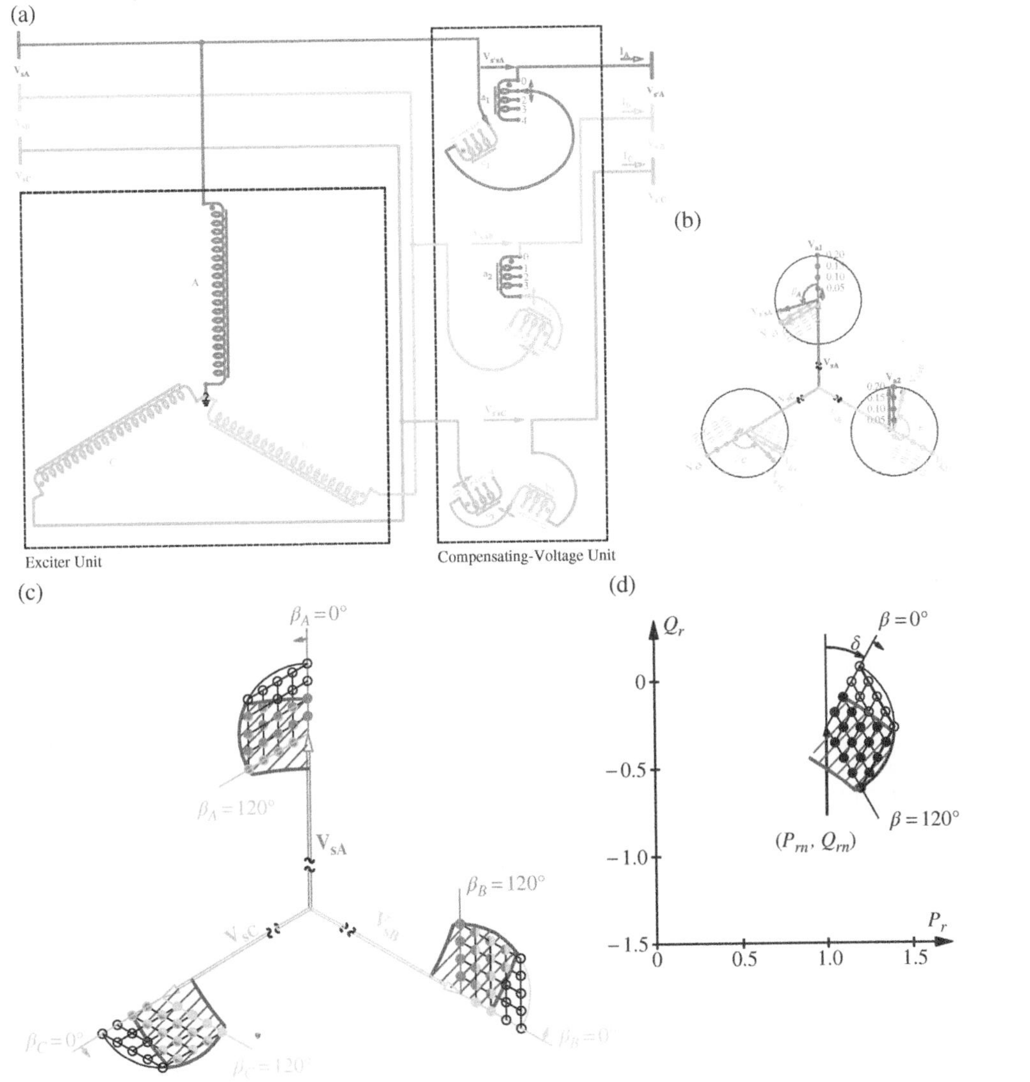

Figure 6-65 (a) ST operating with a compensating voltage in the range of $0 \leq V_{s's} \leq 0.2$ pu and a relative phase angle in the range of $0° \leq \beta \leq 120°$; (b) phasor diagram; (c) modified sending-end voltage points; (d) corresponding receiving-end active and reactive power points for operation within the control range of $0° \leq \beta \leq 120°$; (e) most simplified configuration of the ST for operation in the range of $0° \leq \beta \leq 120°$.

(e)

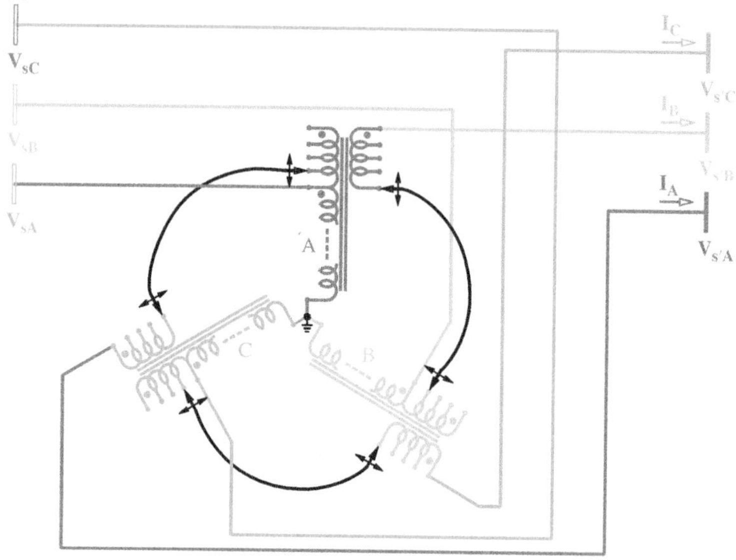

Sen Transformer

Figure 6-65 (Continued)

$$V_{s'sB}\angle\beta_B = V_{sB}\sqrt{k_{b2}^2 + k_{a2}^2 - k_{a2}k_{b2}}\angle \tan^{-1}\frac{\sqrt{3}(k_{a2})}{2k_{b2} - k_{a2}} \tag{6-27a}$$

$$V_{s'sC}\angle\beta_C = V_{sC}\sqrt{k_{c3}^2 + k_{b3}^2 - k_{b3}k_{c3}}\angle \tan^{-1}\frac{\sqrt{3}(k_{b3})}{2k_{c3} - k_{b3}} \tag{6-31a}$$

where $k_{a1} = k_{b2} = k_{c3} = x \times LTC_1$ and $k_{c1} = k_{a2} = k_{b3} = x \times LTC_3$ are the active turns-ratios of the secondary, series-compensating windings in the A, B, and C phases and the corresponding primary windings; x is the voltage (for example 0.05 pu) across each tap in the LTC; $LTC_1 = 1$ and $LTC_3 = 4$ are the LTC tap settings for the three-phase operating point shown in Figure 6-65.

For $N = 4$ in each secondary winding, the total number of possible compensating points (cp) in the 120° control range of operation, is given by

$$cp = N(N + 2) \tag{6-35}$$

and it is 24. The filled dots, in Figure 6-65c, show a total of 16 compensating points of the ST within the allowable voltage range, which is 10% of the nominal voltage in the relative phase angle range of $0° \leq \beta \leq 120°$. Figure 6-65d shows the corresponding P-Q diagram for the power flow at the receiving end.

Figure 6-65a can be further simplified to accommodate the local voltage regulation and independent control of the active and reactive power flows with minimum transformer rating as shown in Figure 6-65e. The compensating windings that provide the in-phase voltages can be constructed to have either more or fewer turns than the rest of the compensating windings. This is to comply with the required local voltage limit. The number of isolated, compensating windings can be reduced by the following action. The compensating winding that provides the in-phase voltage and the corresponding exciting winding can be replaced with a partially tapped autotransformer winding as shown in the Figure 6-65e. As a result, this configuration requires only six windings: three for the Exciter Unit and partially for the Compensating-Voltage Unit and the remaining three for the dedicated Compensating-Voltage Unit.

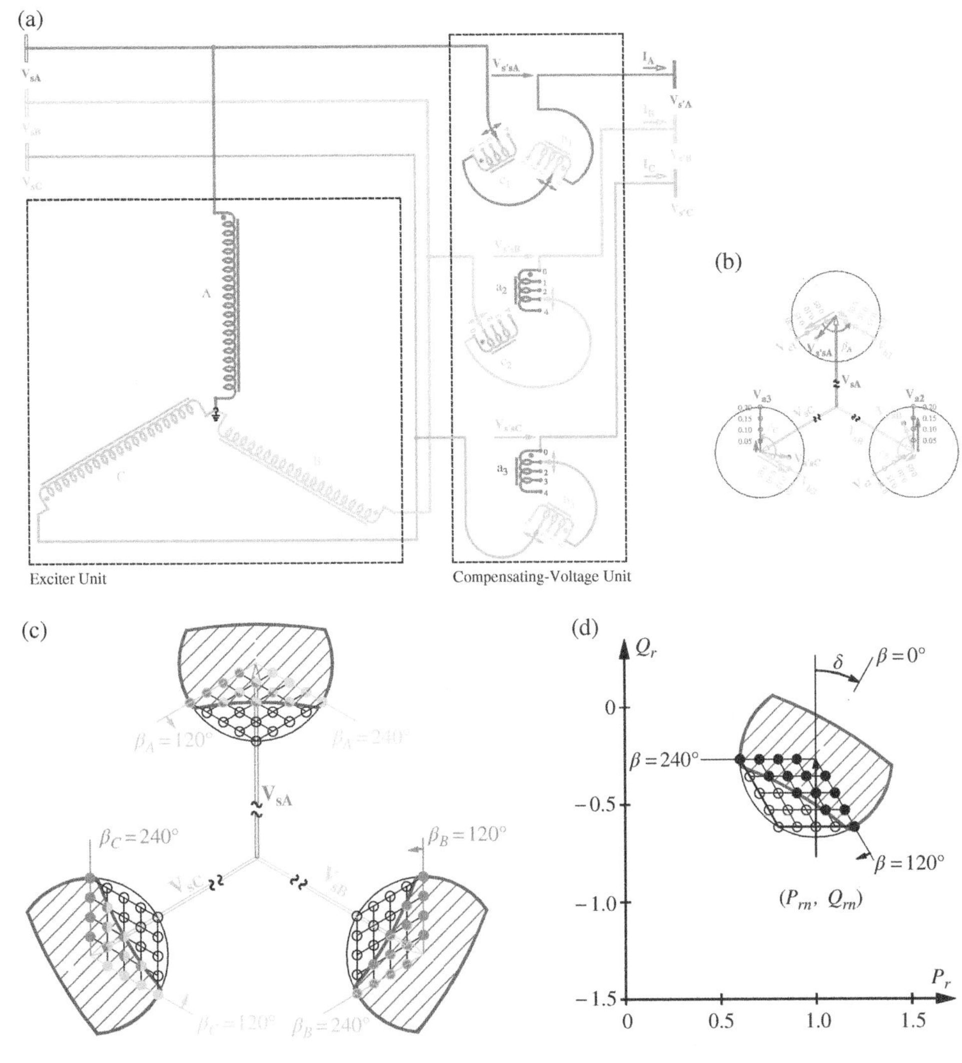

Figure 6-66 (a) ST operating with a compensating voltage in the range of $0 \leq V_{s's} \leq 0.2$ pu and a relative phase angle in the range of $120° \leq \beta \leq 240°$; (b) phasor diagram. (c) Modified sending-end voltage points; (d) corresponding receiving-end active and reactive power points for operation within the control range of $120° \leq \beta \leq 240°$.

Similarly, in an application in which there is a need to vary the relative phase angle (β) of the series-compensating voltage ($\mathbf{V_{s's}}$) between 120 and 240°, a Compensating-Voltage Unit with only six windings, as shown in Figure 6-66a, is needed. This is achieved by constructing the compensating voltage from a combination of two voltages, each of which is induced in a separate winding of a two-phase set (b_1 and c_1 for connection in the A phase, a_2 and c_2 for connection in the B phase, and a_3 and b_3 for connection in the C phase). Figure 6-66b shows the related phasor diagram.

The series-compensating voltages in the A, B, and C phases are defined from Equations (6-8), (6-27), and (6-31) as

$$V_{s'sA}\angle\beta_A = V_{sA}\sqrt{k_{b1}^2 + k_{c1}^2 - k_{b1}k_{c1}}\angle\tan^{-1}\frac{\sqrt{3}(k_{b1} - k_{c1})}{k_{b1} + k_{c1}} \tag{6-8b}$$

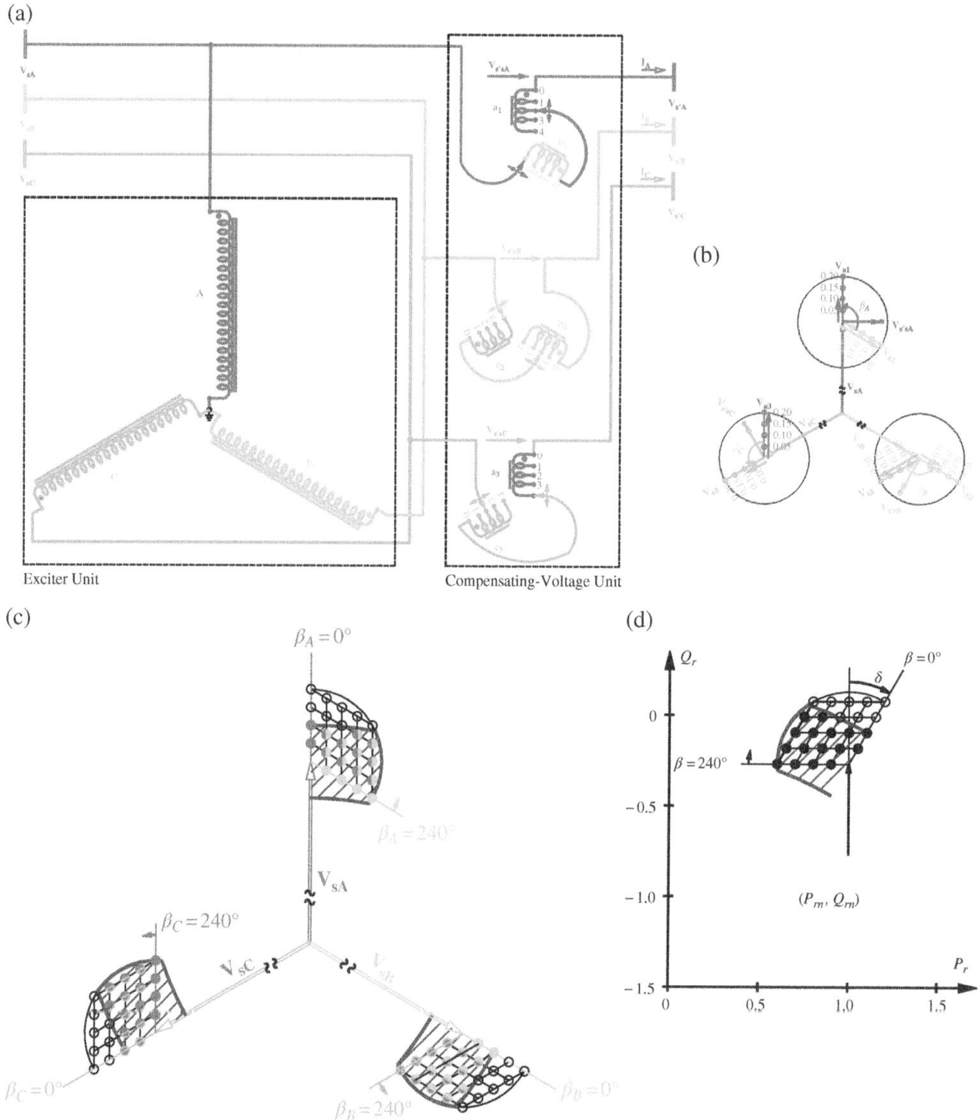

Figure 6-67 (a) ST operating with a compensating voltage in the range of $0 \le V_{s's} \le 0.2$ pu and a relative phase angle in the range of $240° \le \beta \le 360°$; (b) phasor diagram; (c) modified sending-end voltage points; (d) corresponding receiving-end active and reactive power points for operation within the control range of $240° \le \beta \le 360°$.

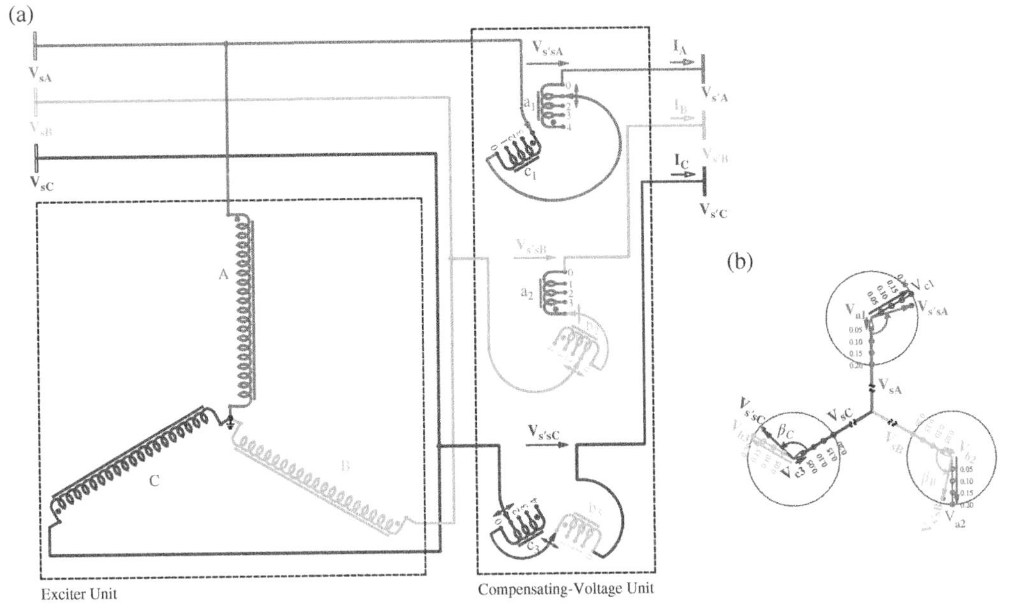

Figure 6-68 (a) ST operating with a compensating voltage in the range of $0 \leq V_{s's} \leq 0.2$ pu and a relative phase angle in the range of $180° \leq \beta \leq 300°$; (b) phasor diagram.

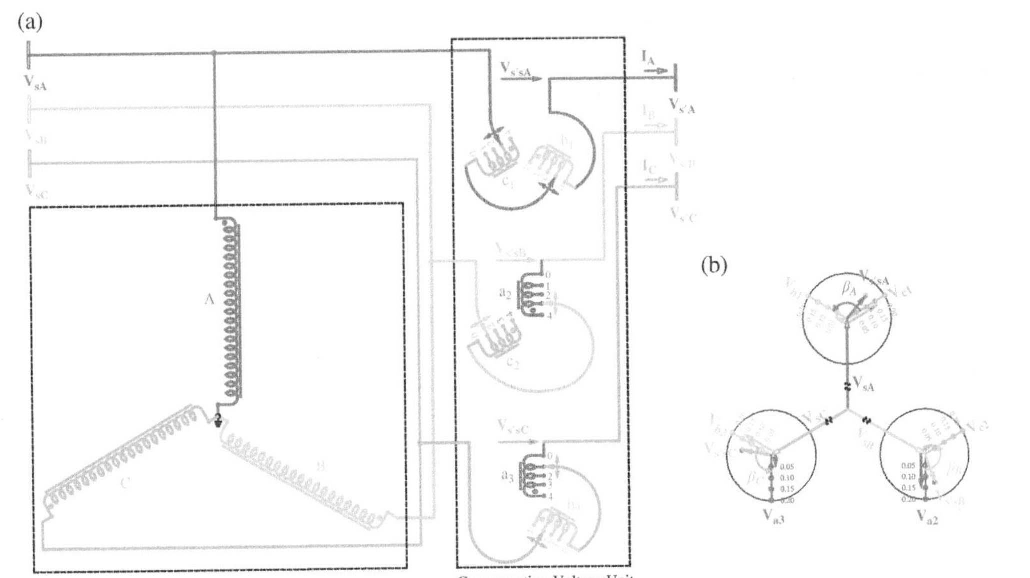

Figure 6-69 (a) ST operating with a compensating voltage in the range of $0 \leq V_{s's} \leq 0.2$ pu and a relative phase angle in the range of $300° \leq \beta \leq 60°$; (b) phasor diagram.

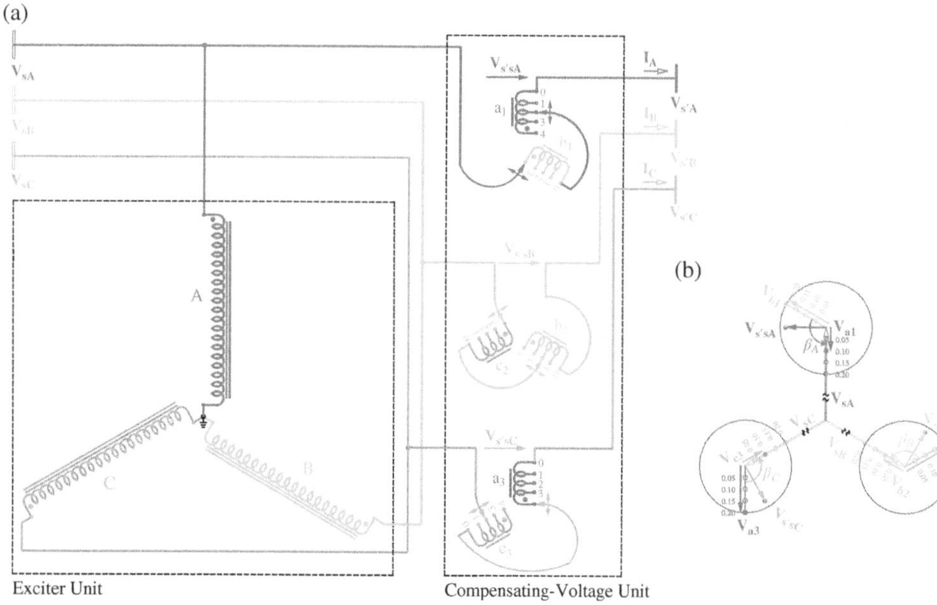

Figure 6-70 (a) ST operating with a compensating voltage in the range of $0 \le V_{s's} \le 0.2$ pu and a relative phase angle in the range of $60° \le \beta \le 180°$; (b) phasor diagram.

$$V_{s'sB}\angle\beta_B = V_{sB}\sqrt{k_{c2}^2 + k_{a2}^2 - k_{c2}k_{a2}}\angle\tan^{-1}\frac{\sqrt{3}(k_{c2} - k_{a2})}{k_{c2} + k_{a2}} \tag{6-27b}$$

$$V_{s'sC}\angle\beta_C = V_{sC}\sqrt{k_{a3}^2 + k_{b3}^2 - k_{a3}k_{b3}}\angle\tan^{-1}\frac{\sqrt{3}(k_{a3} - k_{b3})}{k_{a3} + k_{b3}} \tag{6-31b}$$

where $k_{b1} = k_{c2} = k_{a3} = x \times LTC_2$ and $k_{c1} = k_{a2} = k_{b3} = x \times LTC_3$ are the active turns-ratios of the secondary, series-compensating windings in the A, B, and C phases and the corresponding primary windings; x is the voltage (for example 0.05 pu) across each tap in the LTC; $LTC_2 = 1$ and $LTC_3 = 3$ are the LTC tap settings for the three-phase operating point shown in Figure 6-66.

For $N = 4$ in each secondary winding, the total number of possible compensating points (cp) in the 120° control range of operation is 24 from Equation (6-35). The filled dots, in Figure 6-66c, show a total of 14 compensating points of the ST within the allowable voltage range, which is 10% of the nominal voltage in the relative phase angle range of $120° \le \beta \le 240°$. Figure 6-66d shows the corresponding *P-Q* diagram for the power flow at the receiving end.

Lastly, in an application in which there is a need to vary the relative phase angle (β) of the series-compensating voltage ($\mathbf{V_{s's}}$) in the range of $240° \le \beta \le 360°$, a Compensating-Voltage Unit with only six windings, as shown in Figure 6-67a, is needed. This is achieved by constructing the compensating voltage from a combination of two voltages, each of which is induced in a separate winding of a two-phase set (a_1 and b_1 for connection in the *A* phase, b_2 and c_2 for connection in the *B* phase, and a_3 and c_3 for connection in the *C* phase). Figure 6-67b shows the related phasor diagram.

The series-compensating voltages in the *A*, *B*, and *C* phases are defined from Equations (6-8), (6-27), and (6-31) as

$$V_{s'sA}\angle\beta_A = V_{sA}\sqrt{k_{a1}^2 + k_{b1}^2 - k_{a1}k_{b1}}\angle\tan^{-1}\frac{\sqrt{3}(-k_{b1})}{2k_{a1} - k_{b1}} \tag{6-8c}$$

$$V_{s'sB}\angle\beta_B = V_{sB}\sqrt{k_{b2}^2 + k_{c2}^2 - k_{b2}k_{c2}}\angle\tan^{-1}\frac{\sqrt{3}(-k_{c2})}{2k_{b2} - k_{c2}} \tag{6-27c}$$

$$V_{s'sC}\angle\beta_C = V_{sC}\sqrt{k_{c3}^2 + k_{a3}^2 - k_{c3}k_{a3}}\angle\tan^{-1}\frac{\sqrt{3}(-k_{a3})}{2k_{c3} - k_{a3}} \tag{6-31c}$$

where $k_{a1} = k_{b2} = k_{c3} = x \times LTC_1$ and $k_{b1} = k_{c2} = k_{a3} = x \times LTC_2$ are the active turns-ratios of the secondary, series-connected windings in the A, B, and C phases and the corresponding primary windings; x is the voltage (for example 0.05 pu) across each tap in the LTC; $LTC_1 = 2$ and $LTC_2 = 4$ are the LTC tap settings for the three-phase operating point shown in Figure 6-67.

For $N = 4$, in each secondary, series-compensating winding, the total number of possible compensating points (cp) in the 120° control range of operation is 24 from Equation (6-35). The filled dots, in Figure 6-67c, show a total of 16 compensating points of the ST within the allowable voltage range, which is 10% of the nominal voltage in the relative phase angle range of $240° \leq \beta \leq 360°$. Figure 6-67d shows the corresponding P-Q diagram for the power flow at the receiving end.

Extending the concept just presented, if the polarities of the windings in the Compensating-Voltage Unit in Figure 6-65a are reversed, the relative phase angle (β) of the series-compensating voltage ($\mathbf{V_{s's}}$) can vary between 180° and 300°. The schematic for this configuration is shown in Figure 6-68a. In the same way, if the polarities of the windings in the Compensating-Voltage Units, presented in Figures 6-66a and 6-67a, are reversed, then the relative phase angle (β) of the compensating voltage ($\mathbf{V_{s's}}$) can vary between 300° and 60° and 60° and 180°, respectively. The schematics for these configurations are shown in Figures 6-69a and 6-70a. The related phasor diagrams are shown in Figures 6-68b, 6-69b, and 6-70b, respectively.

The series-compensating voltages in the A, B, and C phases are defined by the Equations (6-8), (6-27), and (6-31), respectively, where $LTC_1 = 1$ and $LTC_3 = 4$ are the LTC tap settings for the three-phase operating point shown in Figure 6-68b; $LTC_2 = 1$ and $LTC_3 = 3$ are the LTC tap settings for the three-phase operating point shown in Figure 6-69b; $LTC_1 = 2$ and $LTC_2 = 4$ are the LTC tap settings for the three-phase operating point shown in Figure 6-70b. Note that the calculated relative phase angles (β_A, β_B, and β_C) using the above-mentioned equations need to be adjusted by a phase angle of π radians, since the polarity of the three secondary, series-compensating windings are reversed in Figures 6-68a, 6-69a, and 6-70a.

In all these cases, the compensating voltage can be at any phase angle with the prevailing line current, which emulates a four-quadrant, series-compensating impedance ($\mathbf{Z_{se}} = R_{se} - jX_{se}$) that consists of a resistance ($R_{se} = +R$ or $-R$) and a reactance ($X_{se} = X_C$ or $-X_L$) in series with the line. The effect is such that the active and reactive power flows in a line can be regulated independently as desired. In addition, the function of a voltage regulator can also be implemented, provided that the final variable magnitude and variable phase angle of the compensating voltage are within the controllable operating range.

So far, the ST configuration is discussed with three primary exciting windings and nine secondary, series-compensating windings for its full 360° operation. The corresponding primary windings are designed for full line voltage and fractional line current. The secondary windings are designed for fractional line voltage and full line current. Therefore, the LTCs carry the full line current and a higher fault current. In an alternate configuration, the LTCs may be moved to the primary side so that the taps carry a fractional amount of line current.

6.2.8 ST Using LTCs with Lower Current Rating

Figure 6-71 shows the generation of a line-frequency, compensating voltage ($\mathbf{V}_{s's}$), through transformer action, that is of variable magnitude and operated at variable phase angle. When this compensating voltage is connected in series with the line, the modified sending-end voltage becomes $\mathbf{V}_{s'}$. The result is that the active and reactive power flows in a line are regulated independently as desired.

Figure 6-71a shows that the three single-phase voltages ($\mathbf{V}_{sA}$, $\mathbf{V}_{sB}$, and $\mathbf{V}_{sC}$) of a three-phase line voltage ($\mathbf{V}_s$) are applied across three-winding sets. The voltage ($\mathbf{V}_{sA}$) is applied across the three windings A_1, A_2, and A_3. The voltage ($\mathbf{V}_{sB}$) is applied across the three windings B_1, B_2, and B_3. The voltage ($\mathbf{V}_{sC}$) is applied across the three windings C_1, C_2, and C_3. Each of the windings (A_1, B_1, and C_1) is tapped at the same number of turns through LTC_1; each of the windings (A_2, B_2, and C_2) is tapped at the same number of turns through LTC_2; and each of the windings (A_3, B_3, and C_3) is tapped at the same number of turns through LTC_3. However, the number of turns in the A1-B1-C1 set, A2-B2-C2 set, and A3-B3-C3 set can be different from each other. The composite voltage from the three windings (A_1, C_2, and B_3) on the primary side (Exciter Unit) is reflected in a single winding on the secondary side (Compensating-Voltage Unit) of the *A* phase. The composite voltage from the three windings (B_1, A_2, and C_3) on the primary side (Exciter Unit) is reflected in a single winding on the secondary side (Compensating-Voltage Unit) of the *B* phase. The composite voltage from the three windings (C_1, B_2, and A_3) on the primary side (Exciter Unit) is reflected in a single winding on the secondary side (Compensating-Voltage Unit) of the C phase. By choosing the number of turns from each of the three windings (A_1, C_2, and B_3), for example, the compensating voltage ($\mathbf{V}_{s'sA}$) can be of variable magnitude and at variable phase angle (β_A) as shown in Figure 6-71b. Similarly, by choosing the number of turns from each of the three windings (B_1, A_2, and C_3), the compensating voltage ($\mathbf{V}_{s'sB}$) can be of variable magnitude and at variable phase angle (β_B) as shown in Figure 6-71b. Finally, by choosing the number of turns from each of the three windings (C_1, B_2, and A_3), the compensating voltage ($\mathbf{V}_{s'sC}$) can be of variable magnitude and at variable phase angle (β_C) as shown in Figure 6-71b.

Figure 6-71c shows the phasor diagram when the exciter voltage is applied across three windings (A_1, A_2, and A_3 in the *A* phase; B_1, B_2, and B_3 in the *B* phase; and C_1, C_2, and C_3 in the *C* phase), each with the same number of turns. Each of the nine primary windings is tapped at the same number of turns. The compensating voltage, which is the phasor sum of three equal voltages with 120° phase difference from one another, is zero. If equal voltages that are induced from three windings (A_1, C_2, and B_3 on the core of the *A* phase) are combined, the composite voltage is always zero. This property makes it possible to vary the compensating voltage from zero to its rated value in a configuration when the LTCs are on the Exciter Unit and the Compensating-Voltage Unit is in its simplest form with only one series insertion winding in each phase. The LTCs operate at a low current and ride through a higher fault current.

Figure 6-71d shows the phasor diagram when the exciter voltage is applied only across one winding in each phase. For example, $\mathbf{V}_{sA}$ is applied only across the winding (A_1), $\mathbf{V}_{sB}$ is applied only across the winding (B_1), and $\mathbf{V}_{sC}$ is applied only across the winding (C_1). In this case, the series-compensating voltage in each phase is in phase with the corresponding phase voltage of the line and its magnitude is dependent on the number of turns in the compensating winding and the exciter winding.

Figure 6-71e shows that the three-phase voltages ($\mathbf{V}_{sA}$, $\mathbf{V}_{sB}$, and $\mathbf{V}_{sC}$) are applied across the windings (A_2, B_2, and C_2), respectively. In this case, the series-compensating voltage in each phase leads the corresponding phase voltage of the line by 120°.

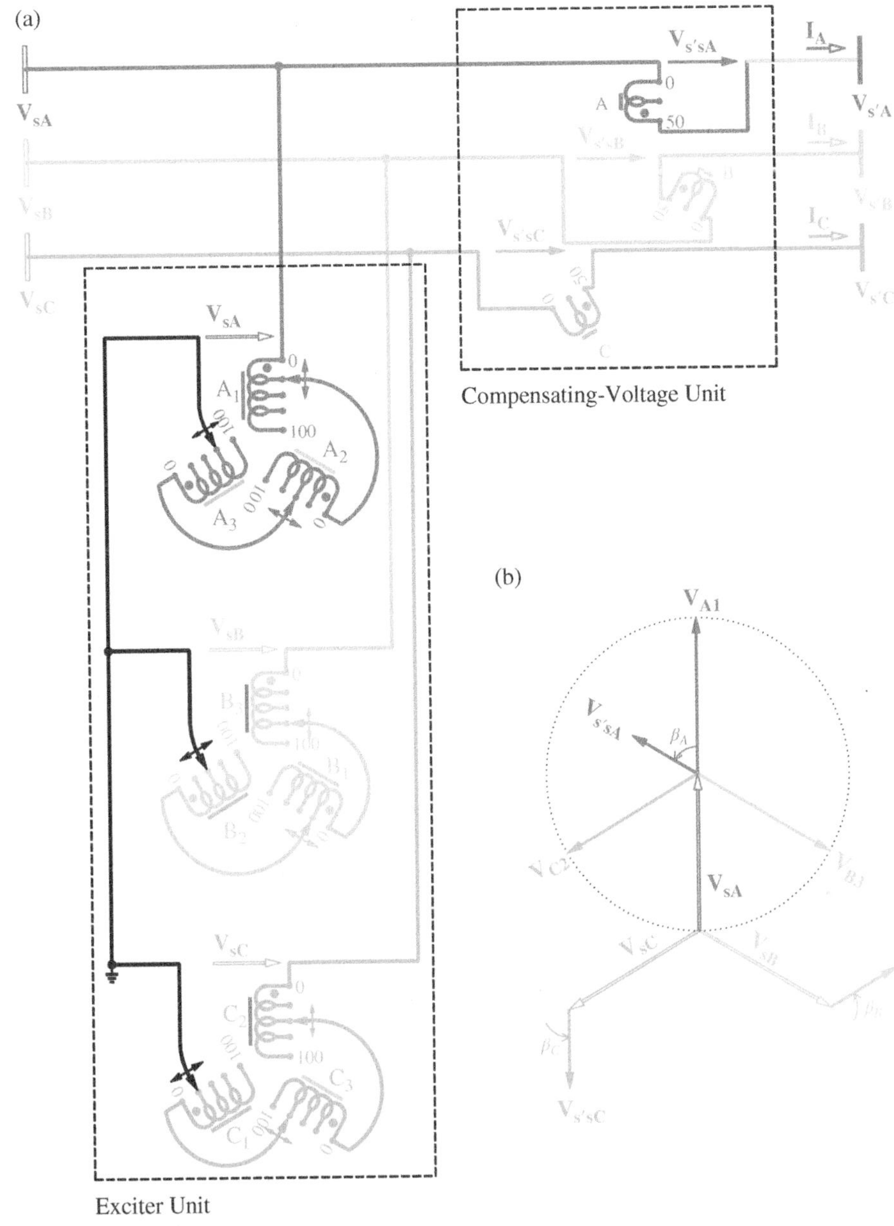

Figure 6-71 (a) ST configuration using LTCs with lower current rating; (b) phasor diagram for compensating voltage with variable magnitude and phase angle. Phasor diagrams for (c) zero compensating voltage, (d) in-phase compensating voltage with variable magnitude, (e) 120° leading compensating voltage with variable magnitude, and (f) 120° lagging compensating voltage with variable magnitude.

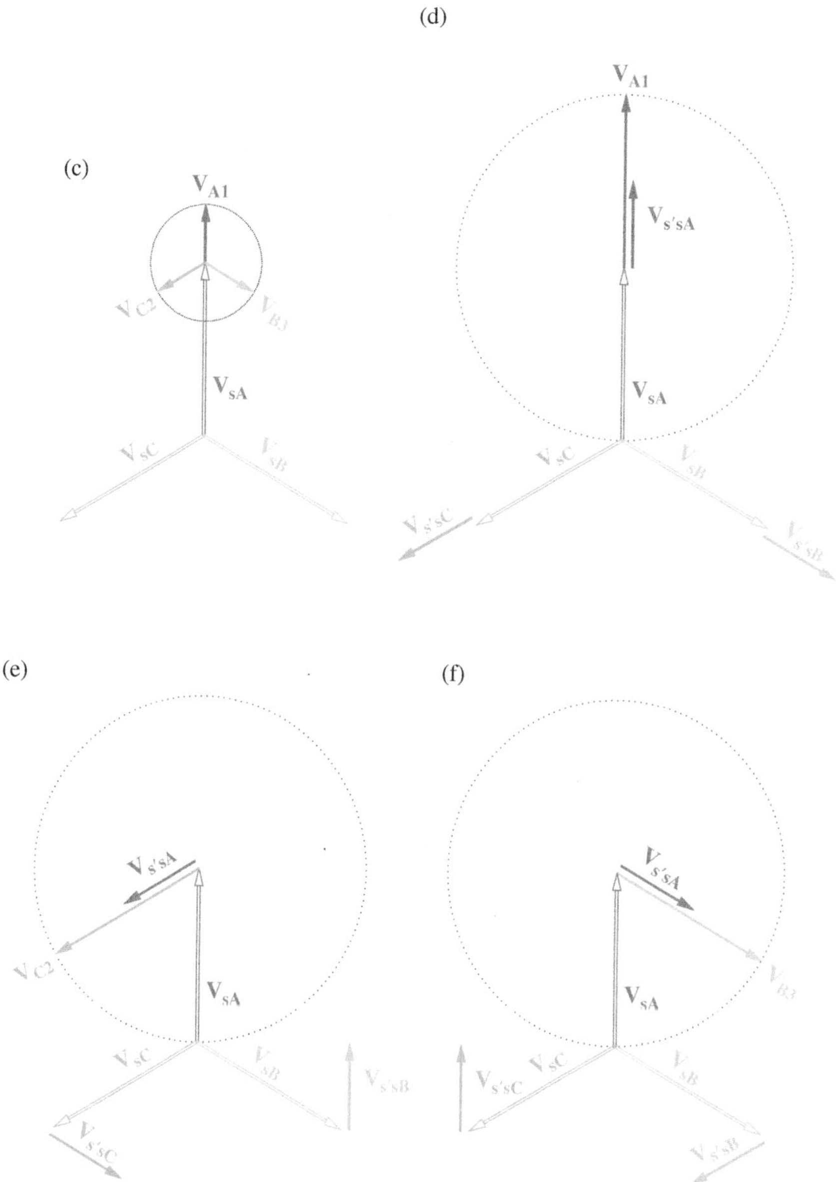

Figure 6-71 (Continued)

Figure 6-71f shows that the three-phase voltages ($\mathbf{V_{sA}}$, $\mathbf{V_{sB}}$, and $\mathbf{V_{sC}}$) are applied across the windings (A_3, B_3, and C_3), respectively, and the compensating voltage in each phase now lags the corresponding phase voltage of the line by 120°.

In summary, the compensating voltage can be at any phase angle with the prevailing line current, which emulates a four-quadrant, series-compensating impedance ($\mathbf{Z_{se}} = R_{se} - jX_{se}$) that consists of a resistance ($R_{se} = +R$ or $-R$) and a reactance ($X_{se} = X_C$ or $-X_L$) in series with the line. The effect is such that the active and reactive power flows in a line can be regulated independently as desired. In addition, the function of a voltage regulator can also be realized.

The ST, in its basic design, is a single-core transformer, consisting of 12 windings – three primary and nine secondary as shown in Figure 6-11. The primary windings are Y-connected. A dedicated three-winding set creates a compensating voltage that is connected in series in each of the three phases. Each of the windings in the three-winding set is excited from a different phase voltage. Therefore, by selecting the number of turns in each of these three windings, a composite voltage, called a compensating voltage that is variable in magnitude ($V_{s's}$) and variable in relative phase angle (β) is created, which in turn, makes the modified line voltage of desired magnitude ($V_{s'}$) and desired phase-shift angle (ψ).

In many applications, the series-compensating voltage is only a few percent of the line voltage. The series winding as well as the LTCs in a single-core design carry high line current and even higher fault current. Sometimes, the LTCs may be required to operate at a low current, so that they can ride through a high fault current. In such cases, three secondary voltages of the Compensating-Voltage Unit, each from a three-winding set, need to be high-voltage and low-current rated. Since the LTCs are exposed to the high-voltage transmission line, the single-core design may be suitable for applications up to a voltage level, beyond which a two-core design must be considered.

6.2.9 ST with a Two-Core Design

Applications that are above some voltage class, for example 230-kV, require a two-core design where the taps are not exposed to a high-voltage transmission line as shown in Figure 6-72. The

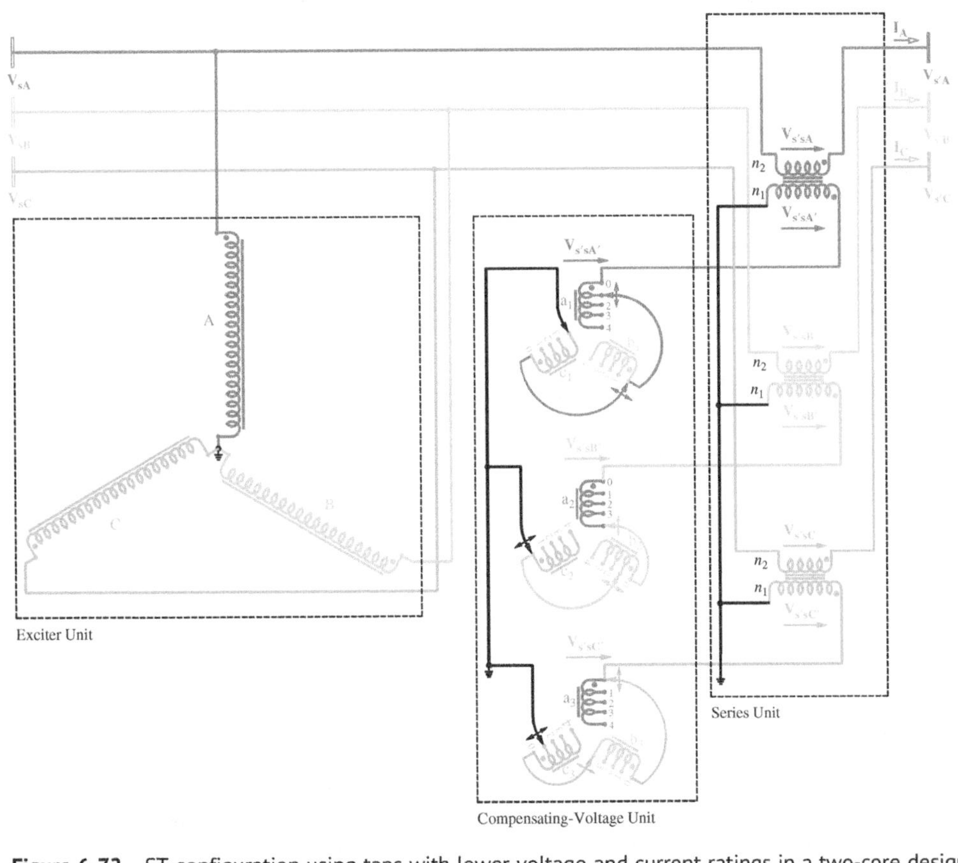

Figure 6-72 ST configuration using taps with lower voltage and current ratings in a two-core design.

Exciter Unit excites the primary windings (A, B, and C). The secondary voltages in the Compensating-Voltage Unit create the following intermediate voltages ($V_{s'sA'}$, $V_{s'sB'}$, and $V_{s'sC'}$) as

$$V_{s'sA'}\angle\beta_A = \mathbf{V_{s'sA'}} = \mathbf{V_{a1}} + \mathbf{V_{b1}} + \mathbf{V_{c1}}$$

or

$$V_{s'sA'}\angle\beta_A = k_{a1}V_{sA} + k_{b1}V_{sB} + k_{c1}V_{sC} = k_{a1}V_{sA} + k_{b1}e^{-j2\pi/3}V_{sA} + k_{c1}e^{j2\pi/3}V_{sA}, \quad (6\text{-}36)$$

$$V_{s'sB'}\angle\beta_B = \mathbf{V_{s'sB'}} = \mathbf{V_{b2}} + \mathbf{V_{c2}} + \mathbf{V_{a2}}$$

or

$$V_{s'sB'}\angle\beta_B = k_{b2}V_{sB} + k_{c2}V_{sC} + k_{a2}V_{sA} = k_{b2}e^{-j2\pi/3}V_{sA} + k_{c2}e^{j2\pi/3}V_{sA} + k_{a2}V_{sA}, \quad (6\text{-}37)$$

and

$$V_{s'sC'}\angle\beta_C = \mathbf{V_{s'sC'}} = \mathbf{V_{c3}} + \mathbf{V_{a3}} + \mathbf{V_{b3}}$$

or

$$V_{s'sC'}\angle\beta_C = k_{c3}V_{sC} + k_{a3}V_{sA} + k_{b3}V_{sB} = k_{c3}e^{j2\pi/3}V_{sA} + k_{a3}V_{sA} + k_{b3}e^{-j2\pi/3}V_{sA}. \quad (6\text{-}38)$$

The intermediate voltages in the A, B, and C phases can be written in terms of active turns-ratios using the Equations (6-8), (6-27), and (6-31) as

$$V_{s'sA'}\angle\beta_A = V_{sA}\sqrt{k_{a1}^2 + k_{b1}^2 + k_{c1}^2 - k_{a1}k_{b1} - k_{b1}k_{c1} - k_{c1}k_{a1}}\angle\tan^{-1}\frac{\sqrt{3}(k_{c1} - k_{b1})}{2k_{a1} - k_{b1} - k_{c1}} \quad (6\text{-}39)$$

$$V_{s'sB'}\angle\beta_B = V_{sB}\sqrt{k_{b2}^2 + k_{c2}^2 + k_{a2}^2 - k_{b2}k_{c2} - k_{c2}k_{a2} - k_{a2}k_{b2}}\angle\tan^{-1}\frac{\sqrt{3}(k_{a2} - k_{c2})}{2k_{b2} - k_{c2} - k_{a2}} \quad (6\text{-}40)$$

$$V_{s'sC'}\angle\beta_C = V_{sC}\sqrt{k_{c3}^2 + k_{a3}^2 + k_{b3}^2 - k_{c3}k_{a3} - k_{a3}k_{b3} - k_{b3}k_{c3}}\angle\tan^{-1}\frac{\sqrt{3}(k_{b3} - k_{a3})}{2k_{c3} - k_{a3} - k_{b3}} \quad (6\text{-}41)$$

where $k_{a1} = k_{b2} = k_{c3} = x \times LTC_1$, $k_{b1} = k_{c2} = k_{a3} = x \times LTC_2$, and $k_{c1} = k_{a2} = k_{b3} = x \times LTC_3$ are the active turns-ratios of the secondary, series-compensating windings in the A, B, and C phases and the corresponding primary windings; x is the voltage (for example 0.05 pu) across each tap in the LTC; $LTC_1 = 1$, $LTC_2 = 0$, and $LTC_3 = 4$ are the LTC tap settings for the three-phase operating point shown in Figure 6-72.

The intermediate series-compensating voltages may be used to excite three additional single-phase step-down transformers (or one three-phase step-down transformer with three isolated secondary windings), called a Series Unit, whose secondary voltages are connected in series with the line. The series-compensating voltages in the A, B, and C phases are defined by the following equations:

$$\frac{V_{s'sA}}{V_{s'sA'}} = \frac{V_{s'sB}}{V_{s'sB'}} = \frac{V_{s'sC}}{V_{s'sC'}} = \frac{n_2}{n_1}. \quad (6\text{-}42)$$

Therefore, the Compensating-Voltage Unit creates the intermediate voltage and the Series Unit connects the compensating voltage in series with the line.

It would be advantageous to use such a scheme, which allows the generation of the required compensating voltage of variable magnitude and variable phase angle by using a three-phase transformer and LTCs and connection of the compensating voltage through three separate

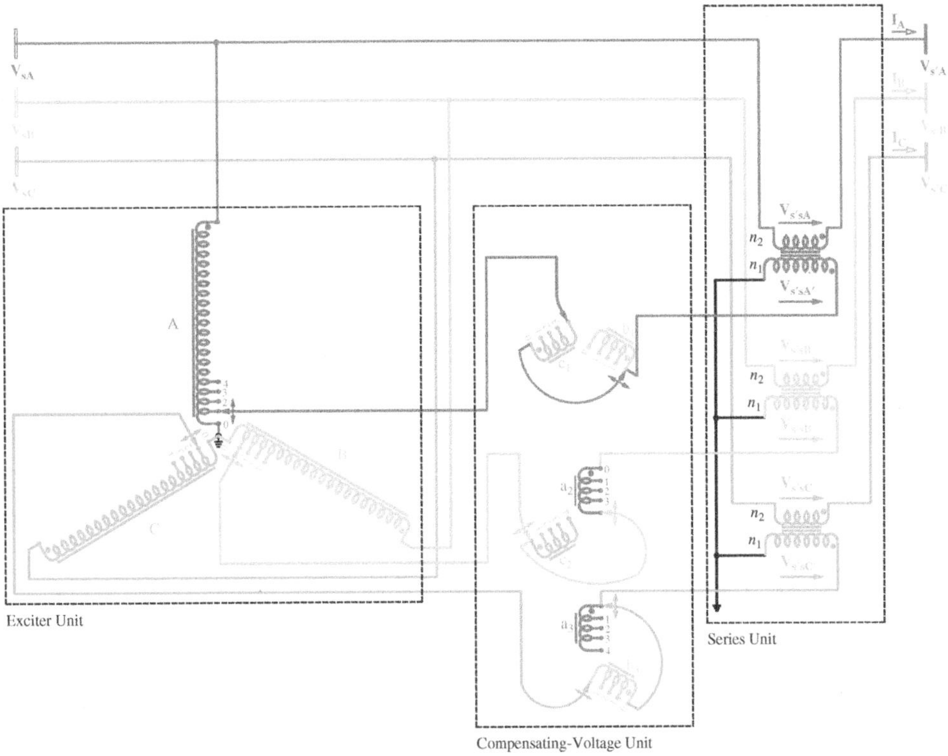

Figure 6-73 ST configuration with six secondary windings in a two-core design.

single-phase transformers (or one three-phase transformer with three isolated secondary windings), thus isolating the taps from any electrical connection with the high-voltage transmission line.

When a two-core design is used, the configuration of the Compensating-Voltage Unit can be further simplified, which requires only six secondary windings, as shown in Figure 6-73, to provide the same hexagonal operating area as shown in Figure 6-12. In this case, the primary windings in the Exciter Unit also induce the in-phase secondary voltages as is done in an autotransformer.

This configuration can be even further simplified, which requires only three secondary windings and two three-phase tap changers (one less than the usual three three-phase tap changers). In this simplest configuration of ST, out of the two parts of the compensating voltage, the first part is induced in the Exciter Unit as an in-phase or an out-of-phase voltage in all three phases.

There are three options in the arrangement of the secondary windings in the Compensating-Voltage Unit to induce the second part of the compensating voltage and achieve the entire range of the relative phase angle of $0° \leq \beta \leq 360°$. Each compensating voltage is designated as (+) if it is an in-phase component or (−) if it is an out-of-phase component. The compensating voltages in Option 1 for all the phases are given in Table 6-13. The required compensating points for the A phase are shown in Figure 6-74 for the entire range of the relative phase angle of $0° \leq \beta \leq 360°$. The required winding configuration is shown in Figures 6-75. Note that the LTC tap settings are

Table 6-13 Components of Option 1 compensating voltages in the A, B, and C phases for the entire range of the relative phase angle of $0° \leq \beta \leq 360°$.

Ranges of relative phase angle (β) of the compensating voltages	Components of *A*-phase compensating voltage	Components of *B*-phase compensating voltage	Components of *C*-phase compensating voltage
0–120°	+A phase and +C phase	+B phase and +A phase	+C phase and +B phase
120–180°	+C phase and −A phase	+A phase and −B phase	+B phase and −C phase
180–300°	−A phase and −C phase	−B phase and −A phase	−C phase and −B phase
300–360°	−C phase and +A phase	−A phase and +B phase	−B phase and +C phase

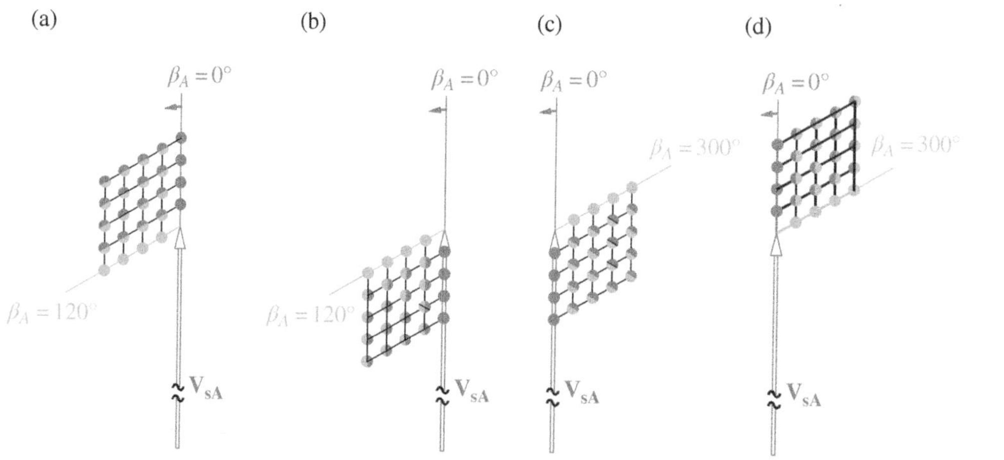

Figure 6-74 The voltage compensating points for the *A* phase in Option 1 configuration using (a) +A and +C secondary windings, (b) −A and +C secondary windings, (c) −A and −C secondary windings, and (d) +A and −C secondary windings.

set up according to Figure 6-74a. The line voltage and power flow operating areas are shown in Figure 6-76.

The operating area, which is a rhombus consists of a total number of possible compensating points (cp) that is defined by

$$cp = 4N(N + 1) \tag{6-43}$$

and it is 80. Therefore, from Equations (6-5) and (6-43), the additional number of compensating points are defined by

$$cp_{additional} = N(N + 1), \tag{6-44}$$

which is one-third more than the number of the compensating points in a hexagon.

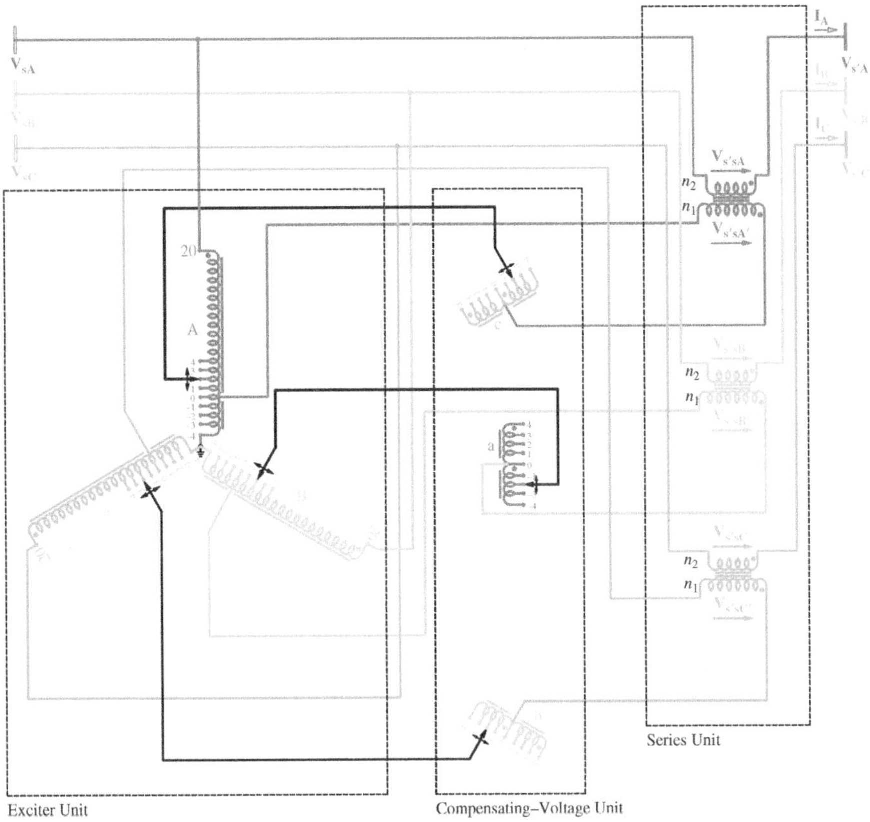

Figure 6-75 ST configuration with three secondary windings in a two-core design: ±A and ±C secondary voltages for compensation in the *A* phase, ±B and ±A secondary voltages for compensation in the *B* phase, and ±C and ±B secondary voltages for compensation in the *C* phase.

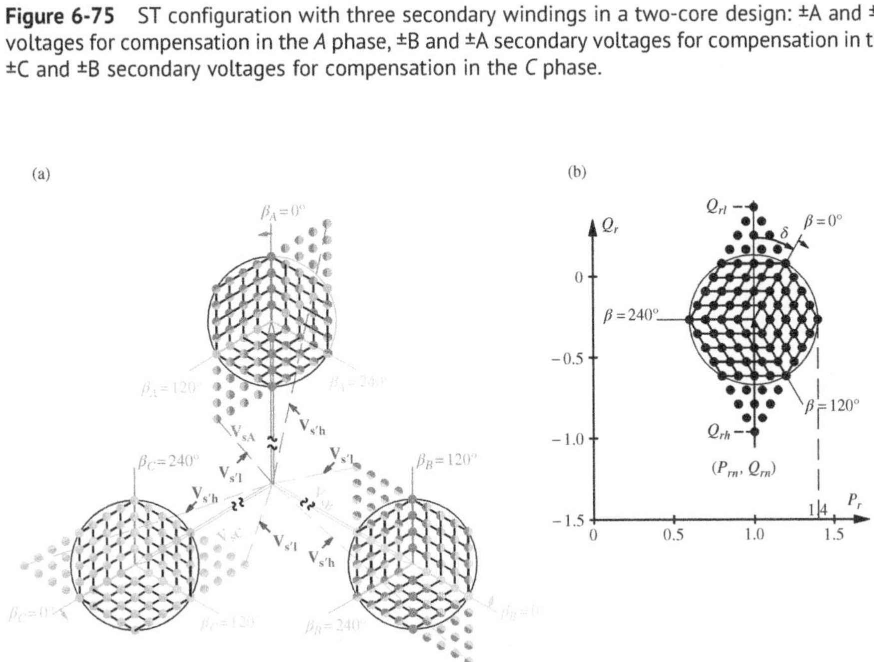

Figure 6-76 Operating points of an ST in Option 1: (a) modified sending-end voltage and (b) active and reactive power flows at the receiving end.

Table 6-14 Components of Option 2 compensating voltages in the A, B, and C phases for the entire range of the relative phase angle of 0° $\leq \beta \leq$ 360°.

Ranges of relative phase angle (β) of the compensating voltages	Components of *A*-phase compensating voltage	Components of *B*-phase compensating voltage	Components of *C*-phase compensating voltage
0–60°	+A phase and −B phase	+B phase and −C phase	+C phase and −A phase
60–180°	−B phase and −A phase	−C phase and −B phase	−A phase and −C phase
180–240°	−A phase and +B phase	−B phase and +C phase	−C phase and +A phase
240–360°	+B phase and +A phase	+C phase and +B phase	+A phase and +C phase

In Option 1, the line voltage varies the most as evidenced in Figure 6-76a. The highest voltage ($\mathbf{V_{s'h}}$) results in the lowest reactive power (Q_{rl}) and lowest voltage ($\mathbf{V_{s'l}}$) results in the highest reactive power (Q_{rh}) at the receiving end while the active power at the receiving end stays at one pu as evidenced in Figure 6-76b.

The compensating voltages in Option 2 for all the phases are given in Table 6-14. The required compensating points for the *A* phase are shown in Figure 6-77 for the entire range of the relative phase angle of 0° $\leq \beta \leq$ 360°. The required winding configuration is shown in Figures 6-78. Note that the LTC tap settings are set up according to Figure 6-77d. The line voltage and power flow operating areas are shown in Figure 6-79.

In Option 2, the line voltage varies the most as evidenced in Figure 6-79a. However, the highest voltage ($\mathbf{V_{s'h}}$) results in the highest active power (P_{rh}) and lowest reactive power (Q_{rl}); lowest voltage ($\mathbf{V_{s'l}}$) results in the lowest active power (P_{rl}) and highest reactive power (Q_{rh}) at the receiving end as evidenced in Figure 6-79b.

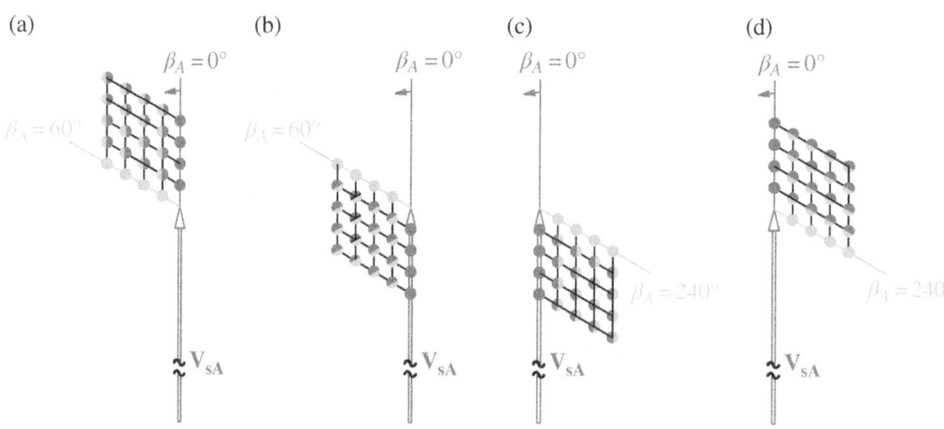

Figure 6-77 The voltage compensating points for the *A* phase in Option 2 configuration using (a) +A and −B secondary windings, (b) −A and −B secondary windings, (c) −A and +B secondary windings, and (d) +A and +B secondary windings.

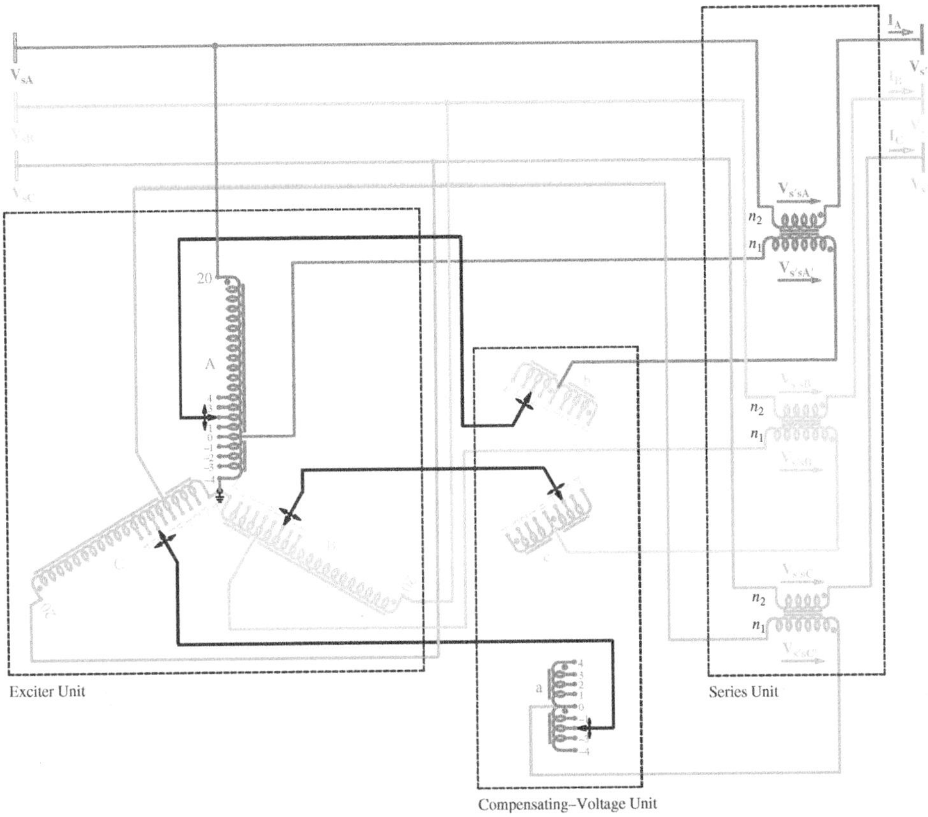

Figure 6-78 ST configuration with three secondary windings in a two-core design: ±A and ±B secondary voltages for compensation in the *A* phase, ±B and ±C secondary voltages for compensation in the *B* phase, and ±C and ±A secondary voltages for compensation in the *C* phase.

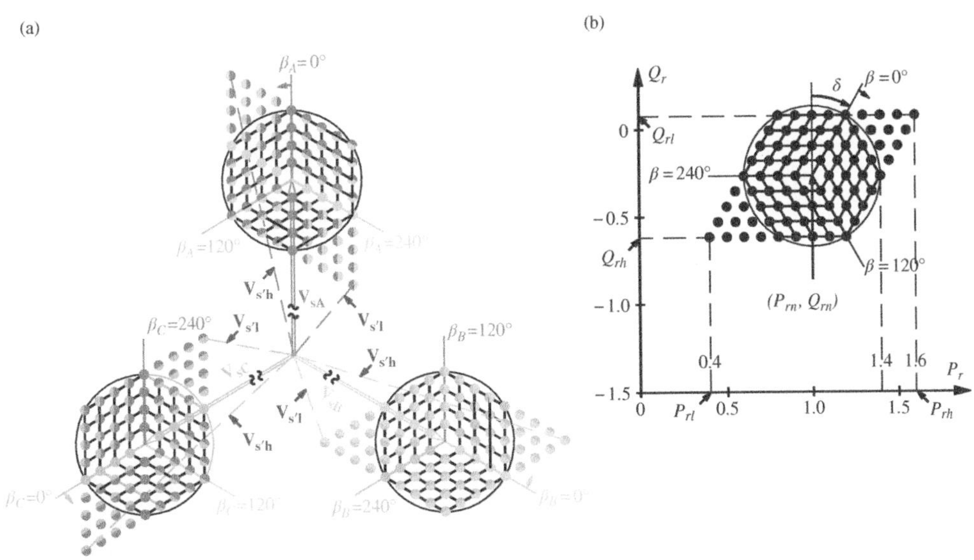

Figure 6-79 Operating points of an ST in Option 2: (a) modified sending-end voltage and (b) active and reactive power flows at the receiving end.

The compensating voltages in Option 3 for all the phases are given in Table 6-15. The required compensating points for the A phase are shown in Figure 6-80 for the entire range of the relative phase angle of $0° \leq \beta \leq 360°$. The required winding configuration is shown in Figures 6-81. Note that the LTC tap settings are set up according to Figure 6-80c. The line voltage and power flow operating areas are shown in Figure 6-82.

In Option 3, the phase-shift angle varies the most as evidenced in Figure 6-82a. The highest modified power angle (δ'_h) results in the highest active power (P_{rh} = 1.6 pu) and the highest reactive power (Q_{rh}); lowest modified power angle (δ'_l) results in the lowest active power (P_{rl} = 0.4 pu) and the lowest reactive power (Q_{rl}) at the receiving end as evidenced in Figure 6-82b.

Each of these alternate configurations must be studied further for its merits and demerits to select the best power flow controller that meets the needs of the utilities worldwide at an affordable price. For example, while Option 1 offers a 40% active power flow enhancement, each of Options 2 and 3 offers a 60% active power flow enhancement. However, in Option 3, for a 60% increase in active power flow, the line voltage increases only ~6%, since $V_{s'} = \sqrt{\left\{1^2 + \left(0.2 \times \sqrt{3}\right)^2\right\}} = 1.058$ (see Figure 6-82a) whereas, in Option 2, for a 60% increase in active power flow, the line voltage increases ~31%, since $V_{s'} = \sqrt{\left\{1.3^2 + \left(0.1 \times \sqrt{3}\right)^2\right\}} = 1.311$ (see Figure 6-79a), which is beyond

Table 6-15 The compensating voltages in Option 3 for all phases.

Ranges of relative phase angle (β) of the compensating voltages	Components of A-phase compensating voltage	Components of B-phase compensating voltage	Components of C-phase compensating voltage
−60° (300°)–60°	−C phase and −B phase	−A phase and −C phase	−B phase and −A phase
60–120°	−B phase and +C phase	−C phase and +A phase	−A phase and +B phase
120–240°	+C phase and +B phase	+A phase and +C phase	+B phase and +A phase
240–300°	+B phase and −C phase	+C phase and −A phase	+A phase and −B phase

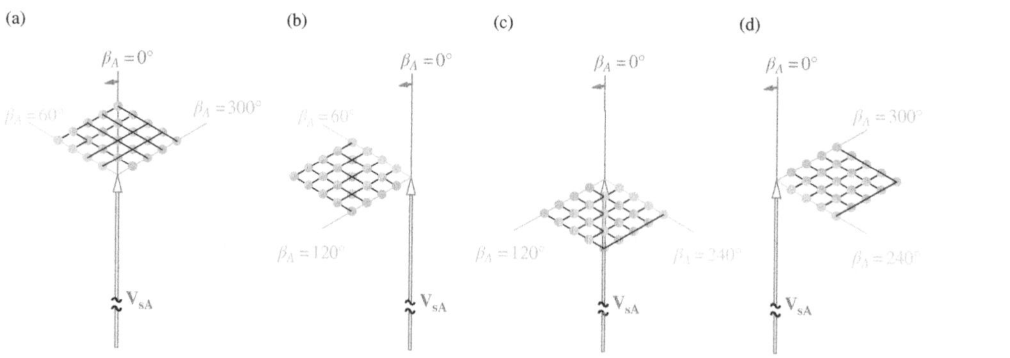

Figure 6-80 The voltage compensating points for the A phase in Option 3 configuration using (a) −B and −C secondary windings, (b) −B and +C secondary windings, (c) +B and +C secondary windings, and (d) +B and −C secondary windings.

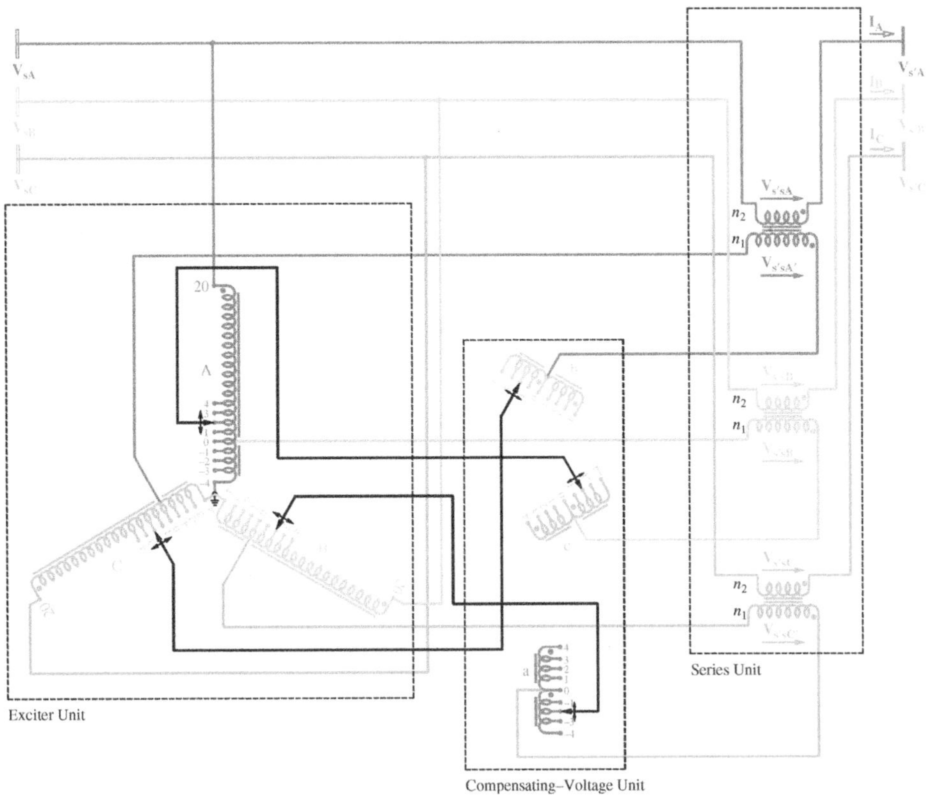

Figure 6-81 An ST configuration with three secondary windings in a two-core design: ±B and ±C secondary voltages for compensation in the *A* phase, ±C and ±A secondary voltages for compensation in the *B* phase, and ±A and ±B secondary voltages for compensation in the *C* phase.

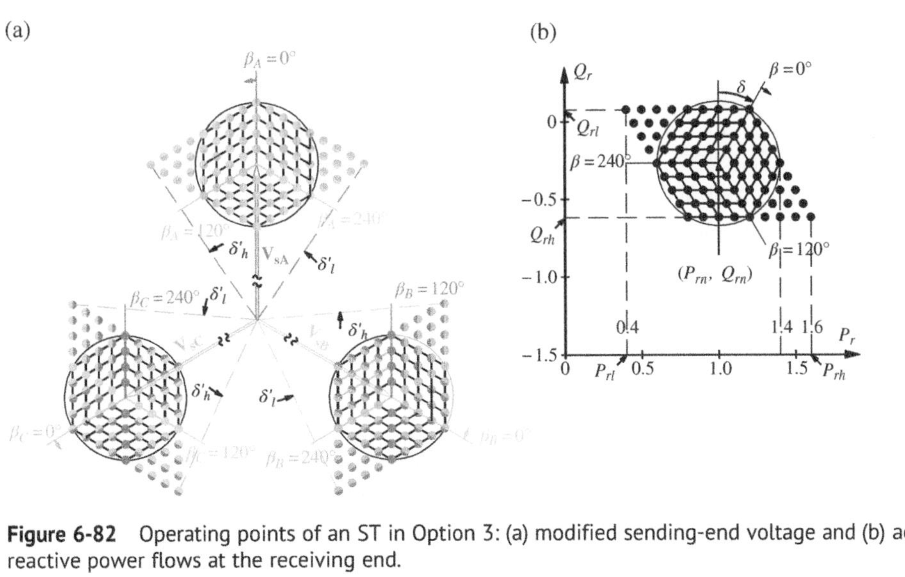

Figure 6-82 Operating points of an ST in Option 3: (a) modified sending-end voltage and (b) active and reactive power flows at the receiving end.

the allowable limit in most utility applications. Therefore, Option 3 provides maximum active power flow at maximum reactive power flow while the line voltage remains within a realistic limit of <10% increase of the nominal voltage. Note that by choosing the number of LTC taps, which is essentially the active turns-ratio of the secondary, series-compensating windings and the exciting primary windings of the ST, the realistic limit of the line voltage can be set at a desired level.

6.3 Comparison Among the VRT, PAR, UPFC, and ST

The comparison between the VSC-based UPFC and the transformer/LTCs-based ST can be done on several key points. The transformer/LTCs-based VRT and PAR are also included for reference.

The ST generates a compensating voltage ($\mathbf{V}_{s's}$) of variable magnitude and variable phase angle through transformer action between the shunt-connected primary, exciting windings and the series-connected secondary, compensating windings. When connected in series with the line, the compensating voltage modifies the magnitude and phase angle of the uncompensated sending-end voltage ($\mathbf{V}_s$) to be the modified sending-end voltage ($\mathbf{V}_{s'}$) and, therefore, control the active and reactive power flows in the line independently.

The ST offers an area of controllability in the *P*-*Q* plane just like the UPFC. Note that the VRT and PAR offer only linear characteristics in the *P*-*Q* plane and, therefore, do not regulate the active and reactive power flows independently as desired.

6.3.1 Power Flow Enhancement

The simulation results for the network example considered in Chapter 3 shows that a 30 MVA-rated Series Unit (Compensating-Voltage Unit in the ST) increases the power flow in the line by 57 MW in the case of the ST and by 58 MW in the case of the UPFC. This increased capacity of the line may be used to redirect power from an overloaded line to an underloaded line that may be very useful during the peak demand period.

In the case of the ST, both active and reactive powers exchanged by the series-connected Compensating-Voltage Unit flow through the shunt-connected Exciter Unit. However, in the case of the UPFC, only the exchanged active power flows through the shared DC link between the series-connected SSSC and the shunt-connected STATCOM. Each VSC is capable of generating or absorbing reactive power at its AC terminal. The series-connected VSC in the UPFC must be designed for the apparent power, which carries the exchanged active and reactive powers with the line. The rating of the shunt-connected VSC must be at least equal to the active power rating of the series-connected VSC. Two VSCs in the UPFC are of equal ratings for simpler housekeeping and practical reasons, such as storing spare parts. Therefore, the remaining rating of the STATCOM can be utilized to regulate the bus voltage by exchanging the reactive power at the POC. In the UPFC configuration, the STATCOM has to let the active power of the series-connected VSC go through it first, thereby reducing its full reactive power-carrying and voltage-regulating capabilities.

Since the shunt- and the series-connected VSCs of the UPFC, considered in Chapter 3, are designed to be of the same rating, a 30 MVA-rated Shunt Unit is able to provide 21 Mvar of reactive power while delivering the maximum 21 MW of active power. If this capability of supplying shunt reactive power is really needed, particularly when the short-circuit impedance at the POC is high, a separate capacitor bank as a shunt reactive compensator may be used along with the ST. Ultimately, the life-cycle cost of a PFC must be the deciding factor for its selection among all the available options.

6.3.2 Speed of Operation

Figure 6-83a shows the test results from the world's first UPFC that consists of a shunt-connected ±160 MVA, 138 kV-rated VSC (STATCOM) and a series-connected ±160 MVA, 13.33 kV-rated VSC (SSSC). The STATCOM is operated in "Voltage Control" mode to hold the bus voltage ($\mathbf{V_s}$)

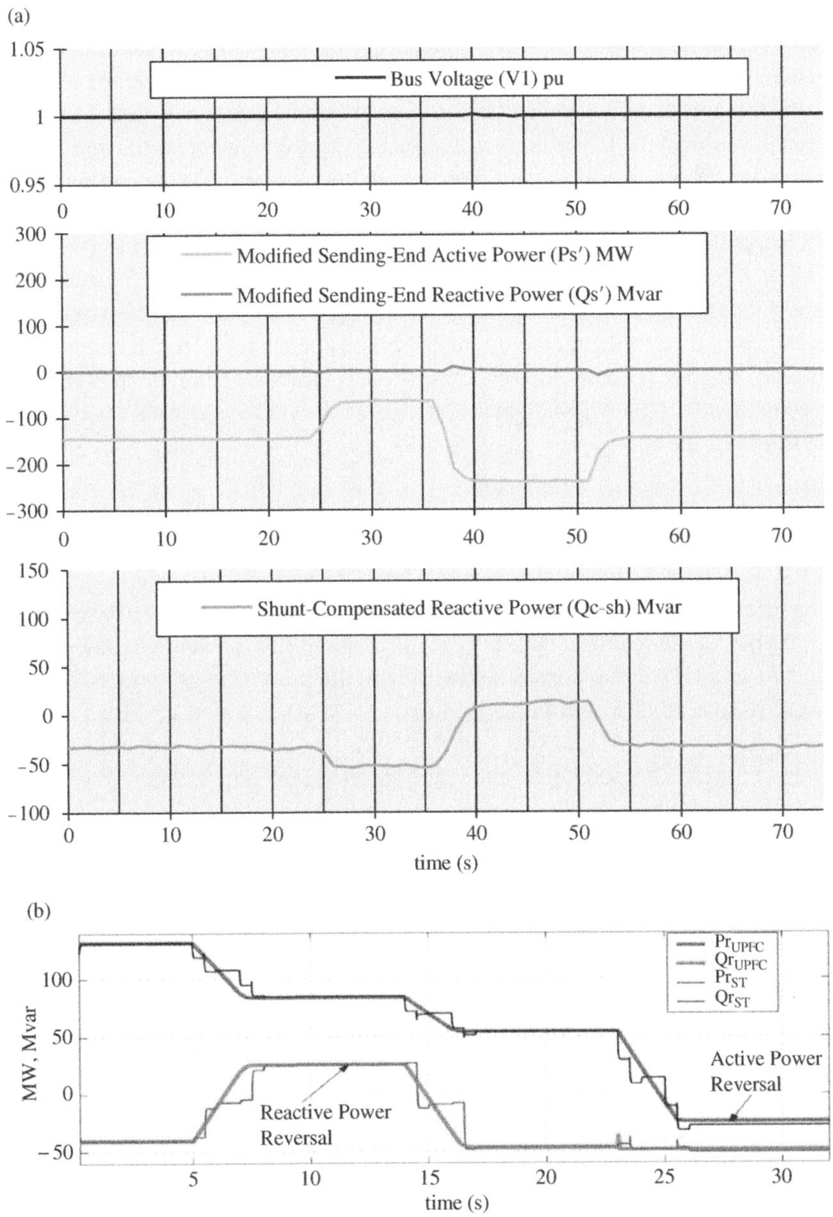

Figure 6-83 (a) UPFC test results for automatic power flow control mode of operation with maintaining unity power factor and 1 pu bus voltage while regulating the line power (Sen and Keri 2002); (b) ST simulation results for open-loop power flow control mode of operation (Faruque and Dinavahi 2007).

at one pu. While maintaining unity power factor load on the line and the bus voltage at one pu level, the active power flow in the line is varied. The active power flow at the modified sending end is regulated at −145, −65, −240, and −145 MW levels while holding the reactive power flow at the same point at zero (*i.e.* unity power factor). The STATCOM exchanges an appropriate amount of reactive power with the line to hold the BUS01 voltage at one pu level. The response time of the UPFC is adjusted to be in several seconds.

The simulated transient responses of P_r and Q_r of the ST and UPFC are shown in Figure 6-83b. The natural power flows between 0 and 5 s. A series-compensating voltage ($\mathbf{V}_{s's}$) of magnitude 0.2 pu is applied at a relative phase angle (β) of 300° at 5 s and 240° at 14 s. While keeping the relative phase angle at 240°, the magnitude of the compensating voltage is increased to 0.4 pu at 23 s. The line power due to the application of the UPFC is changed in a linear fashion, whereas the line power due to the application of the ST is changed in multiple steps. In both cases, the same level of independent power flow control and the line power flow reversal are demonstrated.

The merit of the ST is that it allows independent control of active and reactive power flows in the line, similar to the UPFC. The UPFC can offer a faster response (in ms at a given constant line impedance), but this capability cannot be utilized, as shown in Figure 6-83a, in order to assure continued operation under various contingencies. In most utility applications in which the fast response is not required, the ST presents an economically attractive solution for power flow enhancement when compared with the UPFC, which has higher life-cycle costs (installation, operation, and maintenance) than the ST.

6.3.3 Losses

The power loss in a VSC is defined as the total power consumed by various components of a VSC while carrying the rated current at rated voltage. A Harmonic-Neutralized VSC (HN-VSC) with Gate-Turn Off (GTO) thyristors, switching at once per cycle, has about 1.5% power loss under rated condition. That means for the UPFC that consists of two VSCs, the power loss is about 3% of the VSC rating. The power loss in a single Pulse Width Modulation (PWM)-operated VSC is about 4%. Therefore, for the UPFC that consists of two PWM-operated VSCs, the power loss is about 8%. This makes the operating cost of the VSC-based UPFC to be the highest among the available power flow solutions. With the advancement of semiconductor switches, such as GaN and SiC, the lower turn-on and turn-off times offer lower switching losses; a lower voltage across the switch also offers a lower conduction loss. Thus, the overall loss may be reduced.

6.3.4 Switch Rating

Each VSC that carries its nominal current at its nominal voltage is considered to have one-pu-rated semiconductor devices. In practice, the allowable current through a GTO is rated at 0.5 pu of the controllable current of the device. Note that there are a number of GTOs connected in series and each one of these devices experiences a voltage overshoot at turn-off from the stored energy in the snubber and stray inductances. Considering that the continuous operation is required even with a number of failed-short GTOs in a switch, a conservative design dictates that each device be allowed to operate normally at 0.25 pu of the rated voltage. This means that the installed power rating of the semiconductor devices is eight times the nominal rating in each VSC. In addition, there are two electronic switches: one for clamping the DC link voltage and another one for the electronic bypass. The semiconductor and snubber losses in the VSC require deionized-water cooling and heat exchangers to remove the heat that is generated from the switching of the semiconductor devices.

The first three installed UPFCs in Kentucky, USA, New York, USA and South Korea were built with the use of relatively slow semiconductor devices, such as GTO thyristors that are switched at fundamental frequency. This arrangement requires the use of a HN-VSC. A VSC that is built with relatively faster semiconductor devices, such as Insulated Gate Bipolar Transistors (IGBTs), are operated with PWM techniques with a higher switching frequency. Its magnetic circuit may be limited to just the coupling transformer. However, this arrangement requires the use of a filter at the output of each VSC. In order to reduce the size of the filter, it is desirable to increase the switching frequency, which is limited due to the rising switching losses. The filter may be eliminated in a VSC configuration with multi-modular units where the semiconductor devices reside in the series-connected full bridges or half bridges. A VSC, operated with the PWM techniques, may also need a shield to limit electromagnetic interference (EMI), generated from the high di/dt of the faster semiconductor switches.

6.3.5 Magnetic Circuit Design

Each coupling transformer in the UPFC carries its rated (one pu) power under steady-state condition. The intermediate magnetic circuit, associated with each GTO-based VSC, is rated at 0.5 pu. A one-pu-rated series-coupling transformer permits the compensating voltage to start at its peak value, so that a de-energized transformer can start operating from its zero flux condition and not be saturated. However, this restriction can delay the response of the VSC up to a quarter cycle. If the controller demands the compensating voltage to be applied at its zero value instead of its peak value, the flux in the transformer may ideally build up to the two pu level. Therefore, the series-coupling transformer needs to be designed to be at the two pu level in order to avoid saturation. In this case, the intermediate magnetic circuit associated with the series-connected VSC needs to be rated at one pu instead of 0.5 pu. Therefore, the theoretical rating of the magnetic circuit of the HN-VSC-based UPFC is 4.5 pu (shunt-coupling transformer: one pu; shunt intermediate transformer: 0.5 pu; series-coupling transformer: two pu; and series intermediate transformer: one pu). In practice, the magnetic circuit for each VSC is typically designed to be rated at 15 to 25% more than the theoretical rating.

In most utility applications, it is adequate to apply the series-compensating voltage, calculated by the controller, over the following 10 to 20 cycles. This relaxes the requirement of the rating of the series-coupling transformer. A series-coupling transformer of 1.5 pu rating is considered to be adequate in a utility application. Then the question arises: Why should one invest in a fast-acting VSC when its output voltage cannot be delivered at the same speed of the electronic VSC because the series-coupling transformer is not designed with full rating?

Each of the three primary windings in the Exciter Unit is designed for full line voltage and the rated single-phase power of the ST. Each of the nine secondary windings in the Compensating-Voltage Unit is designed to carry the full line current and the rated single-phase power of the ST. This requires the magnetic rating in Volt-second of the 12 windings (three primary and nine secondary) of the ST to be $12/6 = 2$ pu, considering the three primary and three secondary windings, a total of six windings, make a one pu transformer.

6.3.6 Optimization of Transformer Rating

The UPFC's rated compensating voltage in series with a transmission line in its possible $0° \leq \beta \leq 360°$ range has very unlikely been used. If the rated compensating voltage ($\mathbf{V}_{\mathbf{s}'\mathbf{s}}$) is added to the sending-end voltage ($\mathbf{V}_{\mathbf{s}}$) at 0° or 180°, the modified sending-end voltage ($\mathbf{V}_{\mathbf{s}'} = \mathbf{V}_{\mathbf{s}} + \mathbf{V}_{\mathbf{s}'\mathbf{s}}$) goes out of the allowable range, which may be within the ±10% of the nominal value. Moreover, in the $0° \leq \beta \leq 360°$ range of operation of the UPFC, the maximum power flow enhancement occurs within the first

120° and, therefore, the rest of the operating region is unnecessary in most applications. The ST, in its simpler, customized configuration with a reduced amount of hardware provides the same limited angle (120°) operation, which is adequate in most utility applications.

Whether it is utilized or not, the UPFC is always designed to operate in the $0° \leq \beta \leq 360°$ range, since it does not offer the limited angle design option. In a UPFC with HN-VSCs, the required magnetic rating in Volt-second alone is at least 4.5 pu, whereas, in the ST, the magnetic rating in Volt-second is only two pu for the same $0° \leq \beta \leq 360°$ range of operation. In a practical implementation, only $0° \leq \beta \leq 120°$ range of operation is required for power flow enhancement. Consequently, this requirement is satisfied by constructing the ST as shown in Figure 6-65a. This configuration requires only three primary windings and six secondary windings for a limited angle operation. This results in a 1.5-pu-rated transformer, instead of a regular two-pu-rated transformer for $0° \leq \beta \leq 360°$ range of operation. In comparison, the autotransformer and the PAR, each with one pu rating, can jointly perform in the $0° \leq \beta \leq 90°$ range of operation, requiring a total of two pu rating. Therefore, it can be concluded that the functions of the autotransformer and PAR are combined in a single unit of ST that results in a reduced amount of hardware from what is required for separate autotransformer and PAR. The number of isolated compensating windings in the ST can be reduced by the following action. The compensating winding that provides the in-phase voltage and the corresponding exciting winding can be replaced with a partially tapped autotransformer winding as shown in Figure 6-65e.

Figure 6-84 shows the practical control range of the VRT, PAR, UPFC, and ST. The filled dots, in Figure 6-84a, show a total of 36 compensating points of the ST within the allowable voltage range, which is ±10% of the nominal voltage in the relative phase angle range of $0° \leq \beta \leq 360°$. Figure 6-84b shows that the maximum active power flow enhancement at the receiving end from the natural flow of one pu, within the allowable voltage range, is 0.40 pu with the UPFC, 0.35 pu with the ST, and 0.30 pu with the PAR (asym). Moreover, the ST can provide an active power flow enhancement of 0.3 pu as does the PAR, but along a constant reactive power line (without altering the natural flow of reactive power, Q_{rn}), whereas the PAR increases the reactive power from the

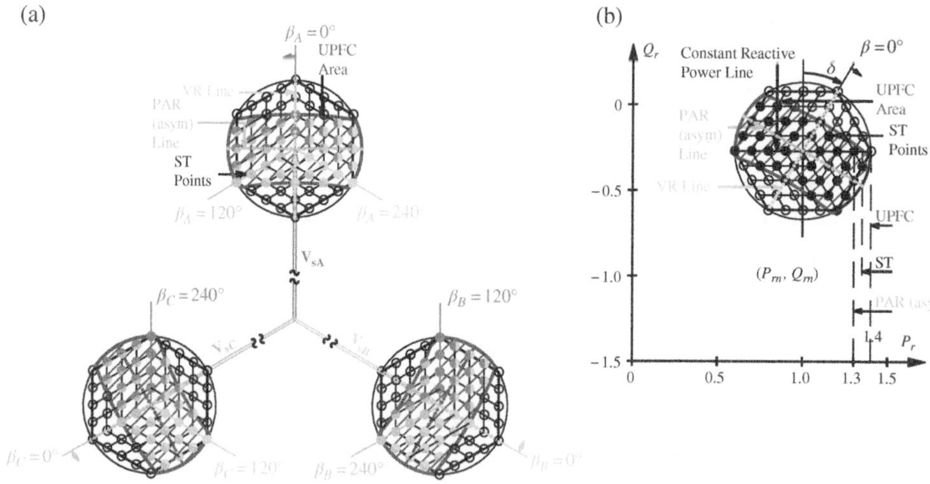

Figure 6-84 Practical voltage compensation range with the use of the VRT, PAR, UPFC, and ST within the entire range of the relative phase angle of $0° \leq \beta \leq 360°$: (a) modified sending-end voltage and (b) active and reactive power flows at the receiving end.

natural flow of 0.268 pu to the simultaneously regulated flow of 0.441 pu. The VRT also regulates the active and reactive power flows simultaneously.

6.3.7 Harmonic Injection into the Power System Network

The ST always generates a sinusoidal compensating voltage of line frequency. Even though the compensating voltage emulates an impedance, this virtual impedance is actually a finite voltage and, therefore, cannot cause resonance with the line inductance or capacitance.

The UPFC always injects harmonic components (voltage and current) into the power system network. The magnitudes of the harmonic components are reduced with a higher leakage reactance in the coupling transformers, but the leakage reactance of the series-coupling transformer increases the overall undesirable inductive compensation range that reduces power flow in the line and reduces the overall desirable capacitive compensation range that increases power flow in the line.

6.3.8 Operation During Line Faults

Generally, to limit a phase-to-neutral fault current, additional inductive impedance in the path of the fault is needed. If, for example, the additional emulated impedance is equal to the equivalent line impedance, the steady-state fault current is reduced to half of the natural fault current. The steady-state faulted voltage across the emulated impedance is half of the phase-to-neutral voltage. In the worst case, the emulated impedance must withstand twice the volt-seconds that are needed to support half of the steady-state phase-to-neutral voltage in order to avoid magnetic saturation. The relatively insignificant (typically 0.05 to 0.15 pu) voltage rating of the series core of the ST would saturate early in the fault cycle. The saturated windings do not provide significant current-limiting effect in the ST, so the ST, designed for power flow regulation, rides through a fault with little effect.

During a fault in the transmission line, the saturated core of the series-coupling transformer does not provide significant current-limiting effect in the UPFC. The current through the series-connected VSC will exceed its rated limit. As a protective measure, the semiconductor switches of a shunt-connected VSC are turned off to ensure zero current flow through it, but the semiconductor switches of the series-connected VSC have to be all turned on to ensure zero compensating voltage, so that the VSC side of the series-coupling transformer is shorted; as a result, the line is not open-circuited. Therefore, to ride through the line faults, the series-connected VSC must be designed to carry the fault current. For example, if the short circuit current is 10 times the nominal line current, the series-connected VSC needs to be designed to carry the fault current. This is an expensive proposition. The alternate solution is to have a nominally rated VSC as a series compensator and an additional electronic bypass-switch that is specifically designed to carry the fault current. Then the question arises as to whether the electronic bypass-switch should be located on the high-voltage line side or on the low-voltage VSC side. If the electronic bypass-switch is located on the high-voltage side, the VSC "sees" an additional inductance from the leakage reactance of the series-coupling transformer in its forward path. Designers prefer to have the electronic bypass-switch on the high-voltage line side because when shorted, all of the leakage reactances of the coupling transformer and intermediate transformer appear across the output terminal of the VSC and a short circuit at its output terminal is prevented. However, it is quite expensive to have electronic switches mounted on an isolated platform and connected to the high-voltage transmission line. It is certainly less expensive to have the electronic bypass-switch located on the low-voltage side (VSC side) of the series-coupling transformer. In this case, when the electronic bypass-switch is operated, the output voltage of the VSC is applied only to the very low leakage reactance of the intermediate transformer. Then the question arises as to whether the VSC is adequately designed to carry this resulting current until the fault current is transferred to the mechanical bypass switch, which takes a few line cycles to operate. At that time, the series-connected VSC can be stopped gating without opening the line. If the electronic bypass-switch is located on the low-voltage VSC side, the VSC should be protected in some way in order to avoid a high current through it during a line fault condition. The obvious solution is to put an additional reactor on the low-voltage VSC side. This additional reactor always stays connected in the forward path of the VSC and, as a result, increases the overall undesirable inductive compensation range and reduces the overall desirable capacitive compensation range.

Note that a HN-VSC that operates at a fundamental switching frequency, provides a positive sequence fundamental voltage. In the case of a line fault, the negative sequence component of the unbalanced line voltage causes the negative sequence current to flow freely into the VSC, which is not desirable. A VSC that uses a PWM modulation technique provides a desirable voltage, not just a positive sequence fundamental voltage, which is effective in operations during unbalance and faults.

The basic differences between the traditional technology of transformer and LTCs versus the VSC are given in Table 6-16.

6.4 Multiline Sen Transformer

A Multiline Sen Transformer (MST) connects a compensating voltage in series with each line in a multiline transmission system and transfers active and reactive powers selectively from one or more transmission lines, referred to as "leader" lines, to the others, referred to as "follower" lines. Each compensating voltage in the "leader" lines is of any magnitude within its allowable limit and at any phase angle with respect to the line voltage, as well as the prevailing line current, so that the active and reactive power flows in that line can be controlled independently as desired. Each series-compensating voltage in the "follower" lines is of specific magnitude and phase angle with respect to the prevailing line current, so that the active and reactive powers from the "leader" lines are transferred bidirectionally to the "follower" lines. This technique provides a desirable power flow management for a multiline transmission system by decreasing the power flow in the overloaded lines and increasing it in the underloaded lines with little impact on the other uncompensated lines.

In the MST configuration, shown in Figure 6-85, the line voltage ($\mathbf{V_s}$) is applied to a shunt-connected three-phase transformer's Y-connected primary windings in the Exciter Unit. In the Compensating-Voltage Unit 1, a total of nine secondary windings (a_{11}, a_{12}, and a_{13} on the core of the *A* phase; b_{11}, b_{12}, and b_{13} on the core of the *B* phase; and c_{11}, c_{12}, and c_{13} on the core of the *C* phase) derive a three-phase compensating voltage, each phase of which is a phasor sum of the voltages induced in a three-winding set (a_{11}, b_{11}, and c_{11} for compensating voltage in the *A* phase; a_{12}, b_{12}, and c_{12} for compensating voltage in the *B* phase; and a_{13}, b_{13}, and c_{13} for compensating voltage in the *C* phase). In the Compensating-Voltage Unit 2, a total of nine secondary windings (a_{21}, a_{22}, and a_{23} on the core of the *A* phase; b_{21}, b_{22}, and b_{23} on the core of the *B* phase; and c_{21}, c_{22}, and c_{23} on the core of the *C* phase) derive a three-phase compensating voltage, each phase of which is a phasor sum of the voltages induced in a three-winding set (a_{21}, b_{21}, and c_{21} for compensating voltage in the *A* phase; a_{22}, b_{22}, and c_{22} for compensating voltage in the *B* phase; and a_{23}, b_{23}, and c_{23} for compensating voltage in the *C* phase). By choosing the number of turns from each of the three windings, and, therefore, the magnitudes of the components of the three 120° phase-shifted

Table 6-16 Comparison among the voltage-regulating transformer (VRT), phase angle regulator (PAR), unified power flow controller (UPFC), and Sen transformer (ST).

	Interdependent active and reactive power flow control		Shunt-series interconnected voltage injection with independent local voltage, active and reactive power flow control		
	VRT	PAR	UPFC with HN-VSCs	UPFC with PWM-VSCs	ST
In-phase voltage injection	Yes	No	Yes	Yes	Yes
Quadrature voltage injection	No	Yes	Yes	Yes	Yes
Independent control of active and reactive power flows in the line	No	No	Yes	Yes	Yes
Number of components/complexity	Low	Low	High	High	Low
Reliability and high availability	Yes	Yes	No	No	Yes
Harmonic voltage injection into the power system	No	No	Yes	Yes	No
Loss of useful capacitive rating due to high series leakage reactance in the coupling transformer	No	No	Yes	Yes	No
Fast fault current bypass-switch needed	No	No	Yes	Yes	No
Adequate dynamic response for utility power flow regulation	Yes	Yes	Yes	Yes	Yes
Compensating voltage steps depend on number of taps installed	Yes	Yes	No	No	Yes
Capability of independent local reactive power generation and absorption	No	No	Yes	Yes	No
VSC output filter needed	No	No	No	Yes[a]	No
VSC EMI shields may be needed	No	No	Yes[b]	Yes[b]	No
Estimated losses at rated power	<1%	<1%	3%	6–8%	<1%
Required power rating of magnetic components (pu)	1	1	4.5	3	1.5
Estimated equipment cost normalized for a unit rated 50 MVA or more	1	1	7.5	6	1.5
Estimated losses/operating cost normalized for a unit rated 50 MVA or more	1	1	5	10	1

[a] Series-connected half- or full-bridge converters may or may not need any passive filter at its output.
[b] Semiconductor switches with fast turn-on and turn-off capabilities may need an EMI shield.

induced voltages, the composite compensating voltage magnitude and the relative phase angle with respect to the line voltage can be selected.

It should be noted that in the Compensating-Voltage Unit 1, each of the windings (a_{11}, b_{12}, and c_{13}) is tapped at the same number of turns; each of the windings (b_{11}, c_{12}, and a_{13}) is tapped at the same number of turns; and each of the windings (c_{11}, a_{12}, and b_{13}) is tapped at the same number of turns. However, the number of turns in the a_{11}-b_{12}-c_{13} set, b_{11}-c_{12}-a_{13} set, and c_{11}-a_{12}-b_{13} set can be different from each other. Also, in the Compensating-Voltage Unit 2, each of the windings (a_{21}, b_{22},

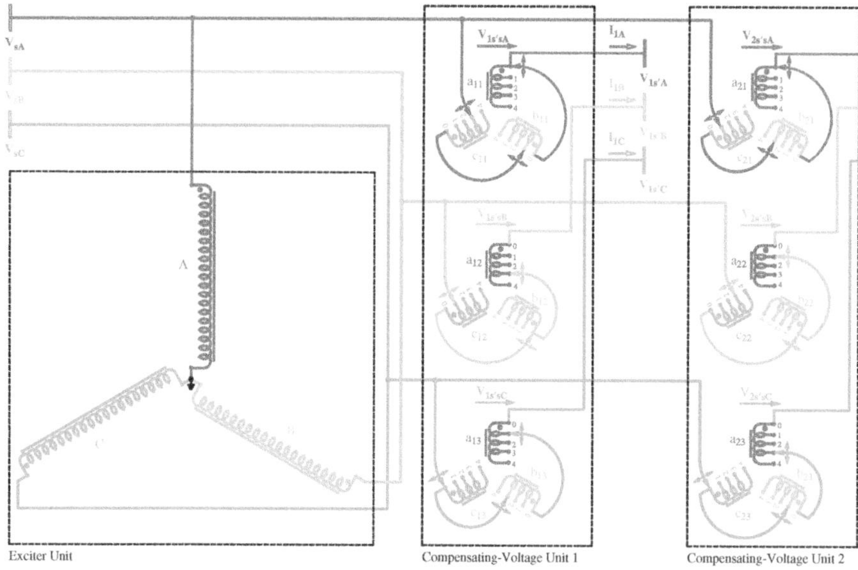

Figure 6-85 Multiline Sen Transformer (MST).

and c_{23}) is tapped at the same number of turns; each of the windings (b_{21}, c_{22}, and a_{23}) is tapped at the same number of turns; and each of the windings (c_{21}, a_{22}, and b_{23}) is tapped at the same number of turns. However, the number of turns in the a_{21}-b_{22}-c_{23} set, b_{21}-c_{22}-a_{23} set, and c_{21}-a_{22}-b_{23} set can be different from each other.

The compensating voltage ($\mathbf{V}_{1s's}$) in the Compensating-Voltage Unit 1 can be at any phase angle with the prevailing line current ($\mathbf{I}_1$), which emulates a four-quadrant, compensating impedance in series with the line 1. The compensating voltage ($\mathbf{V}_{2s's}$) in the Compensating-Voltage Unit 2 can be at any phase angle with the prevailing line current ($\mathbf{I}_2$), which emulates a four-quadrant, compensating impedance in series with the line 2. An emulated impedance can be a combination of a positive resistance ($+R$) or a negative resistance ($-R$) and a capacitive reactance (X_C) or an inductive reactance (X_L).

The series-compensating voltage ($\mathbf{V}_{1s's}$) modifies the sending-end voltage ($\mathbf{V}_s$) to $\mathbf{V}_{1s'}$ in order to independently control the active and reactive power flows in the "leader" line (line 1). The series-compensating voltage ($\mathbf{V}_{2s's}$) modifies the sending-end voltage ($\mathbf{V}_s$) to $\mathbf{V}_{2s'}$ in order to provide the same function for "follower" line (line 2). The compensating voltage in each line is also at any phase angle with the prevailing line current. The active or direct component of the compensating voltage provides the series resistance emulation, whereas the reactive or quadrature component provides the series reactance emulation.

The operations of the Compensating-Voltage Units are such that the compensating voltage in the "leader" line is at any phase angle with the line voltage, as well as the prevailing line current and that in the "follower" line is of specific magnitude and phase angle, so that the active and reactive powers from the "leader" line are transferred bidirectionally to the "follower" line. The basics of this control scheme are discussed next. As discussed earlier, the 120° range of operation of the PFC is often sufficient. The MST, in its simpler customized configuration with a reduced amount of hardware, provides this limited angle (120°) operation.

In the "leader" line (line 1), the controllable range of the active and reactive power flows can easily be determined by applying the series-connected, fully rated, compensating voltage that operates within its entire $0° \leq \beta \leq 360°$ range. As a consequence, the compensating voltage emulates an impedance in series with the line and, in turn, exchanges active and reactive powers with the line. For some compensating voltage ($\mathbf{V_{1s's}}$) in the "leader" line (line 1), the exchanged active and reactive powers are $P_{1se} = -\mathbf{V_{1s's}} \cdot \mathbf{I_1}$ and $Q_{1se} = |-\mathbf{V_{1s's}} \times \mathbf{I_1}|$, respectively. To transfer this exchanged active and reactive powers to the "follower" line, the following relationships must be obeyed, *i.e.*

$$P_{1se} + P_{2se} = 0 \tag{2-302}$$

$$Q_{1se} + Q_{2se} = 0. \tag{6-45}$$

Therefore, the exchanged active and reactive powers in the "follower" line are

$$P_{2se} = -P_{1se} \tag{6-46a}$$

$$Q_{2se} = -Q_{1se}; \tag{6-46b}$$

but the exchanged active and reactive powers in the "follower" line are

$$P_{2se} = I_2^2 R_{2se}^* \tag{6-47a}$$

$$Q_{2se} = I_2^2 X_{2se}^* \tag{6-47b}$$

where R_{2se}^* and X_{2se}^* are the emulated resistance and reactance commands, respectively.

From Equations (6-46) and (6-47), the emulated resistance and reactance commands (R_{2se}^* and X_{2se}^*) for the "follower" line (line 2) are given by

$$R_{2se}^* = -\frac{P_{1se}}{I_2^2} \tag{6-48a}$$

$$X_{2se}^* = -\frac{Q_{1se}}{I_2^2}. \tag{6-48b}$$

The required compensating voltage ($\mathbf{V_{2s's}}$) command can be given to the controller in terms of the emulated resistance and reactance commands (R_{2se}^* and X_{2se}^*). Substituting i with i_2, R_{se} with R_{2se}^*, and X_{se} with X_{2se}^* in Figure 6-13, the same control block diagram can be executed to operate the taps for the Compensating-Voltage Unit 2.

The above concept can be extended to an n-line MST provided the following relationships for the power balance are maintained:

$$P_{1se} + P_{2se} + \ldots + P_{nse} = 0 \tag{6-49a}$$

$$Q_{1se} + Q_{2se} + \ldots + Q_{nse} = 0. \tag{6-49b}$$

Any difference in power transfer between the transmission lines ("leaders" and "followers") will flow through the exciter winding. The exciter winding is rated to carry the differential power between the "leaders" and the "followers" lines. Note that in the implementation of a BTB-SSSC (also referred to as IPFC) for power flow management in a multiline transmission system, a balance of zero net active power at the DC link must be observed.

6.4.1 Basic Differences Between the MST and BTB-SSSC

A VSC-based BTB-SSSC is capable of transferring active and reactive powers selectively from one particular line to another. So can the transformer and the LTC-based MST. Note that in reality, a

BTB-SSSC only transfers the active power from one line to another and, in addition, generates or absorbs the reactive power in each line independently as desired.

Although a properly designed VSC is capable of applying a compensating voltage to the transmission line in sub-cycle time, it is not advisable to do so. System instability may occur during some contingencies. The BTB-SSSC always needs an overrated series-coupling transformer to avoid magnetic saturation. The response time of the MST is adequate in most utility applications, but it can be improved, when needed, by replacing the mechanical LTCs with faster switches. The MST with mechanical LTCs will ride through a line fault. To protect the VSC in a BTB-SSSC from the line fault current, a fast bypass-switch is required.

A BTB-SSSC is capable of connecting its rated compensating voltage in series with a transmission line for a possible $0° \leq \beta \leq 360°$ range of operation. However, in order to maintain the line voltage within an allowable range, the rated compensating voltage is used only over a narrow range of the possible $0° \leq \beta \leq 360°$ phase angle. The MST offers the required operating range with less hardware.

Unlike a VSC, which always injects harmonic components in addition to line frequency voltage into the power system network, the MST only injects a voltage of line frequency. A VSC may need additional filter circuits to reduce the injected harmonic components into the power system network.

The installation cost of the BTB-SSSC is in the order of five times that of the MST. Moreover, depending on the topology of the VSC, the BTB-SSSC with two VSCs has an efficiency in the range of 92–97%, whereas the MST with mechanical LTCs is more than 99% efficient. Therefore, the operating cost of the BTB-SSSC is an order of magnitude higher than that of the MST.

The PFC in a transmission system application is expected to operate reliably 99.9% of the time, except for scheduled maintenance. The reliability of the VSC-based BTB-SSSC concept has not yet been proven, but the components of the MST are transformers and LTCs, which are time-tested and proven to be reliable.

6.5 Flexible Operation of the ST

The basic scheme of the ST can be modified for special applications, such as limited angle operation with a reduced amount of hardware and interconnection of transmission lines of different voltages or phase angles while providing independent control of active and reactive power flows in those lines. Figure 6-86 shows one such scheme in which two transmission lines, one at 345 kV and the other at 138 kV, can be interfaced with the ST when the Compensating-Voltage Unit is connected at the 138 kV line side. The primary windings of the ST are energized with the voltage from the 345 kV line. The magnitude and phase angle of the stepped-down 138 kV voltages are controlled with the use of the Compensating-Voltage Unit.

The above scheme can be extended, as shown in Figure 6-87, to interconnect the transmission lines of different voltage levels while providing independent control of active and reactive power flows in a higher voltage line. The primary windings of the ST are energized from the lower voltage side. The magnitude and phase angle of the stepped up voltages are modified with the use of the Compensating-Voltage Unit.

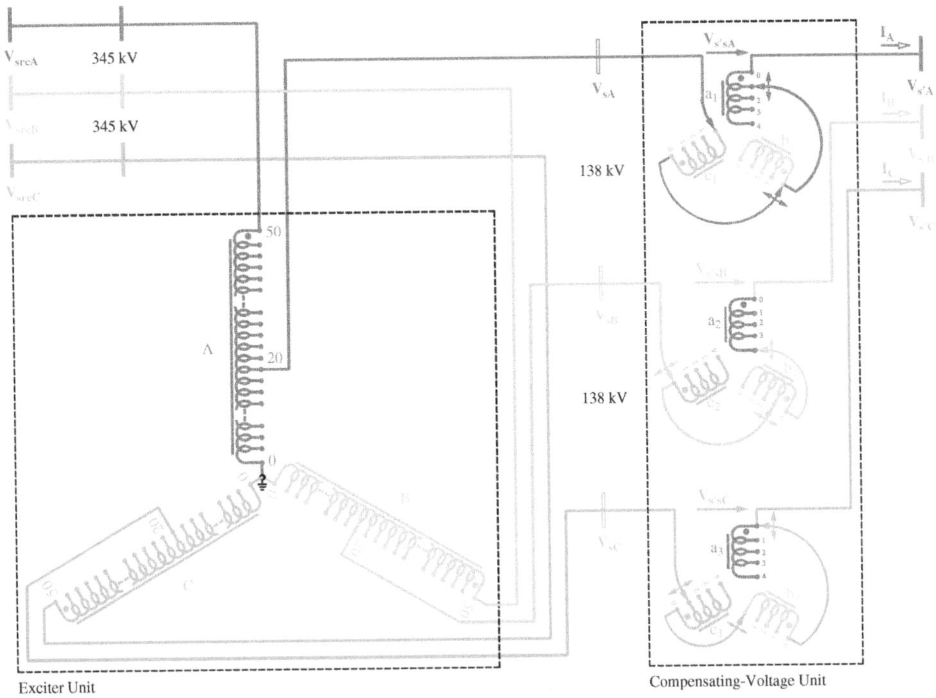

Figure 6-86 ST's Compensating-Voltage Unit is connected to the stepped-down voltage of a transmission line.

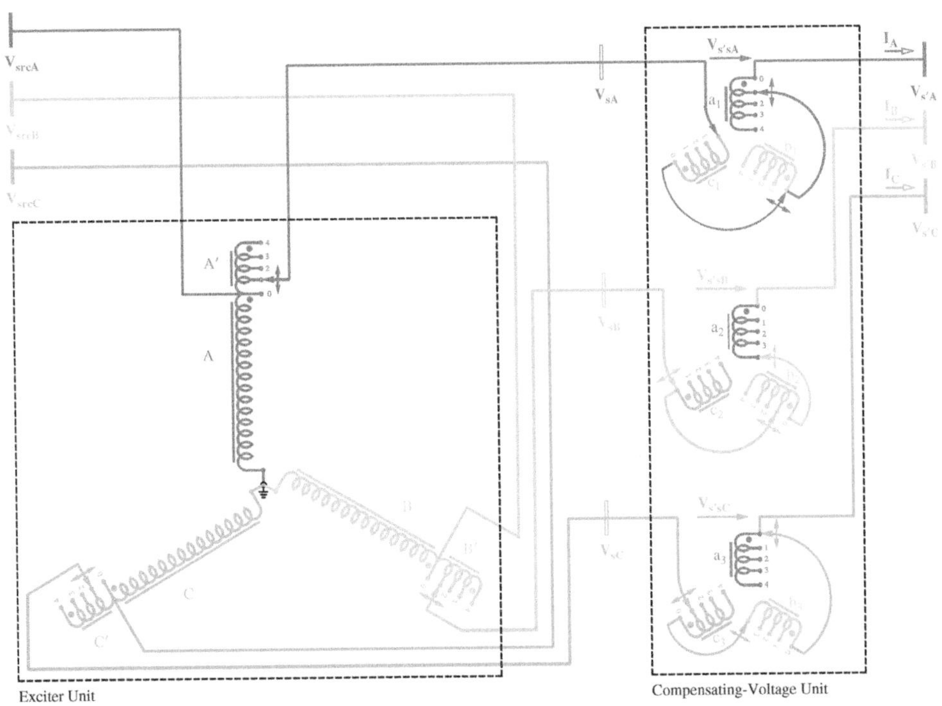

Figure 6-87 ST's Compensating-Voltage Unit is connected to the stepped-up voltage of a transmission line.

6.6 ST with a Shunt-Compensating Voltage

The ST with a Shunt-Series configuration consists of a shunt-connected Exciter Unit (Shunt Unit) and a series-connected Compensating-Voltage Unit (Series Unit) as shown in Figure 6-88a. In an alternate approach, called a Shunt-Shunt configuration, the compensating voltage can be shunt-connected to the line, as shown in Figure 6-88b, providing an electrical isolation between the primary voltage ($\mathbf{V_s}$) of the Exciter Unit (Shunt Unit 1) and the secondary voltage ($\mathbf{V_{s'}}$) of the Compensating-Voltage Unit (Shunt Unit 2). Note that there is no such electrical isolation between the Shunt Unit and the Series Unit in the Shunt-Series configuration. Another important distinction between the Shunt-Series and Shunt-Shunt configurations is that the series-connected Compensating Voltage Unit carries only a fraction of the transmitted power, whereas the shunt-connected Compensating Voltage Unit carries the full transmitted power. The practical application of a shunt-connected Compensating-Voltage Unit might be in a case when two nearby power grids with different voltages or phase angles need to be interconnected.

The circuit configuration, shown in Figure 6-89a, can be used to form a shunt-compensating voltage ($\mathbf{V_{s'}}$) that is at any phase angle with respect to the line voltage ($\mathbf{V_s}$). By choosing the number of turns from each of the three secondary windings in any phase and, therefore, the magnitudes of the components of the three 120° phase-shifted induced voltages, the compensating voltage ($\mathbf{V_{s'}}$) in any phase is derived from the phasor sum of the voltages induced in a three-phase windings set, for example A_1, B_1, and C_1 for compensation in the *A* phase. The corresponding phasor diagram is shown in Figure 6-89b.

The modified sending-end voltage ($\mathbf{V_{s'A}}$), which is the shunt-compensating voltage with magnitude ($V_{s'A}$) and phase-shift angle (ψ_A), is generated for the *A* phase by implementing the following equation:

$$\mathbf{V_{s'A}} = \mathbf{V_{A1}} + \mathbf{V_{B1}} + \mathbf{V_{C1}}$$

or

$$V_{s'A}\angle\psi_A = K_{A1}V_{sA} + K_{B1}V_{sB} + K_{C1}V_{sC} = K_{A1}V_{sA} + K_{B1}e^{-j2\pi/3}V_{sA} + K_{C1}e^{j2\pi/3}V_{sA} \quad (6\text{-}1)$$

where

$\mathbf{V_{A1}}$, $\mathbf{V_{B1}}$, and $\mathbf{V_{C1}}$ are active voltages in the shunt-compensating windings A_1, B_1, and C_1 for the *A* phase,

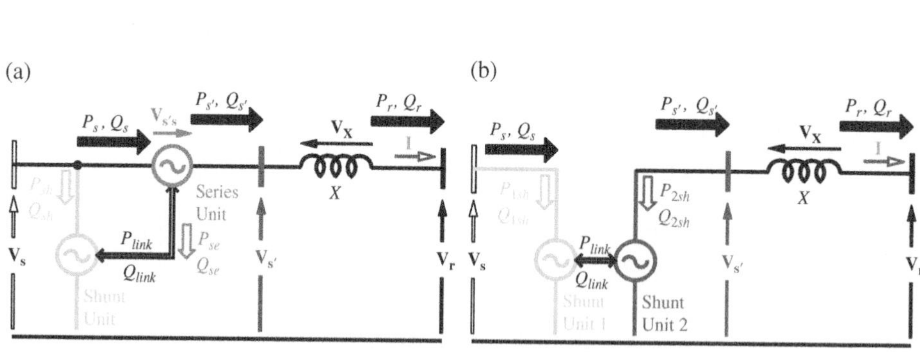

Figure 6-88 Two methods of modifying a transmission line voltage: (a) with a Shunt-Series configuration and (b) with a Shunt-Shunt configuration.

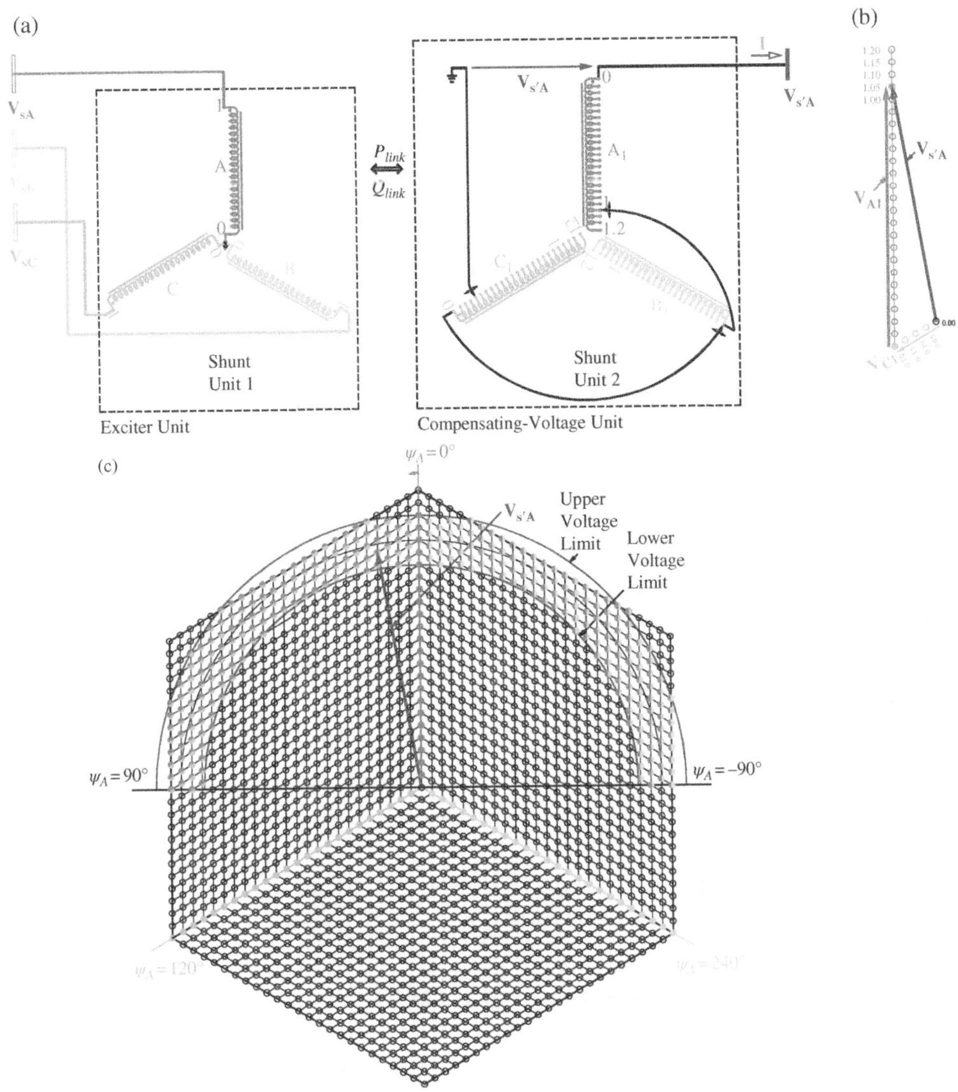

Figure 6-89 (a) Shunt-Shunt configuration of ST for the *A* phase; (b) phasor diagram; (c) operating points of ST for the *A* phase within the upper and lower limits.

$K_{A1} = 1.05$, $K_{B1} = 0$, and $K_{C1} = 0.20$ are the active turns-ratios of the secondary, shunt-compensating windings in the *A* phase and the corresponding primary windings,

$V_{s'A}$ is the magnitude of the shunt-compensating voltage in the *A* phase, which is also the voltage at the modified sending end of the line,

ψ_A is the phase-shift angle for the *A*-phase voltage, and

V_{sA} is the magnitude of the *A*-phase voltage at the sending end of the line.

Note that the *A*-phase shunt-compensating winding is 105% active; the *B*-phase shunt-compensating winding is inactive, since it is bypassed; the *C*-phase shunt-compensating winding is 20% active at the particular operating point, shown in Figure 6-89a. Figure 6-89b shows the corresponding phasor diagram. Note that the filled dots in the phasor diagram represent the end of the A and C compensating phasors, respectively.

The compensating voltage for the A phase and the phase-shift angle in the range of $0° \le \psi_A \le 360°$ is derived from Equation (6-1) as

$$V_{s'A}\cos\psi_A + jV_{s'A}\sin\psi_A = K_{A1}V_{sA} + K_{B1}V_{sA}\left(-1/2 - j\sqrt{3}/2\right) + K_{C1}V_{sA}\left(-1/2 + j\sqrt{3}/2\right)$$

or

$$V_{s'A}\cos\psi_A + jV_{s'A}\sin\psi_A = (K_{A1} - K_{B1}/2 - K_{C1}/2)V_{sA} + j(\sqrt{3}/2)(K_{C1} - K_{B1})V_{sA}.$$

Equating the real and imaginary parts of both sides of the above equation, it can be written that

$$V_{s'}\cos\psi_A = (K_{A1} - K_{B1}/2 - K_{C1}/2)V_{sA} \tag{6-19a}$$

and

$$V_{s'A}\sin\psi_A = \left(\sqrt{3}/2\right)(K_{C1} - K_{B1})V_{sA}. \tag{6-19b}$$

Equations (6-19a) and (6-19b) can be converted into a polar form as

$$V_{s'A}\angle\psi_A = \sqrt{V_{s'A}^2\left(\cos^2\psi_A + \sin^2\psi_A\right)}\angle\tan^{-1}\frac{\sin\psi_A}{\cos\psi_A} \text{ using the identity } \left(\cos^2\psi_A + \sin^2\psi_A = 1\right)$$

or

$$V_{s'A}\angle\psi_A = \sqrt{(V_{s'A}\cos\psi_A)^2 + (V_{s'A}\sin\psi_A)^2}\angle\tan^{-1}\frac{V_{s'A}\sin\psi_A}{V_{s'A}\cos\psi_A}$$

or

$$V_{s'A}\angle\psi_A = V_{sA}\sqrt{K_{A1}^2 + K_{B1}^2 + K_{C1}^2 - K_{A1}K_{B1} - K_{B1}K_{C1} - K_{C1}K_{A1}}\angle\tan^{-1}\frac{\sqrt{3}(K_{C1} - K_{B1})}{2K_{A1} - K_{B1} - K_{C1}}. \tag{6-20}$$

Figure 6-89c shows all the possible operating points of an ST in its Shunt-Shunt configuration. However, most of the operating points are outside any practical operating range, since the line voltage needs to be maintained within an allowable range. The phase-shift angle is also limited to operate in a narrow band in the entire range of $0° \le \psi_A \le 360°$. A voltage variation range of $\pm 10\%$ and a phase-shift angle in the range of $-90° \le \psi_A \le 90°$ is shown in Figure 6-89c. The filled dots represent operating points, determined by the selection of the taps in the LTCs.

The Shunt-Shunt configuration of ST for the A phase can be extended for the B and C phases as shown in Figure 6-90. The windings (A_2, B_2, and C_2) are used for compensation in the B phase and windings (A_3, B_3, and C_3) are used for compensation in the C phase. It should be noted that each of the windings (A_1, B_2, and C_3) is tapped at the same number of turns through LTC_1; each of the windings (B_1, C_2, and A_3) is tapped at the same number of turns through LTC_2; and each of the windings (C_1, A_2, and B_3) is tapped at the same number of turns through LTC_3. However, the number of turns in the A_1-B_2-C_3 set, B_1-C_2-A_3 set, and C_1-A_2-B_3 set can be different from each other.

The modified sending-end voltage ($\mathbf{V_{s'B}}$), which is the shunt-compensating voltage with magnitude ($V_{s'B}$) and phase-shift angle (ψ_B), is generated for the B phase by implementing the following equation:

$$\mathbf{V_{s'B}} = \mathbf{V_{B2}} + \mathbf{V_{C2}} + \mathbf{V_{A2}}$$

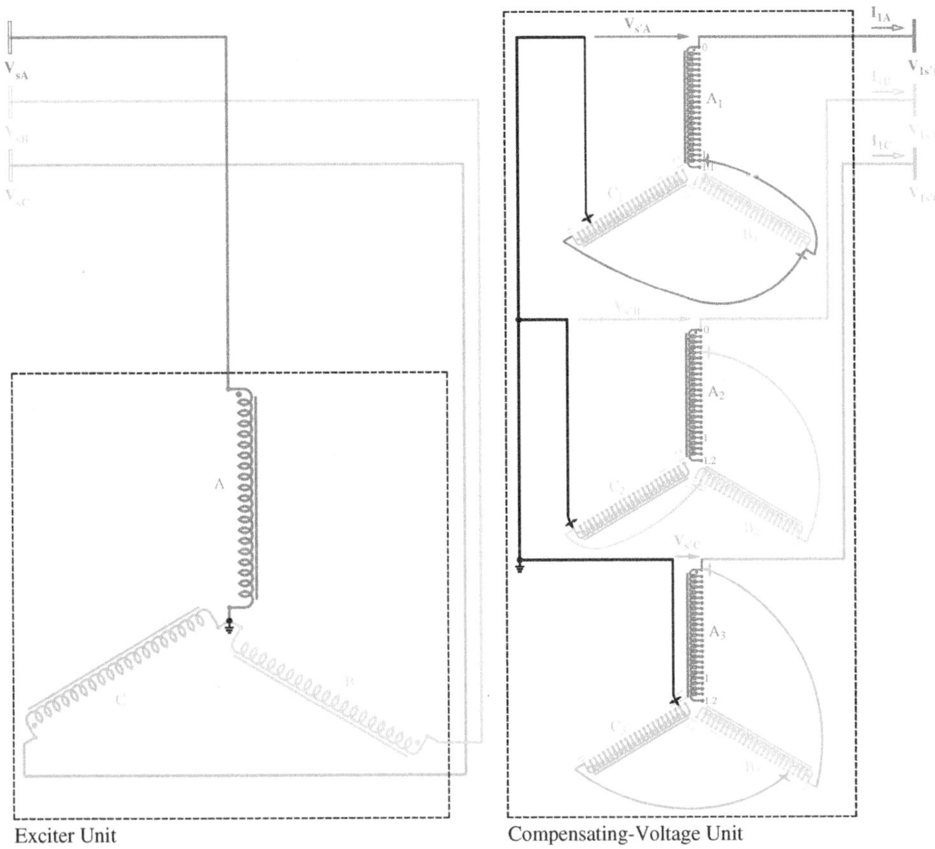

Figure 6-90 Shunt-Shunt configuration of ST for the *A* phase, *B* phase, and *C* phase.

or

$$V_{s'B}\angle\psi_B = K_{B2}V_{sB} + K_{C2}V_{sC} + K_{A2}V_{sA} = K_{B2}e^{-j2\pi/3}V_{sA} + K_{C2}e^{\,j2\pi/3}V_{sA} + K_{A2}V_{sA}. \quad (6\text{-}50)$$

The modified sending-end voltage ($\mathbf{V_{s'C}}$), which is the shunt-compensating voltage with magnitude ($V_{s'C}$) and phase-shift angle (ψ_C), is generated for the *C* phase by implementing the following equation:

$$\mathbf{V_{s'C}} = \mathbf{V_{C3}} + \mathbf{V_{A3}} + \mathbf{V_{B3}}$$

or

$$V_{s'C}\angle\psi_C = K_{C3}V_{sC} + K_{A3}V_{sA} + K_{B3}V_{sB} = K_{C3}e^{\,j2\pi/3}V_{sA} + K_{A3}V_{sA} + K_{B3}e^{-j2\pi/3}V_{sA}. \quad (6\text{-}51)$$

Using Equation (6-20) and replacing K_{A1}, K_{B1}, and K_{C1} with K_{B2}, K_{C2}, and K_{A2}, respectively, it can be written that

$$V_{s'B}\angle\psi_B = V_{sB}\sqrt{K_{B2}^2 + K_{C2}^2 + K_{A2}^2 - K_{B2}K_{C2} - K_{C2}K_{A2} - K_{A2}K_{B2}}\angle\tan^{-1}\frac{\sqrt{3}(K_{A2} - K_{C2})}{2K_{B2} - K_{C2} - K_{A2}}. \quad (6\text{-}52)$$

Using Equation (6-20) and replacing K_{A1}, K_{B1}, and K_{C1} with K_{C3}, K_{A3}, and K_{B3}, respectively, it can be written that

$$V_{s'C}\angle\psi_C = V_{sC}\sqrt{K_{C3}^2 + K_{A3}^2 + K_{B3}^2 - K_{C3}K_{A3} - K_{A3}K_{B3} - K_{B3}K_{C3}}\angle\tan^{-1}\frac{\sqrt{3}(K_{B3} - K_{A3})}{2K_{C3} - K_{A3} - K_{B3}}. \quad (6\text{-}53)$$

The compensating voltage $(\mathbf{V_{s'}} = \mathbf{V_{dq}} = \mathbf{V_d} + \mathbf{V_q})$, being at any phase angle with respect to the prevailing line current (**I**), emulates a four-quadrant, shunt-compensating impedance ($\mathbf{Z_{sh}} = R_{sh} - jX_{sh}$). The active or direct component ($\mathbf{V_d} = V_d$) of the compensating voltage provides the shunt resistance emulation ($R_{sh} = +R$ or $-R$), which can be expressed as

$$\mathbf{V_d} = -R_{sh}\mathbf{I}. \quad (6\text{-}54)$$

The reactive or quadrature component ($\mathbf{V_q} = jV_q$) of the compensating voltage provides the shunt reactance emulation ($X_{sh} = X_C$ or $-X_L$), which can be expressed as

$$\mathbf{V_q} = jX_{sh}\mathbf{I}. \quad (6\text{-}55)$$

The effect of the impedance emulation is such that the line voltage can be modified to a specific magnitude and phase angle.

6.7 Limited Angle Operation of the ST with Shunt-Compensating Voltages

The ST requires the use of a three-phase transformer with three primary windings and nine secondary windings. The shunt-compensating voltage ($\mathbf{V_{s'}}$) is of variable magnitude and variable phase angle. However, in most instances, the capability of connecting a voltage in shunt with a line within its entire range of $0° \leq \psi \leq 360°$ is not needed. The theoretical maximum range is $-90° \leq \psi \leq 90°$. In this case, the circuit configuration for the full $0° \leq \psi \leq 360°$ range of operation can be simplified to the "limited angle" configuration. Figures 6-91 through 6-93 show, through transformer action, the generation of a line frequency, shunt-compensating voltage ($\mathbf{V_{s'}}$), which is of variable magnitude and operated within a limited angle.

Figure 6-91a shows that for the limited angle operation in the range $0° \leq \psi \leq 90°$, the phase angle of the shunt-compensating voltage ($\mathbf{V_{s'}}$) with respect to the line voltage ($\mathbf{V_s}$) can vary between 0° and 90°. The voltage at any point in the electrical system is applied to a shunt-connected three-phase transformer's primary windings. A total of six secondary windings (A_1 and A_2 on the core of the *A* phase, B_2 and B_3 on the core of the *B* phase, and C_3 and C_1 on the core of the *C* phase) constitute the limited angle Compensating-Voltage Unit. By choosing the number of turns from each of the two windings, the compensating voltage in any phase is derived from the phasor sum of the voltages induced in a two-phase winding set (A_1 and C_1 for connection in the *A* phase, A_2 and B_2 for connection in the *B* phase, and B_3 and C_3 for connection in the *C* phase). Figure 6-91b shows the operating points of the ST with a Shunt-Shunt configuration in the *A*, *B*, and *C* phases in the range of $0° \leq \psi \leq 90°$.

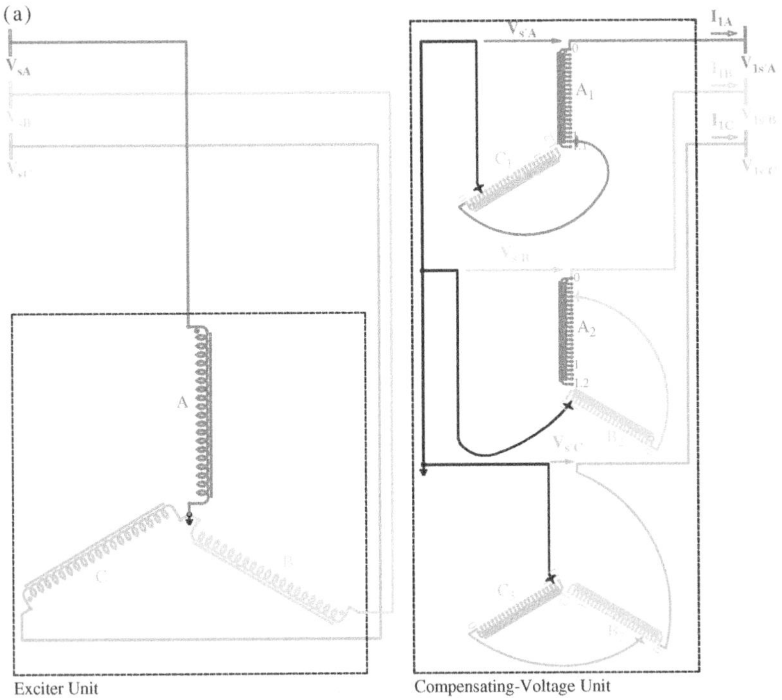

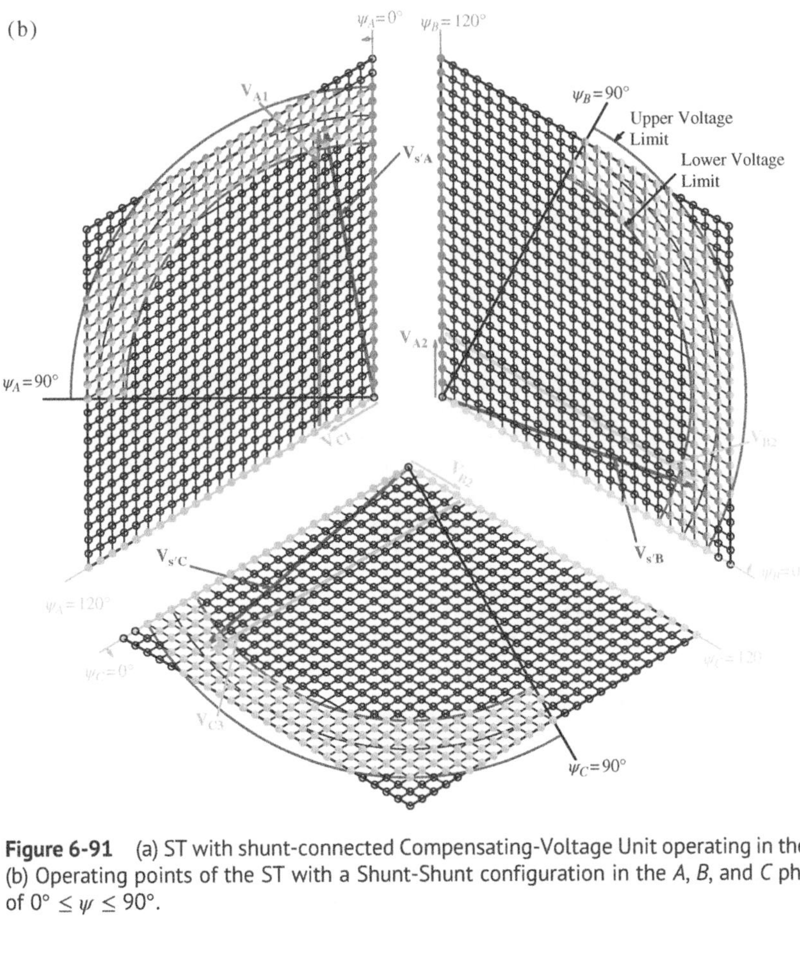

Figure 6-91 (a) ST with shunt-connected Compensating-Voltage Unit operating in the range of $0° \leq \psi \leq 90°$. (b) Operating points of the ST with a Shunt-Shunt configuration in the *A*, *B*, and *C* phases in the range of $0° \leq \psi \leq 90°$.

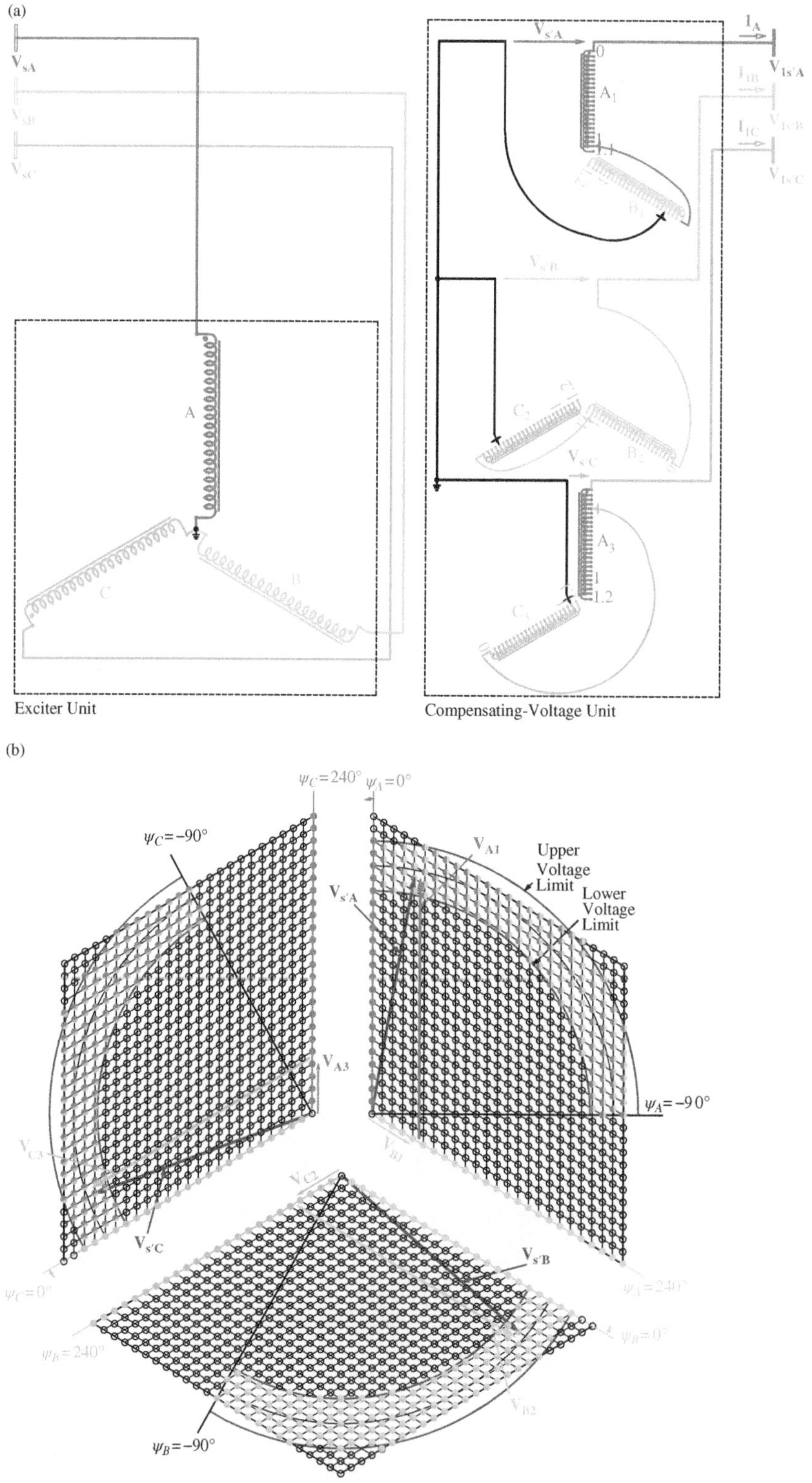

Figure 6-92 (a) ST with shunt-connected Compensating-Voltage Unit operating in the range of $-90° \leq \psi \leq 0°$. (b) Operating points of the ST with a Shunt-Shunt configuration in the *A*, *B*, and *C* phases in the range of $-90° \leq \psi \leq 0°$.

The shunt-compensating voltages in the A, B, and C phases are defined from Equations (6-20), (6-52), and (6-53) as

$$V_{s'A}\angle\psi_A = V_{sA}\sqrt{K_{A1}^2 + K_{C1}^2 - K_{C1}K_{A1}}\angle\tan^{-1}\frac{\sqrt{3}(K_{C1})}{2K_{A1} - K_{C1}} \quad (6\text{-}20a)$$

$$V_{s'B}\angle\psi_B = V_{sB}\sqrt{K_{B2}^2 + K_{A2}^2 - K_{A2}K_{B2}}\angle\tan^{-1}\frac{\sqrt{3}(K_{A2})}{2K_{B2} - K_{A2}} \quad (6\text{-}52a)$$

$$V_{s'C}\angle\psi_C = V_{sC}\sqrt{K_{C3}^2 + K_{B3}^2 - K_{B3}K_{C3}}\angle\tan^{-1}\frac{\sqrt{3}(K_{B3})}{2K_{C3} - K_{B3}} \quad (6\text{-}53a)$$

where $K_{A1} = K_{B2} = K_{C3} = x \times LTC_1$ and $K_{C1} = K_{A2} = K_{B3} = x \times LTC_3$ are the active turns-ratios of the secondary, shunt-compensating windings in the A, B, and C phases and the corresponding primary windings; x is the voltage (for example 0.05 pu) across each tap in the LTC; $LTC_1 = 21$ and $LTC_3 = 4$ are the LTC tap settings for the three-phase operating point shown in Figure 6-91a.

Similarly, in an application in which there is a need to vary the phase-shift angle of the shunt-compensating voltage ($\mathbf{V_{s'}}$) in the range of $-90° \leq \psi \leq 0°$, a Compensating-Voltage Unit with only six windings is needed as shown in Figure 6-92a. This is achieved by constructing the compensating voltage from a combination of two voltages, each of which is induced in a separate winding of a two-phase set (A_1 and B_1 for connection in the A phase, B_2 and C_2 for connection in the B phase, and A_3 and C_3 for connection in the C phase). Figure 6-92b shows the operating points of the ST with a Shunt-Shunt configuration in the A, B, and C phases in the range of $-°90° \leq \psi \leq 0°$.

The shunt-compensating voltages in the A, B, and C phases are defined from Equations (6-20), (6-52), and (6-53) as

$$V_{s'A}\angle\psi_A = V_{sA}\sqrt{K_{A1}^2 + K_{B1}^2 - K_{A1}K_{B1}}\angle\tan^{-1}\frac{\sqrt{3}(-K_{B1})}{2K_{A1} - K_{B1}} \quad (6\text{-}20b)$$

$$V_{s'B}\angle\psi_B = V_{sB}\sqrt{K_{B2}^2 + K_{C2}^2 - K_{B2}K_{C2}}\angle\tan^{-1}\frac{\sqrt{3}(-K_{C2})}{2K_{B2} - K_{C2}} \quad (6\text{-}52b)$$

$$V_{s'C}\angle\psi_C = V_{sC}\sqrt{K_{C3}^2 + K_{A3}^2 - K_{C3}K_{A3}}\angle\tan^{-1}\frac{\sqrt{3}(-K_{A3})}{2K_{C3} - K_{A3}} \quad (6\text{-}53b)$$

where $K_{A1} = K_{B2} = K_{C3} = x \times LTC_1$ and $K_{B1} = K_{C2} = K_{A3} = x \times LTC_2$ are the active turns-ratios of the secondary, shunt-compensating windings in the A, B, and C phases and the corresponding primary windings; x is the voltage (for example 0.05 pu) across each tap in the LTC; $LTC_1 = 21$ and $LTC_2 = 4$ are the LTC tap settings for the three-phase operating point shown in Figure 6-92a.

Perhaps, the most practical range of operation may be obtained by modifying the Shunt-Shunt configuration that operates in the range of $120° \leq \psi \leq 240°$ with the polarities of the windings in the shunt-connected Compensating-Voltage Unit reversed as shown in Figure 6-93a; the phase-shift angle (ψ) of the shunt-compensating voltage ($\mathbf{V_{s'}}$) now varies in the range of $-60° \leq \psi \leq 60°$. The Compensating-Voltage Unit consists of only six windings as shown in Figure 6-93a. The shunt-compensating voltage in each phase is a combination of two voltages, each of which is induced in a separate winding of a two-phase set (B_1 and C_1 for connection in the A phase, C_2 and A_2 for connection in the B phase, and A_3 and B_3 for connection in the C phase). Figure 6-93b shows the operating points of the ST with a Shunt-Shunt configuration in the A, B, and C phases in the range of $-60° \leq \psi \leq 60°$.

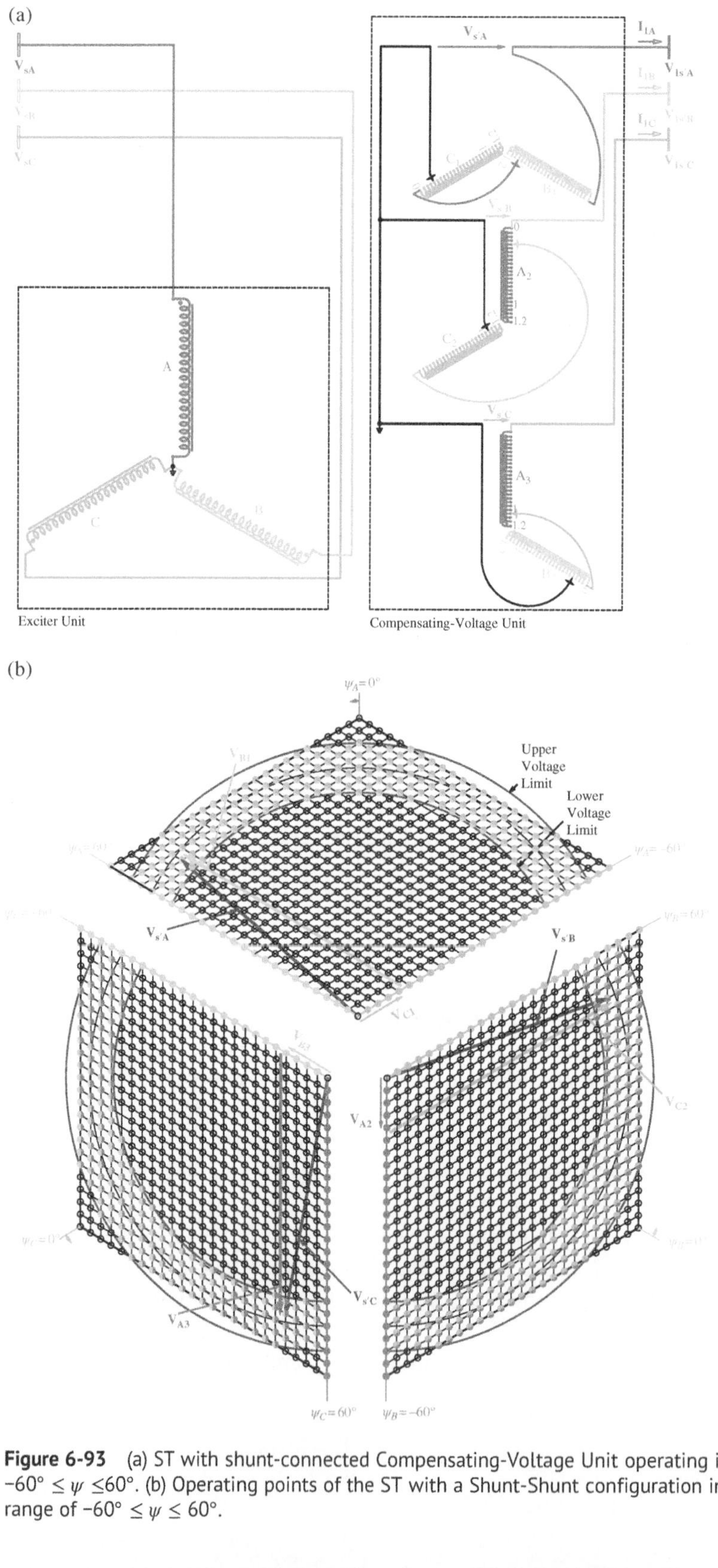

Figure 6-93 (a) ST with shunt-connected Compensating-Voltage Unit operating in the range of $-60° \leq \psi \leq 60°$. (b) Operating points of the ST with a Shunt-Shunt configuration in *A*, *B*, and *C* phases in the range of $-60° \leq \psi \leq 60°$.

The shunt-compensating voltages in the *A*, *B*, and *C* phases are defined from Equations (6-20), (6-52), and (6-53) as

$$V_{s'A}\angle\psi_A = V_{sA}\sqrt{K_{B1}^2 + K_{C1}^2 - K_{B1}K_{C1}}\angle\left\{180^\circ + \tan^{-1}\frac{\sqrt{3}(K_{B1} - K_{C1})}{K_{B1} + K_{C1}}\right\} \quad (6\text{-}20c)$$

$$V_{s'B}\angle\psi_B = V_{sB}\sqrt{K_{C2}^2 + K_{A2}^2 - K_{C2}K_{A2}}\angle\left\{180^\circ + \tan^{-1}\frac{\sqrt{3}(K_{C2} - K_{A2})}{K_{C2} + K_{A2}}\right\} \quad (6\text{-}52c)$$

$$V_{s'C}\angle\psi_C = V_{sC}\sqrt{K_{A3}^2 + K_{B3}^2 - K_{A3}K_{B3}}\angle\left\{180^\circ + \tan^{-1}\frac{\sqrt{3}(K_{A3} - K_{B3})}{K_{A3} + K_{B3}}\right\} \quad (6\text{-}53c)$$

where $K_{B1} = K_{C2} = K_{A3} = x \times LTC_2$ and $K_{C1} = K_{A2} = K_{B3} = x \times LTC_3$ are the active turns-ratios of the secondary, shunt-compensating windings in the *A*, *B*, and *C* phases and the corresponding primary windings; *x* is the voltage (for example 0.05 pu) across each tap in the LTC; $LTC_2 = 21$ and $LTC_3 = 4$ are the LTC tap settings for the three-phase operating point shown in Figure 6-93a.

6.8 MST with Shunt-Compensating Voltages

An MST connects a compensating voltage in shunt with each line in a multiline transmission system. In some special cases for point-to-point transfer of power between two isolated networks with different voltages and/or phase angles, the use of the Shunt-Shunt power conversion may be the preferred choice. In the MST configuration, shown in Figure 6-94, the line voltage ($\mathbf{V_s}$) is applied to a shunt-connected three-phase transformer's Y-connected primary windings in the Exciter Unit. In the Compensating-Voltage Unit 1, a total of nine secondary windings (A_{11}, A_{12}, and A_{13} on the core of the *A* phase; B_{11}, B_{12}, and B_{13} on the core of the *B* phase; and C_{11}, C_{12}, and C_{13} on the core of the *C* phase) derive a three-phase compensating voltage ($\mathbf{V_{1s'A}}$, $\mathbf{V_{1s'B}}$, and $\mathbf{V_{1s'C}}$), each phase of which is the phasor sum of the voltages induced in a three-winding set (A_{11}, B_{11}, and C_{11} for compensating voltage in the *A* phase; A_{12}, B_{12}, and C_{12} for compensating voltage in the *B* phase; and A_{13}, B_{13}, and C_{13} for compensating voltage in the *C* phase). By choosing the number of turns from each of the three windings, and, therefore, the magnitudes of the components of the three 120° phase-shifted induced voltages, the composite compensating voltage's magnitude ($V_{1s'}$) and the phase-shift angle (ψ_1) with respect to the sending-end voltage ($\mathbf{V_s}$) of the line can be selected.

In the Compensating-Voltage Unit 2, a total of nine secondary windings (A_{21}, A_{22}, and A_{23} on the core of the *A* phase; B_{21}, B_{22}, and B_{23} on the core of the *B* phase; and C_{21}, C_{22}, and C_{23} on the core of the *C* phase) derive a three-phase compensating voltage ($\mathbf{V_{2s'A}}$, $\mathbf{V_{2s'B}}$, and $\mathbf{V_{2s'C}}$), each phase of which is the phasor sum of the voltages induced in a three-winding set (A_{21}, B_{21}, and C_{21} for compensating voltage in the *A* phase; A_{22}, B_{22}, and C_{22} for compensating voltage in the *B* phase; and A_{23}, B_{23}, and C_{23} for compensating voltage in the *C* phase). By choosing the number of turns from each of the three windings, and, therefore, the magnitudes of the components of the three 120° phase-shifted induced voltages, the composite compensating voltage's magnitude ($V_{2s'}$) and the phase-shift angle (ψ_2) with respect to the sending-end voltage ($\mathbf{V_s}$) of the line can be selected.

It should be noted that in the Compensating-Voltage Unit 1, each of the windings (A_{11}, B_{12}, and C_{13}) is tapped at the same number of turns; each of the windings (B_{11}, C_{12}, and A_{13}) is tapped at the same number of turns; each of the windings (C_{11}, A_{12}, and B_{13}) is tapped at the same number of turns. However, the number of turns in the A_{11}-B_{12}-C_{13} set, B_{11}-C_{12}-A_{13} set, and C_{11}-A_{12}-B_{13} set can be different from the other.

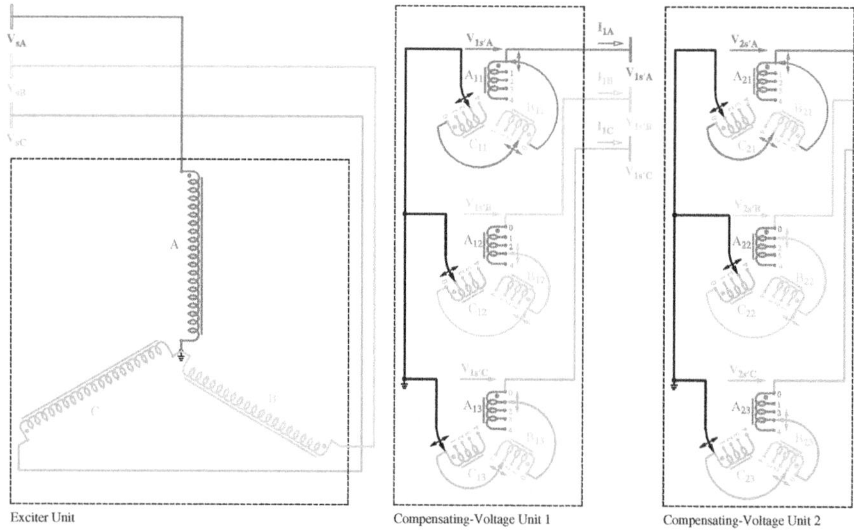

Figure 6-94 MST with shunt-connected Compensating-Voltage Units.

Similarly, in the Compensating-Voltage Unit 2, each of the windings (A_{21}, B_{22}, and C_{23}) is tapped at the same number of turns; each of the windings (B_{21}, C_{22}, and A_{23}) is tapped at the same number of turns; each of the windings (C_{21}, A_{22}, and B_{23}) is tapped at the same number of turns. However, the number of turns in the A_{21}-B_{22}-C_{23} set, B_{21}-C_{22}-A_{23} set, and C_{21}-A_{22}-B_{23} set can be different from the other.

The compensating voltage ($\mathbf{V}_{1s'}$) in the Compensating-Voltage Unit 1 can be at any phase angle with the prevailing line current ($\mathbf{I}_1$), which emulates a four-quadrant, compensating impedance in shunt with the line 1. The compensating voltage ($\mathbf{V}_{2s'}$) in the Compensating-Voltage Unit 2 can be at any phase angle with the prevailing line current ($\mathbf{I}_2$), which emulates a four-quadrant, compensating impedance in shunt with the line 2. An emulated impedance ($\mathbf{Z} = R_{sh} - jX_{sh}$) in each line can be a combination of a resistance ($R_{sh} = +R$ or $-R$) and a reactance ($X_{sh} = X_C$ or $-X_L$).

The operations of the Compensating-Voltage Units are such that the compensating voltage in the "leader" line is at any phase angle with the line voltage, as well as the prevailing line current and that in the "follower" line is of specific magnitude and phase angle, so that the active and reactive powers from the "leader" line are transferred bidirectionally to the "follower" line. As discussed earlier, the 120° range of operation of the PFC is often sufficient. The MST, in its simpler customized configuration with a reduced amount of hardware, provides a limited angle (120°) operation that may be sufficient in the case of a point-to-point power transfer between two isolated networks with possibly, different voltage and phase angle.

6.9 Generalized Sen Transformer

The Generalized Sen Transformer (GST), shown in Figure 6-95, can be used as an extension of the MST that generates both the shunt- and series-compensating voltages in a single unit. The series-compensating voltages serve as the PFCs that control the active and reactive power flows in each line independently and transfer the active and reactive powers from one or more "leader" lines to

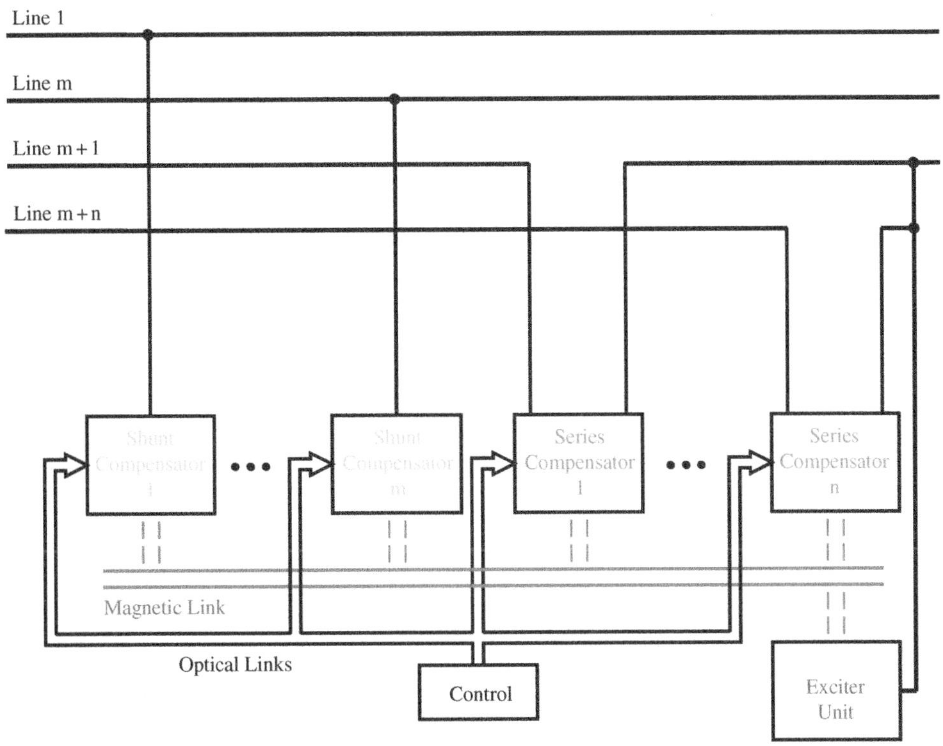

Figure 6-95 Generalized Sen Transformer (GST).

one or more "follower" lines. The shunt-compensating voltages connect the isolated networks with different voltages and phase angles. Each of the compensating voltages is induced from an exciter voltage through transformer action. Therefore, any mismatch of active and reactive powers between various compensating voltages flows to the line that supplies the Exciter Unit of the GST.

Thus, the GST can

1) transfer the active and reactive powers between a number of transmission lines,
2) control the active and reactive power flows in each line selectively, and
3) interconnect transmission lines with different voltages and phase angles.

The dynamic performance of this technology is limited by the operational speed of the LTCs. The mechanical LTCs are likely to be adequate for most utility applications. The dynamic performance can be improved, when needed, by replacing the mechanical LTCs with faster switches.

6.10 Summary

The Sen Transformer (ST) is a power-regulating transformer that modifies the magnitude and phase angle of the transmission line voltage, which results in the same independent control of active and reactive power flows as the Unified Power Flow Controller (UPFC), albeit at a reduced dynamic rate. The ST uses the newest reliable, cost-effective, and proven transformer and LTC-based technology. Since the millisecond-range speed of operation is not essential in most utility power flow applications, the low cost and the high efficiency of the ST becomes attractive when compared with that of the Voltage-Sourced Converter (VSC)-based UPFC solution for the same amount of power flow enhancement. The series-compensating voltage in both the ST and the UPFC exchanges active and reactive powers with the line. In the case of the ST, these exchanged active and reactive powers flow back and forth through the magnetic link to the Shunt Unit to the line. However, in the case of the UPFC, only the exchanged active power by the Series Unit flows back and forth through the DC or AC link to the Shunt Unit to the line. Each of the series- and shunt-connected VSCs in the UPFC is capable of generating or absorbing reactive power independently. If this capability of supplying shunt reactive power is really needed, particularly when the short-circuit impedance at the POC is high, a separate capacitor bank as a shunt reactive compensator may be used along with the ST.

Appendix A

Miscellaneous

A.1 Three-Phase Balanced Voltage, Current, and Power

Consider a three-phase balanced voltage,

$$v_A = V_p \cos(\theta + \varphi_v) \tag{A-1a}$$

$$v_B = V_p \cos(\theta + \varphi_v - 120°) \tag{A-1b}$$

$$v_C = V_p \cos(\theta + \varphi_v + 120°) \tag{A-1c}$$

that supplies a three-phase balanced current,

$$i_A = I_p \cos(\theta + \varphi_i) \tag{A-2a}$$

$$i_B = I_p \cos(\theta + \varphi_i - 120°) \tag{A-2b}$$

$$i_C = I_p \cos(\theta + \varphi_i + 120°) \tag{A-2c}$$

through a three-phase balanced impedance (**Z**) as shown in Figure A-1, where the amplitudes of voltage and current in each phase are V_p and I_p, respectively, and $|\mathbf{Z}| = Z = V_p/I_p$. The system angle ($\theta$), angular frequency ($\omega$), and time ($t$) are related as

$$\theta = \omega t. \tag{A-3}$$

The phase angles of voltage and current from the reference are φ_v and φ_i, respectively. The timing diagrams for one cycle of balanced three-phase voltages and currents and the related phasor diagrams at the end of the cycle are shown in Figures A-2 and A-3, respectively. Note that the vertical projections of the rotating phasors are their instantaneous values.

The instantaneous power in the A, B, and C phases are given by

$$p_A = v_A i_A \tag{A-4a}$$

$$p_B = v_B i_B \tag{A-4b}$$

$$p_C = v_C i_C. \tag{A-4c}$$

From Equations (A-1a), (A-2a), and (A-4a), the A-phase instantaneous power is obtained, using trigonometric identity $\cos(\theta + a)\cos(\theta + b) = \frac{1}{2}[\cos(2\theta + a + b) + \cos(a - b)]$, as

$$p_A = V_p I_p \cos(\theta + \varphi_v)\cos(\theta + \varphi_i) = \frac{V_p I_p}{2}[\cos(2\theta + \varphi_v + \varphi_i) + \cos(\varphi_v - \varphi_i)]. \tag{A-5a}$$

Power Flow Control Solutions for a Modern Grid using SMART Power Flow Controllers, First Edition.
Kalyan K. Sen and Mey Ling Sen.

Companion website: www.wiley.com/go/sen/powerflowcontrol

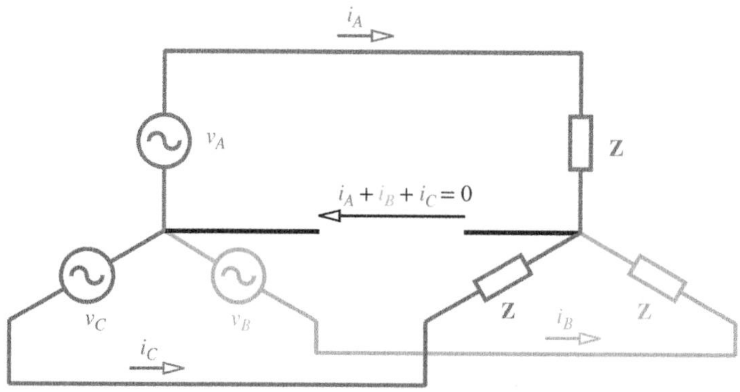

Figure A-1 Three-phase voltage source supplying currents to a balanced three-phase load.

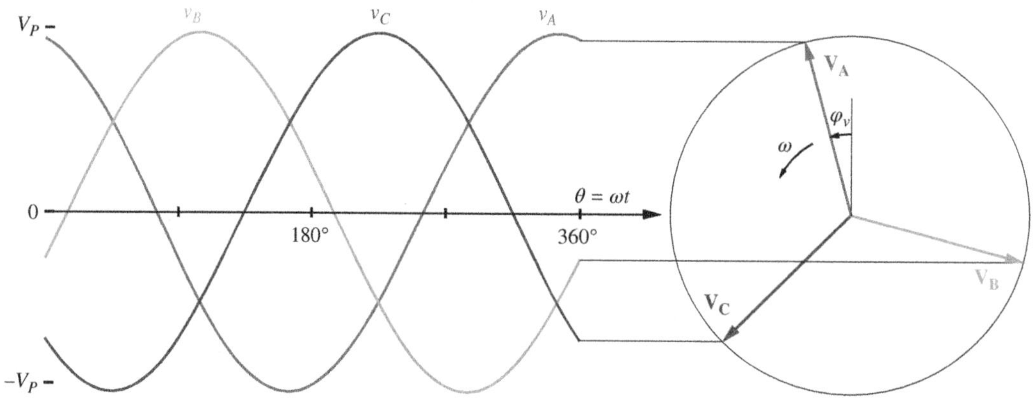

Figure A-2 Balanced three-phase voltages in a timing diagram and the corresponding phasor diagram at the end of the cycle.

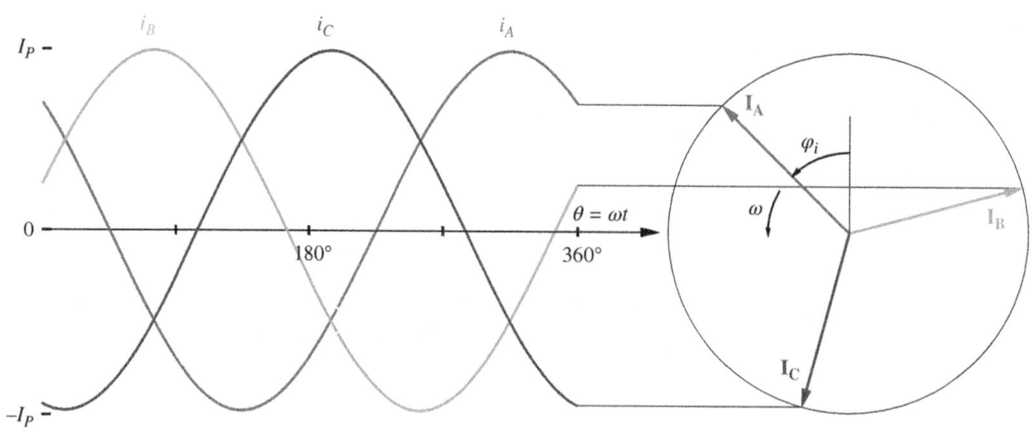

Figure A-3 Balanced three-phase currents in a timing diagram and the corresponding phasor diagram at the end of the cycle.

Similarly, the B-phase instantaneous power is obtained from Equations (A-1b), (A-2b), and (A-4b) as

$$\begin{aligned} p_B &= V_p I_p \cos(\theta + \varphi_v - 120°)\cos(\theta + \varphi_i - 120°) \\ &= \frac{V_p I_p}{2}[\cos(2\theta + \varphi_v + \varphi_i + 120°) + \cos(\varphi_v - \varphi_i)]. \end{aligned} \tag{A-5b}$$

Finally, the C-phase instantaneous power is obtained from Equations (A-1c), (A-2c), and (A-4c) as

$$\begin{aligned} p_C &= V_p I_p \cos(\theta + \varphi_v + 120°)\cos(\theta + \varphi_i + 120°) \\ &= \frac{V_p I_p}{2}[\cos(2\theta + \varphi_v + \varphi_i - 120°) + \cos(\varphi_v - \varphi_i)]. \end{aligned} \tag{A-5c}$$

In each of the three preceding expressions, there exists a double-frequency ($2\theta = 2\omega t$) cosine term whose average value over its one period is zero. There also exists a fixed term, $\frac{V_p I_p}{2}\cos(\varphi_v - \varphi_i)$. Therefore, the three-phase average power (P) is the sum of three single-phase powers or three times the single-phase power, which is given by

$$P = \frac{3}{2}V_p I_p \cos(\varphi_v - \varphi_i) = \frac{3}{2}V_p I_p \cos\varphi \tag{A-6}$$

where the power factor angle is defined as

$$\varphi = \varphi_v - \varphi_i. \tag{A-7}$$

The three-phase average power (P), also referred to as active power (P), has a maximum possible value of term $\frac{3}{2}V_p I_p$, and this maximum value is referred to as apparent power (S). The ratio ($P/S = \cos\varphi$) is referred to as the power factor. The amount of power that can be taken out of the apparent power without altering the active power is the reactive power, and it is given by

$$Q = \sqrt{S^2 - P^2} = \frac{3}{2}V_p I_p \sqrt{1 - \cos^2\varphi} = \frac{3}{2}V_p I_p \sin\varphi. \tag{A-8}$$

Also,

$$\begin{aligned} P &= \frac{3}{2}V_p I_p \cos\varphi = 3\frac{V_p}{\sqrt{2}}\frac{I_p}{\sqrt{2}}\cos\varphi = 3V_{LNrms}I_{rms}\cos\varphi \\ &= \sqrt{3}\left(\sqrt{3}V_{LNrms}\right)I_{rms}\cos\varphi = \sqrt{3}V_{LLrms}I_{rms}\cos\varphi \end{aligned} \tag{A-9}$$

where V_{LNrms} and V_{LLrms} are the line-to-neutral and line-to-line rms voltages, respectively, and I_{rms} is the line rms current.

Using Equations (A-6), (A-7), and (A-8), the expressions for the three-phase active power and reactive power can be written in the form of real (Re) and imaginary (Im) components as

$$P = \frac{3}{2}V_p I_p \cos(\varphi_v - \varphi_i) = \frac{3}{2}V_p I_p \,\mathrm{Re}\left[e^{j(\varphi_v - \varphi_i)}\right] = \frac{3}{2}\mathrm{Re}\left[V_p e^{j\varphi_v} I_p e^{-j\varphi_i}\right] = \frac{3}{2}\mathrm{Re}\,[\mathbf{V}\,\mathbf{I}^*] \tag{A-10}$$

and

$$Q = \frac{3}{2}V_p I_p \sin(\varphi_v - \varphi_i) = \frac{3}{2}V_p I_p \mathrm{Im}\left[e^{j(\varphi_v - \varphi_i)}\right] = \frac{3}{2}\mathrm{Im}\left[V_p e^{j\varphi_v} I_p e^{-j\varphi_i}\right] = \frac{3}{2}\mathrm{Im}[\mathbf{V}\,\mathbf{I}^*]. \tag{A-11}$$

Using Equations (A-10) and (A-11), the expression for apparent power in its complex form can be written as

$$\mathbf{S} = P + jQ = \frac{3}{2}\,\mathrm{Re}\,[\mathbf{V}\,\mathbf{I}^{*}] + j\frac{3}{2}\mathrm{Im}[\mathbf{V}\,\mathbf{I}^{*}] = \frac{3}{2}\mathbf{V}\,\mathbf{I}^{*}. \tag{A-12}$$

Therefore, the expression for instantaneous apparent power in its complex form can be written as

$$\mathbf{s} = p + jq \tag{A-13}$$

(more derivations are given in Equations (A-86) and (A-88)).

The three-phase instantaneous active power is defined as

$$p = p_A + p_B + p_C, \tag{A-14}$$

which is also defined as

$$p = \frac{3}{2} V_p I_p \cos\varphi \tag{A-15}$$

since the double-frequency cosine terms in Equations (A-5a), (A-5b), and (A-5c) are 120° apart and their sum is zero. Using Equations (A-4a), (A-4b), and (A-4c), Equation (A-14) can also be written as

$$p = v_A i_A + v_B i_B + v_C i_C. \tag{A-16}$$

The three-phase instantaneous active power (p) has a maximum possible value of term $\frac{3}{2}V_p I_p$, and this maximum value is referred to as instantaneous apparent power (s). The amount of power that can be taken out of the instantaneous apparent power without altering the instantaneous active power is the instantaneous reactive power, and it is

$$q = \sqrt{s^2 - p^2} = \frac{3}{2} V_p I_p \sqrt{1 - \cos^2\varphi} = \frac{3}{2} V_p I_p \sin\varphi \tag{A-17}$$

(more derivations are given in Equations (A-87), (A-97), and (A-99)).

Note that when $\varphi = 0$, the load is at the unity power factor. In this case, the three-phase reactive power is zero, and the three-phase active power is the same as the three-phase apparent power. When $\varphi = \pm\pi/2$, the load is at the zero power factor. In this case, the three-phase active power is zero, and the three-phase reactive power is the same as the three-phase apparent power.

The analysis of balanced three-phase circuits is straightforward. The analysis of unbalanced three-phase circuits that occur due to unbalanced line impedances, unbalanced loads, unbalanced faults, or short circuits is, however, not so straightforward, but can be simplified with the concept of symmetrical components.

A.2 Symmetrical Components

Any unbalanced three-phase variables, as illustrated in Figure A-4, can be resolved into two sets of balanced three-phase variables and a set of single-phase variables. The three-phase variables in the first set reach their respective peaks in the A-phase, B-phase, and C-phase in counterclockwise sequence; therefore, this set is said to be in positive sequence. The three-phase variables in the second set reach their peaks in the A-phase, C-phase, and B-phase in counterclockwise sequence; therefore, this set is said to be in negative sequence. The variables in the third set overlap with each other; therefore, this set is said to be in zero sequence.

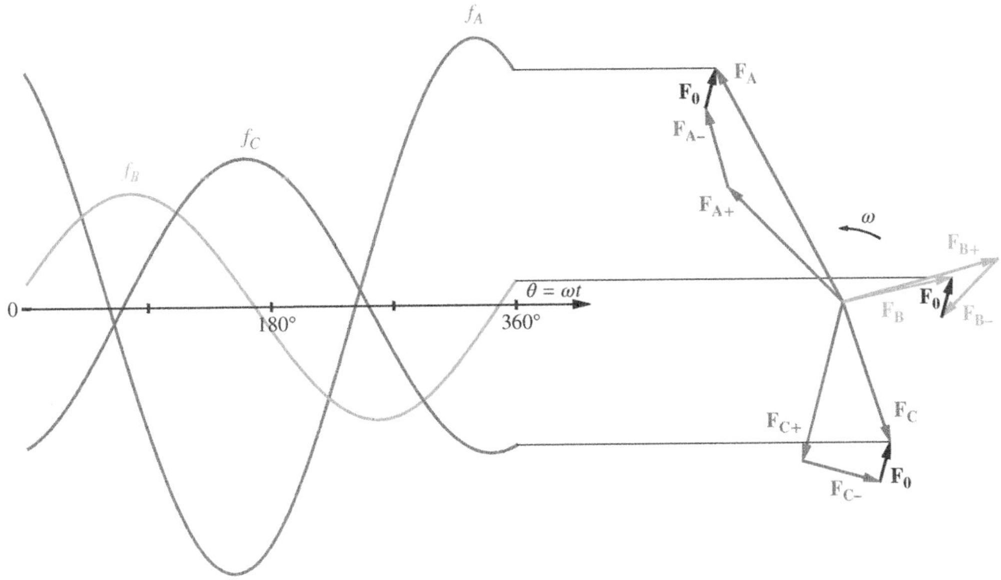

Figure A-4 Unbalanced three-phase variables in a timing diagram and the corresponding phasor diagram at the end of the cycle.

Each phase variable consists of positive-, negative-, and zero-sequence components, such as

$$f_A = f_{A+} + f_{A-} + f_0 \tag{A-18a}$$

$$f_B = f_{B+} + f_{B-} + f_0 \tag{A-18b}$$

$$f_C = f_{C+} + f_{C-} + f_0 \tag{A-18c}$$

where f_{A+}, f_{B+}, and f_{C+} are the positive-sequence components; f_{A-}, f_{B-}, and f_{C-} are the negative-sequence components; and f_0 is the zero-sequence component of the three-phase variables A, B, and C. The phase variable (f) can be voltage, current, flux, and so on.

In phasor form, the above three-phase variables (A-18a) to (A-18c) become

$$\mathbf{F_A} = \mathbf{F_{A+}} + \mathbf{F_{A-}} + \mathbf{F_0} \tag{A-19a}$$

$$\mathbf{F_B} = \mathbf{F_{B+}} + \mathbf{F_{B-}} + \mathbf{F_0} \tag{A-19b}$$

$$\mathbf{F_C} = \mathbf{F_{C+}} + \mathbf{F_{C-}} + \mathbf{F_0}. \tag{A-19c}$$

Figure A-5 shows the balanced, three-phase, positive-sequence components in a timing diagram for one cycle and the corresponding phasor diagram at the end of the cycle. The vertical projections of the three rotating phasors are the instantaneous values of the three-phase variables.

The positive-sequence components can be related to each other with the help of a unit vector operator (a) where $a = e^{j120°}$, $a^2 = e^{j240°} = e^{-j120°}$, and $a^3 = e^{j360°} = 1$. The B- and C-phase positive sequence variables are expressed in terms of the positive sequence A-phase variable as

$$\mathbf{F_{B+}} = a^2\mathbf{F_{A+}} \tag{A-20a}$$

$$\mathbf{F_{C+}} = a\mathbf{F_{A+}}. \tag{A-20b}$$

Figure A-6 shows the balanced, three-phase, negative-sequence components in a timing diagram for one cycle and the corresponding phasor diagram at the end of the cycle. The vertical projections of the three rotating phasors are the instantaneous values of the three-phase variables.

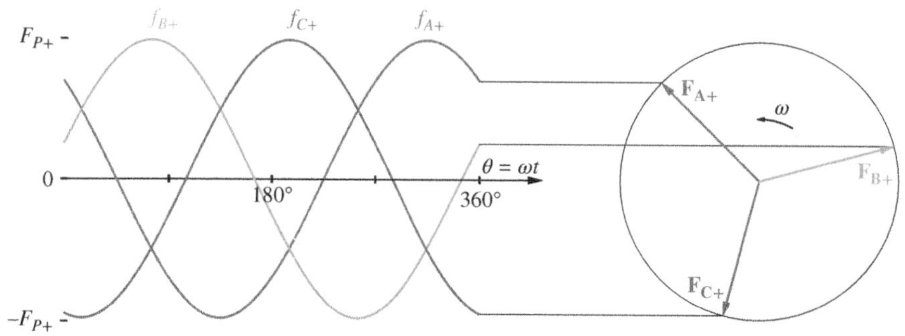

Figure A-5 Balanced three-phase, positive-sequence components in a timing diagram and the corresponding phasor diagram at the end of the cycle.

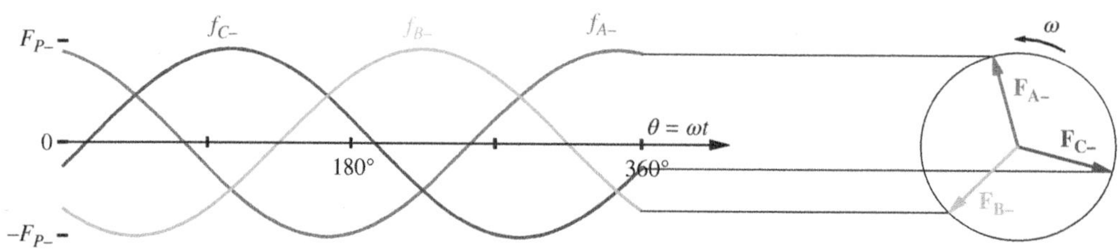

Figure A-6 Balanced three-phase, negative-sequence components in a timing diagram and the corresponding phasor diagram at the end of the cycle.

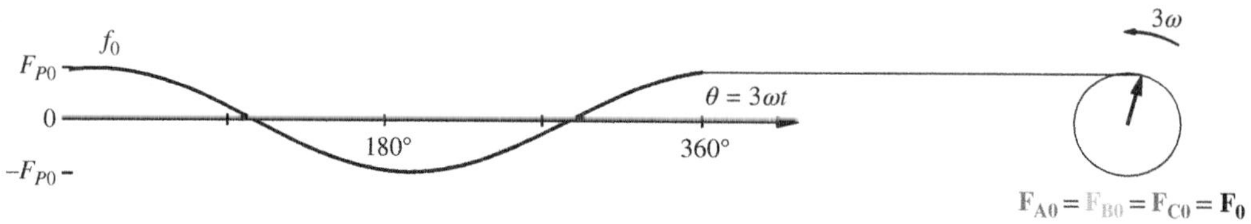

Figure A-7 Zero-sequence component in a timing diagram and the corresponding phasor diagram at the end of the cycle.

The negative-sequence B- and C-phase components can be expressed in terms of the negative-sequence A-phase component as

$$\mathbf{F}_{\mathbf{B}-} = a\mathbf{F}_{\mathbf{A}-} \tag{A-21a}$$

$$\mathbf{F}_{\mathbf{C}-} = a^2\mathbf{F}_{\mathbf{A}-}. \tag{A-21b}$$

Figure A-7 shows the zero-sequence component in a timing diagram for one cycle and the corresponding phasor diagram at the end of the cycle. Three zero-sequence variables from the three phases overlap with each other. The vertical projection of the rotating phasor is the instantaneous value of the zero-sequence component.

Therefore, the three-phase variables defined in Equations (A-19a) to (A-19c) can be rewritten as

$$\mathbf{F_A} = \mathbf{F_{A+}} + \mathbf{F_{A-}} + \mathbf{F_0} \tag{A-22a}$$

$$\mathbf{F_B} = a^2\mathbf{F_{A+}} + a\mathbf{F_{A-}} + \mathbf{F_0} \tag{A-22b}$$

$$\mathbf{F_C} = a\mathbf{F_{A+}} + a^2\mathbf{F_{A-}} + \mathbf{F_0}. \tag{A-22c}$$

Keeping the expression for $\mathbf{F_A}$ as it is and multiplying the expressions for $\mathbf{F_B}$ and $\mathbf{F_C}$ with a and a^2, respectively, Equations (A-22a) to (A-22c) become

$$\mathbf{F_A} = \mathbf{F_{A+}} + \mathbf{F_{A-}} + \mathbf{F_0} \tag{A-23a}$$

$$a\mathbf{F_B} = \mathbf{F_{A+}} + a^2\mathbf{F_{A-}} + a\mathbf{F_0} \tag{A-23b}$$

$$a^2\mathbf{F_C} = \mathbf{F_{A+}} + a\mathbf{F_{A-}} + a^2\mathbf{F_0} \tag{A-23c}$$

since $a^3 = e^{j360^\circ} = 1$.

The addition of the above three equations results in

$$\mathbf{F_A} + a\mathbf{F_B} + a^2\mathbf{F_C} = 3\mathbf{F_{A+}} + \left(\mathbf{F_{A-}} + a^2\mathbf{F_{A-}} + a\mathbf{F_{A-}}\right) + \left(\mathbf{F_0} + a\mathbf{F_0} + a^2\mathbf{F_0}\right) = 3\mathbf{F_{A+}} \tag{A-24}$$

since the sum of the three phasors that are 120° apart is zero.

Therefore, the positive-sequence component is found from the three phase variables as

$$\mathbf{F_{A+}} = \frac{1}{3}\left(\mathbf{F_A} + a\mathbf{F_B} + a^2\mathbf{F_C}\right). \tag{A-25}$$

Keeping the expression for $\mathbf{F_A}$ as it is and multiplying the expressions for $\mathbf{F_B}$ and $\mathbf{F_C}$ with a^2 and a, respectively, Equations (A-22a) to (A-22c) become

$$\mathbf{F_A} = \mathbf{F_{A+}} + \mathbf{F_{A-}} + \mathbf{F_0} \tag{A-26a}$$

$$a^2\mathbf{F_B} = a\mathbf{F_{A+}} + \mathbf{F_{A-}} + a^2\mathbf{F_0} \tag{A-26b}$$

$$a\mathbf{F_C} = a^2\mathbf{F_{A+}} + \mathbf{F_{A-}} + a\mathbf{F_0} \tag{A-26c}$$

since $a^3 = e^{j360^\circ} = 1$.

The addition of the above three equations results in

$$\mathbf{F_A} + a^2\mathbf{F_B} + a\mathbf{F_C} = \left(\mathbf{F_{A+}} + a\mathbf{F_{A+}} + a^2\mathbf{F_{A+}}\right) + 3\mathbf{F_{A-}} + \left(\mathbf{F_0} + a^2\mathbf{F_0} + a\mathbf{F_0}\right) = 3\mathbf{F_{A-}}. \tag{A-27}$$

Therefore, the negative-sequence component is found to be

$$\mathbf{F_{A-}} = \frac{1}{3}\left(\mathbf{F_A} + a^2\mathbf{F_B} + a\mathbf{F_C}\right). \tag{A-28}$$

Simply by adding the expressions for $\mathbf{F_A}$, $\mathbf{F_B}$, and $\mathbf{F_C}$, Equations (A-22a) to (A-22c) become

$$\mathbf{F_A} + \mathbf{F_B} + \mathbf{F_C} = \left(\mathbf{F_{A+}} + a^2\mathbf{F_{A+}} + a\mathbf{F_{A+}}\right) + \left(\mathbf{F_{A-}} + a\mathbf{F_{A-}} + a^2\mathbf{F_{A-}}\right) + 3\mathbf{F_0} = 3\mathbf{F_0}. \tag{A-29}$$

Therefore, the zero-sequence component is found to be

$$\mathbf{F_0} = \frac{1}{3}\left(\mathbf{F_A} + \mathbf{F_B} + \mathbf{F_C}\right). \tag{A-30}$$

The following is an example of decomposing an unbalanced set of phasors, shown in Figure A-4, into three balanced sets of phasors as shown in Figures A-5, A-6, and A-7.

The unbalanced set of phasors, as shown in Figure A-4, are given by

$$\mathbf{F_A} = 1.625\angle 28.35° = 1.43 + j0.7716$$

$$\mathbf{F_B} = 0.6928\angle -77.77° = 0.14676 - j0.6771$$

$$\mathbf{F_C} = 0.9013\angle -161.32° = -0.85382 - j0.2888.$$

From Equation (A-25), the A-phase, positive-sequence component is given by

$$\mathbf{F_{A+}} = \frac{1}{3}\left(\mathbf{F_A} + a\mathbf{F_B} + a^2\mathbf{F_C}\right) \quad \text{or}$$

$$\mathbf{F_{A+}} = \frac{1}{3}\left(1.625\angle 28.35° + e^{j120°}0.6928\angle -77.77° + e^{j240°}0.9013\angle -161.32°\right) \quad \text{or}$$

$$\mathbf{F_{A+}} = \frac{1}{3}\left(1.625e^{j(28.35°)} + 0.6928e^{j(-77.77° + 120°)} + 0.9013e^{j(-161.32° + 240°)}\right) \quad \text{or}$$

$$\mathbf{F_{A+}} = \frac{1}{3}\left(1.625e^{j(28.35°)} + 0.6928e^{j(42.23°)} + 0.9013e^{j(78.68°)}\right)$$

or in Cartesian form

$$\mathbf{F_{A+}} = \frac{1}{3}(1.43 + j0.7716 + (-0.5 + j0.866)(0.14676 - j0.6771) + (-0.5 - j0.866)(-0.85382 - j0.2888)).$$

Therefore,

$$\mathbf{F_{A+}} = \frac{1}{3}\left(\mathbf{F_A} + a\mathbf{F_B} + a^2\mathbf{F_C}\right) = 1\angle 45° = 0.7071 + j0.7071$$

as shown in Figures A-5 and A-8.

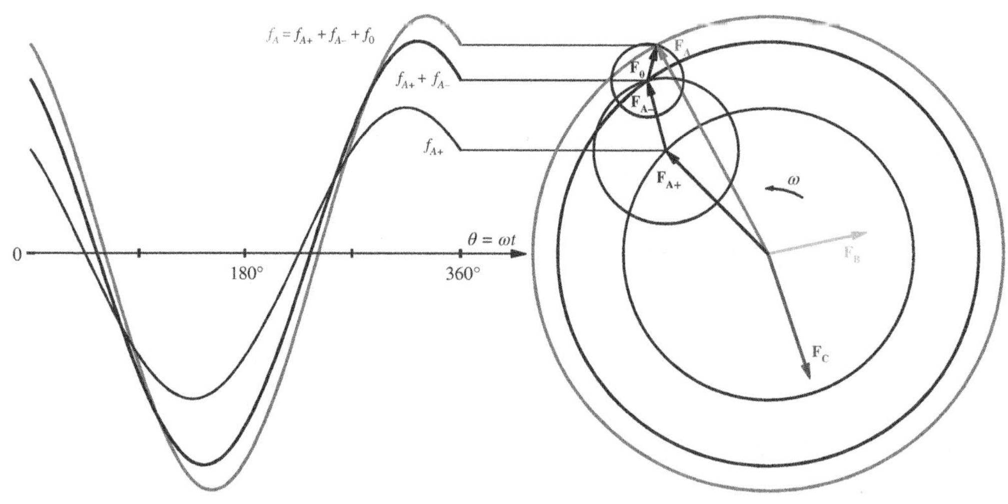

Figure A-8 A-phase, positive-, negative-, and zero-sequence components in a timing diagram and the corresponding phasor diagram at the end of the cycle.

From Equations (A-20a) and (A-20b), the B- and C-phase, positive-sequence components are given by

$$\mathbf{F_{B+}} = a^2\mathbf{F_{A+}} = 1\angle 240° \times 1\angle 45° = 1\angle 285° = 0.2588 - j0.9659$$

as shown in Figures A-5 and A-9 and

$$\mathbf{F_{C+}} = a\mathbf{F_{A+}} = 1\angle 120° \times 1\angle 45° = 1\angle 165° = -0.9659 + j0.2588$$

as shown in Figures A-5 and A-10.

The three phasors are of the same magnitude and 120° apart from each other. Since their sum is zero, they are balanced. Since the A phase leads the B phase and the B phase leads the C phase, the phasors are in positive sequence.

From Equation (A-28), the A-phase, negative-sequence component is given by

$$\mathbf{F_{A-}} = \frac{1}{3}\left(\mathbf{F_A} + a^2\mathbf{F_B} + a\mathbf{F_C}\right) \quad \text{or}$$

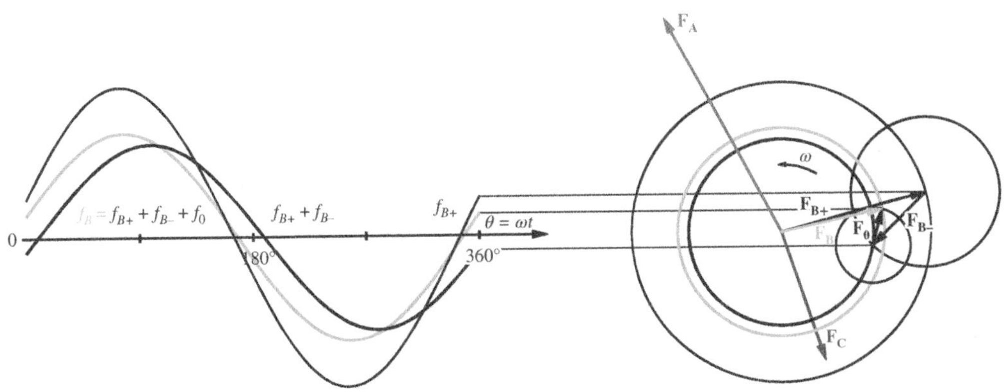

Figure A-9 B-phase positive-, negative-, and zero-sequence components in a timing diagram and the corresponding phasor diagram at the end of the cycle.

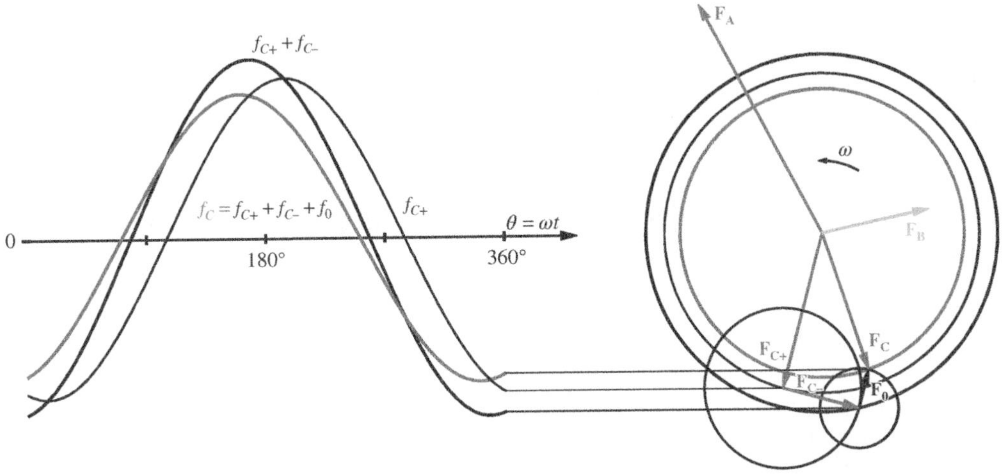

Figure A-10 C-phase positive-, negative-, and zero-sequence components in a timing diagram and the corresponding phasor diagram at the end of the cycle.

$$\mathbf{F}_{A-} = \frac{1}{3}\left(1.625\angle 28.35^\circ + e^{j240^\circ}0.6928\angle -77.77^\circ + e^{j120^\circ}0.9013\angle -161.32^\circ\right) \quad \text{or}$$

$$\mathbf{F}_{A-} = \frac{1}{3}\left(1.625e^{j(28.35^\circ)} + 0.6928e^{j(-77.77^\circ + 240^\circ)} + 0.9013e^{j(-161.32^\circ + 120^\circ)}\right) \quad \text{or}$$

$$\mathbf{F}_{A-} = \frac{1}{3}\left(1.625e^{j(28.35^\circ)} + 0.6928e^{j(162.23^\circ)} + 0.9013e^{j(-41.32^\circ)}\right)$$

or in Cartesian form

$$\mathbf{F}_{A-} = \frac{1}{3}(1.43 + j0.7716 + (-0.5 - j0.866)(0.14676 - j0.6771) + (-0.5 + j0.866)(-0.85382 - j0.2888)).$$

Therefore,

$$\mathbf{F}_{A-} = \frac{1}{3}\left(\mathbf{F}_A + a^2\mathbf{F}_B + a\mathbf{F}_C\right) = 0.5\angle 15^\circ = 0.48296 + j0.1294$$

as shown in Figures A-6 and A-8.

From Equations (A-21a) and (A-21b), the B- and C-phase, negative-sequence components are given by

$$\mathbf{F}_{B-} = a\mathbf{F}_{A-} = 1\angle 120^\circ \times 0.5\angle 15^\circ = 0.5\angle 135^\circ = -0.3535 + j0.3535$$

as shown in Figures A-6 and A-9,

$$\mathbf{F}_{C-} = a^2\mathbf{F}_{A-} = 1\angle 240^\circ \times 0.5\angle 15^\circ = 0.5\angle 255^\circ = -0.1294 - j0.4829$$

as shown in Figures A-6 and A-10.

The three phasors are of the same magnitude and 120° apart from each other. Since their sum is zero, they are balanced. Since the A phase leads the C phase and the C phase leads the B phase, the phasors are in negative sequence.

From Equation (A-30), the A-phase, zero-sequence component is given by

$$\begin{aligned}\mathbf{F}_0 &= \frac{1}{3}(\mathbf{F}_A + \mathbf{F}_B + \mathbf{F}_C)\\ &= \frac{1}{3}(1.43 + j0.7716 + 0.14676 - j0.6771 - 0.85382 - j0.2888)\\ &= 0.241 - j0.064 = 0.25\angle -15^\circ\end{aligned}$$

as shown in Figures A-7, A-8, A-9, and A-10.

The zero-sequence phasors in the three phases are of the same magnitude and they overlap.

A.3 Separation of Positive-, Negative-, and Zero-Sequence Components in a Multiple Frequency Composite Variable

Inverter-Based Resources (IBRs) are an integral part of the modern grid. The IBRs generate multiple frequencies from the Pulse Width Modulation (PWM) operation of the switching devices. The various frequency signals, which are of positive, negative, and zero sequences, appear in the control circuit in the form of feedback signals through the sensors. The various control algorithms that are used in IBRs are implemented in a positive-sequence, fundamental frequency reference frame, which is also called a synchronously rotating reference frame. In this reference frame, any positive-sequence and fundamental frequency variable gets converted into a DC value. Any other positive- or negative-sequence component at any frequency gets converted into an oscillating component. For example, a fundamental frequency, negative-sequence component gets converted into an oscillating component of $-1-1=-2$, *i.e.* 2nd harmonic, rotating clockwise; a 5th harmonic component, which is a negative-sequence component, gets converted into an oscillating component of $-5-1=-6$, *i.e.* 6th harmonic, rotating clockwise; a 7th harmonic component, which is a positive-sequence component, gets converted into an oscillating component of $7-1=+6$, *i.e.* 6th harmonic, rotating counterclockwise; and so on. These non-fundamental components must be filtered out before entering into the control algorithm, since the control algorithm is implemented in a positive-sequence, fundamental frequency reference frame. If not filtered, the non-fundamental components will produce unwanted reference signals. The zero-sequence components also need to be filtered out. If needed, the zero-sequence components need to be treated separately, since their frequencies are odd triple n, meaning $n = 3 \times (2k+1)$ where $k = 0, 1, 2...$, which are different from other frequencies, such as $(6k \pm 1)$ where $k = 0, 1, 2...$ that are present in the PWM waveform.

The zero-sequence component (f_0) is given by

$$f_0 = \frac{1}{3}(f_A + f_B + f_C). \tag{A-31}$$

Therefore, the three-phase variables that consist of positive- and negative-sequence components only are given by

$$f_A' = f_A - f_0 = f_{A+} + f_{A-} \tag{A-32a}$$

$$f_B' = f_B - f_0 = f_{B+} + f_{B-} \tag{A-32b}$$

$$f_C' = f_C - f_0 = f_{C+} + f_{C-}. \tag{A-32c}$$

Consider the previous example with the same frequency positive- and negative-sequence components, but with a triple fundamental frequency, zero-sequence component as shown in Figure A-11.

The timing diagram for one cycle of a set of unbalanced three-phase variables that consist of positive-, negative-, and triple fundamental frequency zero-sequence components and the corresponding phasor diagram at the end of the cycle are shown in Figure A-12.

The A-B-C three-phase variables positive-, negative-, and zero-sequence components in three phases are shown in Figures A-13 through A-15, respectively.

The three-phase variables ($f_{A'}$, $f_{B'}$, and $f_{C'}$) can be used in conjunction with a vector PLL, as discussed in Chapter 3, to calculate the PLL angle (θ) that is the angle of the positive-sequence component (f_{A+}). The amplitude (F) consists of two components: a DC component for the positive-sequence amplitude (F_+) and a double-frequency component for the negative-sequence amplitude (F_-), rotating in the opposite direction as just discussed. The amplitude of the positive-sequence

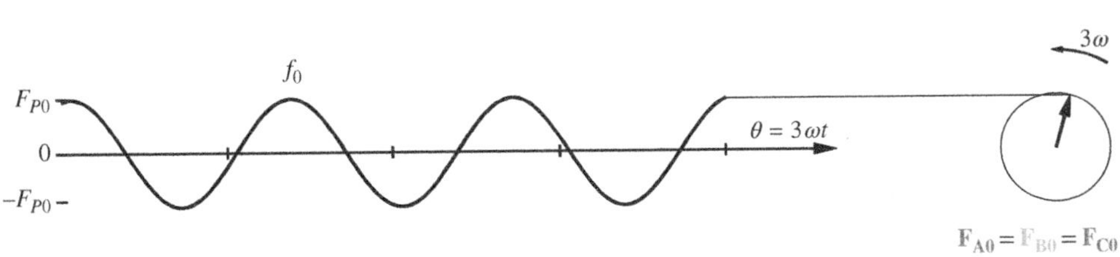

Figure A-11 Triple frequency, zero-sequence component in a timing diagram and the corresponding phasor diagram at the end of the cycle.

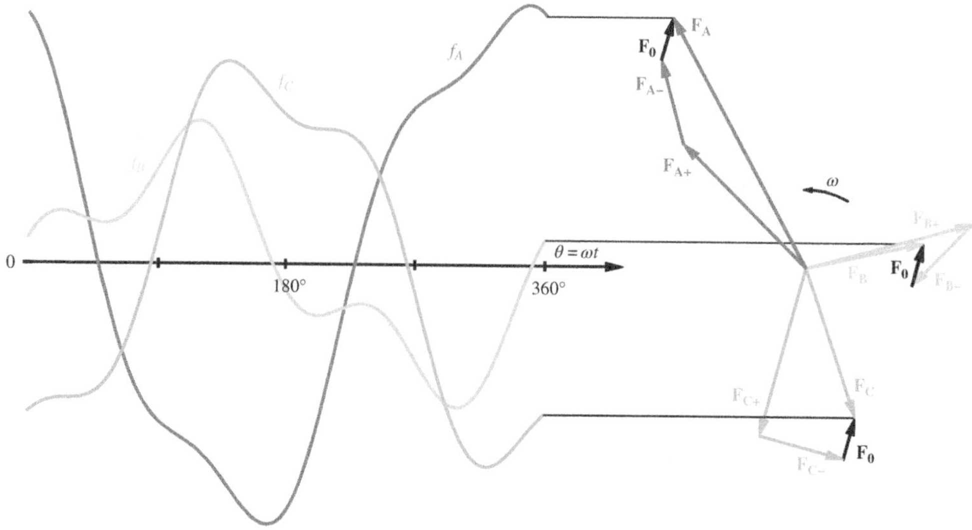

Figure A-12 Unbalanced three-phase variables in a timing diagram and the corresponding phasor diagram at the end of the cycle.

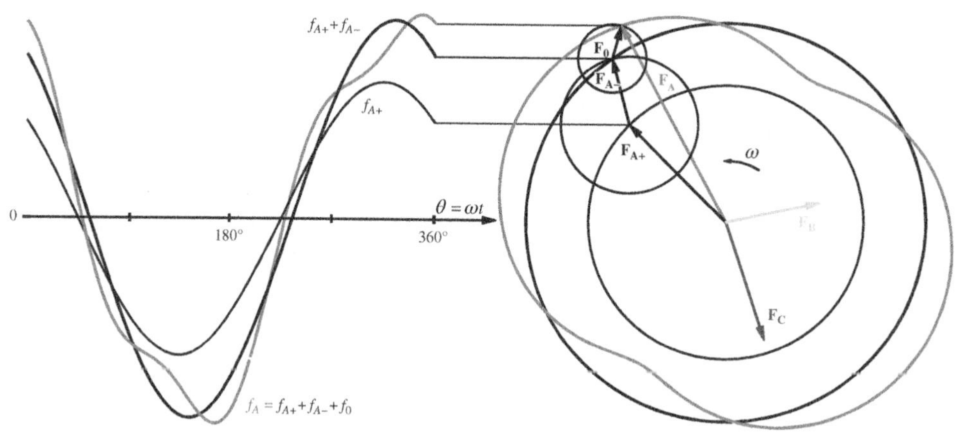

Figure A-13 A-phase positive-, negative-, and zero-sequence components in a timing diagram and the corresponding phasor diagram at the end of the cycle.

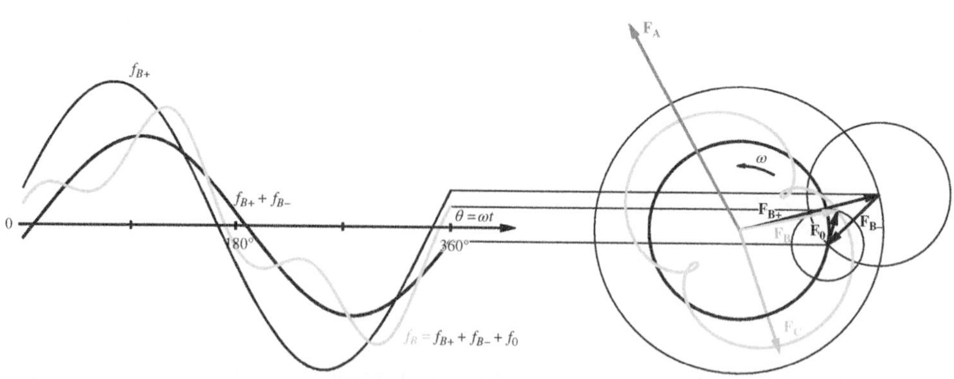

Figure A-14 B-phase positive-, negative-, and zero-sequence components in a timing diagram and the corresponding phasor diagram at the end of the cycle.

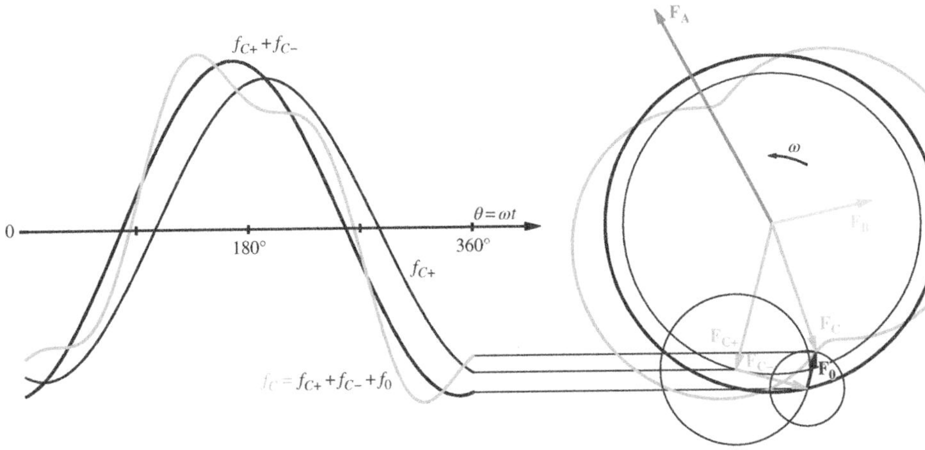

Figure A-15 C-phase positive-, negative-, and zero-sequence components in a timing diagram and the corresponding phasor diagram at the end of the cycle.

component can be segregated using a filter (further discussed later). Once the positive-sequence component is determined from its amplitude and angle, it can be subtracted from the three-phase variables ($f_{A'}$, $f_{B'}$, and $f_{C'}$) to calculate the negative-sequence component. A composite variable that consists of positive- and negative-sequence components with multiple frequencies can be analyzed in various rotating reference frames, such as fundamental negative sequence, 5th harmonic (a negative-sequence component), 7th harmonic (a positive-sequence component), and so on to convert the amplitude of a particular frequency component into DC. All other frequency components can be filtered to isolate the DC component of interest. Each rotating reference frame can isolate one particular frequency component.

A.4 Three-Phase Unbalanced Voltage, Current, and Power

Consider a three-phase unbalanced voltage

$$v_A = v_{A+} + v_{A-} + v_0 = V_{p+} \cos(\theta + \varphi_{v+}) + V_{p-} \cos(\theta + \varphi_{v-}) + v_0 \tag{A-33a}$$

$$v_B = v_{B+} + v_{B-} + v_0 = V_{p+} \cos(\theta + \varphi_{v+} - 120°) + V_{p-} \cos(\theta + \varphi_{v-} + 120°) + v_0 \tag{A-33b}$$

$$v_C = v_{C+} + v_{C-} + v_0 = V_{p+} \cos(\theta + \varphi_{v+} + 120°) + V_{p-} \cos(\theta + \varphi_{v-} - 120°) + v_0 \tag{A-33c}$$

is supplying a three-phase unbalanced current

$$i_A = i_{A+} + i_{A-} + i_0 = I_{p+} \cos(\theta + \varphi_{i+}) + I_{p-} \cos(\theta + \varphi_{i-}) + i_0 \tag{A-34a}$$

$$i_B = i_{B+} + i_{B-} + i_0 = I_{p+} \cos(\theta + \varphi_{i+} - 120°) + I_{p-} \cos(\theta + \varphi_{i-} + 120°) + i_0 \tag{A-34b}$$

$$i_C = i_{C+} + i_{C-} + i_0 = I_{p+} \cos(\theta + \varphi_{i+} + 120°) + I_{p-} \cos(\theta + \varphi_{i-} - 120°) + i_0 \tag{A-34c}$$

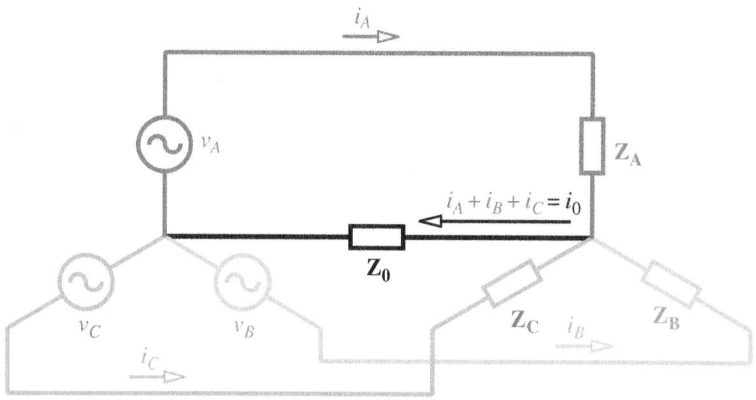

Figure A-16 Three-phase voltage source supplying three-phase unbalanced load.

through a three-phase impedance ($\mathbf{Z_A}$, $\mathbf{Z_B}$, and $\mathbf{Z_C}$) as shown in Figure A-16. The amplitudes of the positive-sequence voltage and current in each phase are V_{p+} and I_{p+}, respectively. The amplitudes of the negative-sequence voltage and current in each phase are V_{p-} and I_{p-}, respectively. The zero-sequence voltage and current in each phase are v_0 and i_0, respectively.

In the case of a three-phase, four-wire system, a zero-sequence current (i_0) can be expressed from Equations (A-34a) to (A-34c) as

$$i_0 = \frac{1}{3}(i_A + i_B + i_C). \tag{A-35}$$

This current flows through a zero-sequence impedance to the neutral of the source. In the case of a three-phase, three-wire system, there is no path for zero-sequence current (i_0) to flow as shown in Figure A-1. Therefore,

$$i_A + i_B + i_C = 0. \tag{A-36}$$

A zero-sequence voltage builds up between the neutral of the load and that of the source. The zero-sequence voltage can be expressed from Equations (A-33a) to (A-33c) as

$$v_0 = \frac{1}{3}(v_A + v_B + v_C). \tag{A-37}$$

The three-phase instantaneous power is defined as

$$p = p_A + p_B + p_C \tag{A-14}$$

where

$$\begin{aligned} p_A &= v_A i_A = (v_{A+} + v_{A-} + v_0) \times (i_{A+} + i_{A-} + i_0) \\ &= \left\{V_{p+} \cos(\theta + \varphi_{v+}) + V_{p-} \cos(\theta + \varphi_{v-}) + v_0\right\} \\ &\quad \times \left\{I_{p+} \cos(\theta + \varphi_{i+}) + I_{p-} \cos(\theta + \varphi_{i-}) + i_0\right\} \end{aligned} \tag{A-38a}$$

from Equations (A-33a) and (A-34a),

$$\begin{aligned} p_B &= v_B i_B = (v_{B+} + v_{B-} + v_0) \times (i_{B+} + i_{B-} + i_0) \\ &= \left\{V_{p+} \cos(\theta + \varphi_{v+} - 120^\circ) + V_{p-} \cos(\theta + \varphi_{v-} + 120^\circ) + v_0\right\} \\ &\quad \times \left\{I_{p+} \cos(\theta + \varphi_{i+} - 120^\circ) + I_{p-} \cos(\theta + \varphi_{i-} + 120^\circ) + i_0\right\} \end{aligned} \tag{A-38b}$$

from Equations (A-33b) and (A-34b),

$$\begin{aligned} p_C &= v_C i_C = (v_{C+} + v_{C-} + v_0) \times (i_{C+} + i_{C-} + i_0) \\ &= \left\{V_{p+} \cos(\theta + \varphi_{v+} + 120^\circ) + V_{p-} \cos(\theta + \varphi_{v-} - 120^\circ) + v_0\right\} \\ &\quad \times \left\{I_{p+} \cos(\theta + \varphi_{i+} + 120^\circ) + I_{p-} \cos(\theta + \varphi_{i-} - 120^\circ) + i_0\right\} \end{aligned} \tag{A-38c}$$

from Equations (A-33c) and (A-34c).

The instantaneous powers in the A phase due to positive-sequence voltage and positive-, negative-, and zero-sequence currents are given by

$$v_{A+} i_{A+} = \frac{V_{p+} I_{p+}}{2} [\cos(2\theta + \varphi_{v+} + \varphi_{i+}) + \cos(\varphi_{v+} - \varphi_{i+})] \tag{A-39a}$$

$$v_{A+} i_{A-} = \frac{V_{p+} I_{p-}}{2} [\cos(2\theta + \varphi_{v+} + \varphi_{i-}) + \cos(\varphi_{v+} - \varphi_{i-})] \tag{A-39b}$$

$$v_{A+} i_0 = V_{p+} \cos(\theta + \varphi_{v+}) i_0. \tag{A-39c}$$

The instantaneous powers in the A phase due to negative-sequence voltage and positive-, negative-, and zero-sequence currents are given by

$$v_{A-} i_{A+} = \frac{V_{p-} I_{p+}}{2} [\cos(2\theta + \varphi_{v-} + \varphi_{i+}) + \cos(\varphi_{v-} - \varphi_{i+})] \tag{A-40a}$$

$$v_{A-} i_{A-} = \frac{V_{p-} I_{p-}}{2} [\cos(2\theta + \varphi_{v-} + \varphi_{i-}) + \cos(\varphi_{v-} - \varphi_{i-})] \tag{A-40b}$$

$$v_{A-} i_0 = V_{p-} \cos(\theta + \varphi_{v-}) i_0. \tag{A-40c}$$

The instantaneous powers in the A phase due to zero-sequence voltage and positive-, negative-, and zero-sequence currents are given by

$$v_0 i_{A+} = v_0 I_{p+} \cos(\theta + \varphi_{i+}) \tag{A-41a}$$

$$v_0 i_{A-} = v_0 I_{p-} \cos(\theta + \varphi_{i-}) \tag{A-41b}$$

$$v_0 i_0 = v_0 i_0. \tag{A-41c}$$

The instantaneous powers in the B phase due to positive-sequence voltage and positive-, negative-, and zero-sequence currents are given by

$$v_{B+} i_{B+} = \frac{V_{p+} I_{p+}}{2} [\cos(2\theta + \varphi_{v+} + \varphi_{i+} + 120^\circ) + \cos(\varphi_{v+} - \varphi_{i+})] \tag{A-42a}$$

$$v_{B+} i_{B-} = \frac{V_{p+} I_{p-}}{2} [\cos(2\theta + \varphi_{v+} + \varphi_{i-}) + \cos(\varphi_{v+} - \varphi_{i-} + 120^\circ)] \tag{A-42b}$$

$$v_{B+} i_0 = V_{p+} \cos(\theta + \varphi_{v+} - 120^\circ) i_0. \tag{A-42c}$$

The instantaneous powers in the B phase due to negative-sequence voltage and positive-, negative-, and zero-sequence currents are given by

$$v_{B-} i_{B+} = \frac{V_{p-} I_{p+}}{2} [\cos(2\theta + \varphi_{v-} + \varphi_{i+}) + \cos(\varphi_{v-} - \varphi_{i+} - 120^\circ)] \tag{A-43a}$$

$$v_{B-} i_{B-} = \frac{V_{p-} I_{p-}}{2} [\cos(2\theta + \varphi_{v-} + \varphi_{i-} - 120^\circ) + \cos(\varphi_{v-} - \varphi_{i-})] \tag{A-43b}$$

$$v_{B-} i_0 = V_{p-} \cos(\theta + \varphi_{v-} + 120^\circ) i_0. \tag{A-43c}$$

The instantaneous powers in the B phase due to zero-sequence voltage and positive-, negative-, and zero-sequence currents are given by

$$v_0 i_{B+} = v_0 I_{p+} \cos(\theta + \varphi_{i+} - 120°) \tag{A-44a}$$

$$v_0 i_{B-} = v_0 I_{p-} \cos(\theta + \varphi_{i-} + 120°) \tag{A-44b}$$

$$v_0 i_0 = v_0 i_0. \tag{A-44c}$$

The instantaneous powers in the C phase due to positive-sequence voltage and positive-, negative-, and zero-sequence currents are given by

$$v_{C+} i_{C+} = \frac{V_{p+} I_{p+}}{2}[\cos(2\theta + \varphi_{v+} + \varphi_{i+} - 120°) + \cos(\varphi_{v+} - \varphi_{i+})] \tag{A-45a}$$

$$v_{C+} i_{C-} = \frac{V_{p+} I_{p-}}{2}[\cos(2\theta + \varphi_{v+} + \varphi_{i-}) + \cos(\varphi_{v+} - \varphi_{i-} - 120°)] \tag{A-45b}$$

$$v_{C+} i_0 = V_{p+} \cos(\theta + \varphi_{v+} + 120°) i_0. \tag{A-45c}$$

The instantaneous powers in the C phase due to negative-sequence voltage and positive-, negative-, and zero-sequence currents are given by

$$v_{C-} i_{C+} = \frac{V_{p-} I_{p+}}{2}[\cos(2\theta + \varphi_{v-} + \varphi_{i+}) + \cos(\varphi_{v-} - \varphi_{i+} + 120°)] \tag{A-46a}$$

$$v_{C-} i_{C-} = \frac{V_{p-} I_{p-}}{2}[\cos(2\theta + \varphi_{v-} + \varphi_{i-} + 120°) + \cos(\varphi_{v-} - \varphi_{i-})] \tag{A-46b}$$

$$v_{C-} i_0 = V_{p-} \cos(\theta + \varphi_{v-} - 120°) i_0. \tag{A-46c}$$

The instantaneous powers in the C phase due to zero-sequence voltage and positive-, negative-, and zero-sequence currents are given by

$$v_0 i_{C+} = v_0 I_{p+} \cos(\theta + \varphi_{i+} + 120°) \tag{A-47a}$$

$$v_0 i_{C-} = v_0 I_{p-} \cos(\theta + \varphi_{i-} - 120°) \tag{A-47b}$$

$$v_0 i_0 = v_0 i_0. \tag{A-47c}$$

The three-phase power due to positive-sequence voltage and positive-, negative-, and zero-sequence currents are given by

$$p_{+} = v_{A+} i_{A+} + v_{B+} i_{B+} + v_{C+} i_{C+} = \frac{3}{2} V_{p+} I_{p+} \cos(\varphi_{v+} - \varphi_{i+}) \tag{A-48a}$$

$$p_{+-} = v_{A+} i_{A-} + v_{B+} i_{B-} + v_{C+} i_{C-} = \frac{3}{2} V_{p+} I_{p-} \cos(2\theta + \varphi_{v+} + \varphi_{i-}) \tag{A-48b}$$

$$p_{+0} = v_{A+} i_0 + v_{B+} i_0 + v_{C+} i_0 = 0. \tag{A-48c}$$

The three-phase power due to negative-sequence voltage and positive-, negative-, and zero-sequence currents are given by

$$p_{-+} = v_{A-} i_{A+} + v_{B-} i_{B+} + v_{C-} i_{C+} = \frac{3}{2} V_{p-} I_{p+} \cos(2\theta + \varphi_{v-} + \varphi_{i+}) \tag{A-48d}$$

$$p_{-} = v_{A-} i_{A-} + v_{B-} i_{B-} + v_{C-} i_{C-} = \frac{3}{2} V_{p-} I_{p-} \cos(\varphi_{v-} - \varphi_{i-}) \tag{A-48e}$$

$$p_{-0} = v_{A-} i_0 + v_{B-} i_0 + v_{C-} i_0 = 0. \tag{A-48f}$$

The three-phase power due to zero-sequence voltage and positive-, negative-, and zero-sequence currents are given by

$$p_{0+} = v_0 i_{A+} + v_0 i_{B+} + v_0 i_{C+} = 0 \quad \text{(A-48g)}$$

$$p_{0-} = v_0 i_{A-} + v_0 i_{B-} + v_0 i_{C-} = 0 \quad \text{(A-48h)}$$

$$p_0 = v_0 i_0 + v_0 i_0 + v_0 i_0 = 3v_0 i_0. \quad \text{(A-48i)}$$

Therefore, the three-phase instantaneous power is given by

$$\begin{aligned} p &= p_+ + p_- + p_{+-} + p_{-+} + p_0 \\ &= \frac{3}{2}V_{p+}I_{p+}\cos(\varphi_{v+} - \varphi_{i+}) + \frac{3}{2}V_{p-}I_{p-}\cos(\varphi_{v-} - \varphi_{i-}) \\ &\quad + \frac{3}{2}V_{p+}I_{p-}\cos(2\theta + \varphi_{v+} + \varphi_{i-}) + \frac{3}{2}V_{p-}I_{p+}\cos(2\theta + \varphi_{v-} + \varphi_{i+}) + 3v_0 i_0. \end{aligned} \quad \text{(A-49)}$$

The first and second terms in Equation (A-49) are time-invariant fixed terms. The third and fourth terms are time-variant, double-frequency cosine terms. The fifth term is time-variant and is contributed by zero-sequence voltage and current. It is important to note that the sufficient condition for a time-variant, double-frequency cosine term to exist is the presence of the negative-sequence term in voltage or current.

A.5 d-q Transformation (3-Phase System, Transformed into d-q axes; d-axis Is the Active Component and q-axis Is the Reactive Component)

A set of three instantaneous phase variables that sum to zero are referred to as a balanced three-phase set of either positive sequence or negative sequence as described before. A positive-sequence set can be represented as the projections of three equal-length rotating phasors that are 120° apart (shown in Figure A-17a) on a stationary reference frame as explained earlier. These instantaneous

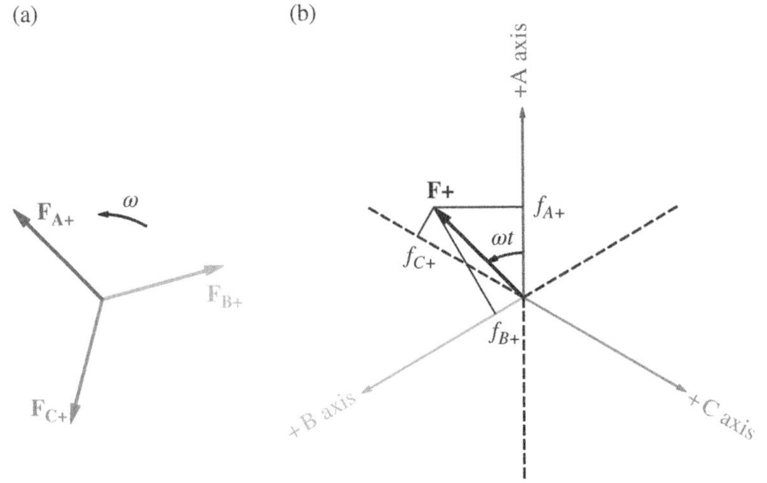

Figure A-17 Representation of three-phase, positive-sequence variables in (a) a phasor diagram, and (b) a vector diagram.

values can also be represented as the projections of a rotating vector upon three stationary axes that are 120° apart as shown in Figure A-17b. Figure A-17b shows **F+** as a positive-sequence spatial vector rotating in the counterclockwise direction. The projections of **F+** on the stationary A, B, and C axes are the positive-sequence instantaneous phase variables.

A negative-sequence set can be represented as the projections of three equal length rotating phasors that are 120° apart (shown in Figure A-18a) on a stationary reference frame. These instantaneous values can also be represented as the projections of a rotating vector upon three stationary axes that are 120° apart as shown in Figure A-18b. Figure A-18b shows **F−** as a negative-sequence spatial vector, rotating in the clockwise direction. The projections of **F−** on the stationary A, B, and C axes are the negative-sequence instantaneous phase variables.

The conversion of the A-B-C stationary variables into the d-q rotating frames is a two-step process. First, the positive and negative-sequence variables are converted into their equivalent d-q stationary variables, denoted by subscript s, assuming that the d axis is aligned with the A axis as shown in Figure A-19a. Second, these d-q stationary variables are rotated counterclockwise from their original position with an arbitrary angular speed of ω_a rad/s as shown in Figure A-19b.

The three-phase set of variables

$$f_A = f_{A+} + f_{A-} + f_0 \tag{A-18a}$$

$$f_B = f_{B+} + f_{B-} + f_0 \tag{A-18b}$$

$$f_C = f_{C+} + f_{C-} + f_0 \tag{A-18c}$$

can be defined as projections of f_{ds} and f_{qs} on the A-B-C axes as

$$f_A = f_{ds} + f_0 \tag{A-50a}$$

$$f_B = \cos(120°) f_{ds} + \cos(30°) f_{qs} + f_0 = -\frac{1}{2} f_{ds} + \frac{\sqrt{3}}{2} f_{qs} + f_0 \tag{A-50b}$$

$$f_C = \cos(120°) f_{ds} + \cos(150°) f_{qs} + f_0 = -\frac{1}{2} f_{ds} - \frac{\sqrt{3}}{2} f_{qs} + f_0. \tag{A-50c}$$

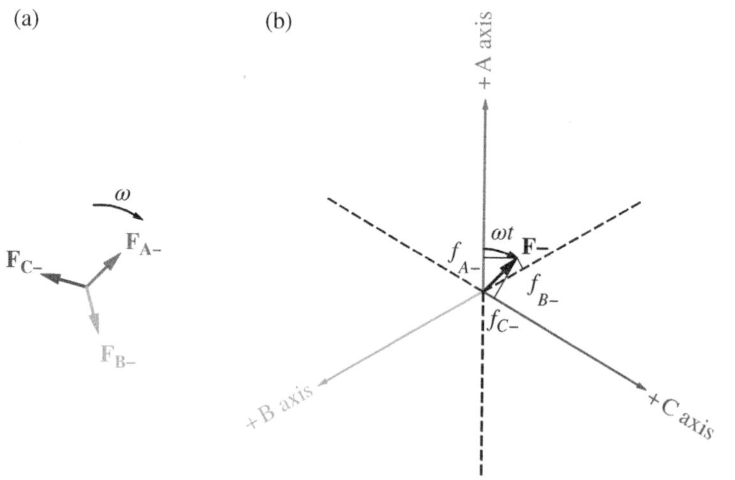

Figure A-18 Representation of three-phase, negative-sequence variables in (a) a phasor diagram and (b) a vector diagram.

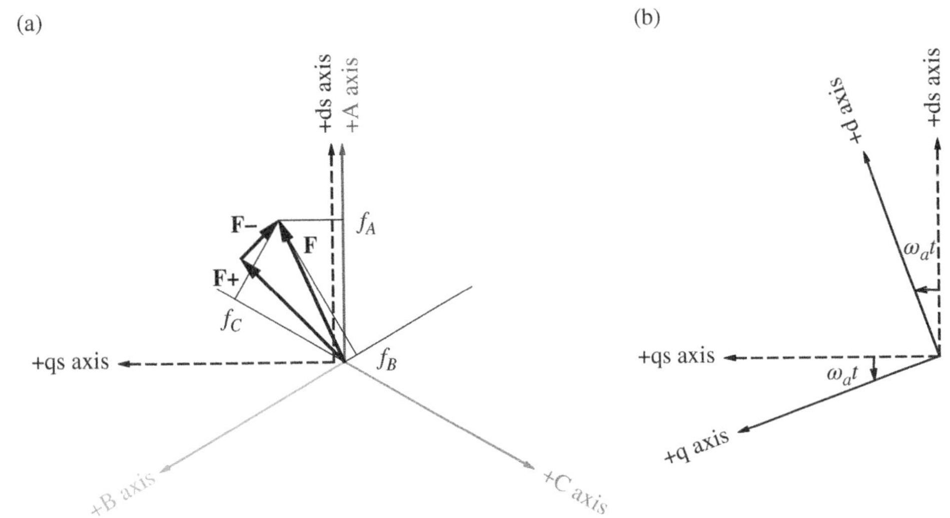

Figure A-19 Conversion of A, B, and C stationary variables into (a) d-q stationary and (b) d-q rotating frames.

In matrix form,

$$\begin{bmatrix} f_A \\ f_B \\ f_C \end{bmatrix} = \begin{bmatrix} 1 & 0 & 1 \\ -\frac{1}{2} & \frac{\sqrt{3}}{2} & 1 \\ -\frac{1}{2} & -\frac{\sqrt{3}}{2} & 1 \end{bmatrix} \times \begin{bmatrix} f_{ds} \\ f_{qs} \\ f_0 \end{bmatrix} \tag{A-51a}$$

and

$$\begin{bmatrix} f_{ds} \\ f_{qs} \\ f_0 \end{bmatrix} = \frac{2}{3} \begin{bmatrix} 1 & -\frac{1}{2} & -\frac{1}{2} \\ 0 & \frac{\sqrt{3}}{2} & -\frac{\sqrt{3}}{2} \\ \frac{1}{2} & \frac{1}{2} & \frac{1}{2} \end{bmatrix} \times \begin{bmatrix} f_A \\ f_B \\ f_C \end{bmatrix}. \tag{A-51b}$$

From Equation (A-51b), it can be written that

$$f_{ds} = \frac{2}{3}\left(f_A - \frac{1}{2}f_B - \frac{1}{2}f_C\right) \tag{A-52a}$$

$$f_{qs} = \frac{2}{3}\left(\frac{\sqrt{3}}{2}f_B - \frac{\sqrt{3}}{2}f_C\right) \tag{A-52b}$$

$$f_0 = \frac{2}{3}\left(\frac{1}{2}f_A + \frac{1}{2}f_B + \frac{1}{2}f_C\right). \tag{A-52c}$$

The above set of equations can be combined as

$$\begin{aligned} f_s &= f_{ds} + jf_{qs} + f_0 \\ &= \frac{2}{3}\left(f_A + \left\{-\frac{1}{2}f_B + j\frac{\sqrt{3}}{2}f_B\right\} + \left\{-\frac{1}{2}f_C - j\frac{\sqrt{3}}{2}f_C\right\}\right) + \frac{1}{3}(f_A + f_B + f_C) \end{aligned}$$

$$f_s = \frac{2}{3}(f_A + af_B + a^2 f_C) + f_0$$

where $a = e^{j120°}$ and $a^2 = e^{j240°} = e^{-j120°}$ or

$$f_s' = f_s - f_0 = \frac{2}{3}\left(f_A + af_B + a^2 f_C\right). \tag{A-53}$$

Note that f_s' is a space vector, such as an airgap flux in an induction machine.

The relationship between the arbitrarily rotating d-q reference frame at ω_a rad/s, denoted by subscript a, and the stationary d-q reference frame, as shown in Figure A-18b, is defined as

$$f_{ds} = f_{da}\cos(\omega_a t) - f_{qa}\sin(\omega_a t) \tag{A-54a}$$

and

$$f_{qs} = f_{da}\sin(\omega_a t) + f_{qa}\cos(\omega_a t). \tag{A-54b}$$

This relationship in matrix form is given by

$$\begin{bmatrix} f_{ds} \\ f_{qs} \end{bmatrix} = \begin{bmatrix} \cos(\omega_a t) & -\sin(\omega_a t) \\ \sin(\omega_a t) & \cos(\omega_a t) \end{bmatrix} \times \begin{bmatrix} f_{da} \\ f_{qa} \end{bmatrix} \tag{A-55a}$$

and

$$\begin{bmatrix} f_{da} \\ f_{qa} \end{bmatrix} = \begin{bmatrix} \cos(\omega_a t) & \sin(\omega_a t) \\ -\sin(\omega_a t) & \cos(\omega_a t) \end{bmatrix} \times \begin{bmatrix} f_{ds} \\ f_{qs} \end{bmatrix}. \tag{A-55b}$$

The complete relationship including the zero-sequence component is given by

$$\begin{bmatrix} f_{da} \\ f_{qa} \\ f_0 \end{bmatrix} = \begin{bmatrix} \cos(\omega_a t) & \sin(\omega_a t) & 0 \\ -\sin(\omega_a t) & \cos(\omega_a t) & 0 \\ 0 & 0 & 1 \end{bmatrix} \times \begin{bmatrix} f_{ds} \\ f_{qs} \\ f_0 \end{bmatrix}. \tag{A-56}$$

From Equations (A-56) and (A-51b), the A-B-C phase variables can be converted into their equivalent d-q variables in an arbitrarily rotating reference frame as

$$\begin{bmatrix} f_{da} \\ f_{qa} \\ f_0 \end{bmatrix} = \begin{bmatrix} \cos(\omega_a t) & \sin(\omega_a t) & 0 \\ -\sin(\omega_a t) & \cos(\omega_a t) & 0 \\ 0 & 0 & 1 \end{bmatrix} \frac{2}{3} \begin{bmatrix} 1 & -\frac{1}{2} & -\frac{1}{2} \\ 0 & \frac{\sqrt{3}}{2} & -\frac{\sqrt{3}}{2} \\ \frac{1}{2} & \frac{1}{2} & \frac{1}{2} \end{bmatrix} \times \begin{bmatrix} f_A \\ f_B \\ f_C \end{bmatrix}.$$

The above equation can be written in another form as

$$\begin{bmatrix} f_{da} \\ f_{qa} \\ f_0 \end{bmatrix} = \frac{2}{3} \begin{bmatrix} \cos(\omega_a t) & \cos(\omega_a t - 120°) & \cos(\omega_a t + 120°) \\ -\sin(\omega_a t) & -\sin(\omega_a t - 120°) & -\sin(\omega_a t + 120°) \\ \frac{1}{2} & \frac{1}{2} & \frac{1}{2} \end{bmatrix} \times \begin{bmatrix} f_A \\ f_B \\ f_C \end{bmatrix}. \tag{A-57}$$

From Equation (A-57), it can be written that

$$f_{da} = \frac{2}{3}\left[f_A\cos(\omega_a t) + f_B\cos(\omega_a t - 120°) + f_C\cos(\omega_a t - 120°)\right] \tag{A-58a}$$

$$f_{qa} = \frac{2}{3}\left[-f_A \sin(\omega_a t) - f_B \sin(\omega_a t - 120°) - f_C \sin(\omega_a t - 120°)\right] \tag{A-58b}$$

$$f_0 = \frac{2}{3}\left(\frac{1}{2}f_A + \frac{1}{2}f_B + \frac{1}{2}f_C\right). \tag{A-58c}$$

The above set of equations can be combined as

$$f_a = f_{da} + jf_{qa} + f_0 = \frac{2}{3}\begin{bmatrix} f_A\{\cos(\omega_a t) - j\sin(\omega_a t)\} + \\ f_B\{\cos(\omega_a t - 120°) - j\sin(\omega_a t - 120°)\} + \\ f_C\{\cos(\omega_a t + 120°) - j\sin(\omega_a t + 120°)\} \end{bmatrix}$$

$$+ \frac{2}{3}\left(\frac{1}{2}f_A + \frac{1}{2}f_B + \frac{1}{2}f_C\right)$$

or

$$f_a = f_{da} + jf_{qa} + f_0 = \frac{2}{3}\begin{bmatrix} f_A\{e^{-j\omega_a t}\} + \\ f_B\left\{e^{-j(\omega_a t - 120°)}\right\} + \\ f_C\left\{e^{-j(\omega_a t + 120°)}\right\} \end{bmatrix} + \frac{1}{3}(f_A + f_B + f_C)$$

or

$$f_a = \frac{2}{3}(f_A + af_B + a^2 f_C)e^{-j\omega_a t} + f_0$$

where $a = e^{j120°}$ and $a^2 = e^{j240°} = e^{-j120°}$, or

$f_a - f_0 = f'_s e^{-j\omega_a t}$ with a substitution from Equation (A-53)

or

$$f'_a = f'_s e^{-j\omega_a t} \tag{A-59a}$$

or

$$\boldsymbol{f_{dq}} = f'_s e^{-j\omega_a t} \tag{A-59b}$$

or

$$\boldsymbol{f_{dq}} = f_d + jf_q \tag{A-59c}$$

where $\boldsymbol{f_{dq}}$ is a space vector, f'_s, which is rotating at ω_a rad/s as defined in Equation (A-59b). This is called d-q transformation. Note that f_d and f_q are orthogonal components in d-q reference frame.

The following matrix is called Park's original transformation matrix. It represents the projections of the stationary A-B-C axes on to the rotating d-q axes:

$$[P] = \frac{2}{3}\begin{bmatrix} \cos(\omega_a t) & \cos(\omega_a t - 120°) & \cos(\omega_a t + 120°) \\ -\sin(\omega_a t) & -\sin(\omega_a t - 120°) & -\sin(\omega_a t + 120°) \\ \frac{1}{2} & \frac{1}{2} & \frac{1}{2} \end{bmatrix}. \tag{A-60}$$

The inverse of Park's original transformation matrix is given by

$$[P]^{-1} = \begin{bmatrix} \cos(\omega_a t) & -\sin(\omega_a t) & 1 \\ \cos(\omega_a t - 120^\circ) & -\sin(\omega_a t - 120^\circ) & 1 \\ \cos(\omega_a t + 120^\circ) & -\sin(\omega_a t + 120^\circ) & 1 \end{bmatrix}. \quad \text{(A-61)}$$

A.5.1 Conversion of a Variable Containing Positive-, Negative-, and Zero-Sequence Components into d-q Frame

Consider a three-phase set of variables that consist of positive-, negative-, and zero-sequence components as

$$f_A = f_{A+} + f_{A-} + f_0 = F_{p+} \cos\left(\theta + \varphi_{f+}\right) + F_{p-} \cos\left(\theta + \varphi_{f-}\right) + f_0 \quad \text{(A-62a)}$$

$$f_B = f_{B+} + f_{B-} + f_0 = F_{p+} \cos\left(\theta + \varphi_{f+} - 120^\circ\right) + F_{p-} \cos\left(\theta + \varphi_{f-} + 120^\circ\right) + f_0 \quad \text{(A-62b)}$$

$$f_C = f_{C+} + f_{C-} + f_0 = F_{p+} \cos\left(\theta + \varphi_{f+} + 120^\circ\right) + F_{p-} \cos\left(\theta + \varphi_{f-} - 120^\circ\right) + f_0. \quad \text{(A-62c)}$$

Then, the relationship between the d-q-0 variables in an arbitrarily rotating reference frame and A-B-C phase variables is given by

$$\begin{bmatrix} f_{da} \\ f_{qa} \\ f_0 \end{bmatrix} = [P] \times \begin{bmatrix} f_A \\ f_B \\ f_C \end{bmatrix} = [P] \times \begin{bmatrix} f_{A+} + f_{A-} + f_0 \\ f_{B+} + f_{B-} + f_0 \\ f_{C+} + f_{C-} + f_0 \end{bmatrix} = [P] \times \begin{bmatrix} f_{A+} \\ f_{B+} \\ f_{C+} \end{bmatrix} + [P] \times \begin{bmatrix} f_{A-} \\ f_{B-} \\ f_{C-} \end{bmatrix} + [P] \times \begin{bmatrix} f_0 \\ f_0 \\ f_0 \end{bmatrix}. \quad \text{(A-63)}$$

But,

$$[P] \begin{bmatrix} f_{A+} \\ f_{B+} \\ f_{C+} \end{bmatrix} = \frac{2}{3} \begin{bmatrix} \cos(\omega_a t) & \cos(\omega_a t - 120^\circ) & \cos(\omega_a t + 120^\circ) \\ -\sin(\omega_a t) & -\sin(\omega_a t - 120^\circ) & -\sin(\omega_a t + 120^\circ) \\ \frac{1}{2} & \frac{1}{2} & \frac{1}{2} \end{bmatrix} \times \begin{bmatrix} F_{p+} \cos\left(\theta + \varphi_{f+}\right) \\ F_{p+} \cos\left(\theta + \varphi_{f+} - 120^\circ\right) \\ F_{p+} \cos\left(\theta + \varphi_{f+} + 120^\circ\right) \end{bmatrix}. \quad \text{(A-64)}$$

Substituting $\theta = \omega t$ from Equation (A-3), it is given by

$$[P]\begin{bmatrix} f_{A+} \\ f_{B+} \\ f_{C+} \end{bmatrix} = \frac{2}{3}F_{p+}\begin{bmatrix} \cos(\omega_a t)\cos\left(\omega t+\varphi_{f+}\right)+\cos(\omega_a t-120^\circ)\cos\left(\omega t+\varphi_{f+}-120^\circ\right)+ \\ \cos(\omega_a t+120^\circ)\cos\left(\omega t+\varphi_{f+}+120^\circ\right) \\ -\sin(\omega_a t)\cos\left(\omega t+\varphi_{f+}\right)-\sin(\omega_a t-120^\circ)\cos\left(\omega t+\varphi_{f+}-120^\circ\right) \\ -\sin(\omega_a t+120^\circ)\cos\left(\omega t+\varphi_{f+}+120^\circ\right) \\ \frac{1}{2}\left\{\cos\left(\omega t+\varphi_{f+}\right)+\cos\left(\omega t+\varphi_{f+}-120^\circ\right)+\cos\left(\omega t+\varphi_{f+}+120^\circ\right)\right\} \end{bmatrix}$$

$$= \frac{1}{3}F_{p+}\begin{bmatrix} \cos\left\{(\omega+\omega_a)t+\varphi_{f+}\right\}+\cos\left\{(\omega-\omega_a)t+\varphi_{f+}\right\}+\cos\left\{(\omega+\omega_a)t+\varphi_{f+}-240^\circ\right\}+ \\ \cos\left\{(\omega-\omega_a)t+\varphi_{f+}\right\}+\cos\left\{(\omega+\omega_a)t+\varphi_{f+}+240^\circ\right\}+\cos\left\{(\omega-\omega_a)t+\varphi_{f+}\right\} \\ -\sin\left\{(\omega+\omega_a)t+\varphi_{f+}\right\}+\sin\left\{(\omega-\omega_a)t+\varphi_{f+}\right\}-\sin\left\{(\omega+\omega_a)t+\varphi_{f+}-240^\circ\right\}+ \\ \sin\left\{(\omega-\omega_a)t+\varphi_{f+}\right\}-\sin\left\{(\omega+\omega_a)t+\varphi_{f+}+240^\circ\right\}+\sin\left\{(\omega-\omega_a)t+\varphi_{f+}\right\} \\ 0 \end{bmatrix}$$

$$= F_{p+}\begin{bmatrix} \cos\left\{(\omega-\omega_a)t+\varphi_{f+}\right\} \\ \sin\left\{(\omega-\omega_a)t+\varphi_{f+}\right\} \\ 0 \end{bmatrix}. \tag{A-65}$$

Therefore,

$$[P]\begin{bmatrix} f_{A-} \\ f_{B-} \\ f_{C-} \end{bmatrix} = \frac{2}{3}\begin{bmatrix} \cos(\omega_a t) & \cos(\omega_a t-120^\circ) & \cos(\omega_a t+120^\circ) \\ -\sin(\omega_a t) & -\sin(\omega_a t-120^\circ) & -\sin(\omega_a t+120^\circ) \\ \frac{1}{2} & \frac{1}{2} & \frac{1}{2} \end{bmatrix}$$

$$\times \begin{bmatrix} F_{p-}\cos\left(\theta+\varphi_{f-}\right) \\ F_{p-}\cos\left(\theta+\varphi_{f-}+120^\circ\right) \\ F_{p-}\cos\left(\theta+\varphi_{f-}-120^\circ\right) \end{bmatrix}. \tag{A-66}$$

Substituting $\theta = \omega t$ from Equation (A-3), it is given by

$$[P]\begin{bmatrix} f_{A-} \\ f_{B-} \\ f_{C-} \end{bmatrix} = \frac{2}{3}F_{p-}\begin{bmatrix} \cos(\omega_a t)\cos\left(\omega t+\varphi_{f-}\right)+\cos(\omega_a t-120^\circ)\cos\left(\omega t+\varphi_{f-}+120^\circ\right)+ \\ \cos(\omega_a t+120^\circ)\cos\left(\omega t+\varphi_{f-}-120^\circ\right) \\ -\sin(\omega_a t)\cos\left(\omega t+\varphi_{f-}\right)-\sin(\omega_a t-120^\circ)\cos\left(\omega t+\varphi_{f-}+120^\circ\right)- \\ \sin(\omega_a t+120^\circ)\cos\left(\omega t+\varphi_{f-}-120^\circ\right) \\ \frac{1}{2}\left\{\cos\left(\omega t+\varphi_{f-}\right)+\cos\left(\omega t+\varphi_{f-}+120^\circ\right)+\cos\left(\omega t+\varphi_{f-}-120^\circ\right)\right\} \end{bmatrix}$$

$$=\frac{1}{3}F_{p-}\begin{bmatrix} \cos\left\{(\omega+\omega_a)t+\varphi_{f-}\right\}+\cos\left\{(\omega+\omega_a)t+\varphi_{f-}\right\}+\cos\left\{(\omega+\omega_a)t+\varphi_{f-}\right\}+ \\ \cos\left\{(\omega-\omega_a)t+\varphi_{f-}+240^\circ\right\}+\cos\left\{(\omega+\omega_a)t+\varphi_{f-}\right\}+\cos\left\{(\omega-\omega_a)t+\varphi_{f-}-240^\circ\right\} \\ -\sin\left\{(\omega+\omega_a)t+\varphi_{f-}\right\}+\sin\left\{(\omega-\omega_a)t+\varphi_{f-}\right\}-\sin\left\{(\omega+\omega_a)t+\varphi_{f-}\right\}+ \\ \sin\left\{(\omega-\omega_a)t+\varphi_{f-}+240^\circ\right\}-\sin\left\{(\omega+\omega_a)t+\varphi_{f-}\right\}+\sin\left\{(\omega-\omega_a)t+\varphi_{f-}-240^\circ\right\} \\ 0 \end{bmatrix}$$

$$=F_{p-}\begin{bmatrix} \cos\left\{(\omega+\omega_a)t+\varphi_{f-}\right\} \\ -\sin\left\{(\omega+\omega_a)t+\varphi_{f-}\right\} \\ 0 \end{bmatrix}. \tag{A-67}$$

Therefore,

$$[P]\begin{bmatrix} f_0 \\ f_0 \\ f_0 \end{bmatrix} = \frac{2}{3}\begin{bmatrix} \{\cos(\omega_a t)+\cos(\omega_a t-120^\circ)+\cos(\omega_a t+120^\circ)\}f_0 \\ -\{\sin(\omega_a t)+\sin(\omega_a t-120^\circ)+\sin(\omega_a t+120^\circ)\}f_0 \\ \frac{1}{2}3f_0 \end{bmatrix} = \begin{bmatrix} 0 \\ 0 \\ f_0 \end{bmatrix}. \tag{A-68}$$

Then, it can be written from Equation (A-60) that

$$\begin{bmatrix} f_{da} \\ f_{qa} \\ f_0 \end{bmatrix} = [P]\times\begin{bmatrix} f_{A+} \\ f_{B+} \\ f_{C+} \end{bmatrix} + [P]\times\begin{bmatrix} f_{A-} \\ f_{B-} \\ f_{C-} \end{bmatrix} + [P]\times\begin{bmatrix} f_0 \\ f_0 \\ f_0 \end{bmatrix}$$

$$= \begin{bmatrix} F_{p+}\cos\left\{(\omega-\omega_a)t+\varphi_{f+}\right\}+F_{p-}\cos\left\{(\omega+\omega_a)t+\varphi_{f-}\right\} \\ F_{p+}\sin\left\{(\omega-\omega_a)t+\varphi_{f+}\right\}-F_{p-}\sin\left\{(\omega+\omega_a)t+\varphi_{f-}\right\} \\ f_0 \end{bmatrix} \tag{A-69}$$

with substitutions from Equations (A-65), (A-67), and (A-68).

When $\omega_a = \omega$, meaning that the rotating reference frame is synchronous with the system angular frequency, Equation (A-69) becomes

$$\begin{bmatrix} f_d \\ f_q \\ f_0 \end{bmatrix} = \begin{bmatrix} F_{p+} \cos\varphi_{f+} + F_{p-} \cos\left(2\omega t + \varphi_{f-}\right) \\ F_{p+} \sin\varphi_{f+} - F_{p-} \sin\left(2\omega t + \varphi_{f-}\right) \\ f_0 \end{bmatrix}. \tag{A-70}$$

Throughout the book, the phase-locked loop is derived from the phase-to-neutral voltage with cosine reference. That means A-phase voltage is at its peak when $\varphi_{v+} = 0$. If the A-B-C line voltages are converted into their equivalent d-q variables in the synchronously rotating reference frame, Equation (A-70) can be rewritten as

$$\begin{bmatrix} v_d \\ v_q \\ v_0 \end{bmatrix} = \begin{bmatrix} V_{p+} + V_{p-} \cos\left(2\omega t + \varphi_{v-}\right) \\ -V_{p-} \sin\left(2\omega t + \varphi_{v-}\right) \\ v_0 \end{bmatrix}. \tag{A-71}$$

If the phase current is at an angle (φ_{i+}) with respect to the phase-to-neutral voltage, Equation (A-70) can be rewritten as

$$\begin{bmatrix} i_d \\ i_q \\ i_0 \end{bmatrix} = \begin{bmatrix} I_{p+} \cos\varphi_{i+} + I_{p-} \cos\left(2\omega t + \varphi_{i-}\right) \\ I_{p+} \sin\varphi_{i+} - I_{p-} \sin\left(2\omega t + \varphi_{i-}\right) \\ i_0 \end{bmatrix}. \tag{A-72}$$

In the synchronously rotating reference frame, the positive-sequence, fundamental component turns into a DC value as described in Section A.3. The d-q components of the fundamental line voltage are V_{p+} and 0, respectively. The rotating d-axis voltage contains a DC component that is the amplitude of the original A-B-C variables. The rotating q-axis voltage contains no DC component under a steady-state condition. The d-q components of the fundamental line current are I_{p+} $\cos(\varphi_{i+})$ and I_{p+} $\sin(\varphi_{i+})$. In addition, the rotating d- and q-axes components contain a double-frequency component whose amplitude is the same as that of the negative-sequence fundamental component in the original variables. The reason for the existence of the double-frequency, negative-sequence component in the d-q frame is because the negative-sequence fundamental component rotates at twice the angular speed in the opposite direction of the synchronously rotating reference frame. It is quite easy to eliminate these negative-sequence terms from the d-q components using a half-cycle running averager. This simply uses the fact that if a variable contains an oscillatory component that is superimposed on a DC component, then averaging the variable over the period of the oscillatory component will result in only the DC component at the output of the averager. In this case, the averager acts as a filter. Once the oscillatory double-frequency component is removed and the DC component is recovered from the d-q variables, the relative angle (φ_{i+}) of the current is then simply

$$\varphi_{i+} = \tan^{-1}\frac{i_q}{i_d}. \tag{A-73}$$

A.5.2 Calculation of Instantaneous Power into d-q Frame

Using Equation (A-16), the instantaneous active power in the three-phase system with phase voltages (v_A, v_B, and v_C) and phase currents (i_A, i_B, and i_C) can be rewritten as

$$p = v_A i_A + v_B i_B + v_C i_C = \begin{bmatrix} v_A & v_B & v_C \end{bmatrix} \times \begin{bmatrix} i_A \\ i_B \\ i_C \end{bmatrix} = \begin{bmatrix} v_A \\ v_B \\ v_C \end{bmatrix}^T \times \begin{bmatrix} i_A \\ i_B \\ i_C \end{bmatrix} = \left[[P]^{-1} \begin{bmatrix} v_d \\ v_q \\ v_0 \end{bmatrix} \right]^T \times [P]^{-1} \begin{bmatrix} i_d \\ i_q \\ i_0 \end{bmatrix}$$

or

$$p = \begin{bmatrix} v_d \\ v_q \\ v_0 \end{bmatrix}^T \left[\left[[P]^{-1} \right]^T [P]^{-1} \right] \begin{bmatrix} i_d \\ i_q \\ i_0 \end{bmatrix} = \begin{bmatrix} v_d & v_q & v_0 \end{bmatrix} \frac{3}{2} \begin{bmatrix} 1 & 0 & 0 \\ 0 & 1 & 0 \\ 0 & 0 & 2 \end{bmatrix} \begin{bmatrix} i_d \\ i_q \\ i_0 \end{bmatrix} \text{ using Equation (A}-61)$$

or

$$p = \frac{3}{2} \left(v_d i_d + v_q i_q \right) + 3 v_0 i_0. \tag{A-74}$$

The instantaneous three-phase power consists of two terms: the first term is due to the positive- and negative-sequence voltage and current and the second term is due to the zero-sequence voltage and current.

A.5.3 Calculation of Instantaneous Power into d-q Frame for a Three-Phase, Three-Wire System

Substituting $i_B = -(i_A + i_C)$ from Equation (A-36), Equation (A-51b) can be written for the current as

$$\begin{bmatrix} i_{ds} \\ i_{qs} \end{bmatrix} = \frac{2}{3} \begin{bmatrix} 1 & -\frac{1}{2} & -\frac{1}{2} \\ 0 & \frac{\sqrt{3}}{2} & -\frac{\sqrt{3}}{2} \end{bmatrix} \times \begin{bmatrix} i_A \\ i_B \\ i_C \end{bmatrix} = \begin{bmatrix} \frac{2}{3} & -\frac{1}{3} & -\frac{1}{3} \\ 0 & \frac{1}{\sqrt{3}} & -\frac{1}{\sqrt{3}} \end{bmatrix} \times \begin{bmatrix} i_A \\ -i_A - i_C \\ i_C \end{bmatrix}$$

$$= \begin{bmatrix} \frac{2}{3} + \frac{1}{3} & \frac{1}{3} - \frac{1}{3} \\ -\frac{1}{\sqrt{3}} & -\frac{1}{\sqrt{3}} - \frac{1}{\sqrt{3}} \end{bmatrix} \times \begin{bmatrix} i_A \\ i_C \end{bmatrix}$$

or

$$\begin{bmatrix} i_{ds} \\ i_{qs} \end{bmatrix} = \begin{bmatrix} 1 & 0 \\ -\frac{1}{\sqrt{3}} & -\frac{2}{\sqrt{3}} \end{bmatrix} \times \begin{bmatrix} i_A \\ i_C \end{bmatrix} \tag{A-75a}$$

and

$$\begin{bmatrix} i_A \\ i_C \end{bmatrix} = \begin{bmatrix} 1 & 0 \\ -\frac{1}{2} & -\frac{\sqrt{3}}{2} \end{bmatrix} \times \begin{bmatrix} i_{ds} \\ i_{qs} \end{bmatrix}. \tag{A-75b}$$

Equation (A-55b) can be rewritten for the current in the synchronously rotating reference frame as

$$\begin{bmatrix} i_d \\ i_q \end{bmatrix} = \begin{bmatrix} \cos(\omega t) & \sin(\omega t) \\ -\sin(\omega t) & \cos(\omega t) \end{bmatrix} \times \begin{bmatrix} i_{ds} \\ i_{qs} \end{bmatrix} = \begin{bmatrix} \cos(\omega t) & \sin(\omega t) \\ -\sin(\omega t) & \cos(\omega t) \end{bmatrix} \times \begin{bmatrix} 1 & 0 \\ -\frac{1}{\sqrt{3}} & -\frac{2}{\sqrt{3}} \end{bmatrix} \begin{bmatrix} i_A \\ i_C \end{bmatrix}$$

with a substitution from Equation (A-75a)

$$= \begin{bmatrix} \cos(\omega t) - \frac{1}{\sqrt{3}}\sin(\omega t) & -\frac{2}{\sqrt{3}}\sin(\omega t) \\ -\sin(\omega t) - \frac{1}{\sqrt{3}}\cos(\omega t) & -\frac{2}{\sqrt{3}}\cos(\omega t) \end{bmatrix} \times \begin{bmatrix} i_A \\ i_C \end{bmatrix}$$

or

$$\begin{bmatrix} i_d \\ i_q \end{bmatrix} = \frac{2}{\sqrt{3}} \begin{bmatrix} \sin(\omega t + 120°) & -\sin(\omega t) \\ \cos(\omega t + 120°) & -\cos(\omega t) \end{bmatrix} \times \begin{bmatrix} i_A \\ i_C \end{bmatrix}. \tag{A-76}$$

Equation (A-76) can be rearranged as

$$\begin{bmatrix} i_A \\ i_C \end{bmatrix} = \begin{bmatrix} \cos(\omega t) & -\sin(\omega t) \\ \cos(\omega t + 120°) & -\sin(\omega t + 120°) \end{bmatrix} \times \begin{bmatrix} i_d \\ i_q \end{bmatrix}. \tag{A-77}$$

Therefore, Park's original transformation matrix for a three-phase, three-wire system is given by

$$[p] = \frac{2}{\sqrt{3}} \begin{bmatrix} \sin(\omega t + 120°) & -\sin(\omega t) \\ \cos(\omega t + 120°) & -\cos(\omega t) \end{bmatrix} \tag{A-78a}$$

and the inverse of Park's original transformation matrix for a three-phase, three-wire system is given by

$$[p]^{-1} = \begin{bmatrix} \cos(\omega t) & -\sin(\omega t) \\ \cos(\omega t + 120°) & -\sin(\omega t + 120°) \end{bmatrix}. \tag{A-78b}$$

This transformation matrix can be used for voltage, current, and flux as well, provided that the zero-sequence component does not exist.

Similarly, in case of voltage,

$$\begin{bmatrix} v_d \\ v_q \end{bmatrix} = \frac{2}{\sqrt{3}} \begin{bmatrix} \sin(\omega t + 120°) & -\sin(\omega t) \\ \cos(\omega t + 120°) & -\cos(\omega t) \end{bmatrix} \begin{bmatrix} v_A \\ v_C \end{bmatrix}. \tag{A-79}$$

In some applications, it may be easier to use the available phase-to-phase voltage. The phase-to-phase voltage is related to the phase-to-neutral voltage as

$$\begin{bmatrix} v_{AB} \\ v_{CB} \end{bmatrix} = \begin{bmatrix} 1 & -1 & 0 \\ 0 & -1 & 1 \end{bmatrix} \times \begin{bmatrix} v_A \\ v_B \\ v_C \end{bmatrix} = \begin{bmatrix} 1 & -1 & 0 \\ 0 & -1 & 1 \end{bmatrix} \times \begin{bmatrix} v_A \\ -v_A - v_C \\ v_C \end{bmatrix} \text{ when } v_A + v_B + v_C = 0.$$

Then,

$$\begin{bmatrix} v_{AB} \\ v_{CB} \end{bmatrix} = \begin{bmatrix} 2 & 1 \\ 1 & 2 \end{bmatrix} \begin{bmatrix} v_A \\ v_C \end{bmatrix} \tag{A-80a}$$

and

$$\begin{bmatrix} v_A \\ v_C \end{bmatrix} = \frac{1}{3}\begin{bmatrix} 2 & -1 \\ -1 & 2 \end{bmatrix}\begin{bmatrix} v_{AB} \\ v_{CB} \end{bmatrix}. \tag{A-80b}$$

Equations (A-79) and (A-80b) can be combined as

$$\begin{aligned}\begin{bmatrix} v_d \\ v_q \end{bmatrix} &= \frac{2}{\sqrt{3}}\begin{bmatrix} \sin(\omega t + 120^\circ) & -\sin(\omega t) \\ \cos(\omega t + 120^\circ) & -\cos(\omega t) \end{bmatrix} \times \begin{bmatrix} v_A \\ v_C \end{bmatrix} \\ &= \frac{1}{3}\frac{2}{\sqrt{3}}\begin{bmatrix} \sin(\omega t + 120^\circ) & -\sin(\omega t) \\ \cos(\omega t + 120^\circ) & -\cos(\omega t) \end{bmatrix} \times \begin{bmatrix} 2 & -1 \\ -1 & 2 \end{bmatrix} \times \begin{bmatrix} v_{AB} \\ v_{CB} \end{bmatrix}\end{aligned}$$

or

$$\begin{bmatrix} v_d \\ v_q \end{bmatrix} = \frac{2}{3}\begin{bmatrix} \cos(\omega t) & \cos(\omega t + 120^\circ) \\ -\sin(\omega t) & -\sin(\omega t + 120^\circ) \end{bmatrix}\begin{bmatrix} v_{AB} \\ v_{CB} \end{bmatrix}. \tag{A-81}$$

Equation (A-81) can be rearranged as

$$\begin{bmatrix} v_{AB} \\ v_{CB} \end{bmatrix} = \sqrt{3}\begin{bmatrix} \sin(\omega t + 120^\circ) & \cos(\omega t + 120^\circ) \\ -\sin(\omega t) & -\cos(\omega t) \end{bmatrix}\begin{bmatrix} v_d \\ v_q \end{bmatrix}. \tag{A-82}$$

Substituting $i_B = -(i_A + i_C)$ from Equation (A-36), Equation (A-16) for instantaneous active power can be rewritten as

$$p = v_A i_A + v_B i_B + v_C i_C = v_A i_A - v_B(i_A + i_C) + v_C i_C = (v_A - v_B)i_A + (v_C - v_B)i_C$$

or

$$p = v_{AB} i_A + v_{CB} i_C = \begin{bmatrix} v_{AB} \\ v_{CB} \end{bmatrix}^T \begin{bmatrix} i_A \\ i_C \end{bmatrix}. \tag{A-83}$$

Substituting Equations (A-77) and (A-82) into Equation (A-83), the expression for instantaneous active power can be derived as

$$\begin{aligned} p &= \begin{bmatrix} v_{AB} \\ v_{CB} \end{bmatrix}^T \begin{bmatrix} i_A \\ i_C \end{bmatrix} \\ &= \begin{bmatrix} \sqrt{3}\{\sin(\omega t + 120^\circ)v_d + \cos(\omega t + 120^\circ)v_q\} \\ \sqrt{3}\{-\sin(\omega t)v_d - \cos(\omega t)v_q\} \end{bmatrix}^T \begin{bmatrix} \cos(\omega t)i_d - \sin(\omega t)i_q \\ \cos(\omega t + 120^\circ)i_d - \sin(\omega t + 120^\circ)i_q \end{bmatrix}\end{aligned}$$

or

$$\begin{aligned} p = {} & \sqrt{3}\{\sin(\omega t + 120^\circ)v_d + \cos(\omega t + 120^\circ)v_q\} \times \{\cos(\omega t)\, i_d - \sin(\omega t)\, i_q\} \\ & + \sqrt{3}\{-\sin(\omega t)v_d - \cos(\omega t)v_q\} \times \{\cos(\omega t + 120^\circ)i_d - \sin(\omega t + 120^\circ)i_q\}\end{aligned}$$

or

$$p = \frac{3}{2}\left(v_d i_d + v_q i_q\right). \tag{A-84}$$

Also, if the voltage and current rotating vectors are represented in their respective d-q reference frames as

$$\mathbf{v} = v_d + jv_q \tag{A-85a}$$

and

$$\mathbf{i} = i_d + ji_q, \tag{A-85b}$$

the instantaneous power is given by

$$\mathbf{s} = \frac{3}{2}\mathbf{v}\,\mathbf{i}^* = \frac{3}{2}\left(v_d + jv_q\right)\left(i_d - ji_q\right) = \frac{3}{2}\left(v_d i_d + v_q i_q\right) + j\frac{3}{2}\left(v_q i_d - v_d i_q\right) = p + jq, \tag{A-86}$$

which is same as Equation (A-13).

Then, the instantaneous active and reactive powers are given by

$$p = \text{Re}\left[\mathbf{s}\right]$$

or

$$p = \frac{3}{2}\left(v_d i_d + v_q i_q\right) \tag{A-84}$$

$$q = \text{Im}\left[\mathbf{s}\right]$$

or

$$q = \frac{3}{2}\left(v_q i_d - v_d i_q\right). \tag{A-87}$$

The instantaneous apparent power (s) is given by

$$\begin{aligned} s &= \sqrt{p^2 + q^2} = \frac{3}{2}\sqrt{\left(v_d i_d + v_q i_q\right)^2 + \left(v_q i_d - v_d i_q\right)^2} = \frac{3}{2}\sqrt{\left(v_d i_d\right)^2 + \left(v_q i_q\right)^2 + \left(v_q i_d\right)^2 + \left(v_d i_q\right)^2} \\ &= \frac{3}{2}\sqrt{\left(v_d^2 + v_q^2\right)i_d^2 + \left(v_d^2 + v_q^2\right)i_q^2} = \frac{3}{2}\sqrt{\left(v_d^2 + v_q^2\right)\left(i_d^2 + i_q^2\right)} = \frac{3}{2}\sqrt{\left(v_{dq}^2\right)\left(i_{dq}^2\right)} = \frac{3}{2}v_{dq}i_{dq} \end{aligned} \tag{A-88}$$

where

$$v_{dq} = \left|\mathbf{v}\right| = \left|v_d + jv_q\right| \tag{A-89}$$

and

$$i_{dq} = \left|\mathbf{i}\right| = \left|i_d + ji_q\right|. \tag{A-90}$$

The instantaneous active and reactive powers, given in Equations (A-84) and (A-87), in engineering units can be rewritten in per unit as

$$p_{pu} = \frac{3}{2}\left(v_d i_d + v_q i_q\right)\frac{1}{P_{base}} = \left(v_d i_d + v_q i_q\right)\frac{1}{V_{base} \times I_{base}} = v_{dpu}i_{dpu} + v_{qpu}i_{qpu} \tag{A-91}$$

$$q_{pu} = \frac{3}{2}\left(v_q i_d - v_d i_q\right)\frac{1}{P_{base}} = \left(v_q i_d - v_d i_q\right)\frac{1}{V_{base} \times I_{base}} = v_{qpu}i_{dpu} - v_{dpu}i_{qpu}. \tag{A-92}$$

Solving the Equations (A-91) and (A-92), it can be written that

$$i_{dpu} = \frac{p_{pu}v_{dpu} + q_{pu}v_{qpu}}{v_{dpu}^2 + v_{qpu}^2} \tag{A-93}$$

$$i_{qpu} = \frac{p_{pu}v_{qpu} - q_{pu}v_{dpu}}{v_{dpu}^2 + v_{qpu}^2}. \tag{A-94}$$

Note that v_{dpu}, v_{qpu}, i_{dpu}, and i_{qpu} are in per units.

From Equation (A-76), it can be written that

$$i_d = \frac{2}{\sqrt{3}}\sin(\omega t + 120°)i_A - \frac{2}{\sqrt{3}}\sin(\omega t)i_C \tag{A-95a}$$

and

$$i_q = \frac{2}{\sqrt{3}}\cos(\omega t + 120°)i_A - \frac{2}{\sqrt{3}}\cos(\omega t)i_C. \tag{A-95b}$$

From Equation (A-81), it can be written that

$$v_d = \frac{2}{3}\cos(\omega t)v_{AB} + \frac{2}{3}\cos(\omega t + 120°)v_{CB} \tag{A-96a}$$

and

$$v_q = -\frac{2}{3}\sin(\omega t)v_{AB} - \frac{2}{3}\sin(\omega t + 120°)v_{CB}. \tag{A-96b}$$

Substituting Equations (A-95a), (A-95b), (A-96a), and (A-96b) into Equation (A-87), the expression for instantaneous reactive power becomes

$$\begin{aligned} q &= \frac{3}{2}\left(v_q i_d - v_d i_q\right) \\ &= \frac{3}{2}\frac{2}{3}\frac{2}{\sqrt{3}}\begin{bmatrix} \{-\sin(\omega t)v_{AB} - \sin(\omega t + 120°)v_{CB}\}\{\sin(\omega t + 120°)i_A - \sin(\omega t)i_C\} - \\ \{\cos(\omega t)v_{AB} + \cos(\omega t + 120°)v_{CB}\}\{\cos(\omega t + 120°)i_A - \cos(\omega t)i_C\} \end{bmatrix} \end{aligned}$$

or

$$q = \frac{2}{\sqrt{3}}\begin{bmatrix} v_{AB}i_A\{-\sin(\omega t)\sin(\omega t + 120°) - \cos(\omega t)\cos(\omega t + 120°)\} + v_{AB}i_C\{\sin^2(\omega t) + \cos^2(\omega t)\} + \\ v_{CB}i_A\{-\sin^2(\omega t + 120°) - \cos^2(\omega t + 120°)\} + \\ v_{CB}i_C\{\sin(\omega t + 120°)\sin(\omega t) + \cos(\omega t + 120°)\cos(\omega t)\} \end{bmatrix}$$

or

$$q = \frac{2}{\sqrt{3}}\left[-v_{AB}i_A\cos(120°) + v_{AB}i_C - v_{CB}i_A + v_{CB}i_C\cos(120°)\right]$$

or

$$q = \frac{2}{\sqrt{3}}\left[\frac{1}{2}v_{AB}i_A + v_{AB}i_C - v_{CB}i_A - \frac{1}{2}v_{CB}i_C\right]$$

or

$$q = \frac{2}{\sqrt{3}}\left[\frac{1}{2}v_{AB}i_A + \frac{1}{2}v_{AB}i_C - \frac{1}{2}v_{CB}i_A - \frac{1}{2}v_{CB}i_C\right] + \frac{2}{\sqrt{3}}\left[\frac{1}{2}v_{AB}i_C - \frac{1}{2}v_{CB}i_A\right]$$

or

$$q=\frac{1}{\sqrt{3}}[v_{AB}i_A+v_{AB}i_C-v_{CB}i_A-v_{CB}i_C]+\frac{1}{\sqrt{3}}[v_{AB}i_C-v_{CB}i_A]$$

or

$$q=\frac{1}{\sqrt{3}}[v_{AB}(i_A+i_C)-v_{CB}(i_A+i_C)]+\frac{1}{\sqrt{3}}[v_{AB}i_C+v_{BC}i_A]$$

or

$$q=\frac{1}{\sqrt{3}}[(v_{AB}-v_{CB})(i_A+i_C)]+\frac{1}{\sqrt{3}}[v_{AB}i_C+v_{BC}i_A]$$

or

$$q=\frac{1}{\sqrt{3}}[v_{AC}(-i_B)]+\frac{1}{\sqrt{3}}[v_{AB}i_C+v_{BC}i_A]$$

or

$$q=\frac{1}{\sqrt{3}}[v_{CA}i_B]+\frac{1}{\sqrt{3}}[v_{AB}i_C+v_{BC}i_A]$$

or

$$q=\frac{1}{\sqrt{3}}[v_{BC}i_A+v_{CA}i_B+v_{AB}i_C]. \tag{A-97}$$

Alternately from Equation (A-97), the instantaneous reactive power is given by

$$q=\frac{1}{\sqrt{3}}[v_{BC}i_A-v_{CA}(i_A+i_C)+v_{AB}i_C]=\frac{1}{\sqrt{3}}[(v_{BC}-v_{CA})i_A+(v_{AB}-v_{CA})i_C]$$

or

$$q=\frac{1}{\sqrt{3}}[(-2v_C+v_B+v_A)i_A+(2v_A-v_B-v_C)i_C]. \tag{A-98}$$

In a three-phase, balanced network, $v_A+v_B+v_C=0$. Substituting $v_A+v_B=-v_C$ and $-v_B-v_C=v_A$ and rewriting Equation (A-98),

$$q=\frac{1}{\sqrt{3}}[(-2v_C-v_C)i_A+(2v_A+v_A)i_C]=\frac{1}{\sqrt{3}}[-3v_Ci_A+3v_Ai_C]$$

or

$$q=\sqrt{3}[v_Ai_C-v_Ci_A]. \tag{A-99}$$

Check:

Consider a three-phase, unity-amplitude voltage, $v_A=\cos\theta$, $v_B=\cos(\theta-120°)$, and $v_C=\cos(\theta+120°)$, driving a three-phase, unity-amplitude current $i_A=\cos(\theta+\varphi)$, $i_B=\cos(\theta+\varphi-120°)$, and $i_C=\cos(\theta+\varphi+120°)$. The instantaneous active and reactive powers are

$$
\begin{aligned}
p &= v_A i_A + v_B i_B + v_C i_C \\
&= \cos\theta\cos(\theta+\varphi) + \cos(\theta-120°)\cos(\theta+\varphi-120°) + \cos(\theta+120°)\cos(\theta+\varphi+120°) \\
&= \frac{1}{2}\left\{\cos 2\left(\theta+\frac{\varphi}{2}\right) + \cos\varphi + \cos 2\left(\theta-120°+\frac{\varphi}{2}\right) + \cos\varphi + \cos 2\left(\theta+120°+\frac{\varphi}{2}\right) + \cos\varphi\right\} \\
&= \frac{3}{2}\cos\varphi + \frac{1}{2}\{\cos(2\theta+\varphi) + \cos(2\theta+120°+\varphi) + \cos(2\theta-120°+\varphi)\} \\
&= \frac{3}{2}\cos\varphi
\end{aligned}
$$

$$
\begin{aligned}
q &= \sqrt{3}[v_A i_C - v_C i_A] \\
&= \sqrt{3}[\cos\theta\cos(\theta+\varphi+120°) - \cos(\theta+120°)\cos(\theta+\varphi)] \\
&= \frac{\sqrt{3}}{2}[\cos(2\theta+\varphi+120°) + \cos(\varphi+120°) - \cos(2\theta+\varphi+120°) - \cos(\varphi-120°)] \\
&= \frac{\sqrt{3}}{2}[\cos(\varphi+120°) - \cos(\varphi-120°)] = -\frac{\sqrt{3}}{2}[2\sin\varphi\sin 120°] \\
&= -\frac{3}{2}\sin\varphi
\end{aligned}
$$

At unity power factor, $\varphi = 0$ and $p = \frac{3}{2}, q = 0.$ (A-100a)

At zero power factor lagging, $\varphi = -90°$ and $p = 0, q = \frac{3}{2}.$ (A-100b)

At zero power factor leading, $\varphi = 90°$ and $p = 0,\ q = -\frac{3}{2}.$ (A-100c)

A.6 Fourier Analysis

Any periodic variable (f) that is a function of angle (θ) can be represented as a Fourier series given by

$$f(\theta) = a_0 + a_1\cos\theta + a_2\cos(2\theta) + a_3\cos(3\theta) + \ldots + b_1\sin\theta + b_2\sin(2\theta) + b_3\sin(3\theta) + \ldots \tag{A-101}$$

$$= a_0 + \sum_{n=1}^{\infty}\{a_n\cos(n\theta) + b_n\sin(n\theta)\} \tag{A-102}$$

$$= a_0 + \sum_{n=1}^{\infty} A_n\sin(n\theta + \alpha_n) \tag{A-103}$$

where

$$a_n = A_n\sin\alpha_n \tag{A-104a}$$

$$b_n = A_n\cos\alpha_n. \tag{A-104b}$$

$$A_n = \sqrt{a_n^2 + b_n^2} \tag{A-104c}$$

$$\alpha_n = \tan^{-1}\frac{a_n}{b_n}. \tag{A-104d}$$

Integrating both sides of Equation (A-101) results in

$$\int_0^{2\pi} f(\theta)d\theta = a_0 \int_0^{2\pi} d\theta + a_1 \int_0^{2\pi} \cos\theta d\theta + a_2 \int_0^{2\pi} \cos(2\theta))d\theta + \ldots$$

$$+ b_1 \int_0^{2\pi} \sin\theta d\theta + b_2 \int_0^{2\pi} \sin(2\theta)d\theta + \ldots$$

$$= a_0\theta|_0^{2\pi} = 2a_0\pi$$

since the integral of a sine or a cosine variable over one period is zero. Then,

$$a_0 = \frac{1}{2\pi}\int_0^{2\pi} f(\theta)d\theta \tag{A-105}$$

is the DC component in the original variable, $f(\theta)$.

The amplitudes (a_n) for $n = 1, 2, 3\ldots$ of the cosine components can be expressed as follows. Multiplying both sides of Equation (A-101) with $\cos(n\theta)$ and integrating over one period results in

$$\int_0^{2\pi} f(\theta)\cos(n\theta)d\theta = a_0 \int_0^{2\pi} \cos(n\theta)d\theta + a_1 \int_0^{2\pi} \cos\theta\cos(n\theta)d\theta + \ldots + a_m \int_0^{2\pi} \cos(m\theta)\cos(n\theta)d\theta + \ldots$$

$$+ b_1 \int_0^{2\pi} \sin\theta\cos(n\theta)d\theta + \ldots + b_m \int_0^{2\pi} \sin(m\theta)\cos(n\theta)d\theta + \ldots$$

Of all the terms on the right-hand side, the only term where $m = n$ results in a nonzero coefficient of a_n and the rest of the terms are zero. Then,

$$\int_0^{2\pi} f(\theta)\cos(n\theta)d\theta = a_n \int_0^{2\pi} \cos^2(n\theta)d\theta = \frac{a_n}{2}\int_0^{2\pi} \{1 + \cos(2n\theta)\}d\theta$$

$$= \frac{a_n}{2}\int_0^{2\pi} d\theta + \frac{a_n}{2}\int_0^{2\pi} \cos(2n\theta)d\theta$$

$$= \frac{a_n}{2}\left[\theta|_0^{2\pi} + \frac{\sin(2n\theta)}{2n}\Big|_0^{2\pi}\right] = \frac{a_n}{2}(2\pi) = a_n\pi$$

and

$$a_n = \frac{1}{\pi}\int_0^{2\pi} f(\theta)\cos(n\theta)\,d\theta. \tag{A-106}$$

The amplitudes (b_n) for $n = 1, 2, 3\ldots$ of the sine components can be expressed following a similar approach. Multiplying both sides of Equation (A-101) by $\sin(n\theta)$ and integrating over one period results in

$$\int_0^{2\pi} f(\theta)\sin(n\theta)\,d\theta = a_0 \int_0^{2\pi} \sin(n\theta)\,d\theta + a_1 \int_0^{2\pi} \cos\theta\sin(n\theta)d\theta + \ldots + a_m \int_0^{2\pi} \cos(m\theta)\sin(n\theta)\,d\theta + \ldots$$

$$+ b_1 \int_0^{2\pi} \sin\theta\sin(n\theta)d\theta + \ldots + b_m \int_0^{2\pi} \sin(m\theta)\sin(n\theta)d\theta + \ldots$$

Of all the terms on the right-hand side, the only term where $m = n$ results in a nonzero coefficient of b_n and the rest of the terms are zero. Then,

$$\int_0^{2\pi} f(\theta)\sin(n\theta)\,d\theta = b_n\int_0^{2\pi}\sin^2(n\theta)d\theta = \frac{b_n}{2}\int_0^{2\pi}\{1-\cos(2n\theta)\}\,d\theta = \frac{b_n}{2}\left[\theta|_0^{2\pi} - \frac{\sin(2n\theta)}{2n}\bigg|_0^{2\pi}\right]$$

$$= \frac{b_n 2\pi}{2} = b_n\pi$$

and

$$b_n = \frac{1}{\pi}\int_0^{2\pi} f(\theta)\sin(n\theta)\,d\theta \tag{A-107}$$

Check:

$$f(\theta) = \frac{1}{2} + \sin\theta + \frac{1}{3}\sin(3\theta) + \frac{1}{5}\sin(5\theta)$$

From Equation (A-105), the DC component is given by

$$a_0 = \frac{1}{2\pi}\int_0^{2\pi}\left[\frac{1}{2} + \sin\theta + \frac{1}{3}\sin(3\theta) + \frac{1}{5}\sin(5\theta)\right]d\theta = \frac{1}{2\pi}\left[\frac{1}{2}\int_0^{2\pi} d\theta + \int_0^{2\pi}\sin\theta d\theta + \frac{1}{3}\int_0^{2\pi}\sin(3\theta)d\theta\right.$$

$$\left. + \frac{1}{5}\int_0^{2\pi}\sin(5\theta)d\theta\right]$$

$$= \frac{1}{2\pi}\left[\frac{1}{2}\theta\bigg|_0^{2\pi} - \cos\theta|_0^{2\pi} - \frac{1}{9}\cos(3\theta)|_0^{2\pi} - \frac{1}{25}\cos(5\theta)|_0^{2\pi}\right] = \frac{1}{2\pi}(\pi) = \frac{1}{2}.$$

From Equation (A-106), the cosine component coefficient is given by

$$a_n = \frac{1}{\pi}\left[\frac{1}{2}\int_0^{2\pi}\cos(n\theta)d\theta + \int_0^{2\pi}\sin\theta\cos(n\theta)d\theta + \frac{1}{3}\int_0^{2\pi}\sin(3\theta)\cos(n\theta)d\theta + \frac{1}{5}\int_0^{2\pi}\sin(5\theta)\cos(n\theta)d\theta\right].$$

For $n = 1$:

$$a_1 = \frac{1}{\pi}\left[\frac{1}{2}\int_0^{2\pi}\cos\theta d\theta + \int_0^{2\pi}\sin\theta\cos\theta d\theta + \frac{1}{3}\int_0^{2\pi}\sin(3\theta)\cos\theta d\theta + \frac{1}{5}\int_0^{2\pi}\sin(5\theta)\cos\theta d\theta\right]$$

$$= \frac{1}{\pi}\left[\frac{1}{2}\int_0^{2\pi}\cos\theta d\theta + \frac{1}{2}\int_0^{2\pi}\sin(2\theta)d\theta + \frac{1}{6}\int_0^{2\pi}\{\sin(4\theta) + \sin(2\theta)\}d\theta\right.$$

$$\left. + \frac{1}{10}\int_0^{2\pi}\{\sin(6\theta) + \sin(4\theta)\}d\theta\right] = 0.$$

For $n = 3$:

$$a_3 = \frac{1}{\pi}\left[\frac{1}{2}\int_0^{2\pi}\cos(3\theta)d\theta + \int_0^{2\pi}\sin\theta\cos(3\theta)d\theta + \frac{1}{3}\int_0^{2\pi}\sin(3\theta)\cos(3\theta)d\theta + \frac{1}{5}\int_0^{2\pi}\sin(5\theta)\cos(3\theta)d\theta\right] = 0.$$

For $n = 5$:

$$a_5 = \frac{1}{\pi}\left[\frac{1}{2}\int_0^{2\pi} \cos(5\theta)d\theta + \int_0^{2\pi} \sin\theta\cos(5\theta)d\theta + \frac{1}{3}\int_0^{2\pi} \sin(3\theta)\cos(5\theta)d\theta + \frac{1}{5}\int_0^{2\pi} \sin(5\theta)\cos(5\theta)d\theta\right] = 0.$$

From Equation (A-107), the sine component coefficient is given by

$$b_n = \frac{1}{\pi}\left[\frac{1}{2}\int_0^{2\pi} \sin(n\theta)d\theta + \int_0^{2\pi} \sin\theta\sin(n\theta)d\theta + \frac{1}{3}\int_0^{2\pi} \sin(3\theta)\sin(n\theta)d\theta + \frac{1}{5}\int_0^{2\pi} \sin(5\theta)\sin(n\theta)d\theta\right].$$

For $n = 1$:

$$b_1 = \frac{1}{\pi}\left[\frac{1}{2}\int_0^{2\pi} \sin\theta d\theta + \int_0^{2\pi} \sin\theta\sin\theta d\theta + \frac{1}{3}\int_0^{2\pi} \sin(3\theta)\sin\theta d\theta + \frac{1}{5}\int_0^{2\pi} \sin(5\theta)\sin\theta d\theta\right]$$

$$= \frac{1}{\pi}\left[0 + \int_0^{2\pi}\left\{\frac{1}{2} - \frac{1}{2}\cos(2\theta)\right\}d\theta + 0 + 0\right] = \frac{1}{\pi}\left[\frac{1}{2}\theta|_0^{2\pi} - \frac{1}{4}\sin(2\theta)|_0^{2\pi}\right] = \frac{1}{\pi}\left[\frac{1}{2}(2\pi - 0) - 0\right] = 1.$$

For $n = 3$:

$$b_3 = \frac{1}{\pi}\left[\frac{1}{2}\int_0^{2\pi} \sin(3\theta)d\theta + \int_0^{2\pi} \sin\theta\sin(3\theta)d\theta + \frac{1}{3}\int_0^{2\pi} \sin(3\theta)\sin(3\theta)d\theta + \frac{1}{5}\int_0^{2\pi} \sin(5\theta)\sin(3\theta)d\theta\right]$$

$$= \frac{1}{\pi}\left[0 + 0 + \frac{1}{6}\int_0^{2\pi}\{1 - \cos(6\theta)\}d\theta + 0\right] = \frac{1}{\pi}\frac{1}{6}\left[\theta|_0^{2\pi} - \frac{1}{6}\sin(6\theta)\Big|_0^{2\pi}\right] = \frac{1}{\pi}\frac{1}{6}[(2\pi - 0) - 0] = \frac{1}{3}.$$

For $n = 5$:

$$b_5 = \frac{1}{\pi}\left[\frac{1}{2}\int_0^{2\pi} \sin(5\theta)d\theta + \int_0^{2\pi} \sin\theta\sin(5\theta)d\theta + \frac{1}{3}\int_0^{2\pi} \sin(3\theta)\sin(5\theta)d\theta + \frac{1}{5}\int_0^{2\pi} \sin(5\theta)\sin(5\theta)d\theta\right]$$

$$= \frac{1}{\pi}\left[0 + 0 + 0 + \frac{1}{10}\int_0^{2\pi}\{1 - \cos(10\theta)\}d\theta\right] = \frac{1}{\pi}\frac{1}{10}\left[\theta|_0^{2\pi} - \frac{1}{10}\sin(10\theta)\Big|_0^{2\pi}\right] = \frac{1}{\pi}\frac{1}{10}[(2\pi - 0) - 0] = \frac{1}{5}.$$

Therefore, the composite variable is given by

$$f(\theta) = a_0 + b_1\sin\theta + b_3\sin(3\theta) + b_5\sin(5\theta) = \frac{1}{2} + \sin\theta + \frac{1}{3}\sin(3\theta) + \frac{1}{5}\sin(5\theta).$$

A.7 Adams-Bashforth Numerical Integration Formula

The exact value of a function (f) at any time (t_{n+1}) can be written using Taylor's series as

$$f(t_{n+1}) = f(t_n) + hDf(t_n) + \frac{h^2}{2}D^2f(t_n) + \frac{h^3}{6}D^3f(t_n) + \frac{h^4}{24}D^4f(t_n) + \varphi(h^5) \tag{A-108}$$

where $f(t_n)$ is the exact value of the function (f) at time (t_n), $Df(t_n)$, $D^2f(t_n)$, $D^3f(t_n)$, and $D^4f(t_n)$ are the exact derivatives of the function (f) at time (t_n), $D \equiv d/dt$, and the time step for numerical integration is $h = t_{n+1} - t_n$. The total time (T) is divided into N equidistant segments where $T = Nh$.

The assumed difference equation for numerical integration is of the form

$$f_{n+1} = \alpha_1 f_n + h\beta_1 Df_n + h\beta_2 Df_{n-1} + h\beta_3 Df_{n-2} + h\beta_4 Df_{n-3} \tag{A-109}$$

where f_{n+1} is the final value of the function (f) at the end of the time step n, f_n is the initial value of the function (f) at the beginning of the time step n, and Df_n, Df_{n-1}, Df_{n-2}, and Df_{n-3} are the derivatives of the function (f) at the beginning of the steps n, $n-1$, $n-2$, and $n-3$, respectively, and α and β are constants. The initial value of the state variable (f) at the beginning of the ($n+1$)th step is the final value of the same state variable (f) at the end of the nth step. It is desired to choose the constants (α and β) to match the highest possible order.

The exact function given in Equation (A-108) is being truncated in Equation (A-109), thereby creating a total truncation error at every step of integration in the order of $\phi(h^5)$. Since, there are N steps in the total duration of integration (T), and $N \sim 1/h$, the global truncation error is in the order of $\phi(h^4)$.

Expanding each term on the both sides of Equation (A-109) using Taylor's series results in

$$\begin{aligned} f_n + hDf_n + \frac{h^2}{2}D^2 f_n + \frac{h^3}{6}D^3 f_n + \frac{h^4}{24}D^4 f_n + \ldots &= \alpha_1 f_n + h\beta_1 Df_n \\ &+ h\beta_2\left(Df_n - hD^2 f_n + \frac{h^2}{2}D^3 f_n - \frac{h^3}{6}D^4 f_n + \ldots\right) \\ &+ h\beta_3\left(Df_n - 2hD^2 f_n + 4\frac{h^2}{2}D^3 f_n - 8\frac{h^3}{6}D^4 f_n + \ldots\right) \\ &+ h\beta_4\left(Df_n - 3hD^2 f_n + 9\frac{h^2}{2}D^3 f_n - 27\frac{h^3}{6}D^4 f_n + \ldots\right). \end{aligned} \tag{A-110}$$

Equating term by term,

$$\begin{aligned} f_n &: 1 = \alpha_1 \\ Df_n &: \ h = h(\beta_1 + \beta_2 + \beta_3 + \beta_4) \Rightarrow \beta_1 + \beta_2 + \beta_3 + \beta_4 = 1 \\ D^2 f_n &: \frac{h^2}{2} = -h^2(\beta_2 + 2\beta_3 + 3\beta_4) \Rightarrow \beta_2 + 2\beta_3 + 3\beta_4 = -\frac{1}{2} \\ D^3 f_n &: \frac{h^3}{6} = \frac{h^3}{2}(\beta_2 + 4\beta_3 + 9\beta_4) \Rightarrow \beta_2 + 4\beta_3 + 9\beta_4 = \frac{1}{3} \\ D^4 f_n &: \frac{h^4}{24} = -\frac{h^4}{6}(\beta_2 + 8\beta_3 + 27\beta_4) \Rightarrow \beta_2 + 8\beta_3 + 27\beta_4 = -\frac{1}{4} \end{aligned} \tag{A-111}$$

From Equation (A-111), it can be found that $\alpha_1 = 1$, $\beta_1 = 55/24$, $\beta_2 = -59/24$, $\beta_3 = 37/24$ and $\beta_4 = -9/24$.

Then, the difference Equation (A-109) becomes

$$f_{n+1} = f_n + \left(\frac{55}{24}Df_n - \frac{59}{24}Df_{n-1} + \frac{37}{24}Df_{n-2} - \frac{9}{24}Df_{n-3}\right)h, \tag{A-112}$$

which is known as Adams-Bashforth's fourth-order integration method.

Appendix B

Power Flow Equations in a Lossy Line

Consider an AC line with a resistance (R), an inductive reactance (X), and $X/R = \tan\varphi = \cot\phi = 7.4$, carrying power from the sending end with a voltage ($\mathbf{V_s} = V_s\angle\delta_s$) to the receiving end with a voltage ($\mathbf{V_r} = V_r\angle\delta_r$) as shown in Figure B-1. Note that φ is the power factor angle of the line impedance; however, ϕ is the phase angle between the natural reactive voltage ($\mathbf{V_{Xn}}$) across the line reactance and the natural voltage ($\mathbf{V_{Rn,Xn}}$) across the line impedance ($\mathbf{Z} = R + jX$), which is also the difference between the sending- and receiving-end voltages. This voltage causes a natural current ($\mathbf{I_n} = I_n\angle\theta_{In}$) to flow through the line impedance ($\mathbf{Z} = R + jX$) as shown in Figure B-1.

Therefore, it can be written from Figure B-1 that

$$\mathbf{V_s} = \mathbf{V_r} + \mathbf{V_{Rn,Xn}} \quad \text{or}$$
$$V_s\angle\delta_s = V_r\angle\delta_r + V_{Rn,Xn}\angle\theta_{VRn,Xn} \tag{B-1a}$$

or

$$V_s\angle\delta_s = V_r\angle(\delta_s - \delta) + V_{Rn,Xn}\angle(\delta_s + \theta_{VRn,Xns}) \text{ using Equation (2-27), or}$$
$$V_{Rn,Xn}\angle(\delta_s + \theta_{VRn,Xns}) = V_s\angle\delta_s - V_r\angle(\delta_s - \delta) \tag{B-1b}$$

and

$$V_s\angle(\delta_r + \delta) = V_r\angle\delta_r + V_{Rn,Xn}\angle(\delta_r + \theta_{VRn,Xnr}) \quad \text{or}$$
$$V_{Rn,Xn}\angle(\delta_r + \theta_{VRn,Xnr}) = V_s\angle(\delta_r + \delta) - V_r\angle\delta_r \tag{B-1c}$$

where

$$\delta = \delta_s - \delta_r. \tag{2-27}$$

Note that $\theta_{VRn,Xns}$ and $\theta_{VRn,Xnr}$ are the phase angles of the voltage across the line impedance with respect to the sending-end and receiving-end voltages, respectively.

Since the line impedance ($\mathbf{Z}$) is inductive, the resulting natural line current ($\mathbf{I_n} = I_n\angle\theta_{In}$) lags the voltage ($\mathbf{V_{Rn,Xn}}$) by a phase angle of 90° – ϕ as shown in Figure B-1.

The natural voltage ($\mathbf{V_{Rn,Xn}}$) across the line impedance can be redefined as

$$V_{Rn,Xn}\angle\theta_{VRn,Xns} = V_s\angle 0° - V_r\angle(-\delta), \tag{B-2a}$$

Power Flow Control Solutions for a Modern Grid using SMART Power Flow Controllers, First Edition.
Kalyan K. Sen and Mey Ling Sen.

Companion website: www.wiley.com/go/sen/powerflowcontrol

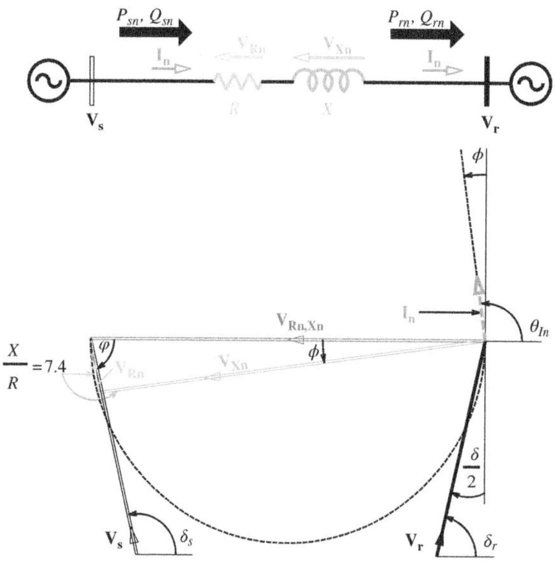

Figure B-1 Two-generator/one-line power system network with a lossy line and its phasor diagram.

considering the sending-end voltage phasor as reference (*i.e.* $\delta_s = 0°$) in Equation (B-1b), or

$$V_{Rn,Xn}\angle\theta_{VRn,Xnr} = V_s\angle\delta - V_r\angle 0°, \tag{B-2b}$$

considering the receiving-end voltage phasor as reference (*i.e.* $\delta_r = 0°$) in Equation (B-1c).

Equation (B-2a) can be rewritten as

$$V_{Rn,Xn}\cos\theta_{VRn,Xns} + jV_{Rn,Xn}\sin\theta_{VRn,Xns} = V_s - V_r\cos\delta + jV_r\sin\delta.$$

Equating the real and imaginary parts of both sides of the above equation, it can be written that

$$V_{Rn,Xn}\cos\theta_{VRn,Xns} = V_s - V_r\cos\delta \tag{B-3a}$$

and

$$V_{Rn,Xn}\sin\theta_{VRn,Xns} = V_r\sin\delta. \tag{B-3b}$$

Therefore, the magnitude of the natural voltage ($\mathbf{V_{Rn,Xn}}$) across the line impedance is derived from Equations (B-3a) and (B-3b) using the identity ($\cos^2\theta_{VRn,Xns} + \sin^2\theta_{VRn,Xns} = 1$) as

$$\begin{aligned} V_{Rn,Xn} &= \sqrt{V_{Rn,Xn}^2\left(\cos^2\theta_{VRn,Xns} + \sin^2\theta_{VRn,Xns}\right)} \\ &= \sqrt{\left(V_{Rn,Xn}\cos\theta_{VRn,Xns}\right)^2 + \left(V_{Rn,Xn}\sin\theta_{VRn,Xns}\right)^2} \end{aligned}$$

or

$$V_{Rn,Xn} = \sqrt{(V_s - V_r\cos\delta)^2 + (V_r\sin\delta)^2} \quad \text{or}$$

$$V_{Rn,Xn} = \sqrt{V_s^2 - 2V_sV_r\cos\delta + V_r^2\left(\cos^2\delta + \sin^2\delta\right)} \quad \text{or}$$

$$V_{Rn,Xn} = \sqrt{V_s^2 - 2V_sV_r\cos\delta + V_r^2} \tag{B-4}$$

using the identity ($\cos^2\delta + \sin^2\delta = 1$).

The phase angle of the voltage ($\mathbf{V_{Rn,Xn}}$) across the line impedance is derived from Equations (B-3a) and (B-3b) as

$$\theta_{VRn,Xns} = \tan^{-1}(\tan\theta_{VRn,Xns}) = \tan^{-1}\frac{\sin\theta_{VRn,Xns}}{\cos\theta_{VRn,Xns}} = \tan^{-1}\frac{V_{Rn,Xn}\sin\theta_{VRn,Xns}}{V_{Rn,Xn}\cos\theta_{VRn,Xns}}$$

or

$$\theta_{VRn,Xns} = \tan^{-1}\frac{V_r\sin\delta}{V_s - V_r\cos\delta}. \tag{B-5}$$

The natural line current ($\mathbf{I_n}$) is defined as

$$\mathbf{I_n} = \frac{\mathbf{V_{Rn,Xn}}}{\mathbf{Z}} \quad \text{or} \quad I_n\angle\theta_{In} = \frac{V_{Rn,Xn}\angle\theta_{VRn,Xn}}{R+jX} \quad \text{or} \quad I_n\angle(\delta_s+\theta_{Ins}) = \frac{V_{Rn,Xn}\angle(\delta_s+\theta_{VRn,Xns})}{R+jX}.$$

Considering the sending-end voltage phasor as reference (*i.e.* $\delta_s = 0°$), it can be written from the above equation that

$$I_n\angle\theta_{Ins} = \frac{V_{Rn,Xn}\angle\theta_{VRn,Xns}}{R+jX} \tag{B-6}$$

or

$$I_n\angle\theta_{Ins} = \frac{V_{Rn,Xn}}{\sqrt{R^2+X^2}}\angle\left(\theta_{VRn,Xns} - \tan^{-1}\frac{X}{R}\right). \tag{B-7}$$

From the above equation, the magnitude of the natural line current ($\mathbf{I_n}$) is given by

$$I_n = \frac{V_{Rn,Xn}}{\sqrt{R^2+X^2}} \tag{B-8}$$

and the phase angle of the line current with respect to the sending-end voltage is given by

$$\theta_{Ins} = \theta_{VRn,Xns} - \tan^{-1}\frac{X}{R} \tag{B-9}$$

where $\theta_{VRn,Xns}$ is given in Equation (B-5).

Alternately, from Equation (B-6), it can be written that

$$I_n\cos\theta_{Ins} + jI_n\sin\theta_{Ins} = \frac{V_{Rn,Xn}\cos\theta_{VRn,Xns} + jV_{Rn,Xn}\sin\theta_{VRn,Xns}}{R+jX} \quad \text{or}$$

$$I_n\cos\theta_{Ins} + jI_n\sin\theta_{Ins} = \frac{V_{Rn,Xn}\cos\theta_{VRn,Xns} + jV_{Rn,Xn}\sin\theta_{VRn,Xns}}{(R+jX)(R-jX)}(R-jX) \quad \text{or}$$

$$I_n\cos\theta_{Ins} + jI_n\sin\theta_{Ins} = \frac{V_{Rn,Xn}\cos\theta_{VRn,Xns} + jV_{Rn,Xn}\sin\theta_{VRn,Xns}}{R^2+X^2}(R-jX) \quad \text{or}$$

$$I_n\cos\theta_{Ins} + jI_n\sin\theta_{Ins} = \frac{RV_{Rn,Xn}\cos\theta_{VRn,Xns} + XV_{Rn,Xn}\sin\theta_{VRn,Xns}}{R^2+X^2} + j\frac{RV_{Rn,Xn}\sin\theta_{VRn,Xns} - XV_{Rn,Xn}\cos\theta_{VRn,Xns}}{R^2+X^2} \quad \text{or}$$

$$I_n \cos\theta_{Ins} + jI_n \sin\theta_{Ins} = \frac{R(V_s - V_r\cos\delta) + XV_r\sin\delta}{R^2 + X^2} + j\frac{RV_r\sin\delta - X(V_s - V_r\cos\delta)}{R^2 + X^2}$$

using Equations (B-3a) and (B-3b).

Equating the real and imaginary parts of both sides of the above equation, it can be written that

$$I_n \cos\theta_{Ins} = \frac{R(V_s - V_r\cos\delta) + XV_r\sin\delta}{R^2 + X^2} \tag{B-10a}$$

and

$$I_n \sin\theta_{Ins} = \frac{RV_r\sin\delta - X(V_s - V_r\cos\delta)}{R^2 + X^2}. \tag{B-10b}$$

Therefore, the magnitude of the natural line current ($\mathbf{I_n}$) is derived from Equations (B-10a) and (B-10b) using the identity $\left(\cos^2\theta_{Ins} + \sin^2\theta_{Ins} = 1\right)$ as

$$I_n = \sqrt{I_n^2\left(\cos^2\theta_{Ins} + \sin^2\theta_{Ins}\right)} = \sqrt{(I_n\cos\theta_{Ins})^2 + (I_n\sin\theta_{Ins})^2} \text{ or}$$

$$I_n = \frac{1}{R^2 + X^2}\sqrt{\{R(V_s - V_r\cos\delta) + XV_r\sin\delta\}^2 + \{RV_r\sin\delta - X(V_s - V_r\cos\delta)\}^2}$$

$$I_n = \frac{1}{R^2 + X^2}\sqrt{(R^2 + X^2)\left\{(V_s - V_r\cos\delta)^2 + V_r^2\sin^2\delta\right\}}$$

$$= \sqrt{\frac{V_s^2 - 2V_sV_r\cos\delta + V_r^2\left(\cos^2\delta + \sin^2\delta\right)}{R^2 + X^2}} \text{ or}$$

$$I_n = \sqrt{\frac{V_s^2 - 2V_sV_r\cos\delta + V_r^2}{R^2 + X^2}} \text{ using the identity } \left(\cos^2\delta + \sin^2\delta = 1\right), \text{ or}$$

$$I_n = \frac{V_{Rn,Xn}}{\sqrt{R^2 + X^2}} \tag{B-8}$$

using Equations (B-4).

The phase angle of the natural line current ($\mathbf{I_n}$) with respect to the sending-end voltage ($\mathbf{V_s}$) is derived from Equations (B-10a) and (B-10b) as

$$\theta_{Ins} = \tan^{-1}(\tan\theta_{Ins}) = \tan^{-1}\frac{\sin\theta_{Ins}}{\cos\theta_{Ins}} = \tan^{-1}\frac{I_n\sin\theta_{Ins}}{I_n\cos\theta_{Ins}}$$

or

$$\theta_{Ins} = \tan^{-1}\frac{RV_r\sin\delta - X(V_s - V_r\cos\delta)}{R(V_s - V_r\cos\delta) + XV_r\sin\delta}. \tag{B-11}$$

Figure B-2 shows the magnitude (I_n) of the natural current ($\mathbf{I_n}$) for the range of power angle ($\delta = \delta_s - \delta_r$) from 0° to 180° when $V_s = V_r = 1$ pu, $X = 0.5$ pu, and $R = 0$ ($X/R = \infty$), 0.2 pu

$(X/R = 2.5)$, and 0.5 pu $(X/R = 1)$. Note that at a given power angle (δ), the magnitude of natural line current (I_n) is less for an increased value of line resistance (R) with a fixed line reactance (X).

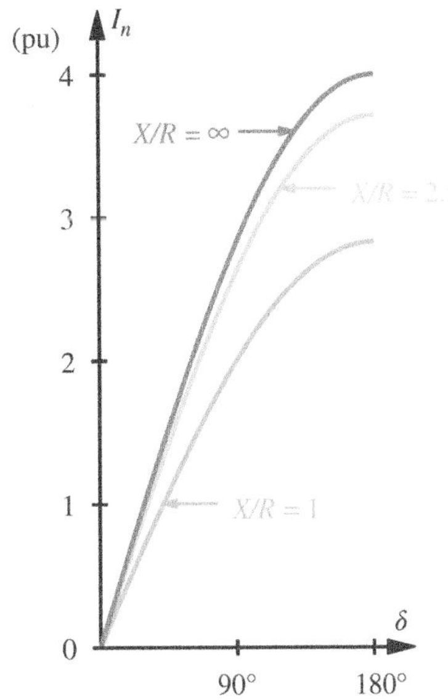

Figure B-2 Natural current (I_n) as a function of power angle ($\delta = \delta_s - \delta_r$) from 0° to 180° when $V_s = V_r = 1$ pu, $X = 0.5$ pu, $R = 0$ $(X/R = \infty)$, 0.2 pu $(X/R = 2.5)$, and 0.5 pu $(X/R = 1)$, respectively.

B.1 Power Flow Equations for a Natural or Uncompensated Line

The natural or uncompensated active and reactive power flows (P_{sn} and Q_{sn}) at the sending end and (P_{rn} and Q_{rn}) at the receiving end of the line are derived as follows. The natural active power flow at the sending end is given by

$$P_{sn} = V_s I_n \cos(\delta_s - \theta_{In})$$

or

$$P_{sn} = V_s I_n \cos\theta_{Ins} \tag{2-39}$$

where θ_{Ins} is the phase angle of the natural line current with respect to the sending-end voltage.

Equation (2-39) can be rewritten using Equation (B-10a) as

$$P_{sn} = \frac{A}{\sqrt{R^2 + X^2}}\left\{R\left(\frac{V_s}{V_r} - \cos\delta\right) + X\sin\delta\right\} \tag{B-12}$$

where

$$A = \frac{V_s V_r}{\sqrt{R^2 + X^2}}. \tag{B-13}$$

The natural reactive power flow at the sending end is given by

$$Q_{sn} = V_s I_n \sin(\delta_s - \theta_{In})$$

or

$$Q_{sn} = -V_s I_n \sin\theta_{Ins} \tag{2-42}$$

where θ_{Ins} is the phase angle of the natural line current with respect to the sending-end voltage.

Equation (2-42) can be rewritten using Equation (B-10b) as

$$Q_{sn} = \frac{A}{\sqrt{R^2 + X^2}}\left\{-R\sin\delta + X\left(\frac{V_s}{V_r} - \cos\delta\right)\right\} \tag{B-14}$$

where A is given in Equation (B-13). For $R = 0$, refer to Equations (2-41), (2-40), and (2-43).

Using Equation (B-13), Equation (B-12) can be rewritten as

$$P_{sn} - \frac{RV_s^2}{R^2 + X^2} = \frac{A}{\sqrt{R^2 + X^2}}[-R\cos\delta + X\sin\delta]. \tag{B-15}$$

Using Equation (B-13), Equation (B-14) can be rewritten as

$$Q_{sn} - \frac{XV_s^2}{R^2 + X^2} = \frac{A}{\sqrt{R^2 + X^2}}[-R\sin\delta - X\cos\delta]. \tag{B-16}$$

Using Equations (B-12), (B-14), and (B-13), the natural apparent power (S_{sn}) at the sending end is given by

$$S_{sn} = \sqrt{P_{sn}^2 + Q_{sn}^2} \tag{B-17}$$

or

$$\begin{aligned} S_{sn} &= \frac{V_s}{R^2 + X^2}\sqrt{\{R(V_s - V_r\cos\delta) + XV_r\sin\delta\}^2 + \{-RV_r\sin\delta + X(V_s - V_r\cos\delta)\}^2} \\ &= \frac{V_s}{R^2 + X^2}\sqrt{(R^2 + X^2)\{(V_s - V_r\cos\delta)^2 + (V_r\sin\delta)^2\}} \\ &= \frac{V_s}{\sqrt{R^2 + X^2}}\sqrt{V_s^2 - 2V_sV_r\cos\delta + V_r^2(\cos^2\delta + \sin^2\delta)} \\ &= \frac{V_s}{\sqrt{R^2 + X^2}}\sqrt{V_s^2 - 2V_sV_r\cos\delta + V_r^2} \text{ using the identity } (\cos^2\delta + \sin^2\delta = 1), \end{aligned}$$

or

$$S_{sn} = \frac{V_sV_{Rn,Xn}}{\sqrt{R^2 + X^2}} \text{ using Equation (B-4), or}$$

$$S_{sn} = V_sI_n \tag{B-18}$$

using Equation (B-8).

Using Equation (B-13), Equations (B-15) and (B-16) can be combined as

$$\begin{aligned} &\left(P_{sn} - \frac{RV_s^2}{R^2 + X^2}\right) + \left(Q_{sn} - \frac{XV_s^2}{R^2 + X^2}\right)^2 \\ &= \left(\frac{V_sV_r}{R^2 + X^2}\right)^2\{(-R\cos\delta + X\sin\delta)^2 + (-R\sin\delta - X\cos\delta)^2\} \\ &= \left(\frac{V_sV_r}{R^2 + X^2}\right)^2(R^2 + X^2)(\cos^2\delta + \sin^2\delta) \\ &= \left(\frac{V_sV_r}{\sqrt{R^2 + X^2}}\right)^2 \text{ using the identity } (\cos^2\delta + \sin^2\delta = 1), \text{ or} \end{aligned}$$

$$\left(P_{sn} - \frac{RV_s^2}{R^2 + X^2}\right)^2 + \left(Q_{sn} - \frac{XV_s^2}{R^2 + X^2}\right)^2 = A^2. \tag{B-19}$$

The Equation (B-19) represents the locus of a circle, centered at $\left[\frac{RV_s^2}{R^2 + X^2}, \frac{XV_s^2}{R^2 + X^2}\right]$ with a radius of A, which is defined in Equation (B-13).

From Equations (B-15) and (B-16), the relationship between the active and reactive power flows at the sending end can be derived as

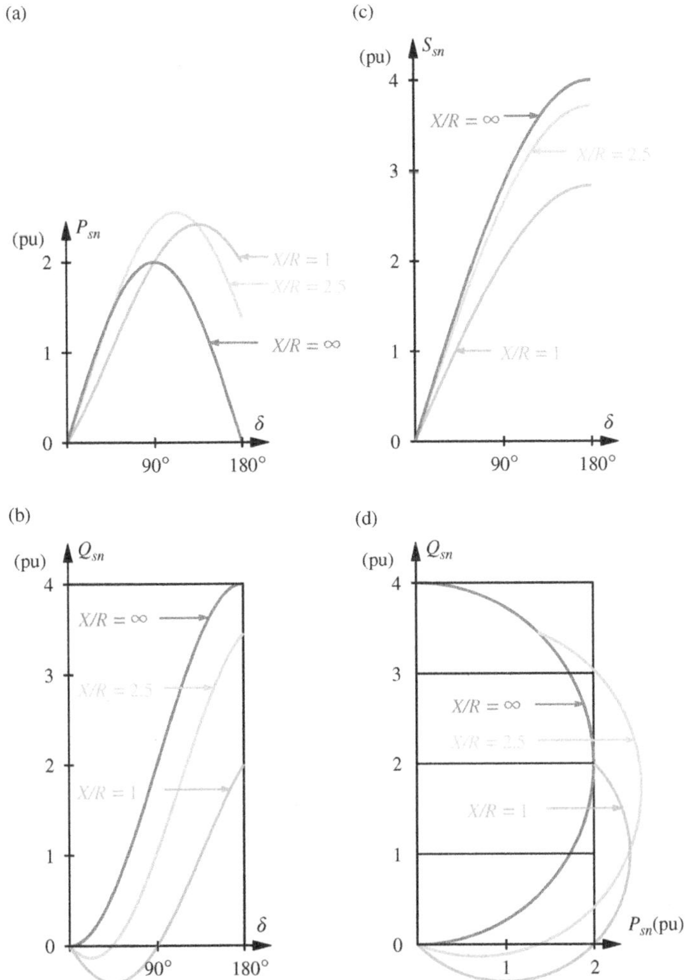

Figure B-3 (a) Natural active power (P_{sn}), (b) natural reactive power (Q_{sn}), and (c) natural apparent power (S_{sn}) as a function of power angle ($\delta = \delta_s - \delta_r$) from 0° to 180° when $V_s = V_r = 1$ pu, $X = 0.5$ pu for three cases: $R = 0$ ($X/R = \infty$), 0.2 pu ($X/R = 2.5$), and 0.5 pu ($X/R = 1$), respectively, and (d) Q_{sn} vs P_{sn} at the sending end of the line.

$$\frac{Q_{sn} - \dfrac{XV_s^2}{R^2 + X^2}}{P_{sn} - \dfrac{RV_s^2}{R^2 + X^2}} = \frac{-R\sin\delta - X\cos\delta}{-R\cos\delta + X\sin\delta} \quad \text{or}$$

$$Q_{sn} = \frac{-R\sin\delta - X\cos\delta}{-R\cos\delta + X\sin\delta}\left(P_{sn} - \frac{RV_s^2}{R^2 + X^2}\right) + \frac{XV_s^2}{R^2 + X^2}. \tag{B-20}$$

At first glance, it seems that if the line is operated at a larger power angle, more current and more power will flow through the line. Let us examine the active and reactive power flows in the line.

Figure B-3 shows the natural active power (P_{sn}), natural reactive power (Q_{sn}), and natural apparent power (S_{sn}) at the sending end, and Q_{sn} vs P_{sn} for the range of power angle ($\delta = \delta_s - \delta_r$) from 0° to 180° when $V_s = V_r = 1$ pu, $X = 0.5$ pu for three cases: $R = 0$ ($X/R = \infty$), 0.2 pu ($X/R = 2.5$), and 0.5 pu ($X/R = 1$), respectively. Note that at a given power angle (δ), the magnitude of apparent power (S_{sn}) at the sending end is less for an increased value of line resistance (R) with a fixed line reactance (X). Typically, the X/R ratio in a transmission network is 20 or higher, sub-transmission network is 5 to 20 and distribution network is 1 to 10, respectively. As X/R ratio decreases, the stability limit increases from 90° to higher values. However, the addition of line resistance (R) also increases line losses. Therefore, it is not a practical way to increase the system stability limit.

The natural active power flow at the receiving end is given by

$$P_{rn} = V_r I_n \cos\left(\delta_r - \theta_{In}\right) \quad \text{or}$$

$$P_{rn} = V_r \cos\delta I_n \cos\theta_{Ins} - V_r \sin\delta I_n \sin\theta_{Ins} \tag{2-45}$$

where θ_{Ins} is the phase angle of the natural line current with respect to the sending-end voltage.

Equation (2-45) can be rewritten using Equations (B-10a) and (B-10b) as

$$P_{rn} = V_r \cos\delta \frac{R(V_s - V_r\cos\delta) + XV_r\sin\delta}{R^2 + X^2} - V_r\sin\delta \frac{RV_r\sin\delta - X(V_s - V_r\cos\delta)}{R^2 + X^2} \quad \text{or}$$

$$P_{rn} = \frac{V_r}{R^2 + X^2}\left[\{R(V_s - V_r\cos\delta) + XV_r\sin\delta\}\cos\delta - \{RV_r\sin\delta - X(V_s - V_r\cos\delta)\}\sin\delta\right] \quad \text{or}$$

$$P_{rn} = \frac{V_r}{R^2 + X^2}\left[RV_s\cos\delta - RV_r\left(\cos^2\delta + \sin^2\delta\right) + XV_s\sin\delta\right] \quad \text{or}$$

$$P_{rn} = \frac{V_r}{R^2 + X^2}\left[R\{V_s\cos\delta - V_r\} + XV_s\sin\delta\right] \text{ using the identity } \left(\cos^2\delta + \sin^2\delta = 1\right), \quad \text{or}$$

$$P_{rn} = \frac{A}{\sqrt{R^2 + X^2}}\left\{R\left(\cos\delta - \frac{V_r}{V_s}\right) + X\sin\delta\right\} \tag{B-21}$$

where A is given in Equation (B-13).

The natural reactive power flow at the receiving end is given by

$$Q_{rn} = V_r I_n \sin\left(\delta_r - \theta_{In}\right) \quad \text{or}$$

$$Q_{rn} = -V_r \sin\delta I_n \cos\theta_{Ins} - V_r \cos\delta I_n \sin\theta_{Ins} \tag{2-47}$$

where θ_{Ins} is the phase angle of the natural line current with respect to the sending-end voltage.

Equation (2-47) can be rewritten using Equations (B-10a) and (B-10b) as

$$Q_{rn} = -V_r\sin\delta \frac{R(V_s - V_r\cos\delta) + XV_r\sin\delta}{R^2 + X^2} - V_r\cos\delta \frac{RV_r\sin\delta - X(V_s - V_r\cos\delta)}{R^2 + X^2} \quad \text{or}$$

$$Q_{rn} = \frac{V_r}{R^2 + X^2}[-\{R(V_s - V_r\cos\delta) + XV_r\sin\delta\}\sin\delta$$

$$-\{RV_r\sin\delta - X(V_s - V_r\cos\delta)\}\cos\delta] \text{ or}$$

$$Q_{rn} = \frac{V_r}{R^2 + X^2}\left[-RV_s\sin\delta + XV_s\cos\delta - XV_r(\cos^2\delta + \sin^2\delta)\right] \text{ or}$$

$$Q_{rn} = \frac{V_r}{R^2 + X^2}[-RV_s\sin\delta + X\{V_s\cos\delta - V_r\}] \text{ using the identity } (\cos^2\delta + \sin^2\delta = 1), \text{ or}$$

$$Q_{rn} = \frac{A}{\sqrt{R^2 + X^2}}\left\{-R\sin\delta + X\left(\cos\delta - \frac{V_r}{V_s}\right)\right\} \tag{B-22}$$

where A is given in Equation (B-13). For $R = 0$, refer to Equations (2-46) and (2-48).

Using Equation (B-13), Equation (B-21) can be rewritten as

$$P_{rn} + \frac{RV_r^2}{R^2 + X^2} = \frac{V_sV_r}{R^2 + X^2}[R\cos\delta + X\sin\delta]. \tag{B-23}$$

Using Equation (B-13), Equations (B-22) can be rewritten as

$$Q_{rn} + \frac{XV_r^2}{R^2 + X^2} = \frac{V_sV_r}{R^2 + X^2}[-R\sin\delta + X\cos\delta]. \tag{B-24}$$

Using Equations (B-21), (B-22), and (B-13), the natural apparent power (S_{rn}) at the receiving end is given by

$$S_{rn} = \sqrt{P_{rn}^2 + Q_{rn}^2} \tag{B-25}$$

or

$$S_{rn} = \frac{V_r}{R^2 + X^2}\sqrt{\{R(V_s\cos\delta - V_r) + XV_s\sin\delta\}^2 + \{-RV_s\sin\delta + X(V_s\cos\delta - V_r)\}^2}$$

$$= \frac{V_r}{R^2 + X^2}\sqrt{(R^2 + X^2)\{(V_s\cos\delta - V_r)^2 + (V_s\sin\delta)^2\}}$$

$$= \frac{V_r}{\sqrt{R^2 + X^2}}\sqrt{V_s^2(\cos^2\delta + \sin^2\delta) - 2V_sV_r\cos\delta + V_r^2}$$

$$= \frac{V_r}{\sqrt{R^2 + X^2}}\sqrt{V_s^2 - 2V_sV_r\cos\delta + V_r^2} \text{ using the identity } (\cos^2\delta + \sin^2\delta = 1), \text{ or}$$

$$S_{rn} = \frac{V_rV_{Rn,Xn}}{\sqrt{R^2 + X^2}} \text{ using Equation (B-4), or}$$

$$S_{rn} = V_rI_n \tag{B-26}$$

using Equation (B-8).

Using Equation (B-13), Equations (B-23) and (B-24) can be combined as

$$\left(P_{rn} + \frac{RV_r^2}{R^2 + X^2}\right)^2 + \left(Q_{rn} + \frac{XV_r^2}{R^2 + X^2}\right)^2$$

$$= \left(\frac{V_sV_r}{R^2 + X^2}\right)^2 \left\{(R\cos\delta + X\sin\delta)^2 + (-R\sin\delta + X\cos\delta)^2\right\}$$

$$= \left(\frac{V_sV_r}{R^2 + X^2}\right)^2 (R^2 + X^2)\left(\cos^2\delta + \sin^2\delta\right)$$

$$= \left(\frac{V_sV_r}{\sqrt{R^2 + X^2}}\right)^2 \text{ using the identity } \left(\cos^2\delta + \sin^2\delta = 1\right), \text{ or}$$

$$\left(P_{rn} + \frac{RV_r^2}{R^2 + X^2}\right)^2 + \left(Q_{rn} + \frac{XV_r^2}{R^2 + X^2}\right)^2 = A^2. \quad \text{(B-27)}$$

The Equation (B-27) represents the locus of a circle, centered at $\left[\frac{-RV_r^2}{R^2 + X^2}, \frac{-XV_r^2}{R^2 + X^2}\right]$ with a radius of A, which is defined in Equation (B-13).

From Equations (B-23) and (B-24), the relationship between the active and reactive power flows at the receiving end can be derived as

$$\frac{Q_{rn} + \dfrac{XV_r^2}{R^2 + X^2}}{P_{rn} + \dfrac{RV_r^2}{R^2 + X^2}} = \frac{-R\sin\delta + X\cos\delta}{R\cos\delta + X\sin\delta} \quad \text{or}$$

$$Q_{rn} = \frac{-R\sin\delta + X\cos\delta}{R\cos\delta + X\sin\delta}\left(P_{rn} + \frac{RV_r^2}{R^2 + X^2}\right) - \frac{XV_r^2}{R^2 + X^2}. \quad \text{(B-28)}$$

Figure B-4 shows the natural active power (P_{rn}), natural reactive power (Q_{rn}), and natural apparent power (S_{rn}) at the receiving end, and Q_{rn} vs P_{rn} for the range of power angle ($\delta = \delta_s - \delta_r$) from 0° to 180° when $V_s = V_r = 1$ pu, $X = 0.5$ pu for three cases: $R = 0$ ($X/R = \infty$), 0.2 pu ($X/R = 2.5$), and 0.5 pu ($X/R = 1$), respectively. Note that at a given power angle (δ), the magnitude of apparent power (S_{rn}) at the receiving end is less for an increased value of line resistance (R) with a fixed line reactance (X).

The difference between the active power at the sending and receiving ends is given by

$$P_{sn} - P_{rn} = \frac{A}{\sqrt{R^2 + X^2}}\left\{R\left(\frac{V_s}{V_r} - \cos\delta\right) + X\sin\delta\right\}$$

$$- \frac{A}{\sqrt{R^2 + X^2}}\left\{R\left(\cos\delta - \frac{V_r}{V_s}\right) + X\sin\delta\right\}$$

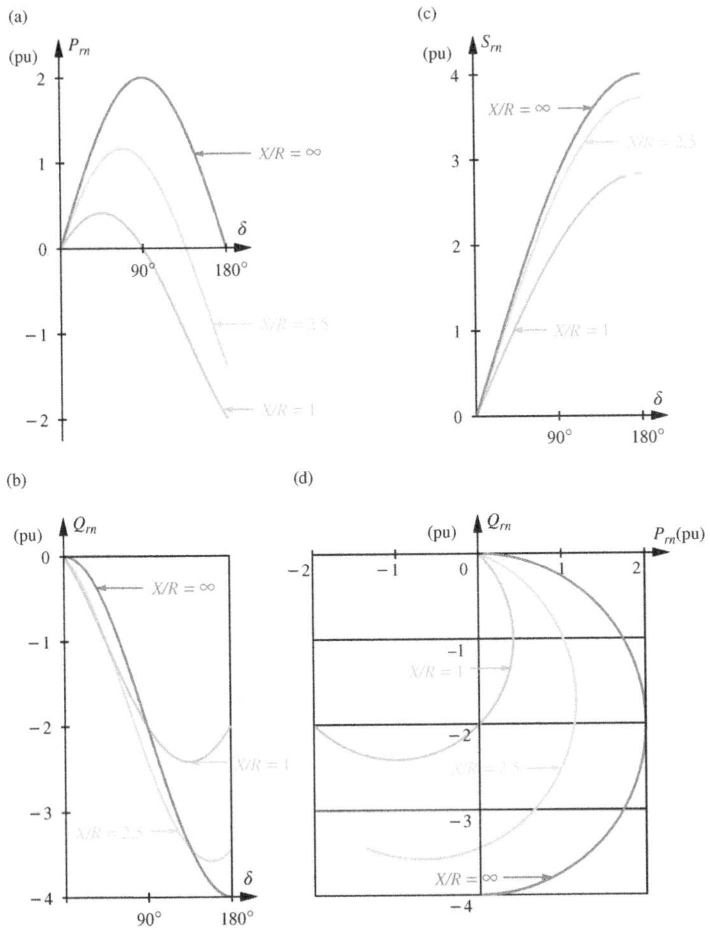

Figure B-4 (a) Natural active power (P_{rn}), (b) natural reactive power (Q_{rn}), and (c) natural apparent power (S_{rn}) as a function of power angle ($\delta = \delta_s - \delta_r$) from 0° to 180° when $V_s = V_r = 1$ pu, $X = 0.5$ pu for three cases: $R = 0$ ($X/R = \infty$), 0.2 pu ($X/R = 2.5$), and 0.5 pu ($X/R = 1$), respectively, and (d) Q_{rn} vs P_{rn} at the receiving end of the line.

using Equations (B-12) and (B-21), or

$$P_{sn} - P_{rn} = \frac{A}{\sqrt{R^2 + X^2}}\left[R\left\{\frac{V_s}{V_r} - \cos\delta - \cos\delta + \frac{V_r}{V_s}\right\} + X\sin\delta - X\sin\delta\right]$$

$$= \frac{R}{R^2 + X^2}\left(V_s^2 + V_r^2\right) - \frac{2RV_sV_r}{R^2 + X^2}\cos\delta \text{ after substituting for } A \text{ from Equation (B-13), or}$$

$$= \frac{R}{R^2 + X^2}\left(V_s^2 - 2V_sV_r\cos\delta + V_r^2\right) = \frac{R}{R^2 + X^2}\left(V_{Rn,Xn}^2\right) \text{ using Equation (B-4), or}$$

$$= \left(\frac{V_{Rn,Xn}}{\sqrt{R^2 + X^2}}\right)^2 R$$

or

$$P_{sn} - P_{rn} = I_n^2 R \tag{B-29a}$$

using Equation (B-8). It is the active power absorbed by the line resistance (R).

Similarly, the difference between the reactive power at the sending and receiving ends is given by

$$\begin{aligned} Q_{sn} - Q_{rn} = & \frac{A}{\sqrt{R^2 + X^2}} \left\{ -R\sin\delta + X\left(\frac{V_s}{V_r} - \cos\delta\right) \right\} \\ & - \frac{A}{\sqrt{R^2 + X^2}} \left\{ -R\sin\delta + X\left(\cos\delta - \frac{V_r}{V_s}\right) \right\} \end{aligned}$$

using Equations (B-14) and (B-22), or

$$\begin{aligned} Q_{sn} - Q_{rn} &= \frac{A}{\sqrt{R^2 + X^2}} \left[-R\sin\delta + X\left\{\frac{V_s}{V_r} - \cos\delta - \cos\delta + \frac{V_r}{V_s}\right\} + R\sin\delta \right] \\ &= \frac{X}{R^2 + X^2}(V_s^2 + V_r^2) - \frac{2XV_sV_r}{R^2 + X^2}\cos\delta \text{ after substituting for } A \text{ from Equation (B-13), or} \\ &= \frac{X}{R^2 + X^2}(V_s^2 - 2V_sV_r\cos\delta + V_r^2) = \frac{X}{R^2 + X^2}(V_{Rn,Xn}^2) \text{ using Equation (B-4), or} \\ &= \left(\frac{V_{Rn,Xn}}{\sqrt{R^2 + X^2}}\right)^2 X \end{aligned}$$

or

$$Q_{sn} - Q_{rn} = I_n^2 X \tag{B-29b}$$

using Equation (B-8). It is the reactive power absorbed by the line reactance (X).

B.2 Power Flow Equations for a Compensated Line

The electric power flow through an AC transmission/distribution line is a function of the line impedance ($\mathbf{Z} = R + jX$), the magnitudes of the sending-end voltage ($\mathbf{V_s}$) and the receiving-end voltage ($\mathbf{V_r}$), and the difference of phase angles (δ_s–δ_r) between these voltages. Figure B-5 shows that the active power at the sending end is the same as that at the receiving end since there is no active power absorbed by the line reactance (X) when $R = 0$, ($X/R = \infty$). However, the reactive power at the sending end is inductive and that at the receiving end is capacitive. Figure B-3 shows that when the line becomes more and more resistive, the value of δ at which the maximum active power at the sending end takes place is higher. Figure B-4 shows that when the line becomes more and more resistive, the value of δ at which the maximum active power at the receiving end takes place is lower. In the case when $X/R = \infty$, the active power at the sending end, as well as at the receiving end, is maximum when $\delta = 90°$. However, for stable operation of the electric power system, δ is maintained at a much lower value. The simulations in EMTP throughout this book are set at $\delta = 20°$ and $X/R = 7.4$.

Assuming that the line impedance does not change in a short period of time, and there is little control over the distant bus voltage (e.g. receiving-end voltage), the only parameter that is able to

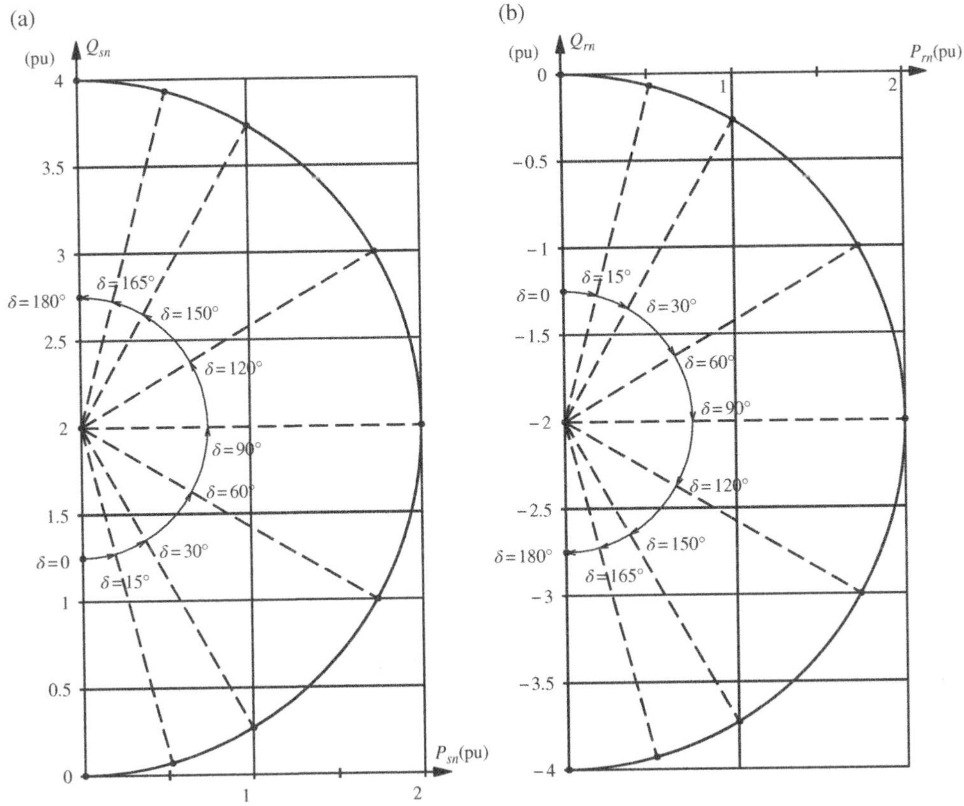

Figure B-5 (a) Q_{sn} vs P_{sn} at the sending end and (b) Q_{rn} vs P_{rn} at the receiving end of the line for the range of power angle ($\delta = \delta_s - \delta_r$) from 0° to 180° when $V_s = V_r = 1$ pu, $X = 0.5$ pu, and $R = 0$ ($X/R = \infty$).

vary the power flow in a line is the modified sending-end voltage ($\mathbf{V_{s'}} = V_{s'}\angle\delta_{s'}$). There are basically two ways to modify the sending-end voltage as discussed below.

B.2.1 Shunt-Compensating Voltage

The line voltage can be modified by connecting a compensating voltage ($\mathbf{V_{s'}} = V_{s'}\angle\delta_{s'}$) with the desired magnitude and phase angle in shunt with the modified sending end of the line as shown in Figure B-6. As an example, the sending-end voltage ($\mathbf{V_s}$) can be used to excite the primary windings of a transformer. The induced secondary voltage ($\mathbf{V_{s'}}$) can be shunt-connected as the modified sending-end voltage. For a given amount of transmitted power, the Shunt-Shunt configuration is rated for the full amount of transmitted power. In certain special cases for point-to-point transfer of power between two isolated networks with different voltages or phase angles, the use of the Shunt-Shunt configuration may be the preferred topology. If a magnetic link-based Shunt-Shunt compensator is used for this purpose, the frequencies of both AC lines must be the same.

When the sending-end voltage is modified to $\mathbf{V_{s'}}$, the line voltage ($\mathbf{V_{R,X}}$), which is the difference between the modified sending- and receiving-end voltages, causes a prevailing current ($\mathbf{I} = I\angle\theta_I$) to flow through the line impedance ($\mathbf{Z} = R + jX$). Therefore, it can be written from Figure B-6 that

$$\mathbf{V_{R,X}} = \mathbf{V_{s'}} - \mathbf{V_r}; \tag{B-30a}$$

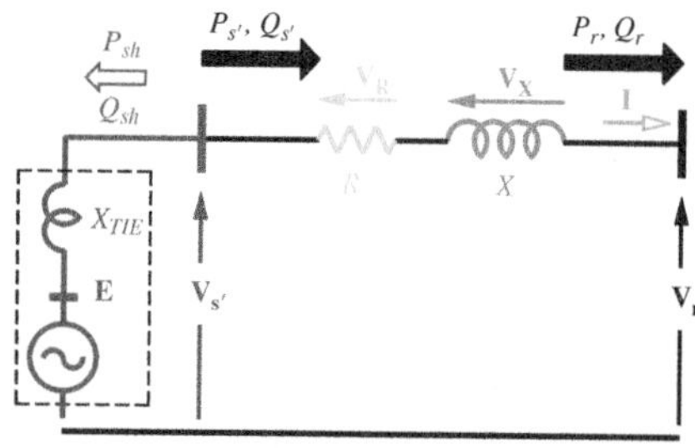

Figure B-6 Modified sending-end voltage ($\mathbf{V_{s'}}$) with a shunt-compensating voltage.

substituting the subscript s with s' in Equation (B-4), it can be written that

$$V_{R,X} = \sqrt{V_{s'}^2 - 2V_{s'}V_r \cos\delta' + V_r^2} \tag{B-30b}$$

where

$$\delta' = \delta_{s'} - \delta_r \tag{2-62}$$

$$\delta' = \delta + \psi. \tag{2-51a}$$

Rewriting Equations (B-8) and (B-11) for the modified sending-end voltage ($\mathbf{V_{s'}} = V_{s'}\angle\delta_{s'}$), the magnitude and phase angle of the prevailing line current are given by

$$I = \frac{V_{R,X}}{\sqrt{R^2 + X^2}} \tag{B-31}$$

and

$$\theta_{Is'} = \tan^{-1}\frac{RV_r \sin\delta' - X(V_{s'} - V_r \cos\delta')}{R(V_{s'} - V_r\cos\delta') + XV_r \sin\delta'}. \tag{B-32}$$

B.2.1.1 Power Flow at the Modified Sending End with a Shunt Compensating Voltage

Rewriting Equation (B-12), replacing $\mathbf{V_s} = V_s\angle\delta$ with $\mathbf{V_{s'}} = V_{s'}\angle\delta'$, the active power flow at the modified sending end of the line is given by

$$P_{s'} = \frac{A'}{\sqrt{R^2 + X^2}}\left\{R\left(\frac{V_{s'}}{V_r} - \cos\delta'\right) + X\sin\delta'\right\} \tag{B-33}$$

where

$$A' = \frac{V_{s'}V_r}{\sqrt{R^2 + X^2}} \tag{B-34}$$

and δ' is defined in Equations (2-62) and (2-51a).

Rewriting Equation (B-14), replacing $\mathbf{V_s} = V_s\angle\delta$ with $\mathbf{V_{s'}} = V_{s'}\angle\delta'$, the reactive power flow at the modified sending end of the line is given by

$$Q_{s'} = \frac{A'}{\sqrt{R^2 + X^2}}\left\{-R\sin\delta' + X\left(\frac{V_{s'}}{V_r} - \cos\delta'\right)\right\}. \tag{B-35}$$

For $R = 0$, refer to Equations (2-59), (2-61), and (2-60).

Following the steps to derive Equation (B-19), Equations (B-33), (B-35), and (B-34) can be combined as

$$\left(P_{s'}-\frac{RV_{s'}^2}{R^2+X^2}\right)^2+\left(Q_{s'}-\frac{XV_{s'}^2}{R^2+X^2}\right)^2=A'^2. \tag{B-36}$$

From Equations (B-34), (B-33), and (B-35), it can be written that

$$\tan\delta'=\frac{-R\left(Q_{s'}-\frac{XV_{s'}^2}{R^2+X^2}\right)+X\left(P_{s'}-\frac{RV_{s'}^2}{R^2+X^2}\right)}{-R\left(P_{s'}-\frac{RV_{s'}^2}{R^2+X^2}\right)-X\left(Q_{s'}-\frac{XV_{s'}^2}{R^2+X^2}\right)}. \tag{B-37}$$

Digress:

$$\frac{P_{s'}-\frac{RV_{s'}^2}{R^2+X^2}}{Q_{s'}-\frac{XV_{s'}^2}{R^2+X^2}}=\frac{-R\cos\delta'+X\sin\delta'}{-R\sin\delta'-X\cos\delta'}\quad\text{or}$$

$$\left(P_{s'}-\frac{RV_{s'}^2}{R^2+X^2}\right)(-R\sin\delta'-X\cos\delta')=\left(Q_{s'}-\frac{XV_{s'}^2}{R^2+X^2}\right)(-R\cos\delta'+X\sin\delta')\quad\text{or}$$

$$-R\left(P_{s'}-\frac{RV_{s'}^2}{R^2+X^2}\right)\sin\delta'-X\left(Q_{s'}-\frac{XV_{s'}^2}{R^2+X^2}\right)\sin\delta'$$
$$=-R\left(Q_{s'}-\frac{XV_{s'}^2}{R^2+X^2}\right)\cos\delta'+X\left(P_{s'}-\frac{RV_{s'}^2}{R^2+X^2}\right)\cos\delta'$$

or

$$\tan\delta'=\frac{-R\left(Q_{s'}-\frac{XV_{s'}^2}{R^2+X^2}\right)+X\left(P_{s'}-\frac{RV_{s'}^2}{R^2+X^2}\right)}{-R\left(P_{s'}-\frac{RV_{s'}^2}{R^2+X^2}\right)-X\left(Q_{s'}-\frac{XV_{s'}^2}{R^2+X^2}\right)}.$$

Recall that for $R=0$,

$$P_{s'}^2+\left(Q_{s'}-\frac{V_{s'}^2}{X}\right)^2=a'^2. \tag{2-64}$$

The Equation (B-36) represents the locus of a circle, centered at $\left[\frac{RV_{s'}^2}{R^2+X^2},\frac{XV_{s'}^2}{R^2+X^2}\right]$ with a radius of A', which is defined in Equation (B-34). The Equation (2-64) represents the locus of a circle, centered at $\left[0,\frac{V_{s'}^2}{X}\right]$ with a radius of a', which is defined in Equation (2-61).

Using Equation (B-34), Equation (B-36) can be rearranged as

$$\left(P_{s'}-\frac{RV_{s'}^2}{R^2+X^2}\right)^2+\left(Q_{s'}-\frac{XV_{s'}^2}{R^2+X^2}\right)^2=A'^2$$

$$\left(\frac{RV_{s'}^2}{R^2+X^2}\right)^2+\left(\frac{XV_{s'}^2}{R^2+X^2}\right)^2-\left(\frac{2RP_{s'}}{R^2+X^2}+\frac{2XQ_{s'}}{R^2+X^2}+\frac{V_r^2}{R^2+X^2}\right)V_{s'}^2+P_{s'}^2+Q_{s'}^2=0$$

or

$$\left(V_{s'}^2\right)^2-\left(2RP_{s'}+2XQ_{s'}+V_r^2\right)V_{s'}^2+\left(R^2+X^2\right)\left(P_{s'}^2+Q_{s'}^2\right)=0. \tag{B-38}$$

Solving the above quadratic equation results in

$$V_{s'}^2=\frac{\left(2RP_{s'}+2X\,Q_{s'}+V_r^2\right)\pm\sqrt{\left(2RP_{s'}+2X\,Q_{s'}+V_r^2\right)^2-4\left(R^2+X^2\right)\left(P_{s'}^2+Q_{s'}^2\right)}}{2}$$

or

$$V_{s'}^2=\left(RP_{s'}+X\,Q_{s'}+\frac{V_r^2}{2}\right)\pm\sqrt{\left(RP_{s'}+X\,Q_{s'}+\frac{V_r^2}{2}\right)^2-\left(R^2+X^2\right)\left(P_{s'}^2+Q_{s'}^2\right)}$$

or

$$V_{s'}=+\sqrt{\left(RP_{s'}+X\,Q_{s'}+\frac{V_r^2}{2}\right)+\sqrt{\left(RP_{s'}+X\,Q_{s'}+\frac{V_r^2}{2}\right)^2-\left(R^2+X^2\right)\left(P_{s'}^2+Q_{s'}^2\right)}} \tag{B-39a}$$

or

$$V_{s'}=+\sqrt{\left(RP_{s'}+X\,Q_{s'}+\frac{V_r^2}{2}\right)-\sqrt{\left(RP_{s'}+X\,Q_{s'}+\frac{V_r^2}{2}\right)^2-\left(R^2+X^2\right)\left(P_{s'}^2+Q_{s'}^2\right)}} \tag{B-39b}$$

or

$$V_{s'}=-\sqrt{\left(RP_{s'}+X\,Q_{s'}+\frac{V_r^2}{2}\right)+\sqrt{\left(RP_{s'}+X\,Q_{s'}+\frac{V_r^2}{2}\right)^2-\left(R^2+X^2\right)\left(P_{s'}^2+Q_{s'}^2\right)}} \tag{B-39c}$$

or

$$V_{s'}=-\sqrt{\left(RP_{s'}+X\,Q_{s'}+\frac{V_r^2}{2}\right)-\sqrt{\left(RP_{s'}+X\,Q_{s'}+\frac{V_r^2}{2}\right)^2-\left(R^2+X^2\right)\left(P_{s'}^2+Q_{s'}^2\right)}}. \tag{B-39d}$$

Out of the four possible values of $V_{s'}$, given in Equations (B-39a through B-39d), the most desired value is the one that provides the lowest Apparent Power Rating (*APR*), thus the lowest cost solution.

From Equation (B-37), the modified power angle is given by

$$\delta'=\tan^{-1}\frac{-R\left(Q_{s'}-\frac{XV_{s'}^2}{R^2+X^2}\right)+X\left(P_{s'}-\frac{RV_{s'}^2}{R^2+X^2}\right)}{-R\left(P_{s'}-\frac{RV_{s'}^2}{R^2+X^2}\right)-X\left(Q_{s'}-\frac{XV_{s'}^2}{R^2+X^2}\right)}, \tag{B-40a}$$

which can be rearranged to obtain the phase-shift angle as

$$\psi = \tan^{-1}\frac{-R\left(Q_{s'} - \frac{XV_{s'}^2}{R^2+X^2}\right) + X\left(P_{s'} - \frac{RV_{s'}^2}{R^2+X^2}\right)}{-R\left(P_{s'} - \frac{RV_{s'}^2}{R^2+X^2}\right) - X\left(Q_{s'} - \frac{XV_{s'}^2}{R^2+X^2}\right)} - \delta, \tag{B-40b}$$

using Equation (2-51b).

For the desired active and reactive power flows ($P_{s'}{}^*$ and $Q_{s'}{}^*$) at the modified sending end within the allowable range, the magnitude ($V_{s'}$) and the phase-shift angle (ψ) of the shunt-compensating voltage are determined by substituting $P_{s'} = P_{s'}^*$ and $Q_{s'} = Q_{s'}^*$ in Equations (B-39a through B-39d) and (B-40b), respectively. A Power Flow Controller (PFC) with a Shunt-Shunt configuration applies a voltage ($\mathbf{V_{s'}} = V_{s'}\angle\psi$) at the modified sending end. For $R = 0$, refer to Equations (2-67a through 2-67d) and (2-68).

Using Equations (B-33), (B-35), and (B-34), the apparent power ($S_{s'}$) at the modified sending end is given by

$$S_{s'} = \sqrt{P_{s'}^2 + Q_{s'}^2} \tag{2-63}$$

or

$$S_{s'} = \frac{V_{s'}}{R^2+X^2}\sqrt{\{R(V_{s'} - V_r\cos\delta') + XV_r\sin\delta'\}^2 + \{-RV_r\sin\delta' + X(V_{s'} - V_r\cos\delta')\}^2}$$

$$= \frac{V_{s'}}{R^2+X^2}\sqrt{(R^2+X^2)\left\{(V_{s'} - V_r\cos\delta')^2 + (V_r\sin\delta')^2\right\}}$$

$$= \frac{V_{s'}}{\sqrt{R^2+X^2}}\sqrt{V_{s'}^2 - 2V_{s'}V_r\cos\delta' + V_r^2(\cos^2\delta' + \sin^2\delta')}$$

$$= \frac{V_{s'}}{\sqrt{R^2+X^2}}\sqrt{V_{s'}^2 - 2V_{s'}V_r\cos\delta' + V_r^2} \text{ using the identity } \left(\cos^2\delta' + \sin^2\delta' = 1\right), \text{or}$$

$$S_{s'} = \frac{V_{s'}V_{R,X}}{\sqrt{R^2+X^2}} \text{ using Equation (B-30b), or using Equation (B-31)}$$

$$S_{s'} = V_{s'}I, \tag{B-41}$$

which is same as Equation (2-69).

Rewriting Equation (B-20) by substituting $\mathbf{V_s} = V_s\angle\delta$ with $\mathbf{V_{s'}} = V_{s'}\angle\delta'$, P_{sn} with $P_{s'}$, and Q_{sn} with $Q_{s'}$, the relationship between the active and reactive power flows at the modified sending end is given by

$$Q_{s'} = \frac{-R\sin\delta' - X\cos\delta'}{-R\cos\delta' + X\sin\delta'}\left(P_{s'} - \frac{RV_{s'}^2}{R^2+X^2}\right) + \frac{XV_{s'}^2}{R^2+X^2}. \tag{B-42}$$

For $R = 0$, refer to Equation (2-70a). For a given line with fixed R, X, and $\mathbf{V_r}$, the locus of ($Q_{s'}$ versus $P_{s'}$) depends on the modified sending-end voltage $\mathbf{V_{s'}} = V_{s'}\angle\delta'$.

B.2.1.2 Power Flow at the Receiving End with a Shunt-Compensating Voltage

Rewriting Equation (B-21) by substituting $\mathbf{V_s} = V_s\angle\delta$ with $\mathbf{V_{s'}} = V_{s'}\angle\delta'$, A with A', and P_{rn} with P_r, the active power flow at the receiving end of the line is given by

$$P_r = \frac{A'}{\sqrt{R^2+X^2}}\left\{R\left(\cos\delta' - \frac{V_r}{V_{s'}}\right) + X\sin\delta'\right\} \tag{B-43}$$

where A' is given in Equation (B-34) and δ' is defined in Equations (2-62) and (2-51a).

Rewriting Equation (B-22) by substituting $\mathbf{V_s} = V_s\angle\delta$ with $\mathbf{V_{s'}} = V_{s'}\angle\delta'$, A with A', and Q_{rn} with Q_r, the reactive power flow at the receiving end of the line is given by

$$Q_r = \frac{A'}{\sqrt{R^2 + X^2}}\left\{-R\sin\delta' + X\left(\cos\delta' - \frac{V_r}{V_{s'}}\right)\right\}. \tag{B-44}$$

For $R = 0$, refer to Equations (2-74), (2-75), and (2-61).

Following the steps to derive Equation (B-27), Equations (B-43) and (B-44) can be combined as

$$\left(P_r + \frac{RV_r^2}{R^2 + X^2}\right)^2 + \left(Q_r + \frac{XV_r^2}{R^2 + X^2}\right)^2 = A'^2. \tag{B-45}$$

From Equations (B-34), (B-43), and (B-44), it can be written that

$$\tan\delta' = \frac{R\left(Q_r + \frac{XV_r^2}{R^2 + X^2}\right) - X\left(P_r + \frac{RV_r^2}{R^2 + X^2}\right)}{-R\left(P_r + \frac{RV_r^2}{R^2 + X^2}\right) - X\left(Q_r + \frac{XV_r^2}{R^2 + X^2}\right)}. \tag{B-46}$$

Digress:

$$\frac{P_r + \frac{RV_r^2}{R^2 + X^2}}{Q_r + \frac{XV_r^2}{R^2 + X^2}} = \frac{R\cos\delta' + X\sin\delta'}{-R\sin\delta' + X\cos\delta'} \quad \text{or}$$

$$\left(P_r + \frac{RV_r^2}{R^2 + X^2}\right)(-R\sin\delta' + X\cos\delta') = \left(Q_r + \frac{XV_r^2}{R^2 + X^2}\right)(R\cos\delta' + X\sin\delta') \quad \text{or}$$

$$-R\left(P_r + \frac{RV_r^2}{R^2 + X^2}\right)\sin\delta' - X\left(Q_r + \frac{XV_r^2}{R^2 + X^2}\right)\sin\delta'$$

$$= R\left(Q_r + \frac{XV_r^2}{R^2 + X^2}\right)\cos\delta' - X\left(P_r + \frac{RV_r^2}{R^2 + X^2}\right)\cos\delta'$$

or

$$\tan\delta' = \frac{R\left(Q_r + \frac{XV_r^2}{R^2 + X^2}\right) - X\left(P_r + \frac{RV_r^2}{R^2 + X^2}\right)}{-R\left(P_r + \frac{RV_r^2}{R^2 + X^2}\right) - X\left(Q_r + \frac{XV_r^2}{R^2 + X^2}\right)}.$$

Recall that for $R = 0$,

$$P_r^2 + \left(Q_r + \frac{V_r^2}{X}\right)^2 = a'^2. \tag{2-79}$$

The Equation (B-45) represents the locus of a circle, centered at $\left[\frac{-RV_r^2}{R^2 + X^2}, \frac{-XV_r^2}{R^2 + X^2}\right]$ with a radius of A', which is defined in Equation (B-34). The Equation (2-79) represents the locus of a circle, centered at $\left[0, -\frac{V_r^2}{X}\right]$ with a radius of a', which is defined in Equation (2-61).

Using Equation (B-34), Equation (B-45) can be rearranged as

$$V_{s'} = \frac{\sqrt{R^2+X^2}}{V_r}\sqrt{\left(P_r + \frac{RV_r^2}{R^2+X^2}\right)^2 + \left(Q_r + \frac{XV_r^2}{R^2+X^2}\right)^2}. \tag{B-47}$$

From Equation (B-46), the modified power angle is given by

$$\delta' = \tan^{-1}\frac{R\left(Q_r + \frac{XV_r^2}{R^2+X^2}\right) - X\left(P_r + \frac{RV_r^2}{R^2+X^2}\right)}{-R\left(P_r + \frac{RV_r^2}{R^2+X^2}\right) - X\left(Q_r + \frac{XV_r^2}{R^2+X^2}\right)}, \tag{B-48a}$$

which can be rearranged to obtain the phase-shift angle as

$$\psi = \tan^{-1}\frac{R\left(Q_r + \frac{XV_r^2}{R^2+X^2}\right) - X\left(P_r + \frac{RV_r^2}{R^2+X^2}\right)}{-R\left(P_r + \frac{RV_r^2}{R^2+X^2}\right) - X\left(Q_r + \frac{XV_r^2}{R^2+X^2}\right)} - \delta \tag{B-48b}$$

using Equation (2-51b).

For the desired active and reactive power flows (P_r^*, Q_r^*) at the receiving end within the allowable range, the magnitude ($V_{s'}$) and the phase-shift angle (ψ) of the shunt-compensating voltage are determined by substituting $P_r = P_r^*$ and $Q_r = Q_r^*$ in Equations (B-47) and (B-48b), respectively. A PFC with a Shunt-Shunt configuration applies a voltage ($\mathbf{V}_{s'} = V_{s'}\angle\psi$) at the modified sending end. For $R = 0$, refer to Equations (2-81) and (2-82a) or (2-82b).

Using Equations (B-43), (B-44), and (B-34), the apparent power (S_r) at the receiving end is given by

$$S_r = \sqrt{P_r^2 + Q_r^2} \tag{2-76}$$

or

$$S_r = \frac{V_r}{R^2+X^2}\sqrt{\{R(V_{s'}\cos\delta' - V_r) + XV_{s'}\sin\delta'\}^2 + \{-RV_{s'}\sin\delta' + X(V_{s'}\cos\delta' - V_r)\}^2}$$

$$= \frac{V_r}{R^2+X^2}\sqrt{(R^2+X^2)\left\{(V_{s'}\cos\delta' - V_r)^2 + (V_{s'}\sin\delta')^2\right\}}$$

$$= \frac{V_r}{\sqrt{R^2+X^2}}\sqrt{V_{s'}^2\left(\cos^2\delta' + \sin^2\delta'\right) - 2V_{s'}V_r\cos\delta' + V_r^2}$$

$$= \frac{V_r}{\sqrt{R^2+X^2}}\sqrt{V_{s'}^2 - 2V_{s'}V_r\cos\delta' + V_r^2} \text{ using the identity } \left(\cos^2\delta' + \sin^2\delta' = 1\right)\text{, or}$$

$$S_r = \frac{V_rV_{R,X}}{\sqrt{R^2+X^2}} \text{ using Equation (B-30b), or using Equation (B-31)}$$

$$S_r = V_rI, \tag{B-49}$$

which is same as Equation (2-77).

Rewriting Equation (B-28) with a substitution $\delta = \delta'$, the relationship between the active and reactive power flows at the receiving end is given by

$$Q_r = \frac{-R\sin\delta' + X\cos\delta'}{R\cos\delta' + X\sin\delta'}\left(P_r + \frac{RV_r^2}{R^2+X^2}\right) - \frac{XV_r^2}{R^2+X^2}. \tag{B-50}$$

For $R = 0$, refer to Equation (2-83a). For a given line with fixed R, X, and $\mathbf{V_r}$, the locus of (Q_r versus P_r) depends on the modified power angle ($\delta' = \delta_{s'} - \delta_r$).

B.2.1.3 Exchanged Power by a Shunt-Compensating Voltage

The shunt-compensating voltage is an AC voltage source that carries a current, which is at any phase angle with its terminal voltage. Therefore, the compensating voltage, essentially, emulates a four-quadrant, shunt-compensating impedance ($\mathbf{Z_{sh}} = R_{sh} - jX_{sh}$) at its terminals. This virtual impedance consists of a resistance ($R_{sh} = +R$ or $-R$) and a reactance ($X_{sh} = X_C$ or $-X_L$) in shunt with the line. A $X_{sh} > 0$ means the compensating reactance is capacitive and $X_{sh} < 0$ means the compensating reactance is inductive.

The active and reactive powers that are exchanged by the shunt-compensating voltage source are P_{sh} and Q_{sh}, respectively. Referring to Figure B-6, for any ($P_{s'}$ and $Q_{s'}$), the active and reactive powers (P_{sh} and Q_{sh}) of the shunt compensator are defined as

$$P_{sh} = -P_{s'} \tag{2-87a}$$

and

$$Q_{sh} = -Q_{s'} \tag{2-87b}$$

where $P_{s'}$ and $Q_{s'}$ are defined in Equations (B-33) and (B-35), respectively.

B.2.1.4 Representation of a Shunt-Compensating Voltage as a Shunt-Compensating Impedance

The active power (P_{sh}) and reactive power (Q_{sh}) of the shunt compensator are defined by

$$P_{sh} = I^2 R_{sh} \tag{2-88a}$$

and

$$Q_{sh} = -I^2 X_{sh} \tag{2-88b}$$

where the line current (I) is defined in Equation (B-31).

Using Equations (2-87a), (2-87b), (2-88a), and (2-88b), the shunt-compensating resistance (R_{sh}) and shunt-compensating reactance (X_{sh}) are given by

$$R_{sh} = \frac{P_{sh}}{I^2} \tag{2-89a}$$

or

$$R_{sh} = -\frac{P_{s'}}{I^2} \tag{2-89b}$$

and

$$X_{sh} = -\frac{Q_{sh}}{I^2} \tag{2-90a}$$

or

$$X_{sh} = \frac{Q_{s'}}{I^2}. \tag{2-90b}$$

The emulated shunt-compensating resistance (R_{sh}), reactance (X_{sh}), and impedance (Z_{sh}) are defined as

$$\mathbf{Z_{sh}} = Z_{sh}\angle\theta_{Zsh} = R_{sh} - jX_{sh} \tag{2-91a}$$

or

$$Z_{sh}\angle\theta_{Zsh} = \sqrt{R_{sh}^2 + X_{sh}^2}\angle\left(\tan^{-1}\frac{-X_{sh}}{R_{sh}}\right) \tag{2-91b}$$

where

$$Z_{sh}\angle\theta_{Zsh} = Z_{sh}\cos(\theta_{Zsh}) + jZ_{sh}\sin(\theta_{Zsh}), \tag{2-92}$$

$$R_{sh} = Z_{sh}\cos(\theta_{Zsh}) \tag{2-93a}$$

and

$$X_{sh} = -Z_{sh}\sin(\theta_{Zsh}). \tag{2-93b}$$

Note that $X_{sh} > 0$ means the shunt-compensating reactance is capacitive and $X_{sh} < 0$ means the shunt-compensating reactance is inductive.

B.2.2 Series-Compensating Voltage as an Impedance Regulator, Voltage Regulator, and Phase Angle Regulator (Asymmetric)

The line voltage can be modified by connecting a compensating voltage ($\mathbf{V_{s's}} = V_{s's}\angle(\delta_s + \beta)$) in series with the line as shown in Figure B-7. As an example, the sending-end voltage ($\mathbf{V_s}$) can be used to excite the primary windings of a transformer. The induced secondary (compensating) voltage modifies the sending-end voltage to be the modified sending-end voltage ($\mathbf{V_{s'}}$). For a given amount of transmitted power, the Shunt-Series configuration needs to be rated for only a fractional amount of transmitted power. Therefore, it is the preferred topology of a PFC.

When the sending-end voltage is modified to $\mathbf{V_{s'}}$, the line voltage ($\mathbf{V_{R,X}}$), which is the difference between the modified sending- and receiving-end voltages, causes a prevailing current ($\mathbf{I} = I\angle\theta_I$) to flow through the line impedance ($\mathbf{Z} = R + jX$). When a series-compensating voltage is used to modify the line voltage, the modified sending-end voltage is given by

$$\mathbf{V_{s'}} = \mathbf{V_s} + \mathbf{V_{s's}}. \quad \text{or} \quad V_{s'}\angle(\delta_s + \psi) = V_s\angle\delta_s + V_{s's}\angle(\delta_s + \beta) \tag{2-94a}$$

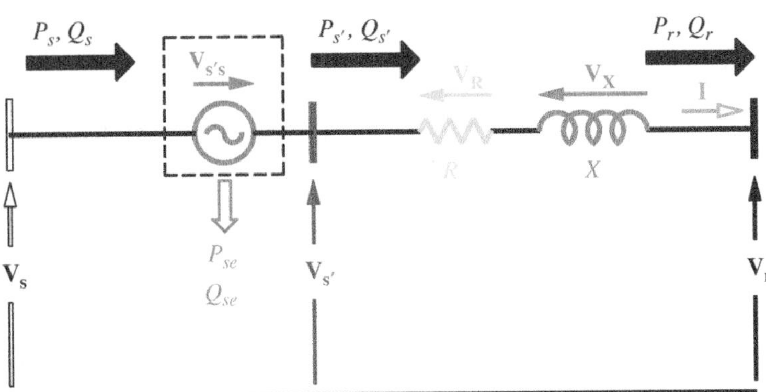

Figure B-7 Modified sending-end voltage ($\mathbf{V_{s'}}$) with a series-compensating voltage.

as shown in Figure B-7, or

$$\mathbf{V_{s's}} = \mathbf{V_{s'}} - \mathbf{V_s} \qquad \text{or} \quad V_{s's}\angle(\delta_s + \beta) = V_{s'}\angle(\delta_s + \psi) - V_s\angle\delta_s \tag{2-94b}$$

where $\mathbf{V_{s's}}$, the difference between the modified sending-end voltage ($\mathbf{V_{s'}}$) and the sending-end voltage ($\mathbf{V_s}$), is the compensating voltage, β is the relative phase angle of the compensating voltage ($\mathbf{V_{s's}}$) with respect to the sending-end voltage ($\mathbf{V_s}$), and ψ is the phase-shift angle of the modified sending end voltage ($\mathbf{V_{s'}}$) with respect to the sending-end voltage ($\mathbf{V_s}$).

Alternatively, the above equations can be rewritten as

$$V_{s'}\angle(\delta_r + \delta') = V_s\angle(\delta_r + \delta) + V_{s's}\angle(\delta_r + \delta + \beta) \tag{2-95a}$$

or

$$V_{s's}\angle(\delta_r + \delta + \beta) = V_{s'}\angle(\delta_r + \delta') - V_s\angle(\delta_r + \delta). \tag{2-95b}$$

For a given compensating voltage ($\mathbf{V_{s's}} = V_{s's}\angle(\delta_s + \beta)$), the magnitude ($V_{s'}$) and modified power angle ($\delta' = \delta_{s'} - \delta_r$) of the modified sending-end voltage ($\mathbf{V_{s'}}$) are derived as

$$V_{s'} = \sqrt{V_s^2 + V_{s's}^2 + 2V_sV_{s's}\cos\beta} \tag{2-103}$$

and

$$\delta' = \tan^{-1}\frac{V_s\sin\delta + V_{s's}\sin(\delta + \beta)}{V_s\cos\delta + V_{s's}\cos(\delta + \beta)}. \tag{2-104}$$

For a given modified sending-end voltage ($\mathbf{V_{s'}} = V_{s'}\angle\delta_{s'}$), the magnitude ($V_{s's}$) and the relative phase angle (β) of the series-compensating voltage ($\mathbf{V_{s's}}$) are derived as

$$V_{s's} = \sqrt{V_{s'}^2 + V_s^2 - 2V_{s'}V_s\cos(\delta' - \delta)} \tag{2-106}$$

and

$$\beta = \tan^{-1}\frac{V_{s'}\sin\delta' - V_s\sin\delta}{V_{s'}\cos\delta' - V_s\cos\delta} - \delta. \tag{2-107}$$

The phase-shift angle ($\psi = \delta' - \delta$) of the modified sending-end voltage ($\mathbf{V_{s'}}$) is derived as

$$\psi = \tan^{-1}\frac{V_{s's}\sin\beta}{V_s + V_{s's}\cos\beta}. \tag{2-98}$$

Alternately, this can be found by substituting Equation (2-51a) into Equation (2-104) and setting $\delta = 0°$ on both sides of the Equation (2-104).

The magnitude ($V_{s's}$) and relative phase angle (β) of the series-compensating voltage ($\mathbf{V_{s's}}$) are derived as

$$V_{s's} = \sqrt{V_{s'}^2 + V_s^2 - 2V_{s'}V_s\cos\psi} \tag{2-100}$$

and

$$\beta = \tan^{-1}\frac{V_{s'}\sin\psi}{V_{s'}\cos\psi - V_s}. \tag{2-101}$$

From Equation (B-30a), it can be written that

$$\mathbf{V_{R,X}} = \mathbf{V_{s'}} - \mathbf{V_r} \qquad \text{or}$$

$$\mathbf{V_{R,X}} = \mathbf{V_s} + \mathbf{V_{s's}} - \mathbf{V_r} \text{ using Equation (2-94a)} \qquad \text{or}$$

$$V_{R,X}\angle\theta_{VR,X} = V_s\angle\delta_s + V_{s's}\angle(\delta_s + \beta) - V_r\angle\delta_r \qquad \text{or}$$

$$V_{R,X}\angle(\delta_s + \theta_{VR,Xs}) = V_s\angle\delta_s + V_{s's}\angle(\delta_s + \beta) - V_r\angle(\delta_s - \delta) \text{ using Equation (2-27)}$$

where $\theta_{VR,Xs}$ is the phase angle of the voltage across the line impedance with respect to the sending-end voltage.

Considering the sending-end voltage phasor as reference (*i.e.* $\delta_s = 0°$), it can be written from the above equation that

$$V_{R,X}\angle\theta_{VR,Xs} = V_s\angle 0° + V_{s's}\angle\beta - V_r\angle(-\delta) \qquad \text{or}$$

$$V_{R,X}\cos\theta_{VR,Xs} + jV_{R,X}\sin\theta_{VR,Xs} = V_s + V_{s's}\cos\beta - V_r\cos\delta + jV_{s's}\sin\beta + jV_r\sin\delta.$$

Equating the real and imaginary parts of both sides of the above equation, it can be written that

$$V_{R,X}\cos\theta_{VR,Xs} = V_s - V_r\cos\delta + V_{s's}\cos\beta \tag{B-51a}$$

and

$$V_{R,X}\sin\theta_{VR,Xs} = V_r\sin\delta + V_{s's}\sin\beta. \tag{B-51b}$$

Therefore, the magnitude of the voltage ($\mathbf{V_{R,X}}$) across the line impedance is derived from Equations (B-51a) and (B-51b) using the identity ($\cos^2\theta_{VR,Xs} + \sin^2\theta_{VR,Xs} = 1$) as

$$V_{R,X} = \sqrt{V_{R,X}^2\left(\cos^2\theta_{VR,Xs} + \sin^2\theta_{VR,Xs}\right)} = \sqrt{\left(V_{R,X}\cos\theta_{VR,Xs}\right)^2 + \left(V_{R,X}\sin\theta_{VR,Xs}\right)^2}$$

or

$$V_{R,X} = \sqrt{(V_s - V_r\cos\delta + V_{s's}\cos\beta)^2 + (V_r\sin\delta + V_{s's}\sin\beta)^2}$$

or

$$V_{R,X} = \sqrt{\begin{aligned}&V_s^2 - 2V_sV_r\cos\delta + V_r^2\left(\cos^2\delta + \sin^2\delta\right) + V_{s's}^2\left(\cos^2\beta + \sin^2\beta\right)\\&+ 2V_sV_{s's}\cos\beta - 2V_{s's}V_r\cos(\delta + \beta)\end{aligned}}$$

or

$$V_{R,X} = \sqrt{V_s^2 - 2V_sV_r\cos\delta + V_r^2 + V_{s's}^2 + 2V_sV_{s's}\cos\beta - 2V_{s's}V_r\cos(\delta + \beta)} \tag{B-52}$$

using the identities ($\cos^2\delta + \sin^2\delta = 1$) and ($\cos^2\beta + \sin^2\beta = 1$).

Using Equation (B-4), Equation (B-52) can be rewritten as

$$V_{R,X} = \sqrt{V_{Rn,Xn}^2 + V_{s's}[V_{s's} + 2\{V_s\cos\beta - V_r\cos(\delta + \beta)\}]}. \tag{B-53}$$

The phase angle of the voltage ($\mathbf{V_{R,X}}$) across the line impedance with respect to the sending-end voltage is derived from Equations (B-51a) and (B-51b) as

$$\theta_{VR,Xs} = \tan^{-1}(\tan\theta_{VR,Xs}) = \tan^{-1}\frac{\sin\theta_{VR,Xs}}{\cos\theta_{VR,Xs}} = \tan^{-1}\frac{V_{R,X}\sin\theta_{VR,Xs}}{V_{R,X}\cos\theta_{VR,Xs}}$$

or

$$\theta_{VR,Xs} = \tan^{-1}\frac{V_r \sin\delta + V_{s's}\sin\beta}{V_s - V_r\cos\delta + V_{s's}\cos\beta}. \tag{B-54}$$

The line current (**I**) is defined as

$$\mathbf{I} = \frac{\mathbf{V_{R,X}}}{\mathbf{Z}} \quad \text{or} \quad I\angle\theta_I = \frac{V_{R,X}\angle\theta_{VR,X}}{R + jX} \quad \text{or} \quad I\angle(\delta_s + \theta_{Is}) = \frac{V_{R,X}\angle(\delta_s + \theta_{VR,Xs})}{R + jX}.$$

Considering the sending-end voltage phasor as reference (*i.e.* $\delta_s = 0°$), it can be written from the above equation that

$$I\angle\theta_{Is} = \frac{V_{R,X}\angle\theta_{VR,Xs}}{R + jX} \tag{B-55a}$$

or

$$I\angle\theta_{Is} = \frac{V_{R,X}}{\sqrt{R^2 + X^2}}\angle\left(\theta_{VR,Xs} - \tan^{-1}\frac{X}{R}\right) \tag{B-55b}$$

where the magnitude of the prevailing line current (**I**) is given by

$$I = \frac{V_{R,X}}{\sqrt{R^2 + X^2}} \tag{B-31}$$

and the phase angle of the line current (**I**) with respect to the sending-end voltage ($\mathbf{V_s}$) is given by

$$\theta_{Is} = \theta_{VR,Xs} - \tan^{-1}\frac{X}{R} \tag{B-56}$$

where $\theta_{VR,Xs}$ is given in Equations (B-54).

Equations (B-31) and (B-53) can be combined as

$$I = \frac{\sqrt{V^2_{Rn,Xn} + V_{s's}[V_{s's} + 2\{V_s\cos\beta - V_r\cos(\delta + \beta)\}]}}{\sqrt{R^2 + X^2}} \tag{B-57a}$$

$$I = \sqrt{\frac{V^2_{Rn,Xn}}{R^2 + X^2} + \frac{V_{s's}[V_{s's} + 2\{V_s\cos\beta - V_r\cos(\delta + \beta)\}]}{R^2 + X^2}} \quad \text{or}$$

$$I = \sqrt{I_n^2 + \frac{V_{s's}[V_{s's} + 2\{V_s\cos\beta - V_r\cos(\delta + \beta)\}]}{R^2 + X^2}} \tag{B-57b}$$

using Equation (B-8).

Equation (B-55a) can be expanded as

$$I\angle\theta_{Is} = \frac{V_{R,X}\angle\theta_{VR,Xs}}{R + jX} \quad \text{or}$$

$$I\cos\theta_{Is} + jI\sin\theta_{Is} = \frac{V_{R,X}\cos\theta_{VR,Xs} + jV_{R,X}\sin\theta_{VR,Xs}}{R + jX} \quad \text{or}$$

$$I\cos\theta_{Is} + jI\sin\theta_{Is} = \frac{V_{R,X}\cos\theta_{VR,Xs} + jV_{R,X}\sin\theta_{VR,Xs}}{(R + jX)(R - jX)}(R - jX) \quad \text{or}$$

$$I\cos\theta_{Is} + jI\sin\theta_{Is} = \frac{V_{R,X}\cos\theta_{VR,Xs} + jV_{R,X}\sin\theta_{VR,Xs}}{R^2 + X^2}(R - jX) \quad \text{or}$$

$$I\cos\theta_{Is} + jI\sin\theta_{Is} = \frac{RV_{R,X}\cos\theta_{VR,Xs} + XV_{R,X}\sin\theta_{VR,Xs}}{R^2 + X^2} + j\frac{RV_{R,X}\sin\theta_{VR,Xs} - XV_{R,X}\cos\theta_{VR,Xs}}{R^2 + X^2} \quad \text{or}$$

$$I\cos\theta_{Is} + jI\sin\theta_{Is} = \frac{R(V_s - V_r\cos\delta + V_{s's}\cos\beta) + X(V_r\sin\delta + V_{s's}\sin\beta)}{R^2 + X^2} + j\frac{R(V_r\sin\delta + V_{s's}\sin\beta) - X(V_s - V_r\cos\delta + V_{s's}\cos\beta)}{R^2 + X^2}$$

using Equations (B-51a) and (B-51b).

Equating the real and imaginary parts of both sides of the above equation, it can be written that

$$I\cos\theta_{Is} = \frac{R(V_s - V_r\cos\delta + V_{s's}\cos\beta) + X(V_r\sin\delta + V_{s's}\sin\beta)}{R^2 + X^2} \tag{B-58a}$$

and

$$I\sin\theta_{Is} = \frac{R(V_r\sin\delta + V_{s's}\sin\beta) - X(V_s - V_r\cos\delta + V_{s's}\cos\beta)}{R^2 + X^2}. \tag{B-58b}$$

Therefore, the magnitude of the prevailing line current (**I**) is derived from Equations (B-58a) and (B-58b) using the identity $\left(\cos^2\theta_{Is} + \sin^2\theta_{Is} = 1\right)$ as

$$I = \sqrt{I^2\left(\cos^2\theta_{Is} + \sin^2\theta_{Is}\right)} = \sqrt{(I\cos\theta_{Is})^2 + (I\sin\theta_{Is})^2} \quad \text{or}$$

$$I = \frac{1}{R^2 + X^2}\sqrt{\begin{array}{l}\{R(V_s - V_r\cos\delta + V_{s's}\cos\beta) + X(V_r\sin\delta + V_{s's}\sin\beta)\}^2 + \\ \{R(V_r\sin\delta + V_{s's}\sin\beta) - X(V_s - V_r\cos\delta + V_{s's}\cos\beta)\}^2\end{array}} \quad \text{or}$$

$$I = \frac{1}{R^2 + X^2}\sqrt{(R^2 + X^2)\left\{(V_s - V_r\cos\delta + V_{s's}\cos\beta)^2 + (V_r\sin\delta + V_{s's}\sin\beta)^2\right\}} \quad \text{or}$$

$$I = \frac{1}{\sqrt{R^2 + X^2}}\sqrt{(V_s - V_r\cos\delta + V_{s's}\cos\beta)^2 + (V_r\sin\delta + V_{s's}\sin\beta)^2} \quad \text{or}$$

$$I = \frac{1}{\sqrt{R^2 + X^2}}\sqrt{V_s^2 - 2V_sV_r\cos\delta + V_r^2 + V_{s's}^2 + 2V_sV_{s's}\cos\beta - 2V_{s's}V_r\cos(\delta + \beta)} \quad \text{or}$$

$$I = \frac{V_{R,X}}{\sqrt{R^2 + X^2}} \tag{B-31}$$

using Equation (B-52).

The phase angle of the line current (**I**) with respect to the sending-end voltage ($\mathbf{V_s}$) is given by

$$\theta_{Is} = \tan^{-1}\frac{V_r\sin\delta + V_{s's}\sin\beta}{V_s - V_r\cos\delta + V_{s's}\cos\beta} - \tan^{-1}\frac{X}{R} \tag{B-59a}$$

using Equations (B-56) and (B-54), or

$$\theta_{Is} = \tan^{-1}(\tan\theta_{Is}) = \tan^{-1}\frac{\sin\theta_{Is}}{\cos\theta_{Is}} = \tan^{-1}\frac{I\sin\theta_{Is}}{I\cos\theta_{Is}}$$

Table B-1 Line current for a VR and a PAR (asym).

	VR	PAR (asym)
I	$\sqrt{I_n^2 + \frac{V_{s's}[V_{s's} \pm 2\{V_s - V_r \cos\delta\}]}{R^2 + X^2}}$	$\sqrt{I_n^2 + \frac{V_{s's}[V_{s's} \pm 2V_r \sin\delta]}{R^2 + X^2}}$
θ_{Is}	$\tan^{-1}\frac{V_r \sin\delta}{V_s - V_r\cos\delta \pm V_{s's}} - \tan^{-1}\frac{X}{R}$	$\theta_{Is} = \tan^{-1}\frac{V_r \sin\delta \pm V_{s's}}{V_s - V_r\cos\delta} - \tan^{-1}\frac{X}{R}$
	$\tan^{-1}\frac{RV_r \sin\delta - X(V_s - V_r\cos\delta \pm V_{s's})}{R(V_s - V_r\cos\delta \pm V_{s's}) + XV_r\sin\delta}$	$\theta_{Is} = \tan^{-1}\frac{R(V_r\sin\delta \pm V_{s's}) - X(V_s - V_r\cos\delta)}{R(V_s - V_r\cos\delta) + X(V_r\sin\delta \pm V_{s's})}$

or

$$\theta_{Is} = \tan^{-1}\frac{R(V_r\sin\delta + V_{s's}\sin\beta) - X(V_s - V_r\cos\delta + V_{s's}\cos\beta)}{R(V_s - V_r\cos\delta + V_{s's}\cos\beta) + X(V_r\sin\delta + V_{s's}\sin\beta)} \tag{B-59b}$$

using Equations (B-58a) and (B-58b).

For a Voltage Regulator (VR), $\beta = 0°$ and 180° represent increasing and decreasing voltages, respectively. For a Phase Angle Regulator (Asymmetric), PAR (asym), $\beta = \pm 90°$ represents increasing and decreasing power flows, respectively. The expressions for the magnitude and phase angle for the line current (**I**) are derived in Equations (B-57b) and (B-59a) or (B-59b), respectively. They are shown in Table B-1 for a VR and a PAR (asym), respectively.

B.2.2.1 Power Flow at the Sending End with a Series-Compensating Voltage

The active power flow at the sending end of the line when the line is series-compensated is given by

$$P_s = V_s I \cos(\delta_s - \theta_I) \quad \text{or}$$
$$P_s = V_s I \cos\theta_{Is} \tag{2-116}$$

where θ_{Is} is the phase angle of the line current with respect to the sending-end voltage.

Equation (2-116) can be rewritten using Equation (B-58a) as

$$P_s = V_s\frac{R(V_s - V_r\cos\delta + V_{s's}\cos\beta) + X(V_r\sin\delta + V_{s's}\sin\beta)}{R^2 + X^2} \quad \text{or}$$

$$P_s = \frac{A}{\sqrt{R^2 + X^2}}\left\{R\left(\frac{V_s}{V_r} - \cos\delta + \frac{V_{s's}}{V_r}\cos\beta\right) + X\left(\sin\delta + \frac{V_{s's}}{V_r}\sin\beta\right)\right\} \quad \text{or}$$

$$P_s = \frac{A}{\sqrt{R^2 + X^2}}\left\{R\left(\frac{V_s}{V_r} - \cos\delta\right) + X\sin\delta + R\frac{V_{s's}}{V_r}\cos\beta + X\frac{V_{s's}}{V_r}\sin\beta\right\} \tag{B-60}$$

or

$$P_s = P_{sn} + \frac{A_s}{\sqrt{R^2 + X^2}}(R\cos\beta + X\sin\beta) \tag{B-61}$$

where P_{sn} is given in Equation (B-12) and

$$A_s = \frac{V_s V_{s's}}{\sqrt{R^2 + X^2}}. \tag{B-62}$$

The reactive power flow at the sending end of the line when the line is series-compensated is given by

$$Q_s = V_s I \sin(\delta_s - \theta_I) \qquad \text{or}$$

$$Q_s = -V_s I \sin\theta_{Is} \tag{2-120}$$

where θ_{Is} is the phase angle of the line current with respect to the sending-end voltage.

Equation (2-120) can be rewritten using Equation (B-58b) as

$$Q_s = -V_s \frac{R(V_r \sin\delta + V_{s's} \sin\beta) - X(V_s - V_r \cos\delta + V_{s's} \cos\beta)}{R^2 + X^2} \qquad \text{or}$$

$$Q_s = \frac{A}{\sqrt{R^2 + X^2}} \left\{ -R\left(\sin\delta + \frac{V_{s's}}{V_r} \sin\beta \right) + X\left(\frac{V_s}{V_r} - \cos\delta + \frac{V_{s's}}{V_r} \cos\beta \right) \right\} \qquad \text{or}$$

$$Q_s = \frac{A}{\sqrt{R^2 + X^2}} \left\{ -R\sin\delta + X\left(\frac{V_s}{V_r} - \cos\delta \right) - R\frac{V_{s's}}{V_r} \sin\beta + X\frac{V_{s's}}{V_r} \cos\beta \right\} \tag{B-63}$$

or

$$Q_s = Q_{sn} + \frac{A_s}{\sqrt{R^2 + X^2}}(-R\sin\beta + X\cos\beta) \tag{B-64}$$

where Q_{sn} and A_s are given in Equations (B-14) and (B-62), respectively. For $R = 0$, the active and reactive power flows at the sending end of the line are given in Equations (2-118) and (2-122), respectively.

Equations (B-61) and (B-64) can be combined as

$$(P_s - P_{sn})^2 + (Q_s - Q_{sn})^2 = \left(\frac{A_s}{\sqrt{R^2 + X^2}} \right)^2 \left[\{R\cos\beta + X\sin\beta\}^2 + \{-R\sin\beta + X\cos\beta\}^2 \right]$$

$$(P_s - P_{sn})^2 + (Q_s - Q_{sn})^2 = \frac{A_s^2}{R^2 + X^2} \left[(R^2 + X^2)\{\cos^2\beta + \sin^2\beta\} \right]$$

$$(P_s - P_{sn})^2 + (Q_s - Q_{sn})^2 = A_s^2 \tag{B-65}$$

using the identity $(\cos^2\beta + \sin^2\beta = 1)$ and

$$\tan\beta = \frac{-R(Q_s - Q_{sn}) + X(P_s - P_{sn})}{R(P_s - P_{sn}) + X(Q_s - Q_{sn})}. \tag{B-66}$$

Digress:

From Equations (B-61) and (B-64), the active and reactive power flows at the sending end are related as

$$\frac{Q_s - Q_{sn}}{P_s - P_{sn}} = \frac{-R\sin\beta + X\cos\beta}{R\cos\beta + X\sin\beta} \qquad \text{or}$$

$$(Q_s - Q_{sn})(R\cos\beta + X\sin\beta) = (P_s - P_{sn})(-R\sin\beta + X\cos\beta)$$

or $\quad R(P_s - P_{sn})\sin\beta + X(Q_s - Q_{sn})\sin\beta = -R(Q_s - Q_{sn})\cos\beta + X(P_s - P_{sn})\cos\beta$

or

$$\tan\beta = \frac{-R(Q_s - Q_{sn}) + X(P_s - P_{sn})}{R(P_s - P_{sn}) + X(Q_s - Q_{sn})}.$$

Using Equation (B-62), Equation (B-65) can be rearranged as

$$V_{s's} = \frac{\sqrt{R^2 + X^2}}{V_s} \sqrt{(P_s - P_{sn})^2 + (Q_s - Q_{sn})^2}. \tag{B-67}$$

Equation (B-66) can be rearranged as

$$\beta = \tan^{-1}\frac{-R(Q_s - Q_{sn}) + X(P_s - P_{sn})}{R(P_s - P_{sn}) + X(Q_s - Q_{sn})}. \tag{B-68}$$

For $R = 0$, $V_{s's}$ and β are given in Equations (2-125) and (2-126), respectively.

For the desired active and reactive power flows (P_s^*, Q_s^*) within the allowable range, the magnitude ($V_{s's}$) and the relative phase angle (β) of the series-compensating voltage ($\mathbf{V_{s's}}$) are determined by substituting $P_s = P_s^*$ and $Q_s = Q_s^*$ in Equations (B-67) and (B-68), respectively. The relative phase angle (β) is the angle between the voltage regulator line (*i.e.* $\beta = 0°$ line) and the line that passes through (P_{sn}, Q_{sn}) and (P_s^*, Q_s^*).

Recall that for $R = 0$,

$$(P_s - P_{sn})^2 + (Q_s - Q_{sn})^2 = a_s^2. \tag{2-123}$$

The Equation (B-65) represents the locus of a circle, centered at $[P_{sn}, Q_{sn}]$ with a radius of A_s, which is defined in Equation (B-62). The Equation (2-123) represents the locus of a circle, centered at $[P_{sn}, Q_{sn}]$ with a radius of a_s, which is defined in Equation (2-119).

For implementation of the VR, the relative phase angle is set to $\beta = 0°$ or $\beta = 180°$ in Equations (B-61) and (B-64). For $\beta = 0°$, the variations of P_s and Q_s are given by

$$P_s = P_{sn} + \frac{A_s}{\sqrt{R^2 + X^2}}\{R\cos(0°) + X\sin(0°)\} \quad \text{or} \quad P_s = P_{sn} + \frac{RA_s}{\sqrt{R^2 + X^2}}$$

and

$$Q_s = Q_{sn} + \frac{A_s}{\sqrt{R^2 + X^2}}\{-R\sin(0°) + X\cos(0°)\} \quad \text{or} \quad Q_s = Q_{sn} + \frac{XA_s}{\sqrt{R^2 + X^2}}.$$

For $\beta = 180°$, the variations of P_s and Q_s are given by

$$P_s = P_{sn} + \frac{A_s}{\sqrt{R^2 + X^2}}\{R\cos(180°) + X\sin(180°)\} \quad \text{or} \quad P_s = P_{sn} - \frac{RA_s}{\sqrt{R^2 + X^2}}$$

and

$$Q_s = Q_{sn} + \frac{A_s}{\sqrt{R^2 + X^2}}\{-R\sin(180°) + X\cos(180°)\} \quad \text{or} \quad Q_s = Q_{sn} - \frac{XA_s}{\sqrt{R^2 + X^2}}.$$

For $\beta = 0°$ or $\beta = 180°$, the variations of P_s and Q_s are then given by

$$P_s = P_{sn} \pm \frac{RA_s}{\sqrt{R^2 + X^2}} \tag{B-69a}$$

or

$$V_{s's} = \pm\frac{R^2 + X^2}{RV_s}(P_s - P_{sn}) \quad \text{or} \quad V_{s's} = \frac{R^2 + X^2}{RV_s}|P_s - P_{sn}| \tag{B-69b}$$

using Equation (B-62) and

$$Q_s = Q_{sn} \pm \frac{XA_s}{\sqrt{R^2 + X^2}} \tag{B-69c}$$

or

$$V_{s's} = \pm\frac{R^2 + X^2}{XV_s}(Q_s - Q_{sn}) \quad \text{or} \quad V_{s's} = \frac{R^2 + X^2}{XV_s}|Q_s - Q_{sn}| \tag{B-69d}$$

using Equation (B-62). For a desired P_s, the required $V_{s's}$ can be determined using Equation (B-69b); for a desired Q_s, the required $V_{s's}$ can be determined using Equation (B-69d).

For implementation of the PAR (asym), the relative phase angle is set to $\beta = 90°$ or $\beta = -90°$ in Equations (B-61) and (B-64). For $\beta = 90°$, the variations of P_s and Q_s are given by

$$P_s = P_{sn} + \frac{A_s}{\sqrt{R^2 + X^2}}\{R\cos(90°) + X\sin(90°)\} \quad \text{or} \quad P_s = P_{sn} + \frac{XA_s}{\sqrt{R^2 + X^2}}$$

and

$$Q_s = Q_{sn} + \frac{A_s}{\sqrt{R^2 + X^2}}\{-R\sin(90°) + X\cos(90°)\} \quad \text{or} \quad Q_s = Q_{sn} - \frac{RA_s}{\sqrt{R^2 + X^2}}.$$

For $\beta = -90°$, the variations of P_s and Q_s are given by

$$P_s = P_{sn} + \frac{A_s}{\sqrt{R^2 + X^2}}\{R\cos(-90°) + X\sin(-90°)\} \quad \text{or} \quad P_s = P_{sn} - \frac{XA_s}{\sqrt{R^2 + X^2}}$$

and

$$Q_s = Q_{sn} + \frac{A_s}{\sqrt{R^2 + X^2}}\{-R\sin(-90°) + X\cos(-90°)\} \quad \text{or} \quad Q_s = Q_{sn} + \frac{RA_s}{\sqrt{R^2 + X^2}}.$$

For $\beta = \pm 90°$, the variations of P_s and Q_s are then given by

$$P_s = P_{sn} \pm \frac{XA_s}{\sqrt{R^2 + X^2}} \tag{B-70a}$$

or

$$V_{s's} = \pm\frac{R^2 + X^2}{XV_s}(P_s - P_{sn}) \quad \text{or} \quad V_{s's} = \frac{R^2 + X^2}{XV_s}|P_s - P_{sn}| \tag{B-70b}$$

using Equation (B-62) and

$$Q_s = Q_{sn} \mp \frac{RA_s}{\sqrt{R^2 + X^2}} \tag{B-70c}$$

or

$$V_{s's} = \mp\frac{R^2 + X^2}{RV_s}(Q_s - Q_{sn}) \quad \text{or} \quad V_{s's} = \frac{R^2 + X^2}{RV_s}|Q_s - Q_{sn}| \tag{B-70d}$$

using Equation (B-62). For a desired P_s, the required $V_{s's}$ can be determined using Equation (B-70b); for a desired Q_s, the required $V_{s's}$ can be determined using Equation (B-70d).

From equations (B-61) and (B-64), the active and reactive power flows at the sending end are related as

$$\frac{Q_s - Q_{sn}}{P_s - P_{sn}} = \frac{-R\sin\beta + X\cos\beta}{R\cos\beta + X\sin\beta} \quad \text{or} \quad Q_s = \frac{-R\sin\beta + X\cos\beta}{R\cos\beta + X\sin\beta}(P_s - P_{sn}) + Q_{sn}. \tag{B-71}$$

For a VR, $\beta = 0°$ or $\beta = 180°$ represents an increase or decrease in voltage, respectively. The relationship between active and reactive power flows at the sending end, derived in Equation (B-71), can be reduced to

$$Q_s = \frac{X}{R}(P_s - P_{sn}) + Q_{sn}. \tag{B-72}$$

For a PAR (asym), $\beta = +90°$ or $\beta = -90°$ represents an increase or decrease in power flow, respectively. The relationship between the active and the reactive power flows at the sending end, derived in Equation (B-71), can be reduced to

$$Q_s = -\frac{R}{X}(P_s - P_{sn}) + Q_{sn}. \tag{B-73}$$

B.2.2.2 Power Flow at the Receiving End with a Series-Compensating Voltage

The active power flow at the receiving end of the line when the line is series-compensated is given by

$$P_r = V_r I \cos(\delta_r - \theta_I) \quad \text{or}$$

$$P_r = V_r \cos\delta I \cos\theta_{Is} - V_r \sin\delta I \sin\theta_{Is} \tag{2-129}$$

where θ_{Is} is the phase angle of the line current with respect to the sending-end voltage.

Equation (2-129) can be rewritten using Equations (B-58a) and (B-58b) as

$$P_r = V_r \cos\delta \frac{R(V_s - V_r\cos\delta + V_{s's}\cos\beta) + X(V_r\sin\delta + V_{s's}\sin\beta)}{R^2 + X^2}$$
$$- V_r \sin\delta \frac{R(V_r\sin\delta + V_{s's}\sin\beta) - X(V_s - V_r\cos\delta + V_{s's}\cos\beta)}{R^2 + X^2}$$

or

$$P_r = \frac{A}{\sqrt{R^2 + X^2}}\left\{R\left(1 - \frac{V_r}{V_s}\cos\delta + \frac{V_{s's}}{V_s}\cos\beta\right)\cos\delta + X\left(\frac{V_r}{V_s}\sin\delta + \frac{V_{s's}}{V_s}\sin\beta\right)\cos\delta\right.$$
$$\left. - R\left(\frac{V_r}{V_s}\sin\delta + \frac{V_{s's}}{V_s}\sin\beta\right)\sin\delta + X\left(1 - \frac{V_r}{V_s}\cos\delta + \frac{V_{s's}}{V_s}\cos\beta\right)\sin\delta\right\}$$

or

$$P_r = \frac{A}{\sqrt{R^2 + X^2}}\left\{R\left(1 - \frac{V_r}{V_s}\cos\delta + \frac{V_{s's}}{V_s}\cos\beta\right)\cos\delta - R\left(\frac{V_r}{V_s}\sin\delta + \frac{V_{s's}}{V_s}\sin\beta\right)\sin\delta\right.$$
$$\left. + X\left(\frac{V_r}{V_s}\sin\delta + \frac{V_{s's}}{V_s}\sin\beta\right)\cos\delta + X\left(1 - \frac{V_r}{V_s}\cos\delta + \frac{V_{s's}}{V_s}\cos\beta\right)\sin\delta\right\}$$

or

$$P_r = \frac{A}{\sqrt{R^2 + X^2}}\left\{R\left(1 - \frac{V_r}{V_s}\cos\delta\right)\cos\delta + R\frac{V_{s's}}{V_s}\cos(\delta + \beta) - R\left(\frac{V_r}{V_s}\sin\delta\right)\sin\delta\right.$$
$$\left. + X\left(\frac{V_r}{V_s}\sin\delta\right)\cos\delta + X\left(1 - \frac{V_r}{V_s}\cos\delta\right)\sin\delta + X\frac{V_{s's}}{V_s}\sin(\delta + \beta)\right\}$$

or

$$P_r = \frac{A}{\sqrt{R^2 + X^2}}\left\{R\left(\cos\delta - \frac{V_r}{V_s}(\cos^2\delta + \sin^2\delta)\right) + X\sin\delta + R\frac{V_{s's}}{V_s}\cos(\delta + \beta)\right.$$
$$\left. + X\frac{V_{s's}}{V_s}\sin(\delta + \beta)\right\}$$

or

$$P_r = \frac{A}{\sqrt{R^2+X^2}}\left\{R\left(\cos\delta - \frac{V_r}{V_s}\right) + X\sin\delta + R\frac{V_{s's}}{V_s}\cos(\delta+\beta) + X\frac{V_{s's}}{V_s}\sin(\delta+\beta)\right\} \tag{B-74}$$

using the identity ($\cos^2\delta + \sin^2\delta = 1$), or

$$P_r = P_{rn} + \frac{A_r}{\sqrt{R^2+X^2}}\{R\cos(\delta+\beta) + X\sin(\delta+\beta)\} \tag{B-75}$$

where P_{rn} is given in Equation (B-21) and

$$A_r = \frac{V_r V_{s's}}{\sqrt{R^2+X^2}}. \tag{B-76}$$

The reactive power flow at the receiving end of the line when the line is series-compensated is given by

$$Q_r = V_r I\sin(\delta_r - \theta_I) \quad \text{or}$$

$$Q_r = -V_r\sin\delta\, I\cos\theta_{Is} - V_r\cos\delta\, I\sin\theta_{Is} \tag{2-133}$$

where θ_{Is} is the phase angle of the line current with respect to the sending-end voltage.

Equation (2-133) can be rewritten using Equations (B-58a) and (B-58b) as

$$Q_r = -V_r\sin\delta\frac{R(V_s - V_r\cos\delta + V_{s's}\cos\beta) + X(V_r\sin\delta + V_{s's}\sin\beta)}{R^2+X^2}$$
$$-V_r\cos\delta\frac{R(V_r\sin\delta + V_{s's}\sin\beta) - X(V_s - V_r\cos\delta + V_{s's}\cos\beta)}{R^2+X^2}$$

or

$$Q_r = \frac{A}{\sqrt{R^2+X^2}}\left\{-R\left(1-\frac{V_r}{V_s}\cos\delta + \frac{V_{s's}}{V_s}\cos\beta\right)\sin\delta - X\left(\frac{V_r}{V_s}\sin\delta + \frac{V_{s's}}{V_s}\sin\beta\right)\sin\delta\right.$$
$$\left.-R\left(\frac{V_r}{V_s}\sin\delta + \frac{V_{s's}}{V_s}\sin\beta\right)\cos\delta + X\left(1-\frac{V_r}{V_s}\cos\delta + \frac{V_{s's}}{V_s}\cos\beta\right)\cos\delta\right\}$$

or

$$Q_r = \frac{A}{\sqrt{R^2+X^2}}\left\{-R\left(1-\frac{V_r}{V_s}\cos\delta + \frac{V_{s's}}{V_s}\cos\beta\right)\sin\delta - R\left(\frac{V_r}{V_s}\sin\delta + \frac{V_{s's}}{V_s}\sin\beta\right)\cos\delta\right.$$
$$\left.-X\left(\frac{V_r}{V_s}\sin\delta + \frac{V_{s's}}{V_s}\sin\beta\right)\sin\delta + X\left(1-\frac{V_r}{V_s}\cos\delta + \frac{V_{s's}}{V_s}\cos\beta\right)\cos\delta\right\}$$

or

$$Q_r = \frac{A}{\sqrt{R^2+X^2}}\left\{-R\left(1-\frac{V_r}{V_s}\cos\delta\right)\sin\delta - R\frac{V_{s's}}{V_s}\sin(\delta+\beta) - R\left(\frac{V_r}{V_s}\sin\delta\right)\cos\delta\right.$$
$$\left.-X\left(\frac{V_r}{V_s}\sin\delta\right)\sin\delta + X\left(1-\frac{V_r}{V_s}\cos\delta\right)\cos\delta + X\frac{V_{s's}}{V_s}\cos(\delta+\beta)\right\}$$

or

$$Q_r = \frac{A}{\sqrt{R^2+X^2}}\left\{-R\sin\delta + X\left(\cos\delta - \frac{V_r}{V_s}\left(\cos^2\delta + \sin^2\delta\right)\right)\right.$$

$$\left. -R\frac{V_{s'S}}{V_s}\sin(\delta+\beta) + X\frac{V_{s'S}}{V_s}\cos(\delta+\beta)\right\}$$

or

$$Q_r = \frac{A}{\sqrt{R^2+X^2}}\left\{-R\sin\delta + X\left(\cos\delta - \frac{V_r}{V_s}\right) - R\frac{V_{s'S}}{V_s}\sin(\delta+\beta) + X\frac{V_{s'S}}{V_s}\cos(\delta+\beta)\right\} \tag{B-77}$$

using the identity ($\cos^2\delta + \sin^2\delta = 1$), or

$$Q_r = Q_{rn} + \frac{A_r}{\sqrt{R^2+X^2}}\{-R\sin(\delta+\beta) + X\cos(\delta+\beta)\} \tag{B-78}$$

where Q_{rn} and A_r are given in Equations (B-22) and (B-76), respectively. For $R = 0$, the active and reactive power flows at the receiving end of the line are given in Equations (2-131) and (2-135), respectively.

Equations (B-75) and (B-78) can be combined as

$$(P_r - P_{rn})^2 + (Q_r - Q_{rn})^2 = \left(\frac{A_r}{\sqrt{R^2+X^2}}\right)^2 \left[\{R\cos(\delta+\beta) + X\sin(\delta+\beta)\}^2 + \{-R\sin(\delta+\beta) + X\cos(\delta+\beta)\}^2\right].$$

or

$$(P_r - P_{rn})^2 + (Q_r - Q_{rn})^2 = \frac{A_r^2}{R^2+X^2}\left[\left(R^2+X^2\right)\left\{\cos^2(\delta+\beta) + \sin^2(\delta+\beta)\right\}\right]$$

or

$$(P_r - P_{rn})^2 + (Q_r - Q_{rn})^2 = A_r^2 \tag{B-79}$$

using the identity $\left\{\cos^2(\delta+\beta) + \sin^2(\delta+\beta) = 1\right\}$ and

$$\tan(\delta+\beta) = \frac{-R(Q_r - Q_{rn}) + X(P_r - P_{rn})}{R(P_r - P_{rn}) + X(Q_r - Q_{rn})}. \tag{B-80}$$

Digress:
From Equations (B-75) and (B-78), the active and reactive power flows at the receiving end are related as

$$\frac{Q_r - Q_{rn}}{P_r - P_{rn}} = \frac{-R\sin(\delta+\beta) + X\cos(\delta+\beta)}{R\cos(\delta+\beta) + X\sin(\delta+\beta)} \quad \text{or}$$

$$(Q_r - Q_{rn})\{R\cos(\delta+\beta) + X\sin(\delta+\beta)\} = (P_r - P_{rn})\{-R\sin(\delta+\beta) + X\cos(\delta+\beta)\}$$

or $$R(P_r - P_{rn})\sin(\delta+\beta) + X(Q_r - Q_{rn})\sin(\delta+\beta) = -R(Q_r - Q_{rn})\cos(\delta+\beta) + X(P_r - P_{rn})\cos(\delta+\beta)$$

or

$$\tan(\delta + \beta) = \frac{-R(Q_r - Q_{rn}) + X(P_r - P_{rn})}{R(P_r - P_{rn}) + X(Q_r - Q_{rn})}.$$

Using Equation (B-76), Equations (B-79) can be rearranged as

$$V_{s's} = \frac{\sqrt{R^2 + X^2}}{V_r}\sqrt{(P_r - P_{rn})^2 + (Q_r - Q_{rn})^2}. \tag{B-81}$$

Equation (B-80) can be rearranged as

$$\beta = \tan^{-1}\frac{-R(Q_r - Q_{rn}) + X(P_r - P_{rn})}{R(P_r - P_{rn}) + X(Q_r - Q_{rn})} - \delta. \tag{B-82}$$

For $R = 0$, $V_{s's}$ and β are given in Equations (2-138) and (2-139), respectively.

For the desired active and reactive power flows (P_r^*, Q_r^*) within the allowable range, the magnitude ($V_{s's}$) and the relative phase angle (β) of the series-compensating voltage ($\mathbf{V_{s's}}$) are determined by substituting $P_r = P_r^*$ and $Q_r = Q_r^*$ in Equations (B-81) and (B-82), respectively. The relative phase angle (β) is the angle between the voltage regulator line (*i.e.* $\beta = 0°$ line) and the line that passes through (P_{rn}, Q_{rn}) and (P_r^*, Q_r^*).

Recall that for $R = 0$,

$$(P_r - P_{rn})^2 + (Q_r - Q_{rn})^2 = a_r^2. \tag{2-136}$$

The Equation (B-79) represents the locus of a circle, centered at $[P_{rn}, Q_{rn}]$ with a radius of A_r, which is defined in Equation (B-76). The Equation (2-136) represents the locus of a circle, centered at $[P_{rn}, Q_{rn}]$ with a radius of a_r, which is defined in Equation (2-132).

For implementation of the VR, the relative phase angle is set to $\beta = 0°$ or $\beta = 180°$ in Equations (B-75) and (B-78). For $\beta = 0°$, the variations of P_r and Q_r are given by

$$P_r = P_{rn} + \frac{A_r}{\sqrt{R^2 + X^2}}\{R\cos(\delta + 0°) + X\sin(\delta + 0°)\} \quad \text{or}$$

$$P_r = P_{rn} + \frac{A_r}{\sqrt{R^2 + X^2}}(R\cos\delta + X\sin\delta)$$

and

$$Q_r = Q_{rn} + \frac{A_r}{\sqrt{R^2 + X^2}}\{-R\sin(\delta + 0°) + X\cos(\delta + 0°)\} \quad \text{or}$$

$$Q_r = Q_{rn} + \frac{A_r}{\sqrt{R^2 + X^2}}(-R\sin\delta + X\cos\delta).$$

For $\beta = 180°$, the variations of P_s and Q_s are given by

$$P_r = P_{rn} + \frac{A_r}{\sqrt{R^2 + X^2}}\{R\cos(\delta + 180°) + X\sin(\delta + 180°)\} \quad \text{or}$$

$$P_r = P_{rn} + \frac{A_r}{\sqrt{R^2 + X^2}}(-R\cos\delta - X\sin\delta)$$

and

$$Q_r = Q_{rn} + \frac{A_r}{\sqrt{R^2 + X^2}}\{-R\sin(\delta + 180°) + X\cos(\delta + 180°)\} \quad \text{or}$$

$$Q_r = Q_{rn} + \frac{A_r}{\sqrt{R^2 + X^2}}(R\sin\delta - X\cos\delta).$$

For $\beta = 0°$ or $\beta = 180°$, the variations of P_r and Q_r are then given by

$$P_r = P_{rn} \pm \frac{A_r}{\sqrt{R^2 + X^2}}(R\cos\delta + X\sin\delta) \tag{B-83a}$$

or

$$V_{s's} = \pm\frac{R^2 + X^2}{V_r(R\cos\delta + X\sin\delta)}(P_r - P_{rn}) \quad \text{or} \quad V_{s's} = \frac{R^2 + X^2}{V_r(R\cos\delta + X\sin\delta)}|P_r - P_{rn}| \tag{B-83b}$$

using Equation (B-76) and

$$Q_r = Q_{rn} \pm \frac{A_r}{\sqrt{R^2 + X^2}}(-R\sin\delta + X\cos\delta) \tag{B-83c}$$

or

$$V_{s's} = \pm\frac{R^2 + X^2}{V_r(-R\sin\delta + X\cos\delta)}(Q_r - Q_{rn}) \quad \text{or} \quad V_{s's} = \frac{R^2 + X^2}{V_r(-R\sin\delta + X\cos\delta)}|Q_r - Q_{rn}| \tag{B-83d}$$

using Equation (B-76). For a desired P_r, the required $V_{s's}$ can be determined using Equation (B-83b); for a desired Q_r, the required $V_{s's}$ can be determined using Equation (B-83d).

For implementation of the PAR (asym), the relative phase angle is set to $\beta = 90°$ or $\beta = -90°$ in Equations (B-75) and (B-78). For $\beta = 90°$, the variations of P_r and Q_r are given by

$$P_r = P_{rn} + \frac{A_r}{\sqrt{R^2 + X^2}}\{R\cos(\delta + 90°) + X\sin(\delta + 90°)\} \quad \text{or}$$

$$P_r = P_{rn} + \frac{A_r}{\sqrt{R^2 + X^2}}(-R\sin\delta + X\cos\delta)$$

and

$$Q_r = Q_{rn} + \frac{A_r}{\sqrt{R^2 + X^2}}\{-R\sin(\delta + 90°) + X\cos(\delta + 90°)\} \quad \text{or}$$

$$Q_r = Q_{rn} + \frac{A_r}{\sqrt{R^2 + X^2}}\{-R\cos\delta - X\sin\delta\}.$$

For $\beta = -90°$, the variations of P_r and Q_r are given by

$$P_r = P_{rn} + \frac{A_r}{\sqrt{R^2 + X^2}}\{R\cos(\delta - 90°) + X\sin(\delta - 90°)\} \quad \text{or}$$

$$P_r = P_{rn} + \frac{A_r}{\sqrt{R^2 + X^2}}(R\sin\delta - X\cos\delta)$$

and

$$Q_r = Q_{rn} + \frac{A_r}{\sqrt{R^2 + X^2}}\{-R\sin(\delta - 90°) + X\cos(\delta - 90°)\} \quad \text{or}$$

$$Q_r = Q_{rn} + \frac{A_r}{\sqrt{R^2 + X^2}}(R\cos\delta + X\sin\delta).$$

For $\beta = \pm 90°$, the variations of P_r and Q_r are then given by

$$P_r = P_{rn} \pm \frac{A_r}{\sqrt{R^2 + X^2}}(-R\sin\delta + X\cos\delta) \tag{B-84a}$$

or

$$V_{s's} = \pm\frac{R^2 + X^2}{V_r(-R\sin\delta + X\cos\delta)}(P_r - P_{rn}) \quad \text{or} \quad V_{s's} = \frac{R^2 + X^2}{V_r(-R\sin\delta + X\cos\delta)}|P_r - P_{rn}| \tag{B-84b}$$

using Equation (B-76) and

$$Q_r = Q_{rn} \mp \frac{A_r}{\sqrt{R^2 + X^2}}(R\cos\delta + X\sin\delta) \tag{B-84c}$$

or

$$V_{s's} = \mp\frac{R^2 + X^2}{V_r(R\cos\delta + X\sin\delta)}(Q_r - Q_{rn}) \quad \text{or} \quad V_{s's} = \frac{R^2 + X^2}{V_r(R\cos\delta + X\sin\delta)}|Q_r - Q_{rn}| \tag{B-84d}$$

using Equation (B-76). For a desired P_r, the required $V_{s's}$ can be determined using Equation (B-84b); for a desired Q_r, the required $V_{s's}$ can be determined using Equation (B-84d).

From Equations (B-75) and (B-78), the active and reactive power flows at the receiving end are related as

$$\frac{Q_r - Q_{rn}}{P_r - P_{rn}} = \frac{-R\sin(\delta + \beta) + X\cos(\delta + \beta)}{R\cos(\delta + \beta) + X\sin(\delta + \beta)}$$

or

$$Q_r = \frac{-R\sin(\delta + \beta) + X\cos(\delta + \beta)}{R\cos(\delta + \beta) + X\sin(\delta + \beta)}(P_r - P_{rn}) + Q_{rn}. \tag{B-85}$$

For a VR, $\beta = 0°$ or $180°$ represents an increase or decrease in voltage, respectively. The relationship between the active and reactive power flows at the receiving end, derived in Equation (B-85), can be reduced to

$$Q_r = \frac{\mp R\sin\delta \pm X\cos\delta}{\pm R\cos\delta \pm X\sin\delta}(P_r - P_{rn}) + Q_{rn} \quad \text{or} \quad Q_r = \frac{-R\sin\delta + X\cos\delta}{R\cos\delta + X\sin\delta}(P_r - P_{rn}) + Q_{rn}. \tag{B-86}$$

For a PAR (asym), $\beta = 90°$ or $-90°$ represents an increase or decrease in power flow, respectively. The relationship between the active and reactive power flows at the receiving end, derived in Equation (B-85), can be reduced to

$$Q_r = \frac{\mp R\cos\delta \mp X\sin\delta}{\mp R\sin\delta \pm X\cos\delta}(P_r - P_{rn}) + Q_{rn} \quad \text{or} \quad Q_r = \frac{R\cos\delta + X\sin\delta}{R\sin\delta - X\cos\delta}(P_r - P_{rn}) + Q_{rn}. \tag{B-87}$$

B.2.2.3 Power Flow at the Modified Sending End with a Series-Compensating Voltage

The active power flow at the modified sending end of the line when the line is series-compensated is given by

$$P_{s'} = V_{s'} I\cos(\delta_{s'} - \theta_I) \quad \text{or}$$
$$P_{s'} = V_{s'} \cos\psi I \cos\theta_{Is} + V_{s'} \sin\psi I \sin\theta_{Is}. \tag{2-147}$$

where θ_{Is} is the phase angle of the line current with respect to the sending-end voltage.

Equation (2-147) can be rewritten using Equations (2-96a), (2-96b), (B-58a), and (B-58b) as

$$P_{s'} = (V_s + V_{s's}\cos\beta)\frac{R(V_s - V_r\cos\delta + V_{s's}\cos\beta) + X(V_r\sin\delta + V_{s's}\sin\beta)}{R^2 + X^2} + V_{s's}\sin\beta\frac{R(V_r\sin\delta + V_{s's}\sin\beta) - X(V_s - V_r\cos\delta + V_{s's}\cos\beta)}{R^2 + X^2}$$

or

$$P_{s'} = R\frac{(V_s + V_{s's}\cos\beta)(V_s - V_r\cos\delta + V_{s's}\cos\beta) + V_{s's}\sin\beta(V_r\sin\delta + V_{s's}\sin\beta)}{R^2 + X^2} + X\frac{(V_s + V_{s's}\cos\beta)(V_r\sin\delta + V_{s's}\sin\beta) - V_{s's}\sin\beta(V_s - V_r\cos\delta + V_{s's}\cos\beta)}{R^2 + X^2}$$

or

$$P_{s'} = \frac{1}{R^2 + X^2}\begin{bmatrix} R\{(V_s + V_{s's}\cos\beta)(V_s - V_r\cos\delta + V_{s's}\cos\beta) + V_{s's}\sin\beta(V_r\sin\delta + V_{s's}\sin\beta)\} \\ + X\{(V_s + V_{s's}\cos\beta)(V_r\sin\delta + V_{s's}\sin\beta) - V_{s's}\sin\beta(V_s - V_r\cos\delta + V_{s's}\cos\beta)\} \end{bmatrix}$$

or

$$P_{s'} = \frac{1}{R^2 + X^2}\begin{bmatrix} R\{V_s^2 + 2V_sV_{s's}\cos\beta + V_{s's}^2 - V_sV_r\cos\delta - V_{s's}V_r\cos(\delta + \beta)\} \\ + X\{V_sV_r\sin\delta + V_{s's}V_r\sin(\delta + \beta)\} \end{bmatrix} \tag{B-88}$$

or

$$P_{s'} = \frac{A}{\sqrt{R^2 + X^2}}\left[R\left(\frac{V_s}{V_r} - \cos\delta\right) + X\sin\delta\right] + \frac{RV_{s's}^2}{R^2 + X^2} + \frac{V_{s's}}{R^2 + X^2}[R\{2V_s\cos\beta - V_r\cos(\delta + \beta)\} + XV_r\sin(\delta + \beta)]$$

or

$$P_{s'} = P_{sn} + \frac{RV_{s's}^2}{R^2 + X^2} + \frac{1}{\sqrt{R^2 + X^2}}[R\{2A_s\cos\beta - A_r\cos(\delta + \beta)\} + XA_r\sin(\delta + \beta)] \tag{B-89}$$

where P_{sn}, A_s, and A_r are given in Equations (B-12), (B-62), and (B-76), respectively.

The reactive power flow at the modified sending end of the line when the line is series-compensated is given by

$$Q_{s'} = V_{s'} I\sin(\delta_{s'} - \theta_I) \quad \text{or}$$
$$Q_{s'} = V_{s'} \sin\psi I \cos\theta_{Is} - V_{s'} \cos\psi I \sin\theta_{Is} \tag{2-150}$$

where θ_{Is} is the phase angle of the line current with respect to the sending-end voltage.

Equation (2-150) can be rewritten using Equations (2-96a), (2-96b), (B-58a), and (B-58b) as

$$Q_{s'} = V_{s's}\sin\beta\frac{R(V_s - V_r\cos\delta + V_{s's}\cos\beta) + X(V_r\sin\delta + V_{s's}\sin\beta)}{R^2 + X^2}$$
$$- (V_s + V_{s's}\cos\beta)\ \frac{R(V_r\sin\delta + V_{s's}\sin\beta) - X(V_s - V_r\cos\delta + V_{s's}\cos\beta)}{R^2 + X^2}$$

or

$$Q_{s'} = R\frac{V_{s's}\sin\beta(V_s - V_r\cos\delta + V_{s's}\cos\beta) - (V_s + V_{s's}\cos\beta)(V_r\sin\delta + V_{s's}\sin\beta)}{R^2 + X^2}$$
$$+ X\frac{V_{s's}\sin\beta(V_r\sin\delta + V_{s's}\sin\beta) + (V_s + V_{s's}\cos\beta)(V_s - V_r\cos\delta + V_{s's}\cos\beta)}{R^2 + X^2}$$

or

$$Q_{s'} = \frac{1}{R^2 + X^2}\begin{bmatrix} R\{V_{s's}\sin\beta(V_s - V_r\cos\delta + V_{s's}\cos\beta) - (V_s + V_{s's}\cos\beta)(V_r\sin\delta + V_{s's}\sin\beta)\} \\ + X\{V_{s's}\sin\beta(V_r\sin\delta + V_{s's}\sin\beta) + (V_s + V_{s's}\cos\beta)(V_s - V_r\cos\delta + V_{s's}\cos\beta)\} \end{bmatrix}$$

or

$$Q_{s'} = \frac{1}{R^2 + X^2}\begin{bmatrix} -R\{V_sV_r\sin\delta + V_{s's}V_r\sin(\delta + \beta)\} \\ + X\{V_s^2 + 2V_sV_{s's}\cos\beta + V_{s's}^2 - V_sV_r\cos\delta - V_{s's}V_r\cos(\delta + \beta)\} \end{bmatrix} \tag{B-90}$$

or

$$Q_{s'} = \frac{A}{\sqrt{R^2 + X^2}}\left[-R\sin\delta + X\left(\frac{V_s}{V_r} - \cos\delta\right)\right] + \frac{XV_{s's}^2}{R^2 + X^2}$$
$$+ \frac{V_{s's}}{R^2 + X^2}[-RV_r\sin(\delta + \beta) + X\{2V_s\cos\beta - V_r\cos(\delta + \beta)\}]$$

or

$$Q_{s'} = Q_{sn} + \frac{XV_{s's}^2}{R^2 + X^2} + \frac{1}{\sqrt{R^2 + X^2}}[-RA_r\sin(\delta + \beta) + X\{2A_s\cos\beta - A_r\cos(\delta + \beta)\}] \tag{B-91}$$

where Q_{sn}, A_s, and A_r are given in Equations (B-14), (B-62), and (B-76), respectively. For $R = 0$, the active and reactive power flows at the modified sending end of the line are given in Equations (2-149) and (2-152), respectively.

Equations (B-89) and (B-91) can be combined as

$$(P_{s'} - P_{sn})^2 + (Q_{s'} - Q_{sn})^2$$
$$= \frac{1}{R^2 + X^2}\left\{\begin{aligned} &\left[\frac{RV_{s's}^2}{\sqrt{R^2 + X^2}} + R\{2A_s\cos\beta - A_r\cos(\delta + \beta)\} + XA_r\sin(\delta + \beta)\right]^2 \\ &+ \left[\frac{XV_{s's}^2}{\sqrt{R^2 + X^2}} - RA_r\sin(\delta + \beta) + X\{2A_s\cos\beta - A_r\cos(\delta + \beta)\}\right]^2 \end{aligned}\right\}$$

or substituting for A_s and A_r,

$$(P_{s'} - P_{sn})^2 + (Q_{s'} - Q_{sn})^2$$

$$= \frac{V_{s's}^2}{(R^2 + X^2)^2} \left\{ \begin{array}{l} [RV_{s's} + R\{2V_s \cos\beta - V_r \cos(\delta + \beta)\} + XV_r \sin(\delta + \beta)]^2 \\ + [XV_{s's} - RV_r \sin(\delta + \beta) + X\{2V_s \cos\beta - V_r \cos(\delta + \beta)\}]^2 \end{array} \right\}$$

or

$$\begin{aligned}(P_{s'} - P_{sn})^2 + (Q_{s'} - Q_{sn})^2 = \frac{V_{s's}^2}{(R^2 + X^2)^2} \{&[RV_{s's}]^2 + [R\{2V_s \cos\beta - V_r \cos(\delta + \beta)\}]^2 \\ &+ [XV_r \sin(\delta + \beta)]^2 + 2RV_{s's}R\{2V_s \cos\beta - V_r \cos(\delta + \beta)\} \\ &+ 2R\{2V_s \cos\beta - V_r \cos(\delta + \beta)\}XV_r \sin(\delta + \beta) + 2XV_r \sin(\delta + \beta)RV_{s's} + [XV_{s's}]^2 \\ &+ [RV_r \sin(\delta + \beta)]^2 + [X\{2V_s \cos\beta - V_r \cos(\delta + \beta)\}]^2 - 2XV_{s's}RV_r \sin(\delta + \beta) \\ &- 2RV_r \sin(\delta + \beta)X\{2V_s \cos\beta - V_r \cos(\delta + \beta)\} + 2X\{2V_s \cos\beta - V_r \cos(\delta + \beta)\}XV_{s's}\}\end{aligned}$$

or

$$\begin{aligned}(P_{s'} - P_{sn})^2 + (Q_{s'} - Q_{sn})^2 = \frac{V_{s's}^2}{R^2 + X^2} \{&[V_{s's}]^2 + [\{2V_s \cos\beta - V_r \cos(\delta + \beta)\}]^2 \\ &+ [V_r \sin(\delta + \beta)]^2 + 2V_{s's}\{2V_s \cos\beta - V_r \cos(\delta + \beta)\}\}\end{aligned}$$

or

$$\begin{aligned}(P_{s'} - P_{sn})^2 + (Q_{s'} - Q_{sn})^2 = \frac{V_{s's}^2}{R^2 + X^2} [&V_{s's}^2 + 2V_{s's}\{2V_s \cos\beta - V_r \cos(\delta + \beta)\} \\ &+ 4V_s^2 \cos^2\beta - 4V_s V_r \cos\beta \cos(\delta + \beta) + V_r^2].\end{aligned} \quad \text{(B-92)}$$

or

$$\begin{aligned}\{(P_{s'} - P_{sn})^2 + (Q_{s'} - Q_{sn})^2\}(R^2 + X^2) = V_{s's}^2 [&V_{s's}^2 + 2V_{s's}\{2V_s \cos\beta - V_r \cos(\delta + \beta)\} \\ &+ 4V_s^2 \cos^2\beta - 4V_s V_r \cos\beta \cos(\delta + \beta) + V_r^2]\end{aligned}$$

or

$$\begin{aligned}&V_{s's}^4 + 2\{2V_s \cos\beta - V_r \cos(\delta + \beta)\}V_{s's}^3 + \{4V_s^2 \cos^2\beta - 4V_s V_r \cos\beta \cos(\delta + \beta) + V_r^2\}V_{s's}^2 \\ &- \{(P_{s'} - P_{sn})^2 + (Q_{s'} - Q_{sn})^2\}(R^2 + X^2) = 0.\end{aligned} \quad \text{(B-93)}$$

Note that the Equation (B-93) is a fourth-order equation that may need to be solved using a numerical method to determine the magnitude ($V_{s's}$) of the series-compensating voltage ($\mathbf{V_{s's}}$). Out of the four possible values of $V_{s'}$, given in Equation (B-93), the most desired value is the one that provides the lowest *APR*, thus the lowest cost solution.

Recall that for $R = 0$,

$$\begin{aligned}&V_{s's}^4 + 2\{2V_s \cos\beta - V_r \cos(\delta + \beta)\}V_{s's}^3 + \{4V_s^2 \cos^2\beta - 4V_s V_r \cos\beta \cos(\delta + \beta) + V_r^2\}V_{s's}^2 \\ &- \{(P_{s'} - P_{sn})^2 + (Q_{s'} - Q_{sn})^2\}X^2 = 0.\end{aligned} \quad \text{(2-154)}$$

Alternately, Equations (B-89) and (B-91) can be combined as

$$\left(P_{s'}-P_{sn}-\frac{RV_{s's}^2}{R^2+X^2}\right)^2+\left(Q_{s'}-Q_{sn}-\frac{XV_{s's}^2}{R^2+X^2}\right)^2$$
$$-\frac{1}{R^2+X^2}\left\{\begin{array}{l}[R\{2A_s\cos\beta-A_r\cos(\delta+\beta)\}+XA_r\sin(\delta+\beta)]^2+\\ [-RA_r\sin(\delta+\beta)+X\{2A_s\cos\beta-A_r\cos(\delta+\beta)\}]^2\end{array}\right\}$$

or

$$\left(P_{s'}-P_{sn}-\frac{RV_{s's}^2}{R^2+X^2}\right)^2+\left(Q_{s'}-Q_{sn}-\frac{XV_{s's}^2}{R^2+X^2}\right)^2 = 4A_s^2\cos^2\beta-4A_sA_r\cos\beta\cos(\delta+\beta)+A_r^2 \quad \text{(B-94)}$$

and

$$\tan\beta=-\frac{\left\{R\left(P_{s'}-P_{sn}-\frac{RV_{s's}^2}{R^2+X^2}\right)+X\left(Q_{s'}-Q_{sn}-\frac{XV_{s's}^2}{R^2+X^2}\right)\right\}V_r(\sin\delta)+\left\{R\left(Q_{s'}-Q_{sn}-\frac{XV_{s's}^2}{R^2+X^2}\right)-X\left(P_{s'}-P_{sn}-\frac{RV_{s's}^2}{R^2+X^2}\right)\right\}\{2V_s-V_r(\cos\delta)\}}{\left\{R\left(P_{s'}-P_{sn}-\frac{RV_{s's}^2}{R^2+X^2}\right)+X\left(Q_{s'}-Q_{sn}-\frac{XV_{s's}^2}{R^2+X^2}\right)\right\}V_r(\cos\delta)+\left\{R\left(Q_{s'}-Q_{sn}-\frac{XV_{s's}^2}{R^2+X^2}\right)-X\left(P_{s'}-P_{sn}-\frac{RV_{s's}^2}{R^2+X^2}\right)\right\}V_r\sin\delta}. \quad \text{(B-95)}$$

Digress:
From Equations (B-89) and (B-91), it can be written that

$$\frac{P_{s'}-P_{sn}-\frac{RV_{s's}^2}{R^2+X^2}}{Q_{s'}-Q_{sn}-\frac{XV_{s's}^2}{R^2+X^2}}=\frac{R\{2A_s\cos\beta-A_r\cos(\delta+\beta)\}+XA_r\sin(\delta+\beta)}{-RA_r\sin(\delta+\beta)+X\{2A_s\cos\beta-A_r\cos(\delta+\beta)\}}. \quad \text{or}$$

$$\left(P_{s'}-P_{sn}-\frac{RV_{s's}^2}{R^2+X^2}\right)(-RA_r\sin(\delta+\beta)+X\{2A_s\cos\beta-A_r\cos(\delta+\beta)\})$$
$$=\left(Q_{s'}-Q_{sn}-\frac{XV_{s's}^2}{R^2+X^2}\right)(R\{2A_s\cos\beta-A_r\cos(\delta+\beta)\}+XA_r\sin(\delta+\beta)) \quad \text{or}$$

$$-R\left(P_{s'}-P_{sn}-\frac{RV_{s's}^2}{R^2+X^2}\right)A_r\sin(\delta+\beta)+X\left(P_{s'}-P_{sn}-\frac{RV_{s's}^2}{R^2+X^2}\right)\{2A_s\cos\beta-A_r\cos(\delta+\beta)\}$$
$$=R\left(Q_{s'}-Q_{sn}-\frac{XV_{s's}^2}{R^2+X^2}\right)\{2A_s\cos\beta-A_r\cos(\delta+\beta)\}+X\left(Q_{s'}-Q_{sn}-\frac{XV_{s's}^2}{R^2+X^2}\right)A_r\sin(\delta+\beta)$$

or

$$-R\left(P_{s'}-P_{sn}-\frac{RV_{s's}^2}{R^2+X^2}\right)A_r\sin(\delta+\beta)-X\left(Q_{s'}-Q_{sn}-\frac{XV_{s's}^2}{R^2+X^2}\right)A_r\sin(\delta+\beta)$$
$$=R\left(Q_{s'}-Q_{sn}-\frac{XV_{s's}^2}{R^2+X^2}\right)\{2A_s\cos\beta-A_r\cos(\delta+\beta)\}$$
$$-X\left(P_{s'}-P_{sn}-\frac{RV_{s's}^2}{R^2+X^2}\right)\{2A_s\cos\beta-A_r\cos(\delta+\beta)\}$$

or

$$-\left\{R\left(P_{s'}-P_{sn}-\frac{RV_{s's}^2}{R^2+X^2}\right)+X\left(Q_{s'}-Q_{sn}-\frac{XV_{s's}^2}{R^2+X^2}\right)\right\}A_r\sin(\delta+\beta)$$
$$=\left\{R\left(Q_{s'}-Q_{sn}-\frac{XV_{s's}^2}{R^2+X^2}\right)-X\left(P_{s'}-P_{sn}-\frac{RV_{s's}^2}{R^2+X^2}\right)\right\}\{2A_s\cos\beta-A_r\cos(\delta+\beta)\}$$

or using intermediate variables (P and Q)

$$-\{R(P)+X(Q)\}A_r\sin(\delta+\beta)=\{R(Q)-X(P)\}\{2A_s\cos\beta-A_r\cos(\delta+\beta)\}$$

or

$$-\{R(P)+X(Q)\}A_r(\sin\delta\cos\beta+\cos\delta\sin\beta)$$
$$=\{R(Q)-X(P)\}\{2A_s\cos\beta-A_r(\cos\delta\cos\beta-\sin\delta\sin\beta)\}$$

or

$$-\{R(P)+X(Q)\}A_r(\cos\delta\sin\beta)-\{R(Q)-X(P)\}A_r\sin\delta\sin\beta$$
$$=\{R(P)+X(Q)\}A_r(\sin\delta\cos\beta)+\{R(Q)-X(P)\}\{2A_s\cos\beta-A_r(\cos\delta\cos\beta)\}$$

or

$$-[\{R(P)+X(Q)\}A_r(\cos\delta)+\{R(Q)-X(P)\}A_r\sin\delta]\sin\beta$$
$$=[\{R(P)+X(Q)\}A_r(\sin\delta)+\{R(Q)-X(P)\}\{2A_s-A_r(\cos\delta)\}]\cos\beta$$

or

$$\tan\beta=-\frac{\left\{R\left(P_{s'}-P_{sn}-\frac{RV_{s's}^2}{R^2+X^2}\right)+X\left(Q_{s'}-Q_{sn}-\frac{XV_{s's}^2}{R^2+X^2}\right)\right\}V_r(\sin\delta)+\left\{R\left(Q_{s'}-Q_{sn}-\frac{XV_{s's}^2}{R^2+X^2}\right)-X\left(P_{s'}-P_{sn}-\frac{RV_{s's}^2}{R^2+X^2}\right)\right\}\{2V_s-V_r(\cos\delta)\}}{\left\{R\left(P_{s'}-P_{sn}-\frac{RV_{s's}^2}{R^2+X^2}\right)+X\left(Q_{s'}-Q_{sn}-\frac{XV_{s's}^2}{R^2+X^2}\right)\right\}V_r(\cos\delta)+\left\{R\left(Q_{s'}-Q_{sn}-\frac{XV_{s's}^2}{R^2+X^2}\right)-X\left(P_{s'}-P_{sn}-\frac{RV_{s's}^2}{R^2+X^2}\right)\right\}V_r\sin\delta}.$$

Recall that for $R=0$,

$$(P_{s'}-P_{sn})^2+\left(Q_{s'}-Q_{sn}-\frac{V_{s's}^2}{X}\right)^2=4a_s^2\cos^2\beta-4a_sa_r\cos\beta\cos(\delta+\beta)+a_r^2. \qquad \text{(2-153a)}$$

The Equation (B-94) represents the locus of a circle, centered at $\left[P_{sn} + \frac{RV_{s's}^2}{R^2+X^2}, Q_{sn} + \frac{XV_{s's}^2}{R^2+X^2}\right]$ with a radius of $\sqrt{4A_s^2\cos^2\beta - 4A_sA_r\cos\beta\cos(\delta+\beta) + A_r^2}$. The Equation (2-153a) represents the locus of a circle, centered at $\left[P_{sn}, Q_{sn} + \frac{V_{s's}^2}{X}\right]$ with a radius of $\sqrt{4a_s^2\cos^2\beta - 4a_sa_r\cos\beta\cos(\delta+\beta) + a_r^2}$. Note that for a given series-compensating voltage ($V_{s's}$), the center of the locus is fixed, but the radius is a function of the relative phase angle (β) and the locus is actually an ellipse.

For the desired active and reactive power flows ($P_{s'}*$ and $Q_{s'}*$) at the modified sending end, the required magnitude ($V_{s's}$) and the relative phase angle (β) of the series-compensating voltage ($\mathbf{V_{s's}}$) can be determined by substituting $P_{s'} = P_{s'}^*$ and $Q_{s'} = Q_{s'}^*$ in Equations (B-93) and (B-95), respectively. The relative phase angle (β) is the angle between the voltage regulator line (*i.e.* $\beta = 0°$ line) and the line that passes through (P_{sn}, Q_{sn}) and ($P_{s'}^*$, $Q_{s'}^*$).

For implementation of the VR, the relative phase angle is set to $\beta = 0°$ or $\beta = 180°$ in Equations (B-89) and (B-91). For $\beta = 0°$, the variations of $P_{s'}$ and $Q_{s'}$ are given by

$$P_{s'} = P_{sn} + \frac{RV_{s's}^2}{R^2+X^2} + \frac{1}{\sqrt{R^2+X^2}}[R\{2A_s\cos 0° - A_r\cos(\delta+0°)\} + XA_r\sin(\delta+0°)] \quad \text{or}$$

$$P_{s'} = P_{sn} + \frac{RV_{s's}^2}{R^2+X^2} + \frac{1}{\sqrt{R^2+X^2}}[R\{2A_s - A_r\cos(\delta)\} + XA_r\sin(\delta)]$$

and

$$Q_{s'} = Q_{sn} + \frac{XV_{s's}^2}{R^2+X^2} + \frac{1}{\sqrt{R^2+X^2}}[-RA_r\sin(\delta+0°) + X\{2A_s\cos 0° - A_r\cos(\delta+0°)\}] \quad \text{or}$$

$$Q_{s'} = Q_{sn} + \frac{XV_{s's}^2}{R^2+X^2} + \frac{1}{\sqrt{R^2+X^2}}[-RA_r\sin(\delta) + X\{2A_s - A_r\cos(\delta)\}].$$

For $\beta = 180°$, the variations of $P_{s'}$ and $Q_{s'}$ are given by

$$P_{s'} = P_{sn} + \frac{RV_{s's}^2}{R^2+X^2} + \frac{1}{\sqrt{R^2+X^2}}[R\{2A_s\cos 180° - A_r\cos(\delta+180°)\} + XA_r\sin(\delta+180°)] \quad \text{or}$$

$$P_{s'} = P_{sn} + \frac{RV_{s's}^2}{R^2+X^2} + \frac{1}{\sqrt{R^2+X^2}}[R\{-2A_s + A_r\cos(\delta)\} - XA_r\sin(\delta)]$$

and

$$Q_{s'} = Q_{sn} + \frac{XV_{s's}^2}{R^2+X^2} + \frac{1}{\sqrt{R^2+X^2}}[-RA_r\sin(\delta+180°) + X\{2A_s\cos 180° - A_r\cos(\delta+180°)\}] \quad \text{or}$$

$$Q_{s'} = Q_{sn} + \frac{XV_{s's}^2}{R^2+X^2} + \frac{1}{\sqrt{R^2+X^2}}[RA_r\sin(\delta) + X\{-2A_s + A_r\cos(\delta)\}].$$

For $\beta = 0°$ or $\beta = 180°$, the variations of $P_{s'}$ and $Q_{s'}$ are then given by

$$P_{s'} = P_{sn} + \frac{RV_{s's}^2}{R^2+X^2} + \frac{1}{\sqrt{R^2+X^2}}[R\{\pm 2A_s \mp A_r\cos(\delta)\} \pm XA_r\sin(\delta)] \qquad \text{(B-96a)}$$

or substituting for A_s and A_r from Equations (B-62) and (B-76)

$$(P_{s'} - P_{sn})(R^2 + X^2) = RV_{s's}^2 \pm [R(2V_s - V_r\cos\delta) + XV_r\sin\delta]V_{s's}$$

or

$$V_{s's}^2 \pm \left[(2V_s - V_r\cos\delta) + \frac{X}{R}V_r\sin\delta\right]V_{s's} - (P_{s'} - P_{sn})\frac{1}{R}(R^2 + X^2) = 0$$

or

$$V_{s's} = \frac{\mp\left[(2V_s - V_r\cos\delta) + \frac{X}{R}V_r\sin\delta\right] \pm\sqrt{\left[(2V_s - V_r\cos\delta) + \frac{X}{R}V_r\sin\delta\right]^2 + \frac{4}{R}(P_{s'} - P_{sn})(R^2 + X^2)}}{2} \quad \text{or}$$

$$V_{s's} = \frac{1}{2}\left|\left[(2V_s - V_r\cos\delta) + \frac{X}{R}V_r\sin\delta\right] - \sqrt{\left[(2V_s - V_r\cos\delta) + \frac{X}{R}V_r\sin\delta\right]^2 + \frac{4}{R}(P_{s'} - P_{sn})(R^2 + X^2)}\right| \tag{B-96b}$$

and

$$Q_{s'} = Q_{sn} + \frac{XV_{s's}^2}{R^2 + X^2} + \frac{1}{\sqrt{R^2 + X^2}}[\mp RA_r\sin(\delta) + X\{\pm 2A_s \mp A_r\cos(\delta)\}] \tag{B-96c}$$

or substituting for A_s and A_r from Equations (B-62) and (B-76)

$$(Q_{s'} - Q_{sn})(R^2 + X^2) = XV_{s's}^2 \pm [-RV_r\sin\delta + X(2V_s - V_r\cos\delta)]V_{s's}$$

or

$$V_{s's}^2 \pm \left[-\frac{R}{X}V_r\sin\delta + (2V_s - V_r\cos\delta)\right]V_{s's} - (Q_{s'} - Q_{sn})\frac{1}{X}(R^2 + X^2) = 0$$

or

$$V_{s's} = \frac{\mp\left[-\frac{R}{X}V_r\sin\delta + (2V_s - V_r\cos\delta)\right] \pm\sqrt{\left[-\frac{R}{X}V_r\sin\delta + (2V_s - V_r\cos\delta)\right]^2 + \frac{4}{X}(Q_{s'} - Q_{sn})(R^2 + X^2)}}{2} \quad \text{or}$$

$$V_{s's} = \frac{1}{2}\left|\left[-\frac{R}{X}V_r\sin\delta + (2V_s - V_r\cos\delta)\right] - \sqrt{\left[-\frac{R}{X}V_r\sin\delta + (2V_s - V_r\cos\delta)\right]^2 + \frac{4}{X}(Q_{s'} - Q_{sn})(R^2 + X^2)}\right|. \tag{B-96d}$$

For implementation of the PAR (asym), the relative phase angle is set to $\beta = 90°$ or $\beta = -90°$ in Equations (B-89) and (B-91). For $\beta = 90°$, the variations of $P_{s'}$ and $Q_{s'}$ are given by

$$P_{s'} = P_{sn} + \frac{RV_{s's}^2}{R^2 + X^2} + \frac{1}{\sqrt{R^2 + X^2}}[R\{2A_s \cos 90° - A_r\cos(\delta + 90°)\} + XA_r\sin(\delta + 90°)] \quad \text{or}$$

$$P_{s'} = P_{sn} + \frac{RV_{s's}^2}{R^2 + X^2} + \frac{1}{\sqrt{R^2 + X^2}}[R\{A_r\sin(\delta)\} + XA_r\cos(\delta)]$$

and

$$Q_{s'} = Q_{sn} + \frac{XV_{s's}^2}{R^2 + X^2} + \frac{1}{\sqrt{R^2 + X^2}}[-RA_r\sin(\delta + 90°) + X\{2A_s\cos(+90°) - A_r\cos(\delta + 90°)\}] \quad \text{or}$$

$$Q_{s'} = Q_{sn} + \frac{XV_{s's}^2}{R^2 + X^2} + \frac{1}{\sqrt{R^2 + X^2}}[-RA_r\cos(\delta) + X\{A_r\sin(\delta)\}].$$

For $\beta = -90°$, the variations of $P_{s'}$ and $Q_{s'}$ are given by

$$P_{s'} = P_{sn} + \frac{RV_{s's}^2}{R^2 + X^2} + \frac{1}{\sqrt{R^2 + X^2}}[R\{2A_s\cos(-90°) - A_r\cos(\delta - 90°)\} + XA_r\sin(\delta - 90°)] \quad \text{or}$$

$$P_{s'} = P_{sn} + \frac{RV_{s's}^2}{R^2 + X^2} + \frac{1}{\sqrt{R^2 + X^2}}[R\{-A_r\sin(\delta)\} - XA_r\cos(\delta)]$$

and

$$Q_{s'} = Q_{sn} + \frac{XV_{s's}^2}{R^2 + X^2} + \frac{1}{\sqrt{R^2 + X^2}}[-RA_r\sin(\delta - 90°) + X\{2A_s\cos(-90°) - A_r\cos(\delta - 90°)\}] \quad \text{or}$$

$$Q_{s'} = Q_{sn} + \frac{XV_{s's}^2}{R^2 + X^2} + \frac{1}{\sqrt{R^2 + X^2}}[RA_r\cos(\delta) + X\{-A_r\sin(\delta)\}].$$

For $\beta = \pm 90°$, the variations of $P_{s'}$ and $Q_{s'}$ are then given by

$$P_{s'} = P_{sn} + \frac{RV_{s's}^2}{R^2 + X^2} \pm \frac{1}{\sqrt{R^2 + X^2}}[R\{A_r\sin(\delta)\} + XA_r\cos(\delta)] \tag{B-97a}$$

or substituting for A_s and A_r from Equations (B-62) and (B-76)

$$(P_{s'} - P_{sn})(R^2 + X^2) = RV_{s's}^2 \pm [RV_r \sin\delta + XV_r \cos\delta]V_{s's}$$

or

$$V_{s's}^2 \pm \left[V_r \sin\delta + \frac{X}{R}V_r \cos\delta\right]V_{s's} - (P_{s'} - P_{sn})\frac{1}{R}(R^2 + X^2) = 0$$

or

$$V_{s's} = \frac{\mp\left[V_r \sin\delta + \frac{X}{R}V_r \cos\delta\right] \pm \sqrt{\left[V_r \sin\delta + \frac{X}{R}V_r \cos\delta\right]^2 + \frac{4}{R}(P_{s'} - P_{sn})(R^2 + X^2)}}{2} \quad \text{or}$$

$$V_{s's} = \frac{1}{2}\left|\left[V_r \sin\delta + \frac{X}{R}V_r \cos\delta\right] - \sqrt{\left[V_r \sin\delta + \frac{X}{R}V_r \cos\delta\right]^2 + \frac{4}{R}(P_{s'} - P_{sn})(R^2 + X^2)}\right|. \tag{B-97b}$$

and

$$Q_{s'} = Q_{sn} + \frac{XV_{s's}^2}{R^2 + X^2} \pm \frac{1}{\sqrt{R^2 + X^2}}[-RA_r \cos(\delta) + XA_r \sin(\delta)]. \tag{B-97c}$$

or substituting for A_s and A_r from Equations (B-62) and (B-76)

$$(Q_{s'} - Q_{sn})(R^2 + X^2) = XV_{s's}^2 \pm [-RV_r \cos\delta + XV_r \sin\delta]V_{s's}$$

or

$$V_{s's}^2 \pm \left[-\frac{R}{X}V_r \cos\delta + V_r \sin\delta\right]V_{s's} - (Q_{s'} - Q_{sn})\frac{1}{X}(R^2 + X^2) = 0$$

or

$$V_{s's} = \frac{\mp\left[-\frac{R}{X}V_r \cos\delta + V_r \sin\delta\right] \pm \sqrt{\left[-\frac{R}{X}V_r \cos\delta + V_r \sin\delta\right]^2 + \frac{4}{X}(Q_{s'} - Q_{sn})(R^2 + X^2)}}{2}$$

$$V_{s's} = \frac{1}{2}\left|\left[-\frac{R}{X}V_r \cos\delta + V_r \sin\delta\right] - \sqrt{\left[-\frac{R}{X}V_r \cos\delta + V_r \sin\delta\right]^2 + \frac{4}{X}(Q_{s'} - Q_{sn})(R^2 + X^2)}\right|. \tag{B-97d}$$

For a desired $P_{s'}$, the required $V_{s's}$ can be determined using Equation (B-97b); for a desired $Q_{s'}$, the required $V_{s's}$ can be determined using Equation (B-97d).

From Equations (B-89) and (B-91), the active and reactive power flows at the modified sending end are related as

$$\frac{Q_{s'} - Q_{sn} - \frac{XV_{s's}^2}{R^2 + X^2}}{P_{s'} - P_{sn} - \frac{RV_{s's}^2}{R^2 + X^2}} = \frac{-RA_r \sin(\delta + \beta) + X\{2A_s \cos\beta - A_r \cos(\delta + \beta)\}}{R\{2A_s \cos\beta - A_r \cos(\delta + \beta)\} + XA_r \sin(\delta + \beta)}$$

or

$$Q_{s'} = \frac{-RA_r \sin(\delta + \beta) + X\{2A_s \cos\beta - A_r \cos(\delta + \beta)\}}{R\{2A_s \cos\beta - A_r \cos(\delta + \beta)\} + XA_r \sin(\delta + \beta)}\left(P_{s'} - P_{sn} - \frac{RV_{s's}^2}{R^2 + X^2}\right) + Q_{sn} + \frac{XV_{s's}^2}{R^2 + X^2}. \tag{B-98}$$

For a VR, $\beta = 0^\circ$ or 180° represents an increase or decrease in voltage, respectively. The relationship between the active and reactive power flows at the modified sending end, derived in Equation (B-98), can be reduced to

$$Q_{s'} = \frac{\mp RA_r \sin\delta \pm X\{2A_s - A_r \cos\delta\}}{\pm R\{2A_s - A_r \cos\delta\} \pm XA_r \sin\delta}\left(P_{s'} - P_{sn} - \frac{RV_{s's}^2}{R^2 + X^2}\right) + Q_{sn} + \frac{XV_{s's}^2}{R^2 + X^2}$$

or

$$Q_{s'} = \frac{-RA_r \sin\delta + X\{2A_s - A_r \cos\delta\}}{R\{2A_s - A_r \cos\delta\} + XA_r \sin\delta}\left(P_{s'} - P_{sn} - \frac{RV_{s's}^2}{R^2 + X^2}\right) + Q_{sn} + \frac{XV_{s's}^2}{R^2 + X^2}. \tag{B-99}$$

For a PAR (asym), $\beta = 90°$ or $-90°$ represents an increase or decrease in power flow, respectively. The relationship between the active and reactive power flows at the modified sending end, derived in Equation (B-98), can be reduced to

$$Q_{s'} = \frac{\mp RA_r \cos\delta \pm XA_r \sin\delta}{\pm RA_r \sin\delta \pm XA_r \cos\delta}\left(P_{s'} - P_{sn} - \frac{RV_{s's}^2}{R^2 + X^2}\right) + Q_{sn} + \frac{XV_{s's}^2}{R^2 + X^2}$$

or

$$Q_{s'} = \frac{-RA_r \cos\delta + XA_r \sin\delta}{RA_r \sin\delta + XA_r \cos\delta}\left(P_{s'} - P_{sn} - \frac{RV_{s's}^2}{R^2 + X^2}\right) + Q_{sn} + \frac{XV_{s's}^2}{R^2 + X^2}. \quad \text{(B-100)}$$

B.2.2.4 Exchanged Power by a Series-Compensating Voltage

The series-compensating voltage is an AC voltage source that carries a current, which is at any phase angle with its terminal voltage. Therefore, the compensating voltage, essentially, emulates a four-quadrant, series-compensating impedance ($\mathbf{Z_{se}} = R_{se} - jX_{se}$) at its terminals. This virtual impedance consists of a resistance ($R_{se} = +R$ or $-R$) and a reactance ($X_{se} = X_C$ or $-X_L$) in series with the line. A $X_{se} > 0$ means the compensating reactance is capacitive and $X_{se} < 0$ means the compensating reactance is inductive.

The exchanged active and reactive powers by the series-compensating voltage with the line are defined as

$$P_{se} = P_s - P_{s'} \quad \text{(2-169a)}$$

$$Q_{se} = Q_s - Q_{s'} \quad \text{(2-179a)}$$

where

$$P_s = P_{sn} + \frac{A_s}{\sqrt{R^2 + X^2}}(R\cos\beta + X\sin\beta) \quad \text{(B-61)}$$

$$Q_s = Q_{sn} + \frac{A_s}{\sqrt{R^2 + X^2}}(-R\sin\beta + X\cos\beta) \quad \text{(B-64)}$$

$$P_{s'} = P_{sn} + \frac{RV_{s's}^2}{R^2 + X^2} + \frac{1}{\sqrt{R^2 + X^2}}[R\{2A_s\cos\beta - A_r\cos(\delta + \beta)\} + XA_r\sin(\delta + \beta)] \quad \text{(B-89)}$$

$$Q_{s'} = Q_{sn} + \frac{XV_{s's}^2}{R^2 + X^2} + \frac{1}{\sqrt{R^2 + X^2}}[-RA_r\sin(\delta + \beta) + X\{2A_s\cos\beta - A_r\cos(\delta + \beta)\}]. \quad \text{(B-91)}$$

Using Equations (B-61) and (B-89), it can be written from Equation (2-169a) that

$$P_{se} = P_s - P_{s'} = \frac{A_s}{\sqrt{R^2 + X^2}}(R\cos\beta + X\sin\beta) - \frac{RV_{s's}^2}{R^2 + X^2} - \frac{1}{\sqrt{R^2 + X^2}}[R\{2A_s\cos\beta - A_r\cos(\delta + \beta)\} + XA_r\sin(\delta + \beta)]$$

or

$$P_{se} = -\frac{RV_{s's}^2}{R^2+X^2} + \frac{1}{\sqrt{R^2+X^2}}[-R\{A_s\cos\beta - A_r\cos(\delta+\beta)\} + X\{A_s\sin\beta - A_r\sin(\delta+\beta)\}] \tag{B-101}$$

where A_s and A_r are given in Equations (B-62) and (B-76), respectively.

Using Equations (B-64) and (B-91), it can be written from Equation (2-179a) that

$$Q_{se} = Q_s - Q_{s'} = \frac{A_s}{\sqrt{R^2+X^2}}(-R\sin\beta + X\cos\beta) - \frac{XV_{s's}^2}{R^2+X^2}$$
$$-\frac{1}{\sqrt{R^2+X^2}}[-RA_r\sin(\delta+\beta) + X\{2A_s\cos\beta - A_r\cos(\delta+\beta)\}]$$

or

$$Q_{se} = -\frac{XV_{s's}^2}{R^2+X^2} + \frac{1}{\sqrt{R^2+X^2}}[-R\{A_s\sin\beta - A_r\sin(\delta+\beta)\} - X\{A_s\cos\beta - A_r\cos(\delta+\beta)\}] \tag{B-102}$$

where A_s and A_r are given in Equations (B-62) and (B-76), respectively.

Equations (B-101) and (B-102) can be combined as

$$\left(P_{se} + \frac{RV_{s's}^2}{R^2+X^2}\right)^2 + \left(Q_{se} + \frac{XV_{s's}^2}{R^2+X^2}\right)^2 = \frac{1}{R^2+X^2} \times$$
$$\begin{pmatrix} [-R\{A_s\cos\beta - A_r\cos(\delta+\beta)\} + X\{A_s\sin\beta - A_r\sin(\delta+\beta)\}]^2 \\ + [-R\{A_s\sin\beta - A_r\sin(\delta+\beta)\} - X\{A_s\cos\beta - A_r\cos(\delta+\beta)\}]^2 \end{pmatrix}$$

or

$$\left(P_{se} + \frac{RV_{s's}^2}{R^2+X^2}\right)^2 + \left(Q_{se} + \frac{XV_{s's}^2}{R^2+X^2}\right)^2 = \frac{1}{R^2+X^2}(R^2+X^2)\begin{bmatrix} \{A_s\cos\beta - A_r\cos(\delta+\beta)\}^2 \\ + \{A_s\sin\beta - A_r\sin(\delta+\beta)\}^2 \end{bmatrix}$$

or

$$\left(P_{se} + \frac{RV_{s's}^2}{R^2+X^2}\right)^2 + \left(Q_{se} + \frac{XV_{s's}^2}{R^2+X^2}\right)^2 = [A_s^2 + A_r^2 - 2A_sA_r\cos\delta]$$

or substituting for A_s and A_r from Equations (B-62) and (B-76)

$$\left(P_{se} + \frac{RV_{s's}^2}{R^2+X^2}\right)^2 + \left(Q_{se} + \frac{XV_{s's}^2}{R^2+X^2}\right)^2 = \frac{V_{s's}^2}{R^2+X^2}[V_s^2 + V_r^2 - 2V_sV_r\cos\delta]$$

or with a substitution from Equations (B-4)

$$\left(P_{se} + \frac{RV_{s's}^2}{R^2+X^2}\right)^2 + \left(Q_{se} + \frac{XV_{s's}^2}{R^2+X^2}\right)^2 = \frac{V_{s's}^2}{R^2+X^2}[V_{Rn,Xn}^2]$$

or

$$\left(P_{se} + \frac{RV_{s's}^2}{R^2+X^2}\right)^2 + \left(Q_{se} + \frac{XV_{s's}^2}{R^2+X^2}\right)^2 = A_{se}^2 \tag{B-103}$$

where

$$A_{se} = V_{s's}\frac{V_{Rn,Xn}}{\sqrt{R^2 + X^2}} \tag{B-104}$$

with a substitution from Equation (B-8) and

$$\tan\beta = \frac{-\left(P_{se} + \frac{RV_{s's}^2}{R^2 + X^2}\right)[RA_r \sin\delta - X\{A_s - A_r\cos\delta\}] + \left(Q_{se} + \frac{XV_{s's}^2}{R^2 + X^2}\right)[-R\{A_s - A_r\cos\delta\} - XA_r\sin\delta]}{\left(P_{se} + \frac{RV_{s's}^2}{R^2 + X^2}\right)[-R\{A_s - A_r\cos\delta\} - XA_r\sin\delta] - \left(Q_{se} + \frac{XV_{s's}^2}{R^2 + X^2}\right)[-RA_r\sin\delta + X\{A_s - A_r\cos\delta\}]}. \tag{B-105}$$

Digress:
From Equations (B-101) and (B-102), the exchanged active and reactive powers by the series-compensating voltage are related as

$$\frac{P_{se} + \frac{RV_{s's}^2}{R^2 + X^2}}{Q_{se} + \frac{XV_{s's}^2}{R^2 + X^2}} = \frac{-R\{A_s\cos\beta - A_r\cos(\delta + \beta)\} + X\{A_s\sin\beta - A_r\sin(\delta + \beta)\}}{-R\{A_s\sin\beta - A_r\sin(\delta + \beta)\} - X\{A_s\cos\beta - A_r\cos(\delta + \beta)\}}$$

or

$$\left(P_{se} + \frac{RV_{s's}^2}{R^2 + X^2}\right)[-R\{A_s\sin\beta - A_r\sin(\delta + \beta)\} - X\{A_s\cos\beta - A_r\cos(\delta + \beta)\}]$$

$$= \left(Q_{se} + \frac{XV_{s's}^2}{R^2 + X^2}\right)[-R\{A_s\cos\beta - A_r\cos(\delta + \beta)\} + X\{A_s\sin\beta - A_r\sin(\delta + \beta)\}].$$

$$\left(P_{se} + \frac{RV_{s's}^2}{R^2 + X^2}\right)[-R\{A_s\sin\beta - A_r\sin\delta\cos\beta - A_r\cos\delta\sin\beta\}$$
$$-X\{A_s\cos\beta - A_r\cos\delta\cos\beta + A_r\sin\delta\sin\beta\}]$$
$$= \left(Q_{se} + \frac{XV_{s's}^2}{R^2 + X^2}\right)[-R\{A_s\cos\beta - A_r\cos\delta\cos\beta + A_r\sin\delta\sin\beta\}$$
$$+X\{A_s\sin\beta - A_r\sin\delta\cos\beta - A_r\cos\delta\sin\beta\}].$$

$$\left(P_{se} + \frac{RV_{s's}^2}{R^2 + X^2}\right)[-R\{A_s - A_r\cos\delta\}\sin\beta + RA_r\sin\delta\cos\beta$$
$$-X\{A_s - A_r\cos\delta\}\cos\beta - XA_r\sin\delta\sin\beta]$$
$$= \left(Q_{se} + \frac{XV_{s's}^2}{R^2 + X^2}\right)[-R\{A_s - A_r\cos\delta\}\cos\beta - RA_r\sin\delta\sin\beta$$
$$+X\{A_s - A_r\cos\delta\}\sin\beta - XA_r\sin\delta\cos\beta].$$

$$\tan\beta = \frac{\begin{aligned}&\left(P_{se} + \frac{RV_{s's}^2}{R^2+X^2}\right)[-R\{A_s - A_r\cos\delta\} - XA_r\sin\delta]\sin\beta \\ &- \left(Q_{se} + \frac{XV_{s's}^2}{R^2+X^2}\right)[-RA_r\sin\delta + X\{A_s - A_r\cos\delta\}]\sin\beta \\ &= -\left(P_{se} + \frac{RV_{s's}^2}{R^2+X^2}\right)[RA_r\sin\delta - X\{A_s - A_r\cos\delta\}]\cos\beta \\ &+ \left(Q_{se} + \frac{XV_{s's}^2}{R^2+X^2}\right)[-R\{A_s - A_r\cos\delta\} - XA_r\sin\delta]\cos\beta \\ &-\left(P_{se} + \frac{RV_{s's}^2}{R^2+X^2}\right)[RA_r\sin\delta - X\{A_s - A_r\cos\delta\}] \\ &+\left(Q_{se} + \frac{XV_{s's}^2}{R^2+X^2}\right)[-R\{A_s - A_r\cos\delta\} - XA_r\sin\delta]\end{aligned}}{\begin{aligned}&\left(P_{se} + \frac{RV_{s's}^2}{R^2+X^2}\right)[-R\{A_s - A_r\cos\delta\} - XA_r\sin\delta] \\ &-\left(Q_{se} + \frac{XV_{s's}^2}{R^2+X^2}\right)[-RA_r\sin\delta + X\{A_s - A_r\cos\delta\}]\end{aligned}}.$$

Recall that for $R = 0$,

$$P_{se}^2 + \left(Q_{se} + \frac{V_{s's}^2}{X}\right)^2 = a_{se}^2. \tag{2-183}$$

The Equation (B-103) represents the locus of a circle, centered at $\left[-\frac{RV_{s's}^2}{R^2+X^2}, -\frac{XV_{s's}^2}{R^2+X^2}\right]$ with a radius of A_{se}, which is defined in Equation (B-104). The Equation (2-183) represents the locus of a circle, centered at $\left[0, -\frac{V_{s's}^2}{X}\right]$ with a radius of a_{se}, which is defined in Equation (2-178).

Using Equation (B-104), Equation (B-103) can be rearranged as

$$P_{se}^2 + 2P_{se}\frac{RV_{s's}^2}{R^2+X^2} + \left(\frac{RV_{s's}^2}{R^2+X^2}\right)^2 + Q_{se}^2 + 2Q_{se}\frac{XV_{s's}^2}{R^2+X^2} + \left(\frac{XV_{s's}^2}{R^2+X^2}\right)^2 - \left(\frac{V_{s's}V_{Rn,Xn}}{\sqrt{R^2+X^2}}\right)^2 = 0$$

or

$$\left(V_{s's}^2\right)^2 - \left(V_{Rn,Xn}^2 - 2RP_{se} - 2XQ_{se}\right)V_{s's}^2 + \left(R^2+X^2\right)\left(P_{se}^2 + Q_{se}^2\right) = 0. \tag{B-106}$$

Solving the above quadratic equation results in

$$V_{s's}^2 = \frac{\left(V_{Rn,Xn}^2 - 2RP_{se} - 2XQ_{se}\right) \pm \sqrt{\left(V_{Rn,Xn}^2 - 2RP_{se} - 2XQ_{se}\right)^2 - 4\left(R^2+X^2\right)\left(P_{se}^2 + Q_{se}^2\right)}}{2}$$

or

$$V_{s's}^2 = \left(\frac{V_{Rn,Xn}^2}{2} - RP_{se} - XQ_{se}\right) \pm \sqrt{\left(\frac{V_{Rn,Xn}^2}{2} - RP_{se} - XQ_{se}\right)^2 - \left(R^2+X^2\right)\left(P_{se}^2 + Q_{se}^2\right)}$$

or

$$V_{s's} = +\sqrt{\left(\frac{V_{Rn,Xn}^2}{2} - RP_{se} - XQ_{se}\right) + \sqrt{\left(\frac{V_{Rn,Xn}^2}{2} - RP_{se} - XQ_{se}\right)^2 - (R^2 + X^2)(P_{se}^2 + Q_{se}^2)}} \quad \text{(B-107a)}$$

$$V_{s's} = +\sqrt{\left(\frac{V_{Rn,Xn}^2}{2} - RP_{se} - XQ_{se}\right) - \sqrt{\left(\frac{V_{Rn,Xn}^2}{2} - RP_{se} - XQ_{se}\right)^2 - (R^2 + X^2)(P_{se}^2 + Q_{se}^2)}} \quad \text{(B-107b)}$$

$$V_{s's} = -\sqrt{\left(\frac{V_{Rn,Xn}^2}{2} - RP_{se} - XQ_{se}\right) + \sqrt{\left(\frac{V_{Rn,Xn}^2}{2} - RP_{se} - XQ_{se}\right)^2 - (R^2 + X^2)(P_{se}^2 + Q_{se}^2)}} \quad \text{(B-107c)}$$

$$V_{s's} = -\sqrt{\left(\frac{V_{Rn,Xn}^2}{2} - RP_{se} - XQ_{se}\right) - \sqrt{\left(\frac{V_{Rn,Xn}^2}{2} - RP_{se} - XQ_{se}\right)^2 - (R^2 + X^2)(P_{se}^2 + Q_{se}^2)}}. \quad \text{(B-107d)}$$

Out of the four possible values of $V_{s's}$, given by Equations (B-107a through B-107d), the most desired value is the one that provides the lowest *APR*, thus the lowest cost solution.

The relative phase angle (β) of the series-compensating voltage ($V_{s's}$) is derived from Equation (B-105) as

$$\beta = \tan^{-1}\frac{-\left(P_{se} + \frac{RV_{s's}^2}{R^2 + X^2}\right)[RA_r \sin\delta - X\{A_s - A_r\cos\delta\}] + \left(Q_{se} + \frac{XV_{s's}^2}{R^2 + X^2}\right)[-R\{A_s - A_r\cos\delta\} - XA_r\sin\delta]}{\left(P_{se} + \frac{RV_{s's}^2}{R^2 + X^2}\right)[-R\{A_s - A_r\cos\delta\} - XA_r\sin\delta] - \left(Q_{se} + \frac{XV_{s's}^2}{R^2 + X^2}\right)[-RA_r\sin\delta + X\{A_s - A_r\cos\delta\}]}. \quad \text{(B-108)}$$

For the desired exchanged active and reactive powers (P_{se}^*, Q_{se}^*) within the allowable range, the required magnitude ($V_{s's}$) and the relative phase angle (β) of the series-compensating voltage ($\mathbf{V_{s's}}$) can be determined by substituting $P_{se} = P_{se}^*$ and $Q_{se} = Q_{se}^*$ in Equations (B-107a through B-107d) and (B-108), respectively. The relative phase angle (β) is the angle between the voltage regulator line (*i.e.* $\beta = 0°$ line) and the line that passes through (P_{rn}, Q_{rn}) and (P_{se}^*, Q_{se}^*). For $R = 0$, refer to Equations (2-186a through 2-186d) and (2-187a or 2-187b).

For implementation of the VR, the relative phase angle is set to $\beta = 0°$ or $\beta = 180°$ in Equations (B-101) and (B-102). For $\beta = 0°$, the variations of P_{se} and Q_{se} are given by

$$P_{se} = -\frac{RV_{s's}^2}{R^2 + X^2} + \frac{1}{\sqrt{R^2 + X^2}}[-R\{A_s\cos 0° - A_r\cos(\delta + 0°)\} + X\{A_s\sin 0° - A_r\sin(\delta + 0°)\}] \quad \text{or}$$

$$P_{se} = -\frac{RV_{s's}^2}{R^2 + X^2} + \frac{1}{\sqrt{R^2 + X^2}}[-R\{A_s - A_r\cos(\delta)\} + X\{-A_r\sin(\delta)\}]$$

and

$$Q_{se} = -\frac{XV_{s's}^2}{R^2+X^2} + \frac{1}{\sqrt{R^2+X^2}}[-R\{A_s \sin 0° - A_r \sin(\delta + 0°)\}$$
$$-X\{A_s \cos 0° - A_r \cos(\delta + 0°)\}] \quad \text{or}$$

$$Q_{se} = -\frac{XV_{s's}^2}{R^2+X^2} + \frac{1}{\sqrt{R^2+X^2}}[-R\{-A_r \sin(\delta)\} - X\{A_s - A_r \cos(\delta)\}].$$

For $\beta = 180°$, the variations of P_{se} and Q_{se} are given by

$$P_{se} = -\frac{RV_{s's}^2}{R^2+X^2} + \frac{1}{\sqrt{R^2+X^2}}[-R\{A_s \cos 180° - A_r \cos(\delta + 180°)\}$$
$$+X\{A_s \sin 180° - A_r \sin(\delta + 180°)\}] \quad \text{or}$$

$$P_{se} = -\frac{RV_{s's}^2}{R^2+X^2} + \frac{1}{\sqrt{R^2+X^2}}[-R\{-A_s + A_r \cos(\delta)\} + X\{+A_r \sin(\delta)\}]$$

and

$$Q_{se} = -\frac{XV_{s's}^2}{R^2+X^2} + \frac{1}{\sqrt{R^2+X^2}}[-R\{A_s \sin 180° - A_r \sin(\delta + 180°)\}$$
$$-X\{A_s \cos 180° - A_r \cos(\delta + 180°)\}] \quad \text{or}$$

$$Q_{se} = -\frac{XV_{s's}^2}{R^2+X^2} + \frac{1}{\sqrt{R^2+X^2}}[-R\{+A_r \sin(\delta)\} - X\{-A_s + A_r \cos(\delta)\}].$$

For $\beta = 0°$ or $\beta = 180°$, the variations of P_{se} and Q_{se} are then given by

$$P_{se} = -\frac{RV_{s's}^2}{R^2+X^2} + \frac{1}{\sqrt{R^2+X^2}}[-R\{\pm A_s \mp A_r \cos(\delta)\} + X\{\mp A_r \sin(\delta)\}] \tag{B-109a}$$

or substituting for A_s and A_r from Equations (B-62) and (B-76)

$$P_{se}(R^2+X^2) = -RV_{s's}^2 + [\mp R(V_s - V_r \cos\delta) \mp XV_r \sin\delta]V_{s's}$$

or

$$V_{s's}^2 \pm \left[V_s - V_r \cos\delta + \frac{X}{R}V_r \sin\delta\right]V_{s's} + P_{se}\frac{1}{R}(R^2+X^2) = 0$$

or

$$V_{s's} = \frac{\mp\left[V_s - V_r\cos\delta + \frac{X}{R}V_r\sin\delta\right] \pm \sqrt{\left[V_s - V_r\cos\delta + \frac{X}{R}V_r\sin\delta\right]^2 - \frac{4}{R}P_{se}(R^2+X^2)}}{2} \tag{B-109b}$$

and

$$Q_{se} = -\frac{XV_{s's}^2}{R^2+X^2} + \frac{1}{\sqrt{R^2+X^2}}[-R\{\mp A_r \sin(\delta)\} - X\{\pm A_s \mp A_r \cos(\delta)\}] \tag{B-109c}$$

or substituting for A_s and A_r from Equations (B-62) and (B-76)

$$Q_{se}(R^2+X^2) = -XV_{s's}^2 \mp [-RV_r \sin\delta + X(V_s - V_r \cos\delta)]V_{s's}$$

or

$$V_{s's}^2 \pm \left[-\frac{R}{X}V_r \sin\delta + V_s - V_r \cos\delta\right]V_{s's} + Q_{se}\frac{1}{X}\left(R^2 + X^2\right) = 0$$

or

$$V_{s's} = \frac{\mp\left[-\frac{R}{X}V_r \sin\delta + V_s - V_r \cos\delta\right] \pm \sqrt{\left[-\frac{R}{X}V_r \sin\delta + V_s - V_r \cos\delta\right]^2 - \frac{4}{X}Q_{se}(R^2 + X^2)}}{2}$$

or

$$V_{s's} = \frac{1}{2}\left|-\frac{R}{X}V_r \sin\delta + V_s - V_r \cos\delta \mp \sqrt{\left[-\frac{R}{X}V_r \sin\delta + V_s - V_r \cos\delta\right]^2 - \frac{4}{X}Q_{se}(R^2 + X^2)}\right|. \tag{B-109d}$$

For $R = 0$, it can be written from Equation (B-109b) that

$$V_{s's} = \mp X\frac{P_{se}}{V_r \sin\delta} \quad \text{or} \quad V_{s's} = \frac{X}{V_r \sin\delta}\left|P_{se}\right|. \tag{B-109e}$$

For $R = 0$, it can be written from Equation (B-109d) that

$$V_{s's} = \frac{\mp(V_s - V_r \cos\delta) \pm \sqrt{(V_s - V_r \cos\delta)^2 - 4XQ_{se}}}{2} \quad \text{or}$$

$$V_{s's} = \left|\frac{V_s - V_r \cos\delta \mp \sqrt{(V_s - V_r \cos\delta)^2 - 4XQ_{se}}}{2}\right|. \tag{B-109f}$$

For implementation of the PAR (asym), the relative phase angle is set to $\beta = 90°$ or $\beta = -90°$ in Equations (B-101) and (B-102). For $\beta = 90°$, the variations of P_{se} and Q_{se} are given by

$$P_{se} = -\frac{RV_{s's}^2}{R^2 + X^2} + \frac{1}{\sqrt{R^2 + X^2}}[-R\{A_s \cos 90° - A_r \cos(\delta + 90°)\} + X\{A_s \sin 90° - A_r \sin(\delta + 90°)\}] \quad \text{or}$$

$$P_{se} = -\frac{RV_{s's}^2}{R^2 + X^2} + \frac{1}{\sqrt{R^2 + X^2}}[-R\{A_r \sin(\delta)\} + X\{A_s - A_r \cos(\delta)\}]$$

and

$$Q_{se} = -\frac{XV_{s's}^2}{R^2 + X^2} + \frac{1}{\sqrt{R^2 + X^2}}[-R\{A_s \sin 90° - A_r \sin(\delta + 90°)\} - X\{A_s \cos 90° - A_r \cos(\delta + 90°)\}] \quad \text{or}$$

$$Q_{se} = -\frac{XV_{s's}^2}{R^2 + X^2} + \frac{1}{\sqrt{R^2 + X^2}}[-R\{A_s - A_r \cos(\delta)\} - X\{A_r \sin(\delta)\}].$$

For $\beta = -90°$, the variations of P_{se} and Q_{se} are given by

$$P_{se} = -\frac{RV_{s's}^2}{R^2 + X^2} + \frac{1}{\sqrt{R^2 + X^2}}[-R\{A_s \cos(-90°) - A_r \cos(\delta - 90°)\} + X\{A_s \sin(-90°) - A_r \sin(\delta - 90°)\}] \quad \text{or}$$

$$P_{se} = -\frac{RV_{s's}^2}{R^2 + X^2} + \frac{1}{\sqrt{R^2 + X^2}}[-R\{-A_r \sin(\delta)\} + X\{-A_s + A_r \cos(\delta)\}]$$

and

$$Q_{se} = -\frac{XV_{s's}^2}{R^2 + X^2} + \frac{1}{\sqrt{R^2 + X^2}}[-R\{A_s\sin(-90°) - A_r\sin(\delta - 90°)\}$$

$$-X\{A_s\cos(-90°) - A_r\cos(\delta - 90°)\}] \quad \text{or}$$

$$Q_{se} = -\frac{XV_{s's}^2}{R^2 + X^2} + \frac{1}{\sqrt{R^2 + X^2}}[-R\{-A_s + A_r\cos(\delta)\} - X\{-A_r\sin(\delta)\}].$$

For $\beta = \pm 90°$, the variations of P_{se} and Q_{se} are then given by

$$P_{se} = -\frac{RV_{s's}^2}{R^2 + X^2} + \frac{1}{\sqrt{R^2 + X^2}}[-R\{\pm A_r\sin(\delta)\} + X\{\pm A_s \mp A_r\cos(\delta)\}] \tag{B-110a}$$

or substituting for A_s and A_r from Equations (B-62) and (B-76)

$$P_{se}(R^2 + X^2) = -RV_{s's}^2 \mp [RV_r\sin\delta + X(V_s - V_r\cos\delta)]V_{s's}$$

or

$$V_{s's}^2 \pm \left[V_r\sin\delta + \frac{X}{R}(V_s - V_r\cos\delta)\right]V_{s's} + P_{se}\frac{1}{R}(R^2 + X^2) = 0$$

or

$$V_{s's} = \frac{\mp\left[V_r\sin\delta + \frac{X}{R}(V_s - V_r\cos\delta)\right] \pm \sqrt{\left[V_r\sin\delta + \frac{X}{R}(V_s - V_r\cos\delta)\right]^2 - \frac{4}{R}P_{se}(R^2 + X^2)}}{2}$$

or

$$V_{s's} = \left|\frac{\left[V_r\sin\delta + \frac{X}{R}(V_s - V_r\cos\delta)\right] - \sqrt{\left[V_r\sin\delta + \frac{X}{R}(V_s - V_r\cos\delta)\right]^2 - \frac{4}{R}P_{se}(R^2 + X^2)}}{2}\right| \tag{B-110b}$$

and

$$Q_{se} = -\frac{XV_{s's}^2}{R^2 + X^2} + \frac{1}{\sqrt{R^2 + X^2}}[-R\{\pm A_s \mp A_r\cos(\delta)\} - X\{\pm A_r\sin(\delta)\}] \tag{B-110c}$$

or substituting for A_s and A_r from Equations (B-62) and (B-76)

$$Q_{se}(R^2 + X^2) = -XV_{s's}^2 \mp [R(V_s - V_r\cos\delta) + XV_r\sin\delta]V_{s's}$$

or

$$V_{s's}^2 \pm \left[\frac{R}{X}(V_s - V_r\cos\delta) + V_r\sin\delta\right]V_{s's} + Q_{se}\frac{1}{X}(R^2 + X^2) = 0$$

or

$$V_{s's} = \frac{\mp\left[\frac{R}{X}(V_s - V_r\cos\delta) + V_r\sin\delta\right] \pm \sqrt{\left[\frac{R}{X}(V_s - V_r\cos\delta) + V_r\sin\delta\right]^2 - \frac{4}{X}Q_{se}(R^2 + X^2)}}{2}$$

or

$$V_{s's} = \left|\frac{\left[\frac{R}{X}(V_s - V_r\cos\delta) + V_r\sin\delta\right] - \sqrt{\left[\frac{R}{X}(V_s - V_r\cos\delta) + V_r\sin\delta\right]^2 - \frac{4}{X}Q_{se}(R^2 + X^2)}}{2}\right|. \tag{B-110d}$$

For $R = 0$, it can be written from Equation (B-110b) that

$$V_{s's} = \mp X\frac{P_{se}}{V_s - V_r\cos\delta} \quad \text{or} \quad V_{s's} = \frac{X}{V_s - V_r\cos\delta}|P_{se}|. \tag{B-110e}$$

For $R = 0$, it can be written from Equation (B-110d) that

$$V_{s's} = \frac{\mp V_r\sin\delta \pm \sqrt{(V_r\sin\delta)^2 - 4XQ_{se}}}{2} \quad \text{or} \quad V_{s's} = \left|\frac{V_r\sin\delta - \sqrt{(V_r\sin\delta)^2 - 4XQ_{se}}}{2}\right|. \tag{B-110f}$$

From Equations (B-62), (B-76), (B-101), and (B-102), the exchanged active and reactive powers by the compensating voltage are related as

$$\frac{Q_{se} + \frac{XV_{s's}^2}{R^2 + X^2}}{P_{se} + \frac{RV_{s's}^2}{R^2 + X^2}} = \frac{-R\{V_s\sin\beta - V_r\sin(\delta+\beta)\} - X\{V_s\cos\beta - V_r\cos(\delta+\beta)\}}{-R\{V_s\cos\beta - V_r\cos(\delta+\beta)\} + X\{V_s\sin\beta - V_r\sin(\delta+\beta)\}}$$

or

$$Q_{se} = \frac{R\{V_s\sin\beta - V_r\sin(\delta+\beta)\} + X\{V_s\cos\beta - V_r\cos(\delta+\beta)\}}{R\{V_s\cos\beta - V_r\cos(\delta+\beta)\} - X\{V_s\sin\beta - V_r\sin(\delta+\beta)\}}\left(P_{se} + \frac{RV_{s's}^2}{R^2 + X^2}\right) - \frac{XV_{s's}^2}{R^2 + X^2}. \tag{B-111}$$

For a VR, $\beta = 0°$ or $180°$ represents an increase or decrease in voltage, respectively. The relationship between the exchanged active and reactive powers by the compensating voltage, derived in Equation (B-111), can be reduced to

$$Q_{se} = \frac{R\{\mp V_r\sin\delta\} + X\{\pm V_s \mp V_r\cos\delta\}}{R\{\pm V_s \mp V_r\cos\delta\} - X\{\mp V_r\sin\delta\}}\left(P_{se} + \frac{RV_{s's}^2}{R^2 + X^2}\right) - \frac{XV_{s's}^2}{R^2 + X^2}$$

or

$$Q_{se} = \frac{-RV_r\sin\delta + X\{V_s - V_r\cos\delta\}}{R\{V_s - V_r\cos\delta\} + XV_r\sin\delta}\left(P_{se} + \frac{RV_{s's}^2}{R^2 + X^2}\right) - \frac{XV_{s's}^2}{R^2 + X^2}. \tag{B-112}$$

For a PAR (asym), $\beta = 90°$ or $-90°$ represents an increase or decrease in power flow, respectively. The relationship between the exchanged active and reactive powers by the compensating voltage, derived in Equation (B-111), can be reduced to

$$Q_{se} = \frac{R\{\pm V_s \mp V_r \cos\delta\} + X\{\pm V_r \sin\delta\}}{R\{\pm V_r \sin\delta\} - X\{\pm V_s \mp V_r \cos\delta\}}\left(P_{se} + \frac{RV_{s's}^2}{R^2 + X^2}\right) - \frac{XV_{s's}^2}{R^2 + X^2}$$

or

$$Q_{se} = \frac{R\{V_s - V_r \cos\delta\} + XV_r \sin\delta}{RV_r \sin\delta - X\{V_s - V_r \cos\delta\}}\left(P_{se} + \frac{RV_{s's}^2}{R^2 + X^2}\right) - \frac{XV_{s's}^2}{R^2 + X^2}. \tag{B-113}$$

B.2.2.5 Additional Series-Compensating Voltages

A series-compensating voltage can be used to implement (1) Phase Angle Regulator (symmetric) and (2) Reactance Regulator. The expressions for the magnitude and phase angle for the line current (**I**) are derived in Equations (B-57b) and (B-59a) or (B-59b), respectively, where the relative phase angle (β) is given by

Equation (2-200): $\beta = \frac{\psi}{2} + 90°$ for $\psi > 0$ or $\beta = \frac{\psi}{2} + 270°$ for $\psi < 0$ for a PAR (sym) and

Equation (B-82): $\beta = \tan^{-1}\frac{-R(Q_r - Q_{rn}) + X(P_r - P_{rn})}{R(P_r - P_{rn}) + X(Q_r - Q_{rn})} - \delta$ for a RR.

B.2.2.5.1 Phase Angle Regulator (Symmetric)

A PAR (sym) connects a compensating voltage ($\mathbf{V_{s's}}$) in series with the line to regulate the phase-shift angle (ψ) of the line voltage ($\mathbf{V_s}$) at the sending end. The magnitude ($V_{s'}$) of the modified sending-end voltage ($\mathbf{V_{s'}}$) and that (V_s) of the sending-end voltage ($\mathbf{V_s}$) are the same, *i.e.* $V_{s'} = V_s$. The magnitude ($V_{s's}$) of the series-compensating voltage ($\mathbf{V_{s's}}$) for a phase-shift angle (ψ) is defined as

$$V_{s's} = 2V_s \sin\left(\frac{\psi}{2}\right). \tag{2-198}$$

The relative phase angle is defined as

$$\beta = \frac{\psi}{2} + 90° \text{ for } \psi > 0 \quad \text{or} \quad \beta = \frac{\psi}{2} + 270° \text{ for } \psi < 0. \tag{2-200}$$

Rewriting Equation (B-12), the active power flow at the modified sending end of the line is given by

$$P_{s'} = \frac{A}{\sqrt{R^2 + X^2}}\left[R\left\{\frac{V_s}{V_r} - \cos\delta'\right\} + X\sin\delta'\right] \tag{B-114}$$

where

$$A = \frac{V_s V_r}{\sqrt{R^2 + X^2}} \tag{B-13}$$

and the modified power angle (δ') is given by

$$\delta' = \delta + \psi. \tag{2-51a}$$

Rewriting Equation (B-14), the reactive power flow at the modified sending end of the line is given by

$$Q_{s'} = \frac{A}{\sqrt{R^2 + X^2}}\left[-R\sin\delta' + X\left\{\frac{V_s}{V_r} - \cos\delta'\right\}\right]. \tag{B-115}$$

Using Equation (B-13), Equations (B-114) and (B 115) can be combined as

$$\left(P_{s'} - \frac{RV_s^2}{R^2 + X^2}\right)^2 + \left(Q_{s'} - \frac{XV_s^2}{R^2 + X^2}\right)^2$$
$$= \frac{A^2}{R^2 + X^2}\left\{(-R\cos\delta' + X\sin\delta')^2 + (-R\sin\delta' - X\cos\delta')^2\right\}$$

or

$$\left(P_{s'} - \frac{RV_s^2}{R^2 + X^2}\right)^2 + \left(Q_{s'} - \frac{XV_s^2}{R^2 + X^2}\right)^2$$
$$= \frac{A^2}{R^2 + X^2}\left\{R^2(\cos^2\delta' + \sin^2\delta') + X^2(\cos^2\delta' + \sin^2\delta')\right\} = A^2$$

using the identity $(\cos^2\delta' + \sin^2\delta' = 1)$, or

$$\left(P_{s'} - \frac{RV_s^2}{R^2 + X^2}\right)^2 + \left(Q_{s'} - \frac{XV_s^2}{R^2 + X^2}\right)^2 = A^2. \tag{B-116}$$

Recall that for $R = 0$,

$$P_{s'}^2 + \left(Q_{s'} - \frac{V_s^2}{X}\right)^2 = a^2. \tag{2-205}$$

The Equation (B-116) represents the locus of a circle, centered at $\left[\frac{RV_s^2}{R^2 + X^2}, \frac{XV_s^2}{R^2 + X^2}\right]$ with a radius of A, which is defined in Equation (B-13). The Equation (2-205) represents the locus of a circle, centered at $\left[0, \frac{V_s^2}{X}\right]$ with a radius of a, which is defined in Equation (2-41).

Rewriting Equation (B-21), the active power flow at the receiving end of the line is given by

$$P_r = \frac{A}{\sqrt{R^2 + X^2}}\left\{R\left(\cos\delta' - \frac{V_r}{V_s}\right) + X\sin\delta'\right\} \tag{B-117}$$

since $V_{s'} = V_s$.

Rewriting Equation (B-22), the reactive power flow at the receiving end of the line is given by

$$Q_r = \frac{A}{\sqrt{R^2 + X^2}}\left\{-R\sin\delta' + X\left(\cos\delta' - \frac{V_r}{V_s}\right)\right\} \tag{B-118}$$

since $V_{s'} = V_s$.

Using Equation (B-13), Equations (B-117) and (B-118) can be combined as

$$\left(P_r + \frac{RV_r^2}{R^2 + X^2}\right)^2 + \left(Q_r + \frac{XV_r^2}{R^2 + X^2}\right)^2$$
$$= \frac{A^2}{R^2 + X^2}\left\{(R\cos\delta' + X\sin\delta')^2 + (-R\sin\delta' + X\cos\delta')^2\right\}$$

or

$$\left(P_r + \frac{RV_r^2}{R^2 + X^2}\right)^2 + \left(Q_r + \frac{XV_r^2}{R^2 + X^2}\right)^2$$
$$= \frac{A^2}{R^2 + X^2}\left\{R^2\left(\cos^2\delta' + \sin^2\delta'\right) + X^2\left(\cos^2\delta' + \sin^2\delta'\right)\right\} = A^2$$

using the identity $\left(\cos^2\delta' + \sin^2\delta' = 1\right)$, or

$$\left(P_r + \frac{RV_r^2}{R^2 + X^2}\right)^2 + \left(Q_r + \frac{XV_r^2}{R^2 + X^2}\right)^2 = A^2 \tag{B-119}$$

and

$$\tan\delta' = \frac{R\left(Q_r + \frac{XV_r^2}{R^2 + X^2}\right) - X\left(P_r + \frac{RV_r^2}{R^2 + X^2}\right)}{-R\left(P_r + \frac{RV_r^2}{R^2 + X^2}\right) - X\left(Q_r + \frac{XV_r^2}{R^2 + X^2}\right)}. \tag{B-46}$$

Recall that for $R = 0$,

$$P_r^2 + \left(Q_r + \frac{V_r^2}{X}\right)^2 = a^2. \tag{2-206}$$

The Equation (B-119) represents the locus of a circle, centered at $\left[\frac{-RV_r^2}{R^2 + X^2}, \frac{-XV_r^2}{R^2 + X^2}\right]$ with a radius of A, which is defined in Equation (B-13). The Equation (2-206) represents the locus of a circle, centered at $\left[0, -\frac{V_r^2}{X}\right]$ with a radius of a, which is defined in Equation (2-41).

Rearranging Equation (B-119), it can be written that

$$\left(P_r + \frac{RV_r^2}{R^2 + X^2}\right)^2 + \left(Q_r + \frac{XV_r^2}{R^2 + X^2}\right)^2 = A^2 \quad \text{or}$$

$$\left(Q_r + \frac{XV_r^2}{R^2 + X^2}\right)^2 = A^2 - \left(P_r + \frac{RV_r^2}{R^2 + X^2}\right)^2 \quad \text{or}$$

$$Q_r + \frac{XV_r^2}{R^2 + X^2} = \sqrt{A^2 - \left(P_r + \frac{RV_r^2}{R^2 + X^2}\right)^2}. \tag{B-120}$$

Rearranging Equation (B-46), it can be written that

$$\tan\delta' = \frac{R\left(Q_r + \frac{XV_r^2}{R^2 + X^2}\right) - X\left(P_r + \frac{RV_r^2}{R^2 + X^2}\right)}{-R\left(P_r + \frac{RV_r^2}{R^2 + X^2}\right) - X\left(Q_r + \frac{XV_r^2}{R^2 + X^2}\right)} \quad \text{or}$$

$$-R\left(P_r + \frac{RV_r^2}{R^2 + X^2}\right)\tan\delta' - X\left(Q_r + \frac{XV_r^2}{R^2 + X^2}\right)\tan\delta'$$
$$= R\left(Q_r + \frac{XV_r^2}{R^2 + X^2}\right) - X\left(P_r + \frac{RV_r^2}{R^2 + X^2}\right) \quad \text{or}$$

$$-R\left(Q_r + \frac{XV_r^2}{R^2+X^2}\right) - X\left(Q_r + \frac{XV_r^2}{R^2+X^2}\right)\tan\delta'$$

$$= R\left(P_r + \frac{RV_r^2}{R^2+X^2}\right)\tan\delta' - X\left(P_r + \frac{RV_r^2}{R^2+X^2}\right) \quad \text{or}$$

$$Q_r + \frac{XV_r^2}{R^2+X^2} = \frac{\left(P_r + \frac{RV_r^2}{R^2+X^2}\right)(-R\tan\delta' + X)}{R + X\tan\delta'}. \tag{B-121}$$

The Equations (B-120) and (B-121) can be combined as

$$\sqrt{A^2 - \left(P_r + \frac{RV_r^2}{R^2+X^2}\right)^2} = \frac{\left(P_r + \frac{RV_r^2}{R^2+X^2}\right)(-R\tan\delta' + X)}{R + X\tan\delta'} \quad \text{or}$$

$$(R + X\tan\delta')\sqrt{A^2 - \left(P_r + \frac{RV_r^2}{R^2+X^2}\right)^2} = \left(P_r + \frac{RV_r^2}{R^2+X^2}\right)(-R\tan\delta' + X) \quad \text{or}$$

$$R\sqrt{A^2 - \left(P_r + \frac{RV_r^2}{R^2+X^2}\right)^2} + X\sqrt{A^2 - \left(P_r + \frac{RV_r^2}{R^2+X^2}\right)^2}\tan\delta'$$

$$= -R\left(P_r + \frac{RV_r^2}{R^2+X^2}\right)\tan\delta' + X\left(P_r + \frac{RV_r^2}{R^2+X^2}\right) \quad \text{or}$$

$$\left(R\left(P_r + \frac{RV_r^2}{R^2+X^2}\right) + X\sqrt{A^2 - \left(P_r + \frac{RV_r^2}{R^2+X^2}\right)^2}\right)\tan\delta'$$

$$= -R\sqrt{A^2 - \left(P_r + \frac{RV_r^2}{R^2+X^2}\right)^2} + X\left(P_r + \frac{RV_r^2}{R^2+X^2}\right) \quad \text{or}$$

$$\tan\delta' = \frac{-R\sqrt{A^2 - \left(P_r + \frac{RV_r^2}{R^2+X^2}\right)^2} + X\left(P_r + \frac{RV_r^2}{R^2+X^2}\right)}{R\left(P_r + \frac{RV_r^2}{R^2+X^2}\right) + X\sqrt{A^2 - \left(P_r + \frac{RV_r^2}{R^2+X^2}\right)^2}} \tag{B-122a}$$

or

$$\delta' = \tan^{-1}\frac{-R\sqrt{A^2 - \left(P_r + \frac{RV_r^2}{R^2+X^2}\right)^2} + X\left(P_r + \frac{RV_r^2}{R^2+X^2}\right)}{R\left(P_r + \frac{RV_r^2}{R^2+X^2}\right) + X\sqrt{A^2 - \left(P_r + \frac{RV_r^2}{R^2+X^2}\right)^2}}. \tag{B-122b}$$

For a desired active power flow (P_r^*), the required modified power angle ($\delta' = \delta_{s'} - \delta_r$) can be determined by substituting $P_r = P_r^*$ in Equation (B-122b). The phase-shift angle (ψ) can be determined using the following equation:

$$\psi = \delta' - \delta. \tag{2-51b}$$

B.2.2.5.2 Reactance Regulator

A Reactance Regulator (RR) connects a variable capacitor/reactor with a compensating reactance ($X_{se} = X_C$ or $-X_L$) either directly or indirectly in series with the line to regulate the effective reactance of the line between its two ends. A $X_{se} > 0$ means the series-compensating reactance is capacitive and $X_{se} < 0$ means the series-compensating reactance is inductive. When a compensating reactance (X_{se}) is connected directly in series with a line, a voltage ($\mathbf{V_q}$), which is in quadrature with the prevailing line current ($\mathbf{I}$), is formed across the reactance as shown in Figure 2-28. A compensating voltage ($\mathbf{V_{s's}} = -\mathbf{V_q}$) may also be used, maintaining the quadrature relationship with the prevailing line current ($\mathbf{I}$) to emulate a capacitor or reactor in series with the line. In the process, the compensating voltage modifies the magnitude ($V_{s'}$) and phase angle ($\delta_{s'}$) of the modified sending-end voltage ($\mathbf{V_{s'}}$) simultaneously.

For a desired series-compensating reactance ($X_{se}{}^*$), the required effective line reactance (X_{eff}) can be determined by substituting $X_{se} = X_{se}^*$ in the following equation:

$$X_{eff} = X - X_{se}. \tag{2-210a}$$

Rewriting Equation (B-21), the active power flow at the receiving end of the line is given by

$$P_r = \frac{A_{eff}}{\sqrt{R^2 + X_{eff}^2}} \left\{ R\left(\cos\delta - \frac{V_r}{V_s} \right) + X_{eff} \sin\delta \right\}. \tag{B-123}$$

Rewriting Equation (B-22), the reactive power flow at the receiving end of the line is given by

$$Q_r = \frac{A_{eff}}{\sqrt{R^2 + X_{eff}^2}} \left\{ -R\sin\delta + X_{eff}\left(\cos\delta - \frac{V_r}{V_s} \right) \right\} \tag{B-124}$$

where

$$A_{eff} = \frac{V_s V_r}{\sqrt{R^2 + X_{eff}^2}}. \tag{B-125}$$

Using Equation (B-125), Equations (B-123) and (B-124) can be combined as

$$\left(P_r + \frac{RV_r^2}{R^2 + X_{eff}^2} \right)^2 + \left(Q_r + \frac{X_{eff} V_r^2}{R^2 + X_{eff}^2} \right)^2 = \frac{A_{eff}^2}{R^2 + X_{eff}^2} \left\{ (R\cos\delta + X_{eff}\sin\delta)^2 + (-R\sin\delta + X_{eff}\cos\delta)^2 \right\}$$

or

$$\left(P_r + \frac{RV_r^2}{R^2 + X_{eff}^2} \right)^2 + \left(Q_r + \frac{X_{eff} V_r^2}{R^2 + X_{eff}^2} \right)^2 = \frac{A_{eff}^2}{R^2 + X_{eff}^2} \left\{ R^2(\cos^2\delta + \sin^2\delta) + X_{eff}^2(\cos^2\delta + \sin^2\delta) \right\} = A_{eff}^2$$

using the identity $(\cos^2\delta + \sin^2\delta = 1)$, or

$$\left(P_r + \frac{RV_r^2}{R^2 + X_{eff}^2} \right)^2 + \left(Q_r + \frac{X_{eff} V_r^2}{R^2 + X_{eff}^2} \right)^2 = A_{eff}^2 \tag{B-126}$$

and

$$\tan\delta = \frac{R\left(Q_r + \frac{X_{eff}V_r^2}{R^2+X_{eff}^2}\right) - X_{eff}\left(P_r + \frac{RV_r^2}{R^2+X_{eff}^2}\right)}{-R\left(P_r + \frac{RV_r^2}{R^2+X_{eff}^2}\right) - X_{eff}\left(Q_r + \frac{X_{eff}V_r^2}{R^2+X_{eff}^2}\right)}, \tag{B-127}$$

which can also be verified by substituting δ' with δ and X with X_{eff} in Equation (B-46).

Combining Equations (B-123) and (B-125), it can be written that

$$P_r = \frac{V_sV_r}{R^2+X_{eff}^2}\left[R\left\{\cos\delta - \frac{V_r}{V_s}\right\} + X_{eff}\sin\delta\right] \tag{B-128}$$

or

$$X_{eff}^2 - \frac{V_sV_r}{P_r}\sin\delta X_{eff} + R^2 - R\frac{V_sV_r}{P_r}\left\{\cos\delta - \frac{V_r}{V_s}\right\} = 0.$$

Solving the above quadratic equation results in

$$X_{eff} = \frac{\frac{V_sV_r}{P_r}\sin\delta \pm \sqrt{\left(\frac{V_sV_r}{P_r}\sin\delta\right)^2 - 4R^2 + 4R\frac{V_sV_r}{P_r}\left\{\cos\delta - \frac{V_r}{V_s}\right\}}}{2}. \tag{B-129}$$

For a desired active power flow (P_r^*), the required effective line reactance (X_{eff}) can be determined by substituting $P_r = P_r^*$ in Equation (B-129). The series-compensating reactance (X_{se}) can be determined using the following equation:

$$X_{se} = X - X_{eff} \tag{2-210b}$$

using Equation (2-210a).

The active power flow at the receiving end of the line when the line is series-compensated is given by

$$P_r = V_rI\cos(\delta_r - \theta_I) = V_rI\cos(\delta_r - \{\delta_r + \theta_{Ir}\}) = V_rI\cos(-\theta_{Ir})$$

where θ_{Ir} is the phase angle of the line current with respect to the receiving-end voltage or

$$P_r = V_rI\cos\theta_{Ir} \quad \text{or} \quad I\cos\theta_{Ir} = \frac{P_r}{V_r}. \tag{2-218}$$

The reactive power flow at the receiving end of the line when the line is series-compensated is given by

$$Q_r = V_rI\sin(\delta_r - \theta_I) = V_rI\sin(\delta_r - \{\delta_r + \theta_{Ir}\}) = V_rI\sin(-\theta_{Ir})$$

where θ_{Ir} is the phase angle of the line current with respect to the receiving-end voltage or

$$Q_r = -V_rI\sin\theta_{Ir} \quad \text{or} \quad I\sin\theta_{Ir} = -\frac{Q_r}{V_r}. \tag{2-219}$$

From Figure B-7, it can be written that

$$\mathbf{V_{s'}} = \mathbf{V_r} + \mathbf{V_R} + \mathbf{V_X}. \tag{B-130}$$

Considering the receiving-end voltage phasor as reference (*i.e.* $\delta_r = 0°$), it can be written from Equation (B-130) that

$$V_{s'}\angle\delta' = V_r\angle 0° + I\angle\theta_{Ir}(R + jX) \quad \text{or}$$

$$V_{s'}\cos\delta' + jV_{s'}\sin\delta' = V_r + (I\cos\theta_{Ir} + jI\sin\theta_{Ir}) \times (R + jX) \quad \text{or}$$

$$V_{s'}\cos\delta' + jV_{s'}\sin\delta' = V_r + RI\cos\theta_{Ir} - XI\sin\theta_{Ir} + jRI\sin\theta_{Ir} + jXI\cos\theta_{Ir}.$$

Equating the real and imaginary parts of the both sides of the above equation, it can be written that

$$V_{s'}\cos\delta' = V_r + RI\cos\theta_{Ir} - XI\sin\theta_{Ir} \tag{B-131a}$$

and

$$V_{s'}\sin\delta' = RI\sin\theta_{Ir} + XI\cos\theta_{Ir}. \tag{B-131b}$$

Therefore, the magnitude ($V_{s'}$) and modified power angle ($\delta' = \delta_{s'} - \delta_r$) of the modified sending-end voltage ($\mathbf{V_{s'}}$) are derived from Equations (B-131a) and (B-131b) using the identity $\left(\cos^2\delta' + \sin^2\delta' = 1\right)$ as

$$V_{s'} = \sqrt{V_{s'}^2\left(\cos^2\delta' + \sin^2\delta'\right)} = \sqrt{(V_{s'}\cos\delta')^2 + \left(V_{s'}\sin\delta'\right)^2}$$

or

$$V_{s'} = \sqrt{(V_r + RI\cos\theta_{Ir} - XI\sin\theta_{Ir})^2 + \left(RI\sin\theta_{Ir} + XI\cos\theta_{Ir}\right)^2}$$

or

$$V_{s'} = \sqrt{V_r^2 + (RI)^2 + (XI)^2 + 2V_rI(R\cos\theta_{Ir} - X\sin\theta_{Ir})}$$

or using Equations (2-218) and (2-219)

$$V_{s'} = \sqrt{V_r^2 + (R^2 + X^2)I^2 + 2(RP_r + XQ_r)} \tag{B-132}$$

and

$$\delta' = \tan^{-1}(\tan\delta') = \tan^{-1}\frac{\sin\delta'}{\cos\delta'} = \tan^{-1}\frac{V_{s'}\sin\delta'}{V_{s'}\cos\delta'} = \tan^{-1}\frac{RI\sin\theta_{Ir} + XI\cos\theta_{Ir}}{V_r + RI\cos\theta_{Ir} - XI\sin\theta_{Ir}}$$

or

$$\delta' = \tan^{-1}\frac{XP_r - RQ_r}{V_r^2 + RP_r + XQ_r}. \tag{B-133}$$

Using Equation (2-51b), Equation (B-133) can be rewritten as

$$\psi = \tan^{-1}\frac{XP_r - RQ_r}{V_r^2 + RP_r + XQ_r} - \delta. \tag{B-134}$$

Recall that for $R = 0$,

$$V_{s'} = \sqrt{V_r^2 + (XI)^2 + 2XQ_r} \tag{2-221}$$

and

$$\delta' = \tan^{-1}\frac{XP_r}{V_r^2 + XQ_r}. \tag{2-222}$$

The apparent power (S_r) at the receiving end is given by

$$S_r = \sqrt{P_r^2 + Q_r^2}. \tag{2-76}$$

The line current (I) is given by

$$I = \frac{S_r}{V_r}. \tag{2-77}$$

The series-compensating voltage ($V_{s's}$) is given by

$$V_{s's} = IX_{se}. \tag{2-263a}$$

Note that only the absolute value is considered to determine the rating of the series-compensating voltage.

The relative phase angle (β) is given by

$$\beta = \tan^{-1}\frac{-R(Q_r - Q_{rn}) + X(P_r - P_{rn})}{R(P_r - P_{rn}) + X(Q_r - Q_{rn})} - \delta. \tag{B-82}$$

The active and reactive power flows ($P_{s'}$ and $Q_{s'}$) at the modified sending end are given by

$$P_{s'} = P_r + P_{line} \tag{B-135}$$

and

$$Q_{s'} = Q_r + Q_{line} \tag{2-78a}$$

where P_{line} and Q_{line} are given in Equations (2-318) and (2-11b), respectively.

Changing one parameter (voltage, $V_{s'}$, or phase angle, $\delta_{s'}$, of the modified sending-end voltage) using a PFC affects both the active and the reactive power flows in the line simultaneously. The key to be able to independently control the active and reactive power flows (P and Q) in a line is to be able to control both the magnitude and phase angle of the line voltage independently. A PFC modifies the line voltage using the series-compensating voltage of a shunt-series power converter.

The UPFC that consists of two VSCs in a shunt-series configuration can provide independent controls of the active and reactive power flows in the line. However, the faster response of the UPFC in ms to subcycle is not needed in most utility applications. The ST brings the right technology for an independent power flow control at a cost that is an order-of-magnitude less than the life-cycle cost (installation, operation, and maintenance) of a UPFC.

B.2.2.6 Representation of a Series-Compensating Voltage as a Series-Compensating Impedance

The series-compensating voltage ($\mathbf{V_{s's}} = V_{s's}\angle(\delta_s + \beta) = -\mathbf{V_{dq}}$), being at any phase angle with the prevailing line current ($\mathbf{I}$), emulates a four-quadrant, series-compensating impedance ($\mathbf{Z_{se}} = R_{se} - jX_{se}$) that consists of a resistance ($R_{se} = +R$ or $-R$) and a reactance ($X_{se} = X_C$ or $-X_L$) in series with the line. The equation defining the series-compensating impedance ($\mathbf{Z_{se}}$) can be written as

$$\mathbf{Z_{se}} = Z_{se}\angle\theta_{Zse} = \frac{\mathbf{V_{dq}}}{I\angle\theta_I} \quad \text{or using Equation (1-4)}, Z_{se}\angle\theta_{Zse} = \frac{-\mathbf{V_{s's}}}{I\angle\theta_I}$$

or

$$Z_{se}\angle\theta_{Zse} = \frac{\mathbf{V_{s's}}\angle 180^\circ}{I\angle\theta_I} \quad \text{or} \quad Z_{se}\angle\theta_{Zse} = \frac{V_{s's}\angle(\delta_s + \beta + 180^\circ)}{I\angle(\delta_s + \theta_{Is})}$$

or

$$Z_{se}\angle\theta_{Zse} = \frac{V_{s's}}{I}\angle(\beta + 180^\circ - \theta_{Is}), \tag{2-245a}$$

which can be further expanded using Equation (B-31) to obtain

$$Z_{se}\angle\theta_{Zse} = \sqrt{R^2 + X^2}\frac{V_{s's}}{V_{R,X}}\angle(\beta + 180^\circ - \theta_{Is}) \tag{B-136}$$

where θ_{Is} is the phase angle of the line current with respect to the sending-end voltage.

Equation (B-136) can be expanded using Equation (B-59a) as

$$Z_{se}\angle\theta_{Zse} = \sqrt{R^2 + X^2}\frac{V_{s's}}{V_{R,X}}\angle(\beta + 180^\circ - \theta_{Is})$$

or

$$Z_{se}\angle\theta_{Zse} = \sqrt{R^2 + X^2}\frac{V_{s's}}{V_{R,X}}\angle\left(\beta + 180^\circ - \tan^{-1}\frac{V_r \sin\delta + V_{s's}\sin\beta}{V_s - V_r\cos\delta + V_{s's}\cos\beta} + \tan^{-1}\frac{X}{R}\right)$$

or

$$Z_{se}\angle\theta_{Zse} = \sqrt{R^2 + X^2}\frac{V_{s's}}{V_{R,X}}\angle\left(\tan^{-1}\frac{-V_s\sin\left(\beta + \tan^{-1}\frac{X}{R}\right) + V_r\sin\left(\delta + \beta + \tan^{-1}\frac{X}{R}\right) - V_{s's}\sin\left(\tan^{-1}\frac{X}{R}\right)}{-V_s\cos\left(\beta + \tan^{-1}\frac{X}{R}\right) + V_r\cos\left(\delta + \beta + \tan^{-1}\frac{X}{R}\right) - V_{s's}\cos\left(\tan^{-1}\frac{X}{R}\right)}\right). \tag{B-137}$$

Digress:

$$F = \beta + 180^\circ - \tan^{-1}\frac{V_r \sin\delta + V_{s's}\sin\beta}{V_s - V_r\cos\delta + V_{s's}\cos\beta} + \tan^{-1}\frac{X}{R}$$

or

$$\tan^{-1}\frac{V_r \sin\delta + V_{s's}\sin\beta}{V_s - V_r\cos\delta + V_{s's}\cos\beta} = \beta + 180^\circ + \tan^{-1}\frac{X}{R} - F$$

or

$$\frac{V_r \sin\delta + V_{s's}\sin\beta}{V_s - V_r\cos\delta + V_{s's}\cos\beta} = \tan\left(\beta + 180^\circ + \tan^{-1}\frac{X}{R} - F\right)$$

or

$$\frac{V_r \sin\delta + V_{s's} \sin\beta}{V_s - V_r \cos\delta + V_{s's} \cos\beta} = \tan\left(\beta + \tan^{-1}\frac{X}{R} - F\right)$$

or

$$\frac{V_r \sin\delta + V_{s's} \sin\beta}{V_s - V_r \cos\delta + V_{s's} \cos\beta} = \frac{\sin\left(\beta + \tan^{-1}\frac{X}{R} - F\right)}{\cos\left(\beta + \tan^{-1}\frac{X}{R} - F\right)}$$

$$= \frac{\sin\left(\beta + \tan^{-1}\frac{X}{R}\right)\cos F - \cos\left(\beta + \tan^{-1}\frac{X}{R}\right)\sin F}{\cos\left(\beta + \tan^{-1}\frac{X}{R}\right)\cos F + \sin\left(\beta + \tan^{-1}\frac{X}{R}\right)\sin F}$$

or

$$(V_r \sin\delta + V_{s's} \sin\beta)\left\{\cos\left(\beta + \tan^{-1}\frac{X}{R}\right)\cos F + \sin\left(\beta + \tan^{-1}\frac{X}{R}\right)\sin F\right\}$$
$$= \left\{\sin\left(\beta + \tan^{-1}\frac{X}{R}\right)\cos F - \cos\left(\beta + \tan^{-1}\frac{X}{R}\right)\sin F\right\}(V_s - V_r \cos\delta + V_{s's} \cos\beta).$$

or

$$\cos\left(\beta + \tan^{-1}\frac{X}{R}\right)\cos F(V_r \sin\delta + V_{s's} \sin\beta)$$
$$- \sin\left(\beta + \tan^{-1}\frac{X}{R}\right)\cos F(V_s - V_r \cos\delta + V_{s's} \cos\beta)$$
$$= - \cos\left(\beta + \tan^{-1}\frac{X}{R}\right)\sin F(V_s - V_r \cos\delta + V_{s's} \cos\beta)$$
$$- \sin\left(\beta + \tan^{-1}\frac{X}{R}\right)\sin F(V_r \sin\delta + V_{s's} \sin\beta)$$

or

$$\cos F\left\{\cos\left(\beta + \tan^{-1}\frac{X}{R}\right)(V_r \sin\delta + V_{s's} \sin\beta)\right.$$
$$\left. - \sin\left(\beta + \tan^{-1}\frac{X}{R}\right)(V_s - V_r \cos\delta + V_{s's} \cos\beta)\right\}$$
$$= - \sin F\left\{\cos\left(\beta + \tan^{-1}\frac{X}{R}\right)(V_s - V_r \cos\delta + V_{s's} \cos\beta)\right.$$
$$\left. + \sin\left(\beta + \tan^{-1}\frac{X}{R}\right)(V_r \sin\delta + V_{s's} \sin\beta)\right\}$$

or

$$\frac{\sin F}{\cos F} = -\frac{\cos\left(\beta + \tan^{-1}\frac{X}{R}\right)(V_r \sin\delta + V_{s's}\sin\beta) - \sin\left(\beta + \tan^{-1}\frac{X}{R}\right)(V_s - V_r\cos\delta + V_{s's}\cos\beta)}{\cos\left(\beta + \tan^{-1}\frac{X}{R}\right)(V_s - V_r\cos\delta + V_{s's}\cos\beta) + \sin\left(\beta + \tan^{-1}\frac{X}{R}\right)(V_r\sin\delta + V_{s's}\sin\beta)}$$

or

$$\tan F = -\frac{V_r\sin\delta\cos\left(\beta + \tan^{-1}\frac{X}{R}\right) + V_{s's}\sin\beta\cos\left(\beta + \tan^{-1}\frac{X}{R}\right) - V_s\sin\left(\beta + \tan^{-1}\frac{X}{R}\right) + V_r\cos\delta\sin\left(\beta + \tan^{-1}\frac{X}{R}\right) - V_{s's}\cos\beta\sin\left(\beta + \tan^{-1}\frac{X}{R}\right)}{V_s\cos\left(\beta + \tan^{-1}\frac{X}{R}\right) - V_r\cos\delta\cos\left(\beta + \tan^{-1}\frac{X}{R}\right) + V_{s's}\cos\beta\cos\left(\beta + \tan^{-1}\frac{X}{R}\right) + V_r\sin\delta\sin\left(\beta + \tan^{-1}\frac{X}{R}\right) + V_{s's}\sin\beta\sin\left(\beta + \tan^{-1}\frac{X}{R}\right)}$$

or

$$\tan F = \frac{-V_s\sin\left(\beta + \tan^{-1}\frac{X}{R}\right) + V_r\sin\left(\delta + \beta + \tan^{-1}\frac{X}{R}\right) - V_{s's}\sin\left(\tan^{-1}\frac{X}{R}\right)}{-V_s\cos\left(\beta + \tan^{-1}\frac{X}{R}\right) + V_r\cos\left(\delta + \beta + \tan^{-1}\frac{X}{R}\right) - V_{s's}\cos\left(\tan^{-1}\frac{X}{R}\right)}$$

or

$$F = \tan^{-1}\frac{-V_s\sin\left(\beta + \tan^{-1}\frac{X}{R}\right) + V_r\sin\left(\delta + \beta + \tan^{-1}\frac{X}{R}\right) - V_{s's}\sin\left(\tan^{-1}\frac{X}{R}\right)}{-V_s\cos\left(\beta + \tan^{-1}\frac{X}{R}\right) + V_r\cos\left(\delta + \beta + \tan^{-1}\frac{X}{R}\right) - V_{s's}\cos\left(\tan^{-1}\frac{X}{R}\right)}$$

The emulated series-compensating resistance (R_{se}) and series-compensating reactance (X_{se}) are defined in Equations (2-244a) and (2-244b), which can be redefined using Equation (B-137) as

$$R_{se} = \sqrt{R^2 + X^2}\frac{V_{s's}}{V_{R,X}}\left\{\frac{-V_s\cos\left(\beta + \tan^{-1}\frac{X}{R}\right) + V_r\cos\left(\delta + \beta + \tan^{-1}\frac{X}{R}\right) - V_{s's}\cos\left(\tan^{-1}\frac{X}{R}\right)}{V_{R,X}}\right\} \tag{B-138a}$$

and

$$X_{se} = \sqrt{R^2 + X^2}\frac{V_{s's}}{V_{R,X}}\left\{\frac{V_s\sin\left(\beta + \tan^{-1}\frac{X}{R}\right) - V_r\sin\left(\delta + \beta + \tan^{-1}\frac{X}{R}\right) + V_{s's}\sin\left(\tan^{-1}\frac{X}{R}\right)}{V_{R,X}}\right\} \tag{B-138b}$$

where $X_{se} > 0$ means that the series-compensating reactance is capacitive and $X_{se} < 0$ means that the series-compensating reactance is inductive.

The magnitude and phase angle of the series-compensating impedance are given by

$$Z_{se}\angle\theta_{Zse} = \sqrt{R_{se}^2 + X_{se}^2}\angle\left(\tan^{-1}\frac{-X_{se}}{R_{se}}\right). \tag{2-242b}$$

Recall for $R = 0$,

$$R_{se} = X\frac{V_{s's}}{V_X}\left\{\frac{V_s\sin\beta - V_r\sin(\delta+\beta)}{V_X}\right\}, \tag{2-247a}$$

$$X_{se} = X\frac{V_{s's}}{V_X}\left\{\frac{V_s\cos\beta - V_r\cos(\delta+\beta) + V_{s's}}{V_X}\right\}, \text{ and} \tag{2-247b}$$

$$Z_{se}\angle\theta_{Zse} = X\frac{V_{s's}}{V_X}\angle\left(\tan^{-1}\frac{-V_s\cos\beta + V_r\cos(\delta+\beta) - V_{s's}}{V_s\sin\beta - V_r\sin(\delta+\beta)}\right). \tag{2-246}$$

B.2.2.6.1 Equivalent Impedance of a Voltage Regulator (VR)

For a VR, the relative phase angle (β) is either 0° or 180°. For $\beta = 0°$, the series-compensating resistance (R_{se}) can be defined from Equation (B-138a) as

$$R_{se} = \sqrt{R^2+X^2}\frac{V_{s's}}{V_{R,X}}\left\{\frac{-V_s\cos\left(\tan^{-1}\frac{X}{R}\right) + V_r\cos\left(\delta + \tan^{-1}\frac{X}{R}\right) - V_{s's}\cos\left(\tan^{-1}\frac{X}{R}\right)}{V_{R,X}}\right\}$$

and the series-compensating reactance (X_{se}) can be defined from Equation (B-138b) as

$$X_{se} = \sqrt{R^2+X^2}\frac{V_{s's}}{V_{R,X}}\left\{\frac{V_s\sin\left(\tan^{-1}\frac{X}{R}\right) - V_r\sin\left(\delta + \tan^{-1}\frac{X}{R}\right) + V_{s's}\sin\left(\tan^{-1}\frac{X}{R}\right)}{V_{R,X}}\right\}.$$

For $\beta = 180°$, the series-compensating resistance (R_{se}) can be defined from Equation (B-138a) as

$$R_{se} = \sqrt{R^2+X^2}\frac{V_{s's}}{V_{R,X}}\left\{\frac{V_s\cos\left(\tan^{-1}\frac{X}{R}\right) - V_r\cos\left(\delta + \tan^{-1}\frac{X}{R}\right) - V_{s's}\cos\left(\tan^{-1}\frac{X}{R}\right)}{V_{R,X}}\right\}$$

and the series-compensating reactance (X_{se}) can be defined from Equation (B-138b) as

$$X_{se} = \sqrt{R^2+X^2}\frac{V_{s's}}{V_{R,X}}\left\{\frac{-V_s\sin\left(\tan^{-1}\frac{X}{R}\right) + V_r\sin\left(\delta + \tan^{-1}\frac{X}{R}\right) + V_{s's}\sin\left(\tan^{-1}\frac{X}{R}\right)}{V_{R,X}}\right\}.$$

In general,

$$R_{se} = \sqrt{R^2 + X^2}\frac{V_{s's}}{V_{R,X}}\left\{\frac{-V_s\cos\left(\tan^{-1}\frac{X}{R}\right) + V_r\cos\left(\delta + \tan^{-1}\frac{X}{R}\right) - V_{s's}\cos\left(\tan^{-1}\frac{X}{R}\right)}{V_{R,X}}\right\} \tag{B-139a}$$

and

$$X_{se} = \sqrt{R^2 + X^2}\frac{V_{s's}}{V_{R,X}}\left\{\frac{V_s\sin\left(\tan^{-1}\frac{X}{R}\right) - V_r\sin\left(\delta + \tan^{-1}\frac{X}{R}\right) + V_{s's}\sin\left(\tan^{-1}\frac{X}{R}\right)}{V_{R,X}}\right\} \tag{B-139b}$$

where $V_{s's} > 0$ represents $\beta = 0°$ and $V_{s's} < 0$ represents $\beta = 180°$.

Using Equation (2-242b), the series-compensating impedance (Z_{se}) can be written from Equations (B-139a) and (B-139b) as

$$Z_{se}\angle\theta_{Zse} = \sqrt{R_{se}^2 + X_{se}^2}\angle\left(\tan^{-1}\frac{-X_{se}}{R_{se}}\right)$$

$$= \sqrt{R^2 + X^2}\frac{V_{s's}}{V_{R,X}}\angle\left(\tan^{-1}\frac{-V_s\sin\left(\tan^{-1}\frac{X}{R}\right) + V_r\sin\left(\delta + \tan^{-1}\frac{X}{R}\right) - V_{s's}\sin\left(\tan^{-1}\frac{X}{R}\right)}{-V_s\cos\left(\tan^{-1}\frac{X}{R}\right) + V_r\cos\left(\delta + \tan^{-1}\frac{X}{R}\right) - V_{s's}\cos\left(\tan^{-1}\frac{X}{R}\right)}\right)$$

or

$$Z_{se}\angle\theta_{Zse} = \sqrt{R^2 + X^2}\frac{V_{s's}}{V_{R,X}}\angle\left(\tan^{-1}\frac{V_s\sin\left(\tan^{-1}\frac{X}{R}\right) - V_r\sin\left(\delta + \tan^{-1}\frac{X}{R}\right) + V_{s's}\sin\left(\tan^{-1}\frac{X}{R}\right)}{V_s\cos\left(\tan^{-1}\frac{X}{R}\right) - V_r\cos\left(\delta + \tan^{-1}\frac{X}{R}\right) + V_{s's}\cos\left(\tan^{-1}\frac{X}{R}\right)}\right) \tag{B-140}$$

where $V_{s's} > 0$ represents $\beta = 0°$ and $V_{s's} < 0$ represents $\beta = 180°$.

Recall that for $R = 0$,

$$R_{se} = -X\frac{V_{s's}}{V_X}\left\{\frac{V_r\sin\delta}{V_X}\right\}, \tag{2-249a}$$

$$X_{se} = X\frac{V_{s's}}{V_X}\left\{\frac{V_s - V_r\cos\delta + V_{s's}}{V_X}\right\}, \text{ and} \tag{2-249b}$$

$$Z_{se}\angle\theta_{Zse} = X\frac{V_{s's}}{V_X}\angle\left(\tan^{-1}\frac{V_s - V_r\cos\delta + V_{s's}}{V_r\sin\delta}\right). \tag{2-250}$$

B.2.2.6.2 Equivalent Impedance of a Phase Angle Regulator (Asymmetric)

For a PAR (asym), the relative phase angle (β) is either +90° or −90°. For $\beta = 90°$, the series-compensating resistance (R_{se}) can be defined from Equation (B-138a) as

$$R_{se} = \sqrt{R^2 + X^2}\frac{V_{s's}}{V_{R,X}}\left\{\frac{V_s\sin\left(\tan^{-1}\frac{X}{R}\right) - V_r\sin\left(\delta + \tan^{-1}\frac{X}{R}\right) - V_{s's}\cos\left(\tan^{-1}\frac{X}{R}\right)}{V_{R,X}}\right\}$$

and the series-compensating reactance (X_{se}) can be defined from Equation (B-138b) as

$$X_{se} = \sqrt{R^2 + X^2}\frac{V_{s's}}{V_{R,X}}\left\{\frac{V_s\cos\left(\tan^{-1}\frac{X}{R}\right) - V_r\cos\left(\delta + \tan^{-1}\frac{X}{R}\right) + V_{s's}\sin\left(\tan^{-1}\frac{X}{R}\right)}{V_{R,X}}\right\}.$$

For $\beta = -90°$, the series-compensating resistance (R_{se}) can be defined from Equation (B-138a) as

$$R_{se} = \sqrt{R^2 + X^2}\frac{V_{s's}}{V_{R,X}}\left\{\frac{-V_s\sin\left(\tan^{-1}\frac{X}{R}\right) + V_r\sin\left(\delta + \tan^{-1}\frac{X}{R}\right) - V_{s's}\cos\left(\tan^{-1}\frac{X}{R}\right)}{V_{R,X}}\right\}$$

and the series-compensating reactance (X_{se}) can be defined from Equation (B-138b) as

$$X_{se} = \sqrt{R^2 + X^2}\frac{V_{s's}}{V_{R,X}}\left\{\frac{-V_s\cos\left(\tan^{-1}\frac{X}{R}\right) + V_r\cos\left(\delta + \tan^{-1}\frac{X}{R}\right) + V_{s's}\sin\left(\tan^{-1}\frac{X}{R}\right)}{V_{R,X}}\right\}.$$

In general,

$$R_{se} = \sqrt{R^2 + X^2}\frac{V_{s's}}{V_{R,X}}\left\{\frac{V_s\sin\left(\tan^{-1}\frac{X}{R}\right) - V_r\sin\left(\delta + \tan^{-1}\frac{X}{R}\right) - V_{s's}\cos\left(\tan^{-1}\frac{X}{R}\right)}{V_{R,X}}\right\} \tag{B-141a}$$

and

$$X_{se} = \sqrt{R^2 + X^2}\frac{V_{s's}}{V_{R,X}}\left\{\frac{V_s\cos\left(\tan^{-1}\frac{X}{R}\right) - V_r\cos\left(\delta + \tan^{-1}\frac{X}{R}\right) + V_{s's}\sin\left(\tan^{-1}\frac{X}{R}\right)}{V_{R,X}}\right\} \tag{B-141b}$$

where $V_{s's} > 0$ represents $\beta = 90°$ and $V_{s's} < 0$ represents $\beta = -90°$.

Using Equation (2-242b), the series-compensating impedance (Z_{se}) can be written from Equations (B-141a) and (B-141b) as

$$Z_{se}\angle\theta_{Zse} = \sqrt{R^2 + X^2}\frac{V_{s's}}{V_{R,X}}\angle\left(\tan^{-1}\frac{-V_s\cos\left(\tan^{-1}\frac{X}{R}\right) + V_r\cos\left(\delta + \tan^{-1}\frac{X}{R}\right) - V_{s's}\sin\left(\tan^{-1}\frac{X}{R}\right)}{V_s\sin\left(\tan^{-1}\frac{X}{R}\right) - V_r\sin\left(\delta + \tan^{-1}\frac{X}{R}\right) - V_{s's}\cos\left(\tan^{-1}\frac{X}{R}\right)}\right) \tag{B-142}$$

where $V_{s's} > 0$ represents $\beta = 90°$ and $V_{s's} < 0$ represents $\beta = -90°$.

Recall that for $R = 0$,

$$R_{se} = X\frac{V_{s's}}{V_X}\left\{\frac{V_s - V_r\cos\delta}{V_X}\right\}, \tag{2-251a}$$

$$X_{se} = X\frac{V_{s's}}{V_X}\left\{\frac{V_r\sin\delta + V_{s's}}{V_X}\right\}, \text{ and} \tag{2-251b}$$

$$Z_{se}\angle\theta_{Zse} = X\frac{V_{s's}}{V_X}\angle\left(\tan^{-1}\frac{-V_r\sin\delta - V_{s's}}{V_s - V_r\cos\delta}\right). \tag{2-252}$$

B.2.2.6.3 *Equivalent Impedance of a Phase Angle Regulator (Symmetric)*

For a PAR (sym), the regulation parameter is the phase-shift angle, ψ. For $\psi > 0$, $\beta = \frac{\psi}{2} + 90°$, the series-compensating resistance (R_{se}) can be defined from Equation (B-138a) as

$$R_{se} = \sqrt{R^2 + X^2}\frac{V_{s's}}{V_{R,X}}\left\{\frac{V_s\sin\left(\frac{\psi}{2} + \tan^{-1}\frac{X}{R}\right) - V_r\sin\left(\delta + \frac{\psi}{2} + \tan^{-1}\frac{X}{R}\right) - V_{s's}\cos\left(\tan^{-1}\frac{X}{R}\right)}{V_{R,X}}\right\}$$

and the series-compensating reactance (X_{se}) can be defined from Equation (B-138b) as

$$X_{se} = \sqrt{R^2 + X^2}\frac{V_{s's}}{V_{R,X}}\left\{\frac{V_s\cos\left(\frac{\psi}{2} + \tan^{-1}\frac{X}{R}\right) - V_r\cos\left(\delta + \frac{\psi}{2} + \tan^{-1}\frac{X}{R}\right) + V_{s's}\sin\left(\tan^{-1}\frac{X}{R}\right)}{V_{R,X}}\right\}.$$

For $\psi < 0$, $\beta = \frac{\psi}{2} + 270°$, the series-compensating resistance (R_{se}) can be defined from Equation (B-138a) as

$$R_{se} = \sqrt{R^2 + X^2}\frac{V_{s's}}{V_{R,X}}\left\{\frac{-V_s\sin\left(\frac{\psi}{2} + \tan^{-1}\frac{X}{R}\right) + V_r\sin\left(\delta + \frac{\psi}{2} + \tan^{-1}\frac{X}{R}\right) - V_{s's}\cos\left(\tan^{-1}\frac{X}{R}\right)}{V_{R,X}}\right\}$$

and the series-compensating reactance (X_{se}) can be defined from Equation (B-138b) as

$$X_{se} = \sqrt{R^2 + X^2}\frac{V_{s's}}{V_{R,X}}\left\{\frac{\begin{array}{c}-V_s\cos\left(\frac{\psi}{2} + \tan^{-1}\frac{X}{R}\right) + V_r\cos\left(\delta + \frac{\psi}{2} + \tan^{-1}\frac{X}{R}\right)\\ + V_{s's}\sin\left(\tan^{-1}\frac{X}{R}\right)\end{array}}{V_{R,X}}\right\}.$$

In general,

$$R_{se} = \sqrt{R^2 + X^2}\frac{V_{s's}}{V_{R,X}}\left\{\frac{+V_s\sin\left(\frac{\psi}{2} + \tan^{-1}\frac{X}{R}\right) - V_r\sin\left(\delta + \frac{\psi}{2} + \tan^{-1}\frac{X}{R}\right) - V_{s's}\cos\left(\tan^{-1}\frac{X}{R}\right)}{V_{R,X}}\right\} \tag{B-143a}$$

and

$$X_{se} = \sqrt{R^2 + X^2}\frac{V_{s's}}{V_{R,X}}\left\{\frac{V_s\cos\left(\frac{\psi}{2} + \tan^{-1}\frac{X}{R}\right) - V_r\cos\left(\delta + \frac{\psi}{2} + \tan^{-1}\frac{X}{R}\right) + V_{s's}\sin\left(\tan^{-1}\frac{X}{R}\right)}{V_{R,X}}\right\} \tag{B-143b}$$

where $V_{s's} > 0$ represents $\beta = \frac{\psi}{2} + 90^\circ$ and $V_{s's} < 0$ represents $\beta = \frac{\psi}{2} + 270^\circ$.

Therefore, the series-compensating impedance (Z_{se}) can be written from Equations (B-143a) and (B-134b) as

$$Z_{se}\angle\theta_{Zse} = \sqrt{R^2 + X^2}\frac{V_{s's}}{V_{R,X}}\angle\left(\tan^{-1}\frac{-V_s\cos\left(\frac{\psi}{2} + \tan^{-1}\frac{X}{R}\right) + V_r\cos\left(\delta + \frac{\psi}{2} + \tan^{-1}\frac{X}{R}\right) - V_{s's}\sin\left(\tan^{-1}\frac{X}{R}\right)}{+V_s\sin\left(\frac{\psi}{2} + \tan^{-1}\frac{X}{R}\right) - V_r\sin\left(\delta + \frac{\psi}{2} + \tan^{-1}\frac{X}{R}\right) - V_{s's}\cos\left(\tan^{-1}\frac{X}{R}\right)}\right) \tag{B-144}$$

where $V_{s's} > 0$ represents $\beta = \frac{\psi}{2} + 90^\circ$ and $V_{s's} < 0$ represents $\beta = \frac{\psi}{2} + 270^\circ$.

Recall that for $R = 0$,

$$R_{se} = X\frac{V_{s's}}{V_X}\left\{\frac{V_s\cos\left(\frac{\psi}{2}\right) - V_r\cos\left(\delta + \frac{\psi}{2}\right)}{V_X}\right\}, \tag{2-253a}$$

$$X_{se} = X\frac{V_{s's}}{V_X}\left\{\frac{-V_s\sin\left(\frac{\psi}{2}\right) + V_r\sin\left(\delta + \frac{\psi}{2}\right) + V_{s's}}{V_X}\right\}, \text{ and} \tag{2-253b}$$

$$Z_{se}\angle\theta_{Zse} = X\frac{V_{s's}}{V_X}\angle\left(\tan^{-1}\frac{+V_s\sin\left(\frac{\psi}{2}\right) - V_r\sin\left(\delta + \frac{\psi}{2}\right) - V_{s's}}{V_s\cos\left(\frac{\psi}{2}\right) - V_r\cos\left(\delta + \frac{\psi}{2}\right)}\right). \tag{2-254}$$

B.2.2.6.4 Equivalent Impedance of a Reactance Regulator

For a RR, the regulation parameter is the series-compensating reactance (X_{se}) while the series-compensating resistance (R_{se}) is set to zero. Therefore,

$$R_{se} = 0 \tag{2-257}$$

$$X_{se} = \sqrt{R^2 + X^2}\frac{V_{s's}}{V_{R,X}}. \tag{B-145}$$

From Equations (B-145) and (B-31), it can be written that

$$\frac{V_{s's}}{X_{se}} = \frac{V_{R,X}}{\sqrt{R^2 + X^2}} = I, \tag{B-146}$$

which is same as Equation (2-234d).

Therefore, the series-compensating impedance (Z_{se}) can be written from Equation (2-257) and (B-145) as

$$Z_{se}\angle\theta_{Zse} = R_{se} - jX_{se} = \sqrt{R^2 + X^2}\frac{V_{s's}}{V_{R,X}}\angle(-90°) \tag{B-147}$$

where $V_{s's} > 0$ represents a capacitive compensating-reactance and $V_{s's} < 0$ represents an inductive compensating-reactance.

Recall that for $R = 0$,

$$X_{se} = X\frac{V_{s's}}{V_X} \tag{2-262}$$

and

$$Z_{se}\angle\theta_{Zse} = X\frac{V_{s's}}{V_X}\angle(-90°). \tag{2-258}$$

The compensating reactance (X_{se}) can be emulated using a series-connected inverter, called Static Synchronous Series Compensator (SSSC). Note that the SSSC inherently emulates a compensating resistance ($R_{se} = +R$) that is equivalent to the active power loss in the inverter and the series-coupling transformer, which is in the range of 2–3% of the rating of the inverter. It is also possible to design a compensating resistance ($R_{se} = +R$ or $-R$) by using an energy sink or source that is coupled with the inverter. The most inexpensive way to emulate a four-quadrant, series-connected reactance/impedance is possible by using a traditional transformer/LTCs-based technology, such as Sen Transformer in which the active power loss is limited to the loss in the transformer in less than 1% of the rating of the Sen Transformer.

B.2.2.7 *RPI*, *LI*, and *APR* of a PFC

The magnitude ($V_{Rn,Xn}$) of the natural voltage ($\mathbf{V_{Rn,Xn}}$) across the line is given by

$$V_{Rn,Xn} = \sqrt{V_s^2 - 2V_sV_r\cos\delta + V_r^2}. \tag{B-4}$$

For the desired series-compensating resistance ($R_{se}{}^*$) and reactance ($X_{se}{}^*$), the required effective line resistance (R_{eff}) and reactance (X_{eff}) can be determined by substituting $R_{se} = R_{se}^*$ and $X_{se} = X_{se}^*$ in the following equations:

$$R_{eff} = R + R_{se} \tag{B-148}$$

and

$$X_{eff} = X - X_{se}. \tag{2-210a}$$

Therefore, the line current is given by

$$I = \frac{V_{Rn,Xn}}{\sqrt{R_{eff}^2 + X_{eff}^2}}. \tag{B-149}$$

Rewriting Equation (B-21) with a substitution from Equation (B-13), the active power flow at the receiving end of the line is given by

$$P_r = \frac{V_s V_r}{R_{eff}^2 + X_{eff}^2}\left[R_{eff}\left\{\cos\delta - \frac{V_r}{V_s}\right\} + X_{eff}\sin\delta\right]. \tag{B-150}$$

The Reactive Power Index (RPI), which is a measure of total reactive power (Q_{line}) needed by the line for a unit of active power (P_r) transmitted, is defined as

$$RPI = \frac{Q_{line}}{P_r}. \tag{2-321}$$

The reactive power absorbed by the line is given by

$$Q_{line} = I^2 X \tag{2-11b}$$

where X is the line reactance.

From Equation (2-321), the RPI per unit of active power (P_r) transmitted is given by

$$\frac{RPI}{P_r} = \frac{Q_{line}}{P_r^2} = \frac{I^2 X}{P_r^2} = \frac{V_{Rn,Xn}^2 X}{\left(R_{eff}^2 + X_{eff}^2\right) V_s^2 V_r^2}\frac{\left(R_{eff}^2 + X_{eff}^2\right)^2}{\left[R_{eff}\left\{\cos\delta - \frac{V_r}{V_s}\right\} + X_{eff}\sin\delta\right]^2}$$

using Equation (2-11b), (B-149), (B-150), or

$$\frac{RPI}{P_r} = \frac{V_{Rn,Xn}^2 X}{V_s^2 V_r^2}\frac{R_{eff}^2 + X_{eff}^2}{\left[R_{eff}\left\{\cos\delta - \frac{V_r}{V_s}\right\} + X_{eff}\sin\delta\right]^2}. \tag{B-151}$$

The Loss Index (LI), which is a measure of line loss as a multiple of baseline loss, is defined as

$$LI = k_R \frac{I^2}{I_n^2} \tag{2-324}$$

where

$$k_R = \frac{R}{R_n} \tag{2-325}$$

where the line resistance is R at line current I and the line resistance is R_n at line current I_n. Note that as the line current (I) increases, the loss increases, which increases the temperature and, in turn, the resistance of the conductor.

From Equations (2-324) and (2-325), the LI per unit of active power (P_r) transmitted is given by

$$\frac{LI}{P_r} = k_R \frac{I^2}{I_n^2 P_r} = k_R \frac{V_{Rn,Xn}^2}{\left(R_{eff}^2 + X_{eff}^2\right) I_n^2}\frac{R_{eff}^2 + X_{eff}^2}{V_s V_r}\frac{1}{R_{eff}\left\{\cos\delta - \frac{V_r}{V_s}\right\} + X_{eff}\sin\delta}$$

using Equations (B-149) and (B-150), or

$$\frac{LI}{P_r} = k_R \frac{V_{Rn,Xn}^2}{V_{Rn,Xn}^2 V_s V_r} \frac{R^2 + X^2}{R_{eff}\left\{\cos\delta - \frac{V_r}{V_s}\right\} + X_{eff}\sin\delta} = k_R \frac{1}{V_s V_r} \frac{R^2 + X^2}{R_{eff}\left\{\cos\delta - \frac{V_r}{V_s}\right\} + X_{eff}\sin\delta} \tag{B-152}$$

using Equation (B-8).

The Apparent Power Rating (*APR*) of the PFC is defined as

$$APR = |V_{s's}|I \tag{2-326}$$

when the PFC is used to increase power flow with a prevailing line current (*I*);

$$APR = |V_{s's}|I_n \tag{2-327}$$

when the PFC is used to decrease power flow from a natural line current (I_n).

From Equation (2-326), the *APR* per unit of active power (P_r) transmitted is given by

$$\frac{APR}{P_r} = \frac{|V_{s's}|I}{P_r} = \frac{\left\{\sqrt{R_{se}^2 + X_{se}^2}\right\}I^2}{P_r} = \frac{V_{Rn,Xn}^2}{R_{eff}^2 + X_{eff}^2} \frac{R_{eff}^2 + X_{eff}^2}{V_s V_r} \frac{\sqrt{R_{se}^2 + X_{se}^2}}{R_{eff}\left\{\cos\delta - \frac{V_r}{V_s}\right\} + X_{eff}\sin\delta}$$

using Equation (2-245a), (2-242b), (B-149), (B-150), or

$$\frac{APR}{P_r} = \frac{|V_{s's}|I}{P_r} = \frac{V_{Rn,Xn}^2}{V_s V_r} \frac{\sqrt{R_{se}^2 + X_{se}^2}}{R_{eff}\left\{\cos\delta - \frac{V_r}{V_s}\right\} + X_{eff}\sin\delta}. \tag{B-153}$$

Equations (B-151), (B-152), and (B-153) apply to a PAR (sym), a PAR (asym), and an IR, since all three PFCs emulate a four-quadrant, series-compensating impedance ($\mathbf{Z_{se}} = R_{se} - jX_{se}$) that consists of a resistance ($R_{se} = +R$ or $-R$) and a reactance ($X_{se} = X_C$ or $-X_L$) in series with the line. However, for a RR, $R_{se} = 0$. Therefore, the Equations (B-151), (B-152), and (B-153) can be rewritten as follows.

The *RPI* per unit of active power (P_r) transmitted is given by

$$\frac{RPI}{P_r} = \frac{Q_{line}}{P_r^2} = \frac{I^2 X}{P_r^2} = \frac{V_{Rn,Xn}^2 X}{V_s^2 V_r^2} \frac{R^2 + X_{eff}^2}{\left[R\left\{\cos\delta - \frac{V_r}{V_s}\right\} + X_{eff}\sin\delta\right]^2}. \tag{B-151a}$$

The *LI* per unit of active power (P_r) transmitted is given by

$$\frac{LI}{P_r} = k_R \frac{I^2}{I_n^2 P_r} = k_R \frac{V_{Rn,Xn}^2}{I_n^2 V_s V_r} \frac{1}{R_{eff}\left\{\cos\delta - \frac{V_r}{V_s}\right\} + X_{eff}\sin\delta}$$

$$= k_R \frac{1}{V_s V_r} \frac{R^2 + X^2}{R\left\{\cos\delta - \frac{V_r}{V_s}\right\} + X_{eff}\sin\delta}. \tag{B-152a}$$

The APR per unit of active power (P_r) transmitted is given by

$$\frac{APR}{P_r} = \frac{|V_{s's}|I}{P_r} = \frac{X_{se}I^2}{P_r} = \frac{V_{Rn,Xn}^2}{V_sV_r}\frac{X_{se}}{R\left\{\cos\delta - \frac{V_r}{V_s}\right\} + X_{eff}\sin\delta}. \tag{B-153a}$$

If the line resistance (R) is neglected, the Equations (B-151), (B-152), and (B-153) can be rewritten as follows.

The RPI per unit of active power (P_r) transmitted is given by

$$\frac{RPI}{P_r} = \frac{Q_{line}}{P_r^2} = \frac{I^2X}{P_r^2} = \frac{V_{Xn}^2X}{V_s^2V_r^2}\frac{R_{se}^2 + X_{eff}^2}{\left[R_{se}\left\{\cos\delta - \frac{V_r}{V_s}\right\} + X_{eff}\sin\delta\right]^2}. \tag{B-151b}$$

The LI per unit of active power (P_r) transmitted is given by

$$\frac{LI}{P_r} = k_R\frac{I^2}{I_n^2P_r} = k_R\frac{V_{Xn}^2}{I_n^2V_sV_r}\frac{1}{R_{se}\left\{\cos\delta - \frac{V_r}{V_s}\right\} + X_{eff}\sin\delta}$$

$$= k_R\frac{1}{V_sV_r}\frac{X^2}{R_{se}\left\{\cos\delta - \frac{V_r}{V_s}\right\} + X_{eff}\sin\delta}. \tag{B-152b}$$

The APR per unit of active power (P_r) transmitted is given by

$$\frac{APR}{P_r} = \frac{|V_{s's}|I}{P_r} = \frac{\sqrt{R_{se}^2 + X_{se}^2}I^2}{P_r} = \frac{V_{Xn}^2}{V_sV_r}\frac{\sqrt{R_{se}^2 + X_{se}^2}}{R_{se}\left\{\cos\delta - \frac{V_r}{V_s}\right\} + X_{eff}\sin\delta}. \tag{B-153b}$$

Equations (B-151b), (B-152b), and (B-153b) apply to a PAR (sym), a PAR (asym), and an IR, since all three PFCs emulate a four-quadrant, series-compensating impedance ($\mathbf{Z_{se}} = R_{se} - jX_{se}$) that consists of a resistance ($R_{se} = +R$ or $-R$) and a reactance ($X_{se} = X_C$ or $-X_L$) in series with the line. However, for a RR, $R_{se} = 0$. Therefore, the Equations (B-151b), (B-152b), and (B-153b) can be rewritten as follows.

The RPI per unit of active power (P_r) transmitted is given by

$$\frac{RPI}{P_r} = \frac{Q_{line}}{P_r^2} = \frac{I^2X}{P_r^2} = \frac{V_{Xn}^2X}{V_s^2V_r^2}\frac{1}{[\sin\delta]^2} = \frac{V_{Xn}^2}{P_{rn}^2}\frac{1}{X} \tag{B-151c}$$

using Equations (2-41) and (2-46).

The LI per unit of active power (P_r) transmitted is given by

$$\frac{LI}{P_r} = k_R\frac{I^2}{I_n^2P_r} = k_R\frac{V_{Xn}^2}{I_n^2V_sV_r}\frac{1}{X_{eff}\sin\delta} = k_R\frac{1}{V_sV_r}\frac{X^2}{X_{eff}\sin\delta} = k_R\frac{1}{P_{rn}}\frac{X}{X_{eff}} \tag{B-152c}$$

using Equations (2-33), (2-41), and (2-46).

The APR per unit of active power (P_r) transmitted is given by

$$\frac{APR}{P_r} = \frac{|V_{s's}|I}{P_r} = \frac{X_{se}I^2}{P_r} = \frac{X_{se}V_{Xn}^2}{X_{eff}^2}\frac{X_{eff}}{V_sV_r\sin\delta} = \frac{V_{Xn}^2}{V_sV_r}\frac{X_{se}}{X_{eff}\sin\delta} \tag{B-153c}$$

using Equations (2-234d), (2-207), and (2-209). Note that $V_{Xn} = V_{Xeff}$.

B.3 Descriptions of the Examples in Chapter 2

A list of all the "Examples" from Chapter 2 is presented below.

Example 2-1 In Figures 2-2a and 2-2b, the magnitude (V_s) of the sending-end voltage and the line reactance (X) are 1 and 0.5 pu, respectively. For maximum power transfer, (a) what is the load resistance (R_{load})? (b) what is the load voltage (V_r)? (c) what is the line current (I)? (d) what is the load power (P_r)? and (e) what is the line reactive power (Q_{line})?

Example 2-2 In Figures 2-4a and 2-4b, (a) what type and how much of shunt compensation (X_{sh}) is needed for maximum amount of power transfer to the resistive load ($R_{load} = 0.5$ pu), assuming the line reactance, $X = 0.5$pu? (b) what is the load voltage (V_r)? (c) what is the line current (I_{load})? (d) what is the load power (P_r)? (e) what is the line reactive power (Q_{line})? and (f) what is the rating of the shunt compensator (Q_{sh})?

Example 2-3 Consider in a near no-load line, the magnitude (V_s) of the sending-end voltage ($\mathbf{V_s}$) in Figure 2-1 is 0.99 pu and $\mathbf{Z} = jX = j0.045$ pu; (a) what is the magnitude (V_r) of the receiving-end voltage ($\mathbf{V_r}$) when $\mathbf{Z_{load}} = jX_L$ where $X_L = 0.945$ pu? (b) what is the magnitude (V_r) when $\mathbf{Z_{load}} = -jX_C$ where $X_C = 1.035$ pu? (c) what is the rating of the shunt compensator?

Example 2-4 In Figure 2-6, the magnitudes (V_s and V_r) of the sending- and receiving-end voltages are 1 pu each, the power angle, $\delta = \delta_s - \delta_r = 30°$, and the line reactance, $X = 0.5$ pu; (a) what is the natural voltage ($\mathbf{V_{Xn}}$) across the line reactance (X)? (b) what is the resulting current ($\mathbf{I_n}$)? (c) plot the voltage magnitude along the line.

Example 2-5 Consider the electrical system data as given in Table 2-1. What are the active and reactive power flows (P_{sn} and Q_{sn}) at the sending end, (P_{mn} and Q_{mn}) at the midpoint, and (P_{rn} and Q_{rn}) at the receiving end of the line?

Examples 2-6 through 2-43, except 2-34, employ the electrical system data as given in Table 2-1.

Example 2-6 (VRT) For $V_{s'} = 1.2$ pu and $V_{s'} = 0.8$ pu, calculate $m_{s'}$ and $c_{s'}$.

Example 2-7 (VRT) For $\beta = 0°$ or $\beta = 180°$, calculate m_r and c_r.

Examples 2-8 through 2-11 are applicable to PFCs with both Shunt-Shunt and Shunt-Series configurations.

Example 2-8 For a given $V_{s'} = 0.967715$ pu and $\delta' = 41.515336°$, calculate $P_{s'}$ and $Q_{s'}$.

Example 2-9 For a given $V_{s'} = 0.967715$ pu and $\delta' = 41.515336°$, calculate P_r and Q_r.

Example 2-10 For a given $P_{s'} = 1.282843$ pu and $Q_{s'} = 0.423737$ pu, calculate $V_{s'}$ and δ'.

Example 2-11 For a given $P_r = 1.282843$ pu and $Q_r = -0.550792$ pu, calculate $V_{s'}$ and δ'.

Example 2-12 Using a Shunt-Shunt configuration, $V_{s'} = 0.967715$ pu, $\delta' = 41.515336°$, $P_{s'} = 1.282843$ pu, and $Q_{s'} = 0.423737$. Calculate V_X, I, R_{sh}, and X_{sh}.

Example 2-13 (IR) For a given $V_{s's} = 0.2$ pu and $\beta = 105°$, calculate $V_{s'}$, ψ, and δ'.

Example 2-14 (IR) For a given $V_{s'} = 0.967715$ pu, $\psi = 11.515336°$, and $\delta' = 41.515336°$, calculate $V_{s's}$ and β.

Example 2-15 (IR) For a given $V_{s's} = 0.2$ pu and $\beta = 105°$, calculate V_X and I.

Example 2-16 VR & PAR(asym) Using a VR, (a) for $P_s = 1.4$pu, what is $V_{s's}$? (b) for $P_s = 0.6$pu, what is $V_{s's}$? Using a PAR (asym), (c) for $Q_s = 0.667949$ pu, what is $V_{s's}$? (d) for $Q_s = -0.132051$ pu, what is $V_{s's}$?

Example 2-17 VR Using a VR, (a) for $P_r = 1.2$ pu, what is $V_{s's}$? (b) for $P_r = 0.8$ pu, what is $V_{s's}$? (c) for $Q_r = 0.078461$ pu, what is $V_{s's}$? (d) for $Q_r = -0.614359$ pu, what is $V_{s's}$?

Example 2-18 PAR(asym)	Using a PAR (asym), (a) for $P_r = 1.34641$ pu, what is $V_{s's}$? (b) for $P_r = 0.65359$ pu, what is $V_{s's}$? (c) for $Q_r = -0.467949$ pu, what is $V_{s's}$? (d) for $Q_r = -0.067949$ pu, what is $V_{s's}$?
Example 2-19 PAR(asym)	For $\beta = 90°$ or $\beta = -90°$, calculate m_r and c_r.
Example 2-20 VR	Using a VR, (a) for $P_{s'} = 1.2$ pu, what is $V_{s's}$? (b) for $P_{s'} = 0.8$ pu, what is $V_{s's}$? (c) for $Q_{s'} = 0.801539$ pu, what is $V_{s's}$? (d) for $Q_{s'} =$ 0.105641 pu, what is $V_{s's}$?
Example 2-21	For $\beta = 0°$ or $\beta = 180°$ and $V_{s's} = 0.2$ pu, calculate $m_{s'}$ and $c_{s'}$.
Example 2-22 PAR(asym)	Using a PAR (asym), (a) for $P_{s'} = 1.34641$ pu, what is $V_{s's}$? (b) for $P_{s'} = 0.65359$ pu, what is $V_{s's}$? (c) for $Q_{s'} = 0.547949$ pu, what is $V_{s's}$? (d) for $Q_{s'} = 0.147949$ pu, what is $V_{s's}$?
Example 2-23 PAR(asym)	For $\beta = 90°$ or $\beta = -90°$ and $V_{s's} = 0.2$, calculate $m_{s'}$ and $c_{s'}$.
Example 2-24 VR	Using a VR, (a) for $P_{se} = -0.2$ pu, what is $V_{s's}$? (b) for $P_{se} = 0.2$ pu, what is $V_{s's}$? (c) for $Q_{se} = -0.13359$ pu, what is $V_{s's}$? (d) for $Q_{se} = -0.02641$ pu, what is $V_{s's}$?
Example 2-25	For $\beta = 0°$ or $\beta = 180°$, calculate m_{se} and c_{se}.
Example 2-26 PAR(asym)	Using a PAR (asym), (a) for $P_{se} = 0.05359$ pu, what is $V_{s's}$? (b) for $P_{se} = -0.05359$ pu, what is $V_{s's}$? (c) for $Q_{se} = -0.28$ pu, what is $V_{s's}$? (d) for $Q_{se} = 0.12$ pu, what is $V_{s's}$?
Example 2-27 PAR(asym)	For $\beta = 90°$ or $\beta = -90°$, calculate m_{se} and c_{se}.
Example 2-28 (IR)	For a given $V_{s's} = 0.2$ pu and $\beta = 105°$, calculate P_s and Q_s.
Example 2-29 (IR)	For a given $V_{s's} = 0.2$ pu and $\beta = 105°$, calculate P_r and Q_r.
Example 2-30 (IR)	For a given $V_{s's} = 0.2$ pu and $\beta = 105°$, calculate $P_{s'}$ and $Q_{s'}$.
Example 2-31 (IR)	For a given $V_{s's} = 0.2$ pu and $\beta = 105°$, calculate P_{se} and Q_{se}
Example 2-32 (IR)	For a given $P_s = 1.386370$ pu and $Q_s = 0.164422$ pu, calculate $V_{s's}$ and β.
Example 2-33 (IR)	For a given $P_r = 1.282843$ pu and $Q_r = -0.550792$ pu, calculate $V_{s's}$ and β
Example 2-34 (IR)	For a given $P_{s'} = 1.282843$ pu and $Q_{s'} = 0.423737$ pu, calculate $V_{s's}$ and β.
Example 2-35 (IR)	For a given $P_{se} = 0.103528$ pu and $Q_{se} = -0.259315$ pu, calculate $V_{s's}$ and β.
Example 2-36 (IR)	For a given $V_{s's} = 0.2$ pu and $\beta = 105°$, calculate R_{se} and X_{se}.
Example 2-37 (IR)	For an IR operating at $V_{s's} = 0.2$ pu and $\beta = 105°$, calculate R_{se} and X_{se}.
Example 2-38 (VR)	For a VR operating at $V_{s's} = 0.2$ pu and $\beta = 0°$, calculate $V_{s'}$, δ', R_{se}, and X_{se}.
Example 2-39 PAR(asym)	For a PAR (asym) operating at $V_{s's} = 0.2$ pu and $\beta = 90°$, calculate $V_{s'}$, δ', R_{se}, and X_{se}.
Example 2-40 PAR(sym)	For a PAR (sym) operating at $\psi = 11.478341°$, calculate $V_{s'}$, δ', R_{se}, and X_{se}.
Example 2-41 (RR)	Consider the magnitudes of the sending- and receiving-end voltages: $V_s = V_r = 1$ pu, the power angle: $\delta = \delta_s - \delta_r = 60°$, and the line reactance: $X = 0.5$ pu. For a RR operating at $V_{s's} = 0.5$ pu (capacitive) and -0.5 pu (inductive), respectively, calculate $V_{s'}$, δ', and X_{se}.

(Continued)

Example 2-42 PAR(sym)	For a PAR (sym) operating at $\psi = -30°$, calculate $P_{s'}$, $Q_{s'}$, P_r, Q_r, R_{se}, and X_{se}.
Example 2-43 PAR(sym)	For a PAR (sym) operating at $\psi = 60°$, calculate $P_{s'}$, $Q_{s'}$, P_r, Q_r, R_{se}, X_{se}, *RPI*, *LI* for $k_R = 1$, and *APR*. Adjusting the control parameter (ψ), tabulate *RPI*, *LI*, and *APR* of a PAR (sym) for power flow increases in three cases: Case 1 (40%), Case 2 (70%), and Case 3 (100%).
Example 2-44 PAR(asym)	For a PAR (asym) operating at $V_{s's} = -0.577350$ pu, calculate $P_{s'}$, $Q_{s'}$, P_r, Q_r, R_{se}, and X_{se}.
Example 2-45 PAR(asym)	For a PAR (asym) operating at $V_{s's} = 0.577350$ pu, calculate $P_{s'}$, $Q_{s'}$, P_r, Q_r, R_{se}, X_{se}, *RPI*, *LI* for $k_R = 1$, and *APR*. Adjusting the control parameter, $V_{s's}$, tabulate *RPI*, *LI*, and *APR* of a PAR (asym) for power flow increases in three cases: Case 1 (40%), Case 2 (70%), and Case 3 (100%).
Example 2-46 (RR)	For a RR operating at $V_{s's} = -0.517638$ pu, calculate P_r, Q_r, $P_{s'}$, $Q_{s'}$, R_{se}, and X_{se}.
Example 2-47 (RR)	For a RR operating at $V_{s's} = 0.5$ pu, calculate P_r, Q_r, $P_{s'}$, $Q_{s'}$, R_{se}, X_{se}, *RPI*, *LI* for $k_R = 1$, and *APR*. Adjusting the control parameter, $V_{s's}$, tabulate *RPI*, *LI*, and *APR* of a RR for power flow increases in three cases: Case 1 (40%), Case 2 (70%), and Case 3 (100%).
Example 2-48 (IR)	For an IR operating at $V_{s's} = 0.5$ pu and $\beta = 240°$, calculate $P_{s'}$, $Q_{s'}$, P_r, Q_r, R_{se}, and X_{se}.
Example 2-49 (IR)	For an IR operating at $V_{s's} = 0.5$ pu and $\beta = 60°$, calculate $P_{s'}$, $Q_{s'}$, P_r, Q_r, R_{se}, X_{se}, *RPI*, *LI* for $k_R = 1$, and *APR*. Adjusting the control parameter ($V_{s's}$) while $\beta = 60°$, tabulate *RPI*, *LI*, and *APR* of an IR for power flow increases in three cases: Case 1 (40%), Case 2 (70%), and Case 3 (100%).
Example 2-50 (BTB-SSSC)	Consider two identical lines that are connected in parallel as shown in Figure 2-68. Using the electrical system data for each line as given in Table 2-1, use a BTB-SSSC with a series-compensating voltage ($\mathbf{V_{1s's}}$) to make $P_{1r} = 2 \times P_{1rn}$ and $Q_{1r} = Q_{1rn}$. Calculate $V_{1s's}$, β_1, P_{1se}, and Q_{1se}. For the second line with a given series-compensating voltage ($\mathbf{V_{2s's}}$) when $P_{link} = 0$ and $Q_{2se} = 0$, calculate $V_{2s's}$ and β_2.

Appendix C

Modeling of the Sen Transformer in PSS®E

SIEMENS

Siemens Industry, Inc.,
Siemens Power Technologies International
400 State Street
Schenectady, New York 12305 USA
Tel: +1 (518)395-5000 • Fax: +1 (518) 346-2777
www.siemens.com/power-technotogies

Memorandum

Date: May 2. 2019
To: Bruce Fardanesh, NYPA
CC:
From: Daniel Feltes, Carlos Grande-Moran, Dinemayer Silva, and James Feltes
RE: **Modeling of the Sen Transformer in PSS®E**

The Sen single-core three-phase transformer is a power flow-regulating transformer that performs independent active and reactive power flow control. The Sen transformer includes wye-grounded shunt primary windings (exciter unit) electrically connected to secondary windings in series with the transmission circuit where active and reactive flows will be controlled (by the compensating voltage unit). Active and reactive flow control is carried out by changes in taps available on the Sen transformer's series windings. This change in tap position results in changes in both voltage magnitude and phase angle across the compensating voltage unit.

Currently, there is no mathematical model for representing the Sen transformer for steady-state analysis in commercial-grade software such as PSS®E. A series connection of a phase angle regulator (PAR) transformer and a voltage regulating transformer (LTC) is proposed for the modeling of the Sen transformer. This model was tested in the Chilean transmission grid to evaluate if it emulates the expected performance of a Sen transformer.

C.1 Sen Transformer

The Sen transformer can be represented by the asymmetrical PAR transformer model shown in Figure C-1. This transformer model has a complex voltage ratio $te^{j\psi}$, where t is associated to the magnitudes of the voltage ratio and the angle ψ is associated with the voltage phase angle difference

Power Flow Control Solutions for a Modern Grid using SMART Power Flow Controllers, First Edition.
Kalyan K. Sen and Mey Ling Sen.

Companion website: www.wiley.com/go/sen/powerflowcontrol

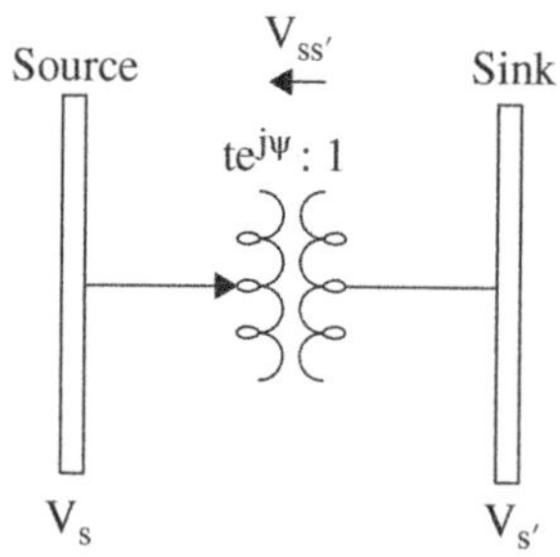

Figure C-1 Sen transformer representation.

between phasors V_s and $V_{s'}$. An important component of the transformer is the series compensating voltage $V_{ss'} \angle\beta$ introduced by the compensating voltage unit. The Q-component (i.e. quadrature) of the compensating voltage is used to control the throughput reactive power flow. Whereas the P-component (i.e. in-phase) is used to adjust the phase shift angle. ψ, so that the throughput active flow can be controlled. These parameters are mutually dependent as shown in the phasor diagram in Figure C-2.

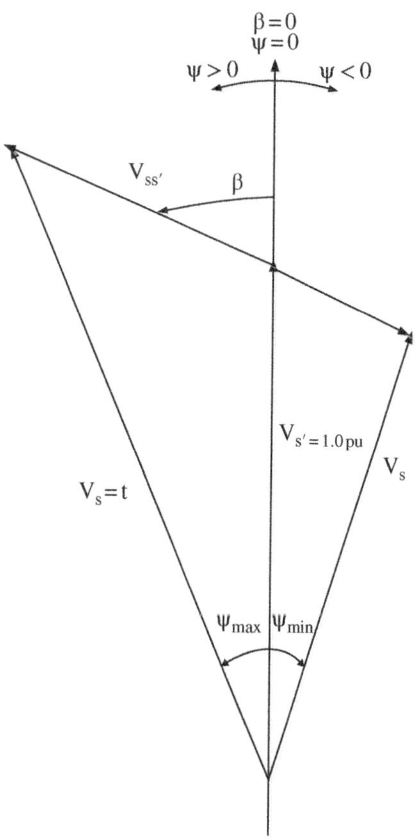

Figure C-2 Phasor diagram of the Sen transformer.

This document was prepared by Siemens Industry. Inc., Siemens Power Technologies International (Siemens PTI), solely for the benefit of the recipient named in this memorandum. Siemens PTI nor any party acting on its behalf (a) makes any warranty, expressed or implied, with respect to the use of any information or methods disclosed m this document; or (b) assumes any liability with respect to the use of any information or methods disclosed in this document.

Any party other than the named recipient of this memorandum, by their acceptance or use of this document, releases Siemens PTI from any liability for direct, indirect, consequential or special loss or damage whether arising in contract, warranty, express or implied, tort or otherwise, and irrespective of fault negligence, and strict liability.

The phasor equation to represent the sending-end voltage to the receiving-end voltage is shown below.

$$\overrightarrow{V_{s'}} = \frac{1}{t}\overrightarrow{V_s}e^{-j\psi} \tag{C-1}$$

C.2 Modeling with Two Transformers in Series

Using the no-load complex voltage ratio shown in eqn. C-1, a suggested representation for the Sen transformer in load flow analysis is one in which a symmetrical PAR on MW control mode is connected in series with a load tap changer transformer (LTC) on voltage control mode, as shown in Figure C-3.

The PAR controls the MW flow and the LTC regulates the phasor V_s voltage, although it may also regulate the receiving end phasor voltage $V_{s'}$ or a remote bus voltage. The winding one of the symmetrical PAR transformer model is connected to the sending-end bus and the winding one of the LTC transformer model is connected to the midpoint bus.

The active power-regulating transformer uses the continuous phase angle adjustment control mode in PSS®E and the voltage-regulating transformer uses the discrete tap adjustment control mode. It is also suggested that the PAR leakage impedance be set to a small value, for example,

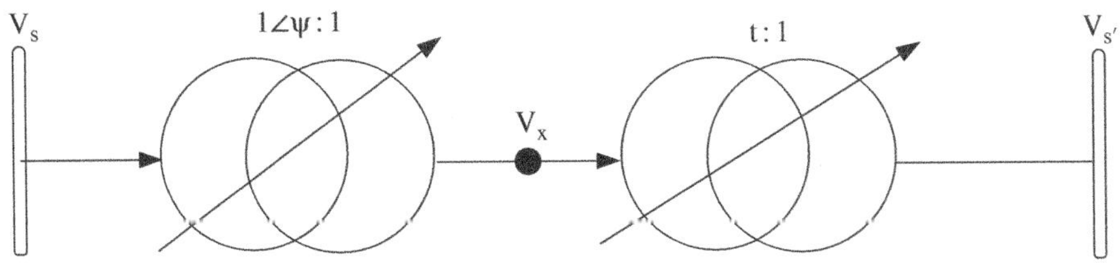

Figure C-3 Proposed Sen transformer representation in PSS®E load flow.

5 to 10 times the default value for zero impedance in PSS®E (default value is 0+j0.0001 pu). The LTC leakage impedance will be equal to the difference between the Sen transformer leakage impedance and the PAR leakage impedance.

The per-unit positive sequence leakage impedance values can be specified on either winding MVA base or system MVA base, and winding voltage base.

The following quantities need to be specified for the PAR and the LTC transformers:

PAR transformer:

- Phase angle range (**ψ**max, **ψ**min)
- Desired MW control band (**MW**max, **MW**min)
- Control mode "**3**" (symmetrical PAR transformer)
- MVA ratings (same as LTC transformer model)

LTC transformer:

- Tap range (**t**max, **t**min)
- Desired voltage control band (**V**max, **V**min) at the controlled bus voltage
- Control mode "**1**" (LTC voltage-regulating transformer)
- MVA ratings (same as PAR transformer model)

The desired MW flow would be adjusted by the PAR phase angle and MW control band settings and the desired voltage would be adjusted by the LTC tap range and voltage control band settings. Note that when solving the load flow in PSS®E, the options "Tap Adjustment" and "Adjust phase shift" should be activated.

The phasor equation to solve is shown below. This is the same phasor equation that is used for the Sen transformer above.

$$\overrightarrow{V_{S'}} = \frac{1}{t}\,\vec{V}_S e^{-j\psi} \tag{C-2}$$

C.3 Relating the Sen Transformer with the PSS®E Model

Because the Sen transformer model in PSS®E is not a one-to-one match to the original model of the Sen transformer, a set of equations were derived to relate the two models. The equations below can be used to compute the PSS®E quantities of off-nominal tap t and phase angle ψ from the given magnitude voltage change $\Delta V(V_{SS'})$ and phase angle β from the Sen transformer.

$$t = \sqrt{\Delta V^2 + 2\Delta V\cos\beta + 1} \tag{C-3}$$

$$\psi = \sin^{-1}\left(\frac{\Delta V}{t} * \sin\beta\right) \tag{C-4}$$

The inverse equations are given below. These equations would be used to compute $\boldsymbol{\Delta V}(|\boldsymbol{V_{ss'}}|)$ and $\boldsymbol{\beta}$ from the PSS®E quantities $\boldsymbol{t}$ and $\boldsymbol{\psi}$.

$$\Delta V = \sqrt{t^2 - 2t\cos\psi + 1} \tag{C-5}$$

$$\beta = \sin^{-1}\left(\frac{t}{\Delta V} * \sin\psi\right) \tag{C-6}$$

These equations are important for exchanging parameters between the two models and checking the validity of the results.

C.4 Chilean Case Study

To test the performance of the two series transformer representation, a three-phase 300 MVA, 154/154 kV Sen transformer was added to a PSS®E load flow base case of the Chilean grid, supplied by for use in these investigations by Transelec of Chile. This transformer was placed in one of the parallel paths between the Southern and Central control areas of the system. The proposed PSS®E model was added between the Punta de Cortes and the Tilcoco 154 kV substations. These buses and the surrounding area are shown in Figure C-4.

The study area after adding the Sen transformer is shown in Figure C-5. The PAR component of the two-transformer model is placed between buses 4074 TAP PUNTA DE and 10000 SEN. This transformer has a small of impedance of 0.1% on a 300 MVA base. This transformer is also in control mode 3 or MW flow control. The LTC portion of the two-transformer model is placed between bus 10000 SEN and 4130 TAP TILCOCO. This transformer has a leakage impedance of 6% on a 300 MVA and 154 kV base. The control mode for the LTC transformer is control mode 1 or voltage control and is controlling the voltage at bus 4074 TAP PUNTA DE. The base settings for the model are to control 90 MW (within a range of ±1 MW) from TAP PUNTA DE to TAP TILCOCO and to control the voltage at TAP PUNTA DE to a desired magnitude of 1.0 pu (within a rangc of ± 0.005 pu).

Several steady-state simulations were performed with the model controlling different MW and voltage setpoints. In these simulations, the MW flow was varied between 70 MW and 120 MW and three different voltage setpoints of 0.95 pu, 1.0 pu, and 1.05 pu were tested. For all simulations tested, the model was able to control to its desired control point. The equivalent voltage magnitude $\boldsymbol{\Delta V}$ and phase angle $\boldsymbol{\beta}$ were also calculated for easy comparison to actual Sen transformer values. These results are shown in Table C-1.

Simulations were also performed to test the model under contingency conditions. For these tests, the model was set to control the flow of 90 MW from TAP PUNTA DE to TAP TILCOCO and to control voltage at TAP PUNTA DE at 1.0 pu. This was the base case pre-contingency conditions. N-1 and N-2 outages of transmission lines in the study area were simulated with both PAR adjustment and LTC adjustment solution options enabled. Table C-2 below shows the results of this analysis with the first two columns showing the results of the case without the Sen transformer model and the other columns showing the results with the Sen transformer included in study area. This comparison shows that without the Sen transformer model, the controlled voltage and power flow are below the desired control points, but the desired control points are satisfied when the Sen transformer is included.

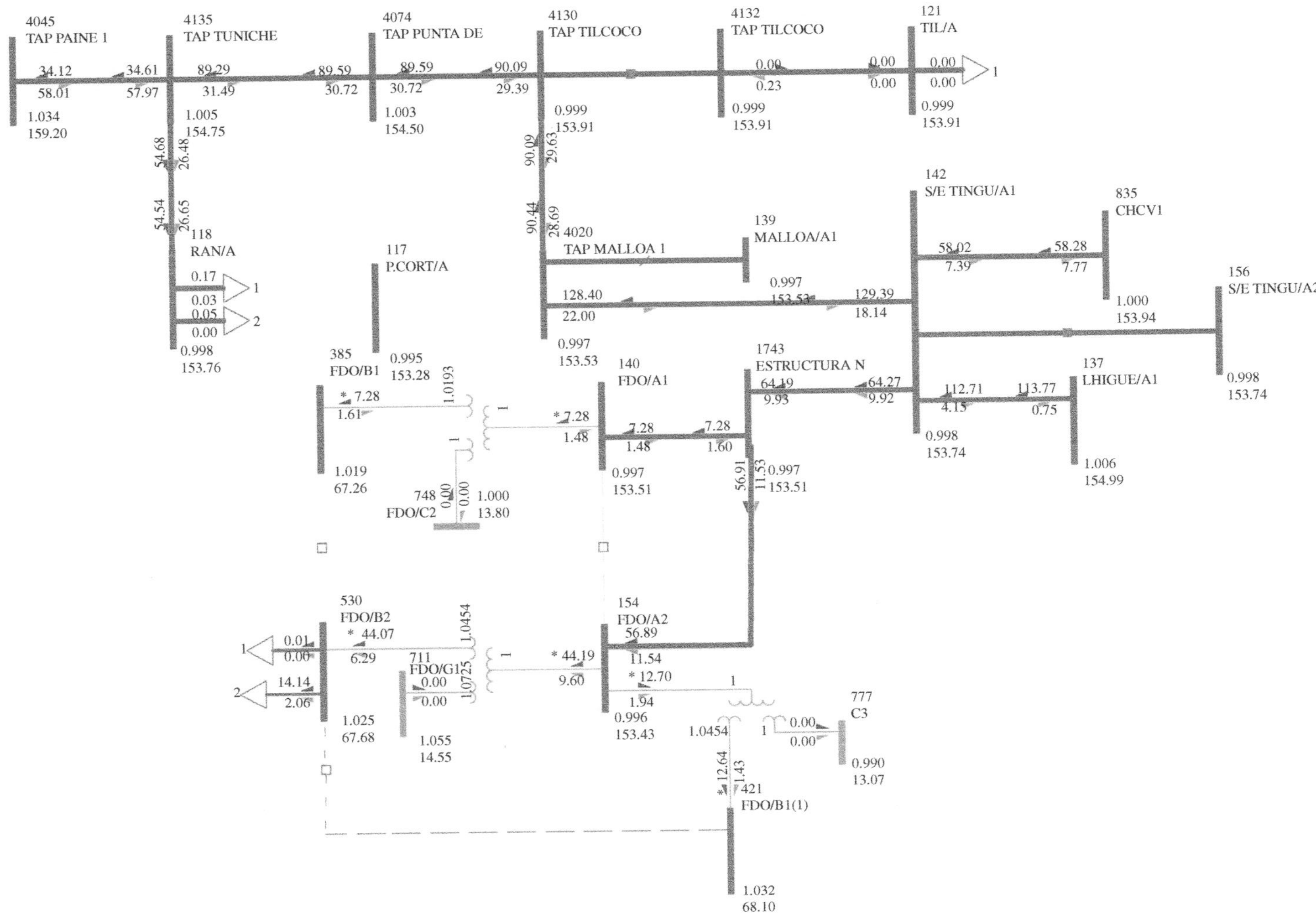

Figure C-4 Chilean power system – study area.

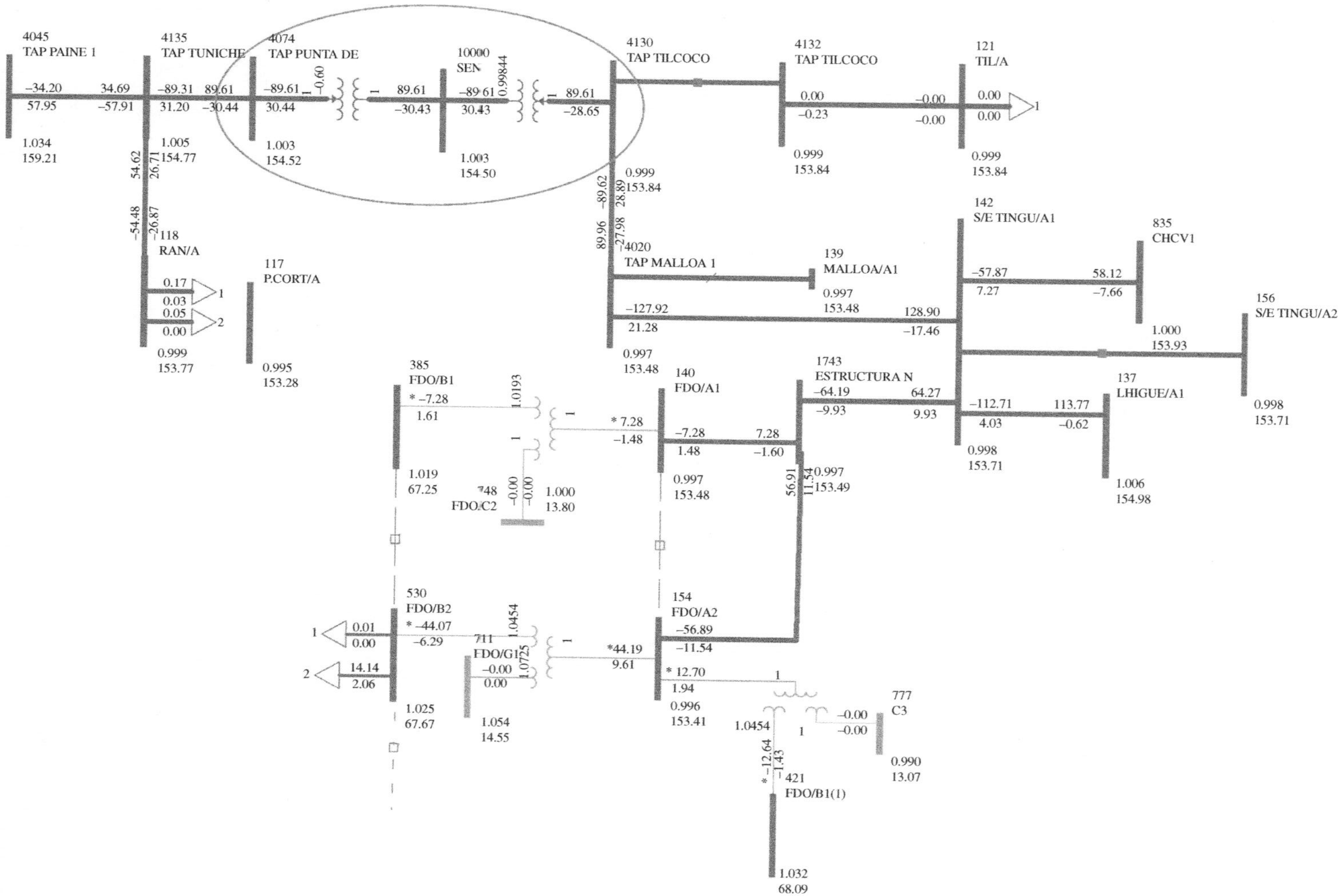

Figure C-5 Chilean system post Sen transformer.

Table C-1 Chilean grid simulation results.

Voltage controlled to 0.95 pu					
Pr (MW)	**Measured Vs**	**Tap(t = t1/t2)**	**PSA ψ(deg)**	**ΔV_{ssp}**	**WindingAng β(deg)**
70	0.954	0.9000	−1.800	0.10	195.719
80	0.954	0.9000	−0.790	0.10	187.068
90	0.953	0.9000	0.220	0.10	178.021
100	0.953	0.9000	1.240	0.10	169.001
110	0.952	0.9000	2.260	0.11	160.585
120	0.952	0.9000	3.270	0.11	153.163
Voltage controlled to 1.0 pu					
Pr (MW)	**Measured Vs**	**Tap(t = t1/t2)**	**PSA ψ(deg)**	**ΔV_{ssp}**	**WindingAng β(deg)**
70	1.004	1.0000	−2.600	0.05	268.700
80	1.004	1.0000	−1.590	0.03	269.205
90	1.004	1.0000	−0.550	0.01	269.725
100	1.004	1.0000	0.450	0.01	90.225
110	1.003	1.0000	1.470	0.03	90.735
120	1.003	1.0000	2.500	0.04	91.250
Voltage controlled to 1.05 pu					
Pr (MW)	**Measured Vs**	**Tap(t = t1/t2)**	**PSA ψ(deg)**	**ΔV_{ssp}**	**WindingAng β(deg)**
70	1.048	1.1156	−3.330	0.13	209.677
80	1.048	1.1156	−2.290	0.12	201.236
90	1.048	1.1156	−1.250	0.12	191.915
100	1.048	1.1156	−0.210	0.12	182.026
110	1.047	1.1156	0.840	0.12	171.939
120	1.047	1.1156	1.890	0.12	162.257

The most constraining case of the tested contingencies was the loss of the154 kV lines S/E TINGU/A to LHIGUE/A1 154 kV and the TAP PAINE1 to TAP TUNICHE. Without the Sen transformer model in the power flow case, the MW flow from TAP PUNTA DE to TAP TILCOCO dropped to 77 MW and the voltage at TAP PUNTA DE dropped to 0.973 pu. With the Sen

Table C-2 Chilean contingency scenario results.

Base case with no Sen transformer		PAR set to control 90 MW and Voltage to 1.0 pu					
Power flow (MW)	Voltage at 4074 TAP PUNTA DE	Pr (MW)	Measured Vs	Tap (t = t1/t2)	PSA ψ(deg)	ΔV_{ssp}	WindingAng β(deg)
N-0		N-0					
89	1.003	90	1.003	0.99844	−0.6	0.01	261.221
LOSS of 4045 TAP PAINE 1 to 4135 TAP TUNICHE		LOSS of 4045 TAP PAINE 1 to 4135 TAP TUNICHE					
78	0.981	90	0.997	1.0297	0.68	0.03	157.585
LOSS of 4135 TAP TUNICHE to 118 RAN/A		LOSS of 4135 TAP TUNICHE to 118 RAN/A					
66	1.017	90	1.001	0.97969	3.4	0.06	110.769
LOSS of 142 S/E TINGU/A to 137 LHIGUE/A1		LOSS of 142 S/E TINGU/A to 137 LHIGUE/A1					
89	0.998	90	0.998	0.99844	−0.46	0.01	258.766
LOSS of 142 S/E TINGU/A to 137 LHIGUE/A1 and LOSS of 4045 TAP PAINE 1 to 4135 TAP TUNICHE		LOSS of 142 S/E TINGU/A to 137 LHIGUE/A1 LOSS Of 4045 TAP PAINE 1 to 4135 TAP TUNICHE					
77	0.973	90	0.997	1.0484	0.65	0.05	166.176

transformer model added in the power flow case, the MW flow from TAP PUNTA DE to TAP TILCOCO was controlled to 90 MW and the voltage at TAP PUNTA DE was also controlled to 1.0 pu. Thus, the Sen transformer was able to control the MW flow and bus voltage to their pre-contingency values under this contingent scenario.

C.5 Limitations – PSS®E Two-Transformer Model

In PSS®E, the phase angle limits of a PAR transformer are defined by ψ_{max} and ψ_{min} and similarly for an LTC voltage-regulating transformer, its limits are defined by R_{max} and R_{min}. PSS®E then controls the phase angle ψ continuously between the limits to maintain the desired MW flow and adjusts the taps of the LTC discretely between the limits to control voltage. The Sen transformer, on the other hand, is controlled by adjusting its taps and thus the parameters: voltage magnitude $V_{ss'}$ and phase angle β. This control method is constrained by a hexagon that represents the allowable operating area.

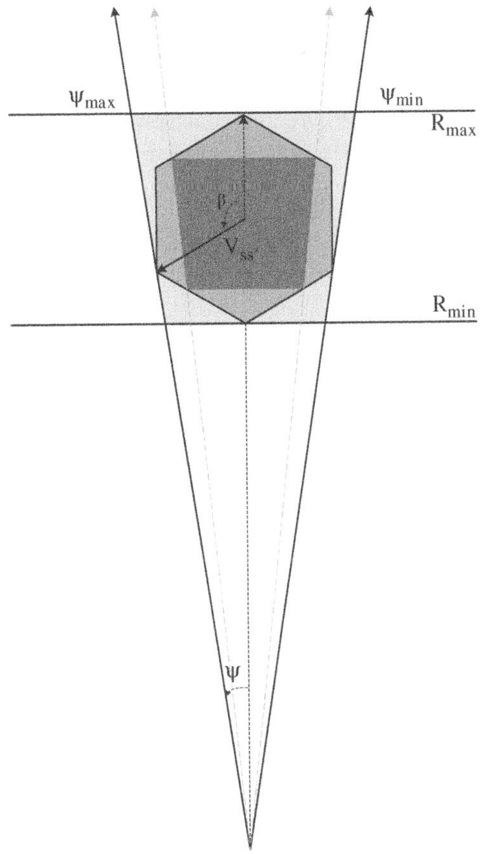

Figure C-6 Operating limits of the two-transformer model and the Sen transformer.

The limits used in the PSS®E model and the Sen transformer are shown in Figure C-6. The ψ_{max} and ψ_{min} and R_{max} and R_{min} limits used in PSS®E result in a control area having a trapezoidal shape, rather than the hexagonal shape of the control area of the Sen transformer. It is important to note that if the PSS®E limits (the blue trapezoidal area) are set to the bottom edges of the hexagonal operating range of the Sen transformer (shown in yellow), then the model in PSS®E may reach an operating point where the Sen transformer cannot operate as shown by the blue region of Figure C-6. A solution to this problem is to limit the PSS®E model to only operate within the region bound by the hexagon. This operating range is shown by the green trapezoidal region in Figure C-6. With this solution, the model is conservative and always operates within the equipment capabilities, although the model will not use the full capability available of the Sen transformer.

Another solution to this problem is to develop a python script that would adjust the PSS®E model to perform additional checks. The script would solve the load flow using the above limits of the Sen transformer. If we are using the recommended conservative approach (the green area in Figure C-6), then the script would check to see if the Sen transformer was on limits. If the Sen transformer was on limits, the script could then adjust the phase angle of the PAR and the tap of the LTC to implement the full range of the Sen transformer. This would be an iterative approach, likely requiring several load flow solutions to converge into either a .solution that meets the flow and voltage settings or alternately is on the actual Sen transformer limits. But, we again note that if the conservative green area limits are used and the script is not run, the Sen transformer is operating within its limits in the load flow solution, although not necessarily optimally as some of its limits are modeled as too limiting in the load flow model. The importance of this "lost area" in modeling of the available control range is hard to quantify without analyzing an actual application of the Sen transformer.

C.6 Conclusion

The two-transformer representation using standard PSS®E transformer models is a reasonable way to represent the Sen transformer. Use of this representation in the steady-state simulations with the Chilean transmission system showed that it was able to control to the desired MW and voltage control points. Also, under selected N-1 and N-2 contingency conditions, the proposed Sen transformer model was able to control MW flow and bus voltage magnitude to the desired set points and provide some voltage support in the study area.

When using the proposed PSS®E two-transformer representation, the phase angle and tap range limits must be carefully considered as to not overestimate the operating range of the Sen transformer. This can be done either by setting the PSS®E model limits within the hexagonal operating zone of the Sen transformer limits or by the use of a python script to perform more detailed calculations incorporating the Sen transformer's full control range, which was also discussed.

References

H. Barot, L. Singh and R. K. Varma, "Operational Experience with Static Var Compensators in Ontario, Canada," *IEEE Power & Energy Society General Meeting*, Paper No. PESGM2014-1756, Jul. 2014, Washington, DC, USA.

M. O. Faruque and V. Dinavahi, "A tap-changing algorithm for the implementation of Sen transformer," *IEEE Trans. Power Delivery*, vol. 22, no. 3, pp. 1750–1757, Jul. 2007.

Leuven EMTP Center, *ATP Rule Book*, Ed. Leuven, Belgium: Leuven EMTP Center, 1990.

K. K. Sen, "Overview of Voltage Regulation Schemes for Utility and Industrial Applications," *How2Power.com e-magazine*, Sep. 2015. http://www.how2power.com/newsletters/1509/articles/H2PToday1509_design_Sen.pdf?NOREDIR=1

K. K. Sen and M. L. Sen, *Introduction to FACTS Controllers: Theory, Modeling, and Applications*. New York: IEEE Press and John Wiley & Sons, Inc., 2009.

K. K. Sen and A. J. F. Keri, "Comparison of field results and digital simulation results of voltage-sourced converter-based FACTS controllers," *IEEE Trans. Power Delivery*, vol. 18, no. 1, pp. 300–306, Jan. 2002.

Further Reading

Books

1 IEEE Working Group (15.05.07) on HVDC Bibliography and Records, Bonneville Power Administration, Western Area Power Administration, *An Annotated Bibliography of HVDC Transmission and FACTS Devices 1991–1993*. Portland, OR: Bonneville Power Administration, Western Area Power Administration, 1993.
2 IEEE Working Group (15.05.07) on HVDC Bibliography and Records, Bonneville Power Administration, Western Area Power Administration, *An Annotated Bibliography of HVDC Transmission and FACTS Devices 1994–1995*. Portland, OR: Bonneville Power Administration, Western Area Power Administration, 1995.
3 IEEE Working Group (15.05.07) on HVDC Bibliography and Records, Bonneville Power Administration, Western Area Power Administration, *An Annotated Bibliography of HVDC Transmission and FACTS Devices 1996–1997*. Portland, OR: Bonneville Power Administration, Western Area Power Administration, 1998.
4 Y. H. Song and A. T. Johns, *Flexible AC Transmission Systems (FACTS)*. London, UK: IEE, 1999.

Power Flow Control Solutions for a Modern Grid using SMART Power Flow Controllers, First Edition.
Kalyan K. Sen and Mey Ling Sen.

Companion website: www.wiley.com/go/sen/powerflowcontrol

5 N. G. Hingorani and L. Gyugyi, *Understanding FACTS: Concepts and Technology of Flexible AC Transmission Systems*. New York: IEEE Press, 2000.
6 R. M. Mathur and R. K. Varma, *Thyristor-Based FACTS Controllers for Electrical Transmission Systems*. New York: IEEE Press and John Wiley & Sons, Inc., 2002.
7 E. Acha, C. R. Fuerte-Esquivel, H. Ambriz-Pérez and C. Angeles-Camacho, *FACTS: Modelling and Simulation in Power Networks*. New York: John Wiley & Sons, Inc., 2004.
8 K. H. Sueker, *Power Electronics Design: A Practitioner's Guide*, 1st Edition. Amsterdam: Elsevier, 2005.
9 K. R. Padiyar, *Facts: Controllers in Power Transmission and Distribution*. Bangalore, India: Anshan, 2007.
10 S. Bhowmick, *Flexible AC Transmission Systems (FACTS)*. Boca Raton, FL: CRC Press, 2016.
11 M. Eremia, C.-C. Liu and A.-A. Edris, *Advanced Solutions in Power Systems: HVDC, FACTS and Artificial Intelligence*. Piscataway, NJ and Hoboken, NJ: IEEE Press and John Wiley & Sons, Inc., 2016.

General

12 Westinghouse Electric Corporation (Ed.), "AC voltage regulator," in *Electrical Transmission and Distribution Reference Book*. East Pittsburgh, PA: Westinghouse Electric & Manufacturing Company, 1942.
13 L. Gyugyi and M. B. Brennen, "Method and control apparatus for static VAR generator and compensator," U.S. Patent 3,999,117, 1976.
14 W. S. Meyer, "EMTP data modularization and sorting by class: a foundation upon which EMTP data bases can be built," *EMTP Newsletter*, vol. 4, no. 2, pp. 28–40, Nov. 1983.
15 E. J. Stacey and M. B. Brennen, "Active power conditioner system," U.S. Patent 4,651,265, 1987.
16 T. T. Nguyen and W. D. Humpage, "TCR and SR compensators in composite electromagnetic transient simulation," *IEE Proc. C*, vol. 136, no. 3, pp. 195–205, May 1989.
17 L. J. Bohmann and R. H. Lasseter, "Stability and harmonics in thyristor controlled reactors," *IEEE Trans. Power Delivery*, vol. 5, no. 2, pp. 1175–1181, Apr. 1990.
18 A. M. Gole and V. K. Sood, "A static compensator model for use with electromagnetic transients simulation programs," *IEEE Trans. Power Delivery*, vol. 5, no. 3, pp. 1398–1407, Jul. 1990.
19 C. K. K. Tang and D. R. Turner, "Modeling of large nonlinear loads and thyristor controlled shunt compensators," *IEE Proc. C*, vol. 138, no. 6, pp. 505–517, Nov. 1991.
20 A. N. Vasconcelos, A. J. P. Ramos, J. S. Monteiro, M. B. V. C. Lima, H. D. Silva and L. R. Lins, "Detailed modeling of an actual static VAR compensator for electromagnetic transients studies," *IEEE Trans. Power Syst.*, vol. 7, no. 1, pp. 11–19, Feb. 1992.
21 J. A. Martinez, "How to adopt the EMTP for classroom instruction," *IEEE Trans. Power Syst.*, vol. 7, no. 1, pp. 351–358, Feb. 1992.
22 S. Y. Lee, S. Bhattacharya, T. Lejonberg, A. E. Hammad and S. Lefebvre, "Detailed modeling of static VAR compensators using the electromagnetic transients program (EMTP)," *IEEE Trans. Power Delivery*, vol. 7, no. 2, pp. 836–847, Apr. 1992.
23 S. Lefebvre and L. Gerin-Lajoie, "A static compensator model for the EMTP," *IEEE Trans. Power Syst.*, vol. 7, no. 2, pp. 477–486, May 1992.
24 A. J. F. Keri, B. J. Ware, R. A. Byron and A. S. Mehraban, "Improving Transmission System Performance Using Controlled Series Capacitors," *CIGRE*, Paper No. 14/37/38-07, 1992.
25 B. Pilvelait, T. H. Ortmeyer and D. Maratukulam, "Advanced series compensation for transmission system using a switched capacitor module," *IEEE Trans. Power Delivery*, vol. 8, no. 2, pp. 584–590, Apr. 1993.

26 A. T. Johns and Q. Y. Xuan, "Digital study of thyristor controlled series capacitor compensated EHV transmission systems," *IEEE Proc. 10th IEEE Regional Conf. Computer, Communication, Control and Power Engineering, TENCON*, pp. 263–265.

27 L. Gyugyi, C. D. Schauder, C. W. Edwards and M. Sarkozi, "Apparatus and method for dynamic voltage restoration of utility distribution networks," U.S. Patent 5,329,222, 1994.

28 S. G. Helbing and G. G. Karady, "Investigations of an advanced form of series compensation," *IEEE Trans. Power Delivery*, vol. 9, no. 2, pp. 939–947, Apr. 1994.

29 M. R. Iravani, P. L. Dandeno, K. H. Nguyen, D. Zhu and D. Maratukulam, "Applications of static phase shifters in power systems," *IEEE Trans. Power Delivery*, vol. 9, no. 3, pp. 1600–1608, Jul. 1994.

30 M. B. Brennen and B. Banerjee, "Low cost, high performance active power line conditioners," *Third Int. Conf. Power Quality: End–Use Applications and Perspectives*, EPRI, Amsterdam, Oct. 1994.

31 M. R. Iravani and D. J. Maratukulam, "Review of semiconductor controlled (static) phase shifters for power system application," *IEEE Trans. Power Syst.*, vol. 9, no. 4, pp. 1833–1839, Nov. 1994.

32 T. F. Godart, A. F. Imece, J. C. McIver and E. A. Chebli, "Feasibility of thyristor controlled series capacitor for distribution substation enhancements," *IEEE Trans. Power Delivery*, vol. 10, no. 1, pp. 203–209, Jan. 1995.

33 M. Kezunovic, M. Aganagic, S. McKenna and D. Hamai, "Computing responses of series compensation capacitors with MOV protection in real-time," *IEEE Trans. Power Delivery*, vol. 10, no. 1, pp. 244–251, Jan. 1995.

34 A. Ghosh and G. Ledwich, "Modeling and control of thyristor-controlled series capacitors," *IEE Proc. Gener. Transm. Distrib.*, vol. 142, no. 3, pp. 297–304, May 1995.

35 J. A. Martinez, "EMTP simulation of a digitally-controlled static VAr system for optimal load compensation," *IEEE Trans. Power Delivery*, vol. 10, no. 3, pp. 1408–1415, Jul. 1995.

36 W. Zhu, R. Spee, R. R. Mohler, G. C. Alexander, W. A. Mittelstadt and D. Maratukulam, "An EMTP study of SSR mitigation using the thyristor controlled series capacitor," *IEEE Trans. Power Delivery*, vol. 10, no. 3, pp. 1479–1485, Jul. 1995.

37 D. J. Nyati, S. Eitzmann, M. Kappenman, J. Van House, D. Mohan and A. N. Edris, "Design issues for a single core transformer thyristor controlled phase-angle regulator," *IEEE Trans. Power Delivery*, vol. 10, no. 4, pp. 2013–2019, Oct. 1995.

38 S. Lefebvre, A. M. Gole, J. Reeve, L. Piliotto, N. Martins and S. Bhattacharya, "Working group on dynamic performance and modeling of DC systems and power electronics for transmission systems: report on test systems for AC/DC interaction studies," *IEEE Trans. Power Delivery*, vol. 10, no. 4, pp. 2027–2034, Oct. 1995.

39 G. E. Lee and D. L. Goldsworthy, "BPA's Pacific AC intertie series capacitors: experience, equipment and protection," *IEEE Trans. Power Delivery*, vol. 11, no. 1, pp. 253–259, Jan. 1996.

40 M. Junya, I. Keiichi, T. Takeshi and A. Minoru, "Dynamic performance of a thyristor controlled series capacitor system," *Proc. 31st Universities Power Engineering Conf.*, vol. 3, pp. 961–964, Iraklio, Greece, 1996.

41 H. A. Othman and L. Angquist, "Analytical modeling of thyristor controlled series capacitors for SSR studies," *IEEE Trans. Power Syst.*, vol. 11, no. 1, pp. 119–127, Feb. 1996.

42 S. G. Jalali, R. A. Hedin, M. Pereira and K. Sadek, "Stability model for the advanced series compensator (ASC)," *IEEE Trans. Power Delivery*, vol. 11, no. 2, pp. 1128–1137, Apr. 1996.

43 A. R. M. Tenorio, J. B. Ekanayake and N. Jenkins, "Modeling of FACTS devices," *Proc. IEE Conf. AC and DC Power Transmission*, pp. 340–345, London, UK, Apr./May 1996.

44 A. C. S. de Lima, S. S. Wanderley and R. M. Stephan, "Modeling a static VAR compensator using EMTP," *Proc. 38th Midwest Symp. Circuits and Systems*, vol. 2, pp. 219–222, Rio de Janeiro, Brazil, 1996.

45 J. Ndayizamba, S. Lefebvre and D. D. Xuan, "Efficient and flexible thyristor controlled series capacitor simulation in EMTP-TACS," *Proc. 12th Power Systems Computation Conf.*, vol. 2, pp. 1122–1127, Dresden, Germany, 1996.

46 J. Ndayizamba, S. Lefebvre and D. D. Xuan, "Torsional interaction in flexible AC transmission systems with a thyristor controlled series capacitor," *Proc. 5th Electrimacs, Modeling and Simulation of Electric Machines, Converters & Systems*, pp. 1187–1192, Saint Nazaire, France, 1996.

47 B. Das, A. Ghosh and A. Sachchidanand, "A new control scheme for dynamic brake," *Proc. Ninth Nat. Power Systems Conf., NPSC*, vol. 2, pp. 501–507, Kanpur, India, 1996.

48 A. M. Gole, A. Keri, C. Kwankpa, E. W. Gunther, H. W. Dommel, I. Hassan, J. R. Marti, J. A. Martinez, K. G. Fehrle, L. Tang, M. F. McGranaghan, O. B. Nayak, P. F. Ribeiro, R. Iravani and R. Lasseter, "Guidelines for modeling power electronics in electric power engineering applications," *IEEE Trans. Power Delivery*, vol. 12, no. 1, pp. 505–514, Jan. 1997.

49 V. Kaura and V. Blasko, "Operation of a phase locked loop system under distorted utility conditions," *IEEE Trans. Ind. Appl.*, vol. 33, no. 1, pp. 58–63, Jan./Feb. 1997.

50 P. Parihar and G. G. Karady, "Characterization of a thyristor controlled reactor," *Electric Power Syst. Res.*, vol. 41, no. 2, pp. 141–149, May 1997.

51 L. Wang, "Simulations of prefiring NGH damping scheme on suppressing torsional oscillations using EMTP," *IEEE Trans. Power Syst.*, vol. 12, no. 2, pp. 882–888, May 1997.

52 L. Prikler and G. Banfai, "EMTP models for simulation of shunt reactor switching transients," *Int. J. Electr. Power Energy Syst.*, vol. 19, no. 4, pp. 235–240, May 1997.

53 M. R. Sultan, J. Reeve and R. Adapa, "Transient Behavior of Systems Containing FACTS Devices: Modeling of the Thyristor Controlled Series Capacitor," EPRI Final Report TR-108191, 2149-15, Jul. 1997.

54 P. Mattavelli, G. C. Verghese and A. M. Stankovic, "Phasor dynamics of thyristor controlled series capacitor systems," *IEEE Trans. Power Syst.*, vol. 12, no. 3, pp. 1259–1267, Aug. 1997.

55 E. A. Leonidaki, B. C. Papadias, N. D. Hatziargyriou and S. Giorgas, "A physical laboratory model of a thyristor controlled series capacitor," *Proc. 33rd Universities Power Engineering Conf.*, Edinburgh, UK, 1998.

56 O. M. Neto, P. C. A. Leao and A. E. A. Araujo, "Controlled series compensation: dynamic performance under fault conditions," *Proc. 32nd Universities Power Engineering Conf., UPEC*, vol. 1, pp. 5–8, Manchester, UK, 1997.

57 O. M. Neto, R. A. Meneguim, F. D. Oliveira and J. M. Oliveira, "Modeling of a thyristor controlled series capacitor in MATLAB environment," *Proc. 32nd Univertsities Power Engineering Conf., UPEC*, vol. 1, pp. 14–21, Manchester, UK, 1997.

58 J. A. Martinez and N. Mohan, "ATP simulation of power electronics systems using a data-module approach," *Proc. 32nd Universities Power Engineering Conf., UPEC*, vol. 1, pp. 495–498, Manchester, UK, 1997.

59 E. A. Leonidaki, N. D. Hatziargyriou and G. J. Georgantzis, "Simulation of thyristor controlled series capacitors and subsynchronous studies," *Proc. 32nd Universities Power Engineering Conf., UPEC*, vol. 2, pp. 883–886, Manchester, UK, 1997.

60 H.-G. Han, J.-K. Park and B.-H. Le, "Analysis of thyristor controlled series compensator dynamics using the state variable approach of a periodic system model," *IEEE Trans. Power Delivery*, vol. 12, no. 4, pp. 1744–1750, Oct. 1997.

61 B. K. Perkins and M. R. Iravani, "Dynamic modeling of a TCSC with application to SSR analysis," *IEEE Trans. Power Syst.*, vol. 12, no. 4, pp. 1619–1625, Nov. 1997.

62 A. A. Girgis, A. A. Sallam and A. K. El-Din, "An adaptive protection scheme for advanced series compensated (ASC) transmission lines," *IEEE Trans. Power Delivery*, vol. 13, no. 2, pp. 414–420, Apr. 1998.

63 L. Palav and A. M. Gole, "On using the solid-state breaker in distribution systems," *Proc. IEEE Canad. Conf. Electrical and Computer Engineering*, vol. 1, pp. 693–700, Waterloo, Ontario, 1998.

64 A. R. M. Tenorio and N. Jenkins, "Use of EMTP-ATP MODELS language for modeling thyristor controlled series capacitors," *Proc. 33rd Universities Power Engineering Conf., UPEC*, vol. 2, pp. 679–682, Edinburgh, UK, 1998.

65 M. Sultan, J. Reeve and R. Adapa, "Combined transient and dynamic analysis of HVDC and FACTS systems," *IEEE Trans. Power Delivery*, vol. 13, no. 4, pp. 1271–1277, Oct. 1998.

66 P. S. S. Holenarsipur, D. J. Christofersen and N. Mohan, "Thyristor switched arrester for improved protection of power system equipment," *Proc. IEEE PES Winter Meeting*, vol. 2, pp. 1009–1014, New York, 1999.

67 H. Xueqiang-Zhao and H. Chen-Chen, "Circuit analysis of a thyristor controlled series compensation," *Proc. IEEE PES Winter Meeting*, vol. 2, pp. 1067–1072, New York, 1999.

68 Y. Tanaka, H. Taniguchi, M. Egawa, H. Fujita, M. Watanabe and H. Konishi, "Using a miniature model and EMTP simulations to evaluate new methods to control and protect a thyristor controlled series compensator," *Proc. IEEE PES Winter Meeting*, vol. 2, pp. 1196–1201, New York, 1999.

69 P. Mattavelli, G. C. Verghese and A. M. Stankovic, "SSR analysis with dynamic phasor model of thyristor controlled series capacitor," *IEEE Trans. Power Syst.*, vol. 14, no. 1, pp. 200–208, Feb. 1999.

70 Z. Xu and C. Chen, "Damping subsynchronous resonance using an improved NGH SSR damping," *Proc. IEEE PES Summer Meeting*, vol. 2, pp. 780–785, Edmonton, AB, Canada, 1999.

71 D. D. Xueqiang-Zhao, "Study of TCSC model and prospective application in the power systems of China," *Proc. Int. Conf. Power Electronics and Drive Systems, PEDS*, vol. 2, pp. 688–691, Kowloon, Hong Kong, 1999.

72 C. Gama and R. Tenorio, "Improvements for power system performance: modeling, analysis and benefits of TCSCs," *Proc. IEEE PES Winter Meeting*, vol. 2, pp. 1462–1467, 2000.

73 L. F. W. De Souza, E. H. Watanabe and M. Aredes, "GTO controlled series capacitor," *Proc. IEEE PES Winter Meeting*, vol. 4, pp. 2520–2525, 2000.

74 L. F. W. De Souza, E. H. Watanabe and M. Aredes, "GTO controlled series capacitors: multi-module and multi-pulse arrangements," *IEEE Trans. Power Delivery*, vol. 15, no. 2, pp. 725–731, Apr. 2000.

75 P. K. Muttik, P. Wang and M. W. H. Minchin, "Detailed simulation of SVC transient performance using PSCAD," *Proc. Int. Conf. Power System Technology, PowerCon*, Perth, Australia, 2000.

76 R. J. Davalos and J. M. Ramirez, "A review of a quasistatic and a dynamic TCSC model," *IEEE Power Eng. Rev.*, vol. 20, no. 11, pp. 63–65, Nov. 2000.

77 B. K. Johnson, "Simulation of TSC and TSSC," *Proc. IEEE PES Winter Meeting*, vol. 2, pp. 637–638, 2001.

78 L. L. Grigsby, *The Electric Power Engineering Handbook*, vol. 3. Boca Raton, FL: CRC Press, 2001, pp. 191–200.

79 Manitoba HVDC Research Centre, *PSCAD/EMTDC Manual*. Manitoba HVDC Research Centre, 2001.

80 B. K. Johnson, "Benchmark system for the simulation of TCSC and SVC," *Proc. IEEE PES Winter Meeting*, pp. 484–487, New York, 2002.

81 P. Irwin, "Transmission at a crossroads," *Electrical T&D World*, vol. 216, no. 1, pp. 13—23, first quarter 2002.

82 S. Al-Senaidi, "An application of thyristor controlled series capacitor to selective pole switching," *Electric Power Compon. Syst.*, vol. 30, no. 8, pp. 14–20, Aug. 2002.

83 W. Geraldes, E. H. Rose, C. Machado, J. T. Honda and P. M. Vendramini, "Static VAR compensator from Coxipo substation (Brazil), comparison between computer simulations and field tests," *Proc. 5th Int. Conf. AC and DC Power Transmission*, pp. 301–306, London, England, 1991.

84 U.S.-Canada Power System Outage Task Force, Final Report on the August 14, 2003 Blackout in the United States and Canada: Causes and Recommendations," Apr. 2004. https://www.energy.gov/sites/prod/files/oeprod/DocumentsandMedia/BlackoutFinal-Web.pdf, Accessed on Nov. 5, 2020.

STATCOM

85 Y. Sumi, Y. Harumoto, T. Hasegawa, M. Yano, K. Ikeda and T. Matsuura, "New static var control using force-commutated inverters," *IEEE Trans. Power Syst.*, vol. PAS-1OO, no. 9, pp. 4216–4224, Sep. 1981

86 1 MVAR Advanced Static VAR Generator Development Program, *ESEERCO* Report, EP 84-30, Apr. 1987.

87 L. Gyugyi, N. G. Hingorani, P. R. Nannery and N. Tai, "Advanced Static VAR Compensator Using Gate Turn-Off Thyristors for Utility Applications," *CIGRE,* Paper No. 23-203, 1990.

88 C. Schauder and H. Mehta, "Vector analysis and control of advanced static VAR compensators," *IEE Proc. Gener. Transm. Distrib.*, vol. 140, no. 4, pp. 299–306, Jul. 1993.

89 C. Schauder, M. Gernhardt, E. Stacey, T. Lemak, L. Gyugyi, T. W. Cease and A. Edris, "Development of a ±100 MVAr static condenser for voltage control of transmission systems," *IEEE Trans. Power Delivery*, vol. 10, no. 3, pp. 1486–1496, Jul. 1995.

90 Y. Ni and L. Snider, "STATCOM power frequency model with VSC charging dynamics and its application in the power system stability analysis," *Proc. Fourth Int. Conf. Advances in Power System Control, Operation and Management, APSCOM*, vol. 1, pp. 119–124, Nov. 1997.

91 C. Hochgraf and R. H. Lasseter, "STATCOM controls for operation with unbalanced voltages," *IEEE Trans. Power Delivery*, vol. 13, no. 2, pp. 538–544, Apr. 1998.

92 K. V. Patil, J. Senthil, J. Jiang and R. M. Mathur, "Application of STATCOM for damping torsional oscillations in series compensated AC systems," *IEEE Trans. Energy Conversion*, vol. 13, no. 3, pp. 237–243, Sep. 1998.

93 P. W. Lehn and M. R. Iravani, "Experimental evaluation of STATCOM closed loop dynamics," *IEEE Trans. Power Delivery*, vol. 13, no. 4, pp. 1378–1384, Oct. 1998.

94 H. Okuno, T. Tominaka, S. Fujishima, T. Mitsumoto, T. Kubo, T. Kawaguchi, J. Kim, K. Ikegami, N. Sakamoto, N. Yokouchi, T. Morikawa, T. Tanaka, A. Goto and Y. Yano, "Physical experiments on a 10 kVAr STATCOM model," *Proc. IEEE PES Winter Meeting*, vol. 2, pp. 1075–1077, 1999.

95 K. K. Sen, "STATCOM-STATic synchronous COMpensator: theory, modeling, and applications," *Proc. IEEE PES Winter Meeting*, vol. 2, pp. 1177–1183, 1999.

96 C. Schauder, "STATCOM for compensation of large electric arc furnace installations," *Proc. IEEE PES Summer Meeting*, vol. 2, pp. 1109–1112, Jul. 1999.

97 H. F. Wang, "Phillips-Heffron model of power systems installed with STATCOM and applications," *IEE Proc. Gener. Transm. Distrib.*, vol. 146, no. 5, pp. 521–527, Sep. 1999.

98 H. Yonezawa, T. Shimato, M. Tsukada, K. Matsuno, I. Iyoda, J. J. Paserba and G. F. Reed, "Study of a STATCOM application for voltage stability evaluated by dynamic PV curves and time simulations," *Proc. IEEE PES Winter Meeting*, vol. 2, pp. 1471–1476, 2000.

99 Y. Yu, C. Jianye and H. Yingduo, "STATCOM modeling and analysis in damping power system oscillations," *Proc. 35th Intersociety Conference and Exhibition on Energy Conversion Engineering, IECEC*, vol. 2, pp. 756–762, Jul. 2000.

100 P. Garica-Gonzalez and A. Garcia-Cerrada, "Control system for a PWM-based STATCOM," *IEEE Trans. Power Delivery*, vol. 15, no. 4, pp. 1252–1257, Oct. 2000.

101 P. Rao, M. L. Crow and Z. Yang, "STATCOM control for power system voltage control applications," *IEEE Trans. Power Delivery*, vol. 15, no. 4, pp. 1311–1317, Oct. 2000.

102 J. A. Martinez, "Modeling of custom power equipment using electromagnetic transients programs," *Proc. Ninth Int. Conf. Harmonics and Quality of Power*, vol. 3, pp. 769–774, Oct. 2000.

103 L. Chun, J. Qirong and W. Zhonghong, "Study of STATCOM control for power swings damping improvement," *Proc. Int. Conf. Power System Technology, PowerCon*, vol. 1, pp. 535–540, Dec. 2000.

104 S. Chen, G. Joos and L. T. Moran, "Dynamic performance of PWM STATCOMs operating under unbalance and fault conditions in distribution systems," *Proc. IEEE PES Winter Meeting*, vol. 2, pp. 950–955, 2001.

105 P. S. Sensarma, K. R. Padiyar and V. Ramanarayanan, "Analysis and performance evaluation of a distribution STATCOM for compensating voltage fluctuations," *IEEE Trans. Power Delivery*, vol. 16, no. 2, pp. 259–264, Apr. 2001.

106 Z. Yang, C. Shen, L. Zhang, M. L. Crow and S. Atcitty, "Integration of STATCOM and battery energy storage," *IEEE Trans. Power Systems*, vol. 16, no. 2, pp. 254–260, May 2001.

107 V. Dinavahi, R. Iravani and R. Bonert, "Real-time digital simulation of power electronic apparatus interfaced with digital controllers," *IEEE Trans. Power Delivery*, vol. 16, no. 4, pp. 775–781, Oct. 2001.

108 B. K. Keshavan and N. Prabhu, "Damping of subsynchronous oscillations using STATCOM- a FACTS device," *Proc. IEEE/PES Transmission and Distribution Conference and Exposition*, vol. 1, pp. 1–7, 2001.

109 P. W. Lehn, "Exact modeling of the voltage source converter," *IEEE Trans. Power Delivery*, vol. 17, no. 1, pp. 217–222, Jan. 2002.

110 O. Anaya-Lara and E. Acha, "Modeling and analysis of custom power systems by PSCAD/EMTDC," *IEEE Trans. Power Delivery*, vol. 17, no. 1, pp. 266–272, Jan. 2002.

111 L. Ran, L. Holdsworth and G. A. Putrus, "Dynamic selective harmonic elimination of a three-level inverter used for static VAr compensation," *IEE Proc. Gener. Transm. Distrib.*, vol. 149, no. 1, pp. 83–89, Jan. 2002.

112 P. W. Lehn, "A benchmark system for simulation of the DSTATCOM," *Proc. IEEE PES Winter Meeting*, vol. 1, pp. 496–498, 2002.

113 C. Qian and M. L. Crow, "A cascaded converter-based STATCOM with energy storage," *Proc. IEEE PES Winter Meeting*, vol. 1, pp. 544–549, 2002.

114 C. K. Sao, P. W. Lehn, M. R. Iravani and J. A. Martinez, "A benchmark system for digital time-domain simulation of a pulse-width modulated D-STATCOM," *IEEE Trans. Power Delivery*, vol. 17, no. 4, pp. 1113–1120, Oct. 2002.

115 L. Cong and Y. Wang, "Coordinated control of generator excitation and STATCOM for rotor angle stability and voltage regulation enhancement of power systems," *IEE Proc. Gener. Transm. Distrib.*, vol. 149, no. 6, pp. 659–666, Nov. 2002.

116 A. B. Arsoy, Y. Liu, P. F. Ribeiro and F. Wang, "STATCOM-SMES," *IEEE Ind. Appl. Mag.*, vol. 9, no. 2, pp. 21–28, Mar.–Apr. 2003.

117 V. Dinavahi, R. Iravani and R. Bonert, "Design of a real-time digital simulator for a D-STATCOM system," *IEEE Trans. Ind. Electron.*, vol. 51, no. 5, pp. 1001–1008, Oct. 2004.

SSSC

118 L. Gyugyi, C. D. Schauder and K. K. Sen, "Static synchronous series compensator: a solid-state approach to the series compensation of transmission lines," *IEEE Trans. Power Delivery*, vol. 12, no. 1, pp. 406–417, Jan. 1997.

119 K. K. Sen, "SSSC-static synchronous series compensator: theory, modeling, and application," *IEEE Trans. Power Delivery*, vol. 13, no. 1, pp. 241–246, Jan. 1998.

120 R. Mihalic, "Power flow control with controllable reactive series elements," *IEE Proc. Gener. Transm. Distrib.*, vol. 145, no. 5, pp. 493–498, Sep. 1998.

121 H. Okuno, T. Tominaka, S. Fujishima, T. Mitsumoto, T. Kubo, T. Kawaguchi, J.-W. Kim, K. Ikegami, N. Sakamoto, S. Yokouchi, T. Morikawa, T. Tanaka, A. Goto and Y. Yano, "Switching-level simulation model for SSSC with EMTP," *Proc. IEEE PES Winter Meeting*, vol. 2, pp. 1075–1077, 1999.

122 R. Mihalic and I. Papic, "Mathematical models and simulation of a static synchronous series compensator," *Proc. IEEE Int. Conf. Electric Power Engineering, PowerTech*, Budapest, Hungary, 1999.

123 L. Sunil Kumar and A. Ghosh, "Modeling and control design of a static synchronous series compensator," *IEEE Trans. Power Delivery*, vol. 14, no. 4, pp. 1448–1453, Oct. 1999.

124 K. K. Sen, "A power flow controller with a stable reversing capability," *Proc. IEEE PES Summer Meeting*, Seattle, USA, Jul. 2000.

125 B. Han, S. Moon, J. Park and G. Karady, "Dynamic characteristic analysis of multi-bridge PWM inverter for SSSC," *Proc. IEEE PES Summer Meeting*, vol. 3, pp. 1618–1623, 2000.

126 J. Seo, S. Moon, J. Park and J. Choe, "Design of a robust SSSC supplementary controller to suppress the SSR in the series compensated system," *Proc. IEEE PES Winter Meeting*, vol. 3, pp. 1283–1288, 2001.

127 B. Han, S. Baek, H. Kim and G. Karady, "Dynamic characteristic analysis of SSSC based on multi-bridge inverter," *IEEE Trans. Power Delivery*, vol. 17, no. 2, pp. 623–629, Apr. 2002.

128 G. N. Pillai, A. Ghosh and A. Joshi, "Torsional interaction between an SSSC and a PSS in a series compensated power system," *IEE Proc. Gener. Transm. Distrib.*, vol. 149, no. 6, pp. 653–658, Nov. 2002.

129 M. O. Faruque, V. Dinavahi, S. Santoso and R. Adapa, "Review of electromagnetic transient models of non-VSC FACTS," *IEEE Trans. Power Delivery*, vol. 20, no. 2, pp. 1065–1079, Apr. 2005.

UPFC

130 L. Gyugyi, "A unified power flow control concept for flexible AC transmission systems," *IEE Proc. C*, vol. 139, no. 4, p. 323, Jul. 1992.

131 L. Gyugyi, C. D. Schauder, S. L. Williams, T. R. Rietman, D. R. Torgerson and A. Edris, "The unified power flow controller: a new approach to power transmission control," *IEEE Trans. Power Delivery*, vol. 10, no. 2, pp. 1085–1097, Apr. 1995.

132 R. Mihalic, P. Zunko and D. Povh, "Improvement of transient stability using unified power flow controller," *IEEE Trans. Power Delivery*, vol. 11, no. 1, pp. 485–492, Jan. 1996.

133 A. Nabavi-Niaki and M. R. Iravani, "Steady-state and dynamic models of unified power flow controller (UPFC) for power system studies," *IEEE Trans. Power Systems*, vol. 11, no. 4, pp. 1937–1943, Nov. 1996.

134 K.S. Smith, L. Ran and J. Penman, "Dynamic modelling of a unified power flow controller," *IEE Proc. Gener. Transm. Distrib.*, vol. 144, no. 1, pp. 7–12, Jan. 1997.

135 X. Lombard and P. G. Therond, "Control of unified power flow controller: comparison of methods on the basis of a detailed numerical model," *IEEE Trans. Power Systems*, vol. 12, no. 2, pp. 824–830, May 1997.

136 A. M. Kulkarni and K. R. Padiyar, "Performance evaluation of unified power flow controller using transient simulation," *Proc. Int. Conf. Power Electronics and Drive Systems*, vol. 2, pp. 734–739, Singapore, May 1997.

137 J. Bian, D. G. Ramey, R. J. Nelson and A. Edris, "A study of equipment sizes and constraints for a unified power flow controller," *IEEE Trans. Power Delivery*, vol. 12, no. 3, pp. 1385–1391, Jul. 1997.

138 I. Papic, P. Zunko, D. Povh and M. Weinhold, "Basic control of unified power flow controller," *IEEE Trans. Power Systems*, vol. 12, no. 4, pp. 1734–1739, Nov. 1997.

139 M. Rahman, M. Ahmed, R. Gutman, R. J. O'Keefe, R. J. Nelson and J. Bian, "UPFC application on the AEP system: planning considerations," *IEEE Trans. Power Systems*, vol. 12, no. 4, pp. 1695–1701, Nov. 1997.

140 C. D. Schauder, L. Gyugyi, M. R. Lund, D. M. Hamai, T. R. Rietman, D. R. Torgerson and A. Edris, "Operation of the unified power flow controller (UPFC) under practical constraints," *IEEE Trans. Power Delivery*, vol. 13, no. 2, pp. 630–639, Apr. 1998.

141 A. S. Mehraban, A. Edris, C. D. Schauder and J. H. Provanzana, "Installation, commissioning, and operation of the world's first UPFC on the AEP system," *Proc. Int. Conf. Power System Technology, POWERCON*, vol. 1, pp. 323–327, Beijing, China, Aug. 1998.

142 K.R. Padiyar and A.M. Kulkarni, "Control design and simulation of unified power flow controller," *IEEE Trans. Power Delivery*, vol. 13, no. 4, pp. 1348–1354, Oct. 1998.

143 K. K. Sen and E. J. Stacey, "UPFC-unified power flow controller: theory, modeling, and applications," *IEEE Trans. Power Delivery*, vol. 13, no. 4, pp. 1453–1460, Oct. 1998.

144 C. Schauder, E. Stacey, M. Lund, L. Gyugyi, L. Kovalsky, A. Keri, A. Mehraban and A. Edris, "AEP UPFC project: installation, commissioning and operation of the 160 MVA STATCOM (phase I)," *IEEE Trans. Power Delivery*, vol. 13, no. 4, pp. 1530–1535, Oct. 1998.

145 Y. Ni, Z. Huang, S. Chen and B. Zhang, "Incorporating UPFC model into the power system toolbox of the MATLAB for transient stability study," *Proc. IEEE Region 10 Int. Conf. Global Connectivity in Energy, Computer, Communication and Control, TENCON*, vol. 2, pp. 506–509, New Delhi, India, Dec. 1998.

146 H. Okuno, T. Tominaka, S. Fujishima, T. Mitsumoto, T. Kubo, T. Kawaguchi, J. W. Kim, K. Ikegami, N. Sakamoto, S. Yokouchi, T. Morikawa, T. Tanaka, A. Goto and Y. Yano, "Unified power flow controller on the AEP system: commissioning and operation," *Proc. IEEE PES Winter Meeting*, vol. 2, 31, pp. 1075–1077, New York, 1999.

147 A. J. F. Keri, A. S. Mehraban, X. Lombard, A. Eiriachy and A. A. Edris, "Unified power flow controller (UPFC): modeling and analysis," *IEEE Trans. Power Delivery*, vol. 14, no. 2, pp. 648–654, Apr. 1999.

148 Y. Morioka, M. Kato, Y. Mishima, Y. Nakachi, M. Asada and K. Tokuhara, "Implementation of unified power flow controller and verification for transmission capability improvement," *IEEE Trans. Power Systems*, vol. 14, no. 2, pp. 575–581, May 1999.

149 H. F. Wang, "Applications of modelling UPFC into multimachine power systems," *IEE Proc. Gener. Transm. Distrib.*, vol. 146, no. 3, pp. 306–312, May 1999.

150 T. Makombe and N. Jenkins, "Investigation of a unified power flow controller," *IEE Proc. Gener. Transm. Distrib.*, vol. 146, no. 4, pp. 400–408, Jul. 1999.

151 R. Caldon, A. Mari, A. Paolucci and N. Turri, "An efficient modular model for static and dynamic analysis of FACTS performances on electric power systems," *Proc. Int. Conf. Electric Power Engineering, PowerTech*, Budapest, Hungary, 1999.

152 B. A. Renz, A. Keri, A. S. Mehraban, C. Schauder, E. Stacey, L. Kovalsky, L. Gyugyi and A. Edris, "AEP unified power flow controller performance," *IEEE Trans. Power Delivery*, vol. 14, no. 4, pp. 1374–1381, Oct. 1999.

153 H. Fujita, Y. Watanabe and H. Akagi, "Control and analysis of a unified power flow controller," *IEEE Trans. Power Electronics*, vol. 14, no. 6, pp. 1021–1027, Nov. 1999.

154 S. Zheng, Y. L. Tan, S. Cheng and Y. Zhu, "Dynamic characteristic study of UPFC based on a detailed simulation model," *Proc. IEEE PES Winter Meeting*, vol. 2, pp. 1360–1364, 2000.

155 C.A. Canizares, "Power flow and transient stability models of FACTS controllers for voltage and angle stability studies," *Proc. IEEE PES Winter Meeting*, vol. 2, pp. 1447–1454, 2000.

156 Z. J. Meng and P. L. So, "A current injection UPFC model for enhancing power system dynamic performance," *Proc. IEEE PES Winter Meeting*, vol. 2, pp. 1544–1549, 2000.

157 S. Arabi, P. Kundur and R. Adapa, "Innovative techniques in modeling UPFC for power system analysis," *IEEE Trans. Power Systems*, vol. 15, no. 1, pp. 336–341, Feb. 2000.

158 Z. Huang, Y. Ni, C. M. Shen, F. F. Wu, S. Chen and B. Zhang, "Application of unified power flow controller in interconnected power systems-modeling, interface, control strategy, and case study," *IEEE Trans. Power Systems*, vol. 15 no. 2, pp. 817–824, May 2000.

159 H. Wang, "A unified model for the analysis of FACTS devices in damping power system oscillations. III. Unified power flow controller," *IEEE Trans. Power Delivery*, vol. 15, no. 3, pp. 978–983, Jul. 2000.

160 D. J. Won and S. I. Moon, "A study on the dynamic responses of a PWM based UPFC and its application to power system," *Proc. IEEE PES Summer Meeting*, vol. 3, pp. 1590–1595, 2000.

161 K. Schoder, A. Hasanovic and A. Feliachi, "Load-flow and dynamic model of the unified power flow controller (UPFC) within the power system toolbox (PST)," *Proc. 43rd IEEE Midwest Symp. Circuits and Systems*, vol. 2, pp. 634–637, Lansing, MI, Aug. 2000.

162 K. Sen, "Simulation of UPFC," *Proc. IEEE PES Winter Meeting*, vol. 2, pp. 641–642, 2001.

163 J. Seo, S. Moon, J. Park and J. Choe, "Design of a robust UPFC controller for enhancing the small signal stability in the multimachine power systems," *Proc. IEEE PES Winter Meeting*, vol. 3, pp. 1197–1202, 2001.

164 Y. L. Kang, G. B. Shrestha and T. T. Lie, "Consideration of DC capacitor and line inductance in the design of UPFC to improve transient stability," *Proc. IEEE PES Winter Meeting*, vol. 3, pp. 1277–1282, 2001.

165 K. Sedraoui, K. Al-haddad and G. Olivier, "A new approach for the dynamic control of unified power flow controller (UPFC)," *Proc. IEEE PES Summer Meeting*, vol. 2, pp. 955–960, 2001.

166 Y. L. Kang, G. B. Shresta and T.T. Lie, "Application of an NLPID controller on a UPFC to improve transient stability of a power system," *IEE Proc. Gener. Transm. Distrib.*, vol. 148, no. 6, pp. 523–529, 2001.

167 P. C. Stefanov and A. M. Stankovic, "Modeling of UPFC operation under unbalanced conditions with dynamic phasors," *IEEE Trans. Power Systems*, vol. 17, no. 2, pp. 395–403, May 2002.

168 H. Fujita, H. Akagi and Y. Watanabe, "Dynamic performance of a unified power flow controller for stabilizing AC transmission systems," *Proc. 33rd Annual IEEE Power Electronics Specialists Conf.*, vol. 1, pp. 81–87, 2002.

169 C. T. Chang and Y. Y. Hsu, "Design of UPFC controllers and supplementary damping controller for power transmission control and stability enhancement of a longitudinal power system," *IEE Proc. Gener. Transm. Distrib.*, vol. 149, no. 4, pp. 463–471, Jul. 2002.

170 J. B. Choo, B. H. Chang, H. S. Lee, H. S. Shin and K. K. Koh, "Development of FACTS Operation Technology to the KEPCO Power Network – Installation & Operation," *Proc. IEEE Power Engineering Society Transmission & Distribution Conf.*, Yokohama, Japan, Paper No. 231, Oct. 2002.

171 P. Garcia-Gonzalez and A. Garcia-Cerrada, "Detailed analysis and experimental results of the control system of a UPFC," *IEE Proc. Gener. Transm. Distrib.*, vol. 150, no. 2, pp. 147–154, Mar. 2003.

172 B. H. Chang, S. Y. Kim, J. S. Yoon, S. P. Moon, D. H. Baek, B. M. Kwak and J. B. Choo, "Control Strategies Study for Kepco UPFC Operation Automation in Korean Sub-Transmission System," *CIGRE*, Paper No. B4-306, 2006.

IPFC

173 B. Fardanesh, M. Henderson, B. Shperling, S. Zelingher, L. Gyugyi, C. Schauder, B. Lam, J. Mounford, R. Adapa and A. Edris, "Convertible Static Compensator Application to the New York Transmission System," *CIGRE*, Paper No. 14-103, 1998.

174 L. Gyugyi, K. K. Sen and C. D. Schauder, "The interline power flow controller concept: a new approach to power flow management in transmission systems," *IEEE Trans. Power Delivery*, vol. 14, no. 3, pp. 1115–1123, Jul. 1999.

175 B. Fardanesh, B. Shperling, E. Uzunovic and S. Zelingher, "Multiconverter FACTS devices: the generalized unified power flow controller (GUPFC)," *Proc. IEEE PES Summer Meeting*, vol. 2, pp. 1020–1025, Jul. 2000.

176 J. Chen, T. T. Lie and D. M. Vilathgamuwa, "Basic control of interline power flow controller," *Proc. IEEE PES Winter Meeting*, vol. 1, pp. 521–525, 2002.

177 X.-P. Zhang, "Modelling of the interline power flow controller and the generalised unified power flow controller in Newton power flow," *IEE Proc. Commun.*, vol. 150, no. 3, pp. 268–274, May 2003.

178 B. Fardanesh, "Optimal utilization, sizing, and steady-state performance comparison of multiconverter VSC based FACTS controllers," *IEEE Trans. Power Delivery*, vol. 19, no. 3, pp. 1321–1327, Jul. 2004.

ST

179 A. Krämer, "On-Load Tap-Changers for Power Transformers: Operation, Principles, Applications and Selection," Maschinenfabrik Reinhausen, 2000.

180 K. K. Sen and M. L. Sen, "Versatile power flow transformers for compensating power flow in a transmission line," U.S. Patents 6,335,613, 6,384,581, 6,396,248, and 6,420,856, 2002.

181 K. K. Sen and M. L. Sen, "Introducing the family of Sen transformers: a set of power flow controlling transformers," *IEEE Trans. Power Delivery*, vol. 18, no. 1, pp. 149–157, Jan. 2003.

182 K. K. Sen and M. L. Sen, "Comparison of the Sen transformer with the unified power flow controller," *IEEE Trans. Power Delivery*, vol. 18, no. 3, pp. 1523–1533, Oct. 2003.

183 K. K. Sen and M. L. Sen, "Multiline power flow transformer for compensating power flow among transmission lines," U.S. Patent 6,841,976, 2005.

184 B. Asghari, M. O. Faruque and V. Dinavahi, "Detailed real-time transient model of Sen transformer," *IEEE Trans. Power Delivery*, vol. 23, no. 3, pp. 1513–1521, Jul. 2008.

185 D. Fentie, "Transient Modeling and Protection of Sen Transformer," M.S. Thesis, University of Saskatchewan, 2010.

186 M. L. Sen and K. K. Sen, "Introducing the SMART Power Flow Controller – An Integral Part of Smart Grid," *Electrical Power and Energy Conference*, Paper No. 103, London, Canada, Oct. 2012.

187 K. Ashwani and K. Jitendra, "Comparison of UPFC and SEN transformer for ATC enhancement in restructured electricity markets," *Int. J. Electr. Power Energy Syst.*, vol. 41, no. 1, pp. 96–104, Oct. 2012.

188 K. Ashwani and W. Gao, "Power flow model of Sen transformer for loadability enhancement and comparison with UPFC in hybrid electricity markets," *Electric Power Compon. Syst.*, vol. 37, no. 2, pp. 189–209, Feb. 2012.

189 G. Maddela, "Transmission Line Loss Minimization Using Power Flow Controllers," M.Sc. Thesis, University of Saskatchewan, Aug. 2013.

190 K. K. Sen, "Practical Power Flow Controller Brings Benefits of Power Electronics to the Grid," *How2Power.com e-magazine*, Mar. 2015. http://www.how2power.com/newsletters/1503/articles/H2PToday1503_design_Sen.pdf?NOREDIR=1

191 D. Fentie, J. C. Garcia, R. Gokaraju and S. O. Faried, “EMT model of the ‘Sen Transformer’ for fault analysis studies,” *Int. Conf. Power Systems Transients* (IPST 2015), Cavtat, Croatia, Jun. 2015

192 K. K. Sen and M. L. Sen, “SMART Power Flow Controller for Smarter Grid Applications,” *IEEE Power & Energy Society General Meeting*, Paper No. 15PESGM1910, Denver, USA, Jul. 2015.

193 K. K. Sen and M. L. Sen, “Comparison of Operational Characteristics Between the Sen Transformer and Phase Angle Regulator,” *IEEE Power & Energy Society General Meeting*, Paper No. 16PESGM0959, Boston, USA, Jul. 2016.

194 K. K. Sen and M. L. Sen, “Unique Capabilities of Sen Transformer: A Power Flow Regulating Transformer,” *IEEE Power & Energy Society General Meeting*, Paper No. 16PESGM0961, Boston, USA, Jul. 2016.

195 J. Liu and V. Dinavahi, “Nonlinear magnetic equivalent circuit-based real-time Sen transformer electro-magnetic transient model on FPGA for HIL emulation,” *IEEE Trans. Power Delivery*, vol. 31, no. 6, pp. 2483–2493, Dec. 2016.

196 K. K. Sen and M. L. Sen, “Phase Angle Regulation Versus Impedance Regulation: Which Offers Greater Control of Power Flow on the Grid?,” *How2Power.com e-magazine*, Feb. 2017. http://www.how2power.com/newsletters/1702/articles/H2PToday1702_design_Sen.pdf?NOREDIR=1

197 K. K. Sen and M. L. Sen, “Modeling of the Sen Transformer Using an Electromagnetic Transients Program,” *How2Power.com e-magazine*, Mar. 2018. http://www.how2power.com/newsletters/1803/articles/H2PToday1803_design_Sen.pdf?NOREDIR=1

198 B. Chen, W. Fei, C. Tian, L. Yang and J. Gu, “A high-voltage ‘Sen’ transformer: configuration, principles, and applications,” *Energies*, vol. 11, no. 4, pp. 1–18, 2018. http://www.mdpi.com/1996-1073/11/4/918/htm

199 Modeling of the Sen Transformer in PSS®E, http://files.midphasesitebuilder.com/ad/33/ad335f60-d07b-4b8b-a9a3-ef0b6ab450ee.pdf, Accessed on Aug. 12, 2020.

Index

Power Flow Control Solutions for a Modern Grid using SMART Power Flow Controllers, First Edition.
Kalyan K. Sen and Mey Ling Sen.

Companion website: www.wiley.com/go/sen/powerflowcontrol

Books in the IEEE Press Series on Power and Engineering

Series Editor: ME El-Hawary, Dalhousie University, Halifax, Nova Scotia, Canada

The mission of IEEE Press Series on Power Engineering is to publish leading-edge books that cover the broad spectrum of current and forward-looking technologies in this fast-moving area. The series attracts highly acclaimed authors from industry/academia to provide accessible coverage of current and emerging topics in power engineering and allied fields. Our target audience includes the power engineering professional who is interested in enhancing their knowledge and perspective in their areas of interest.

1. *Electric Power Systems: Design and Analysis, Revised Printing*
Mohamed E. El-Hawary

2. *Power System Stability*
Edward W. Kimbark

3. *Analysis of Faulted Power Systems*
Paul M. Anderson

4. *Inspection of Large Synchronous Machines: Checklists, Failure Identification, and Troubleshooting*
Isidor Kerszenbaum

5. *Electric Power Applications of Fuzzy Systems*
Mohamed E. El-Hawary

6. *Power System Protection*
Paul M. Anderson

7. *Subsynchronous Resonance in Power Systems*
Paul M. Anderson, B.L. Agrawal, J.E. Van Ness

8. *Understanding Power Quality Problems: Voltage Sags and Interruptions*
Math H. Bollen

9. *Analysis of Electric Machinery*
Paul C. Krause, Oleg Wasynczuk, and S.D. Sudhoff

10. *Power System Control and Stability, Revised Printing*
Paul M. Anderson, A.A. Fouad

11. *Principles of Electric Machines with Power Electronic Applications, Second Edition*
Mohamed E. El-Hawary

12. *Pulse Width Modulation for Power Converters: Principles and Practice*
D. Grahame Holmes and Thomas Lipo

13. *Analysis of Electric Machinery and Drive Systems, Second Edition*
Paul C. Krause, Oleg Wasynczuk, and S.D. Sudhoff

14. *Risk Assessment for Power Systems: Models, Methods, and Applications*
Wenyuan Li

15. *Optimization Principles: Practical Applications to the Operations of Markets of the Electric Power Industry*
Narayan S. Rau

16. *Electric Economics: Regulation and Deregulation*
Geoffrey Rothwell and Tomas Gomez

17. *Electric Power Systems: Analysis and Control*
Fabio Saccomanno

18. *Electrical Insulation for Rotating Machines: Design, Evaluation, Aging, Testing, and Repair*
Greg C. Stone, Edward A. Boulter, Ian Culbert, and Hussein Dhirani

19. *Signal Processing of Power Quality Disturbances*
Math H. J. Bollen and Irene Y. H. Gu

20. *Instantaneous Power Theory and Applications to Power Conditioning*
Hirofumi Akagi, Edson H. Watanabe and Mauricio Aredes

21. *Maintaining Mission Critical Systems in a 24/7 Environment*
Peter M. Curtis

22. *Elements of Tidal-Electric Engineering*
Robert H. Clark

23. *Handbook of Large Turbo-Generator Operation and Maintenance, Second Edition*
Geoff Klempner and Isidor Kerszenbaum

24. *Introduction to Electrical Power Systems*
Mohamed E. El-Hawary

25. *Modeling and Control of Fuel Cells: Distributed Generation Applications*
M. Hashem Nehrir and Caisheng Wang

26. *Power Distribution System Reliability: Practical Methods and Applications*
Ali A. Chowdhury and Don O. Koval

27. *Introduction to FACTS Controllers: Theory, Modeling, and Applications*
Kalyan K. Sen, Mey Ling Sen

28. *Economic Market Design and Planning for Electric Power Systems*
James Momoh and Lamine Mili

29. *Operation and Control of Electric Energy Processing Systems*
James Momoh and Lamine Mili

30. *Restructured Electric Power Systems: Analysis of Electricity Markets with Equilibrium Models*
Xiao-Ping Zhang

31. *An Introduction to Wavelet Modulated Inverters*
S.A. Saleh and M.A. Rahman

32. *Control of Electric Machine Drive Systems*
Seung-Ki Sul

33. *Probabilistic Transmission System Planning*
Wenyuan Li

34. *Electricity Power Generation: The Changing Dimensions*
Digambar M. Tagare

35. *Electric Distribution Systems*
Abdelhay A. Sallam and Om P. Malik

36. *Practical Lighting Design with LEDs*
Ron Lenk, Carol Lenk

37. *High Voltage and Electrical Insulation Engineering*
Ravindra Arora and Wolfgang Mosch

38. *Maintaining Mission Critical Systems in a 24/7 Environment, Second Edition*
Peter Curtis

39. *Power Conversion and Control of Wind Energy Systems*
Bin Wu, Yongqiang Lang, Navid Zargari, Samir Kouro

40. *Integration of Distributed Generation in the Power System*
Math H. Bollen, Fainan Hassan

41. *Doubly Fed Induction Machine: Modeling and Control for Wind Energy Generation Applications*
Gonzalo Abad, Jesús López, Miguel Rodrigues, Luis Marroyo, and Grzegorz Iwanski

42. *High Voltage Protection for Telecommunications*
Steven W. Blume

43. *Smart Grid: Fundamentals of Design and Analysis*
James Momoh

44. *Electromechanical Motion Devices, Second Edition*
Paul Krause, Oleg Wasynczuk, Steven Pekarek

45. *Electrical Energy Conversion and Transport: An Interactive Computer-Based Approach, Second Edition*
George G. Karady, Keith E. Holbert

46. *ARC Flash Hazard and Analysis and Mitigation*
J.C. Das

47. *Handbook of Electrical Power System Dynamics: Modeling, Stability, and Control*
Mircea Eremia, Mohammad Shahidehpour

48. *Analysis of Electric Machinery and Drive Systems, Third Edition*
Paul C. Krause, Oleg Wasynczuk, S.D. Sudhoff, Steven D. Pekarek

49. *Extruded Cables for High-Voltage Direct-Current Transmission: Advances in Research and Development*
Giovanni Mazzanti, Massimo Marzinotto

50. *Power Magnetic Devices: A Multi-Objective Design Approach*
S.D. Sudhoff

51. *Risk Assessment of Power Systems: Models, Methods, and Applications, Second Edition*
Wenyuan Li

52. *Practical Power System Operation*
Ebrahim Vaahedi

53. *The Selection Process of Biomass Materials for the Production of Bio-Fuels and Co-Firing*
Najib Altawell

54. *Electrical Insulation for Rotating Machines: Design, Evaluation, Aging, Testing, and Repair, Second Edition*
Greg C. Stone, Ian Culbert, Edward A. Boulter, and Hussein Dhirani

55. *Principles of Electrical Safety*
Peter E. Sutherland

56. *Advanced Power Electronics Converters: PWM Converters Processing AC Voltages*
Euzeli Cipriano dos Santos Jr., Edison Roberto Cabral da Silva

57. *Optimization of Power System Operation, Second Edition*
Jizhong Zhu

58. *Power System Harmonics and Passive Filter Designs*
J.C. Das

59. *Digital Control of High-Frequency Switched-Mode Power Converters*
Luca Corradini, Dragan Maksimoviæ, Paolo Mattavelli, Regan Zane

60. *Industrial Power Distribution, Second Edition*
Ralph E. Fehr, III

61. *HVDC Grids: For Offshore and Supergrid of the Future*
Dirk Van Hertem, Oriol Gomis-Bellmunt, Jun Liang

62. *Advanced Solutions in Power Systems: HVDC, FACTS, and Artificial Intelligence*
Mircea Eremia, Chen-Ching Liu, Abdel-Aty Edris

63. *Operation and Maintenance of Large Turbo-Generators*
Geoff Klempner, Isidor Kerszenbaum

64. *Electrical Energy Conversion and Transport: An Interactive Computer-Based Approach*
George G. Karady, Keith E. Holbert

65. *Modeling and High-Performance Control of Electric Machines*
John Chiasson

66. *Rating of Electric Power Cables in Unfavorable Thermal Environment*
George J. Anders

67. *Electric Power System Basics for the Nonelectrical Professional*
Steven W. Blume

68. *Modern Heuristic Optimization Techniques: Theory and Applications to Power Systems*
Kwang Y. Lee, Mohamed A. El-Sharkawi

69. *Real-Time Stability Assessment in Modern Power System Control Centers*
Savu C. Savulescu

70. *Optimization of Power System Operation*
Jizhong Zhu

71. *Insulators for Icing and Polluted Environments*
Masoud Farzaneh, William A. Chisholm

72. *PID and Predictive Control of Electric Devices and Power Converters Using MATLAB®/Simulink®*
Liuping Wang, Shan Chai, Dae Yoo, Lu Gan, Ki Ng

73. *Power Grid Operation in a Market Environment: Economic Efficiency and Risk Mitigation*
Hong Chen

74. *Electric Power System Basics for Nonelectrical Professional, Second Edition*
Steven W. Blume

75. *Energy Production Systems Engineering*
Thomas Howard Blair

76. *Model Predictive Control of Wind Energy Conversion Systems*
Venkata Yaramasu, Bin Wu

77. *Understanding Symmetrical Components for Power System Modeling*
J.C. Das

78. *High-Power Converters and AC Drives, Second Edition*
Bin Wu, Mehdi Narimani

79. *Current Signature Analysis for Condition Monitoring of Cage Induction Motors: Industrial Application and Case Histories*
William T. Thomson, Ian Culbert

80. *Introduction to Electric Power and Drive Systems*
Paul Krause, Oleg Wasynczuk, Timothy O'Connell, Maher Hasan

81. *Instantaneous Power Theory and Applications to Power Conditioning, Second Edition*
Hirofumi, Edson Hirokazu Watanabe, Mauricio Aredes

82. *Practical Lighting Design with LEDs, Second Edition*
Ron Lenk, Carol Lenk

83. *Introduction to AC Machine Design*
Thomas A. Lipo

84. *Advances in Electric Power and Energy Systems: Load and Price Forecasting*
Mohamed E. El-Hawary

85. *Electricity Markets: Theories and Applications*
Jeremy Lin, Jernando H. Magnago

86. *Multiphysics Simulation by Design for Electrical Machines, Power Electronics and Drives*
Marius Rosu, Ping Zhou, Dingsheng Lin, Dan M. Ionel, Mircea Popescu, Frede Blaabjerg, Vandana Rallabandi, David Staton

87. *Modular Multilevel Converters: Analysis, Control, and Applications*
Sixing Du, Apparao Dekka, Bin Wu, Navid Zargari

88. *Electrical Railway Transportation Systems*
Morris Brenna, Federica Foiadelli, Dario Zaninelli

89. *Energy Processing and Smart Grid*
James A. Momoh

90. *Handbook of Large Turbo-Generator Operation and Maintenance, Third Edition*
Geoff Klempner, Isidor Kerszenbaum

91. *Advanced Control of Doubly Fed Induction Generator for Wind Power Systems*
Dehong Xu, Dr. Frede Blaabjerg, Wenjie Chen, Nan Zhu

92. *Electric Distribution Systems, 2nd Edition*
Abdelhay A. Sallam, Om P. Malik

93. *Power Electronics in Renewable Energy Systems and Smart Grid: Technology and Applications*
Bimal K. Bose

94. *Distributed Fiber Optic Sensing and Dynamic Rating of Power Cables*
Sudhakar Cherukupalli, and George J Anders

95. *Power System and Control and Stability, Third Edition*
Vijay Vittal, James D. McCalley, Paul M. Anderson, and A. A. Fouad

96. *Electromechanical Motion Devices: Rotating Magnetic Field-Based Analysis and Online Animations, Third Edition*
Paul Krause, Oleg Wasynczuk, Steven D. Pekarek, and Timothy O'Connell

97. *Applications of Modern Heuristic Optimization Methods in Power and Energy Systems*
Kwang Y. Lee and Zita A. Vale

98. *Handbook of Large Hydro Generators: Operation and Maintenance*
Glenn Mottershead, Stefano Bomben, Isidor Kerszenbaum, and Geoff Klempner

99. *Advances in Electric Power and Energy: Static State Estimation*
Mohamed E. El-Hawary

100. *Arc Flash Hazard Analysis and Mitigation, Second Edition*
J.C. Das

101. *Maintaining Mission Critical Systems in a 24/7 Environment, Third Edition*
Peter M. Curtis

102. *Real-Time Electromagnetic Transient Simulation of AC-DC Networks*
Venkata Dinavahi and Ning Lin

103. *Probabilistic Power System Expansion Planning with Renewable Energy Resources and Energy Storage Systems,*
Jaeseok Choi and Kwang Y. Lee

104. *Power Magnetic Devices: A Multi-Objective Design Approach, Second Edition.*
Scott D. Sudhoff.

105. *Optimal Coordination of Power Protective Devices with Illustrative Examples*
Ali R. Al-Roomi

106. *Alternative Liquid Dielectrics for High Voltage Transformer Insulation Systems: Performance Analysis and Applications,*
U. Mohan Rao, I. Fofana, and R. Sarathi

107. *Power System Protection: Fundamentals and Applications.*
John Ciufo and Aaron Cooperberg

108. *Power Flow Control Solutions for a Modern Grid using SMART Power Flow Controllers*
Kalyan K. Sen and Mey Ling Sen

Printed and bound by CPI Group (UK) Ltd, Croydon, CR0 4YY